SECOND EDITION

Microbiology

An Introduction

Of Related Interest from the Benjamin/Cummings Series in the Life Sciences

SECOND EDITION

Microbiology
An Introduction

Gerard J. Tortora
Bergen Community College

Berdell R. Funke
North Dakota State University

Christine L. Case
Skyline College

The Benjamin/Cummings Publishing Company, Inc.

Menlo Park, California • Reading, Massachusetts • Don Mills, Ontario • Wokingham, U.K. • Amsterdam • Sydney • Singapore • Tokyo • Mexico City • Bogota • Santiago • San Juan

Sponsoring Editor: Jane R. Gillen
Developmental Editor: Jo Andrews
Production Editor: Pat Waldo
Copy Editor: Fannie Toldi
Book and Cover Designer: John Edeen
Artist: Barbara Haynes, Carla Simmons, and Carol Verbeeck
Illustrator: Michael Fornalski
Photographs: Biological Photo Service

About the cover: Glowing in the flask is newly synthesized cytosine, one of the building blocks of DNA, the genetic material. Modern technology has made it possible not only to synthesize the building blocks, but also to assemble completely synthetic DNA. Many of the new techniques were developed through research using microorganisms (see Chapter 8 and Microview 2).

Figure acknowledgments begin on page 795.

Library of Congress Cataloging-in-Publication Data

Tortora, Gerard J.
Microbiology : an introduction.

(Benjamin/Cummings series in the life sciences)
Includes bibliographies and index.
1. Microbiology. I. Funke, Berdell R.
II. Case, Christine L., 1948– III. Title.
IV. Series. [DNLM: 1. Microbiology. QW 4 T712m]
QR41.2.T67 1986 576 85–22962
ISBN 0–8053–9315–3

defghij-DO-8987

The Benjamin/Cummings Publishing Company, Inc.
2727 Sand Hill Road
Menlo Park, California 94025

Preface

Microbiology: An Introduction is a comprehensive text for students in a wide variety of programs, including allied-health sciences, biological sciences, environmental studies, animal science, forestry, agriculture, home economics, and liberal arts. It is a beginning text, assuming no previous study of biology or chemistry.

In the four years since the publication of the first edition, this book has been used at more than 600 colleges and universities, by over 150,000 students. We have been gratified to hear from instructors and students alike that the book has become a favorite among their textbooks—a learning tool both effective and enjoyable. So, in planning the second edition, we strove to integrate the suggestions made by users of the first edition, yet maintain our clear coverage of the basic principles of microbiology.

FEATURES OF THE REVISION

Our primary goal in revising Microbiology: An Introduction was to update the book throughout, to make sure it reflected the important new discoveries of the last few years. We also wanted to address several issues of organization and coverage that had been brought up by users of the first edition. At the same time, we wanted to keep the book at a manageable length and to maintain the emphasis on *fundamental* facts and principles. Every page of the revised manuscript was scrutinized with those goals in mind. The following list highlights the major changes that ensued:

- Almost all of the special topic boxes—both *Microbiology in the News* and *MMWR*—have been replaced with more timely ones.
- *Microbial Metabolism* (Chapter 5) has been carefully revised to make the basic principles of this important but difficult topic clearer and more accessible.
- *Microbial Genetics* (Chapter 8) has been expanded to include slightly more detail on molecular biology and recombinant DNA technology.
- *Bacteria* (Chapter 10) has been revised to reflect the organizational scheme of the new *Bergey's Manual of Systematic Bacteriology.*
- The chapter on *eucaryotic microorganisms* (Chapter 11) now includes algae, and fungi are covered in greater depth than previously.
- The chapters on *immunology* (Chapters 16 and 17) have been heavily revised to reflect the many recent advances in this fast-moving field.
- Part Four, *Microorganisms and Human Disease,* has been made more consistent in organization. The major change has been the reassignment of nosocomial diseases and diseases associated with wounds and bites (formerly covered in Chapter 20) to other chapters according to the host organ systems these diseases affect.
- Three full-color photo essays, called "Microviews," are included. *Microview 1* (after page 90), a revision of the color section in the first edition, covers microbial morphology, stains, and biochemical tests. *Microview 2* (after page 250) concerns genetic engineering and includes a tour of Cetus, a leading biotechnology company. *Microview 3* (after page 522) is a tour of hospital laboratories at the University of California at San Francisco. We think students will find these photoessays informative and interesting.

FEATURES RETAINED FROM THE FIRST EDITION

As mentioned earlier, we have retained the features that made the first edition so popular. These include:

- **An appropriate balance between microbiological fundamentals and applications, and between medical applications and other applied areas.** As before, basic principles are given greater emphasis than applications, and health-related applications are emphasized. Nevertheless—as before—applications are integrated throughout the text, and considerable attention is devoted to microorganisms in habitats outside the human body. The two new photoessays demonstrate several different applications of microbiology, and show how a student might work with microbes in his or her career. We hope that all students will gain an appreciation of the fascinating diversity of microbial life, the central roles of microorganisms in nature, and the importance of microorganisms in our daily lives.
- **Illustration program carefully developed to support the text.** Included are both state-of-the art micrographs that dramatically show microbial structure and good examples of conventional micrographs that more closely resemble what is usually seen in a microbiology laboratory. About a quarter of the photos in this second edition are new.
- **Special topic boxes that relate microbiology to real-life issues and applications.** There are two main types of boxes: *Microbiology in the News* items and reprints from the *Morbidity and Mortality Weekly Report* (*MMWR*). The *MMWR* boxes are particularly valuable as optional reading for allied-health students; abridged but not simplified from the original reports written for health professionals, these articles are easily comprehensible by beginning microbiology students, often to their amazement.

IN-TEXT LEARNING AIDS

A major goal of writing this text was to create a book that would be an effective tool for learning. Therefore, we have included many student aids in each chapter.

- **Objectives** provide students with guidelines for what they should know after studying the chapter.
- **Tables** summarize, organize, and complement the text discussions.
- **Study Outlines** at the end of each chapter aid review. Each heading is keyed to specific pages in the text.
- **Study Questions** test recall of information presented in the chapter (*Review Questions*) and provide an opportunity to apply knowledge in problem-solving and interpretation (*Challenge Questions*).
- **Further Reading** suggestions give sources for further investigation of the topics in the chapter.

Several **appendices** at the end of the book heighten its usefulness. The first appendix is the *Classification of Bacteria According to Bergey's Manual of Systematic Bacteriology*. This is followed by a guide to *Pronunciation of Scientific Names,* which provides basic rules of pronunciation and phonetic pronunciations of genus and species names used in the text. Also included is a guide to *Word Roots Used in Microbiology,* a *Most Probable Numbers Table,* and a brief description of *Methods for Taking Clinical Samples*. A **Glossary** provides definitions of all important terms used in the text.

COURSE SEQUENCES

We have organized the book in what we feel is a useful fashion, but we recognize that there are a number of alternative sequences in which the material might be effectively presented. For those who wish to follow another sequence, we have made each chapter as independent as possible and have included numerous cross-references. Thus, the survey of the microbial world, Part Two, could be studied at the beginning of a course, immediately after Chapter 1. Or environmental and industrial microbiology, Part Five, could follow Parts One and Two. Since Chapters 7 and 18 both deal with the control of microbial growth, they could be covered together. The material on microbes and human disease, Part Four, readily lends itself to rearrangement or selective coverage. The various diseases are organized into chapters according to host organ system affected. However, the *Instructor's Guide* provides detailed guidelines for organizing the disease material in several alternative ways.

SUPPLEMENTARY MATERIALS

All the ancillaries to the text have been revised by their original authors.

- The ***Study Guide,*** by Berdell Funke, is available to help students master and review major concepts and facts from the text. Each chapter of the study guide begins with a chapter summary organized under the text headings. Important terms are printed in boldface and defined, and important figures and tables from the text are included. Following the summary is an extensive self-testing section containing matching questions, fill-in questions, and an answer key.

- The ***Instructor's Guide,*** by Christine Case, includes many practical suggestions for using the text in a course. *Suggested course outlines* are provided. For presentation of microbial diseases by microbial agent (taxonomic group), mode of transmission, or portal of entry, sequences of topics and pertinent pages are listed. Also included are the *answers to Study Questions,* in a format that can be used for grading, or reproduced and distributed to students for self-study. The final section of the Instructor's Guide is devoted to *test items;* each chapter has two objective tests, each containing 15 questions. The tests can be reproduced and used directly from the Guide.

- **Acetate overhead transparencies** of 100 two-color line drawings from the text are available from the publisher upon written request.

- ***Laboratory Experiments in Microbiology: Brief Edition*** (1986, ISBN 0-8053-9316-1), by Ted Johnson and Christine Case, is an abridgement of the more comprehensive manual by the same authors. It is specifically designed to accompany the Second Edition of *Microbiology: An Introduction*. A detailed *Intructor's Guide* for the lab manual facilitates preparation for laboratory sessions.

ACKNOWLEDGMENTS

In the preparation of this textbook, we have benefited from the guidance and advice of a large number of microbiology instructors across the country. Six contributors provided early drafts of chapters or portions of chapters. Reviewers offered constructive criticism and suggestions at various stages of manuscript preparation. Contributors and reviewers are listed on the following page. We gratefully acknowledge our debt to these individuals. In addition, we would like to thank the staff members of the UCSF Hospitals Laboratories and Cetus who helped in the preparation of Microviews 2 and 3.

Gerard J. Tortora
Berdell R. Funke
Christine L. Case

Contributors and Reviewers of *Microbiology: An Introduction*, second edition

Contributors:

Richard Bernstein, San Francisco State University
Frank Binder, Marshall University
Cynthia Moffet, San Francisco
Roger Nichols, Weber State College
J. Dennis O'Malley, University of Arkansas
Leleng To, Hardin-Simmons University

Reviewers of the First and Second Editions:

Lucia Anderson, Queensborough Community College
Kenneth L. Anderson, California State University, Los Angeles
Barry Batzing, SUNY College at Cortland
Jeff Becker, University of Tennessee
Lois Beishir, Antelope Valley College
Harold Bendigkeit, De Anza College
Richard Bernstein, San Francisco State University
L. J. Berry, University of Texas
Russell F. Bey, University of Minnesota
Frank Binder, Marshall University
Robert B. Boley, University of Texas
Russell J. Centanni, Boise State University
J. John Cohen, University of Colorado, Denver
Richard Davis, West Valley College
Norman Epps, University of Guelph
Cindy Erwin, San Francisco City College
David Filmer, Purdue University
Roger Furbee, Middlesex County College
David Gabrielson, North Dakota State University
James G. Garner, C. W. Post College
Joseph J. Gauthier, University of Alabama
Joan Handley, University of Kansas
Bettina Harrison, University of Massachusetts, Boston
Diane S. Herson, University of Delaware
Ronald Hochede, City College of San Francisco
John G. Holt, Iowa State University
Robert Janssen, University of Arizona
Ted R. Johnson, St. Olaf College
Diana Kaftan, Diablo Valley College
Alan Konopka, Purdue University
Walter Koostra, University of Montana
Robert I. Krasner, Providence College
John Lammert, Gustavus Adolphus College
John Lewis, San Bernardino Valley College
Peter Ludovici, University of Arizona
John C. Makemson, Florida International University
Eleanor K. Marr, Dutchess Community College
William Matthai, Tarrant County Junior College
Frank Mittermeyer, Elmhurst College
Henry Mulcahy, Suffolk University
Roger Nichols, Weber State College
Elinor O'Brien, Boston College
J. Dennis O'Malley, University of Arkansas
Dennis Opheim, Texas A & M University
Joseph R. Powell, Florida Junior College
Violet Schirone, Suffolk County College
Louis Shainberg, Mount San Antonio College
Gregory L. Shipley, Texas A & M University
Bernice Stewart, Prince Georges Community College
Roberta Wald, Portland Community College
Brian J. Wilkinson, Illinois State University

Brief Table of Contents

PART ONE

FUNDAMENTALS OF MICROBIOLOGY 1

Chapter 1 The Microbial World and You 2
Chapter 2 Chemical Principles 24
Chapter 3 Observing Microorganisms Through a Microscope 59
Chapter 4 Functional Anatomy of Procaryotic and Eucaryotic Cells 79
Chapter 5 Microbial Metabolism 113
Chapter 6 Microbial Growth 151
Chapter 7 Control of Microbial Growth 182
Chapter 8 Microbial Genetics 210

PART TWO

SURVEY OF THE MICROBIAL WORLD 253

Chapter 9 Classification of Microorganisms 254
Chapter 10 Bacteria 275
Chapter 11 Fungi, Algae, Protozoans, and Multicellular Parasites 301
Chapter 12 Viruses 339

PART THREE

INTERACTION BETWEEN MICROBE AND HOST 377

Chapter 13 Principles of Disease and Epidemiology 377
Chapter 14 Mechanisms of Pathogenicity 399
Chapter 15 Nonspecific Defenses of the Host 416
Chapter 16 Specific Defenses of the Host: The Immune Response 437
Chapter 17 Harmful Aspects of the Immune Response 471
Chapter 18 Antimicrobial Drugs 493

PART FOUR

MICROORGANISMS AND HUMAN DISEASE 523

Chapter 19 Microbial Diseases of the Skin and Eyes 524
Chapter 20 Microbial Diseases of the Nervous System 548
Chapter 21 Microbial Diseases of the Cardiovascular and Lymphatic Systems 574
Chapter 22 Microbial Diseases of the Respiratory System 608
Chapter 23 Microbial Diseases of the Digestive System 640
Chapter 24 Microbial Diseases of the Urinary and Genital Systems 674

PART FIVE

MICROBIOLOGY, THE ENVIRONMENT, AND HUMAN AFFAIRS 699

Chapter 25 Soil and Water Microbiology 700
Chapter 26 Food and Industrial Microbiology 732

Appendices 756
Glossary 771
Acknowledgments 775
Index 798

Detailed Table of Contents

PART ONE

FUNDAMENTALS OF MICROBIOLOGY 1

Chapter 1 The Microbial World and You 2

MICROBES IN OUR LIVES 4
A BRIEF HISTORY OF MICROBIOLOGY 5
- The First Observations 5
- Fermentation 5
- The Debate Over Spontaneous Generation 6
- The Germ Theory of Disease 8
- Vaccination 10
- The Birth of Modern Chemotherapy: Dreams of a "Magic Bullet" 11

NAMING AND CLASSIFYING MICROORGANISMS 12
THE DIVERSITY OF MICROORGANISMS 13
- Bacteria 13
- Fungi 13
- Protozoans 15
- Algae 15
- Multicellular Animal Parasites 15
- Viruses 15

MICROBES AND HUMAN WELFARE 15
- Recycling Vital Elements 16
- Sewage Treatment: Using Microbes to Recycle Water 17
- Insect Control by Microorganisms 18
- Modern Industrial Microbiology and Genetic Engineering 18

MICROBES AND HUMAN DISEASE 19
BOX WHAT MAKES SOURDOUGH BREAD DIFFERENT? 7
STUDY OUTLINE 20
STUDY QUESTIONS 22
FURTHER READING 23

Chapter 2 Chemical Principles 24

STRUCTURE OF ATOMS 25
- Chemical Elements 25
- Electronic Configurations 26

HOW ATOMS FORM MOLECULES: CHEMICAL BONDS 28
- Ionic Bonds 28
- Covalent Bonds 31
- Hydrogen Bonds 31
- Molecular Weight and Moles 32

CHEMICAL REACTIONS 32
- Synthesis Reactions 32
- Decomposition Reactions 32
- Exchange Reactions 33
- The Reversibility of Chemical Reactions 33
- How Chemical Reactions Occur 33
- Energy of Chemical Reactions 35
- Forms of Energy 35

IMPORTANT BIOLOGICAL MOLECULES 35
INORGANIC COMPOUNDS 35
- Water 35
- Acids, Bases, and Salts 36
- Acid–Base Balance 36

ORGANIC COMPOUNDS 38
- Functional Groups 40
- Carbohydrates 42
- Lipids 43
- Proteins 44
- Nucleic Acids 51
- Adenosine Triphosphate (ATP) 52

BOX EXPONENTS, EXPONENTIAL NOTATION, AND LOGARITHMS 39
STUDY OUTLINE 53
STUDY QUESTIONS 56
FURTHER READING 58

Chapter 3 Observing Microorganisms Through a Microscope 59

UNITS OF MEASUREMENT 60
MICROSCOPY: THE INSTRUMENTS 60
Compound Light Microscopy 61
Darkfield Microscopy 61
Phase-Contrast Microscopy 64
Differential Interference Contrast (DIC) Microscopy 65
Fluorescence Microscopy 65
Electron Microscopy 67
PREPARATION OF SPECIMENS FOR LIGHT MICROSCOPY 68
Hanging-Drop Preparation 68
Staining and Preparing Smears 69
Simple Stains 69
Differential Stains 70
Special Stains 74
Flagella Staining 75
BOX *BDELLOVIBRIO*—PREDATOR EXTRAORDINAIRE 71
STUDY OUTLINE 75
STUDY QUESTIONS 76
FURTHER READING 78

Chapter 4 Functional Anatomy of Procaryotic and Eucaryotic Cells 79

PROCARYOTIC AND EUCARYOTIC CELLS 80
THE PROCARYOTIC CELL 81
SIZE, SHAPE, AND ARRANGEMENT OF BACTERIAL CELLS 81
STRUCTURES EXTERNAL TO THE CELL WALL 82
Glycocalyx 82
Flagella 84
Axial Filaments 84
Pili 84
CELL WALL 86
Composition and Characteristics 86
Damage to the Cell Wall 89
STRUCTURES INTERNAL TO THE CELL WALL 90
Plasma (Cytoplasmic) Membrane 91
Movement of Materials Across Membranes 92
Cytoplasm 95
Endospores 96
THE EUCARYOTIC CELL 100
FLAGELLA AND CILIA 101
CELL WALL 102
PLASMA (CYTOPLASMIC) MEMBRANE 103
CYTOPLASM 103
ORGANELLES 103
Nucleus 103
Endoplasmic Reticulum 105
Ribosomes 105
Golgi Complex 105
Mitochondria 106
Lysosomes 106
Centrosome and Centrioles 107
Chloroplasts 107
CYTOPLASMIC INCLUSIONS 107
BOX MICROBIAL MAGNETS 97
STUDY OUTLINE 108
STUDY QUESTIONS 110
FURTHER READING 112

Chapter 5 Microbial Metabolism 113

ENZYMES 115
Parts of an Enzyme 115
Mechanisms of Enzymatic Action 117
Naming Enzymes 118
Factors Influencing Enzymatic Activity 118
Feedback Inhibition 121
ENERGY PRODUCTION METHODS 122
The Concept of Oxidation-Reduction 122
Generation of ATP: Types of Phosphorylation 122
Chemiosmotic Mechanism of ATP Generation 123
Nutritional Patterns Among Organisms 125
Summary of Energy Production Methods 128
BIOCHEMICAL PATHWAYS OF ENERGY PRODUCTION 129
CARBOHYDRATE CATABOLISM 130
Glycolysis 130
Respiration 131
Fermentation 135
Lipid Catabolism 138
Protein Catabolism 139
BIOCHEMICAL PATHWAYS OF ENERGY UTILIZATION 140
Biosynthesis of Polysaccharides 140
Biosynthesis of Lipids 140
Biosynthesis of Amino Acids 142
Biosynthesis of Purines and Pyrimidines 143

INTEGRATION OF METABOLISM 143
BOX PHOTOSYNTHESIS WITHOUT CHLOROPHYLL 127
BOX WHAT IS FERMENTATION? 137
STUDY OUTLINE 145
STUDY QUESTIONS 147
FURTHER READING 150

Chapter 6 Microbial Growth 151

REQUIREMENTS FOR GROWTH 152
Physical Requirements 152
Chemical Requirements 154
CULTURE MEDIA 159
Chemically Defined Media 160
Complex Media 161
Anaerobic Growth Media and Methods 161
Special Culture Techniques 162
Selective and Differential Media 163
Mixed and Pure Cultures 167
Isolating Pure Cultures 167
INOCULATION 168
PRESERVING BACTERIAL CULTURES 168
GROWTH OF BACTERIAL CULTURES 168
Bacterial Division 168
Generation Time 168
Phases of Growth 171
Measurement of Microbial Growth 172
BOX MAKING FOOD WITHOUT LIGHT—CHEMOSYNTHESIS ON THE OCEAN FLOOR 155
STUDY OUTLINE 177
STUDY QUESTIONS 179
FURTHER READING 181

Chapter 7 Control of Microbial Growth 182

CONDITIONS INFLUENCING MICROBIAL CONTROL 183
Temperature 183
Kind of Microbe 183
Physiological State of the Microbe 184
Environment 184
ACTIONS OF MICROBIAL CONTROL AGENTS 185
Alteration of Membrane Permeability 185
Damage to Proteins and Nucleic Acids 185
PATTERN OF MICROBIAL DEATH 185
PHYSICAL METHODS OF MICROBIAL CONTROL 186
Heat 186
Filtration 189
Low Temperature 189
Dessication 189
Osmotic Pressure 190
Radiation 191
CHEMICAL METHODS OF MICROBIAL CONTROL 192
Qualities of an Effective Disinfectant 192
Principles of Effective Disinfection 192
Evaluating a Disinfectant 194
Kinds of Disinfectants 194
BOX *MMWR PSEUDOMONAS AERUGINOSA* PERITONITIS ATTRIBUTED TO A CONTAMINATED IODOPHOR SOLUTION 197
STUDY OUTLINE 204
STUDY QUESTIONS 207
FURTHER READING 209

Chapter 8 Microbial Genetics 210

STRUCTURE AND FUNCTION OF THE GENETIC MATERIAL 211
GENOTYPE AND PHENOTYPE 211
DNA AND CHROMOSOMES 212
DNA REPLICATION 213
RNA AND PROTEIN SYNTHESIS 218
Transcription 218
Translation 221
THE GENETIC CODE 222
REGULATION OF GENE EXPRESSION IN BACTERIA 223
INDUCTION AND REPRESSION 224
Induction 224
Repression 224
Operon Model 224
MUTATION: CHANGE IN THE GENETIC MATERIAL 226
TYPES OF MUTATIONS 226
MUTAGENESIS 228
Chemical Mutagens 228
Radiation 230
Frequency of Mutation 230
IDENTIFYING MUTATIONS 230
THE AMES TEST FOR IDENTIFYING CHEMICAL CARCINOGENS 233
GENETIC TRANSFER AND RECOMBINATION 234
TRANSFORMATION IN BACTERIA 234
CONJUGATION IN BACTERIA 236
TRANSDUCTION IN BACTERIA 237
RECOMBINATION IN EUCARYOTES 239
PLASMIDS 240

GENETIC ENGINEERING: RECOMBINANT DNA 241
GENES AND EVOLUTION 244
BOX *MICROBIOLOGY IN THE NEWS* THE PROMISE AND PERIL OF GENETIC ENGINEERING 244
STUDY OUTLINE 246
STUDY QUESTIONS 249
FURTHER READING 251

PART TWO

SURVEY OF THE MICROBIAL WORLD 253

Chapter 9 Classification of Microorganisms 254

FIVE-KINGDOM SYSTEM OF CLASSIFICATION 255
NAMING ORGANISMS 255
CLASSIFICATION OF EUCARYOTIC ORGANISMS 257
CLASSIFICATION OF BACTERIA 257
CLASSIFICATION OF VIRUSES 260
CRITERIA FOR CLASSIFICATION AND IDENTIFICATION OF MICROORGANISMS 261
- Morphological Characteristics 262
- Differential Staining 262
- Biochemical Tests 263
- Serology 263
- Phage Typing 264
- Amino Acid Sequencing 264
- Protein Analysis 264
- Base Composition of Nucleic Acids 266
- Nucleic Acid Hybridization 267
- Genetic Recombination 269
- Numerical Taxonomy 269

BOX THE ARCHAEOBACTERIA—NEITHER PROCARYOTES NOR EUCARYOTES? 270
STUDY OUTLINE 270
STUDY QUESTIONS 272
FURTHER READING 273

Chapter 10 Bacteria 275

BERGEY'S MANUAL: BACTERIAL TAXONOMY 275
BACTERIAL GROUPS 276
- Spirochetes 276
- Helical/Vibrioid, Gram-Negative Bacteria 278
- Gram-Negative, Aerobic Rods and Cocci 280
- Facultatively Anaerobic, Gram-Negative Rods 281
- Anaerobic, Gram-Negative, Straight, Curved or Helical Rods 286
- Dissimilatory Sulfate- or Sulfur-Reducing Bacteria 286
- Anaerobic, Gram-Negative Cocci 287
- Rickettsias and Chlamydias 287
- Mycoplasmas 289
- Gram-Positive Cocci 289
- Endospore-Forming, Gram-Positive Rods and Cocci 291
- Regular, Non-Sporing, Gram-Positive Rods 291
- Irregular, Non-Sporing, Gram-Positive Rods 291
- Mycobacteria 292
- Nocardioforms 292
- Gliding, Fruiting, Sheathed, and Budding and/or Appendaged Bacteria 293
- Chemoautotrophic Bacteria 295
- Archaeobacteria 295
- Phototrophic Bacteria 295
- Actinomycetes 297

BOX HOW *E. COLI* GOT ITS NAME 283
STUDY OUTLINE 298
STUDY QUESTIONS 299
FURTHER READING 300

Chapter 11 Fungi, Algae, Protozoans, and Multicellular Parasites 301

FUNGI 302
VEGETATIVE STRUCTURES 302
- Yeasts 303
- Dimorphic Fungi 304

REPRODUCTIVE STRUCTURES 305
- Asexual Spores 305
- Sexual Spores 306

MEDICALLY IMPORTANT PHYLA OF FUNGI 307
- Zygomycota 307
- Ascomycota 307
- Basidiomycota 307
- Deuteromycota 307

FUNGAL DISEASES 308
UNDESIRABLE ECONOMIC EFFECTS OF FUNGI 309
ALGAE 310

STRUCTURE AND REPRODUCTION 313
ROLES OF ALGAE IN NATURE 313
LICHENS 315
SLIME MOLDS 317
PROTOZOANS 317
PROTOZOAN BIOLOGY 318
Nutrition 318
Reproduction 318
Encystment 319
MEDICALLY IMPORTANT PHYLA OF PROTOZOANS 319
Sarcodina 319
Mastigophora 319
Ciliata 321
Sporozoa 321
HELMINTHS 325
HELMINTH BIOLOGY 325
Reproduction 325
Life Cycle 325
PLATYHELMINTHES 325
Trematodes 325
Cestodes 325
ASCHELMINTHES 329
Nematodes 329
ARTHROPODS 331
BOX *MICROBIOLOGY IN THE NEWS* PARASITE OUTWITS THE IMMUNE SYSTEM, INTRIGUES SCIENTISTS 322
STUDY OUTLINE 333
STUDY QUESTIONS 336
FURTHER READING 338

Chapter 12 Viruses 339

GENERAL CHARACTERISTICS OF VIRUSES 340
Definition 340
Host Range 340
Size 341
VIRAL STRUCTURE 341
Nucleic Acid 341
Capsid and Envelope 342
General Morphology 342
CLASSIFICATION OF VIRUSES 344
ISOLATION, CULTIVATION, AND IDENTIFICATION OF VIRUSES 347
Growth of Bacteriophages in the Laboratory 349
Growth of Animal Viruses in the Laboratory 349
VIRAL MULTIPLICATION 352
Multiplication of Bacteriophages 352
Multiplication of Animal Viruses 357
EFFECTS OF ANIMAL VIRAL INFECTION ON HOST CELLS 365
VIRUSES AND CANCER 365
Transformation of Normal Cells Into Tumor Cells 366
DNA-Containing Oncogenic Viruses 368
RNA-Containing Oncogenic Viruses 368
Activation of Oncogenes 368
LATENT VIRAL INFECTIONS 369
SLOW VIRAL INFECTIONS 369
PRIONS 369
PLANT VIRUSES AND VIROIDS 370
BOX *MICROBIOLOGY IN THE NEWS* ESTABLISHING A LINK BETWEEN EB VIRUS AND CANCER 367
STUDY OUTLINE 371
STUDY QUESTIONS 374
FURTHER READING 376

PART THREE

INTERACTION BETWEEN MICROBE AND HOST 377

Chapter 13 Principles of Disease and Epidemiology 378

PATHOLOGY, INFECTION, AND DISEASE 379
NORMAL FLORA 379
Functions 380
Relationships Between Normal Flora and Host 381
CAUSES OF DISEASE 382
Main Categories of Disease 382
Koch's Postulates 383
SPREAD OF INFECTION 385
Reservoirs 385
Nocosomial (Hospital-Acquired) Infections 385
Transmission of Disease 388
KINDS OF DISEASES 390
SIGNALS OF DISEASE 392
EPIDEMIOLOGY 393
BOX *MMWR* ENDOTOXIC REACTIONS ASSOCIATED WITH THE REUSE OF CARDIAC CATHETERS 387
STUDY OUTLINE 394

STUDY QUESTIONS 396
FURTHER READING 398

Chapter 14 Mechanisms of Pathogenicity 399

ENTRY OF A MICROORGANISM INTO THE HOST 400
Mucous Membranes 400
Skin 400
Parenteral Route 400
Preferred Portal of Entry 400
Numbers of Invading Microbes 400
ADHERENCE 402
HOW PATHOGENS RESIST HOST DEFENSES 402
Capsules 402
Components of the Cell Wall 403
Enzymes 403
Other Factors 404
DAMAGE TO HOST CELLS 404
Direct Damage 405
Toxins 405
PATHOGENIC PROPERTIES OF OTHER MICROORGANISMS 407
Viruses 407
Fungi, Protozoans, Helminths, and Algae 408
PATTERN OF DISEASE 410
PORTALS OF EXIT 411
BOX *MMWR* FATAL DIPHTHERIA 407
STUDY OUTLINE 413
STUDY QUESTIONS 414
FURTHER READING 415

Chapter 15 Nonspecific Defenses of the Host 416

SKIN AND MUCOUS MEMBRANES 417
Mechanical Factors 417
Chemical Factors 419
PHAGOCYTOSIS 420
Formed Elements of the Blood 421
Kinds of Phagocytic Cells 421
Mechanism of Phagocytosis 424
INFLAMMATION 424
Vasodilation and Increased Permeability of Blood Vessels 425
Phagocyte Migration 426
Repair 427
FEVER 427
ANTIMICROBIAL SUBSTANCES 428
Interferon (IFN) 428
Complement and Properdin 429
FACTORS THAT LOWER RESISTANCE 432
STUDY OUTLINE 433
STUDY QUESTIONS 435
FURTHER READING 436

Chapter 16 Specific Defenses of the Host: The Immune Response 437

THE DUALITY OF THE IMMUNE SYSTEM 438
ACQUIRED IMMUNITY 438
Naturally Acquired Immunity 438
Artificially Acquired Immunity 438
ANTIGENS 440
ANTIBODIES 441
Structure of Antibodies 441
Classes of Immunoglobulins 441
MECHANISMS OF THE IMMUNE RESPONSE 444
B Cells and Humoral Immunity 445
Antibody Responses to Antigens 445
T Cells and Cell-Mediated Immunity 447
SEROLOGY 451
Precipitation Reactions 452
Agglutination Reactions 452
Complement Fixation Reactions 457
Neutralization Reactions 457
Immunofluorescence and Fluorescent Antibody Techniques 458
Radioimmunoassay 459
Enzyme-Linked Immunosorbent Assay (ELISA) 459
VACCINATIONS AND PUBLIC HEALTH 462
Vaccines 462
BOX *MICROBIOLOGY IN THE NEWS* NOBEL PRIZE-WINNING IMMUNOLOGY—THEORY AND PRACTICE 448
STUDY OUTLINE 465
STUDY QUESTIONS 468
FURTHER READING 470

Chapter 17 Harmful Aspects of the Immune Response 471

HYPERSENSITIVITY 472
Type I (Anaphylaxis) Reactions 472
Type II (Cytotoxic) Reactions 475

Type III (Immune Complex) Reactions 476
Type IV (Cell-Mediated) Reactions 477
AUTOIMMUNITY 478
Loss of Immunologic Tolerance 478
Major Histocompatibility Complex 479
Transplantation 480
Natural Immune Deficiencies 481
IMMUNE RESPONSE TO CANCER 481
Immunologic Surveillance 481
Immunologic Escape 484
Immunotherapy 485
TRANFUSION REACTIONS AND RH INCOMPATIBILITY 485
The ABO Blood Group System 486
The Rh Blood Group System 486
BOX *MICROBIOLOGY IN THE NEWS* ACQUIRED IMMUNE DEFICIENCY SYNDROME (AIDS) UPDATE: A MEDICAL DETECTION REPORT 482
STUDY OUTLINE 488
STUDY QUESTIONS 490
FURTHER READING 492

Chapter 18 Antimicrobial Drugs 493

HISTORICAL DEVELOPMENT 494
CRITERIA FOR ANTIMICROBIAL DRUGS 495
SPECTRUM OF ACTIVITY 497
ACTION OF ANTIMICROBIAL DRUGS 497
Inhibition of Cell Wall Synthesis 497
Inhibition of Protein Synthesis 499
Injury to the Cell Membrane 500
Inhibition of Nucleic Acid Synthesis 500
Inhibition of Enzymatic Activity 500
SURVEY OF COMMONLY USED ANTIMICROBIAL DRUGS 501
Synthetic Antimicrobial Drugs 501
Antibiotics 502
Antifungal Drugs 508
Antiviral Drugs 508
Antiprotozoan and Antihelminthic Drugs 509
TESTS FOR MICROBIAL SUSCEPTIBILITY TO CHEMOTHERAPEUTIC AGENTS 511
Disk Diffusion Method 512
Tube Dilution and Agar Dilution Tests 513
Automated Tests 514
DRUG RESISTANCE 514
BOX *MICROBIOLOGY IN THE NEWS* CONTROVERSY SURROUNDS THE USE OF ANTIBIOTICS IN FEED 496
BOX *MMWR* PENICILLIN-RESISTANT GONOCOCCI AND NATURAL SELECTION 515
STUDY OUTLINE 517
STUDY QUESTIONS 520
FURTHER READING 521

PART FOUR

MICROORGANISMS AND HUMAN DISEASE 523

Chapter 19 Microbial Diseases of the Skin and Eyes 524

STRUCTURE AND FUNCTION OF THE SKIN 525
NORMAL FLORA OF THE SKIN 525
BACTERIAL DISEASES OF THE SKIN 527
Staphylococcal Skin Infections 527
Streptococcal Skin Infections 529
Infections by Pseudomonads 530
Acne 531
VIRAL DISEASES OF THE SKIN 531
Warts 531
Smallpox (Variola) 532
Chickenpox (Varicella) and Shingles (Herpes Zoster) 533
Measles (Rubeola) 534
German Measles (Rubella) 535
Cold Sores (Herpes Simplex) 536
FUNGAL DISEASES OF THE SKIN 537
Superficial Mycoses 537
Cutaneous Mycoses 537
Subcutaneous Mycoses 537
Candidiasis 538
INFECTIONS OF THE EYE 539
Bacterial Infections 539
Viral Infections 542
BOX SCABIES—AN ITCHY INFECTION BY A PARASITIC ARTHROPOD 538
STUDY OUTLINE 542
STUDY QUESTIONS 545
FURTHER READING 547

Chapter 20 Microbial Diseases of the Nervous System 548

ORGANIZATION OF THE NERVOUS SYSTEM 549
MENINGITIS 550
Neisseria Meningitis (Meningococcal Meningitis) 551

Hemophilus influenzae Meningitis 552
Streptococcus pneumoniae Meningitis (Pneumococcal Meningitis) 553
Cryptococcus neoformans Meningitis (Cryptococcosis) 553
Tetanus 554
BOTULISM 555
LEPROSY 557
POLIOMYELITIS 558
RABIES 560
ARTHROPOD-BORNE ENCEPHALITIS 565
AFRICAN TRYPANOSOMIASIS 566
NAEGLERIA MICROENCEPHALITIS 567
SLOW VIRUS DISEASES 567
BOX *MMWR* SYSTEMIC ALLERGIC REACTIONS FOLLOWING IMMUNIZATION WITH HUMAN DIPLOID CELL RABIES VACCINE 563
STUDY OUTLINE 569
STUDY QUESTIONS 572
FURTHER READING 573

Chapter 21 Microbial Diseases of the Cardiovascular and Lymphatic Systems 574

STRUCTURE AND FUNCTION OF THE CARDIOVASCULAR SYSTEM 575
STRUCTURE AND FUNCTION OF THE LYMPHATIC SYSTEM 575
BACTERIAL DISEASES OF THE CARDIOVASCULAR AND LYMPHATIC SYSTEMS 577
Septicemia 577
Puerperal Sepsis 577
Bacterial Endocarditis 579
Rheumatic Fever 579
Tularemia 580
Brucellosis (Undulant Fever) 581
Anthrax 582
Listeriosis 584
Gangrene 584
SYSTEMIC DISEASES CAUSED BY ANIMAL BITES AND SCRATCHES 585
Plague 585
Relapsing Fever 587
Lyme Disease 587
RICKETTSIAL DISEASES OF THE CARDIOVASCULAR AND LYMPHATIC SYSTEMS 587
Typhus 587
Rocky Mountain Spotted Fever 589
VIRAL DISEASES OF THE CARDIOVASCULAR AND LYMPHATIC SYSTEMS 590
Myocarditis 590
Infectious Mononucleosis 590
Burkitt's Lymphoma 571
Yellow Fever 592
Dengue 592
Viral Hemorrhagic Fevers 592
PROTOZOAN AND HELMINTHIC DISEASES OF THE CARDIOVASCULAR AND LYMPHATIC SYSTEMS 593
Toxoplasmosis 593
American Trypanosomiasis (Chagas' Disease) 595
Malaria 595
Schistosomiasis 598
BOX *MMWR* BACTEREMIA AMONG AORTIC-VALVE SURGERY PATIENTS 578
STUDY OUTLINE 601
STUDY QUESTIONS 605
FURTHER READING 607

Chapter 22 Microbial Diseases of the Respiratory System 608

STRUCTURE AND FUNCTION OF THE RESPIRATORY SYSTEM 609
NORMAL FLORA OF THE RESPIRATORY SYSTEM 611
BACTERIAL DISEASES OF THE UPPER RESPIRATORY SYSTEM 611
Streptococcal Pharyngitis ("Strep" Throat) 612
Scarlet Fever 612
Diphtheria 613
Cutaneous Diphtheria 614
Otitis Media 614
VIRAL DISEASE OF THE UPPER RESPIRATORY SYSTEM 615
Common Cold (Coryza) 615
BACTERIAL DISEASES OF THE LOWER RESPIRATORY SYSTEM 615
Whooping Cough (Pertussis) 615
Tuberculosis 618
Bacterial Pneumonias 621
VIRAL DISEASES OF THE LOWER RESPIRATORY SYSTEM 625
Viral Pneumonia 625
Influenza ("Flu") 625
FUNGAL DISEASES OF THE LOWER RESPIRATORY SYSTEM 627
Histoplasmosis 627

Coccidioidomycosis 628
Blastomycosis (North American Blastomycosis) 629
Other Fungi Involved in Respiratory Disease 630
PROTOZOAN AND HELMINTHIC DISEASES OF THE LOWER RESPIRATORY SYSTEM 630
BOX *MICROBIOLOGY IN THE NEWS* A CURE FOR THE COMMON COLD? 616
STUDY OUTLINE 633
STUDY QUESTIONS 637
FURTHER READING 639

Chapter 23 Microbial Diseases of the Digestive System 640

STRUCTURE AND FUNCTION OF THE DIGESTIVE SYSTEM 641
ANTIMICROBIAL FEATURES OF THE DIGESTIVE SYSTEM 642
NORMAL FLORA OF THE DIGESTIVE SYSTEM 642
BACTERIAL DISEASES OF THE DIGESTIVE SYSTEM 643
Diseases of the Mouth 643
Diseases of the Lower Digestive System 646
VIRAL DISEASES OF THE DIGESTIVE SYSTEM 654
Mumps 654
Cytomegalovirus (CMV) Inclusion Disease 656
Hepatitis 657
Viral Gastroenteritis 660

MYCOTOXINS 660
PROTOZOAN DISEASES OF THE DIGESTIVE SYSTEM 660
Giardiasis 660
Balantidiasis (Balantidial Dysentery) 661
Amoebic Dysentery (Amoebiasis) 661
Cryptosporidiosis 662
HELMINTHIC DISEASES OF THE DIGESTIVE SYSTEM 662
Tapeworm Infestation 662
Trichinosis 665
BOX *MMWR* MILK-BORNE SALMONELLOSIS 649
STUDY OUTLINE 668
STUDY QUESTIONS 672
FURTHER READING 673

Chapter 24 Microbial Diseases of the Urinary and Genital Systems 674

STRUCTURE AND FUNCTION OF THE URINARY SYSTEM 675
STRUCTURE AND FUNCTION OF THE GENITAL SYSTEM 675
NORMAL FLORA OF THE URINARY AND GENITAL SYSTEMS 676
BACTERIAL DISEASES OF THE URINARY SYSTEM 677
Cystitis 678
Pyelonephritis 678
Leptospirosis 678
Glomerulonephritis 679
BACTERIAL DISEASES OF THE GENITAL SYSTEM 679
Gonorrhea 680
Syphilis 682
Nongonococcal Urethritis (NGU) 686
Gardnerella Vaginalis 686
Lymphogranuloma Venereum (LGV) 687
Chancroid (Soft Chancre) and Granuloma Inguinale 687
VIRAL DISEASE OF THE GENITAL SYSTEM 688
Genital Herpes 688
FUNGAL DISEASE OF THE GENITAL SYSTEM 689
Candidiasis 689
PROTOZOAN DISEASE OF THE GENITAL SYSTEM 692
Trichomoniasis 692
BOX *MICROBIOLOGY IN THE NEWS* NEW HOPE FOR TREATING HERPES 689
STUDY OUTLINE 693
STUDY QUESTIONS 695
FURTHER READING 697
FURTHER READING FOR PART FOUR 697

PART FIVE

MICROBIOLOGY, THE ENVIRONMENT, AND HUMAN AFFAIRS 699

Chapter 25 Soil and Water Microbiology 700

SOIL MICROBIOLOGY AND CYCLES OF THE ELEMENTS 701
THE COMPONENTS OF SOIL 701

Rock Particles and Minerals 701
Soil Water 701
Soil Gases 701
Organic Matter 701
Organisms 701
MICROORGANISMS AND BIOGEOCHEMICAL CYCLES 703
Carbon Cycle 703
Nitrogen Cycle 705
Other Biogeochemical Cycles 711
DEGRADATION OF PESTICIDES AND OTHER SYNTHETIC CHEMICALS 711
AQUATIC MICROBIOLOGY AND SEWAGE TREATMENT 711
FRESHWATER MICROBIAL FLORA 713
SEAWATER MICROBIAL FLORA 713
EFFECTS OF POLLUTION 714
Transmission of Infectious Diseases 714
Chemical Pollution 714
TESTS FOR WATER PURITY 716
WATER TREATMENT 718
SEWAGE TREATMENT 720
Primary Treatment 721
Biochemical Oxygen Demand 721
Secondary Treatment 722
Sludge Digestion 723
Septic Tanks 725
Oxidation Ponds 725
Tertiary Treatment 726
BOX *MICROBIOLOGY IN THE NEWS* WAGING A BACTERIAL WAR ON PESTS 704
STUDY OUTLINE 726
STUDY QUESTIONS 729
FURTHER READING 731

Chapter 26 Food and Industrial Microbiology 732

FOOD SPOILAGE AND PRESERVATION 733
SPOILAGE OF CANNED FOOD 735
ASEPTIC PACKAGING 736
LOW-TEMPERATURE PRESERVATION 736
RADIATION AND FOOD PRESERVATION 737
PASTEURIZATION 738
CHEMICAL PRESERVATIVES 738
FOOD-BORNE INFECTIONS AND MICROBIAL INTOXICATIONS 739
ROLE OF MICROORGANISMS IN FOOD PRODUCTION 740
Cheese 740
Other Dairy Products 741
Nondairy Fermentations 741
Alcoholic Beverages 741
FOOD FROM MICROORGANISMS 742
ALTERNATIVE ENERGY SOURCES FROM MICROORGANISMS 746
GENETIC ENGINEERING AND INDUSTRIAL MICROBIOLOGY 747
Genetic Engineering 747
Industrial Microbiology 749
BOX *MICROBIOLOGY IN THE NEWS* SINGLE-CELL PROTEIN MAY HELP EASE WORLD FOOD PROBLEM 744
STUDY OUTLINE 752
STUDY QUESTIONS 754
FURTHER READING 755

APPENDIX A: CLASSIFICATION OF BACTERIA ACCORDING TO *BERGEY'S MANUAL OF SYSTEMATIC BACTERIOLOGY* 756
APPENDIX B: PRONUNCIATION OF SCIENTIFIC NAMES 760
APPENDIX C: WORD ROOTS USED IN MICROBIOLOGY 764
APPENDIX D: MOST PROBABLE NUMBERS (MPN) TABLE 768
APPENDIX E: METHODS FOR TAKING CLINICAL SAMPLES 769
GLOSSARY 771
ACKNOWLEDGMENTS 775
INDEX 798
MICROVIEW 1: BACTERIAL STAINS AND MORPHOLOGY after 90
MICROVIEW 2: RECOMBINANT DNA AT WORK after 250
MICROVIEW 3: INSIDE A CLINICAL LAB after 522

SECOND EDITION

Microbiology

An Introduction

PART ONE

FUNDAMENTALS OF MICROBIOLOGY

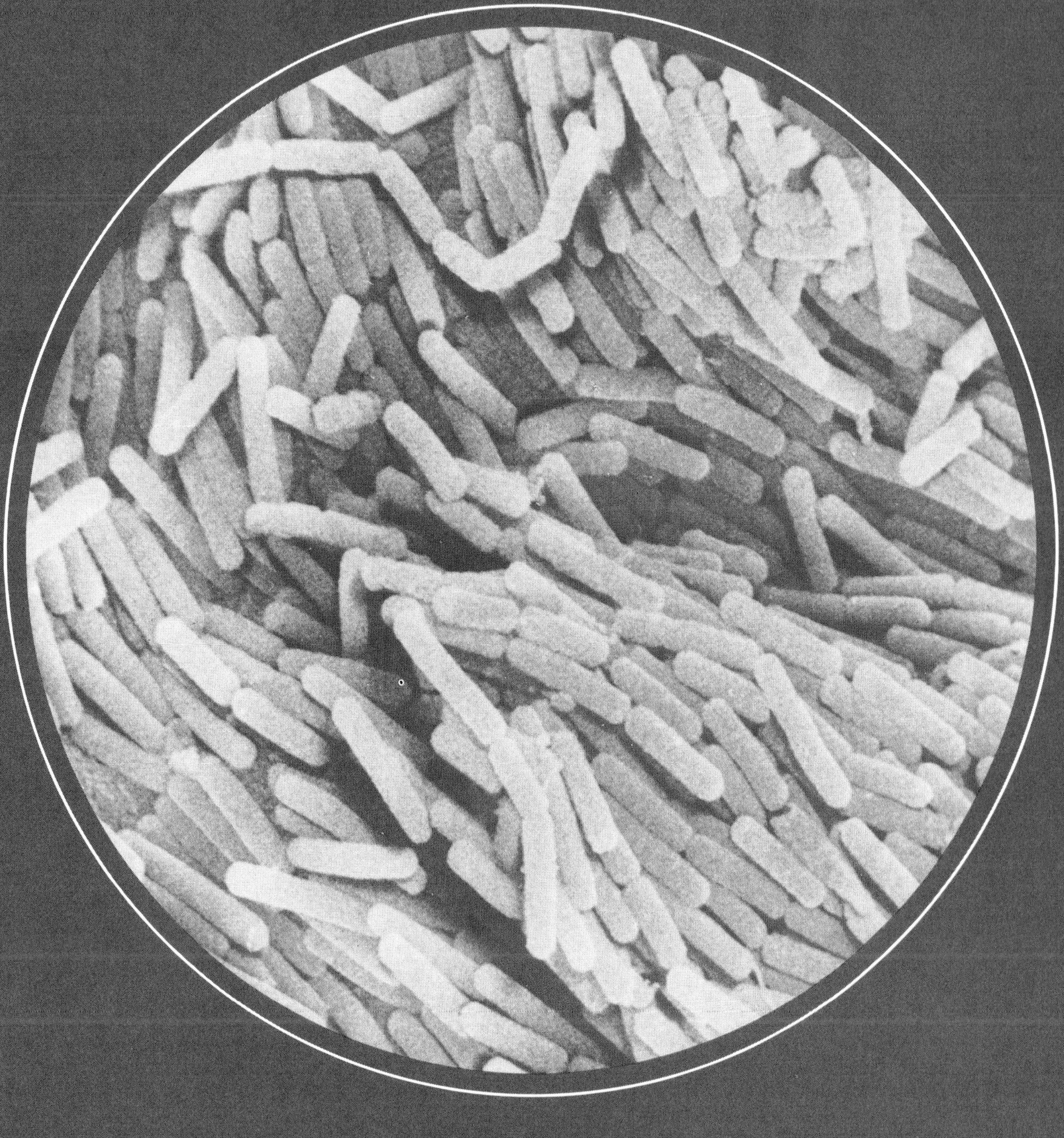

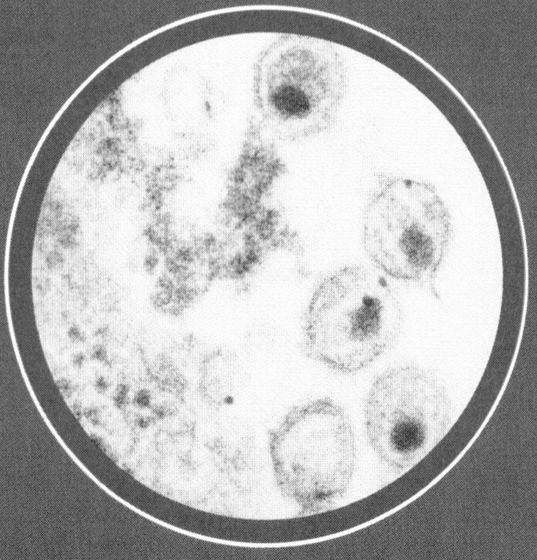

CHAPTER 1

The Microbial World and You

OBJECTIVES

After completing this chapter you should be able to

- Identify the contributions to microbiology made by
  Anton van Leeuwenhoek
  Robert Hooke
  Louis Pasteur
  Robert Koch
  Joseph Lister
  Paul Ehrlich
  Alexander Fleming
  Edward Jenner
- Compare the theories of spontaneous generation and biogenesis.
- Provide the rationale for the system of scientific names.
- List the major groups of organisms studied in microbiology.
- List at least four beneficial activities of microorganisms.
- Define normal flora.

One hot summer in 1976, 182 guests at an elegant Philadelphia hotel became mysteriously ill. Within a few weeks, 29 had died. All the victims suffered fever, coughing, and pneumonia, yet the hospital tests could not point to a microorganism or poison as the culprit. Thus, **Legionnaires' disease,** named after the American Legion Convention held at the hotel during the outbreak, became the subject of an intense investigation. Across the nation, scientists teamed up to find the source of the disease and how it was spread.

Modern microbiological techniques eventually helped scientists reveal that the disease was caused by an airborne microorganism—a tiny bacterium

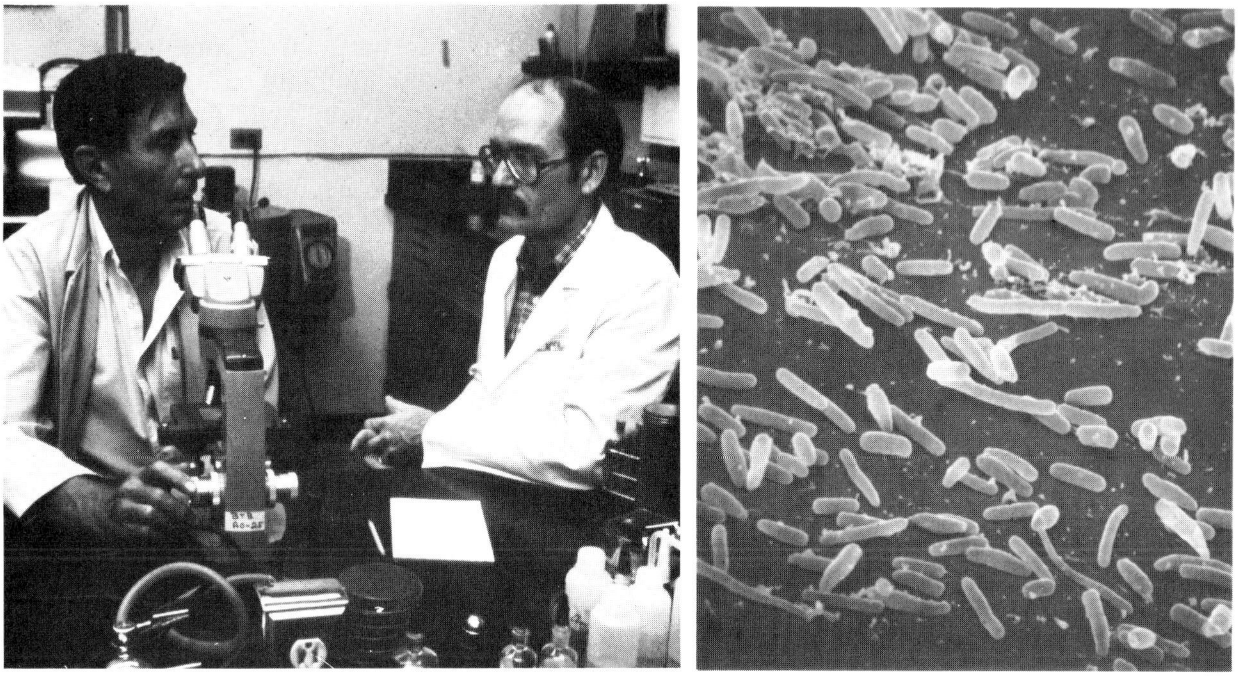

(a) (b)

Figure 1–1 **(a)** Joseph E. McDade and Charles C. Shepard, the microbiologists who first isolated *Legionella pneumophila* at the Centers for Disease Control in Atlanta, Georgia. **(b)** *Legionella pneumophila*, the bacterium that causes Legionnaires' disease.

not detectable by standard laboratory tests (Figure 1–1 and Microview 1–1C). Researchers now believe that guests exposed to the hotel's air conditioning system inhaled the bacterium, which had penetrated the system's ventilation filters. The bacterium was not new, but was unfamiliar to microbiologists. It had, in fact, caused several earlier unexplained epidemics of pneumonia.

After it had been carefully documented, the bacterium of Legionnaires' disease was assigned the scientific name *Legionella pneumophila* (lē-jä-nel'lä nü-mō' fi lä),* and the disease itself was later renamed **legionellosis.** About 10% of pneumonias previously classified as "viral" are actually caused by the *Legionella* bacterium. The drug erythromycin has subsequently successfully treated patients with legionellosis.

In the past decade, another previously unrecognized disease has caught the attention of the public—**toxic shock syndrome (TSS)**. TSS usually strikes otherwise healthy, young, menstruating females who use tampons, particularly the "super-absorbent" kind. The most frequent symptoms were often not immediately worrisome to the women affected: fever, sore throat, headache, fatigue, a sunburn-like rash, and peeling skin on the palms and soles. Untreated, however, the disease often progressed to more serious conditions, including respiratory and circulatory distress; in a few cases, patients died. Microbiologists and public health officials naturally focused attention on the problem and within months had shown that the disease is caused by toxins (poisonous substances) produced by the common bacterium *Staphylococcus aureus* (staf-i-lō-kok'kus ô'rē-us) (see Figure 10–1). In 1985, a Harvard research team found that certain tampon fibers soak up large amounts of magnesium normally found in the vagina. When the amount of magnesium in the vagina is reduced, *S. aureus* produces large amounts of toxin. The association of TSS with a particular brand of tampon resulted in the recall of that brand and the publication of cautions about the use of tampons in general.

Most recently, **acquired immune deficiency syndrome (AIDS)** has become a source of alarm to

* Pronunciations follow the first major mention of an organism within each part.

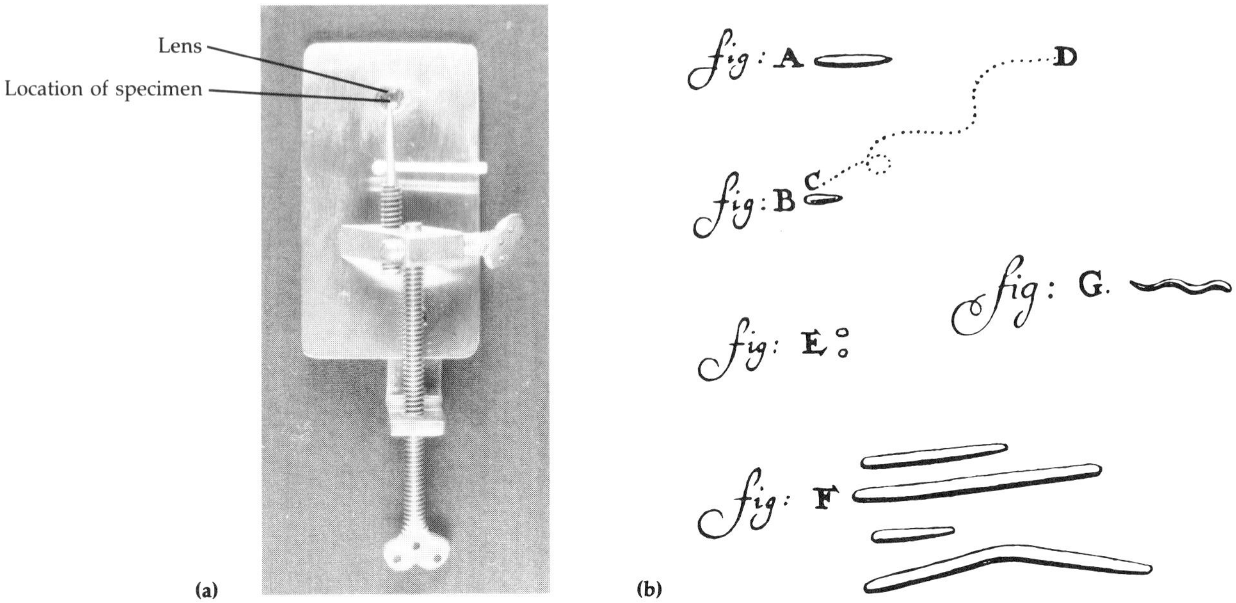

Figure 1–2 **(a)** A replica of a simple microscope made by Anton van Leeuwenhoek to observe living organisms too small to be seen with the naked eye. The specimen was placed on the tip of the adjustable point and viewed from the other side through the tiny round lens. The highest magnification possible with his lenses was about 300 × (times). **(b)** Some of Leeuwenhoek's drawings of bacteria, made in 1683. He was the first to see the microorganisms we now call bacteria and protozoans.

microbiologists and to the public at large. There is as yet no cure for this fatal disease. AIDS is the only known infectious disease in which the microbe responsible (a virus) acts only to lower the victim's immunity. Secondary, unrelated diseases produce the symptoms that result in death. Among these secondary diseases are a type of pneumonia and a form of cancer. AIDS is caused by a virus (HTLV-III) that is related to viruses that cause leukemia. Symptoms begin insidiously: a person suffering from AIDS might first experience a low-grade fever, shortness of breath, or muscle aches.

This book will show you how microbiologists use specific techniques and procedures to track down the agents that cause diseases such as legionellosis, TSS, and AIDS. You will also learn about the methods used to study microbes, the roles microbes play in the world around us, the body's responses to microbial infection, and the ways in which certain drugs combat microbial diseases.

MICROBES IN OUR LIVES

Most of us have had plenty of contact with the subject of this text—and not just from our last cold or bout with influenza. Any cheese or yogurt we eat has already been processed by the controlled activity of microorganisms.

For many people, words like *germ, microbe,* or *microorganism* bring to mind a group of tiny creatures that do not quite fit any of the categories in that old question, "Is it animal, vegetable, or mineral?" Microorganisms are minute living things, individually too small to be seen with the naked eye. The term includes bacteria, fungi (yeasts and molds), protozoans, and microscopic algae. It also includes viruses, those noncellular entities sometimes regarded as being at the border of life and nonlife.

We tend to associate these small organisms with uncomfortable infections, major diseases such as AIDS, or such common inconveniences as spoiled food. The truth, however, is that the majority of microorganisms make crucial contributions to the welfare of the world's inhabitants. Many act as miniature benefactors, helping to maintain the balance of living organisms and chemicals in our global environment. Marine and freshwater microorganisms form the basis of the food chain in oceans, lakes, and rivers. Soil microbes help to break down wastes and incorporate nitrogen gas from the air into organic compounds; chemical elements in the land and air are thus recycled. Certain bacteria and microscopic algae play important roles in photosynthesis, a food- and oxygen-generating process critical to our life on earth. Such microbes are essential to the survival of life on the planet.

Practical knowledge of microbiology is needed for the fields of medicine and the related health sciences. For example, hospital workers must be able to protect patients from bacteria that normally flourish around our homes and work places, but which are more dangerous to the sick and injured. Yet, an antiseptic world would be undesirable—even uncomfortable. We humans (and many other animals) depend on the bacteria in our intestines to synthesize some vitamins that the body requires.

Microorganisms provide not only natural aids to health, but also commercial benefits. Microbiological techniques are used every day by industrial chemists and technicians. Manufacturers can often culture (grow) microorganisms to be used in the synthesis of their products easier and cheaper than chemists can synthesize them. Acetone, glycerin, organic acids, enzymes, alcohols, and many drugs are produced this way. And our food industry frequently relies upon microbes to perform crucial steps in food processing. Microorganisms help produce vinegar, sauerkraut, pickles, alcoholic beverages, green olives, soy sauce, buttermilk, cheese, yogurt, and bread (see Box on p. 7).

Today, we casually accept the fact that microorganisms are found almost everywhere. Yet there was a time, not long ago, when microbes could not be seen because there were no microscopes, when cells were unknown to scientists, and when people believed that life sprang up spontaneously from nonliving matter. In the not-so-distant past, food spoilage often could not be controlled, and entire families died because vaccinations and antibiotics were not available to fight an infection.

Perhaps we can get an idea of how our current concepts of microbiology developed if we look at a few of the breakthroughs that have changed our lives.

A BRIEF HISTORY OF MICROBIOLOGY

The First Observations

Astronomers had been using telescopes to look at the stars for centuries before anyone thought to turn an optic lens upon a drop of water. The Dutch merchant and amateur scientist Anton van Leeuwenhoek was one of the first to observe microorganisms through magnifying lenses. In 1673, he wrote the first of a series of letters to the Royal Society of London describing the "animalcules" he saw through his simple, single-lens microscope. Leeuwenhoek's detailed drawings of "animalcules" in rain water, in peppercorn infusions, and in material taken from teeth scrapings have since been identified as representations of bacteria and protozoans (Figure 1–2).

While Leeuwenhoek was observing the microbial world, the Englishman Robert Hooke was using a microscope to observe thin slices of cork—which is composed of the walls of dead plant cells. Hooke called the pores between the walls "little boxes" or "cells." His discovery of this structure, reported in 1665, marked the beginning of a cell theory.

By 1838–39, two German biologists, botanist Matthias Schleiden and zoologist Theodor Schwann, had consolidated the current observations about cells in living matter. Their work clearly stated the theory that *all living things are composed of cells.* Subsequent investigations into the structure and functions of cells have been based on this cell theory, which is one of the most important generalizations of modern biology. We shall discuss the structure of cells in Chapter 4.

Fermentation

Because Louis Pasteur had a reputation as a solver of microbiological problems, a group of French merchants asked him to find out why wine and beer soured. They hoped to develop a method that would prevent the spoiling of those beverages shipped long distances. At the time, many scientists believed that the sugars in these fluids were

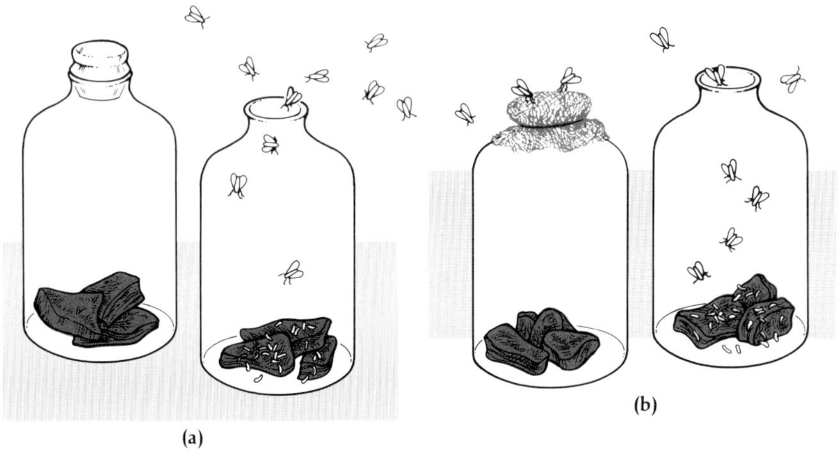

Figure 1–3 Redi's experiments to demonstrate that maggots do not arise spontaneously from decaying meat. **(a)** In his first experiment, Redi showed that maggots (fly larvae) would appear on uncovered meat but not on the meat in a sealed jar. **(b)** In his second experiment, Redi used gauze to cover half the jars, in order to allow the free entrance of air (but not flies). Again, the meat in the covered jars showed no maggots.

converted by air into alcohol. By 1860, Pasteur found instead that microorganisms called yeasts convert the sugars to alcohol in the absence of air. This process is called **fermentation** and is used to make wine or beer. Souring and spoiling, which occur later, are caused by a different group of microorganisms, called bacteria. In the presence of air, bacteria change the alcoholic beverage into a sour waste product known as acetic acid (vinegar).

Pasteur's solution to spoilage was to heat the alcohol just enough to kill most of the bacteria; this process does not greatly affect the flavor of wine or beer. We call this process **pasteurization,** and it is used for milk as well as for some alcoholic drinks to kill bacteria.

The Debate over Spontaneous Generation

After van Leeuwenhoek discovered the "invisible" world of microorganisms, the scientific community became interested in the origins of these tiny, living things. Until the second half of the nineteenth century, it was generally believed that some forms of life could arise spontaneously from nonliving matter; this process was known as **spontaneous generation.** People thought that toads, snakes, and mice could be born of moist soil, that flies could emerge from manure, and that maggots could arise from decaying corpses.

Experiments pro and con

A strong opponent of spontaneous generation, the Italian physician Francesco Redi, set out in 1668 to demonstrate that maggots, the larvae of flies, do not arise spontaneously from decaying meat, as was commonly believed. Redi filled three jars with decaying meat and sealed them tightly. Then he similarly arranged three other jars and left them open. Maggots appeared in the open vessels after flies entered the jars and laid their eggs. But the sealed containers showed no forms of life (Figure 1–3a). Still, his antagonists were not convinced, but claimed that fresh air was needed for spontaneous generation. So Redi set up a second experiment in which three jars were covered with a fine net instead of being sealed. No worms appeared in the gauze-covered jars, even though air was present. Maggots appeared only if flies were allowed to leave their eggs on the meat (Figure 1–3b).

Redi's results were a serious blow to the long-held belief that large forms of life could arise from nonlife. However, many scientists still believed that small organisms such as Leeuwenhoek's "animalcules" were simple enough to be generated by nonliving materials.

What Makes Sourdough Bread Different?

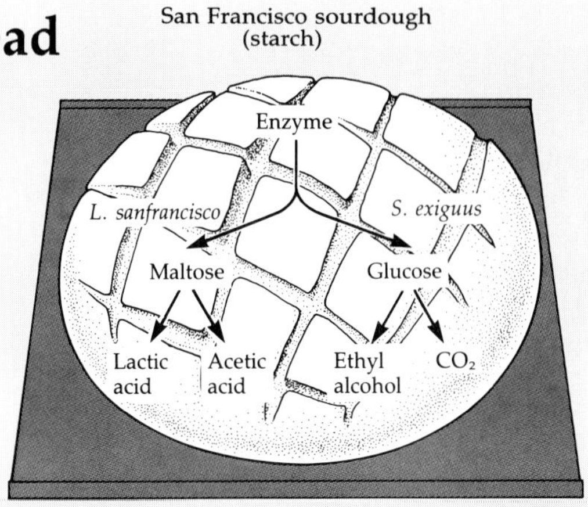

Imagine being a miner during the California gold rush. You've just made bread dough from your last supplies of flour and salt when someone yells, "Gold!" Temporarily forgetting your hunger, you run off to the gold fields. Many hours later you return. Before cooking the dough, you save a spoonful as a "starter" for your next batch of bread. The dough has been rising longer than usual, but you are too cold, tired, and hungry to start a new batch. Later you find that your bread tastes different than previous batches; it is slightly sour. During the gold rush, miners baked so many sour loaves that they were nicknamed "sourdoughs." Perhaps some of the sour batches of bread tasted like San Francisco's famous sourdough bread of today.

Conventional bread is made from flour, water, sugar, salt, shortening, and a living microbe, yeast. The yeast belongs to the kingdom Fungi and is named *Saccharomyces cerevisiae.* When flour is mixed with water, an enzyme in the flour breaks its starch into two sugars, maltose and glucose. After the ingredients for the bread are mixed, the yeast metabolizes the sugars and produces ethanol and carbon dioxide as waste products. This metabolic process is called fermentation. The dough rises as carbon dioxide bubbles get trapped in the sticky matrix. The alcohol evaporates during baking. It and the carbon dioxide gas form spaces that remain in the bread.

Originally, breads were leavened by wild yeast from the air, which had been trapped in the dough. Later, bakers kept a starter culture of yeast—dough from the last batch of bread—to leaven each new batch of dough. Sourdough bread is made with a special sourdough starter culture that is added to flour, water, and salt. Perhaps the most famous sourdough bread made today comes from San Francisco, where a handful of bakeries have continuously cultivated their starters for more than 100 years, and meticulously maintained the starters to keep out unwanted microbes that can produce different and undesirable flavors. After bakeries in other areas made several unsuccessful attempts to match the unique flavor of San Francisco sourdough, rumors arose of a unique local climate or contamination from bakery walls. T. F. Sugihara and L. Kline from the United States Department of Agriculture set out to debunk these rumors and to determine the microbiological reason for the breads' different taste so that it could be made in other areas.

The U.S.D.A. workers found that sourdough is eight to ten times more acidic than conventional bread because of the presence of lactic and acetic acids. These acids account for the sour flavor of the bread. The workers isolated and identified the yeast in the starter: *Saccharomyces exiguus,* a unique yeast that does not ferment maltose and thrives in the acidic environment of this dough. Because the yeast did not produce the acids and did not use maltose, Sugihara and Kline searched the starter for a bacterium capable of fermenting maltose and producing the acids. The bacterium, so carefully guarded all those years, was isolated and was classified into the genus *Lactobacillus.* Many members of this genus are used in dairy fermentations and are part of the normal flora of humans and other mammals.

Analyses of the cell structure and genetic composition of this bacterium showed that it is genetically different from other previously characterized lactobacilli. It has been given the name *Lactobacillus sanfrancisco.*

The case for spontaneous generation of microorganisms was strengthened in 1745 when John Needham found that even after he heated nutrient fluids (chicken broth and corn infusions) before pouring them into covered flasks, the cooled solutions were soon teeming with microorganisms. Needham claimed that the microbes developed spontaneously from the fluids. Twenty years later, Lazzaro Spallanzani suggested that microorganisms from the air probably had entered Needham's solutions after they were boiled, and showed that nutrient fluids heated *after* being sealed in a flask did not develop microbial growth. Needham responded by claiming that the "vital force" necessary for spontaneous generation had been destroyed by the heat, and kept out of the flasks by the seals.

The theory of biogenesis

The issue was still unresolved in 1858, when the German scientist Rudolf Virchow challenged spontaneous generation with the concept of **biogenesis.** An extension of the cell theory, biogenesis is the claim that living cells can arise only from preexisting living cells. But arguments about spontaneous generation continued for a few years before the issue was resolved experimentally.

Pasteur settles the argument

In 1861, Pasteur designed a series of ingenious and persuasive experiments that finally ended the debate. He demonstrated that microorganisms are indeed present in the air and that they can contaminate seemingly sterile solutions, but that air itself does not give rise to microbial life.

Pasteur began these experiments by filling several short-necked flasks with beef broth and boiling them. Some were then left open and allowed to cool. In a few days, these flasks were found to be contaminated with microbes. The other flasks, sealed after boiling, remained free of microorganisms. From these results, Pasteur reasoned that microbes in the air were the agents responsible for contaminating nonliving matter such as the broths in Needham's flasks.

Pasteur's next step was to place the broth in long-necked flasks; he then bent the necks into S-shaped curves (Figure 1–4). The contents of these flasks were then boiled and cooled. The broth in the flasks did not decay, and showed no signs of life after days, weeks, and even months. (Some of these original vessels are on display at the Pasteur Institute in Paris. They still show no sign of contamination, more than 100 years later.) Pasteur's unique design allowed air to pass into the flask, but the neck trapped any airborne microorganisms that might contaminate the broth.

Pasteur showed that microorganisms can be present in nonliving matter—on solids, in liquids, and in the air. Further, he demonstrated conclusively that microbial life can be destroyed by heat and that methods can be devised to block the access of airborne microorganisms to nutrient environments. These discoveries form the basis of the **aseptic techniques** (techniques to prevent contamination by unwanted microorganisms) that are now standard practice in laboratory and many medical procedures. Modern aseptic techniques are among the first and most important things that a beginning microbiologist learns.

Pasteur's work provided evidence that microorganisms cannot originate from mystical forces present in nonliving materials. Rather, any appearance of "spontaneous" life in nonliving solutions can be attributed to microorganisms that were already present in the air or in the fluids themselves. (Keep in mind that although a form of spontaneous generation probably did occur on primitive earth when life first began, scientists now agree that this does not happen under our present environmental conditions.)

The Germ Theory of Disease

The realization that yeasts play a crucial role in fermentation was the first concept to link a microorganism's activity to physical and chemical changes in organic materials. This discovery alerted scientists to the possibility that microorganisms might have similar relationships with plants and animals—specifically that microorganisms might cause disease. This idea was called the **germ theory of disease.**

The germ theory was a difficult concept for many to accept because for centuries people had believed that disease was punishment for an individual's crimes or misdeeds. When the inhabitants of an entire village became ill, foul odors from sewage or poisonous vapors from swamps were often blamed for the sickness. Most people born in Pasteur's time found it inconceivable that "invisible" microbes could travel through the air to infect plants and animals, or remain on clothing and bedding to be transmitted from one person to another. But gradually, scientists accumulated the information needed to support the unorthodox germ theory.

In 1836, Agostino Bassi, an amateur microscopist, made the first association between a microorganism and a disease when he proved that a silkworm disease was caused by a fungus. In 1865, Pasteur was called upon to help stop another silkworm disease, which was ruining the silk industry throughout Europe. Pasteur found that the second infection was caused by a protozoan, and he developed a method for recognizing afflicted silkworm moths.

In the 1860s, Joseph Lister, an English surgeon, applied the germ theory to medical issues. Disinfectants were unknown at the time, but Lister had heard of Pasteur's work connecting microbes to animal diseases. Lister knew that carbolic acid (now called phenol) kills bacteria, so he began soaking surgical dressings in a mild solution of it. The practice so reduced the incidence of infections and deaths that other surgeons quickly adopted it. Lister's technique was one of the earliest medical attempts to control the infections caused by microorganisms.

The first proof that bacteria actually cause disease was given by Robert Koch in 1876, just a century ago. Koch, a brilliant German physician, was

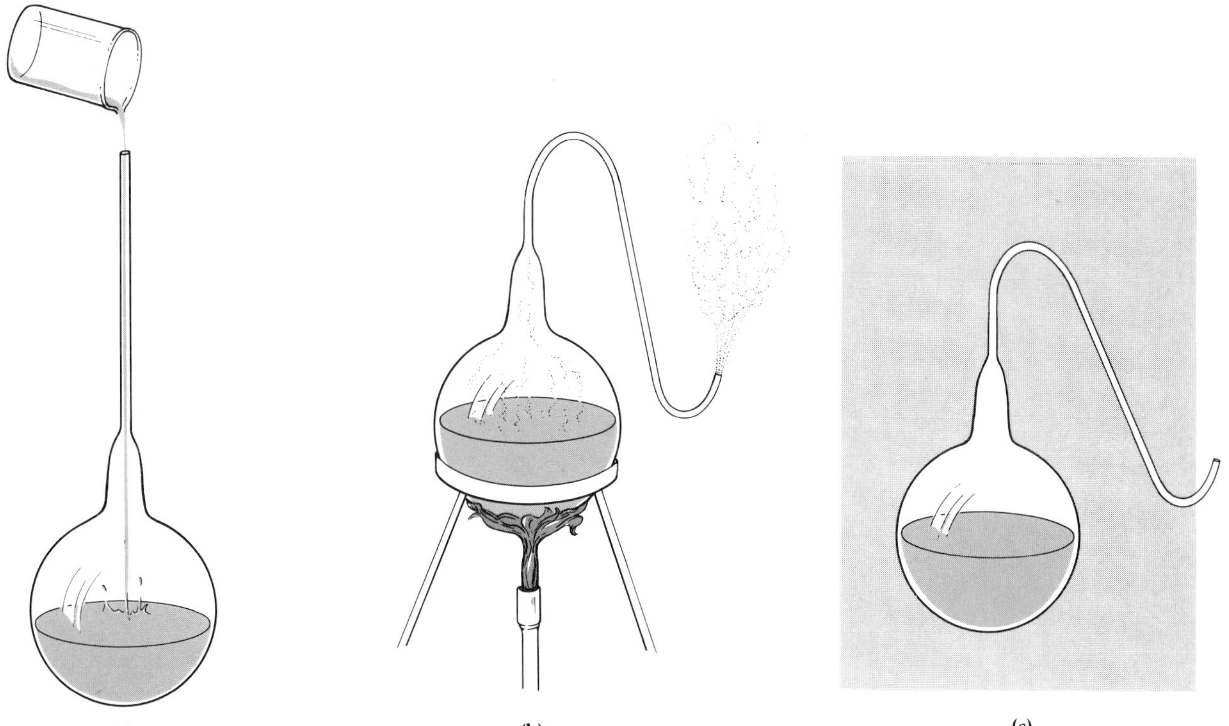

(a) (b) (c)

Figure 1–4 Pasteur's experiment that disproved the theory of spontaneous generation. **(a)** Pasteur first poured nutrient broth into a long-necked flask. **(b)** Next he heated the neck of the flask and bent it into an S-shaped curve; then he boiled the solution for several minutes. **(c)** Microorganisms did not appear in the cooled broth, even after long periods of time.

Pasteur's young rival in the race to discover the cause of anthrax, a disease that was destroying cattle and sheep in Europe.

Koch discovered rod-shaped bacteria in the blood of cattle that had died of anthrax. He cultured the bacteria on artificial media and then injected samples of the culture into healthy animals. When these animals became sick and died, Koch isolated the bacteria in their blood and compared them to the bacteria originally isolated. He found that the two sets of blood cultures contained the same bacteria.

Koch thus established a sequence of experimental steps for directly relating a specific microbe to a specific disease. These steps are actually a set of criteria, known today as **Koch's Postulates.** Over the past hundred years, these same criteria have been invaluable in investigations proving that specific microorganisms cause many diseases. Koch's Postulates, their limitations, and their application to Legionnaires' disease will be discussed in greater detail in Chapter 13.

Vaccination

Often a treatment or preventive procedure is developed before scientists know why it works. The smallpox vaccine is an example of preventive medicine jumping ahead of research. In 1798, almost 70 years before Koch established that a specific microorganism causes anthrax, Edward Jenner, a daring young British physician, embarked on an experiment to find a protection from smallpox.

Smallpox epidemics were greatly feared. The disease periodically swept through Europe killing thousands, and it wiped out 90% of the American Indians on the East Coast when European settlers first brought the infection to the New World. The disease usually starts with the appearance of small red spots (pox) all over the body. The pox become hard pimples that break down into blisters filled with pus.

When a young milkgirl told Jenner that she couldn't get smallpox because she already had been sick from cowpox—a much milder disease—he decided to put the girl's story to a test. First Jenner collected scrapings from cowpox blisters. Then he made inoculations with the cowpox material by scratching the patient's arm with a pox-infected needle. The scratch would turn into a raised bump, in a few days the patient would become mildly sick and recover, and would then never contract either cowpox or smallpox again. The process was called **vaccination** from the Latin word "vaca" for cow. The protection from disease provided by vaccination (or by recovery from the disease itself) is called **immunity.** We will discuss the mechanisms of immunity in Chapter 16.

We now know that cowpox and smallpox are caused by viruses. Thus, Jenner's experiment was the first time in a Western culture that a living viral agent—the cowpox virus—was used to produce immunity. In ancient times, the Chinese and the Indians rubbed material from smallpox victims into the skin of healthy persons, or breathed it into their nostrils, attempting to contract a mild form of the disease to protect themselves. But this technique would sometimes backfire by initiating an epidemic. The Royal Society of London had warned Jenner against risking his reputation with his experiment, but fortunately, Jenner's use of cowpox virus produced favorable results. Because the cowpox virus closely resembles the smallpox virus, it induces immunity without giving humans a deadly infection.

Years later, around 1880, Pasteur discovered why vaccinations work. He found that the bacterium that causes a disease called chicken cholera lost its ability to cause disease (lost its *virulence,* or became avirulent) after it was grown in the laboratory for long periods. He discovered that microorganisms with decreased virulence still retained their ability to induce immunity against subsequent infections by their virulent counterparts. The discovery of this phenomenon provided a clue to Jenner's successful use of cowpox virus. Even though it is not a laboratory-produced derivative of smallpox virus, cowpox virus can induce immunity to both viruses because it is so closely related to the smallpox virus. Pasteur called cultures of avirulent microorganisms used for preventive inoculation **vaccines.**

Some vaccines are still produced from avirulent microbial strains that stimulate immunity to the related virulent strain. Other vaccines are made from killed virulent cells or from isolated components of virulent microorganisms. Vaccines are now available to protect us against such diseases as polio, whooping cough, measles, mumps, and rubella. Researchers have recently applied recombinant DNA techniques (to be discussed in Chapter 8) to the development of vaccines against

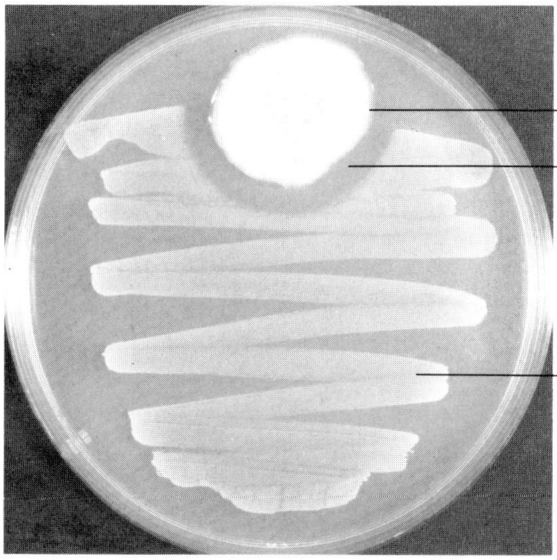

Figure 1–5 Inhibition of bacterial growth around the mold *Penicillium chrysogenum*. Colonies of the bacteria *Staphylococcus aureus* do not grow in the vicinity of the contaminating *Penicillium* colony near the top of the photo.

a variety of viruses, including those that cause herpes, hepatitis B, and influenza.

The Birth of Modern Chemotherapy: Dreams of a "Magic Bullet"

After the relationship between microorganisms and disease was established, the next major focus for medical microbiologists was the search for substances that could destroy **pathogenic** (disease-causing) microorganisms without damaging the infected animal or human. The treatment of disease by chemical substances is called **chemotherapy.** Chemotherapeutic agents prepared from chemicals in the laboratory are called **synthetic drugs.** Agents produced naturally by bacteria and fungi are called **antibiotics.** The success of chemotherapy is based on the fact that some chemicals are more poisonous to microorganisms than to their hosts.

The first synthetic drugs

Paul Ehrlich, A German physician, is recognized as the imaginative thinker who fired the first shot in the chemotherapy revolution. As a medical student, Ehrlich speculated about a "magic bullet" that could hunt down and destroy a pathogen without harming the infected host. Ehrlich launched a search for such a bullet, and in 1910 he found a chemotherapeutic agent called *salvarsan,* an arsenic derivative effective against syphilis. Prior to this discovery, the only known chemical in Europe's medical arsenal was an extract from South American tree bark, *quinine,* which had been used by Spanish conquistadors to treat malaria.

By the late 1930s, researchers had developed several other synthetic drugs that could kill microorganisms. Most of these drugs were derivatives of dyes. The *sulfa drugs* were one important group discovered at that time.

A fortunate accident—antibiotics

In contrast to the sulfa drugs, which were deliberately developed from a series of industrial chemicals, the first antibiotic was discovered by accident. When culturing bacteria, Alexander Fleming almost tossed out some culture plates that had been contaminated by mold. Fortunately, he took a second look at the curious pattern of growth on the contaminated plates. There was a clear area around the mold, where the bacterial culture had stopped growing (Figure 1–5). Fleming, a Scottish physician and bacteriologist, was looking at a mold that could inhibit the growth of a bacterium. The mold was later identified as *Penicillium notatum* (pe-ni-sil'lē-um nō-tä'-tum), and in 1928, Fleming named its active inhibitor *penicillin.*

Penicillin is an antibiotic produced by a fungus. The enormous usefulness of penicillin was not apparent until the 1940s, when it was finally mass-produced and tested clinically. Since then, many other antibiotics have been found.

Unfortunately, antibiotics and other antimicrobial drugs are not without problems. Many antimicrobial chemicals are too toxic to humans for practical use; they kill the pathogenic microbe, but also have damaging side effects on the infected host. Toxicity for humans is a particular problem in the development of drugs for treating viral diseases, for reasons we shall discuss later; there are as yet very few successful antiviral drugs.

The quest for better targeting

Antimicrobial drugs are sent to a diseased body by a variety of delivery systems: pills, salves, syrups, and syringes. A major drawback of current chemotherapy is that most such systems provide only general deliveries. Rarely can special messengers be directed to a particular, infected site; the exceptions are messengers to some localized infections.

Microbiologists, pharmacologists, and clinicians are now working on a variety of techniques to target antimicrobial drugs. The idea is to attach a drug molecule to a so-called bullet, which then travels like a homing device to its target, a specific site of infected cells. The bullet then releases the drug, which "attacks" without hurting nearby normal cells. Such a method would permit the use of drugs now regarded as too toxic for general use.

Thus, the search for a better-targeted magic bullet continues—perhaps with more success soon to come. Advances in cell biology and immunology have provided researchers with information about several instruments that eventually might turn chemotherapy into a more potent medical weapon. Future bullets might include the "ghosts," or membranes, of red blood cells; a group of highly specific molecules called *monoclonal antibodies* (see Chapter 16); *immunotoxins,* monoclonal antibodies attached to toxins; and artificial fatty globules known as *liposomes.*

The inadequate targeting of drugs and the lack of antiviral drugs are only two limitations of modern chemotherapy. An equally important problem is the emergence and spread of new varieties of microorganisms that are resistant to antibiotics. The quest to solve these problems requires sophisticated research techniques and correlated studies never dreamed of in the days of Koch and Pasteur. But before we continue our discussion of medical techniques that control microbes, we need to know more about the microbes themselves. For example, how do microorganisms fit into the scheme of life? The next section is a short introduction to this important aspect of microbiology.

NAMING AND CLASSIFYING MICROORGANISMS

We name the things around us in order to communicate more easily. The common names we assign are often limited to a geographical area. You might call one type of lumber tree, for example, by one of its more than twenty common names, depending on where you live in the United States. Sometimes, one common name refers to two different things. For example, "crown-of-thorns" is the name of both a plant and an animal. To avoid such confusion, scientists have developed a standardized system of scientific names.

The system of naming (nomenclature) that we use was established in 1735 by Carolus Linnaeus, whose name is the latinized form of Carl von Linné. Scientific names are latinized because Latin was the language traditionally used by scholars. Scientific nomenclature assigns each organism two names: the **genus** is the first name and is always capitalized; the **specific epithet** follows and is not capitalized. The species is called by both the genus and specific epithet and is underlined or italicized.

Scientific names can, among other things, describe the organism, honor a researcher, or identify the habitat of the species. For example, a bacterium commonly found on the skin of humans is *Staphylococcus aureus.* "Staphylo-" describes the clustered arrangement of the cells; "coccus" indicates that they are shaped like spheres. The specific epithet, "aureus," is Latin for "golden," the color of this bacterium. As you recall, *Legionella pneumophila* was named after the famous epidemic that first brought it to the attention of physicians and researchers. The genus of the bacterium *Escherichia coli* (esh-ėr-i'kē-ä kō'lē) is named for a scientist, whereas its specific epithet, "coli," reminds us that *E. coli* live in the colon, or large intestine. Notice that after a scientific name has been mentioned once, it can be abbreviated with the initial of the genus followed by the specific epithet.

Similar species are included in the same genus; similar genera are placed in a **family;** related families are put in an **order;** related orders are in a **class;** related classes are part of a **phylum**; and related phyla constitute a **kingdom.**

Before the existence of microbes was known, all organisms were grouped into either the animal kingdom or the plant kingdom. That system of classification was developed by Aristotle during the fourth century B.C. When microscopic organisms with mixed characteristics were discovered late in the seventeenth century, a new system of classification was needed. In 1866, Ernst Heinrich Haeckel, a German zoologist, proposed a classification system with three kingdoms: Animalia, Plantae, and Protista. The Protista consisted of single-celled microorganisms such as bacteria, some fungi, algae, and protozoa.

In 1969, H. R. Whittaker of Cornell University devised a five-kingdom classification system that is now widely accepted. It groups all organisms into the following kingdoms: Procaryotae or Monera (bacteria and cyanobacteria), Protista (funguslike slime molds, animallike protozoans, and some algae), Fungi (unicellular yeasts, multicellular molds, and macroscopic mushrooms), Plantae (some algae, and all mosses, ferns, conifers, and flowering plants), and Animalia (such as sponges, worms, insects, and vertebrates). Noncellular infectious agents including viruses, viroids, and prions are not placed in any kingdom at the present time. In fact, it appears that such organisms may be more closely related to their host cells than to each other. As researchers continue to work on these groups and acquire more information, a classification scheme that includes noncellular agents may be adopted. The Whittaker five–kingdom classification system is based on the cellular organization and nutritional patterns of organisms; it will be discussed in more detail in Part Two. For now, however, let us simply review the major groups of microorganisms covered in this text. These descriptions are only a brief overview; details and explanations will follow in later chapters.

THE DIVERSITY OF MICROORGANISMS

Bacteria

Bacteria are very small, relatively simple, single-celled organisms whose genetic material is not enclosed in a special nuclear membrane (Figure 1–6a). For this reason, bacteria are called **procaryotes,** from the Greek words meaning "pre-nucleus." Bacteria make up the kingdom Whittaker calls Monera; others, including microbiologists, call it Procaryotae. Criteria for classifying bacteria will be discussed in Part Two.

Bacteria cells generally have one of three shapes: **bacillus** (rodlike), **coccus** (spherical or ovoid), and **spirillum** (spiral or corkscrew) (see Figures 4–1, 4–2, and 4–3). (There is also a genus *Bacillus,* but it includes only some of the rod-shaped bacteria.) Individual bacteria may form pairs, chains, clusters, or other groupings; such formations are usually the same within a particular species.

Bacteria are enclosed in cell walls that are largely composed of a substance called *peptidoglycan.* (Cellulose is the main substance of plant cell walls.) Because of differences in the components of their cell walls, different bacteria show characteristic reactions to a variety of stains (dyes). Bacteria generally reproduce by dividing into two equal daughter cells; this process is called *binary fission.* For nutrition, most bacteria use organic material, which in nature can be derived from other, dead organisms or from a living host. Some bacteria can carry out photosynthesis, and some can derive nutrition from inorganic substances. Many bacteria can "swim" by means of moving propellerlike appendages called *flagella.*

Fungi

Fungi are **eucaryotes**—organisms whose cells have a distinct nucleus, which contains the cell's genetic material and which is surrounded by a special envelope called the nuclear membrane. They make up the kingdom Fungi. Fungi may be unicellular or multicellular. Large, multicellular fungi, such as mushrooms, may look somewhat like plants, but they cannot carry out photosynthesis, as most plants can. True fungi have cell walls composed primarily of a substance called *chitin.* The unicellular forms of fungi, *yeasts,* are ovoid microorganisms larger than bacteria. The most typical fungi are *molds.* Molds form *mycelia,* which are long filaments that branch and intertwine (Figure 1–6b). The cottony growths sometimes found on bread and fruit are mold mycelia. The crosswalls separating the nuclei in the mycelia of most molds are not complete, making these fungi exceptions to the strict version of the cell theory; in a sense, such a mycelium is one giant, multinucleate cell.

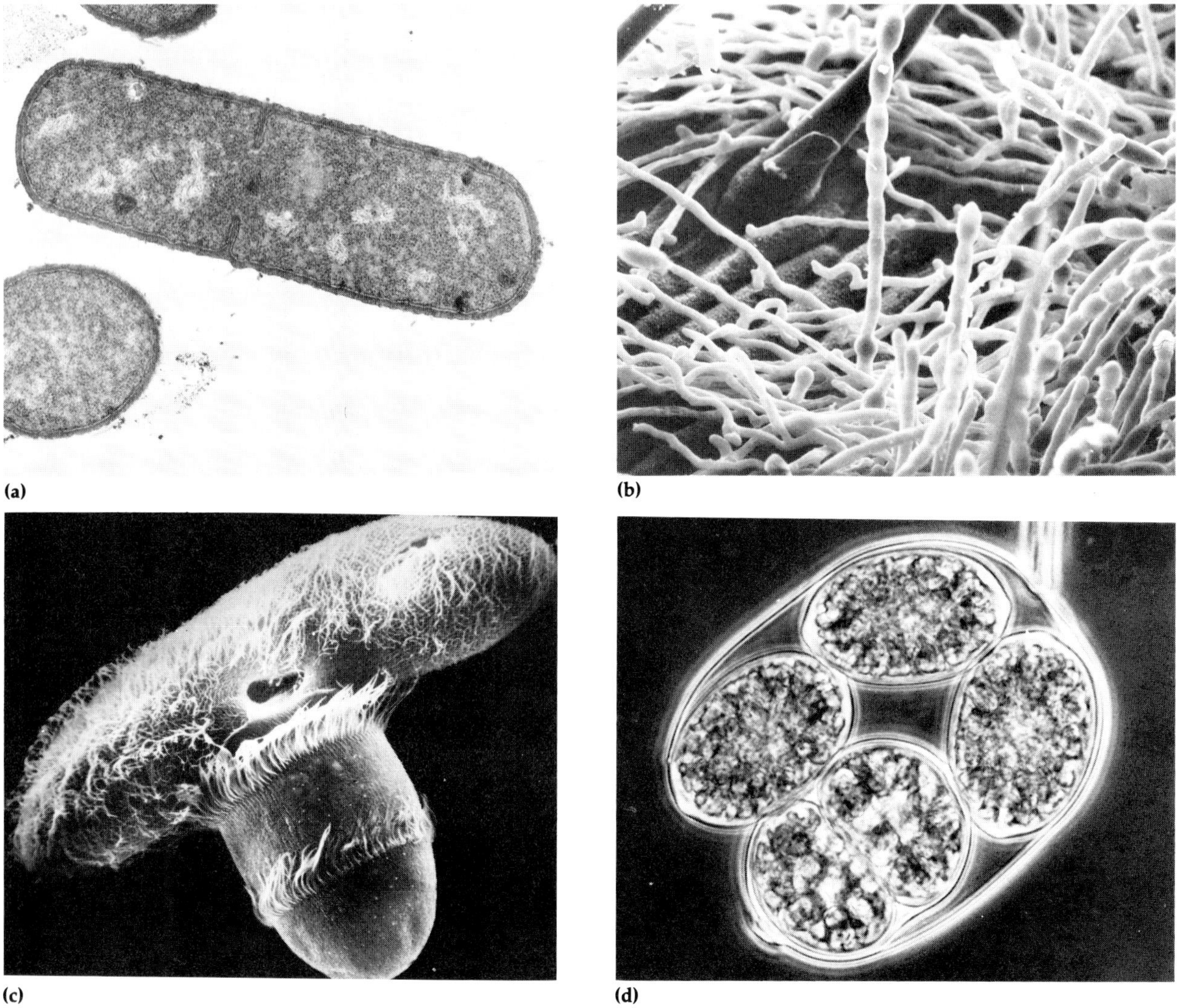

Figure 1–6 Types of microorganisms. **(a)** The rod-shaped bacterium *Bacillus cereus* (×22,000). This cell is in the process of dividing. **(b)** Mycelia from the fungus *Erysiphe grammis*, infecting the surface of a grass plant (×700). **(c)** The protozoan *Didinium nasutum* (lower, bullet-shaped cell) eating one of its fellow protozoans, *Paramecium multimicronucleatum* (×500). **(d)** The freshwater green alga *Oocystis* (×840). **(e)** Several T2 bacteriophages, viruses that attack bacteria.

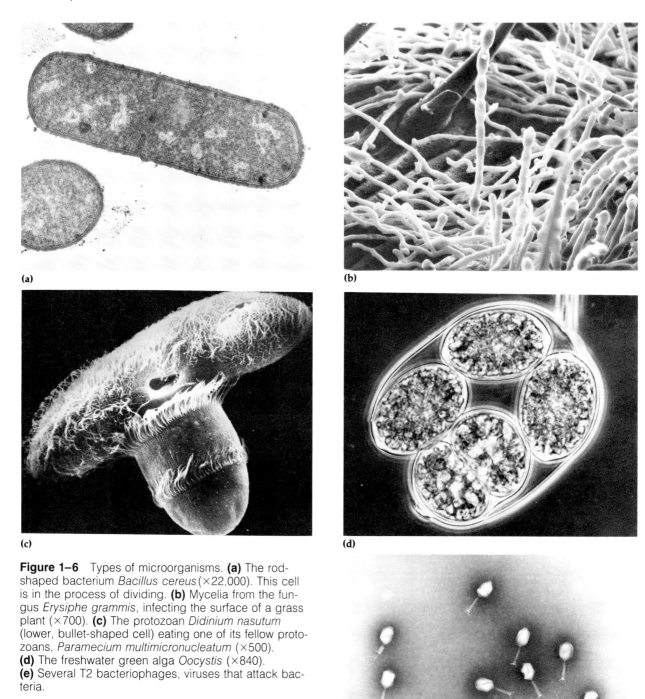

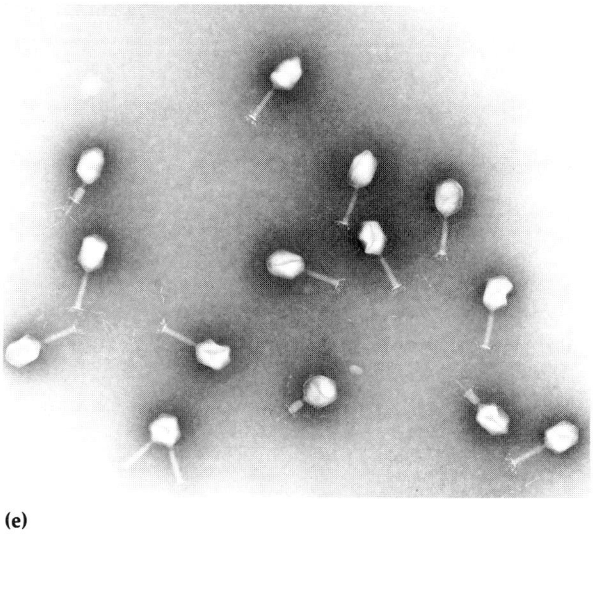

Fungi can reproduce sexually or asexually. They obtain nourishment by absorbing solutions of organic material from their environment—soil, water, or an animal or plant host. The classification of fungi is also discussed in Part Two.

Protozoans

Protozoans are unicellular, eucaryotic microbes that belong to the kingdom Protista (Figure 1–6c). Protozoans are classified according to their means of locomotion. Amoebas move using extensions of their cytoplasm called *pseudopods* ("false feet"). Other protozoans have flagella or numerous, shorter appendages called *cilia.* Protozoans have a flexible outer covering called a *pellicle,* rather than a rigid cell wall. They have a variety of shapes, and live as free entities or parasites, who absorb or ingest organic compounds from their environment. Often they have a number of tiny, specialized, subcellular structures that perform separate functions analogous to those of animal and plant organs. *Paramecium* (par-ä-mē'sē-um) is a protozoan with cilia, separate nuclei for growth regulation and sexual reproduction, an *oral groove* for the ingestion of food, and an *anal pore* for the excretion of wastes.

Algae

Algae are photosynthetic eucaryotes with a wide variety of shapes and both sexual and asexual reproductive forms (Figure 1–6d). The algae of interest to microbiologists are members of the kingdom Protista and are usually unicellular. Algae are classified according to the biochemical characteristics of their pigments, storage products, cell walls, and by the kind of flagella they have. They are abundant in fresh and salt water, soil, plants, and other media. As photosynthesizers, algae need light and air for food production and growth, but do not generally require organic compounds from the environment. By producing oxygen, which is utilized by other organisms including animals, algae play an important role in the balance of nature.

Multicellular Animal Parasites

Although they are not strictly microorganisms, multicellular parasites of medical importance will be discussed in this text. The two major groups of parasites are the flatworms and roundworms, collectively called **helminths.** During some stages of their life cycle, the helminths are microscopic in size, and laboratory identification of these organisms includes many techniques used for the identification of microbes that are traditionally considered microorganisms.

Viruses

Viruses (Figure 1–6e) are very different from the other microbial groups mentioned here. They are so small that most can be seen only with an electron microscope, and they are not cellular. Structurally very simple, the virus particle contains a core made of only one type of nucleic acid, either DNA (deoxyribonucleic acid) or RNA (ribonucleic acid). This core is surrounded by a protein coat. Sometimes the coat is encased by an additional layer, a lipid membrane called an envelope. All living cells have RNA *and* DNA, can carry out chemical reactions, and can reproduce as self-sufficient units. Viruses, however, have no "machinery" for metabolism and can reproduce only inside the cells of other organisms. Thus, viruses are parasites of other biological groups.

In addition to viruses, which contain a core of nucleic acid surrounded by a protein coat, there is a group of infectious molecules (capable of causing infection) that lack a protein coat. These molecules are called **viroids.** They are short pieces of RNA that are known to cause disease in plants (See Figure 12–24). It is suspected but not yet proven that viroids are implicated in human diseases.

Another group of infectious molecules called **prions** appear to lack nucleic acids, and consist only of protein. At present, prions are known to cause two diseases: scrapie and Creutzfeldt–Jakob disease. Scrapie, named for the tendency of sheep and goats infected with the disease to scrape off large clumps of their wool, is a neurological disorder. Creutzfeldt–Jakob disease is a rare form of dementia that affects humans. There is also strong evidence that prions cause two other human diseases that involve the central nervous system, and there is circumstantial evidence that prions are involved in Alzheimer's disease.

MICROBES AND HUMAN WELFARE

As we mentioned earlier, only a minority of microorganisms are pathogenic (disease-producing).

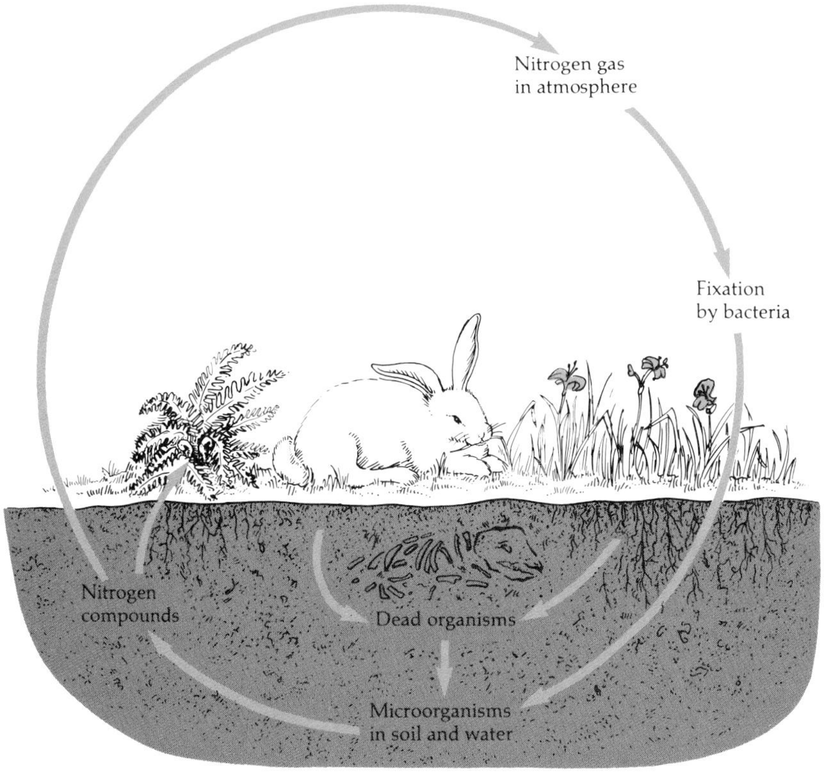

Figure 1–7 The nitrogen cycle. Microorganisms in the soil and water play key roles in recycling chemical elements essential to life, as shown here and in Figure 1–8.

Fortunately, too, there are many methods for controlling the growth of these pathogens outside the body and destroying them once they enter the body. For example, various forms of temperature control, filtration, dehydration, radiation, and disinfection are used to inhibit the growth of microbes outside the body. Antibiotics can be used to kill microbes inside the body.

Microbes that result in food spoilage, such as soft spots on fruits and vegetables, putrefaction of meats, rancidity of fats and oils, and ropiness of bread, are also a minority of microbes. The growth of these microorganisms can be prevented by the use of various techniques, such as boiling, pasteurizing, canning, refrigerating, freezing, irradiating, drying, and adding chemical preservatives.

The vast majority of microbes benefits humans as well as other plants and animals in numerous ways. The following sections will outline some of these beneficial activities. In later chapters we will discuss these activities in greater detail.

Recycling Vital Elements

The chemical elements nitrogen, carbon, oxygen, sulfur, and phosphorus are essential for life and are available only in limited, though large, amounts. Pools of nitrogen, oxygen, and carbon (as carbon dioxide) exist as gases in the atmosphere. Sulfur and phosphorus are stored in the earth's crust. For the most part, it is microorganisms that convert these elements into forms that can be used or stored by plants and animals.

The **nitrogen cycle** is one example of how microorganisms recycle an element (Figure 1–7). Nitrogen is a major constituent of all living cells. It is present in proteins and in nucleic acids—DNA and RNA—which contain the genetic codes for the function and reproduction of living cells. Although

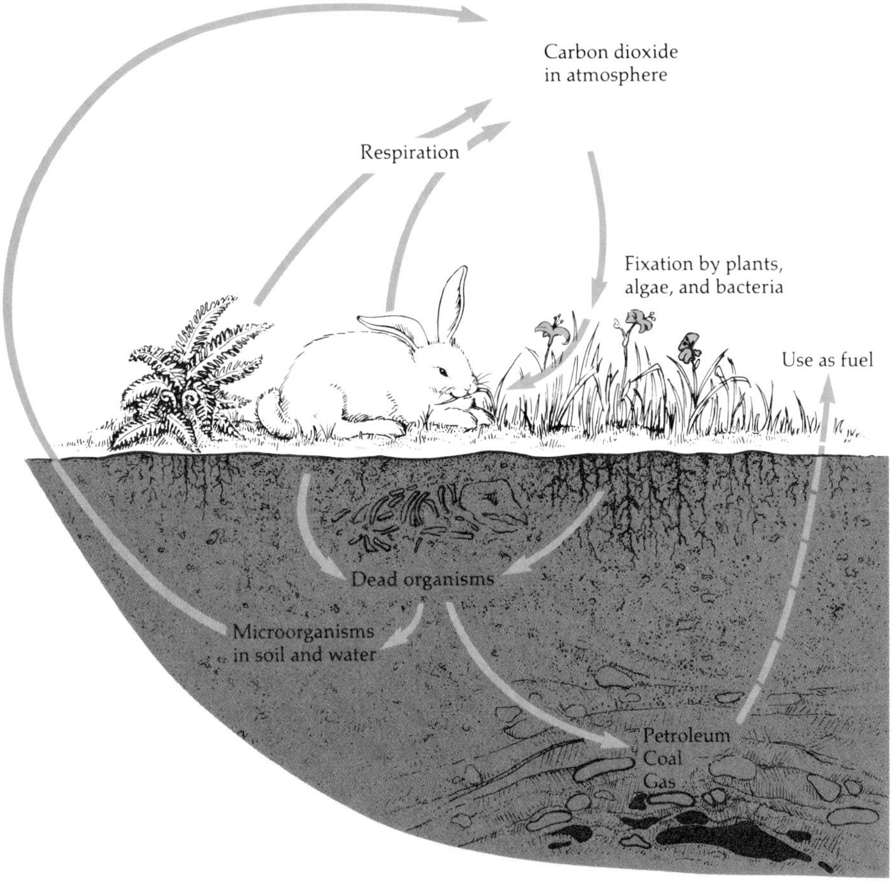

Figure 1–8 The carbon cycle.

79% of the atmosphere is nitrogen, animals and plants cannot use it in that gaseous form. Certain soil bacteria, and *cyanobacteria* (Monera formerly called blue-green algae) in the oceans combine gaseous nitrogen with other elements (fix it to those elements) so that it can be incorporated into living cells and enter the food chain. When other bacteria help to decompose dead plants and animals, nitrogen-containing chemicals are released into the soil; additional microbes then convert the soil nitrogen back into nitrogen gas.

In the **carbon cycle,** green plants and algae remove carbon dioxide from the air and by the process of photosynthesis, use sunlight to convert the carbon dioxide into food (Figure 1–8). When this food is consumed by other organisms, including humans, carbon dioxide is produced in the process called respiration. Microorganisms, mostly bacteria, play another key role in the carbon cycle by also returning carbon dioxide to the atmosphere, when they decompose organic wastes and dead plants and animals. Algae, higher plants, and some bacteria also participate in the **oxygen cycle** by recycling oxygen to the air during photosynthesis. The nitrogen, carbon, and oxygen cycles, as well as those involving sulfur and phosphorus, will be discussed in greater detail in Chapters 5 and 25.

Sewage Treatment: Using Microbes to Recycle Water

With our growing awareness of the need to preserve the environment, we are conscious of our responsibility to recycle precious water and prevent the pollution of rivers and oceans. One major pollutant is **sewage,** which consists of human and animal excrement, wash water, industrial wastes, and ground, surface, and rain water. Sewage is about 99.9% water with a few hundredths of one percent suspended solids. The remainder is a variety of dissolved materials.

These undesirable materials and harmful microorganisms are removed by sewage plant treatments. These combine various physical and chemical processes with treatment by beneficial microbes. First, large solids like paper, wood, glass, gravel, and plastic are removed from sewage; left behind are liquid and organic materials that bacteria convert into byproducts such as carbon dioxide, nitrates, phosphates, sulfates, ammonia, hydrogen sulfide, and methane. We shall discuss sewage treatment in detail in Chapter 25.

Insect Control by Microorganisms

Insect control is important for agricultural reasons, as well as for the prevention of human disease. The fact is, each year devastating crop damage is caused by insects. Although more research into the use of microorganisms to control insects is needed, several types of bacteria are already being used to reduce insect infestations.

The bacterium *Bacillus thuringiensis* (bä-sil'lus thür-in-jē-en'sis) has been used extensively in the United States to control pests, including alfalfa caterpillars, bollworms, corn borers, cabbageworms, tobacco budworms, and fruit tree leaf rollers. This species of bacteria produces protein crystals that are toxic to the digestive systems of the insects. Enormous quantities of the bacteria are grown, dried, and incorporated into a dusting powder that is applied to the crops upon which the insects feed.

Microbial insect control can have important effects on the environment. Many chemical insecticides, such as DDT, are not easily degraded by soil organisms. They remain in the soil and air as toxic pollutants. After rain showers, they can be carried to streams and rivers, and eventually incorporated into the food chain. It has been shown that some insecticides ingested by fish and small animals are poisonous to them. Also, some of these chemicals might be responsible for birth defects in larger animals.

On the other hand, bacteria and other microorganisms used for insect control do not permanently disturb the ecology of the farm, forestland, or public water supply. And so far, insects have not developed a resistance to microbial insecticides. In contrast, many chemicals used to control insects have been needed in larger concentrations, because new generations of resistant insects are no longer killed by the original levels of poison sprayed on crops.

Modern Industrial Microbiology and Genetic Engineering

Earlier in this chapter, we touched on the commercial use of microorganisms to produce several common foods and chemicals. Two recent extensions of industrial microbiology have been attracting considerable attention: **single-cell protein (SCP),** a microbe-made food substitute, and **genetic engineering,** a new technology that greatly expands the potential of bacteria as miniature biochemical factories.

The demand for food increases with the world's growing population and the continuous consumption of decreasing energy supplies. In response to escalating needs for food, scientists are investigating several new food sources. One such source is SCP—the protein (and other nutrients) produced by microorganisms cultivated on industrial wastes.

The advantage of SCP is that microorganisms grow rapidly and can produce a high yield, estimated to be 15 times greater than the amount of soybeans and 50 times greater than the amount of corn grown in the same time. For example, one-half metric ton of yeast can produce 50 metric tons of protein daily. However, SCP is not very tasty and in some cases might need to be supplemented with essential amino acids to provide a balanced protein source. At present, SCP is used primarily in animal feed.

Besides growing food in tremendous quantities, microorganisms might soon be used to manufacture large amounts of human hormones and other badly needed medical substances. With the help of new techniques developed by biochemists, fragments of human or animal DNA that code for important proteins can be attached to bacterial DNA. The resulting hybrid DNA is called **re-**

combinant DNA. When recombinant DNA is inserted into bacteria, the bacteria can make large quantities of the desired protein.

Recombinant DNA techniques have been used thus far to produce a number of proteins of tremendous potential for medical use, including human insulin, human growth hormone, and some types of interferon, and factor VIII.

It was noted earlier that scientists are using recombinant DNA techniques to develop vaccines against several viruses, including those that cause herpes, hepatitis B, influenza, and malaria. Just recently, scientists using recombinant DNA techniques have produced monoclonal antibodies to use in the diagnosis of liver cancer in its early stages. It is possible that within a decade certain genetic diseases like diabetes will be eliminated by the insertion of correct DNA sequences directly into unhealthy human cells. In the meantime, recombinant DNA techniques provide scientists with the means to cultivate large amounts of natural proteins and other substances, such as insulin, that are otherwise very expensive and difficult to isolate and purify.

An even more recently discovered area is **recombinant RNA** technology. This technique permits scientists to make almost limitless amounts of RNA molecules that are not obtainable by other current methods of biotechnology. It will also provide new opportunities to study and alter viruses and viroids, which contain RNA as the genetic informational molecule.

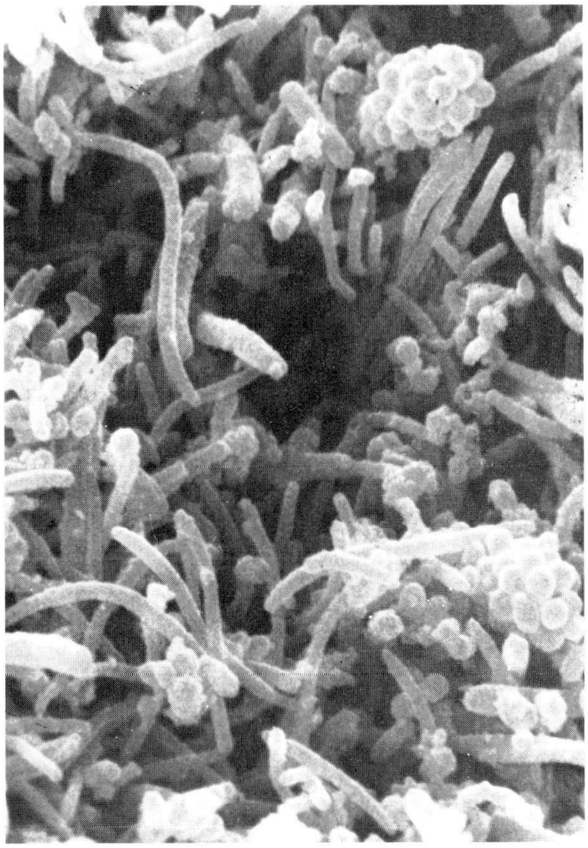

Figure 1–9. Several bacteria found as normal flora inside the human mouth (×3000).

MICROBES AND HUMAN DISEASE

We all live in a microbial world, from birth until death. And we all have a normal group of microorganisms, or **flora,** on and inside our bodies (Figure 1–9). Although these normal flora do not usually disturb us, and often benefit us, they can under some circumstances cause us illness or infect people we contact.

When is a microbe a welcome part of the healthy human, and when is it a harbinger of disease? The distinction between health and disease is in large part a balance between the natural defenses of the body and the disease-producing properties of microorganisms. Microorganisms can enter our bodies whenever we breathe, swallow, cut our skin, or have close contact with other people and animals. Some microorganisms cause disease by invading body tissues; others release poisons called toxins. Usually, large numbers of microorganisms are needed to cause disease.

Whether our bodies overcome the offensive tactics of a particular microbe depends on our **resistance.** Important natural resistance is provided by skin, mucous membranes, cilia, stomach acid, antimicrobial chemicals (such as interferon), destruction of microbes by white blood cells, inflammation, fever, and our powerful immune systems. Sometimes our natural defenses have to be supplemented by antibiotics or other drugs.

People may become particularly susceptible to infection when their natural defenses are weakened by age, diet, or prior illnesses. This was shown in the original outbreak of legionellosis. Not everyone at the convention developed the disease. And of those who contracted legionellosis, a higher number of elderly persons and those with prior respiratory ailments became seriously ill and suffered complications. Researchers still do not completely understand why some victims had mild cases and others became seriously ill. Further information about the bacterium's natural habitat and growth requirements may provide clues.

One of the ongoing tasks of microbiologists is to explore such questions as why an exposure to pathogenic microbes does not always lead to an infection. In the chapters that follow, you will become acquainted with the principles and tools that microbiologists use to understand the many interactions between microbes and their hosts.

STUDY OUTLINE

MICROBES IN OUR LIVES (pp. 4–5)

1. Living things too small to be seen with the naked eye are called microorganisms.
2. Microorganisms play important roles in the maintenance of an ecological balance on earth.
3. Some microorganisms live in humans and other animals and are needed to maintain the animal's health.
4. Some microorganisms are used to produce foods and useful chemicals.
5. Some microorganisms cause disease.

A BRIEF HISTORY OF MICROBIOLOGY (pp. 5–12)

The First Observations (p. 5)

1. Anton van Leeuwenhoek, using a simple microscope, was the first to observe microorganisms (1673).
2. Robert Hooke observed that plant material was composed of "little boxes"; he introduced the term "cell" (1665).
3. Matthias Schleiden and Theodor Schwann introduced the cell theory: the concept that all living things are composed of cells (1838–39).

Fermentation (pp. 5–6)

1. Pasteur found that yeasts and bacteria are responsible for converting fruit sugars into alcohol and then acids.
2. Bacteria in some alcoholic beverages and milk are killed by pasteurization.

The Debate over Spontaneous Generation (pp. 6–8)

1. Until the mid-1880s, people believed in spontaneous generation, the idea that living organisms could arise from nonliving matter.
2. Francesco Redi demonstrated that maggots appear on decaying meat only when flies are able to lay eggs on the meat (1668).
3. John Needham claimed that microorganisms could arise spontaneously from heated nutrient broth (1745).
4. Lazzaro Spallanzani repeated Needham's experiments and suggested that Needham's results were due to microorganisms in the air entering his broth (1765).
5. Rudolf Virchow introduced the concept of biogenesis: living cells can arise only from preexisting cells (1858).
6. Louis Pasteur demonstrated that microorganisms are in the air everywhere and offered proof of biogenesis (1861).
7. Aseptic techniques, based on Pasteur's discoveries, are methods used in laboratory and medical procedures to prevent contamination by microorganisms that are in the air.

The Germ Theory of Disease (pp. 8–10)

1. Agostino Bassi (1834) and Pasteur (1865) showed a relationship between microorganisms and disease.
2. Joseph Lister introduced the use of a disinfectant to clean surgical dressings in order to control infections in humans (1860s).
3. Robert Koch proved that microorganisms transmit disease. He used a sequence of procedures called Koch's Postulates (1876), which are used today to prove that a particular microorganism causes a particular disease.

Vaccination (pp. 10–11)

1. In a vaccination, immunity (resistance to a particular disease) is conferred by inoculation with a vaccine.
2. In 1798, Edward Jenner demonstrated that inoculation with cowpox material provides humans with immunity from smallpox.
3. About 1880, Pasteur discovered that avirulent bacteria could be used as a vaccine for chicken cholera; he coined the word *vaccine*.
4. Modern vaccines are prepared from living avirulent microorganisms or killed pathogens, from isolated components of pathogens, or by recombinant DNA techniques.

The Birth of Modern Chemotherapy: Dreams of a "Magic Bullet" (pp. 11–12)

1. Chemotherapy is chemical treatment of a disease.
2. Two types of chemotherapeutic agents are synthetic drugs (chemically prepared in the laboratory) and antibiotics (substances produced naturally by bacteria and fungi that inhibit the growth of other microorganisms).
3. Paul Ehrlich introduced an arsenic-containing chemical called salvarsan to treat syphilis (1910).
4. Alexander Fleming observed that the mold *Penicillium* inhibited the growth of a bacterial culture. He named the active ingredient penicillin (1928).
5. Penicillin is an antibiotic produced by a fungus.
6. Researchers are continually looking for new antimicrobial drugs that harm pathogens but do not harm the host.
7. Scientists are developing procedures for delivering drugs directly to diseased tissues without harming healthy cells.

NAMING AND CLASSIFYING MICROORGANISMS (pp. 12–13)

1. In a nomenclature system designed by Carolus Linnaeus (1735), each living organism is assigned two names.
2. The two names consist of a genus and specific epithet, which must be underlined or italicized.
3. All organisms are classified into five kingdoms: Procaryotae (or Monera), Protista, Fungi, Plantae, and Animalia.

THE DIVERSITY OF MICROORGANISMS (pp. 13–15)

Bacteria (p. 13)

1. Bacteria are one-celled organisms. Because they have no nucleus, the cells are described as procaryotic.
2. Bacteria can use a wide range of substances for their nutrition.

Fungi (pp. 13–15)

1. Fungi (mushrooms, molds, and yeasts) have eucaryotic cells (with a true nucleus). Most fungi are multicellular.
2. Fungi obtain nutrients by absorbing organic material from their environment.

Protozoans (p. 15)

1. Protozoans are unicellular eucaryotes, and are classified according to their means of locomotion.
2. Protozoans obtain nourishment by absorption or ingestion through specialized structures.

Algae (p. 15)

1. Algae are unicellular or multicellular eucaryotes that obtain nourishment by photosynthesis.
2. Most algae are of no medical importance.
3. Algae produce most of the oxygen in the atmosphere.

Multicellular Animal Parasites (p. 15)

1. The principal groups of multicellular animal parasites are flatworms and roundworms, collectively called helminths.
2. Microscopic stages of helminths are identified by traditional microbiologic procedures.

Viruses (p. 15)

1. Viruses are noncellular entities that are parasites of cells.
2. Viruses consist of a nucleic acid core (DNA or RNA) surrounded by a protein coat. An envelope may surround the coat.
3. Viroids are infectious pieces of RNA known to cause certain plant diseases.
4. Prions are infectious proteins known to cause scrapie and Creutzfeldt–Jakob disease.

MICROBES AND HUMAN WELFARE (pp. 15–19)

1. Microorganisms degrade dead plants and animals and recycle chemical elements to be used by living plants and animals.
2. Bacteria are used to decompose organic matter in sewage.
3. Bacteria that cause diseases in insects are being used as biological controls of insect pests. Biological controls are specific for the pest and do not harm the environment.
4. Microorganisms can be used to help produce foods. They are also food sources (single-cell protein) themselves.
5. Using recombinant DNA, bacteria can produce important proteins such as insulin, beta-endorphin, and interferon.

MICROBES AND HUMAN DISEASE (pp. 19–20)

1. Everyone has normal flora in and on the body.
2. The quantity of microbes and their disease-producing properties, and the host's resistance and general health are important factors in whether a person will contract a disease.

STUDY QUESTIONS

REVIEW

1. Match the following people to their contribution toward the advancement of microbiology.

____ Ehrlich	(a) First to observe bacteria
____ Fleming	(b) First to observe cells in plant material and name them
____ Hooke	(c) Disproved spontaneous generation
____ Koch	(d) Proved that microorganisms can cause disease
____ Lister	(e) Discovered penicillin
____ Pasteur	(f) Used the first synthetic chemotherapeutic agent
____ Leeuwenhoek	(g) First to employ disinfectants in surgical procedures

2. How did the idea of spontaneous generation come about?
3. Some proponents of spontaneous generation believed that air is necessary for life. They felt that Spallanzani did not really disprove spontaneous generation because he hermetically sealed his flasks to keep air out. How did Pasteur's experiments address the air question without allowing the microbes in the air to ruin his experiment?
4. Briefly explain why scientific names are more useful than common names.
5. Match the following microorganisms to their descriptions.

____ Algae	(a) Not composed of cells
____ Bacteria	(b) Cell walls made of chitin
____ Fungi	(c) Cell walls made of peptidoglycan
____ Protozoans	(d) Cell walls made of cellulose; photosynthetic
____ Viruses	(e) Complex cell structure lacking cell walls

6. Compare and contrast the following terms and name one disease caused by each: virus, viroid, prion.
7. Briefly state the role played by microorganisms in each of the following.
(a) Biological control of pests
(b) Recycling of elements
(c) Normal flora
(d) Sewage treatment
(e) Human insulin production

CHALLENGE

1. How did the theory of biogenesis lead the way for the germ theory of disease?
2. The genus name of a bacterium is "erwinia" and the species name is "carotovora." Write the scientific name of this organism correctly. Using this name as an example, explain how scientific names are chosen.
3. Find at least three supermarket products made by microorganisms. (*Hint:* The label will state the scientific name of the organism or include the word *culture.*)

FURTHER READING

Brock, T. D., ed. 1971. *Milestones in Microbiology.* Washington, D. C.: American Society for Microbiology. The writings of Jenner, Lister, Fleming, and others, with instructive editorial comments.

De Kruif, P. 1953, *Microbe Hunters.* New York: Harcourt, Brace, and World. The stories of Leeuwenhoek, Koch, Pasteur, and others are presented in an interesting narrative.

Dixon, B. 1976. *Magnificent Microbes.* New York: Atheneum. A best-selling account of our dependence on microbes.

Fraser, D. W. and J. E. McDade. October 1979. Legionellosis. *Scientific American 241:82–99.* A summary of the mysterious epidemic that led to the discovery of a new bacterium.

Margulis, L. and K. V. Schwartz. 1982. *Five Kingdoms: An Illustrated Guide to the Phyla of Life on Earth.* San Francisco: W. H. Freeman. A discussion of the classification of living organisms and descriptions of phyla.

Noble, W. C. and J. Naidoo. 1979. *Microorganisms and Man.* Baltimore: University Park Press. A short and easy way to get acquainted with microorganisms.

Rosebury, T. 1969. *Life on Man.* New York: Viking Press. A humorous yet scientific account of the role of microbes on the human body.

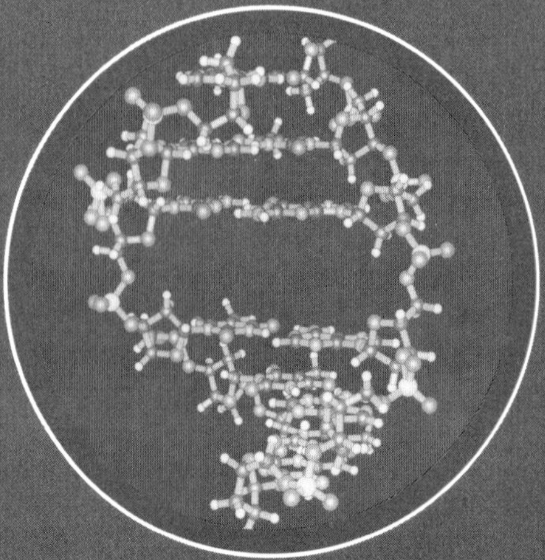

CHAPTER 2

Chemical Principles

OBJECTIVES

After completing this chapter you should be able to

- Discuss the structure of an atom and its relation to the chemical properties of elements.
- Define the following terms: ionic bond; covalent bond; hydrogen bond; molecular weight; mole.
- Diagram three basic types of chemical reactions.
- Identify the role of enzymes in chemical reactions.
- List several properties of water that are important to living systems.
- Distinguish between inorganic and organic molecules.
- Define the following terms: acid; base; salt; pH.
- Name the building blocks of the following macromolecules: carbohydrates; simple lipids; phospholipids; proteins; nucleic acids.
- Identify the role of ATP in cellular activities.

Why study chemistry in a microbiology course? Microorganisms are so small that any examination of their functions is quickly directed to the level of molecules. Most of their activities are, in fact, nothing more than a series of chemical reactions. Although we can see a tree rot and smell milk when it sours, we might not realize what is happening on a microscopic level. In both cases, microbes are conducting chemical operations. The tree rots when microorganisms induce the hydrolysis (breakdown by water) of wood. And the milk turns sour due to fermentation of lactose sugar to lactic acid by bacteria.

Like all organisms, microorganisms must use nutrients to make chemical building blocks that are used for growth and for all the other functions essential to life. For most microorganisms, synthesizing these building blocks requires that nutrient substances be broken down and the resulting energy and molecular fragments be assembled into new substances. This complicated biochemistry takes place daily in countless microbial "laboratories"—any of which could fit on the head of a pin.

All matter—whether air, rock, or living organism—is made up of small units called **atoms.** Atoms interact with each other and in certain combinations to form **molecules.** Living cells are made up of molecules, some of which are very complex. The interaction of atoms and molecules is called **chemistry.**

The chemistry of microbes is the microbiologist's concern. To understand the changes that occur in microorganisms, and the changes they make to the world around us, you will need to know how molecules are formed and how they interact.

STRUCTURE OF ATOMS

Atoms are the smallest units of matter that enter into chemical reactions. How big are atoms? It has been estimated that the largest atoms in nature are less than 0.00000005 centimeters in diameter, and the smallest, the atoms of hydrogen, are less than 0.00000001 cm in diameter. In other words, a line of 50 million of the largest atoms placed end to end would be, at most, 2.5 cm (1 inch) long.

It was once believed that atoms were the smallest particles of matter, but physicists have since demonstrated that huge amounts of energy can split atoms into even smaller particles. However, in the chemical reactions of our everyday world, atoms are the smallest units.

All atoms have a centrally located **nucleus** and particles called **electrons** that move around the nucleus in arrangements called electronic configurations (Figure 2–1). The nuclei of most atoms are stable—that is, they do not change spontaneously—and nuclei do not participate in chemical reactions. The nucleus is made up of positively (+) charged particles called **protons** and uncharged (neutral) particles called **neutrons.** The nucleus, therefore, bears a net positive charge. Neutrons and protons have approximately the same weight, which is about 1840 times that of an electron. The charge on electrons is negative (−), and in all atoms the number of protons is equal to the number of electrons. Because the total positive charge of the nucleus equals the total negative charge of the electrons, each atom is electrically neutral.

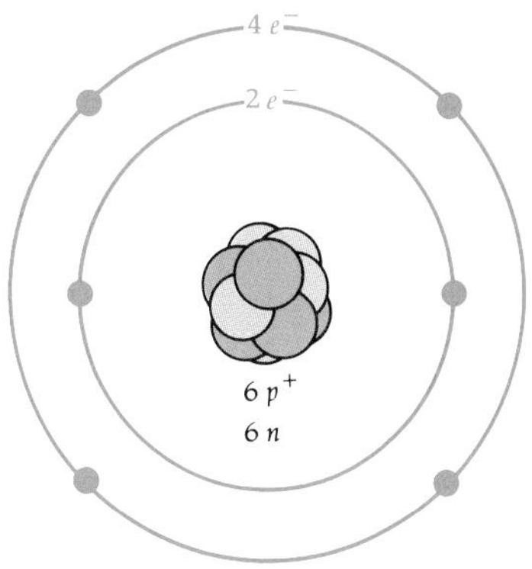

Figure 2–1 Atomic structure. Illustrated here is a very simplified version of a carbon atom (C). It contains six protons (p^+, dark color) and six neutrons (n, light color) in its centrally located nucleus, and six electrons (e^-, black) orbiting the nucleus.

The number of protons in an atomic nucleus ranges from one (in a hydrogen atom) to more than 100, in the largest known atoms. Atoms are often listed by their **atomic number,** the number of protons in the nucleus (see Table 2–1). The total number of protons and neutrons in an atom is its **atomic weight.**

Chemical Elements

All atoms with the same atomic number behave the same way chemically and are classified as the same **chemical element.** Each element has its own name and a one- or two-letter symbol, usually derived from the English or Latin name for the element. For example, the symbol for the element hydrogen is H, and the symbol for carbon is C. The symbol for sodium is Na—the first two letters of its Latin name *natrium*—to distinguish it from nitrogen, N, and sulfur, S. There are 92 naturally occurring elements. However, only about 26 elements are commonly found in living things; these are listed in Table 2–1. The elements most abundant in living matter—hydrogen, carbon, nitrogen, and oxygen—are shown in color in the table. The next most abundant elements appear in gray.

Most elements have several **isotopes,** atoms with differing numbers of neutrons in their nuclei.

All isotopes of an element have the same number of protons in their nuclei, but their atomic weights differ because of the difference in the number of neutrons. For example, in a natural sample of oxygen, all the atoms will contain eight protons. However, 99.76% of the atoms will have eight neutrons, 0.04% will contain nine neutrons, and the remaining 0.2% will contain ten neutrons. Therefore, the three isotopes composing a natural sample of oxygen will have atomic weights of 16, 17, and 18, although all will have the atomic number 8. Atomic numbers are written as a subscript to the left of an element's chemical symbol. Atomic weights are written as a superscript above the atomic number. Thus, natural oxygen isotopes are designated as ${}^{16}_{8}O$, ${}^{17}_{8}O$, and ${}^{18}_{8}O$. Rare isotopes of certain elements are extremely useful in biological research, medical diagnosis, and treatment of some disorders.

Electronic Configurations

In the atom, electrons are arranged in **electron shells,** which are regions corresponding to different **energy levels.** The arrangement is called an **electronic configuration.** Layered outward from the nucleus, each shell can hold a characteristic maximum number of electrons: two electrons in the innermost shell (lowest energy level), eight electrons in the second shell, and eight electrons in the third shell, if it is the atom's outermost (valence) shell. The fourth, fifth, and sixth electron shells can each accommodate 18 electrons, although there are some exceptions to this generalization. Table 2–2 shows the electronic configurations for atoms of some elements found in living organisms.

There is a tendency for the outermost shell to be filled with the maximum number of electrons.

Table 2–1 The Elements of Life

Element	Symbol	Approximate Atomic Weight	Atomic Number
Hydrogen	H	1	1
Carbon	C	12	6
Nitrogen	N	14	7
Oxygen	O	16	8
Fluorine	F	19	9
Sodium	Na	23	11
Magnesium	Mg	24	12
Silicon	Si	28	14
Phosphorus	P	31	15
Sulfur	S	32	16
Chlorine	Cl	35	17
Potassium	K	39	19
Calcium	Ca	40	20
Vanadium	V	51	23
Chromium	Cr	52	24
Manganese	Mn	55	25
Iron	Fe	56	26
Cobalt	Co	59	27
Nickel	Ni	59	28
Copper	Cu	64	29
Zinc	Zn	65	30
Arsenic	As	75	33
Selenium	Se	79	34
Molybdenum	Mo	96	42
Tin	Sn	119	50
Iodine	I	127	53

Table 2–2 Electronic Configurations for the Atoms of Some Elements Found in Living Systems

Element	Electronic Configuration: First Electron Shell	Second Electron Shell	Third Electron Shell	Diagram	Number of Valence (Outermost) Shell Electrons
Hydrogen	1			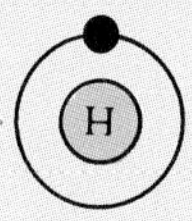	1
Carbon	2	4		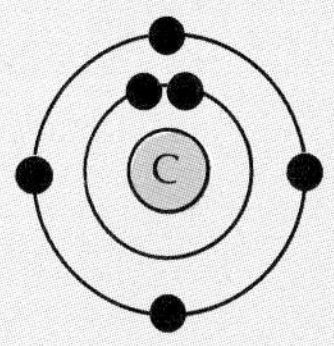	4
Nitrogen	2	5		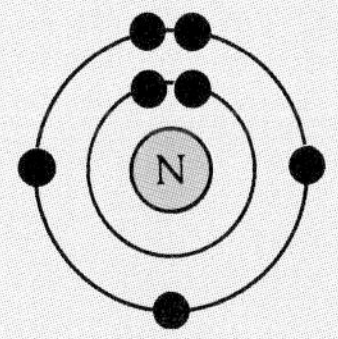	5
Oxygen	2	6		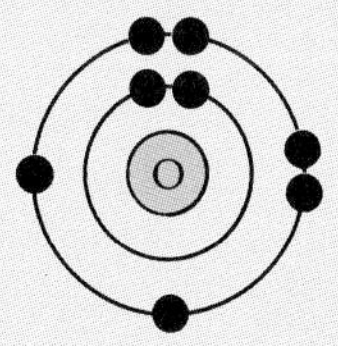	6
Magnesium	2	8	2	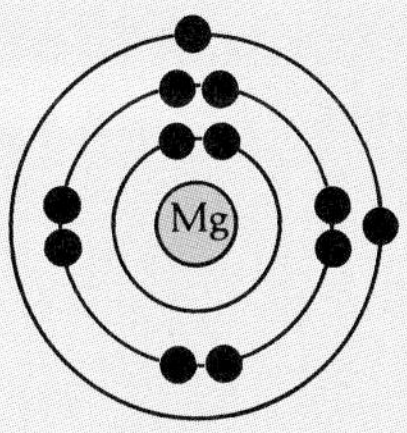	2
Phosphorus	2	8	5	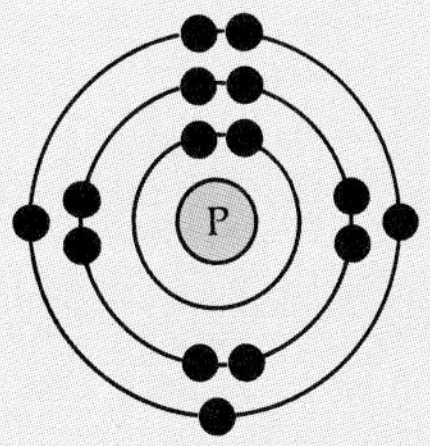	5
Sulfur	2	8	6	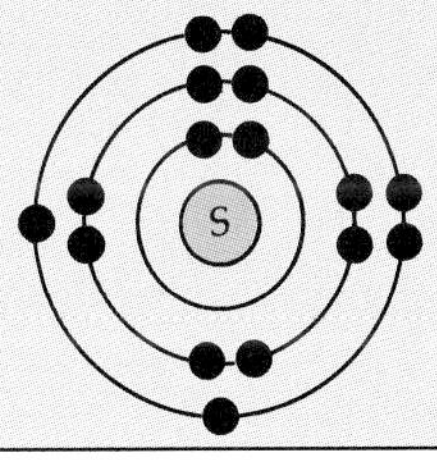	6

An atom can give up, accept, or share electrons with other atoms to fill this shell.

The chemical properties of atoms are largely a function of the number of electrons in the outermost electron shell. When its outer shell is filled, the atom is stable, or inert: it does not tend to react with other atoms. Helium (atomic number 2) and neon (atomic number 10) are examples of atoms that have filled outer shells. Helium has two electrons in the first and only shell, and neon has two electrons in the first shell and eight electrons in the second, its outermost, shell.

When an atom's outer electron shell is only partially filled, the atom is unstable. Such an atom reacts with other atoms to become more stable, and this reaction depends, in part, on the degree to which the outer energy levels are filled. Note the number of electrons in the outer energy levels for the atoms in Table 2–2.

HOW ATOMS FORM MOLECULES: CHEMICAL BONDS

When the outermost energy level of an atom is not completely filled by electrons, it may be thought of as having either unfilled spaces or "extra" electrons in that energy level. For example, an atom of oxygen, with two electrons in the first energy level and six in the second, has two unfilled spaces in the second electron shell; an atom of potassium has one extra electron in its outermost shell. The most stable configuration for any atom is to have its outermost shell filled, as do the inert gases; therefore, for these two atoms to attain that stability, oxygen has to gain two electrons and potassium has to lose one electron. Atoms tend to combine such that the extra electrons in the outermost shell of one atom fill the spaces of the outermost shell of the other atom; the outermost shell of each atom would then have eight electrons.

The **valence,** or combining capacity, of an atom is the number of extra, or deficient, electrons in its outermost electron shell. For example, hydrogen has a valence of 1 (one unfilled space or one extra electron), oxygen a valence of 2 (two unfilled spaces), carbon a valence of 4 (four unfilled spaces), and magnesium a valence of 2 (two extra electrons). Valence may also be viewed as the bonding capacity of an element: hydrogen can form one chemical bond with another atom, oxygen can form two chemical bonds with various atoms, carbon can form four chemical bonds, and magnesium can form two chemical bonds.

Basically, atoms gain stability by completing the full complement of electrons in their outermost energy shells. They do this by combining to form **molecules,** which can be made up of atoms of one or more elements. A molecule that contains at least two different kinds of atoms, such as H_2O, the water molecule, is called a **compound.** In H_2O, the subscript 2 indicates that there are two atoms of hydrogen to one atom of oxygen. Molecules of a compound hold together because the valence electrons of the combining atoms form attractive forces, called **chemical bonds,** between the atomic nuclei. Because energy is required for chemical bond formation, each chemical bond possesses a certain amount of potential chemical energy. Atoms of a given element can form only a specific number of chemical bonds, based on their configurations.

In general, atoms form bonds in one of two ways: (1) by gaining or losing electrons from their outer electron shell or (2) by sharing outer electrons. When atoms gain or lose outer electrons, the chemical bond is called an ionic bond. When outer electrons are shared, the bond is called a covalent bond. Although we will discuss ionic and covalent bonds separately, the kinds of bonds actually found in molecules do not entirely belong to either category. Instead, molecules range from the highly ionic to the highly covalent.

Ionic Bonds

Atoms are electrically neutral when the number of positive charges (protons) equals the number of negative charges (electrons). But when an isolated atom gains or loses electrons, this balance is upset. If the atom gains electrons, it acquires an overall negative charge; if the atom loses electrons, it acquires an overall positive charge. Such a negatively or positively charged particle or group of particles is called an **ion.**

Consider the following examples. Sodium (Na) has 11 protons and 11 electrons, with one electron in its outer electron shell. Sodium tends to lose the single outer electron; it is an *electron donor.* When sodium donates an electron, it is left with 11 protons and only 10 electrons and so has an overall

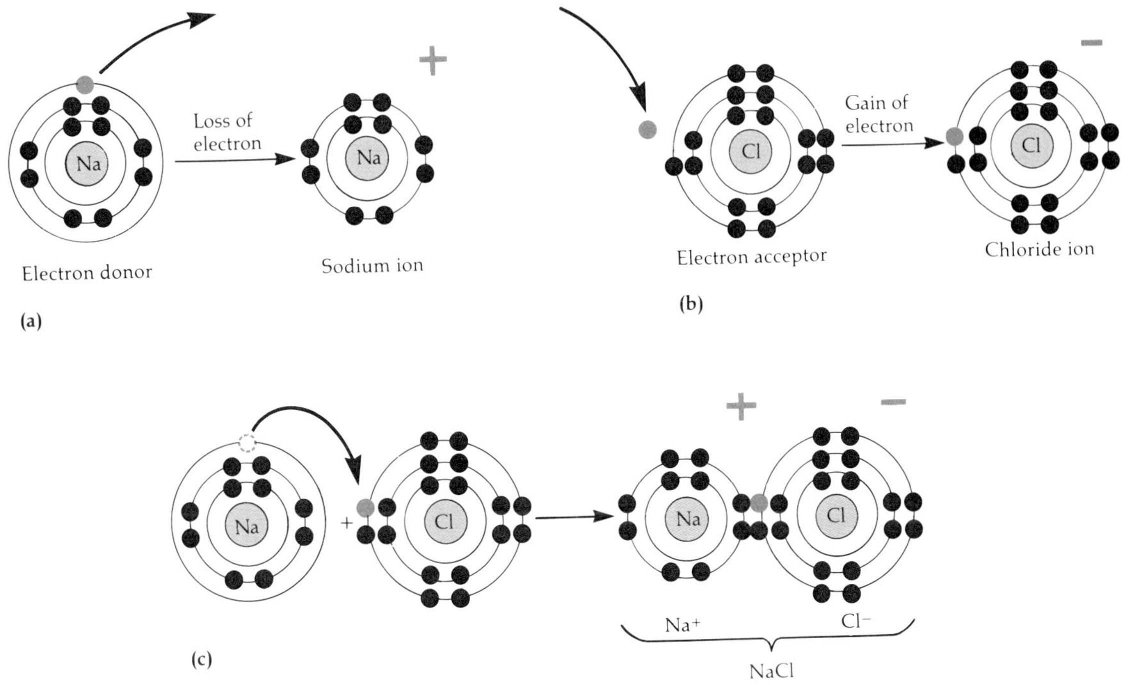

Figure 2–2 Ionic bond formation. **(a)** A sodium atom (Na) loses one electron to an electron acceptor and forms a sodium ion (Na^+). **(b)** A chlorine atom (Cl) accepts one electron from an electron donor to become a chloride ion (Cl^-). **(c)** When the sodium and chloride ions are held together by an ionic bond, a molecule of sodium chloride (NaCl) is formed.

charge of +1. This positively charged sodium atom is called a sodium ion and is written as Na^+ (Figure 2–2a). Chlorine (Cl) has a total of 17 electrons, 7 of them in the outer electron shell. Because this outer shell can hold 8 electrons, chlorine tends to pick up an electron that has been lost by another atom; it is an *electron acceptor*. By accepting an electron, chlorine acquires a total of 18 electrons. However, it still has only 17 protons in its nucleus. The chloride ion therefore has a charge of −1 and is written as Cl^- (Figure 2–2b).

The opposite charges of the sodium ion (Na^+) and chloride ion (Cl^-) attract each other. The attraction, an ionic bond, holds the two atoms together, and a molecule is formed (Figure 2–2c). The formation of this molecule, called sodium chloride (NaCl) or table salt, is a common example of ionic bonding. Thus, an **ionic bond** is an attraction between atoms in which one atom loses electrons and another atom gains electrons. Strong ionic bonds, such as those that hold Na^+ and Cl^- ions together in salt crystals, have only limited importance in living cells. But the weaker ionic bonds formed in aqueous solutions are important in biochemical reactions of microbes and other organisms. For example, weaker ionic bonds assume a role in certain antigen-antibody reactions.

In general, an atom whose outer electron shell is less than half-filled will lose electrons and form positively charged ions, called **cations.** Examples of cations are the potassium ion (K^+), calcium ion (Ca^{2+}), and sodium ion (Na^+). But when an atom's outer electron shell is more than half-filled, the atom will gain electrons and form negatively charged ions, called **anions.** Examples are the iodide ion (I^-), chloride ion (Cl^-), and sulfur ion

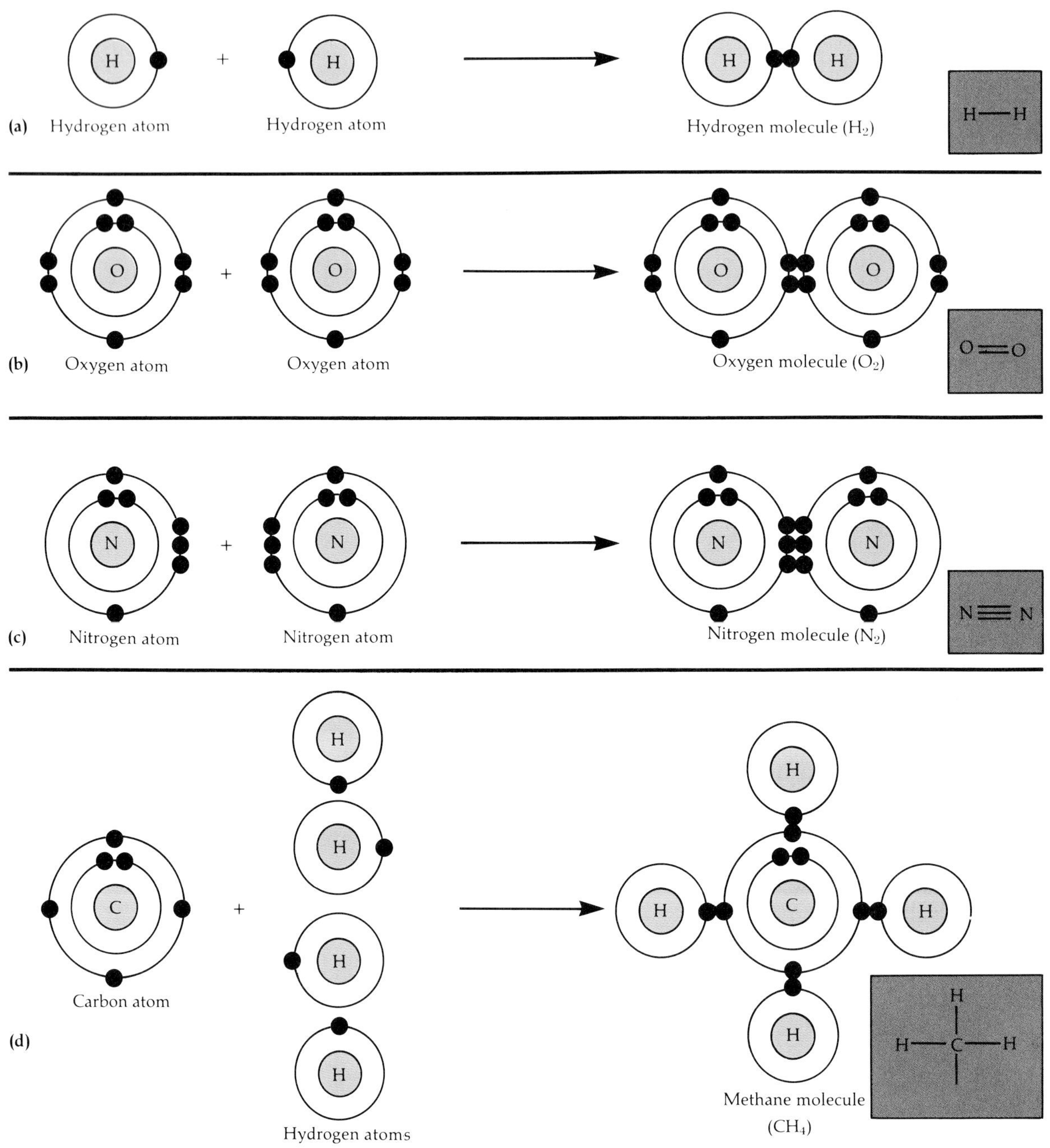

Figure 2–3 Covalent bond formation. **(a)** Single covalent bond between two hydrogen atoms. **(b)** Double covalent bond between two oxygen atoms. **(c)** Triple covalent bond between two nitrogen atoms. **(d)** Single covalent bonds between four hydrogen atoms and a carbon atom.

(S^{2-}). Notice that the symbol for an ion is the chemical abbreviation followed by the ion's number of positive (+) or negative (−) charges.

Hydrogen is an example of an atom whose outer level is exactly half-filled. The first energy level can hold two electrons, but the first energy level of a hydrogen atom contains only one electron. Hydrogen can lose its electron and become a positive ion (H^+), which is precisely what happens when a hydrogen ion combines with a chloride ion to form hydrochloric acid (HCl). Hydrogen can gain an electron and become a negative ion (H^-), which can form ionic bonds with positive ions, but H^- ions are not very important in living systems. More importantly, hydrogen atoms can form an altogether different kind of bond: a covalent bond.

Covalent Bonds

A **covalent bond** is a chemical bond formed by two atoms sharing one or more pairs of electrons. Covalent bonds are stronger and far more common in organisms than true ionic bonds. In the hydrogen molecule, H_2, two hydrogen atoms share a pair of electrons. Each hydrogen atom has its own electron plus one electron from the other atom (Figure 2–3a). The shared pair of electrons actually orbits the nuclei of both atoms. Therefore, the outer electron shells of both atoms are filled. When only one pair of electrons is shared between atoms, a *single covalent bond* is formed. For simplicity, a single covalent bond is expressed as a single line between the atoms (H—H). When two pairs of electrons are shared between atoms, a *double covalent bond* is formed, expressed as two single lines (=) (Figure 2–3b). A *triple covalent bond,* expressed as three single lines (≡), occurs when three pairs of electrons are shared (Figure 2–3c).

The principles of covalent bonding that apply to atoms of the same element also apply to atoms of different elements. Methane (CH_4) is an example of covalent bonding between atoms of different elements (Figure 2–3d). The outer electron shell of the carbon atom can hold eight electrons but has only four. Each hydrogen atom can hold two electrons but has only one. Consequently, in the methane molecule, the carbon atom gains four hydrogen electrons to complete its outer shell, while each hydrogen atom completes its pair by sharing one electron from the carbon atom. Each outer electron of the carbon atom orbits both the carbon nucleus and a hydrogen nucleus. Each hydrogen electron orbits both its own nucleus and the carbon nucleus.

Elements such as hydrogen and carbon, whose outer electron shells are half-filled, form covalent bonds quite easily. In fact, in living organisms, carbon always forms covalent bonds; it never becomes an ion. However, many atoms whose outer electron shells are almost full will also form covalent bonds. An example is oxygen. We will not discuss the reasons why some atoms tend to form covalent bonds rather than ionic bonds. But it is important to understand the basic principles of bond formation, because chemical reactions are nothing more than the making or breaking of bonds between atoms. *Remember:* Covalent bonds are formed by the sharing of electrons between atoms. Ionic bonds are formed by attractions between atoms that have lost or gained electrons.

Hydrogen Bonds

Another chemical bond of special importance to all organisms is the **hydrogen bond,** in which a hydrogen atom that is covalently bonded to one oxygen or nitrogen atom is attracted to another oxygen or nitrogen atom in a different compound. Such bonds are weak and do not bind atoms into molecules. However, they do serve as bridges between different molecules or between various portions of the same molecule.

When hydrogen combines with atoms of oxygen or nitrogen, the relatively large positive nucleus of these larger atoms attracts the hydrogen electron more strongly than does the small hydrogen nucleus. Thus, in a molecule of water (H_2O), all the electrons tend to be closer to the oxygen nucleus than to the hydrogen nuclei. The oxygen portion of the molecule thus has a slightly negative charge, and the hydrogen portion of the molecule has a slightly positive charge (Figure 2–4a). When the positively charged end of one molecule is attracted to the negative end of another molecule, a hydrogen bond is formed (Figure 2–4b). This attraction can occur between hydrogen and other atoms of the same molecule. Because nitrogen and oxygen have unshared pairs of electrons, they are the elements most frequently involved in hydrogen bonding.

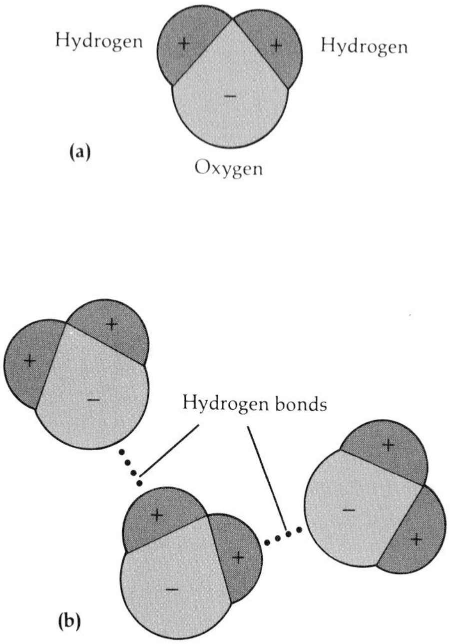

Figure 2–4 Hydrogen bond formation in water. **(a)** In a water molecule, the electrons of the hydrogen atoms are strongly attracted to the oxygen atom. Therefore, the part of the water molecule containing the oxygen atom has a slightly negative charge and the part containing hydrogen atoms has a slightly positive charge. **(b)** In a hydrogen bond between water molecules, the hydrogen of one water molecule is attracted to the oxygen of another water molecule. Many water molecules may be attracted to each other in hydrogen bonds (black dots).

Hydrogen bonds are considerably weaker than either ionic or covalent bonds; they have only about 5% the strength of covalent bonds. Consequently, hydrogen bonds are formed and broken relatively easily. This property accounts for the temporary bonding that occurs between certain atoms of large and complex molecules, such as proteins and nucleic acids. Even though hydrogen bonds are relatively weak, large molecules containing several hundred of these bonds have considerable strength and stability.

Molecular Weight and Moles

The **molecular weight** of a molecule is the sum of the atomic weights of all its atoms. To relate the molecular level to the laboratory level, we use a unit called the mole. One **mole** of a substance is the number of grams equal to the molecular weight of the substance. For example, one mole of water weighs 18 grams, because the molecular weight of H_2O is 18 ($2 \times 1 + 16$). The word mole can also be applied to atoms, or even ions. Thus, one mole of hydrogen ions is equal to 1 gram.

CHEMICAL REACTIONS

As we said earlier, **chemical reactions** are the making or breaking of bonds between atoms. After a chemical reaction, the total number of atoms remains the same, but because they are rearranged, there are new molecules with new properties. In this section, we will look at three basic chemical reactions common to all living cells. By becoming familiar with these reactions, you will be able to understand the specific chemical reactions we will discuss later, particularly in Chapter 5.

Synthesis Reactions

When two or more atoms, ions, or molecules combine to form new and larger molecules, the reaction is called a **synthesis reaction.** To synthesize means "to put together," and a synthesis reaction *forms new bonds.* Synthesis reactions can be expressed in the following way:

A	+	B	$\longrightarrow$	AB
Atom, ion, or molecule A		Atom, ion, or molecule B	Combine to form	New molecule AB

The combining substances, A and B, are called the *reactants;* the substance formed by the combination is the *product.* The arrow indicates the direction in which the reaction proceeds.

Synthesis reactions in living organisms are collectively called anabolic reactions, or simply **anabolism.** The combining of sugar molecules to form starch, and of amino acids to form protein, are two examples of anabolism.

Decomposition Reactions

The reverse of a synthesis reaction is a **decomposition reaction.** To decompose means "to break down into smaller parts," and in a decomposition reaction, *bonds are broken.* Typically, decomposition reactions split large molecules into smaller molecules, ions, or atoms. A decomposition reaction occurs in this way:

$$\underset{\text{Molecule AB}}{AB} \xrightarrow[\text{down into}]{\text{Breaks}} \underset{\text{Atom, ion, or molecule A}}{A} + \underset{\text{Atom, ion, or molecule B}}{B}$$

Decomposition reactions that occur in living organisms are collectively called catabolic reactions, or simply **catabolism.** The breakdown of food molecules in digestion is an example of catabolism.

Exchange Reactions

All chemical reactions are based on synthesis and decomposition. Many reactions, such as **exchange reactions,** are actually part synthesis and part decomposition. An exchange reaction works in this way:

$$AB + CD \xrightarrow[\text{to form}]{\text{Recombine}} AD + BC$$

First the bonds between A and B and between C and D are broken in a decomposition process. New bonds are then formed between A and D and between B and C in a synthesis process. Exchange reactions are involved in the conversion of organic acids into amino acids.

The Reversibility of Chemical Reactions

All chemical reactions are, in theory, reversible; that is, they can occur in either direction. In practice, however, some reactions do this more easily than others. A chemical reaction that is readily reversible (meaning, the end product can revert to the original molecules) is termed a **reversible reaction,** and is indicated by two arrows, as shown here:

$$A + B \underset{\text{Breaks down into}}{\overset{\text{Combines to form}}{\rightleftharpoons}} AB$$

Some reversible reactions occur because neither the reactants nor the end products are very stable. Other reactions will reverse only under special conditions:

$$A + B \underset{\text{Water}}{\overset{\text{Heat}}{\rightleftharpoons}} AB$$

Whatever is written above or below the arrows indicates the special condition under which the reaction in that direction occurs. In this case, A and B react to produce AB only when heat is applied, and AB breaks down into A and B only in the presence of water.

How Chemical Reactions Occur

The **collision theory** explains how chemical reactions occur and how certain factors affect the rates of those reactions. The basis of the collision theory is that all atoms, ions, and molecules are continuously moving and are thus continuously colliding with one another. The energy transferred by the particles in the collision might disrupt their electron structures enough that chemical bonds are broken or new ones are formed.

Factors affecting chemical reactions

Several factors determine whether a collision will cause a chemical reaction: the velocities of the colliding particles, their energy, and their specific chemical configurations. Up to a point, the higher the particles' velocities, the greater the probability that their collision will cause a reaction. Also, each chemical reaction requires a specific level of energy. But even if colliding particles possess the minimum energy needed for reaction, no reaction will take place unless the particles are properly oriented toward each other.

Let us assume that molecules of substance X (the reactant) are to be converted to molecules of substance Y (the product). In a given population of molecules of substance X, at a specific temperature, some molecules will possess relatively little energy; the majority of the population will possess an average amount of energy; and a small portion of the population will have high energy. Only the energy-rich X molecules are able to react and be converted to Y molecules. Therefore, only relatively few molecules at any one time possess enough energy to react in a collision. The collision energy required for a chemical reaction is its **activation energy,** which is the amount of energy needed to disrupt the stable electronic configuration of a specific molecule so that the electrons can be rearranged.

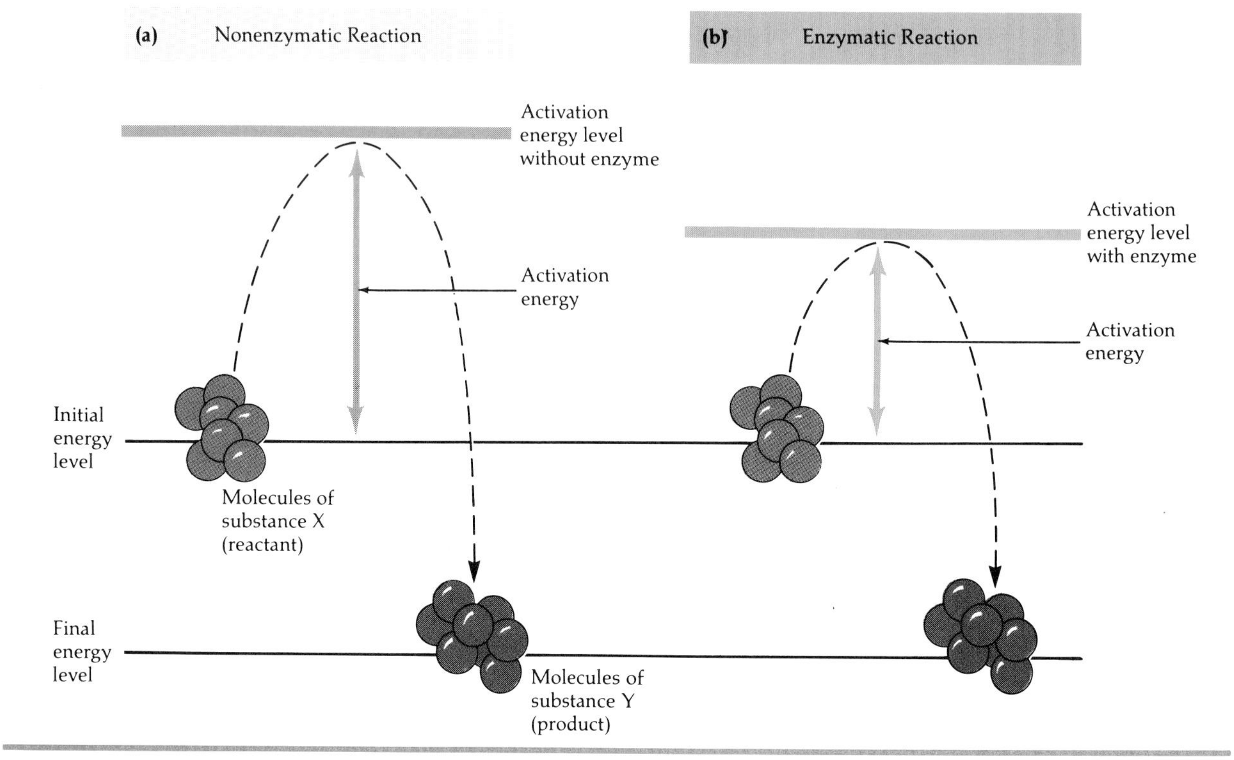

Figure 2–5 Energy requirements of a chemical reaction **(a)** without an enzyme and **(b)** with an enzyme. In **(b)**, the enzyme lowers the activation energy level. Thus, more molecules of X are converted to Y because more molecules of substance X possess the activation energy needed for the reaction.

The *reaction rate,* the frequency of collisions containing sufficient energy, depends on the number of reacting molecules at the activation energy level. One way to increase the reaction rate of a substance is to raise its temperature. By causing the molecules to move faster, heat increases both the frequency of collisions and the number of molecules that attain activation energy. The number of collisions is also increased when the pressure and the concentrations of the reactants are increased, and the distance between molecules is thereby decreased.

Enzymes and chemical reactions

Living systems contain large protein molecules called **enzymes** that act as catalysts. A **catalyst** is a substance that changes the rate of a chemical reaction, usually by increasing it. As catalysts, enzymes typically accelerate chemical reactions. The three-dimensional enzyme molecule has an *active site,* a place that will interact with a specific molecule, or **substrate** (see Figure 5–3).

The enzyme orients the substrate into a position that increases the probability of a reaction. The **enzyme–substrate complex** formed by the temporary binding of enzyme and reactants enables the collisions to be more effective and lowers the activation energy of the reaction (Figure 2–5). In this way, many more molecules of X can participate in the reaction than would without the enzyme. The enzyme, therefore, speeds up the reaction, by increasing the number of X molecules that attain sufficient activation energy (Figure 2–5). (Heat, high pressure, and high concentration of reactants are all nonenzymatic agents of acceleration.)

An enzyme is able to accelerate a reaction without an increase in temperature. This ability is crucial to living systems, because a significant temperature increase would destroy cellular proteins. The crucial function of enzymes, therefore, is to speed up biochemical reactions at a temperature that is compatible with the normal functioning of the cell.

Energy of Chemical Reactions

Some change of energy occurs whenever bonds between atoms are formed or broken during chemical reactions. When a chemical bond is formed, energy is required. Such a chemical reaction that requires energy is called an **endergonic reaction.** When a bond is broken, energy is released. Such a chemical reaction that releases energy is an **exergonic reaction.** Thus, synthesis reactions are endergonic (require energy), whereas decomposition reactions are exergonic (give off energy). The building processes of organisms are achieved through synthesis reactions. The breakdown of foods, on the other hand, occurs through decomposition reactions, which release the energy to be used by organisms for building processes.

Forms of Energy

The ability to use energy and change it into different forms is basic to the survival of any living cell. Besides **chemical energy,** the energy of chemical reactions, there are three other forms of energy that are fundamental in biological systems: mechanical, radiant, and electrical energy. These forms can exist as **potential energy** (stored or inactive energy) or **kinetic energy** (the energy of motion). Cells require **mechanical energy** for movement, reorganization of structures inside the cell, and changes in cell shape. **Radiant energy** is heat or light. By agitating the molecules of any system, heat promotes chemical activity. Light is the primary source of energy for green plants, algae, and bacteria that use photosynthesis to make organic compounds for food. **Electrical energy** is produced whenever electrons move from one place to another. Although electrons do not usually flow through living cells, cells do have electrical currents associated with the ionic flows that occur where charged particles or molecules cross cell membranes.

IMPORTANT BIOLOGICAL MOLECULES

Biologists and chemists divide compounds into two principal classes: inorganic and organic. **Inorganic compounds** are molecules, usually small, in which ionic bonds may play an important role. Inorganic compounds include water and many salts, acids, and bases.

Organic compounds always contain hydrogen and carbon, a unique element in the chemistry of life. Because carbon has four electrons in its outer shell, it can combine with a variety of atoms, including other carbon atoms, to form straight or branched chains and rings. Carbon chains are the backbone for many substances of living cells; organic compounds include sugars, amino acids, and vitamins. Organic compounds are held together mostly or entirely by covalent bonds. Some organic molecules, such as polysaccharides, proteins, and nucleic acids, are very large, usually containing thousands of atoms. Such molecules are referred to as *macromolecules.*

INORGANIC COMPOUNDS

Water

All living organisms require a wide variety of inorganic compounds for growth, repair, maintenance, and reproduction. **Water** is one of the most important, as well as one of the most abundant, of these compounds, and it is particularly vital to microorganisms. Outside the cell, nutrients are dissolved in water, and can therefore pass through cell membranes. And inside the cell, water is the medium for most chemical reactions. In fact, water is by far the most abundant component of almost all living cells, and comprises 5 to 95% or more of each cell, the average being 65 to 75%. Simply stated, no organism can survive without water.

Water has structural and chemical properties that make it particularly suitable for its role in living cells. As we discussed, the total charge on the water molecule is neutral, but the oxygen region of the molecule has a slightly negative charge and the hydrogen region has a slightly positive charge (see Figure 2–4a). Any molecule having such an unequal distribution of charges is called a *polar mole-*

cule. The polar nature of water gives it four characteristics that make it a useful medium for living cells.

First, every water molecule is capable of forming four hydrogen bonds with nearby water molecules. This property results in a strong attraction between water molecules. Because of this strong attraction, a great deal of heat is required to separate water molecules from each other and form water vapor; thus, water has a relatively high boiling point (100°C). With such a high boiling point, water exists in the liquid state on most of the earth's surface. Furthermore, the hydrogen bonding between water molecules makes its crystalline structure (ice) less dense than liquid water at 4°C. Hence, ice floats and can serve as an insulating layer on the surfaces of lakes and streams that harbor living organisms.

Second, the polarity of water makes it an excellent dissolving medium or *solvent.* Many polar substances dissociate (dissolve) into individual molecules in water, because the negative part of the water molecule is attracted to the positive part of the molecules in the *solute* (dissolving substance), while the positive part of the water molecules is attracted to the negative part of solute molecules. Substances, such as salts, that are composed of atoms (or groups of atoms) held together by ionic bonds, tend to dissociate into separate cations and anions in water. Thus, the polarity of water allows molecules of many different substances to separate and become surrounded by water molecules (Figure 2–6).

Third, polarity accounts for water's characteristic role as a reactant or product in many chemical reactions. Its polarity facilitates the splitting and rejoining of hydrogen (H^+) and hydroxyl (OH^-) groups. Water is a key reactant in the digestive processes of organisms—processes in which larger molecules are broken down into smaller ones. Water molecules are also involved in synthetic reactions: water is an important source of the hydrogen and oxygen that are incorporated into numerous organic compounds in living cells.

Finally, the relatively strong hydrogen bonding between water molecules (see Figure 2–4b) makes water an excellent temperature buffer. A given quantity of water, compared with the same quantity of many other substances, requires a great amount of heat to increase its temperature and a great loss of heat to decrease its temperature. Normally, heat absorption by molecules increases their kinetic energy and thus their rate of motion, and increases reactivity. In water, however, heat absorption first breaks hydrogen bonds, rather than increasing molecular motion. Therefore, much more heat must be applied to raise the temperature of water than to raise the temperature of a nonhydrogen-bonded liquid. The reverse is true as water cools. Thus, water more easily maintains a constant temperature than do other solvents and tends to protect a cell from fluctuations in environmental temperatures.

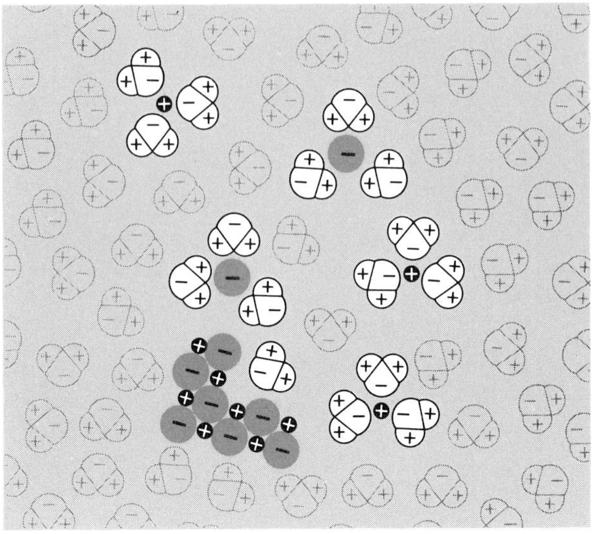

Figure 2–6 How water acts as a solvent for NaCl. The negative part of the water molecule is attracted to the positive sodium ion (Na^+) (small black circles), while the positive part of the water molecule is attracted to the negative chloride ion (Cl^-) (large colored circles).

Acids, Bases, and Salts

As we saw in Figure 2–6, when inorganic salts such as sodium chloride (NaCl) are dissolved in water, they undergo **dissociation,** or **ionization.** That is, they break apart into ions. Substances called acids and bases show similar behavior.

An **acid** can be defined as a substance that dissociates into one or more hydrogen ions (H^+) and one or more negative ions (anions). Thus, an acid can also be defined as a proton (H^+) donor. A **base,** on the other hand, dissociates into one or more positive ions (cations) plus one or more negatively charged groups that can accept, or combine with, protons. Thus, sodium hydroxide (NaOH) is a base because it dissociates to release hydroxyl ions (OH^-), which have a strong attraction for protons, and are among the most important proton acceptors. And finally, a **salt** is a substance that dissociates in water into cations and anions, neither of which are H^+ or OH^-. Figure 2–7 shows common examples of each type of compound and how they dissociate in water.

Acid–Base Balance

An organism must maintain a fairly constant balance of acids and bases to remain healthy. In the aqueous environment within organisms, acids dissociate into hydrogen ions (H^+) and anions. Bases, on the other hand, dissociate into hydroxyl ions (OH^-) and cations. The more hydrogen ions that are free in a solution, the more acid the solution. Conversely, the more hydroxyl ions that are free in a solution, the more basic, or alkaline, it is.

Biochemical reactions, that is, reactions in living systems, are extremely sensitive to even small changes in the acidity or alkalinity of the environments in which they occur. In fact, H^+ and OH^- ions are involved in almost all biochemical processes, and the functions of a cell are modified greatly by any deviation from its narrow band of normal H^+ and OH^- concentrations. For this reason, the acids and bases that are continuously formed in an organism must be kept in balance.

It is convenient to express the amount of H^+ in a solution by a logarithmic pH scale, which ranges from 0 to 14 (Figure 2–8). The **pH** of a solution is the negative logarithm to the base 10 of the hydrogen ion concentration, determined in moles per liter $[H^+]$, or $-\log_{10} [H^+]$.

For example, if the H^+ concentration of a solution is 1.0×10^{-4} moles/liter, or 10^{-4}, its pH equals $-\log_{10}10^{-4} = -(-4) = 4$; this is about the pH of wine. You might refer to the box on Exponents, Exponential Notation, and Logarithms to help you understand this calculation. In the laboratory, however, you will usually measure the pH of a solution with a pH meter or with chemical test papers, and will not need calculations.

Acidic solutions contain more H^+ ions than OH^- ions and have a pH lower than 7. If a solution has more OH^- ions than H^+ ions, it is **basic,** or **alkaline.** In pure water, a small percentage of the

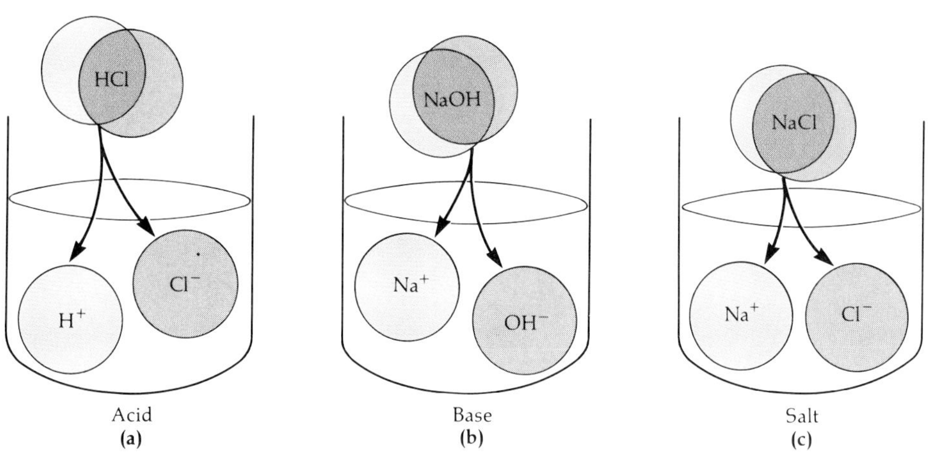

Figure 2–7 Acids, bases, and salts. **(a)** In water, hydrochloric acid (HCl) dissociates into H^+ ions and Cl^- ions. **(b)** Sodium hydroxide (NaOH), a base, dissociates into OH^- ions and Na^+ ions in water. **(c)** In water, table salt (NaCl) dissociates into positive ions (Na^+) and negative ions (Cl^-), neither of which are H^+ or OH^- ions.

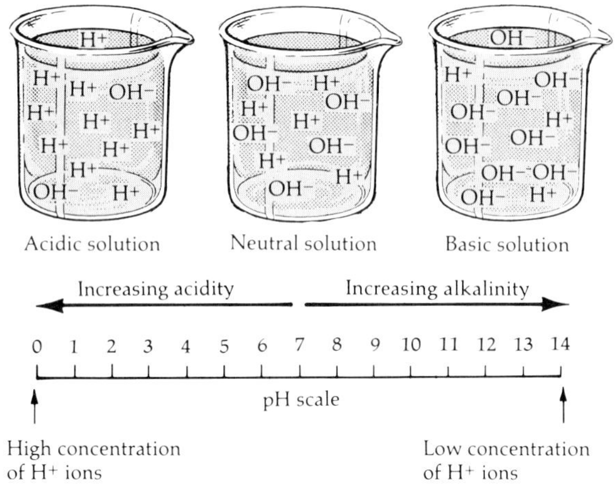

Figure 2–8 The pH scale. The concentrations of H^+ and OH^- ions are equal at pH 7, which is the neutral point. As the pH values decrease from 14 to 0, the H^+ concentration increases. Thus, the lower the pH is, the more acidic the solution; the higher the pH is, the more basic the solution. If the pH value of a solution is below 7, the solution is acidic; if the pH is above 7, the solution is basic (alkaline).

Table 2–3 pH Values of Some Human Bodily Fluids and Common Substances

Fluid or Substance	pH
Stomach acid	1.5–2.5
Lemon juice	2.2–2.4
Carbonated soft drinks	3–3.5
Tomato juice	4.2
Urine	5.0–7.8
Saliva	6.35–6.85
Milk	6.6–6.9
Distilled (pure) water	7.0
Blood	7.35–7.45
Eggs	7.6–8.0
Milk of magnesia	10.0–11.0
Limewater	12.3

molecules are dissociated into H^+ and OH^- ions, so it has a pH of 7. Because the concentrations of H^+ and OH^- are equal, this pH level is said to be **neutral.**

The pH scale is logarithmic, so a change of one whole number represents a tenfold change from the previous concentration. Thus, a solution of pH 1 has 100 times more H^+ ions than a solution with a pH 3, and 10 times more H^+ ions than one with a pH 2.

The pH values of some human body fluids and other common substances are shown in Table 2–3.

Keep in mind that the pH of a solution can be changed. We can increase its acidity by adding substances that will increase the concentration of hydrogen ions. As a living organism takes up nutrients, carries out chemical reactions, and excretes wastes, its balance of acids and bases tends to change, and the pH fluctuates. Fortunately, organisms possess natural **pH buffers,** compounds that help keep the pH from changing drastically. But the pH in our environment's water and soil can be altered by waste products from organisms, pollutants from industry, or fertilizers used in agricultural fields or gardens. When bacteria are grown in laboratory medium, they excrete waste products such as acids that can alter the pH of the medium. If this effect were to continue, the pH of the medium would become acidic enough to inhibit bacterial enzymes and cause death of the bacteria. In order to prevent this problem, pH buffers are added to the culture medium. One very effective pH buffer for some culture media uses a mixture of K_2HPO_4 and KH_2PO_4 (See Table 6–3).

Different microbes function best within different pH ranges, but most organisms grow best in environments with pH between 6.5 and 8.5. The microbes most able to tolerate acid conditions are fungi, whereas the procaryotes called cyanobacteria (blue-green algae) tend to do well in alkaline habitats. *Propionibacterium acnes* (prō-pē-on-ē-bak-ti'rē-um ak'nēz), a bacterium that contributes to acne, has its natural environment on human skin, which tends to be slightly acidic, with pH around 4. *Thiobacillus thiooxidans* (thī-ō-bä-sil'lus thī-ō-oks' i-danz) is a bacterium that grows on elemental sulfur and produces sulfuric acid (H_2SO_4). Its optimum growth pH is 1–3.5. The sulfuric acid produced by this bacterium in mine water is important in dissolving uranium and copper from deposits in the ground.

ORGANIC COMPOUNDS

Inorganic compounds, excluding water, constitute about 1 to 1.5% of living cells. These components,

Exponents, Exponential Notation, and Logarithms

Very large and very small numbers—such as 4,650,000,000 and 0.00000032—are cumbersome to work with. It is more convenient to express such numbers in exponential notation, that is, as a power of 10. For example, 4.65×10^9 is in **standard exponential notation,** or **scientific notation.** 4.65 is the **coefficient,** and 9 is the power or **exponent.** In standard exponential notation, the coefficient is always a number between 1 and 10 and the exponent can be positive or negative.

To change a number into exponential notation, follow two steps. First, determine the coefficient by moving the decimal point so there is only one nonzero digit to the left of it. For example,

0.0000003 2

The coefficient is 3.2. Second, determine the exponent by counting the number of places you moved the decimal point. If you moved it to the left, the exponent is positive. If you moved it to the right, the exponent is negative. In the example, you moved the decimal point 7 places to the right, so the exponent is −7. Thus

$$0.00000032 = 3.2 \times 10^{-7}$$

Now suppose we are working with a large number instead of a very small number. The same rules apply, but our exponential value will be positive rather than negative. For example,

$$4{,}650{,}000{,}000. = 4.65 \times 10^{+9}$$
$$= 4.65 \times 10^9$$

To multiply numbers written in exponential notation, multiply the coefficients and *add* the exponents. For example,

$$(3 \times 10^4) \times (2 \times 10^3) =$$
$$(3 \times 2) \times 10^{4+3} = 6 \times 10^7$$

To divide, divide the coefficients and *subtract* the exponents. For example,

$$\frac{3 \times 10^4}{2 \times 10^3} = \frac{3}{2} \times 10^{4-3} = 1.5 \times 10^1$$

Microbiologists use exponential notation in many situations. For instance, exponential notation is used to describe the number of microorganisms in a population. Such numbers are often very large (see Chapter 6). Another application of exponential notation is to express concentrations of chemicals in a solution—chemicals such as media components (see Chapter 6), disinfectants (see Chapter 7), or antibiotics (see Chapter 18). Such numbers are often very small. Converting from one unit of measurement to another in the metric system requires multiplying or dividing by a power of 10, which is easiest to carry out in exponential notation.

A **logarithm** (log) is the power to which a base number is raised to produce a given number. Usually we work with logarithms to the base 10, abbreviatied $\mathbf{log_{10}}$. The first step in finding the $\log_{10}$ of a number is to write the number in standard exponential notation. If the coefficient is exactly 1, the $\log_{10}$ is simply equal to the exponent. For example,

$$\log_{10} 0.00001 = \log_{10}(1 \times 10^{-5})$$
$$= -5$$

If the coefficient is not 1, as is often the case, a logarithm table or calculator must be used to determine the logarithm.

Microbiologists use logs for calculating pH levels and for graphing the growths of microbial populations in culture (see Chapter 6).

whose molecules have only a few atoms, cannot provide the chemical specificity that cells need to perform complicated biological functions. Such specificity is possible with organic molecules, whose carbon atoms can combine in an enormous variety of ways with other carbon atoms and atoms of other elements.

In the formation of organic molecules, carbon's four outer electrons can participate in up to four covalent bonds, and carbon atoms can bond to each other, to form straight-chain, branched-chain, or ring structures (Figure 2–9).

When single bonds hold a carbon atom to four other atoms, those four atoms form the skeleton of a three-dimensional shape called a **tetrahedron** (Figure 2–10). Carbon atoms can also bond with double or triple bonds; the molecular shape then is not tetrahedral.

In addition to carbon, the most frequently found elements in organic compounds are hydrogen (which can form one bond), oxygen (two bonds), and nitrogen (three bonds). Sulfur (two bonds) and phosphorus (five bonds) appear less often. Other elements are found, but only in a relatively few organic compounds. If you look back at Table 2–1, you will see that the elements most abundant in organic compounds are the same as those most abundant in living cells.

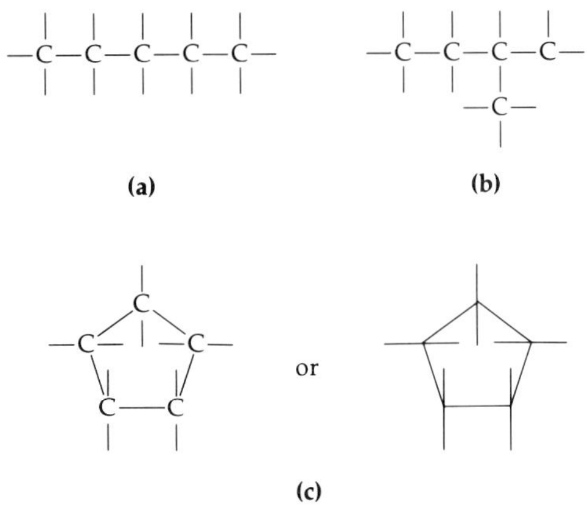

Figure 2–9 Various bonding patterns of carbon. **(a)** Straight-chain structures. **(b)** Branched-chain structures. **(c)** Ring. In ring structures, the carbon atoms at the corners are often not shown.

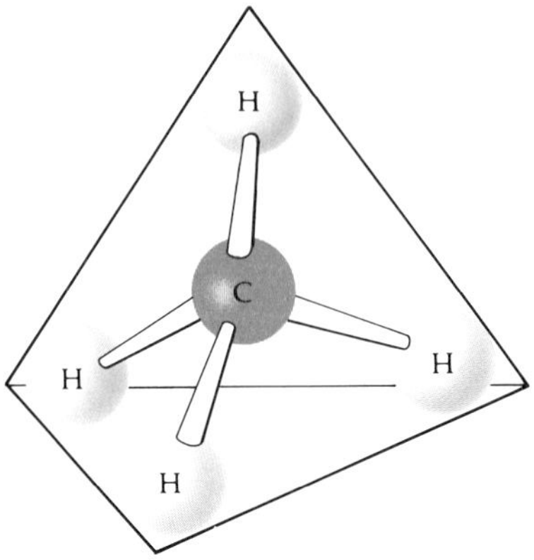

Figure 2–10 A tetrahedral carbon compound. Organic molecules containing carbon atoms with four single bonds have this distinctive shape due to the directions of the four bonds. Here the carbon atom is bonded to four hydrogen atoms to form the tetrahedral methane molecule (CH_4).

Functional Groups

The basic chain of carbon atoms in a molecule is called the **carbon skeleton;** there is a huge number of combinations possible for carbon skeletons. Most of these carbons are bonded to hydrogen atoms. The bonding of other elements with carbon and hydrogen forms characteristic **functional groups** (Table 2–4). Functional groups are responsible for most of the characteristic chemical properties and many of the physical properties of a particular organic compound. So functional groups help us to classify organic compounds. For example, the —OH group is present in each of the following molecules.

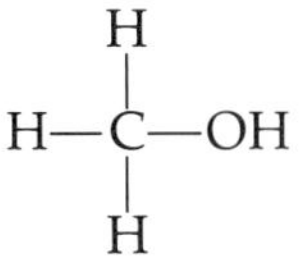

Methanol

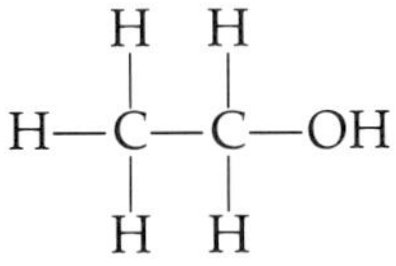

Ethanol

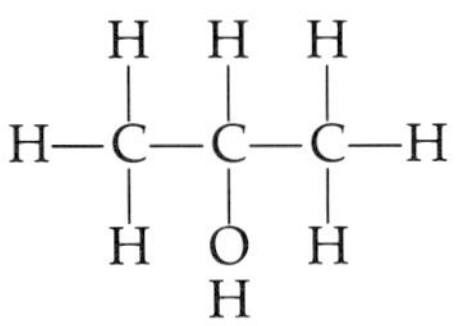

Isopropyl alcohol

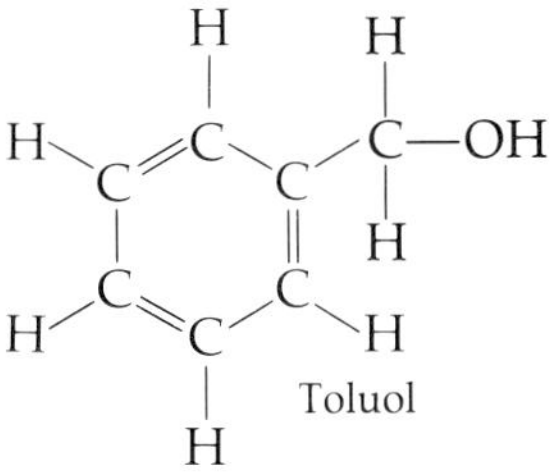

Because the characteristic reactivity of the molecules is based on the —OH group, they are grouped together in a class called alcohols. The —OH group is called the *hydroxy group* and is not to be confused with the *hydroxyl ion* (OH⁻) of bases. The hydroxy group of alcohols does not ionize; it is covalently bonded to a carbon atom.

When a class of compounds is characterized by a certain functional group, the letter R is usually substituted for the remainder of the molecule. For example, alcohols in general may be written R—OH.

Frequently, more than one functional group is found in a single molecule. For example, an amino

Table 2–4 Representative Functional Groups and the Compounds in Which They Are Found

Functional Group	Name of Group	Class of Compounds
R—O—H	Hydroxy	Alcohol
R—C(=O)—H	Carbonyl (terminal)	Aldehyde
R—C(=O)—R	Carbonyl (internal)	Ketone
R—CH₂—NH₂	Amino	Amine
R—C(=O)—O—R′	Ester	Ester
R—CH₂—O—CH₂—R′	Ether	Ether
R—CH₂—SH	Sulfhydryl	Sulfhydryl
R—C(=O)—OH	Carboxyl	Organic acid

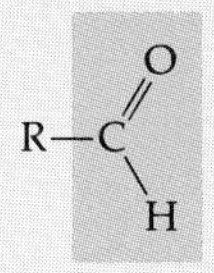

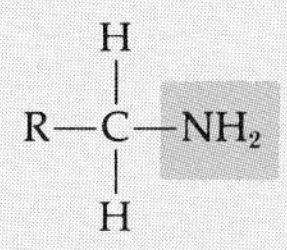

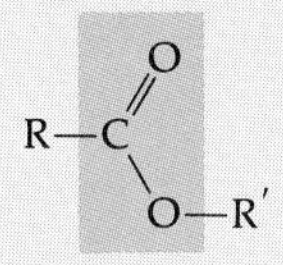

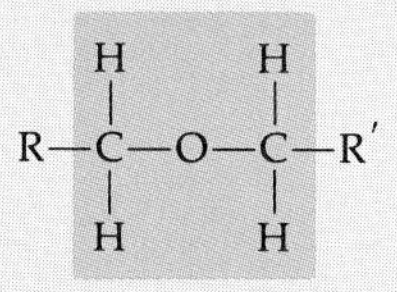

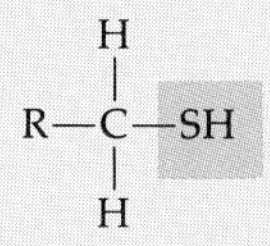

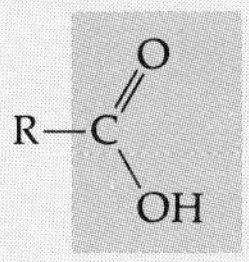

acid molecule contains both amino and carboxyl groups (glycine is shown below).

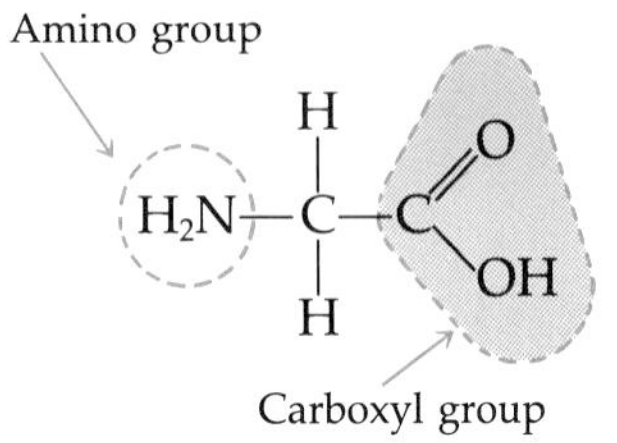

Most of the organic compounds found in living organisms are quite complex; a large number of carbon atoms form the skeleton and many functional groups are attached. In organic and biochemical molecules, it is important to note that each of the four bonds of carbon is satisfied (attached to another atom) and each of the attaching atoms has its characteristic number of bonds satisfied. Because of this, such molecules are stable.

Small organic molecules can be combined into the very large molecules called macromolecules. Macromolecules are usually **polymers,** that is, large molecules formed by covalent bonding of many repeating small molecules called **subunits** or **monomers.** When two monomers join together, the reaction usually involves the elimination of a hydrogen atom from one monomer and a hydroxy group from the other; these combine to produce water:

$$X{-}OH + H{-}Y \longrightarrow X{-}Y + H_2O$$

This type of reaction is called **condensation** or **dehydration** because a molecule of water is released.

The synthesis of macromolecules such as carbohydrates, lipids, proteins, and nucleic acids is a dominant activity of cell chemistry.

Carbohydrates

A large and diverse group of organic compounds are the **carbohydrates,** which include sugars and macromolecules such as starches. The carbohydrates perform a number of major functions in living systems. For instance, one type of sugar (deoxyribose) is a building block of DNA, the molecule that carries hereditary information. Other sugars help to form the cell walls of bacterial cells. Macromolecular carbohydrates, meanwhile, function as food reserves. Simple carbohydrates are utilized in the synthesis of amino acids and fats or fatlike substances, which are used to build structures and provide an emergency source of energy (see Figure 5–22). The principal function of carbohydrates, however, is to fuel cell activities with a ready source of energy.

Carbohydrates are made up of carbon, hydrogen, and oxygen atoms. The ratio of hydrogen to oxygen atoms is always 2 to 1 in carbohydrates. This ratio can be seen in the formulas for the carbohydrates ribose ($C_5H_{10}O_5$), glucose ($C_6H_{12}O_6$), and sucrose ($C_{12}H_{22}O_{11}$). Although there are exceptions, the general formula for carbohydrates is $(CH_2O)_n$, where n symbolizes that there are three or more CH_2O units. Carbohydrates can be divided into three major groups on the basis of size: monosaccharides, disaccharides, and polysaccharides.

Monosaccharides

Simple sugars are called **monosaccharides;** each molecule contains three to seven carbon atoms. The number of carbon atoms in the molecule of a simple sugar is indicated by the prefix of its name: simple sugars with three carbons are called trioses. There are also tetroses (four-carbon sugars), pentoses (five-carbon sugars), hexoses (six-carbon sugars), and heptoses (seven-carbon sugars). Pentoses and hexoses are extremely important to living organisms. Deoxyribose is a pentose found in DNA, the genetic material of the cell. And glucose, a common hexose, is the main energy-supplying molecule of living cells.

Disaccharides

Disaccharides are formed when two monosaccharides bond in a dehydration synthesis (condensation) reaction. For example, molecules of two monosaccharides, glucose and fructose, combine to form a molecule of the disaccharide sucrose (table sugar):

$$\underset{\substack{\text{Glucose}\\ \text{(Monosaccharide)}}}{C_6H_{12}O_6} + \underset{\substack{\text{Fructose}\\ \text{(Monosaccharide)}}}{C_6H_{12}O_6} \longrightarrow \underset{\substack{\text{Sucrose}\\ \text{(Disaccharide)}}}{C_{12}H_{22}O_{11}} + \underset{\text{Water}}{H_2O}$$

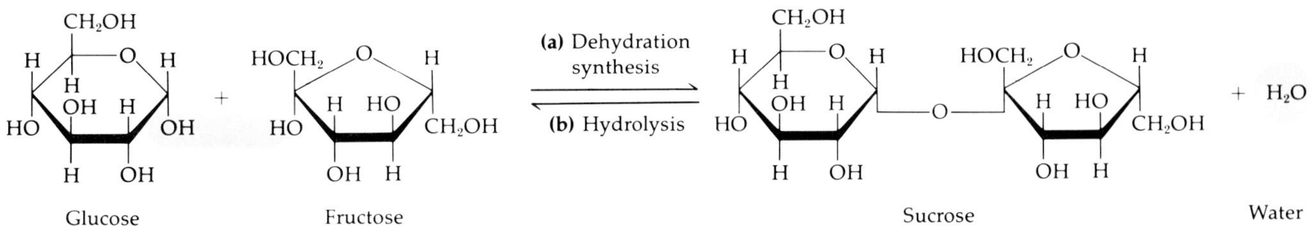

Figure 2–11 **(a)** In dehydration synthesis (left to right), the monosaccharides glucose and fructose combine to form a molecule of the disaccharide sucrose. A molecule of water is lost in the reaction. **(b)** In hydrolysis (right to left), the sucrose molecule breaks down into smaller molecules, glucose and fructose. For the hydrolysis reaction to proceed, water must be added to the sucrose.

The formula for sucrose is $C_{12}H_{22}O_{11}$ and not $C_{12}H_{24}O_{12}$ because a molecule of water (H_2O) is lost in the process of disaccharide formation (Figure 2–11a). Similarly, the dehydration synthesis of the monosaccharides glucose and galactose forms the disaccharide lactose (milk sugar).

It may seem odd that glucose and fructose should have the same chemical formula. Actually, they are different monosaccharides. The positions of the oxygens and carbons vary in the two different molecules (Figure 2–11) and consequently, the molecules have different physical and chemical properties. Two molecules with the same chemical formula but different structures and properties are called **isomers.**

Disaccharides can be broken down into smaller, simpler molecules when water is added. This chemical reaction, the reverse of dehydration synthesis, is called **digestion** or **hydrolysis,** which means "to split by using water." A molecule of sucrose, for example, may be digested into its components of glucose and fructose by the addition of water. This reaction is represented in Figure 2–11b.

Polysaccharides

Carbohydrates in the third major group, the **polysaccharides,** consist of eight or more monosaccharides (glucose is a common sugar unit) joined together through dehydration synthesis; they often have side chains branching off the main structure. Polysaccharides are macromolecules. Like disaccharides, polysaccharides can be split apart into their constituent sugars through hydrolysis. Unlike monosaccharides and disaccharides, however, they usually lack the characteristic sweetness of sugars like fructose and sucrose and usually are not soluble in water.

One important polysaccharide is *glycogen,* which is composed of glucose subunits and is synthesized as a storage material by animals and some bacteria. *Cellulose,* another important glucose polymer, is the main component of the cell walls of plants and most algae. The polysaccharide *dextran,* which is produced as a sugary slime by certain bacteria, is used in a blood plasma substitute.

Lipids

Lipids are a second major group of organic compounds found in living matter. Like carbohydrates, they are composed of atoms of carbon, hydrogen, and oxygen, but lipids lack the 2 to 1 ratio between hydrogen and oxygen atoms. Lipids are a very diverse group of compounds. Most are insoluble in water, but dissolve readily in nonpolar solvents such as ether, chloroform, and alcohol.

If lipids were suddenly to disappear from the earth, all living cells would collapse in a pool of fluid, because lipids are essential to the structure and function of membranes that separate living cells from their environment. In this role, lipids are indispensable to a cell's survival. Within living organisms, they also function as fuel reserves: examples are fats in animals and poly-β-hydroxybutyric acid in bacteria.

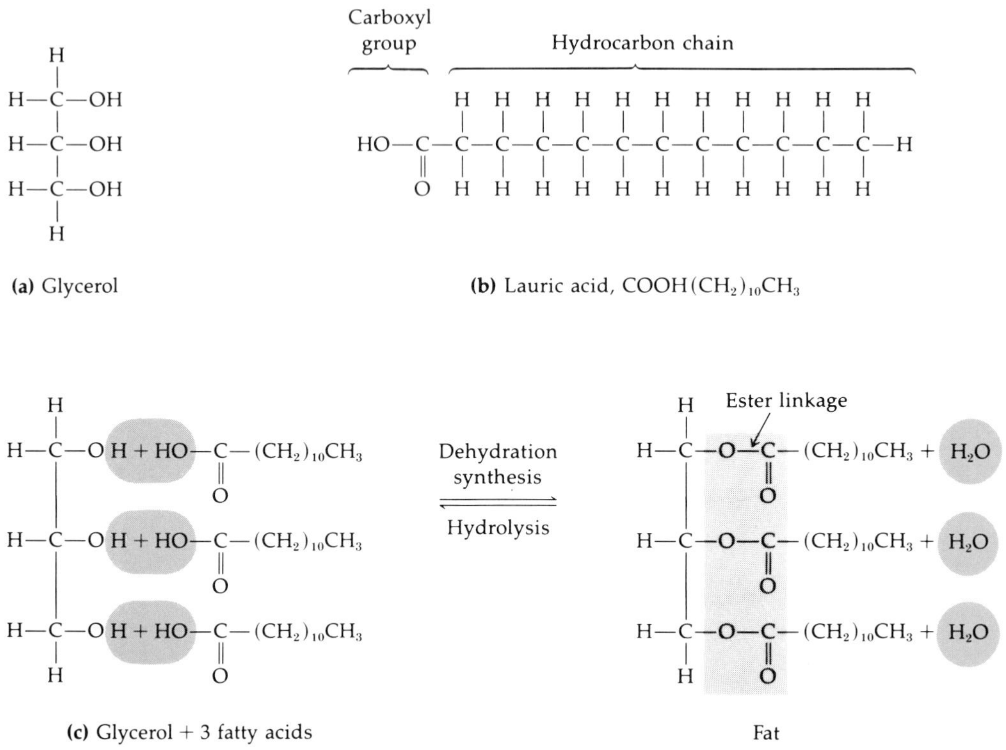

Figure 2–12 Structural formulas for **(a)** glycerol and **(b)** lauric acid, a fatty acid. In the condensed formula for lauric acid, $COOH(CH_2)_{10}CH_3$, the subscript 10 means that the molecule has ten CH_2 units. **(c)** The chemical combination of a molecule of glycerol with three fatty acid molecules forms one molecule of fat and three molecules of water in a dehydration synthesis reaction. The addition of three water molecules to a fat forms glycerol and three fatty acid molecules in a hydrolysis reaction.

Simple lipids

Simple lipids, or *fats,* contain an alcohol called *glycerol* and a group of compounds known as *fatty acids.* Glycerol molecules have three carbon atoms to which are attached three hydroxy (—OH) groups. The structural formula for glycerol is shown in Figure 2–12a. Fatty acids consist of long hydrocarbon chains (composed only of carbon and hydrogen atoms) ending in a carboxyl (—COOH) (organic acid) group (Figure 2–12b). Most common fatty acids contain an even number of carbon atoms.

A fat molecule is formed when a molecule of glycerol combines with one to three fatty acid molecules to form a monoglyceride, diglyceride, or triglyceride (Figure 2–12c). In the reaction, one to three molecules of water are formed (dehydration), depending on the number of fatty acid molecules reacting. The chemical bond formed where the water molecule is removed is an *ester linkage.* In the reverse reaction, hydrolysis, a fat molecule is broken down into its component fatty acid and glycerol molecules.

Because the fatty acids that form lipids have different structures, there is a wide variety of lipids. For example, three molecules of fatty acid A might combine with a glycerol molecule; or one molecule each of fatty acids A, B, and C might unite with a molecule of glycerol. Simple lipids serve as storage materials in higher organisms, but are not found in bacteria.

Complex lipids

Complex lipids contain elements such as phosphorus, nitrogen, and sulfur, in addition to the

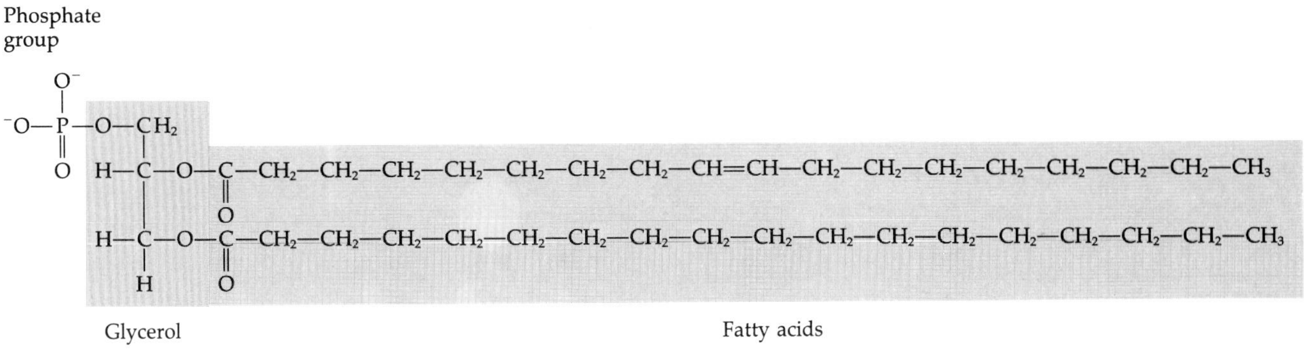

Figure 2–13 Structure of a phospholipid.

carbon, hydrogen, and oxygen found in simple lipids. The complex lipids called *phospholipids* are made up of glycerol, two fatty acids, and, in place of a third fatty acid, a phosphate group bonded to one of several organic groups (Figure 2–13). Phospholipids are the lipids that build membranes, and are thus essential to all procaryotic and eucaryotic cells. The structure of membranes will be discussed in detail in Chapter 4.

Steroids

Structurally very different from the lipids described previously are the lipids called *steroids*. Figure 2–14 shows the structure of the steroid cholesterol. The —OH group in cholesterol makes it a *sterol*. Sterols are important constituents of the plasma membranes of animal cells and of one group of bacteria (mycoplasmas), and they are also found in fungi and plants. Animals synthesize steroid hormones and vitamin D from sterols.

The cell wall of *Mycobacterium tuberculosis* (mī-kō-bak-ti'rē-um tü-ber-kū-lō'sis), the bacterium that causes tuberculosis, is distinguished by its lipid-rich content. The wall contains waxes and glycolipids (carbohydrates joined to lipids) that give the bacterium its distinctive staining characteristics. Other members of the genus *Mycobacterium* also have lipid-rich cell walls.

Proteins

Proteins are organic molecules that contain carbon, hydrogen, oxygen, and nitrogen. Some also contain sulfur. If you were to separate out and weigh all the groups of organic compounds in a living cell, the proteins would tip the scale. Hundreds of different proteins can be found in any single cell, and together they make up 50% or more of the cell's dry weight. They are essential ingredients in all aspects of cell structure and function. We have already mentioned enzymes, the proteins that catalyze biochemical reactions. But proteins have other vital functions as well. Carrier proteins help to transport certain chemicals into and out of cells. Other proteins, such as the bacteriocins produced by many bacteria, serve to kill other bacteria. Toxins produced by certain disease-causing microorganisms also fall into the protein category.

Some proteins play a key role in cell contraction and locomotion. Others, such as the hormones of eucaryotic organisms, have regulatory functions. Then there are proteins that play a key

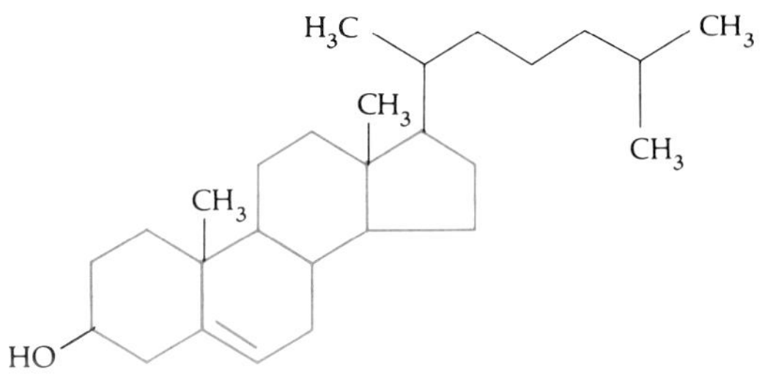

Figure 2–14 Cholesterol, a steroid. Note the four "fused" carbon rings, which are characteristic of steroid molecules. The hydrogen atoms attached to the carbons at the corners of the rings have been omitted.

role in vertebrate immune systems: antibodies, for instance, are proteins. And some proteins are simply structural, as is the collagen of animal skin and bones.

Amino acids

Just as the monosaccharides are the building blocks of larger carbohydrate molecules, and fatty acids and glycerol are the building blocks of fats, **amino acids** are the building blocks of proteins. Amino acids contain at least one carboxyl (—COOH) group and one amino (—NH_2) group attached to the same carbon atom, called an alpha-carbon (α-carbon) (Figure 2–15). Such amino acids are called *alpha-amino acids.* Also attached to the alpha-carbon is a side group (R group), which is the amino acid's distinguishing factor. The side group can be a hydrogen atom, an unbranched or branched chain, or a ring structure that is cyclic (all carbon) or heterocyclic (an atom other than carbon is included in the ring). The side group can contain functional groups such as the sulfhydryl group (—SH), the hydroxy group (—OH), or additional carboxyl or amino groups. These side groups and the carboxyl and alpha-amino groups affect the total structure of a protein, as will be described later. The structures and standard abbreviations of the 20 amino acids naturally found in proteins, are shown in Table 2–5.

Amino acids exist in either of two configurations called *stereoisomers,* designated by D and L. These two configurations are mirror images, corresponding to "right-handed" (D) and "left-handed" (L) three-dimensional shapes (Figure 2–16). The amino acids found in proteins are always the L-isomers (except for glycine, the simplest amino acid, which does not have stereoisomers). However, D-amino acids occasionally occur in nature, for example, in certain bacterial cell walls and antibiotics. (Many other kinds of organic molecules also can exist in D and L forms. One example is the sugar glucose, which occurs in nature as D-glucose.)

Although only 20 kinds of amino acids occur naturally in proteins, the total number of amino acids in a single protein molecule can vary from 50 to hundreds or more, and they can be arranged in an infinite number of ways. Thus, proteins have

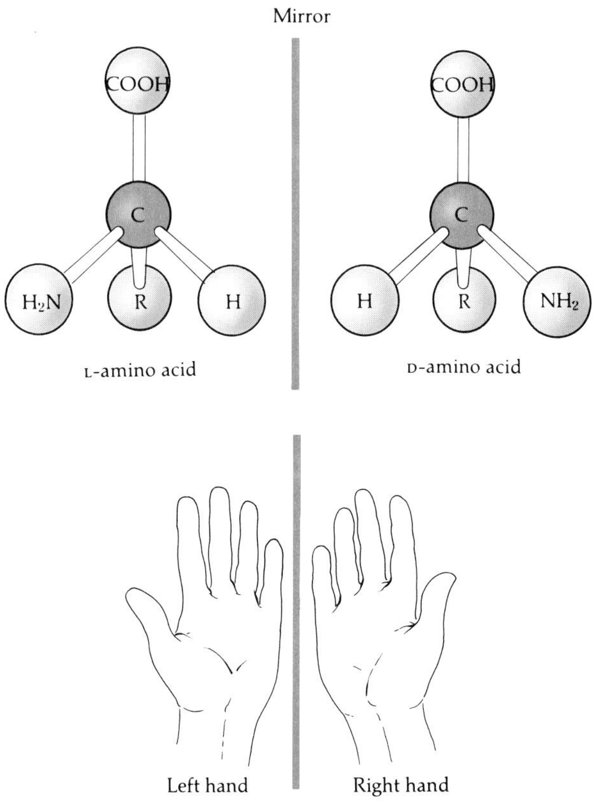

Figure 2–16 The L- and D-isomers of an amino acid, shown with ball-and-stick models. The two isomers, like left and right hands, are mirror images of each other and cannot be superimposed on one another (try it!).

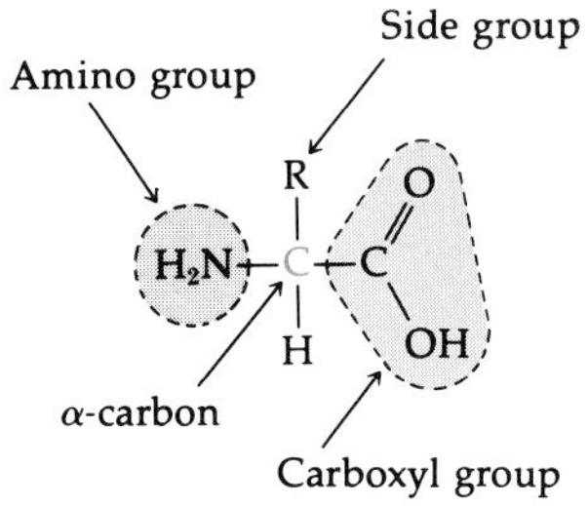

Figure 2–15. General structural formula for an amino acid. The alpha-carbon is shown in color. The letter R can stand for any of a number of groups of atoms. Different amino acids have different R groups.

Table 2–5 Twenty Amino Acids Found in Proteins

Amino Acid (Abbreviation)	Structural Formula	Characteristic of R Group
Glycine (Gly)	$H-\underset{\underset{NH_2}{\vert}}{\overset{\overset{H}{\vert}}{C}}-COOH$	Hydrogen atom
Alanine (Ala)	$CH_3-\underset{\underset{NH_2}{\vert}}{\overset{\overset{H}{\vert}}{C}}-COOH$	Unbranched chain
Valine (Val)	$H_3C-\underset{\underset{CH_3}{\vert}}{CH}-\underset{\underset{NH_2}{\vert}}{\overset{\overset{H}{\vert}}{C}}-COOH$	Branched chain
Leucine (Leu)	$H_3C-\underset{\underset{CH_3}{\vert}}{CH}-CH_2-\underset{\underset{NH_2}{\vert}}{\overset{\overset{H}{\vert}}{C}}-COOH$	Branched chain
Isoleucine (Ileu)	$H_3C-CH_2-\underset{\underset{CH_3}{\vert}}{CH}-\underset{\underset{NH_2}{\vert}}{\overset{\overset{H}{\vert}}{C}}-COOH$	Branched chain
Serine (Ser)	$HO-CH_2-\underset{\underset{NH_2}{\vert}}{\overset{\overset{H}{\vert}}{C}}-COOH$	Hydroxy (—OH) group
Threonine (Thr)	$H_3C-\underset{\underset{OH}{\vert}}{CH_2}-\underset{\underset{NH_2}{\vert}}{\overset{\overset{H}{\vert}}{C}}-COOH$	Hydroxy (—OH) group
Cysteine (Cys)	$HS-CH_2-\underset{\underset{NH_2}{\vert}}{\overset{\overset{H}{\vert}}{C}}-COOH$	Sulfhydryl (—SH) group

continued

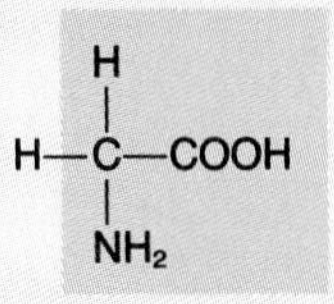

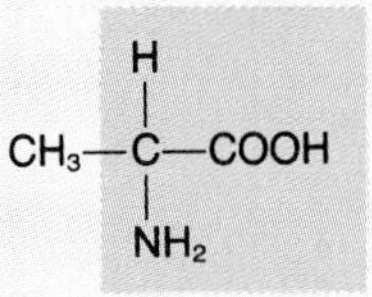

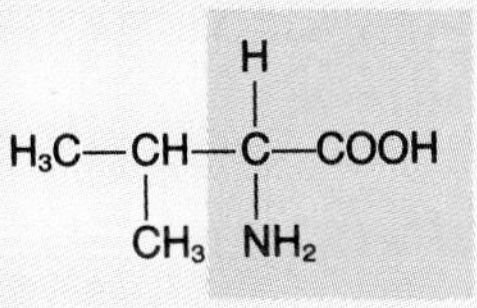

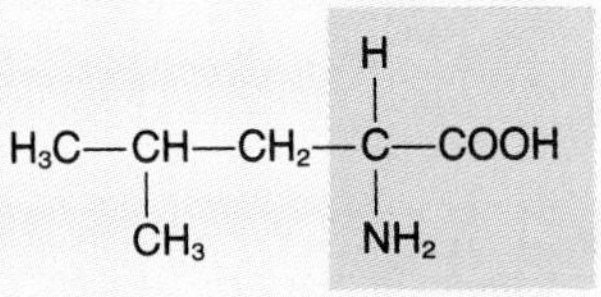

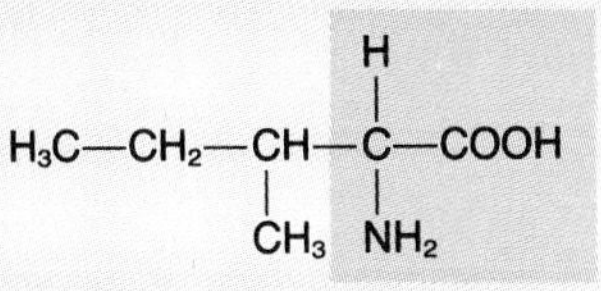

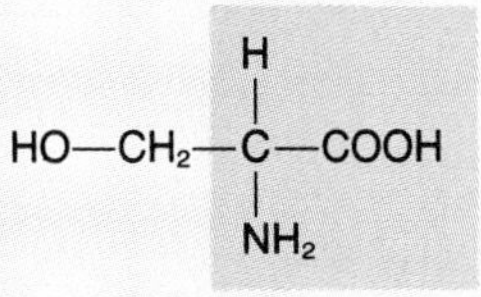

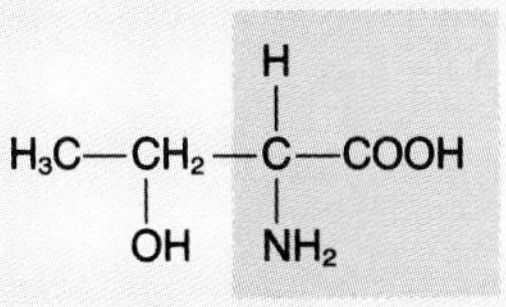

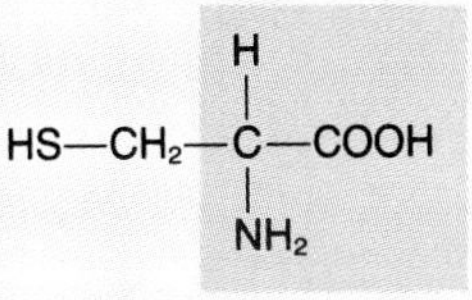

Table 2–5 Twenty Amino Acids Found in Proteins, *continued*

Amino Acid (Abbreviation)	Structural Formula	Characteristic of R Group
Methionine (Met)	$H_3C-S-CH_2-CH_2-\overset{\displaystyle H}{\underset{\displaystyle NH_2}{C}}-COOH$	Sulfhydryl (—SH) group
Glutamic acid (Glu)	$HOOC-CH_2-CH_2-\overset{\displaystyle H}{\underset{\displaystyle NH_2}{C}}-COOH$	Additional carboxyl (—COOH) group, acidic
Aspartic acid (Asp)	$HOOC-CH_2-\overset{\displaystyle H}{\underset{\displaystyle NH_2}{C}}-COOH$	Additional carboxyl (—COOH) group, acidic
Lysine (Lys)	$H_2N-CH_2-CH_2-CH_2-CH_2-\overset{\displaystyle H}{\underset{\displaystyle NH_2}{C}}-COOH$	Additional amino (—NH₂) group, basic
Arginine (Arg)	$H_2N-\overset{\displaystyle }{\underset{\displaystyle NH}{C}}-NH-CH_2-CH_2-CH_2-\overset{\displaystyle H}{\underset{\displaystyle NH_2}{C}}-COOH$	Additional amino (—NH₂) group, basic
Asparagine (Asn)	$H_2N-\overset{\displaystyle }{\underset{\displaystyle O}{C}}-CH_2-\overset{\displaystyle H}{\underset{\displaystyle NH_2}{C}}-COOH$	Additional amino (—NH₂) group, basic
Glutamine (Gln)	$H_2N-\overset{\displaystyle }{\underset{\displaystyle O}{C}}-CH_2-CH_2-\overset{\displaystyle H}{\underset{\displaystyle NH_2}{C}}-COOH$	Additional amino (—NH₂) group, basic
Phenylalanine (Phe)	$\langle\text{ring}\rangle-CH_2-\overset{\displaystyle H}{\underset{\displaystyle NH_2}{C}}-COOH$	Cyclic
Tyrosine (Tyr)	$HO-\langle\text{ring}\rangle-CH_2-\overset{\displaystyle H}{\underset{\displaystyle NH_2}{C}}-COOH$	Cyclic

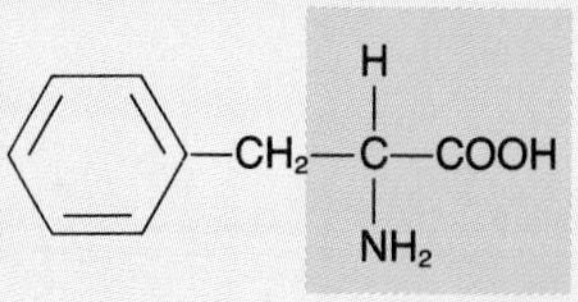

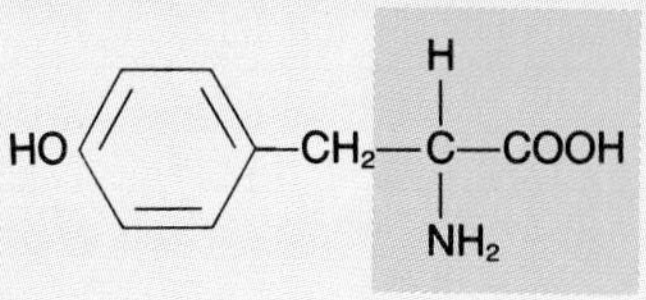

Table 2–5 Twenty Amino Acids Found in Proteins, *continued*

Amino Acid (Abbreviation)	Structural Formula	Characteristic of R Group
Histidine (His)	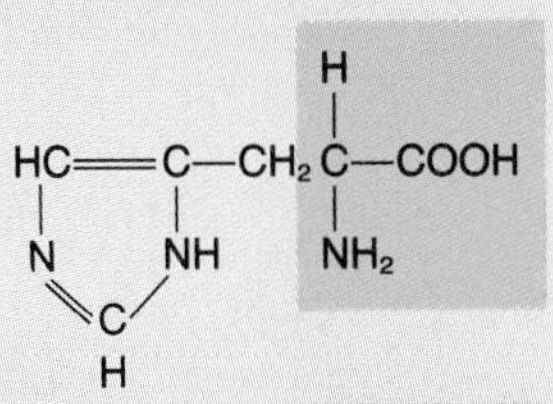	Heterocyclic
Trytophan (Trp)	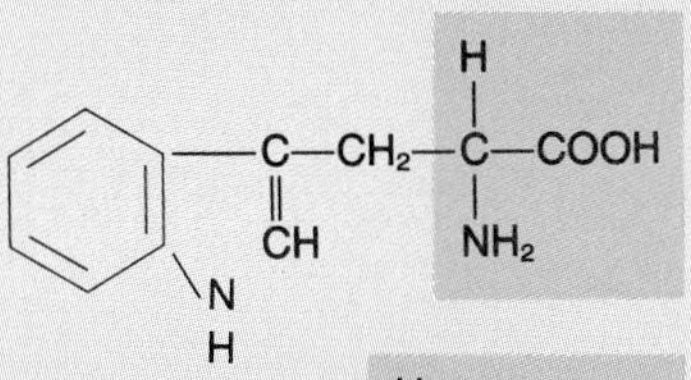	Heterocyclic
Proline (Pro)	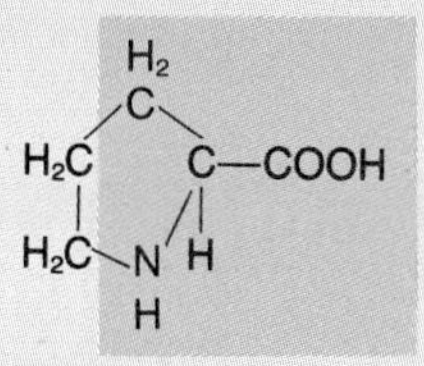	Heterocyclic

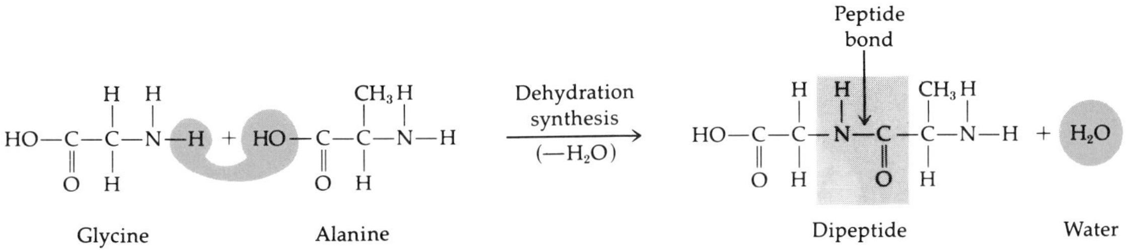

Figure 2–17 Peptide bond formation by dehydration synthesis. The amino acids glycine and alanine combine to form a dipeptide. The newly formed bond between the nitrogen atom of glycine and the carbon atom of alanine is called a peptide bond.

different lengths, different quantities of the various amino acid units, and different specific sequences in which the amino acids are bonded. The number of proteins is practically endless, and every living cell produces many different proteins.

Peptide bonds

The amino acids bond between the carboxyl (—COOH) group of one amino acid and the amino (—NH_2) group of another. For every bond between two amino acids, one water molecule is released (dehydration). The bonds between amino acids are called **peptide bonds.**

In the example in Figure 2–17, a peptide bond between two amino acids is formed by dehydration synthesis. The carboxyl group of one amino acid supplies an OH^- and the amino group of the other releases an H^+ for the formation of water. The peptide bond is between the carbon atom of the carboxyl group of one amino acid and the nitrogen atom of the amino group of another amino acid. The resulting compound, because it consists of two amino acids joined via a peptide bond, is called a *dipeptide.* Adding another amino acid to a dipeptide would form a *tripeptide.* Further additions of amino acids would produce a long, chainlike molecule called a *polypeptide,* a large protein molecule.

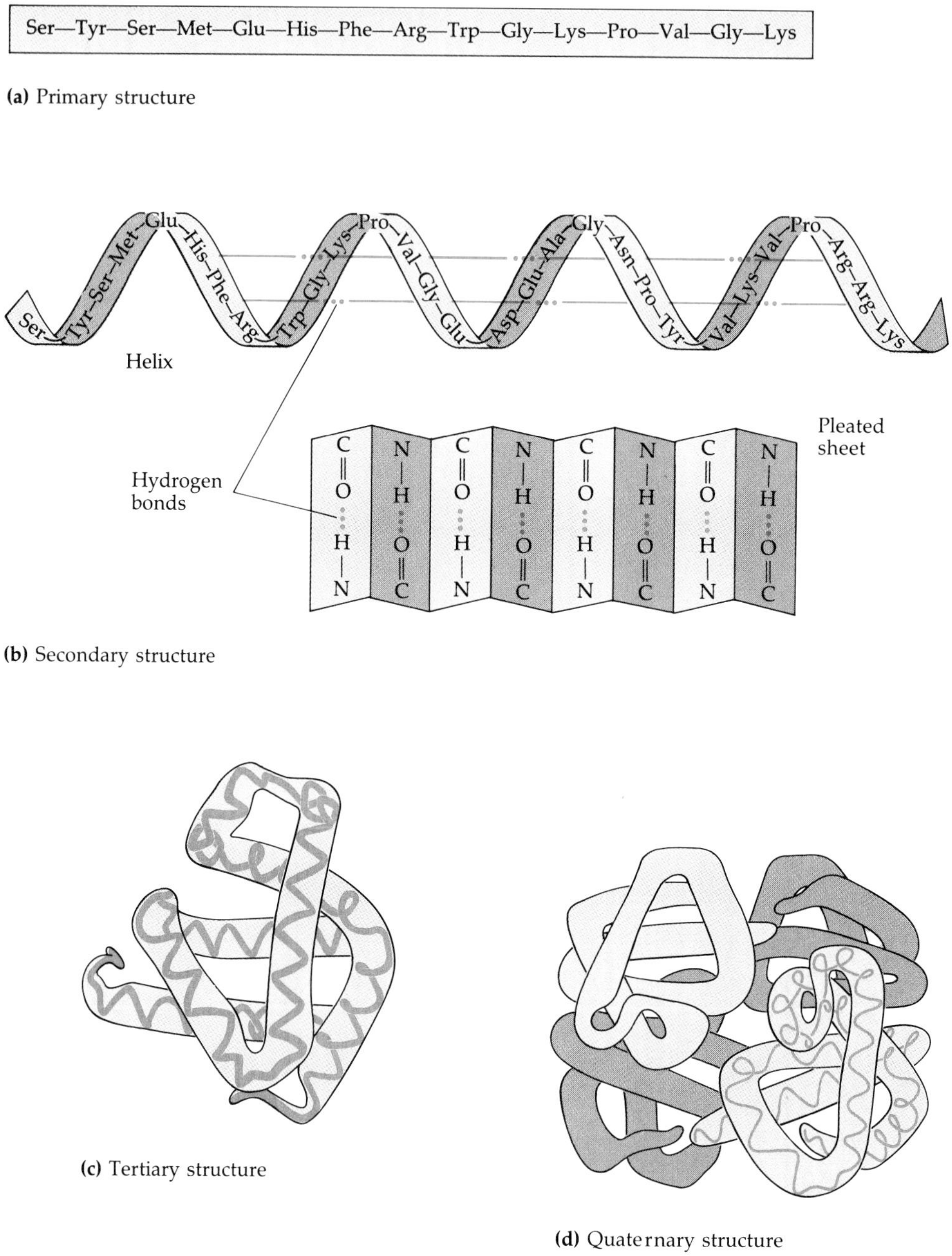

Figure 2–18 Protein structure. **(a)** Primary structure, the sequence of amino acids. Each amino acid is designated by its three-letter abbreviation. **(b)** Secondary structure, a helix and a pleated sheet. **(c)** Tertiary structure, the folding of a polypeptide chain. **(d)** Quaternary structure, the relationship between several polypeptide chains that make up a protein. Shown here is the quaternary structure of a protein with four chains, two of one type and two of another.

Levels of protein structure

Proteins can have three or four levels of organization: primary, secondary, tertiary, and quaternary. The *primary structure* is the order (the sequence) in which the amino acids are linked together (Figure 2–18a). Alterations in sequence can have profound metabolic effects. For example, a single incorrect amino acid in a blood protein can produce the deformed hemoglobin molecule characteristic of sickle-cell anemia.

Proteins are not simple linear chains of amino acids. A protein's *secondary structure* is any regular coil or zigzag arrangement along one dimension. This aspect of a protein's shape generally comes from hydrogen bonds joining the atoms of peptide bonds at different locations along the polypeptide chain. Common secondary protein structures are clockwise spirals called helixes and pleated sheets (Figure 2–18b), which form from roughly parallel portions of the chain. Both structures are held together by hydrogen bonds.

Tertiary structure refers to how the protein is bent or folded into a three-dimensional shape (Figure 2–18c). This nonregular looping and twisting is the result of interactions between various functional groups or branches of specific amino acids. Hydrogen bonds and other relatively weak interactions play an important role in tertiary structure. In addition, sulfhydryl groups (—SH) on two amino acid subunits can form a covalent, disulfide link (—S—S—) by removing the hydrogen atoms (oxidation).

Some proteins have a *quaternary structure,* which results from an aggregation of two or more individual polypeptide units (Figure 2–18d). For example, an enzyme called phosphorylase *a* consists of four identical polypeptide subunits. These units alone do not exhibit enzymatic activity, but they become active when in groups of four. Certain other proteins with quaternary structure have two or more different kinds of polypeptide subunits.

The proteins we have been discussing are *simple proteins,* which contain only amino acids. *Conjugated proteins* are combinations of amino acids with other organic or inorganic components. Conjugated proteins are named by their non-amino-acid component. Thus, glycoproteins contain sugars, nucleoproteins contain nucleic acids, metalloproteins contain metal atoms, lipoproteins contain lipids, and phosphoproteins contain phosphate groups. An example of a microbial phosphoprotein is phospholipase C. Manufactured by the bacterium *Clostridium perfringens* (klôs-tri'dē-um pėr-frin'jens), this enzyme breaks down red blood cells and induces some of the symptoms of gas gangrene.

Nucleic Acids

First discovered in the nuclei of cells, **nucleic acids** are exceedingly large organic molecules containing carbon, hydrogen, oxygen, nitrogen, and phosphorus. Whereas the basic structural units of proteins are amino acids, the basic units of nucleic acids are *nucleotides.* Nucleic acids are of two principal kinds: *deoxyribonucleic acid (DNA)* and *ribonucleic acid (RNA).*

DNA

A molecule of DNA is a chain of many nucleotide units. Each nucleotide of DNA has three parts: a nitrogen-containing base, a pentose (five-carbon sugar), and a phosphate group (phosphoric acid molecule) (Figure 2–19a). The nitrogen-containing base is either adenine, guanine, cytosine, or thymine. All the nitrogen-containing groups are ring structures containing atoms of carbon, hydrogen, oxygen, and nitrogen. Adenine and guanine are double-ring structures, collectively referred to as *purines.* Thymine and cytosine are smaller, single-ring structures, called *pyrimidines.* The pentose in DNA is *deoxyribose.*

Nucleotides are named according to their nitrogenous base. Thus, a nucleotide containing thymine is called a *thymine nucleotide.* One containing adenine is called an *adenine nucleotide,* and so on. The term *nucleoside* refers to the combination of a purine or pyrimidine plus a pentose sugar; it does not contain a phosphate group.

Although the chemical composition of the DNA molecule was known before 1900, it was not until 1953 that the organization of the chemical subunits was modeled. J. D. Watson and F. H. C. Crick proposed this model on the basis of data from many investigations. Figure 2–19b shows the following structural characteristics of the DNA molecule:

1. The molecule consists of two strands with crossbars. The strands twist around each other

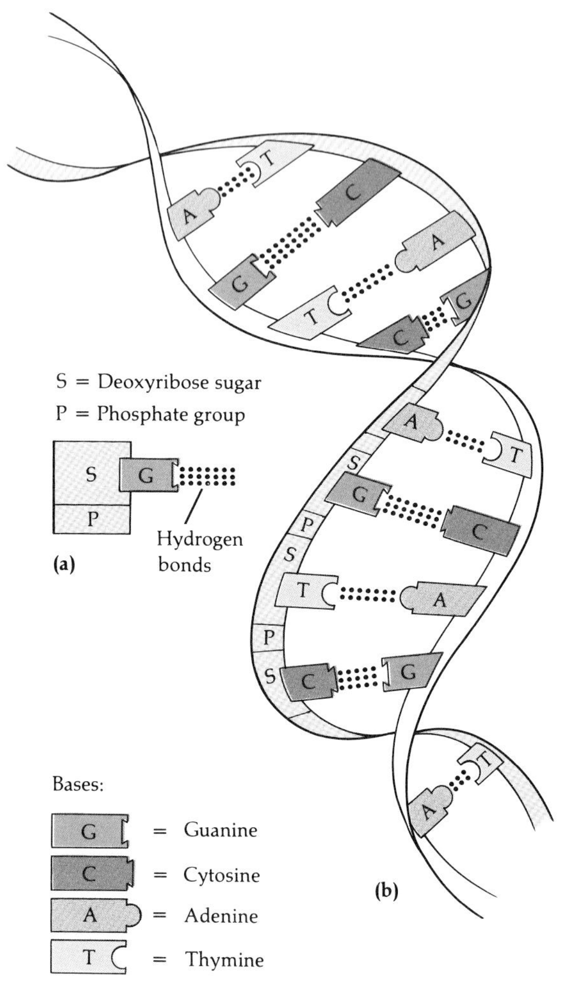

Figure 2–19 Structure of DNA. **(a)** Diagram of a guanine nucleotide. **(b)** Part of an assembled DNA molecule. The colored dots represent hydrogen bonds between nitrogenous bases.

in the form of a *double helix,* so that the shape resembles a twisted ladder.

2. The uprights of the DNA ladder, called the backbone, consist of alternating phosphate groups and the deoxyribose (sugar) portions of the nucleotides.

3. The rungs of the ladder contain nitrogenous bases in pairs joined by hydrogen bonds. As shown, a purine always pairs with a pyrimidine; that is, adenine always pairs with thymine, and cytosine always pairs with guanine.

Cells contain hereditary material called genes, each of which is a segment of a DNA molecule. Genes determine all hereditary traits, and they control all the activities that take place within cells. When a cell divides, its hereditary information is passed on to the next generation. This transfer of information is possible because of DNA's unique structure, and will be discussed further in Chapter 8.

RNA

RNA, the second principal kind of nucleic acid, differs from DNA in several respects. Whereas DNA is double-stranded, RNA is usually single-stranded. The five-carbon sugar in the RNA nucleotide is ribose, which has one more oxygen atom than deoxyribose. And one of RNA's bases is uracil instead of thymine. At least three different kinds of RNA have been identified in cells. These are referred to as messenger RNA, ribosomal RNA, and transfer RNA. As we shall see in Chapter 8, each type of RNA has a specific role in protein synthesis.

Adenosine Triphosphate (ATP)

A molecule that is indispensable to the life of the cell is **adenosine triphosphate (ATP).** ATP is the principal energy-carrying molecule of all cells. It stores the chemical energy released by some chemical reactions, and it provides the energy for energy-requiring reactions. ATP consists of an adenosine unit, composed of adenine and ribose, and three phosphate groups (Figure 2–20). In other words, it is an adenine nucleotide with two extra phosphate groups. ATP is a high-energy molecule because it releases a large amount of usable energy when a water molecule is added and it loses a terminal phosphate group ($—PO_3$ or Ⓟ), to become **adenosine diphosphate (ADP).** This reaction can be represented as follows:

$$\underset{\text{Adenosine triphosphate}}{\text{ATP}} \rightleftharpoons \underset{\text{Adenosine diphosphate}}{\text{ADP}} + \underset{\text{Phosphate}}{Ⓟ} + \underset{\text{Energy}}{\text{E}}$$

Because the supply of ATP at any particular time is limited, a mechanism exists to replenish it: the addition of a phosphate group to ADP manufactures more ATP. Because we can assume that

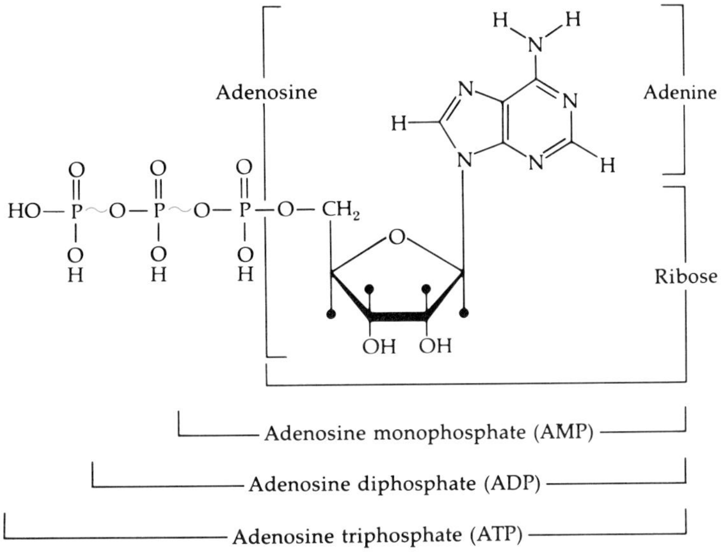

Figure 2–20 Structure of ATP. High-energy phosphate bonds are shown in colored wavy lines. When ATP breaks down to ADP and inorganic phosphate, a large amount of chemical energy is released for use in other chemical reactions.

energy is required to manufacture ATP, the reaction can be represented as follows:

ADP	+	Ⓟ	+	E	⇌	ATP
Adenosine diphosphate		Phosphate		Energy		Adenosine triphosphate

The energy required to attach a phosphate group to ADP is supplied by the cell's various decomposition reactions, particularly the decomposition of glucose. ATP can be stored in every cell, where its potential energy is not released until needed.

STUDY OUTLINE

WHY STUDY CHEMISTRY IN A MICROBIOLOGY COURSE? (pp. 24–25)

1. The interaction of atoms and molecules is called chemistry.
2. The metabolic activities of microorganisms involve complex chemical reactions.
3. Nutrients are broken down by microbes to obtain energy and molecules that are used to make new cells.

STRUCTURE OF ATOMS (pp. 25–28)

1. Atoms are the smallest units of chemical elements that enter into chemical reactions.
2. Atoms consist of a nucleus, which contains protons and neutrons, and electrons that orbit around the nucleus.
3. Atomic number refers to the number of protons in the nucleus; the total number of protons and neutrons is the atomic weight.

Chemical Elements (pp. 25–26)

1. Atoms with the same atomic number and same chemical behavior are classified as the same chemical element.
2. Chemical elements are designated by letter abbreviations called chemical symbols.

3. There are about 26 elements commonly found in living cells.
4. Atoms that are of the same element and have the same atomic number but different atomic weights are called isotopes.

Electronic Configurations (pp. 26–28)

1. In an atom, electrons are arranged around the nucleus in electron shells.
2. Each shell can hold a characteristic number of electrons.
3. Chemical properties of an atom are largely due to the number of electrons in its outermost shell: to fill the outermost shell, it reacts with other atoms.

HOW ATOMS FORM MOLECULES: CHEMICAL BONDS (pp. 28–32)

1. Atoms form molecules in order to fill their outermost electron shells.
2. Attractive forces that bind the atomic nuclei of two atoms are called chemical bonds.
3. The combining capacity of an atom is its valence, the number of chemical bonds the atom can form with other atoms.

Ionic Bonds (pp. 28–31)

1. A positively or negatively charged atom or group of atoms is called an ion.
2. A chemical attraction between ions of opposite charge is called an ionic bond.
3. Ionic bonds have only limited importance in living cells.

Covalent Bonds (p. 31)

1. In a covalent bond, atoms share pairs of electrons.
2. Covalent bonds are stronger than ionic bonds and are far more common in organisms.

Hydrogen Bonds (pp. 31–32)

1. A hydrogen bond exists when a hydrogen atom covalently bonded to one oxygen or nitrogen atom is attracted to another.
2. Hydrogen bonds form weak links between different molecules or between parts of the same large molecule.

Molecular Weight and Moles (p. 32)

1. Molecular weight is the sum of the atomic weights of all the atoms in a molecule.
2. A mole of an atom, ion, or molecule is equal to its weight divided by the molecular weight.

CHEMICAL REACTIONS (pp. 32–35)

1. Chemical reactions are the making or breaking of chemical bonds between atoms.
2. Three basic types of chemical reactions are synthesis, decomposition, and exchange reactions. In theory, all chemical reactions are reversible.

How Chemical Reactions Occur (pp. 33–35)

1. For a chemical reaction to take place, the reactants must collide with each other (collision theory).
2. The minimum energy of collision that can produce a chemical reaction is called activation energy.
3. Specialized proteins called enzymes accelerate chemical reactions in living systems by lowering the activation energy.

Energy of Chemical Reactions (p. 35)

1. When a chemical bond is formed, energy is used.
2. When a bond is broken, energy is released.
3. Endergonic reactions require energy; exergonic reactions release energy.

Forms of Energy (p. 35)

1. Chemical, mechanical, radiant, and electrical energy all are important to living cells.
2. These forms of energy can exist as either potential or kinetic energy.

IMPORTANT BIOLOGICAL MOLECULES (pp. 35–53)

INORGANIC COMPOUNDS (pp. 35–38)

1. Inorganic compounds are usually small, ionically bonded molecules.
2. Water, acids, bases, and salts are examples of inorganic compounds.

Water (pp. 35–36)

1. Water is the most abundant substance in cells.
2. Because water is a polar molecule, it is an excellent solvent.
3. Water is an excellent temperature buffer.

Acids, Bases, and Salts (pp. 36–37)

1. An acid dissociates into H^+ ions and anions.
2. A base dissociates into OH^- ions and cations.
3. A salt dissociates into negative and positive ions, neither of which are H^+ or OH^-.

Acid–Base Balance (pp. 37–38)

1. The term pH refers to the concentration of H^+ in a solution.
2. A solution with a pH of 7 is neutral; a pH below 7 indicates acidity; a pH above 7 indicates alkalinity.
3. A pH buffer, which stabilizes the pH inside a cell, can be used in culture media.

ORGANIC COMPOUNDS (pp. 38–53)

1. Carbon is the characteristic element of organic compounds.
2. Carbon atoms form four bonds with carbon and other atoms.
3. Organic compounds are mostly or entirely covalently bonded, and many of them are large molecules.
4. Small organic molecules may combine into very large molecules called macromolecules.
5. Monomers usually bond together by condensation reactions that form water and a polymer.
6. Polymers may be broken down by hydrolysis, a reaction requiring the splitting of a water molecule.
7. Functional groups are responsible for most of the properties of organic molecules.

Carbohydrates (pp. 42–43)

1. Carbohydrates consist of atoms of carbon, hydrogen, and oxygen, with hydrogen and oxygen in a 2:1 ratio.
2. Carbohydrates include sugars and starches.
3. Carbohydrates are divided into three groups: monosaccharides, disaccharides, and polysaccharides.
4. Monosaccharides may form disaccharides and polysaccharides by dehydration synthesis.
5. Polysaccharides and disaccharides may be broken down by hydrolysis.

Lipids (pp. 43–45)

1. Lipids are a diverse group of compounds distinguished by their insolubility in water.
2. Simple lipids (fats) consist of a molecule of glycerol and three molecules of fatty acids.
3. Phospholipids are complex lipids consisting of glycerol, two fatty acids, and phosphate.
4. Steroids are carbon-ring systems with functional hydroxyl and carbonyl groups.

Proteins (pp. 45–51)

1. Amino acids are the building blocks of proteins.
2. Amino acids consist of carbon, hydrogen, oxygen, nitrogen, and sometimes sulfur.
3. By linking amino acids, peptide bonds (formed by hydrolysis) form polypeptide chains.
4. Twenty amino acids occur naturally.
5. Proteins have four levels of structure: primary (linear chain of amino acids); secondary (regular coils or pleats); tertiary (irregular folds); and quaternary (two or more polypeptide chains).
6. Conjugated proteins consist of amino acids and other organic or inorganic compounds.

Nucleic Acids (pp. 51–52)

1. Nucleic acids—DNA and RNA—are macromolecules consisting of repeating nucleotides.
2. A nucleotide is composed of a pentose, phosphoric acid, and a nitrogenous base. A nucleoside is composed of a pentose and a nitrogenous base.
3. A DNA nucleotide consists of deoxyribose (pentose) and one of these nitrogenous bases: thymine or cytosine (called pyrimidines), adenine or guanine (called purines).
4. DNA consists of two strands of nucleotides wound in a double helix. The strands are held together by hydrogen bonds between a purine and a pyrimidine.
5. An RNA nucleotide consists of ribose (pentose) and one of these nitrogenous bases: cytosine, guanine, adenine, or uracil.

Adenosine Triphosphate (ATP) (pp. 52–53)

1. ATP stores chemical energy for various cellular activities.
2. When ATP's terminal phosphate group is broken, energy is released.
3. The energy from decomposition reactions is used to regenerate ATP from ADP.

STUDY QUESTIONS

REVIEW

1. What is a chemical element?
2. Diagram the electronic configuration of a carbon atom.
3. How does $^{14}_{6}C$ differ from $^{12}_{6}C$?
4. What type of bonds will hold the following atoms together?
   (a) Li^+ and Cl^- ions in LiCl
   (b) Carbon and oxygen atoms in CO_2
   (c) Oxygen atoms in O_2
   (d) A hydrogen atom from glutamic acid and a nitrogen atom from lysine in

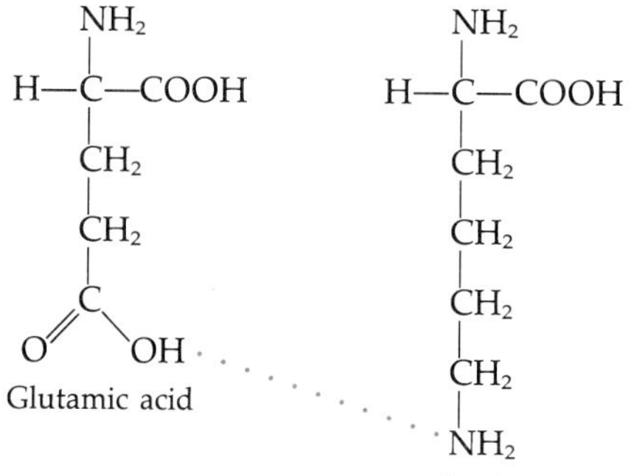

Glutamic acid
Lysine

   (e) A hydrogen atom of one nucleotide to nitrogen or oxygen atoms of another nucleotide in

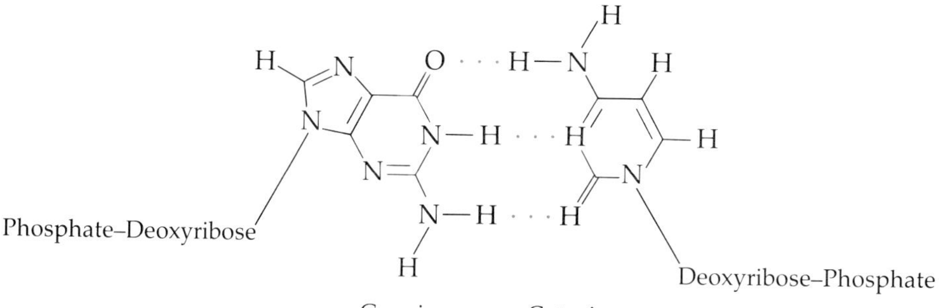

Guanine Cytosine

5. What type of bonding exists between water molecules in a beaker of water?
6. Classify the following inorganic molecules as an acid, base, or salt. Their dissociation products are shown to help you.
   (a) $HNO_3 \rightarrow H^+ + NO_3^-$
   (b) $H_2SO_4 \rightarrow 2H^+ + SO_4^{2-}$
   (c) $NaOH \rightarrow Na^+ + OH^-$
   (d) $Mg_2SO_4 \rightarrow 2Mg^{2+} + SO_4^{2-}$

7. Vinegar, pH 3, is how many times more acidic than pure water?

8. Calculate the molecular weight of $C_6H_{12}O_6$.

9. How many moles are in 360 grams of glucose?

10. Classify the following types of chemical reactions.
 (a) Glucose + Fructose → Sucrose
 (b) Lactose → Glucose + Galactose
 (c) $NH_4Cl + H_2O → NH_4OH + HCl$
 (d) ATP ⇌ ADP + Ⓟ

11. The following reaction requires an enzyme. What purpose does the enzyme serve in this reaction?

 Lactose → Glucose + Galactose

12. Classify the following as subunits of either a carbohydrate, lipid, protein, or nucleic acid.

 (a) CH_3—$(CH_2)_7$—CH=CH—$(CH_2)_7$—COOH
 　　　　　　　Oleic acid

 (b)　　　NH_2
 　　　　　|
 　　H—C—COOH
 　　　　　|
 　　　　CH_2
 　　　　　|
 　　　　OH
 　　　Serine

 (c) $C_6H_{12}O_6$
 (d) Thymine nucleotide

13. Water plays an important role in these reactions:

 HO—C—C—N—C—C—N—H + H_2O ⇌ HO—C—C—N—H + HO—C—C—N—H

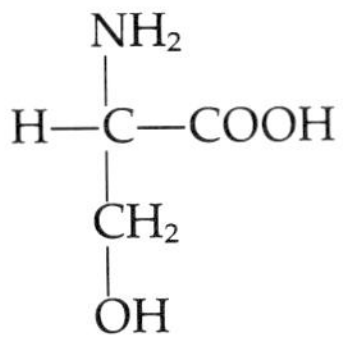

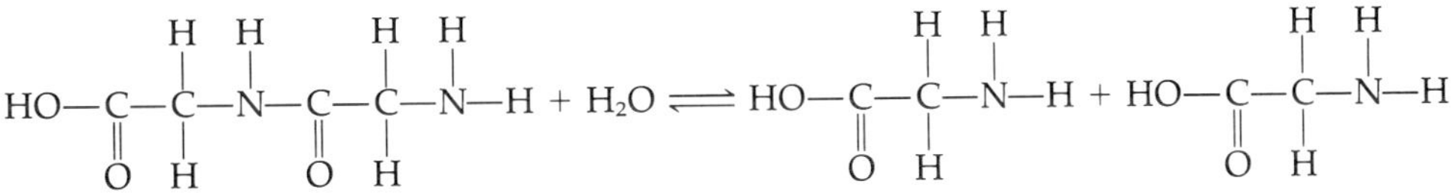

 (a) What direction is the hydrolysis reaction (left to right or right to left)?
 (b) What direction is the dehydration synthesis reaction?
 (c) Circle the atoms involved in the formation of water.
 (d) Identify the peptide bond.

14. This diagram shows a protein molecule. Indicate the regions of primary, secondary, and tertiary structure. Does this protein have quaternary structure?

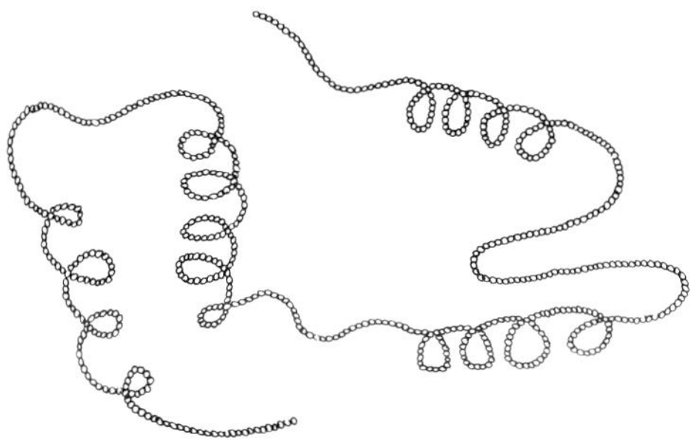

15. Draw a simple lipid and show how it could be modified to a phospholipid.

16. ATP is an energy-storage compound. Where does it get this energy from?

17. The energy-carrying property of the ATP molecule is due to energy dynamics that favor the breaking of bonds between ______________. These bonds are (ionic/covalent/hydrogen) bonds.

CHALLENGE

1. When you blow bubbles into a glass of water, the following reactions take place:

$$H_2O + CO_2 \xrightarrow{A} H_2CO_3 \xrightarrow{B} H^+ + HCO_3^-$$

(a) What type of reaction is *A*?
(b) What does reaction *B* tell you about the type of molecule H_2CO_3 is?

2. What are the common structural characteristics of ATP and DNA molecules?

3. *Thiobacillus ferrooxidans* was responsible for destroying buildings in the midwest by causing changes in the earth. The original rock, which contained lime ($CaCO_3$) and pyrite (FeS_2), expanded as the bacteria's metabolism caused gypsum ($CaSO_4$) crystals to form. How did *Thiobacillus* bring about the change from lime to gypsum?

FURTHER READING

Asimov, I. 1962. *The World of Carbon.* New York: Collier Books. Designed for the nonchemist, an introduction to organic chemistry.

Dickerson, R. E., H. B. Gray, M. Y. Darensbourg, and D. J. Darensbourg. 1984. *Chemical Principles,* 4th ed. Menlo Park, CA: Benjamin/Cummings. An introductory college chemistry textbook with a chapter on the "special role of carbon."

Lechtman, M. D., B. Roohk, and R. J. Egan. 1979. *The Games Cells Play: Basic Concepts of Cellular Metabolism.* Menlo Park, CA: Benjamin/Cummings. An innovative guide to the basics of chemistry for the biology student; contains chapter self-tests.

Luria, S. E., S. J. Gould, and S. Singer, 1981. *A View of Life.* Menlo Park, CA: Benjamin/Cummings. Chapter 2 provides an introduction to chemistry for the biology student.

Stryer, L. 1981. *Biochemistry,* 2nd ed. San Francisco, CA: W. H. Freeman. A technical survey of biochemistry, with a cellular approach.

Watson, J. D. *The Double Helix,* ed. Gunther S. Stent, 1980. New York: W. W. Norton. Original papers and Watson's account of the struggle to decipher the structure of DNA give insight to creative scientific processes. Includes commentaries and reviews by other scientists.

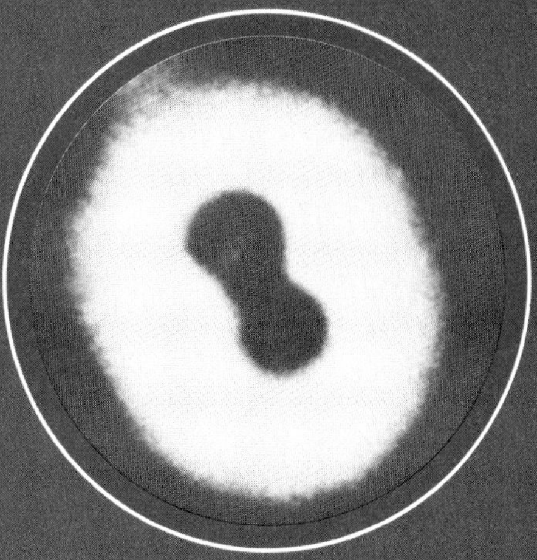

CHAPTER 3

Observing Microorganisms Through a Microscope

OBJECTIVES

After completing this chapter you should be able to

- List the units of measurement used for microorganisms, and know their equivalents.
- Diagram the path of light through a compound microscope.
- Define resolution and total magnification.
- Cite an advantage of each of the following types of microscopy and compare each to brightfield illumination: (a) phase-contrast, (b) darkfield, (c) differential interference contrast, (d) fluorescence.
- Explain how electron microscopy differs from light microscopy.
- Differentiate between an acid dye and a basic dye.
- Compare simple, differential, and special stains.
- List the steps in a Gram stain and describe the appearance of a gram-positive and gram-negative cell after each step.

Microorganisms are too small to be seen with the naked eye, so they must be observed with a microscope (*micro* means "small," *skopein* means "to see"). In Leeuwenhoek's day, looking through a microscope meant looking through rainbow rings and shadows and multiple images. Modern microbiologists, however, have access to microscopes that produce, with great clarity, magnifications anywhere from 10 to thousands of times more powerful than those of Leeuwenhoek's simple single lens (see Figure 1–2a). A major part of this chapter will describe how different microscopes function and the advantages of each.

Some microbes are more visible than others. Many have to be escorted through several staining procedures before cell walls, membranes, and structures lose their opaque or colorless natural state. The last part of this chapter will explain the methods of preparing specimens for light microscopy.

With the prospect of microbial images dancing around in your head, you may be wondering just how we are going to sort out, measure, and count the specimens we will be studying. Therefore, we open this chapter with a short discussion of how to use the metric system for measuring microbes.

UNITS OF MEASUREMENT

Because microorganisms and their component parts are so very small, they are measured in units that are unfamiliar to many of us in everyday life. When measuring microorganisms, we use the metric system. The standard unit of length in the metric system is the **meter (m).** A major advantage of the metric system is that the units are related to each other by factors of 10. Thus, 1 m is the same as 10 decimeters (dm) or 100 centimeters (cm) or 1000 millimeters (mm). (To convert units within the English system of length we would have to use 3 feet or 36 inches to equal 1 yard.)

Microorganisms and their structural components are measured in even smaller units, such as micrometers, nanometers, and angstroms. A **micrometer (μm),** formerly known as a micron (μ), is equal to 0.000001 m. The prefix *micro* indicates that the unit following it should be divided by one million (10^6) (see Box on Exponential Notation, in Chapter 2). A **nanometer (nm),** formerly known as a millimicron (mμ), is equal to 0.000000001 m. *Nano* tells us that the unit after it should be divided by one billion (10^9). An **angstrom (Å)** is equal to 0.0000000001 m, or 10^{-10} m. The angstrom is no longer an official unit of measure; however, because of its widespread presence in the literature you should be familiar with it. Its accepted equivalent is 0.1 nm. Table 3–1 presents the basic metric units of length and some of their English equivalents. Here you can compare the microscopic units of measurement with the commonly known macroscopic units of measurement such as centimeters, meters, and kilometers.

MICROSCOPY: THE INSTRUMENTS

The simple microscopes used by Leeuwenhoek and Hooke had only one lens, similar to a magnifying glass. A combination of lenses was later used by Joseph Jackson Lister (the father of Joseph Lister) in 1830. This development led to the mod-

Table 3–1 Metric Units of Length and English Equivalents

Metric Unit	Meaning of Prefix	Metric Equivalent	English Equivalent
1 kilometer (km)	*kilo* = 1000	1000 m = 10^3 m	3280.84 ft or 0.62 mi; 1 mi = 1.61 km
1 meter (m)		Standard unit of length	39.37 inches or 3.28 ft or 1.09 yd
1 decimeter (dm)	*deci* = 1/10	0.1 m = 10^{-1} m	3.94 inches
1 centimeter (cm)	*centi* = 1/100	0.01 m = 10^{-2} m	0.394 inch; 1 inch = 2.54 cm
1 millimeter (mm)	*milli* = 1/1000	0.001 m = 10^{-3} m	
1 micrometer (μm) (formerly micron (μ))	*micro* = 1/1,000,000	0.000001 m = 10^{-6} m	
1 nanometer (nm) (formerly millimicron (mμ))	*nano* = 1/1,000,000,000	0.000000001 m = 10^{-9} m	
1 angstrom (Å)		0.0000000001 m = 10^{-10} m	

ern compound microscope, the kind used today in a microbiology laboratory.

Compound Light Microscopy

A modern **compound light microscope** has two sets of lenses, objective and ocular, and uses visible light as its source of illumination (Figure 3–1a). Using the compound light microscope, we can examine very small specimens, as well as their fine detail or *ultrastructure* in some cases. A series of finely ground lenses form a clearly focused image that is many times larger than the specimen itself. This magnification is achieved when light rays from an *illuminator,* the light source, are passed through a *condenser,* which directs the light rays through the specimen. From here, light rays pass into the *objective lens,* the lens closest to the specimen. The image of the specimen is magnified again by the *ocular lens,* or *eyepiece* (Figure 3–1b).

We calculate the total magnification of a specimen by multiplying the objective lens magnification (power) by the ocular lens magnification (power). Most microscopes used in microbiology have several objective lenses, including 10 times (low power), 40 times (high power), and 100 times (oil immersion). Most oculars magnify specimens by 10 times. Multiplying the magnification of a specific objective lens with that of the ocular, we see that the total magnifications would be 100 times for low power, 400 times for high power, and 1000 times for oil immersion. Some compound light microscopes can achieve a magnification of 2000 times with oil immersion.

A general principle of microscopy is that the shorter the wavelength of light used in the instrument, the greater the resolution. **Resolution,** or **resolving power,** is the ability of the lenses to distinguish fine detail and structure. Specifically, it refers to the ability of the lenses to distinguish between two points set at a specified distance apart. For example, if a microscope has a resolving power of 0.4 nm (4 Å), it is capable of distinguishing two points as separate objects if they are at least 0.4 nm (4 Å) apart. The white light used in a compound light microscope has a relatively long wavelength and cannot resolve structures smaller than 0.3 μm. This fact and practical considerations limit the magnification achieved by even the best compound light microscopes to about 2000 times.

It is usually necessary to stain a specimen for examination under a compound light microscope, because the refractive index of the specimen is nearly the same as the refractive index of the medium that surrounds it. The **refractive index** is the relative velocity at which light passes through a material. Light rays traveling in a single medium usually move in a straight line. But when light rays pass through two materials with different refractive indexes, the rays change direction (refract) at the boundary between the materials, and increase the image's contrast between the specimen and the medium.

It is important to preserve the direction of light rays after they have passed through the stained specimen; this is accomplished by the use of immersion oil between the glass slide and the objective lens. Because immersion oil has the same refractive index as glass, the oil becomes part of the optics of the glass of the microscope. Unless the oil is used, light rays are refracted as they enter the air from the slide, and the objective would have to be increased in diameter to capture them. The oil has the same effect as increasing the objective diameter and, therefore, of improving the resolving power of the lenses. If oil is not used with an oil immersion objective, the image becomes fuzzy with poor resolution (Figure 3–1c).

Under usual operating conditions, the field of vision in a compound light microscope is brightly illuminated. By focusing the light, the condenser produces a **brightfield** illumination (Figure 3–2a). An unstained cell has little contrast with its surroundings and is therefore difficult to see. Unstained cells are more easily observed in the modified compound microscopes described in the next section.

Darkfield Microscopy

A **darkfield microscope** is used for examining certain microorganisms that either are invisible in the living state in the ordinary light microscope, cannot be stained by standard methods, or are so distorted by staining that their characteristics then cannot be identified. A darkfield microscope uses, instead of the normal condenser, a darkfield condenser that does not permit light to transmit

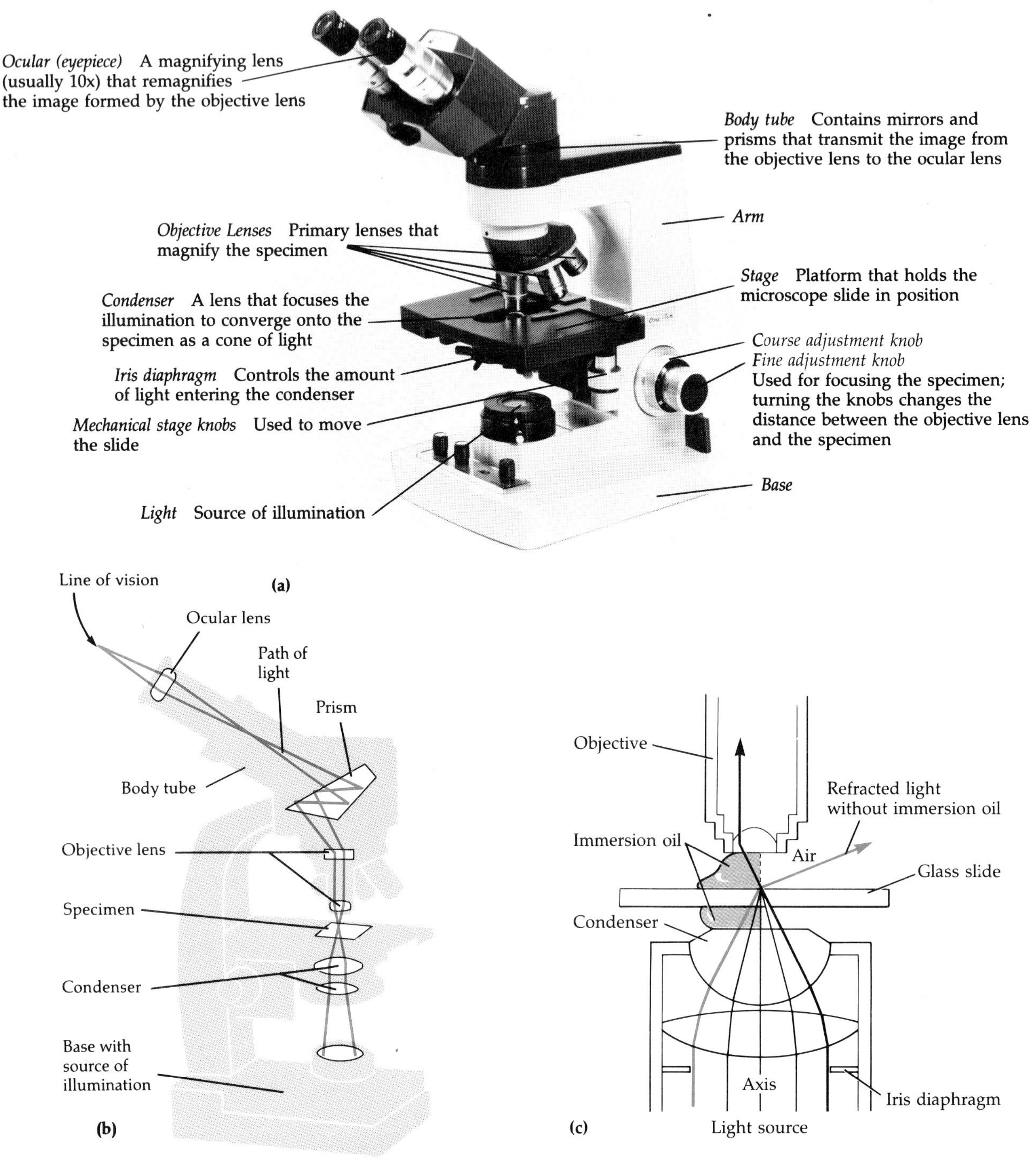

Figure 3–1 The compound light microscope. **(a)** Principal parts and their functions. **(b)** The path of light. **(c)** Refractive index. Because the refractive indexes of the glass microscope slide and immersion oil are the same, the light rays do not refract when passing from one to the other.

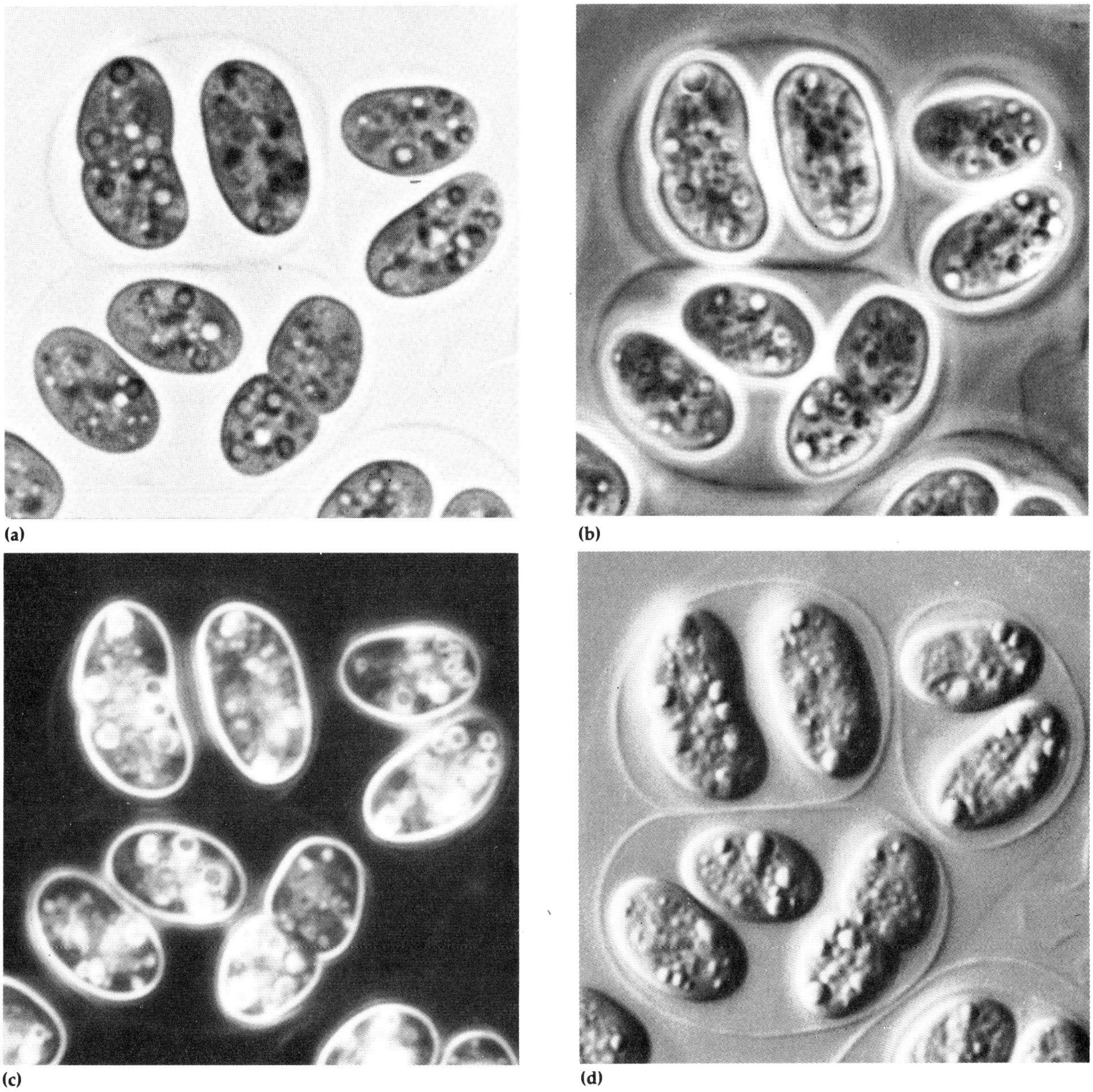

Figure 3–2 Types of images produced by a compound light microscope. Shown here are the same cells of a cyanobacterium (also known as a blue-green alga) in the genus *Gloeocapsa*, as seen using different microscopic techniques. All cells are printed at a magnification of × 2800. **(a)** Brightfield illumination shows internal structures in the cells and the outline of the transparent sheath that often surrounds two or four recently divided cells. **(b)** Phase-contrast microscopy shows greater differentiation among the internal structures and also shows the sheath well. The wide light band around each cell is an artifact resulting from this type of microscopy—a "phase halo." **(c)** Against the dark background seen with darkfield microscopy, edges of the cell are bright, some internal structures seem to "sparkle," and the sheath is almost invisible. **(d)** Differential interference contrast microscopy results in a three-dimensional appearance for the cells and their inclusions. The surrounding sheaths are sharply defined.

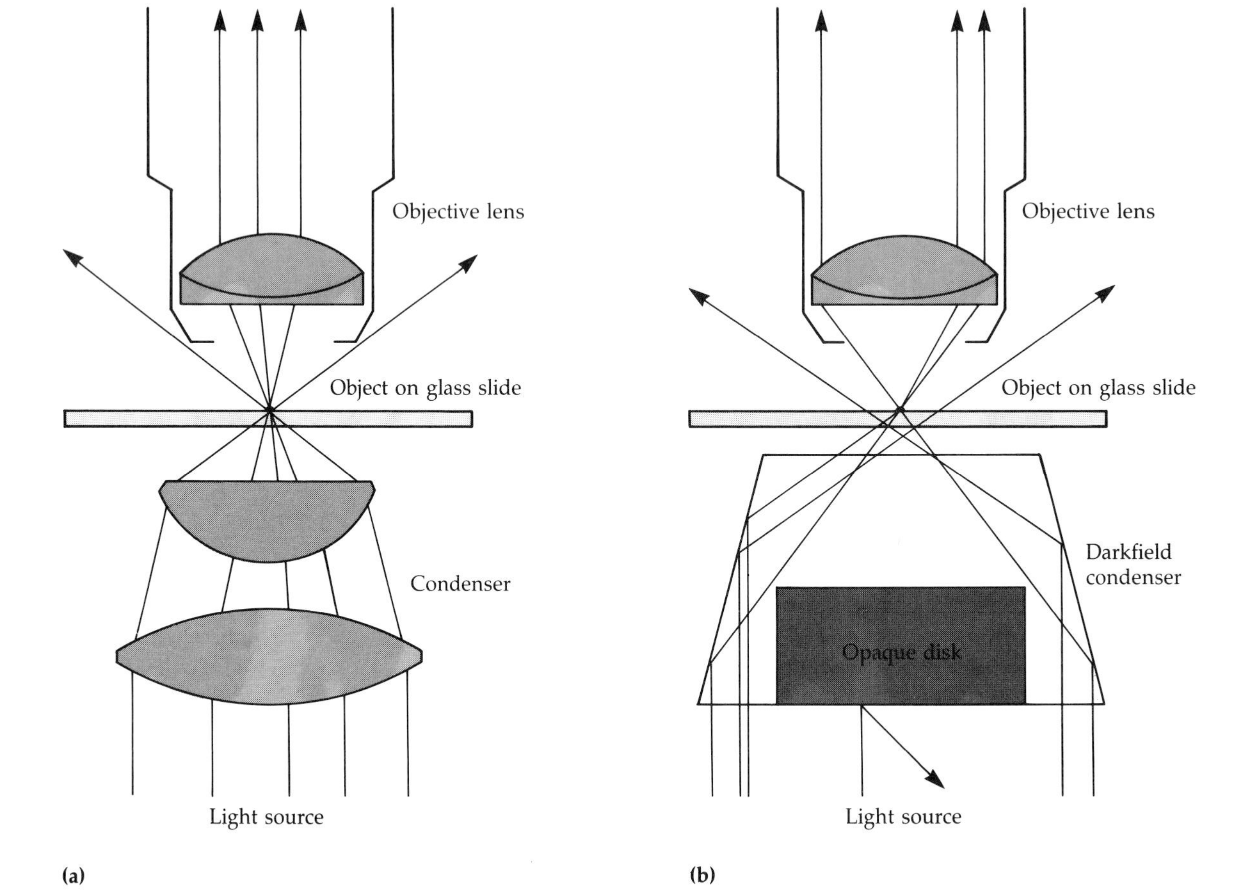

Figure 3–3 Comparison between **(a)** brightfield and **(b)** darkfield microscopy.

through the specimen into the objective lens (Figure 3–3). The darkfield condenser reflects light on the specimen at an oblique angle. Light that does not reflect off the specimen does not enter the objective lens. Because there is no direct background light, the specimen appears light against a black background—the dark field (Figure 3–2b). Internal structures of the cell are often not visible. This technique is frequently used to examine unstained microorganisms suspended in liquid. One use for darkfield microscopy is the examination of very thin spirochetes, such as *Treponema pallidum* (trep-pō-nē′ mä pal′li-dum), the causative agent of syphilis.

Phase-Contrast Microscopy

Another way to observe microorganisms is with a **phase-contrast microscope.** This technique is especially useful because it permits the detailed examination of internal structures in *living* microorganisms. In this microscope, a special kind of condenser focuses light on the specimen. Light is transmitted through a ring-like diffraction plate built into the objective lens. Light rays passing through different parts of the specimen are diffracted (retarded) differently and travel different pathways through or around the ring-like diffraction plate. Light rays traveling different paths will be slightly out of phase with one another when they reach the eye of the viewer. These phase differences and those subtle differences between different parts of a living cell are detected by the observer's eye as differences in contrast between the specimen and its background, and between details of the internal structures of the specimen. The internal details of a cell appear as degrees of brightness against a dark background (Figure 3–2c). The advantage of this technique is that the microorganism can be studied in its natural state; it does not have to be killed or stained.

Differential Interference Contrast (DIC) Microscopy

Differential interference contrast (DIC) microscopy is a relatively new technique for viewing transparent living microscopic entities. In DIC microscopy, beams of light are split and recombined by specially designed prisms. The result is a spectacular three-dimensional view of the object (Figure 3–2d).

Fluorescence Microscopy

Fluorescence microscopy takes advantage of the fluorescence of substances. Fluorescent substances give off light of one color when light of another color strikes them. Fluorescence microscopy uses an ultraviolet or near-ultraviolet light source. If the specimen to be viewed does not naturally fluoresce, it is stained with one of a group of fluorescent dyes called *fluorochromes*. Microorganisms stained with a fluorochrome and examined under a microscope with ultraviolet light appear as luminescent, bright objects against a dark background (Figure 3–4b).

Fluorochromes have special attractions for different microorganisms. For example, the fluorochrome auramine O, which glows yellow when

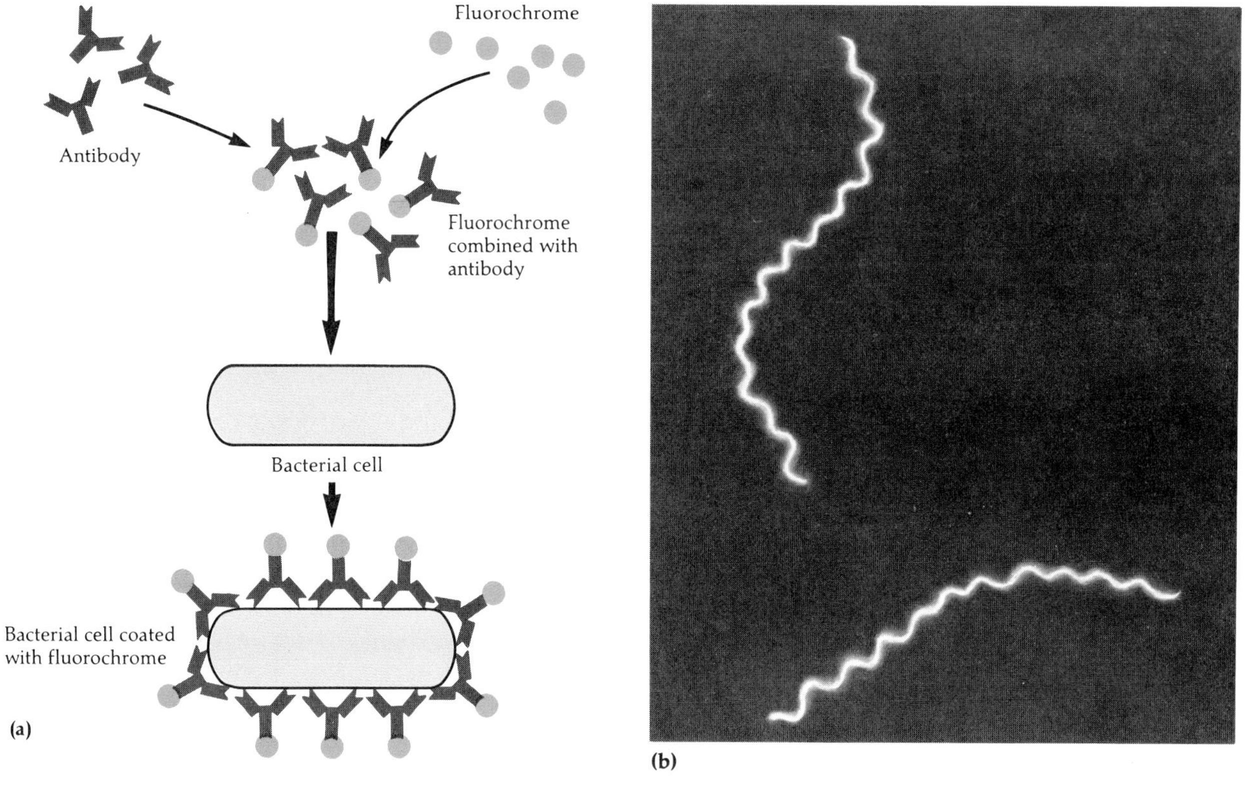

Figure 3–4 **(a)** The principle of immunofluorescence. A fluorochrome is combined with a specific bacterial antibody. When the preparation is added to bacterial cells on a microscope slide, the antibody attaches to the bacterial cells and the cells fluoresce. **(b)** In the fluorescent-antibody technique, the structures to which the specific antibodies attach show up as light regions against a darker background.

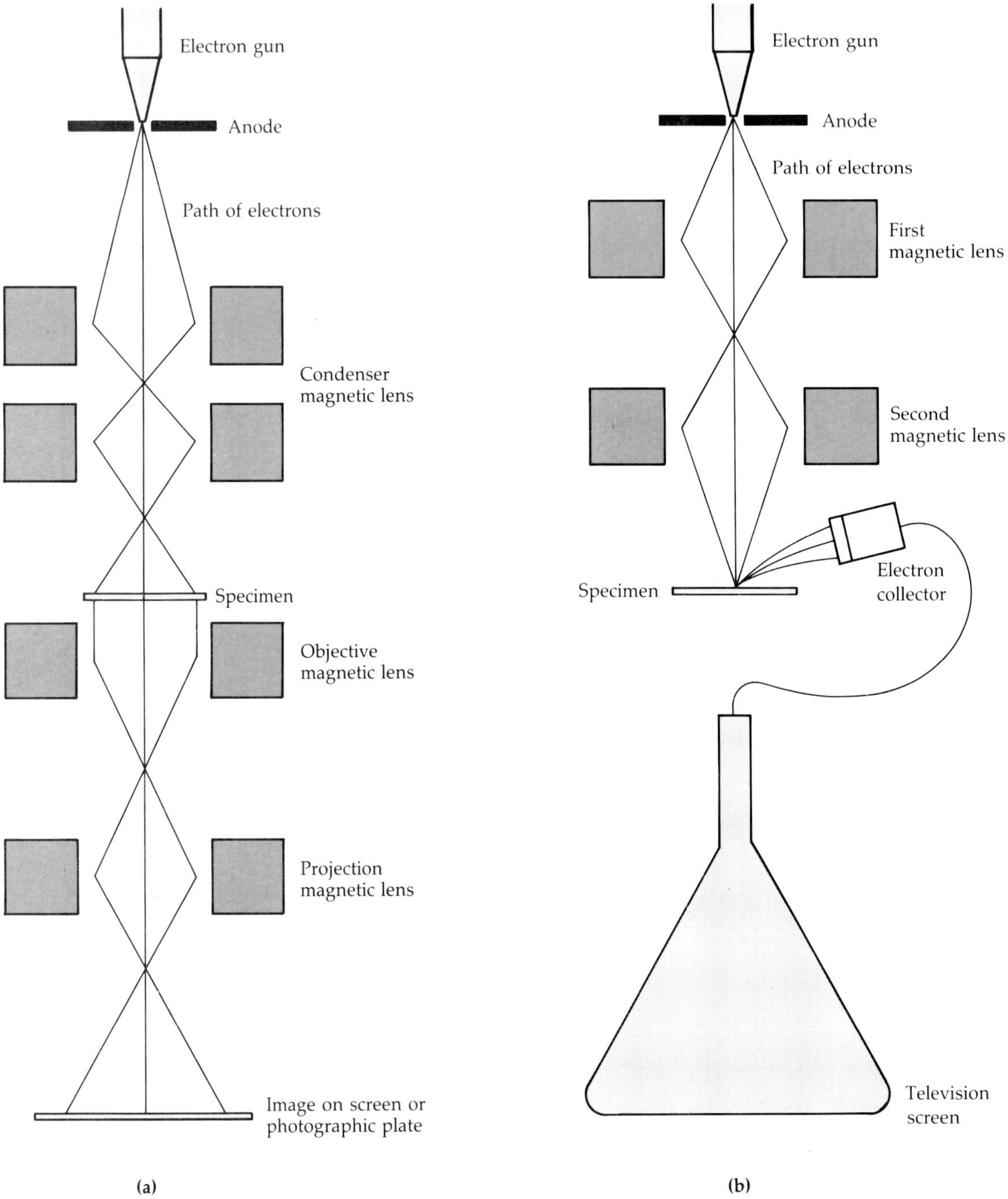

Figure 3–5 Electron microscopes. **(a)** In a transmission electron microscope, electrons pass through the specimen and are scattered. Magnetic lenses focus the image onto a screen or photographic plate. **(b)** In a scanning electron microscope, the electrons sweep across the specimen and its surface releases electrons. The electrons are picked up by a detector and transmitted onto a viewing screen.

exposed to ultraviolet light, is strongly absorbed by *Mycobacterium tuberculosis,* the bacterium that causes tuberculosis. When the dye is applied to a sample of material suspected of containing the bacterium, the bacterium is detected by the appearance of bright yellow organisms against a dark background. *Bacillus anthracis* (bä-sil'lus an-thrā' sis), the causative agent of anthrax, appears apple-green when stained with another fluorochrome, fluorescein isothiocyanate (FITC).

The principal use of fluorescence microscopy is a diagnostic technique called the **fluorescent-antibody technique** or **immunofluorescence. Antibodies** are natural defense molecules that are produced by humans and many animals in reaction to a foreign substance, or **antigen.** Fluorescent antibodies for a particular antigen are obtained as follows: an animal is injected with a specific antigen, such as a bacterium; the animal then begins to produce specific antibodies against that antigen. After a sufficient time, the antibodies are removed from the serum of the animal. Next, a fluorochrome is chemically combined with the antibodies. These fluorescent antibodies are then added to a microscope slide containing an unknown bacterium. If this unknown bacterium is the same bacterium that was injected into the animal, the fluorescent antibodies bind to the surface of the bacteria, causing them to fluoresce (Figure 3–4a). This technique can detect bacteria or other disease-producing microorganisms even within cells, tissues, or other clinical specimens (Figure 3–4b and Microview 3–4). It is especially useful in diagnosing syphilis and rabies. We will say more about antigen-antibody reactions and immunofluorescence in Chapter 16.

Electron Microscopy

Objects smaller than 0.3 μm, such as viruses or the internal structures of cells, must be examined with an **electron microscope.** Its resolving power is far greater than that of the other microscopes mentioned. In electron microscopy, a beam of electrons is used instead of light. Free electrons travel in waves, just as light does. Instead of glass lenses, magnets are used in an electron microscope to focus a beam of electrons through a vacuum tube onto a specimen. Because the wavelengths of electrons are about 1/100,000 that of visible light, the resolving power of a very sophisticated electron microscope can come close to 0.25 nm (2.5 Å), and therefore can resolve small molecules. Conventional electron microscopes have a resolving power of about 1 nm (10 Å) and can magnify objects up to 200,000 times.

Transmission electron microscopy

There are two types of electron microscopes: the transmission electron microscope and the scanning electron microscope. In the **transmission electron microscope,** a finely focused beam of electrons passes through a specially prepared, ultrathin section of the specimen (Figure 3–5a). The beam is then refocused by magnetic lenses, and the image reveals the effect of the specimen on the transmitted electron beam. The image is called a **transmission electron micrograph** (Figure 3–6a). Because the density of most microscopic specimens is very low, the contrast between their ultrastructures and the background is weak. Contrast can be greatly enhanced by use of a "stain," salts of various heavy metals such as lead, osmium, tungsten, and uranium. These metals can be fixed onto the specimen (positive staining) or used to increase the electron opacity of the surrounding field (negative staining). Negative staining is useful for the study of the very smallest specimens, such as virus particles, bacterial flagella, and protein molecules.

Scanning electron microscopy

Until recently, a disadvantage of electron microscopes has been the necessity of sectioning (cutting into thin slices) the specimen before examination. Although transmission microscopy of thin sections is an extremely valuable technique for examining different layers of specimens, it does not provide a three-dimensional effect. The **scanning electron microscope** overcomes this difficulty. This instrument provides striking three-dimensional views of specimens at magnifications of about 10,000 times (Figure 3–6b). In scanning electron microscopy, a finely focused beam of electrons is directed over the specimen and a secondary set of electrons is *ejected from the surface* of the specimen onto a televisionlike screen or photographic plate (Figure 3–5b). The scanning electron microscope is especially useful in studying the surface structures of intact cells and viruses.

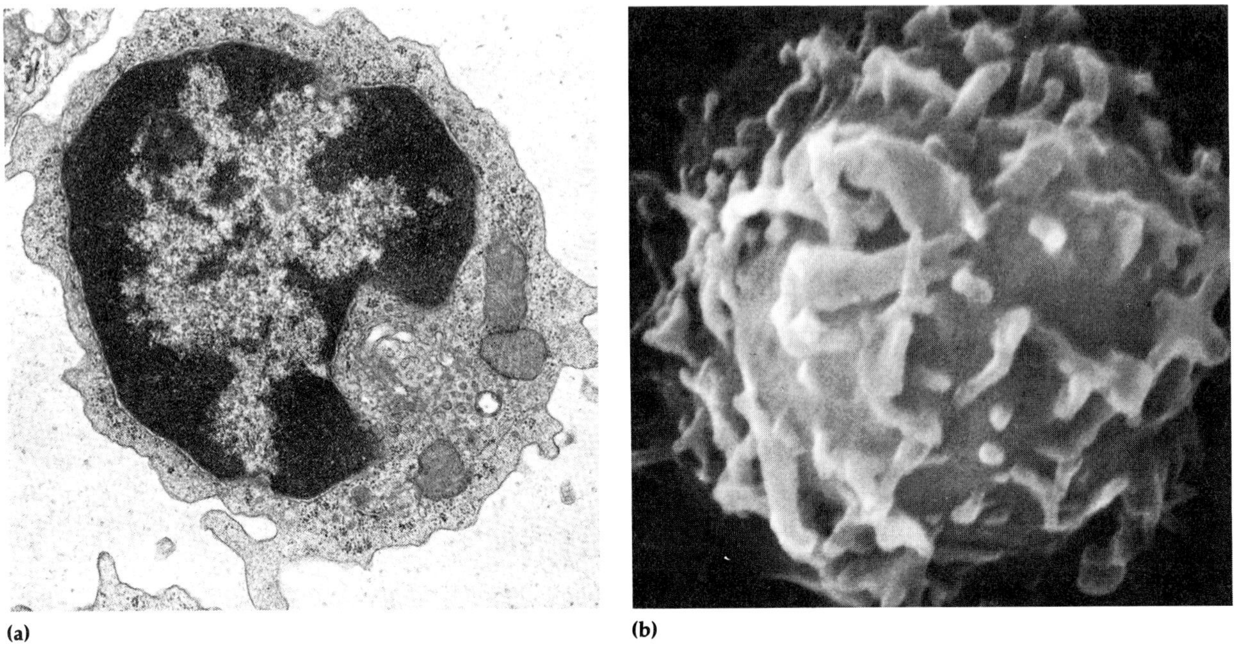

Figure 3–6 The same type of cell viewed through a transmission electron microscope and a scanning electron microscope. **(a)** This transmission electron micrograph shows a thin slice of a lymphocyte, a type of white blood cell. This type of microscopy allows one to see internal structures—at least those structures present in the slice. **(b)** In a scanning electron micrograph, surface structures can be seen, as demonstrated in this view of a lymphocyte. Note the three-dimensional appearance of this cell in contrast to the two-dimensional appearance of the cell in (a). (Both cells × 18,000).

PREPARATION OF SPECIMENS FOR LIGHT MICROSCOPY

Because most microorganisms appear almost colorless through a standard light microscope, one often must prepare them for observation by fixing and staining them. However, when it is important to observe microorganisms that are still alive, stains cannot be used. We therefore begin with a brief discussion of a hanging-drop preparation.

Hanging-Drop Preparation

In a **hanging-drop preparation,** a drop of a live microbial suspension is placed on the center of a cover glass. Then the edges of the cover glass are coated with petroleum jelly, such as Vaseline (Figure 3–7a). Next, a special microscope slide with a concave center (depression slide) is pressed against the petroleum jelly (Figure 3–7b) and quickly inverted. If this is done correctly, the drop of microorganisms hangs from the cover glass into the concavity of the slide. The slide is then examined through a microscope (Figure 3–7c). The petroleum jelly seals the cover glass to the slide and prevents excessive evaporation of the solution in which the microorganisms are suspended.

One's primary reason for using a hanging-drop preparation is to observe microbial motility. **Motility** is the ability of an organism to move by itself. It should not be confused with **Brownian movement,** the movement of suspended particles (and microorganisms) due to their bombardment by moving molecules in the suspension. When they are showing Brownian movement, the particles or microbes all vibrate at about the same rate and maintain a fairly consistent spatial relationship with one another. An organism's motility is continuous in a given direction. Because the organism is free to move in a hanging-drop suspension, its motility can be easily observed. Using a hanging-drop preparation, unique microbial dramas can be observed as they happen; see Box on p. 71.

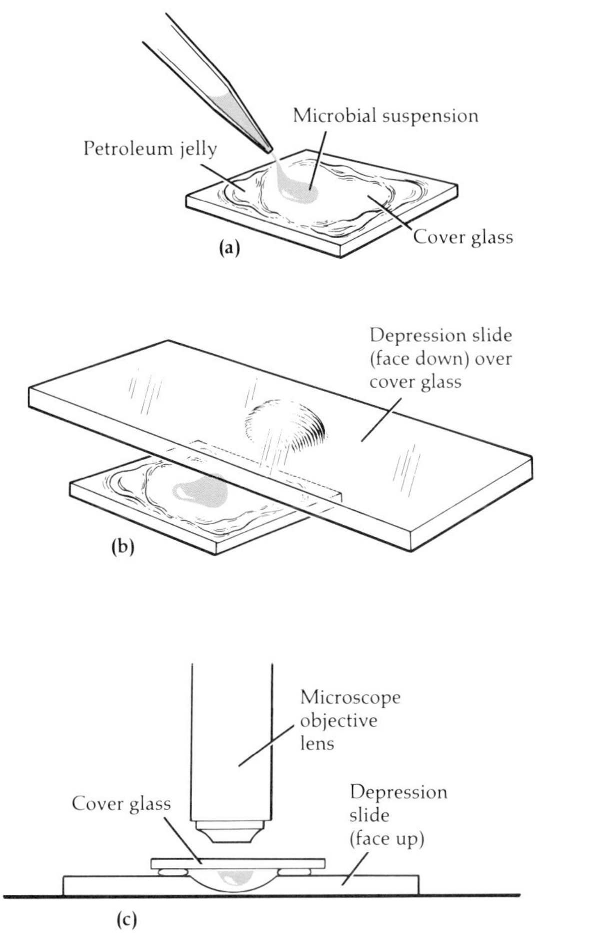

Figure 3–7 Procedure for making a hanging-drop preparation.

Staining and Preparing Smears

Most studies of the shapes and cellular arrangements of microorganisms are made from stained preparations. **Staining** simply means coloring the microorganisms with a dye that emphasizes certain structures. Before they can be stained, however, the microorganisms must be attached, or **fixed,** to the microscope slide; otherwise, the stain might wash them from the slide.

When a specimen is fixed, a thin film of material containing the microorganisms is spread over the surface of the slide (Figure 3–8a). This film is called a **smear.** After it is allowed to air dry the slide is slowly passed through the flame of a Bunsen burner several times, smear side up (Figure 3–8b). Flaming coagulates the microbial proteins and fixes the microorganisms to the slide. (This fixing procedure usually kills them.) Stain is applied, then washed off with water (Figure 3–8c and d), and the slide is blotted with absorbent paper. The stained microorganisms are now ready for microscopic examination.

Stains are salts composed of a positive and a negative ion, one of which is colored and is known as the chromophore. The color of so-called **basic dyes** is in the positive ion. In **acidic dyes,** it is in the negative ion. Bacteria are slightly negatively charged at a pH of 7.0. Thus, the colored positive ion in a basic dye is attracted to the negatively charged bacterial cell. Examples of basic dyes are crystal violet, methylene blue, and safranin. Acid dyes are not attracted to most types of bacteria because the dye's negative ions are repelled by the negatively charged bacterial surface. So the stain is repelled by the bacteria and instead colors the background. This preparation of colorless bacteria against a colored background is called **negative staining.** It is valuable in the observation of overall cell shapes, sizes, and capsules, because the cells are made highly visible against a contrastingly dark background. Distortions of cell size and shape are minimized because heat fixing is not necessary and the cells do not pick up the stain. Examples of acidic dyes are eosin and nigrosin.

To apply acidic or basic dyes, microbiologists use three kinds of staining techniques: simple, differential, and special.

Simple Stains

A **simple stain** is an aqueous or alcohol solution of a single basic dye. Although different dyes bind specifically to different parts of cells, the primary purpose of a simple stain is to highlight the entire microorganism, so that cellular shapes and arrangements are visible. The stain is applied to the fixed smear for a certain amount of time, then washed off, and the slide is dried and examined (Figure 3–8d). Occasionally, a chemical is added to the solution to intensify the stain. Such an additive is called a **mordant.** Some of the simple stains commonly used in the laboratory are methylene blue, carbolfuchsin, gentian violet, and safranin.

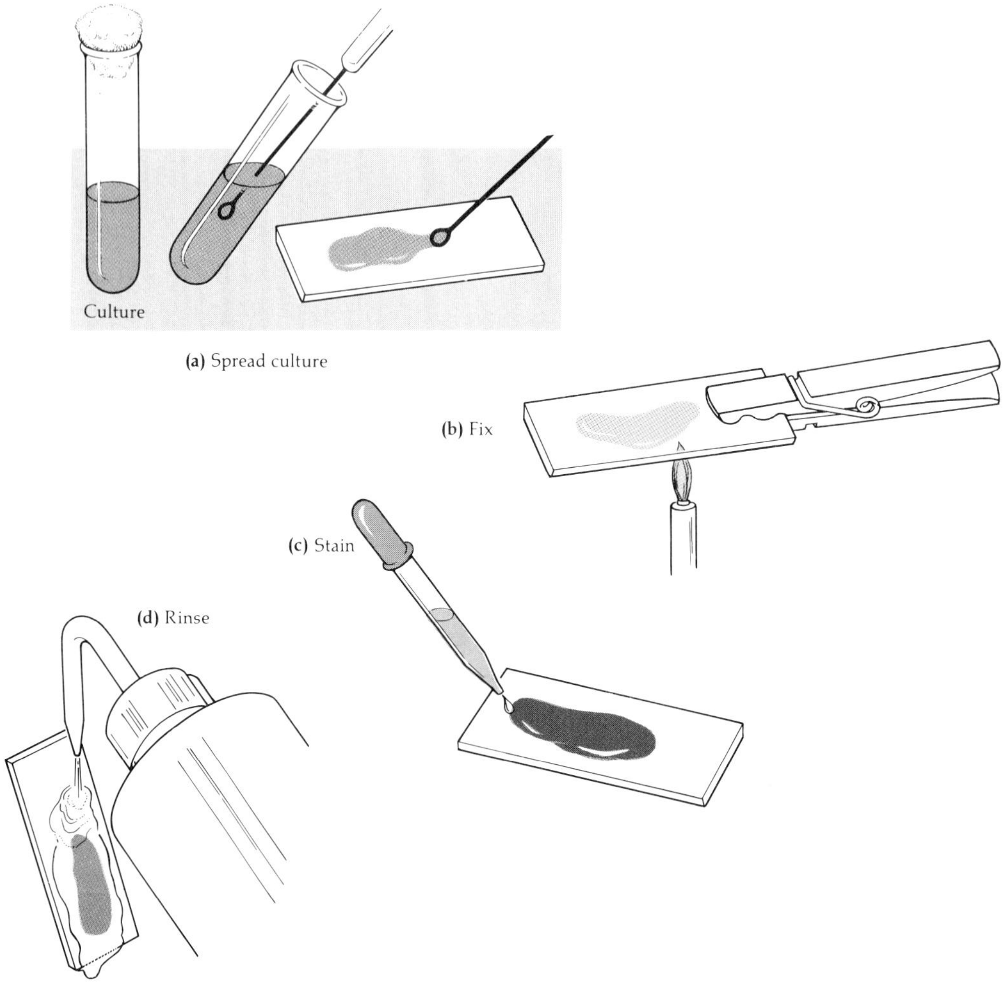

Figure 3–8 Smear preparation and staining procedures. **(a)** A suspension of microorganisms is spread in a thin film over a slide and air dried. **(b)** The smear is fixed by flaming. **(c)** The fixed smear is flooded with stain, then **(d)** rinsed. After drying, the slide is ready for microscopic examination.

Differential Stains

Unlike simple stains, **differential stains** react differently with different kinds of bacteria, and thus distinguish between different bacteria. The differential stains most frequently used for bacteria are the Gram stain and the acid-fast stain.

Gram stain

The **Gram stain** was developed in 1884 by the Danish bacteriologist Hans Christian Gram. It is one of the most useful staining procedures because it divides bacteria into two large groups: gram-positive (+) and gram-negative (−).

In this procedure, the heat-fixed smear is covered with a basic purple dye, usually crystal violet (Figure 3–9a). After a short time, the dye is washed off and the smear is covered with iodine (a mordant) (Figure 3–9b). When the iodine is washed off, both gram-positive and gram-negative bacteria appear dark violet or purple. Next, the slide is washed with an ethanol or ethanol–acetone solution (Figure 3–9c). This solution is a decolorizing

Bdellovibrio, Predator Extraordinaire

At first glance, the 1 μm-long, curved, gram-negative rods of *Bdellovibrio* do not seem unusual. *Bdellovibrio* exhibit the normal runs and tumbles of flagellated bacteria; however, they move about ten times faster than other bacteria of the same size. Patient observation or use of a microscope equipped with a video camera reveals that *Bdellovibrio*'s behavior is unlike that of any other known bacteria. A unique microbial drama occurs when *Bdellovibrio* are placed in a suspension with other bacteria such as *E. coli.* Apparently drawn by chemical attractants, *Bdellovibrio* speeds towards and rams the slower *E. coli.* The two cells become atttached and rotate rapidly. As the two bacteria whirl about, the *Bdellovibrio* paralyzes the *E. coli* by use of an unknown mechanism that most likely involves one of *Bdellovibrio*'s protein products. *Bdellovibrio* kills *E. coli* without breaking down the cell's contents.

Next, in preparation for moving into its newly acquired home, *Bdellovibrio* adds structural components to the outer membrane of *E. coli.* This activity, which stabilizes the outer membrane, is another ability unique to *Bdellovibrio.* After penetrating *E. coli*'s cell wall and detaching its flagellum, *Bdellovibrio* snuggles into the periplasmic space (that is, the space between the plasma membrane and the cell wall), and prepares for a feast (see the figure).

Macromolecules leak from *E. coli*'s cytoplasm into the periplasmic space. *Bdellovibrio* digests the macromolecules and grows into a long, coiled filament. In about two to three hours, the cytoplasmic contents of the host (including its DNA) are digested, and the *Bdellovibrio* filament fragments into as many as 20 individual curved rods. Each rod develops a flagellum and is released when the host cell is broken open by a *Bdellovibrio* enzyme. The new *Bdellovibrio* scurry away to feed off other bacteria.

Bdellovibrio was once thought to be a parasite. However, thorough studies have shown that *Bdellovibrio* does not merely siphon off its host's resources and energy, it actually digests the cell's contents as food, so *Bdellovibrio* is now considered to be a bacterial predator. The ecological importance of *Bdellovibrio* is that it, along with protozoans and bacteriophages, provides a means of control on the population growth of soil bacteria.

Some researchers have suggested a use for *Bdellovibrio* in industrial microbiology. Genetically engineered bacteria can be made to produce large quantities of a desired chemical; however, these bacteria tend to store the chemical product rather than release it. This characteristic makes it very difficult to harvest the chemical product in large quantities. Researchers would like to adapt the mechanism, used by *Bdellovibrio,* that causes large molecules to leave the host cell. If genetically engineered bacteria could be made to secrete rather than store their metabolic products, the mass production of a desired chemical would be greatly facilitated.

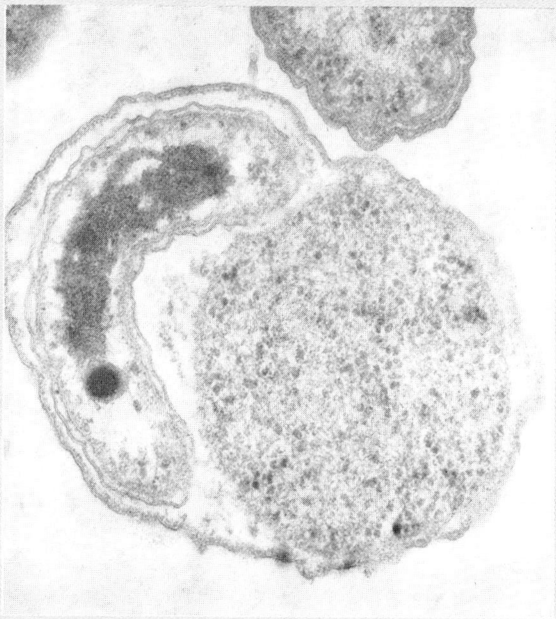

A *Bdellovibrio* enters an *E. coli* cell and forms a bdelloplast. The *Bdellovibrio*, having killed the prey cell, utilizes the *E. coli*'s constituents for its elongation and multiplication.

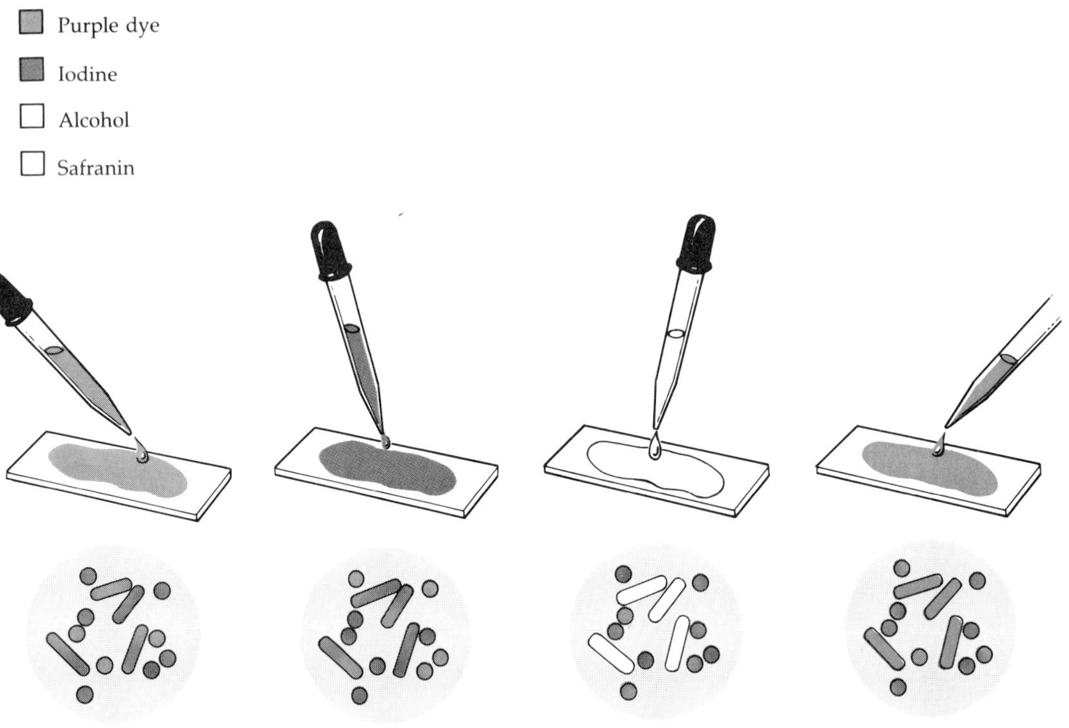

(a) Application of crystal violet (purple dye)

(b) Application of iodine (mordant)

(c) Alcohol wash (decolorization)

(d) Application of safranin (counterstain)

Figure 3–9 Gram-staining procedure. Carefully note the key; it indicates which colors replace the bacteria's actual colors. **(a)** A heat-fixed smear of gram-positive cocci and gram-negative rods is first covered with a basic purple dye such as crystal violet, then the dye is washed off. **(b)** Then the smear is covered with iodine (a mordant) and washed off. At this time, both gram-positive and gram-negative bacteria are purple. **(c)** The slide is washed with ethanol or an alcohol and acetone solution (a decolorizer), and then water. Now the gram-positive cells are purple and the gram-negative cells are colorless. **(d)** In the final step, safranin is added as a counterstain, and the slide is washed, dried, and examined microscopically. Gram-positive bacteria retain the purple dye, even through the alcohol wash. Gram-negative bacteria appear pink because they pick up the safranin counterstain.

agent, which removes the purple from the cells of some species but not from others. The alcohol is rinsed off and the slide is then stained with safranin, a basic red dye (Figure 3–9d). The smear is washed again, blotted dry, and examined microscopically.

The purple dye and the iodine combine with each bacterium and color it dark violet or purple. Bacteria that retain this color after the alcohol has attempted to decolorize them are classified as **gram-positive** (see Microview 1–1A). Bacteria that lose the dark violet or purple color after decolorization are classified as **gram-negative** (see Microview 1–1B). Because gram-negative bacteria are colorless after the alcohol wash, they are no longer visible. It is for this reason that the safranin is applied; it turns the gram-negative bacteria pink. Stains such as safranin that are used to give contrast in color are called **counterstains.** Because gram-positive bacteria retain the original purple stain, they are not affected by the safranin counterstain.

As you will see in the next chapter, different kinds of bacteria react differently to the Gram stain, probably because chemical differences in their cell walls affect the retention or escape of the crystal violet–iodine (CV–I) complex. The alcohol rinse disrupts the plasma membranes of both gram-positive and gram-negative cells, so that the CV–I complex passes into them. However, the intact, thick peptidoglycan layers of the gram-positive cells' walls prevent the complex from leaving the cells, so the gram-positive bacteria retain the color of the crystal violet dye. The gram-negative bacteria have a much thinner peptidoglycan layer, probably only a monolayer (see Figure 4–11), and an outer membrane composed mostly of lipopolysaccharide. This outer membrane is mostly removed by the ethanol rinse, and the monolayer of peptidoglycan has discontinuities that allow the CV–I complex to escape. In summary, the gram-positive cells retain the dye and remain purple. The gram-negative cells do not retain the dye. They are colorless until counterstained with a red dye, after which they appear pink.

The Gram method is one of the most important staining techniques in medical microbiology. But Gram staining results are not absolute, because some bacterial cells stain poorly or not at all. The Gram reaction is most consistent when it is used on young, growing bacteria. In many cases, the Gram reaction of a bacterium provides valuable information for the treatment of disease. For example, gram-positive bacteria tend to be killed easily by penicillin and sulfonamide drugs. Gram-negative bacteria resist these drugs, but are much more susceptible to drugs such as streptomycin, chloramphenicol, and tetracycline. Thus, Gram stain identification of a bacterium can help determine which drug will be most effective against a disease.

Some of the features of gram-positive and gram-negative bacteria are compared in Table 3–2.

Acid-fast stain

Another important differential stain (one that divides bacteria into distinctive groups) is the **acid-fast stain.** Microbiologists use this stain to identify members of bacteria belonging to the genera *Mycobacterium* and *Nocardia* (nō-kär'dē-ä). Although many bacteria in these groups are nonpathogenic, two are important disease producers, *Mycobacterium tuberculosis* and *Mycobacterium leprae* (lep'rī) (see Microview 1–1D).

In the acid-fast staining procedure, the red dye carbolfuchsin is applied to a fixed smear, and the slide is gently steamed over a flame for several minutes. Then water is used to wash the stain from the slide. The bacteria are then red. The smear is next treated with acid–alcohol, a decolorizer, which removes the red stain from bacteria that are not acid-fast. The acid-fast microorganisms retain the red color. The smear is then stained with a methylene blue counterstain. The acid-fast bacteria retain the carbolfuchsin and repel the counterstain, so they remain red. But the non-acid-fast bacteria take up the counterstain and appear blue. The acid-fast bacteria probably retain the carbolfuchsin because waxy chemical components in their cell walls and cell membranes attract carbolfuchsin.

Table 3–2 Some Comparative Characteristics of Gram-Positive and Gram-Negative Bacteria

Characteristic	Gram-Positive	Gram-Negative
Gram reaction	Stain dark violet or purple	Stain Pink
Peptidoglycan layer	Thick	Thin
Lipopolysaccharide content	Low	High
Ratio of RNA to DNA in cell	8 : 1	About 1 : 1
Nutritional requirements	More complex	Less complex
Resistance to physical disruption	High	Low
Cell wall disruption by lysozyme	High	Low (requires pretreatment)
Susceptibility to penicillin and sulfonamide	Marked	Much less marked
Susceptibility to streptomycin, chloramphenicol, and tetracycline	Much less marked	Marked
Inhibition by basic dyes	Marked	Much less marked
Susceptibility to anionic detergents	Marked	Much less marked
Resistance to sodium azide	Marked	Much less marked

Special Stains

Special stains are used to color and isolate specific parts of microorganisms, such as endospores and flagella, and to reveal the presence of capsules.

Negative staining for capsules

Many microorganisms contain a gelatinous covering called a **capsule,** which we will discuss in our examination of the procaryotic cell. In medical microbiology, demonstrating the presence of a capsule is a means of determining the organism's identity and/or **pathogenicity,** that is, its ability to cause disease. To demonstrate the presence of capsules, a microbiologist can mix the bacteria in a solution containing a fine colloidal suspension of colored particles (usually India ink), and then stain the bacteria with a simple stain such as safranin. Because of their chemical composition, capsules do not accept most biological dyes, and thus appear as halos surrounding each stained bacterial cell. This is called a negative staining technique; negative stains do not penetrate the cell capsule, and thus contrast the capsule to the surrounding medium (Figure 3–10a and Microview 1–1E).

Endospore (spore) stain

Although relatively uncommon in bacterial cells, endospores can be formed by six genera of bacteria. Formed within a cell, an **endospore** protects the microorganism from adverse environmental conditions. Endospores cannot be stained by ordinary methods such as simple staining and Gram staining because the dyes do not penetrate the wall

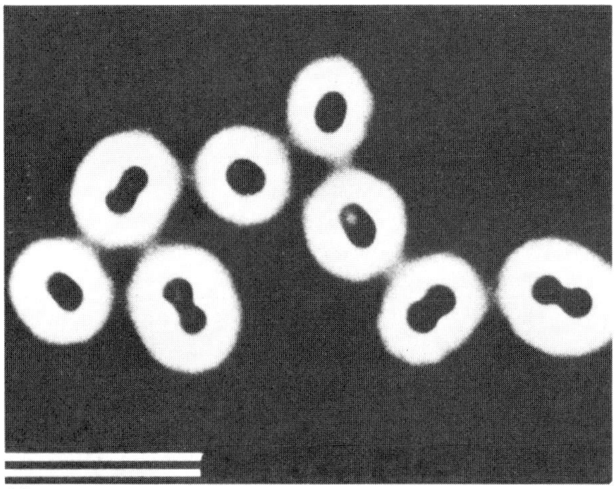

(a)

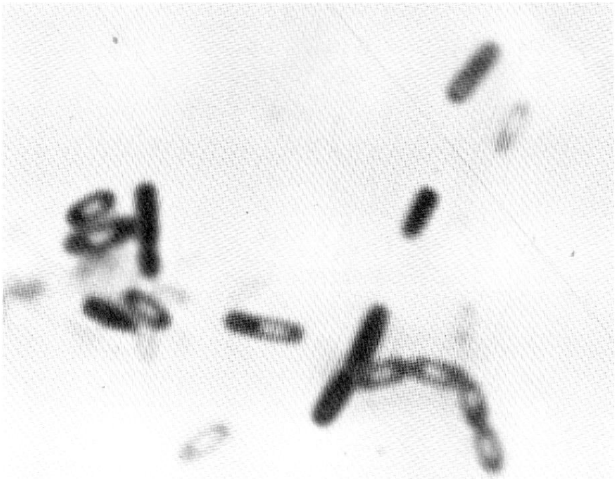

(b)

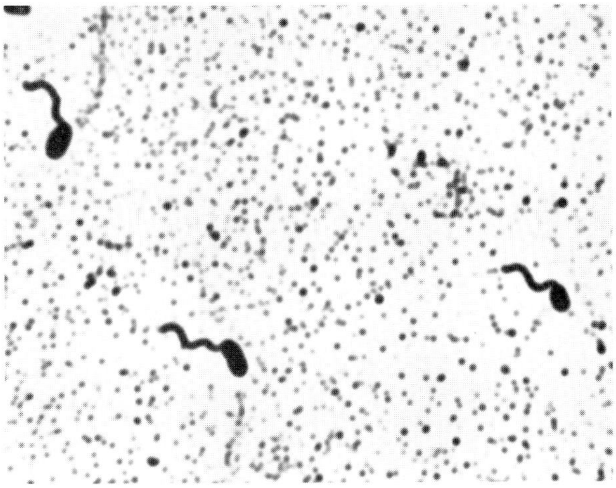

(c)

Figure 3–10 Negative staining of capsules and staining of endospores and flagella. **(a)** India ink provides a dark background, so the capsules of these bacteria *(Thiocapsa floridana)* show up as light areas surrounding the stained cells. **(b)** Endospores are seen as light areas in the centers of these rod- to oval-shaped cells of the bacterium *Clostridium bifermentans.* **(c)** Flagella are shown as thick wavy extensions from one end of these cells of the bacterium *Vibrio cholerae.* In relation to the bodies of the cells, the flagella are much thicker than they normally are, because of the accumulated layers of stain.

of the endospore. But because endospores are highly refractive, they are readily detected under the light microscope (Figure 3–10b).

Most commonly used is the *Schaeffer–Fulton endospore stain.* Malachite green, the primary stain, is applied to a heat-fixed smear and heated to steaming for about 5 minutes. The heat helps the stain to penetrate the endospore wall. Then the preparation is washed for about 30 seconds with water to remove the malachite green stain from all of the cells' parts except the endospores. Next, safranin, a counterstain, is applied to the smear to displace any residual malachite green in those cell parts. In a properly prepared smear, the endospores appear green within red or pink cells.

Flagella staining

Bacterial **flagella** are structures of locomotion too small to be seen by light microscopes. So that they can be viewed, a tedious and delicate staining procedure uses a mordant to build up the diameters of the flagella until they become visible under the light microscope. Medical microbiology uses the number and arrangement of flagella as diagnostic aids. Figure 3–10c depicts a flagella stain.

STUDY OUTLINE

UNITS OF MEASUREMENT (p. 60)

1. The standard unit of length is the meter (m).
2. Microorganisms are measured in micrometers, μm (10^{-6} m), nanometers (10^{-9} m), and angstroms (10^{-10} m).

MICROSCOPY: THE INSTRUMENTS (pp. 60–61)

Compound Light Microscopy (p. 61)

1. The most common microscope used in microbiology is the compound light microscope, which uses two sets of lenses (ocular and objective).
2. We calculate the total magnification of an object by multiplying the magnification of the objective lens by the magnification of the ocular lens.
3. The compound light microscope uses visible light.
4. The maximum resolving power (ability to distinguish two points) is 0.3 μm; maximum magnification is 2000×.
5. Brightfield illumination is used for stained smears.
6. Unstained cells are more productively observed in the following four types of microscopy, which use modified compound microscopes.

Darkfield Microscopy (pp. 61–64)

1. The darkfield microscope shows a light silhouette of an organism against a dark background.
2. It is most useful to detect the presence of extremely small organisms.

Phase-Contrast Microscopy (pp. 64–65)

1. The phase-contrast microscope uses a special condenser to enhance differences in the refractive indexes of the cell's parts and its surroundings.
2. It allows the detailed observation of living organisms.

Differential Interference Contrast (DIC) Microscopy (p. 65)

1. The DIC microscope provides a three-dimensional image of the object being observed.
2. In the microscope, beams of light are split and recombined by prisms.

Fluorescence Microscopy (pp. 65–67)

1. In fluorescence microscopy, specimens are first stained with fluorochromes, then viewed through a compound microscope using an ultraviolet (or near-ultraviolet) light source.
2. The microorganisms appear as bright objects against a dark background.
3. Fluorescence microscopy is primarily used in a diagnostic procedure called fluorescent antibody technique.

Electron Microscopy (p. 67)

1. A beam of electrons, instead of light, is used with an electron microscope.
2. Conventional electron microscopes have resolving power down to about 1 nm (1000 μm) and achieve a magnification of about 200,000×.

3. Thin sections of organisms can be seen in an electron micrograph produced with a transmission electron microscope.
4. Three-dimensional views of the surfaces of whole microorganisms can be obtained with a scanning electron microscope.

PREPARATION OF SPECIMENS FOR LIGHT MICROSCOPY (pp. 68–75)

Hanging-Drop Preparation (p. 68)

1. This procedure is used to study microbial motility, the ability of an organism to move by itself.
2. When displaying Brownian movement, particles or microorganisms are moved about by random bombardment of molecules moving in the suspension. Such particles maintain a fairly constant spatial relationship with one another.

Staining and Preparing Smears (p. 69)

1. Staining means to color with a dye, to make certain structures more visible.
2. Fixing uses heat to attach microorganisms to a slide.
3. A smear is a thin film of material used for microscopic examination.
4. Bacteria are negatively charged, and the colored positive ion of a basic dye will stain bacterial cells.
5. The colored negative ion of an acid dye will stain the background of a bacterial smear; a negative stain is produced.

Simple Stains (p. 69)

1. A simple stain is an aqueous or alcohol solution of a single basic dye.
2. It is used to make cellular shapes and arrangements visible.

Differential Stains (pp. 70–73)

1. Differential stains, such as the Gram stain and acid-fast stain, divide bacteria into groups according to their reactions to the stains.
2. The Gram-stain procedure uses a purple stain, iodine as a mordant, and a red counterstain.
3. Gram-positive bacteria retain the purple stain after the decolorization step; gram-negative bacteria do not, and thus appear red from the counterstain.
4. Acid-fast bacteria, such as members of the genera *Mycobacterium* and *Nocardia* retain carbolfuchsin after decolorization and appear red; non-acid-fast bacteria take up the methylene blue counterstain and appear blue.

Special Stains (pp. 74–75)

1. Stains such as the endospore stain and flagella stain color only certain parts of microbes.
2. Negative staining is used to make microbial capsules visible.

STUDY QUESTIONS

REVIEW

1. Fill in the following blanks.
   1 μm = ______________ m
   1 ______________ = 10^{-9} m
   1 Å = ______________ m
   1 μm = ______________ nm = ______________ Å
2. Label the parts of the compound light microscope (at right).

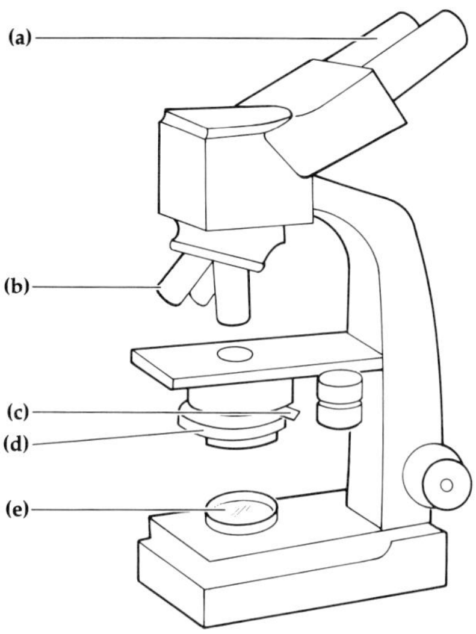

3. Calculate the total magnification of the nucleus of a cell being observed through a compound light microscope using a 10× ocular lens and the oil-immersion lens.

4. Which type of microscope would be best to use to observe the following?
(a) A stained bacterial smear
(b) Unstained bacterial cells where the cells are small and no detail is needed
(c) Unstained live tissue, where it is desirable to see some intracellular detail
(d) A sample that emits light when illuminated with ultraviolet light
(e) Intracellular detail of a cell that is 1 μm long

5. An electron microscope differs from a light microscope in that ______________ focused by ______________ is used instead of light and the image is viewed not through the ocular lenses but on ______________.

6. The maximum magnification of a compound microscope is ______________; of an electron microscope, ______________. The maximum resolution of a compound microscope is ______________; of an electron microscope, ______________.

7. One advantage of a scanning electron microscope over a transmission electron microscope is ______________.

8. Acid dyes stain the (cells/background) in a smear and are used for (negative/simple) stains.

9. Basic dyes stain the (cells/background) in a smear and are used for (negative/simple) stains.

10. Why do basic dyes stain bacterial cells? Why don't acid dyes stain bacterial cells?

11. When is it most appropriate to use a simple stain? Differential stain? Negative stain? Flagella stain?

12. Why is a mordant used in the Gram stain? In the flagella stain?

13. What is the purpose of a counterstain in the acid-fast stain?

14. What is the purpose of a decolorizer in the Gram stain? In the acid-fast stain?

15. Fill in the following table regarding the Gram stain.

	Appearance after this step of	
Steps	Gram (+) cells	Gram (−) cells
crystal violet		
iodine		
alcohol–acetone		
safranin		

CHALLENGE

1. In a Gram stain, one step could be omitted, and you could still differentiate between gram-positive and gram-negative cells. What is that one step?

2. Using a good compound light microscope with a resolving power of 0.3 μm, a 10× ocular lens, and a 100× oil-immersion lens, would you be able to discern two objects separated by 3 μm? 0.3 μm? 300 nm? 30,000 Å?

3. Why isn't the Gram stain used on acid-fast bacteria? If you did Gram stain acid-fast bacteria, what would their Gram reaction be? What is the Gram reaction of non-acid-fast bacteria?

FURTHER READING

Branson, D. 1972. *Methods in Clinical Bacteriology.* Springfield, IL: C. C. Thomas. Good stepwise directions for staining procedures and preparation of stains. Also see *Procedure Manual for Clinical Bacteriology* (1982).

Gerhardt, P., *et al*, eds. 1981. *Manual of Methods for General Bacteriology.* Washington, D. C.: American Society for Microbiology. Section II covers microscopy and preparation of specimens.

Gray, P., ed. 1981. *Encyclopedia of Microscopy and Microtechniques.* New York: Van Nostrand. An illustrated encyclopedia of terminology and techniques.

Lennette, E. H., A. Balows, W. J. Hausler, and J. Shadomy, eds. 1985. *Manual of Clinical Microbiology,* 4th ed. Washington, D. C.: American Society for Microbiology. Contains a section on microscopy, specimen preparation, and staining.

Sieburth, J. M. 1975. *Microbial Seascapes.* Baltimore, MD: University Park Press. An atlas of electron micrographs of marine microbes. Introduction explains preparation techniques for transmission electron microscopy and scanning electron microscopy.

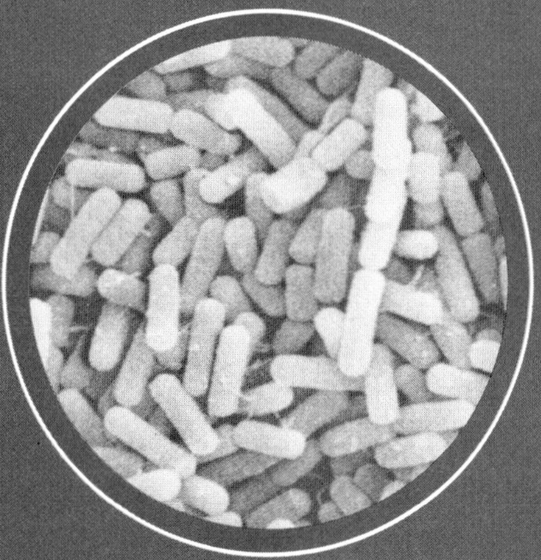

CHAPTER 4

Functional Anatomy of Procaryotic and Eucaryotic Cells

OBJECTIVES

After completing this chapter you should be able to

- Identify the three basic shapes of bacteria.
- Explain the differences between gram-positive and gram-negative cell walls.
- Describe the structure, chemistry, and functions of the procaryotic plasma membrane.
- Define the following: simple diffusion; osmosis; facilitated diffusion; active transport.
- Identify the functions of procaryotic cell structures.
- Compare and contrast the following features of procaryotes and eucaryotes: overall cell structure; flagella; nucleus.
- Distinguish between eucaryotic phagocytosis and pinocytosis.
- Explain what an organelle is.
- Describe the functions of the following: endoplasmic reticulum; Golgi complex; mitochondria; lysosomes; chloroplasts.

In studying human anatomy, we must look carefully at the physical arrangement of organs and tissues in the body. Because most microorganisms are unicellular, the study of microbial anatomy brings us to inspect a single cell's construction—how each cellular component contributes to the structure of the cell as a whole.

Despite their complexity and variety, all living cells can be divided into two groups: procaryotes and eucaryotes. On the macroscopic level, plants and animals are entirely composed of eucaryotic cells. In the microbial world, bacteria and cyanobacteria (formerly known as blue-green algae) are procaryotes. Most other microbes are eucaryotes: fungi (yeasts and molds), protozoans, and true algae.

As noncellular elements with some cell-like properties, viruses do not fit into any organization scheme of living cells. They are genetic particles that reproduce but are unable to perform the usual chemical activities of living cells. Viral structure and activity will be discussed in Chapter 12. For now, we will concentrate on procaryotic and eucaryotic cells.

PROCARYOTIC AND EUCARYOTIC CELLS

Procaryotes and eucaryotes are chemically similar: they both contain nucleic acids, proteins, lipids, and carbohydrates. They use the same kinds of chemical reactions to metabolize food, build proteins, and store energy. It is primarily the *structure* of cell walls, membranes, and various internal compartments (organelles) that distinguish procaryotes from eucaryotes.

Because they have so many chemical properties in common, the origins of procaryotes and eu-

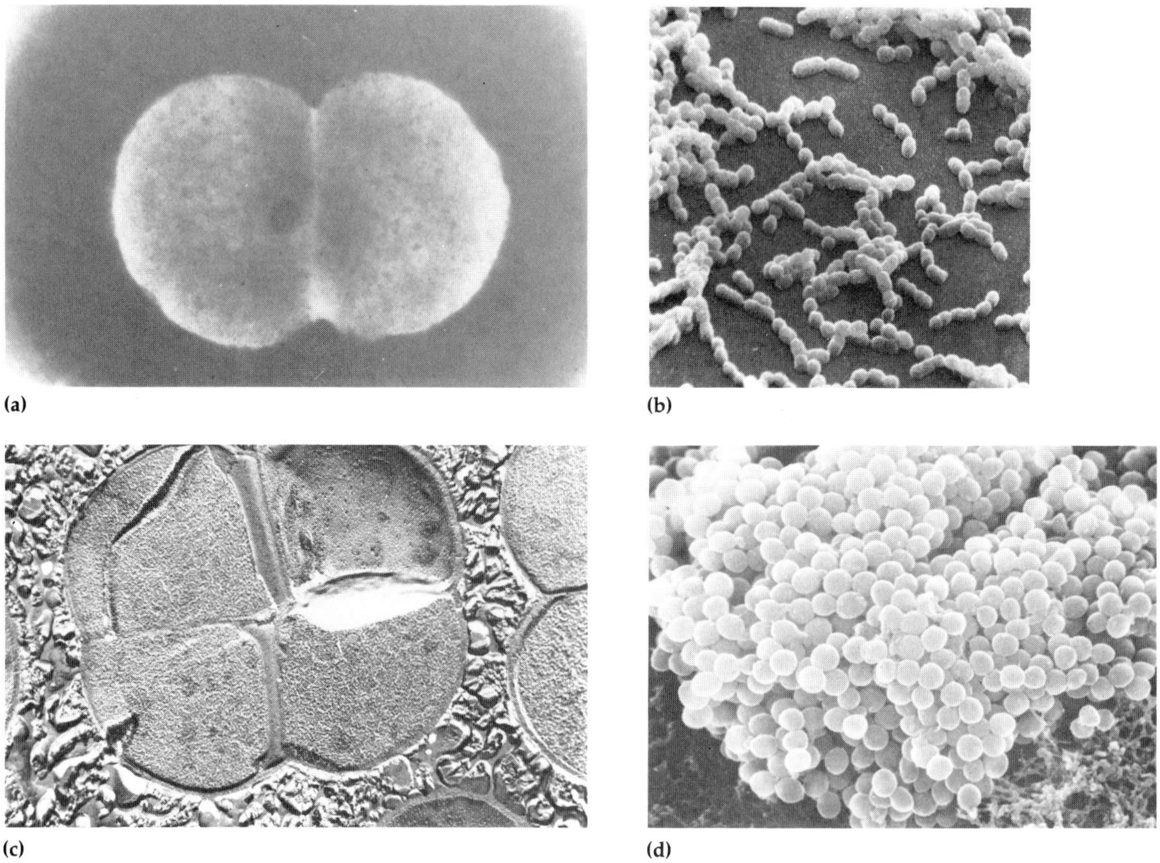

Figure 4–1 Arrangements of cocci. **(a)** Diplococci. *Neisseria gonorrhoeae* (×35,000). **(b)** Streptococci. *Streptococcus mutans* shown in short chains (×2500). **(c)** Tetrads. Freeze-fractured and etched tetrad of *Sporosarcina urea*. **(d)** Staphylococci. Scanning electron micrograph of *Staphylococcus aureus*.

caryotes are clearly related. Scientists think that procaryotic cells evolved first, and that the larger, more complex eucaryotic cells might have developed from a group of relatively large procaryotes that were parasitized by smaller bacteria. Some of the organelles in eucaryotes might be derived from these ancient remnants of bacteria.

The chief distinguishing characteristics of **procaryotic** (from the Greek for *pre-nucleus*) **cells** are as follows:

1. Their genetic material (DNA) is not enclosed within a membrane.
2. They lack other membrane-bounded organelles common to eucaryotic cells.
3. Their DNA is not associated with histone proteins.
4. Their cell walls almost always contain peptidoglycan, which is unique to procaryotes.

Eucaryotic (from the Greek for *true nucleus*) **cells** have linear structures of DNA called chromosomes; these are found in the cell's nucleus, which is separated from the cytoplasm by a nuclear membrane. The DNA of eucaryotic chromosomes is consistently associated with special chromosomal proteins, called histones and nonhistones. Eucaryotes also have a mitotic apparatus and a number of organelles, including mitochondria, endoplasmic reticulum, and, sometimes, chloroplasts. We will go over the particular characteristics and functions of these organelles later.

THE PROCARYOTIC CELL

The members of the procaryotic world, the bacteria and cyanobacteria, comprise a vast heterogeneous group of unicellular organisms. The thousands of species of bacteria are differentiated by many factors, including morphology (shape), chemical composition (often detected by staining reactions), nutritional requirements, and biochemical activities.

These general features of bacteria can be observed with the light microscope:

1. Almost all bacteria have a semirigid cell wall.
2. If bacteria are motile, their motility is usually achieved by flagella.
3. Bacteria are unicellular; although they are frequently found in characteristic groupings, each cell carries out all the functions of the organism.
4. Most bacteria multiply by binary fission, a process by which a single cell divides into two identical daughter cells.

SIZE, SHAPE, AND ARRANGEMENT OF BACTERIAL CELLS

There are a great many sizes and shapes among bacteria. Most bacteria fall within a range of 0.20 to 2.0 μm in diameter, and have one of three basic shapes: the spherical **coccus** (plural *cocci,* meaning "berries"); the rod-shaped **bacillus** (plural *bacilli,* meaning "little staffs"); and the **spiral.**

Cocci are usually round, but can be oval, elongated, or flattened on one side. When cocci divide to reproduce, the cells can remain attached to one another; they might not separate. Cocci that remain in pairs after dividing are called **diplococci** (Figure 4–1a and Microview 1–1F); those that divide and remain attached in chainlike patterns are **streptococci** (Figure 4–1b); those that divide in two planes and remain in groups of four are known as **tetrads** (Figure 4–1c); those that divide in three regular planes and remain attached in cubelike groups of eight are called **sarcinae;** and those that divide at random planes and form grapelike clusters or broad sheets are **staphylococci** (Figure 4–1d). These groups are frequently helpful in the identification of certain cocci.

Bacilli divide only across their short axis, so there are fewer groupings of bacilli than of cocci. **Diplobacilli** appear in pairs after division and **streptobacilli** occur in chains (Figure 4–2a and Microview 1–1G). Some bacilli look like cigarettes. Others have tapered ends like cigars. Still others are oval and look so much like cocci that they are called **coccobacilli** (Figure 4–2d). However, most bacilli appear as single rods (Figure 4–2c).

Spiral bacteria have one or more twists; they are never straight. Curved rods that look like commas are called **vibrios** (Figure 4–3a). Others, called **spirilla,** have a distinctive helical shape like a corkscrew and fairly rigid bodies (Figure 4–3b). Yet another group of spirals are called **spirochetes** (Figure 4–3c). Unlike the spirilla, which have flagella, spirochetes move by means of an axial filament,

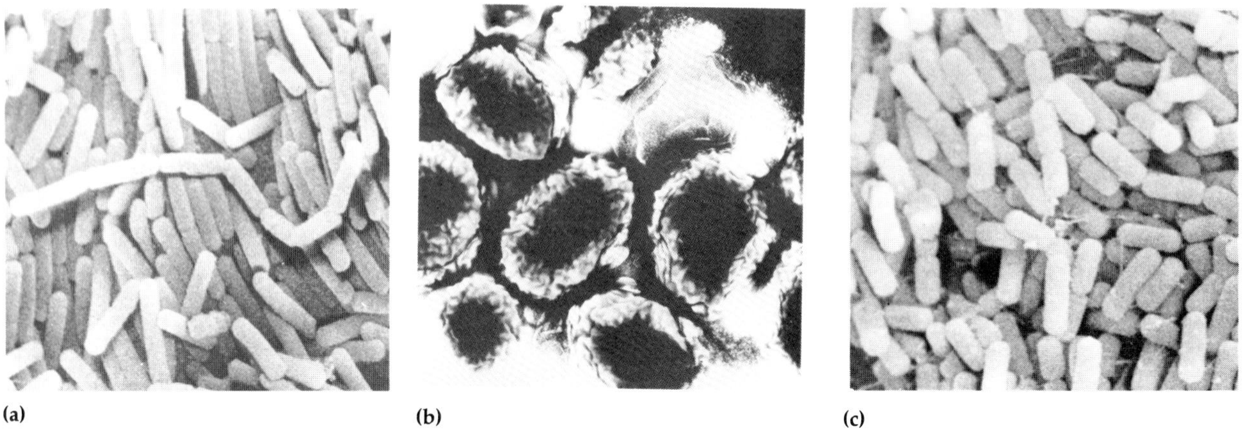

Figure 4–2 Bacilli. **(a)** Streptobacilli. **(b)** Coccobacilli. **(c)** Single bacilli. A few joined pairs of bacilli serve as examples of diplobacilli.

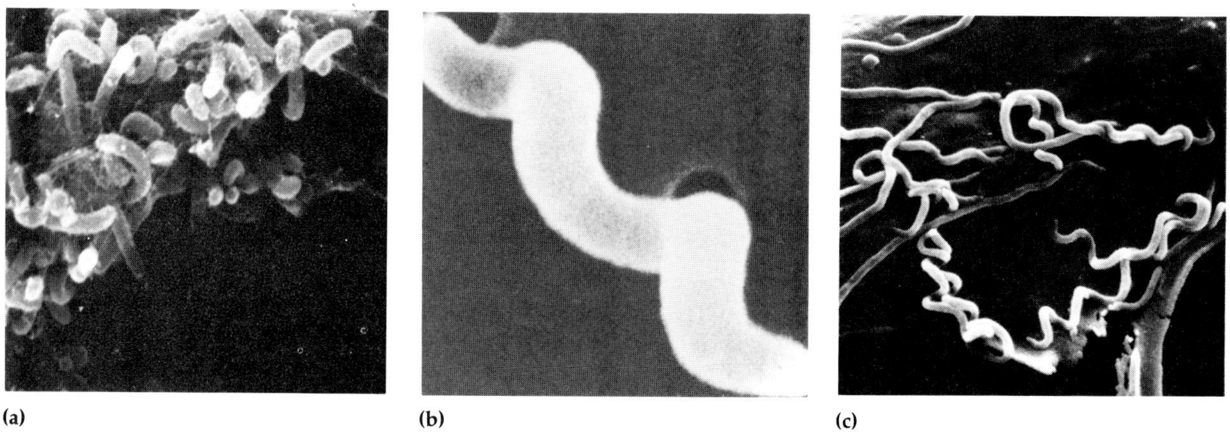

Figure 4–3 Spiral bacteria. **(a)** Vibrios. Several cells of *Vibrio cholerae* (×4800). **(b)** Spirillum. *Campylobacter jejuni* (×17,000). **(c)** Spirochetes. Several cells of *Treponema pallidum* (×4000).

which resembles a flagellum but is contained under an external flexible sheath. As you will see later, some genera of bacteria have more complex shapes and arrangements.

The shape of a bacteria is determined by heredity. However, a number of environmental conditions can alter that shape. If this happens, identification becomes even more difficult. Moreover, some bacteria are genetically **pleomorphic,** which means they can have many shapes, not just one.

Figure 4–4 shows the structure of a typical procaryotic (bacterial) cell. We will discuss its components according to the following organization: structures external to the cell wall; the cell wall; and structures internal to the cell wall.

STRUCTURES EXTERNAL TO THE CELL WALL

Among the structures external to the procaryotic cell wall are the glycocalyx, which can take the form of a capsule or a slime layer; flagella; axial filaments; and pili (fimbriae).

Glycocalyx

The **glycocalyx** is the general term used for the polymeric substances that surround bacterial cells. (An alternative term for the glycocalyx is **extracellular polymeric substance** or **EPS.**) The glycocalyx is composed of a gelatinous polymer of poly-

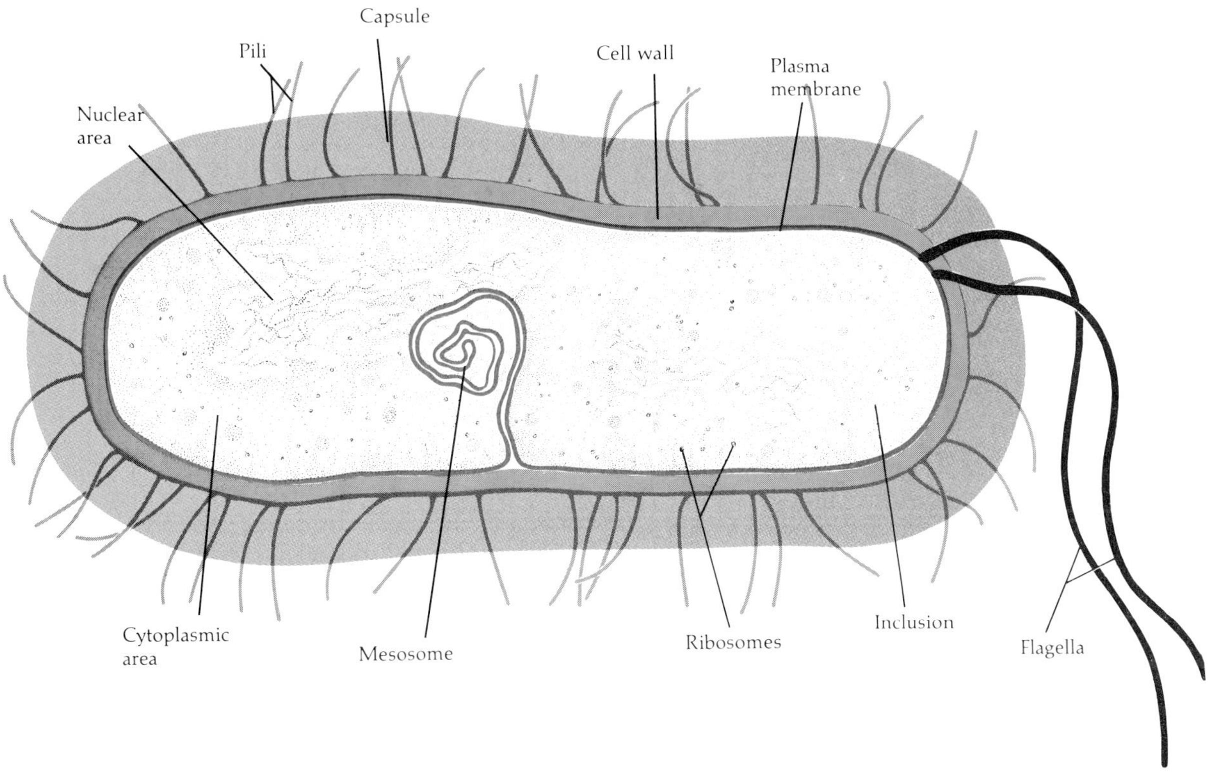

Figure 4–4 Structure of a typical procaryotic (bacterial) cell as seen in longitudinal section.

saccharide, polypeptide, or both. The chemical complexity of different species of bacteria varies widely. The material of the glycocalyx is viscous (sticky). For the most part, it is made inside the cell and excreted to the cell surface. If the substance is organized and is firmly attached to the cell wall, the glycocalyx is described as a **capsule** (see Figure 4–4). The presence of a capsule can be determined by use of negative staining, such as the India ink method mentioned in Chapter 3 (see Figure 3–10a and Microview 1–1E). If the substance is unorganized and only loosely attached to the cell wall, the glycocalyx is described as a **slime layer.**

It is clear that a major function of the sticky glycocalyx is **adherence,** that is, the bacterium's attaching to various surfaces in order to survive in its natural environment. Through adherence, bacteria can attach to diverse surfaces such as rocks in fast-moving streams, plant roots, human teeth and tissues, and even other bacteria. In some cases adhesion is quite specific because of the possible selectivity of the glycocalyx.

The glycocalyx can have functions other than adherence. Capsules are an important mechanism of bacterial virulence (the degree to which a pathogen causes disease). Capsules often protect pathogenic bacteria from phagocytosis by cells of the host. Phagocytosis, which is discussed in Chapter 15, is a process by which certain white blood cells engulf and destroy microbes. The identification of capsular material is often an important step in the identification of certain pathogenic bacteria. *Streptococcus pneumoniae* (strep-tō-kok′kus nu-mō′nē-ī), *Klebsiella pneumoniae* (kleb-sē-el′la), and *Bacillus anthracis* are three important species of encapsulated bacteria that can be identified by the swelling of their capsules when exposed to specific chemicals (antibodies). This swelling of bacterial capsules is

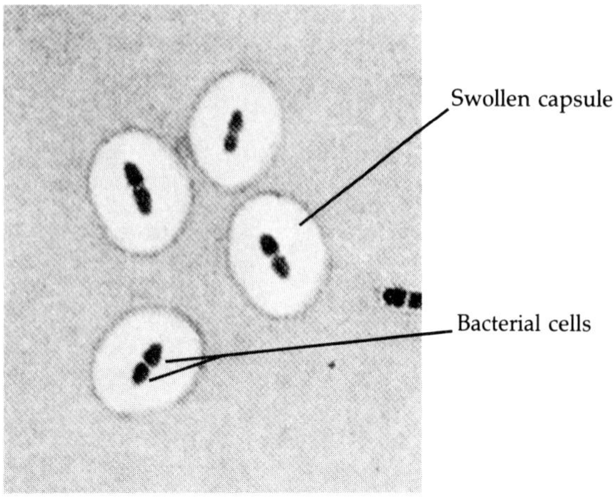

Figure 4–5 In this drawing of the quellung reaction, the capsules around pairs of *Streptococcus pneumoniae* cells appear swollen because of their reaction with specific antibodies.

called the **quellung reaction.** Figure 4–5 shows the quellung reaction of *Streptococcus pneumoniae.*

The glycocalyx may also aid the cell in securing nutrients.

Flagella

A second structural component of some procaryotic cells is flagella (see Figure 4–4). **Flagella** (singular, *flagellum,* meaning "whip") are long filamentous appendages that propel the bacteria. The flagellum has three basic parts (Figure 4–6a).The outermost region, the *filament,* is constant in diameter and contains the globular (roughly spherical) protein *flagellin* arranged in several chains that form a helix around a hollow core. The filament is attached to a slightly wider *hook,* which consists of a different protein. The third portion of a flagellum is the *basal body,* which anchors the flagellum to the cell wall and cytoplasmic membrane.

The basal body is composed of a small, central rod inserted into a series of rings. Gram-negative bacteria contain two pairs of rings. The outer rings are anchored to the cell wall, and the inner rings are anchored to the plasma membrane. In gram-positive bacteria, only the inner pair is present. As you will see later in the chapter, the flagella (and cilia) structures of eucaryotic cells are more complex than those of procaryotic cells.

Bacteria with flagella are motile. That is, they have the ability to move on their own. Motility can be seen microscopically in hanging-drop preparations (see Figure 3–7) or in bacteria grown in a semisolid "motility medium," in which movement is seen outward from the inoculum.

The flagellar mechanism by which bacteria are able to propel themselves is known. Each procaryotic flagellum is a semirigid, helical rotor that pushes the cell by spinning either clockwise or counterclockwise around its axis (Figure 4–6b). Although the exact mechano-chemical basis for this biological "motor" is not completely understood, we know that it depends on the cell's continuous generation of energy. Eucaryotic flagella undulate in a wavelike motion.

Bacterial cells have four arrangements of flagella (Figure 4–6c): **monotrichous** (single polar flagellum); **amphitrichous** (single flagella at both ends of the cell); **lophotrichous** (two or more flagella at one or both poles of the cell); and **peritrichous** (flagella distributed over the entire cell).

Flagellated bacteria can move in one direction, tumble, and reverse movement. Some species of bacteria endowed with large numbers of flagella (*Proteus,* for example) can "swarm" or show rapid, wavelike growth across a culture medium (see Microview 1–2C). Flagellar proteins serve to identify certain pathogenic bacteria.

Axial Filaments

Spirochetes are a group of bacteria that have unique structure and motility. One of the best known spirochetes is *Treponema pallidum,* the causative agent of syphilis. Spirochetes move by means of **axial filaments,** which arise at the poles of the cell within the cell wall and spiral around the cell (Figure 4–7). Axial filaments, which are anchored at one end of the spirochete, have a structure similar to that of flagella. The rotation of the filaments causes the rigid helical cell to rotate in the opposite direction and move as a corkscrew. Eucaryotic cells do not contain axial filaments.

Pili

Pili, or **fimbriae,** are hairlike appendages attached to bacterial cells in much the same way as are flagella. But pili are considerably shorter and thinner than flagella (see Figure 4–4). Like flagella, pili consist of a protein (called *pilin*) arranged helically

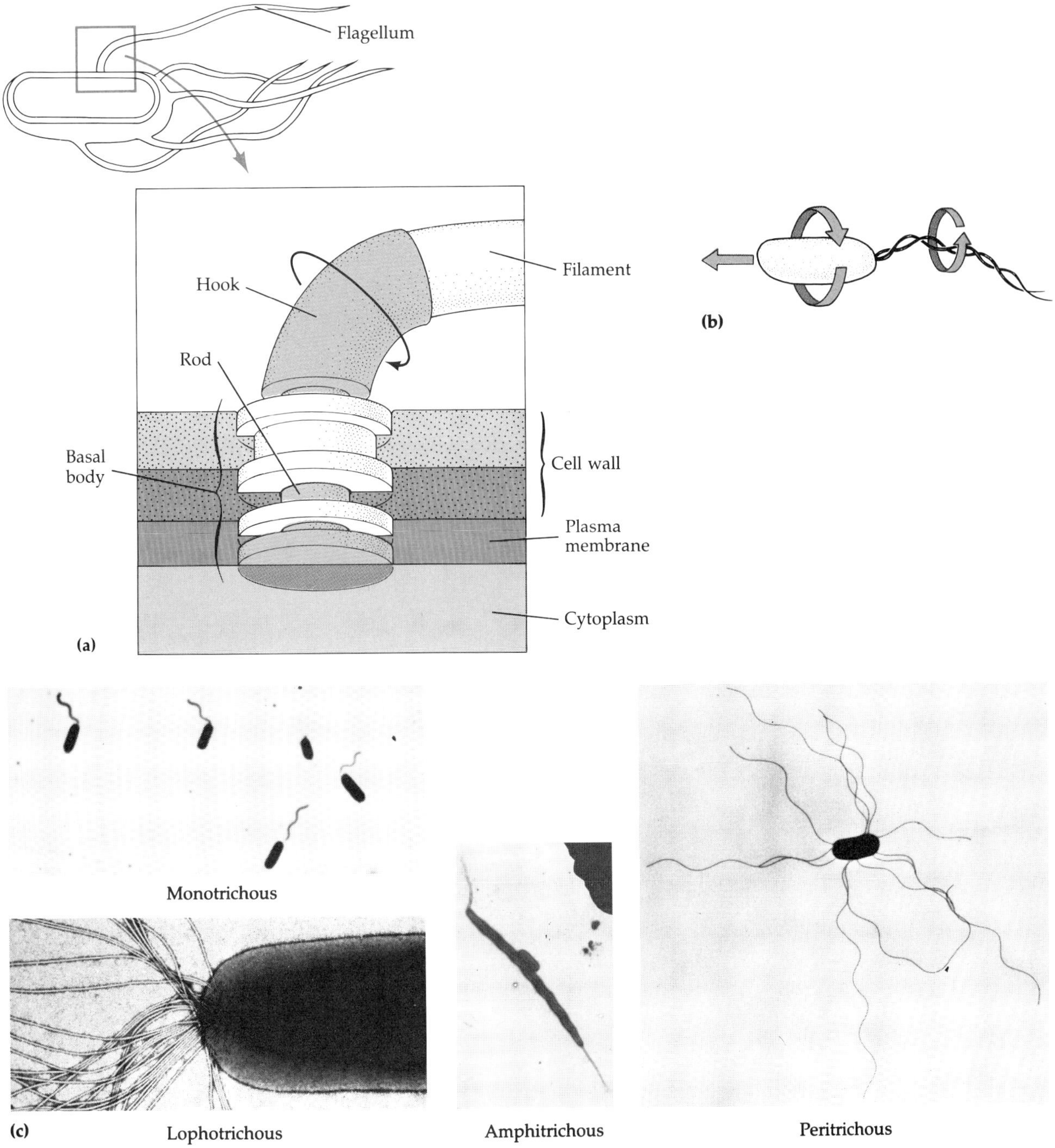

Figure 4–6 Flagella. **(a)** Parts and attachment of a flagellum of a gram-negative bacterium. **(b)** Mechanism of flagellar motion. **(c)** The four basic types of flagellar arrangements: monotrichous (*Pseudomonas aeruginosa*); lophotrichous (*Aquaspirillum graniferum*; ×108,000); amphitrichous (*Spirillum ostreae*; ×3500); peritrichous (*Proteus vulgaris;* ×4200).

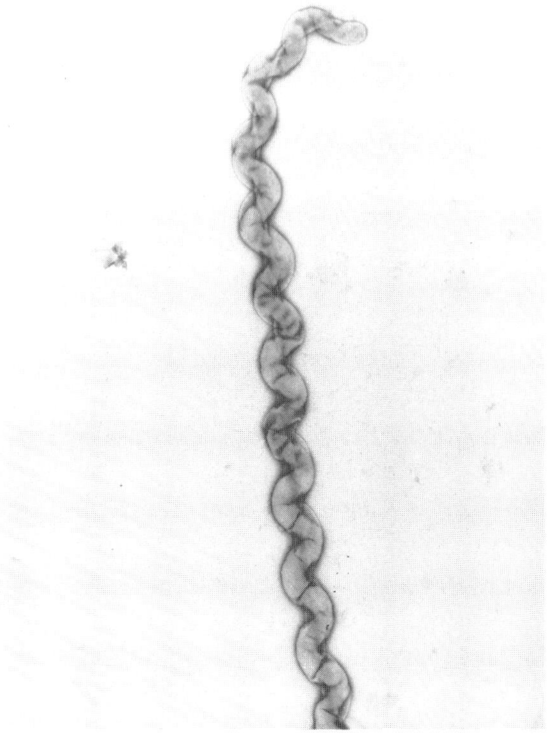

Figure 4–7 Axial filament of the spirochete, *Leptospira* sp.

around a central core. Pili can occur at the poles of the bacterial cell, or they can be evenly distributed over the entire surface of the cell. There can be anywhere from a few to several hundred pili per cell (Figure 4–8). Many gram-negative bacteria have pili.

Pili have two main functions. The first is to adhere to surfaces, including the surfaces of other cells. This function is similar in the pili and the glycocalyx. Pili associated with the bacterium *Neisseria gonorrhoeae* (nī-se´-rē-ä go-nôr-rē´ä), the causative agent of gonorrhea, help the microbe to colonize mucous membranes. Once colonization occurs, the bacteria are capable of causing the disease. When pili are absent (because of genetic mutation), colonization cannot occur and disease does not occur.

A second function of pili is to join bacterial cells prior to the transfer of DNA from one cell to another. Such pili are called **sex pili** and are discussed in detail in Chapter 8. (Eucaryotic cells do not contain pili.)

CELL WALL

The **cell wall** of the bacterial cell is a complex, semirigid structure that is responsible for the characteristic shape of the cell. The cell wall surrounds the underlying, fragile plasma (cytoplasmic) membrane and protects it and the internal parts of the cell from adverse changes in the surrounding environment (see Figure 4–4).

The major function of the cell wall is to prevent bacterial cells from rupturing when the osmotic pressure inside the cell is greater than that outside the cell. It also serves as a point of anchorage for flagella, helps determine which substances enter and leave the cell, and produces symptoms of disease in some species. Clinically, the cell wall is important because it is the site of action of some antibiotics.

Although some eucaryotes, such as algae and fungi, contain cell walls, those cell walls differ chemically from those of procaryotes, are simpler in structure, and are less rigid.

Composition and Characteristics

The bacterial cell wall is composed of a macromolecular network called *peptidoglycan (murein)*. Peptidoglycan consists of two sugars, which are related to glucose, called N-acetylglucosamine (NAG) and N-acetylmuramic acid (NAM) *(murus,*

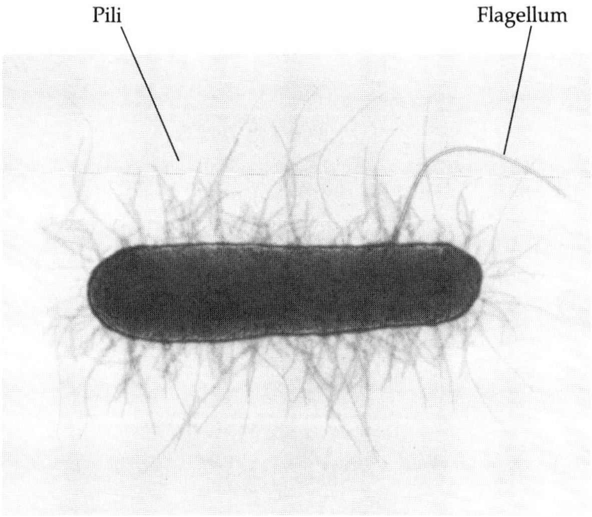

Figure 4–8 Hairlike pili seem to bristle from this cell of *Escherichia coli*. Note the flagellum.

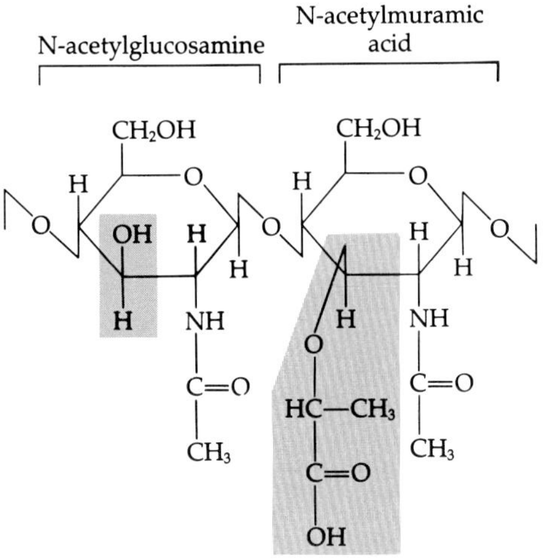

Figure 4–9 N-acetylglucosamine (NAG) and N-acetylmuramic acid (NAM) joined together as in peptidoglycan. The linkage between them is called a β–1,4 linkage.

meaning "wall") and chains of four or five amino acids. The structural formulas for NAG and NAM are shown in Figure 4–9.

The various components of peptidoglycan are assembled in the cell wall as follows. The N-acetylglucosamine and N-acetylmuramic acid alternate in rows, each row forming a carbohydrate "backbone" (Figure 4–10). There are 10 to 65 sugars in each row. To each molecule of N-acetylmuramic acid is attached a *tetrapeptide side chain,* consisting of four amino acids (alternating D and L forms). Adjacent tetrapeptide side chains may be directly bonded to each other or linked by a *peptide cross bridge* consisting of one to five amino acids.

In Chapter 3 we discussed the classification of bacteria by use of a differential staining method called the Gram stain. In most gram-positive bacteria, the cell wall consists of several layers of peptidoglycan connected by peptide side chains and cross bridges, the structure shown in Figure 4–9. This arrangement provides a very rigid framework.

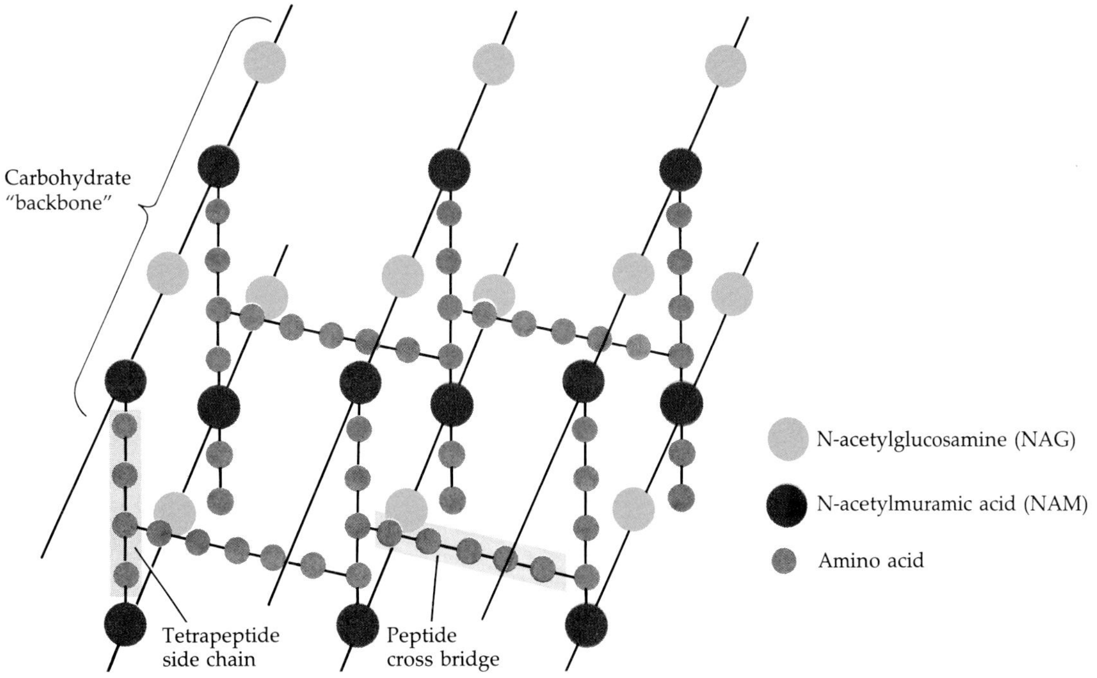

Figure 4–10 Chemical structure and organization of peptidoglycan of the bacterial cell wall. The frequency of peptide cross bridges and the number of amino acids in these bridges varies with the species of bacterium. In addition to the bridges between chains in each peptidoglycan sheet, there are bridges between different sheets. Thus, a peptidoglycan macromolecule might be quite thick.

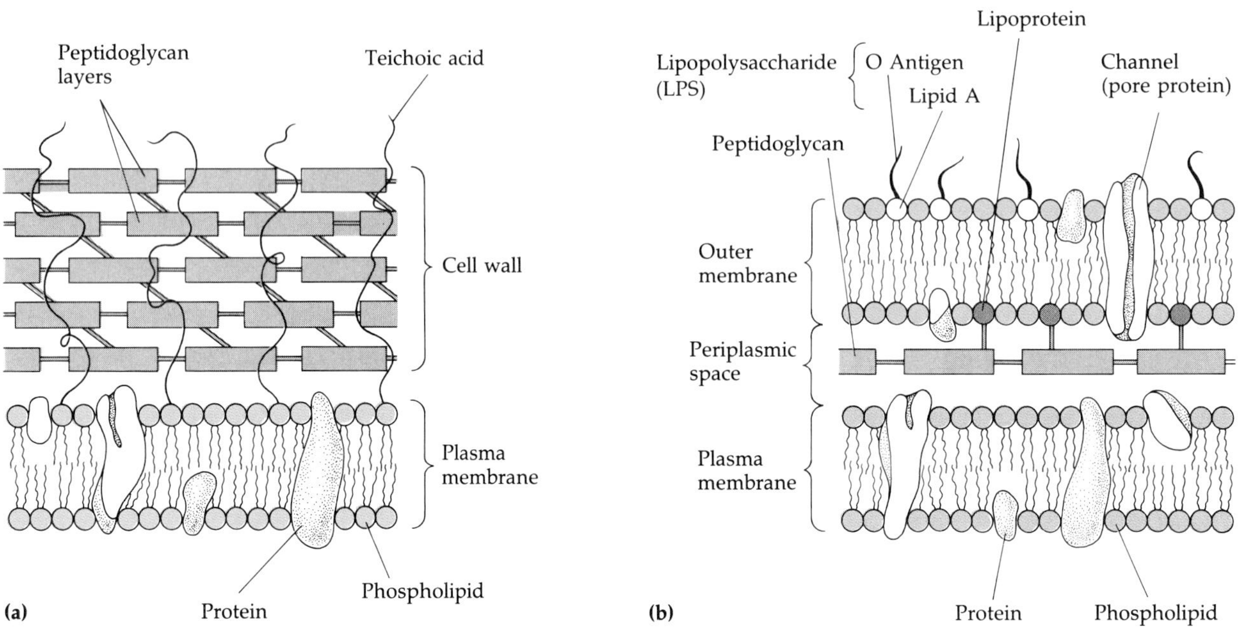

Figure 4–11 Comparison of the structures and chemical components of **(a)** gram-positive and **(b)** gram-negative cell walls.

The layers of peptidoglycan are considerably thicker in gram-positive than in gram-negative bacteria (Figure 4–11a). The cell walls of many gram-positive bacteria contain polysaccharides called *teichoic acids,* which make it possible for the bacteria to be identified by immunological means, to be described in Chapter 16. Teichoic acids are bonded to the peptidoglycan layers or to the plasma membrane. They might aid the transport of ions into and out of the cell, and they also might play a role in normal cell division.

Gram-negative bacteria also contain peptidoglycan, but in very small amounts, and they contain no teichoic acids at all. The peptidoglycan is found in the *periplasmic space* (described in the next section), a space between the plasma membrane and outer membrane, and is covalently bonded to lipoproteins in the outer membrane. Because they contain only a small amount of peptidoglycan, the cell walls of gram-negative bacteria are susceptible to mechanical breakage. The peptidoglycan layer is surrounded by an *outer membrane,* which consists of lipoproteins, lipopolysaccharides (LPS), and phospholipids (Figure 4–11b). This outer layer is an important barrier between the cell's interior and certain substances from the environment, including antibiotics such as penicillins, certain dyes, bile salts, and heavy metals.

The polysaccharide portion of the LPS is composed of sugars, called O antigens, that help to distinguish species of gram-negative bacteria (for example, *Salmonella* species) by immunological means. This role is comparable to that of teichoic acids in gram-positive cells. The lipid portion of the LPS, called *lipid A,* is referred to as *endotoxin,* and is toxic when in the host's bloodstream. It causes fever and intravascular hemolysis (disintegration of red blood cells). The nature and importance of these and other toxins will be discussed in Chapter 14.

Among procaryotes, there are cells that naturally have no walls or have very little wall material. These include members of the genus *Mycoplasma* (mī-kō-plaz'ma) and related organisms. Mycoplasmas are the smallest known bacteria that can grow and reproduce outside of living host cells. Because they have no cell walls, they pass through most bacterial filters and were first mistaken for viruses. Their plasma membranes are unique among bacteria in having lipids called *sterols,* which are thought to help protect them from osmotic lysis.

Other atypical bacterial cells are the L **forms** (named after the Lister Institute, where they were discovered). These are tiny mutant bacteria with defective cell walls. Certain chemicals and antibiotics like penicillin induce many bacteria to produce L forms. Although some L forms can revert to the original bacterial form, others are stable. L forms tend to contain just enough cell wall material to prevent lysis from occurring in dilute solutions.

Recall from Chapter 3 that the mechanism for the Gram stain is related to the structure and chemical composition of the cell wall. It might serve you well to review that mechanism now.

Damage to the Cell Wall

The cell wall is a good target for certain antimicrobial drugs, because the wall is made of chemicals unlike those in eucaryotic cells. Thus, chemicals that will damage bacterial cell walls, or interfere with their synthesis, often will not harm the cells of an animal host. One way that the cell wall can be damaged is by exposure to the enzyme *lysozyme*. This enzyme occurs naturally in some eucaryotic cells and is a constituent of tears, mucus, and saliva. In the cell wall of many types of gram-positive bacteria, lysozyme makes it vulnerable to rupture, or lysis. What happens is this: the lysozyme catalyzes the hydrolysis of the bonds between the sugars in the polysaccharide chain of peptidoglycan. This act is analogous to cutting the steel supports of a bridge with a cutting torch. The gram-positive bacterium's cell wall is completely destroyed by lysozyme. However, the cellular contents that remain surrounded by the plasma membrane may remain intact; this structure is termed a **protoplast.** Typically, the protoplast is spherical and still capable of carrying on metabolism.

When lysozyme is similarly applied to gram-negative cells, usually the wall is not completely destroyed; much of the outer membrane remains. In this case, the cellular contents, plasma membrane, and remaining outer wall layer are a **spheroplast,** also a spherical structure. Gram-negative

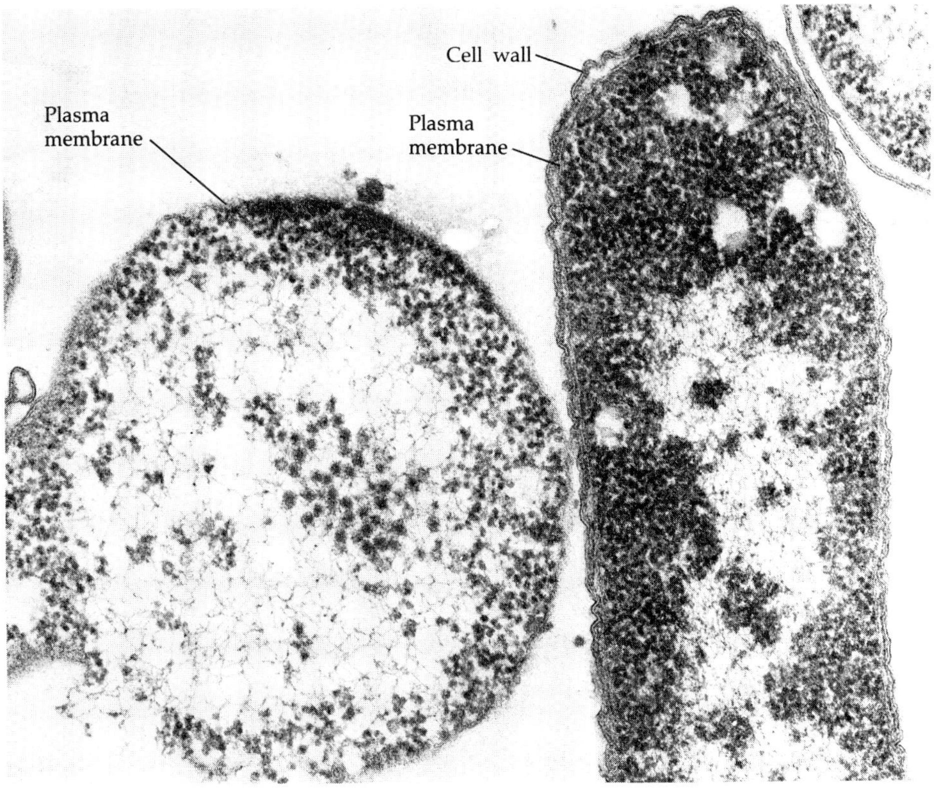

Figure 4–12 Protoplast. On the left is a rounded protoplast of the bacterium *Escherichia coli;* it is bounded only by the plasma membrane. To the right is one end of a normal, rod-shaped cell of *E. coli* with its cell wall intact (×61,000).

cells that lose all of their wall material are referred to as protoplasts (Figure 4–12). Protoplasts and spheroplasts burst in pure water or very dilute salt or sugar solutions because the water molecules from the surrounding fluid rapidly diffuse into and enlarge the cell, which has a much lower concentration of water. This rupturing is called **osmotic lysis,** and will soon be discussed in detail.

Antibiotics destroy bacteria by interfering with the formation of the peptide bridges of peptidoglycan, thus preventing the formation of a functional cell wall. In general, the antibiotic penicillins are more effective against gram-positive bacteria than against gram-negative. The outer membrane of gram-negative bacteria seems to interfere with the entry of penicillins into the cells. Organisms with deficient cell walls, such as mycoplasmas and L forms, are also resistant to antibiotics that interfere with cell wall synthesis.

STRUCTURES INTERNAL TO THE CELL WALL

Thus far, we have discussed the procaryotic cell wall and structures external to it. We will now look inside the procaryotic cell and discuss the structures and functions of the plasma membrane and other components within the cytoplasm of the cell.

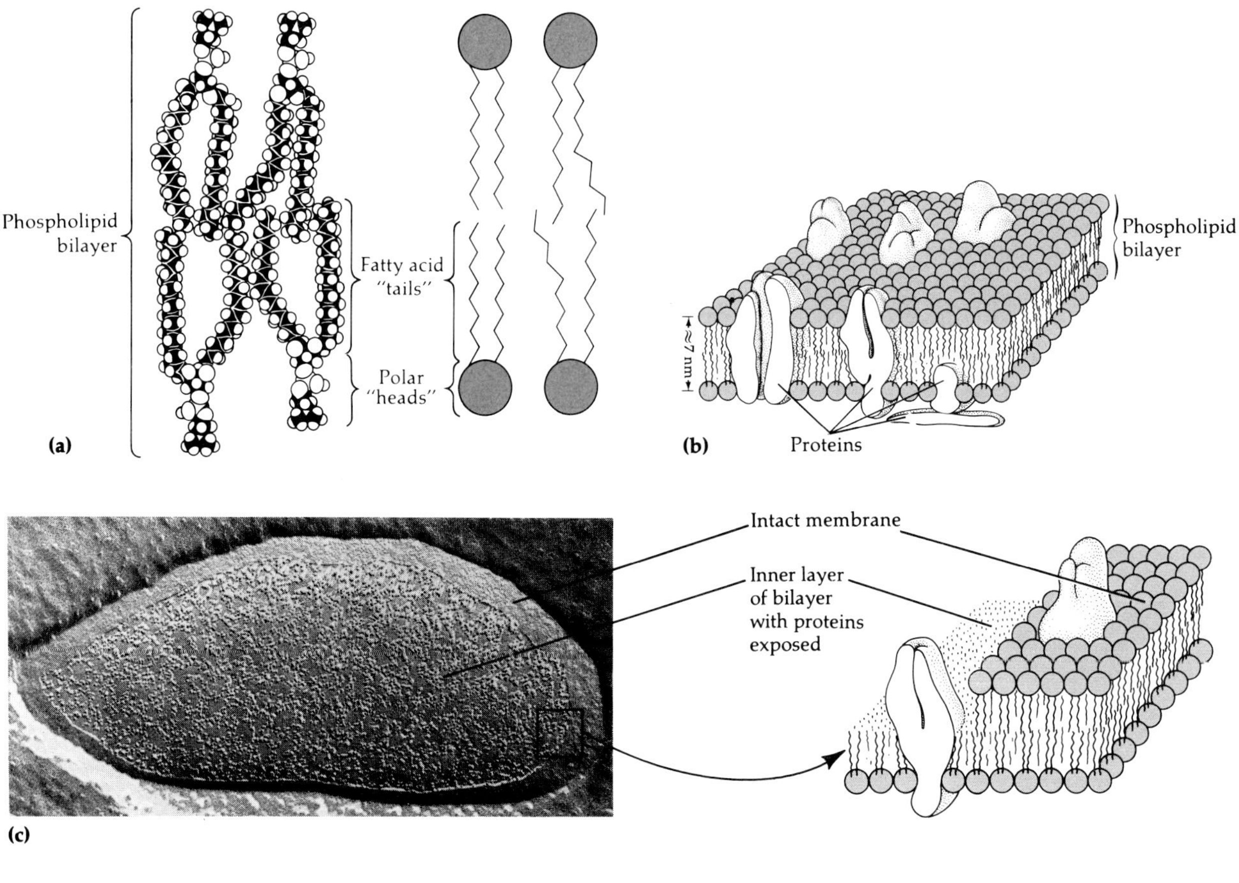

Figure 4–13 Plasma membrane. **(a)** Phospholipid bilayer. Space-filling models of several molecules next to a schematic drawing. **(b)** Drawing of membrane showing phospholipid bilayer and proteins. **(c)** An electron micrograph showing a freeze-etched membrane of a red blood cell (×20,000). The outer layer of the phospholipid bilayer has been chipped away, leaving proteins sticking up through the inner layer. The bilayer is still intact around the edge of the cell.

Bacterial Stains and Morphology

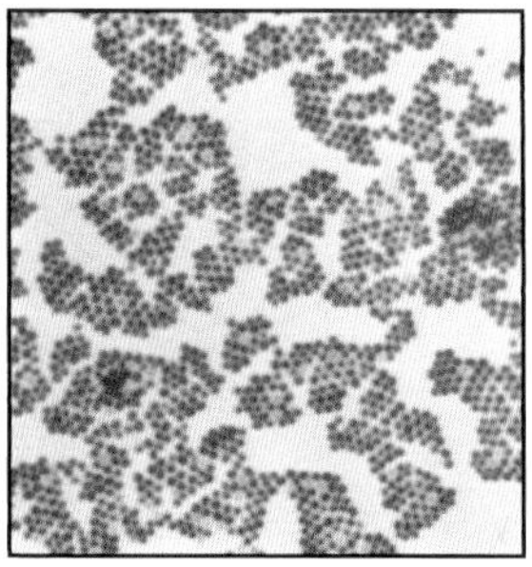

A. Clusters of *Staphylococcus aureus* demonstrate a gram-positive stain.

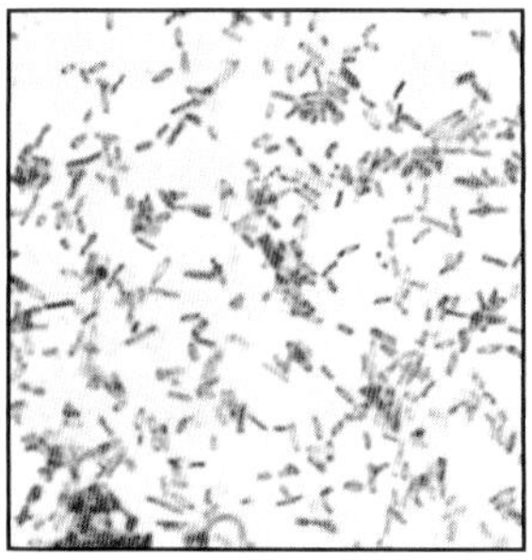

B. Gram-negative rods of the intestinal bacterium *Escherichia coli*.

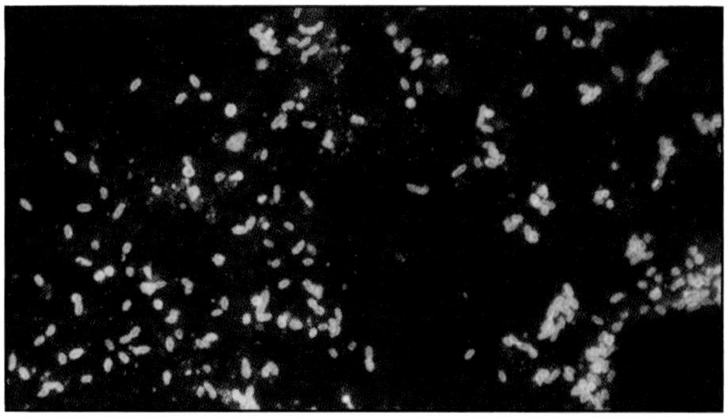

C. Fluorescent-antibody stain of *Legionella pneumophila* in lung tissue from a fatal case of Legionnaires' disease.

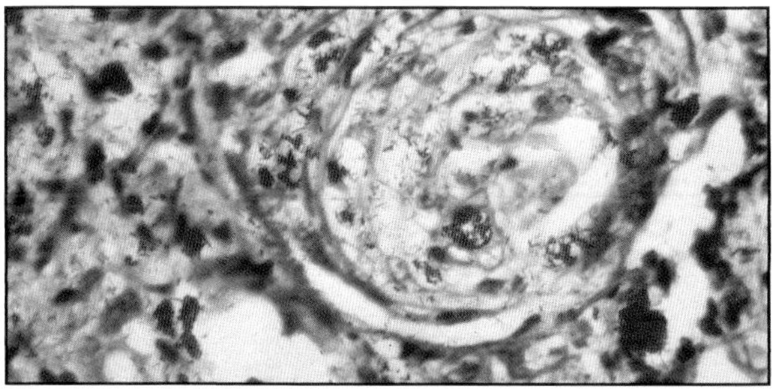

D. Section of a cutaneous nerve showing red, acid-fast bacteria of the leprosy-causing species *Mycobacterium leprae*.

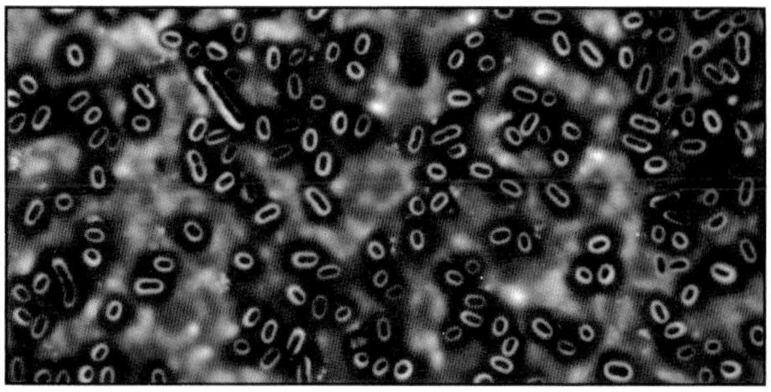

E. A capsule stain of *Klebsiella pneumoniae*, clearly showing the light, unstained capsules around the dark cell bodies.

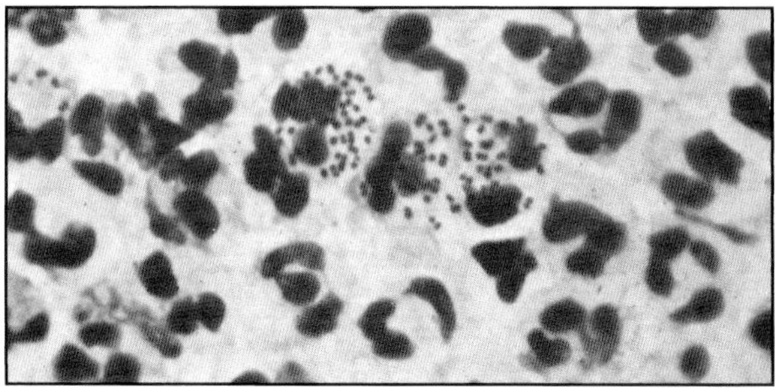

F. Numerous intracellular diplococci of *Neisseria gonorrhoeae* are in this urethral discharge smear from a case of gonorrhea. (Large, irregular red bodies are nuclei of white blood cells.)

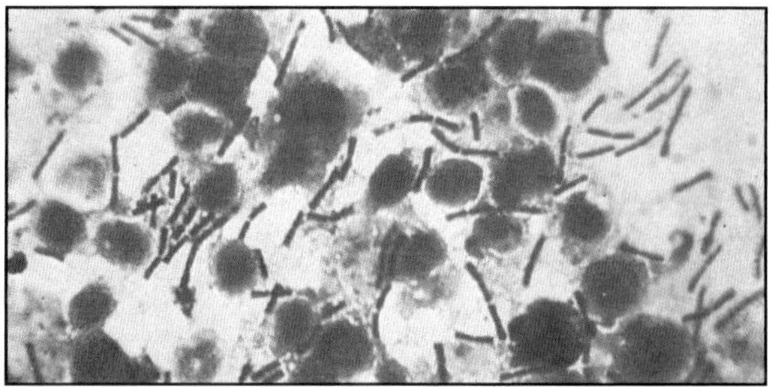

G. Short chains of thick rods of *Bacillus anthracis* as seen among tissue cells from a case of anthrax.

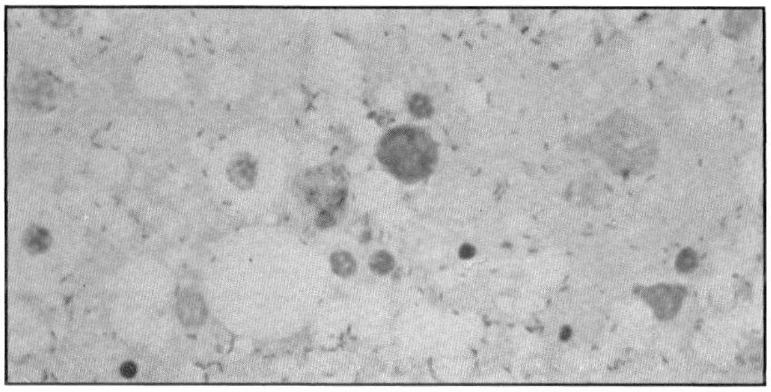

H. Rickettsias of the species *Rickettsia rickettsii* show up as tiny, red, intracellular forms in the Giménez stain of a yolk sac smear.

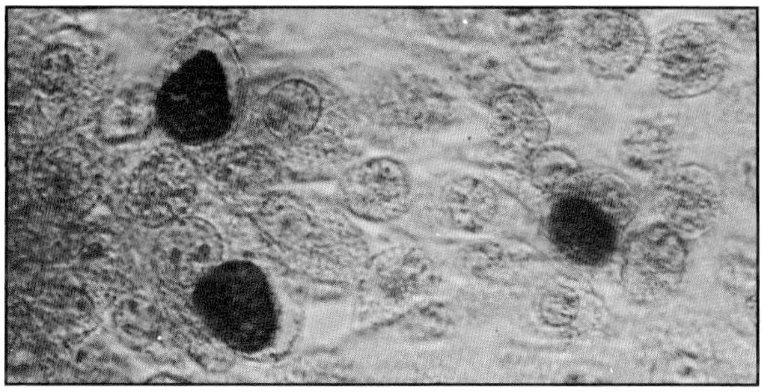

I. Dark, reddish-brown inclusion bodies of *Chlamydia trachomatis* are seen in several cells of a cultured monolayer of McCoy cells.

Eucaryotic Pathogens

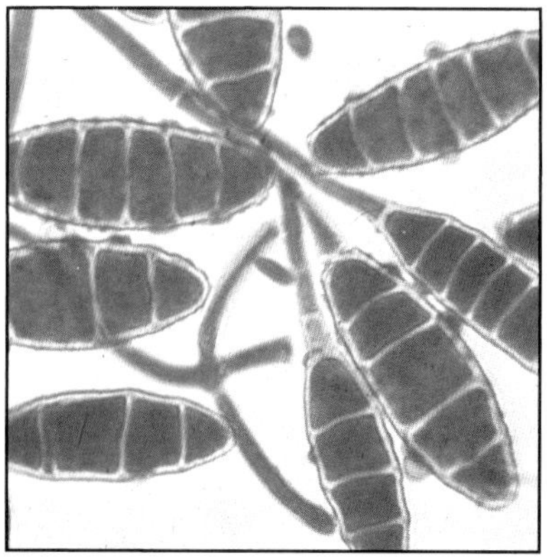

A.

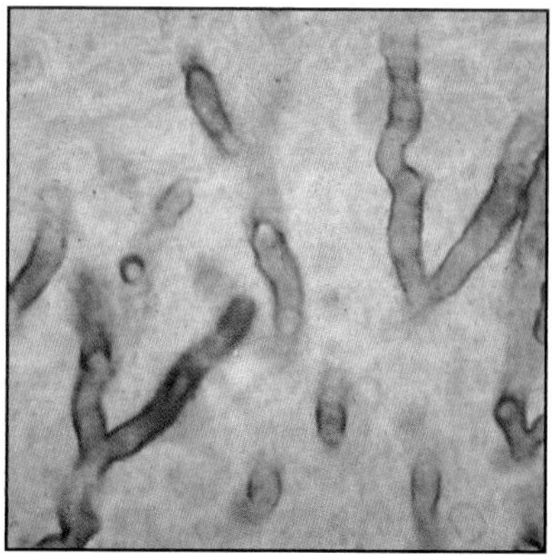

B.

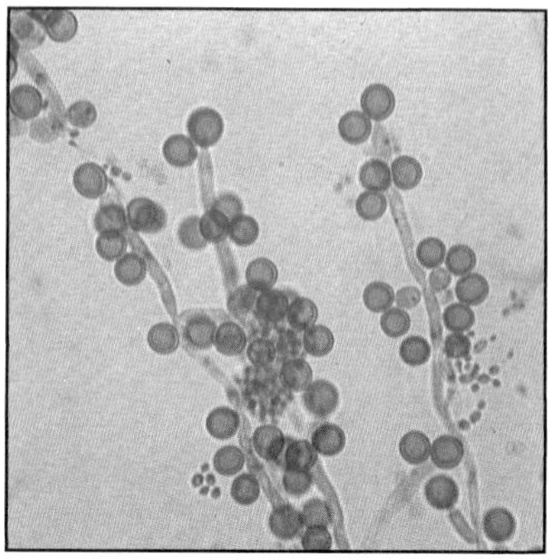

C.

Fungi A. Macroconidia of the fungal dermatophyte *Microsporum gypseum*. **B.** Hyphae of *Aspergillus fumigatus* in lung tissue from a case of aspergillosis. **C.** Appearance of *Candida albicans* grown on corn meal agar; the large spherical cells are chlamydospores, the much smaller rounded cells are blastoconidia, and the elongated strands of cells are pseudohyphae.

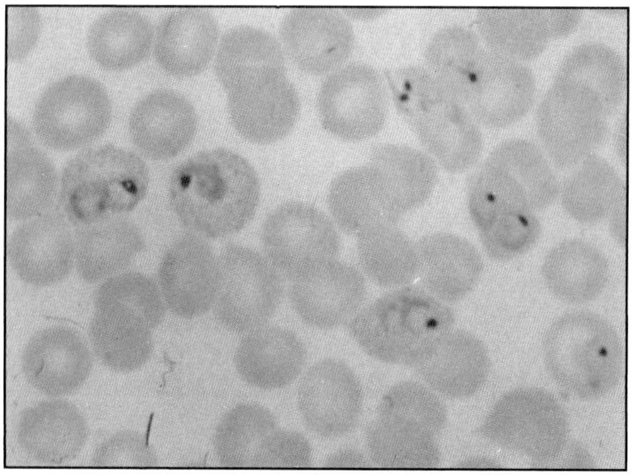

D.

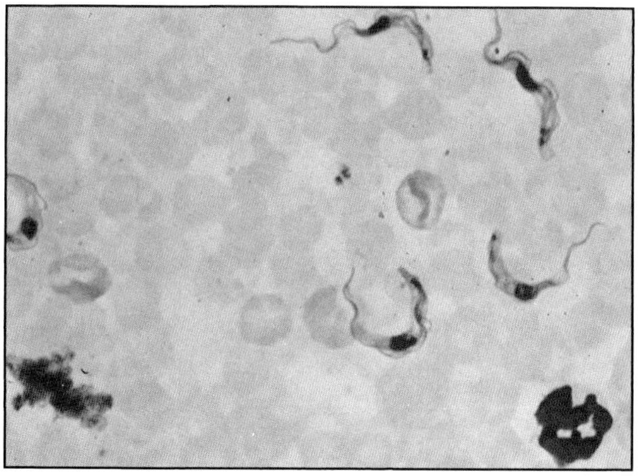

E.

Protozoans D. Blood smear showing intracellular rings and trophozoites of the malarial parasite *Plasmodium vivax*. **E.** Trypanomastigotes of *Trypanosoma brucei* in a blood smear from an experimentally infected animal; note the undulating membrane along the body of each trypanosome.

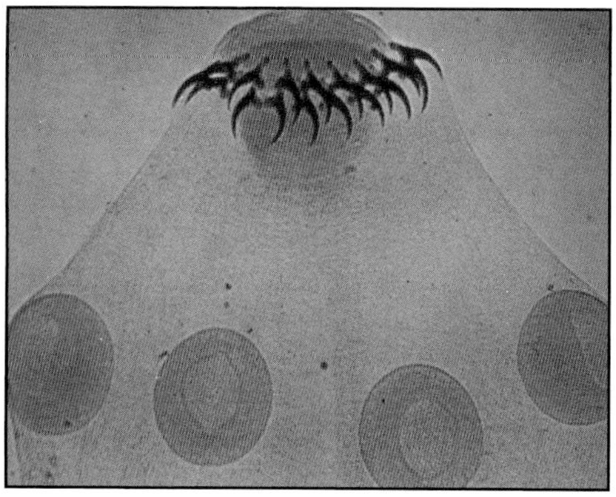

F.

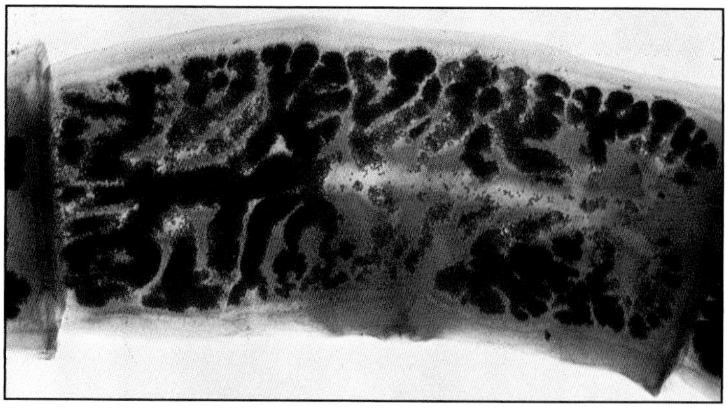

G.

Helminths F. Scolex of the pork tapeworm *Taenia solium*; note the sucking cups and hooks used to attach to the intestinal wall. **G.** A gravid proglottid of *Taenia solium*.

Plasma (Cytoplasmic) Membrane

The **plasma (cytoplasmic) membrane** is a thin structure lying inside the cell wall and enclosing the cytoplasm of the cell (see Figure 4–4). The chemical composition of the procaryotic plasma membrane is about 60% protein and 40% lipid, most of which is phospholipid. (Eucaryotic plasma membranes also contain carbohydrates and sterols.)

Structure

In electron micrographs, procaryotic and eucaryotic cell membranes both look like two-layered structures: there are two dark lines with a light space between the lines (Figure 4–13a). The phospholipid molecules are arranged in two parallel rows, called a *phospholipid bilayer* (Figure 4–13b). Each phospholipid molecule contains a charged polar head (the phosphate end), which is soluble in water, and an uncharged nonpolar tail (the hydrocarbon end), which is insoluble in water. The polar ends are on the outside of the phospholipid bilayer, and the nonpolar ends are on the inside.

The plasma membrane also contains protein molecules, which can be arranged in a variety of ways (Figure 4–13c). Some lie at or near the inner and outer surfaces of the membrane. Others penetrate the membrane partway or completely. Recent studies by physicists and biologists have demonstrated that the phospholipid and protein molecules in membranes are not static, but seem to move quite freely within the membrane surface. This movement is most likely associated with the many functions performed by the plasma membrane. This dynamic arrangement of phospholipid and protein is referred to as the *fluid mosaic model.*

Functions

The most important function of the plasma membrane is to serve as a selective barrier through which materials enter and exit the cell. In this function, plasma membranes are **selectively permeable** (sometimes called **semipermeable**). This term indicates that certain molecules and ions pass through the membrane, but others are restricted. The permeability of the membrane depends on several factors. Although large molecules (such as proteins) with molecular weights over several hundred cannot pass through the membrane, smaller molecules (such as water, amino acids, and some simple sugars) usually pass through easily if they are uncharged. Ions penetrate only very slowly. Substances that dissolve easily in lipids enter and exit more easily than do other substances, because the membrane consists mostly of phospholipids. The movement of materials across plasma membranes also depends on carrier molecules, which will be described shortly.

Plasma membranes are also important to the breakdown of foods and the production of energy. The plasma membranes of bacteria contain enzymes capable of catalyzing the chemical reactions that break down nutrients and produce energy. In some bacteria, pigments and enzymes involved in photosynthesis (the conversion of light energy into chemical energy) are found in infoldings of the plasma membrane that extend into the cytoplasm. These membrane layers are called **chromatophores** (Figure 4–14).

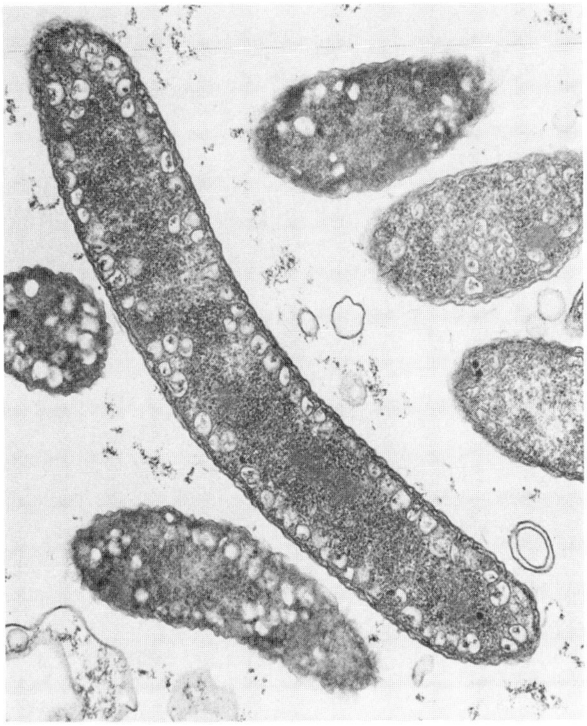

Figure 4–14 Chromatophores, membranous structures that are sites of photosynthesis, can be seen in this transmission electron micrograph of *Rhodospirillum rubrum* (×30,000).

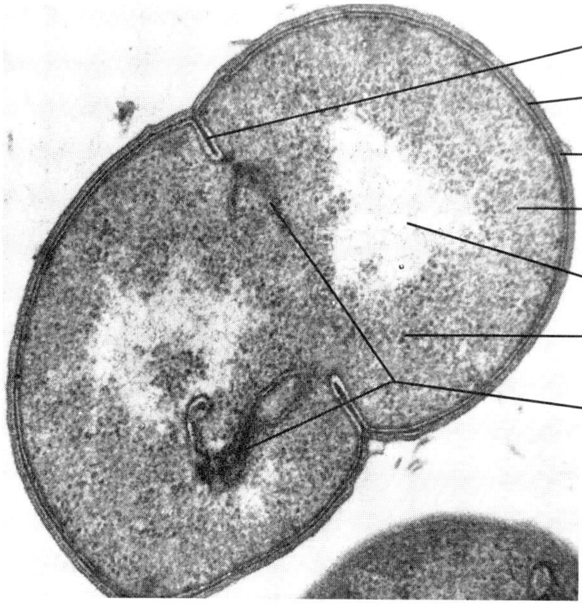

Figure 4–15 Transverse septum in a dividing cell of *Sporosarcina ureae* (×72,000). Two mesosomes extending into the cytoplasm from the innermost edges of the septa can be seen.

Mesosomes

The plasma membranes of gram-positive bacteria often contain one or more large, irregular folds called **mesosomes** (see Figure 4–4). In gram-negative cells, the mesosomes are considerably smaller. Generally, mesosomes are layered structures. Mesosomes are not present in eucaryotic cells.

Although the exact role of mesosomes is unknown, they might play a role in reproduction and metabolism. When a bacterial cell divides (binary fission), a cross wall called a *transverse septum* forms and partitions the genetic material of the parent cell into each of the two identical daughter cells (see Figure 6–11). Mesosomes might begin the formation of the transverse septum and attach bacterial DNA to the plasma membrane (Figure 4–15). Mesosomes might also help to separate DNA into each daughter cell following binary fission. And the ability of cells to concentrate nutrients might be enhanced by the presence of mesosomes, because their folds increase the surface area of the plasma membrane.

Destruction of plasma membrane by antimicrobial agents

Because the plasma membrane is vital to the bacterial cell, it is not surprising that it is the site at which several antimicrobial agents exert their effect. In addition to the chemicals that damage the cell wall and hence indirectly expose the membrane to injury, many compounds specifically damage plasma membranes. These compounds include certain alcohols and quaternary ammonium compounds, which are used as disinfectants. By disintegrating the membrane's phospholipids, a group of antibiotics known as the polymyxins cause leakage of intracellular contents and subsequent cell death. This mechanism is discussed in Chapter 18.

Movement of Materials Across Membranes

When a concentration of a substance is stronger on one side of a membrane than on another, a **concentration gradient** (difference) exists. If the substance can cross the membrane, it will move to the more dilute side until the concentrations are equal—or until other forces stop its movement.

Materials move across plasma membranes of both procaryotic and eucaryotic cells by two kinds of processes: passive and active. In **passive processes,** substances cross the membrane from an area of high concentration to an area of low concentration, without any expenditure of energy by the cell. Examples of passive processes are simple diffusion, osmosis, and facilitated diffusion. In **active processes,** the cell must use energy to move substances from areas of low concentration to areas of high concentration. An example of an active process is active transport.

Simple diffusion

Simple diffusion is the net (overall) movement of molecules or ions from an area of high concentration to an area of low concentration (Figure 4–16). The movement from areas of high to low concentration continues until the molecules or ions are evenly distributed. The point of even distribution is called *equilibrium.* Cells rely on diffusion to transport certain small molecules such as oxygen and carbon dioxide across their cell membranes.

Figure 4–16 The principle of simple diffusion. The molecules of dye in the pellet are diffusing into the water, from an area of high dye concentration to areas of low dye concentration.

Osmosis

Osmosis is the net movement of solvent molecules across a selectively permeable membrane, from an area in which the molecules are highly concentrated to an area of low concentration. In living systems, the chief solvent is water.

Osmosis may be demonstrated with the apparatus shown in Figure 4–17. A tube constructed

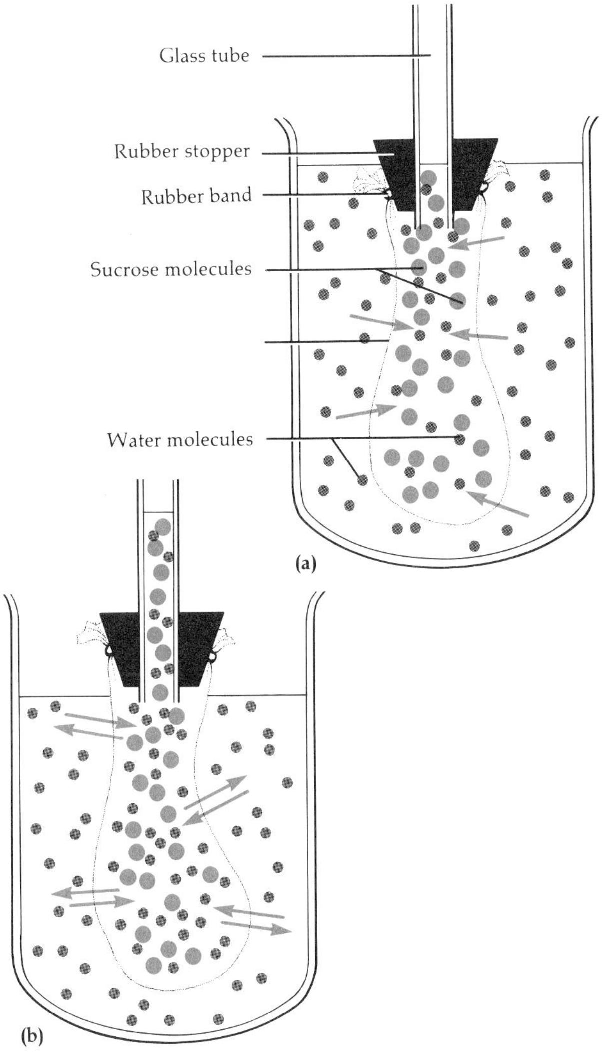

Figure 4–17 The principle of osmosis. **(a)** Setup at the beginning of the experiment. **(b)** Setup at equilibrium. The final height of solution in the glass tube in **(b)** is a measure of the osmotic pressure.

from cellophane, which is a selectively permeable membrane, is filled with a colored, aqueous solution of 20% sucrose. The opening of the cellophane tube is plugged with a rubber stopper through which a glass tubing is fitted. The cellophane tube is placed into a beaker containing distilled water. Initially, the concentrations of water on either side of the membrane are different. Because of the glucose molecules, the concentration of water is lower inside the cellophane tube than outside. To eliminate this difference, water moves from the beaker into the cellophane tube. The force with which the water moves is called osmotic pressure.

Very simply, **osmotic pressure** is the force with which a solvent moves from a solution of low solute concentration to a solution of high solute concentration when the solutions are separated by a semipermeable membrane. In our example, there is no movement of sugar out of the cellophane tube into the beaker, however, because the cellophane is impermeable to molecules of sugar—the sugar molecules are too large to go through the pores of the membrane. As water moves into the cellophane tube, the sugar solution becomes increasingly dilute, and because the cellophane tube is at maximum expansion due to its increased volume of water, water begins to move up the glass tubing. In time, the water that has accumulated in the cellophane tube and the glass tubing exerts a downward pressure that forces water molecules out of the cellophane tube and back into the beaker. When water molecules leave and enter the cellophane tube at the same rate, equilibrium is reached.

A bacterial cell may be subjected to three kinds of osmotic solutions: isotonic, hypotonic, or hypertonic. An **isotonic solution** (*iso* means "equal") is one in which the overall concentrations of solutes are the same on both sides of the membrane. Neither water nor solutes tend to flow into or out of the cell; the cell's contents are in equilibrium with the solution outside the cell wall.

Earlier we mentioned that lysozyme and certain antibiotics damage bacterial cell walls by causing the cells to rupture, or lyse. Such rupturing occurs because bacterial cytoplasm usually contains such a strong concentration of solutes that when the wall is weakened or removed, water enters the cell by osmosis. The damaged (or removed) cell wall cannot constrain the swelling of the cytoplasmic membrane, and the membrane bursts. This is an example of osmotic lysis caused by immersion in a hypotonic solution. A **hypotonic solution** (*hypo* means "under" or "less") outside the cell is a medium whose concentration of solutes is lower than that inside the cell. Most bacteria live in hypotonic solutions, and swelling is contained by the cell wall.

A **hypertonic solution** (*hyper* means "above") is a medium having a higher concentration of solutes than the cell has. Most bacterial cells placed in a hypertonic solution shrink and collapse because water leaves the cells by osmosis.

Keep in mind that the terms isotonic, hypotonic, and hypertonic describe the concentration of solutions outside the cell *relative* to the concentration inside the cell.

Facilitated diffusion

When **facilitated diffusion** occurs, the substance to be transported combines with a *carrier protein* in the plasma membrane. Such carriers are sometimes called *permeases.* The carrier can transport a substance across the membrane from an area of high concentration to one of low concentration. Facilitated diffusion is similar to simple diffusion in that because the substance moves from a high to a low concentration, the cell does not expend energy. The process differs from simple diffusion in its use of carriers (Figure 4–18a).

In some cases, molecules required by bacteria are too large to be transported into the cells by the methods just described. Most bacteria, however, produce enzymes that can break down large molecules like proteins into small amino acids, and polysaccharides into simple sugars. Such enzymes, which are released by the bacteria into the surrounding medium, are appropriately called *extracellular enzymes.* Once the enzymes degrade the large molecules, the subunits are transported by permeases into the cell. For example, specific carriers retrieve DNA bases, such as the purine guanine, from extracellular media, and bring them into the cell's cytoplasm.

Active transport

In performing **active transport,** the cell *uses energy* in the form of adenosine triphosphate (ATP) to

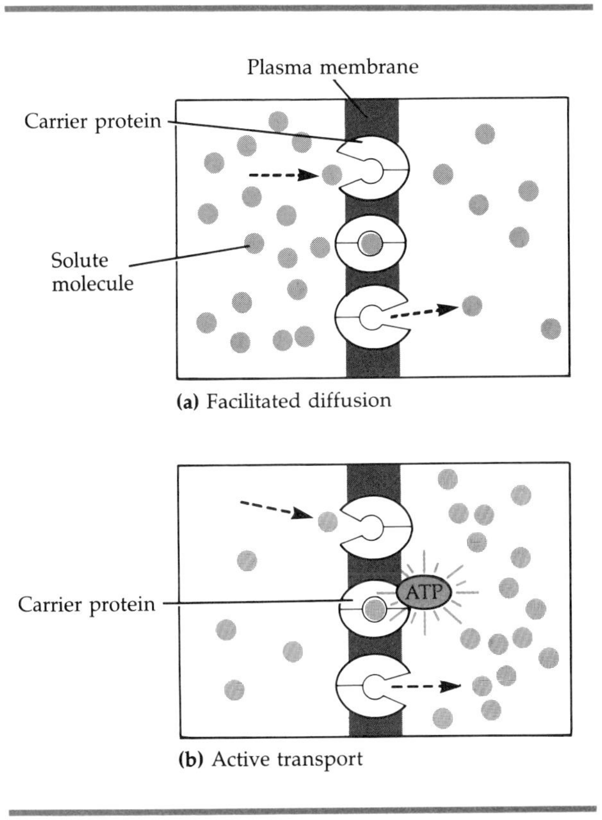

Figure 4–18 Facilitated diffusion compared with active transport. **(a)** In facilitated diffusion, carrier proteins in the membrane transport molecules across the membrane from an area of high concentration to one of low concentration (with the concentration gradient). This process does not require ATP. **(b)** For active transport, ATP supplies the energy needed for a carrier protein to transport a molecule across the membrane against a concentration gradient.

move substances across the plasma membrane. The movement is usually from outside to inside, even though the concentration might be much higher inside the cell. Like facilitated diffusion, active transport depends on carrier proteins in the plasma membrane (Figure 4–18b). There appears to be a different carrier for each transported substance or group of closely related transported substances.

Simple diffusion and facilitated diffusion are useful mechanisms for transporting substances into cells when their concentration is greater outside the cell. However, when a bacterial cell is in an environment in which nutrients are in low concentration, the cell must utilize active transport to accumulate the needed substances. For example, carrier proteins called galactoside permease enable some bacteria to acquire certain sugars from growth media.

Eucaryotic cells can use two additional active transport processes called phagocytosis and pinocytosis. Both processes are described in later sections.

Cytoplasm

For a procaryotic cell, the term **cytoplasm** refers to everything inside the plasma membrane (see Figures 4–4 and 4–15). Cytoplasm is about 80% water and contains nucleic acids, proteins, carbohydrates, lipids, inorganic ions, and many low-molecular-weight compounds. It includes a fluid component (matrix), particles with various functions, and a nuclear region. The fluid component is water, in which a vast array of molecules are dissolved in high concentration. Inorganic ions are present in much higher concentrations in cytoplasm than in most media. The fluid is thick, semitransparent, and elastic.

Cytoplasm is the substance in which the chemical reactions of the cell occur. When the cytoplasm receives raw materials from the external environment, enzymatic reactions within the cytoplasm degrade them to yield usable energy. The cytoplasm is also the place in which new substances are synthesized for cellular use.

Procaryotic cytoplasm lacks certain features of eucaryotic cytoplasm, such as a cytoskeleton and cytoplasmic streaming. These features will be described later.

Nuclear area

The **nuclear area,** or **nucleoid,** of a bacterial cell (see Figure 4–15) contains a single, long, circular molecule of DNA, the **bacterial chromosome** (see Figure 8–2a). This is the cell's genetic information. Unlike the chromosomes of eucaryotic cells, bacterial chromosomes are not surrounded by a nuclear envelope. Nucleoids can be spherical, elongated, or dumbbell shaped.

Bacteria often contain, in addition to the bacterial chromosome, small cyclic DNA molecules called **plasmids** (see Figure 8–23a). These are extrachromosomal genetic elements; that is, they are not connected to the main bacterial chromosome and replicate autonomously. Plasmids contain genes that are generally not crucial for the bacteria to grow under the usual environmental conditions, and plasmids may be gained or lost without harming the cell. Under certain conditions, however, plasmids are an advantage to cells. We shall discuss plasmids further in Chapter 8.

Ribosomes

All eucaryotic and procaryotic cells contain **ribosomes.** The cytoplasm of a procaryotic cell contains thousands of these very small structures, which give the cytoplasm a granular appearance (see Figure 4–15). These structures are composed of RNA and protein and are somewhat smaller and less dense than ribosomes of eucaryotic cells. As you will see in Chapter 8, ribosomes are factories that guide the synthesis of proteins needed for cellular activities. Several antibiotics, such as streptomycin, neomycin, and tetracyclines, exert their antimicrobial effects by inhibiting protein synthesis in the ribosomes.

Inclusions

Within the cytoplasm of procaryotic (and eucaryotic) cells are several kinds of reserve deposits, known as **inclusions.** Some inclusions are common to a wide variety of bacteria, whereas others are limited to a small number of species and therefore serve as a basis for identification. Among the more prominent bacterial inclusions are the following.

Metachromatic granules These inclusions stain red with certain blue dyes, such as methylene blue, and are collectively known as **volutin.** A stored form of phosphate, they are generally formed by cells that grow in phosphate-rich environments. Metachromatic granules are found in algae, fungi, and protozoans, as well as in bacteria. These granules are quite large and are characteristic of *Corynebacterium diphtheriae* (kô-rī-nē-bak-ti′rē-um dif-thi′rē-ī), the causative agent of diphtheria, so they do have diagnostic significance.

Polysaccharide granules These inclusions typically consist of glycogen and starch, and their presence can be demonstrated when iodine is applied to the cells. In the presence of iodine, glycogen granules appear reddish brown, and starch granules appear blue.

Lipid inclusions Lipid inclusions appear in various species of *Mycobacterium, Bacillus, Azotobacter* (*ä-zō-tō-bak′ter*), *Spirillum* (spī-ril′lum), and other genera. (A lipid commonly found as a storage material and unique to bacteria is the polymer *poly-β-hydroxybutyric acid.*) Lipid inclusions are revealed by use of fat-soluble dyes, such as Sudan dyes.

Sulfur granules Certain bacteria known as the "sulfur bacteria," which belong to the genus *Thiobacillus,* derive energy by oxidizing sulfur and sulfur-containing compounds. These bacteria may deposit sulfur granules in the cell, where they serve as an energy reserve.

Carboxysomes These are polyhedral or hexagonal inclusions that contain the enzyme ribulose 1,5-diphosphate carboxylase. Bacteria that use carbon dioxide as their sole source of carbon require this enzyme for photosynthesis. Among the bacteria containing carboxysomes are nitrifying bacteria, cyanobacteria, and thiobacilli.

Gas vacuoles These are hollow cavities found in many aquatic procaryotes, including cyanobacteria, photosynthetic bacteria, and halobacteria. Each vacuole consists of rows of several individual *gas vesicles,* which are hollow cylinders constructed of protein. The function of gas vacuoles is to maintain buoyancy so that the cells can remain at the depth in the water appropriate to their receiving sufficient amounts of oxygen, light, and nutrients.

For a discussion of inclusions in magnetotactic bacteria, see Box.

Endospores

When essential nutrients are depleted, certain gram-positive bacteria, those of the genera *Clostridium* and *Bacillus,* form specialized "resting" cells called **endospores.** Endospores are highly durable, dehydrated bodies with a thick wall. They are formed *inside* the bacterial cell wall. Released into the environment, they can survive extreme heat, lack of water, and exposure to many toxic chemicals.

The process of endospore formation within a

Microbial Magnets

For nearly 100 years biologists have suspected that birds can use the earth's magnetic field for orientation during flight. During the 1970's, magnetotaxis (response to a magnetic field) was demonstrated in some birds, as well as in honey bees, and some bacteria. In 1975, R. P. Blakemore made an observation that has led to a more complete understanding of the possible mechanism behind magnetotaxis. He reported,

"I casually observed that these [sulfide-rich mud] enrichments also contained highly motile bacteria that migrated nearly unidirectionally across the microscope field of view. These . . . microorganisms became dislodged from sediment particles in the course of [our] making preparations for phase contrast microscopy. They then persistently swam toward and accumulated at one edge of drops of sediment transferred to depression slides. They swam in the same geographic direction even when the microscope was turned around, moved to another location, or covered with a pasteboard box. Thus, it was evident their swimming direction was influenced not by light, as I had at first supposed, but by some pervasive stimulus. That the cells were, in fact, magnetically responsive was vividly demonstrated when a magnet was brought near the microscope. To my astonishment, the hundreds of swimming cells instantly turned and rushed away from the end of the magnet! They were always attracted by the end that also attracted the North-seeking end of a compass needle and they were repelled by its opposite end. Their swimming speed was very fast, on the order of 100 μm per second, and the entire population consisting of hundreds of freely and independently swimming cells swerved in unison as the magnet was moved about nearby."[1]

Magnetotactic bacteria synthesize magnetite (iron oxide) (Fe_3O_4) and store it in inclusions called magnetosomes, as shown in the figure. The magnetosomes act like magnets and cause the bacterial cells to align on the earth's geomagnetic field. All bacteria with

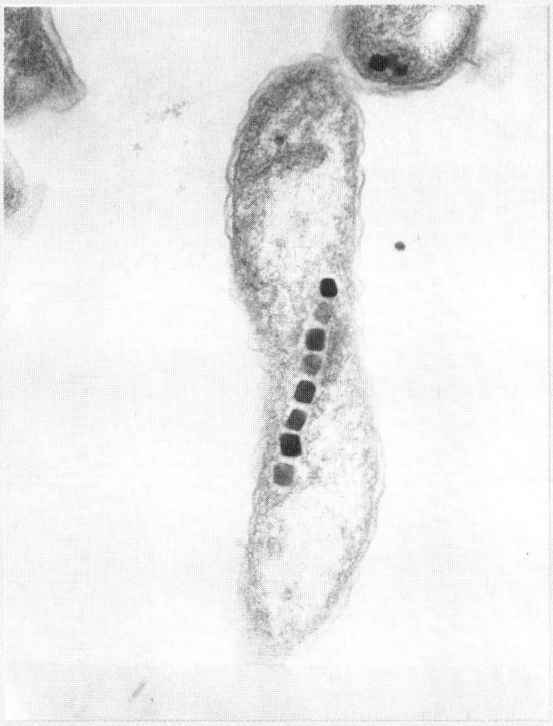

Aquaspirillum magnetotacticum showing chain of magnetosomes. Outer membrane of gram-negative wall is also visible.

magnetosomes are gram-negative, have pili or a polysaccharide glycocalyx to provide attachment, and may be either rods or cocci. These bacteria are found in sediments throughout the world. In the northern hemisphere, North-seeking bacteria predominate, and in the southern hemisphere, South-seeking bacteria predominate.

Only *Aquaspirillum magnetotacticum* (see figure) has been studied in pure culture. This bacterium metabolizes organic molecules for carbon and energy and requires a microaerophilic environment. It was found that when *A. magnetotacticum* are separated from an attachment site, these bacteria swim along what appear to be geomagnetic lines. They move downward, toward either the North or South pole, until reaching a suitable attachment site.

(continued on next page)

How these bacteria synthesize magnetite is presently being studied; however, the function of magnetite in the cell's metabolism is not yet clear. The formation of ferric ion (Fe^{3+}) from ferrous ion (Fe^{2+}) could provide ATP for the cell, but the estimated amount of ATP is so small that the magnetite must serve another function. In vitro, magnetosomes can decompose hydrogen peroxide, which forms in cells in the presence of oxygen and is usually degraded by the enzyme catalase before it reaches toxic levels. Researchers speculate that magnetosomes protect the cell against hydrogen peroxide accumulation in vivo.

The synthesis and function of magnetosomes are of interest to microbiologists to further our understanding of cells and may provide a model to explain the formation of similar iron oxides in birds, insects, and other animals. The ecology of magnetic microbes may also provide a tool for scientists studying the earth's magnetic field. Evidence suggests that the earth's magnetic field may have reversed or shifted several times. Examination of the orientation of fossils of magnetotactic bacteria preserved in rock or sediment may shed light on the movement of the continents and magnetic poles in relation to one another. Magnetotactic bacteria might also help us learn about evolution. Normally, the earth's magnetic field repels a portion of the ionized gases that continually emanate from the sun. During a magnetic reversal, more of this radiation strikes the earth, which may lead to an increase in mutations. The orientation of fossilized magnetotactic bacteria in relation to mutant fossils might help us learn whether magnetic reversals affect evolution.

[1]Reproduced, with permission, from the *Annual Review of Microbiology*, Volume 36, © 1982 by Annual Reviews Inc. and Richard P. Blakemore. (Richard P. Blakemore. 1982. Magnetotactic Bacteria. *Annual Review of Microbiology* 36:217–238.)

vegetative (parent) cell is known as **sporulation** or **sporogenesis** (Figure 4–19a and b). It is not clear what biochemical events trigger this process. In the first observable stage of sporogenesis, a newly replicated bacterial chromosome and a small portion of cytoplasm are isolated by an ingrowth of the plasma membrane called a *spore septum*. The spore septum becomes a double-layered membrane that surrounds the chromosome and cytoplasm. This structure, entirely enclosed within the original cell, is called a *forespore*. Thick layers of peptidoglycan are laid down between the two membrane layers. Then a thick *spore coat* of protein forms around the outside membrane. It is this coat that is responsible for the resistance of endospores to many harsh chemicals.

The diameter of the endospore may be the same as, smaller than, or larger than the diameter of the vegetative cell. Depending on the species, the endospore might be located *terminally* (at one end), *subterminally* (near one end), or *centrally* inside the vegetative cell (Figure 4–19c). When the endospore matures, the vegetative cell wall dissolves (lyses) and the endospore is freed.

Most of the water present in the forespore cytoplasm is eliminated by the time sporogenesis is complete, and endospores do not carry out metabolic reactions. The highly dehydrated endospore core contains only DNA, small amounts of RNA, ribosomes, enzymes, and a few important small molecules. The latter include a strikingly large amount of an organic acid called *dipicolinic acid*, which is accompanied by a large amount of calcium ion. These cellular components will be essential for resuming metabolism later.

Endospores can remain dormant for a long time, even hundreds of years. However, an endospore returns to its vegetative state by a process called **germination.** Germination is triggered by physical or chemical damage to the endospore's coat. The endospore's enzymes then break down the extra layers surrounding the endospore, water enters, and metabolism resumes. Because one vegetative cell forms a single endospore, which after germination remains one cell, sporogenesis in bacteria is *not* a means of reproduction. There is no increase in the number of cells.

Endospores are important from a clinical viewpoint, because they are resistant to processes that normally kill vegetative cells. Such processes include heating, freezing, desiccation, use of chemicals, and radiation. Whereas most vegetative cells are killed by temperatures above 70°C, endospores can survive in boiling water for an hour or more.

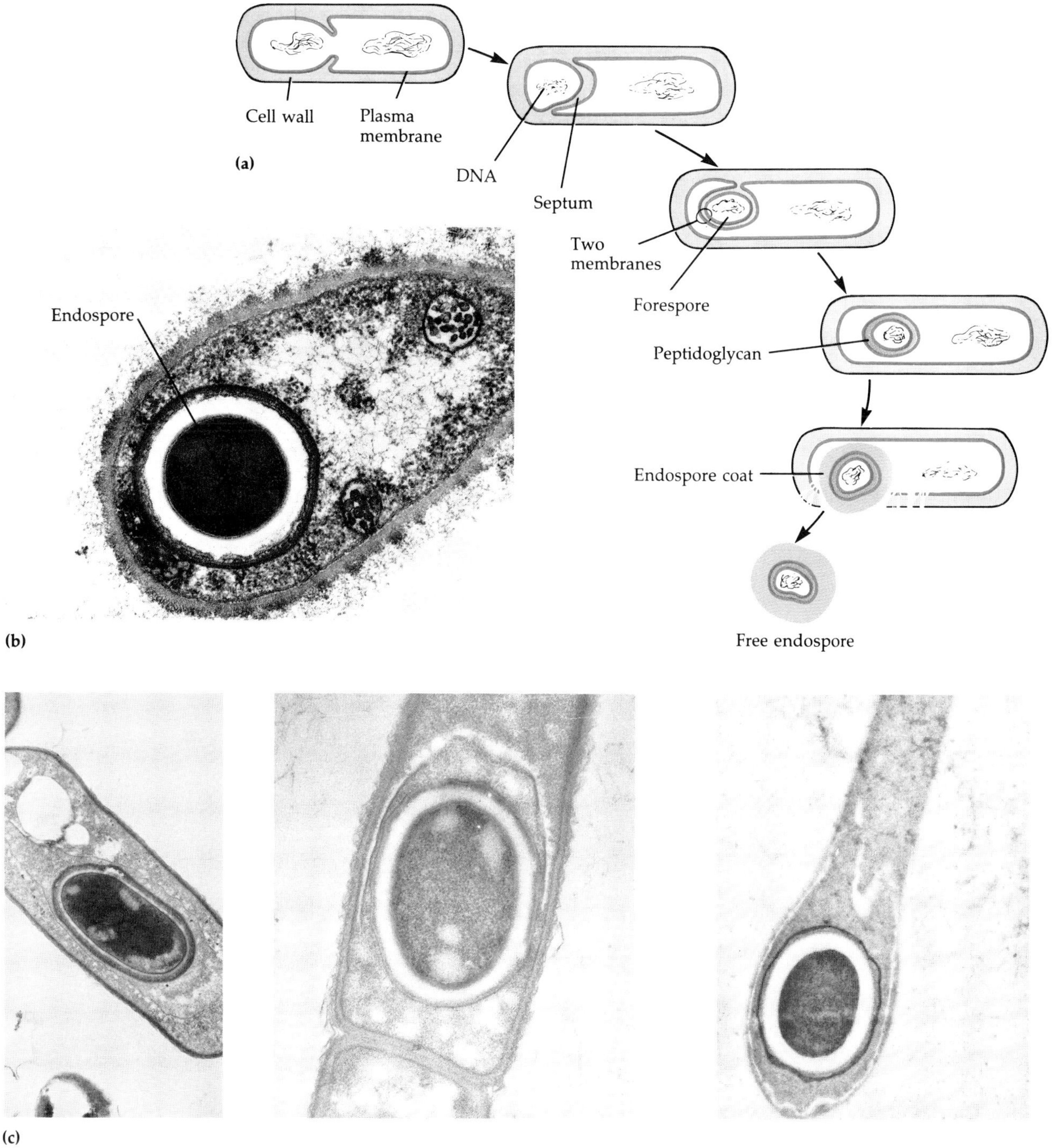

Figure 4–19 Endospores. **(a)** Sporogenesis, the process of endospore formation. **(b)** Electron micrograph of an endospore in *Bacillus sphaericus*. **(c)** Various locations of endospores. *Left:* central endospore in *Bacillus megaterium* (×40,500); *center:* subterminal endospore in *B. anthracis* (×38,000); *right:* round, terminal endospore at the ends of a cell of *C. tetani* (approx. ×14,000).

Endospore-forming bacteria are a problem in the food industry, because they are likely to survive any underprocessing, and because some species produce toxins and disease. Special methods used to control organisms that produce endospores are discussed in Chapter 7.

As noted in Chapter 3, endospores are difficult to stain for detection. Thus, a specially prepared stain must be used along with heat. (The Schaeffer–Fulton endospore stain is commonly used.)

Having examined the functional anatomy of the procaryotic cell, we will now look at the functional anatomy of the eucaryotic cell.

THE EUCARYOTIC CELL

As mentioned earlier, eucaryotic organisms include algae, protozoans, fungi, higher plants, and animals. The eucaryotic cell (Figure 4–20) is typically larger and structurally more complex than the procaryotic cell. By comparing the structure of the procaryotic cell in Figure 4–4 with that of the eucaryotic cell, we can see the differences between the two types of cells.

To review briefly, the genetic material (DNA) of procaryotic cells is not membrane-bounded or associated in a regular way with protein. The genetic material of eucaryotic cells *is* membrane-bounded, organized into chromosomes, and closely associated with histones and other proteins. Eucaryotic cells also contain membrane-bounded organelles; procaryotic cells do not. **Organelles** are specialized structures that perform specific functions. Although both types of cells carry on the same basic functions, procaryotic cells do not localize these functions in specific organelles.

The following discussion of eucaryotic cell anatomy will parallel our discussion of procaryotic cell anatomy by starting with structures that extend beyond the body of the cell. At the end of the discussion, the principal differences between procaryotic and eucaryotic cells will be summarized in Table 4–1.

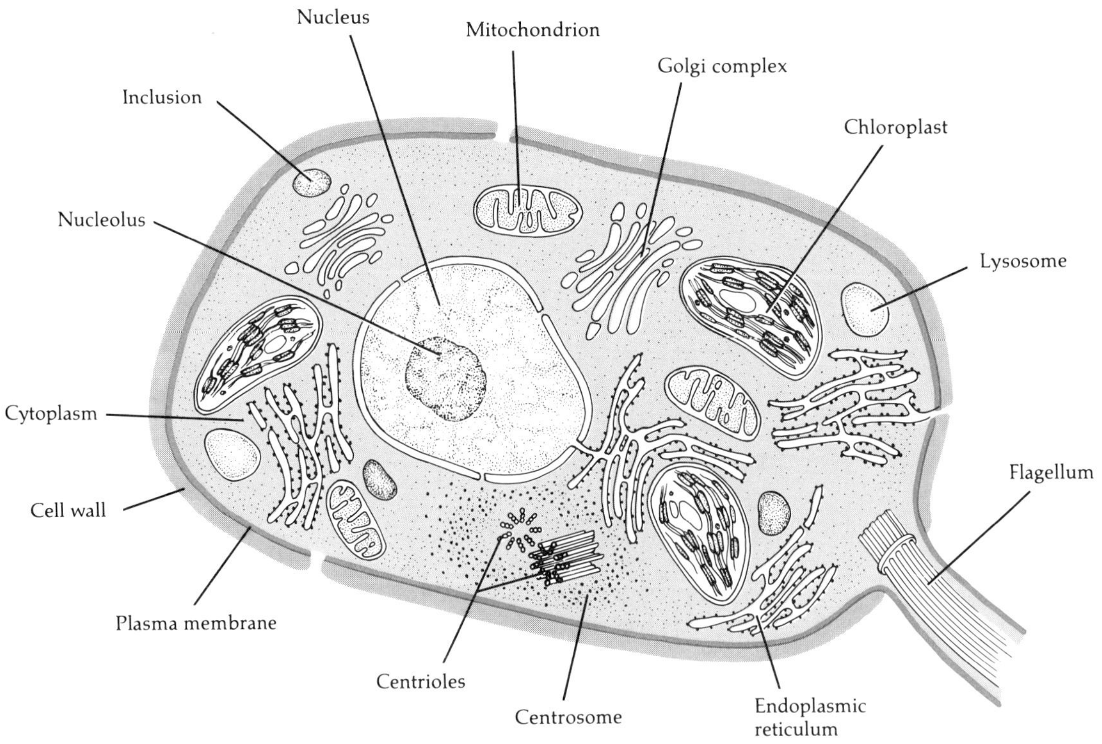

Figure 4–20 Highly schematic diagram of a composite eucaryotic cell.

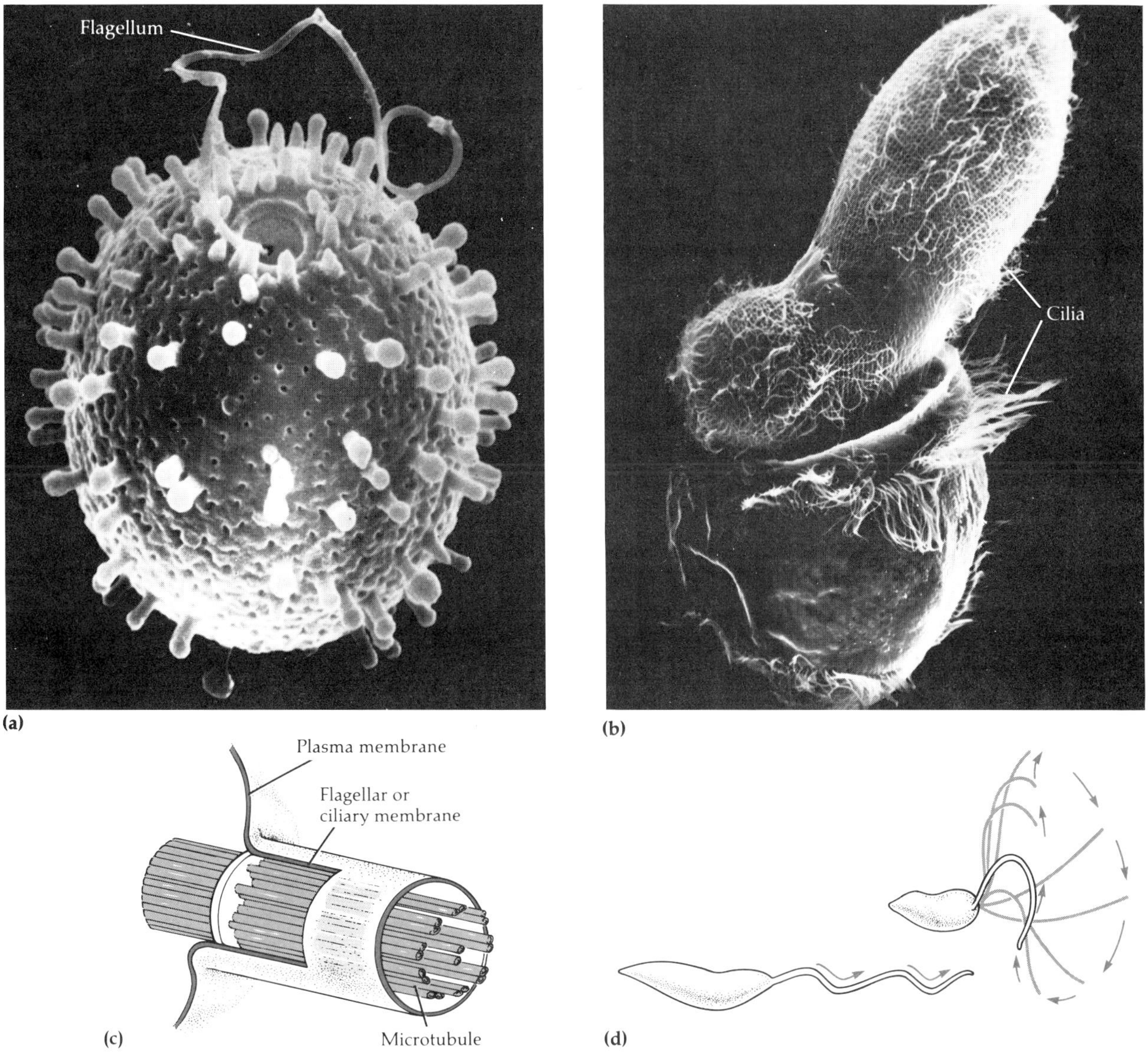

Figure 4–21 Eucaryotic flagella and cilia. **(a)** The freshwater euglenoid *Trachelomonas* sp. (×5000). Its long, whiplike flagellum is emerging from the cell body at the top of this scanning electron micrograph. **(b)** Ciliate eats ciliate. *Didinium nasutum,* the lower cell with two rows of cilia around its body, has begun to engulf a cilia-covered cell of *Paramecium multimicronucleatum* (×1500). **(c)** Structure of a flagellum or cilium. **(d)** Movement of a eucaryotic flagellum.

FLAGELLA AND CILIA

Many types of eucaryotic cells have projections that are used for cellular locomotion or for moving substances along the surface of the cell. These projections contain cytoplasm and are enclosed by the plasma membrane. If the projections are few and are long in relation to the size of the cell, they are called **flagella.** If the projections are numerous and short, resembling hairs, they are **cilia.** Euglenoid algae use a flagellum for locomotion, whereas protozoans, such as *Paramecium,* utilize cilia for locomotion (Figure 4–21a and b). In humans, the tail

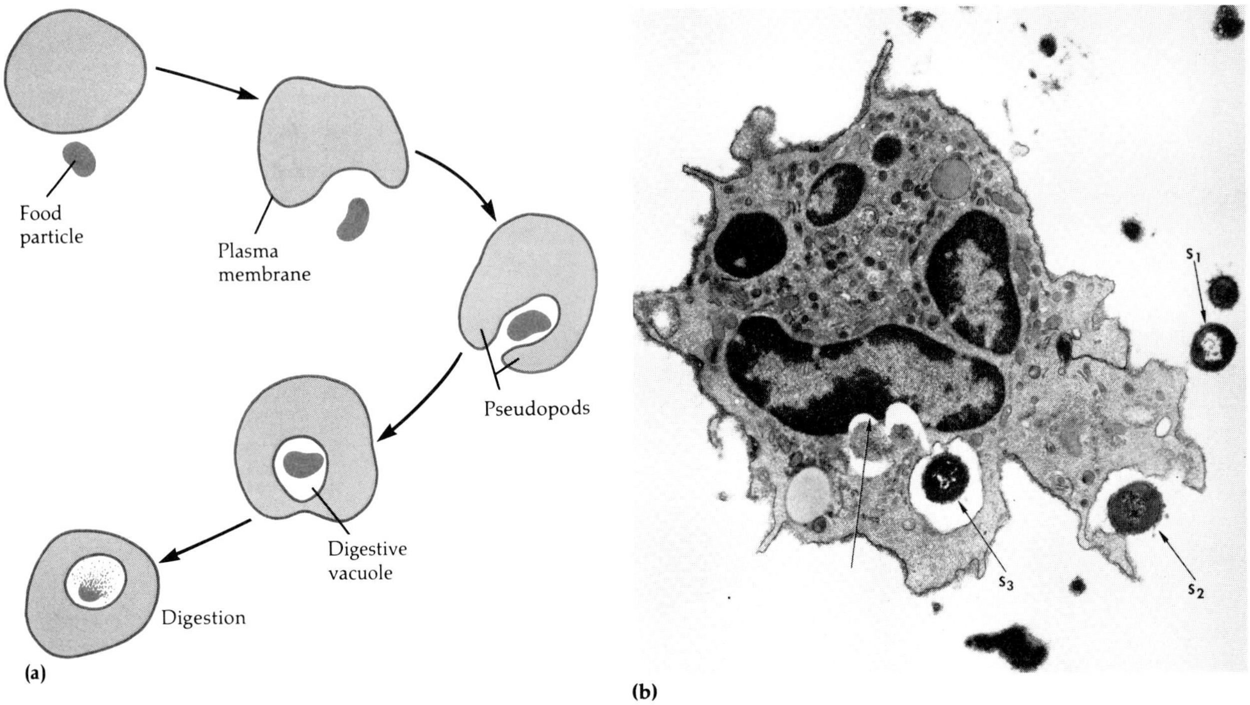

Figure 4–22 Phagocytosis. **(a)** Diagram of ingestion of a solid particle by phagocytosis. **(b)** Phagocytosis of bacteria (*Streptococcus pyogenes*) by a white blood cell in the lung of a mouse (×23,500). S_1 is a free bacterium, S_2 is a bacterium that has been partially engulfed, and S_3 is a bacterium completely engulfed in a clear vacuole. (The unlabeled arrow points to an area of the nucleus that appears to have been digested away.)

of a sperm cell is a flagellum that propels the sperm cell through the female reproductive system. Whereas a procaryotic flagellum rotates, a eucaryotic flagellum moves in a wavelike manner. To help keep foreign material out of the lungs, ciliated cells of the human respiratory system move the material along the surface of the cells in the bronchial tubes and trachea toward the throat and mouth (see Figure 15–3).

Eucaryotic flagella and cilia are structurally more complex than procaryotic flagella. The eucaryotic structures are composed of small tubules of protein called *microtubules* (Figure 4–21c). Both flagella and cilia consist of nine pairs of microtubules that encircle an inner central pair. The central microtubules arise from a plate near the surface of the cell, and the outer pairs arise from a structure called a centriole.

CELL WALL

In general, the eucaryotic **cell wall** is considerably simpler than that of the procaryotic cell. Most algae have cell walls consisting of the polysaccharide *cellulose*. Cell walls of some fungi also contain cellulose, but in most fungi, the principal structural component of the cell wall is the polysaccharide *chitin*, a polymer of N-acetylglucosamine units. (Chitin is also the main structural component of the exoskeleton of crustaceans and insects.) The cell wall of yeasts contains the polysaccharides *glucan* and *mannan*. In eucaryotes that lack a cell wall, the plasma membrane may be the outer covering; however, cells such as slime molds and sarcodines, which have direct contact with the environment, may have coatings outside the plasma membrane (see Microview 1–3E). Protozoans do not have a typical cell wall; instead, they have a flexible outer covering called a *pellicle*.

An important clinical consideration is that eucaryotic cells do not contain peptidoglycan, the framework of the procaryotic cell wall. This is important medically because antibiotics, such as penicillins and cephalosporins, act against peptidoglycan and therefore do not affect human eucaryotic cells.

PLASMA (CYTOPLASMIC) MEMBRANE

In eucaryotic cells that lack a cell wall, the **plasma membrane** is the external covering of the cell (see Figure 4–20). In function and structure, the eucaryotic and procaryotic plasma membranes are very similar. There are, however, differences in the proteins and in the carbohydrates attached to the proteins of some eucaryotic membranes. Another difference is that eucaryotic plasma membranes contain *sterols,* complex lipids not found in procaryotic plasma membranes (with the exception of the mycoplasmas). Sterols seem to be associated with the ability of the membranes to resist lysis due to increased osmotic pressure. (The chemical activity of sterols can be altered by a group of antimicrobial drugs called polyenes.)

Substances can cross eucaryotic and procaryotic plasma membranes by simple diffusion, osmosis, facilitated diffusion, or active transport. Eucaryotic cells can use two additional mechanisms called phagocytosis and pinocytosis. During **phagocytosis,** or "cell eating," cellular projections called *pseudopods* engulf solid particles external to the cell. Once the particle is surrounded, the pseudopod's membrane folds inward, forming a membrane sac around the particle (Figure 4–22a). This newly formed sac, called a *digestive vacuole,* breaks off from the outer cell membrane, and the solid material inside the vacuole is digested. Indigestible particles and cell products are removed from the cell by a reverse phagocytosis. This process is important because molecules and particles of material that would normally be restricted from crossing the plasma membrane can be brought into or removed from the cell. Human phagocytic white blood cells constitute a vital defensive mechanism. Using phagocytosis, the white blood cells destroy bacteria and other foreign substances (Figure 4–22b).

In **pinocytosis,** or "cell drinking," the engulfed material is liquid rather than solid. Moreover, no cytoplasmic projections are formed. Instead, the membrane folds inward, surrounds the liquid, and detaches from the rest of the intact membrane. Few cells are capable of phagocytosis, but many cells can carry on pinocytosis.

When phagocytosis and pinocytosis involve the inward movement of materials, they are together referred to as **endocytosis.** However, because phagocytosis and pinocytosis can also work in reverse, expelling materials from the cells, they are also referred to as **exocytosis.**

CYTOPLASM

The **cytoplasm** of the eucaryotic cell encompasses everything inside the cell's membrane and outside the nucleus (see Figure 4–20). The cytoplasm is the matrix, or ground substance, in which various cellular components are found. Physically and chemically, the fluid component of the cytoplasm of eucaryotes is similar to that of procaryotes. A major difference, however, is that eucaryotic cytoplasm has a complex internal structure, consisting of a series of exceedingly small rods called *microfilaments* and cylinders called *microtubules.* Together, they form the *cytoskeleton.* The cytoskeleton provides support and shape, and is involved in the movement of structures in the cytoplasm and even the entire cell, as occurs in phagocytosis. The movement of eucaryotic cytoplasm from one part of the cell to another, which helps to distribute nutrients and move the cell over a surface, is called *cytoplasmic streaming.* Another difference between procaryotic and eucaryotic cytoplasm is that many of the important enzymes found in the cytoplasmic fluid of procaryotes are sequestered in the organelles of eucaryotes.

ORGANELLES

Nucleus

The most characteristic eucaryotic organelle is the nucleus (see Figure 4–20). The **nucleus** is usually spherical or oval (Figure 4–23), is frequently the largest structure in the cell, and contains almost all of the cell's hereditary information (DNA). Some DNA is found in mitochondria and chloroplasts (of photosynthetic organisms).

The nucleus is separated from the cytoplasm by a double membrane called the *nuclear envelope.*

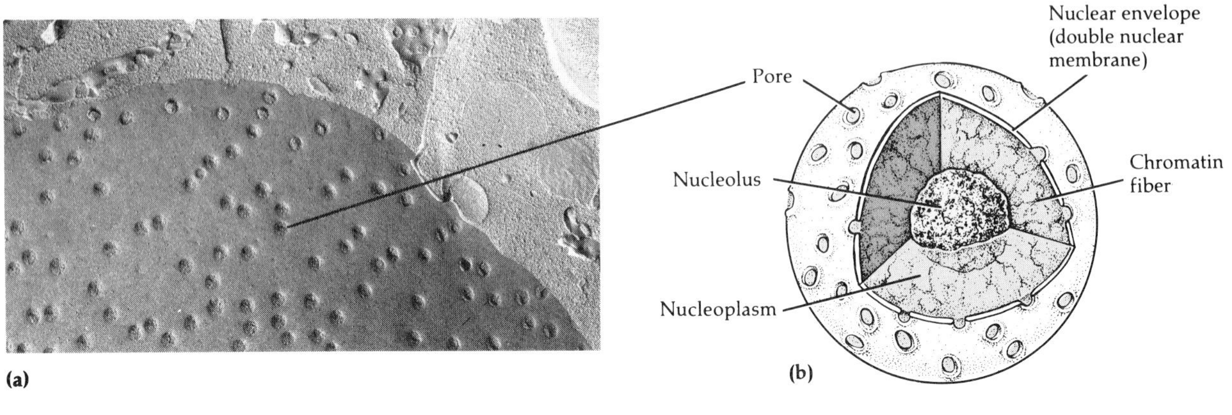

Figure 4–23 Nucleus. **(a)** Electron micrograph of a freeze-etched section, part of a nucleus in an onion root tip cell (×17,000). **(b)** Diagram of a nucleus.

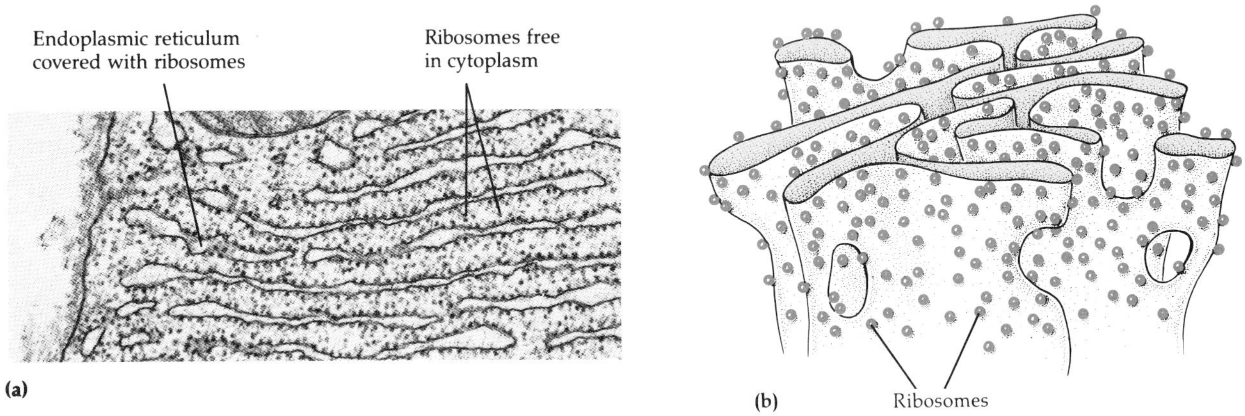

Figure 4–24 Endoplasmic reticulum and ribosomes. **(a)** Transmission electron micrograph of endoplasmic reticulum and ribosomes in cross section (×24,000). **(b)** Three-dimensional drawing of endoplasmic reticulum and ribosomes.

Each of the two membranes resembles the plasma membrane in structure. Minute *pores* in the nuclear membrane allow the nucleus to communicate with the membranous network in the cytoplasm, called the endoplasmic reticulum (see next section). Substances entering and exiting the nucleus are believed to pass through the tiny pores. Within the nuclear envelope is a gelatinous fluid called *nucleoplasm.* Spherical bodies called the *nucleoli* are also present. These structures are a center for the synthesis of ribosomal RNA, an essential constituent of ribosomes (see next section). Finally, there is the DNA, which is combined with a number of proteins, including several basic proteins called *histones* and *nonhistones*. When the cell is not reproducing, the DNA and its associated proteins are a threadlike mass called **chromatin.** Prior to nuclear division, the chromatin coils into shorter and thicker rodlike bodies called **chromosomes.** Procaryotic "chromosomes" do not undergo this process, do not have histones, and are not enclosed in a nuclear envelope.

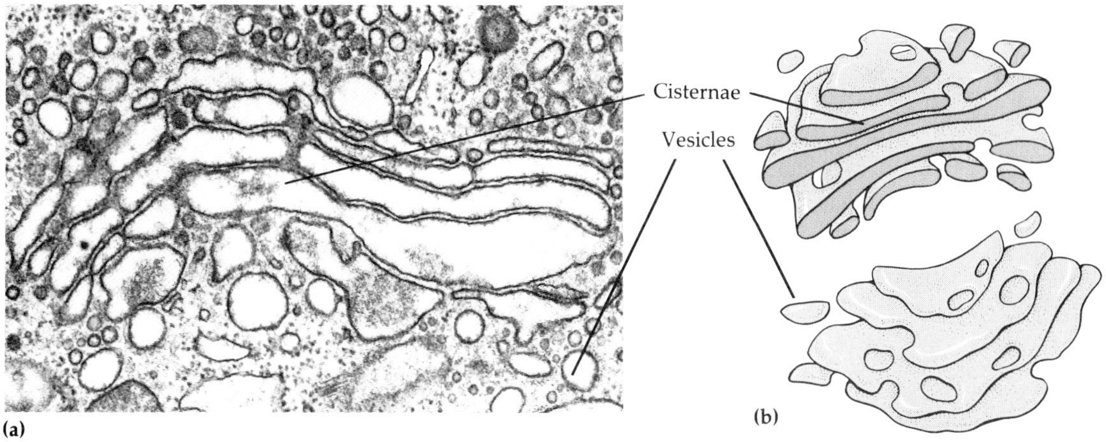

Figure 4–25 Golgi complex. **(a)** Transmission electron micrograph of a Golgi complex in cross section (×26,000). **(b)** Three-dimensional drawing of a Golgi complex.

Endoplasmic Reticulum

Within the cytoplasm, there is a system consisting of pairs of parallel membranes that enclose narrow cavities of varying shapes. This system, not present in procaryotes, is known as the **endoplasmic reticulum,** or **ER** (Figure 4–24).

The ER is a network of canals running throughout the cytoplasm. These canals are continuous with both the plasma membrane and the nuclear membrane (see Figure 4–20). It is thought that the ER provides a surface area for chemical reactions, a pathway for the transportation of molecules within the cell, and a storage area for synthesized molecules. The ER plays a role in both lipid synthesis and protein synthesis.

Ribosomes

Attached to the outer surface of some of the ER are exceedingly small, dense, spherical bodies called **ribosomes,** which are also found free in the cytoplasm. Ribosomes are the sites of protein synthesis in the cell.

The ribosomes of the eucaryotic ER and cytoplasm are somewhat larger and more dense than those of procaryotic cells. Accordingly, these eucaryotic ribosomes are called 80S ribosomes, and those of procaryotic cells are known as 70S ribosomes. Chloroplasts and mitochondria contain 70S ribosomes. The letter S refers to Svedberg units, which indicate the relative rate of sedimentation during ultra-high-speed centrifugation. The structure of ribosomes and their role in protein synthesis will be discussed in more detail in Chapter 8.

Golgi Complex

Another organelle found in the cytoplasm of eucaryotic cells is the **Golgi complex.** This structure usually consists of four to eight flattened discs called *cisternae,* which have expanded areas at their ends from which bud spherical *vesicles* (Figure 4–25). The Golgi complex is sometimes connected to the ER (see Figure 4–20). Its function is to package and secrete (release from the cell) certain proteins, lipids, and carbohydrates.

Lipids synthesized by the ER and proteins synthesized by the ribosomes that are associated with ER are transported from the ER tubules to the Golgi complex. As these substances accumulate in the Golgi complex, *vesicles* form and pinch off. Termed a *secretory granule,* the protein and its associated vesicle move toward the surface of the cell, where the contents of the vesicle are secreted.

The Golgi complex also functions in the synthesis of carbohydrates, which combine with proteins to form complexes called *glycoproteins.* Glycoproteins are also secreted from the cell in vesicles.

Mitochondria

Spherical, rod-shaped, or filamentous organelles called **mitochondria** appear throughout the cytoplasm of eucaryotic cells (see Figure 4–20). Sectioned and viewed under an electron microscope, each of these small organelles is revealed to have an elaborate internal organization (Figure 4–26). A mitochondrion consists of a double membrane similar in structure to the plasma membrane. The outer mitochondrial membrane is smooth, but the inner membrane is arranged in a series of folds called *cristae.* The center of the mitochondrion is called the *matrix.* Because of the nature and arrangement of the cristae, the inner membrane provides an enormous surface area on which chemical reactions can occur. Enzymes involved in energy-releasing reactions that form ATP are located on the cristae. Mitochondria are frequently called the "powerhouses of the cell" because of their central role in the production of ATP (see Chapter 5).

Lysosomes

As viewed under the electron microscope, **lysosomes,** which are formed from Golgi complexes, appear to be membrane-enclosed spheres in eucaryotic cytoplasm (Figure 4–27). Unlike mitochondria, lysosomes have only a single membrane and lack detailed structure (see Figure 4–20). But they contain powerful digestive enzymes capable of breaking down many kinds of molecules. Moreover, these enzymes can also digest bacteria that enter the cell. Human white blood cells, which use phagocytosis to ingest bacteria, contain large numbers of lysosomes. Scientists have wondered why these powerful enzymes do not also destroy their own cells. Perhaps the lysosome membrane in a healthy cell is impermeable to enzymes, and so prevents the enzymes from moving out into the cytoplasm. However, when a cell is injured and the lysosomes do release their enzymes, the enzymes then promote reactions that break the cell down

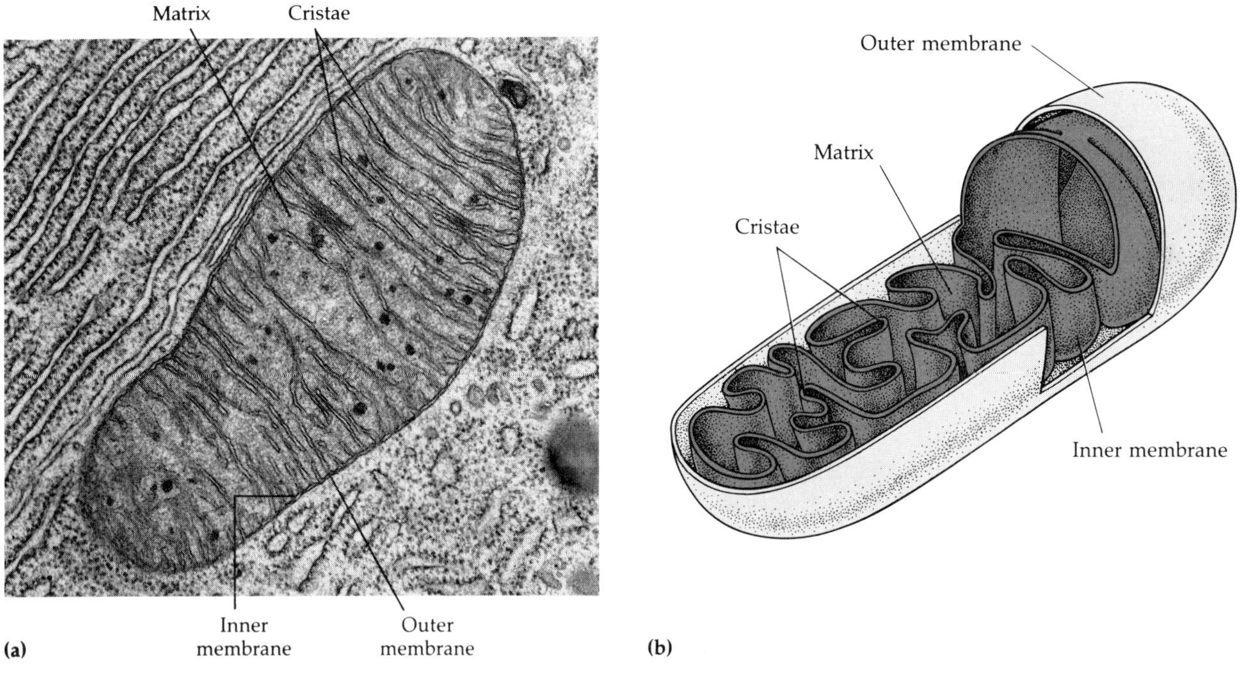

Figure 4–26 Mitochondria. **(a)** Transmission electron micrograph of a mitochondrion in longitudinal section from a rat pancreas cell (×34,000). **(b)** Three-dimensional drawing of a mitochondrion.

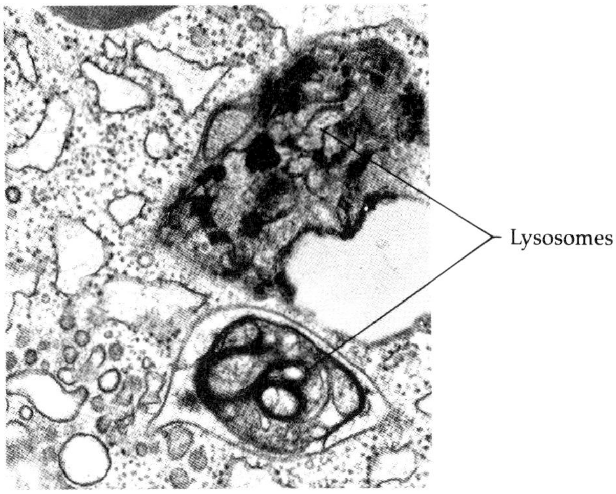

Figure 4–27 Lysosomes. Transmission electron micrograph of lysosomes in a rat pancreas cell (×26,000).

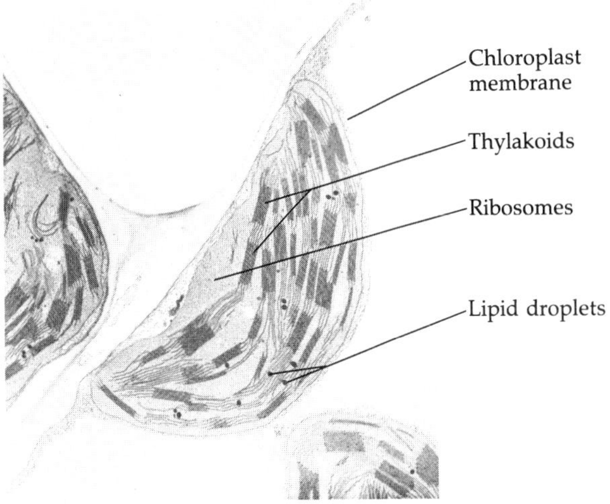

Figure 4–28 Transmission electron micrograph of a chloroplast from a cell in a corn leaf (×8,000). The light-trapping pigments are located on the thylakoids.

into its chemical constituents. The chemical remains are either reused by the body or excreted. Because of this function, lysosomes have been called "suicide packets."

Centrosome and Centrioles

There is a dense area of eucaryotic cytoplasm, generally spherical and located near the nucleus, that is called the **centrosome.** Within the centrosome is a pair of cylindrical structures, the **centrioles** (see Figure 4–20). Each centriole is a ring of nine evenly spaced bundles, each of which, in turn, consists of three microtubules. The two centrioles are situated so that the long axis of one is at right angles to the long axis of the other. Centrioles play a role in eucaryotic cell division.

Chloroplasts

Algae (and green plants) contain a unique organelle called a **chloroplast** (Figure 4–28). A chloroplast is a membrane-bounded structure that contains both the pigment chlorophyll and the enzymes required for the light-gathering phases of photosynthesis. The chlorophyll is contained in stacks of membranes called *thylakoids.* Like mitochondria, chloroplasts contain 70S ribosomes, DNA, and enzymes involved in protein synthesis. They are capable of multiplying by binary fission within the cell. Some of the needed genes are carried in the nuclear DNA; others are in the cytoplasm.

CYTOPLASMIC INCLUSIONS

Eucaryotic cells contain cytoplasmic inclusions that have various influences on cellular function. The size and composition of these inclusions vary widely among the various species of Protista. *Zymogen* (an enzyme storage bank), *fat inclusions, vacuoles* (raw food forms), and *glycogen* (complex carbohydrates) are examples of eucaryotic inclusions.

The principal differences between procaryotic and eucaryotic cells are presented in Table 4–1. Most of the differences have been discussed in this chapter; a few will be treated in greater detail in subsequent chapters.

Our next concern is to examine microbial metabolism. In Chapter 5 you will learn the importance of enzymes to microbes, and the ways in which microbes produce and utilize energy.

Table 4–1 Principal Differences Between Procaryotic and Eucaryotic Cells

Characteristic	Procaryotic	Eucaryotic
Nucleus	No nuclear membrane or nucleoli	True nucleus consisting of nuclear envelope and nucleoli
Membrane-bounded organelles	Absent	Present; examples include lysosomes, Golgi complex, endoplasmic reticulum, mitochondria, and chloroplasts
Flagella	Submicroscopic; simple	Microscopic; complex
Glycocalyx	Extracellular polymeric capsule or slime layer	Absent
Cytoplasm	No cytoskeleton or cytoplasmic streaming	Cytoskeleton; cytoplasmic streaming
Mitotic apparatus	Absent	Present during nuclear division
Chromosome (DNA) arrangement	Single circular chromosome; lacks histones	Several or many linear chromosomes with histones
Cell membrane	Generally lacks sterols	Sterols present
Ribosomes	Small size (70S)	Large (80S) and small sizes
Cell Wall	When present, chemically complex	When present, chemically simple
Vacuoles	Atypical	Typical
Sexual reproduction	Fragmentary; no meiosis, only portions of chromosomes are reassorted	Regular; meiosis occurs, whole chromosome is reassorted

STUDY OUTLINE

PROCARYOTIC AND EUCARYOTIC CELLS (pp. 80–81)

1. Procaryotic and eucaryotic cells are similar in their chemical composition and chemical reactions.
2. Procaryotic cells lack membrane-bounded organelles (including a nucleus).
3. Peptidoglycan is found in procaryotic cell walls and not in eucaryotic cell walls.

THE PROCARYOTIC CELL (pp. 81–100)

1. Bacteria have a semirigid cell wall. If they are motile, it is achieved by flagella. They are unicellular, and most multiply by binary fission.
2. Bacterial species are differentiated by morphology, chemical composition, nutritional requirements, and biochemical activities.

SIZE, SHAPE, AND ARRANGEMENT OF BACTERIAL CELLS (pp. 81–82)

1. Most bacteria are between 0.20 and 2.0 μm in diameter.
2. The three basic bacterial shapes are coccus (spheres), bacillus (rods), and spiral.
3. Pleomorphic bacteria can assume several shapes.

STRUCTURES EXTERNAL TO THE CELL WALL (pp. 82–86)

1. The glycocalyx (capsule or slime layer) is a gelatinous polysaccharide and/or polypeptide covering. It allows cells to adhere to surfaces and might protect pathogens from phagocytosis.

2. Flagella are relatively long filamentous appendages consisting of a filament, hook, and basal body. They are used for motility.
3. Motility of spirochetes is by axial filaments, which wrap around the cell in a spiral fashion.
4. Pili (fimbriae) are appendages found on gram-negative bacteria. They help the cells attach to surfaces and transfer genetic material.

CELL WALL (pp. 86–90)

Composition and Characteristics (pp. 86–89)

1. The cell wall surrounds the plasma membrane and protects the cell from changes in osmotic pressure.
2. The cell wall consists of peptidoglycan (sugars and amino acids).
3. Many gram-positive bacteria also contain teichoic acids.
4. Gram-negative bacteria have a lipopolysaccharide-phospholipid-lipoprotein outer membrane surrounding the peptidoglycan. This layer keeps some injurious chemicals out of the cell and causes the gram-negative reaction.
5. The (LPS) polysaccharide is composed of O antigens. Lipid A (LPS) is an endotoxin.

Damage to the Cell Wall (pp. 89–90)

1. In the presence of lysozyme, the gram-positive bacteria's cell wall is destroyed, and the remaining cellular contents are referred to as a protoplast.
2. In the presence of lysozyme, the gram-negative bacteria's cell wall is not completely destroyed, and the remaining cellular contents are referred to as a spheroplast.
3. Protoplasts and spheroplasts are subject to osmotic lysis.
4. The mycoplasmas are bacteria that naturally lack cell walls. L forms are bacteria that temporarily lack cell walls or have very little wall material.
5. Antibiotics like penicillin interfere with cell wall synthesis.

STRUCTURES INTERNAL TO THE CELL WALL (pp. 90–100)

Plasma (Cytoplasmic) Membrane (pp. 91–92)

1. The plasma membrane encloses the cytoplasm and is a phospholipid bilayer with protein (fluid mosaic).
2. The plasma membrane is selectively permeable.
3. Plasma membranes carry enzymes for metabolic reactions such as nutrient breakdown, energy production, and photosynthesis.
4. Mesosomes, irregular infoldings of the plasma membrane, function in cell division and metabolism.
5. Plasma membranes can be destroyed by alcohols and polymyxins.

Movement of Materials Across Membranes (pp. 92–95)

1. Movement across the membrane may be by passive processes, in which materials move from higher to lower concentration and no energy is expended by the cell.
2. In simple diffusion, molecules and ions move until equilibrium is reached.
3. Osmosis is the movement of water from high to low concentrations across a semipermeable membrane until equilibrium is reached.
4. In facilitated diffusion, substances are transported by permeases across membranes from high to low concentration.
5. In active transport, materials move from low to high concentrations by permeases and the cell must expend energy.

Cytoplasm (pp. 95–96)

1. Cytoplasm, in which the chemical reactions of the cell occur, is the fluid component (cytoplasmic area) and nuclear area inside the plasma membrane.
2. The fluid component contains mostly water and high concentrations of molecules and inclusions (metachromatic granules, polysaccharide granules, lipid inclusions, sulfur granules, carboxysomes, and gas vacuoles).
3. The nuclear area contains the DNA of the main bacterial chromosome. Bacteria can also contain plasmids, which are extrachromosomal DNA circles.
4. The cytoplasmic area contains numerous ribosomes.

Endospores (pp. 96–100)

1. Endospores are resting structures formed by some bacteria for survival during adverse environmental conditions.
2. The process of endospore formation is called sporulation (sporogenesis); the return of an endospore to its vegetative state is called germination.

THE EUCARYOTIC CELL (pp. 100–108)

FLAGELLA AND CILIA (pp. 101–102)

1. Whereas flagella are few and long in relation to cell size, cilia are numerous and short.
2. Flagella and cilia are used for motility, and cilia also move substances along the surface of the cells.
3. Both flagella and cilia consist of nine pairs + two pairs of microtubules.

CELL WALL (pp. 102–103)

1. The cell walls of most algae and some fungi consist of cellulose.
2. The main material of fungal cell walls is chitin.

PLASMA (CYTOPLASMIC) MEMBRANE (p. 103)

1. Like the procaryotic plasma membrane, the eucaryotic plasma membrane is a phospholipid bilayer containing proteins.
2. Eucaryotic plasma membranes contain carbohydrates attached to the proteins and sterols not found in procaryotic cells.
3. Besides the methods used by procaryotic cells to move substances across the plasma membrane, eucaryotic cells use phagocytosis ("cell eating") and pinocytosis ("cell drinking").

CYTOPLASM (p. 103)

1. The cytoplasm of eucaryotic cells includes everything inside the plasma membrane and external to the nucleus.
2. The physical and chemical characteristics of the cytoplasm of eucaryotic cells resemble that of the cytoplasm of procaryotic cells.
3. Eucaryotic cytoplasm has a cytoskeleton and exhibits cytoplasmic streaming.

ORGANELLES (pp. 103–107)

1. Organelles are specialized membrane-bounded structures in the cytoplasm.
2. They are characteristic of eucaryotic cells.
3. The nucleus, which contains DNA in the form of chromosomes, is the most characteristic eucaryotic organelle.
4. The nuclear membrane is connected to a system of parallel membranes in the cytoplasm, called the endoplasmic reticulum.
5. The endoplasmic reticulum provides a surface for chemical reactions, serves as a transporting network, and stores synthesized molecules.
6. Ribosomes are found in the cytoplasm or attached to the endoplasmic reticulum.
7. Eucaryotic ribosomes are 80S, and procaryotic ribosomes are 70S.
8. The Golgi complex consists of cisternae. It functions in secretion, carbohydrate synthesis, and glycoprotein formation.
9. Mitochondria are the primary sites of ATP production. They contain small amounts of ribosomes and DNA, and multiply by fission.
10. Lysosomes are formed from Golgi complexes. They store powerful digestive enzymes and are referred to as "suicide packets."
11. The centrosome is a dense area of cytoplasm near the nucleus. It contains a pair of cylindrical structures called centrioles that are involved in cell division.
12. Chloroplasts contain chlorophyll and enzymes for photosynthesis. Like mitochondria, they contain some ribosomes and DNA and multiply by fission.

CYTOPLASMIC INCLUSIONS (pp. 107–108)

1. Cytoplasmic inclusions vary in size and composition.
2. Examples are zymogen, fat, vacuoles, and glycogen.

STUDY QUESTIONS

REVIEW

1. Draw each of the following bacterial shapes:
   (a) Spiral
   (b) Bacillus
   (c) Coccus

2. List three differences between procaryotic and eucaryotic cells.

3. Diagram each of the following flagellar arrangements:
(a) Lophotrichous
(b) Monotrichous
(c) Peritrichous

4. Match the structures to their functions.

______ Axial filament	(a) Motility
______ Capsule	(b) Motility in spirochetes
______ Cell wall	(c) Transfer of genetic material
______ Endospore	(d) Protection from phagocytes
______ Flagella	(e) Cell shape
______ Glycocalyx	(f) Selectively permeable
______ Mesosomes	(g) Resting
______ Pili	(h) Cell wall formation
______ Plasma membrane	(i) Attachment to surfaces

5. Identify eight structures/inclusions found in the cytoplasm of procaryotic cells.

6. Endospore formation is called ______. It is initiated by ______. Formation of a new cell from an endospore is called ______. This process is triggered by ______.

7. Why is an endospore called a resting structure? Of what advantage is an endospore to a bacterial cell?

8. Explain what would happen in the following experiments.
(a) A suspension of bacteria is placed in distilled water.
(b) A suspension of bacteria is placed in distilled water with lysozyme.
(c) A suspension of bacteria is placed in an aqueous solution of lysozyme and 10% sucrose.

9. A cell requires an amino acid from the environment, where it is present in a higher concentration than within the cell. How could this amino acid be brought into the cell?

10. Describe the process called active transport.

11. Why are mycoplasmas resistant to antibiotics that interfere with cell wall synthesis?

12. Answer the following question using these diagrams (below), which represent cross sections of bacterial cell walls.
(a) Which diagram represents a gram-positive bacterium? How can you tell?
(b) Explain how the Gram stain works to distinguish between these two types of cell walls.
(c) Why does penicillin have no effect on most gram-negative cells?

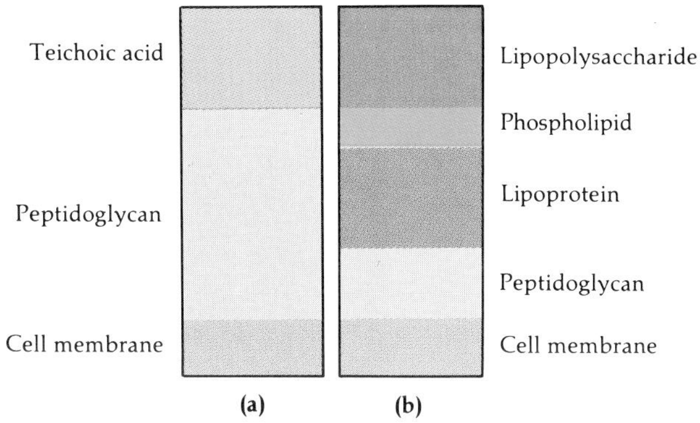

13. Match the following characteristics of eucaryotic cells with their functions.

______ Chloroplasts	(a) Intracellular transport
______ Endoplasmic reticulum	(b) Photosynthesis
______ Golgi complex	(c) ATP production
______ Lysosomes	(d) Digestive enzyme storage
______ Mitochondria	(e) Secretion

14. What process would a eucaryotic cell use to ingest a procaryotic cell?

CHALLENGE

1. Eucaryotic cells might have evolved from early procaryotic cells living in close association. What do you know about eucaryotic organelles that would support this theory?
2. Patients with *Enterobacter* and *Pseudomonas* infections exhibit similar symptoms regardless of the site of infection (e.g., digestive or circulatory system). On the basis of cell structure, explain this phenomenon.
3. Why can procaryotic cells be smaller than eucaryotic cells and still carry on all of the functions for life?

FURTHER READING

Adler, J. April 1976. The sensing of chemicals by bacteria. *Scientific American* 235:40–47. Describes the chemotactic response of bacteria using molecules that detect the presence of chemicals in the environment.

Beveridge, T. J. 1981. Ultrastructure, chemistry, and function of the bacterial wall. *International Review of Cytology* 72:229–317. An overview of properties of procaryotic cell walls.

Costerton, J. W., R. T. Irvin, and K.-J. Cheng. 1981. The bacterial glycocalyx in nature and disease. *Annual Review of Microbiology* 35:299–324. A thorough discussion of the functions of the glycocalyx.

Doetsch, R. N. and R. D. Sjoblad. 1980. Flagella structure and function in eubacteria. *Annual Review of Microbiology* 34:69–108. A detailed description of procaryotic flagellar ultrastructure and mechanics.

Ferris, F. G. and T. J. Beveridge. 1985. Functions of Bacterial Cell Surface Structures. *BioScience* 35:172–177. Includes ultrastructures and functions of the cell wall, capsules, flagella, pili, and a newly discovered structure, spinae.

Hurst, A. and G. W. Gould. 1984. *The Bacterial Spore.* New York: Academic Press. Volume II includes current information on endospore structures, sporulation, and germination.

Unwin, N. and R. Henderson. February 1984. The structure of proteins in biological membranes. *Scientific American* 250:78–94. Electron micrographs illustrate membrane-bounded proteins including bacteriorhodopsin.

Walsy, A. E. August 1977. The gas vacuoles of blue-green algae. *Scientific American* 237:90–97. Summarizes the structure of gas vacuoles and how they are used to regulate buoyancy in cyanobacteria.

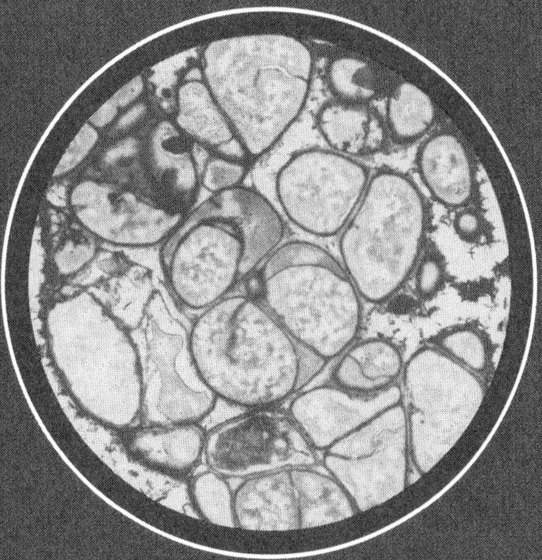

CHAPTER 5

Microbial Metabolism

OBJECTIVES

After completing this chapter you should be able to

- Define metabolism and contrast the fundamental differences between anabolism and catabolism.
- Describe the mechanism of enzymatic action.
- List factors that influence enzymatic activity.
- List and provide examples of three types of phosphorylation reactions that generate ATP.
- Describe the chemiosmotic model for ATP generation.
- Explain what is meant by oxidation–reduction.
- Categorize the various nutritional patterns among organisms.
- Describe the chemical reactions of glycolysis.
- Compare and contrast aerobic and anaerobic respiration.
- Explain the products of the Krebs cycle.
- Describe the chemical reactions and list the products of fermentation.

Metabolism refers to the sum of all chemical reactions within a living organism. Because chemical reactions either release or require energy, metabolism can be visualized as an energy-balancing act. Accordingly, metabolic activity can be divided into two classes of chemical reactions: **anabolic (synthetic) reactions** and **catabolic (degradative) reactions.**

In living cells, the chemical reactions that combine simple substances into more complex molecules are collectively known as **anabolism.** Overall, an anabolic process requires energy. Examples of

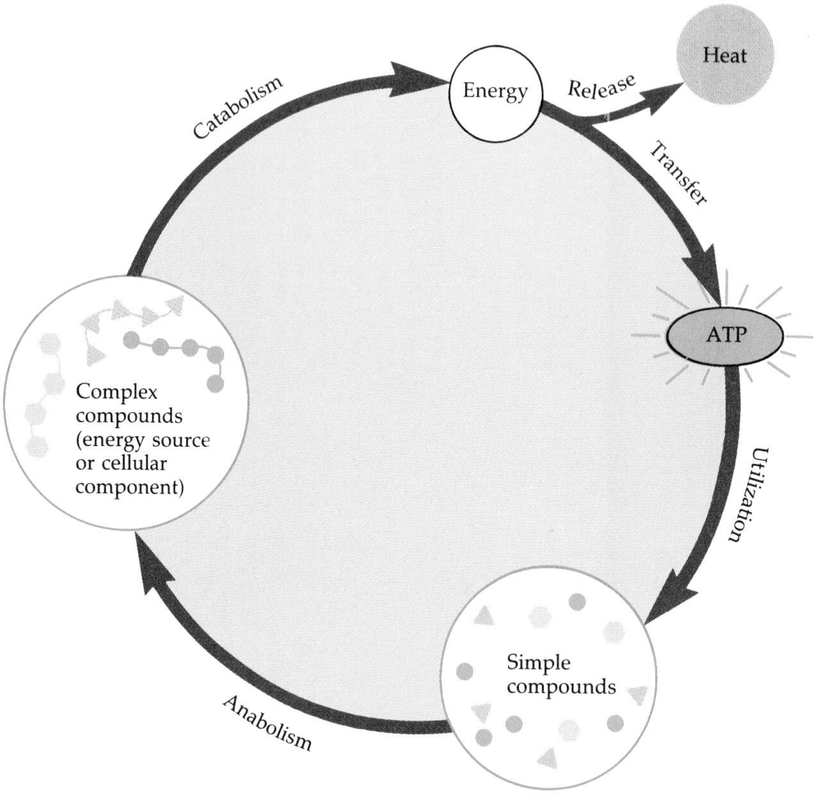

Figure 5–1 Relationship between anabolism, catabolism, and ATP. When simple compounds are combined to form complex compounds (anabolism), ATP provides the energy for synthesis. When large compounds are split apart (catabolism), much of the energy is given off as heat. Some is transferred to and trapped in ATP and then utilized to drive anabolic reactions.

anabolic processes are the formation of proteins from amino acids, of nucleic acids from mononucleotides, and of polysaccharides from simple sugars. These biosynthetic reactions generate the materials for cell growth.

The chemical reactions that break down complex organic compounds into simple ones are collectively known as **catabolism.** Catabolic reactions release chemical energy stored in organic molecules. This energy is then stored in the *energy-rich bonds* of molecules produced during catabolism. The high-potential energy of these bonds carries the energy to drive energy-requiring reactions. In this way, catabolic reactions release energy that can be used to drive anabolic reactions. However, only a small part of the energy released is actually available for cellular functions. The most common energy carrier in all biological systems is adenosine triphosphate (ATP); its structure can be reviewed in Figure 2–20. The role of ATP in the relationship between catabolic and anabolic processes is shown in Figure 5–1.

A major part of the energy released in catabolism is lost to the environment as heat. Thus, there is a continuous need of new external sources of energy for the cell. The chemical composition of a living cell is constantly changing: some molecules are being broken down while others are being synthesized. This balanced flow of chemicals and energy maintains the life of a cell.

This chapter will later examine some representative chemical reactions that produce energy (catabolic reactions) or use energy (anabolic reactions) in microorganisms. We will then look at how these various reactions are integrated within the cell. But first let us consider the principal properties of a group of proteins involved in almost all biologically important chemical reactions. These proteins, the enzymes, were described briefly in Chapter 2. Although it is beyond the scope of this text to name and discuss the actions of individual enzymes, you should be aware of the central role of enzymes in metabolic reactions. It is important to understand that a cell's metabolic pathways are determined by its enzymes, which are, in turn, determined by its genetic makeup.

ENZYMES

We indicated in Chapter 2 that chemical reactions occur when chemical bonds are made or broken. In order for reactions to take place, atoms, ions, or molecules must collide with each other. The effectiveness of the collision depends on the velocity of the particles, the amount of energy required for the reaction to occur (*activation energy*), and the specific configuration of the particles. Any change in condition that increases the frequency of collisions also increases the rate of the chemical reaction. Raising the temperature, pressure, and the amounts of the reacting molecules are such changes.

Substances that can speed up a chemical reaction by increasing the frequency of collisions or lowering the activation energy requirement, without themselves being altered, are called **catalysts.** In living cells, **enzymes** serve as biological catalysts. They can speed up chemical reactions by increasing the frequency of collisions, lowering the activation energy, and properly orienting the colliding molecules. They do this without increasing the temperature or pressure, changes that could disrupt or kill living cells. As catalysts, enzymes are specific: each particular enzyme will affect only specific **substrates**—reactant molecules in that particular chemical reaction. The specificity of enzymes is made possible by their structures. Generally large globular proteins, enzymes range in molecular weight from about 10,000 to somewhere in the millions. Of the thousand or more known enzymes, each has a three-dimensional characteristic shape with a specific surface configuration due to its primary, secondary, and tertiary structures (see Figure 2–18).

Enzymes are extremely efficient. Under optimum conditions, they can catalyze reactions at rates that are 10^8 to 10^{10} times (up to 10 billion times) more rapid than those of comparable reactions occurring without enzymes. The *turnover number* (number of substrate molecules metabolized per enzyme molecule per second) is generally between 1 and 10,000 and can be as high as 500,000.

Yet, as previously mentioned, enzymes are specific in the reactions they catalyze, as well as in the substrates they act upon. From the large number of diverse molecules in the cell, an enzyme must "find" the correct substrate. Moreover, the enzyme-catalyzed reactions tend to take place in aqueous solutions and at relatively low temperatures—conditions that otherwise would not favor rapid movement of molecules or rapid chemical reactions.

Enzymes are also subject to various cellular controls. Their rate of synthesis and their concentration at any given time are under the control of the cell's genes and are influenced by various other molecules in the cell, as will be discussed in Chapter 8. Many enzymes in both active and inactive forms are in the cell. The rate at which the inactive form becomes active or the active form becomes inactive is determined by the cellular environment.

Parts of an Enzyme

Some enzymes consist entirely of proteins. But many enzymes contain a protein called an **apoenzyme** that is inactive without a nonprotein component called the **cofactor.** Together, the apoenzyme and cofactor are an activated **holoenzyme,** or whole enzyme. If the cofactor is removed, the apoenzyme will not function (Figure 5–2). The cofactor can be a metal ion or a complex organic molecule called a **coenzyme.**

Coenzymes assist in the enzyme's activity by accepting atoms removed from the substrate or donating atoms required by the substrate. Many coenzymes are derived from vitamins (Table 5–1; for the sake of completeness, vitamins not used by microorganisms are also included). Two of the coenzymes most important in cellular metabolism are *NAD* (nicotinamide adenine dinucleotide) and *NADP* (nicotinamide adenine dinucleotide phosphate). Both compounds contain derivatives of the B vitamin nicotinic acid (niacin) and function with their respective enzymes in removing and transferring hydrogen ions and electrons (hydrogen

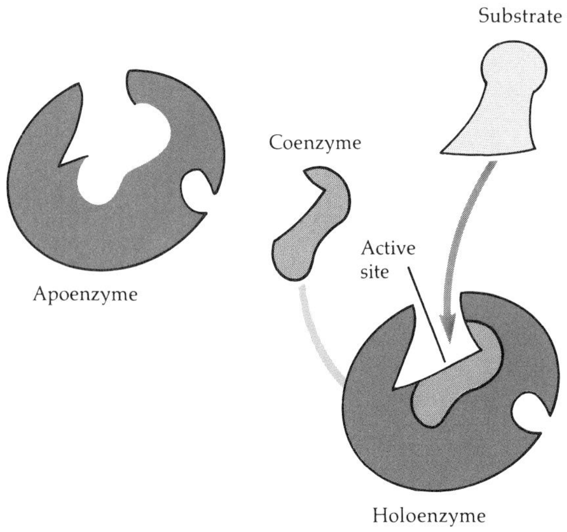

Figure 5–2 Components of a holoenzyme. Many enzymes require both an apoenzyme (protein portion) and a cofactor (nonprotein portion) to become active. The cofactor can be a metal ion or an organic molecule, called a coenzyme (as shown here). The apoenzyme and cofactor together make up the holoenzyme, or complete enzyme. The substrate is the substance with which the enzyme reacts.

atoms) from substrates. Enzymes that remove hydrogen atoms from a substance are called *dehydrogenases* and the process is called *dehydrogenation.*

The flavin coenzymes, such as *FMN* (flavin mononucleotide) and *FAD* (flavin adenine dinucleotide), contain derivatives of the B vitamin riboflavin and are important in photosynthetic reactions. Like NAD and NADP, these coenzymes also function in hydrogen transfer reactions with dehydrogenase enzymes.

Another important coenzyme, *coenzyme A (CoA),* contains a derivative of pantothenic acid, another B vitamin. This coenzyme plays an important role in the synthesis and breakdown of fats and in a series of oxidizing reactions called the *Krebs cycle.* Coenzyme A is used in decarboxylation reactions (removal of CO_2) and is associated with a useful fragment of cellular catabolism, the acetyl group:

$$CH_3{-}\overset{\overset{\displaystyle O}{\|}}{C}{-}$$

When coenzymes are tightly bonded to their apoenzymes, they are called **prosthetic groups.** One example is the heme (iron-containing) group of an enzyme called cytochrome *c.* **Cytochromes** are enzymes that function as electron carriers in

Table 5–1 Vitamins and Their Functions

Vitamin	Function
Water-soluble vitamins:	
Vitamin B_1 (thiamine)	Part of coenzyme cocarboxylase; has many functions, including the metabolism of pyruvic acid
Vitamin B_2 (riboflavin)	Coenzyme in flavoproteins; active in electron transfers
Niacin	Part of NAD molecule; active in electron transfers
Vitamin B_6 (pyridoxine)	Coenzyme in amino acid metabolism
Vitamin B_{12} (cyanocobalamin)	Active in red blood cell formation, amino acid metabolism
Pantothenic acid	Part of coenzyme A molecule; involved in metabolism of pyruvic acid and lipids
Biotin	Involved in carbon dioxide fixation reactions, fatty acid synthesis
Folic acid	Coenzyme used in synthesis of purines and pyrimidines
Vitamin C (ascorbic acid)	Involved in collagen deposition in connective tissue
Lipid-soluble vitamins:	
Vitamin A	Active in formation of visual pigment, bone and teeth growth
Vitamin D	Involved in intestinal absorption of calcium and phosphorus
Vitamin E	Needed for cellular and macromolecular syntheses
Vitamin K	Coenzyme used in formation of blood-clotting proteins

respiration and photosynthesis, as we shall see later in this chapter. These compounds contain pigments and metal ions and are structurally related to hemoglobin and chlorophyll.

As noted earlier, some cofactors are metal ions, including iron, copper, magnesium, manganese, zinc, calcium, and cobalt. It is believed that such cofactors bridge the enzyme and the substrate, thus binding them to facilitate substrate transformation. For example, magnesium (Mg^{2+}) is required by many phosphorylating enzymes that act together with ATP. The Mg^{2+} can form a bond between the enzyme and the ATP molecule. Most trace elements required by living cells are probably used to activate cellular enzymes.

Mechanism of Enzymatic Action

Although scientists do not completely understand how enzymes lower activation energy, the general sequence of events is believed to be as follows (Figure 5–3):

1. The surface of the substrate contacts a specific region on the surface of the enzyme molecule, called the *active site.*
2. A temporary intermediate compound called an *enzyme–substrate complex* forms.
3. The substrate molecule is transformed (by rearrangement of existing atoms, a breakdown of the substrate molecule, or the combination of several substrate molecules).
4. The transformed substrate molecules, the products of the reaction, move away from the enzyme molecule.
5. The recovered enzyme, now freed, reacts with other substrate molecules.

As mentioned earlier, enzymes have *specificity* for particular substrates. For example, a specific enzyme may be capable of hydrolyzing a peptide bond only between two specific amino acids. And other enzymes are capable of hydrolyzing starch, but not cellulose; even though both starch and cellulose are polysaccharides composed of glucose subunits, the orientations of the subunits in the two polysaccharides differ. Enzymes have this specificity because the three-dimensional shape of the active site fits the substrate somewhat as a lock with its key. The substrate is usually much smaller than the enzyme, and relatively few of the enzyme's amino acids make up the active site.

A certain compound can be a substrate for a number of different enzymes that catalyze different reactions, so the fate of a reactant (substrate) depends on the enzyme that reacts upon it. For example, glucose-6-phosphate, a molecule important in cell metabolism, can be acted upon by at least four different enzymes, and each reaction will give a different product.

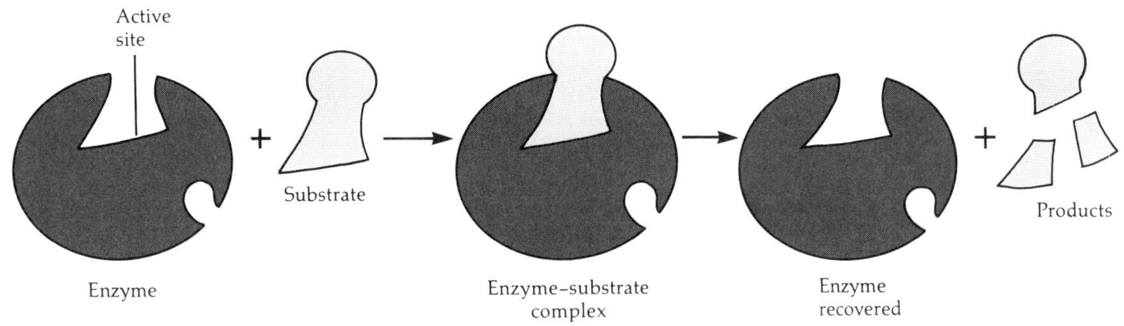

Figure 5–3. Mechanism of enzymatic action. The substrate contacts the active site on the enzyme to form an enzyme–substrate complex. The substrate is then transformed into products and the enzyme is recovered.

Table 5–2 Enzyme Classification Based on Type of Chemical Reaction

Class	Type of Chemical Reaction	Examples
Oxidoreductase	Oxidation–reduction in which oxygen and hydrogen are gained or lost	Cytochrome oxidase, D-lactate dehydrogenase
Transferase	Transfer of functional groups, such as an amino group, acetyl group, or phosphate group	Acetate kinase, alanine deaminase
Hydrolase	Hydrolysis (addition of water)	Lipase, sucrase
Lyase	Without hydrolysis, removal of groups of atoms	Oxalate decarboxylase, isocitrate lyase
Isomerase	Rearrangement of atoms within a molecule	Glucose-phosphate isomerase, alanine racemase
Ligase	Joining of two molecules in which ATP is broken down	Acetyl-CoA synthetase, methionyl-tRNA synthetase

Naming Enzymes

The names of enzymes usually end in the suffix *-ase.* All enzymes can be grouped into six classes according to the type of chemical reactions they catalyze (Table 5–2). Enzymes within each of the major classes are named according to specific reactions. For example, the class called oxidoreductases is involved with oxidation–reduction reactions (described shortly). During some oxidation reactions a substance loses hydrogen; in others, a substance gains oxygen. Enzymes that remove hydrogen are called *dehydrogenases;* those that add oxygen are called *oxidases.* As you will see later, even dehydrogenase and oxidase enzymes have more specific names, such as D-lactate dehydrogenase and cytochrome oxidase, depending on the specific substrates with which they react.

Factors Influencing Enzymatic Activity

Several factors influence the activity of an enzyme. Among the more important are temperature, pH, substrate concentration, and inhibitors.

Temperature

The speed of most chemical reactions increases as the temperature rises. For enzymatic reactions, however, elevation beyond a certain temperature drastically reduces the reaction's rate (Figure 5–4a). This decrease in the reaction's rate is due to the enzyme's denaturation, a phenomenon common to all proteins. **Denaturation** (Figure 5–5) usually involves breakage of the hydrogen bonds and other noncovalent bonds that hold the enzyme in its characteristic three-dimensional structure (tertiary configuration). A common example of denaturation is the transformation of uncooked egg white (albumin) to a hardened state after it is heated. As might be expected, this alteration in structure changes the arrangement of the amino acids in the active site and causes the enzyme to lose its catalytic ability and vital biological activity. In some cases, denaturation is partially or fully reversible. But if denaturation continues to the point of the enzyme losing its solubility and coagulating, denaturation becomes irreversible, because the enzyme cannot regain its original properties. Enzymes can also be denatured by concentrated acids, bases, heavy metal ions (such as copper, zinc, silver, arsenic, or mercury), alcohol, and ultraviolet radiation.

pH

Most enzymes have a pH at which their activity is characteristically maximal; this level is the **pH optimum.** Above or below this pH value, enzyme activity, and therefore reaction rates, decline (Figure 5–4b). When the ion concentrations (pH) in the medium are changed, the many ionizable groups essential in stabilizing the enzymatic three-dimensional structure are affected. Extreme changes in pH can cause denaturation.

Substrate concentration

There is a maximum rate at which a certain amount of enzyme can catalyze a specific reaction. Only when the concentration of substrate(s) is extremely high can this maximum rate be attained. Under

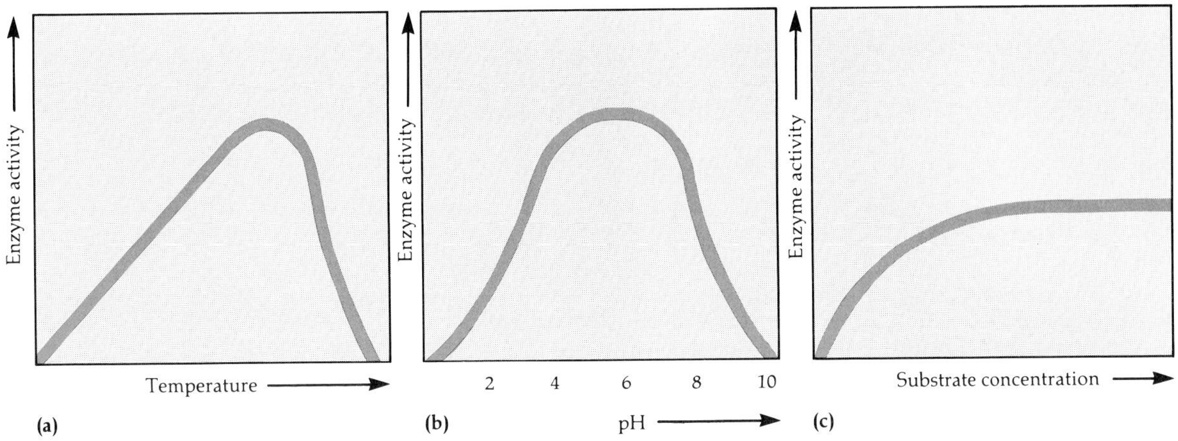

Figure 5–4 Factors that influence enzymatic activity, plotted for a hypothetical enzyme. **(a)** Temperature. The enzymatic activity (rate of reaction catalyzed by the enzyme) increases with increasing temperature until the enzyme, a protein, is denatured by heat and inactivated. At this point, the reaction rate falls steeply. **(b)** pH. The level at which this enzyme is most active is around pH 5. **(c)** Substrate concentration. With increasing concentration of substrate molecules, the rate of reaction increases until the active sites on all the enzyme molecules are filled, at which point the maximum rate of reaction is reached.

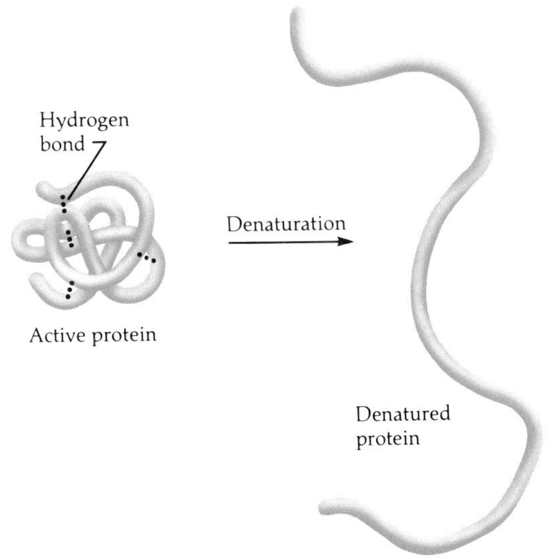

Figure 5–5 Denaturation of a protein comes about by the breakage of the noncovalent bonds (such as hydrogen bonds) that hold the active protein in its three-dimensional shape. The denatured protein is no longer functional.

conditions of high substrate concentration, the enzyme is said to be **saturated;** that is, its active site is always occupied by substrate or product molecules. In this condition, a further increase in substrate concentration will not affect the reaction rate because all active sites are already in use. (See Figure 5–4c). If a substrate's concentration exceeds a cell's saturation level, the rate of reaction can be increased only if the cell produces additional enzyme molecules. But under normal cellular conditions, enzymes are not saturated with substrate(s). At any given time, many of the enzyme molecules are inactive for lack of substrate; thus, the rate of reaction is likely to be determined (restricted) by the substrate concentration.

Inhibitors

As you will see in later chapters, an effective way to control the growth of bacteria is to control their enzymes. Certain poisons such as cyanide, arsenic, mercury, and nerve gas combine with enzymes and prevent their functioning. This results in the inhibition or death of a cell. Many antimicrobial drugs used in medicine are enzyme inhibitors.

According to their mechanism of action, enzyme inhibitors are classified as competitive inhibitors and noncompetitive inhibitors. **Competitive inhibitors** compete with the normal substrate for the active site of the enzyme. The competitive inhibitor is able to do this because its shape and chemical structure are very similar to those of the normal substrate (Figure 5–6a and b). One good example of a competitive inhibitor is sulfanilamide (a sulfa drug), which inhibits the enzyme whose normal substrate is para-aminobenzoic acid (PABA).

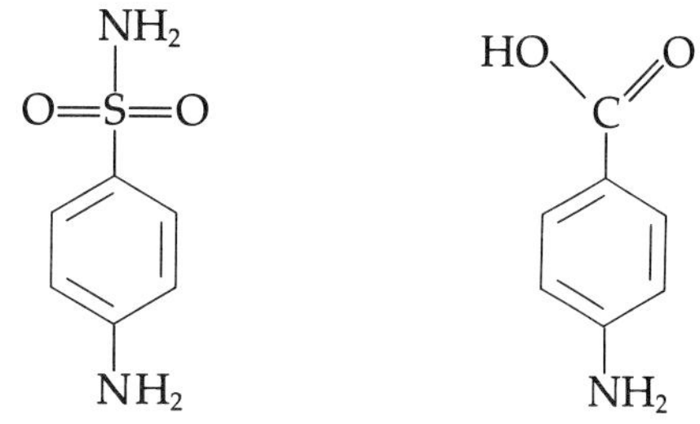

PABA is an essential nutrient used by many bacteria in the synthesis of folic acid, a vitamin that functions as a coenzyme. When sulfanilamide is administered to bacteria, the enzyme that normally converts PABA to folic acid combines instead with the sulfanilamide. The folic acid is not synthesized, and the bacteria cannot grow. Human cells do not use PABA to make their folic acid, so sulfanilamide selectively kills the bacteria but does not harm human cells.

Noncompetitive inhibitors do not compete with the substrate for the enzyme's active site; instead, they act on a part of the enzyme other than the active site so as to alter the active site and prevent substrates from binding there. The site of the noncompetitive inhibitor's binding is called an **allosteric site** ("other space") and is specific for the inhibitor (Figure 5–6c). This process is an **allosteric transition:** an enzyme's activity is changed because of selective binding at a site that does not overlap with the substrate's binding site. The change to the enzyme can be either reversible or irreversible. In some cases, allosteric mechanisms can activate an enzyme, rather than inhibit it.

Another type of noncompetitive inhibition can operate on enzymes that require metal ions for their activity. Certain chemicals can bind or tie up the metal-ion activators and thus prevent an enzymatic reaction. Cyanide can bind the iron in iron-containing enzymes, and fluoride can bind the calcium or magnesium in enzymes that contain those ions.

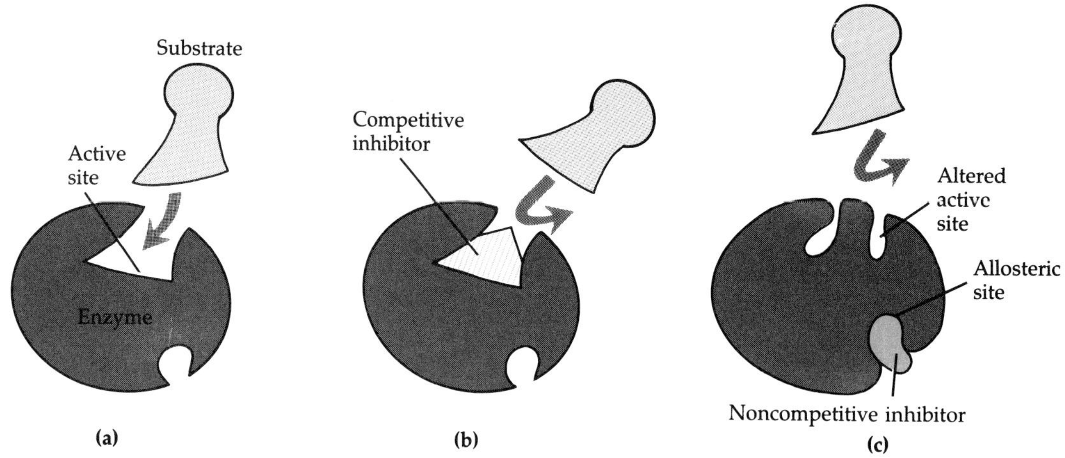

Figure 5–6 Enzyme inhibitors. **(a)** Uninhibited enzyme and its normal substrate. **(b)** Competitive inhibitor. **(c)** One type of noncompetitive inhibitor, causing allosteric inhibition.

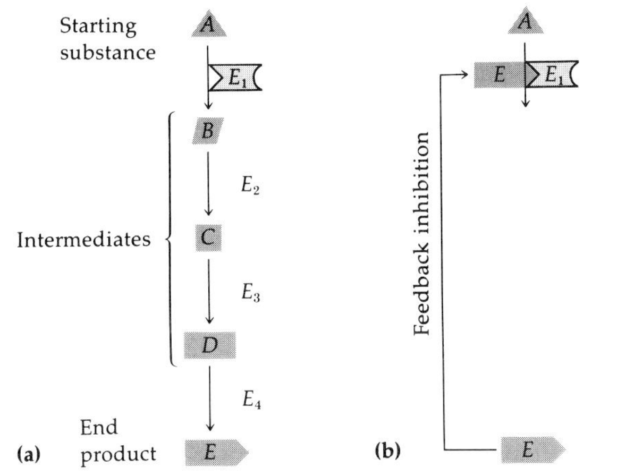

Figure 5–7 Feedback inhibition. **(a)** Synthesis of the end product. **(b)** Inhibition of an early step in the pathway by the end product. E_1 through E_4 represent different enzymes.

Feedback Inhibition

Allosteric noncompetitive inhibitors play a role in a biochemical control called **feedback inhibition** (or *end product inhibition*). In some metabolic reactions, several steps are required for the synthesis of a particular chemical compound, called the end product. The cell makes this end product by converting one substance to another in the presence of specific enzymes (Figure 5–7a). In many biosynthetic pathways, the final product can inhibit activity of one of the enzymes earlier in the pathway. This phenomenon is feedback inhibition (Figure 5–7b). This mechanism regulates the cell from making excessive end product.

Feedback inhibition acts on enzymes that have already been synthesized; it does not affect the synthesis of these enzymes. A single enzyme, usually the first one in a metabolic pathway, is allosterically inhibited by the end product in a reversible reaction. Because the enzyme is inhibited, the product of the first enzymatic reaction in the pathway is not synthesized. Because that unsynthesized product would normally be the substrate for the second enzyme in the pathway, the second reaction stops immediately too. Thus, even though only the first enzyme in the pathway is inhibited, the entire pathway shuts down and no new end product is formed. As the existing end product is used up by the cell, the first enzyme's allosteric site will more often remain unbound, and the pathway will resume activity.

The bacterium *Escherichia coli* (esh-er-i'kē-ä kō'lē) can demonstrate feedback inhibition in the synthesis of the amino acid isoleucine. In this metabolic pathway, five steps are taken to enzymatically convert the amino acid threonine to isoleucine (Figure 5–8). If isoleucine is added to the *E. coli* culture medium, the bacteria no longer synthesize isoleucine. This condition is maintained until the supply of isoleucine is depleted. This type of feedback inhibition is also involved in regulating the cell's production of other amino acids, vitamins, purines, and pyrimidines.

Before we look at the actual chemical reactions by which organisms produce energy, let us examine a few basic concepts of energy production by cells.

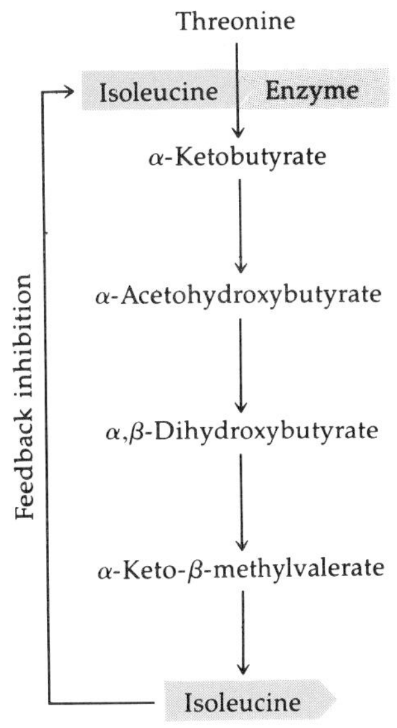

Figure 5–8 Feedback inhibition in *E. coli*. Shown here is a sequence of reactions in which threonine is converted to isoleucine. If isoleucine is added to the culture medium, *E. coli* no longer synthesizes isoleucine, because the isoleucine combines with the enzyme that catalyzes the first step of the synthesis. Thus, isoleucine inhibits the synthesis of more isoleucine.

ENERGY PRODUCTION METHODS

Nutrient molecules, like all molecules, have energy stored in the bonds between their atoms. But when this energy is spread throughout the molecule, it is difficult for the cell to use. Various reactions in catabolic pathways, however, concentrate the released energy into the high-energy bonds of ATP, which serves as a convenient energy carrier. To consider how this is done, we will focus on three important areas of energy production: oxidation–reduction, ATP generation, and nutritional patterns of organisms.

The Concept of Oxidation–Reduction

Oxidation is the addition of oxygen to a molecule, or, more generally, the removal of electrons (e^-) from a molecule. In many cellular oxidations, two electrons and two hydrogen ions (H^+) are removed at the same time; this is equivalent to the removal of two hydrogen atoms. Because most biological oxidations involve the loss of hydrogen atoms, they are called **dehydrogenation** reactions. For example, when lactic acid oxidizes to form pyruvic acid, the lactic acid loses two hydrogen atoms.

$$\text{COOH}-\text{CHOH}-\text{CH}_3 \xrightarrow{\text{2H (oxidation)}} \text{COOH}-\text{C}=\text{O}-\text{CH}_3$$

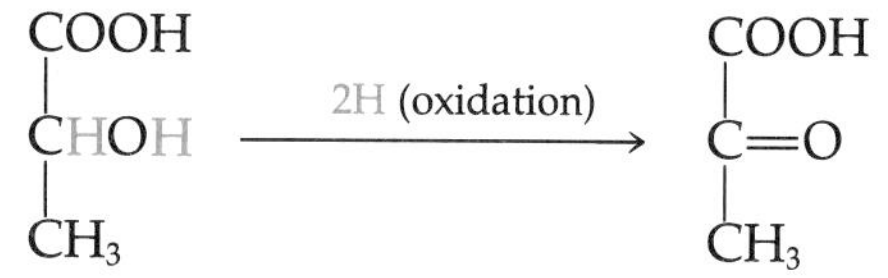

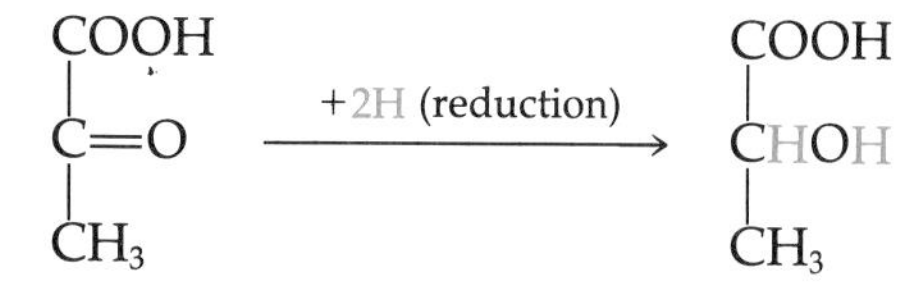

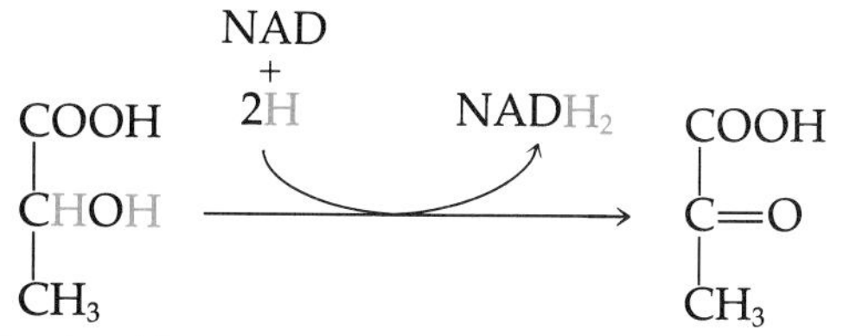

Lactic acid → Pyruvic acid

When a compound has picked up electrons or hydrogen atoms, it is said to be reduced. Thus, **reduction** is a gain of electrons, and is the opposite of oxidation. Consider the reduction of pyruvic acid to form lactic acid.

$$\text{COOH}-\text{C}=\text{O}-\text{CH}_3 \xrightarrow{\text{+2H (reduction)}} \text{COOH}-\text{CHOH}-\text{CH}_3$$

Pyruvic acid → Lactic acid

In reactions within a cell, oxidations and reductions are always coupled. In other words, each time one substance is oxidized, another is almost simultaneously reduced. The pairing of these reactions is **oxidation–reduction.**

When a substance is oxidized in this way, the freed hydrogen atoms do not remain free in the cell, but are transferred immediately by coenzymes to another compound. Two coenzymes commonly used by living cells to carry hydrogen atoms are derivatives of the vitamin niacin. They are nicotinamide adenine dinucleotide (NAD) and nicotinamide adenine dinucleotide phosphate (NADP). The oxidation and reduction states of NAD and NADP can be symbolized as follows:

$$\text{NAD} \underset{-2H}{\overset{+2H}{\rightleftharpoons}} \text{NADH}_2$$

$$\underset{\text{Oxidized}}{\text{NADP}} \underset{-2H}{\overset{+2H}{\rightleftharpoons}} \underset{\text{Reduced}}{\text{NADPH}_2}$$

Thus, when lactic acid is *oxidized* to form pyruvic acid, the two hydrogen atoms removed in the reaction are used to *reduce* NAD. This coupled, oxidation–reduction reaction may be written as follows:

$$\underset{\text{Lactic acid}}{\text{COOH}-\text{CHOH}-\text{CH}_3} \xrightarrow[\text{NAD} + 2H \quad \text{NADH}_2]{} \underset{\text{Pyruvic acid}}{\text{COOH}-\text{C}=\text{O}-\text{CH}_3}$$

An important point to remember about oxidation–reduction reactions is that oxidation is usually an energy-producing reaction. Cells take foodstuffs (energy sources) and degrade them from highly reduced compounds (with many hydrogen atoms) to highly oxidized compounds (with many oxygen atoms or multiple bonds). For example, when a cell oxidizes a molecule of glucose ($C_6H_{12}O_6$), the energy in the glucose molecule is removed in a stepwise manner and ultimately trapped by ATP, which is then an energy source for energy-requiring reactions. Compounds such as glucose that have many hydrogen atoms are highly reduced compounds, containing more potential energy than oxidized compounds.

Generation of ATP: Types of Phosphorylation

The energy released during oxidation reactions is trapped within the cell and ATP is formed. As described in Chapter 2, a phosphate group (symbolized Ⓟ) is added to ADP to form ATP.

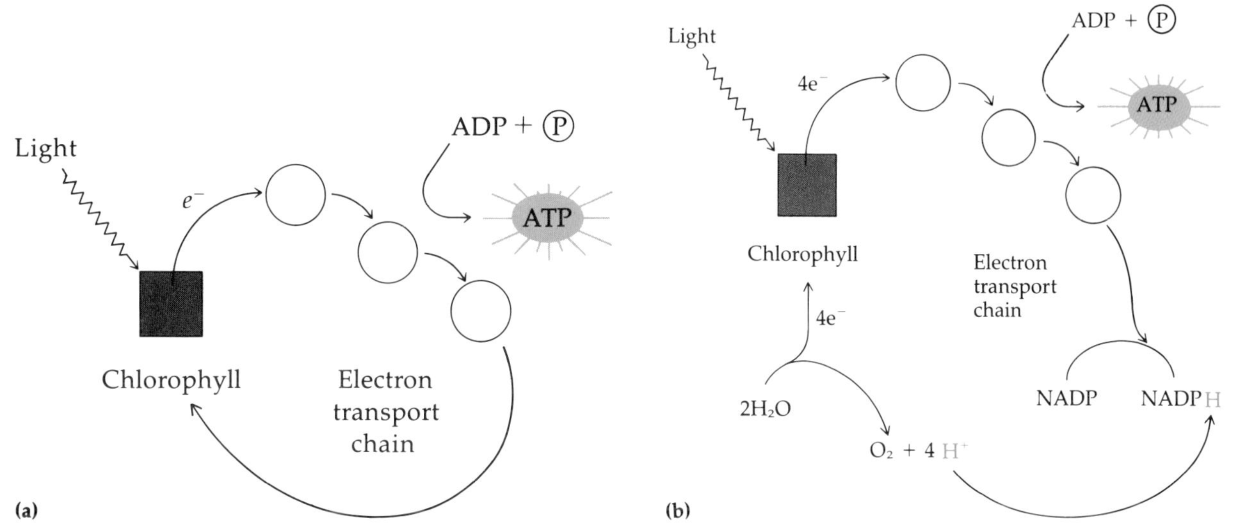

Figure 5–9 Photophosphorylation. **(a)** Cyclic photophosphorylation. **(b)** Noncyclic photophosphorylation.

$$\text{Adenosine—}\textcircled{P} \sim \textcircled{P}\text{(ADP)} + \text{Energy} + \textcircled{P} \longrightarrow$$
$$\text{Adenosine—}\textcircled{P} \sim \textcircled{P} \sim \textcircled{P}\ \text{(ATP)}$$

The symbol ~ designates a high-energy bond, one that can readily be broken to release usable energy. The high-energy bond that attaches the third phosphate contains the energy stored in this reaction. The addition of a phosphate to a chemical compound is called **phosphorylation.** Organisms use three mechanisms of phosphorylation to generate ATP.

In **substrate-level phosphorylation,** ATP is generated when a high-energy phosphate group is directly transferred from an intermediate metabolic compound to ADP. The following example shows only the carbon skeleton and the phosphate group of the metabolic compound.

$$\text{C—C—C} \sim \textcircled{P} + \text{ADP} \longrightarrow \text{C—C—C} + \text{ATP}$$

When **oxidative phosphorylation** occurs, electrons are removed from organic compounds (usually by NAD) and passed through a series of electron acceptors to molecules of oxygen (O_2) or other inorganic molecules. This process occurs in the plasma membrane of procaryotes and in the inner mitochondrial membrane of eucaryotes. The series of electron acceptors used in oxidative phosphorylation is called the **electron transport chain** (see Figure 5–15). The transfers of electrons from one electron acceptor to the next release energy, which is used to generate ATP from ADP through a process called chemiosmosis.

The third mechanism of phosphorylation, **photophosphorylation,** occurs only in photosynthetic cells, which contain light-trapping pigments such as the chlorophylls. In this mechanism, light energy liberates an electron, which is passed along a series of electron acceptors. Each transfer releases energy used in chemiosmosis to convert ADP to ATP. In **cyclic phosphorylation**, the electron returns to chlorophyll; in **noncyclic phosphorylation**, it becomes incorporated in the coenzyme nicotinamide adenine dinucleotide phosphate (NADP) and NADPH is formed (Figure 5–9).

The significance of these three forms of phosphorylation will become clearer later in this chapter, as we discuss in more detail their role in energy production.

Chemiosmotic Mechanism of ATP Generation

The chemiosmotic mechanism of ATP synthesis—**chemiosmosis**—was first proposed by the British biochemist Peter Mitchell in 1961. Although it took a number of years to gain full acceptance, it is now recognized as a major milestone in the history of biochemistry.

What is chemiosmosis, and how does it drive the energy-requiring synthesis of ATP from ADP? To understand the answers to these questions, we need to recall several concepts that were introduced in Chapter 4, as part of the discussion of "Movement of Materials Across Membranes." Recall that substances diffuse passively across membranes from regions of high to regions of low concentration; this diffusion yields energy. Recall also that the movement of substances against such a concentration gradient *requires* energy, and that in such an active transport of molecules or ions across biological membranes the required energy is provided by ATP. In one sense, chemiosmosis is the opposite of active transport, that is, the energy released when a substance moves along a gradient is used to *synthesize* ATP. The "substance" in this case refers to protons (H^+). The steps in chemiosmosis are as follows (Figure 5–10):

1. As electrons pass down an electron transport chain, some of the carrier molecules pump protons from one side of the membrane to the other. The carrier molecules in the membrane are arranged so that they take up protons on one side of the membrane and release them on the other side.
2. The phospholipid membrane is normally impermeable to protons, so this pumping establishes a proton gradient. The proton gradient has potential energy, called the *proton motive force.*
3. The protons on the outside of the membrane can be transported in by an enzyme in the membrane called **adenosine triphosphatase (ATPase)**. When this movement occurs, energy is released and is used by the enzyme to convert ADP to ATP.

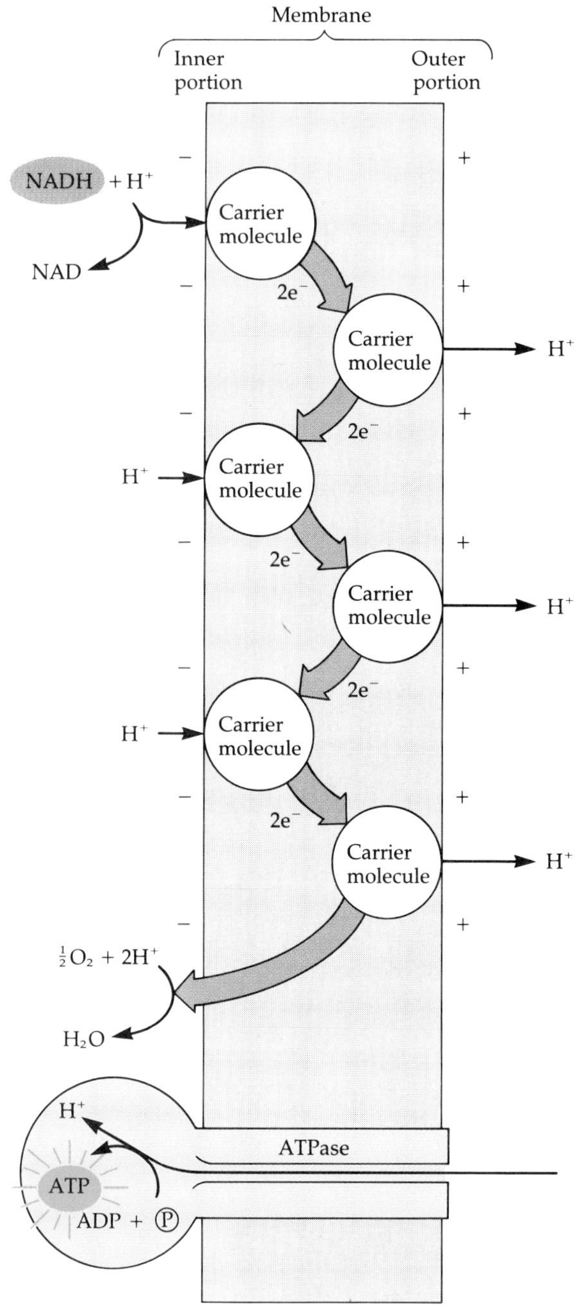

Figure 5–10 Chemiosmosis.

Both procaryotic and eucaryotic cells use the chemiosmotic mechanism to generate energy for ATP production in both oxidative phosphorylation and photophosphorylation. The membrane that contains the electron transport carriers and the ATPase may be the plasma membrane of procaryotes, the inner mitochondrial membrane, or the thylakoid membrane.

Table 5–3 Nutritional Classification of Organisms

Nutritional Type	Energy Source	Carbon Source	Examples
Photoautotroph	Light	Carbon dioxide (CO_2)	Photosynthetic bacteria (green sulfur and purple sulfur bacteria), cyanobacteria, algae, plants
Photoheterotroph	Light	Organic compounds	Purple nonsulfur and green nonsulfur bacteria
Chemoautotroph	Inorganic compounds	Carbon dioxide (CO_2)	Hydrogen, sulfur, iron, and nitrifying bacteria
Chemoheterotroph	Organic compounds	Organic compounds	Most bacteria, fungi, protozoans, and all animals

Nutritional Patterns Among Organisms

To acquire the energy for various metabolic activities, microorganisms need raw materials—mainly a carbon source for new organic compounds. Nutritional patterns among organisms can be distinguished on the basis of two criteria: energy source and principal source of carbon. First considering the energy source, we can generally classify organisms as phototrophs or chemotrophs. **Phototrophs** use light as their primary energy source, whereas **chemotrophs** depend on oxidation–reduction reactions for energy. For a principal carbon source, **autotrophs** (self-feeders) use carbon dioxide and **heterotrophs** (feeders on others) require an organic carbon source.[†]

If we combine the energy and carbon sources, we derive the following nutritional classifications for organisms: **photoautotrophs**, **photoheterotrophs**, **chemoautotrophs**, and **chemoheterotrophs**. The nutritional classification of microbes is summarized in Table 5–3. Almost all of the medically important microorganisms discussed in this book are chemoheterotrophs.

Photoautotrophs

Photoautotrophs use light as a source of energy, and carbon dioxide as their chief source of carbon. They include photosynthetic bacteria (green sulfur and purple sulfur bacteria, and cyanobacteria), algae, and green plants. The process by which photoautotrophs transform carbon dioxide and water into carbohydrates and oxygen gas is called **photosynthesis.** Essentially, photosynthesis is the conversion of light energy into chemical energy; chlorophyll molecules trap the light. The photosynthesis carried out by cyanobacteria, algae, and green plants can be represented as follows:

$$\underbrace{\underset{\text{Carbon dioxide}}{6CO_2} + \underset{\text{Water}}{12H_2O}}_{\text{Raw materials}} \quad \underbrace{\xrightarrow[\text{Chlorophyll, Enzymes}]{\text{Light}}}_{\text{Necessary conditions}}$$

$$\underbrace{\underset{\text{Sugar}}{C_6H_{12}O_6} + \underset{\text{Oxygen}}{6O_2} + \underset{\text{Water}}{6H_2O}}_{\text{Products}}$$

In the synthesized sugar, the carbon (C) and the oxygen (O) come from carbon dioxide (CO_2), and the hydrogen (H) comes from water (H_2O). The oxygen (O) in water (H_2O) is eventually given off as oxygen gas (O_2). The energy for the photosynthetic synthesis of sugar is derived from ATP, which was generated by photophosphorylation, which in turn used energy from the chemiosmotic mechanism.

In these photosynthetic reactions of cyanobacteria, algae, and green plants, the hydrogen atoms of water are used to reduce carbon dioxide, and oxygen gas is given off. Because this photosynthetic process uses O_2, it is sometimes called

[†]Autotrophs may also be referred to as *lithotrophs* (rock eating), and heterotrophs may also be referred to as *organotrophs.*

oxygenic. Some cyanobacteria can carry on photosynthesis without oxygen. The light-trapping pigment principally used by green plants, algae, and cyanobacteria is *chlorophyll a.* It is located in the thylakoids of chloroplasts in algae and green plants, and in the thylakoids that form a part of an elaborate internal membrane structure in cyanobacteria.

In addition to the cyanobacteria, there are several other families of photosynthetic procaryotes, who are classified according to the way they reduce carbon dioxide. These bacteria cannot use water to reduce carbon dioxide, and cannot carry on photosynthesis when oxygen is present (they must have an *anaerobic* environment). Consequently, their photosynthetic process does not produce oxygen gas and is called nonoxygenic. Two of the families are photoautotrophs: the green sulfur and purple sulfur bacteria. The **green sulfur bacteria** use sulfur, sulfur compounds, or hydrogen gas to reduce carbon dioxide and form organic compounds. Applying the energy from light to chlorophyll and the appropriate enzymes, these bacteria oxidize sulfur to sulfuric acid, hydrogen sulfide to sulfur, or hydrogen gas to water. The **purple sulfur bacteria** also use sulfur, sulfur compounds, or hydrogen gas to reduce carbon dioxide. They are distinguished from the green sulfur bacteria on biochemical and morphological grounds.

The chlorophyll used by these photosynthetic bacteria is a group of pigments called *bacteriochlorophylls,* which absorb light of longer wavelengths than that absorbed by chlorophyll *a.* Bacteriochlorophylls of green sulfur bacteria are found in vesicles underlying and attached to the plasma membrane called *chlorosomes* (*chlorobium vesicles*). In the purple sulfur bacteria, the bacteriochlorophylls are located in the plasma membrane.

Several characteristics that distinguish eucaryotic photosynthesis from procaryotic photosynthesis are presented in Table 5–4. See Box for a discussion of an exceptional photosynthetic system that exists in *Halobacterium.* The system does not use chlorophyll.

Photoheterotrophs

Photoheterotrophs use light as a source of energy but cannot convert carbon dioxide to sugar; rather, they use organic compounds as sources of carbon. These organic compounds include alcohols, fatty acids, other organic acids, and carbohydrates. Among the photoheterotrophs are green nonsulfur and purple nonsulfur bacteria.

Chemoautotrophs

Chemoautotrophs use inorganic compounds as a source of energy and carbon dioxide as their principal source of carbon. (See Figure 26–1, the carbon cycle). Inorganic sources of energy for these organisms include hydrogen sulfide, H_2S (for *Beggiatoa,* bej-jē-ä-tō'ä); elemental sulfur, S (*Thiobacillus*); ammonia, NH_3 (*Nitrosomonas,* nī-trō-sō-mō'näs); nitrites, NO_2^- (*Nitrobacter,* nī-trō-bak'tėr); hydrogen gas, H_2 (*Hydrogenomonas,* hī-drō-je-nō-mō'näs); and iron, Fe^{2+} (*Thiobacillus ferrooxidans,* fer-rō-oks'i-danz). The energy derived from the oxidation of these inorganic compounds is eventually

Table 5–4 Comparison of Eucaryotic and Procaryotic Photosynthesis

Characteristic	Eucaryotes: Algae, plants	Procaryotes: Cyanobacteria	Procaryotes: Green sulfur bacteria	Procaryotes: Purple sulfur bacteria
Substance that reduces CO_2	H atoms of H_2O	H atoms of H_2O	Sulfur, sulfur compounds, H_2 gas	Sulfur, sulfur compounds, H_2 gas
Oxygen production	Oxygenic	Oxygenic (and nonoxygenic)	Nonoxygenic	Nonoxygenic
Light-trapping pigment	Chlorophyll *a*	Chlorophyll *a*	Bacteriochlorophylls	Bacteriochlorophylls
Site of photosynthesis	Chloroplasts	Thylakoids	Chlorosomes	Plasma membrane
Environment	Aerobic	Aerobic (and anaerobic)	Anaerobic	Anaerobic

Photosynthesis Without Chlorophyll

The strange archaeobacterium, *Halobacterium*, lives where very little else can grow. This bacterium is found in salt lakes, in the salt licks on ranches, in salt flats—any environment with a concentration of salt 5–7 times greater than that of the ocean. Halobacteria are easy to detect, because they turn their environment red. Because these bacteria cannot ferment carbohydrates and do not contain chlorophyll, it was assumed that all their energy comes from oxidative phosphorylation. The exciting discovery of a new system of photophosphorylation arose through the study of the cell membrane of *Halobacterium halobium*.

Researchers found that the plasma membrane of *H. halobium* fragments into two fractions (red and purple) when the cell is broken down and its components sorted. The red fraction, which comprises most of the membrane, contains cytochromes, flavoproteins, and other parts of the electron transport system that carries out oxidative phosphorylation. The purple fraction is more interesting. This purple membrane occurs in distinct patches of hexagonal lattices within the plasma membrane. The purple color comes from a protein that comprises 75% of the purple membrane fraction. This protein is similar to the retinal pigment in the rod cells of the human eye, rhodopsin, so the protein was named bacteriorhodopsin. At the time it was discovered, its function was not known.

In the 1970s, further study of *H. halobium* suggested some startling explanations. When starved cells in anaerobic, dark environments were exposed to either light or oxygen, ATP synthesis increased. The table compares the effects of light and oxygen on anaerobic photosynthetic bacteria, photosynthetic green algae, a facultative anaerobe (*E. coli*), and *H. halobium*.

	Synthesis of ATP			
	Oxygen		No Oxygen	
Microorganism	Light	Dark	Light	Dark
Photosynthetic bacteria	–	–	+	–
Green algae	+	–	–	–
Facultative anaerobe (*E. coli*)	+	+	+	+
H. halobium	+	+	+	–

E. coli can grow in any environment because it requires neither oxygen nor light. The photosynthetic bacteria and green algae synthesize ATP only when they are in the appropriate concentration of oxygen and only in the presence of light. *H. halobium* can grow in the presence of either light or oxygen but cannot grow when neither is present.

This unexpected result suggested that *Halobacterium* can obtain energy by using either of two systems: one that operates in the presence of oxygen (oxidative phosphorylation) and one that operates in the presence of light (some kind of photophosphorylation). It was found that the rate of ATP synthesis by *H. halobium* is highest when the cells receive light between 550–600nm in wavelength; this range exactly corresponds to the absorption spectrum of bacteriorhodopsin. This and other clues suggested that the mechanism of photophosphorylation is tied to the pigment bacteriorhodopsin.

Researchers hypothesized that bacteriorhodopsin, like chlorophyll, acts as a proton pump to create a proton gradient across a cell membrane; in this case, the gradient is created across the purple membrane. To test this theory, researchers isolated pure purple membrane and used a special technique to form the membrane into vesicles. Illumination caused the pH level of the medium inside and surrounding the vesicles to change. This change indicated that protons were being transported across the purple membrane. According to the chemiosmotic theory, this transport of protons in intact cells creates a proton gradient that can drive the synthesis of ATP (see figure).

Halobacterium synthesizes the purple membrane only when it is in low oxygen concentrations and in the presence of light. It is clear

(*continued on next page*)

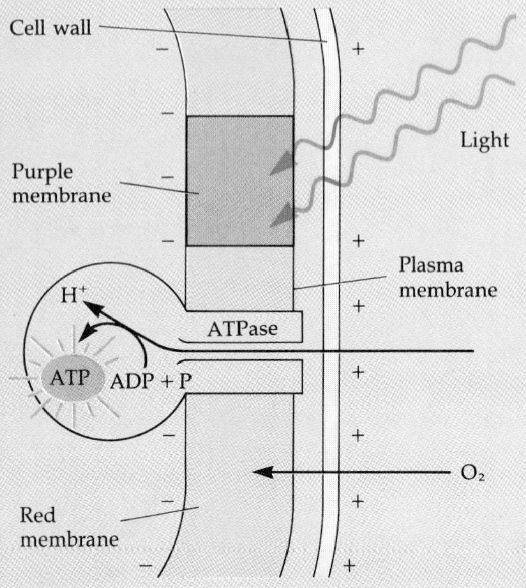

that the membrane provides the bacterium with energy when oxygen concentrations in its highly salinic environment become too low to support oxidative phosphorylation; that condition occurs frequently.

Until the discovery of the halobacterial system, chlorophyll-containing systems were the only ones known to generate ATP using sunlight. Use of the pigment bacteriorhodopsin by *Halobacterium* is the simplest form of photophosphorylation known, and provides a useful model for the study of chemiosmosis. Bacteriorhodopsin is itself an intriguing discovery. Its similarity to rhodopsin, the pigment of animal vision, is the basis of much conjecture about both the mechanism and evolution of sight (the perception of light) and its relationship to the generation of energy.

stored in ATP, which is produced by oxidative phosphorylation.

Chemoheterotrophs

When we discuss photoautotrophs, photoheterotrophs, and chemoautotrophs, it is easy to categorize energy source and carbon source because they occur as separate entities. However, in **chemoheterotrophs** the distinction is not so clear, because both energy source and carbon source are usually the same organic compound—glucose, for example.

Heterotrophs are classified according to their source of organic molecules: **saprophytes** live on dead organic matter, and **parasites** derive nutrients from a living host. The vast majority of bacteria, and all fungi, protozoans, and animals, are chemoheterotrophs.

Summary of Energy Production Methods

In living systems, energy is passed from one source to another in the form of electrons. To produce energy, a cell must have an energy source that functions as an electron donor. Energy sources can be as diverse as glucose, elemental sulfur, ammonia, hydrogen gas, or light in conjunction with certain pigments (Figure 5–11). Electrons removed from these energy sources are next transferred to electron carriers, such as the coenzymes NAD and FAD. This is an oxidation–reduction reaction: The initial energy source is oxidized as the electron carrier is reduced. In the third stage, electrons are transferred from electron carriers to final electron acceptors, in further oxidation–reduction reactions. As we will discuss in later sections, the process is called aerobic respiration when oxygen (O_2) serves as the final electron acceptor; in the process called anaerobic respiration, inorganic substances other than oxygen, such as nitrate ions (NO_3^-) or sulfate ions (SO_4^{2-}), serve as the final electron acceptors; and in the process called fermentation, organic compounds serve as the final electron acceptors. A series of electron carriers called an electron transport chain releases energy that is used by the mechanism of chemiosmosis to synthesize ATP. Regardless of their energy sources, all organisms use similar oxidation–reduction reactions to transfer electrons and similar mechanisms to use the energy released to produce ATP.

We will now consider the specific biochemical pathways organisms use to produce energy.

(a) ENERGY SOURCE
(glucose, elemental sulfur, hydrogen gas, light)

Electrons

ATP

(b) ELECTRON CARRIER
(NAD, FAD)

ATP

(c) FINAL ELECTRON ACCEPTOR

O_2 (aerobic respiration)	NO_3^-, SO_4^{2-} (anaerobic respiration)	Organic compound (fermentation)

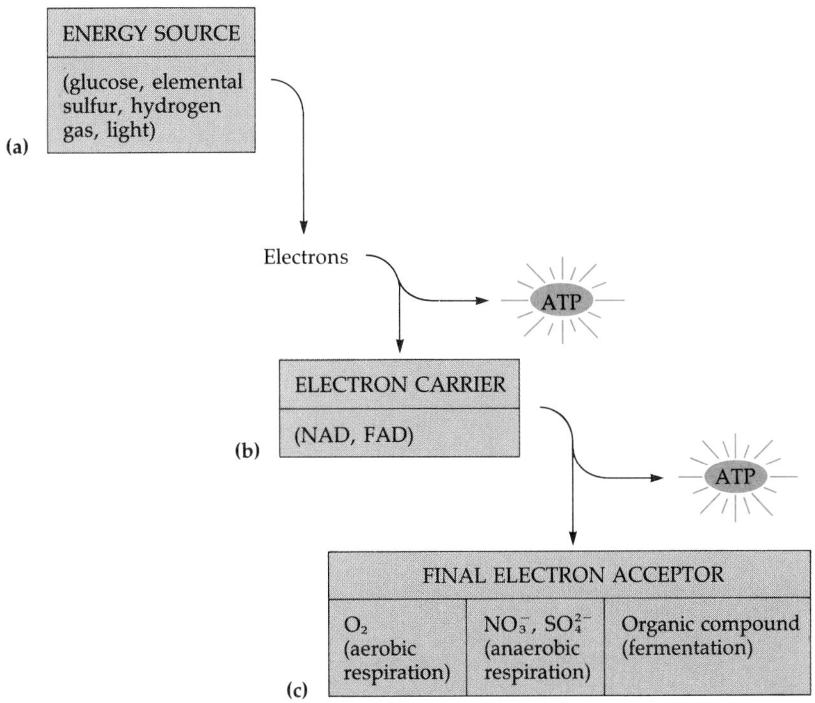

Figure 5–11 The production of ATP requires **(a)** a source of energy as an electron donor, **(b)** the transfer of electrons to an electron carrier, and **(c)** a final electron acceptor.

BIOCHEMICAL PATHWAYS OF ENERGY PRODUCTION

Organisms that produce and store energy from organic molecules do so in a series of controlled reactions rather than in a single burst. If the energy were released all at once, it would be released as a large amount of heat, which could not be readily used to drive chemical reactions and would, in fact, damage the cell. To extract energy from organic compounds and store it in cells, organisms pass electrons from one compound to another through a series of oxidation–reduction reactions.

A sequence of enzymically catalyzed chemical reactions occurring in a cell is called a **biochemical pathway.** Pathways are usually written as follows:

$$H_2O \qquad ATP \quad ADP + \textcircled{P} \qquad CO_2$$

$$A \longrightarrow B \longrightarrow C \longrightarrow D \rightleftharpoons E$$

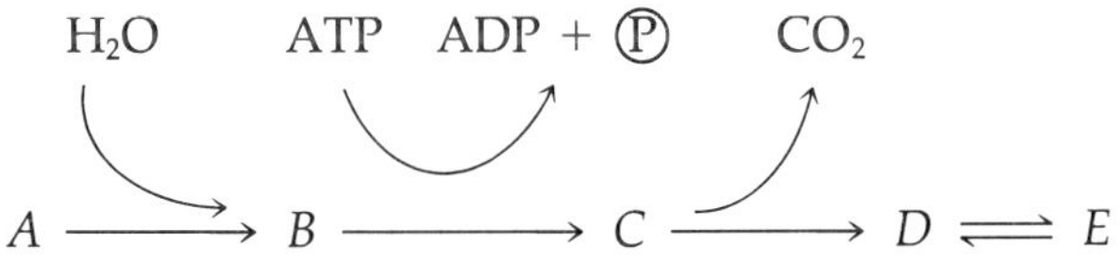

Note that pathways shown in this way are not usually written as balanced chemical equations. This hypothetical pathway converts starting material A into the end product E in a series of four steps. (Of course, if this were an actual pathway, chemical structures or names would replace letters A through E.)

The curved arrow originating at H_2O indicates that water is a reactant in the reaction converting A to B. Similarly, the curved arrow pointing to CO_2 indicates that carbon dioxide is a product of the reaction converting C to D. To drive the reaction converting B to C, a molecule of ATP must be broken down to ADP and a phosphate in a coupled reaction; that is, the two reactions occur simultaneously, each depending on the other. The reaction converting D to E is readily reversible, as indicated by the double arrow. Keep in mind that almost every reaction in a biochemical pathway is catalyzed by a specific enzyme; sometimes the name of the enzyme is printed near the arrow.

CARBOHYDRATE CATABOLISM

Most microorganisms oxidize carbohydrates to provide most of the cell's energy. **Carbohydrate catabolism,** the breakdown of carbohydrate molecules to produce energy, is therefore of great importance in cell metabolism. Glucose is the carbohydrate energy source most used by cells. Microorganisms, as you will see later, can also catabolize various lipids and proteins for energy production.

To produce energy from glucose, microorganisms use two general processes, respiration and fermentation. The first stage of both respiration and fermentation is the oxidation of glucose to pyruvic acid. This is most commonly accomplished by a process called glycolysis.

Glycolysis

Glycolysis, usually the first stage in carbohydrate catabolism, is the oxidation of glucose to pyruvic acid. Glycolysis is also called the **Embden–Meyerhof pathway.** Most microorganisms, as well as higher organisms, use this pathway, which is a series of ten chemical reactions, each catalyzed by a different enzyme. The principal reactions are shown in Figure 5–12 and summarized below:

1. First, two molecules of ATP are used when a six-carbon glucose molecule is phosphorylated, rearranged, and split into two three-carbon compounds, glyceraldehyde-3-phosphate and dihydroxyacetone phosphate. These two compounds are interconvertible.
2. Next, the two three-carbon molecules are oxidized to two molecules of pyruvic acid. In these reactions, four molecules of ATP are formed by substrate-level phosphorylation, and two molecules of $NADH_2$ are produced.
3. Because two molecules of ATP are needed to get glycolysis started, and four molecules of ATP are generated when the process is completed, *there is a net gain of two molecules of ATP for each molecule of glucose that is oxidized.*

Pentose phosphate pathway

Many bacteria have, in addition to glycolysis, an alternate pathway for the oxidation of glucose. This is called the **pentose phosphate pathway (hexose monophosphate shunt),** and it operates simultaneously with glycolysis. The pentose phosphate pathway is a cyclic pathway that provides a means for the breakdown of five-carbon sugars (pentoses), as well as glucose. A key feature of this pathway is that it produces important intermediate pentoses that act as precursors in the synthesis of (1) nucleic acids, (2) glucose from carbon dioxide in photosynthesis, and (3) certain amino acids. The

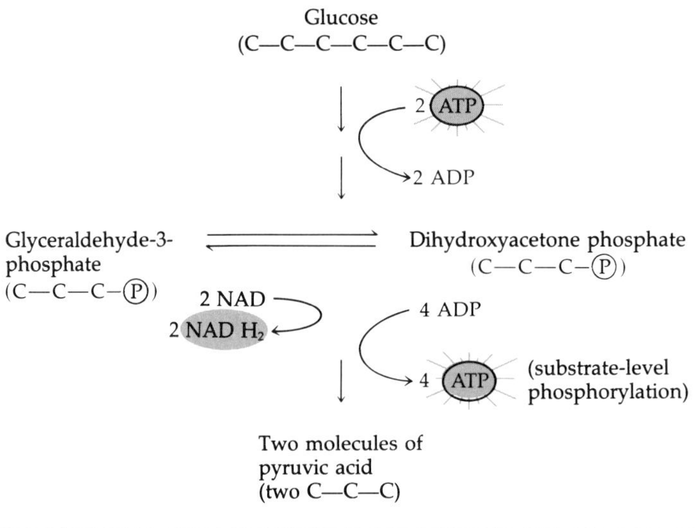

Figure 5–12 Outline of the principal reactions of glycolysis.

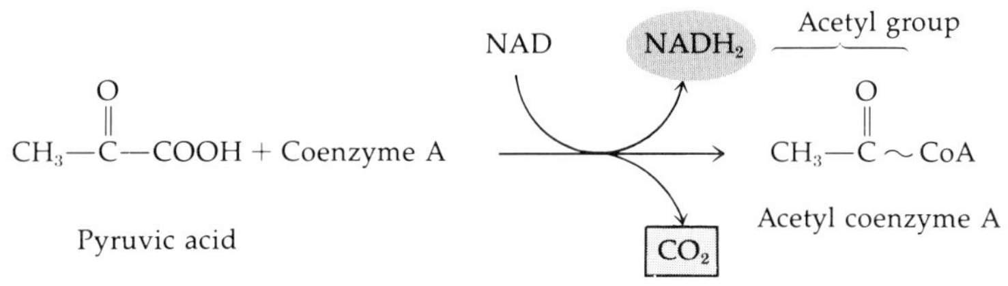

Figure 5–13 Formation of acetyl CoA from pyruvic acid.

pathway is an important producer of $NADPH_2$ from NADP; $NADPH_2$ is used in various biosynthetic reactions in the cell. Unlike glycolysis, the pentose phosphate pathway produces only one ATP. Bacteria that use the pentose phosphate pathway include *Bacillus subtilis* (su'til-us), *Escherichia coli, Leuconostoc mesenteroides* (lü-kō-nos'tok mes-en-ter-oi'dēz), and *Streptococcus faecalis* (strep-tō-kok'kus fē-kāl'is).

Entner–Doudoroff pathway (EDP)

The **Entner–Doudoroff pathway (EDP)** is still another pathway for the oxidation of glucose to pyruvic acid. From each molecule of glucose, two molecules of $NADPH_2$ are produced for use in cellular biosynthetic reactions. Bacteria that have the enzymes for the EDP can metabolize without either glycolysis or the pentose phosphate pathway. The only bacteria known to possess the EDP are some species of the genus *Rhizobium* (rī-zō'bē-um) and several of the genus *Pseudomonas* (sū-dō-mō'nas). Tests for the ability to oxidize glucose by this pathway are sometimes used to identify *Pseudomonas* in the clinical laboratory.

Respiration

After glucose has been broken down to pyruvic acid, the pyruvic acid can undergo further fermentation, described later, or it can undergo respiration. **Respiration** is an ATP-generating process in which chemical compounds are oxidized and the final electron acceptor is almost always an inorganic molecule. Pyruvic acid, produced by glycolysis, is split and a fragment of it is attached to a coenzyme molecule called coenzyme A or CoA; that combination is acetyl CoA. Acetyl CoA next enters the Krebs cycle, a series of reactions that release electrons and protons (H^+). Electrons released during glycolysis, the formation of acetyl CoA, and the Krebs cycle are passed to NAD or FAD molecules. These molecules carry the electrons to an electron transport chain, where ATP is formed. A major feature of respiratory processes is the operation of an electron transport chain. As we discussed before, electrons move through the chain in the chemiosmosis process, which produces the energy to create ATP. This is the process called oxidative phosphorylation.

Any naturally occurring organic molecule can be degraded by some microbe during respiration. And, as you will see, the organic molecule is usually oxidized completely to carbon dioxide. Consequently, the yield of ATP per molecule of substrate is much greater in respiration than in fermentation (to be described shortly).

In most respirations, the final hydrogen (electron) acceptor is free oxygen (O_2). This type of respiration is called **aerobic respiration.** A few bacteria carry on **anaerobic respiration,** in which the final hydrogen (electron) acceptor is an inorganic molecule other than free oxygen. Examples of such inorganic acceptors are nitrates (NO_3^-), sulfates (SO_4^{2-}), and carbonates (CO_3^-). Anaerobic respiration can produce an energy yield almost as high as that of aerobic respiration.

We will first consider aerobic respiration.

Aerobic respiration

Before pyruvic acid can enter the processes that completely oxidize it to carbon dioxide and water, it must lose one molecule of carbon dioxide, and become a two-carbon compound. This process is called **decarboxylation.** The two-carbon compound, called an *acetyl group,* then attaches to the coenzyme known as *coenzyme A (CoA)*; the resulting complex is known as *acetyl coenzyme A (acetyl CoA)* (Figure 5–13.) During this reaction,

NAD is reduced to $NADH_2$. Remember that the oxidation of one glucose molecule produces two molecules of pyruvic acid, so for each molecule of glucose, two molecules of carbon dioxide are lost, and two molecules of $NADH_2$ are produced. Each molecule of $NADH_2$ will later yield 6 molecules of ATP in the electron transport chain. Once the pyruvic acid has undergone decarboxylation and its derivative (the acetyl group) has attached to CoA, the resulting compound (acetyl CoA) is ready to enter the Krebs cycle.

Krebs Cycle The **Krebs cycle,** also called the **tricarboxylic acid cycle (TCA)** or **citric acid cycle,** is a series of chemical reactions in which the large amount of potential chemical energy stored in intermediate compounds derived from pyruvic acid is released step by step. In this cycle, a series of oxidations and reductions transfers that potential energy, in the form of electrons, to a number of coenzymes. The pyruvate derivatives are oxidized; the coenzymes are reduced.

As acetyl CoA enters the Krebs cycle, CoA detaches from the acetyl group and can pick up more acetyl groups for the next Krebs cycle. Meanwhile, the acetyl group combines with a substance called oxaloacetic acid to form citric acid. At this point the Krebs cycle begins. The major chemical reactions of this cycle are outlined in Figure 5–14.

As the various acids move through the cycle, they undergo a number of changes, all controlled by specific enzymes. One of these changes is decarboxylation. Isocitric acid, a six-carbon compound, is decarboxylated to the five-carbon compound called α-ketoglutaric acid. By losing a molecule of CO_2 and picking up a molecule of CoA, the five-carbon compound, α-ketoglutaric acid, is then decarboxylated to form the four-carbon compound called succinyl CoA. Once again, remember that the oxidation of one molecule of glucose produces two molecules of acetyl CoA that enter the Krebs cycle. This means that from each molecule of glucose, two molecules of CO_2 are produced when pyruvic acid is decarboxylated, and four molecules of CO_2 are produced in the Krebs cycle.

As the acids go through the cycle, they are also involved in a series of oxidation–reduction reactions. In the conversion of the six-carbon isocitric acid to a five-carbon compound, two hydrogen atoms are lost. In other words, the six-carbon compound is oxidized. The hydrogen atoms released in the Krebs cycle are picked up by the coenzymes NAD and FAD. When NAD picks up hydrogen atoms, it is reduced and may be represented as $NADH_2$; similarly, FAD is reduced to $FADH_2$. Notice the points in the synthesis of acetyl CoA and in the Krebs cycle where coenzymes pick up hydrogen atoms (see Figure 5–14).

If we look at the Krebs cycle as a whole, we see that for each two molecules of acetyl CoA that enter the cycle, four molecules of carbon dioxide are liberated by decarboxylation; six molecules of $NADH_2$ and two molecules of $FADH_2$ are produced by oxidation–reduction reactions; and two molecules of GTP (the equivalent of ATP) are generated. Many of the intermediates in the Krebs cycle also play a role in other pathways, especially in amino acid biosynthesis. In the electron transport chain, the six molecules of $NADH_2$ will later yield 18 ATP molecules, and the two molecules of $FADH_2$ will yield 4 ATP molecules.

The carbon dioxide produced in the Krebs cycle is liberated into the atmosphere as a gaseous by-product of aerobic respiration. (Humans produce carbon dioxide from the Krebs cyle in each cell of the body and discharge it through the lungs during exhalation.) The reduced coenzymes ($NADH_2$ and $FADH_2$) are the most important outcome of the Krebs cycle, because they contain the energy originally stored in glucose and then in pyruvic acid. During the next phase of aerobic respiration, a series of reductions transfer the energy stored in the coenzymes to ATP. These reactions are collectively called the electron transport chain.

Electron Transport Chain The **electron transport chain** consists of a sequence of carrier molecules that are capable of oxidation and reduction. As electrons are passed through the chain, there is a stepwise release of energy, and ATP is generated from ADP and a free phosphate molecule. In aerobic respiration, the terminal electron acceptor of the chain is oxygen (O_2); in anaerobic respiration, the final acceptor is an inorganic molecule other than oxygen (or, rarely, an organic compound). The final oxidation is irreversible in both respirations. In eucaryotic cells, the electron transport chain is contained in membranes in mitochondria; in procaryotic cells, it is found in the plasma membrane.

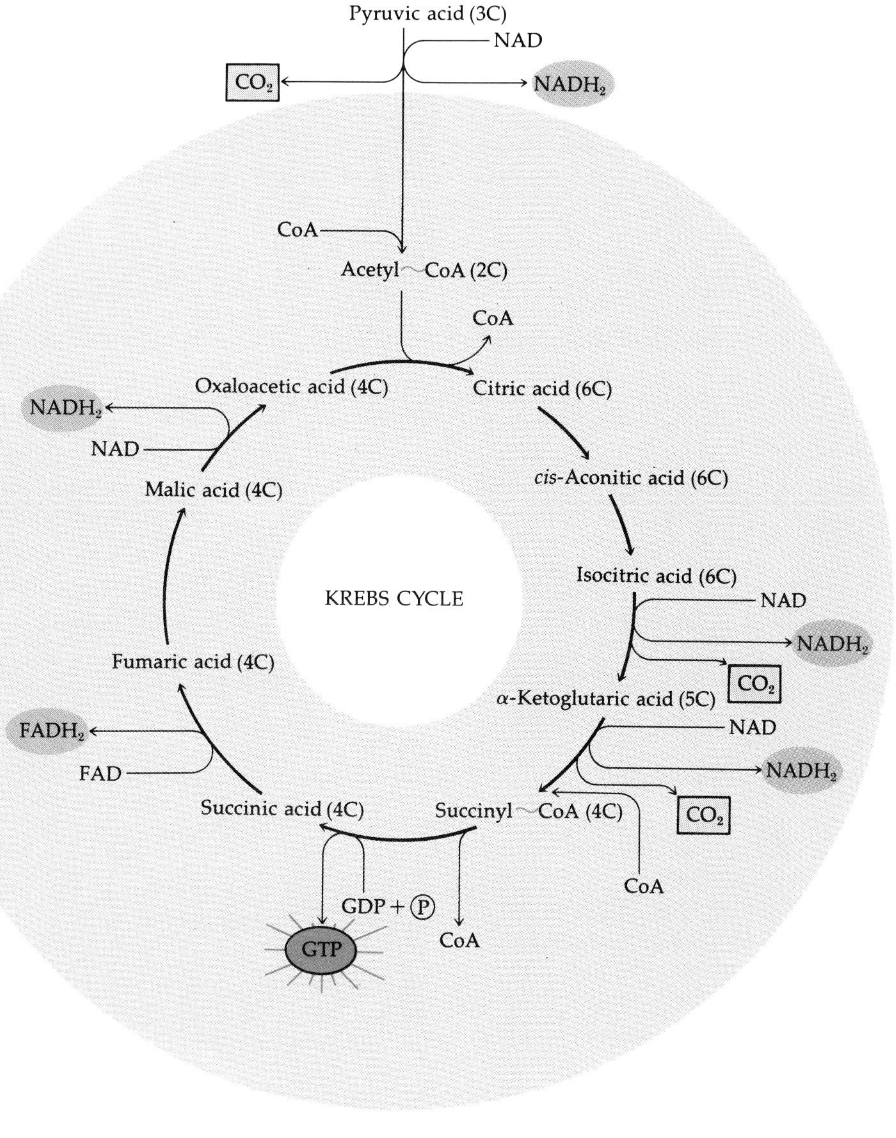

Figure 5–14 The Krebs cycle. Note the points where CO_2 is produced, $NADH_2$ and $FADH_2$ are formed, and GTP is generated. GTP is used directly in certain cell processes and can also be used to convert ADP to ATP.

The respiration of organic compounds involves three classes of carrier molecules in the electron transport chain: **flavoproteins,** proteins that contain a coenzyme derived from riboflavin (vitamin B_2), and that are capable of performing alternating oxidations and reductions; **cytochromes,** proteins that contain an iron-containing group (heme) capable of existing alternately as a reduced form (Fe^{2+}) and an oxidized form (Fe^{3+}); and **quinones,** nonprotein carriers of low molecular weight.

Much is known about the electron transport chain in the mitochondria of eucaryotic cells, so this is the one we will describe. The electron transport chains of different bacterial systems are quite diverse, and their particular carrier molecules may differ markedly from those of other bacteria and from those of eucaryotic mitochondrial systems. However, keep in mind that they achieve the same basic goal.

The first step in the electron transport chain (shown in Figure 5–15) involves the transfer of hydrogen atoms from $NADH_2$ to FAD. In this transfer, $NADH_2$ is oxidized to NAD, and FAD is reduced to $FADH_2$. The importance of the hydrogen transfer from $NADH_2$ to FAD is that it releases energy. This energy is used to produce ATP from ADP and phosphoric acid (phosphate). The hydrogen atoms are then used to reduce a quinone. After that reduction, however, the hydrogen atoms do not stay intact. They ionize into hydrogen ions (H^+) and electrons (e^-) according to the following reaction:

$$\underset{\text{Hydrogen atom}}{H} \longrightarrow \underset{\text{Hydrogen ion}}{H^+} + \underset{\text{Electron}}{e^-}$$

In the next step of the electron transport chain, the electrons (e^-) from the hydrogen atoms are passed to cytochrome *b*. (At the same time, the H^+ ions are released into solution.) The electrons are passed from one cytochrome to another—from cytochrome *b* to cytochrome *c* to cytochrome *a*, and finally to cytochrome oxidase. Each cytochrome in the electron transport system is reduced as it picks up electrons and oxidized as it gives up electrons. (The electron transport chains of some bacteria, such as *Escherichia coli*, lack cytochrome *c*. The presence of cytochrome *c* can be determined by the diagnostically useful **oxidase test.**)

At several steps in the electron transport chain, protons are actively pumped across the membrane, and a proton gradient and the resulting proton motive force are created. Moving the protons back across the membrane releases energy used to synthesize ATP from ADP. At the end of the electron transport system, the electrons are passed to oxygen, which becomes negatively charged. The oxygen then combines with the free H^+ ions to form water. As you may recall, the transfer of electrons from reduced coenzymes to oxygen is known as oxidative phosphorylation, and the mechanism by which this generates ATP is called chemiosmosis. The various electron transfers in the electron transport chain generate 34 ATP molecules from each molecule of glucose oxidized.

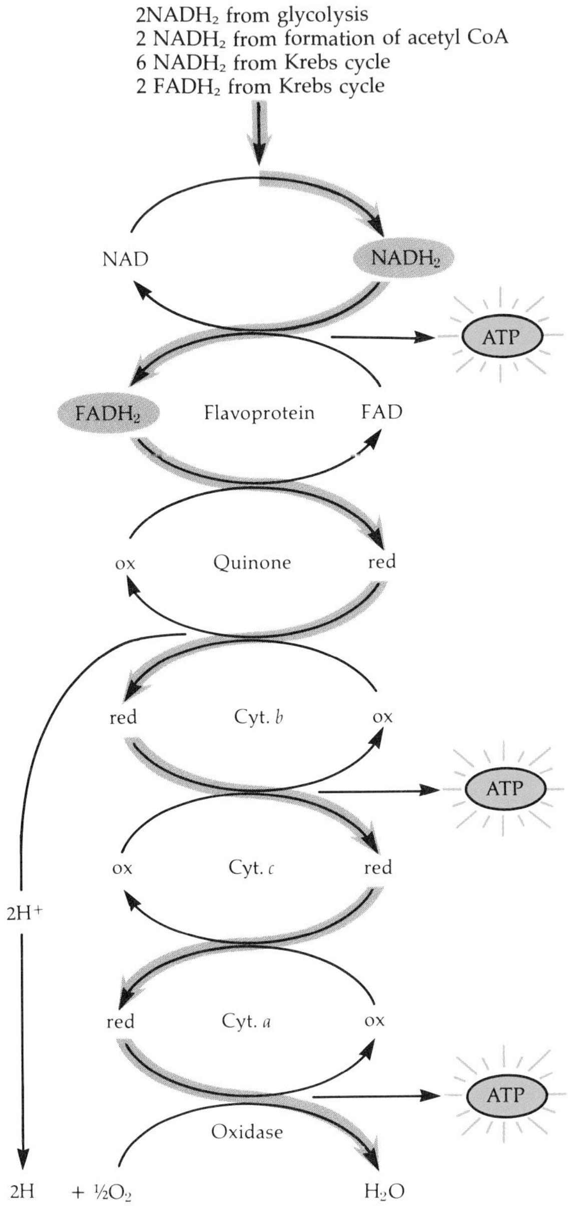

Figure 5–15 An electron transport chain. The heavy, colored arrows show the path of the electrons as they pass from one carrier to the next. ATP is produced at three points along the chain.

Table 5–5 provides a detailed accounting of the ATP yield during aerobic respiration. We can now summarize the overall reaction for aerobic respiration as follows:

$$\underset{\text{Glucose}}{C_6H_{12}O_6} + \underset{\text{Oxygen}}{6O_2} + 38\ ADP + 38\ \text{Ⓟ} \longrightarrow$$

$$\underset{\text{Carbon dioxide}}{6CO_2} + \underset{\text{Water}}{6H_2O} + 38\ ATP$$

A summary of the various stages of aerobic respiration is presented in Figure 5–16.

Anaerobic respiration

In the process called **anaerobic respiration,** the final electron acceptor is usually an inorganic substance other than oxygen (O_2). Some bacteria, such as *Pseudomonas* and *Bacillus,* can use a nitrate ion (NO_3^-) as a final electron acceptor; it is reduced to nitrite ion (NO_2^-), nitrous oxide (N_2O), or nitrogen gas (N_2). Other bacteria, such as *Desulfovibrio* (dē-sul-fō-vib′rē-ō), use sulfate (SO_4^{2-}) as the final electron acceptor, to form hydrogen sulfide (H_2S). Still other bacteria use carbonate (CO_3^{2-}) to form methane (CH_4). (A few microbes that carry on anaerobic respiration use organic compounds such as fumaric acid as the final electron acceptor.) From the oxidation of an organic molecule, anaerobic respiration yields varying amounts of ATP, which depend on the organism and the pathway.

Fermentation

After glucose is broken down to pyruvic acid, the pyruvic acid can undergo respiration, as previously described, or it can undergo fermentation. **Fermentation** can be defined several ways (see Box), but we will define it as a process that

1. Releases energy from sugars or other organic molecules such as amino acids, organic acids, purines, and pyrimidines.
2. Does not require oxygen (but sometimes can occur in its presence).
3. Does not require use of an electron transport chain.
4. Uses an organic molecule as the final electron acceptor.
5. Produces small amounts of ATP (only 2 ATP molecules).

During fermentation, electrons (hydrogen) are transferred from $NADH_2$ to pyruvic acid, which is turned into various end products; substrate-level phosphorylation converts ADP to ATP. Various microorganisms are able to ferment various substrates; the end products depend on the particular microorganism, the substrate, and the enzymes that are present and active. Chemical analyses of these end products are useful in the identification of microorganisms. The products of various

Table 5–5 Summary of ATP Produced During Aerobic Respiration of One Glucose Molecule

Source	ATP Yield
Glycolysis	
(1) oxidation of glucose to pyruvic acid	2 ATP
(2) production of 2 $NADH_2$	6 ATP (in electron transport chain)
Formation of acetyl CoA produces 2 $NADH_2$	6 ATP (in electron transport chain)
Krebs Cycle	
(1) oxidation of succinyl CoA to succinic acid	2 GTP (equivalent of ATP)
(2) production of 6 $NADH_2$	18 ATP (in electron transport chain)
(3) production of 2 $FADH_2$	4 ATP (in electron transport chain)
TOTAL	38 ATP

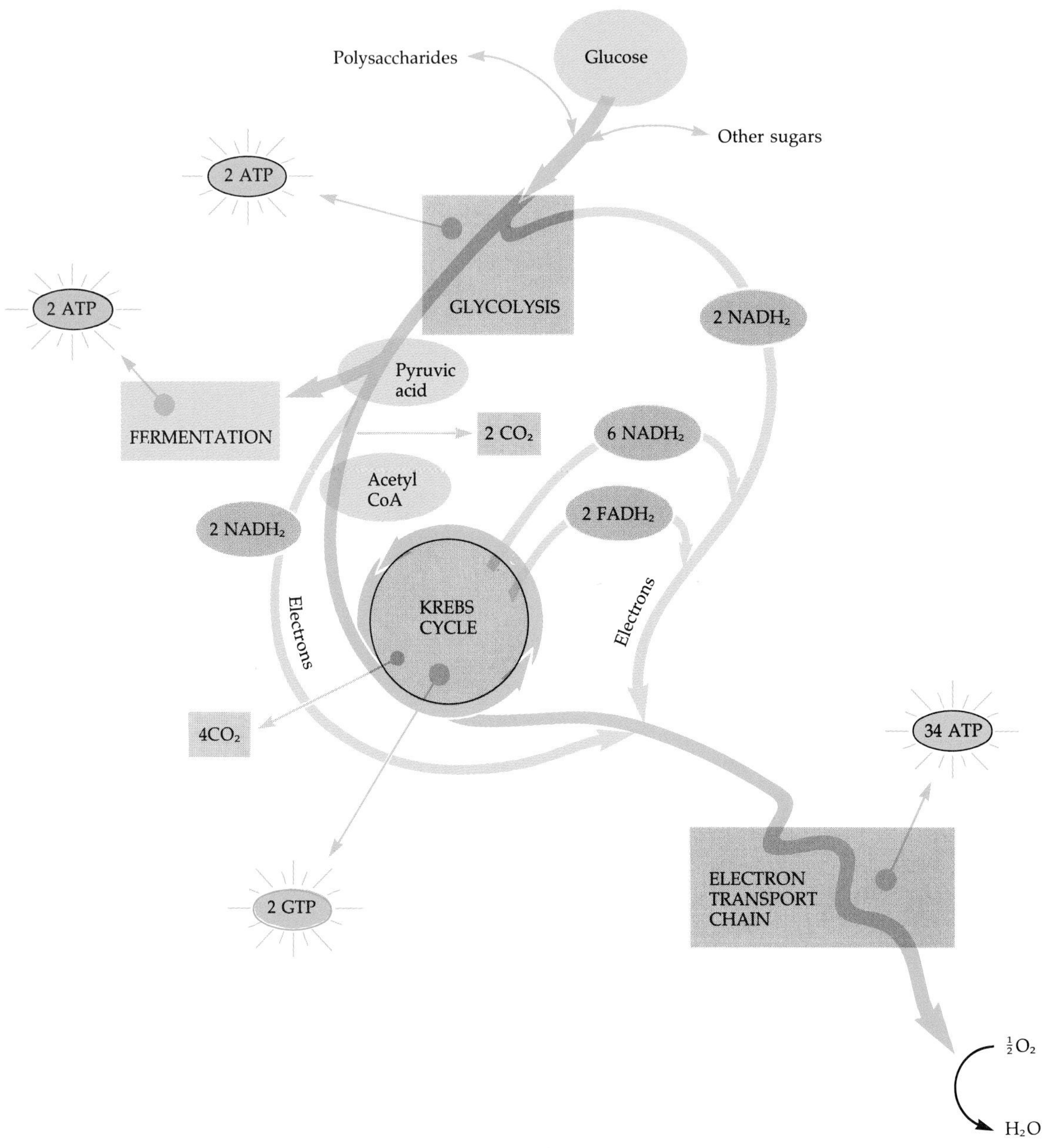

Figure 5–16 Overview of carbohydrate catabolism: Glucose is broken down completely to carbon dioxide and water and ATP is generated. This process has three major phases: glycolysis, the Krebs cycle, and the electron transport chain. The key event in the process is that electrons are removed from intermediates of glycolysis and the Krebs cycle and are carried by NAD or FAD to the electron transport chain. By oxidative phosphorylation, the electron transport chain produces most of the ATP. The process of fermentation uses only the glycolysis portion of this sequence, and generates much less ATP by substrate-level phosphorylation.

What Is Fermentation?

To many people, fermentation simply means the production of alcohol. Grains and fruits ferment into beer and wine. If a food soured, you might say it was "off" or fermented. Here are some definitions of fermentation. They range from informal, general usage to more scientific definitions that you will need to know for this course.

Fermentation is:

1. Any process that produces alcoholic beverages (general use).
2. Any spoilage of food by microorganisms (general use).
3. Any large-scale microbial process, occurring with or without air (common definition used in industry).
4. Any energy-releasing metabolic process that takes place only under anaerobic conditions (becoming more scientific).
5. All metabolic processes that release energy from a sugar or other organic molecule, do not require oxygen or an electron transport system, and use an organic molecule as the final electron acceptor (the definition we will use in this text).

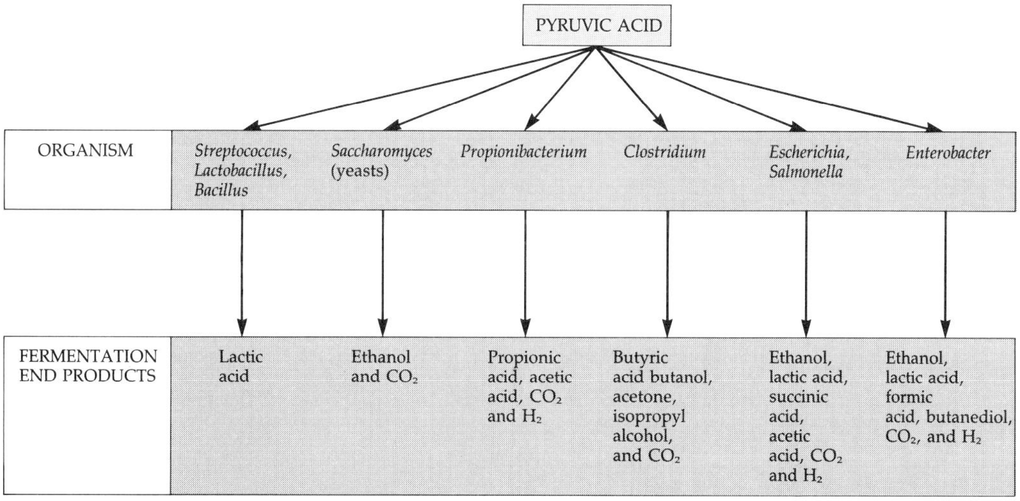

Figure 5–17 End products of various microorganisms' fermentations.

microorganisms' fermentations of pyruvic acid are shown in Figure 5–17.

In discussing the process of fermentation, we will consider two of the more important processes, lactic acid fermentation and alcoholic fermentation. A key feature of all fermentations is that they produce small amounts of ATP (only 2 ATP molecules) because much of the energy remains in the chemical bonds of the organic end products, such as lactic acid or ethanol. Another key feature to remember is that fermentations do not utilize a Krebs cycle or electron transport chain.

Lactic acid fermentation

During the first step of **lactic acid fermentation,** glycolysis, a molecule of glucose is oxidized to two molecules of pyruvic acid (see Figure 5–12). This oxidation generates the energy with which the two molecules of ATP are formed. In the next step, the

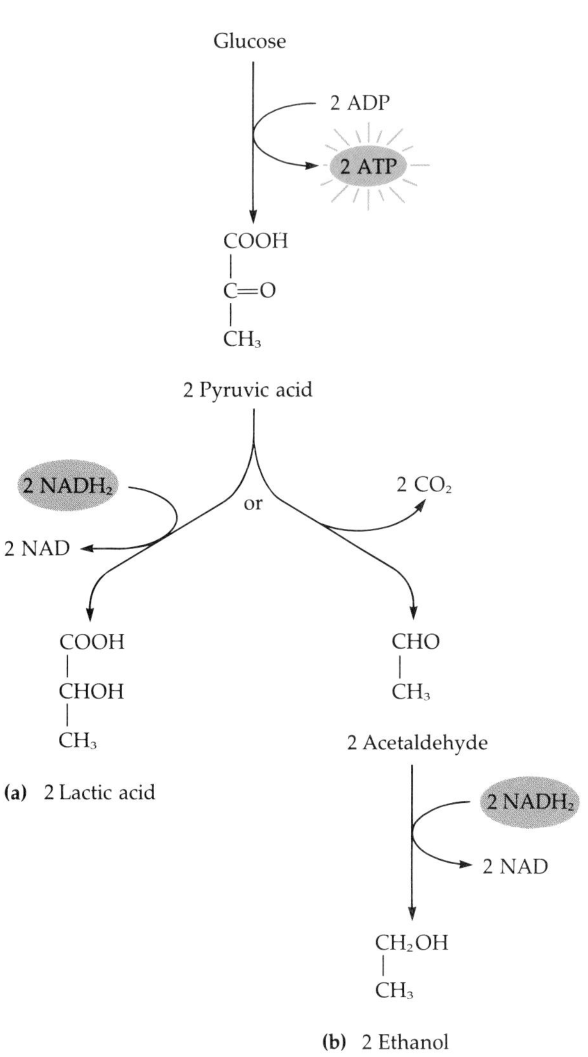

Figure 5–18 Fermentation. **(a)** Lactic acid fermentation. **(b)** Alcoholic fermentation.

two molecules of pyruvic acid are reduced by two molecules of $NADH_2$ to form two molecules of lactic acid (Figure 5–18a). Because lactic acid is the end product of the reaction, it undergoes no further oxidation, and most of the energy produced by the reaction is stored in the lactic acid. Thus fermentation yields only low energy.

Three important genera of lactic acid bacteria are *Streptococcus*, *Lactobacillus* (lak-tō-bä-sil'lus), and *Leuconostoc*. Because these microbes produce only lactic acid, they are referred to as *homofermentative* (*homolactive*). Lactic acid is economically important because it can cause food spoilage. However, the process is also important in food production. Depending on the bacterium used, fermentation can produce yogurt from milk, sauerkraut from fresh cabbage, and pickles from cucumbers.

Alcoholic fermentation

Alcoholic fermentation also begins with the glycolysis of glucose and the yield of pyruvic acid and two molecules of ATP. In the next reaction, the two molecules of pyruvic acid are converted to two molecules of acetaldehyde and two molecules of CO_2 (Figure 5–18b). The two molecules of acetaldehyde are next reduced by two molecules of $NADH_2$ to form two molecules of ethanol (ethyl alcohol). Again, alcoholic fermentation is a low-energy-yield process because most of the energy remains in the ethanol, the end product. Alcoholic fermentation is carried out by *Saccharomyces* (sak-ä-rō-mī'sēs), a type of yeast, and is the basis of wine and beer production. Organisms that produce lactic acid as well as other acids, alcohols, or acetone are known as *heterofermentative* (*heterolactic*) and often use the pentose phosphate pathway.

The ethanol and carbon dioxide produced by yeasts are waste products for yeast cells but useful to humans. Ethanol is used by the brewing and distilling industries to make alcoholic beverages, and bakers need carbon dioxide's help in making bread dough rise. Table 5–6 lists some of the various microbial fermentations used by industry to convert inexpensive raw materials into useful end products.

A summary of aerobic respiration, anaerobic respiration, and fermentation is given in Table 5–7.

Lipid Catabolism

Thus far in our discussion of energy production, we have emphasized the oxidation of glucose, the principal energy-supplying carbohydrate. But microbes also oxidize lipids and proteins, and the oxidations of all these nutrients are related.

Fats, you might recall, are lipids consisting of fatty acids and glycerol. Microbes produce extracellular enzymes called *lipases* that break fats down

Table 5–6 Some Industrial Fermentations

Fermentation Product	Commercial Use	Starting Material	Microorganism
Ethanol	Beer	Malt extract	*Saccharomyces cerevisiae* (yeast, a fungus)
	Wine	Grape or other fruit juices	*Saccharomyces ellipsoideus* (yeast, a fungus)
Acetic acid	Vinegar	Alcohol	*Acetobacter* (bacterium)
Lactic acid	Cheese, yogurt	Milk	*Lactobacillus, Streptococcus* (bacteria)
	Rye bread	Grain, sugar	*Lactobacillus bulgaricus* (bacterium)
	Summer sausage	Meat	*Pediococcus* (bacterium)
Propionic acid and carbon dioxide	Swiss cheese	Milk	*Propionibacterium freudenreichii* (bacterium)
Acetone and butanol	Pharmaceutical, industrial uses	Molasses	*Clostridium acetobutylicum* (bacterium)
Glycerol	Pharmaceutical, industrial uses	Molasses	*Saccharomyces cerevisiae* (yeast, a fungus)
Citric acid	Flavoring	Molasses	*Aspergillus* (fungus)
Methanol	Fuel	Agricultural wastes	*Clostridium* (bacteria)

Table 5–7 Comparison of Aerobic Respiration, Anaerobic Respiration, and Fermentation

Energy-Producing Process	Growth Conditions	Final Hydrogen (Electron) Acceptor	Type of Phosphorylation Used to Build ATP	Number of ATP Molecules Produced
Aerobic respiration	Aerobic	Free oxygen (O_2)	Substrate-level and oxidative	38
Anaerobic respiration	Anaerobic	Usually an inorganic substance (such as NO_3^-, SO_4^{2-}, or CO_3^{2-}), but not free oxygen (O_2)	Oxidative	Variable
Fermentation	Aerobic or Anaerobic	An organic molecule	Substrate-level	2

into their fatty acid and glycerol components. Each component is then metabolized separately.

Many types of microbes can convert glycerol into dihydroxyacetone phosphate, one of the intermediates formed during glycolysis (see Figure 5–12). The dihydroxyacetone phosphate is then metabolized according to the mechanism shown in Figure 5–19.

Fatty acids are catabolized somewhat differently. The mechanism of fatty acid oxidation is called **beta oxidation.** In this process, carbon fragments of a long chain of fatty acid are removed two at a time and acetyl CoA is formed. As the molecules of acetyl CoA form, they enter the Krebs cycle, as do the acetyl CoA molecules formed by the oxidation of pyruvic acid (see Figure 5–14). Thus, the oxidation of glycerol and fatty acids is directly linked to the oxidation of glucose by the Krebs cycle.

Protein Catabolism

Proteins are too large to pass through the plasma membranes of microbes. Microbes produce extracellular *proteases* and *peptidases,* which break down proteins into their component amino acids. However, before amino acids can be catabolized, they must be converted to substances that can enter the Krebs cycle. During one such conversion, called **deamination,** the amino group of an amino acid is removed and converted to an ammonium ion (NH_4^+), which can be excreted from the cell. The remaining organic acid can enter the Krebs cycle. Other conversions involve decarboxylation (removal of —COOH) and dehydrogenation. Although these conversions are complex, the important thing to remember is that various methods can prepare amino acids to enter the Krebs cycle.

BIOCHEMICAL PATHWAYS OF ENERGY UTILIZATION

Up to now we have been considering energy production. Through the oxidation of organic molecules, organisms produce energy by aerobic respiration, anaerobic respiration, and fermentation. Much of this energy is used up as heat. In fact, although the complete metabolic oxidation of glucose to carbon dioxide and water is considered a very efficient process, about 60% of the energy of glucose is lost as heat. Cells use the remaining energy, which is trapped in molecules of ATP, in a variety of ways. Microbes use ATP to provide energy for the transport of substances across plasma membranes—a process called active transport, discussed in Chapter 4. Certain microbes also use some of their ATP for flagellar motion (also discussed in Chapter 4). Most of the ATP, however, is used in the biosynthesis of new cellular components. We will now consider the biosynthesis of a few representative classes of biological molecules: carbohydrates, lipids, amino acids, purines, and pyrimidines.

Biosynthesis of Polysaccharides

Nucleotides carry sugars for the biosynthesis of more complex carbohydrates. Bacteria synthesize glycogen from *adenosine diphosphoglucose (ADPG),* which is the molecule formed when ADP is joined to a glucose molecule and a phosphate is released. Animals synthesize glycogen and many other carbohydrates from *uridine diphosphoglucose (UDPG).* A compound related to UDPG called *UDP-N-acetyl glucosamine (UDPNAG)* is a key starting material in the biosynthesis of peptidoglycan, the substance that forms bacterial cell walls (Figure 5–20).

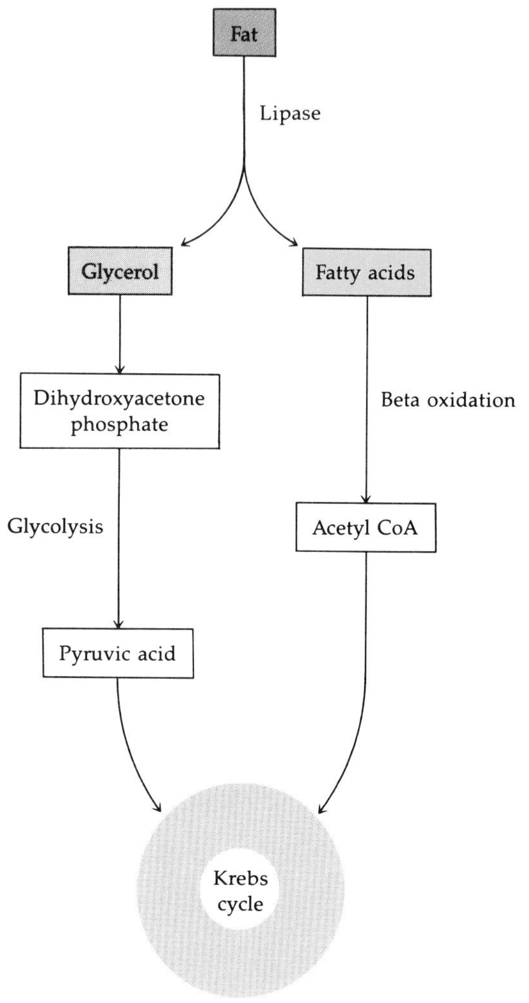

Figure 5–19 Lipid catabolism.

Biosynthesis of Lipids

Microbes synthesize lipids, such as fats, by uniting glycerol and fatty acids. The glycerol portion of the lipid is derived from dihydroxyacetone phosphate, an intermediate formed during glycolysis. Fatty acids, which are long-chained hydrocarbons (hydrogen linked to carbon), are built up when two carbon fragments of acetyl CoA are successively added to each other (Figure 5–21). The enzymatic linkage of glycerol to fatty acids is the biosynthesis of a variety of simple fats or lipids. As mentioned

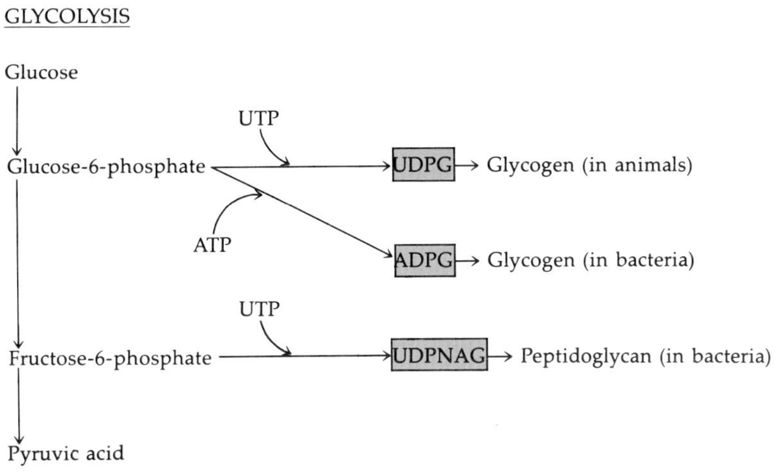

Figure 5–20 Biosynthesis of carbohydrates.

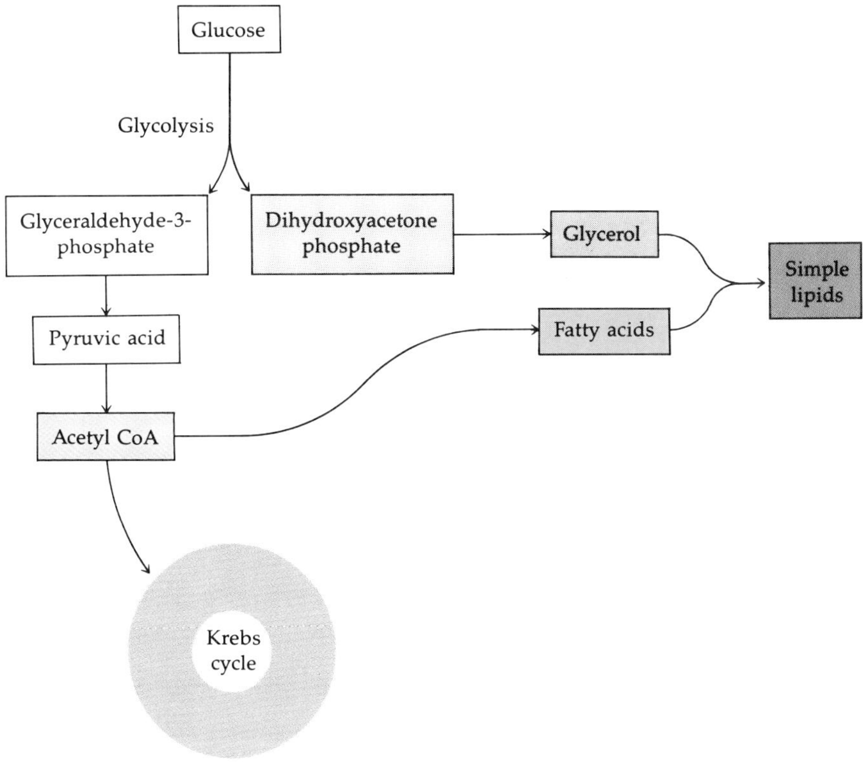

Figure 5–21 Biosynthesis of simple lipids.

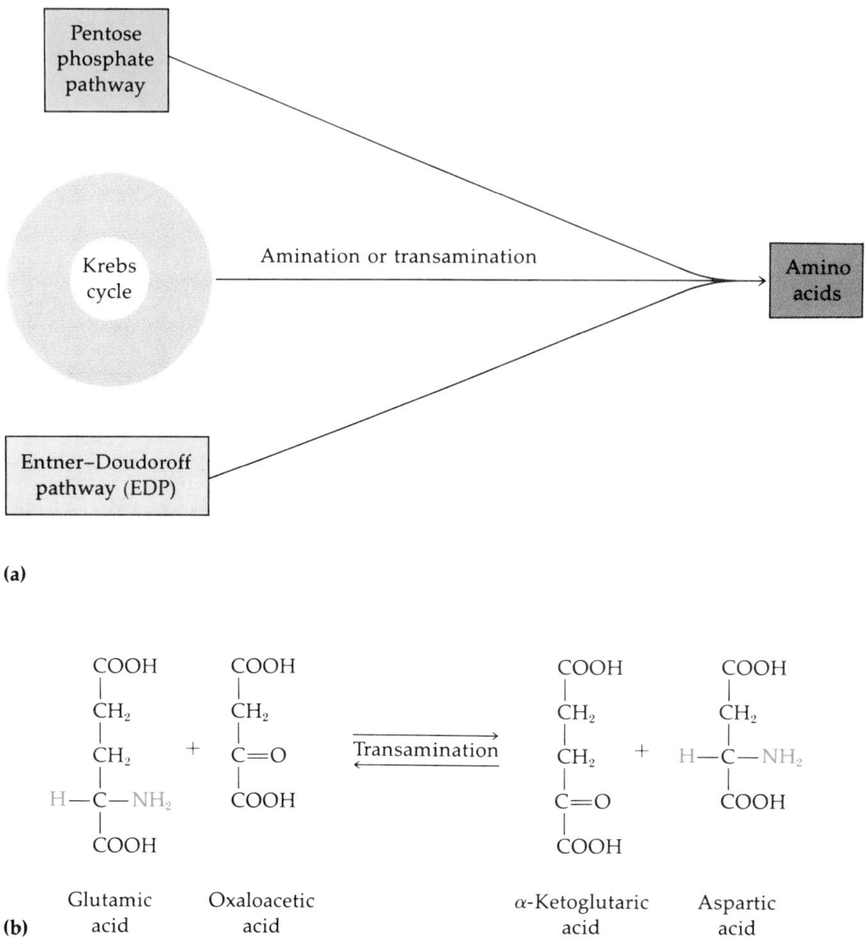

Figure 5–22 Biosynthesis of amino acids. **(a)** Pathways of amino acid biosynthesis. **(b)** Transamination, a process by which new amino acids are made with the amine groups from old amino acids. Glutamic acid and aspartic acid are both amino acids. The other two compounds are intermediates in the Krebs cycle.

previously, the most important role of lipids is as structural components of plasma membranes. Lipids also help to form the cell wall of gram-negative bacteria, and, like carbohydrates, they can also serve as storage forms of energy. Recall that the breakdown products after biological oxidation feed into the Krebs cycle.

Biosynthesis of Amino Acids

Amino acids are required for protein biosynthesis. Some microbes, such as *Escherichia coli,* contain the enzymes necessary to use starting materials such as glucose and inorganic salts in the syntheses of all the amino acids they need. Organisms with the necessary enzymes can synthesize all amino acids directly or indirectly from intermediates of carbohydrate metabolism (Figure 5–22a). Other microbes require that the environment provide some preformed amino acids.

One important source of these precursors (intermediates) used in amino acid synthesis is the Krebs cycle. In the Krebs cycle, adding an amine group to pyruvic acid or an appropriate organic

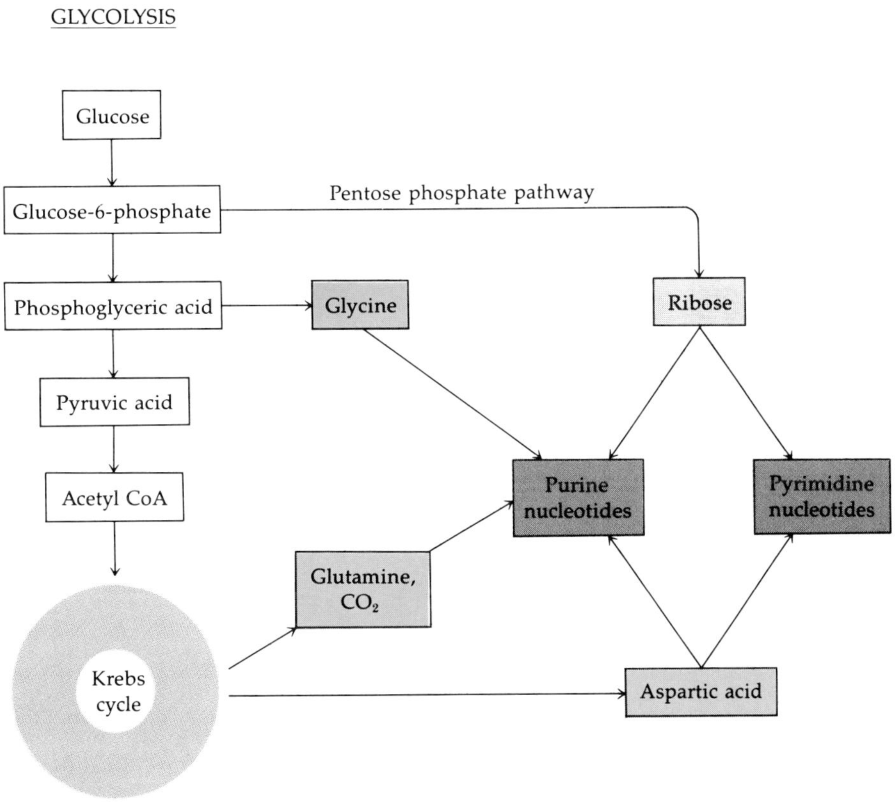

Figure 5–23 Biosyntheses of purine and pyrimidine nucleotides.

acid (*amination*) converts the acid into an amino acid. If this amine group comes from a preexisting amino acid, the process is called *transamination* (Figure 5–22b). Other precursors for amino acid biosynthesis are derived from the pentose phosphate pathway and the Entner–Doudoroff pathway. The main role of amino acids is as building blocks for protein synthesis. The mechanism of protein synthesis will be discussed in Chapter 8.

Biosynthesis of Purines and Pyrimidines

As you may recall from Chapter 2, the informational molecules DNA and RNA consist of repeating units called *nucleotides,* which consist of a purine or pyrimidine, a five-carbon sugar, and a phosphate group. The sugars composing nucleotides are derived from either the pentose phosphate pathway or the Entner–Doudoroff pathway. Certain amino acids—aspartic acid, glycine, and glutamine—play essential roles in the biosyntheses of purines and pyrimidines (Figure 5–23). The carbon and nitrogen atoms derived from these amino acids form the backbones of the purines and pyrimidines. The synthesis of DNA and RNA from nucleotides will be discussed in Chapter 8.

INTEGRATION OF METABOLISM

We have seen thus far that the metabolic processes of microbes produce energy from light, inorganic compounds, and organic compounds. Reactions also occur in which energy is utilized for biosynthesis. With such an array of activity, it might appear that anabolic and catabolic reactions occur independently of each other in space and time.

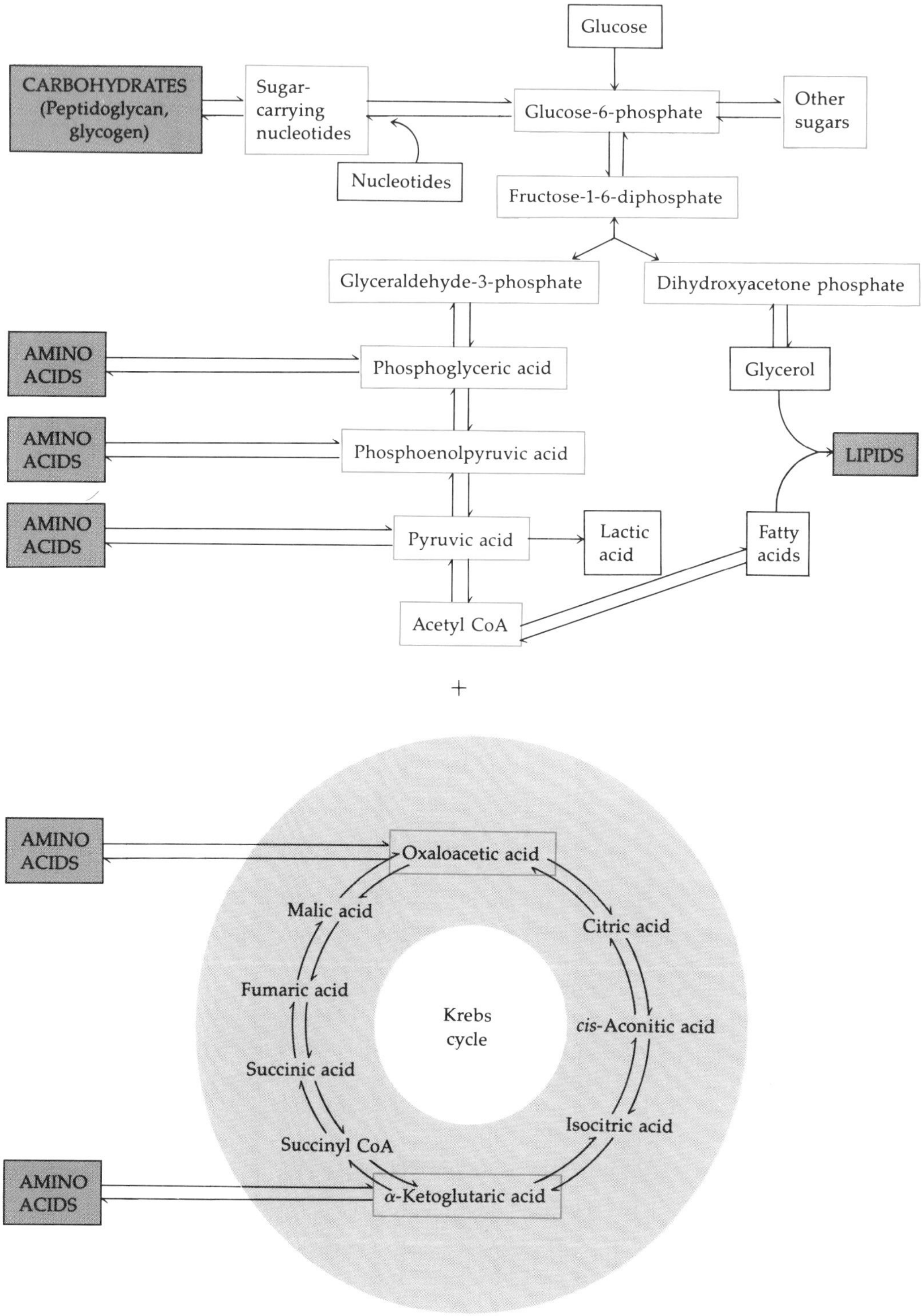

Figure 5–24 Integration of metabolism. Key intermediates are outlined in color. Although not indicated in the figure, amino acids and ribose are used in the synthesis of purine and pyrimidine nucleotides (see Figure 5–23).

Actually, anabolic and catabolic reactions are joined through a group of common intermediates (Figure 5–24). Both anabolic and catabolic reactions also have common biochemical pathways such as the Krebs cycle. Reactions in the Krebs cycle not only form pyruvic acid and acetyl CoA as part of the oxidation of glucose, they also produce intermediates that lead to the synthesis of amino acids, fatty acids, and glycerol. Such pathways are called **amphibolic pathways.**

STUDY OUTLINE

INTRODUCTION (pp. 113–115)

1. The sum of all chemical reactions within a living organism is known as metabolism.
2. Anabolism refers to chemical reactions in which simpler substances are combined to form more complex molecules. Anabolic reactions usually require energy.
3. Catabolism refers to chemical reactions that result in the breakdown of more complex organic molecules into simpler substances. Catabolic reactions usually release energy.
4. The energy of catabolic reactions is used to drive anabolic reactions.
5. The energy for chemical reactions is stored in ATP.

ENZYMES (pp. 115–121)

1. Proteins produced by living cells, enzymes catalyze chemical reactions.
2. Enzymes are generally globular proteins with characteristic three-dimensional shapes.

Parts of an Enzyme (pp. 115–117)

1. Most enzymes are holoenzymes, consisting of a protein portion (apoenzyme) and a nonprotein portion (cofactor or coenzyme).
2. The cofactor can be a metal ion (iron, copper, magnesium, manganese, zinc, calcium, or cobalt) or a complex organic molecule known as a coenzyme (NAD, NADP, FMN, FAD, and coenzyme A).

Mechanism of Enzymatic Action (p. 117)

1. Enzymes catalyze chemical reactions by lowering the activation energy.
2. When an enzyme and substrate combine, the substrate is transformed and the enzyme is recovered.
3. Enzymes are characterized by specificity, a function of their active sites.
4. Enzymes are efficient, able to operate at relatively low temperatures, and subject to various cellular controls.

Naming Enzymes (p. 118)

1. Enzymes' names usually end in *-ase.*

Factors Influencing Enzyme Activity (pp. 118–121)

1. At high temperatures, enzymes undergo denaturation and lose their catalytic properties.
2. The pH at which enzymatic activity is maximal is known as the pH optimum.
3. Within limits, enzymatic activity increases as substrate concentration increases.
4. Competitive inhibitors compete with the normal substrate for the active site of the enzyme. Noncompetitive inhibitors act on other parts of the apoenzyme or on the cofactor and decrease the enzyme's ability to combine with the normal substrate.
5. Feedback inhibition occurs when the end product of a pathway inhibits an enzyme's activity in the pathway.

ENERGY PRODUCTION METHODS (pp. 122–128)

The Concept of Oxidation–Reduction (p. 122)

1. Biological oxidation of a substance means either the addition of oxygen or the removal of a pair of electrons from it (H^+ ions are usually removed with the electrons).
2. Reduction of a substance refers to its gain of a pair of electrons or hydrogen atoms.
3. Each time a substance is oxidized, another is simultaneously reduced.

Generation of ATP: Types of Phosphorylation (pp. 122–123)

1. Energy released during certain metabolic reactions

can be trapped to form ATP from ADP and Ⓟ. Addition of a phosphate to a chemical is called phosphorylation.

2. During substrate-level phosphorylation, a high-energy phosphate group from an intermediate in catabolism is added to ADP.
3. During oxidative phosphorylation, energy is released as electrons are passed to a series of electron acceptors and finally to oxygen or another inorganic compound.
4. During photophosphorylation, energy from light is trapped by chlorophyll, and an electron that passes to a series of electron acceptors is liberated. The electron transfer releases energy used for the synthesis of ATP.

Chemiosmotic Mechanism of ATP Generation (pp. 123–125)

1. Protons being pumped across the membrane generate a proton motive force, as electrons move through a series of acceptors or carriers.
2. Energy produced from movement of the protons back across the membrane is used by ATPase to make ATP from ADP and Ⓟ.

Nutritional Patterns Among Organisms (pp. 125–128)

1. Photoautotrophs use light as an energy source and carbon dioxide as their principal carbon source.
2. Photoheterotrophs use light as an energy source and an organic compound for their carbon source.
3. Chemoautotrophs use inorganic compounds as their energy source and carbon dioxide as their carbon source.
4. Chemoheterotrophs use complex organic molecules as their carbon and energy sources.

Summary of Energy Production Concepts (p. 128)

1. In oxidation–reduction reactions, energy is derived from the transfer of electrons.
2. To produce energy, a cell needs an (organic or inorganic) electron donor, a system of electron carriers, and a final (organic or inorganic) electron acceptor.

BIOCHEMICAL PATHWAYS OF ENERGY PRODUCTION. (p. 129)

1. Enzymatically catalyzed chemical reactions called biochemical pathways release energy from organic molecules.

CARBOHYDRATE CATABOLISM (pp. 130–140)

1. Most of a cell's energy is produced from the oxidation of carbohydrates.
2. Glucose is the most commonly used carbohydrate.

Glycolysis (pp. 130–131)

1. The most common pathway for the oxidation of glucose is glycolysis. Pyruvic acid is the end product.
2. Two ATP and 2 $NADH_2$ molecules are produced from one glucose molecule.
3. Some organisms also use the pentose phosphate pathway or the Entner–Doudoroff pathway to oxidize glucose.

Respiration (pp. 131–135)

1. During respiration, organic molecules are oxidized. Energy is generated from the electron transport chain.
2. The highest energy yield requires the Krebs cycle, which produces CO_2, $NADH_2$, $FADH_2$, and GTP.
3. In aerobic respiration, the Krebs cycle and electron transport chain are used to oxidize a single molecule of glucose to carbon dioxide and water; the yield is 38 ATP molecules. Oxygen is the final electron acceptor.
4. In anaerobic respiration, the final electron acceptor can be nitrate, sulfate, or carbonate. Depending on the organism and other pathways employed, varying amounts of ATP are produced.

Fermentation (pp. 135–138)

1. Fermentation releases energy from sugars or other organic molecules.
2. Oxygen is not required.
3. ATP is produced by substrate-level phosphorylation; the final electron acceptor is an organic molecule.
4. Reduction of pyruvic acid can produce lactic acid or ethanol.

Lipid Catabolism (pp. 138–139)

1. Glycerol is catabolized by conversion to dihydroxyacetone phosphate, and fatty acids are catabolized by beta oxidation.

2. Catabolic products can be further broken down in glycolysis and the Krebs cycle.

Protein Catabolism (pp. 139–140)

1. Before amino acids can be catabolized, they must be converted to various substances that enter the Krebs cycle.
2. Transamination, decarboxylation, and dehydrogenation reactions convert the amino acids to be catabolized.

BIOCHEMICAL PATHWAYS OF ENERGY UTILIZATION (pp. 140–143)

Biosynthesis of Polysaccharides (p. 140)

1. Glycogen is formed from ADPG.
2. UDPG is the starting material for the biosynthesis of peptidoglycan.

Biosynthesis of Lipids (pp. 140–142)

1. Lipids are synthesized from fatty acids and glycerol.
2. Glycerol is derived from dihydroxyacetone phosphate, and fatty acids are built from acetyl CoA.

Biosynthesis of Amino Acids (pp. 142–143)

1. Amino acids are required for protein biosynthesis.
2. All amino acids can be synthesized either directly or indirectly from intermediates of carbohydrate metabolism, particularly from the Krebs cycle.

Biosynthesis of Purines and Pyrimidines (p. 143)

1. The sugars composing nucleotides are derived from either the pentose phosphate pathway or Entner–Doudoroff pathway.
2. Carbon and nitrogen atoms from certain amino acids form the backbones of the purines and pyrimidines.

INTEGRATION OF METABOLISM (pp. 143–145)

1. Anabolic and catabolic reactions are integrated through a group of common intermediates.
2. Such integrated pathways are called amphibolic pathways.

STUDY QUESTIONS

REVIEW

1. Define metabolism.
2. Distinguish between catabolism and anabolism. How are these processes related?
3. Using the diagrams below, show
   (a) Where the substrate will bind
   (b) Where the competitive inhibitor will bind
   (c) Where the noncompetitive inhibitor will bind.

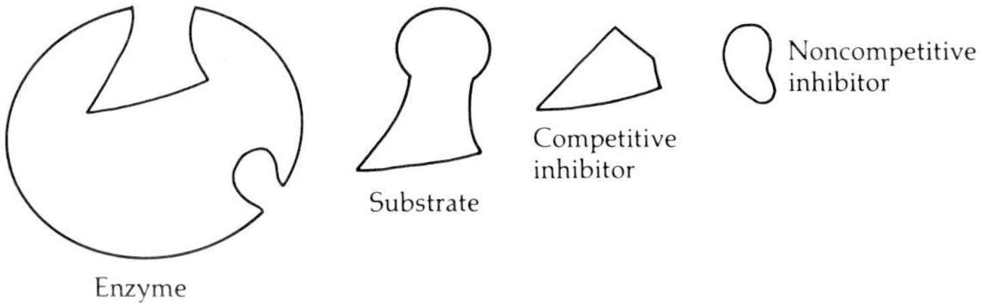

   (d) Which of those four elements could be the inhibitor in feedback inhibition?
4. What will the effect of the reactions in question 3 be?
5. Why are most enzymes active at one particular temperature? Why are enzymes less active below this temperature? What happens above this temperature?

6. Which substances in each of these reactions are being oxidized? Which are being reduced?

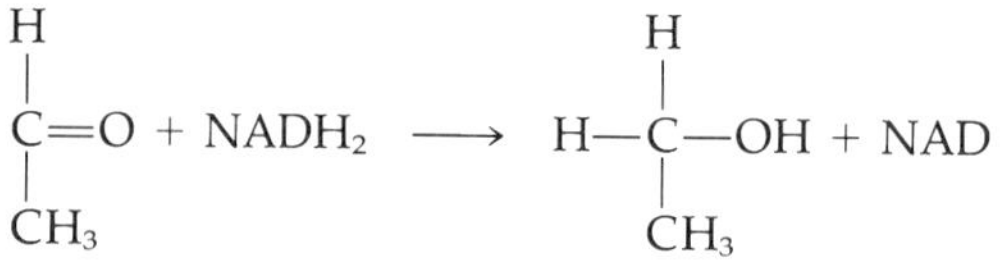

$$\underset{\text{(a) Acetaldehyde}}{\overset{\displaystyle H}{\underset{\displaystyle CH_3}{\overset{|}{\underset{|}{C=O}}}}} + NADH_2 \longrightarrow \underset{\text{Ethanol}}{\overset{\displaystyle H}{\underset{\displaystyle CH_3}{\overset{|}{\underset{|}{H-C-OH}}}}} + NAD$$

$$\underset{\text{(b) Succinic acid}}{\overset{\displaystyle COOH}{\underset{\displaystyle COOH}{\overset{|}{\underset{|}{\overset{H-C-H}{\underset{H-C-H}{|}}}}}}} + FAD \longrightarrow \underset{\text{Fumaric acid}}{\overset{\displaystyle COOH}{\underset{\displaystyle COOH}{\overset{|}{\underset{|}{\overset{C-H}{\underset{C-H}{\|}}}}}}} + FADH_2$$

7. There are three mechanisms for the phosphorylation of ADP to produce ATP. Write the name of the mechanism that describes each of the following reactions.

ATP generated by	Reaction				
	An electron, liberated from chlorophyll by light, is passed down an electron transport chain.				
	Cytochrome c passes two electrons to cytochrome a.				
	$$\underset{\text{Phosphoenolpyruvic acid}}{\overset{\displaystyle CH_2}{\underset{\displaystyle COOH}{\overset{	}{\underset{	}{C-O\sim\text{℗}}}}}} \longrightarrow \underset{\text{Pyruvic acid}}{\overset{\displaystyle CH_3}{\underset{\displaystyle CH_3}{\overset{	}{\underset{	}{C=O}}}}}$$

8. What is the fate of pyruvic acid in an organism that uses aerobic respiration? Fermentation?

9. List four compounds that can be made from pyruvic acid by an organism that uses fermentation only.

10. Fill in the table with the carbon and energy source of each type of organism.

Organism	Carbon Source	Energy Source
Photoautotroph		
Photoheterotroph		
Chemoautotroph		
Chemoheterotroph		

11. Define respiration and differentiate between aerobic and anaerobic respiration.

12. An enzyme and substrate are combined. The rate of reaction begins as shown in this graph. To complete the graph, show the effect of increased substrate concentration on a constant enzyme concentration. Show the effect of increased temperature.

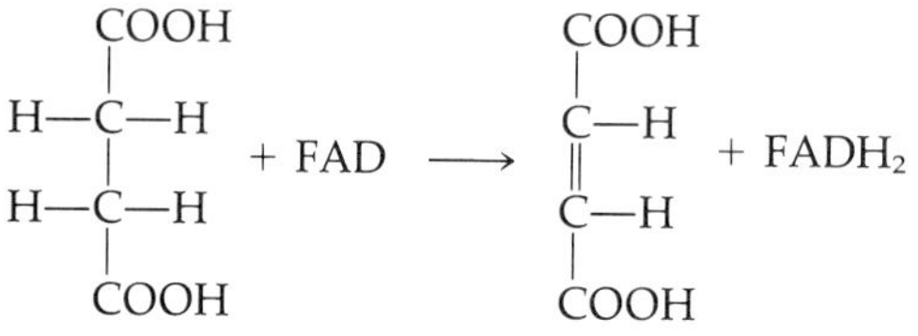

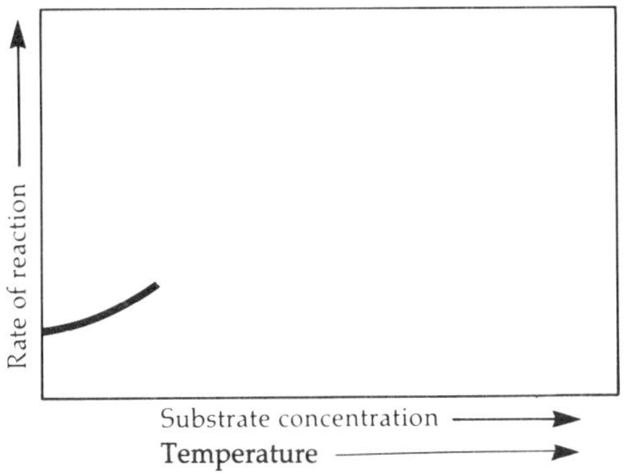

Use the following diagrams for questions 13–17:

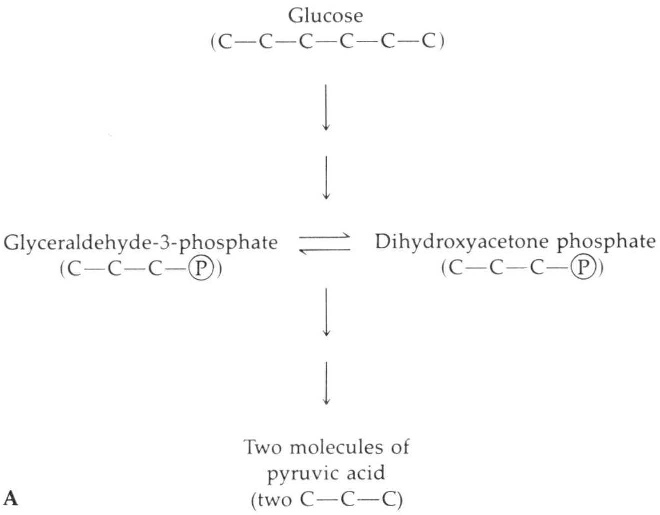

A

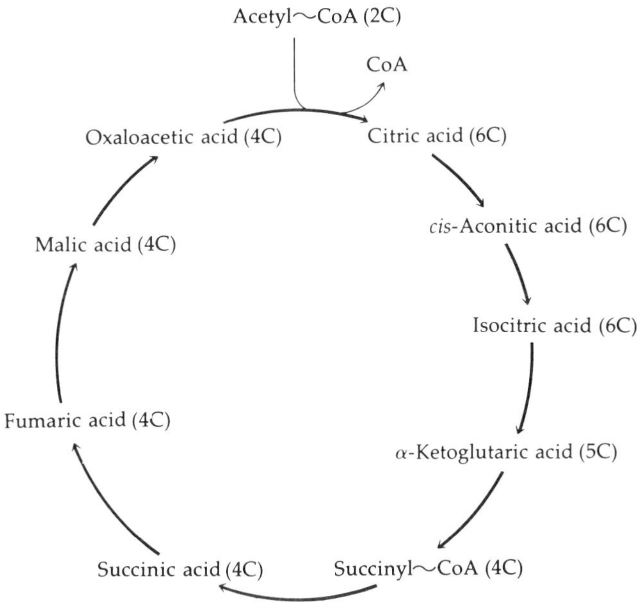

B

13. Name the pathway diagrammed in A. Name pathway B.
14. Glycerol is catabolized by pathway ______________; fatty acids are catabolized by pathway ______________.
15. Identify four places where anabolic and catabolic pathways become integrated.

16. Where is ATP required in pathways A and B?
17. Where is CO_2 released in pathways A and B?
18. The pentose phosphate pathway produces only one ATP. List the cell's four advantages of this pathway.

CHALLENGE

1. Write your own definition of the chemiosmotic mechanism of ATP generation. Look at Figure 5–10 and mark the following. **(a)** The acidic side of the membrane. **(b)** The side with a + electrical charge. **(c)** Sites of potential energy. **(d)** Sites of kinetic energy.
2. Explain why, even under ideal conditions, *Streptococcus* grows slowly.
3. Why must $NADH_2$ be reoxidized? How does this happen in an organism that uses respiration? Fermentation?
4. This graph shows the normal rate of reaction of an enzyme and its substrate (black), and the rate when an excess of competitive inhibitor is present (color). Explain why the graph appears as it does.

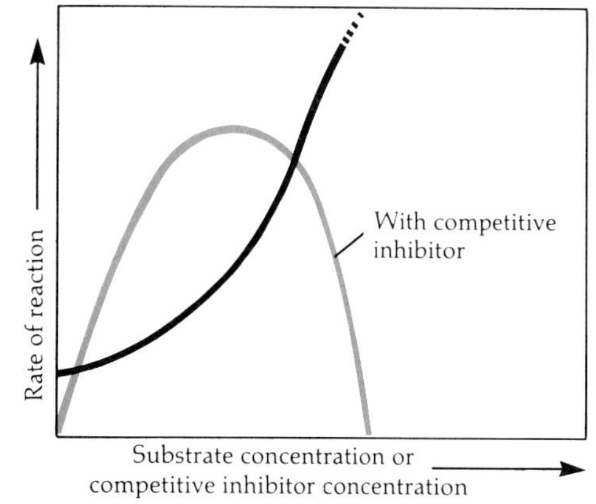

FURTHER READING

Blakemore, R. P. and R. B. Frankel. December 1981. Magnetic navigation in bacteria. *Scientific American* 245:58–65. Describes synthesis and function of magnets in bacterial cells.

Clayton, R. K. and W. R. Sistrom, eds. 1978. *The Photosynthetic Bacteria*. New York: Plenum Press. Discussions of structure and metabolism are included in this treatise.

Doell, H. W., ed. 1974. *Microbial Metabolism*. Stroudsberg, PA: Dowden, Hutchinson, and Ross. A collection of papers on various aspects of metabolism, including fermentation and oxidation of carbohydrates and the metabolism of inorganic compounds.

Glenn, A. R. 1976. Production of extracellular proteins by bacteria. *Annual Review of Microbiology* 30:41–62. A summary of the processes by which exoproteins are synthesized and excreted from cells, and their medical and industrial importance.

Gottschalk, G. 1979. *Bacterial Metabolism*. New York: Springer–Verlag. A detailed description of metabolic pathways in bacteria.

Lehninger, A. L. 1971. *Bioenergetics: The Molecular Basis of Biological Energy Transformations*, 2nd ed. Menlo Park, CA: Benjamin/Cummings. This authoritative reference provides clear explanations of bioenergetics.

Nicholls, D. G. 1982. *Bioenergetics: An Introduction to the Chemiosmotic Theory*. New York: Academic Press. An introductory text on this new theory.

Proton pumping. 1985. *BioScience* 35:14–48. The January issue includes seven articles, with excellent illustrations, describing proton chemistry and photosynthetic, halobacterial, and eucaryotic membrane systems.

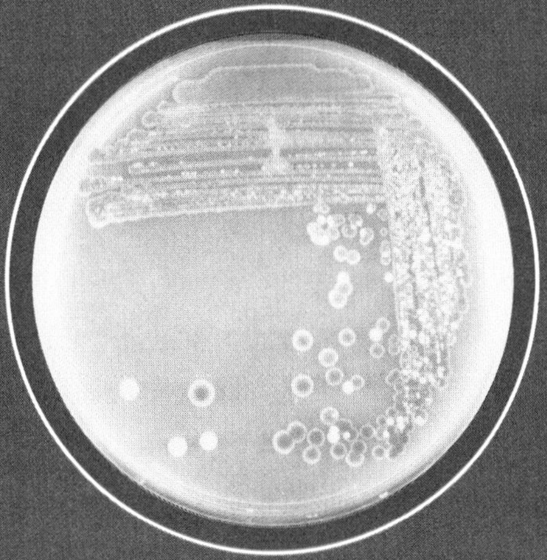

CHAPTER 6

Microbial Growth

OBJECTIVES

After completing this chapter you should be able to

- Define bacterial growth, including transverse fission.
- Classify microbes into three principal groups on the basis of preferred temperature range.
- Explain the importance of pH and osmotic pressure to microbial growth.
- In general terms, list the chemical requirements for growth.
- Explain how microbes are classified on the basis of oxygen requirements.
- Distinguish between chemically defined and complex media.
- Justify the use of the following culture methods: selective and differential media; anaerobic incubators; living host cells; candle jars.
- Describe how pure cultures can be isolated.
- Compare the phases of microbial growth and their relation to generation time.
- Describe several direct and indirect measurements of microbial growth.

When we talk about microbial growth, we are really referring to the *number* of cells, not the *size* of the cells. Microbes that are "growing" are increasing in number, accumulating into clumps of hundreds, colonies of hundreds of thousands, or populations of billions. For the most part, we are not concerned with the size of an individual cell, because it does not vary much during the cell's lifetime. Thus, the total mass of a bacterial population is usually roughly proportional to the number of cells in the population; the rate of increase in a population's mass is often a measure of its reproductive rate.

By understanding the conditions necessary for microbial growth, we can predict how quickly microorganisms will grow in various situations, and determine how to control their growth. Microbial populations can become very large in a very short time, and unchecked microbial growth can cause serious disease and food spoilage.

We will examine in this chapter the physical and chemical requirements for microbial growth, the various kinds of culture media, bacterial division, the phases of microbial growth, and the methods of measuring microbial growth. Once you understand these concepts, you will have a better idea of how microbial growth can be retarded or inhibited.

REQUIREMENTS FOR GROWTH

The requirements for microbial growth can be divided into two categories, physical and chemical. Physical aspects include temperature, pH, and osmotic pressure. Chemical requirements include water, sources of carbon and nitrogen, minerals, oxygen, and organic growth factors.

Physical Requirements

Temperature

Microorganisms grow well at rather ordinary temperatures, not much different from those favored by higher animals. However, certain bacteria are capable of growing in extreme cold or extreme heat, at temperatures that would certainly hinder the survival of most higher organisms.

Microorganisms are divided into three groups on the basis of their preferred range of temperature. These are the **psychrophiles** (cold-loving microbes), **mesophiles** (moderate-temperature-loving microbes), and **thermophiles** (heat-loving microbes). Most bacteria grow only within a limited range of temperatures, and their maximum and minimum growth temperatures are about 30°C apart. They grow poorly at some temperatures within the range and well at others. Each bacterial species grows at particular minimum, optimum, and maximum temperatures. The **minimum growth temperature** is the lowest temperature at which the species will grow. The **optimum growth temperature** is the temperature at which the species grows best. The **maximum growth temperature** is the highest temperature at which growth is possible. By graphing the growth response over a temperature range, we can see that the optimum growth temperature is usually near the top of the range; above that temperature, the rate of growth drops off rapidly (Figure 6–1). This is presumably because the high temperature has inactivated necessary enzymatic systems of the cell.

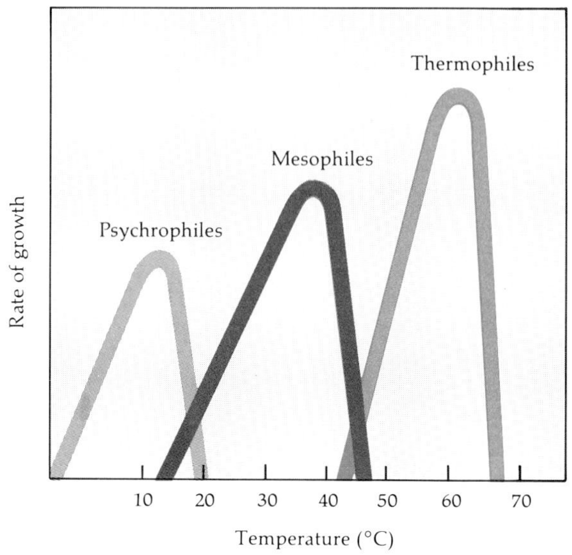

Figure 6–1 Typical responses to temperatures for psychrophile, mesophile, and thermophile growth.

The ranges and maximum growth temperatures that define bacteria as psychrophile, mesophile, or thermophile are not rigidly defined and vary from reference to reference. Psychrophiles, for example, were originally considered simply as organisms capable of growing at 0°C. However, there seem to be two fairly distinct groups of organisms capable of growth at that temperature. It has therefore been suggested that the category of psychrophile be reserved for organisms with an optimum growth temperature of about 15°C and a maximum growth temperature of about 20°C; these requirements would indicate that they cannot grow in reasonably warm rooms (20°C is 68°F). Such organisms are found mostly in the oceans' depths or in certain arctic regions and seldom present concerns in human pursuits, such as the preservation of food. On the other hand, a common and often troublesome group of psychrophiles will grow at refrigerator temperatures, or even below 0°C, but can also grow at temperatures above 20°C. However, few will grow well at temperatures above 30°C, and their optimum growth temperatures tend to be relatively low (often in the upper teens or low twenties). It has been proposed that these organisms be called **psychrotrophs,** and many sources use this term to describe bacteria capable of growth at a common refrigerator temperature (4°C). Other sources apply the term psychrophile to both types of organisms: those that grow best between 15°C and 20°C, and those whose range is greater, between 0°C and 30°C.

Refrigeration is the most common method of preserving household food supplies. It is based on the principle that microbial reproductive rates decrease at low temperatures. Although microbes usually survive even subfreezing temperatures (they might become entirely dormant), they gradually decline in numbers; some species decline faster than others. Psychrophiles, or psychrotrophs, actually do not grow well at low temperatures, except in comparison with other organisms, but given time they are able to slowly deteriorate the quality of food. This deterioration might take the form of mold mycelium, slime on food surfaces, or off tastes or colors in foods. The temperature inside a properly set refrigerator will entirely prevent the growth of most spoilage organisms and prevent growth of almost all the pathogenic bacteria.

The mesophile, with an optimum growth temperature of between 25° and 40°C, is the most common type of microbe. Organisms that have adapted to live in the bodies of higher animals usually have an optimum temperature close to that of their host. The optimum temperature for many pathogenic bacteria is about 37°C, and incubators for clinical cultures are usually set at about this temperature. The mesophiles include most of our common spoilage and disease organisms.

A very interesting group of microorganisms are the thermophiles, those capable of growth at high temperatures. Many of these organisms have an optimum growth temperature of between 50° and 60°C. This is about the temperature of the water from a hot-water tap. Such temperatures can also be reached in soil on which sunlight falls and in thermal waters, such as the hot springs in Yellowstone Park. Some of these organisms grow at temperatures well above 90°C, near the boiling point of water. Remarkably, many thermophiles are incapable of growth at temperatures below about 45°C. (Thermophilic fungi usually have an upper limit of 55° to 60°C, somewhat lower than that of most thermophilic bacteria.) Endospores formed by thermophilic bacteria are usually heat resistant. They survive the usual heat treatment given canned goods, but will not grow at normal storage temperatures. Although elevated temperatures may cause surviving endospores to germinate and grow, these thermophilic bacteria are not a public health problem. Thermophiles are also important in organic compost piles, in which the temperature can rise rapidly to 50° or 60°C.

pH

As you recall from Chapter 2, pH refers to the acidity or alkalinity of a solution. Most bacteria grow best in a narrow range of pH near neutrality, between pH 6.5 and 7.5. Very few bacteria grow at an acid pH below about 4.0. That is the reason why a number of foods, such as sauerkraut, pickles, and many cheeses, are preserved by the acids of bacterial fermentation. Nonetheless, some bacteria are remarkably tolerant of acidity. One autotrophic bacterium, which was found in the drainage water from coal mines and which oxidizes sulfur to form sulfuric acid, can survive at a pH of 1. Molds and yeasts will grow over a greater pH range than bacteria will, but the optimum pH of molds and yeasts is generally below that of bacteria, usually about pH 5 to 6. Alkalinity also inhibits microbial growth, but is rarely used to preserve foods.

When bacteria are cultured in the laboratory, they produce acids that can interfere with desired bacterial growth. To neutralize the acids, chemicals called **buffers** are included in the growth medium. The peptones and amino acids in some media act as buffers, and many media also contain phosphate salts. Phosphate salts have the advantage of exhibiting their buffering effect in the pH growth range of most bacteria. They are also nontoxic; in fact, they provide an essential nutrient element.

Osmotic pressure

Microbes require water for growth and are actually about 80 to 90% water. When a microbial cell is in solution, and the concentration of solutes is higher

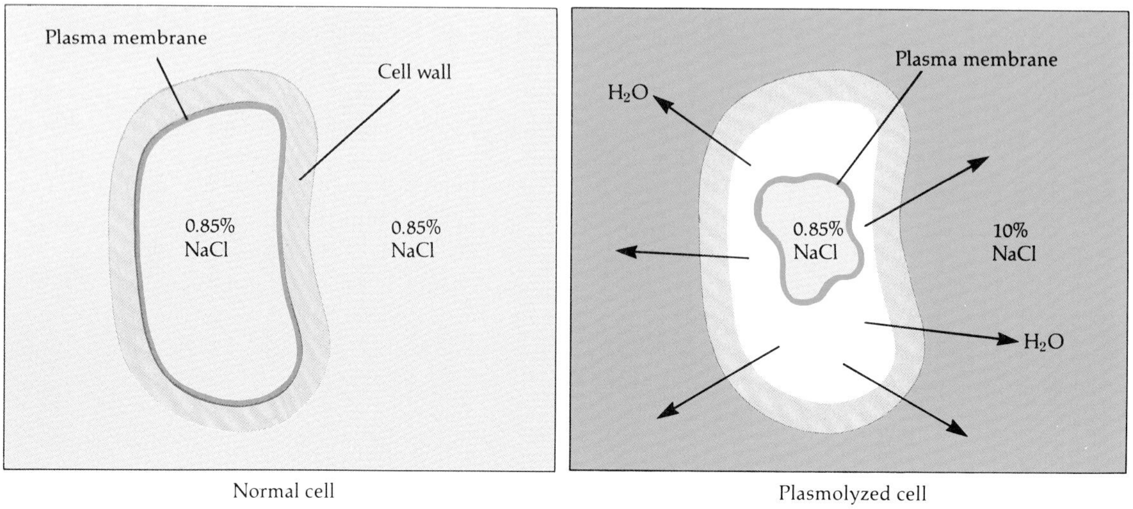

Figure 6–2 Plasmolysis. When the concentration of solutes in the medium exceeds the concentration within the cell (hypotonic), the cellular water leaves the cell by osmosis and the plasma membrane shrinks away from the cell wall.

in the solution than in the cell, the cellular water passes through the cytoplasmic membrane to the high salt concentration. (See the discussion of osmosis in Chapter 4.) This osmotic loss of water is called **plasmolysis** (Figure 6–2). What is so important about this phenomenon is that the growth of the cell is inhibited as the cytoplasmic membrane pulls away from the cell wall. Thus, the addition of salts (or other solutes) to a solution, and the resulting increase in osmotic pressure, can be used to preserve foods. Salted fish, honey, and sweetened condensed milk are preserved largely by this mechanism; the high salt or sugar concentrations draw water out of any microbial cells that are present, and thus prevent their growth.

Some bacteria have adapted so well to an environment of high salt concentrations that they actually require them for growth. These bacteria are called **extreme halophiles.** To grow, isolates from such saline waters as the Dead Sea often require nearly 30% salt, and the inoculating loop used to transfer them must first be dipped into a saturated salt solution. More common are **facultative halophiles,** which do not require high salt concentrations but are able to grow at salt concentrations up to 2%, a concentration that inhibits the growth of many other bacteria. A few species of facultative halophiles can even tolerate 15% salt.

The concentration of agar (a complex polysaccharide isolated from algae) used to solidify microbial growth media is usually about 1.5%. If markedly higher concentrations are used, the growth of some bacteria can be inhibited by the increased osmotic pressure. Because the effects of osmotic pressure are roughly related to the numbers of molecules in a volume of solution, a low-molecular-weight compound such as sodium chloride has a greater antimicrobial effect than does, for example, sucrose, which has a somewhat higher molecular weight.

Chemical Requirements

Carbon

Besides water, one of the most important requirements for microbial growth is a source of carbon, needed for all the organic compounds that make up a living cell. Carbon is the structural backbone of living matter. As we saw in Chapter 2, its four valences allow it to be used in constructing extended and complicated organic molecules (see

Making Food Without Light—Chemosynthesis on the Ocean Floor

One of the earth's most extreme environments exists under one-and-a-half miles of water, at the cold and completely dark bottom of the ocean. Water at this depth contains less than half the oxygen concentration of water near the surface. Some organic matter, such as dead fish and plankton, makes its way to the bottom. However, until humans actually explored the ocean bottom, it was believed that only a few organisms could grow under these harsh conditions. In 1977, the first manned vehicle able to penetrate to the ocean's bottom carried two scientists 2600 meters below the ocean's surface at the Galápagos Rift, about 350 km northeast of the Galápagos Islands. There, amid the vast expanse of barren basalt pillows, the scientists found unexpectedly rich oases of life.

Ecosystem of the Hydrothermal Vents

At the surface of the world's oceans, and in most habitats on earth, ecosystems depend on photosynthetic organisms, such as bacteria, plankton, and green plants, to harness the sun's energy and make carbohydrates. At the bottom of the ocean, where no light penetrates, the so-called primary producers must be nonphotosynthetic. Many are chemoautotrophic bacteria, which do not require preformed organic molecules for nourishment.

(continued on next page)

Chemoautotrophs use inorganic compounds, such as hydrogen sulfide (H_2S), as a source of energy and carbon dioxide as their principal source of carbon.

Along major fractures of the earth's crust, where crustal plates are slowly separating, superheated water (water hotter than 100 C) rises through the sea floor at areas called hydrothermal vents. Because at a depth of 2600 meters the boiling point of water is 450 C, the water does not vaporize, but rises as high as 5 meters in a superhot pressurized plume. The superheated water reacts with surrounding rock and dissolves metal ions, sulfides, and carbon dioxide (CO_2). The subsequent chemical reactions release H^+, which makes the plume of water acidic. Due to these reactions, the concentration of sulfide (S^{2-}) is three times greater than the concentration of molecular oxygen in and around the vents. The ecosystems of these vents depend on the abundance of sulfur compounds.

Mats of bacteria grow along the sides of the vents, where temperatures can be as high as 350 C, the highest temperature that any organism is known to tolerate. Above the vent, where temperatures are around 20 C, the concentration of bacteria is about four times greater than in the surrounding water, and the growth rate of the bacteria is equal to that found in productive, sunlit coastal waters. The bacteria in and around the vent create an environment in which other consumers and decomposers, such as clams, mussels, and worms, can exist.

A Rare Endosymbiosis

One of the most astonishing discoveries in the hydrothermal ecosystem is of animals that live with some species of chemoautotrophic bacteria in mutually beneficial relationships, called endosymbioses. One such relationship involves *Riftia,* a worm that can grow up to 1 meter in length and has no mouth or gut. The photo shows hundreds of such tube worms at a hydrothermal vent on the Galapagos Rift. *Riftia* exists in a situation parallel to those few endosymbiotic relationships involving photosynthetic bacteria or algae—the host receives food from its parasite. *Riftia's* tentacles absorb metal ions, sulfides, CO_2, and O_2 from the plume of hot water. These elements are transported through the worm's circulatory system so that the bacteria throughout its body are supplied with O_2 and H_2S. (Although hydrogen sulfide is usually toxic to animals, *Riftia* appears to have in its blood a protein that binds the sulfide so that it cannot inhibit metabolism.) The bacteria fix the carbon from CO_2 into organic compounds using energy obtained from oxidation of sulfide to sulfate: $CO_2 + H_2S \rightarrow$ carbohydrate + SO_4. These carbohydrates are food for *Riftia.* Thus, *Riftia* obtains its energy ultimately not from the sun but from chemical synthesis—the oxidation of H_2S by its parasite bacteria.

Figures 2–9 and 2–10). One-half of the dry weight of a typical bacterial cell is carbon.

Chemoheterotrophs get most of their carbon from the source of their energy, namely, organic materials such as proteins, carbohydrates, and lipids. Chemoautotrophs derive some carbon from carbon dioxide. And photoautotrophs get their carbon entirely from carbon dioxide.

Because the utilization of carbon is such a basic trait of life on Earth, one way to test for life on another planet is to look for a sign that carbon is being metabolized. If carbon compounds are being either formed or broken down, then living organisms might be at work.

Nitrogen, sulfur, and phosphorus

In addition to carbon, other important elements are needed by microbes for the synthesis of cellular material. For example, protein synthesis requires considerable amounts of nitrogen, as well as some sulfur. The syntheses of DNA and RNA also require nitrogen and some phosphorus, as does the synthesis of ATP, the molecule so important for storage and transfer of chemical energy within the cell. Together, nitrogen, sulfur, and phosphorus constitute about 18% of the dry weight of the cell; of that, 15% is nitrogen. Important natural sources of sulfur include sulfate ion (SO_4^{2-}), hydrogen sulfide (H_2S), and the sulfur-containing amino acids. The box on page 155 describes an unusual ecosystem that is based on a supply of H_2S that would be toxic to most organisms. An important source of phosphorus is phosphate ion (PO_4^{3-}).

Nitrogen is used primarily to form the amino group of the amino acids of proteins. Many bacteria meet this requirement by decomposing proteinaceous material and reincorporating the amino acids into newly synthesized proteins and other nitrogen-containing compounds. Other bacteria use nitrogen from ammonium ions (NH_4^+), which are already in the reduced form and are usually found in organic cellular material. Still other bacteria are able to derive nitrogen from nitrates (compounds that dissociate to the nitrate ion, NO_3^-, in solution).

Some important bacteria and cyanobacteria (blue-green algae) use gaseous nitrogen (N_2) directly from the atmosphere. The process by which they absorb N_2 is called **nitrogen fixation.** Some organisms that can use this method are free-living, mostly in the soil, but others live in symbiosis with the roots of certain plants. *Symbiosis* is a close association of two or more different kinds of organisms. The most important symbiotic nitrogen-fixing bacteria belong to the genus *Rhizobium* and are associated with legumes such as clover, soybeans, alfalfa, beans, and peas. The nitrogen fixed in the symbiosis is used by both the plant and the bacterium. This reaction greatly increases soil fertility and is therefore very important in agriculture.

Trace elements

Microbes require very small amounts of other mineral elements such as iron, copper, molybdenum, and zinc; these are referred to as **trace elements.** Although these components are sometimes added to a laboratory medium, they are usually assumed to be naturally present in tap water or other components of media. Even most distilled waters contain adequate amounts, but tap water is sometimes specified to ensure that these minerals will be present in culture media.

Oxygen

We are accustomed to thinking of oxygen as a necessity of life, but it is actually a rather corrosive gas and did not exist in the atmosphere during most of our planet's history. It is speculated that life could not have first arisen if oxygen had been present. However, many current forms of life have metabolic systems that require oxygen (for aerobic respiration). As we have seen, the oxygen eventually combines with hydrogen atoms stripped from organic compounds and forms water. This process yields a lot of energy, while neutralizing a potentially toxic gas—a very neat solution all in all. When oxygen is not available, the yield is much lower, because the organic material is only partially degraded to compounds such as acids or alcohols. These compounds still contain considerable amounts of unreleased chemical energy.

Microbes that use oxygen, called **aerobes,** produce more energy from nutrients than do microbes that do not use oxygen. Organisms that require oxygen to live are called **obligate aerobes** (Figure 6–3a).

Obligate aerobes are at a disadvantage in that oxygen is poorly soluble in water, and much of the environment is oxygen poor. Therefore, many of the aerobic bacteria have developed, or retained, the ability to continue growing in the absence of oxygen. Such organisms, including some that exist in your intestinal tract, are called **facultative anaerobes** (Figure 6–3b). In other words, facultative anaerobes can use oxygen when it is present, but are able to continue growth by using fermentation or anaerobic respiration when oxygen is not available. However, their efficiency in producing energy decreases in the absence of oxygen. Examples of facultative anaerobes include the familiar *Escherichia coli* and many yeasts.

Obligate anaerobes (Figure 6–3c) are bacteria that are unable to use molecular oxygen (O_2) for energy-yielding reactions. In fact, most are harmed by it. When such a bacterium has an electron transport chain, it cannot use molecular oxygen as a final electron acceptor. These bacteria do use oxygen atoms in cellular materials; the atoms are usually obtained from water.

Understanding how organisms can be harmed by oxygen requires a brief discussion of the toxic forms of oxygen. They are as follows:

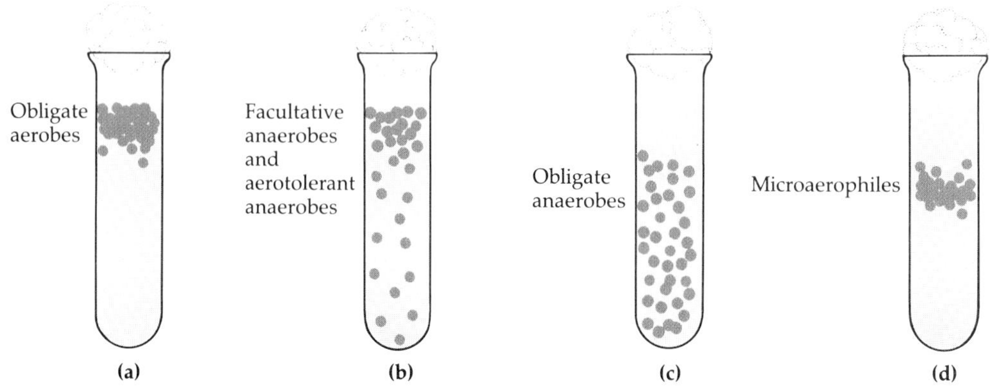

Figure 6–3 Effect of oxygen concentration on the growth of various bacteria in a tube of solid medium. Because oxygen diffuses only a limited distance from the atmosphere into the solid medium, the amount of oxygen available to the bacteria varies.

1. **Singlet oxygen** is normal molecular oxygen (O_2) that has been boosted into a higher energy state and is extremely reactive. It is formed most commonly by the action of visible light. Present in phagocytic cells (Chapter 15), it plays a role in the destruction of foreign cells after they are ingested by the phagocytes.

2. **Superoxide free radicals** (O_2^-) are formed in small amounts during the normal respiration of organisms that use oxygen as a final electron acceptor in water production. In the presence of oxygen, obligate anaerobes, such as bacteria of the genus *Clostridium* (klôs-tri'dē-um), also appear to form some superoxide free radicals. These radicals are so toxic to cellular components that all organisms attempting to grow in atmospheric oxygen must produce an enzyme, **superoxide dismutase,** to neutralize the radicals. Aerobic bacteria, facultative anaerobes growing aerobically, and aerotolerant anaerobes (discussed later) produce this enzyme (Table 6–1). Using superoxide dismutase, they convert the superoxide free radical into molecular oxygen (O_2) and hydrogen peroxide (H_2O_2). The H_2O_2 is then immediately converted into oxygen and water by another enzyme, **catalase.** Catalase is easily detected by its action on hydrogen peroxide. When a drop of hydrogen peroxide is added to a suspension of bacterial cells with catalase, oxygen bubbles are released. This result is designated as catalase+. Anyone who has put hydrogen peroxide on a wound will recognize that cells in human tissue also contain catalase. The actions of the two enzymes, catalase and superoxide dismutase, neutralize toxic forms of oxygen and permit the cell to grow while using oxygen as a final electron acceptor. Another enzyme that breaks down hydrogen peroxide is **peroxidase;** it differs from catalase in that its reaction does not produce oxygen.

Table 6–1 Uses of Oxygen-Related Enzymes by Bacteria

Type of Microbe	Catalase	Superoxide dismutase
Aerobes and facultative aerobes	+	+
Aerotolerant anaerobes	–	+
Obligate anaerobes	–	–

3. **Peroxide** (O^{2-}) is a toxic form of oxygen that is probably generated in small amounts during normal respiration. But, it is better known as the active principle of the antimicrobials hydrogen peroxide and benzoyl peroxide; these are discussed in the next chapter.

4. The **hydroxyl free radical** (OH·) is another intermediate form of oxygen and probably the most reactive. It is formed in the cellular cytoplasm by ionizing radiation. It is also produced by a reaction between superoxide free radicals and peroxide—so most aerobic respiration produces some hydroxyl radicals.

Obligate anaerobes, such as most members of the genus *Clostridium,* produce neither superoxide dismutase nor catalase. Therefore, they are extremely sensitive to oxygen, since the presence of O_2 probably leads to an accumulation of superoxide free radicals in the cell's cytoplasm. The ability to form endospores has probably aided the survival of these oxygen-sensitive bacteria.

Aerotolerant anaerobes cannot use oxygen for growth, but tolerate it fairly well. On the surface of a solid medium, they will grow without the special techniques required by anaerobes that are less oxygen-tolerant (Figure 6–3b). Many of the aerotolerant bacteria characteristically ferment carbohydrates to lactic acid. As lactic acid accumulates, it inhibits the growth of aerobic competitors, and establishes an ecological niche for the lactic acid producers. A common example of lactic acid-producing aerotolerant anaerobes are the lactobacilli used in the production of many acidic fermented foods, such as pickles and cheese. In the laboratory, they are handled and grown much like any other bacterium, but they make no use of the oxygen in the air. These bacteria can tolerate oxygen because they possess superoxide dismutase (See Table 6–1), or an equivalent system, that neutralizes the toxic forms of oxygen discussed above.

A few bacteria are **microaerophilic,** meaning that they grow only in oxygen concentrations lower than those of air. However, they are aerobic in that they require oxygen. In a test tube of solid nutrient medium, they grow only at a depth where small amounts of oxygen have diffused into the medium; they do not grow at the oxygen-rich surface (Figure 6–3d). This limited tolerance is probably due to their sensitivity to superoxide free radicals and peroxides, which they produce in lethal concentrations under oxygen-rich conditions.

Organic growth factors

Organic growth factors are essential organic compounds that the organism is unable to synthesize; they must be directly obtained from the environment. A group of organic growth factors for humans is vitamins. Most vitamins function as coenzymes, the accessories required by certain enzymes in order to function. Many bacteria can synthesize all their own vitamins and are not dependent on outside sources. However, to bacteria that lack the enzymes needed for the synthesis of certain vitamins, those vitamins are organic growth factors. Other organic growth factors required by some bacteria are amino acids, purines, and pyrimidines.

CULTURE MEDIA

A nutrient material prepared for the growth of microorganisms in a laboratory is called a **culture medium.** Some bacteria can grow well on just about any culture medium, others require special media, and still others cannot grow on any nonliving medium yet developed. The microbes that grow and multiply in or on a culture medium are referred to as a **culture.**

Suppose we want to grow a culture of a particular microorganism, perhaps the microbes from a particular clinical specimen. What criteria must the culture medium meet? First of all, it must contain the right nutrients for the particular microorganism we want to grow. It should also contain sufficient moisture and oxygen, and a properly adjusted pH. So that the culture will contain only the microorganisms we add to the medium (and their offspring), the medium must initially be **sterile;** that is, it must initially contain no living microorganisms. Finally, the growing culture should be incubated at the proper temperature.

There is a wide variety of media available for the growth of microorganisms in the laboratory. Most of these, which are available from commercial sources, have premixed components and require only the addition of water and sterilization. For media used in the isolation and identification of bacteria interesting to researchers in such areas as food, water, and clinical microbiology, there are constant revisions of old formulas and introductions of new ones.

When it is desirable to grow bacteria on a solid medium, a solidifying agent such as agar is added to the medium. **Agar** is a complex polysaccharide derived from a marine alga, and it has had a long history as a thickener in foods such as jellies, soups, and ice cream. In fact, it is still used for that purpose.

Agar has some very important properties that make it valuable to microbiology, and no really satisfactory substitute has ever been found. Few microbes can degrade agar, so it remains a solid. Also important is the fact that it melts at about the boiling point of water but remains in a liquid state until the temperature drops to about 40°C. For laboratory use, agar is therefore held in water baths at about 50°C. At this temperature, it does not injure bacteria when it is poured over a bacterial inoculum in a Petri plate. Bacterial suspensions can also be mixed with the melted agar in the water bath so that a uniform suspension of bacteria is made; this can be used for tests of antibiotic susceptibility and other purposes. Once the agar has solidified, it can be incubated at temperatures approaching 100°C before it liquefies; this property is particularly useful when thermophilic bacteria are being used.

Chemically Defined Media

When a medium is being prepared for microbial growth, consideration must be given to the providing of an energy source, as well as sources of carbon, nitrogen, sulfur, phosphorus, and any necessary growth factors, which the organism is unable to synthesize. A **chemically defined medium** is one in which the exact chemical composition is known. Table 6–2 shows a chemically defined medium used for the growth of an organism capable of extracting energy from the oxidation of ammonium ions to nitrite ions.

As Table 6–3 demonstrates, many organic growth factors must be provided in the chemically defined medium used to cultivate species of *Lactobacillus*. Organisms that require many growth factors are described as *fastidious*. Organisms of this type, such as *Lactobacillus*, are sometimes used in tests that determine (assay) the concentration of a

Table 6–2 Chemically Defined Medium for the Growth of a Chemoautotrophic Bacterium That is Able to Use Ammonium Ions for Energy

Constituent	Concentration (Grams/Liter)
$(NH_4)_2SO_4$	0.5
$NaHCO_3$*	0.5
Na_2HPO_4	13.5
KH_2PO_4	0.7
$MgSO_4 \cdot 7H_2O$	0.1
$FeCl_3 \cdot 6H_2O$	0.014
$CaCl_2 \cdot 2H_2O$	0.18
Water	1000 ml (= 1 liter)

*The $NaHCO_3$ is a source of carbon dioxide (the carbon source) in solution.

Source: From D. Pramer and E. L. Schmidt, *Experimental Soil Microbiology*, Burgess Publishing Company, Minneapolis, MN, 1964.

Table 6–3 Chemically Defined Medium for the Growth of a Fastidious Heterotrophic Bacterium, Such as *Lactobacillus* Species

Constituent	Concentration (Grams/Liter)
Carbon and Energy Sources:	
Glucose	10.0
Sodium acetate	6.0
Salts (added as two solutions of mixed salts):	
NH_4Cl	2.5
$(NH_4)_2SO_4$	2.5
KH_2PO_4	0.50
K_2HPO_4	0.50
NaCl, $FeSO_4$, and $MnSO_4$, each	0.005
Amino Acids:	
Arginine, cystine, glycine, histidine, hydroxyproline, proline, tryptophan, and tyrosine, each	0.20
Alanine, aspartic and glutamic acids, isoleucine, leucine, lysine, methionine, norleucine, phenylalanine, serine, threonine, and valine, each	0.10
Glutamine	0.025
Purines:	
Adenine, guanine, and uracil, each	0.010
Vitamins:	
Pyridoxamine-HCl	0.0004
Riboflavin, thiamine, niacin, and pantothenic acid, each	0.0002
Para-aminobenzoic acid	0.00004
Folic acid	0.00002
Biotin	0.0000002
Water	1000 ml (= 1 liter)

Source: From K. Thimann, *The Life of Bacteria*, 2nd ed., The Macmillan Company, New York, 1963.

Table 6–4 Chemically Defined Medium for the Growth of a Less Fastidious Heterotroph, Such as *Escherichia coli*

Constituent	Concentration (Grams/Liter)
Glucose	5.0
$NH_4H_2PO_4$	1.0
NaCl	5.0
$MgSO_4 \cdot 7H_2O$	0.2
K_2HPO_4	1.0
Water	1000 ml (= 1 liter)

Table 6–5 Composition of Nutrient Agar, a Complex Medium for the Growth of Heterotrophic Bacteria

Constituent	Concentration (Grams/Liter)
Peptone	5.0
Beef extract	3.0
Sodium chloride	8.0
Agar	15.0
Water	1000 ml (= 1 liter)

particular vitamin in a substance. For such *microbiological assays* the growth medium to which the test substance and the bacterium will be added contains all the growth requirements of the bacterium except the vitamin being assayed. When the medium, test substance, and bacterium are combined, the growth of bacteria will be proportional to the amount of vitamin in the test substance. This growth is reflected by the amount of lactic acid produced. The more lactic acid, the more the *Lactobacillus* has been able to grow, so the more vitamin is present.

A much simpler chemically defined medium is required for the growth of *Escherichia coli,* a less fastidious bacterium. The composition of this medium is shown in Table 6–4. The simple composition reflects the fact that *E. coli* has a greater synthesizing ability than does the *Lactobacillus* species.

Complex Media

Most heterotrophic bacteria and fungi are routinely grown on **complex media,** media for which the exact chemical composition varies slightly from batch to batch. These complex media are made up of nutrients such as extracts from yeasts, beef, or plants, or digests of proteins from these and other sources. Table 6–5 gives one widely used recipe.

In complex media such as this one, the energy, carbon, nitrogen, and sulfur requirements of the growing microorganisms are met largely by protein. Protein is a large, relatively insoluble molecule that few microorganisms can utilize directly. But a partial digestion by acids or enzymes reduces the protein to shorter chains of amino acids, called *peptones.* These small, soluble fragments can be digested by the bacteria.

The vitamins and other organic growth factors are provided by meat extracts or yeast extracts. The soluble vitamins and minerals from the meats or yeasts are dissolved in the extracting water, which is then evaporated so that these factors are concentrated. (These extracts also supplement the organic nitrogen and carbon compounds.) Yeast extracts are particularly rich in the B vitamins. If this type of medium is in liquid form, it is called **nutrient broth.** When agar is added, it is called **nutrient agar.** (This terminology sometimes confuses students into thinking, wrongly, that agar itself is a nutrient—it is not.)

Anaerobic Growth Media and Methods

The cultivation of anaerobic bacteria poses a special problem. Because anaerobes might be killed by exposure to oxygen, special media called **reducing media** must be used. These media contain ingredients such as sodium thioglycolate that chemically combine with dissolved oxygen and deplete the oxygen of the culture medium. To routinely grow and maintain pure cultures of obligate anaerobes, microbiologists use reducing media stored in ordinary, tightly capped test tubes. These media are heated shortly before use, so that any absorbed oxygen is driven off.

When the culture must be grown in Petri plates so that individual colonies can be observed, special jars that hold several Petri plates in an oxygen-free atmosphere are used. The oxygen is removed by this process: a packet of chemicals (sodium bicarbonate and sodium borohydride) in the jar is moistened with a few milliliters of water and then the jar is sealed. Hydrogen and carbon dioxide are produced by the reaction of the chemicals with the water. A palladium catalyst in the jar com-

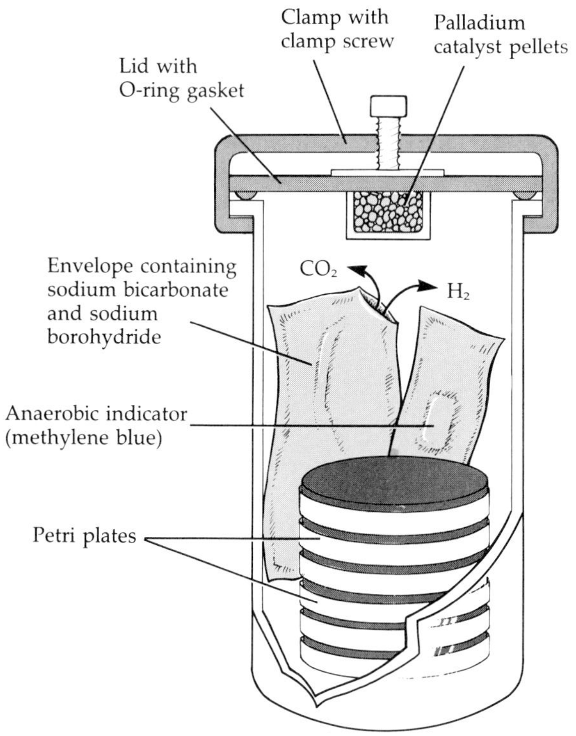

Figure 6–4 Anaerobic container used in cultivation of anaerobic bacteria on Petri plates. When water is mixed with the chemical packet containing sodium bicarbonate and sodium borohydride, hydrogen and carbon dioxide are generated. Reacting on the surface of a palladium catalyst in a screened reaction chamber, the hydrogen and the atmospheric oxygen in the jar combine to form water. The oxygen is thus removed. An anaerobic indicator is also in the container. It contains methylene blue, which is blue when oxidized, and which turns colorless when the oxygen is removed from the container.

bines the oxygen in the jar with the hydrogen produced by the chemical reaction, and water is formed. As a result, the oxygen disappears in a short time. Moreover, the carbon dioxide produced aids the growth of many anaerobic bacteria (Figure 6–4).

Researchers working with anaerobes on a regular basis use more elaborate techniques. Anaerobic glove boxes, which are small, transparent chambers equipped with air locks and filled with inert gases, are sometimes used. The technicians are able to manipulate the equipment by inserting their hands in airtight rubber gloves fitted to the wall of the chamber (Figure 6–6 and Microview 3–3).

Colonies of anaerobes can also be grown in deep test tubes instead of Petri plates. The natural atmosphere in the test tube is replaced with an inert gas such as nitrogen. The bacterial inoculum is then mixed with melted nutrient medium in the tube, and the tube, tightly capped, is rolled on horizontal rollers. The nutrient medium solidifies against the interior walls of the test tube, and colonies that appear there can be counted and picked much as can be done on a shallow layer of agar in a Petri plate.

Special Culture Techniques

Some bacteria have never been successfully grown on artificial laboratory media. *Mycobacterium leprae* (lep'rē), the leprosy bacterium, was grown on the foot pads of mice, and is now usually grown in

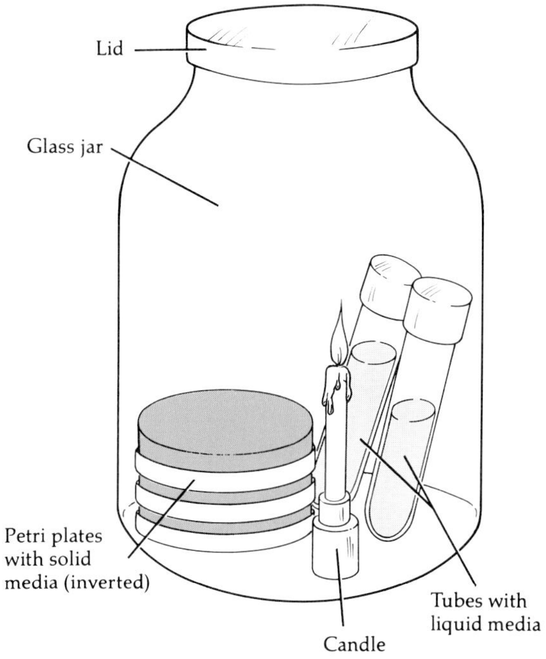

Figure 6–5 Candle jar technique. Plates and tubes inoculated with *Neisseria meningitidis* are placed in a jar with a lighted candle. The cap is placed on the jar. The candle will extinguish when the atmosphere contains approximately 10% carbon dioxide.

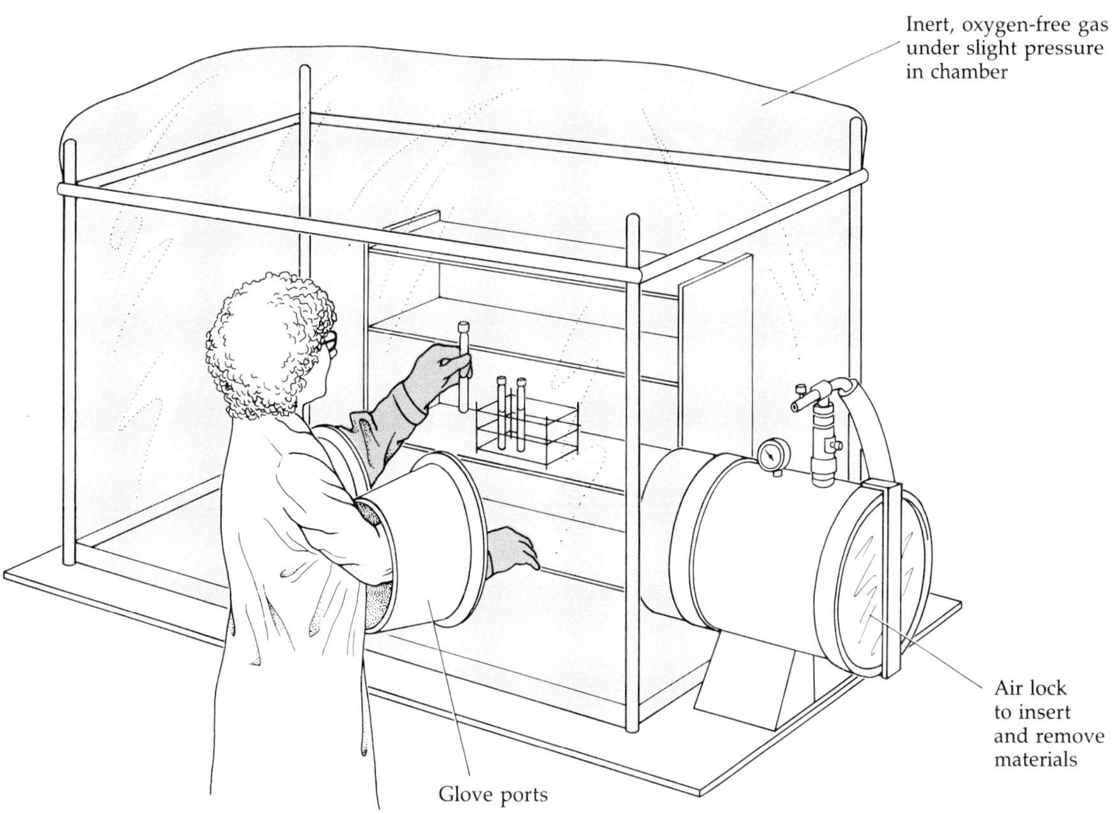

Figure 6–6 Anaerobic chamber. Microbiologists who work regularly with oxygen-sensitive anaerobes often use an anaerobic chamber. The transparent enclosure is filled with an inert, oxygen-free gas. Organisms and materials enter and leave through an air lock and manipulations are made by means of glove ports.

armadillos. Another example is the syphilis spirochete, although certain specialized strains of the latter have been grown on laboratory media. With few exceptions, the obligate intracellular parasites such as the rickettsias and the chlamydia bacteria do not grow on artificial media. They, like viruses (which will be discussed later), can grow only in a living host cell.

Many clinical laboratories have special *carbon dioxide* (CO_2) *incubators* in which to grow bacteria that require concentrations of CO_2 higher or lower than that found in the atmosphere. Desired carbon dioxide levels are maintained by electronic controls. High levels of carbon dioxide are also obtained with simple *candle jars* (Figure 6–5). Petri dish cultures are placed in a large sealed jar containing a lighted candle. The candle stops burning when the atmosphere in the jar has a low concentration of oxygen (but one that is still adequate for the growth of aerobic bacteria) and the high CO_2 concentration required by certain clinically important bacteria, such as those causing brucellosis, gonorrhea, and meningococcal meningitis.

Selective and Differential Media

In clinical and public health microbiology, it is frequently necessary to detect the presence of specific microorganisms associated with disease or poor sanitation. For this task, selective and differential media are used. **Selective media** are designed to suppress the growth of unwanted bacteria and encourage the growth of the desired microbes. For example, bismuth sulfite agar is used to isolate the typhoid bacterium from feces. Sabouraud's glucose agar, which has a pH of 5.6, is used to isolate fungi. Dyes such as brilliant green selectively inhibit gram-positive bacteria, and this dye is the basis of a medium called Brilliant Green agar that is used to isolate the gram-negative rod *Salmonella* (sal-mōn-el'lä).

Differential media make it easier to distinguish colonies of the desired organism from other colonies growing on the same plate. Similarly, pure cultures of microorganisms have identifiable reactions with differential media in tubes or plates. Blood agar (which contains red blood cells) is a dark, reddish brown medium that microbiologists often use to identify bacterial species that destroy blood cells. These species, such as *Streptococcus pyogenes* (strep-tō-kok'kus pī-äj'en-ēz), the bacterium that causes strep throat, show a clear ring around their colonies where they have lysed the surrounding blood cells (Figure 6–7).

Sometimes selective and differential media are used together. Suppose we want to isolate the common bacterium *Staphylococcus aureus,* found in the nasal passages. One of the characteristics of this organism is its tolerance for high concentrations of sodium chloride; another characteristic is its ability to ferment the carbohydrate mannitol to form acid. There is an isolation medium called mannitol salt (MS) agar medium that combines features of both selective and differential media. MS agar contains 7.5% sodium chloride, which will discourage the growth of competing organisms and thus *select* for (favor the growth of) *Staphylococcus.* This salty medium also contains a pH indicator that changes color if the mannitol in the medium is fermented to acid; the mannitol-fermenting colonies of *S. aureus* are thus differentiated from colonies of bacteria that do not ferment mannitol. Bacteria that grow at the high salt concentration *and* ferment mannitol to acid can be readily identified by the color change. These are probably colonies of *S. aureus,* and their identification can be confirmed with additional tests.

Another medium that is both selective and differential is MacConkey agar. This medium contains bile salts and crystal violet, which inhibit the growth of gram-positive bacteria. Because it also contains lactose, gram-negative bacteria that can grow on this disaccharide can be differentiated from similar bacteria that cannot. The ability to distinguish between lactose fermenters (red or pink colonies) and nonfermenters (colorless colonies) is useful in distinguishing between the pathogenic *Salmonella* bacteria and other, related bacteria.

Table 6–6 summarizes the uses for a few differential and selective media.

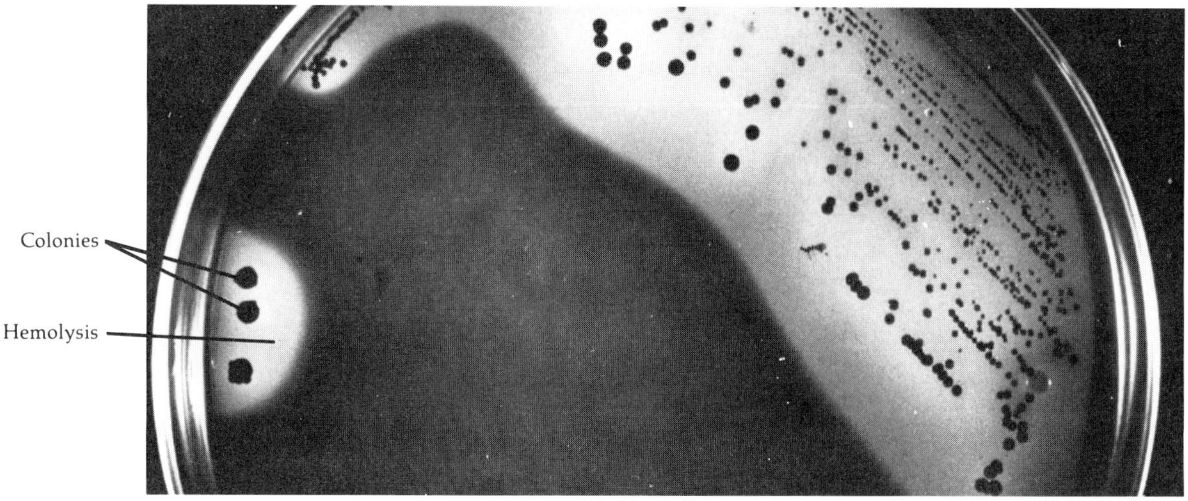

Figure 6–7 Differential medium containing red blood cells. The organism has lysed the red blood cells (hemolysis), so a clear area has formed around the colonies.

Table 6–6 Typical Selective and Differential Media

Medium	Application	Comments
SPS agar	To detect *Clostridium perfringens* in foods.	Under anaerobic incubation, the drugs contained in it (sulfadiazine, polymyxin sulfate) inhibit other clostridia. *C. perfringens* produces black colonies from hydrogen sulfide.
Simmons citrate agar	To differentiate coliform enteric bacteria on their ability to metabolize citric acid.	Medium is neutral; pH indicator in it is green. If citrate (citric acid salt), the nutrient component in the medium, is metabolized, the medium becomes alkaline and the indicator turns blue.
Phenylethyl alcohol agar	To isolate gram-positive organisms such as streptococci.	Phenylethyl alcohol inhibits growth of gram-negative bacteria. Blood is usually added to detect hemolysis.
CTA agar	Differentiates *Corynebacterium diphtheriae* cultures.	Will grow many fastidious bacteria nonselectively. Differentiation primarily based on reaction to paper discs containing dextrose, sucrose, maltose, trehalose, or lactose. A yellow (acidic) zone around disc indicates fermentation.
Bismuth sulfite agar	Isolates *Salmonella typhi* and other enteric bacteria.	Brilliant green dye suppresses gram-positive organisms. Colonies that produce hydrogen sulfide are turned black.
Brilliant green agar	Isolates *Salmonella* in fecal samples.	Brilliant green dye suppresses gram-positive bacteria. Lactose-fermenting or sucrose-fermenting colonies are yellow or greenish. *Salmonella* (except *S. typhi*) form reddish to white colonies.
Desoxycholate agar	Differentiates gram-negative enterics in mixed cultures.	Sodium desoxycholate suppresses gram-positive bacteria. Lactose fermenters form red colonies; lactose-negative colonies are colorless.
Triple sugar iron (TSI) agar	Differentiates isolates of gram-negative enterics.	Differentiates on basis of ability to ferment lactose, sucrose, and glucose. Producers of hydrogen sulfide cause black reaction. Results may be: alkaline slant, acid butt; alkaline slant and butt; acid slant and butt.
XLD agar	Isolates enteric pathogens such as *Shigella*.	Sodium desoxycholate suppresses gram-negative bacteria. Differentiation based on fermentation of xylose, lactose, or sucrose. Producers of hydrogen sulfide form black colonies. Lysine may be decarboxylated to neutralize acidic reaction of *Salmonella*.
Baird–Parker agar	Isolates and identifies *Staphylococcus aureus*.	Staphylococci form black, shiny colonies from reduction of tellurite, surrounded by clear zone in egg yolk component of medium.
SF broth	Detects fecal streptococci in food or water.	Sodium azide inhibits oxygen-using bacteria, but does not affect streptococci. Lactic acid production causes color change in pH indicator.

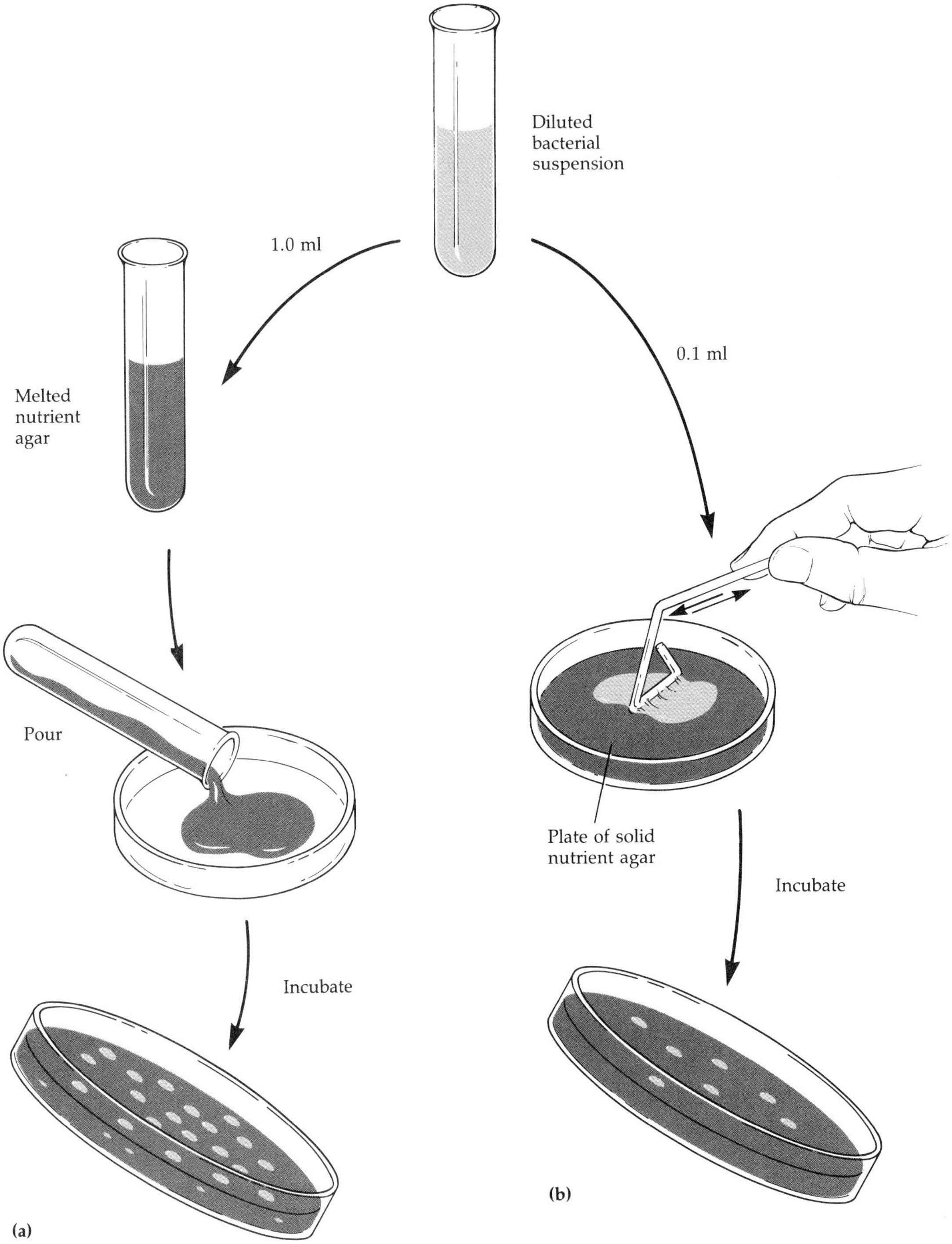

Figure 6–8 Isolating pure cultures. **(a)** Pour plate method. **(b)** Spread plate method.

Mixed and Pure Cultures

Most infectious materials, like pus, sputum, and urine, contain several kinds of bacteria. A culture including more than one kind of microorganism is a **mixed culture;** if grown on solid medium, it will often show several visibly different kinds of colonies. If we wish to study a particular microorganism from a mixed culture, it is necessary to take a sterile needle, touch it to the microbial colony we are interested in, and transfer sample cells from that colony to a separate, sterile medium. The microbial cells will multiply to form new colonies in a **pure culture**; that is, a culture with only one kind of microorganism.

Isolating Pure Cultures

Microbiologists use several procedures for isolating and growing individual colonies of microorganisms.

Pour plate method

The use of solid media makes the isolation of bacteria in pure culture a fairly simple procedure. In the **pour plate method,** dilutions of bacteria suspended in liquid are mixed with melted nutrient agar. Then the agar is poured into a Petri plate, cooled to solidify, and incubated. In such a pour plate, individual bacteria grow into colonies under and on the surface of the agar (Figure 6–8a). Isolated colonies can then be picked (touched) with a sterile inoculating needle and transferred to individual containers. In those containers will grow pure cultures.

Spread plate method

In an alternative method of isolating bacterial colonies, a bacterial suspension is diluted in liquid and then *spread* on a nutrient agar medium. This is called a **spread plate method.** A sterilized bent glass rod is used to spread the bacteria (Figure 6–8b).

Streak plate method

The isolation method most commonly used is the **streak plate method** (Figure 6–9). A sterile inoculating needle is dipped into a mixed culture and

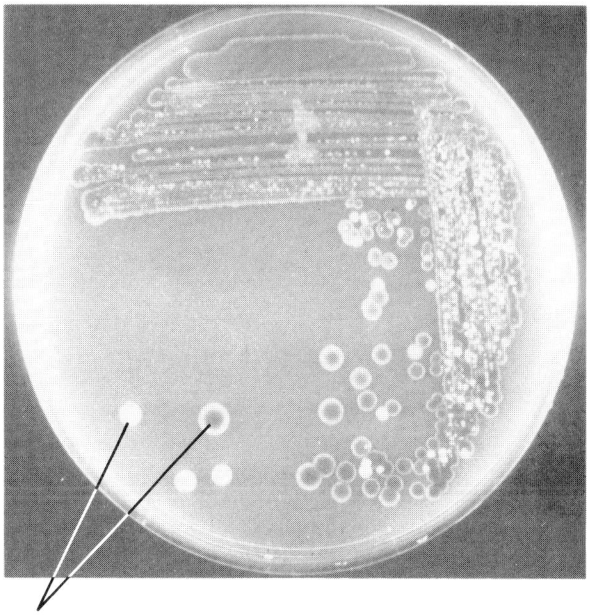

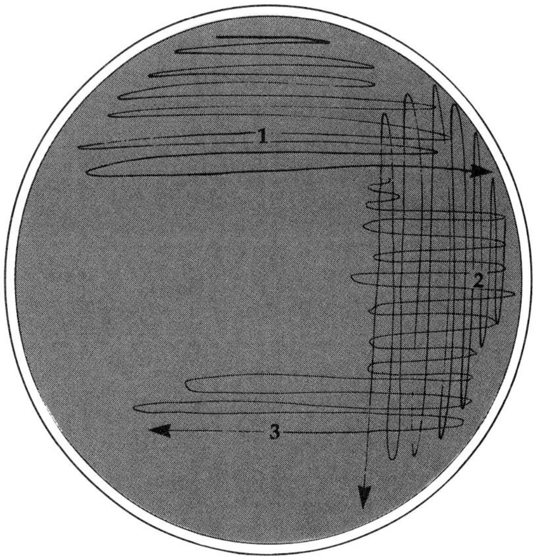

Figure 6–9 Streak plate procedure for pure culture isolation of bacteria. The direction of streaking is indicated by the arrows. Between each streak, the needle is sterilized and then reinoculated with bacteria in the previous section. In this way, the concentration of bacteria is diluted, and well-isolated colonies are obtained. Shown here are isolated colonies of two different bacteria.

streaked in a pattern over the surface of the nutrient medium. As the pattern is traced, bacteria are rubbed off the needle onto the medium, in paths of increasingly fewer cells. The last cells to rub off are wide enough apart that they grow into isolated colonies.

Enrichment culture

Because bacteria present in small numbers can be missed by the previous three methods, and because the bacterium to be isolated might be an unusual physiological type, it is sometimes necessary to resort to an **enrichment culture.** The medium of such a culture is usually liquid and provides nutrients and environmental conditions that favor the growth of a particular microbe but are not suitable for the growth of other types of microbes.

Let us assume that we want to isolate from a soil sample a microbe capable of growth on phenol, a standard disinfectant. If the soil sample is placed in an enrichment culture in which phenol is the only source of carbon and energy, microbes unable to metabolize phenol will not grow. The culture medium is allowed to incubate for a few days, and then a small amount of it is transferred into another culture medium of the same composition. After a series of such transfers, the surviving population will be bacteria capable of metabolizing phenol. The bacteria are given time to grow in the medium between transfers: this is the enrichment stage. Any nutrients in the original inoculum are rapidly diluted out with the successive transfers. When the last dilution is streaked onto a solid medium of the same composition, only colonies of organisms capable of using phenol should grow.

INOCULATION

The **inoculating loop,** or a modification of it, the **inoculating needle,** is made of platinum or stainless steel wire and is used for inoculation and several other microbiological procedures. **Inoculation** is the transfer of microbes to a previously sterilized growth medium. After the medium is inoculated, it is placed in an incubator, usually for 24 to 48 hours. During this time, the bacteria actively grow and reproduce.

If a sufficiently dilute culture is inoculated onto a solid medium, the microorganisms that grow from each cell in the original sample form separate clumps, or masses, that are visible to the naked eye. These masses (each the progeny of one or a few cells) are called **colonies** (Figure 6–10). Colonies have characteristics such as texture, size, shape, color, and adherence to the medium. Such characteristics are fairly constant for each species and are useful in the distinguishing of one microbe from another.

PRESERVING BACTERIAL CULTURES

Refrigeration can be used for short-term storage of bacterial cultures. To preserve microbial cultures for long periods of time, two common methods are deep-freezing and lyophilization. **Deep-freezing** is a process in which a pure culture of microbes is placed in a suspending liquid and quick-frozen at temperatures ranging from −50° to −95°C. The culture can usually be thawed and used several years later.

During **lyophilization (freeze-drying),** a suspension of microbes is quickly frozen at temperatures ranging from −54° to −72°C, and the water is removed by a high vacuum (sublimation). While under vacuum, the container is sealed by a high-temperature torch. The remaining powder-like residue that contains the surviving microbes can be stored for years. The microbes can be revived at any time by hydration with suitable liquid nutrient medium.

GROWTH OF BACTERIAL CULTURES

Bacterial Division

Bacteria normally reproduce by **transverse fission** (Figure 6–11). In preparing for fission, the cell elongates slightly. This is called the *I (initiation) period.* The end of *I* signals the initiation of the *C period,* in which the bacterial chromosome replicates (see chapter 8). Proteins that are needed for division are also synthesized during the C period. When the nuclear material is evenly distributed, the *D (division) period* begins. The plasma membrane near the center of the cell pinches inward. The cell wall thickens and grows inward at this same point, forming the transverse wall.

There are minor variations on this process; the fissions of gram-negative and gram-positive cells differ slightly. A few bacterial species **bud;** that is,

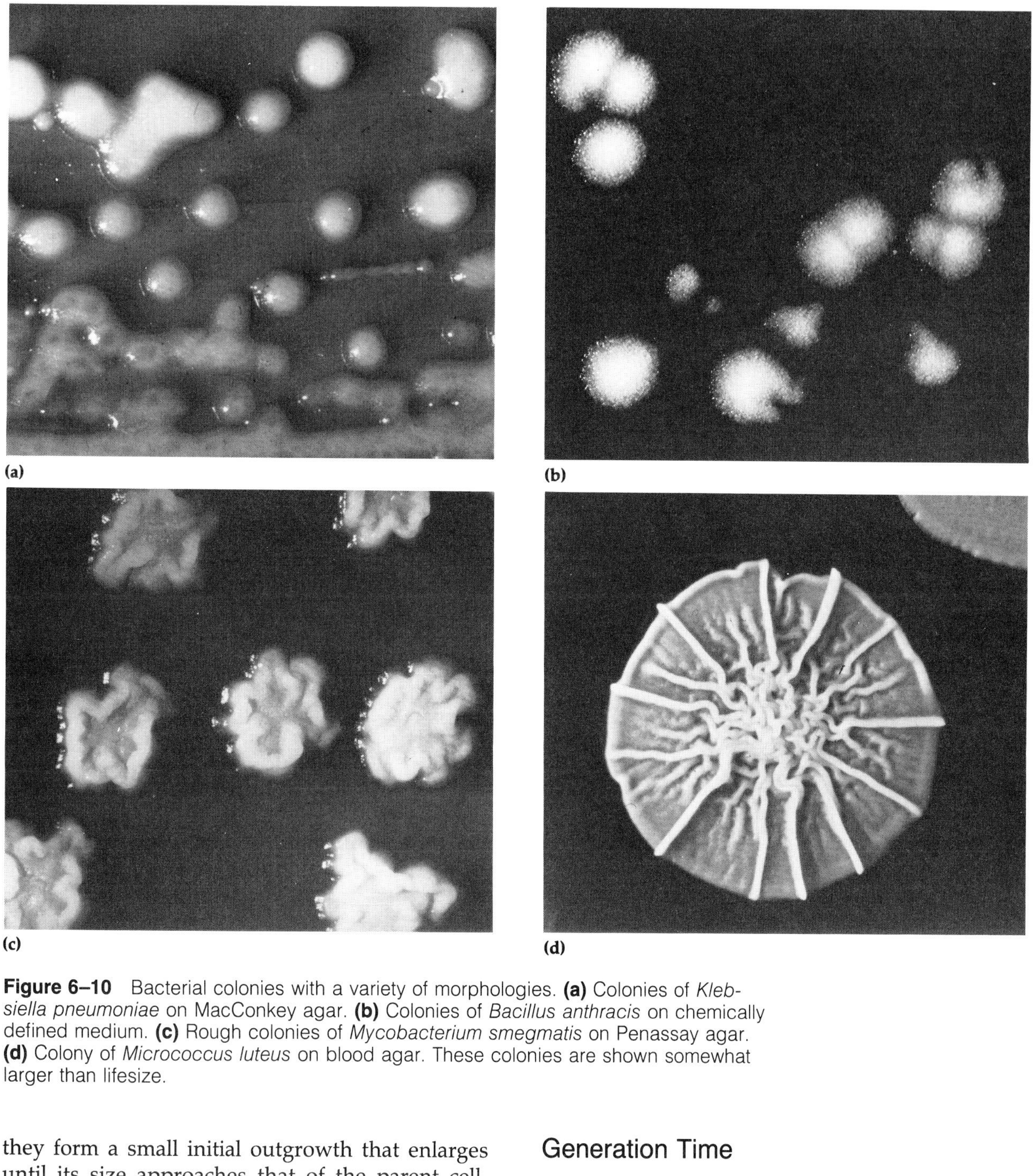

Figure 6–10 Bacterial colonies with a variety of morphologies. **(a)** Colonies of *Klebsiella pneumoniae* on MacConkey agar. **(b)** Colonies of *Bacillus anthracis* on chemically defined medium. **(c)** Rough colonies of *Mycobacterium smegmatis* on Penassay agar. **(d)** Colony of *Micrococcus luteus* on blood agar. These colonies are shown somewhat larger than lifesize.

they form a small initial outgrowth that enlarges until its size approaches that of the parent cell, then it separates. Then there are the filamentous bacteria (actinomycetes) that reproduce by producing chains of spores carried externally at the tips of the filaments. A few filamentous species simply fragment, and the fragments initiate the growth of new cells.

Generation Time

Because of transverse fission, one cell's division produces two cells, the two cells' division produces four cells, and so on. Thus, a single bacterium divides, and transverse fission continues unchecked, an enormous number of cells will be produced in geometric progression (exponential growth):

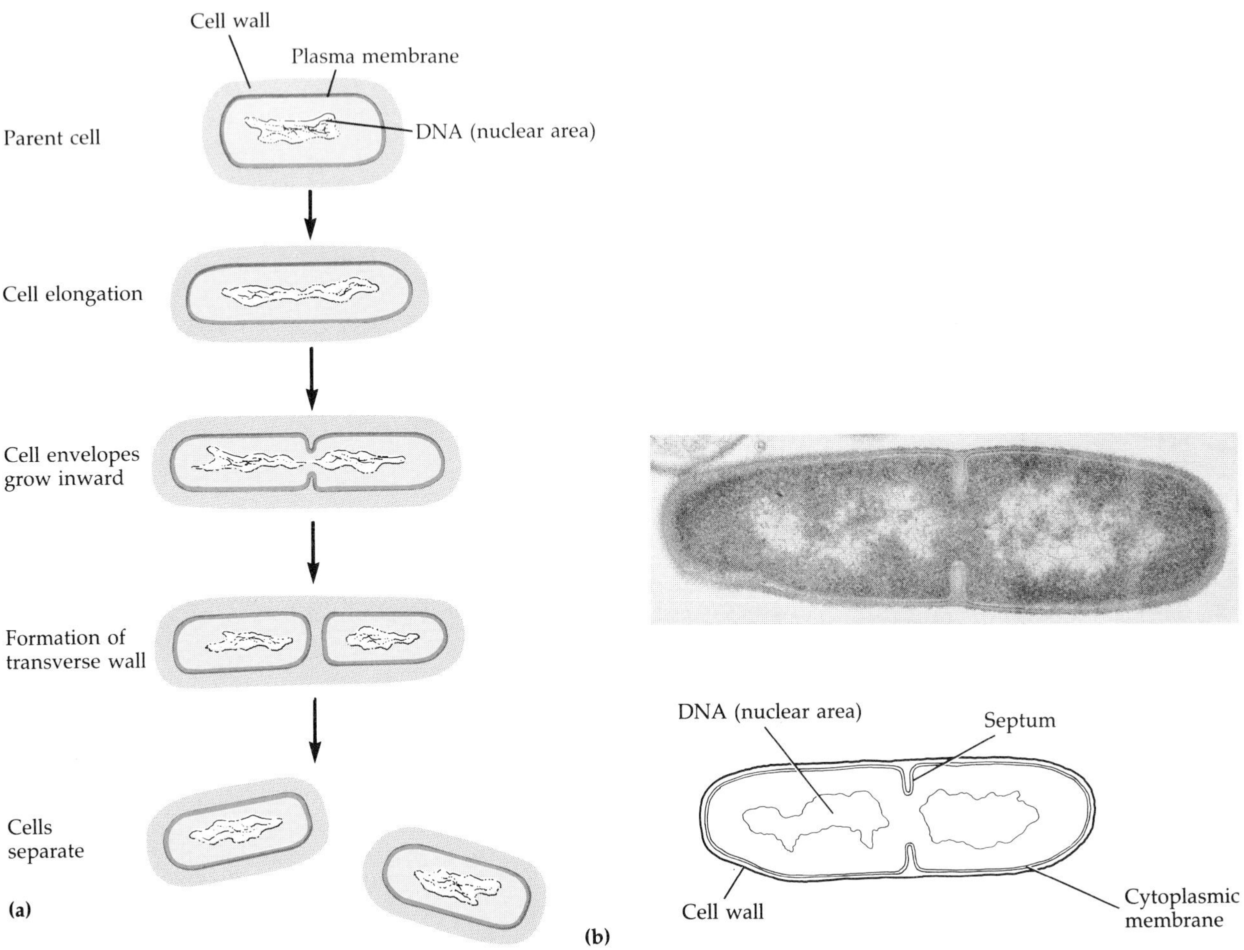

Figure 6–11 Transverse fission in bacteria. **(a)** Diagram of sequence. **(b)** Electron micrograph of a thin section of a cell of *Bacillus licheniformis* starting to divide (×34,600), and drawing of the main structures.

$$\begin{array}{cccccccccccc} 1 & & 2 & & 4 & & 8 & & 16 & & 32 & \ldots \\ & \rightarrow & & \rightarrow & & \rightarrow & & \rightarrow & & \rightarrow & & \\ 2^0 & & 2^1 & & 2^2 & & 2^3 & & 2^4 & & 2^5 & \ldots \end{array}$$

When the number of cells in each generation is expressed as a power of 2, the exponent tells the number of doublings that have occurred. (Before reading on, you might wish to review the box on exponential notation in Chapter 2.)

The time required for a cell to divide and its population to therefore double is called the **generation (doubling) time.** It varies considerably among organisms. For example, although most bacteria have a generation time of 1 to 3 hours, others require over 24 hours per generation. Under favorable conditions, an *E. coli* cell will divide to form two cells every 20 minutes. After 4 hours (12 generations), 4096 cells will have been produced.

There are practical reasons to study the generation times of microbes. A urine sample might be collected, left in a laboratory, and tested for *E. coli* several hours later. If the sample initially contained a few *E. coli* cells (as it normally would), then during the time between collecting and testing, the *E.*

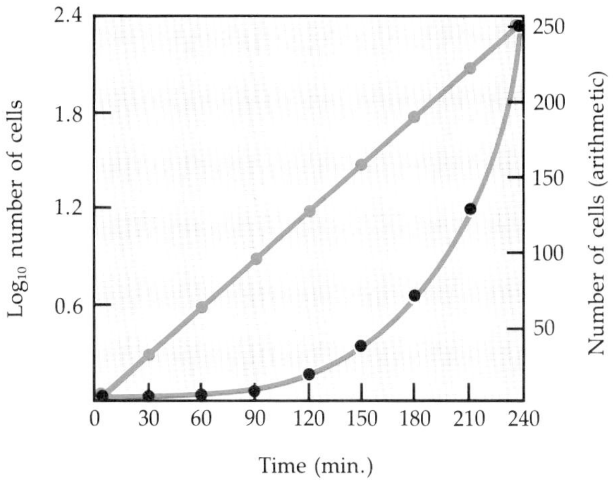

Figure 6–12 Growth curve for an exponentially increasing bacterial population, plotted logarithmically and arithmetically.

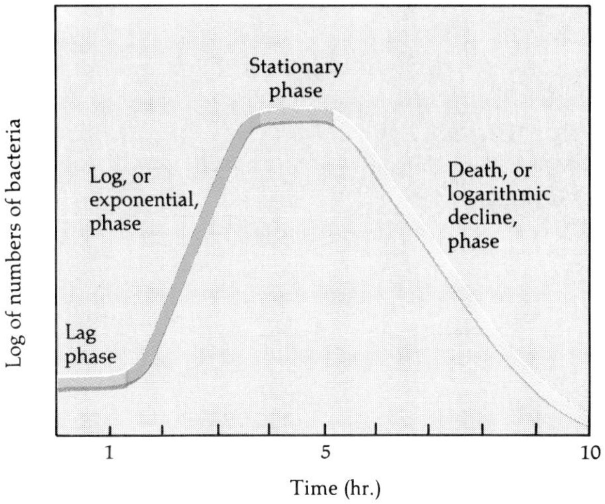

Figure 6–13 Bacterial growth curve, showing the four typical phases of growth.

coli might have reproduced in such quantities that the test results indicate a serious infection.

The easiest way of graphing immense increases in a microbial population is to use a logarithmic, rather than arithmetic, scale (Figure 6–12). Note that if the size of an exponentially increasing microbial population is plotted logarithmically, the growth forms a straight line. The decline and death of a microbial population can be plotted the same way.

Phases of Growth

When a few bacteria are inoculated into a liquid growth medium, and the population is counted at intervals, it is possible to plot a typical **bacterial growth curve** that shows the growth of cells over time (Figure 6–13). There are four basic phases of growth.

Lag phase

For a while, there is little or no change in the number of cells because they do not immediately reproduce in a new medium. This period of little or no cell division is called the **lag phase,** and it can last for an hour or several days. But during this time, the cells are not dormant. The microbial population has intense metabolic activity, in particular, enzyme synthesis. The population is analogous to a factory newly equipped to produce automobiles: there is considerable initial activity, but no immediate increase in the automobile population. Near the end of the lag phase, some cells will double or triple in size, in preparation for reproduction.

Log phase

Eventually, the cells begin to divide and enter a period of growth, or logarithmic increase, called the **log phase,** or **exponential growth phase.** Cellular reproduction is most active during this period, and their generation time reaches a constant minimum. There is apparently a characteristic minimum generation time—maximum rate of doubling—genetically determined for each species. Because the doubling time is constant, a logarithmic plot of growth during the log phase is a straight line. During the log phase, cells show their visible characteristics: the shape, color, density, and groupings of their colonies. The log phase is also the time when cells are most active metabolically, and in industrial production, this is the peak period of activity and efficiency.

On the other hand, during their log phase of growth, microorganisms are more sensitive to adverse conditions than they usually are. Radiation and many antimicrobial drugs—for example, the antibiotic penicillin—exert their effect by interfering with some important step in the growth process, and are therefore most harmful to cells during this phase.

Stationary phase

If exponential growth continues unchecked, some startling numbers of cells can arise. For example, a single bacterium dividing every 20 minutes for two days can theoretically produce an enormously large population of cells (2^{144}, a 44–digit number!). But this does not happen. Eventually, growth slows down, and sooner or later the number of microbial deaths balances the number of new cells, and the population stabilizes. The metabolic activities of individual surviving cells also slow at this stage. This period of equilibrium is called the **stationary phase.**

The reason for cessation of exponential growth is not always clear. The accumulation of waste products toxic to cells and the exhaustion of certain required nutrients are usually involved, along with related changes in pH and temperature. In a specialized apparatus called a *chemostat,* it is possible to keep a population in the exponential growth phase indefinitely by continually draining off spent medium and adding fresh medium.

Death phase

Usually, the number of deaths soon exceeds the number of new cells formed, and the population enters the **death phase,** or **logarithmic decline phase.** This continues until the population is diminished to a tiny fraction of the more resistant cells, or the population might die out entirely. Some species pass through the entire series of phases in only a few days; others retain some surviving cells almost indefinitely. Microbial death will be discussed further in the next chapter.

Measurement of Microbial Growth

There are a number of ways in which bacterial growth can be measured. Some methods measure cell number; other methods measure the population's total mass, which is often directly proportional to cell number. Population numbers are usually recorded as the number of cells in a milliliter of liquid or in a gram of solid material. Because bacterial populations are usually very large, most methods of counting them are based on direct or indirect counts of very small samples; calculations then determine the size of the total population. Assume, for example, that a millionth of a milliliter of sour milk is found to contain 70 bacterial cells. Then there must be 70 times one million cells, or 70 million cells, per milliliter.

The problem in this method is that it is not practical to measure out a millionth of a milliliter of liquid or a millionth of a gram of food. Therefore, the procedure is done indirectly, in a series of dilutions. For example, if we add one milliliter of milk to 99 milliliters of water, each milliliter of this dilution now has one-hundredth as many bacteria as had each milliliter of the original sample. By making a series of such dilutions, we can readily estimate the number of bacteria in our original sample. (This principle is illustrated in Figure 6–15b.) So that microbial populations in foods (such as hamburger) can be counted, a slurry of one part food to nine parts water is finely ground in a food blender. Samples can then be pipetted.

As we now approach the various ways to measure bacterial growth, let us first examine two methods used to determine bacterial numbers in highly concentrated solutions: the direct microscopic count and the standard plate count.

Direct microscopic count

In the method known as the **direct microscopic count,** a measured volume of a bacterial suspension must be inside a defined area on a microscope slide. For example, in the Breed count method for milk bacteria, a 0.01 ml sample is spread over a marked square centimeter of slide and stained. The area of the viewing field of an oil immersion objective can be determined. After many microscope fields have been inspected, the number of bacteria seen in each field can be averaged. This number can be multiplied by a factor that will produce the estimated bacterial count for an entire milliliter of milk.

A specially designed slide called a *Petroff–Hausser counter* is also used in direct microscopic counts (Figure 6–14). A shallow well of known vol-

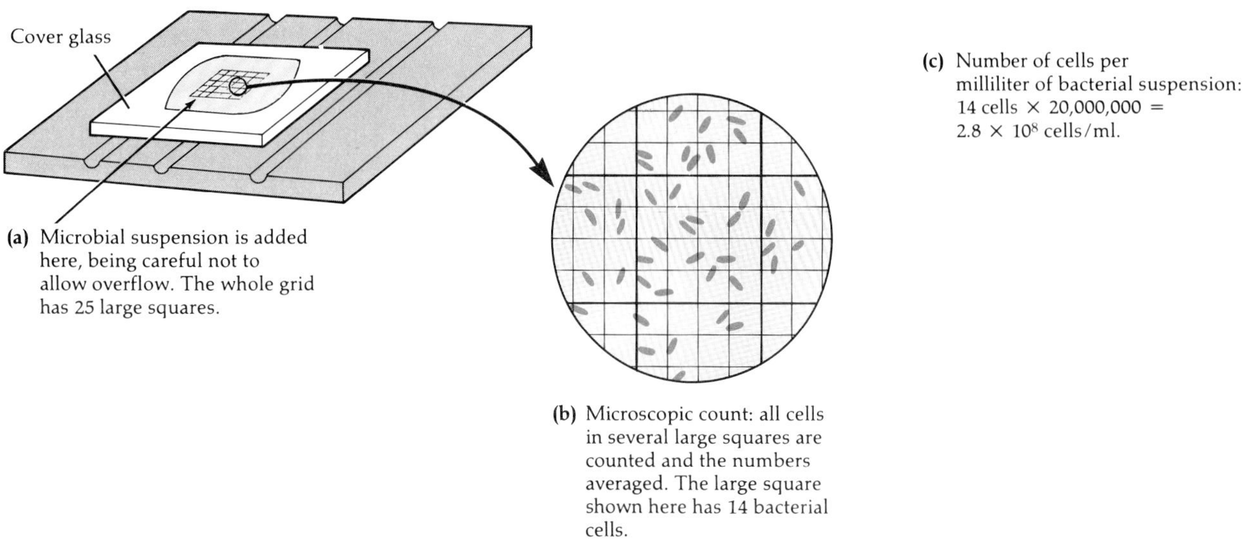

Cover glass
(a) Microbial suspension is added here, being careful not to allow overflow. The whole grid has 25 large squares.
(b) Microscopic count: all cells in several large squares are counted and the numbers averaged. The large square shown here has 14 bacterial cells.
(c) Number of cells per milliliter of bacterial suspension: 14 cells × 20,000,000 = 2.8 × 10⁸ cells/ml.

Figure 6–14 Direct microscopic count of bacteria, with a Petroff–Hausser cell counter. The average number of bacteria within a large square is multiplied by a factor of 20 million to give the number of bacteria per milliliter.

ume is indented into the surface of a microscope slide and covered with a thin glass inscribed with squares of known area. The well is filled with the microbial suspension. The number of bacteria in each of a series of these squares are averaged, and multiplied by the factor that produces the count per milliliter. Motile bacteria are difficult to count by this method, and, as happens with other microscopic methods, dead cells are about as likely as live ones to be counted. In addition to these disadvantages, a rather high concentration of cells, about 10 million bacteria per milliliter, is required. The chief advantage of microscopic counts is that no incubation time is required, and their use is usually reserved for applications having this as the primary consideration. This advantage also holds for *electronic cell counters,* which automatically count the number of cells in a measured volume of liquid. These instruments are available in some research laboratories.

Standard plate count

A second method for the measurement of bacterial growth in highly concentrated solutions is the **standard plate count,** the most frequently used method for counting bacteria. A principal advantage of this technique over the direct microscopic count is that it reflects the number of viable (living) cells. Another advantage is that even small numbers of cells can be detected, because a relatively large volume of microbial suspension can be used. In one procedure, a known volume of 0.1 to 1.0 ml of inoculum is introduced into a Petri plate (Figure 6–15). Then melted nutrient agar is poured into the plate and the two are thoroughly mixed by rotations of the plate. When the agar solidifies, the plate is incubated. The standard plate count is based on the assumption that each bacterium grows and divides to produce a single colony. It is also assumed that the original inoculum is homogeneous and that no aggregates of cells are present.

When a standard plate count is performed, it is important that only a limited number of colonies develop in the plate. When too many colonies are present, some cells are overcrowded and do not develop; these conditions cause inaccuracies in the count. Generally, only those plates with 30 to 300 colonies are counted. To ensure that some colony counts will be within this range, the original inoculum is diluted several times in a process called **serial dilution** (Figure 6–15b).

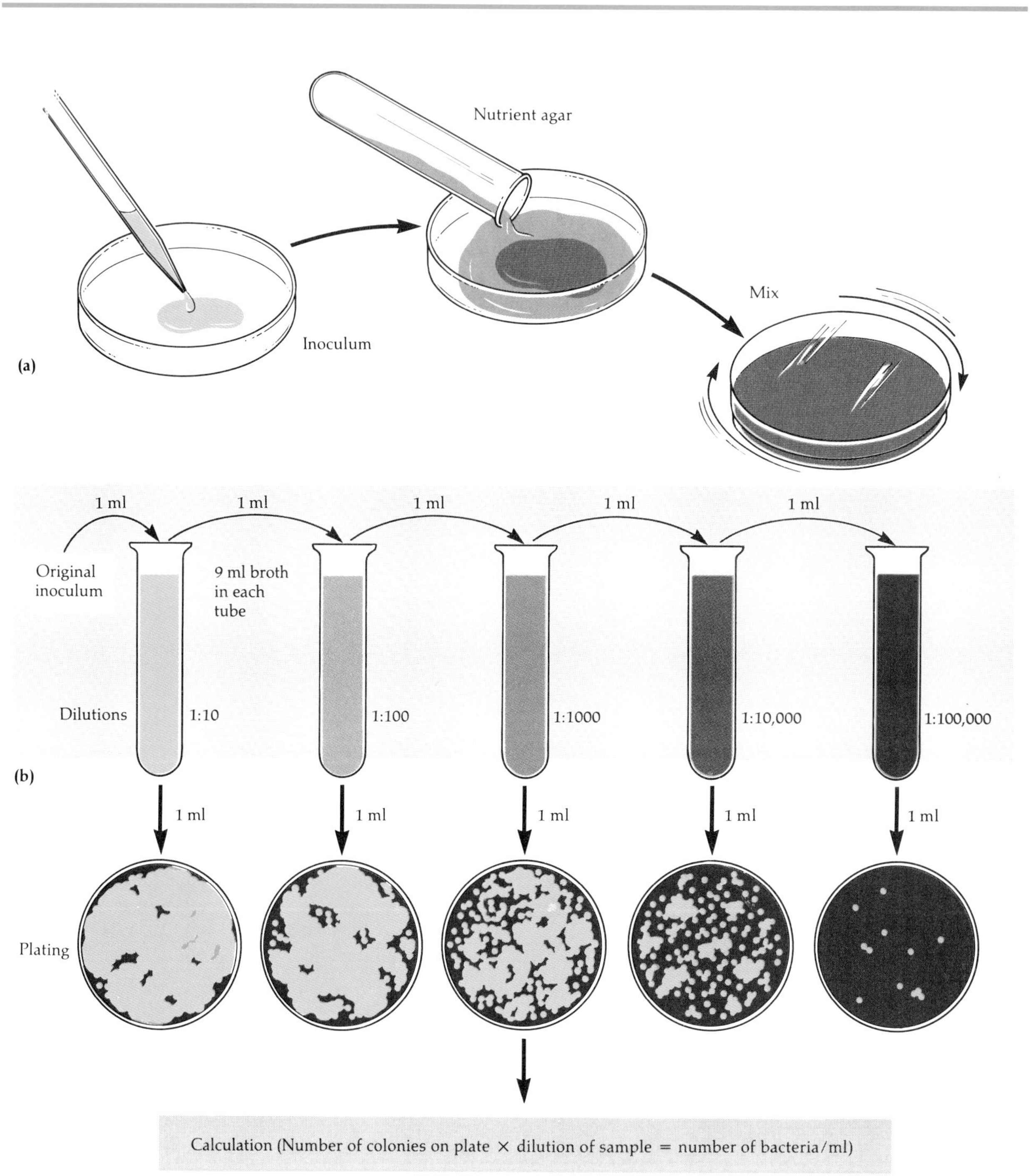

Figure 6–15 Standard plate count. **(a)** Pour plate. Inoculum is pipetted into a Petri plate, liquefied nutrient agar is added, and the agar is mixed with the inoculum by gentle swirling of the plate. **(b)** Serial dilutions. If the Petri plate at the 1:10,000 dilution has 50 colonies, there are 5.0×10^5 bacteria in the original inoculum.

The standard plate count method has a number of shortcomings. As already noted, some bacteria are grouped tightly in aggregations, and unless these are broken up, which is not always possible, each colony on the plate will represent a group of cells rather than an individual cell in the original population. Technically, then, counts are sometimes referred to as **colony-forming units.** Another disadvantage is that the plate must be incubated long enough for a visible colony to form. This takes about 24 hours, sometimes longer. For perishable foods and in health-related bacteriology, this delay is often a problem.

Another disadvantage is that when foods or soils containing a wide variety of microorganisms are plated out, the nutrients and other growth conditions are not always suitable for the growth of all the organisms present. Soil is an extreme example: it is estimated that as few as 1% of the organisms present in soil will grow on the usual test plate. In foods, in which the organisms' nutritional and growth requirements are usually uniform, results are much better. And when a pure culture of bacteria suspended in liquid is counted, considerable accuracy can be achieved by the standard plate count.

Filtration

Where the quantity of bacteria is very low, as in lakes or relatively pure streams, bacteria can be counted by **filtration** methods. One hundred milliliters or more of water is passed through a thin membrane filter whose pores are too small to pass bacteria. Thus, the bacteria are sieved out and retained on the surface of the filter. This filter is then transferred to a pad of nutrient media, where colonies arise from the bacteria on the filter's surface (Figure 6–16). This method is applied frequently to coliform bacteria, indicators of fecal pollution of food or water. The colonies formed by these bacteria are distinct when a differential nutrient medium is used.

Most probable number (MPN)

Another method in the determination of the number of bacteria in a population is the **most probable number (MPN)** method. It is a statistical estimating method based on the fact that the greater the number of bacteria, the more dilution will be needed to

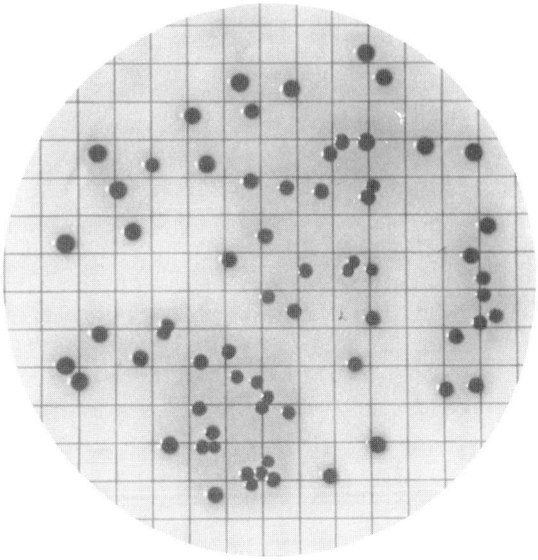

(a)

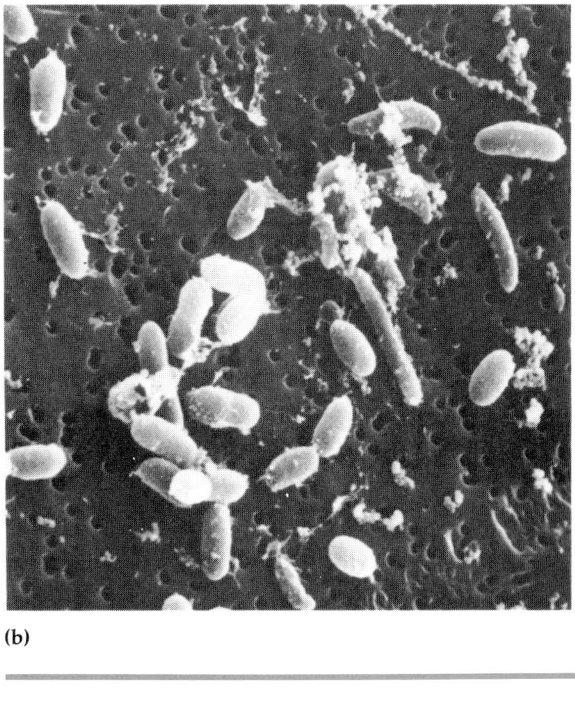

(b)

Figure 6–16 Bacteria on filters. **(a)** Fecal coliform bacteria were collected on this filter and grew into visible colonies when the filter was incubated on a pad moistened with a nutrient medium. A differential medium was used and the fecal coliforms formed distinctive colonies, which appear dark in this photograph. **(b)** Scanning electron micrograph showing cells collected on the surface of a Nalge filter. (Both courtesy of the Millipore Corporation.)

reduce the density to no more than one cell per measured sample. The number of cells in the original medium is first estimated, based on the appearance of the medium, and a range of dilutions is chosen for testing. A number of tubes are inoculated with identical portions of the last few dilutions. Some portions will contain a cell, others will not. The tubes showing growth are counted, and that number is converted to a fraction of the total tubes. The figure is compared to statistical tables that have already been developed, and a cell count is estimated. The MPN technique is often used in issues of food and water sanitation—for example, the determination of water purity—and is used when numbers of bacteria are lower than can be reliably counted by the standard plate count. The MPN states only that there is a 95% chance that the bacterial population falls within a certain range, and that the MPN is statistically the most probable number. A 5-tube MPN table is provided in Appendix D.

We will now look at three ways in which measurements of microbial growth are made with total cell mass. These methods measure turbidity, metabolic activity, and dry weight.

Turbidity

For some types of experimental work, estimating **turbidity** is a practical way of following bacterial growth. As bacteria multiply in a liquid medium, the medium becomes turbid (cloudy) with cells. Because something like 10–100 million cells per milliliter are needed to make a suspension turbid enough to be read, this method has its limitations. It is worth emphasizing here that more than a million cells per milliliter must be present for the first traces of turbidity to be visible. Therefore, visible turbidity is not a useful measure of contamination of liquids by relatively small numbers of bacteria.

In turbidity estimation, a beam of light is transmitted through a bacterial suspension to a photoelectric cell (Figure 6–17). As bacteria increase, less light will reach the photoelectric cell. This change of light will register on the instrument's scale as the percentage of transmission. Also printed on the instrument's scale is a logarithmic expression called the *absorbance* (sometimes *Optical Density* or *OD*), a value derived from the percentage of transmission received. The absorbance is used to plot bacterial growth. When the bacteria are in logarithmic growth or decline, the plot of the absorbance versus time will form an approximately straight line. If absorbance readings are matched with plate counts of the population and other conditions are kept the same, this correlation can be used in future estimations made directly by turbidity. The instrument used for turbidity measurements is usually an electrically operated *spectrophotometer* or *colorimeter*.

Metabolic activity

Another indirect way in which bacterial numbers are estimated is by the measurement of the **metabolic activity** of the population. Rather than attempting to count bacterial colonies or estimate turbidity, we measure, say, the amount of a certain metabolic product and assume it to be in direct proportion to the number of bacteria present. The metabolic product might be acid or carbon dioxide (CO_2). Bacterial numbers can also be estimated with a reduction test, which measures oxygen uptake directly or indirectly. To a medium such as milk, a dye that changes color in the presence or absence of oxygen is added, and the filled tube of milk is tightly capped. Methylene blue, for example, is blue in the presence of oxygen and colorless in its absence. The bacteria use the oxygen as they metabolize the milk. The faster the dye loses color, the faster the oxygen is being depleted and the more bacteria are presumed to be present in the milk.

Dry weight

For filamentous organisms, such as molds, the usual measuring methods are less satisfactory. Molds grow by elongating their vegetative hyphae. A plate count would not measure this increase in biomass. In plate counts of molds, the number of asexual spores are counted instead, and even this is not necessarily a measure of growth. One of the better measurements of growth of filamentous organisms is by **dry weight.** In this procedure, the fungus is removed from the growth medium, filtered to remove extraneous material, placed in a weighing bottle, and dried in a desiccator. For bacteria, the same basic procedure is followed. It is

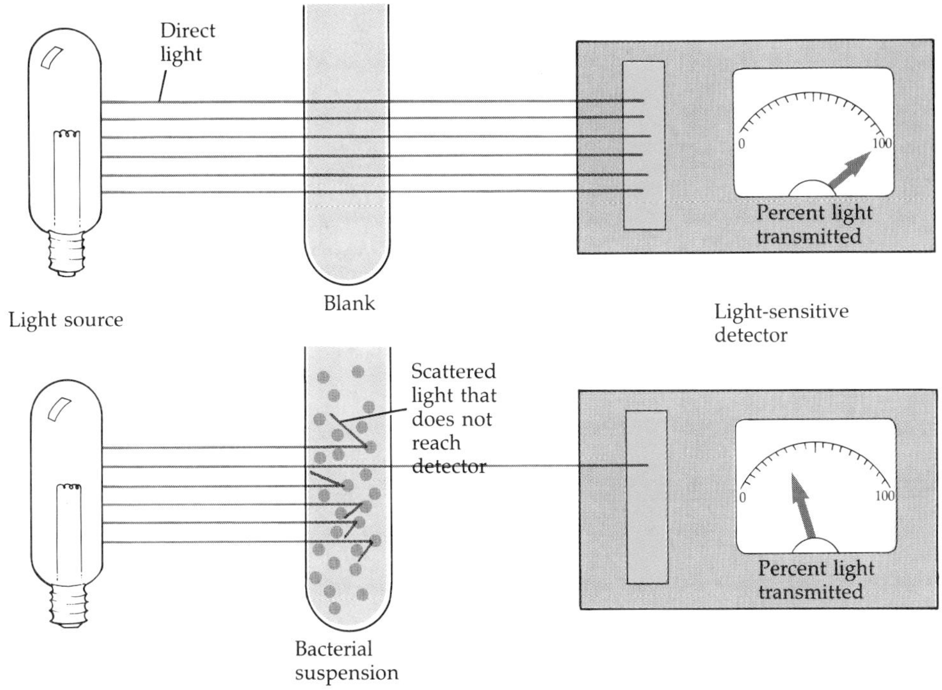

Figure 6–17 Turbidity estimation of bacterial numbers. The amount of light picked up by the light-sensitive detector is inversely proportional to the number of bacteria, under standardized conditions. The less light transmitted, the more bacteria are in the sample.

customary to remove the bacteria from the culture medium by centrifugation.

You now have a basic understanding of the requirements for and measurements of microbial growth. In Chapter 7, we will look at ways in which this growth is controlled in laboratories, hospitals, industry, and our homes.

STUDY OUTLINE

REQUIREMENTS FOR GROWTH (pp. 152–159)

1. Growth of a population is an increase in the number of cells or in mass.
2. Requirements for microbial growth are divided into physical and chemical needs.

Physical Requirements (pp. 152–154)

1. On the basis of growth range of temperatures, microbes are classified as psychrophiles (cold-loving), mesophiles (moderate-temperature-loving), and thermophiles (heat-loving).
2. The minimum growth temperature is the lowest temperature at which a species will grow; the optimum growth temperature is the temperature at which it grows best; and the maximum growth temperature is the highest temperature at which growth is possible.
3. Most bacteria grow best at a pH between 6.5 and 7.5.
4. In a hypertonic solution, microbes undergo plasmolysis; halophiles can tolerate high salt concentrations.
5. On the basis of oxygen requirements, organisms are

classified as obligate aerobes, facultative anaerobes, obligate anaerobes, aerotolerant anaerobes, and microaerophiles.

6. Aerobes, facultative anaerobes, and aerotolerant anaerobes must have the enzymes superoxide dismutase ($2O_2^{2-} + 2H^+ \rightarrow O_2 + H_2O_2$) and catalase ($2H_2O_2 \rightarrow 2H_2O + O_2$), or peroxidase ($H_2O_2 + 2H^+ \rightarrow 2H_2O$).

Chemical Requirements (pp. 154–159)

1. All organisms require a carbon source; chemoheterotrophs use an organic molecule and chemoautotrophs typically use carbon dioxide.
2. Other chemicals required for microbial growth include nitrogen, sulfur, phosphorus, trace elements, and, for some microorganisms, organic growth factors.

CULTURE MEDIA (pp. 159–168)

1. Any material prepared for the growth of bacteria in a laboratory is referred to as a culture medium.
2. Agar is a common solidifying agent for a culture medium.
3. Microbes that grow and multiply in or on a culture medium are known as a culture.

Chemically Defined and Complex Media (pp. 160–161)

1. A chemically defined medium is one in which the exact chemical composition is known.
2. A complex medium is one in which the exact chemical composition is not known.

Anaerobic Growth Media and Methods (pp. 161–162)

1. Reducing media chemically remove molecular oxygen (O_2) that might interfere with the growth of anaerobes.
2. Petri plates can be incubated in an anaerobic jar or anaerobic incubator.

Special Culture Techniques (pp. 162–163)

1. Some parasitic and fastidious bacteria must be cultured in living animals or in tissue cultures.
2. CO_2 incubators or candle jars are used to grow bacteria requiring an increased CO_2 concentration.

Selective and Differential Media (pp. 163–167)

1. By inhibiting unwanted organisms with salts, dyes, or other chemicals, selective media allow growth of only the desired microbe.
2. Differential media are used to distinguish between different organisms.

Mixed and Pure Cultures (p. 167)

1. A mixed culture contains several different species of bacteria. A pure culture contains only one species.

Isolating Pure Cultures (pp. 167–168)

1. Pure cultures may be isolated with any of several methods.
2. Discussed are the pour plate method, spread plate method, streak plate method, and enrichment culture.

INOCULATION (p. 168)

1. The transfer of microbes to a previously sterilized culture medium is called inoculation.
2. It is performed with an inoculating loop or needle.
3. Colonies are the progeny of a single cell that have grown in or on a culture medium.

PRESERVING BACTERIAL CULTURES (p. 168)

1. Microbes can be preserved for long periods of time.
2. This may be accomplished by deep-freezing or lyophilization (freeze-drying).

GROWTH OF BACTERIAL CULTURES (pp. 168–177)

Bacterial Division (pp. 168–169)

1. The normal reproductive method of bacteria is transverse fission, in which a single cell divides into two identical cells.
2. Some bacteria reproduce by budding, aerial spore formation, or fragmentation.

Generation Time (p. 169–171)

1. The time required for a cell or population to divide is known as generation time.
2. Bacterial division occurs according to a logarithmic progression (2 cells, 4 cells, 8 cells, etc.).

Phases of Growth (pp. 171–172)

1. During the lag phase, there is little or no change in the numbers of cells, but metabolic activity is high.
2. During the log phase, the bacteria multiply at the fastest rate possible under the conditions provided.
3. During the stationary phase, there is an equilibrium between cell division and death.
4. During the death phase, the number of deaths exceeds the number of new cells formed.

Measurement of Microbial Growth (pp. 172–177)

1. In a direct microscopic count, the microbes in a measured volume of a bacterial suspension are counted with use of a specially designed slide.
2. A standard plate count reflects the number of viable microbes and assumes that each bacterium grows into a single colony.
3. In filtration, bacteria are retained on the surface of a membrane filter and then transferred to culture medium to grow and subsequently be counted.
4. The most probable number method can be used for microbes that will grow in a liquid medium; it is a statistical estimation.
5. A spectrophotometer is used in turbidity estimates by measuring the amount of light that passes through a suspension of cells.
6. An indirect way of estimating bacterial numbers is by measurement of the metabolic activity of the population, for example, acid production or oxygen consumption.
7. For filamentous organisms such as fungi, measuring dry weight is a convenient method of growth measurement.

STUDY QUESTIONS

REVIEW

1. Describe transverse fission.
2. What physical and chemical properties of agar make it useful in culture media?
3. Draw a typical bacterial growth curve. Label and define each of the four phases.
4. Which of these curves best depicts the log phase of
   (a) A thermophile incubated at room temperature?
   (b) *Listeria monocytogenes* growing in a human?
   (c) A psychrophile when incubated at 9°C?

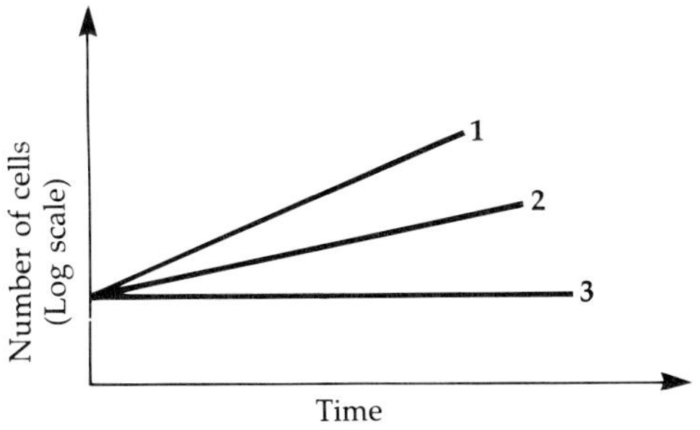

5. Macronutrients (needed in relatively large amounts) are often listed as CHONPS. What does each of these letters indicate, and why are they needed by the cell?
6. Most bacteria grow best at pH ______________.
7. Why can high concentrations of salt or sugar be used to preserve food?

8. Explain five bacterial categories based on their requirements for oxygen.

9. Define the following terms and explain the importance of each enzyme. Catalase, hydrogen peroxide, peroxidase, superoxide free radical, superoxide dismutase.

10. *Clostridium* can be cultured in an anaerobic incubator or in the presence of atmospheric oxygen if thioglycolate is added to the nutrient broth. Compare these two techniques. Using terms from question 9, explain why elaborate culture techniques are used for *Clostridium.*

11. A pastry chef inoculated a cream pie with six *S. aureus* cells. If *S. aureus* has a generation time of 60 minutes, how many cells would be in the cream pie after seven hours?

12. Seven methods of measuring microbial growth were explained in this chapter. Categorize each as a direct or indirect method.

13. Flask *A* contains yeast in glucose-minimal salts broth, incubated at 30°C with aeration. Flask *B* contains yeast in glucose-minimal salts broth, incubated at 30°C in an Anaerobic Jar.
(a) Which culture produced the most ATP?
(b) Which culture produced the most alcohol?
(c) Which culture had the shortest generation time?

14. A normal fecal sample has a high concentration of coliform bacteria. A patient with typhoid fever will excrete *Salmonella* along with normal bacteria. Feces from a suspected case of typhoid fever are inoculated on an S-S agar plate. After proper incubation, coliform bacteria form red colonies and *Salmonella* form colorless colonies with a black center. Is S-S agar selective? Differential? How can you tell?

15. Differentiate between complex and chemically defined media.

16. By deep-freezing, bacteria can be stored without harm for extended periods of time. Why then do refrigeration and freezing preserve foods?

CHALLENGE

1. *E. coli* was incubated with aeration in a nutrient medium containing two carbon sources, and the growth curve below was made from this culture.
(a) Explain what happened at the time marked *x*.
(b) Which substrate provided "better" growth conditions for the bacteria? How can you tell?

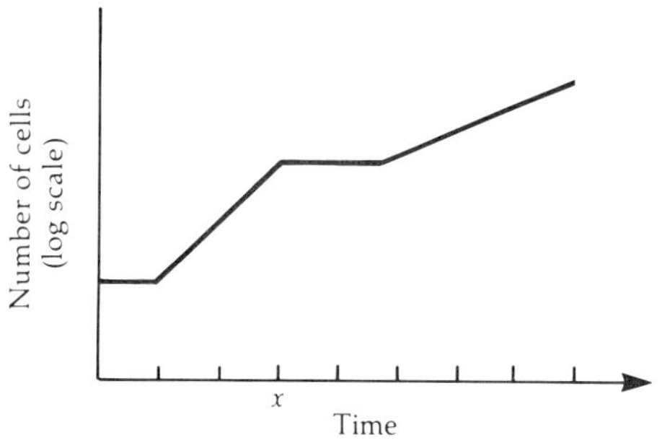

2. *Clostridium* and *Streptococcus* are both catalase-negative. *Streptococcus* grows by fermentation. Why is *Clostridium* killed by oxygen while *Streptococcus* is not?

3. Design an enrichment medium and procedure for growing an endospore-forming bacterium that fixes nitrogen and uses cellulose for its carbon source.

4. Most laboratory media contain a fermentable carbohydrate and peptone because the majority of bacteria require carbon, nitrogen, and energy sources in these forms. How are these three needs met by glucose-minimal salts medium? (See Table 6–4.)

FURTHER READING

Atlas, R. M. and R. Bartha. 1981. *Microbial Ecology: Fundamentals and Applications.* Reading, MA: Addison-Wesley. The interrelationships of microbes with their environments; includes enumeration and isolation techniques.

Difco Manual: Dehydrated Culture Media and Reagents for Microbiology, 10th ed. 1984. Detroit. MI: Difco Laboratories. Formulae and applications of media from the manufacturer.

Ingraham, J. L., O. Maaløe, and F. C. Neidhardt. 1983. *Growth of the Bacterial Cell.* Sunderland, MA: Sinauer Associates. Comprehensive coverage of cell growth.

Kushner, D. J. 1978. *Microbial Life in Extreme Environments.* New York: Academic Press. Includes chapters on physiological adaptations of microbes exposed to extreme temperature, pressure, salt, pH, and radiation conditions in nature.

Payne, W. J. and W. J. Wiebe. 1978. Growth yield and efficiency in chemosynthetic microorganisms. *Annual Review of Microbiology* 32:155–183. A review of the effects of substrate and physical conditions on microbial growth, and techniques to determine growth rate.

Sutter, V. et al. 1980. *Wadsworth Anaerobic Bacteriology Manual,* 3rd ed. St. Louis, MO: C. V. Mosby. Methods for isolation and cultivation of anaerobes.

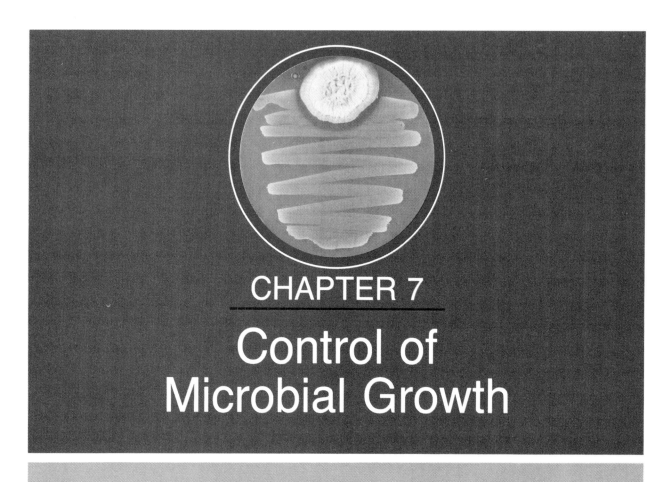

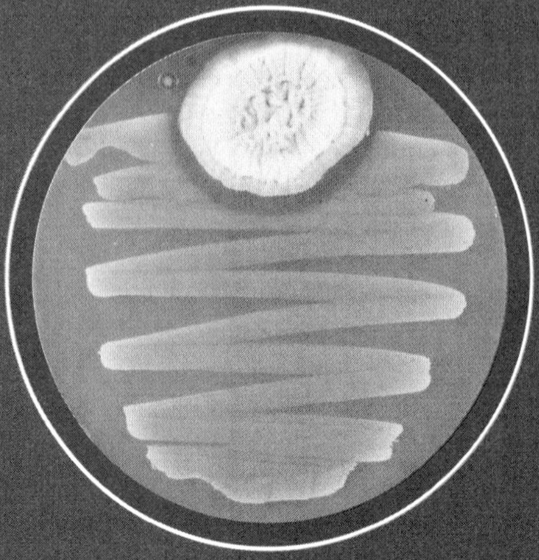

CHAPTER 7

Control of Microbial Growth

OBJECTIVES

After completing this chapter you should be able to

- Define terms related to the destruction or suppression of microbial growth.
- Explain how microbial growth is affected by the type of microbe, its physiological state, and the environmental conditions.
- Describe the effects of microbial control agents on cellular structures.
- Describe the pattern of microbial death caused by treatment with a microbial control agent.
- Describe the physical methods of microbial control.
- Describe the factors related to effective disinfection and their effect on the evaluation of a disinfectant.
- Describe the methods of action and preferred uses for disinfectants.

The scientific control of microbial growth began only about one hundred years ago. Prior to that time, it was not uncommon for massive epidemics to kill thousands of people. In some hospitals 25% of delivering mothers died of infections carried by the hands and instruments of attending nurses and physicians. And, during the American Civil War, a surgeon might have cleaned his scalpel on his bootsole between incisions.

Two persons who introduced the concept of microbial control were Ignatz Semmelweis (1816–1865), a Hungarian physician working in Vienna, and Joseph Lister (1827–1912), an English physician. At the obstetrics ward in the Vienna General Hospital, Semmelweis required all personnel to wash their hands in chlorinated lime; that procedure significantly lowered the infection rate. Lister, meanwhile, read about Pasteur's work with microbes, and assumed that the number of infected surgical wounds (sepsis) could be decreased through use of procedures that prevented the access of microbes to the wound. This system, known as **aseptic surgery,** included the heat sterilization of surgical instruments and, following surgery, the application of phenol (carbolic acid) to wounds.

We have come a long way in controlling microbial growth since the time of Semmelweis and Lister. Today's procedures, far more sophisticated and effective, are used not only to control disease organisms, but also to curb microbial growth that results in food spoilage. This chapter will discuss how microbial growth can be controlled by physical methods and chemical agents. Physical methods include the use of heat, low temperatures, desiccation, osmotic pressure, filtration, and radiation. Chemical agents include several groups of substances that destroy or limit microbial growth on body surfaces or inanimate objects. This discussion will be limited to agents that prevent microbial infection. Later, Chapter 18 will discuss methods for the control of microbes once infection has occurred. Chemotherapy—the use of antimicrobial drugs—is one such method.

Before we begin our survey of physical and chemical methods used to control microbial growth, it would be useful to first define the important terms related to control. Microbial control is needed to prevent the transmission of infection, contamination, and spoilage. But as you will see, it is not always necessary to kill the microbes; some situations require that they simply be inhibited or removed. A variety of agents and procedures are available, and, for different situations, each has its own application and range of controlling effects. Table 7–1 contains a list of terms related to the destruction or suppression of microbial growth. (Note: In this chapter, we use the term *disinfection* to include antisepsis.)

CONDITIONS INFLUENCING MICROBIAL CONTROL

Usually, if we want to sterilize an object, say, a laboratory flask, we do not bother to identify the species of microbes on it. Instead, we launch a general attack calculated to be strong enough to kill the most resistant microbial forms that could possibly be present—endospores. At any time, the flask could carry a multitude of microorganisms in different phases of their growth cycles, with different metabolic states, and even with different microenvironments. Thus, it is possible that an antimicrobial agent will not kill all of the microbial population immediately. The time it takes to reduce or eliminate microbial growth on any object depends, in part, on the number and species of microbes dwelling on the object. It also depends on a number of other factors.

Temperature

Everyone is familiar with the use of ice or mechanical refrigeration to control microbial growth. Like all chemical reactions, the biochemical reactions required for growth are slowed considerably by low temperatures. Chemical disinfectants are also often inhibited by low temperatures. Because their activity is due to temperature-dependent chemical reactions, disinfectants tend to work somewhat better in warm solutions. You might have noticed that directions on disinfectant containers frequently specify that a warm solution be used.

Kind of Microbe

Many disinfectants and antiseptics tend to have a greater effect on gram-positive bacteria, as a group, than on gram-negative bacteria, although this differentiation is by no means as clear cut as it is with antibiotic activity. A certain group of gram-negative bacteria, the pseudomonads (genus *Pseudomonas*), is unusually resistant to chemical activity and will even grow actively *in* some disinfectants and antiseptics. These bacteria are able to maintain themselves in such unlikely substrates as simple saline (salt) solutions, and they are also resistant to many antibiotics. These properties have made these common, and normally harmless, bacteria very troublesome to hospitals. Pseudomonads are **opportunistic pathogens,** that is, microbes that do not ordinarily cause disease, but become pathogenic in the absence of normal competitive flora, as when antibiotics suppress the growth of other microbes but do not adversely affect the pathogen. *Mycobacterium tuberculosis* (mī-kō-bak-ti'rē-um tū-ber-kū-lō'sis), the microbe that causes tuberculosis, is another non-endospore-forming bacterium that exhibits greater than normal chemical resistance.

Special tuberculocidal tests evaluate the effectiveness of chemical antimicrobials acting on the bacterium. The endospores of bacteria and cysts of protozoans are also highly resistant to chemical agents, as are some viruses. Their resistance to chlorine disinfection is particularly important.

Physiological State of the Microbe

Microorganisms that are actively growing tend to be more susceptible to chemical agents than older cells are. This might be because an actively metabolizing and reproducing cell has more points of vulnerability than does an older, possibly dormant cell. Moreover, when microorganisms have formed endospores, the endospores are generally more resistant than the vegetative cells. The endospores of the resistant strains of *Clostridium botulinum* (bo-tū-li'num), for example, can withstand boiling for $5\frac{1}{2}$ hours.

Environment

Organic matter frequently interferes with the action of chemical agents. In hospitals, the presence of organic matter in vomit and feces influences the selection of a disinfectant, because some disinfectants are more effective than others under these conditions. Even bacteria found in food tend to be protected by the organic matter of the food. For example, bacteria in cream are protected by proteins and fats, and can survive more heat than that

TABLE 7–1 Terminology Related to the Control of Microbial Growth

Term	Definition
Terms Related to Destruction:	
Sterilization	The process of destroying all forms of microbial life on an object or in a material. This includes the destruction of endospores—the most resistant form of microbial life. Sterilization is absolute; there are no degrees of sterilization.
Disinfection	The process of destroying vegetative pathogens, but not necessarily endospores or viruses. Usually, a disinfectant is a chemical applied to an object or a material. Disinfectants tend to reduce or inhibit growth; they usually do not sterilize.
Antisepsis	Chemical disinfection of the skin, mucous membranes, or other living tissues.
Germicide (*cide* = killer)	A chemical agent that rapidly kills microbes, but not necessarily their endospores. A *bactericide* kills bacteria, a *sporicide* kills endospores, a *fungicide kills* fungi, a *virucide* kills viruses, and an *amoebicide* kills amoebas and other protozoans.
Terms Related to Suppression:	
Bacteriostasis (*stasis* = halt)	A condition in which bacterial growth and multiplication are inhibited, but the bacteria are not killed. If the bacteriostatic agent is removed, bacterial growth and multiplication may resume. *Fungistasis* refers to the inhibition of fungal growth.
Asepsis (*asepsis* = without infection)	The absence of pathogens from an object or area. Aseptic techniques are designed to prevent the entry of pathogens into the body. Whereas *surgical asepsis* is designed to exclude all microbes, *medical asepsis* is designed to exclude microbes associated with communicable diseases.
Degerming (degermation)	The removal of transient microbes from the skin by mechanical cleansing or by the use of an antiseptic.
Sanitization	The reduction of pathogens on objects to safe public health levels by mechanical cleansing or chemicals.

which would kill bacteria in skim milk. Heat is much more effective under acidic conditions than at a neutral pH. This is also true of a few chemicals.

Let us now examine the mechanisms of action of control agents, that is, the way various agents of control actually kill or inhibit microbes.

ACTIONS OF MICROBIAL CONTROL AGENTS

Alteration of Membrane Permeability

The plasma membrane, located just inside the cell wall, is the target of many control agents. This membrane actively controls the passage of nutrients into the cell and the elimination of wastes from the cell. Damage to the lipids or proteins of the plasma membrane by antimicrobial agents, such as quaternary ammonium compounds, typically causes cellular contents to leak into the surrounding medium and interferes with the growth of the cell. Several forms of chemical agents and antibiotics work, at least in part, in this manner.

Damage to Proteins and Nucleic Acids

You might recall that the functional properties of proteins are due to their three-dimensional shape. This shape is maintained by chemical bonds that hold together adjoining portions of the amino acid chain as it folds back and forth upon itself. Some of those are hydrogen bonds, which are susceptible to breakage by heat or certain chemicals; breakage results in denaturation of the protein. Covalent bonds, which are stronger, are also subject to attack. For example, disulfide bridges, which play an important role in protein structure by joining amino acids with exposed sulfhydryl (—SH) groups, can be broken by certain chemicals or sufficient heat.

Because DNA and RNA carry the genetic message, damage to them by radiation or chemicals, for example, is frequently lethal to the cell; this damage prevents both replication and normal functioning.

PATTERN OF MICROBIAL DEATH

When bacterial populations are heated or treated with antimicrobial chemicals, they die off usually at a constant rate. For example, a population of 1 million microbes has been treated for one minute, and 90% of the population has died. We are now left with 100,000 microbes. If the population is treated for another minute, 90% of *those* microbes die, and we are left 10,000 survivors. In other words, for each minute that the treatment is applied, 90% of the remaining population is killed (Table 7–2). If the death curve is plotted logarithmically, the death rate is seen to be constant (Figure 7–1). Obviously, the more microbes there are at the beginning, the longer it will take to kill them all.

Now that you have an understanding of the basic principles of microbial control, we will examine some representative physical and chemical methods.

Table 7–2 Pattern of Microbial Death

Time (in Minutes)	Deaths per Minute	Number of Survivors
0	0	1,000,000
1	900,000	100,000
2	90,000	10,000
3	9,000	1,000
4	900	100
5	90	10
6	9	1

Source: O. Rahn, *Physiology of Bacteria*, McGraw-Hill, New York, 1932.

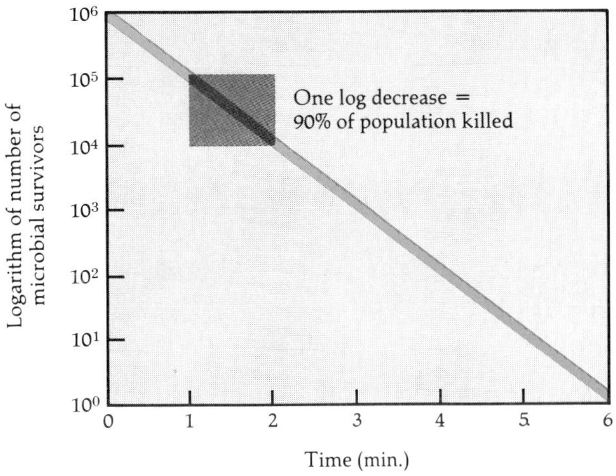

Figure 7–1 Microbial death curve plotted logarithmically.

PHYSICAL METHODS OF MICROBIAL CONTROL

Heat

Probably the most common method by which microbes are killed is **heat.** A visit to any supermarket will demonstrate that canned goods, in which microorganisms have been killed by moist heat, easily outnumber the foods preserved by other methods. Laboratory equipment, such as media and glassware, and hospital instruments are also usually sterilized by heat.

Heat is the most widely applicable and effective agent for sterilization. It is also the most economical and easily controlled. Heat appears to kill microbes by destroying their enzymes by denaturation. In sterilization by heat, the degree of the bacteria's heat resistance must be considered. Heat resistance varies among different microbes; these differences can be expressed through the concept of thermal death point. **Thermal death point (TDP)** is the lowest *temperature* required to kill all of the microorganisms in a liquid suspension in 10 minutes.

Another factor to be considered in sterilization is the length of time required for the material to be rendered sterile. This is expressed as **thermal death time (TDT),** the minimal length of *time* in which all bacteria in a liquid culture will be killed at a given temperature. Both TDP and TDT are useful guidelines that indicate the severity of treatment required to kill a given population of bacteria.

Decimal reduction time is a third concept related to the bacteria's degree of heat resistance. It is the time, in minutes, in which 90% of the population of bacteria at a given temperature will be killed (see Figure 7–1). Decimal reduction time is especially useful in the canning industry.

The heat used in sterilization can be applied in the form of moist heat or dry heat. Dry heat kills by oxidation effects. A simple analogy is the slow charring of paper in a heated oven, when the temperature is below the ignition point of paper. Moist heat kills microorganisms because the water hastens the breaking of hydrogen bonds that hold proteins in their three-dimensional structure.

Moist heat

One type of moist heat sterilization is boiling. At sea level, **boiling** (100°C) kills vegetative forms of bacterial pathogens, many viruses, and fungi within about 10 minutes. Free-flowing (unpressurized) steam is the equivalent in temperature to boiling water. Endospores and some viruses, however, are not destroyed quickly. The hepatitis virus can survive up to 30 minutes of boiling, and some bacterial endospores have resisted boiling temperatures for more than 20 hours. Boiling is therefore not always a reliable sterilization procedure. However, a few minutes of boiling temperatures will kill most pathogens and generally make food or water safe to eat or drink.

To reliably sterilize with moist heat, temperatures above that of boiling water are needed. This high temperature is most commonly achieved by steam under pressure, in an **autoclave** (Figure 7–2). The preferred method of sterilization, autoclaving is used unless the material to be sterilized would be damaged by heat or moisture. The higher the pressure in the autoclave, the higher the temperature. For example, when free-flowing steam, at a temperature of 100°C, is placed under a pressure of 1 atmosphere above sea level pressure, that is, about 15 pounds per square inch, the temperature is increased to 121°C. Increasing the pressure to 20 pounds per square inch raises the temperature to 126°C. The relationship between temperature and pressure is shown in Table 7–3. In an autoclave, steam at a pressure of about 15 pounds per square inch (121°C) will kill *all* organisms and their endospores in about 15 minutes, or perhaps a bit longer, depending on the type and volume of material being sterilized.

Autoclaving is used to sterilize culture media, instruments, dressings, intravenous equipment,

Table 7–3 Relationship Between Pressure and Temperature of Steam at Sea Level*

Pressure in Pounds per Square Inch (psi) (in Excess of Atmospheric Pressure)	Temperature in °C
0 psi	100
5 psi	110
10 psi	116
15 psi	121
20 psi	126
30 psi	135

*At higher altitudes the gauge pressure would be higher for a particular temperature.

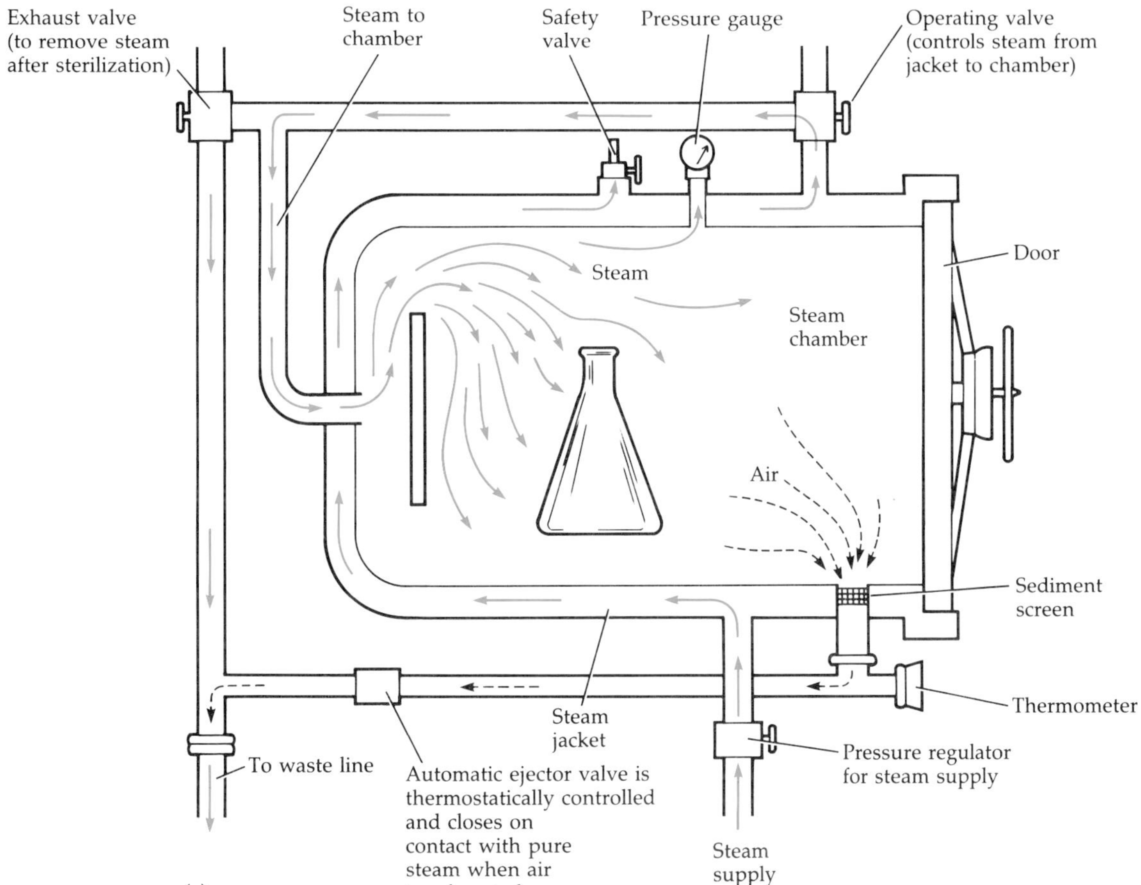

(b)

Figure 7–2 (a) Autoclave. Parts of an autoclave and path of steam flow (solid arrows) within it. The autoclave is a steam chamber capable of withstanding more than 1 atmosphere of pressure (15 pounds per square inch). As steam enters the chamber, air is forced out through a vent in the bottom of the chamber (broken arrows). The automatic ejector valve remains open as long as an air-steam mixture is passing out of the waste line. When all of the air has been ejected, the higher temperature of the remaining pure steam closes the valve. As steam continues to enter, the pressure in the chamber increases. **(b)** Front views of two modern autoclaves.

Table 7–4 Effect of Container Size on Sterilization Times for Liquid Solutions*

Container Size	Liquid Volume	Sterilization Time in Minutes
Test tube: 18 × 150 mm	10 ml	15
Erlenmeyer flask: 125 ml	95 ml	15
Erlenmeyer flask: 2000 ml	1500 ml	30
Fermentation bottle: 9000 ml	6750 ml	70

*Larger containers require additional time to reach sterilizing temperatures. For a test tube, this additional time might be only 5 minutes, but for a 9000-ml bottle it might be 60 minutes. A container is usually not filled past 75% of its capacity.

applicators, solutions, syringes, transfusion equipment, and numerous other items that can withstand high temperature and pressure. Large industrial autoclaves are called *retorts,* but the same principle is used for the common household *pressure cooker,* in the home canning of foods.

Heat requires extra time to reach the center of solid materials such as canned meats, in which heat is conducted without the efficient convection currents in liquids. Extra time is also needed for the heating of large containers. Table 7–4 shows the different time requirements of various container sizes. It is important to remember that to sterilize a surface, steam has to actually contact it. With liquids this is no problem, but with dry glassware, bandages, and the like, care must be taken to ensure that contact is made. For example, aluminum foil is impervious to steam and should not be used for wrapping; paper should be used instead. One should also avoid trapping air in the bottom of a dry container, because trapped air will not be replaced by steam, which is lighter than air. The trapped air is the equivalent of a small hot-air oven, which, as we shall see shortly, requires a vigorous combination of heat and time. Such containers should be tipped so that the air can be displaced by steam. Products that do not permit penetration by moisture, such as mineral oil or petroleum jelly, are not sterilized by the same treatments that would sterilize aqueous solutions.

Steam under pressure commonly fails to sterilize because the air is not completely exhausted. This is usually due to the premature closing of the autoclave's automatic ejector valve (Figure 7–2). Anyone familiar with home canning knows that the steam must flow vigorously out of the valve in the lid for several minutes, to entrain and remove all the air before the pressure cooker is sealed. If the air is not completely exhausted, the container will not reach the temperature expected for a given pressure.

Dry heat sterilization

One of the simplest methods of dry heat sterilization is **direct flaming.** You will use this procedure many times in the laboratory when you sterilize inoculating loops and needles. To sterilize the inoculating loop or needle, all you have to do is heat the wire to a red glow. This method is 100% effective. A similar principle is used in **incineration.** This is an effective way to sterilize and dispose of contaminated paper cups, bags, and dressings.

Another form of dry heat sterilization is **hot air sterilization.** Items to be sterilized by this procedure are placed in an ovenlike apparatus. Generally, a temperature of about 170°C maintained for nearly two hours ensures sterilization. The longer period of time and higher temperature, relative to moist heat, is required because the heat of water is more readily transferred to a cool body than is the heat in air. (Compare the effects of briefly immersing your hand in hot water and of holding it in a hot-air oven at the same temperature for the same time.)

In addition to moist and dry heat, another heating procedure, called pasteurization, is used to control microbes.

Pasteurization

You may recall from Chapter 1 that Louis Pasteur, in the early days of microbiology, found a practical method of preventing the "sickness" or spoilage of beer and wine. Pasteur used mild heating, which was sufficient to kill the organisms that caused the particular spoilage problem without seriously damaging the taste of the product. The same principle was later applied to milk to produce what we now call pasteurized milk. Milk was first pasteurized to eliminate the tuberculosis bacterium. Many relatively heat-resistant (*thermoduric*) organisms survive pasteurization, but these are unlikely to cause disease, nor will they cause refrigerated milk to spoil in a reasonable time.

In the classic **pasteurization** treatment of milk, the milk was exposed to a temperature of about 63°C for 30 minutes. Most pasteurization done today uses higher temperatures, at least 72°C, for about 15 seconds. This treatment, known as **high-temperature, short-time (HTST) pasteurization,** is applied as the milk flows continuously past a heat exchanger. Temperatures higher than are needed to kill pathogens and slightly longer times are common today; the aim is lower bacterial counts so milk keeps well under refrigeration. Milk can be sterilized so that it can be stored without refrigeration. It is heated to about 138°C for a few seconds and then cooled rapidly to minimize the scalded taste. Modern methods of aseptic packaging (Chapter 25) are often used following sterilization.

The heat treatments we have just discussed employ the concept of **equivalent treatments.** For example, the destruction of highly resistant endospores might take 70 minutes at 115°C, whereas only 7 minutes would be needed at 125°C. Both treatments yield the same result. In other words, as the temperature is increased, less time is needed for the same number of microbes to be killed.

Filtration

Filtration is the passage of a liquid or gas through a screenlike material with pores small enough to retain microorganisms (see Figure 6–16). A vacuum that is created in the receiving flask aids gravity in pulling the liquid through the filter. Recall from Chapter 6 that filtration is sometimes used to concentrate microorganisms from a liquid source where small numbers of microbes are found. Filtration is used to sterilize heat-sensitive materials, such as some culture media, enzymes, vaccines, and antibiotic solutions.

In the early days of microbiology, hollow, cylindrical candle-shaped filters of unglazed porcelain were used. The long and indirect passageways through the walls of the filter adsorbed the bacteria. Unseen pathogens that passed through such filters and caused diseases such as rabies were called *filterable viruses.* (The term *virus* was then equated with poison.)

In recent years, **membrane filters,** composed of substances such as cellulose esters or plastic polymers, have become popular for industrial and laboratory use (Figure 7–3). These filters are only 0.1 mm thick. The pores of membrane filters are closely matched in size. In some brands, plastic film is irradiated so that very uniform holes, where the radiation particles have passed, are etched in the plastic. The pores of membrane filters include 0.22 and 0.45 μm sizes, intended for bacteria, and range down to only 0.01 μm, a size that will retain viruses, and even some large protein molecules.

Low Temperature

The effect of **low temperatures** on microbes depends on the particular microbe and the intensity of the application. For example, at temperatures of ordinary refrigerators (0° to 7°C), the metabolic rate of most microbes is so reduced that they do not reproduce or elaborate products. In other words, ordinary refrigeration has a bacteriostatic effect. Yet, psychrophilic (cold-loving) species do grow slowly at refrigerator temperatures and will alter the appearance and taste of foods after a considerable time. Surprisingly, some bacteria can grow at temperatures several degrees below freezing. One-third of the population of some vegetative bacteria might survive a year of freezing; other species might have very few survivors. Rapidly attained subzero (°C) temperatures render microbes dormant, but do not reliably kill them. Freezing is most harmful to bacteria when done slowly; the ice crystals that then form and grow disrupt the cellular and molecular structure of the bacteria.

Desiccation

To grow and multiply, microbes require water. In the absence of water, a condition known as **desiccation,** microbes are not capable of growth or reproduction, but can remain viable for years. Then, when water is made available to them, they can resume their growth and division. This ability is

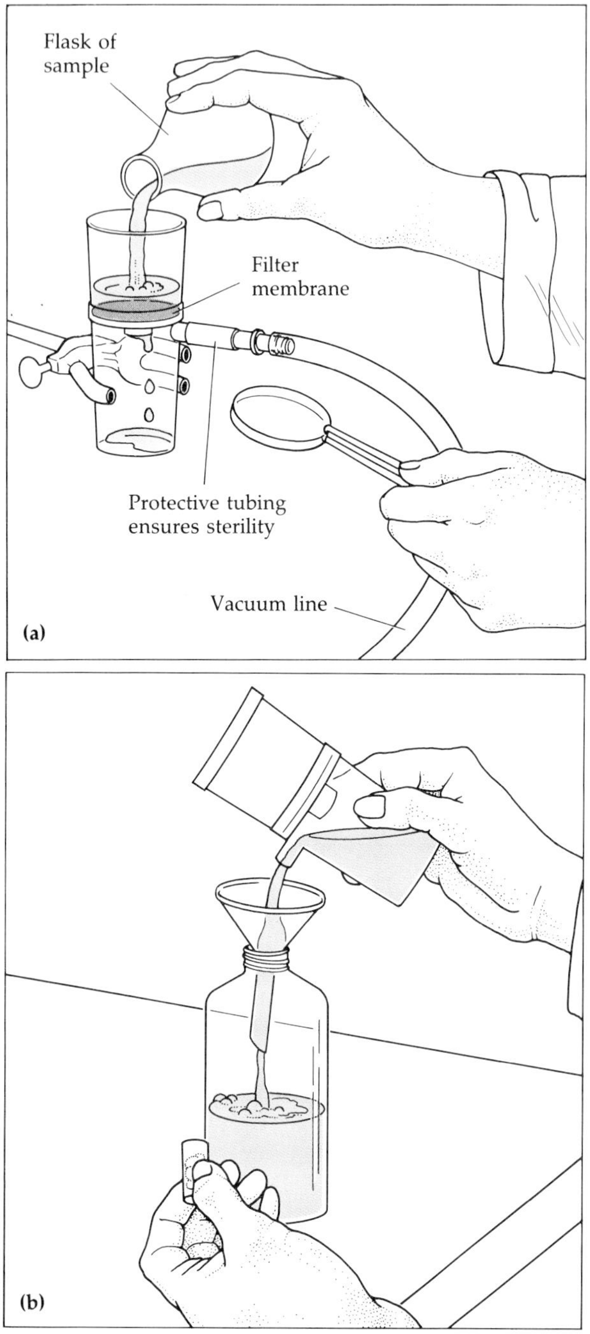

Figure 7–3 Cold sterilization with a disposable, presterilized plastic unit. **(a)** The sample is placed into the upper chamber and forced through the membrane filter by the vacuum in the lower chamber. Pores in the membrane filter are smaller than most bacteria. **(b)** The sterilized sample is decanted and the filter discarded. These assemblies can also be used to count bacteria and are widely used to detect coliforms in water (see Figure 25-10).

used in the laboratory when microbes are preserved by lyophilization, a process described in Chapter 6.

Vegetative cells' resistance to desiccation varies with the species and the organism's environment. For example, the gonorrhea organism can withstand dryness for only about an hour or so, but the tuberculosis bacterium can remain viable for months. A normally susceptible bacterium is much more resistant if it is embedded in mucus, pus, or feces. Viruses are generally resistant to desiccation, but not as resistant as endospores. This ability of certain dried microbes and endospores to remain viable is important in a hospital setting. Dust, clothing, bedding, and dressings might contain infectious microbes in dried mucus, urine, pus, and feces.

Nomadic peoples such as some native Americans used desiccation to preserve foods. If meat is sliced thin and sun dried (made into jerky), it can be preserved for long periods of time. Raisins and similarly dried fruits are other examples.

Osmotic Pressure

The use of high concentrations of salts and sugars to preserve food is based on the effects of **osmotic pressure.** High concentrations of these substances create a hypertonic environment that causes water to leave the microbial cell (see Chapter 4). This process resembles preservation by desiccation in that both methods deny the cell the moisture it needs for growth. As water leaves the microbial cell, the plasma membrane shrinks away from the cell wall (plasmolysis), and the cell stops growing, although it might not immediately die. As mentioned in Chapter 4, the principle of osmotic pressure is used in the preservation of foods. For example, concentrated salt solutions are used to "cure" meats, and thick sugar solutions to preserve fruits.

As a general rule, molds and yeasts are much more capable of growing in materials with low moisture or high osmotic pressures than are bacteria. This property of molds, sometimes combined with their ability to grow under acidic conditions, is the reason fruits and grains are spoiled by molds rather than bacteria. It is also part of the reason why molds are able to form "mildew" growth on such unlikely places as a damp wall or a shower curtain.

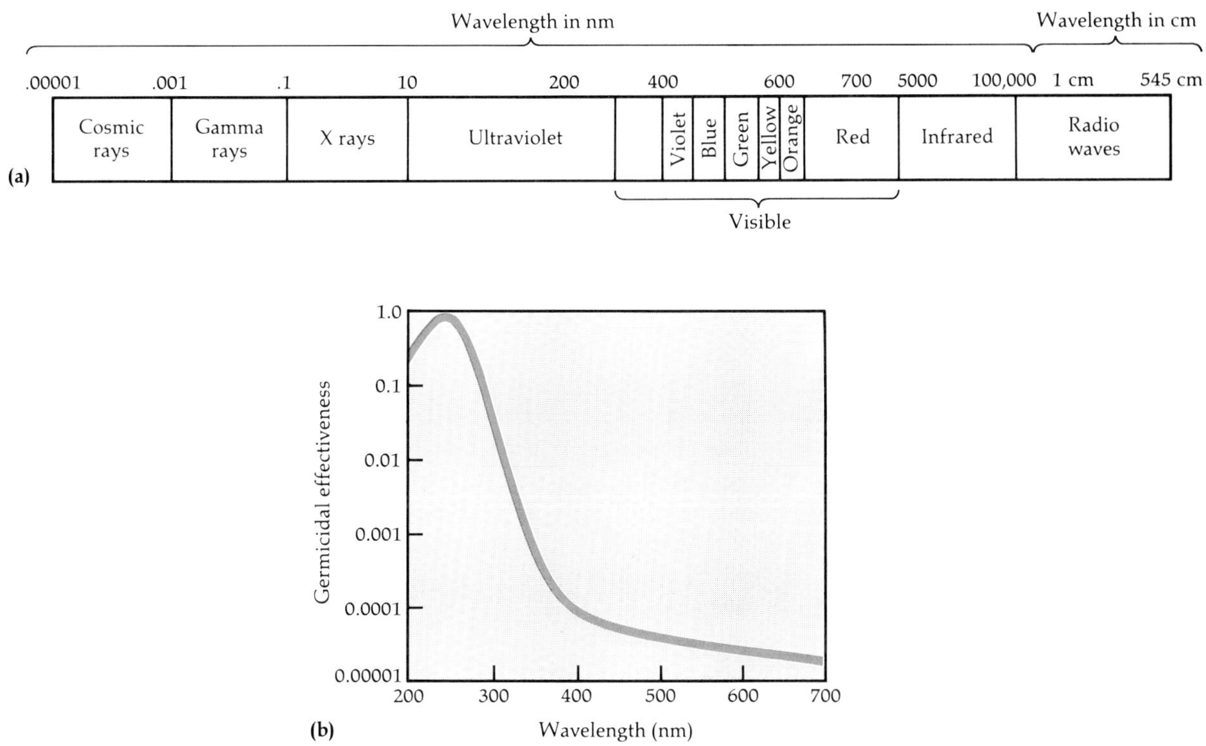

Figure 7–4 Radiant energy. **(a)** Radiant energy spectrum. **(b)** Radiant energy has its maximum germicidal effect at short wavelengths, particularly around 260 nm. Those wavelengths are selectively absorbed by DNA.

Radiation

Depending on its wavelength, intensity, and duration, **radiation** has various effects on cells. There are two types of radiation: ionizing and nonionizing.

Ionizing radiation, such as **X rays, gamma rays,** or **high-energy electron beams,** has a wavelength shorter than that of nonionizing radiation, and therefore carries much more energy (Figure 7–4a). Gamma rays are emitted by radioactive cobalt, and electron beams result from electrons that are accelerated to high energies in special machines. Ionizing radiation has a high degree of penetrating power, and damages microbes when directed at them: Its principal effect is the ionization of water, which forms highly reactive hydroxyl radicals (see discussion of toxic forms of oxygen in Chapter 6).

Interest in the use of radiation for food preservation has recently renewed. It is possible to commercially sterilize foods by radiation with acceptable results. Probably the greatest obstacle to this sterilization method is public apprehension about radiation. In the near future, the most likely applications of radiation are the elimination of insect pests, pork treatment to control trichinosis, and pasteurization-type treatments short of commercial sterilization. Radiation is now used for such purposes in Europe.

Ionizing radiation does not use heat or moisture and has great penetrability. It is therefore very useful for the sterilization of pharmaceuticals and disposable dental and medical supplies, such as plastic syringes, surgical gloves, suturing materials, and catheters. For sterilizing these items, it has recently been replacing sterilizing gases. These gases will be discussed later in the chapter.

Nonionizing radiation has a wavelength longer than that of ionizing radiation and therefore carries less energy. A good example of nonionizing radiation is **ultraviolet (UV) light.** UV light damages the DNA of exposed cells. It causes bonds to form between adjacent thymines in DNA chains (see Figure 8–16). These *thymine dimers* inhibit correct replication of the DNA during reproduction of the cell. The UV wavelengths most effective for killing microorganisms are around 260 nm; these wavelengths are specifically absorbed by cellular DNA (Figure 7–4b). UV radiation is also used to control microbes in the air. A UV or "germicidal" lamp is commonly found in hospital rooms, nurseries, operating rooms, and cafeterias. UV light is also used to sterilize vaccines, sera and toxins, and food. A major drawback of UV light as a sterilizer is that is has low penetrability, so organisms to be killed must be directly exposed to the rays. Organisms protected by solids and coverings such as paper, glass, and textiles are not affected. Another potential problem is that UV light can damage the eyes, and prolonged exposure to UV light can cause burns and skin cancer.

Sunlight contains some UV light, but the most effective (short) wavelengths are screened out by the atmosphere. The antimicrobial effect of sunlight is due almost entirely to its forming singlet oxygen in the cytoplasm (see Chapter 6). Many pigments produced by bacteria provide demonstrated protection from sunlight.

Microwaves do not have much direct effect on microorganisms, but kill indirectly by heating the medium, e.g. food. The physical methods of microbial control are summarized in Table 7–5.

We will now turn our attention to a discussion of several chemical methods of microbial control.

CHEMICAL METHODS OF MICROBIAL CONTROL

Chemical agents are used to control microbes on living tissue and inanimate objects. Unfortunately, few chemical agents achieve sterility; most of them merely reduce microbial populations to safe levels or remove vegetative forms of pathogens from objects. A common problem in disinfection is the selection of an agent that will kill all organisms in the shortest period of time without damaging the contaminated material. Just as there is no single physical method of microbial control that can be used in every situation, there is no one disinfectant that will be appropriate for all circumstances.

Qualities of an Effective Disinfectant

The more of the following qualities a disinfectant has, the more effective it is:

1. Acts rapidly.
2. Attacks all, or a wide range, of microbes.
3. Is able to penetrate thoroughly the material that is contaminated.
4. Readily mixes with water to form a stable solution or emulsion.
5. Is not hampered by organic matter on the substance to be disinfected.
6. Is not likely to decompose and thereby lose its activity after exposure to light, heat, or unfavorable weather.
7. Does not stain, corrode, or destroy the object being disinfected.
8. Is harmless to animals if it is to be used as an antiseptic and does not destroy body tissues or act as a toxin if inhaled or swallowed.

The ideal disinfectant should also have a pleasant odor, and be economical to use and safe to transport.

Principles of Effective Disinfection

In order to select a disinfectant for a particular job, we must first understand its action. For example, what are the properties of the disinfectant? What is it designed to do? Simply by reading the label, we can learn a great deal about a disinfectant's properties. We should also remember that the concentration of a disinfectant will affect its action. A disinfectant should always be diluted exactly as suggested by the manufacturer. Solutions that are too weak may be ineffective, or bacteriostatic instead of bactericidal. On the other hand, solutions that are too strong can be dangerous to humans who come in contact with them.

Table 7–5 Summary of Physical Methods Used to Control Microbial Growth

Method	Mechanism of Action	Comment	Preferred Use
Heat:			
1. Moist heat sterilization			
a. Boiling or flowing steam	Denaturation	Kills vegetative bacterial and fungal pathogens and many viruses within 10 minutes; less effective on endospores	Dishes, basins, pitchers, various equipment
b. Autoclaving	Denaturation	Very effective method of sterilization; at about 15 pounds of pressure (121°C), all vegetative cells and their endospores are killed in about 15 minutes	Microbiological media, solutions, linens, utensils, dressings, equipment, and other items that can withstand temperature and pressure
2. Dry heat sterilization			
a. Direct flaming	Burning to ashes	Very effective method of sterilization	Inoculating loops and needles
b. Incineration	Burning to ashes	Very effective method of sterilization	Paper cups, dressings, animal carcasses, bags, and wipes
c. Hot-air sterilization	Denaturation	Very effective method of sterilization, but requires temperature of 170°C for about 2 hours	Empty glassware, instruments, needles, and syringes
3. Pasteurization	Denaturation	Heat treatment for milk (72°C for about 15 seconds) that kills all pathogens and some nonpathogens	Milk, cream, and certain alcoholic beverages (beer and wine)
Filtration	Separation of bacteria from suspending liquid	Passage of a liquid or gas through a screenlike material that traps microbes; most filters in use consist of cellulose acetate	Useful for sterilizing liquids (toxins, enzymes, vaccines) that are destroyed by heat
Low Temperature:			
1. Refrigeration	Decreased chemical reactions and possible changes in proteins	Has a bacteriostatic effect	Food, drug, and culture preservation
2. Deep-freezing (see Chapter 6)	Decreased chemical reactions and possible changes in proteins	An effective method for *preserving* microbial cultures in which cultures are quick-frozen between −50° and −95°C	Food, drug, and culture preservation
3. Lyophilization (see Chapter 6)	Decreased chemical reactions and possible changes in proteins	Most effective method for long-term preservation of microbial cultures; water removed by high vacuum at low temperature	Food, drug, and culture preservation
Desiccation	Disruption of metabolism	Involves removing water from microbes; primarily bacteriostatic	Food preservation
Osmotic Pressure	Plasmolysis	Results in microbial cells losing water	Food preservation
Radiation:			
1. Ionizing	Destruction of DNA by X rays, gamma rays, and high-energy electron beams	Not widespread in routine sterilization	Used for sterilizing pharmaceuticals and medical and dental supplies
2. Nonionizing	Damage to DNA by UV light	Radiation not very penetrating	Practical application is the UV (germicidal) lamp

Also to be considered is the nature of the material being disinfected. Are organic materials present that might interfere with the action of the disinfectant? The presence of organic materials and the pH of the medium in which the microbes are present might determine whether a chemical control agent is only inhibitory or is lethal to the microbes.

Another very important consideration is whether the disinfectant will easily make contact with the microbes. An area might have to be scrubbed and rinsed before the disinfectant is applied. In general, disinfection is a gradual process. Thus, to be effective, a disinfectant might have to be left on a surface for several hours.

A final factor to keep in mind is that the higher the temperature at which the disinfectant is applied, the more effective it usually is.

Evaluating a Disinfectant

Phenol coefficient

There is an obvious need to compare the effectiveness of disinfectants and antiseptics. Phenol, which was once a widely used disinfectant, has been widely used as a standard for these comparisons. However, some antimicrobials cannot be compared with phenol, particularly if their action is bacteriostatic or if they have a prolonged residual activity on the skin. Nonetheless, the **phenol coefficient test** is still widely used and illustrates the general principles involved in all such comparisons. For this test, three test organisms are used: *Staphylococcus aureus* (a gram-positive bacterium), *Salmonella typhi* (a gram-negative bacterium) and *Pseudomonas aeruginosa* (a gram-negative organism often resistant to antimicrobials). In broth cultures under standard conditions, all three test organisms are exposed to the test chemical for a specified time. If the test chemical must be used at a greater concentration or for a longer period of time than phenol is used, to achieve comparable effects, the test chemical's phenol coefficient will be less than one. This rating indicates that the test chemical is weaker than phenol. A coefficient greater than one indicates that the chemical is more active than phenol. For example, if Brand X at a dilution concentration of 1 : 200 is effective under the specified conditions, and phenol is equally effective at a concentration of 1 : 100, then Brand X is more efficient than phenol by a factor of 200/100—and has a phenol coefficient of 2. The result of a phenol coefficient test is usually confirmed by the use-dilution test.

The **use-dilution test** is sometimes considered the official evaluation of a disinfectant. In this test, standardized preparations of several test bacteria are added to a series of tubes containing increasingly strong concentrations of the test disinfectant. The tubes are then incubated, and growth, or lack of it, is recorded. The more highly diluted the chemical can be and still be effective, the higher its rating.

Many specialized tests can be used to determine specific effects of disinfectants, for example, their sporicidal, tuberculocidal, or fungicidal effectiveness.

Filter paper method

The **filter paper method** is often used in teaching laboratories as a demonstration evaluation of a chemical agent. A disk of filter paper is soaked with a chemical and placed on the surface of agar plates that have been previously inoculated and incubated with the test organism. If the chemical agent is effective, a clear zone, representing inhibition, can be observed around the disk (Figure 7–5).

Kinds of Disinfectants

Phenol and phenolics

Phenol (carbolic acid), the substance first used by Lister in his operating room, is no longer used as an antiseptic and seldom as a disinfectant because it irritates the skin and has a disagreeable odor. It is often used in throat lozenges for its local anesthetic effect but has no antimicrobial effect at the low concentrations used. As mentioned previously, it is also still used as a standard of disinfectant effectiveness. The structure of a phenol molecule is shown in Figure 7–6a.

Derivatives of phenol, called **phenolics,** contain a molecule of phenol that has been chemically altered to reduce its irritating qualities or increase its antibacterial activity in combination with a soap or detergent. Phenolics exert antimicrobial activity

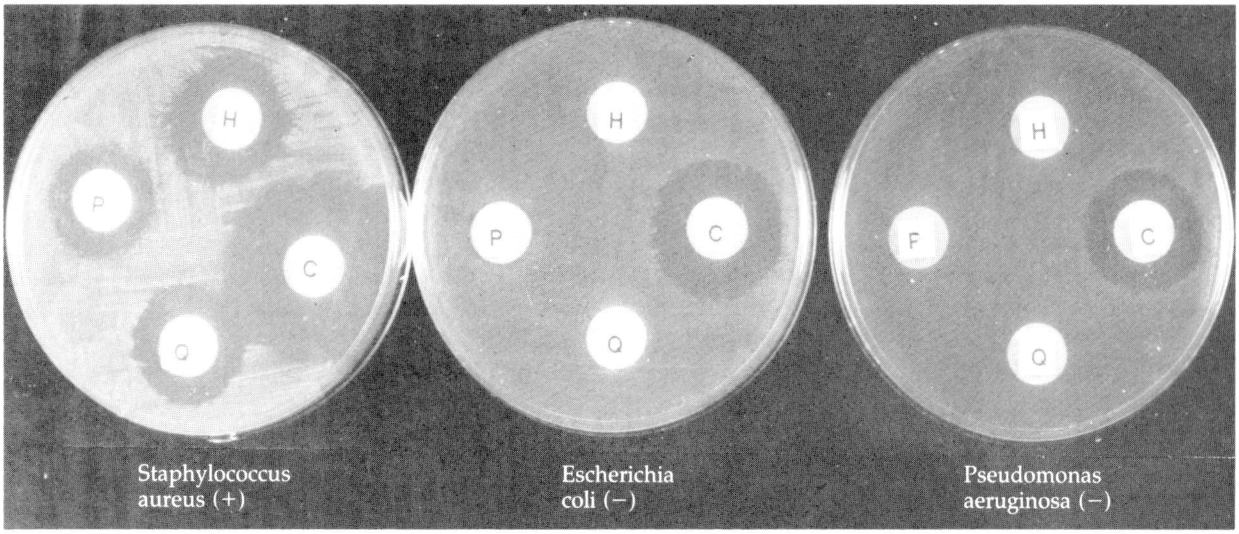

Figure 7–5 Evaluation of disinfectants by the filter paper method. The paper disks are soaked in a solution of the disinfectant and placed on the surface of a nutrient medium, on which a culture of test bacteria has been streaked. The disinfectants, (H) hexachlorophene, (C) sodium hypochlorite (chlorine), (Q) benzalkonium chloride (a quaternary ammonium compound), and (P) O-phenylphenol, are not equally effective against all bacteria. Note that the gram-positive bacterium *Staphylococcus aureus* was the only one susceptible to all of the disinfectants. Note too that *Pseudomonas aeruginosa* is affected only by chlorine (C), but that its zone of inhibition is significantly smaller than those of the other two bacteria. Hexachlorophene (H), which is rather specific for gram-positive organisms, has little effect on gram-negatives, such as *Escherichia coli.* O-phenylphenol (P) and benzalkonium chloride (Q) can be effective against *Escherichia coli,* but their concentrations would have to be much higher than those used effectively on *S. aureus*. The zones of inhibition produced by these concentrations are too small to show.

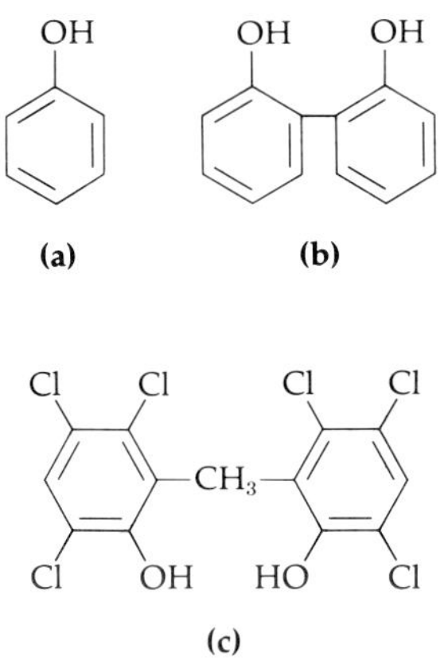

Figure 7–6 Structure of phenol and phenolics. **(a)** Phenol. **(b)** O-phenylphenol. **(c)** Hexachlorophene.

by injuring plasma membranes, inactivating enzymes, and denaturing proteins. Phenolics are frequently used as disinfectants because they remain active in the presence of organic compounds, they are stable, and they persist for long periods of time after application. For these reasons, phenolics are suitable agents for disinfecting pus, saliva, and feces. The addition of halogens such as chlorines to phenolics usually increases their antimicrobial activity.

One of the most frequently encountered phenolics is derived from coal tar, a group of chemicals called *cresols*. A very important cresol is *O-phenylphenol* (Figure 7–6b), the main ingredient in most formulations of Lysol. Cresols are very good environmental disinfectants.

Another phenolic, used in the past much more widely than now, is *hexachlorophene*, consisting of two molecules of phenol (a bis-phenol) joined together (Figure 7–6c). This phenolic was originally in soaps and lotions (pHisoHex) used for surgical scrubs and hospital microbial control procedures, cosmetic soaps, deodorants, feminine hygiene sprays, and even toothpaste. As a bacteriostatic agent, hexachlorophene is effective, especially against gram-positive staphylococcal and streptococcal bacteria, which can cause infections of the skin. Moreover, the hexachlorophene persists on the skin for long periods of time. For these reasons, it was once heavily used to control staphylococcal and streptococcal infections in hospital nurseries. However, in 1972, it was found that *excessive* use of hexachlorophene, such as bathing infants with it several times a day, could lead to neurological damage. Hexachlorophene is still used in hospitals to control nosocomial infections and as a "scrub" for hospital personnel. However, it is no longer casually used by the general public. At present, a prescription is required for the purchase of a 3% or stronger solution of hexachlorophene.

Halogens

The **halogen** elements, particularly iodine and chlorine, are effective antimicrobial agents both alone (as I_2 or Cl_2 in solution) and as constituents of inorganic or organic compounds. *Iodine* (I_2) is not only one of the oldest antiseptics, but also one of the most effective. It is also used to disinfect water, air, and food utensils. Iodine is effective against all kinds of bacteria, many endospores, various fungi, and some viruses. A proposed mechanism for the activity of iodine is that it combines with the amino acid tyrosine, a common component of many enzymes and other cellular proteins (Figure 7–7). As a result, microbial protein function is inhibited. Some of iodine's antimicrobial activity might be due to its powerful oxidizing ability.

Iodine is available as a *tincture,* that is, in solution in aqueous alcohol, and as an iodophor. An *iodophor* is a combination of iodine and an organic molecule, usually a detergent in which the iodine is released slowly. Iodophors have the antimicrobial activity of iodine, but they do not stain and are less irritating than iodine. Some commercially available iodophors include Wescodyne and Ioclide, which are used for environmental disinfection, and Betadine and Isodine, which are used as antiseptics for skin. Although several iodophor antiseptic preparations are widely used in hospitals as disinfectants, they are FDA-approved only for antisepsis of skin and mucous membranes (see Box).

Chlorine, as a gas or in combination with other chemicals, is another widely used disinfectant. Its germicidal action is due to the hypochlorous acid that forms when chlorine is added to water.

$$\text{(1)}\quad \underset{\text{Chlorine}}{Cl_2} + \underset{\text{Water}}{H_2O} \rightleftharpoons \underset{\text{Hydrogen ion}}{H^+} + \underset{\text{Chloride ion}}{Cl^-} + \underset{\text{Hypochlorous acid}}{HOCl}$$

$$\text{(2)}\quad \underset{\text{Hypochlorous acid}}{HOCl} \rightleftharpoons \underset{\text{Hydrogen ion}}{H^+} + \underset{\text{Hypochlorite ion}}{OCl^-}$$

Exactly how hypochlorous acid exerts its killing power is not completely known. Probably it

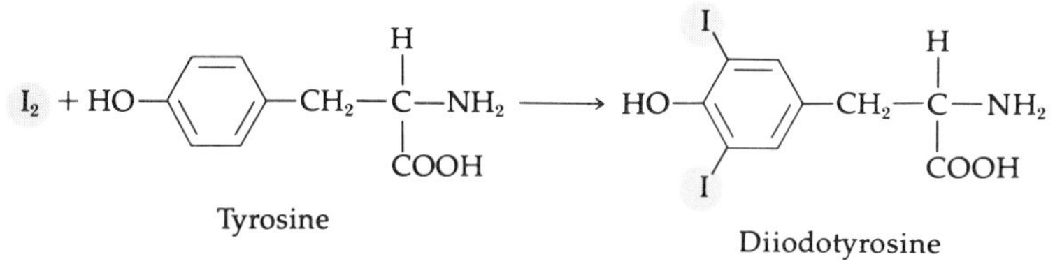

Figure 7–7 Proposed mechanism for the activity of iodine. Tyrosine is an amino acid found in protein. Iodine probably also reacts with other cellular constituents.

Centers for Disease Control
MMWR
Morbidity and Mortality Weekly Report

Pseudomonas aeruginosa Peritonitis Attributed to a Contaminated Iodophor Solution

Five infections involving patients at a hospital have been attributed to use of contaminated Prepodyne Solution.* Intrinsic contamination of this iodophor product with *Pseudomonas aeruginosa* has been confirmed by CDC.

In the period March 9-April 12, 1982, 5 chronic peritoneal dialysis patients at a municipal hospital in Atlanta became infected with *P. aeruginosa.* Four patients developed peritonitis, and 1 developed a skin infection at the catheter insertion site. All 4 patients with peritonitis had low-grade fever, cloudy peritoneal fluid, and abdominal pain. Three of these patients used an automatic peritoneal dialysis machine; 1 used only a bottle cycling machine. All patients had permanent indwelling peritoneal catheters, which were wiped with a 4×4 gauze soaked with an iodophor, Prepodyne Solution, each time the catheter was connected to or disconnected from machine tubing. Aliquots of Prepodyne Solution were transferred from stock bottles to smaller in-use bottles.

Infection-control personnel at the hospital obtained cultures of the dialysate concentrate, internal areas of the dialysis machines, and a small in-use plastic container that had been filled with Prepodyne Solution. All cultures were negative except the Prepodyne Solution, which yielded a pure culture of *P. aeruginosa.* Subsequently, 2 of 8 unopened 1-gallon containers of Prepodyne Solution, lot C109756, one obtained from the dialysis center and a second from another hospital area, were culture-positive; they, too, yielded a pure culture of *P. aeruginosa.* The antimicrobial susceptibility patterns of the isolates obtained from the Prepodyne Solution were identical to those of 2 available isolates from patients; these organisms were sensitive when tested in the hospital to amikacin, carbenicillin, gentamicin, kanamycin, tetracycline, tobramycin, and trimethoprim-sulfamethoxazole, and were resistant to ampicillin and cephalothin.

Editorial Note: This is the second report of contamination of an iodophor solution (*1, 2*) and the first of a poloxamer-iodine solution. Poloxamer-iodine and povidone-iodine are the most commonly used iodophor preparations in hospitals; both preparations have now been demonstrated to be vulnerable to intrinsic contamination. Pseudobacteremia has been described in association with the use of a contaminated iodophor preparation. (In pseudobacteremia, the patient appears to have viable bacteria circulating in his/her blood; that is, bacteremia. The contamination is actually due to a fomite, and the patient's blood is sterile.) The report above demonstrates that contaminated solutions may lead to true infections. Although several iodophor antiseptic preparations, of which Prepodyne Solution is one, are widely used in hospitals for disinfectant purposes, they are FDA-approved only for antisepsis of skin and mucous membranes.

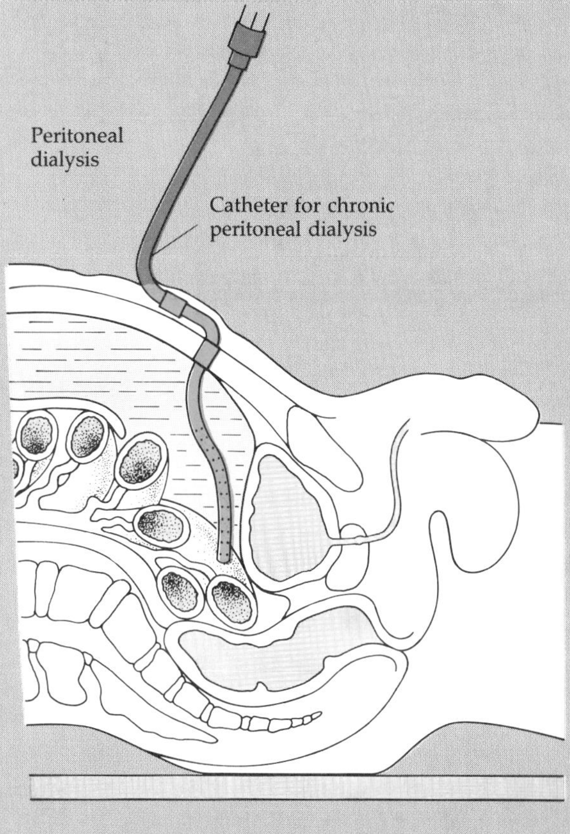

*Use of trade names is for identification only.
Source: MMWR 31:197 (4/23/82).

releases a reactive form of oxygen that combines with cellular protoplasm. The hypochlorite ion (equation 2) also has some killing potential.

A liquid form of compressed chlorine gas is used extensively for disinfecting municipal drinking water, water in swimming pools, and sewage. Several compounds of chlorine are also effective disinfectants. For example, solutions of *calcium hypochlorite* ($Ca{\cdot}(OCl)_2$) are used to disinfect dairies, barns, slaughterhouses, and restaurants' eating utensils. Another chlorine compound, *sodium hypochlorite* (NaOCl), is used as a household disinfectant and a bleach (Clorox), as well as a disinfectant in dairies, food-processing establishments, and hemodialysis systems. When the quality of drinking water is in question, household bleach can achieve a rough equivalent of municipal chlorination. After two drops of bleach are added to a liter of water (4 drops if the water is cloudy) and the mixture has sat for 30 minutes, the water is considered safe for drinking under emergency conditions.

Another group of chlorine compounds, the *chloramines*, consist of chlorine and ammonia. They are used as disinfectants, antiseptics, or sanitizing agents. Chloramines are very stable compounds that release chlorine over long periods of time. They are effective in organic matter, but have the disadvantages of acting more slowly and being less effective purifiers than many other chlorine compounds are. Chloramines are used to sanitize glassware and eating utensils and to treat dairy and food-manufacturing equipment. They may also be used for the emergency disinfection of small amounts of drinking water. Halazone is a chloramine used for this emergency purpose.

Alcohols

Alcohols effectively kill bacteria and fungi but not endospores and nonenveloped viruses. The mechanism of action of alcohol is protein denaturation, so alcohol can dissolve many lipids, including the lipid component of enveloped viruses. Alcohols are not effective against nonenveloped viruses because these viruses can have especially stable proteins. Alcohols are not very effective, for example, against the hepatitis virus, which is not enveloped and is spread by syringes and other skin-penetrating instruments. Alcohols have the advantages of acting and evaporating rapidly and leaving no residue. In a quick swabbing of the skin before an injection (degerming), most of the antiseptic action comes from a simple *wiping away* of dirt and microorganisms. Using an iodine-alcohol solution will increase the number of microbes killed on skin. Alcohols are also used to disinfect clinical oral thermometers.

Two of the most commonly used alcohols are ethanol and isopropanol. The recommended optimum concentration of *ethanol* (CH_3CH_2OH) is 70%, but concentrations between 60 and 95% seem to kill as fast (Table 7–6). Pure alcohol is less effective than aqueous solutions (alcohol mixed with water) because denaturation requires water. *Isopropanol* ($(CH_3)_2CHOH$) often sold as rubbing alcohol, is slightly superior to ethanol as an antiseptic and disinfectant. Moreover, it is less volatile, less expensive, and more easily obtainable than ethanol.

Alcohols are also used to enhance the effectiveness of other chemical agents. For example, an aqueous solution of Zephiran (described shortly) kills about 40% of the population of a test organism in two minutes, whereas a tincture of Zephiran kills about 85% in the same period of time. You can compare the effectiveness of tinctures and aqueous solutions in Figure 7–10 (see page 201).

Table 7–6 Germicidal Action of Various Concentrations of Ethanol Against *Streptococcus pyogenes* (*Note:* + = Growth; − = No Growth)

Percentage of Ethanol	Seconds				
	10	20	30	40	50
100	+	+	+	+	+
95	−	−	−	−	−
90	−	−	−	−	−
80	−	−	−	−	−
70	−	−	−	−	−
60	−	−	−	−	−
50	+	+	−	−	−
40	+	+	+	+	+

Source: H. E. Morton, *Ann. N.Y. Acad. Sci. 53:*191–196, (1950).

Heavy metals and their compounds

Several **heavy metals,** such as silver, mercury, and copper, can be germicidal or antiseptic. The ability of very small amounts of heavy metals, especially silver and copper, to exert antimicrobial activity is referred to as **oligodynamic action** (*oligo* = few). This action can be seen when we place a coin or other clean piece of metal containing silver or copper on an inoculated Petri plate. Extremely small amounts of metal diffuse from the coin and inhibit the growth of bacteria for some distance around the coin (Figure 7–8). This effect is produced by the known action of heavy metal ions on microbes. When the metal's ions combine with the—SH groups on cellular proteins, denaturation results.

Silver is used as an antiseptic in a 1% *silver nitrate* solution. The solution is bactericidal for most organisms. Many states require that the eyes of newborns be treated with a few drops of silver nitrate, to guard against a gonococcal infection of the eyes called gonorrheal ophthalmia neonatorum, which the infants might have contracted as they passed through the birth canal. In recent years, antibiotics have been replacing silver nitrate for this purpose.

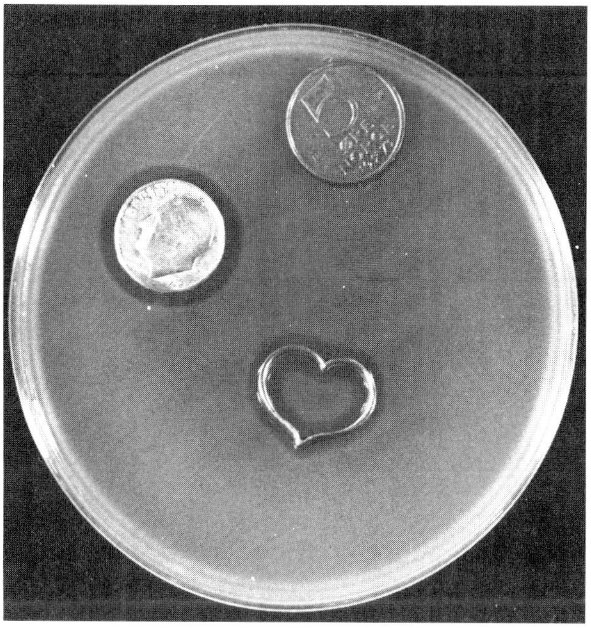

Figure 7–8 Oligodyamic action of silver. Clear zones where bacterial growth has been inhibited are seen around the silver coin and heart. The nonsilver coin has been pushed aside to show the growth beneath it.

Inorganic mercury compounds, such as *mercuric chloride* probably have the longest history of use as a disinfectant. They have a very broad spectrum of activity; their effect is primarily bacteriostatic. However, their use is now limited because of their toxicity, corrosiveness, and ineffectiveness in organic matter. Organic mercury compounds, such as Mercurochrome and Merthiolate are less irritating and less toxic than inorganic mercury compounds. Mercurochrome and Merthiolate are antiseptics used on skin and mucous membranes. Unfortunately, their effects are reversed if they are removed (washed away).

Copper in the form of *copper sulfate* is used chiefly to destroy green algae (algicide) that grow in reservoirs, swimming pools, and fish tanks. If the water does not contain excessive organic matter, copper sulfate is effective in concentrations of 1 part per million (ppm) of water. To prevent mildew, copper compounds are sometimes included in paint.

Another metal used as an antimicrobial is zinc. *Zinc chloride* is a common ingredient in mouthwashes, and *zinc undecylenate* has been used for many years as an antifungal in treatment of athlete's foot. *Zinc oxide* is probably the most widely used antifungal in paints.

Surface-active agents

Surface-active agents, or **surfactants,** can decrease surface tension between molecules of a liquid. Such agents include soaps and detergents. Soap has little value as an antiseptic. However, soaps do have an important function in the mechanical removal of microbes through scrubbing. The skin normally contains dead cells, dust, dried sweat, microbes, and oily secretion from oil glands. Soap breaks the oily film into tiny droplets, a process called **emulsification,** and the water and soap together lift up the emulsified oil and debris and float them away as the lather is washed off. In this sense, soaps are good degerming and emulsifying agents. Many so-called deodorant soaps contain compounds, such as triclocarban, that strongly inhibit gram-positive bacteria.

Acid-anionic surface-active sanitizers are very important in the dairy industry's cleaning of utensils and equipment. Their cleansing ability is related to the negatively charged portion (anion) of the molecule. These sanitizers act on a wide spectrum of microbes including troublesome thermoduric bacteria. They are non-toxic, non-corrosive, and fast acting.

Quaternary ammonium compounds

The most widely used surface-active agents are the cationic detergents (their cleansing ability is related to the positively charged portion—cation—of the molecule), especially the **quaternary ammonium compounds.** Their name comes from the fact that they are modifications of the four-valence ammonium ion (NH_4^+) (Figure 7–9). Quaternary ammonium compounds, or **quats,** as they are commonly called, are strongly bactericidal against gram-positive bacteria, and somewhat less strong against gram-negative bacteria. They also have excellent bacteriostatic action. Quats are also fungicidal, amoebicidal, and virucidal against enveloped viruses. Quats' mode of action is unknown, but they most likely affect the plasma membrane. They change the cell's permeability and cause the loss of essential cytoplasmic constituents such as potassium.

Three popularly used quats are *Zephiran* (benzalkonium chloride), *Phemerol* (benzethonium chloride), and *Ceepryn* (cetylpyridinium chloride). Their popularity is due not only to their strong antimicrobial activity, but also to their being colorless, odorless, tasteless, stable, easily diluted, and non-toxic except at high concentrations. If your mouthwash bottle fills with foam when shaken, the mouthwash probably contains a quat. However, organic matter interferes with their activity, and they are rapidly neutralized by soaps and anionic detergents.

Anyone associated with medical applications of quats should remember that certain bacteria, such as some species of *Pseudomonas,* not only survive in quaternary ammonium compounds but actively grow in them. This resistance occurs not only to the disinfectant solution, but also to the moistened gauze and bandages, the fibers of which tend to neutralize the quats. This resistance is another example of the peculiar and troublesome characteristics of the *pseudomonads.* Their ability to metabolize unusual substrates for growth permits *Pseudomonas* to grow in unlikely places such as simple saline solutions, traces of soap residues, and cap-liner adhesives.

Before we move on to the next group of chemical agents, refer to Figure 7–10, which compares the effectiveness of some of the antimicrobials discussed thus far.

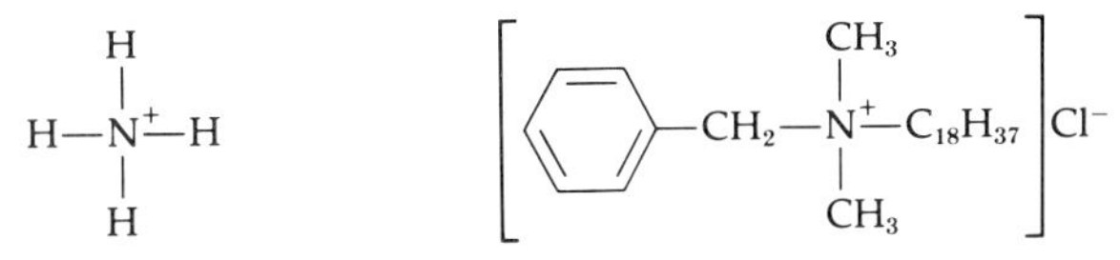

Ammonium ion Benzalkonium chloride

Figure 7–9 The ammonium ion and a quaternary ammonium compound, benzalkonium chloride (Zephiran).

Organic acids

A number of **organic acids** are used as preservatives to control mold growth. *Sorbic acid* (often as the salt *potassium sorbate*) is widely used to inhibit mold growth in acidic foods such as cheese. *Benzoic acid* (or its salt *sodium benzoate*) is an antifungal that is effective at low pH levels and has wide use in soft drinks and other acidic foods. The *parabens,* such as *methylparaben* and *propylparaben,* are often used to inhibit mold growth in liquid cosmetics and shampoos. The parabens are derivatives of benzoic acid but work at a neutral pH. *Calcium propionate* prevents mold growth in bread. The activity of these organic acids is not related to their acidity, but to their ability to inhibit enzymic and metabolic activity. Organic acids are listed on the labels of many foods and a few cosmetic preparations. By and large, the body metabolizes these organic acids readily, and so their use is considered quite safe.

Aldehydes

Aldehydes are among the most effective antimicrobials. Two examples are formaldehyde and glutaraldehyde. They inactivate proteins by forming covalent crosslinks with a number of organic functional groups on proteins (—NH_2, —OH, —COOH, and —SH). *Formaldehyde* gas is an excellent disinfectant. However, it is stable only in high concentrations and at high temperatures. The more commonly used form of formaldehyde is *for-*

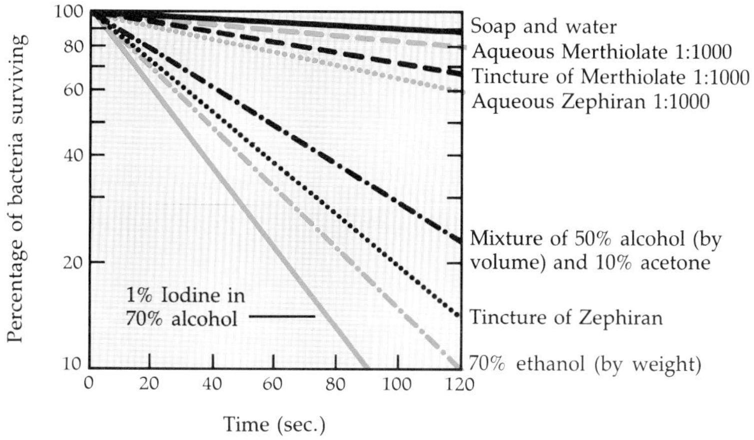

Figure 7–10 Comparison of the effectiveness of various antiseptics. The steeper the downward slope of an antiseptic, the more effective it is. For example, a 1% iodine in 70% alcohol solution is the most effective antiseptic shown here; soap and water is the least effective. Note that tinctures are more effective than aqueous solutions of the same antiseptic.

malin, a 37% aqueous solution of formaldehyde gas. Formalin was once used extensively to preserve biological specimens, embalm corpses, and inactivate bacteria and viruses in vaccines. Formalin is sporicidal at high concentrations. Its sterilizing and disinfecting applications are limited by its tissue-irritating qualities, poor penetration, slow action, unpleasant odor, and its property of leaving a white residue on treated materials.

Glutaraldehyde is a chemical relative of formaldehyde that is less irritating and more pleasant to handle than formaldehyde. Glutaraldehyde is used to sterilize hospital instruments, including respiratory-therapy equipment. When used in a 2% solution (Cidex), it is bactericidal, tuberculocidal, and virucidal in 10 minutes and sporicidal in 3–10 hours. It is probably the only liquid chemical disinfectant that can be considered a possible sterilizing agent. That a lengthy exposure time is required for sporicidal activity emphasizes the fact that chemical agents cannot be relied upon as sterilants.

Gaseous chemosterilizers

Gaseous chemosterilizers are chemicals that sterilize in a closed chamber similar to an autoclave. A gas suitable for this method is *ethylene oxide*

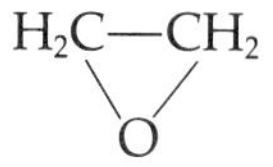

Its activity depends on the denaturation of proteins: the proteins' labile hydrogens, such as —SH, —COOH, or —OH, are replaced by alkyl groups such as $—CH_2CH_2OH$. Ethylene oxide kills all microbes and endospores. It is toxic and explosive in pure form, so it is usually mixed with an inert gas such as carbon dioxide or nitrogen. One of its principal advantages is that it is highly penetrative. Although materials to be sterilized with ethylene oxide must be exposed to it for 4 to 18 hours, its remarkable penetrating power is one reason why ethylene oxide was chosen to sterilize space craft sent to land on the moon and planets. Using heat to sterilize the electronic gear on these vehicles was not practical.

Because of their ability to sterilize without heat, these gases are also widely used on medical supplies and equipment. Examples include disposable sterile plasticware, such as syringes and Petri plates; textiles; sutures; lensed instruments; artificial heart valves; heart-lung machines; and mattresses. Many large hospitals have ethylene oxide chambers, some large enough to sterilize mattresses, as part of their sterilizing equipment. Propylene oxide and beta-propiolactone are also used for gaseous sterilization.

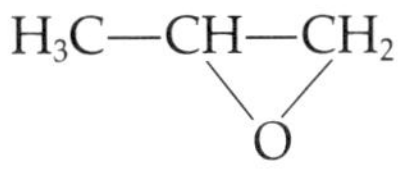

$$H_3C-CH-CH_2$$
$$\backslash \; / $$
$$O$$

Propylene oxide

$$H_2C-CH_2$$
$$| \qquad |$$
$$O-----C=O$$

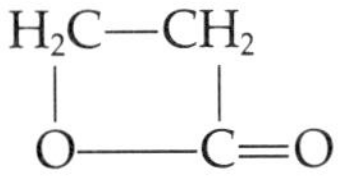

Beta-propiolactone

The second gas is the more hazardous, and is considered a possible carcinogen.

Oxidizing agents

Oxidizing agents exert antimicrobial activity by oxidizing cellular components of the treated microbes. Examples of oxidizing agents are ozone and hydrogen peroxide. *Ozone* (O_3) is a highly reactive form of oxygen that is generated by high-voltage electrical discharges. It is responsible for the air's rather fresh odor after a lightning storm, in the vicinity of electric sparking, or around an ultraviolet light. There is considerable interest in using ozone to replace chlorine in the disinfection of water. Although ozone is an effective killing agent, its residual activity is difficult to maintain in water, and it is more expensive than chlorine.

Hydrogen peroxide is an antiseptic found in many households' medicine cabinets and in hospitals' supply rooms. It is not a good antiseptic for open wounds because it is quickly broken down to hydrogen and oxygen by the action of the enzyme catalase, which is present in human cells (Chapter 6). Although its usefulness as an antiseptic is limited, hydrogen peroxide is effectively used to disinfect inanimate objects. On a nonliving surface, the normally protective enzymes of aerobic bacteria and facultative anaerobes are overwhelmed by the high concentrations of peroxide used. The food industry is increasing its use of hydrogen peroxide for aseptic packaging. The packaging materials pass through a hot solution of the chemical before being assembled into a container. Many wearers of contact lenses are familiar with disinfection by hydrogen peroxide. A platinum catalyst in the lenses' disinfecting kit destroys residual hydrogen peroxide so it does not persist on the lens, where it might be an irritant.

Zinc peroxide, like hydrogen peroxide, is useful in irrigation of deep wounds, where it makes an environment that inhibits the growth of anaerobic bacteria.

Benzoyl peroxide is sometimes a useful treatment of wounds infected by anaerobic pathogens, but is probably much more familiar as the main ingredient in over-the-counter acne medications.

A summary of chemical agents that control microbial growth is presented in Table 7–7.

Antibiotics will be discussed later, in Chapter 18. The compounds discussed in this chapter are not generally useful in the treatment of diseases; because antibiotics are used in chemotherapy, the discussions of antibiotics and the pathogens on which they work have been positioned together.

Table 7–7 Summary of Chemical Agents Used to Control Microbial Growth

Chemical Agent	Mechanism of Action	Preferred Use	Comment
Phenol and phenolics:			
1. Phenol	Disruption of plasma membrane, denaturation, inactivation of enzymes	Still used as a standard for the effectiveness of other disinfectants (phenol coefficient)	No longer used as a disinfectant or antiseptic because of its irritating qualities and disagreeable odor
2. Phenolics	Disruption of plasma membrane, denaturation, inactivation of enzymes	Environmental surfaces, instruments, skin surfaces, and mucous membranes	Derivatives of phenol that are reactive even in the presence of organic material; examples include O-phenylphenol and hexachlorophene

Table 7–7 Summary of Chemical Agents Used to Control Microbial Growth *continued*

Chemical Agent	Mechanism of Action	Preferred Uses	Comment
Halogens	Iodine inhibits protein function and is a strong oxidizing agent; chlorine forms the strong oxidizing agent hypochlorous acid, which alters cellular components	Effective antiseptic available as a tincture and an iodophor; chlorine gas is used to disinfect water; chlorine compounds are used to disinfect dairy equipment, eating utensils, household items, and glassware	Iodine and chlorine may act alone or as components of inorganic and organic compounds
Alcohols	Denaturation and lipid dissolution	Thermometers and other instruments; in a quick swabbing of the skin with alcohol before an injection, most of the disinfecting action probably comes from a simple wiping away (cleansing) of dirt and some microbes; usually another antiseptic is used in addition	Bactericidal and fungicidal, but not effective against endospores and nonenveloped viruses; commonly used alcohols are ethanol (ethyl alcohol) and isopropanol (isopropyl alcohol)
Heavy Metals and Their Compounds	Denature enzymes and other essential proteins	Silver nitrate is used to prevent gonococcal eye infections; Mercurochrome and Merthiolate disinfect skin and mucous membrane; copper sulfate is an algicide	Heavy metals like silver and mercury are germicidal or antiseptic
Surface-active Agents:			
1. Soaps and acid-anionic detergents	Mechanical removal of microbes through scrubbing	Skin degerming and emulsification of debris	Many cosmetic soaps contain antimicrobials such as triclocarban
2. Acid-anionic detergents	Not certain; may involve enzyme inactivation or disruption	Sanitizers in dairy and food-processing industry	Wide spectrum of activity, nontoxic, noncorrosive, fast-acting
3. Cationic detergents (quaternary ammonium compounds)	Enzyme inhibition, protein denaturation, and disruption of plasma membranes	Antiseptic for skin, instruments, utensils, rubber goods	Bactericidal, bacteriostatic, fungicidal, amoebicidal, and virucidal against enveloped viruses; examples of 'quats' are Zephiran, Phemerol, and Ceepryn
Organic acids	Metabolic inhibitors, mostly affecting molds; action not related to their acidity	Sorbic acid and benzoic acid effective at low pH; parabens much used in cosmetics, shampoos; calcium propionate used in bread; all mainly antifungals	Widely used to control molds and some bacteria in foods and cosmetics
Gaseous sterilants	Denaturation	Excellent sterilizing agent, especially for objects that would be damaged by heat	Ethylene oxide is most commonly used

(*continued on next page*)

Table 7–7 Summary of Chemical Agents Used to Control Microbial Growth *continued*

Chemical Agent	Mechanism of Action	Preferred Uses	Comment
Aldehydes	Protein inactivation	Formalin (37% aqueous solution) or formaldehyde preserve specimens and inactivate microbes for vaccines; glutaraldehyde is less irritating than formaldehyde and is used for sterilization of medical equipment (Cidex), and for embalming corpses	Very effective antimicrobials; glutaraldehyde considered a liquid sterilant
Oxidizing Agents	Oxidation	Contaminated surfaces; some deep wounds in which they are very effective against oxygen-sensitive anaerobes	Ozone is gaining attention as a potential replacement for chlorination; hydrogen peroxide a poor antiseptic but a good disinfectant

STUDY OUTLINE

INTRODUCTION (pp. 182–183)

1. Ignatz Semmelweis introduced handwashing for hospital personnel to reduce infections.
2. Joseph Lister introduced antiseptic surgery, including heat sterilization of instruments and disinfection of wounds.
3. Microbial growth can be controlled by physical and chemical methods.
4. Control of microbial growth can prevent infections and food spoilage.
5. Sterilization is the process of destroying all microbial life on an object.
6. Disinfection is the process of reducing or inhibiting microbial growth.

CONDITIONS INFLUENCING MICROBIAL CONTROL (pp. 183–185)

1. A general rule of disinfection is to try to kill the most resistant microbes that can be found on the object to be disinfected.
2. The number and species of microorganisms present will affect the rate of disinfection.

Temperature (p. 183)

1. Biochemical reactions occur more rapidly at warm temperatures.
2. Disinfectant activity is enhanced by warm temperatures.

Kind of Microbe (pp. 183–184)

1. Gram-positive bacteria are generally more susceptible to disinfectants than are gram-negative bacteria.
2. Pseudomonads can even grow in some disinfectants and antiseptics.

Physiological State of the Microbe (p. 184)

1. An actively growing microorganism tends to be less resistant to chemical agents than an older microorganism.
2. Endospores are resistant to chemical agents and physical methods.

Environment (pp. 184–185)

1. Organic matter and pH level frequently interfere with the actions of chemical control agents.
2. Examples of such organic matter include vomit, feces, pus, and food.

ACTIONS OF MICROBIAL CONTROL AGENTS (p. 185)

Alteration of Membrane Permeability (p. 185)

1. The susceptibility of the plasma membrane is due to its lipid and protein components.
2. Certain chemical control agents damage the plasma membrane by altering its permeability.

Damage to Proteins and Nucleic Acids (p. 185)

1. Some microbial control agents damage cellular proteins by breaking hydrogen and covalent bonds.
2. Other interfere with DNA and RNA replication, and protein synthesis.

PATTERN OF MICROBIAL DEATH (p. 185)

1. Bacterial populations subjected to heat or antimicrobial chemicals die at a constant rate.
2. Such a death curve can be plotted logarithmically.

PHYSICAL METHODS OF MICROBIAL CONTROL (pp. 186–192)

Heat (pp. 186–189)

1. Heat is frequently employed to eliminate microorganisms; it is economical and easily controlled.
2. The mechanism of action of heat is denaturation of enzymes.
3. Thermal death point is the lowest temperature at which all the bacteria in a liquid culture will be killed in 10 minutes.
4. Thermal death time is the length of time required to kill all bacteria in a liquid culture at a given temperature.
5. Decimal reduction time is the length of time in which 90% of a bacterial population will be killed at a given temperature.
6. Boiling (100°C) kills many vegetative cells and viruses within 10 minutes.
7. Autoclaving (steam under pressure) is the most effective method of moist heat sterilization. The steam must directly contact the material to be sterilized.
8. Methods of dry heat sterilization include direct flaming, incineration, and hot-air sterilization.
9. In pasteurization, a high temperature is used for a short time (72°C for 15 seconds) to destroy pathogens without altering the flavor of the food.
10. Equivalent treatments are different methods that produce the same effect (reduction in microbial growth).

Filtration (p. 189)

1. Filtration is the passage of a liquid or gas through a filter with pores small enough to retain microbes.
2. Membrane filters composed of nitrocellulose or cellulose acetate are commonly used to filter out bacteria, viruses, and even large proteins.

Low Temperature (p. 189)

1. The effectiveness of low temperatures depends on the particular microorganism and the intensity of the application.
2. Most microorganisms do not reproduce at ordinary refrigerator temperatures (0° to 7°C).
3. Many microbes survive (but do not grow) at the subzero temperatures used to store foods.

Desiccation (pp. 189–190)

1. In the absence of water, microorganisms cannot grow, but can remain viable.
2. Viruses and endospores can resist desiccation.

Osmotic Pressure (p. 190)

1. Microorganisms in high concentrations of salts and sugars undergo plasmolysis.
2. Molds and yeasts are more capable of growing in materials with low moisture or high osmotic pressure than are bacteria.

Radiation (pp. 191–192)

1. The effects of radiation depend on its wavelength, intensity, and duration.
2. Ionizing radiation (X rays, gamma rays and high-energy electron beams) has a high degree of penetration and exerts its effect primarily by ionizing water and forming highly reactive hydroxyl radicals.
3. Ultraviolet radiation, a form of nonionizing radiation, has a low degree of penetration and causes cell damage by making DNA thymine dimers that interfere with the DNA replication; the most effective germicidal wavelength is 260 nm.
4. Microwaves can kill microbes indirectly as materials get hot.

CHEMICAL METHODS OF MICROBIAL CONTROL (pp. 192–204)

1. Chemical agents are used on living tissue (as antiseptics) and on inanimate objects (as disinfectants).

2. Few chemical agents achieve sterility.

Qualities of an Effective Disinfectant (p. 192)

1. Among the qualities of a good disinfectant are rapid action, wide range of action, good penetration, capability of mixing with water, activity in organic matter, and resistance to decomposition.
2. An effective disinfectant is also nonstaining, noncorrosive, and odorless, and exhibits low toxicity to body tissues.

Principles of Effective Disinfection (pp. 192–194)

1. Careful attention should be paid to the properties and concentration of the disinfectant to be used.
2. The presence of organic matter, degree of contact with microorganisms, and temperature should also be considered.

Evaluating a Disinfectant (p. 194)

1. The phenol coefficient is the comparison of one chemical's disinfecting action with that of phenol, applied for the same length of time on the same organism under identical conditions.
2. In the use-dilution test, a series of tubes contain increasing concentrations of disinfectant; the more the chemical can be diluted and still be effective, the higher its rating.
3. In the filter paper method, a disk of filter paper is soaked with a chemical and placed on an inoculated agar plate; a clear zone of inhibition indicates effectiveness.

Kinds of Disinfectants (pp. 194–204)

Phenol and phenolics (pp. 194–196)

1. Phenolics exert their action by injuring plasma membranes, inactivating enzymes, and denaturing proteins.
2. Common phenolics are cresols and hexachlorophene.

Halogens (pp. 196–198)

1. Some halogens are iodine and chlorine alone or as components of inorganic or organic molecules.
2. To inactivate enzymes and other cellular proteins, iodine combines with the amino acid tyrosine.
3. Iodine is available as a tincture (in solution with alcohol) or as an iodophor (combined with an organic molecule).
4. The germicidal action of chlorine is based on the formation of hypochlorous acid when chlorine is added to water.
5. Chlorine is used as a disinfectant in gaseous form (Cl_2), or in the form of a compound, such as calcium hypochlorite, sodium hypochlorite, and chloramines.

Alcohols (p. 198)

1. Alcohols exert their action by protein denaturation and lipid solvation.
2. In tinctures, they enhance the effectiveness of other antimicrobial chemicals.
3. 60–95% aqueous ethanol and isopropanol are used as disinfectants.

Heavy metals and their compounds (p. 199)

1. Silver, mercury, copper, and zinc are used as germicidals or antiseptics.
2. They exert their antimicrobial action through oligodynamic action. When icons combine with —SH groups, proteins are denatured.

Surface-active agents (pp. 199–200)

1. Surface-active agents decrease the tension between molecules that lie on the surface of a liquid; included are soaps and detergents.
2. Soaps and anionic detergents have limited germicidal action but assist in the removal of microorganisms through scrubbing.

Quaternary ammonium compounds (p. 200)

1. Quats are cationic detergents attached to NH_4^+.
2. By disrupting plasma membranes, they allow cytoplasmic constituents to leak out of the cell.
3. They are most effective against gram-positive bacteria.

Organic acids (p. 200)

1. Sorbic acid, benzoic acid, and propionic acid inhibit fungal metabolism. They are used as food preservatives.

Aldehydes (pp. 200–201)

1. Aldehydes such as formaldehyde and glutaraldehyde exert their antimicrobial effect by inactivating proteins.
2. They are among the most effective chemical disinfectants.

Gaseous chemosterilizers (pp. 201–202)

1. Ethylene oxide is the gas most frequently used for sterilization.
2. It penetrates most materials and kills all microorganisms by protein denaturation.

Oxidizing agents (p. 202)

1. Ozone and peroxide are used as antimicrobial agents.
2. They exert their effect by oxidizing molecules inside cells.

STUDY QUESTIONS

REVIEW

1. Name the cause of cell death resulting from damage to the following:
   (a) Cell wall
   (b) Plasma membrane
   (c) Proteins
   (d) Nucleic acids
2. A bacterial culture was in log phase. At time x, a bactericidal compound was added to the culture medium. Explain why the viable count does not immediately drop to zero.

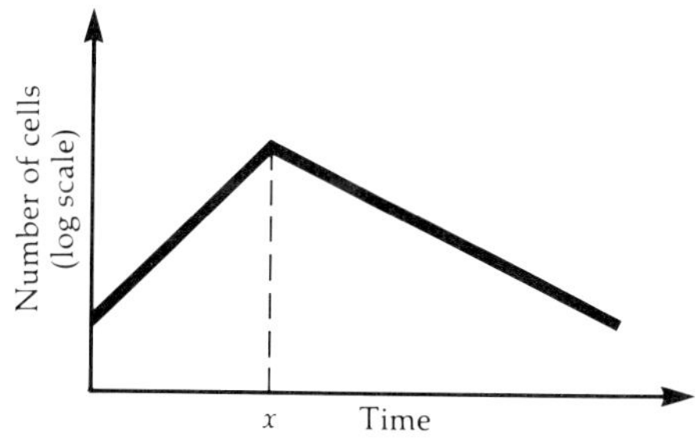

3. The antimicrobial activity of heat is due to ______________.
4. Fill in the following table.

Method of Sterilization	Temperature	Time	Type of Heat	Preferred Use
Autoclaving				
Hot air				

5. The thermal death time for a suspension of *B. subtilis* endospores in dry heat is 30 minutes, and in an autoclave, less than 10 minutes. Which type of heat is more effective? Why?
6. (a) Pasteurization employs what temperature of heat, for how long?
   (b) If pasteurization does not achieve sterilization, why is food treated by pasteurization?
7. Heat-labile solutions such as glucose-minimal salts broth are best sterilized by ______________.

8. Thermal death point is not considered an accurate measure of the effectiveness of heat sterilization. List three factors that can alter the thermal death point.
9. (a) The antimicrobial effect of gamma radiation is due to ________.
   (b) The antimicrobial effect of ultraviolet radiation is due to ________.
10. How do salts and sugars preserve foods? Why are these considered physical rather than chemical methods of control? Name one food that is preserved with sugar and one preserved with salt. How do you account for the occasional growth of *Penicillium* in jelly (which is 50% sucrose)?
11. List ten factors to consider before selecting a disinfectant.
12. Give the method of action and at least one standard use of each of the following types of disinfectants:
    (a) Phenolics (b) Iodine
    (c) Chlorine (d) Alcohol
    (e) Heavy metals (f) Aldehydes
    (g) Ethylene oxide (h) Oxidizing agents
13. The phenol coefficients against *S. aureus* for ethanol and isopropanol are 0.039 and 0.054, respectively. Which is the more effective antimicrobial agent? What is a phenol coefficient?
14. The use-dilution values for two disinfectants tested under the same conditions are given below. If both disinfectants are designed for the same purpose, which would you select? Disinfectant A—1:2; Disinfectant B—1:10,000.
15. A large hospital washes burned patients in a stainless steel tub. After each patient, the tub is cleaned with a quat. It was noticed that 14 out of 20 burn patients acquired *Pseudomonas* infections after being bathed. Provide an explanation for this high rate of infection.

CHALLENGE

1. The filter paper method was used to evaluate three disinfectants. The results were as follows:

Disinfectant	Zone of Inhibition
X	0 mm
Y	5 mm
Z	10 mm

   (a) Which disinfectant was the most effective against the organism?
   (b) Can you determine whether compound *Y* was bactericidal or bacteriostatic?

2. A bacterial culture was in log phase. At time *x*, a bacteriostatic compound was added to the culture. Explain why this graph of the number of viable organisms does not look like the graph in review question 2.

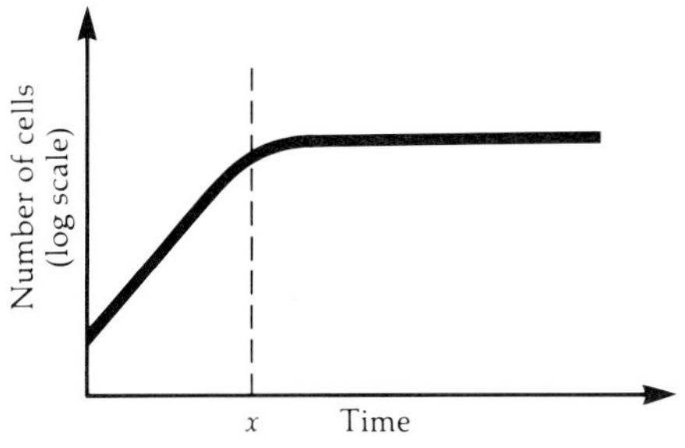

3. *Entamoeba histolytica* and *Giardia lamblia* were isolated from the stool sample of a 45-year-old teacher, and *Shigella sonnei* was isolated from the stool sample of an 18-year-old woman. Both patients experienced diarrhea and severe abdominal cramps, and, prior to onset of digestive symptoms, both had been treated by the same chiropractor.

The chiropractor had administered colonic irrigation to these patients. The device used for this treatment is a gravity-dependent apparatus using 12 liters of tap water. There are no check valves to prevent backflow, so all parts of the apparatus can become contaminated with feces during each colonic treatment. The chiropractor provided colonic treatment to four or five patients per day. Between patients, the adaptor piece that is inserted into the rectum is placed in a "hot water sterilizer."

What two errors does the chiropractor routinely make?

4. Between March 9 and April 12, five chronic peritoneal dialysis patients at one hospital became infected with *Pseudomonas aeruginosa.* Four patients developed peritonitis and one developed a skin infection at the catheter insertion site. All patients with peritonitis had low-grade fever, cloudy peritoneal fluid, and abdominal pain. All patients had permanent indwelling peritoneal catheters, which were wiped with a 4×4 gauze soaked with an iodophor (Prepodyne^R) solution each time the catheter was connected to or disconnected from machine tubing. Aliquots of the iodophor were transferred from stock bottles to small in-use bottles.

   Cultures from the dialysate concentrate and the internal areas of the dialysis machines were negative; iodophor from a small in-use plastic container yielded a pure culture of *P. aeruginosa.*

   What improper technique led to this infection?

FURTHER READING

Block, S. S., ed. 1983. *Disinfection, Sterilization and Preservation,* 3rd ed. Philadelphia, PA: Lea and Febiger. A comprehensive reference on principles and procedures for disinfection and sterilization.

Board of Education and Training. 1976. *Infection Control in the Hospital Environment.* Washington, D.C.: American Society for Microbiology (Proceedings). Lectures and articles on disinfection and environmental monitoring in the hospital.

Borick, P. M., ed. 1973. *Chemical Sterilization.* New York: Van Nostrand Reinhold. A reference on applications and theory of common chemical disinfectants and sterilizers.

Brewer, J. H., ed. 1973. *Lectures on Sterilization.* Durham, NC: Duke University Press. Lectures addressed to hospital personnel who use supplies and are responsible for maintaining sterility and asepsis.

Castle, M. 1980. *Hospital Infection Control.* New York: John Wiley. An excellent reference for practical surveillance and maintenance of asepsis.

Perkins, J. J. 1982. *Principles and Methods of Sterilization in Health Sciences,* 2nd ed. Springfield, IL: C. C. Thomas. A comprehensive reference on theories and applications of sterilization methods.

Phillips, G. B. and W. S. Miller, eds. 1973. *Industrial Sterilization.* Durham, NC: Duke University Press. Methods and applications for the use of gas, formaldehyde, and radiation sterilization.

Russell, A. 1982. *The Destruction of Bacterial Spores.* New York: Academic Press. A discussion of physical and chemical control of endospores in industry and hospitals.

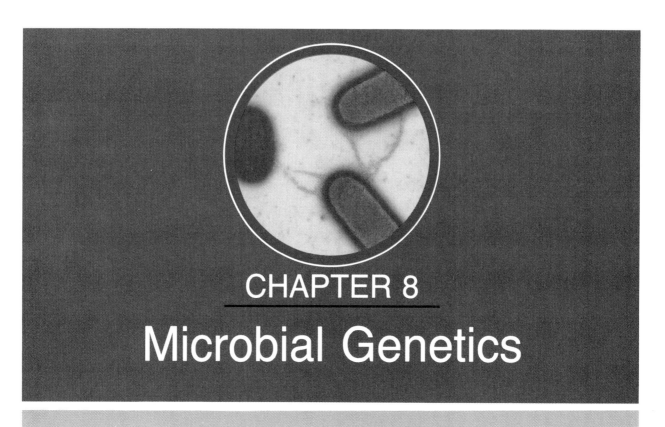

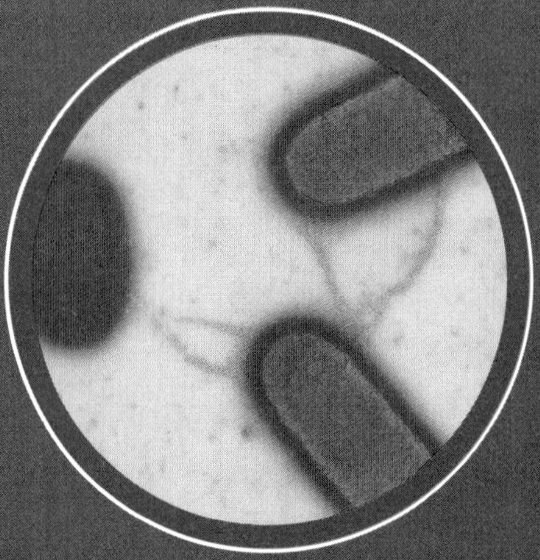

CHAPTER 8

Microbial Genetics

OBJECTIVES

After completing this chapter you should be able to

- Define the following: genetics; chromosome; gene; genetic code; genotype; phenotype; mutagen; genetic recombination.
- Describe how DNA serves as genetic information.
- Describe DNA replication.
- Describe protein synthesis.
- Explain the regulation of gene expression by induction and repression.
- Classify mutations by type.
- Outline direct and indirect selection of mutants.
- Compare mechanisms of genetic transfer in bacteria.
- Define plasmids and discuss their functions.
- Outline how recombinant DNA is produced in vitro and list its potential uses.

All the characteristics of microorganisms, including their morphology, metabolism, behavior, and pathogenicity, are inherited characteristics. Individual organisms transmit these characteristics to their offspring through genes, which are hereditary units that contain the information that determines these characteristics.

Chromosomes are the cellular structures made up of genes that physically carry hereditary information. **Genetics,** the science of heredity, is the study of what genes are, how they carry information, how they are replicated and passed to further generations of cells or passed between organisms, and how the expression of their information within an organism determines the particular characteristics of that organism.

STRUCTURE AND FUNCTION OF THE GENETIC MATERIAL

Genes consist of DNA, deoxyribonucleic acid. We saw in Chapter 2 that DNA is a macromolecule composed of repeating units called *nucleotides.* You may recall that each nucleotide consists of a nitrogenous base (adenine, thymine, cytosine, or guanine), deoxyribose (a pentose sugar), and a phosphate group. The arrangement of the nucleotides in a DNA molecule can be reviewed in Figure 2–19.

The DNA within a cell exists as two long strands of nucleotides twisted together to form a double helix. Each strand is a string of alternating sugar and phosphate groups, and to each sugar is attached a nitrogenous base. The two strands are held together by hydrogen bonds between their nitrogenous bases. The bases are paired in a specific way. Adenine (A) always pairs with thymine (T), and cytosine (C) with guanine (G). The genetic language uses an alphabet with only four letters: the bases A, T, G, and C. But 1000 of these four bases (the number contained in an average gene) can be arranged in 4^{1000} different ways. This enormous flexibility explains how genes can carry so much information. One obvious consequence of the specificity of base pairing is that, if the base sequence of one DNA strand is known, then the sequence of the other strand is also known. The two strands of DNA are *complementary.* A good analogy of complementary information is a photograph and its negative. The complementary structure of DNA provides a clue to its role in the storage and transmission of genetic information, as we shall see later.

Even though the two strands of DNA are held together by weak hydrogen bonds, the double helix is a stable structure. The numerous hydrogen bonds along even a short stretch of DNA provide significant bonding energy. In spite of this stabilizing feature, the two strands of DNA do separate under the influence of special, unwinding proteins. The two strands of separated DNA serve as templates for replication events that lead to the production of two daughter DNA molecules.

DNA carries genetic information from cell to cell and from generation to generation because the specific sequence of nucleotides contained in the DNA duplicates accurately each time the cell divides. And, because the two DNA strands can open up and rejoin easily, genetic information is accessible to the rest of the cell. The information locked within DNA's base sequences determines the characteristics of the cell and transfers these characteristics to subsequent generations of cells. Later in this chapter, we will describe several important experiments that clearly identified DNA as the genetic material of cells.

A **gene** can be defined as a segment of DNA (a sequence of nucleotides in DNA) that codes for a functional product. The final product can be a molecule of ribosomal RNA, for example. Usually, however, the final product is a protein. Each gene has a unique base sequence that codes for a unique protein. The entire DNA molecule specifies the codes for a unique collection of proteins.

Proteins are so important that much of a cell's machinery is concerned with translating the DNA of genes into specific proteins. The genetic information in a region of DNA is transcribed (copied) to produce a specific molecule of RNA. The information encoded in the RNA is translated into a specific sequence of amino acids that form a protein. When a molecule that a gene codes for has been produced, we say that the gene has been "expressed." Thus, gene expression involves the transcription of DNA to produce a specific molecule of RNA. The information encoded in RNA is translated into protein. This process can be symbolized as follows:

$$\text{DNA} \xrightarrow{\text{Transcription}} \text{RNA} \xrightarrow{\text{Translation}} \text{Protein}$$

The flow of genetic information within a cell and between generations of cells is summarized in Figure 8–1.

Before proceeding further, we should mention that DNA can change, or *mutate.* Although some genetic mutations can kill the cell, others permit the offspring a better chance of survival should they be subjected to new environments or stressful conditions. Thus, DNA's ability to mutate is in the long run an advantage that contributes to the successful evolution of the organism.

GENOTYPE AND PHENOTYPE

The **genotype** of an organism is its genetic makeup, the information that codes for all the particular

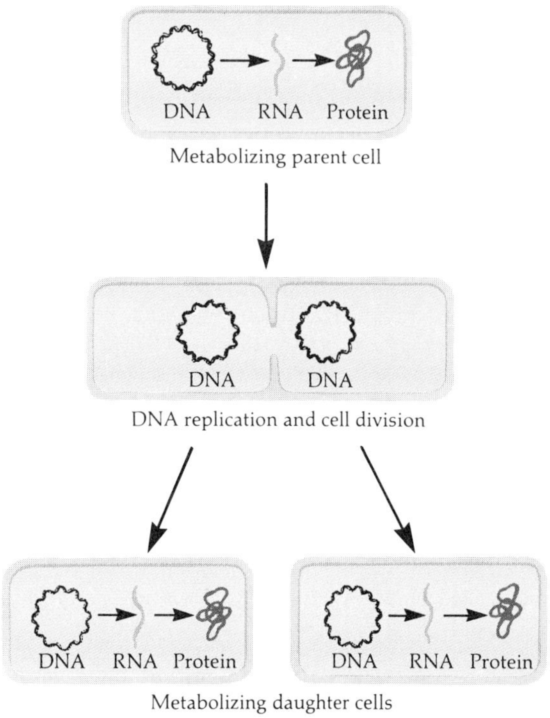

Figure 8–1 Flow of genetic information between generations of cells and within a cell.

characteristics of the organism. The genotype represents the *potential* properties, but not the properties themselves. **Phenotype** refers to the *actual, expressed* properties, such as the organism's ability to perform a particular chemical reaction. The phenotype, then, is the manifestation of the genotype.

In molecular terms, an organism's genotype is its collection of genes, its entire DNA, its blueprint that determines its unique characteristics. What, then, constitutes the organism's phenotype in molecular terms? To put it simply, an organism's phenotype is its collection of proteins. Most of a cell's properties derive from the structures and functions of its proteins. In microorganisms, most proteins are either enzymatic, catalyzing particular reactions, or structural, participating in large functional complexes like membranes or ribosomes. Even those phenotypes that depend on structural macromolecules other than protein rely indirectly on proteins. For instance, the structure of a complex lipid or a polysaccharide molecule is derived from the catalytic activities of enzymes that synthesize, process, and degrade them. Although it is not completely accurate to say that phenotypes are due only to proteins, it is a useful simplification.

DNA AND CHROMOSOMES

Although the structure of DNA is fairly well understood, the packaging of DNA in chromosomes is not. Evidence available so far suggests that the DNA in each chromosome—even in the large, complex chromosomes of eucaryotes—is one long, double helix associated with various proteins that regulate genetic activity.

The DNA of *Escherichia coli,* (esh-ėr-i'kē-ä kō'lē), the common bacterium of the human large intestine, is contained in a typical procaryotic chromosome: a single, long molecule of DNA with no nuclear membrane to enclose it (Figure 8–2a). The chromosome is attached at one or several points to the plasma membrane. The DNA of *E. coli* has about 4 million base pairs and is about 1 mm long—1000 times longer than the entire cell. However, DNA is also very thin and is tightly packed inside the cell, so that this twisted, coiled macromolecule takes up only about 10% of the cell's volume. Bacterial DNA is circular—a closed loop with no free ends. Some proteins (including RNA polymerase and regulatory proteins) are associated with bacterial DNA. Because it is a very long molecule, it can contain large amounts of information, essentially the blueprint for the entire cell.

Eucaryotic chromosomes (Figure 8–2b) contain DNA that is even more highly coiled (condensed) than procaryotic DNA, and eucaryotic chromosomes contain much more protein than procaryotic chromosomes do. In eucaryotic cells, a group of proteins known as histone proteins form complexes around which DNA is wound. The structure and function of a eucaryotic chromosome are also due to diverse nonhistone proteins, which determine many tissue-specific and species-specific phenotypes. The detailed structure of the eucaryotic chromosome, and the precise arrangement of DNA with proteins, is still under investigation. Researchers believe that an understanding of the physical and chemical arrangements of the protein-wrapped DNA is likely to reveal how genes are turned on and off to produce crucial proteins when needed. This regulation of gene expression gov-

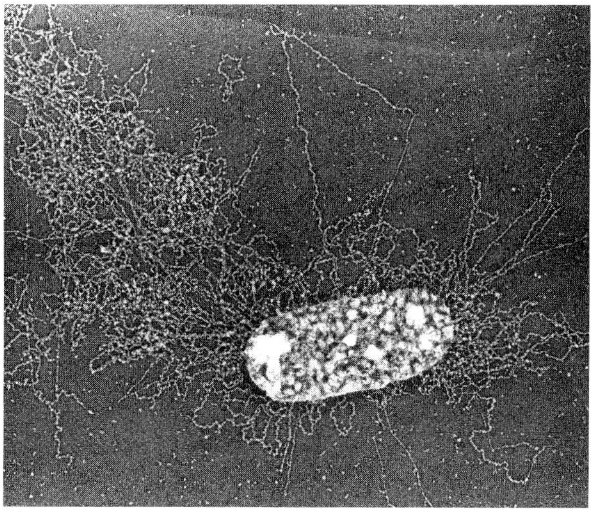
(a)

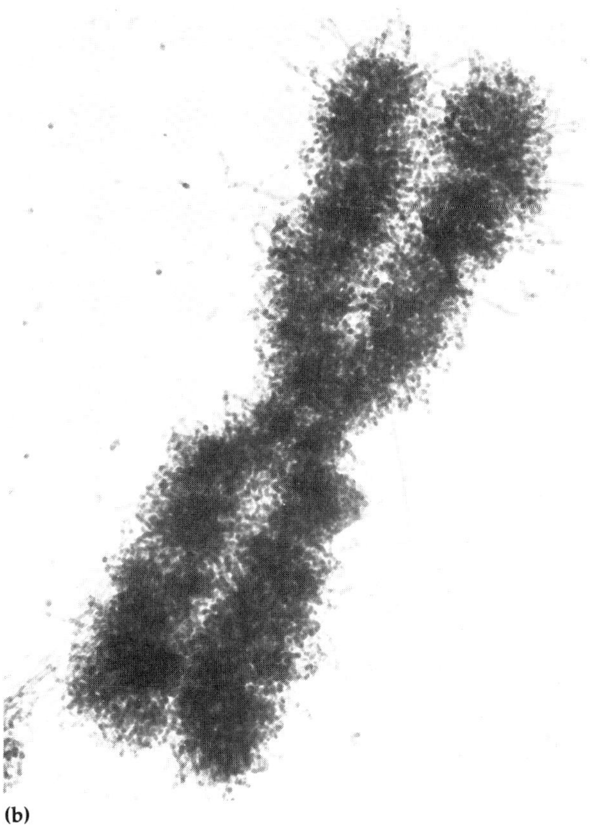
(b)

Figure 8–2 **(a)** Electron micrograph of a procaryotic chromosome. The tangled mass and looping strands of DNA emerging from this disrupted cell of *Escherichia coli* are part of its single chromosome (×6000). **(b)** Electron micrograph of a eucaryotic chromosome. This human chromosome is but one of 46 chromosomes found in a normal human cell. Note the individual strands of protein–DNA fibers (×28,000).

erns the differentiation of eucaryotic cells into the different types of cells found in multicellular organisms.

We will now examine how DNA functions as the genetic material in replication and protein synthesis.

DNA REPLICATION

Replication of DNA results in the synthesis of a duplicate double-stranded DNA molecule. Thus, from one parental DNA molecule, two daughter DNA molecules are produced. The complementary structure of the nitrogenous base sequences in the DNA molecule provides the key to the understanding of DNA replication. Because these bases, which comprise the two strands of double-helical DNA, are complementary, one strand can act as a template for the production of the other strand. When DNA replicates, the two strands of parental DNA temporarily separate from each other (Figure 8–3). The helix unwinds. Free nucleotides present in the cytoplasm of the cell match up to the exposed bases of the single-stranded parental DNA. Where thymine is present on the original strand, only adenine can fit into place on the new strand; where guanine is present on the original strand, only cytosine can fit into place; and so on. Once aligned, the nucleotides are joined by enzymes called *DNA polymerases.* The point at which the parental strands separate is the point at which new strands will be synthesized; this point is called the **replication fork.**

As the site of the replication fork moves along the parental DNA, the DNA unwinds and the unwound single strands each combine with new nucleotides. The original strand and this newly synthesized daughter strand then rewind. Because each new double-stranded DNA molecule contains one original strand (conserved) and one new strand, the process of replication is referred to as **semiconservative replication.**

Before discussing DNA replication in more detail, let us take a closer look at the structure of DNA. Although the two strands of DNA are complementary, their backbones have different chemical senses of direction. Each phosphate group in the sugar–phosphate backbone of DNA attaches the 5′ carbon of one sugar to the 3′ carbon of the next sugar (Figure 8–4). (The base attaches to the 1′

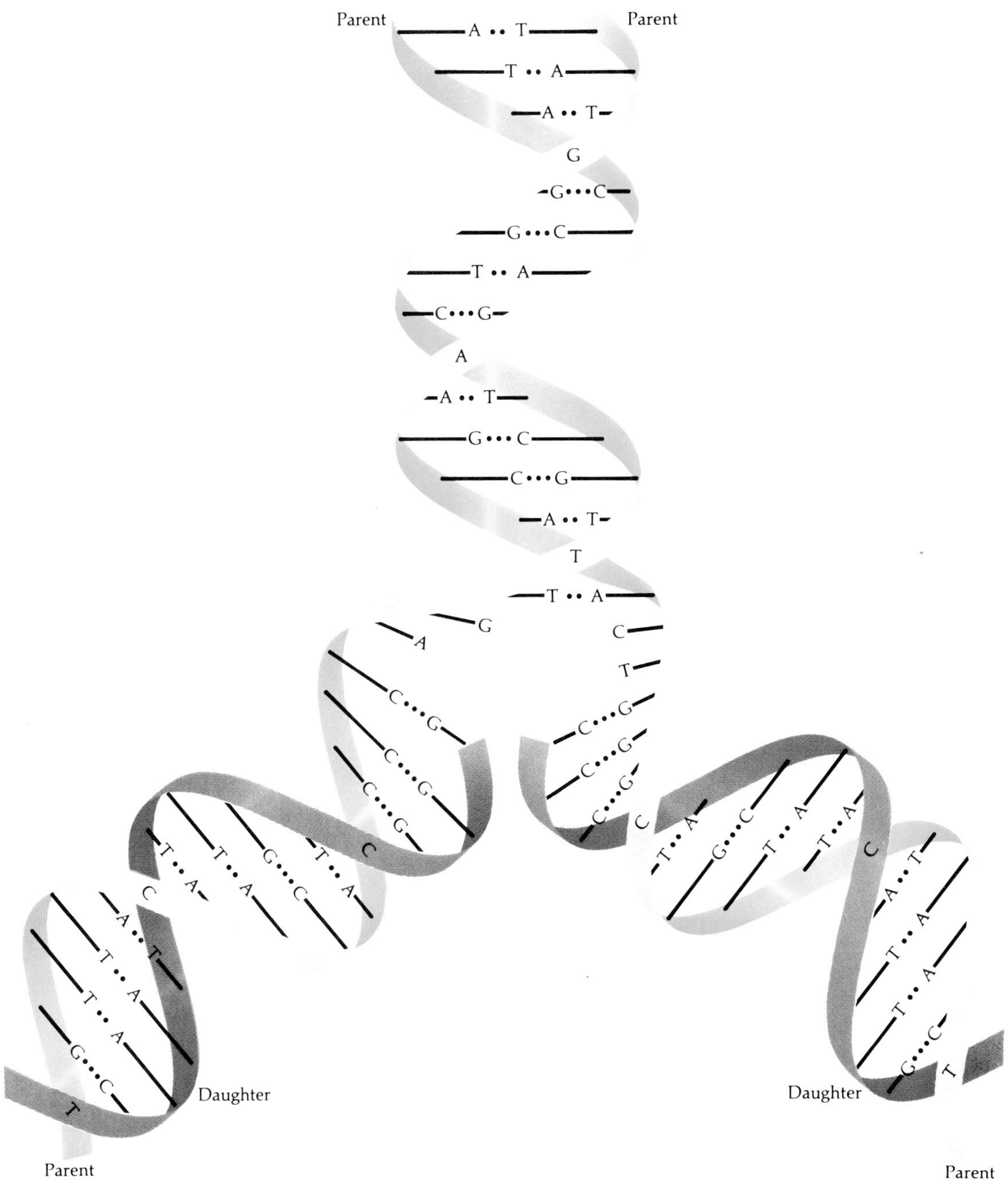

Figure 8–3 DNA replication. The double helix of the parent separates as weak hydrogen bonds between the nucleotides on opposite strands break. Next, hydrogen bonds join new complementary nucleotides with the parental template, to form new base pairs. Enzymes catalyze the formation of sugar–phosphate bonds between sequential nucleotides on each resulting daughter strand.

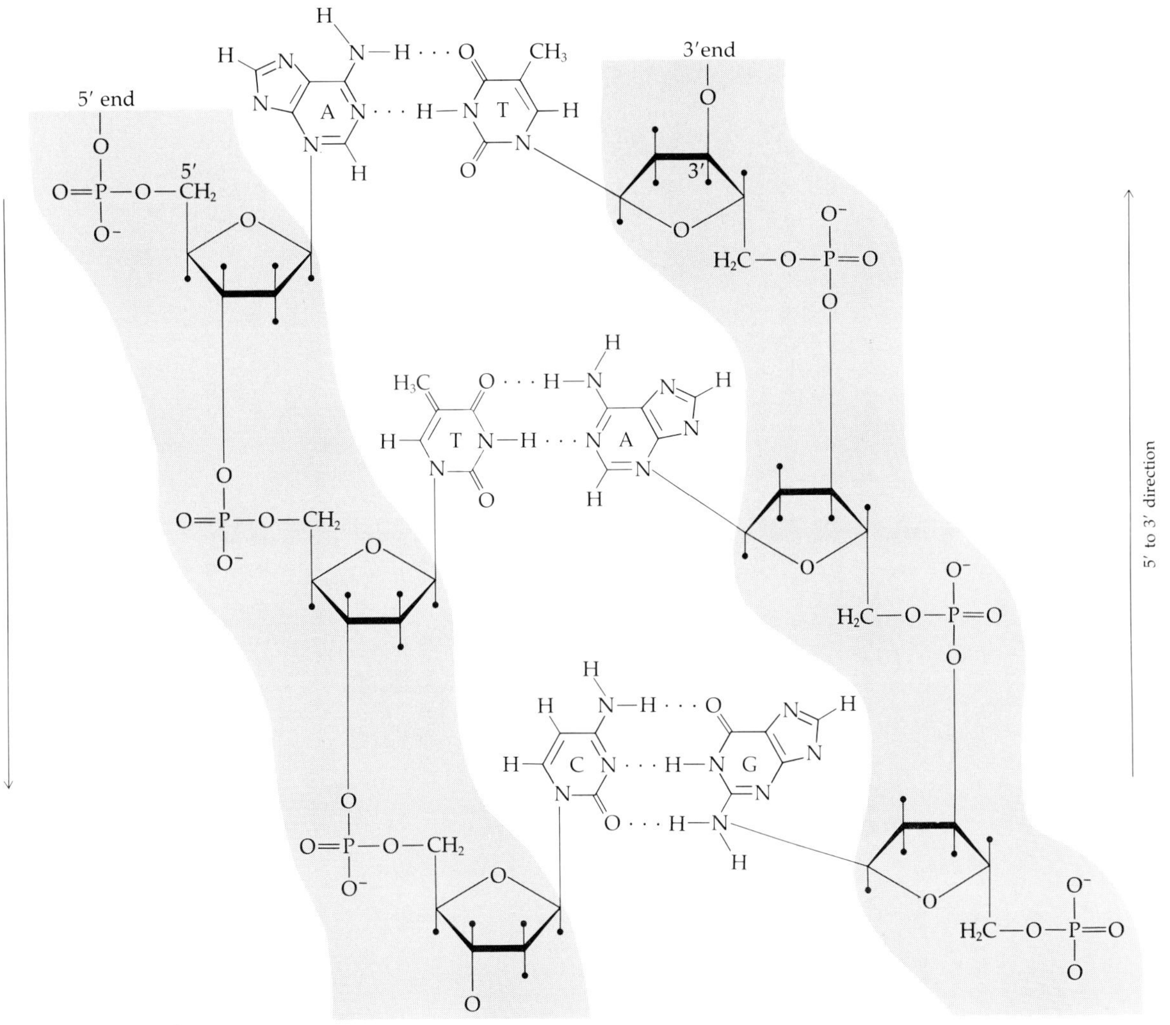

Figure 8–4 5′→3′ orientation of the sugar–phosphate bonds in DNA.

carbon.) The sense of direction is called *5′→3′*, named for the bonds in the sugar–phosphate backbone. If one strand of DNA has the chemical direction *5′→3′*, the other must be reversed, *3′→5′*. The two strands of the DNA's double helix are said to be *antiparallel*.

The antiparallel structure of DNA affects the replication process. DNA polymerases can join new nucleotides to the parental strand only in the *5′→3′* direction. As the site of the replication fork moves along the parental DNA, two product DNA strands are synthesized. One product DNA strand, called the *leading strand,* is synthesized continuously in the *5′→3′* direction (from a template parental strand running *3′→5′*). In contrast, the so-called *lagging strand* of product DNA is synthesized *discontinuously* in fragments of about 1000 nucleotides, which must later be joined to make the complete product DNA.

The complex process of DNA synthesis requires many special proteins and enzymes, only one of which is DNA polymerase (Figure 8–5). At the site of the replication fork, the parental double helix is unwound by enzymes. Stabilizing proteins help maintain the single-stranded parental template DNA. While the leading strand of DNA is

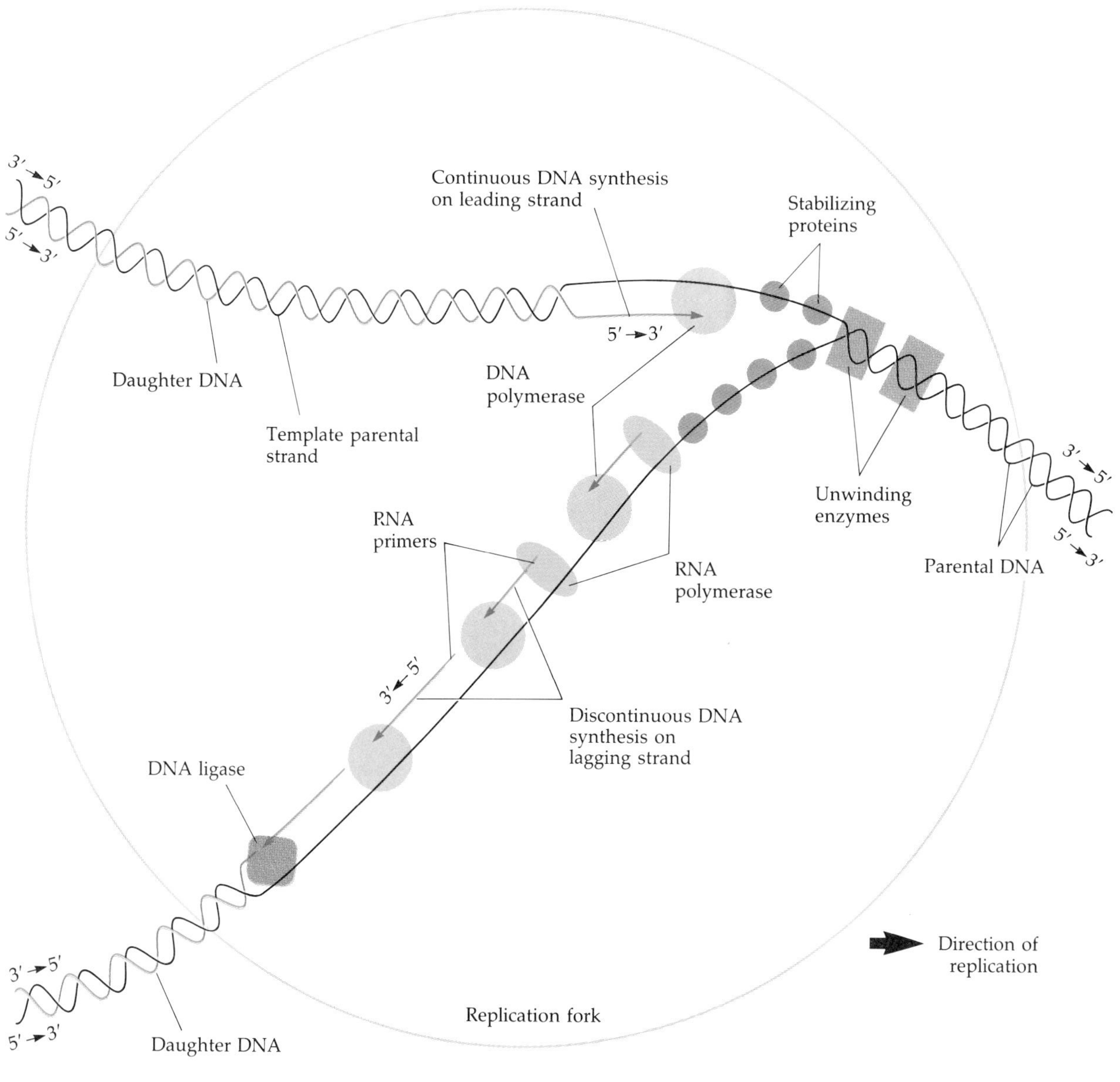

Figure 8–5 Details of the DNA replication fork.

being synthesized by DNA polymerase, an RNA polymerase starts each DNA fragment of the lagging strand with a stretch of about 20 nucleotides called an *RNA primer*. DNA synthesis is continued by DNA polymerase. Another enzyme, *DNA ligase*, joins the fragments together to make a continuous DNA product strand on the parental lagging strand.

In bacteria the replication process begins at a unique site on the bacterial chromosome. It is known that DNA replication of some bacteria, e.g. *E. coli*, goes *bidirectionally* around the chromosome (Figure 8–6). Two replication forks move in opposite directions away from the origin of replication. Because the bacterial chromosome is a closed loop, the replication forks eventually meet

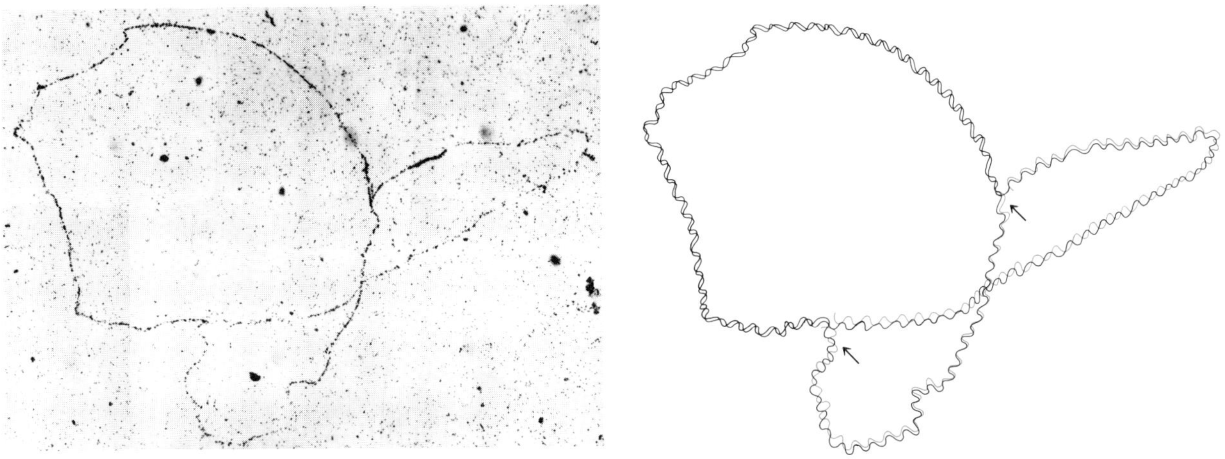

(a)

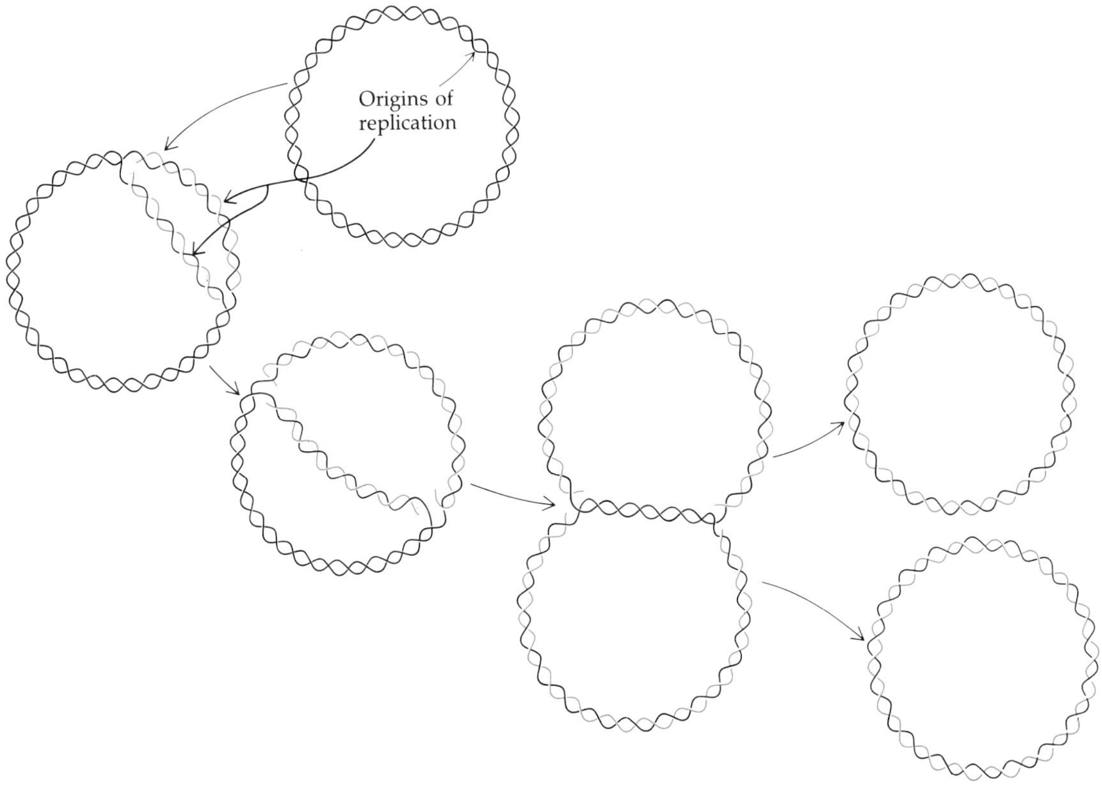

Figure 8–6 Replication of bacterial DNA. **(a)** Autoradiograph of the replication of an *E. coli* chromosome (corresponding diagram in the upper right). The arrows point to the two replication forks. The chromosome is about one-third replicated. Note that one of the new helices is crossed over the other one. **(b)** Diagrammatic representation of the bidirectional replication of a circular DNA molecule. The new strand is shown in color.

when the replication of the chromosome is completed. Much evidence exists to show an association between the bacterial cell membrane and each replication fork, with its unwinding proteins, stabilizing proteins, DNA polymerases, and other enzymes. The DNA–membrane complex of procaryotes might work to ensure that each daughter cell receives one copy of molecule DNA; that is, one complete chromosome.

DNA synthesis is a surprisingly fast process: around 1000 nucleotides per second in *E. coli* growing at 37°C. At first glance this speed seems improbable, considering that nucleotide substrates must be synthesized and then diffuse to the replication fork. Furthermore, several attempts are probably made by the wrong nucleotides to pair at each position before the correct bases pair up. Nevertheless, the speed and specificity of DNA replication is understood to be a process governed by chemical principles only, not requiring unknown forces or unlikely events.

Under some conditions, namely log phase growth in a nutritious medium (Chapter 6; see Figure 6–12), *E. coli* can grow faster than the two replication forks can complete the circular chromosome. Under these conditions the cell initiates *multiple replication forks* at the origin on the chromosome; a new pair of forks begin before the last pair has finished. In this way the total rate of DNA synthesis matches the rate at which the cell divides. Likewise, when the cell's growth greatly slows, the initiation of DNA synthesis at the origin of replication may be delayed. The rate at which DNA nucleotides are synthesized is generally constant (at a stable temperature), but the cell regulates its total rate of DNA replication to match its rate of growth and cell division. The cell does not regulate the elongation or termination of new DNA strands, but regulates the rate at which replication forks on the chromosome are initiated. By regulating the initiation of DNA synthesis, the cell conserves energy because no energy is wasted on the synthesis of incomplete products that have no function.

RNA AND PROTEIN SYNTHESIS

Although the previous section explains how DNA stores genetic information in nucleotide sequences and replicates by a process involving base pairing, it does not explain how the information in DNA is translated into the proteins that control cell activities. In a process called transcription, much of the genetic information in DNA is copied, or encoded, by a complementary base sequence of RNA. The cell then uses this RNA message to synthesize specific proteins through the process of translation. We will now take a closer look at these two processes.

Transcription

A strand of RNA called **messenger RNA (mRNA)** is synthesized in the process of **transcription,** in which a specific portion of the cell's DNA serves as a template (Figure 8–7). In other words, the genetic information stored in the sequence of nitrogenous bases of DNA is rewritten so that the same information appears in the base sequence of mRNA. As in DNA replication, a G in the DNA template dictates a C in the mRNA being made; a C in the DNA template dictates a G in the mRNA; and a T in the DNA template dictates an A in the mRNA. However, an A in the DNA template dictates a uracil (U) in the mRNA, because RNA contains U instead of T. If, for example, the template portion of DNA has the base sequence ATGCAT, then the newly synthesized mRNA strand will have the complementary base sequence UACGUA. This complementary sequence can be represented as follows:

Template DNA		mRNA
A		U
T		A
G	Transcription →	C
C		G
A		U
T		A

The process of transcription requires an enzyme called *RNA polymerase* and a supply of RNA nucleotides. Because of the antiparallel nature of DNA and the fact that mRNA can be synthesized only in the $5' \rightarrow 3'$ direction, only one of the two DNA strands serves as the template for RNA synthesis. This is called the *sense strand.* The other strand, the *antisense strand,* is the complement of the sense strand. The region where RNA polymerase binds tightly to DNA is known as the **promoter** site; this is where transcription begins. The

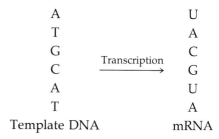

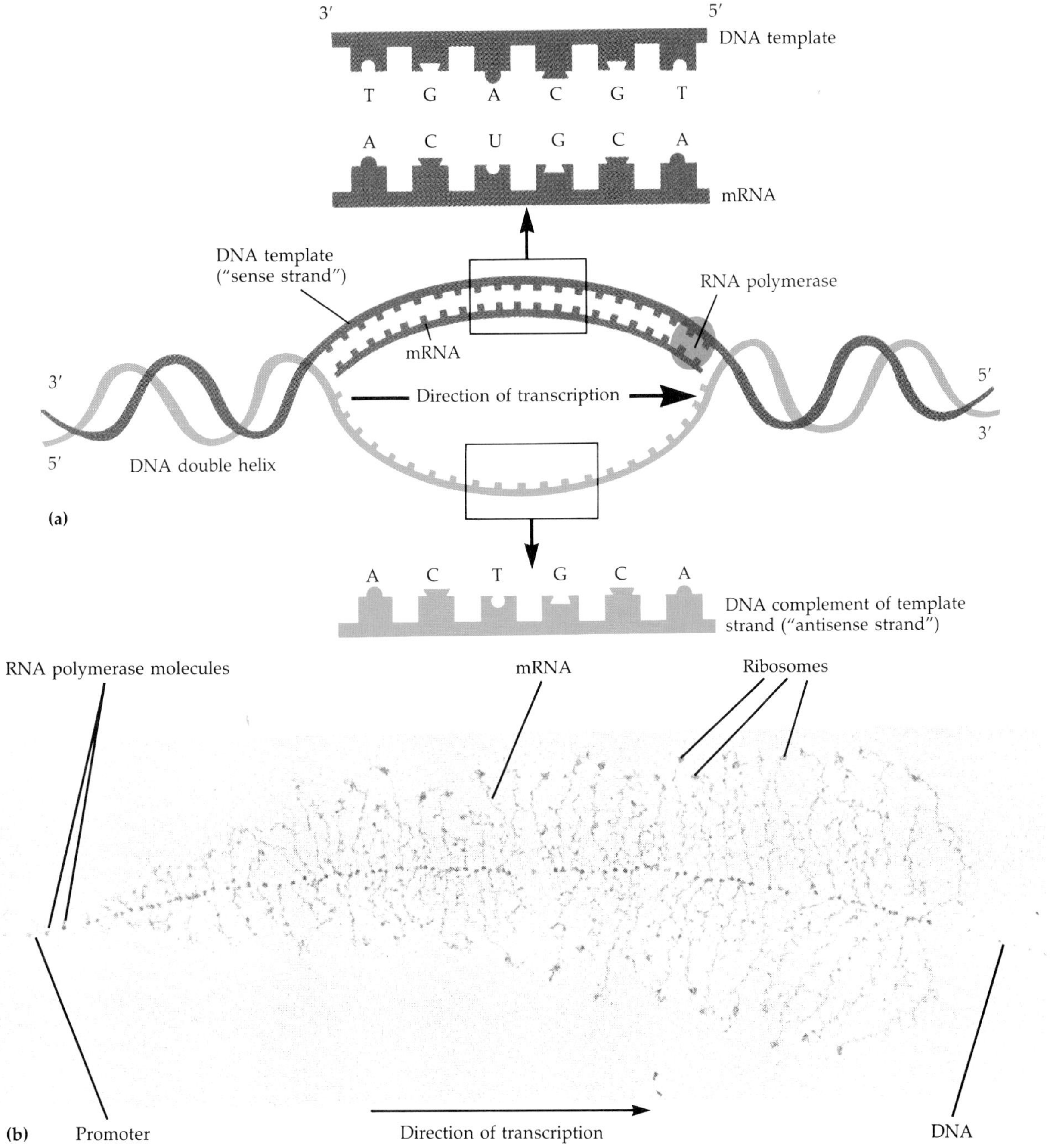

Figure 8–7 Transcription. **(a)** Diagram of a partly uncoiled DNA double helix. Transcription occurs in the unwound region of DNA. The upper box shows temporary base-pairing that occurs during transcription. **(b)** Electron micrograph of transcription from a single gene (×51,000). Many molecules of mRNA are being synthesized simultaneously, starting at the promoter site. The longest mRNA molecules were the first to begin at the promoter. Note the many RNA polymerase molecules along the DNA, and the ribosomes attached to the newly forming mRNA.

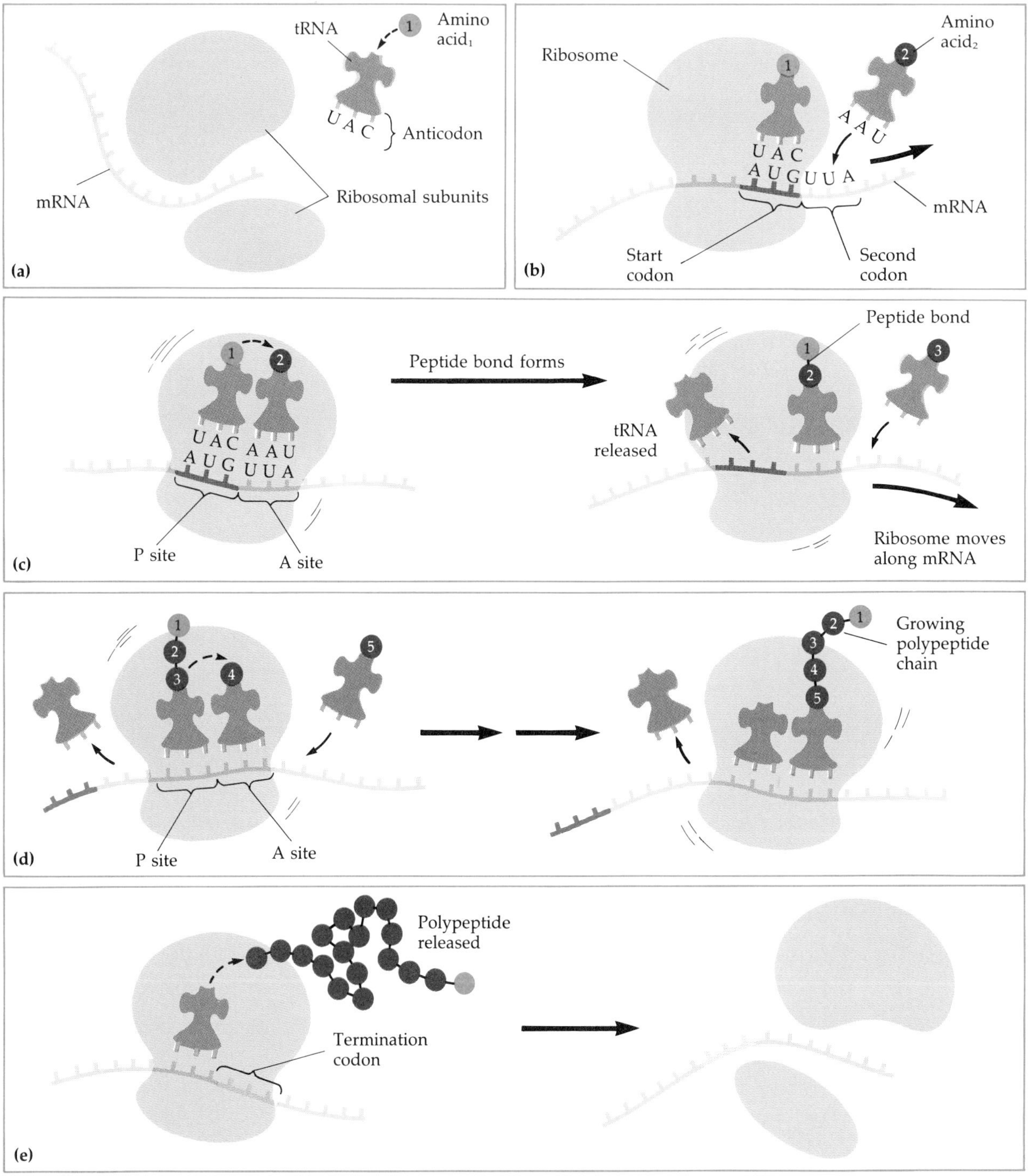

Figure 8–8 Translation. **(a)** Components needed for translation. **(b)** A tRNA carrying the first amino acid pairs with the start codon on the mRNA; this takes place at the A site on the ribosome. **(c)** The ribosome then moves along mRNA until the first tRNA is in the P site. In the A site, the second codon of the mRNA pairs with the tRNA carrying the second amino acid. The first amino acid is joined to the second by a peptide bond, and the first tRNA is released. **(d)** The ribosome moves along the mRNA until the second tRNA is in the P site, and the process continues. (Nucleotide bases are shown only for the first two codons.) The chain of amino acids (polypeptide) grows. **(e)** When the ribosome reaches the termination codon, the polypeptide is released, the last tRNA is released, and the ribosome comes apart.

region of DNA that acts as the end point for transcription is the **terminator** site; at this site the RNA polymerase and newly formed mRNA are released from the DNA. The DNA double helix then reforms.

In summary, then, the genetic information for protein synthesis is stored in DNA and passed to mRNA during transcription.

Translation

The process in which the nitrogenous-base sequence of mRNA dictates the amino acid sequence of a protein is called **translation.**The events involved in translation are described in this section and shown in Figure 8–8.

First, one end of the mRNA molecule becomes associated with a ribosome, the site of protein synthesis (Figure 8–8a). Ribosomes (see Chapter 4) consist of a special type of RNA, called **ribosomal RNA** or **rRNA,** and protein. Each ribosome consists of two subunits.

In solution in the cytoplasm are 20 different amino acids that may participate in protein synthesis (see Table 2–5). These amino acids can be synthesized by the cell or taken up from the external medium. However, before the appropriate amino acids can be joined together to form a protein, they must be *activated.* Another type of RNA, called **transfer RNA** or **tRNA,** participates in this step (Figure 8–9a). For each different amino acid, there is a different type of tRNA. During amino acid activation, a specific amino acid is attached to its specific type of tRNA. This attachment is accomplished with an enzyme and energy from ATP (Figure 8–9b).

One part of the tRNA molecule has a sequence of three nitrogenous bases that matches to its complementary triplet on an mRNA strand. Each set of three nitrogenous bases on mRNA is called a **codon.** Their complement, the three nucleotides in tRNA, is the **anticodon** (Figure 8–9c). During translation, the anticodon of a molecule of tRNA attaches to its complementary codon on mRNA. For example, a tRNA with anticodon UAC pairs with the mRNA codon AUG (Figure 8–8b). The tRNA molecule carries its specific amino acid. The pairing of codon and anticodon occurs only where mRNA is attached to a ribosome.

After the first tRNA, with its amino acid, attaches to mRNA, the ribosome moves along the mRNA strand and the second tRNA molecule, with its amino acid, is moved into position (Figure 8–8b). The two amino acids are joined by a peptide bond, and the first tRNA molecule detaches itself from the mRNA strand (Figure 8–8c). The detached tRNA can now pick up another molecule of the specific amino acid it carries. Each tRNA can be used over and over again. As the proper amino acids are brought into line one by one, peptide bonds form between the amino acids, and a polypeptide chain is formed (Figure 8–8d).

A special termination codon in the mRNA (called a **nonsense codon** because it does not specify any amino acid) signals the end of a polypeptide chain and its release from the ribosome (Figure 8–8e). The ribosome then comes apart into its two subunits.

Before the ribosome moving along the mRNA completes translation of that gene, another ribosome, then another, can attach and begin translation of the same mRNA molecule; there can be a number of ribosomes attached at different positions to the same mRNA molecule (see Figure 8–7b). In this way, a single mRNA strand can be translated simultaneously into several identical protein molecules. An mRNA strand with several attached ribosomes is called a **polyribosome.**

In summary, the synthesis of a protein requires that the genetic information in DNA first be transferred to a molecule of mRNA by a process called transcription. Then, in a process called translation, the mRNA attaches to a ribosome, tRNA activates amino acids, and the amino acids are delivered to mRNA by tRNA according to the complementary base pairings of codons and anticodons. The amino acids are then linked to form polypeptides. Proteins can be composed of one or more polypeptide chains, each of which may range in length from 50 to several hundred amino acids.

Because a typical polypeptide has about 300 amino acids, the typical mRNA must have about 300 codons and thus be at least 900 bases long. As explained earlier, DNA is internally marked off in regions called genes, and each region codes for a different kind of mRNA. A gene, therefore, can also be defined as a DNA base sequence that codes for a functional product—and that product is almost always a protein, the other products being rRNA and tRNA. The DNA of *E. coli* is long enough to contain about 4000 genes and can therefore specify about 4000 different kinds of mRNA.

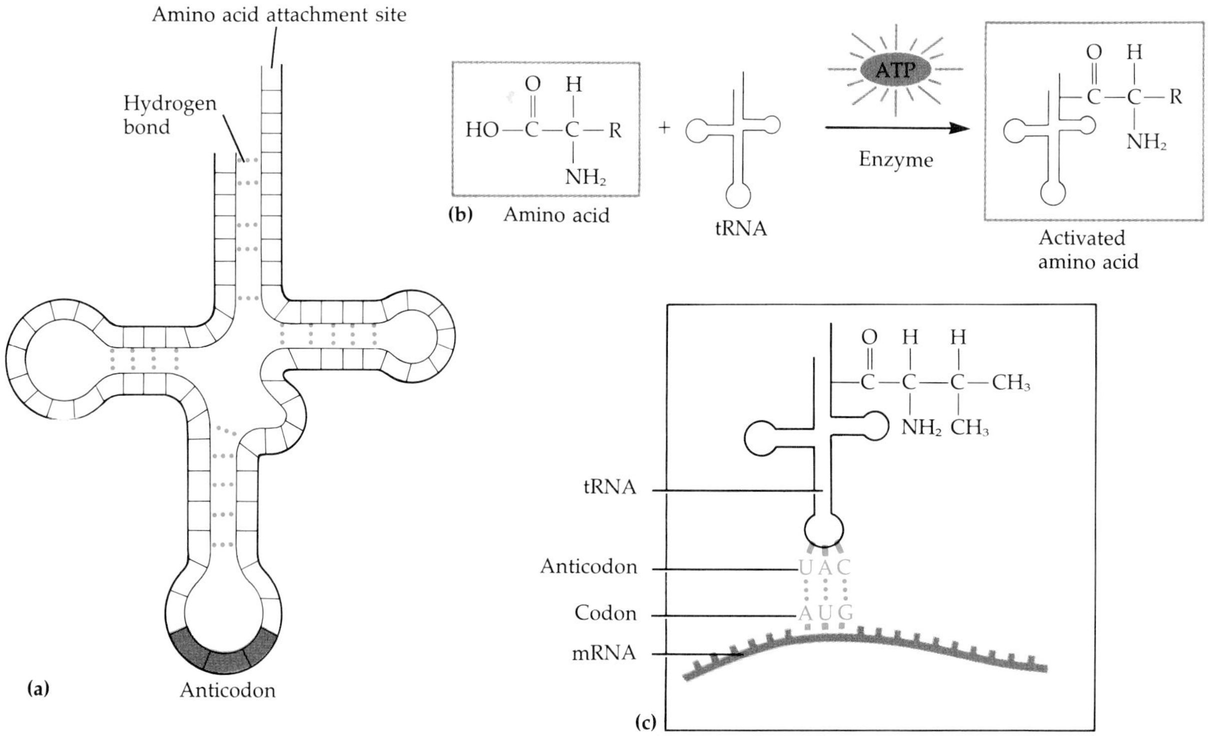

Figure 8–9 Transfer RNA. **(a)** Structure of tRNA. Each "box" represents a nucleotide. Note the regions of hydrogen bonding between base pairs, and the loops of unpaired bases. **(b)** Amino acid activated for attachment of an amino acid to tRNA. **(c)** The anticodon of tRNA pairs with its complementary codon on an mRNA strand.

In fact, *E. coli* has about 4000 different proteins. There are only a few copies of some proteins, but there are several thousand copies of other proteins.

THE GENETIC CODE

The **genetic code** refers to the relationship between the nitrogenous base sequence of DNA, the corresponding codons of mRNA, and the amino acids that the codons stand for (Figure 8–10). Note that there are 64 possible codons but only 20 amino acids. This means that most amino acids are signaled by several codons; this situation is referred to as the **degeneracy** of the code. For example, leucine has 6 codons and alanine has 4 codons.

Of the 64 codons, 61 are sense codons and 3 are nonsense codons. **Sense codons** code for amino acids, while **nonsense codons** do not. Rather, they signal the end of the protein molecule's synthesis. These nonsense codons are UAA, UAG, and UGA. The initiator codon that starts the synthesis of the protein molecule is AUG, which is also the sense codon for methionine. The terminal methionine is usually removed, so all proteins do not begin with

Second Position

First Position	U		C		A		G		Third Position
	UUU	Phe	UCU		UAU	Tyr	UGU	Cys	U
	UUC		UCC		UAC		UGC		C
U				Ser					
	UUA	Leu	UCA		UAA	End	UGA	End	A
	UUG		UCG		UAG	End	UGG	Trp	G
	CUU		CCU		CAU	His	CGU		U
	CUC		CCC		CAC		CGC		C
C		Leu		Pro				Arg	
	CUA		CCA		CAA	Gln	CGA		A
	CUG		CCG		CAG		CGG		G
	AUU		ACU		AAU	Asn	AGU	Ser	U
	AUC	Ile	ACC		AAC		AGC		C
A				Thr					
	AUA		ACA		AAA	Lys	AGA	Arg	A
	AUG	Met	ACG		AAG		AGG		G
	GUU		GCU		GAU	Asp	GGU		U
	GUC		GCC		GAC		GGC		C
G		Val		Ala				Gly	
	GUA		GCA		GAA	Glu	GGA		A
	GUG		GCG		GAG		GGG		G

Figure 8–10 The genetic code. The three nucleotides in an mRNA codon are designated, respectively, as the first position, second position, and third position of the codon. Each set of three nucleotides specifies a particular amino acid, represented by a three-letter abbreviation (see Table 2–5). The codon AUG (which specifies the amino acid methionine) is the start signal for protein synthesis. The word *End* stands for the nonsense codons that serve as signals to terminate protein synthesis.

methionine. The base sequence that specifies the promoter site on DNA is not translated, nor is the terminator.

REGULATION OF GENE EXPRESSION IN BACTERIA

From Chapter 5 we learned that the bacterial cell is involved in an enormously large number of metabolic reactions. Some of these reactions are concerned with biosynthesis and are referred to as anabolic reactions. Others are concerned with degradation and are known as catabolic reactions. The common feature of all metabolic reactions is that they are regulated by enzymes.

We have seen that genes, through transcription and translation, direct the synthesis of proteins, many of which serve as enzymes. These are the very enzymes used for cellular metabolism. Therefore, the genetic machinery and metabolic machinery of a cell are integrated and interdependent. We will now take a look at how gene expression, and therefore metabolism, is regulated. Because protein (enzyme) synthesis requires the cell to spend a tremendous expenditure of energy (ATP), the regulation of protein synthesis is important to the cell's energy economy. Synthesizing only the proteins needed at a particular time is an excellent way to conserve energy.

INDUCTION AND REPRESSION

In the control mechanisms called induction and repression, the synthesis, not the activity, of enzymes in a metabolic pathway is regulated. End-product repression, for example, does not act upon preexisting enzymes. Instead, the end product inhibits the *synthesis* of new enzymes. Induction and repression involve control over enzymic activity at the genetic level; that is, genes are turned on or off so that the synthesis of specific enzymes is increased or decreased.

Induction

The genes required for lactose metabolism in *E. coli* are a well-known example of an inducible system. One of these genes codes for the enzyme β-galactosidase, which splits the substrate lactose into two simple sugars, glucose and galactose. If *E. coli* is placed in a medium in which no lactose is present, the organism is found to contain almost no β-galactosidase enzyme. However, when lactose is added to the medium, the bacterial cells produce a large quantity of the enzyme. In other words, the presence of lactose has induced the cells to synthesize more enzyme. A substance such as lactose, which brings about an increased amount of an enzyme, is called an **inducer,** and enzymes that are synthesized in the presence of inducers are **inducible enzymes.** This response, which is under genetic control, is termed **enzyme induction.**

Repression

The regulation that decreases the synthesis of enzymes is called **enzyme repression.** Repression occurs if cells are exposed to a particular end product of a metabolic pathway; there is a decrease in the rate of synthesis of all the enzymes leading to the formation of that end product.

For example, cells of *E. coli* grown in a medium lacking any amino acids contain the enzymes necessary for the synthesis of all the amino acids contained in the protein molecules of the bacteria. But the introduction of an amino acid to the culture medium greatly decreases the synthesis of the enzymes necessary for the production of that amino acid. Enzymes whose production is reduced by the presence of the end product of a metabolic pathway are called **repressible enzymes,** and the molecule (end product) that brings about repression is termed the **corepressor.**

The control of gene function by induction and repression is explained by a concept called the operon model.

Operon Model

In 1961, François Jacob and Jacques Monod, who had been studying induction and repression, formulated a general model to account for the regulation of enzyme synthesis in bacteria. This model, called the **operon model,** was based on studies of enzymic systems in *E. coli.* Let us examine the components of the operon model and their roles in regulating enzyme synthesis.

Each *E. coli* cell contains all the genetic information required for its metabolic reactions, growth, and reproduction. Some of the enzymes needed to conduct these reactions are present at all times. Others, however, are produced only when they are needed by the cell; that is, they are **induced.** Included in this second group are certain enzymes needed for uptake and catabolism of the disaccharide lactose and other galactosides. These enzymes are β-galactosidase, which breaks lactose into glucose and galactose; β-galactoside permease, which is involved in the transport of lactose into the cell; and thiogalactoside transacetylase, which is involved in the utilization of galactosides other than lactose.

Structural genes are genes that specify the amino acid sequence responsible for the structure of particular enzymes. The structural genes for enzymes involved in lactose uptake and utilization turn out to be next to each other on the bacterial chromosome (Figure 8–11a). These genes produce their enzymes rapidly and simultaneously when lactose is introduced into the culture medium. We will now see how this regulation occurs.

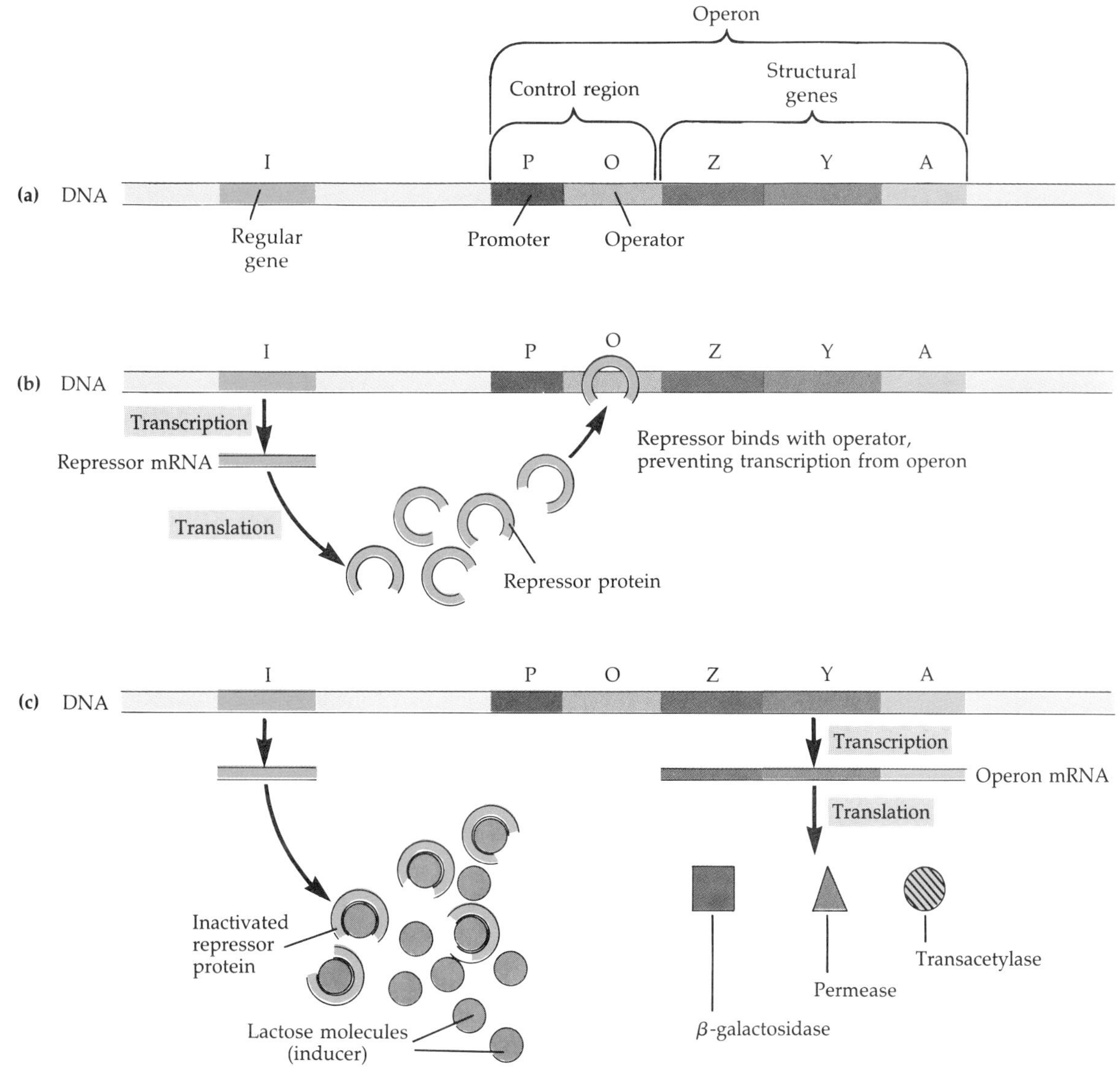

Figure 8–11 The lactose operon. **(a)** This segment of the DNA molecule of *E. coli* shows the genes involved in lactose catabolism. **(b)** Control of the operon in the absence of lactose (repression). **(c)** Control of the operon in the presence of lactose (induction).

A region of DNA called the **regulator gene** codes for a protein called a **repressor.** Another region of DNA called the **operator** is located adjacent to the structural genes. The operator controls the transcription of the structural genes. And the **promoter site** is the region of DNA where RNA polymerase binds to initiate transcription. When lactose is absent (Figure 8–11b), the repressor protein binds tightly to the operator site. This binding prevents the RNA polymerase enzyme from transcribing the adjacent structural genes; consequently, no mRNA is made and no enzymes are synthesized. Thus, in the absence of lactose, no enzymes are produced. But, when lactose, the inducer, is present (Figure 8–11c), it binds to the repressor and alters the repressor so that it cannot bind to the operator site. In the absence of operator-bound repressor, the RNA polymerase enzyme can transcribe the structural genes into mRNA, which is then translated into enzymes. This is why in the presence of lactose, enzymes are produced. Lactose is said to **induce** enzyme synthesis.

The operator and promoter sites and the structural genes they control are referred to as an **operon**—the basis for the term **operon model.**

In repressible operons, the repressor requires the presence of a small molecule called a **corepressor,** which helps it bind to the operator gene and promoter site. Without the corepressor, the repressor is unable to bind. The enzymes involved in the synthesis of many amino acids are regulated in this manner. In such cases, the corepressor is an amino acid produced by a set of enzymes. The cell stops synthesizing these enzymes when the amino acid is already present in the medium.

It should be remembered that many genes in bacteria are not subject to repression. Such genes produce enzymes at certain necessary levels, regardless of how much nutrient is present in the medium. Enzymes so produced are called **constitutive enzymes,** and genes responsible for their production are known as **constitutive genes.** Usually, these genes code for enzymes that the cell always needs in fairly constant amounts for its major life processes. The enzymes of glycolysis and most repressor proteins are examples.

Both induction and repression of enzymes are differential genetic activities in which the synthesis of genetic products is sensitive to a given set of environmental conditions. In this regard, induction and repression are adaptive mechanisms that appear to be of survival value to the organism. When these mechanisms are in operation, cells do not expend large amounts of energy in synthesizing enzymes not immediately required; rather, such mechanisms allow the synthesis of enzymes to respond to conditions in the environment.

Keep in mind that enzyme repression is similar to feedback inhibition (see Chapter 5) because in both processes, the end product acts to prevent its own synthesis. However, repression differs from feedback inhibition in that the end product of enzyme repression regulates enzyme *synthesis,* whereas the end product of feedback inhibition regulates enzyme *activity.*

Although it is fairly common in bacteria for groups of genes to be close together on a chromosome when they govern a sequence of related steps, this is not typical of eucaryotes. In eucaryotes, genes with similar functions tend to be scattered throughout various chromosomes. Scientists do not yet have much information on the regulatory mechanisms that control genes in higher organisms, but it is apparent that natural selection favors those organisms that can regulate their gene expression.

MUTATION: CHANGE IN THE GENETIC MATERIAL

A **mutation** is a change in the base sequence of DNA. We can reasonably expect that a change in the base sequence of a gene will cause a change in the product coded by that gene. When a gene mutates, the enzyme coded by the gene may become inactive or less active because its amino acid sequence has changed. Such a change in genotype may be disadvantageous or even lethal, if the cell loses a phenotypic trait it needs. Yet, a mutation can be beneficial if, for instance, the altered enzyme coded by the mutant gene has a new activity that benefits the cell.

Many simple mutations are neutral: the change in DNA base sequence causes no change in the activity of the product coded by the gene. A common example of a neutral mutation is the substitution of one nucleotide for another in the DNA, especially at a location corresponding to the third position of mRNA codon. Because of the degeneracy of the genetic code, the resulting new codon might still code for the same amino acid. Even if the amino acid is changed, there may be no change in protein function if the amino acid is in a nonvital portion of the protein.

TYPES OF MUTATIONS

The most common type of mutation involving single base pairs is **base substitution** or **point mutation,** in which a single base at one point in the DNA is replaced with a different one. Then, when the DNA replicates, the result is a substituted base pair (Figure 8–12). For example, A–T might be substituted for G–C or C–G for G–C. If a base substitution occurs in a portion of the DNA molecule that codes for a protein, then the mRNA transcribed from the gene will carry an incorrect base at some

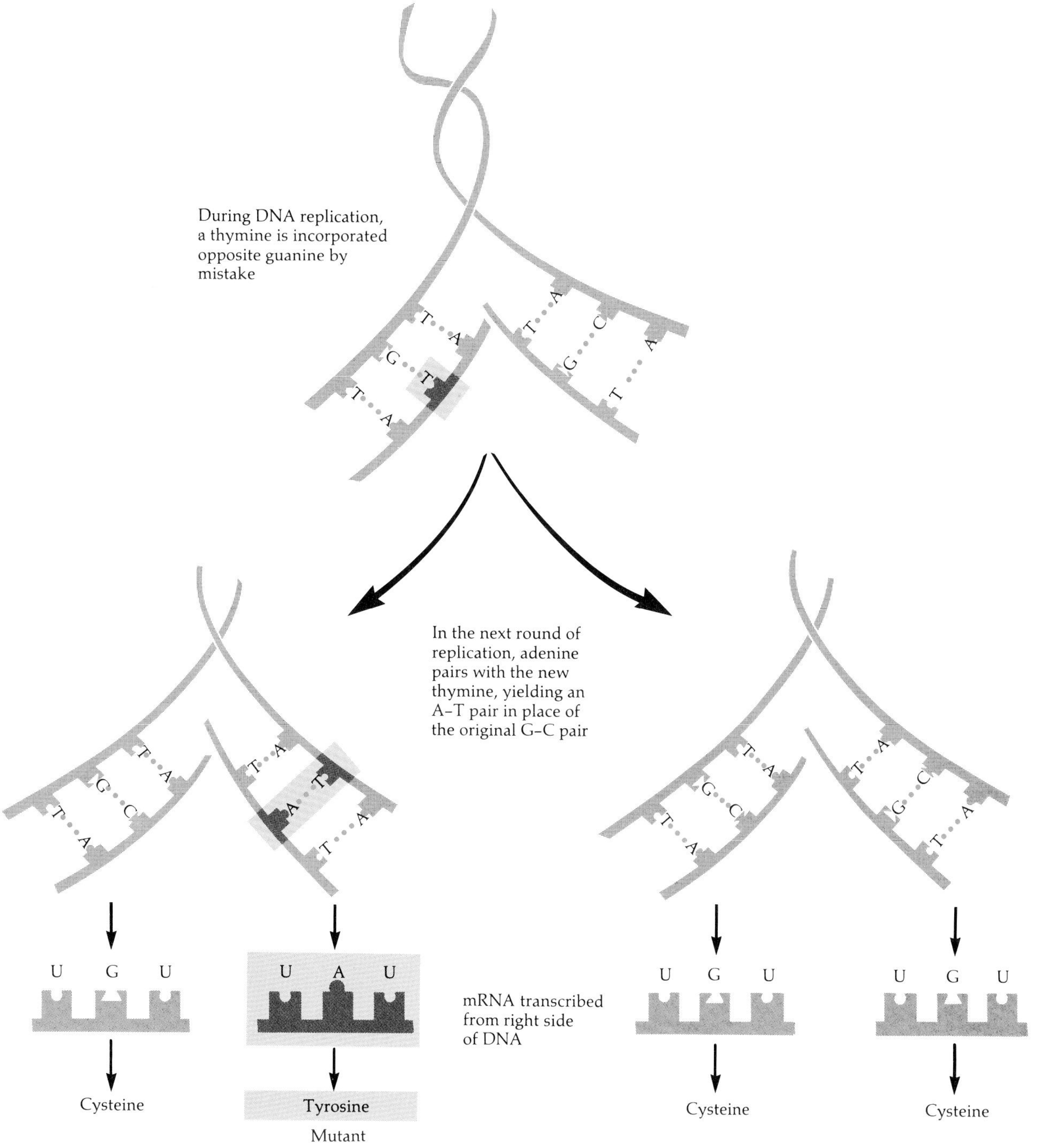

Figure 8–12 Base substitution, leading to base pair substitution.

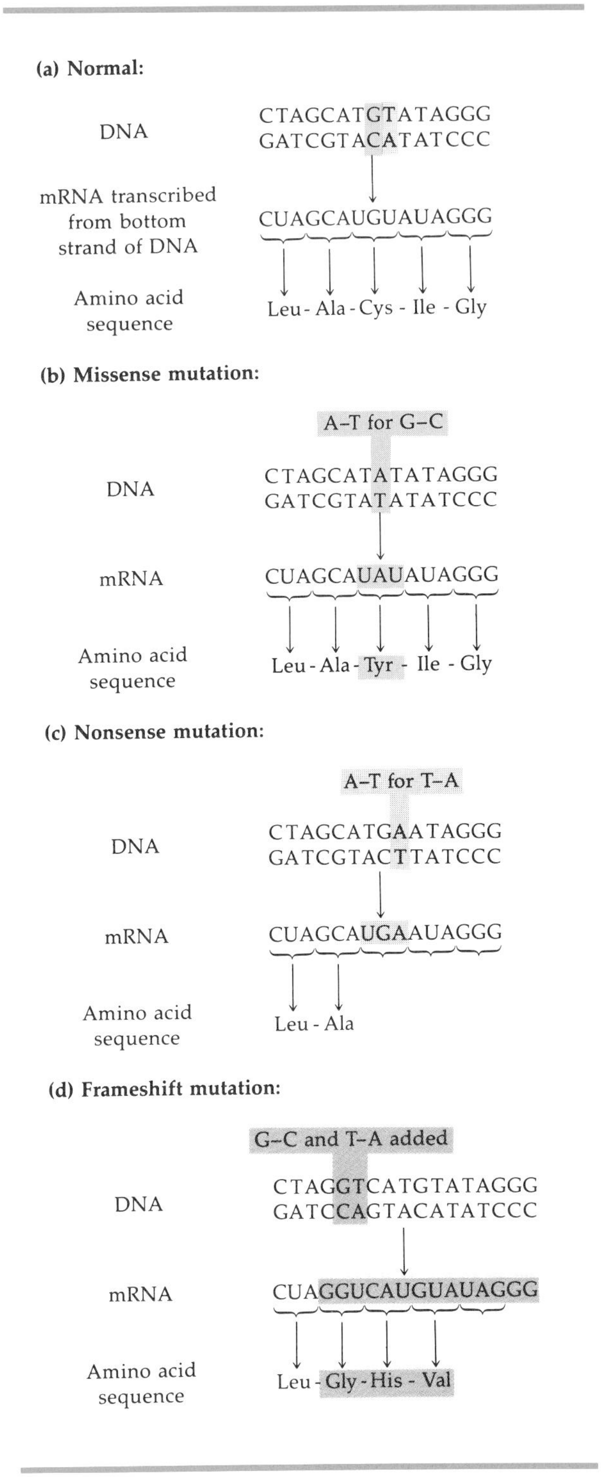

Figure 8–13 Types of mutations. **(a)** Normal DNA molecule. **(b)** Missense mutation. **(c)** Nonsense mutation. **(d)** Frameshift mutation.

position. When the mRNA is translated into protein, the incorrect base can cause the insertion of an incorrect amino acid in the protein. Thus, the base substitution in DNA can result in an amino acid substitution in the synthesized protein. This is known as **missense mutation** (Figure 8–13a and b).

By creating a terminator (nonsense) codon in mRNA before the protein is synthesized, some base substitutions effectively prevent the synthesis of a functional protein. Only a fragment of the protein is synthesized. A base substitution thus resulting in a nonsense codon is called a **nonsense mutation** (Figure 8–13c).

Besides base pair mutations, there are also changes in DNA called **frameshift mutations.** Here, one or a few base pairs are deleted or added to DNA (Figure 8–13d). This can shift the "translational reading frame," that is, the three-by-three grouping of nucleotides recognized by the tRNAs during translation. For example, inserting one base pair in the middle of a gene causes many amino acids downstream from the site of the original mutation to change. Frameshift mutations almost always result in an inactive protein product for the mutated gene. They lead to a long stretch of missense, and in most cases a nonsense codon will eventually be generated and thereby terminate translation.

MUTAGENESIS

Base substitutions and frameshift mutations may occur spontaneously because of occasional mistakes made during DNA replication. These **spontaneous mutations** are mutations that occur without known intervention of mutation-causing agents. Agents in the environment, such as certain chemicals and radiation, that directly or indirectly bring about mutations are called **mutagens.**

Chemical Mutagens

Among the many chemicals known to be mutagenic is *nitrous acid.* Figure 8–14 shows how exposure of DNA to nitrous acid can convert the base adenine (A) to a form that no longer pairs with thymine (T) but with cytosine (C). When DNA containing such modified adenines replicates, one daughter DNA molecule will have a different base-pair sequence than that of the parent DNA. Even-

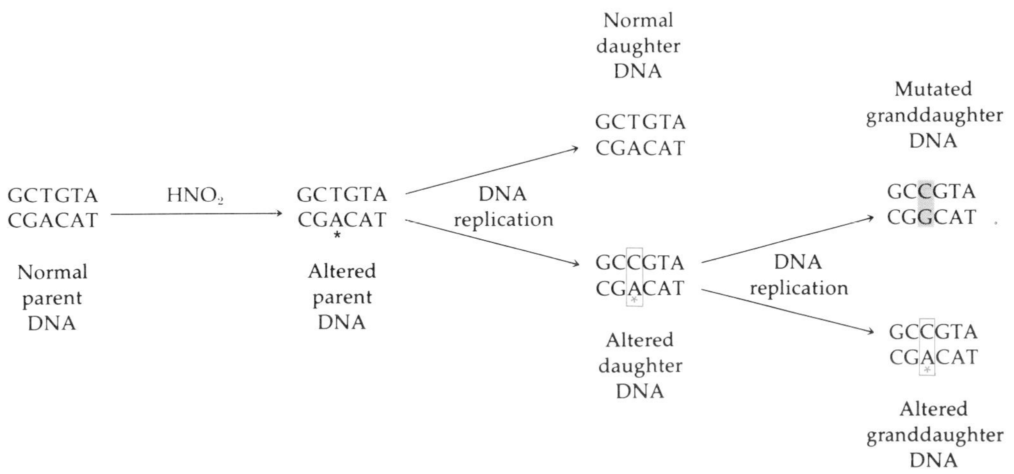

Figure 8–14 Mutagenesis by nitrous acid (HNO_2). The nitrous acid alters an adenine (asterisk) with the result that it pairs with cytosine instead of thymine. In the next round of replication, the new C will pair with G, and a new C—G pair will be introduced into the mutated DNA.

tually, some A–T base pairs of the parent will have been changed to G–C base pairs in a granddaughter cell. Nitrous acid is thus an effective **base pair mutagen,** which makes a specific mutational change in DNA. Like all mutagens, it does not select which gene it will mutate, but alters DNA at random locations.

Other chemical mutagens are **base analogs,** like *2-aminopurine* and *5-bromouracil.* These molecules are structurally similar to normal nitrogenous bases, but have slightly altered base-pairing properties. The 2-aminopurine (Figure 8–15a) is incorporated into DNA in place of adenine (A) but can sometimes pair with cytosine (C). The 5-bromouracil (Figure 8–15b) is incorporated into DNA in place of thymine (T) but often pairs with guanine (G). When base analogs are given to growing cells, the analogs are metabolized and randomly incorporated into cellular DNA in place of the normal bases. Then, during DNA replication, the analogs cause mistakes in base pairing. The wrongly paired bases can be faithfully copied during further DNA replication, and a base pair substitution results. Some antiviral drugs are base analogs.

Still other mutagens can cause small deletions or insertions instead of substitutions. For instance,

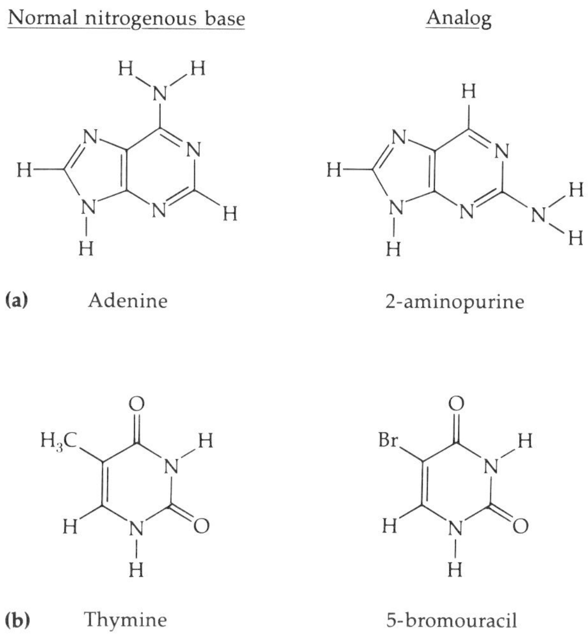

Figure 8–15 Base analogs and the nitrogenous bases they replace. **(a)** Adenine and 2-aminopurine. **(b)** Thymine and 5-bromouracil.

under certain conditions, *benzpyrene,* which is present in smoke and soot, is an effective **frameshift mutagen.** Likewise, *aflatoxin,* produced by *Aspergillus flavus* (a-spėr-jil'lus flā'vus), a mold that grows on peanuts and grain, is a frameshift mutagen, as are the *acridine dyes,* used experimentally against herpesvirus infections. Frameshift mutagens usually have the right size and chemical properties to slip between the stacked base pairs of the DNA double helix. Though there is no direct proof, it is believed that a frameshift mutagen works by slightly offsetting the two strands of DNA, leaving a gap in one strand or a bulge in the other. When the offset or staggered DNA strands are copied during DNA synthesis, one or more base pairs can be inserted or deleted in the new double-stranded DNA. Interestingly, frameshift mutagens are often potent carcinogens.

Radiation

X rays, gamma rays, and other forms of ionizing radiation are potent mutagens. The penetrating rays of ionizing radiation cause electrons to pop out of their usual electron shell locations. The affected electrons bombard other molecules, and cause more damage, and many of the resulting ions and free radicals (molecular fragments with unpaired electrons) are very reactive. Some of these ions can combine with bases in the DNA. Such a combination results in mistakes in base pairing during replication and leads ultimately to base substitutions. An even more serious outcome is the breaking of covalent bonds in the sugar–phosphate backbone of DNA. These cause physical breaks in chromosomes.

Another form of mutagenic radiation is *ultraviolet light (UV),* a nonionizing component of ordinary sunlight. The most important effect of UV light on DNA is the formation of harmful covalent bonds among the bases. Adjacent thymines in a DNA strand can crosslink to form thymine dimers (Figure 8–16a). Such dimers, unless repaired, may cause serious damage or death to the cell, because the cell cannot properly transcribe or replicate such DNA.

Bacteria and other organisms can repair radiation damage with processes involving enzymes. Some components of the repair processes also participate in recombination. Some repair processes are dependent on visible light and some are not. One method of light-independent repair of UV damage is illustrated in Figure 8–16b. Enzymes cut out the distorted cross-linked thymines, widen the gap, and then fill in the gap with synthesized DNA complementary to the undamaged strand, so that the original base pair sequence is restored. The last step is the sealing of the gap with DNA ligase. Occasionally this repair process, as well as other repair processes in the cell, makes errors, and the original base pair sequence is not properly restored. The result of this error is a mutation.

FREQUENCY OF MUTATION

The **mutation rate** is the probability that a gene will mutate when a cell divides. The rate is usually stated as a power of 10, and because mutations are very rare, the exponent is always a negative number. For example, if there is one chance in 10,000 that a gene will mutate when the cell divides, the mutation rate is 1/10,000, which is expressed as 10^{-4}. Spontaneous mistakes in DNA replication occur at a very low rate because the replication machinery is remarkably faithful. Perhaps only once in 10^{9} replicated base pairs does an error occur. Because the average gene has about 10^{3} base pairs, the spontaneous rate of mutation is about once in 10^{6} (a million) replicated genes. Mutations usually occur more or less randomly along a chromosome. No particular gene is singled out, and within a mutated gene, no particular A–T or G–C base pair is likely to be mutated. Having random mistakes at low frequency is an essential aspect in the design of living organisms, for evolution requires that genetic diversity be generated at random and at a low rate. Typically, a mutagen will increase the spontaneous rate from 10 to 1000 times; that is, a mutagen will produce a mutation rate of 10^{-5} to 10^{-3} per gene per cell per generation.

IDENTIFYING MUTATIONS

Mutations can be detected by selecting or testing for an altered phenotype. Whether or not a mutagen is used, mutant cells with specific mutations are always rare compared with other cells in the population. The problem is one of detecting a rare event.

Experiments are done more easily with bacteria than with other organisms, because bacteria reproduce rapidly and large numbers of organisms (more than 10^9 per milliliter of broth or per Petri plate) can easily be used. Furthermore, because bacteria generally have only one version of each gene per cell (they are haploid), the effects of a mutated gene are not masked by the presence of a normal version of the gene.

Positive, or direct, selection involves the detection of mutant cells but rejection of the unmutated parent cells. For example, suppose we were trying to find mutant bacteria that are resistant to penicillin. When the bacterial cells are plated on a medium containing penicillin, the mutant can be identified directly. The few cells in the population that are resistant (mutants) will grow and form colonies. The normal, penicillin-sensitive, parental cells cannot grow.

For identifying mutations in other kinds of genes, it may be necessary to use **negative, or indirect, selection.** This is made possible by the **replica plating technique.** In replica plating, a method that will separate individual cells is used to inoculate a number of bacteria onto an agar plate. This plate,

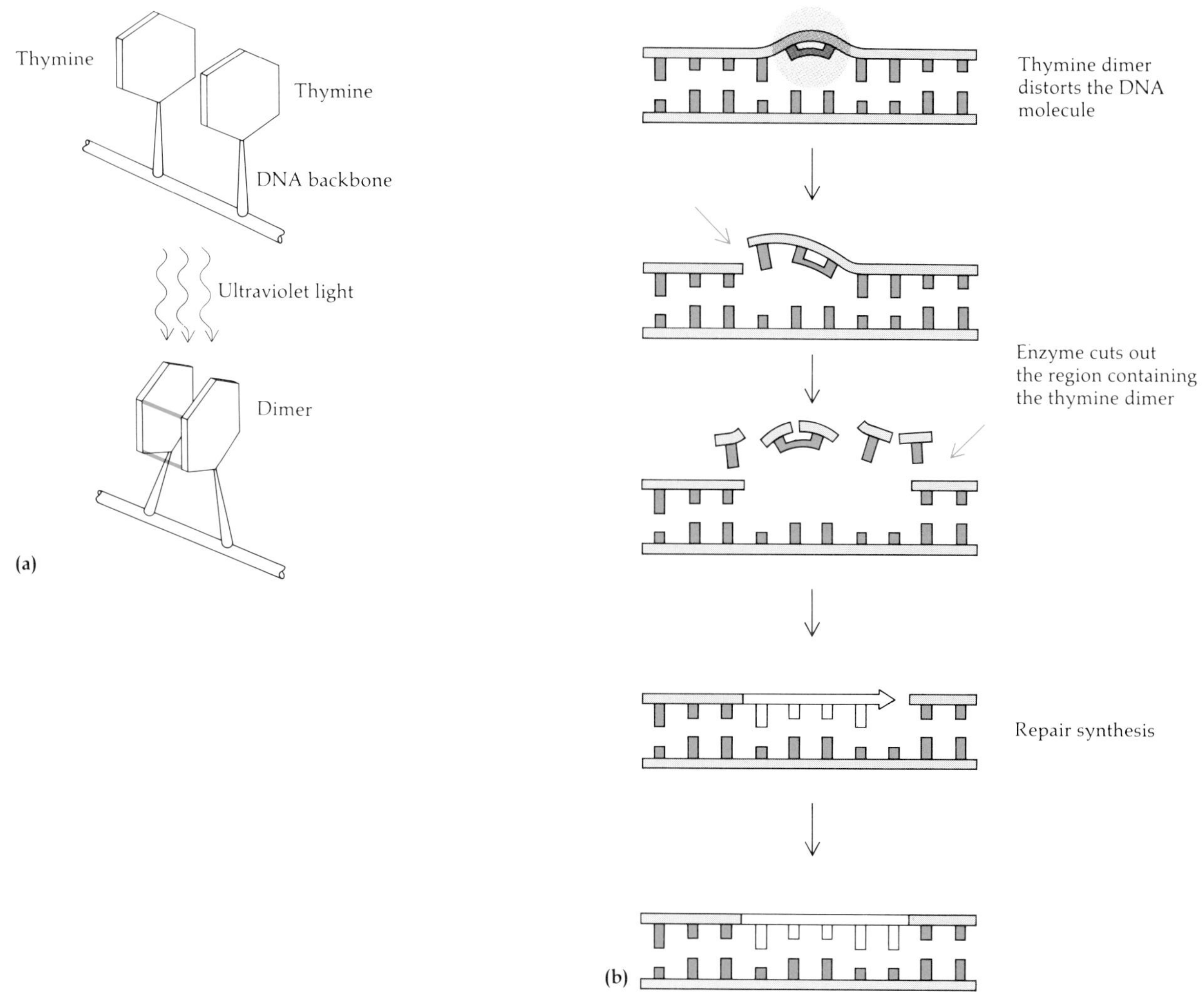

Figure 8–16 Mutagenesis by ultraviolet light. **(a)** Effect of ultraviolet light on adjacent thymines. **(b)** Mechanism of repair of cross-linked thymines. This mechanism does not require light.

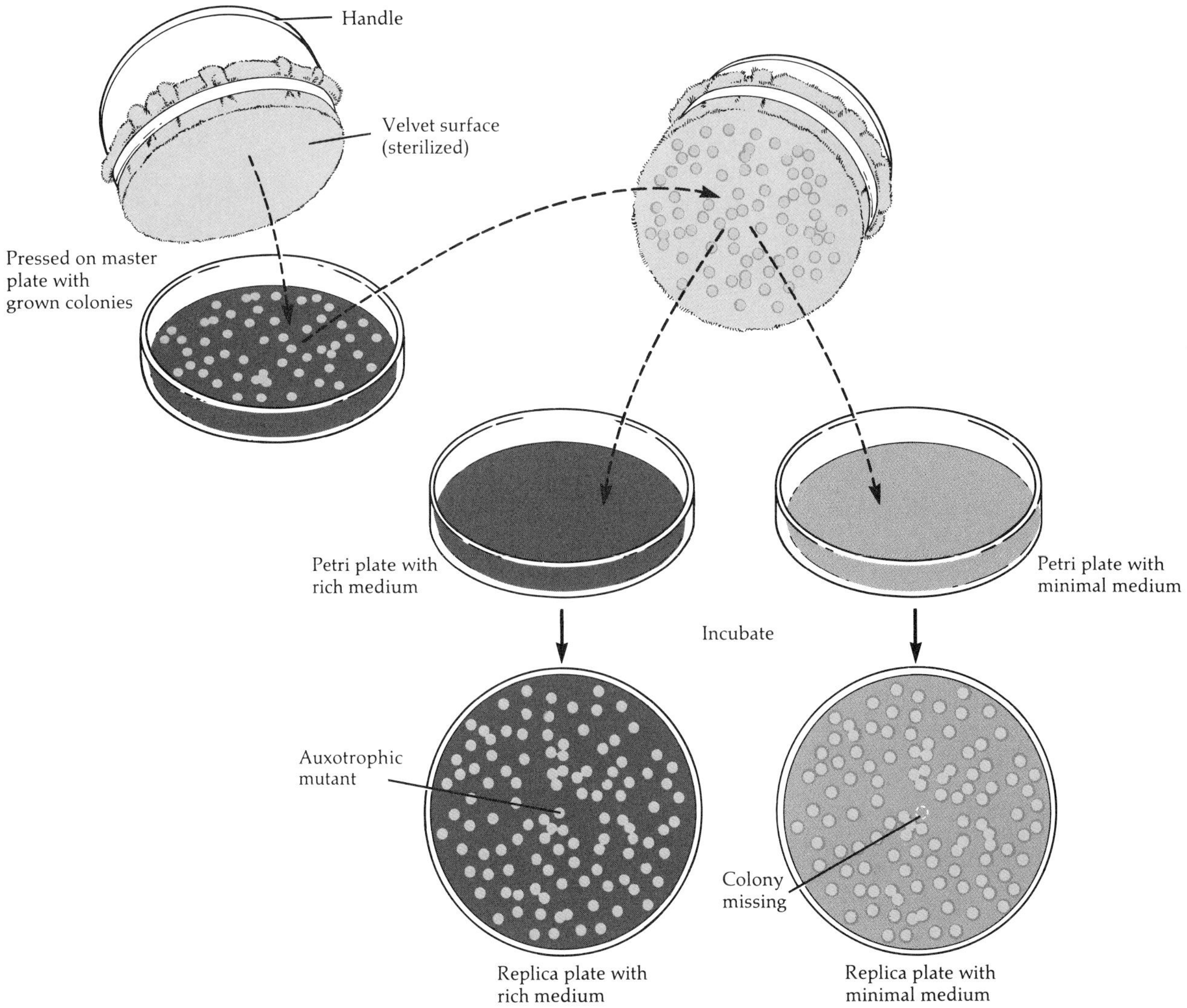

Figure 8–17 Replica plating. This technique transfers bacteria quickly and easily from colonies on a master plate (Petri plate) to a different medium in another plate. The procedure permits the identification of auxotrophic mutants, which form colonies on the supplemented medium of the master plate but are unable to grow on the minimal medium of the replica plate.

called the master plate, contains a rich, nonselective medium on which all cells will grow. After several hours of incubation, each cell reproduces to form a colony. Then a pad of sterile velvet is pressed over the master plate (Figure 8–17). Some of the cells from each colony adhere to the velvet. Next, the velvet is pressed down onto two (or more) sterile plates. One plate contains a rich medium and one contains a minimal medium on which the original, nonmutant bacteria can grow (for example, glucose and inorganic salts). A mutant colony that grew on the rich medium of the master plate, but has a new requirement for a growth factor, will not grow on the minimal me-

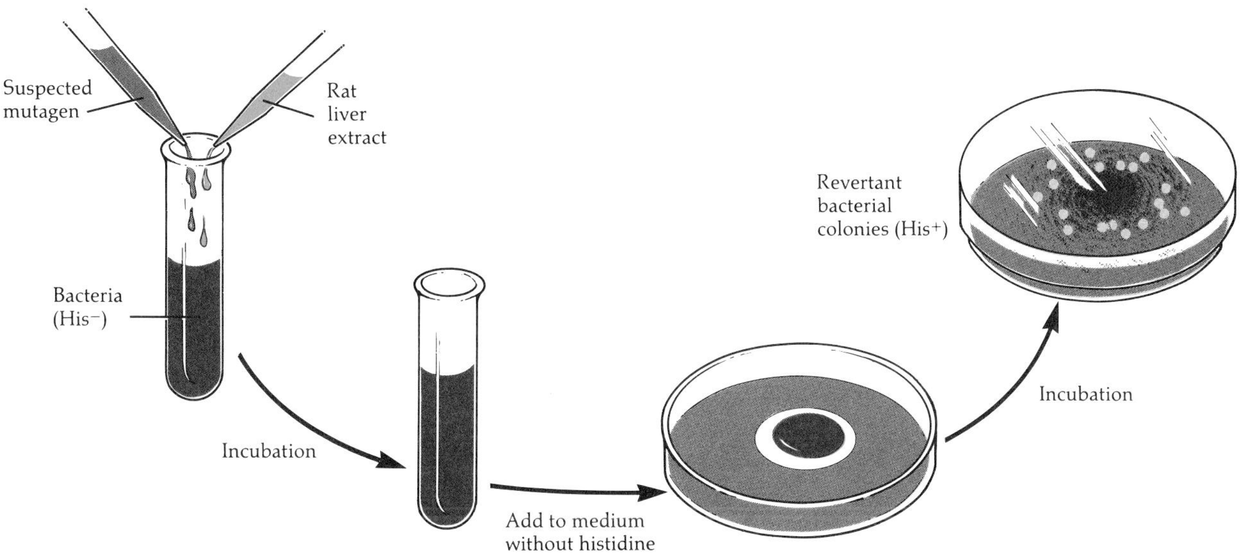

Figure 8–18 Ames test. Mutant *Salmonella* bacteria unable to synthesize histidine (His$^-$) are mixed with a suspected mutagen and rat liver extract and inoculated onto a medium lacking histidine. Only bacteria that have further mutated (reverted) to His$^+$ (able to synthesize histidine) will grow into colonies. The liver extract "activates" the suspected mutagen.

dium, which lacks that growth factor. Cells taken from the mutant colony on the master plate can then be tested on minimal media supplemented with various growth factors. These tests determine precisely which factor is required.

The replica plating technique is a very effective means of isolating mutants that require one or more new growth factors. Any mutant microorganism possessing a nutritional requirement (such as an amino acid) not possessed by the parent is known as an **auxotroph.**

THE AMES TEST FOR THE IDENTIFICATION OF CHEMICAL CARCINOGENS

Many known mutagens have been found to be carcinogens. A **carcinogen** is any substance that causes cancer in animals, including humans. In recent years chemicals in the environment, the work place, and the diet have been implicated as causes of cancer in humans. The usual subjects of tests to determine potential carcinogens are animals, and the testing procedures are both time consuming and expensive. Now there are faster and less expensive procedures for the preliminary screening of potential carcinogens. One of these, called the *Ames test,* employs bacteria as carcinogen indicators.

The Ames test is based on the fact that a mutated cell, by undergoing *another* or *back* mutation, can revert to a cell that resembles the original, nonmutant cell. The test measures the reversion of histidine auxotrophs of *Salmonella* (mutants that have lost the ability to synthesize the amino acid histidine) to nonmutant, histidine-synthesizing cells in both the presence and absence of the substance being tested (Figure 8–18). Many chemicals must be activated (transformed chemically) by animal enzymes in order for mutagenic or carcinogenic activity to be exhibited. The chemical to be tested and the mutant bacteria are incubated together with rat liver extract, a rich source of activation enzymes. If the substance being tested is mutagenic, it will cause a reversion rate higher than the spontaneous reversion rate. The number of observed revertants provides an indication of the degree to which a substance is mutagenic. About 90% of the substances found by the Ames test to be mutagenic have been shown also to be carcinogenic in animals. Furthermore, the more mutagenic substances are generally found to be more carcinogenic as well.

GENETIC TRANSFER AND RECOMBINATION

Genetic recombination is the rearrangement of genes to form new combinations. When we use the term, we are usually referring to rearrangement between genes from two individuals of the same species. Exceptions to this generalization will be discussed later.

Figure 8–19 shows one type of recombinational event occurring between two pieces of DNA, which we shall regard as chromosomes for the sake of simplicity. We have called one chromosome *A* and the other *B*. If these two chromosomes break and rejoin as shown, a process called **crossing over,** some of the genes carried by these chromosomes are shuffled. Thus, each original chromosome has been *recombined* so that it now carries a portion of the other chromosome's genes.

If *A* and *B* represent DNA from different individuals, how are they brought close enough to recombine? Genes of higher organisms are recombined during meiosis, when chromosomes from the two parents pair (a process called synapsis). Thus, the gametes resulting from meiosis can contain recombined DNA. For eucaryotes, genetic recombination is an ordered process that usually occurs as part of the sexual cycle of the organism. In bacteria, genetic recombination can happen in a number of ways, which we will discuss in the following sections.

Like mutation, genetic recombination contributes to a population's genetic diversity, which is the source of variation in evolution. Recombination is more likely than mutation to be beneficial, because recombination brings together different gene groups that may enable the organism to carry out a new function.

Genetic material can be transferred between bacteria in several ways. In all of the mechanisms, the transfer involves a **donor cell** that gives only a portion of its total DNA to a different **recipient cell.** Once transferred, part of the donor's DNA is usually incorporated into the recipient's DNA; the remainder is degraded by cellular enzymes. If the new DNA *replaces* DNA in the recipient's chromosome, as often occurs, there will be no net change in the length of the recipient's DNA. The recipient cell that has DNA from the donor added to its own DNA is called a **recombinant.** The transfer of genetic material between bacteria is by no means a regular event, but occurs in perhaps only 1% or less of an entire population. Let us now examine the specific types of genetic transfer in detail.

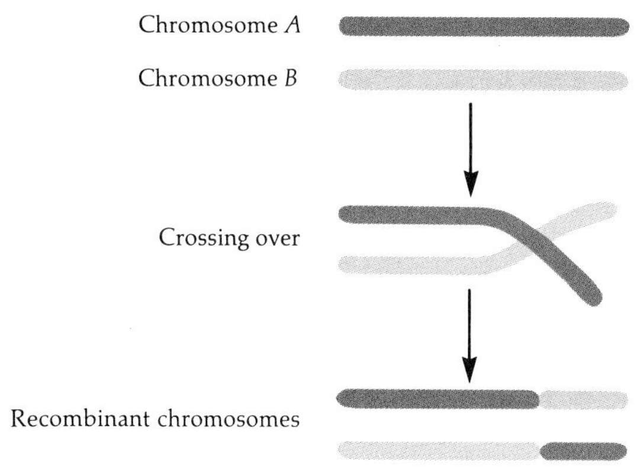

Figure 8–19 Simplified version of one type of genetic recombination between two chromosomes, *A* and *B*. The chromosomes cross over by breaking and rejoining. The result is two recombinant chromosomes, each of which carries genes originally present on two parent chromosomes.

TRANSFORMATION IN BACTERIA

During the process of **transformation,** genes transferred from one bacterium to another are "naked" DNA in solution. This process was first described more than 50 years ago, although it was not understood at that time. Not only did transformation demonstrate that genetic material could be transferred from one bacterial cell to another, but study of this event eventually led to the conclusion that DNA is the genetic material. The initial experiment on transformation was performed by F. Griffith in

England in 1928 while he was working with two strains of *Streptococcus pneumoniae.* One, a virulent strain, has a polysaccharide capsule and causes pneumonia; the second, an avirulent strain, lacks the capsule and does not cause disease.

Griffith was interested in determining whether injections of heat-killed (60°C) encapsulated pneumococci could be used to vaccinate mice against pneumonia. He found that the injected mice did not become ill from the heat-killed strain. However, when these dead cells were mixed with live avirulent cells and injected into the mice, many of the mice died. In the blood of the dead mice, Griffith found living, encapsulated, virulent bacteria. It was presumed that hereditary material from the dead bacteria had entered the live cells and changed them genetically so that their offspring were encapsulated, pathogenic forms (Figure 8–20).

Subsequent investigations based on Griffith's research revealed that bacterial transformation could be duplicated by use of standard microbiological culturing techniques in broth and on agar in Petri plates. A broth (liquid culture medium) was inoculated with live bacteria without capsules. Dead bacteria with capsules were then added to the broth, which was incubated for some hours. A drop of the culture was smeared over agar (solid culture medium) in Petri plates. After a period of incubation, it was found that significantly more of the bacterial colonies than would be expected from spontaneous back mutation or reversion, had the

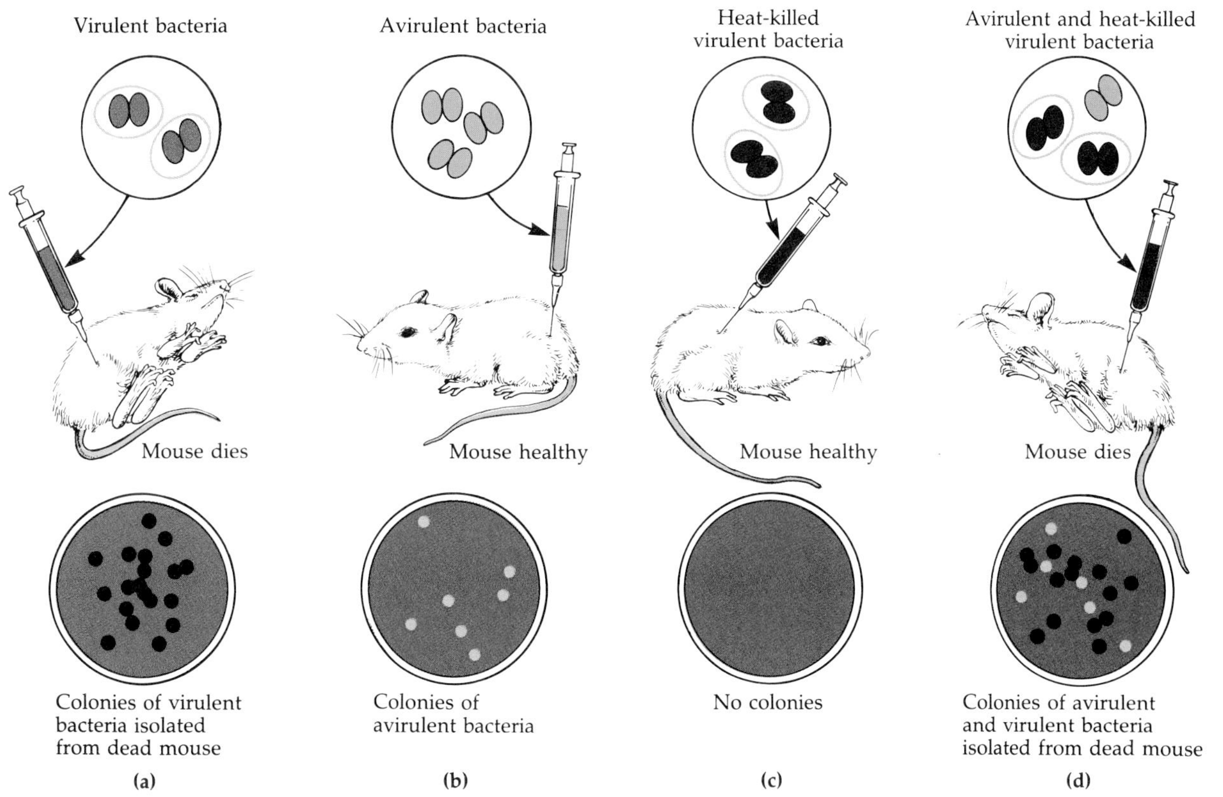

Figure 8–20 Griffith's experiment demonstrating genetic transformation. Some material from the heat-killed virulent bacteria transformed the living avirulent bacteria into virulent bacteria, which killed the mouse. Avirulent bacteria are readily destroyed by the host; therefore, few show up as colonies on the medium.

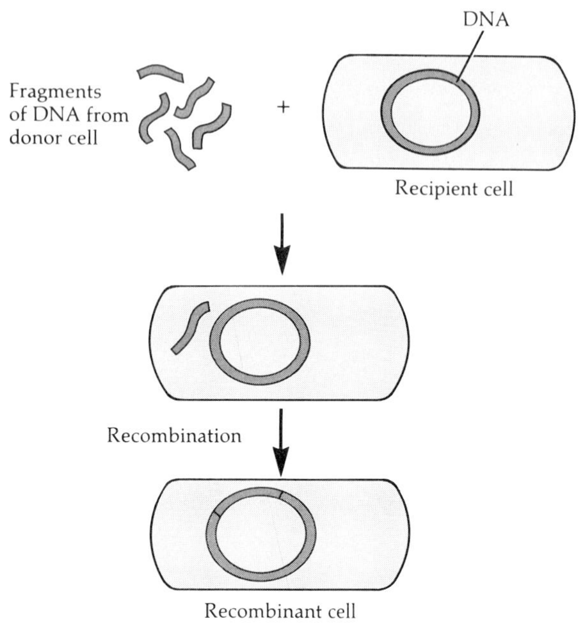

Figure 8–21 Mechanism of transformation.

smooth appearance typical of encapsulated bacteria—that is, they had been transformed. The bacteria had acquired a new hereditary trait.

The next step was to extract various chemical components from the killed cells and determine whether any of these chemicals cause the transformation. These crucial experiments were performed by O. T. Avery and his associates at the Rockefeller Institute in the United States. After years of research, they announced in 1944 that the chemical responsible for transforming harmless pneumococci into virulent strains was DNA.

Since Griffith's experiment, considerable information has been gathered about transformation. In nature, some bacteria, perhaps after death and cell lysis, release their DNA into the environment. Other bacteria can then encounter the DNA, and depending on the particular species and growth conditions, take up fragments of DNA into their cytoplasm and recombine the DNA into their own chromosomes. A recipient cell having this new combination of genes is a kind of hybrid—or recombinant—cell (Figure 8–21). Because its genes have been changed, all the descendants of such a recombinant cell will be identical to it. Transformation occurs naturally among very few genera of bacteria. These include *Bacillus, Hemophilus, Neisseria, Acinetobacter, Rhizobium,* and certain strains of *Streptococcus* and *Staphylococcus.*

Transformation works best when the donor and recipient cells are very closely related and when the recipient cells are in the late log phase of growth. Even though only a small portion of a cell's DNA is transferred to the recipient, it is still a very large piece of DNA that must pass through the recipient cell wall and membrane. When a recipient cell is in a physiological state in which it can take up and incorporate the donor DNA, it is said to be **competent.** Competence may be related to alterations in the cell wall that make it permeable to the large DNA molecule, or to the synthesis of specific receptor sites on the bacterial cell's surface. Among the traits whose transformation has been studied are capsule formation (already noted), nutritional needs, pathogenicity, pigment formation, and drug resistance.

Escherichia coli is not naturally competent for transformation. However, a simple laboratory treatment enables *E. coli* to readily accept its own or a foreign DNA. The discovery of this treatment, along with our recently developed ability to manipulate DNA in the laboratory virtually as a reagent, has sped the recombinant DNA–genetic engineering revolution (see later section in this chapter).

CONJUGATION IN BACTERIA

Conjugation is another mechanism by which genetic material is transferred from one bacterium to another. It differs from transformation in that contact between living cells is required.

Because much experimental work on conjugation has been done with *E. coli,* we will describe the process in this organism (Figure 8–22).

Conjugation requires not only that there be direct cell-to-cell contact, but also that the cells be of opposite mating types. One cell, the donor, is called an **F^+ cell.** The other cell, the recipient, is called an **F^- cell.** The cytoplasm of an F^+ cell contains extra pieces of DNA **F factors.** These are double-stranded pieces of DNA arranged in a closed circle and are not part of the bacterial chromosome. As you may recall from Chapter 4, F factors are one type of **plasmid**—a free genetic element that is not necessarily attached to, or part of, the bacterium's main chromosome. We shall de-

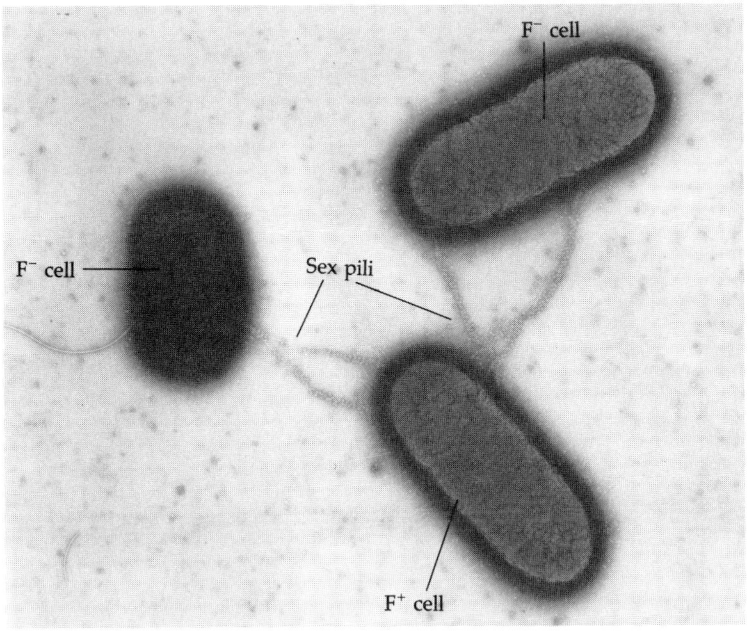

Figure 8–22 Conjugation in *E. coli* (×25,000).

scribe some other types of plasmids later in this chapter.

When F^+ and F^- cells are mixed together, the F^+ cells attach to the F^- cells by means of sex pili. (F factors contain genetic coding for the synthesis of sex pili.) During conjugation, the F factor is duplicated within the F^+ cell, and the new copy is transferred across an intercellular bridge from the F^+ to the F^- cell. This transfer converts the F^- cell into an F^+ cell. Now an F^+, it is capable of transferring its F^+ factor to another F^- cell. Remember, the F factor is a plasmid, separate from the bacterial chromosome. When an F factor is transferred, the bacterial chromosome of the F^+ cell is not passed to the F^- cell. Thus, this event produced no recombinants (Figure 8–23a).

Some F^+ cells have the F factor integrated into their chromosome (Figure 8–23b); they are called **Hfr (high frequency of recombination) cells.** During conjugation between an Hfr and an F^- cell, the Hfr cell's chromosome, with its integrated F factor, replicates, and the new copy of the chromosome is transferred to the recipient cell (Figure 8–23c). Only Hfr cells can transfer a copy of their chromosome. Replication of the Hfr chromosome begins within the F factor, and a small piece of the F factor leads the chromosomal genes into the F^- cell. Most of the integrated F factor enters the recipient cell last, if at all. Usually, the chromosome breaks before it is completely transferred. Once within the recipient cell, some donor DNA can recombine with the recipient's DNA, as occurs in transformation. Donor DNA that is not integrated is degraded.

TRANSDUCTION IN BACTERIA

A third mechanism of genetic transfer between bacteria is **transduction.** In this process, naked DNA is not passed as it is in transformation, or during cell-to-cell contact, as in conjugation. Rather, DNA is passed inside a bacterial virus, called a **bacteriophage** (or simply **phage**) from the donor to the recipient cell. (Phages will be discussed further in Chapter 12.)

To understand how transduction works, we shall consider the life cycle of one type of transducing phage of *E. coli;* this phage carries out a

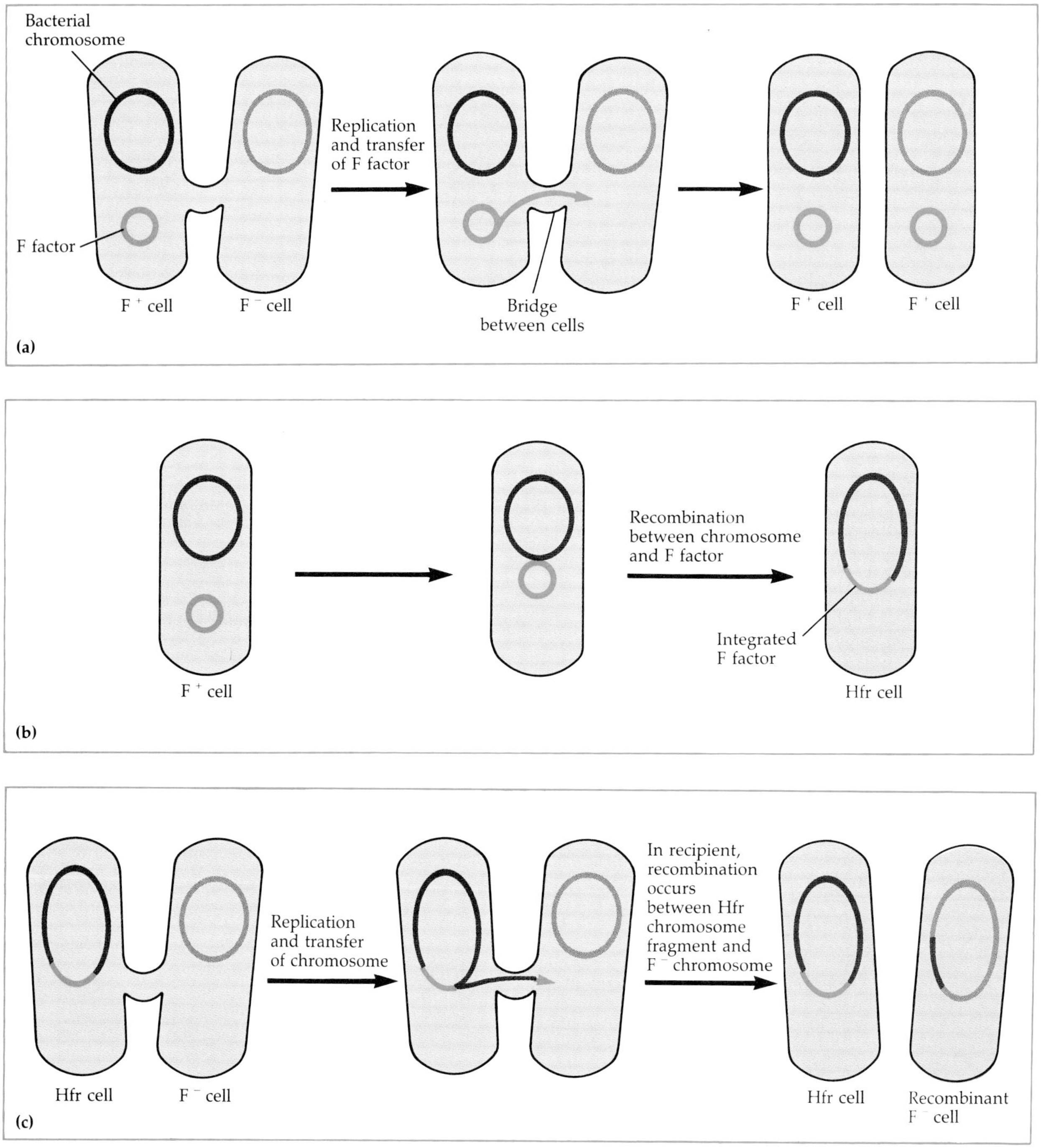

Figure 8–23 Mechanism for conjugation. **(a)** When an F factor is transferred from donor (F^+) to recipient (F^-), the F^- cell is converted into an F^+ cell. **(b)** Integration of an F factor into the bacterial chromosome makes the cell an Hfr. **(c)** An Hfr donor passes a portion of its chromosome into an F^- cell. A recombinant F^- cell results.

process called **generalized transduction.** In the process of infection, the phage attaches to the bacterial cell wall and injects its DNA into the bacterium (Figure 8–24). The viral DNA acts as a template for the synthesis of new viral DNA and also directs the synthesis of viral protein coats to envelop new viral particles. During a viral infection of a bacterial cell, fragments of the bacterial chromosome may occasionally be accidentally incorporated into the viral genome. These particles now carry bacterial DNA in addition to viral DNA, and when the infected bacterial cell releases the newly synthesized viral particles, some bacterial DNA enclosed in viral coats is also released. If the viral-coated bacterial DNA infects a new population of bacteria, that new population will thus receive several bacterial genes. Viral-coated bacterial DNA cannot cause cell lysis, as does a phage. Transduction of the viral-coated bacterial DNA usually leads to recombination between the DNA of the first host and the DNA of the new host cell. This recombination is one way that bacteria acquire new genotypes.

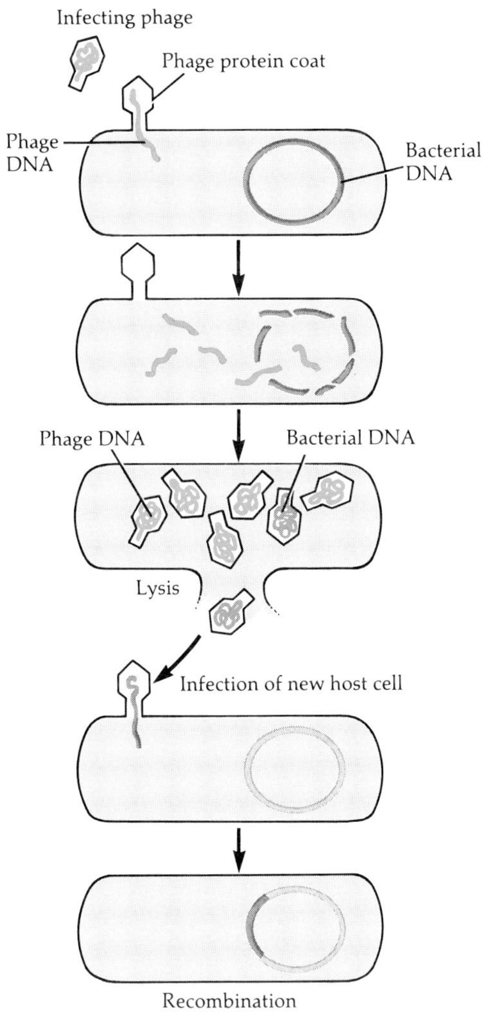

Figure 8–24 Transduction by a bacteriophage. Shown here is generalized transduction, in which any bacterial gene can be transferred.

All genes contained within a bacterium undergoing generalized transduction are equally likely to be picked up in a phage coat and transferred. There is another type of transduction, called **specialized (restricted) transduction,** in which only certain bacterial genes are transferred. Specialized transduction will be discussed in Chapter 12.

RECOMBINATION IN EUCARYOTES

Transformation, conjugation, and transduction all result in genetic recombination and involve only the DNA of the donor cell entering the recipient cell. In eucaryotes, genetic recombination is the result of a sexual reproductive process. In contrast to the mechanisms in bacteria, sexual reproduction is a regular event for most eucaryotes and is necessary for the survival of many species. Sexual reproduction in eucaryotes involves the fusion of the haploid nuclei of two parent cells (Figure 8–25). A **haploid** cell is one that contains only one of each type of chromosome; the number of chromosomes in such a cell is called the **haploid number,** symbolized n. The haploid cells that fuse are called **gametes.** (In animals, the male gametes are sperm and the female gametes are ova.) Fusion of the gametes produces a fertilized ovum, or **zygote,** which receives half its chromosomes from one parent and half from the other.

Each species of eucaryotic organism has a characteristic **chromosome number,** that is, the number of chromosomes in each ordinary body cell (called *somatic* cells). The human chromosome number is 46 (23 pairs). Each parent contributes one set of 23 chromosomes, which carries genes for all the activities of human cells. The other set of 23 chromosomes is more or less a duplicate set; it contains the same or a different version of each gene.

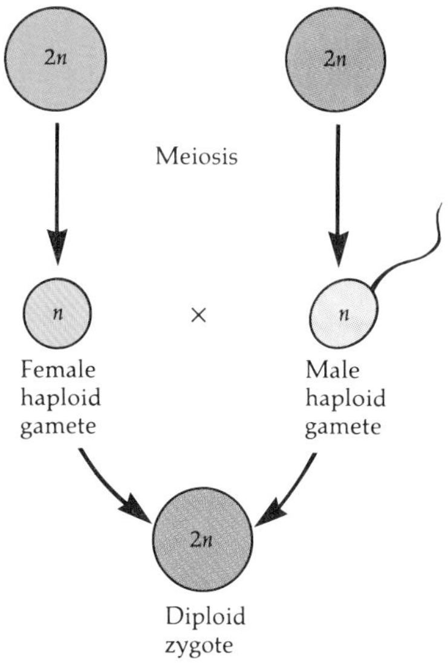

Figure 8–25 Sexual reproduction in eucaryotes.

A **diploid** cell contains two of each type of chromosome; the number of chromosomes in such a cell is called the **diploid number,** symbolized *2n*. The paired chromosomes are known as **homologous chromosomes.**

For eucaryotes to reproduce sexually by fusing their gametes to form zygotes, the chromosome number must be reduced from diploid to haploid in the gametes. This is accomplished by a process called **meiosis,** in which certain cells undergo division to yield haploid gametes. In the process of meiosis, portions of chromosomes can be exchanged with one another by **crossing over** (see Figure 8–19). Crossing over in eucaryotic cells can often be observed microscopically; it is in fact the process of genetic recombination. Once haploid gametes are produced by meiosis, the gamete's nucleus can fuse with another gamete's nucleus and form a diploid zygote. Subsequent growth and development of a zygote produces a mature organism again capable of sexual reproduction.

PLASMIDS

Plasmids are small, self-replicating circles of DNA found in many bacteria. They carry genes that the cell needs only under special conditions, and certain plasmids, such as the F plasmid, can be integrated into the main chromosome. Plasmids are usually dispensable in that the cell can survive under most conditions without the use of the proteins coded by plasmid genes. The F factor, for instance, is useful in gene transfer, which is usually not essential for survival. Many plasmids are **conjugative plasmids;** that is, they have genes for carrying out their transfer to another cell (and for transferring the bacterial chromosome or other plasmids).

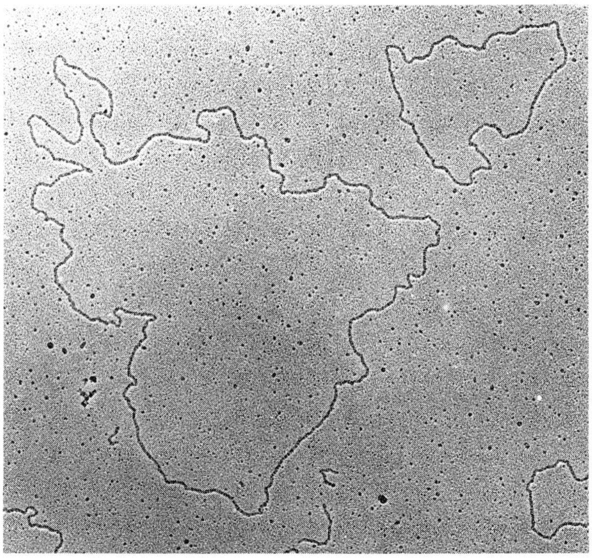

(a)

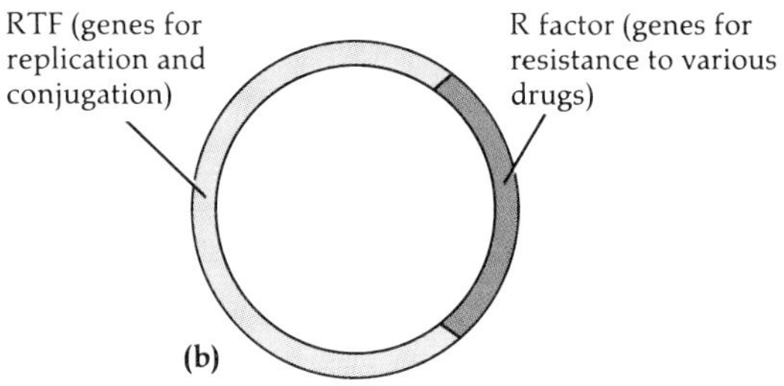

Figure 8–26 R factors, a type of plasmid. **(a)** Electron micrograph showing two different resistance (R) factors (×15,000). The larger circle of DNA is a plasmid isolated from the bacterium *Bacteroides fragilis* and encodes resistance to the antibiotic clindamycin. The smaller loop of DNA is a plasmid from *Escherichia coli* and provides resistance to tetracycline. **(b)** Diagram of an R factor.

Plasmids often carry genes that have a more direct use to the cell than that of simple gene transfer. **Dissimilation plasmids** have genetic coding for enzymes that catalyze the catabolism of certain unusual sugars and hydrocarbons. Some species of *Pseudomonas* can actually use such substances as toluene, camphor, and the high-molecular-weight hydrocarbons of petroleum as primary carbon and energy sources.

Bacteriocinogenic plasmids contain genes for the synthesis of *bacteriocins,* toxic proteins that kill other bacteria. These plasmids have been found in many genera, and they are useful markers for the identification of certain bacteria in clinical laboratories.

Resistance factors (R factors) are plasmids of great medical importance. R factors carry genes that determine the resistance of their host cell to antibiotics, heavy metals, or cellular toxins. Many R factors are conjugative plasmids that contain two groups of genes. One group has the **resistance transfer factor (RTF)** and includes genes for replication and conjugation. The other group, the **R genes,** codes for resistance and contains the genetic instructions for the production of enzymes that inactivate certain drugs (Figure 8–26). Different R factors, when present in the same cell, readily recombine to produce R factors with new combinations of R genes.

R factors present very serious problems for the treatment of infectious diseases by antibiotics. The widespread use of antibiotics in medicine and agriculture (many animal feeds contain antibiotics) has led to the preferential survival (selection) of bacteria that have R factors. The transfer of resistance between bacterial cells of a population, and even between different genera, also contributes to the problem.

GENETIC ENGINEERING: RECOMBINANT DNA

Recombination, as discussed above, is a natural biological process in most organisms. In recent years, however, the term *recombination* has taken on a special, somewhat different meaning when applied to experiments involving genetic engineering. Our detailed knowledge of the plasmids and enzymes involved in DNA metabolism has made it possible for us to isolate genes from various organisms and incorporate them into bacterial cells. This procedure of manufacturing and manipulating genetic material is referred to as **genetic engineering** (see Microview 2). **Recombinant DNA** results from the joining together of DNA from different sources, often from different species, in vitro (Figure 8–27). (*In vitro* means "in glass," that is, not within a living organism.) When the recombinant DNA is introduced into an appropriate bacterium, the bacterium can synthesize the products of the new genes it has acquired.

In the making of recombinant DNA, special bacterial enzymes called **restriction enzymes** are used to cut open bacterial plasmids (see Figure 8–27). These same enzymes are used to cut segments of DNA from some other organism. Scientists have cut segments from frog, rat, and human cells, and from yeast and bacteria. If the DNA segment is small and its base sequence is known, the segment can be chemically synthesized instead of extracted from an organism.

But by whatever method a foreign DNA is obtained, the next step is the insertion of the DNA into a plasmid (Figure 8–27). Restriction enzymes usually cut staggered ends on DNA molecules, so the base pairs can be matched as two different pieces of DNA are combined. The DNA fragment is inserted into the bacterial plasmid, and the junctions are covalently sealed by *DNA ligase.*

The recombinant plasmid is then introduced into bacterial cells by transformation. Chemical treatment of the cell wall of the recipient bacteria removes structural barriers and increases its permeability to DNA during transformation. The resulting cells are hybrids with genotypes that never before existed in nature. Each time the hybrid bacteria divide, so do its plasmids carrying the recombinant DNA. Often such recombinant bacteria can produce an enormous amount of whatever substance is programmed by their new genotype.

Most bacterial cells do not readily secrete the substances produced. After cultivation the cells are collected and lysed to harvest their product. Current efforts in genetic engineering are focused on the culturing of recombinant cells that secrete their product naturally.

Special techniques are required if the DNA of a particular eucaryotic gene is made up of exons

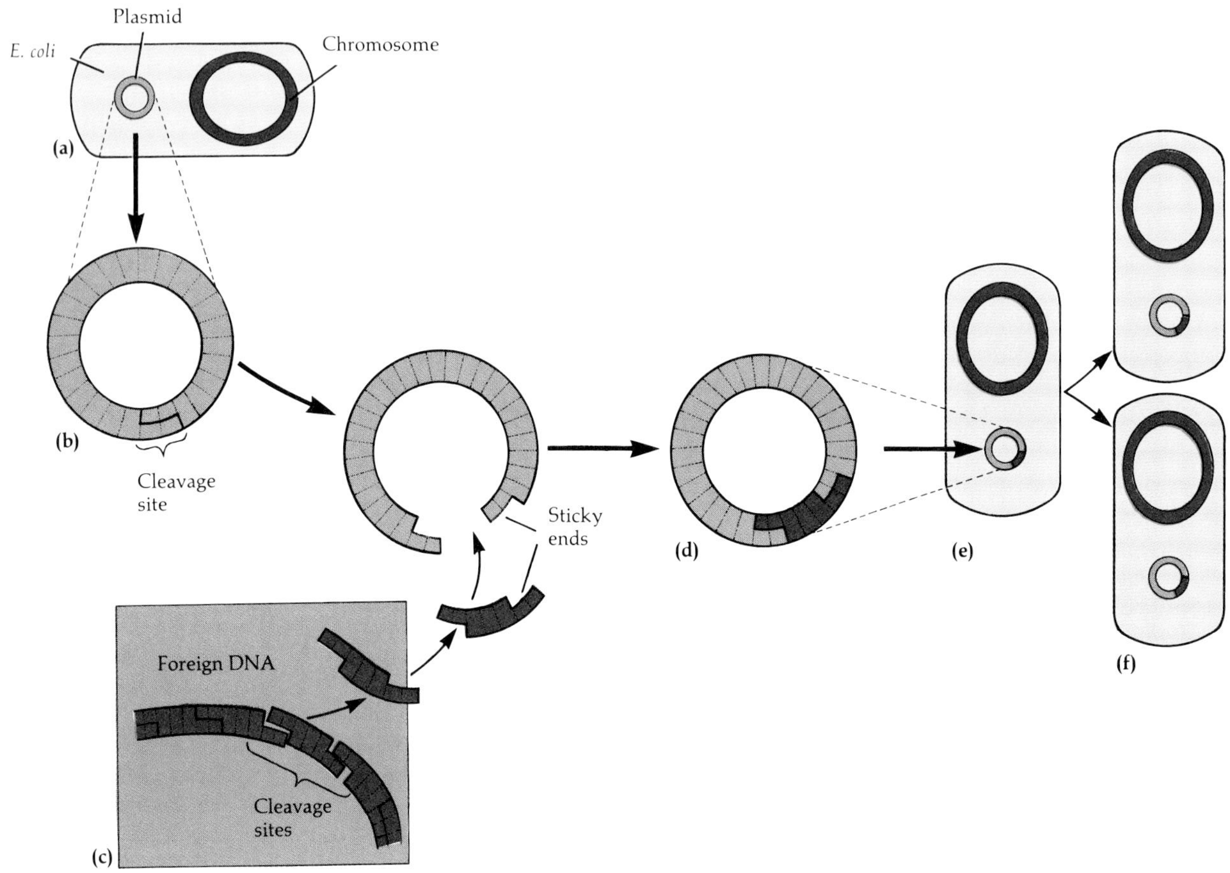

Figure 8–27 Method for in vitro construction of recombinant DNA. **(a)** A plasmid is first isolated from *E. coli*. **(b)** Restriction enzymes split the plasmid at a known cleavage site. **(c)** A segment of foreign DNA from another organism is split by restriction enzymes. Many such enzymes make staggered cuts, as shown on both the plasmid and foreign DNA, resulting in single-strand stretches at the new ends. These short polynucleotide stretches are called "sticky ends." **(d)** The foreign DNA is inserted into the plasmid and joined covalently by the enzyme DNA ligase. **(e)** The recombinant plasmid is introduced into bacterial cells by transformation. **(f)** Successive generations of the hybrid bacteria maintain the recombinant plasmid.

and introns (Figure 8–28). *Exons* are regions of DNA that code for proteins; *introns* are intervening regions that do not code for proteins. The eucaryotic cell transcribes the whole DNA including both exons and introns, removes the intron RNA, and finally splices together the remaining pieces to form mRNA. Bacterial cells do not contain the enzymes to remove the intron RNA; therefore, a bacterial recombinant cannot make the mRNA or protein. The DNA of eucaryotic cells that contain exons and introns is not used for recombination; instead, mRNA is collected. An exact DNA transcript of the mRNA is then made in vitro, with viral reverse transcriptase as a reagent (see Figure 12–20). This *cDNA* (copy DNA) represents the entire correct coding sequence of DNA for the gene of interest.

Once a eucaryotic gene is isolated, it is often useful to make multiple copies of it. These copies are called clones and are produced by recombinant

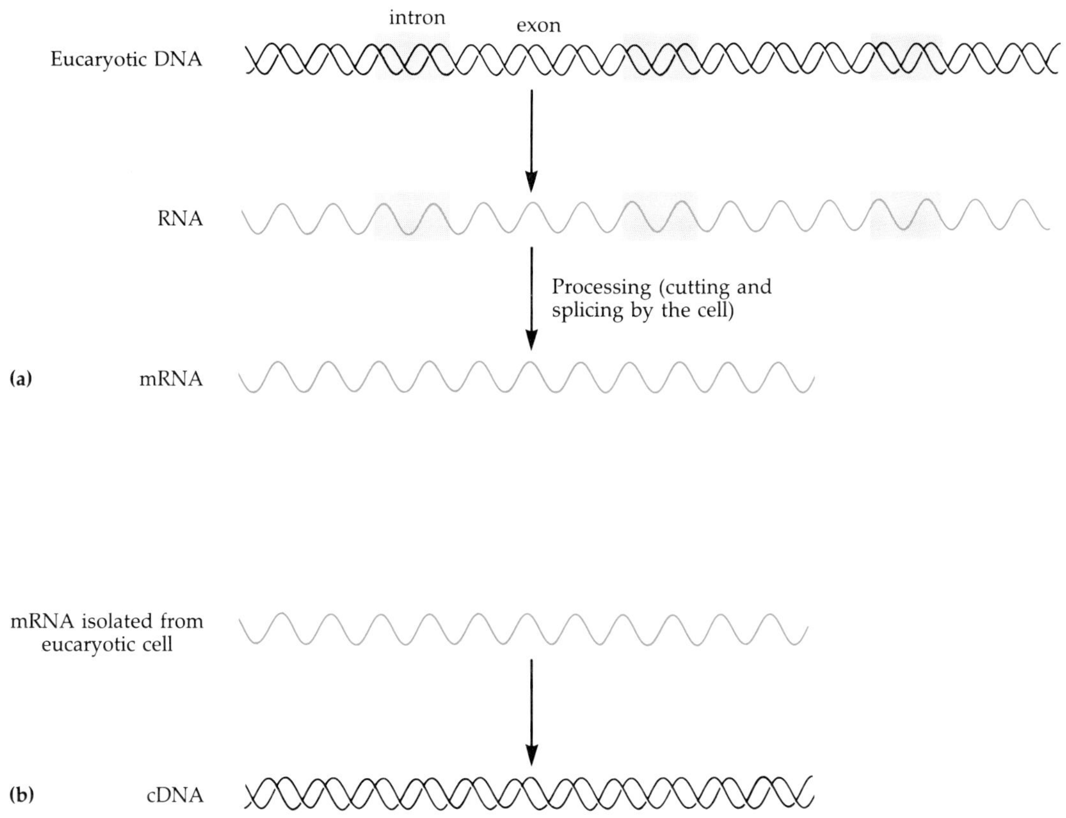

Figure 8–28 **(a)** Transcription and processing of a gene containing exons and introns. **(b)** Reverse transcription of mRNA into DNA in vitro is brought about by the viral enzyme reverse transcriptase (see Chapter 12 and Figure 12–20). The new DNA is called "cDNA."

DNA methods. When a gene of interest is cloned, it must first be incorporated into a plasmid. The plasmid (recombinant DNA) is then inserted into a selected bacterium. The plasmid replicates as the bacterium multiplies, and so multiple copies of the gene are produced.

Although one important goal of recombinant-DNA research is to increase our understanding of how genes are regulated, the technique also has practical applications. First, animal genes can be inserted into bacterial DNA so that bacteria produce human substances of therapeutic importance. For example, bacteria have already been programmed to synthesize *somatostatin,* a brain hormone related to growth; *insulin,* a hormone used by diabetics; *human growth hormone,* a hormone that promotes growth in children; and *interferon,* an important antiviral (and possible anticancer) substance that will be discussed in Chapter 15.

Recombinant-DNA techniques are also being used to produce certain proteins that appear on the surface of pathogenic microorganisms. Large amounts of these proteins are produced by harmless bacteria altered by recombinant DNA. When such a protein is given as a vaccine to an animal, it causes the animal to produce natural antibodies.

A third application of recombinant-DNA research is to improve existing fermentation processes for making microbial products such as antibiotics. And a fourth application is its use by the chemical and energy industries. For example, ways of programming bacteria to convert ethylene to ethylene glycol or to ferment plants into ethanol for use as fuel are now being rigorously pursued.

Genetic engineering is not without dangers. There is always the possibility of creating a lethal organism or a highly resistant one that might cause widespread disease. In view of this possibility, scientists have agreed that any potentially dangerous genetic engineering experiments should be performed only with genetically weakened bacterial strains, and in laboratories from which the newly created recombinants cannot escape and contaminate the general public. In the accompanying box, we discuss some of the positive and negative aspects of genetic engineering currently being debated.

GENES AND EVOLUTION

We have now seen how genes can be controlled by the cell's internal regulatory mechanisms and altered by mutation and recombination. All these processes provide diversity in the descendants of cells, and diversity is the driving force of evolution. Without genetic diversity, there can be no natural selection by environmental factors. The different kinds of microorganisms that exist today, all with relatively stable properties, are the result of a long history of evolution. During that process, microorganisms have continuously changed by altering their genetic properties, adapting, as they did so, to many different habitats.

Microbiology in the News

The Promise and Peril of Genetic Engineering

No technological advance offers as much hope for a better future as genetic engineering, and none promises to revolutionize so much of our society—how we treat diseases, how we grow plants and rear livestock, even how we mine for minerals and drill for oil.

When the promises of genetic engineering were first unfurled in the 1970s, public alarm grew over possible dangers from life forms accidentally or intentionally released into the environment. Warnings of environmental disruption, especially from new diseases that might decimate animal populations, were fairly common. Some argued that manipulating the human genome was philosophically and morally unacceptable. Still others warned of potential misuses—a laboratory Hitler cloning superhumans for subterfuge and diversion.

The cloud of confusion and distrust grew. Would genetic engineering expand resources and feed the hungry, or would it seriously endanger the human race? More than a decade after its inception, the long-term answers to these questions still remain unanswered.

The Promise

Progress in genetic engineering to date shows that the promises for this new technology are very real, and the potential is even much greater than many once hoped. Consider these accomplishments and predictions for genetic engineering:

- Medical researchers have cloned the genes that code for human insulin and one form of interferon, a potentially powerful tool against viral diseases and cancer. Both chemical substances are now produced by genetically engineered bacteria and are commercially available in relatively large quantities.
- Genetic engineering is producing effective vaccines against influenza and hepatitis, and may yield vaccines against herpes, AIDS, and other dangerous viruses.
- Scientists are implanting the gene for human growth hormone into livestock embryos in hopes of producing marketable livestock much faster than by conventional means.
- Medical researchers are developing ways to replace defective human genes with normal genes, with the goal of curing previously incurable genetic disorders such as Lesch-Nyhan syndrome.
- Genes that render certain bacteria toxic to insects have been isolated and transplanted to a bacterium that

grows on plant roots, giving some plants a new defense against pests.

- Genetic researchers have developed a mutant bacterium that retards frost development on plants and promises to save millions of dollars each year.
- Genes that allow plants to fix atmospheric nitrogen are being developed for use in nonleguminous plants, thus reducing the need for fertilizer and the water pollution created by fertilizers.
- Genetic engineers are developing a special strain of bacteria that produces ethanol from cellulose. This bacterium could help convert forest and crop residues into valuable liquid fuel as a substitute for increasingly scarce oil.

The successes and the not-too-distant promises of genetic engineering have spawned a great deal of enthusiasm in the world business community. Dozens of new companies have formed in recent years, and billions of dollars have already been invested in the fledgling industry.

The Questions

Nevertheless, the safety questions remain. Many scientists still are actively trying to decipher the risks posed by genetic engineering, particularly the possibility of creating new forms of disease. Scientists agree that modified microorganisms may behave in unpredictable ways and that, once unleashed, they and their effects could not be recalled.

Another major concern of the scientific community is the ways genetic engineering allow human beings to tamper with the evolutionary process itself. Deliberate genetic manipulations may be different from anything that ordinarily occurs during "natural" evolution.

On the other hand, some evidence suggests that there might already be much more exchange of genes among different groups of organisms in nature than was previously supposed. Furthermore, recent work suggests that the dangers of genetic engineering have been blown out of proportion by a public that is legitimately concerned but not adequately informed, and that genetically engineered bacteria are not generally a threat to ecosystem stability. In short, much of the suspicion seems unfounded.

The Reactions

Further work that would help scientists learn more about the hazards of genetic engineering is now tied up in costly legal battles. For example, Jeremy Rifkin, of the Foundation of Economic Trends, and Michael W. Fox, of the Humane Society of the United States, have filed suit to stop the transfer of human growth hormone genes to animals, work currently spearheaded by the U.S. Department of Agriculture. Rifkin and Fox are concerned on many levels about the mixing of human and animal genes. Perhaps most importantly, they worry about the environmental disruptions created by the loss of genetic traits that have been a part of an animal's genome for centuries.

Many supporters of genetic engineering feel quite the opposite on this question and the general question of halting research in genetic engineering. To them, the promises of this revolutionary technology far outweigh any potential damage. Feeding the world's hungry with a genetically engineered strain of supercattle, they argue, far exceeds the potential threats. Likewise, gene therapy, in which defective human genes are replaced by normal ones, offers too much hope for victims of crippling diseases to be abandoned because of problems that are not proven to exist. Even leading environmentalists, who initially joined Rifkin at the outset of the battle in the 1970s, have switched sides.

The Future

In the midst of the controversy, it is important to remember that scientists themselves are very aware of the potential hazards of genetic engineering and are actively working to make it as safe as possible. Groups such as the Biohazards Conference at Asilomar, The National Academy of Sciences, and The Recombinant DNA Advisory Committee (RAC) of the National Institutes of Health, as well as individual scientists, have debated the questions and suggested guidelines that can function as safeguards for genetic engineering research. The public, in turn, must become better educated so that it can respond not out of panic, but from an informed position.

Genetic engineering stands at a threshold. Still in its infancy, this revolutionary science offers great promise and, at the same time, possible unprecedented peril. Which the legacy will be only time will tell.

STUDY OUTLINE

STRUCTURE AND FUNCTION OF THE GENETIC MATERIAL (pp. 211–223)

1. Genetics is the study of what genes are, how they carry information, how their information is expressed, and how they are replicated and passed to subsequent generations or other organisms.
2. DNA in cells exists as a double-stranded helix; the two strands are held together by hydrogen bonds between specific nitrogenous base pairs: A–T and C–G.
3. A gene is a segment of DNA, a sequence of nucleotides, that codes for a functional product, usually a protein.
4. The genetic information in a region of DNA is transcribed to produce RNA; mRNA is translated into proteins.

GENOTYPE AND PHENOTYPE (pp. 211–212)

1. Genotype is the genetic composition of an organism—its entire DNA.
2. Phenotype is the expression of the genes—the proteins of the cell and their ability to perform particular chemical reactions.

DNA AND CHROMOSOMES (pp. 212–213)

1. The DNA in a chromosome exists as one long, double helix associated with various proteins that regulate genetic activity.
2. Bacterial DNA is circular; the chromosome of *E. coli* contains about 4 million base pairs and is approximately 1000 times longer than the cell.
3. The DNA of eucaryotic chromosomes is more condensed and is associated with much protein, including histones.

DNA REPLICATION (pp. 213–218)

1. During DNA replication, the two strands of the double helix separate at the replication fork, and each strand is used as a template by DNA polymerases to synthesize two new strands of DNA according to the rules of nitrogenous base pairing.
2. The result of DNA replication is two new strands of DNA, each having a base sequence complementary to one of the original strands.
3. Because each double-stranded DNA molecule contains one original and one new strand, the replication process is called semiconservative.
4. DNA is synthesized in one chemical direction called 5′→3′. At the replication fork, one daughter strand is synthesized continuously, the other discontinuously.
5. DNA synthesis of new fragments on the lagging or discontinuous strand moves in the opposite direction from the movement of the replication fork. Completion of the lagging strand requires RNA primers, RNA polymerase, and DNA ligase, as well as DNA polymerase.

RNA AND PROTEIN SYNTHESIS (pp. 218–222)

Transcription (pp. 218–221)

1. During transcription, the enzyme RNA polymerase synthesizes a strand of mRNA from one strand of double-stranded DNA, which serves as a template.
2. The starting point for transcription where RNA polymerase binds to DNA is the promoter site; the region of DNA that is the end point of transcription is the terminator site.
3. The product of transcription is mRNA. This copy of the genetic information is a template for protein synthesis.

Translation (pp. 221–222)

1. The process in which the information in the nucleotide-base sequence of mRNA is used to dictate the amino acid sequence of a protein is called translation.
2. The mRNA associates with ribosomes, which consist of rRNA and protein.
3. Three-base segments of mRNA are called codons.
4. Specific amino acids are attached to molecules of tRNA. Another portion of the tRNA has a base triplet called an anticodon.
5. The interaction of codon and anticodon at the surface of the ribosome results in specific amino acids being brought to the site of protein synthesis.
6. The ribosome moves along an mRNA strand as amino acids are joined to form a growing polypeptide.

7. A polyribosome is an mRNA with several ribosomes attached to it.

THE GENETIC CODE (pp. 222–223)

1. The genetic code refers to the relationship between the nucleotide-base sequence of DNA, the corresponding codons of mRNA, and the amino acids for which the codons code.
2. The genetic code is degenerate; that is, most amino acids are coded for by more than one codon.
3. Of the 64 codons, 61 are sense codons (which code for amino acids) and 3 are nonsense codons (which do not code for amino acids and are stop signals for translation).

REGULATION OF GENE EXPRESSION IN BACTERIA (pp. 223–226)

1. Gene regulation of protein synthesis is energy efficient, because synthesis occurs only when proteins (enzymes) are needed.
2. For gene regulatory mechanisms, the control is aimed at mRNA synthesis, which then controls enzyme synthesis.

INDUCTION AND REPRESSION (pp. 223–226)

Induction (p. 224)

1. In the presence of certain chemicals (inducers), cells are induced to synthesize more enzymes. This process is called induction.
2. An example is that β-galactosidase is produced by *E. coli* in the presence of lactose, so lactose can be metabolized.

Repression (p. 225)

1. Repression controls the synthesis of one or several (repressible) enzymes.
2. When cells are exposed to a particular end product, the synthesis of enzymes related to that product decreases. Biosynthesis of amino acids such as histidine and tryptophan is controlled in this way.

Operon Model (pp. 224–226)

1. The formation of enzymes is controlled by structural genes. These genes and the promoter and operator sites that control their transcription are called an operon.
2. In the operon model, a regulator gene codes for repressor protein.
3. When the inducer is absent, the repressor binds to the operator and no mRNA is synthesized.
4. When the inducer is present, it binds to the repressor so that it cannot bind to the operator; thus, mRNA is made and enzyme synthesis is induced.
5. In some repressible systems, the repressor requires a corepressor in order to bind to the operator site; thus the corepressor controls enzyme synthesis.
6. Some genes, called constitutive genes, produce enzymes regardless of how much substrate is present. Examples are genes for most of the enzymes of glycolysis.

MUTATION: CHANGE IN THE GENETIC MATERIAL (pp. 226–234)

1. A mutation is a change in the nitrogenous-base sequence of DNA; that change causes a change in the gene product coded for by the mutated gene.
2. Many mutations are neutral, some are disadvantageous, and some are beneficial.

TYPES OF MUTATIONS (pp. 226–228)

1. A base occurs when one base pair in DNA is replaced with a different base pair.
2. In a frameshift mutation, one or a few base pairs are deleted or added to DNA.
3. Alterations in DNA can result in missense mutations (amino acid substitutions) or nonsense mutations (which create stop codons).

MUTAGENESIS (pp. 228–230)

1. Mutagens are agents in the environment that cause permanent changes in DNA.
2. Some mutations occur without the presence of a mutagen; these are called spontaneous mutations.

Chemical Mutagens (pp. 228–230)

1. Chemical mutagens include base-pair mutagens (e.g., nitrous acid), base analogs (e.g., 2-aminopurine and 5-bromouracil), and frameshift mutagens (e.g., benzpyrene).

Radiation (p. 230)

1. Ionizing radiation causes the formation of ions and free radicals that react with DNA; base substitutions or breakage of the sugar-phosphate backbone result.
2. Ultraviolet radiation is nonionizing; it causes bonding between adjacent thymines.
3. In the presence of light, radiation-induced damage can be repaired by photoreactivation; it can be repaired in the dark by enzymes that cut out and replace the damaged portion of DNA. Errors in the repair process result in mutations.

FREQUENCY OF MUTATION (p. 230)

1. Mutation rate is the probability that a gene will mutate when a cell divides; the rate is expressed as 10 to a negative power.
2. Mutations usually occur randomly along a chromosome.
3. A low rate of spontaneous mutations is beneficial in providing the genetic diversity needed for evolution.

IDENTIFYING MUTATIONS (pp. 230–233)

1. Mutations can be detected by selecting or testing for an altered phenotype.
2. Positive selection involves the selection of mutant cells and rejection of unmutated cells.
3. Replica plating is used for negative selection, to detect, for example, auxotrophs that have nutritional requirements not possessed by the parent (nonmutated) cell.

THE AMES TEST FOR THE IDENTIFICATION OF CHEMICAL CARCINOGENS (pp. 233–234)

1. This is a relatively inexpensive and rapid test for identifying possible chemical carcinogens.
2. The test assumes that a mutant cell can revert to a normal cell in the presence of a mutagen, and that many mutagens are carcinogens.
3. Histidine auxotrophs of *Salmonella* are exposed to an enzymatically treated potential carcinogen, and reversions to the nonmutant (prototroph) are selected.

GENETIC TRANSFER AND RECOMBINATION (pp. 234–245)

1. Genetic recombination, which refers to the rearrangement of genes from separate groups of genes, usually originates from different organisms; it contributes to genetic diversity.
2. In crossing over, genes from two chromosomes are recombined into one chromosome containing some genes from each original chromosome.
3. Mechanisms for recombination in bacteria involve a portion of the cell's DNA being transferred from donor to recipient.
4. When some of the donor's DNA has been integrated into the recipient's DNA, the resultant cell is called a recombinant.

TRANSFORMATION IN BACTERIA (pp. 234–236)

1. During this process, genes are transferred from one bacterium to another as "naked" DNA in solution.
2. This process was first demonstrated in *Streptococcus pneumoniae*.
3. It occurs naturally among a few genera of bacteria.

CONJUGATION IN BACTERIA (pp. 236–237)

1. This process requires contact between living cells.
2. One type of genetic donor cell is an F^+; recipient cells are F^-. F^+ cells contain cytoplasmic DNA (plasmids) called F factors; these are transferred to the F^- cell during conjugation.
3. When the plasmid becomes incorporated into the chromosome, the cell is called an Hfr (high-frequency recombinant).
4. During conjugation, an Hfr can transfer its entire chromosome to an F^-. Usually, the Hfr chromosome breaks before it is fully transferred.

TRANSDUCTION IN BACTERIA (pp. 237–239)

1. In this process, DNA is passed from one bacterium to another in a bacteriophage.
2. The new bacterial genes can be incorporated into the recipient cell after it is infected by the bacteriophage.

RECOMBINATION IN EUCARYOTES (pp. 239–240)

1. In eucaryotes, genetic recombination is associated with sexual reproduction.
2. Sexual reproduction involves the fusion of haploid gametes to form a diploid zygote.
3. Haploid gametes are produced through meiosis, in which portions of chromosomes can be exchanged in the process of genetic recombination.

PLASMIDS (pp. 240–241)

1. Plasmids, self-replicating circular strands of DNA, are found in the cytoplasm of bacteria.
2. Plasmid genes are not usually essential for survival of the cell.
3. There are several types of plasmids including conjugative, dissimilative, toxigenic, and resistance plasmids.

GENETIC ENGINEERING: RECOMBINANT DNA (p. 241–244)

1. The procedure of manufacturing or manipulating genetic material in the laboratory is called genetic engineering.
2. Recombinant DNA results from the union of DNA from different sources, often different species; it uses restriction enzymes and DNA ligase, as well as other enzymes and biochemical procedures.
3. Foreign DNA is inserted into a bacterial plasmid, and the process of transformation introduces the plasmid into bacterial cells.
4. Through genetic engineering, bacteria can produce human substances of therapeutic importance, carry out improved fermentation processes, and synthesize chemicals for industry.

GENES AND EVOLUTION (p. 244)

1. Diversity is the driving force of evolution.
2. Genetic mutation and recombination provide a diversity of organisms, and the process of natural selection allows the growth of those best adapted for a given environment.

STUDY QUESTIONS

REVIEW

1. Briefly describe the components of DNA and explain its functional relationship to RNA and protein.
2. Draw a diagram showing a portion of a chromosome undergoing replication.
   (a) Identify the replication fork.
   (b) What is the role of DNA polymerase? Of RNA polymerase?
   (c) What is the role of DNA ligase?
   (d) How does this process represent semiconservative replication?
3. Below is a strand of DNA.

(3′)	A	T	A	T	T	A	C	T	A	C	T	T	T	G	C	T	T	G	G	C	A	C	T	G	A	A	C	T	(5′)
	1	2	3	4	5	6	7	8	9	10	11	12	13	14	15	16	17	18	19	20	21	22	23	24	25	26	27	28	

TATAATG = Promoter sequence

   (a) Write the complementary strand of DNA.
   (b) Using the genetic code provided in Figure 8–10, identify the sequence of amino acids coded for by mRNA transcribed from this strand of DNA (the complementary strand is the antisense strand).
   (c) What would the effect be if C was substituted for T at base 13?
   (d) What would the effect be if A was substituted for G at base 14?
   (e) What would the effect be if G was substituted for A at base 21?
   (f) What would the effect be if C was inserted between bases 12 and 13?
   (g) How would ultraviolet radiation affect this strand of DNA?
   (h) Identify a nonsense sequence in this strand of DNA.
4. Describe translation, and be sure to include the following terms: ribosome, rRNA, amino acid activation, tRNA, anticodon, and codon.
5. Contrast the structures of a bacterial chromosome and a eucaryotic chromosome.
6. Explain the selection of an antibiotic-resistant mutant by direct selection, and the selection of an antibiotic-sensitive mutant by indirect selection.

7. Match the following examples of mutagens.

___ A mutagen that is incorporated into DNA in place of a normal base	(a) Frameshift mutagen
___ A mutagen that causes the formation of highly reactive ions	(b) Base analog
___ A mutagen that alters adenine so that it base-pairs with cytosine	(c) Base-pair mutagen
___ A mutagen that causes insertions	(d) Ionizing radiation
___ A mutagen that causes the formation of pyrimidine dimers	(e) Nonionizing radiation

8. Describe the principle of the Ames test for identifying chemical carcinogens.
9. Differentiate between transformation and transduction.
10. Define plasmids and explain the relationship between F plasmids and conjugation.
11. Outline the production of recombinant DNA in vitro.
12. List three practical applications of genetic engineering.
13. Use this metabolic pathway to answer the questions below:

$$\text{Substrate } A \xrightarrow{\text{enzyme } a} \text{Intermediate } B \xrightarrow{\text{enzyme } b} \text{End product } C$$

(a) If enzyme *a* is inducible and is not being synthesized at present, a ________ protein must be bound tightly to the ________ site. When the inducer is present, it will bind to the ________ so that ________ can occur.
(b) If enzyme *a* is repressible, end product *C*, called a ________, causes the ________ to bind to the ________. What causes derepression?
(c) If enzyme *a* is constitutive, what effect, if any, will the presence of *A* or *C* have on it?

14. Define the following terms:
(a) Genotype
(b) Phenotype
(c) Recombination
15. Why are mutation and recombination important in the process of natural selection and the evolution of organisms?
16. You are provided with cultures with the following characteristics.

Culture 1: F^+, genotype $A^+ B^+ C^+$
Culture 2: F^-, genotype $A^- B^- C^-$

(a) Indicate the possible genotypes of a recombinant cell resulting from the conjugation of cultures 1 and 2.
(b) Indicate the possible genotypes of a recombinant cell resulting from conjugation of the two cultures after the F^+ has become an Hfr.

CHALLENGE

1. Base analogs and ionizing radiation are used in the treatment of cancer. These are mutagens, so how do you suppose they are used to treat diseases?
2. Why are semiconservative replication and degeneracy of the genetic code advantageous to survival?

Recombinant DNA at Work

Since ancient times people have harnessed the metabolic abilities of microorganisms for human use—primarily for the making of foods such as bread, cheese, yogurt, wine, and beer. In the past half century, the range of microbial products has expanded to include chemicals for medical use (antibiotics) and industrial use. Within the past 10 years, however, biotechnology has entered a whole new phase. Laboratory techniques for making recombinant DNA—DNA containing genes from two or more sources—have made it possible to "engineer" microorganisms to make products heretofore unique to the cells of animals and plants.

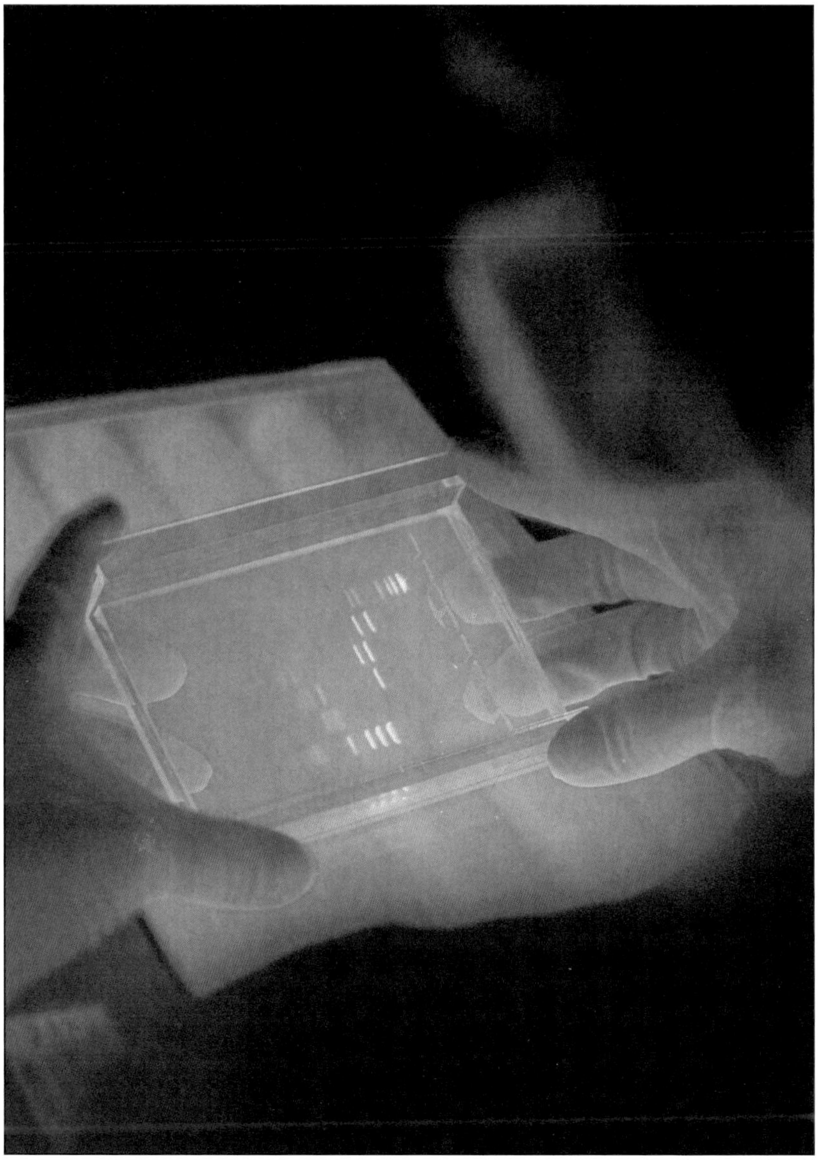

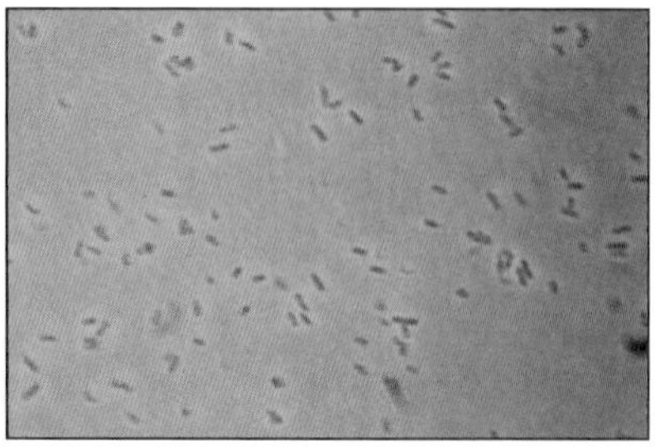

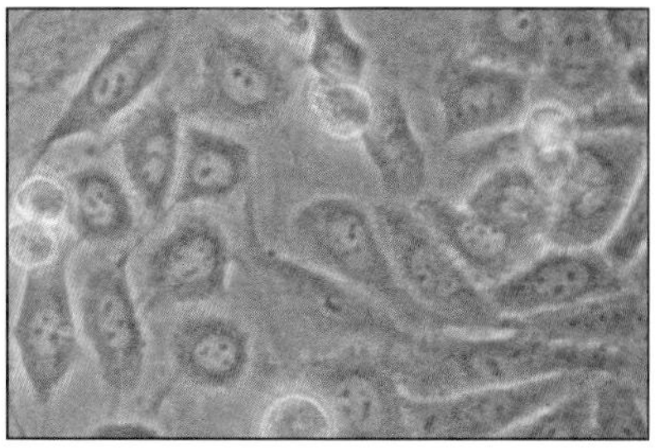

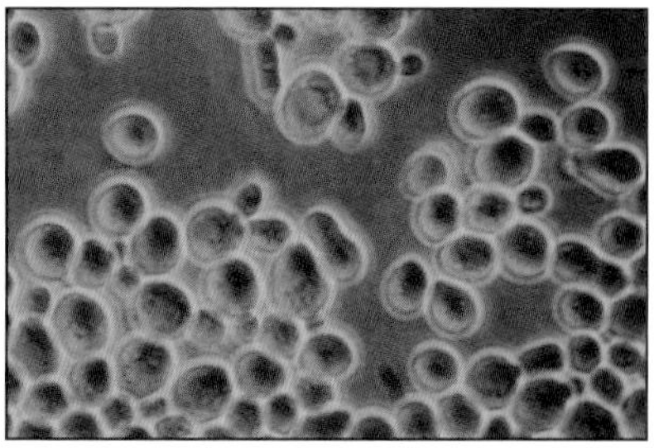

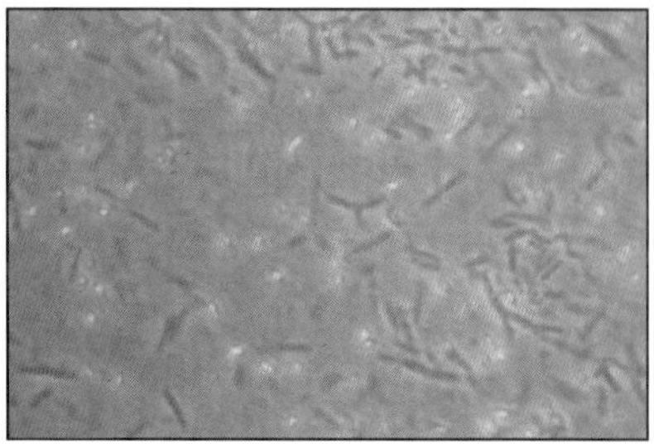

■ The micrographs above show four types of cells used in the production of recombinant-DNA products. From the top, they are *E. coli*, cultured mammalian cells, yeast, and *Bacillus*. (Magnification is approximately 1500×.) The micrographs on this page are from Genentech, a biotechnology company. ■ One of the techniques used in the making of recombinant DNA is slab gel electrophoresis. The slab gel in the picture at the left has been treated with a DNA-binding dye that fluoresces pink in UV light. The pink bands are DNA molecules of different sizes. On the next three pages, you will learn more about this and other processes of the new biotechnology.

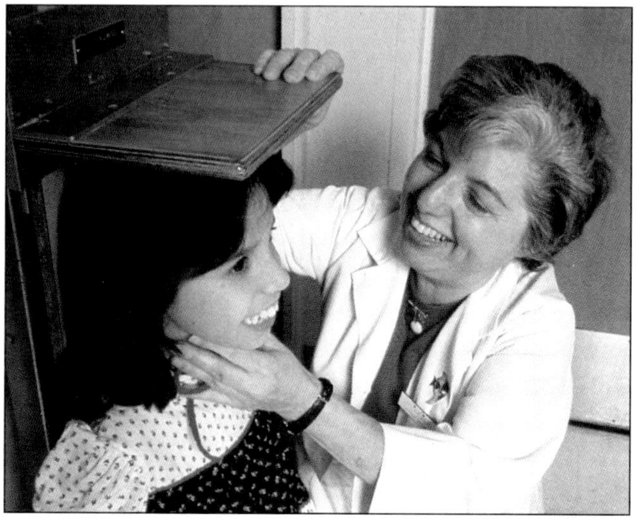

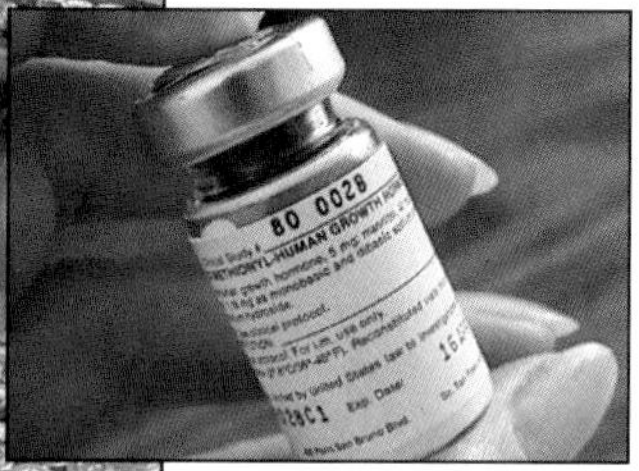

Benefits of Biotechnology

■ **Medicine.** In the photo above, taken at Genentech, bottles of human growth hormone produced by recombinant-DNA techniques are in the final stages of manufacture. Children with growth-hormone deficiencies are already benefiting from this new product (upper right). ■ **Agriculture.** At the near right, a research assistant at Cetus tests seeds coated with genetically engineered microbial treatments, products designed to enhance crop yield. The corn at the far right is a result of a research program at the University of California, Davis. Genetically engineered crop plants with improved characteristics is an ultimate goal of such research.

3. Replication of *E. coli*'s chromosome takes 40 to 45 minutes, and the cell has a generation time of 26 minutes. How does the cell have complete chromosomes for each daughter cell? For each granddaughter cell?

FURTHER READING

Ayala, F. J. and J. A. Kiger, Jr. 1984. *Modern Genetics*, 2nd ed. Menlo Park, CA: Benjamin/Cummings. A recent genetics textbook with excellent coverage of bacterial and viral genetics.

Daven, C. I. (Introduction). 1981. *Genetics: Readings from Scientific American*. San Francisco, CA: W. H. Freeman. A collection of papers covering the chemical structure of DNA, the genetic code, and applied genetics.

Devoret, R. August 1979. Bacterial tests for potential carcinogens. *Scientific American* 241:40–49. Short tests using bacteria instead of animals to detect carcinogenic chemicals and the effects of carcinogens on DNA.

Freifelder, D. 1983. Molecular Biology: *A Comprehensive Introduction to Procaryotes and Eucaryotes*. Boston, MA; Science Books International. An excellent molecular genetics text that discusses topics of this chapter in more detail.

Gilbert, W. and L. Villa-Komaroff. April 1980. Useful proteins from recombinant bacteria. *Scientific American* 242:74–94. How genes are spliced into bacteria so the bacteria can produce nonbacterial proteins such as insulin and interferon.

Howard-Flanders, P. November 1981. Inducible repair of DNA. *Scientific American* 245:72–80. Describes the roles of two enzymes that respond to an "SOS" for DNA damage.

Watson, J. D., J. Tooze, and D. T. Kurtz. 1983. *Recombinant DNA: A Short Course*. New York: W. H. Freeman. Describes laboratory techniques used to identify, isolate, and transfer genes in procaryotic and eucaryotic cells; includes a chapter on the recombinant DNA industry.

PART TWO

SURVEY OF THE MICROBIAL WORLD

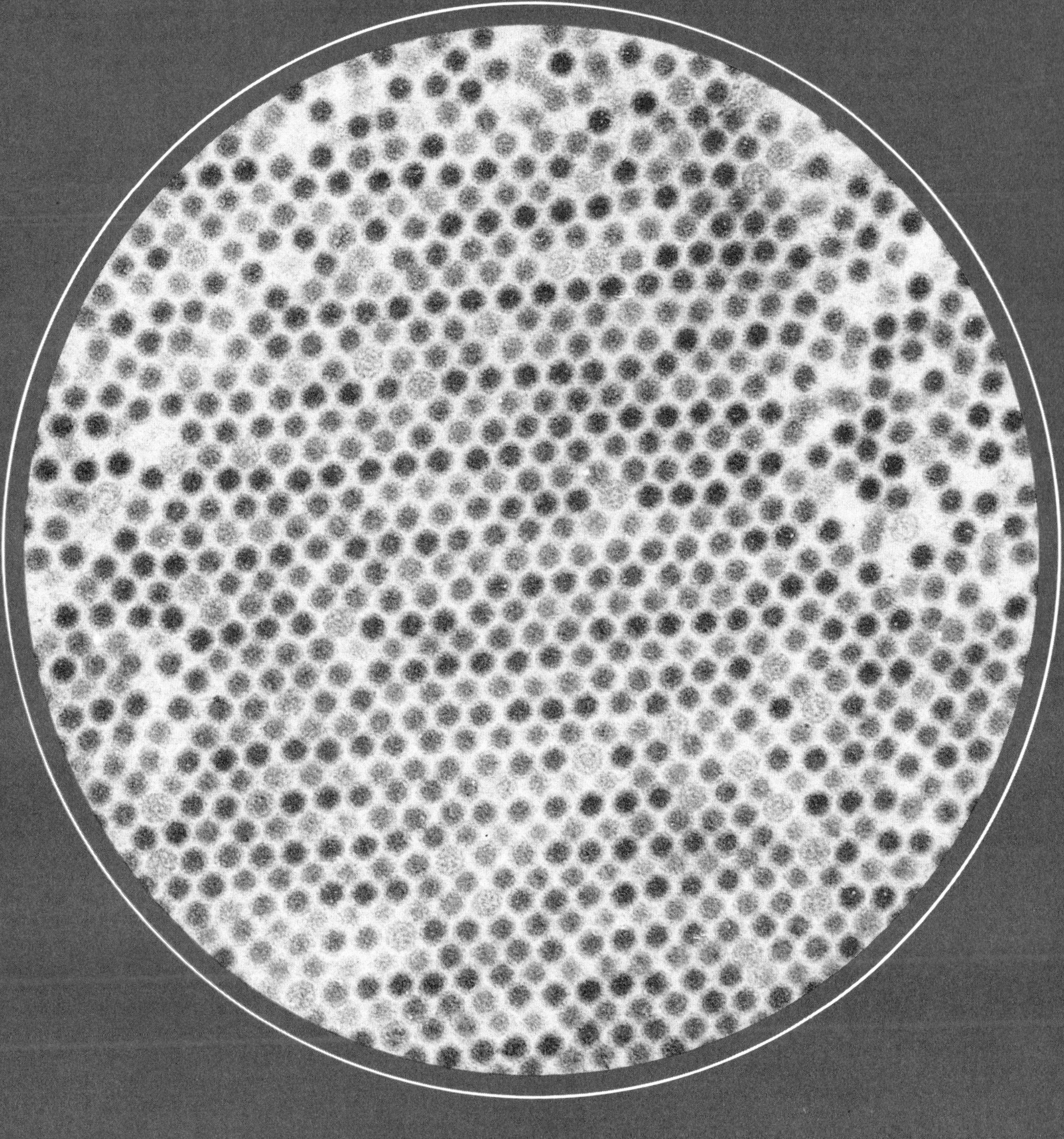

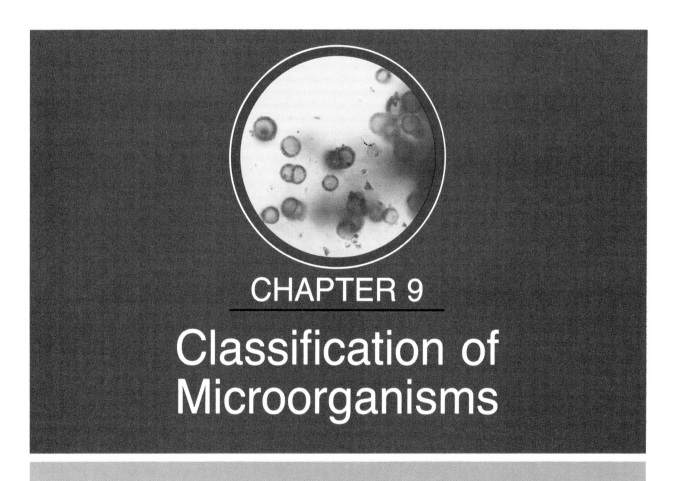

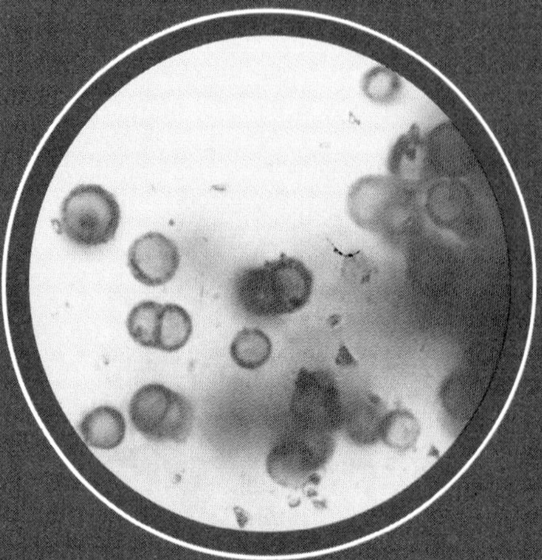

CHAPTER 9

Classification of Microorganisms

OBJECTIVES

After completing this chapter you should be able to

- Define taxonomy.
- List the major characteristics used to differentiate the five kingdoms of living organisms.
- Identify the characteristics of the kingdom Procaryotae that differentiate it from other kingdoms.
- Define binomial nomenclature and discuss its purpose.
- Compare and contrast classification and identification.
- Describe ten methods of identification and classification of microorganisms.

The science of classification, especially of living forms, is called **taxonomy.** The objects of taxonomy are to establish the relationship between one group of organisms and another and to differentiate between them. Until the end of the nineteenth century, all organisms were divided into two kingdoms: plant and animal. Then, when microscopes were developed and biologists came to understand the structures and physiological characteristics of microorganisms, it became apparent that microorganisms did not really belong to either the plant or animal kingdom, although many microbes have both plant and animal characteristics. In 1866, Ernst Haeckel proposed a third kingdom, the Protista, to include all microorganisms—bacteria, fungi, protozoans, and algae. But subsequent research indicated that among the Protista could be distinguished two basic cell types: procaryotic and eucaryotic. The morphological differences between these two cell types were described in detail in Chapter 4.

FIVE-KINGDOM SYSTEM OF CLASSIFICATION

In 1969, H. R. Whittaker proposed the **five-kingdom system** of biological classification, which has been widely accepted. Recall from Chapter 1 that in this system, there is a basic division between organisms with procaryotic cells and those with eucaryotic cells (Figure 9–1). The bacteria and the cyanobacteria (formerly called blue-green algae) are the procaryotic organisms, called **Monera** by Whittaker and **Procaryotae** by others, including *Bergey's Manual of Systematic Bacteriology.* As discussed in Chapter 4, procaryotes do not have a nucleus separated from the cytoplasm by a membrane. Instead, they have a nuclear region in which the genetic material is more or less localized. Fossil evidence for procaryotic organisms indicates their presence on earth more than 3.5 billion years ago.

Higher organisms are based on the structurally much more complicated eucaryotic cell, in which a membrane separates the nucleus from the cytoplasm. This type of cell apparently evolved more recently, about 1.4 billion years ago.

In 1982, Margulis and Schwartz proposed a revision of the five-kingdom classification scheme. Their revised scheme distinguishes among the four eucaryotic kingdoms according to nutritional requirements, patterns of development, tissue differentiation, and possession of complex flagella. (Recall that eucaryotic flagella and cilia are composed of nine outer pairs plus an inner pair of microtubules, whereas procaryotic flagella are composed of repeating flagellin subunits.)

Simple eucaryotic organisms, mostly unicellular, are grouped as the **Protista.** Protists generally have flagella at some time during their life cycle. The multicellular protists lack tissue organization. The kingdom Protista includes funguslike water molds, slime molds, animallike protozoans, and the primitive eucaryotic algae.

Plants, animals, and fungi comprise three kingdoms of complex eucaryotic organisms, most of which are multicellular.

Fungi are eucaryotic organisms such as the unicellular yeasts, multicellular molds, and macroscopic varieties such as mushrooms. To obtain raw materials for vital functions, a fungus absorbs dissolved organic matter through the membranes of cells joined together to form thin tubes called hyphae. These cells lack complete walls, so they form continuous, multinucleated cytoplasm rather than discrete cells. Most fungi lack flagella. Fungi develop from spores or fragments of hyphae.

Plantae are multicellular eucaryotes. This kingdom includes some algae and all mosses, ferns, conifers, and flowering plants. To obtain energy, a plant photosynthesizes food; this process converts carbon dioxide from the air to organic molecules used by the cell.

Animalia are multicellular animals composed of eucaryotic cells. This kingdom includes sponges, various worms, insects, and animals with backbones (vertebrates). Animals obtain carbon and energy by ingesting organic matter through a mouth of some kind.

NAMING ORGANISMS

In a world inhabited by millions of living organisms, biologists must be sure they know exactly which organism is being discussed. We cannot use common names because the same name is often used for many different organisms in different regions. For example, there are two different plants with the common name Spanish Moss, and neither one is a moss. And three different animals are referred to as a gopher. Because common names are rarely specific and can often be misleading, a system of scientific names, referred to as **scientific nomenclature,** was developed in the eighteenth century by Carolus Linnaeus.

This system gives every organism two names, usually derived from either Latin or Greek. These are the **genus** name and **specific epithet,** and both names are printed underlined or italicized. The genus name is always capitalized and is always a noun. The species name is written in lowercase and is usually an adjective. Because this system gives two names to each organism, the system is called **binomial nomenclature.**

Let us consider a few examples. Your own genus and specific epithet are *Homo sapiens* (hō'mō sā' pē-ens); one bacterium that causes pneumonia

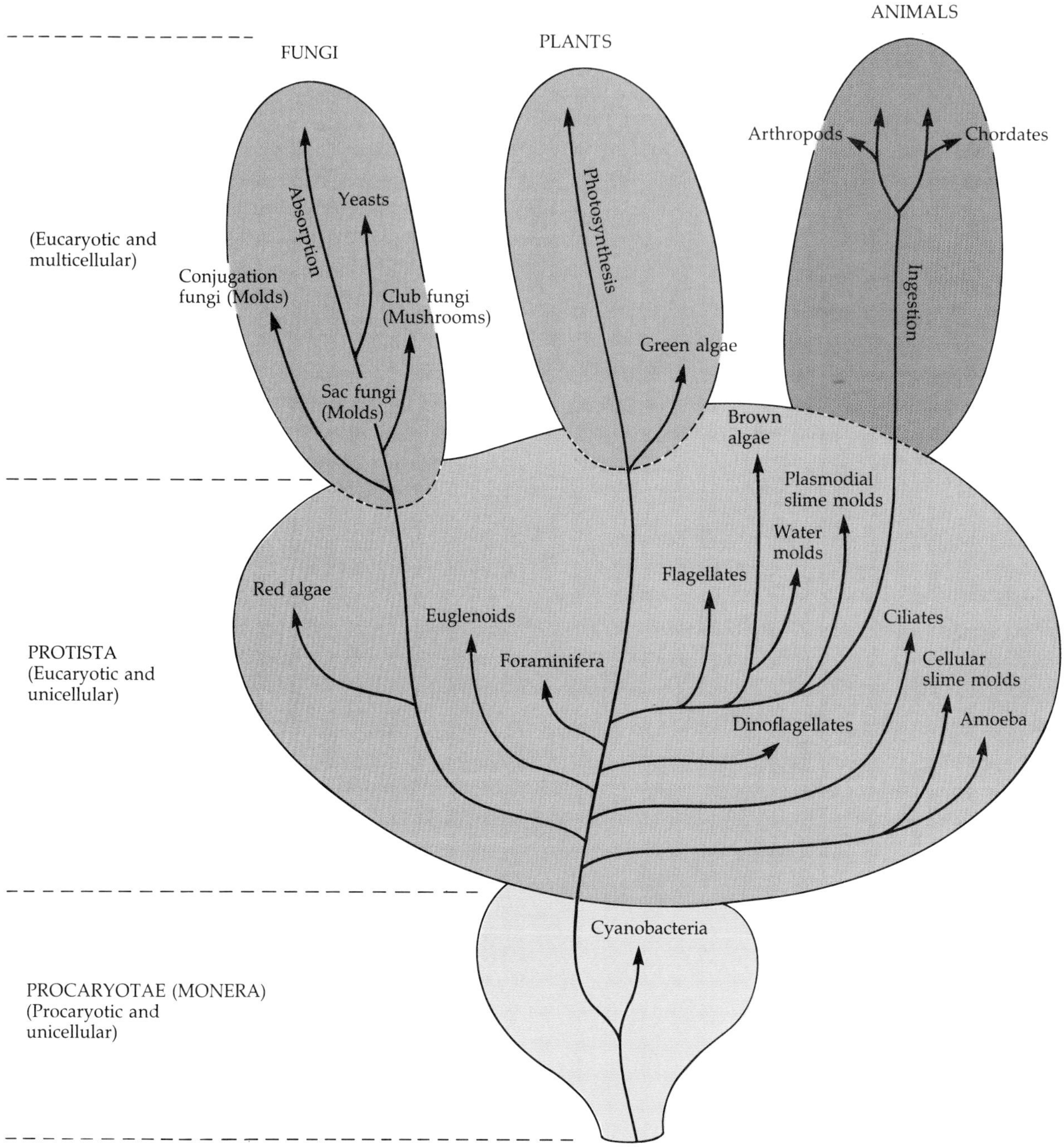

Figure 9–1 A commonly accepted five-kingdom system of classification.

is called *Klebsiella pneumoniae* (kleb-sē-el'lä nü-mō'nē-í); and a mold that contaminates bread is called *Rhizopus nigricans* (rī-zō'pu̇s nī' gri-kans). The genus and specific epithet are spelled out the first time they are used (*Klebsiella pneumoniae*), but thereafter, the genus name may be abbreviated (*K. pneumoniae*).

CLASSIFICATION OF EUCARYOTIC ORGANISMS

For eucaryotic organisms, a **species** is a group of closely related organisms that have a limited geographical distribution and that breed among themselves (interbreed), but do not breed with other species. Consequently, members of one species have morphological characteristics that distinguish them from members of another species. A **genus** (plural, **genera**) consists of species that differ from each other in certain ways but are related by descent. For example, *Quercus* (kwer'kus), the genus name for oak, consists of all types of oak trees (white oak, red oak, bur oak, velvet oak, and so on). Even though each species of oak differs from every other species, they are all related genetically. Just as a number of species make up a genus, related genera make up a **family.** A group of similar families constitutes an **order,** and a group of similar orders makes up a **class.** Related classes, in turn, comprise a **division.** (In zoology, the comparable term **phylum** is used.) Thus, a particular organism has a genus name and specific epithet, and belongs to a family, order, class, and phylum or division. All phyla or divisions that are related to each other make up a **kingdom** (Figure 9–2).

Charles Darwin's (1809–1882) theory of evolution explains that the many similarities and differences among organisms are derived from their descent from common ancestors. The arrangement of organisms into taxonomic categories, collectively called **taxa,** reflects degrees of relatedness among organisms. That is, the hierarchy of taxa shows evolutionary or **phylogenetic** (from a common ancestor) relationships.

CLASSIFICATION OF BACTERIA

The classification of eucaryotic organisms and the classification of bacteria differ in several ways. For example, application of the term *species* to higher organisms is based on geographical distribution and interbreeding that result in distinctive morphological characteristics. Among bacteria, however, morphological traits are limited, and sexuality involves unilateral transfer of genetic material rather than the fusion of two cells such as an egg and a sperm. A **bacterial species** is accordingly defined as a population of cells with similar characteristics. (The types of characteristics will be discussed later in this chapter.) The members of a bacterial species are essentially undistinguishable from each other but are distinguishable from members of other species, usually on the basis of several features. In some cases, not all pure cultures of the same species are identical in all ways. Each such group in a species is called a **strain,** which is a group of cells all derived from a single cell. Strains are identified by numbers, letters, or names that follow the specific epithet. It is therefore possible to define a bacterial species as a collection of closely related strains.

As eucaryotic organisms are grouped, so do several species of bacteria constitute a genus, several genera comprise a family, families are grouped into orders, orders into classes, and classes into divisions. The most recent *Bergey's Manual* (see below) proposes to divide bacteria into seven classes, which fall into four divisions. Proposed higher taxa are listed in Table 9–1.

Some of the information employed to determine evolutionary relationships in higher organisms comes from fossils. Bones, shells, or stems that contain mineral matter or have left imprints in rock that was once mud are examples of fossils. The structures of most microorganisms are not readily fossilized. One exception is a marine protist whose fossilized colonies form the white cliffs of Dover, England. Another exception is the fossilized cyanobacteria found in 3- to 3.5-billion-year-old rocks in Western Australia. These are the oldest known fossils. Some younger fossils of procaryotes and eucaryotes are shown in Figure 9–3.

The taxonomic classification scheme for bacteria may be found in a reference book called *Bergey's Manual of Systematic Bacteriology.* This book classifies bacteria according to phylogenetic information recently provided by advances in molecular biology. A scheme of these suggested evolutionary relationships in shown in Figure 9–4. Attempting to separate microorganisms by evolutionary group-

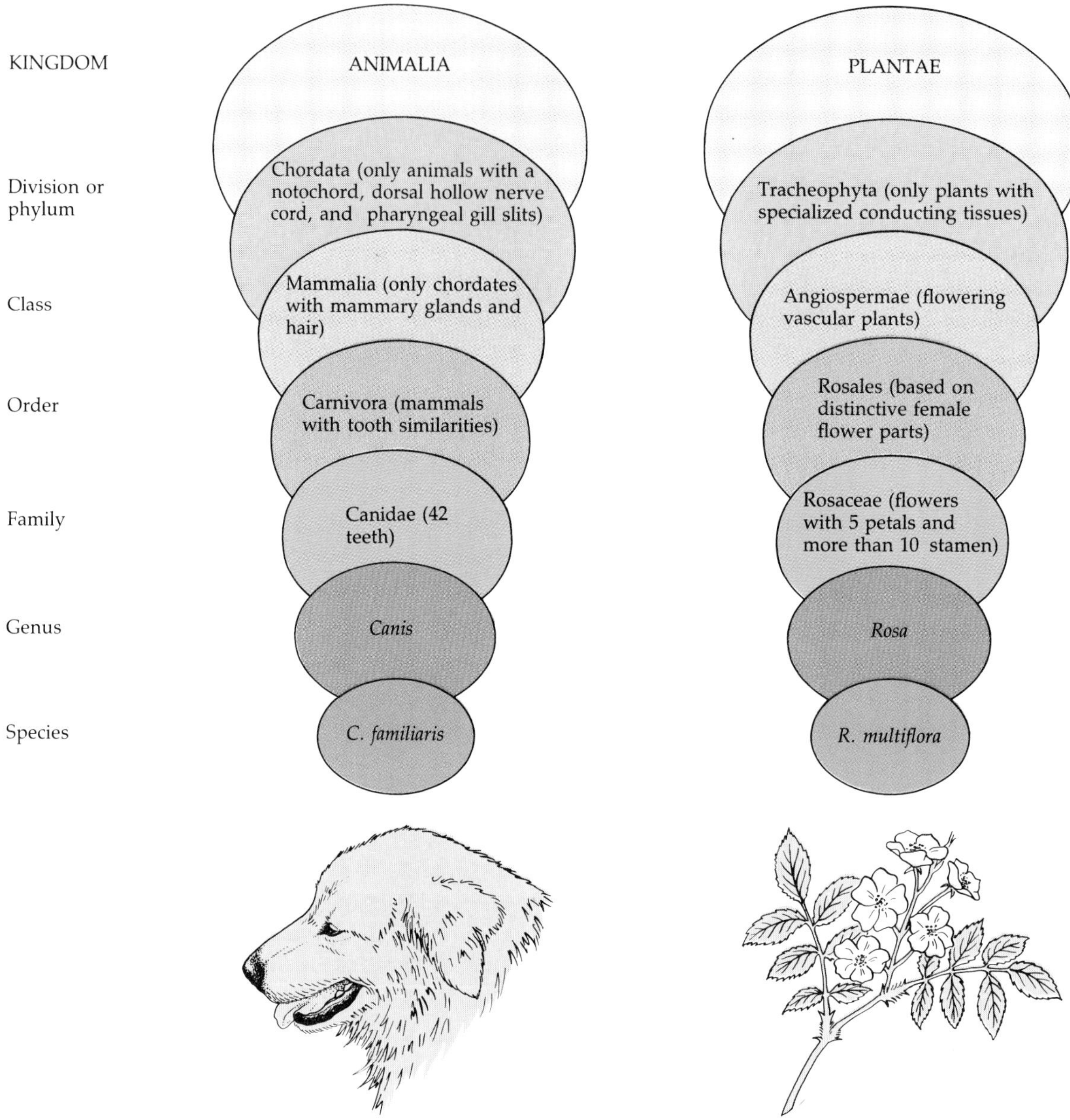

Figure 9–2 Classification of living organisms.

ings—as one would separate whales from fish or opossums from rats—is a new science. Information used to build (and modify) the phylogenetic model comes from analyses of the compositions of nucleic acid bases, from DNA hybridizations, and from chemical analyses of cellular components.

Not all bacteria have yet been surveyed, and those bacteria needing further investigation are temporarily placed in descriptive categories or sections such as "Endosymbionts" and "Dissimilatory Sulfate- or Sulfur-Reducing Bacteria."

Rules for the assignments of names to newly classified bacteria and for the assignments of bacteria to taxa are established by the International Committee on Systematic Bacteriology and are published in the *Bacteriological Code*. Descriptions

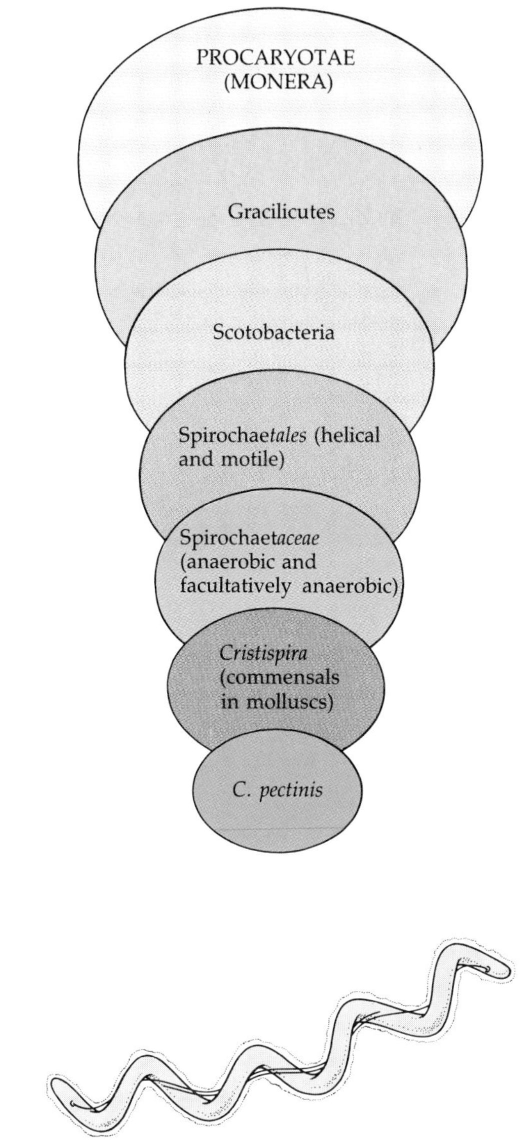

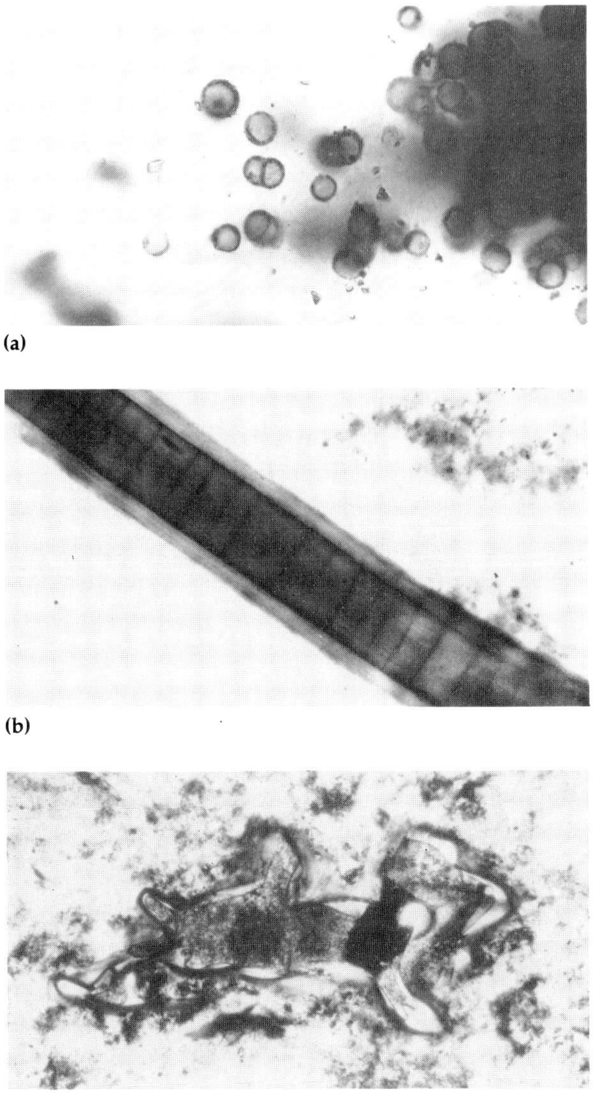

(a)

(b)

(c)

Figure 9–3 Fossilized cells. All are from the Late Proterozoic (ca. 700 Ma) period in China. Shown here are **(a)** solitary and paired coccoids, **(b)** a filamentous cyanobacteria, and **(c)** a complex eucaryotic alga or protozoan.

Table 9–1 Divisions and Classes in the Kingdom Procaryotae (Monera) Identified by Common Names

Division	Class
Typical gram-negative cell wall	Nonphotosynthetic bacteria Anaerobic photosynthetic bacteria Cyanobacteria
Typical gram-positive cell wall	Rods and cocci Actinomycetes and related organisms
Wall-less procaryotes	Mycoplasmas
Unusual walls	Archaeobacteria

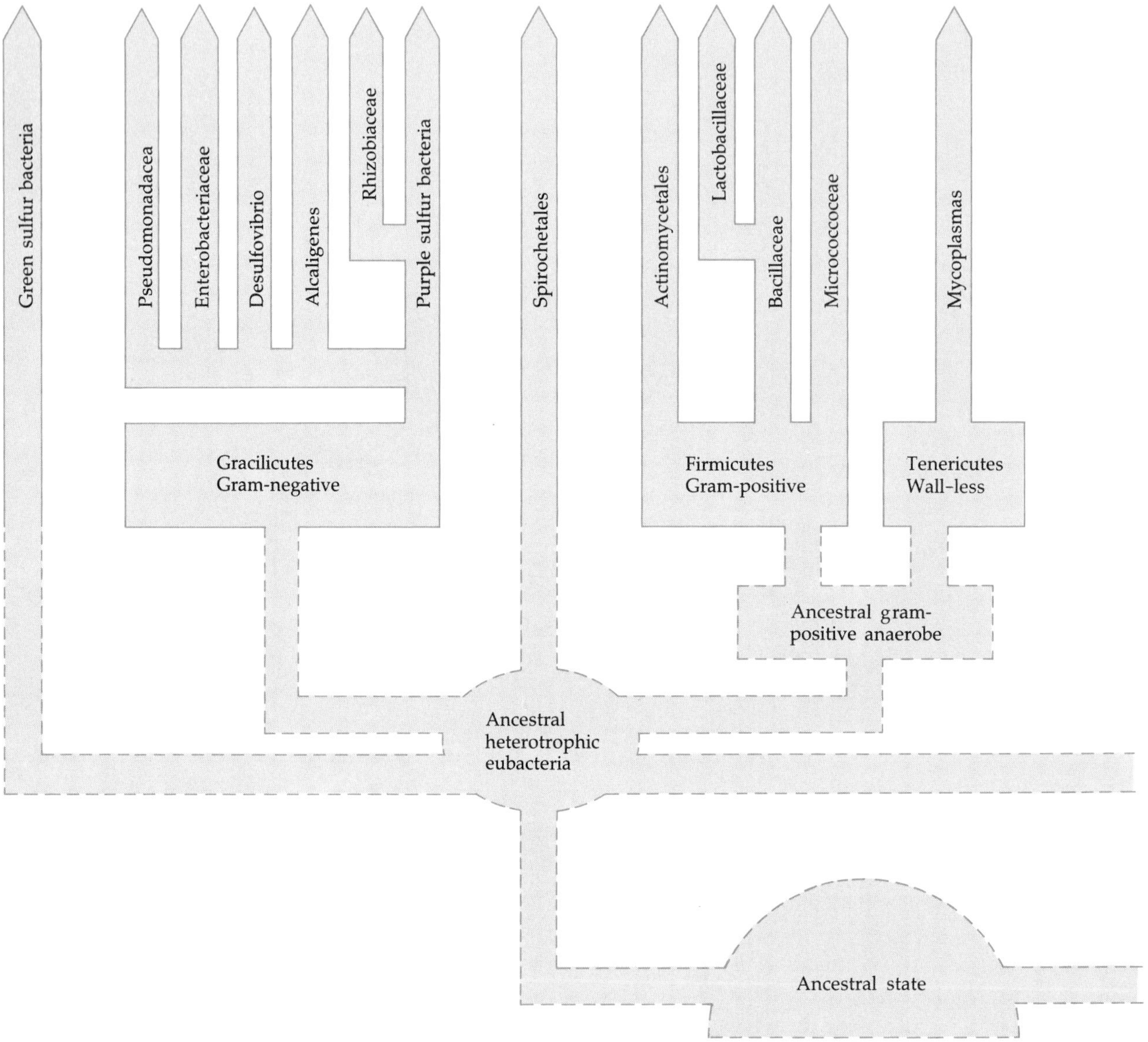

Figure 9–4 Phylogenetic relationships of bacteria.

of bacteria and evidence for their classifications are published in the *International Journal of Systematic Bacteriology* before being incorporated into *Bergey's Manual*. According to the *Bacteriological Code*, scientific names are to be taken from Latin (a genus name can be taken from Greek) or latinized by the addition of the appropriate suffix. Suffixes for order and family are specified as *-ales* and *-aceae*, respectively. (Refer to Figure 9–2.)

CLASSIFICATION OF VIRUSES

Viruses are not classified under any of the five kingdoms because they are not composed of cells and they use the anabolic machinery of host cells to multiply inside living cells. Because a viral genome can direct biosynthesis inside a host cell and some viral genomes can become incorporated into the host genome, it is possible that viruses might be

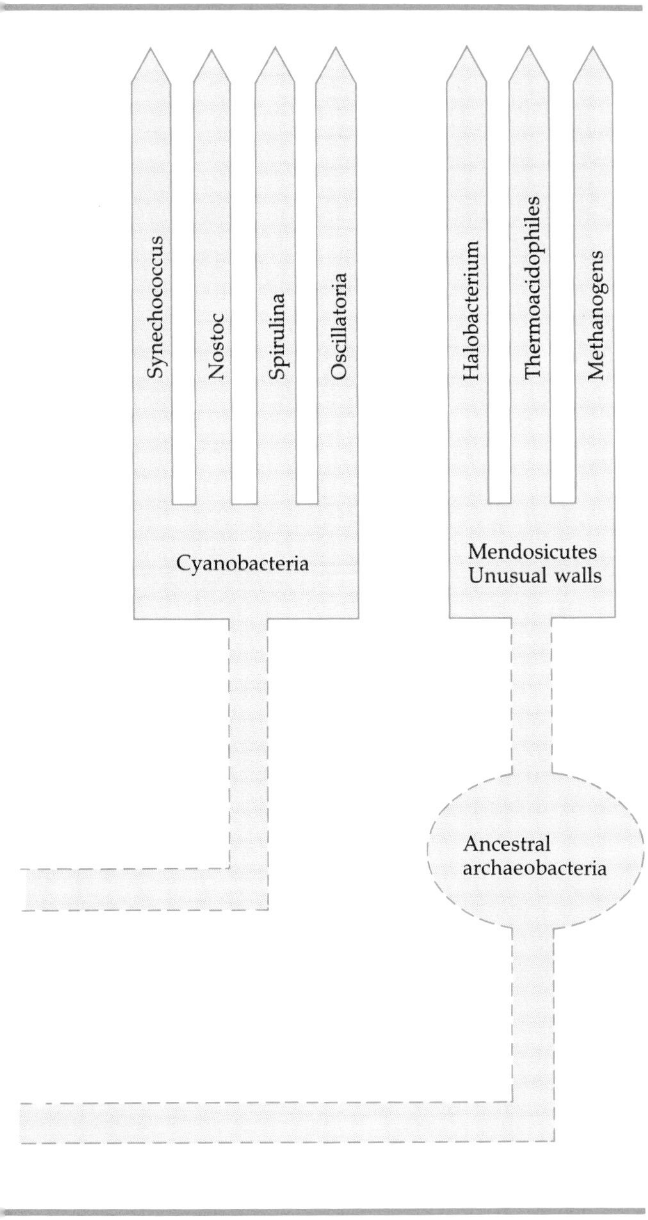

more closely related to their hosts than to other viruses.

Viruses are obligatory intracellular parasites so they must have evolved after a host cell had evolved. The origin of viruses has two hypotheses: (1) they arose from independently replicating strands of nucleic acids (such as plasmids); and (2) they developed from degenerative cells that through many generations gradually lost the ability to survive independently but could survive when associated with another cell. Viruses will be discussed in Chapter 12.

CRITERIA FOR CLASSIFICATION AND IDENTIFICATION OF MICROORGANISMS

Classification provides us with a reference for identification. Microorganisms are identified for practical purposes and not necessarily by the same techniques with which they are classified. Once an organism is identified, it can be placed into a previously devised classification scheme. A classification scheme provides a list of characteristics and means for comparison. Most identification procedures are easily performed in a laboratory and require as few procedures or tests as possible. *Bergey's Manual of Systematic Bacteriology, Volume 1,* includes identifying characteristics for some bacteria. Until the remaining three volumes are completed over the next few years, *Bergey's Manual of Determinative Bacteriology,* 8th edition, can be used for laboratory identification. The latter does not classify bacteria according to evolutionary relatedness, but provides identification schemes based on such criteria as cell wall composition, morphology, differential staining, oxygen requirements, and biochemical testing.

Bergey's Manual of Determinative Bacteriology has been a widely used reference for identification of bacteria since publication of the first edition in 1923. In recent years, microbiologists have been acquiring information on relationships among bacteria and, consequently, are constructing identification schemes that reflect the evolutionary relationships used in *Bergey's Manual of Systematic Bacteriology.*

Medical microbiology (the branch of microbiology dealing with human pathogens) has dominated man's interest in microbes, and this dominance is reflected in many identification schemes. To put the pathogenic properties of bacteria in perspective, it should be noted that although more than 1800 species are listed in the *Approved Lists of Bacterial Names,* less than 200 of these are human pathogens.

We will now discuss several criteria and methods for the classification of microorganisms and the routine identification of some. In addition to these

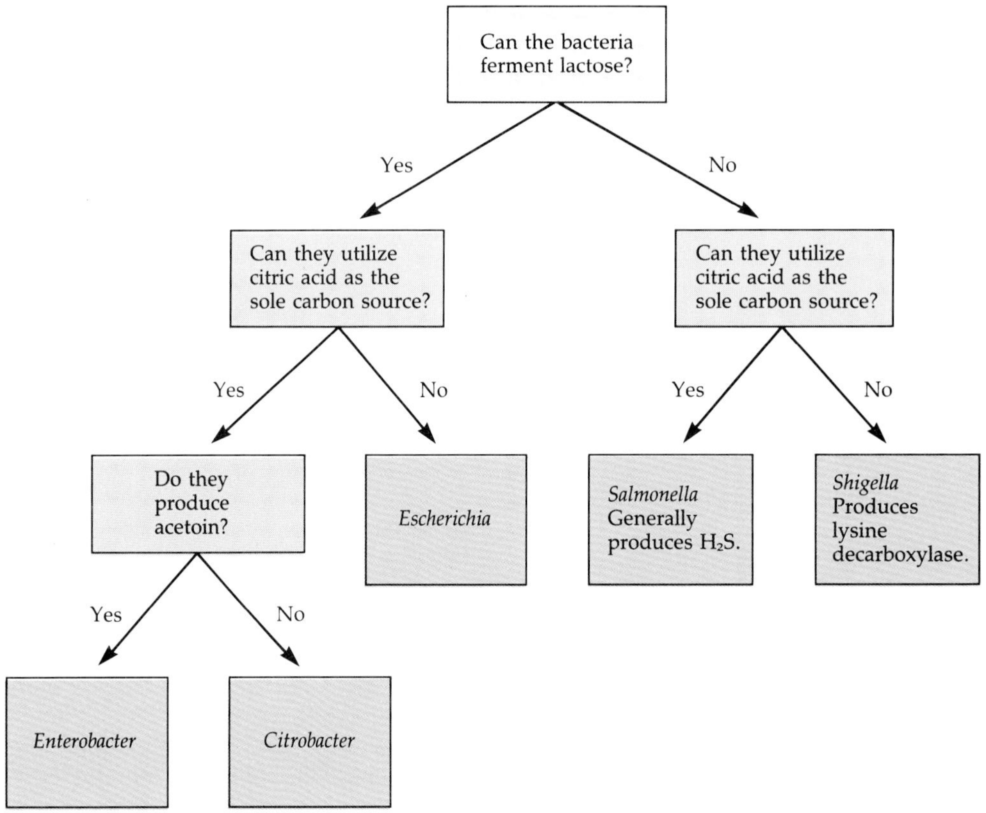

Figure 9–5 Using metabolic characteristics to identify selected genera of enteric bacteria.

methods, the source and habitat of a bacterial isolate will be considered as part of the classification and identification processes.

Morphological Characteristics

Higher organisms are frequently classified according to observed anatomical detail. However, many microorganisms appear too similar to be classified by their structures. Through a microscope, there is a great deal of similarity between organisms that might differ in metabolic or physiological properties. There are literally hundreds of bacterial species that are small rods or small cocci.

Cell morphology tells us little about phylogenetic relationships. However, morphological characteristics are still useful in identifying bacteria. For example, differences in structures such as endospores or flagella can be helpful.

Differential Staining

One of the first steps in the identification of bacteria is differential staining. As we have seen from earlier chapters, just about all bacteria are either gram-positive or gram-negative. Other differential stains, such as the acid-fast stain, can be useful for a more limited group of microorganisms. Recall that these stains are based on the chemical composition of the cell's walls and are therefore not useful in identifying either the wall-less bacteria or the archaeobacteria.

Biochemical Tests

Enzymatic activities are widely used to differentiate among bacteria. Even closely related bacteria can usually be separated into distinct species by testing their ability to ferment an assortment of selected carbohydrates or by subjecting them to other biochemical tests.

Enteric, gram-negative bacteria are a large heterogeneous group of microbes whose natural habitat is the intestinal tract of humans and animals. Among the enteric bacteria are members of the genera *Escherichia* (esh-ėr-i'kē-ä), *Enterobacter* (en-te-rō-bak'ter), *Shigella* (shi-gel'lä), and *Salmonella* (sal-mōn-el'lä). *Escherichia* and *Enterobacter* can be distinguished from *Salmonella* and *Shigella* according to their ability to ferment lactose: the former ferment lactose to produce acid and gas whereas the latter do not. Further biochemical testing as shown in Figure 9–5 can differentiate between the genera.

Microbial identification and classification often include the use of selective or differential media. Recall from Chapter 6 that selective media contain ingredients that suppress the growth of competing organisms and encourage the growth of desired ones, and differential media allow the desired organism to form a colony that is somehow distinctive.

Serology

Serology is the science that deals with the blood serum, and in particular, with immune responses evident in serum (see Chapter 16). Microorganisms are antigenic; that is, injected microorganisms are capable of stimulating antibody formation in animals. For example, the immune system of a rabbit injected with killed typhoid bacteria will respond by producing antibodies against typhoid bacteria. The antibodies are proteins that circulate in the blood and combine in a highly specific way with the bacteria that caused their production. **Antisera** (singular, **antiserum**) are commercially available solutions of such antibodies used in the identification of many medically important bacteria. If an unknown bacterium is isolated from a patient, it can be tested against known antisera and often is identified.

In a procedure called a **slide agglutination test,** the unknown bacterium is placed in a drop of salt solution on a slide and mixed with a drop of known antiserum. The same bacterium is mixed with different antisera on different slides. The bacteria will agglutinate or clump when mixed with antibodies produced to combine with that strain or species. A positive test is therefore determined by agglutination. A slide agglutination test is illustrated in Figure 9–6.

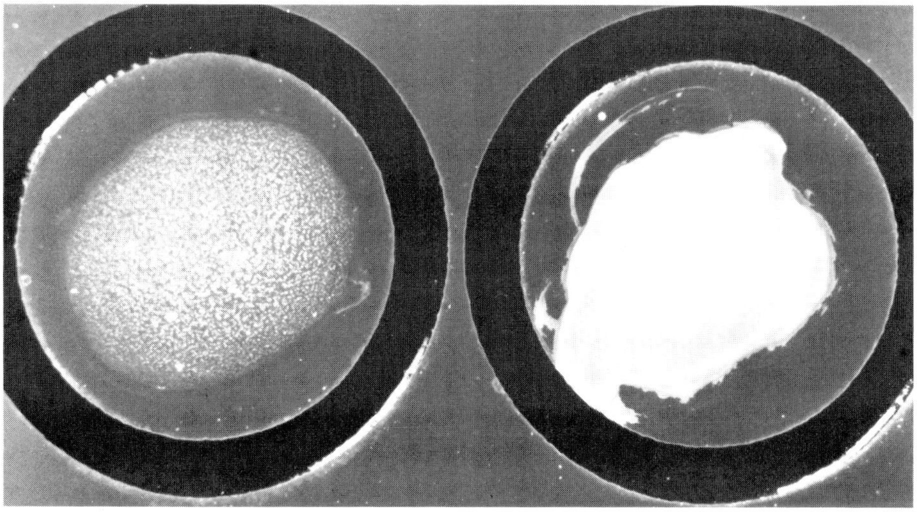

(a) (b)

Figure 9–6 Slide agglutination test. **(a)** Positive test: the grainy appearance is due to the clumping (agglutination) of the bacteria. Agglutination results when the bacteria are mixed with antibodies produced to react with the same strain. **(b)** Negative test in which the bacteria are still evenly distributed in the liquid of the salt solution and antiserum.

These serological techniques can differentiate not only microbial species, but even strains within species can be differentiated. Because closely related bacteria produce some of the same antigens, serological testing is useful in screening bacterial isolates for possible similarities. If an antiserum reacts with proteins from different bacterial strains or species, these bacteria can be tested further for relatedness.

Phage Typing

A concept similar to that of serological testing is the basis for phage typing. Phages are bacterial viruses, and they are highly specialized in that they usually attack only members of a particular species, or even strains within a species. Both serological testing and phage typing are particularly useful in tracing the origin and course of a disease outbreak.

To see how phage typing can help researchers trace the origin and course of a specific strain of a bacterium, let us consider the following example. Years ago there was a very high incidence of hospital-associated (nosocomial) staphylococcal infections. Moreover, the chain of transmission was unknown in many cases. Although such outbreaks still occur, they are less frequent because their sources can be traced by phage typing. In this procedure, a plate of staphylococci growing on an agar medium is marked off in small squares and a drop of each different phage type is placed on the bacteria (Figure 9–7). In squares in which the phage causes lysis, plaques appear. Such a test might show that staphylococci isolated from a surgical wound have the same pattern as those isolated from the operating surgeon or surgical nurses.

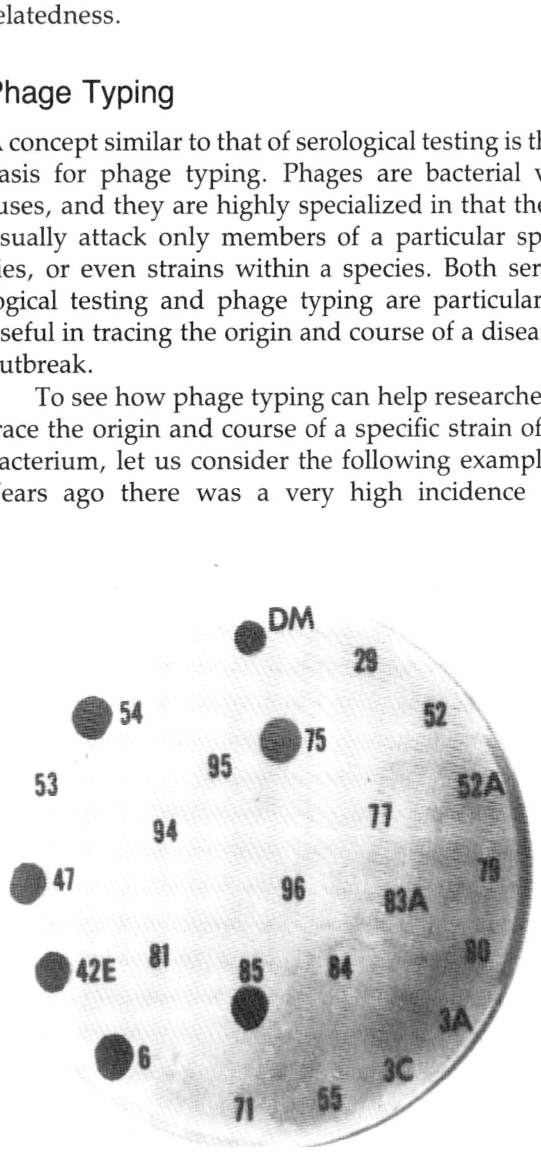

Figure 9–7 Phage typing of a strain of *Staphylococcus aureus*. The tested strain was grown over the entire plate and was lysed by bacteriophages 6, 42E, 47, 54, 75, and 85. Accordingly, the strain is identified as phage type 6/42E/47/54/75/85. (DM is a dye marker.)

Amino Acid Sequencing

A great deal of information can be derived from the study of the amino acid sequences of proteins. The sequence of amino acids in a protein directly reflects the base sequence of the encoding gene. Thus, the comparison of amino acid sequences from proteins of two different organisms can determine their relatedness. The more similar the proteins, the more closely related are the organisms. The phylogenetic tree derived for animals from traditional methods such as fossils and comparative anatomy is reinforced now by such a comparison of amino acid sequences.

The sequencing of amino acids in several important proteins, such as cytochrome *c*, has been useful in the study of phylogenetic relationships among many different organisms (Figure 9–8).

Protein Analysis

Protein profiles are obtained by **polyacrylamide gel electrophoresis (PAGE)**. In this process, proteins from disintegrated cells are dissolved in a detergent, and a drop of the resulting solution is placed on a thin layer of polyacrylamide gel. In a process called **electrophoresis,** an electric current is then passed through the gel. While the charge is applied, different proteins migrate through the gel at different rates, which depend on the size and charge of each protein. After it is stained, the gel shows a "fingerprint" of the cells' proteins (Figure 9–9). Related organisms have identical protein profiles, the profiles of closely related organisms show only minor differences, and profiles of unrelated organisms show major differences. As pro-

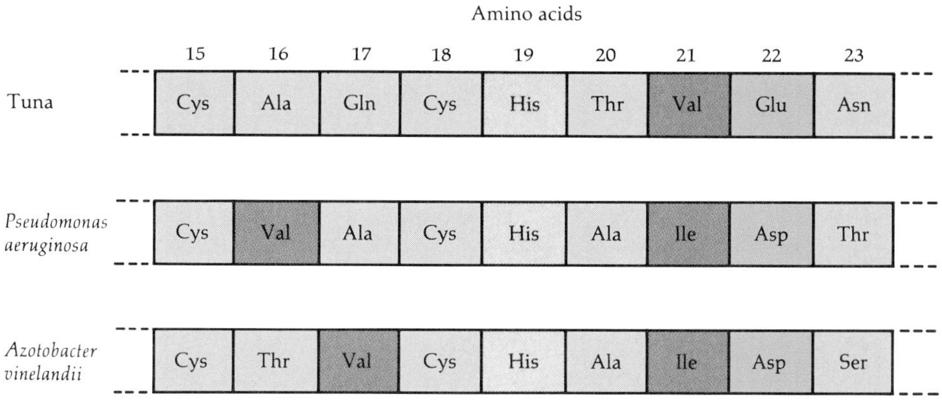

Figure 9–8 Amino acid sequences for a fragment of cytochrome *c* from tuna and from two species of bacteria. Three-letter abbreviations indicate the amino acids, and amino acids of the same chemical type are shown in the same color (e.g., Glu and Asp are both "acidic amino acids," having an extra—COOH group). The similarity among the three sequences is striking, although the two bacteria are clearly more closely related to each other than to tuna.

cedures and materials are developed further, this method may provide a useful tool for identification as well as classification. By comparing the protein profile of the unknown cell to a set of standard profiles, a laboratory technician might be able to identify an organism within 24 hours of its isolation.

One PAGE analysis that is being pursued concerns the identification of cytochromes present in an organism. Although the presence of certain cytochromes is affected by the environment, analysis of bacterial cytochromes has disclosed some patterns that may be useful for classifying organisms. The following table shows the cytochromes that are common to three metabolic categories of organism. The letters correspond to the five cytochromes present in these cells.

Organisms	Cytochromes
Gram-positive chemoheterotrophs	**b, c, aa_3, o**
Gram-negative chemoheterotrophs	**b, c, d, o, a**
Gram-negative chemoautotrophs	**b, c, aa_3, o, a**

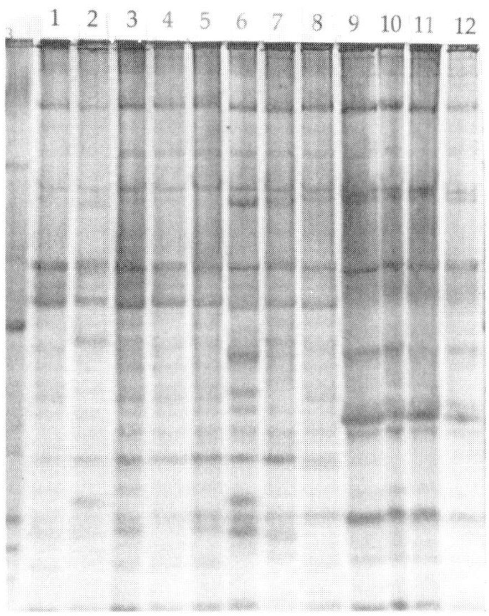

Figure 9–9 PAGE patterns of 12 strains of *Bifidobacterium* isolated from humans and other mammals. All 12 strains produce identical results from biochemical tests. The PAGE patterns, however, reveal that there are two species. Lanes 1–8 are a new species, *B. globosum;* lanes 9–12 are *B. pseudolongum.*

Base Composition of Nucleic Acids

A classification technique that has come into wide use among taxonomists because it has a good possibility of at least suggesting evolutionary relationships is the determination of the nitrogenous-base composition of the DNA. This base composition is usually expressed as percentage of guanine plus cytosine. The base composition of a single species is theoretically a fixed property and thus reveals the degree of species relatedness. As we saw in Chapter 8, each guanine (G) in DNA has a complementary cytosine (C). Similarly, each adenine (A) in the DNA has a complementary thymine (T). Therefore, the percentage of DNA bases that are G–C pairs also tells us the percentage that are A–T pairs (subtracted from 100%). Two organisms that are closely related and hence have many identical or similar genes will naturally have similar amounts of the various bases in their DNA. On the other hand, if there is a difference of more than 10% in their percentage of G–C pairs (for example, if one bacterium's DNA contains 40% G–C and another bacterium has 60% G–C) then these two organisms are probably not related. It should be remembered, however, that two organisms that have the same percentage of G–C are not necessarily closely related; other supporting data are needed to draw conclusions about evolutionary relationships.

Actually determining the entire *sequence* of bases in an organism's DNA is theoretically possible with modern biochemical methods but is impractical for all but the smallest microorganisms (viruses). However, the sequencing of bases in ribosomal RNA has provided useful information on evolutionary relationships among bacteria. Closely related genera have a similar base sequence that differs from that of nonrelated genera. An indirect determination of the degree of similarity between

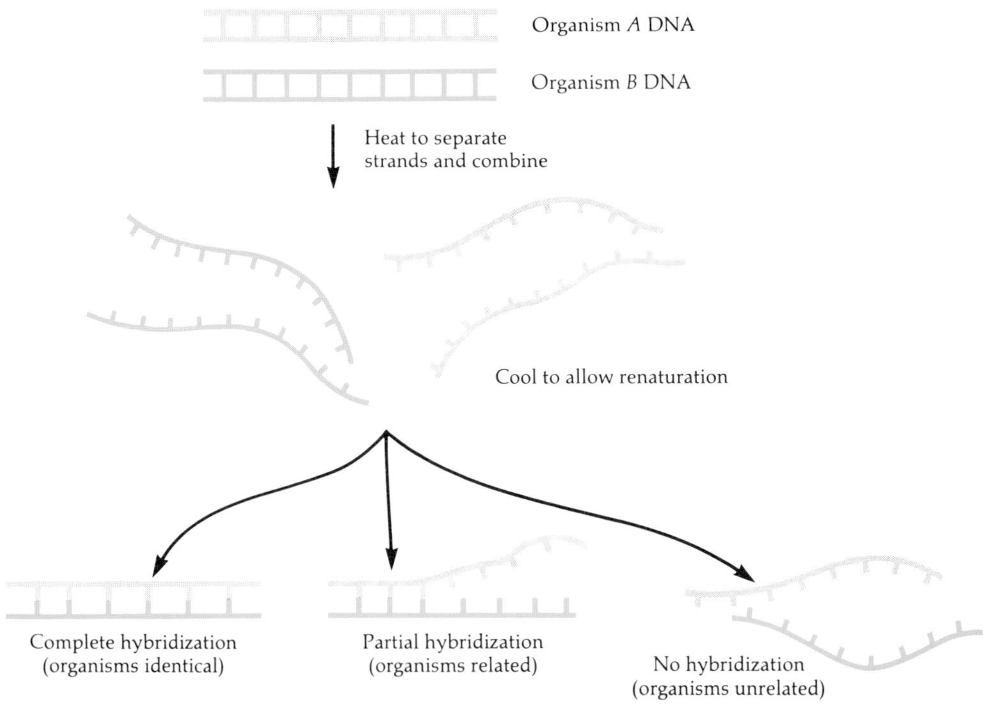

Figure 9–10 DNA hybridization. The greater the amount of pairing between DNA strands from different organisms (hybridization), the more closely related are the organisms.

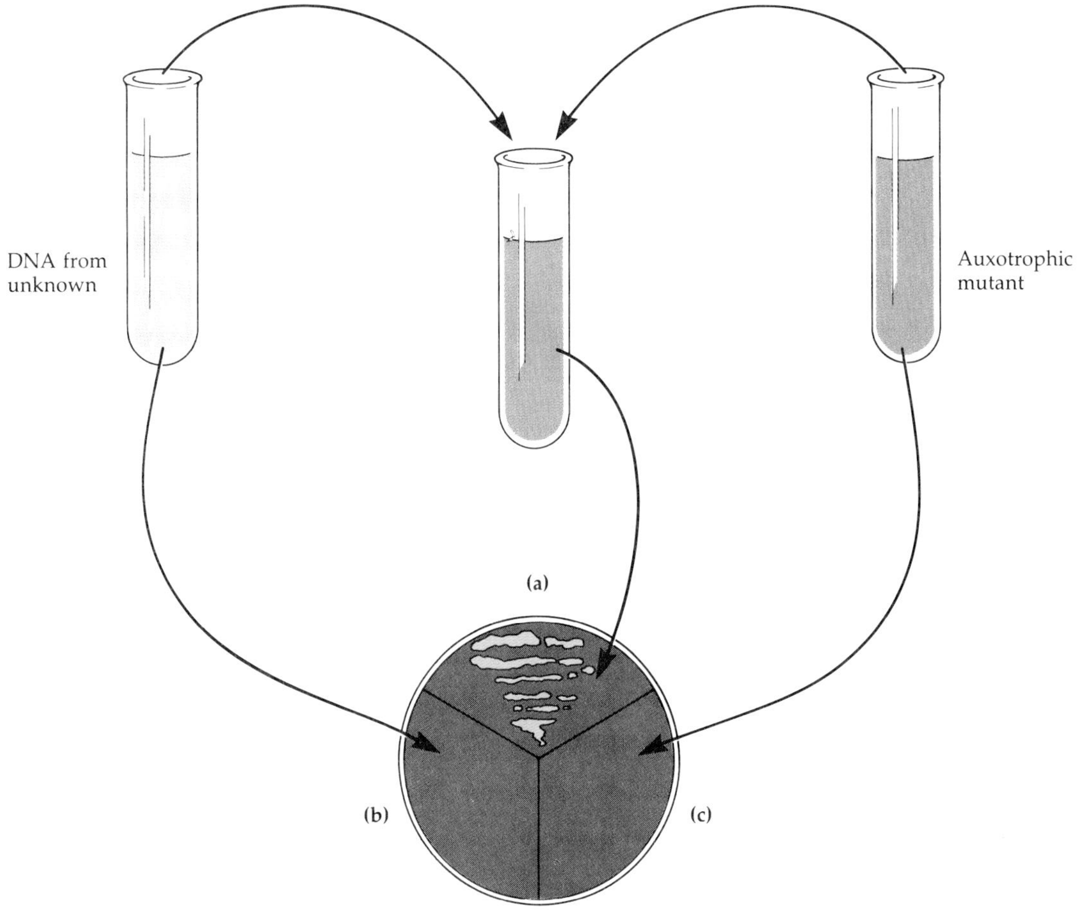

Figure 9–11 Transformation is used to identify bacteria. If the unknown species and the auxotroph are the same species, transformation can occur and **(a)** the resultant recombinant cell will grow. Lack of growth of **(b)**, the DNA alone, and **(c)**, the auxotroph, serve as controls.

the DNA nucleotide sequences of two organisms can be made with the method of nucleic acid hybridization.

Nucleic Acid Hybridization

If a double-stranded molecule of DNA is subjected to heat, the complementary strands will separate as the hydrogen bonds between nitrogenous bases break. If the complementary single strands are then cooled slowly, they will reunite to form a double-stranded molecule identical to the original double strand. (This reunion occurs because the single strands are complementary.) When this technique is applied to separated strands from two different organisms, it is possible to determine the extent of similarity between the base sequences of the two organisms. This method is known as **nucleic acid hybridization.** The procedure assumes that if two species are similar or related, a major portion of their nucleic acid sequences will also be similar. The procedure measures the ability of DNA strands from one organism to hybridize (bind through complementary base pairing) with the DNA strands of another organism (Figure 9–10). The greater the degree of hybridization, the greater the degree of relatedness.

As we learned in Chapter 8, RNA is single-stranded and transcribed from one strand of DNA; a particular strand of RNA therefore is comple-

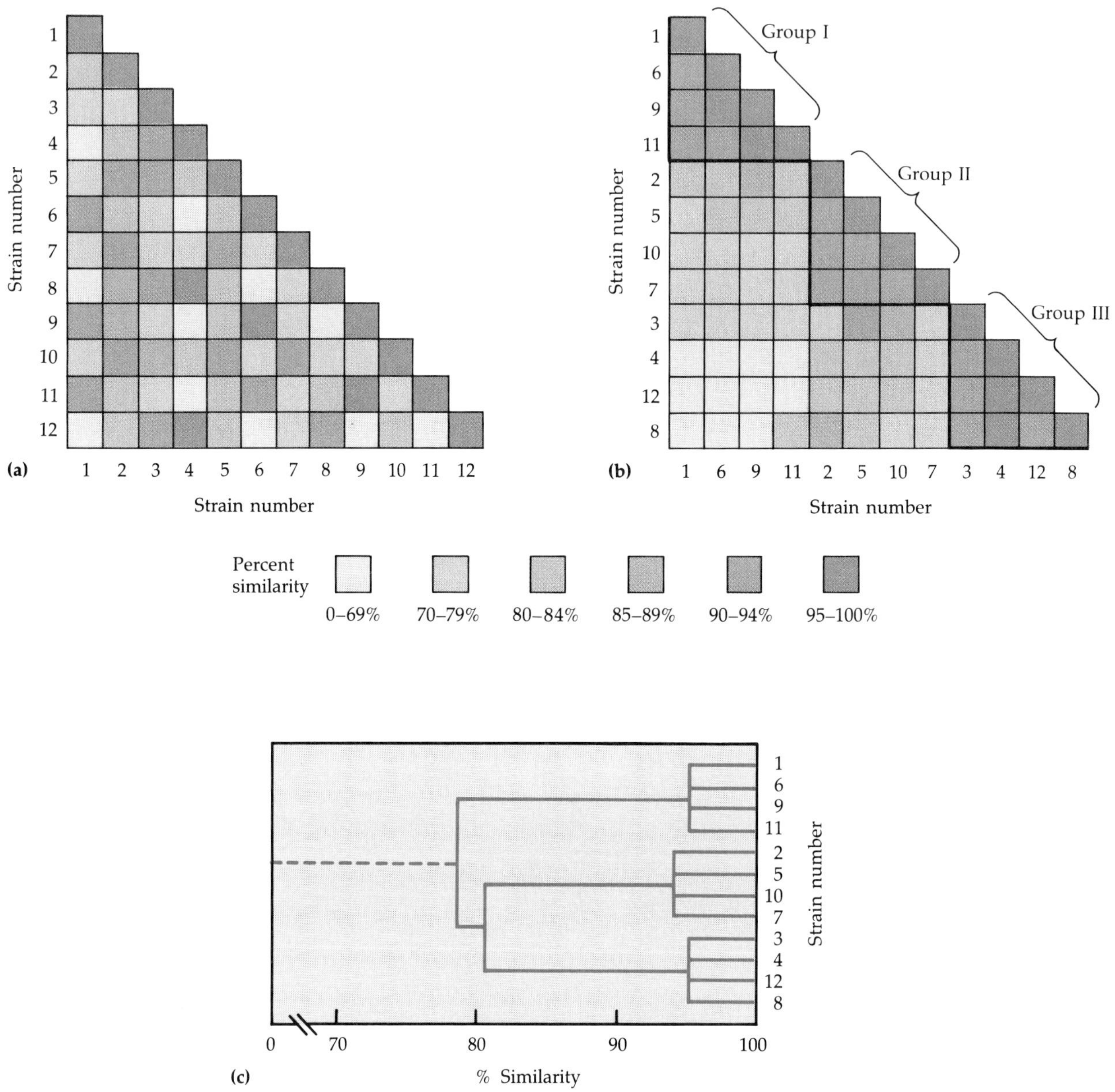

Figure 9–12 Numerical taxonomy. With the aid of a computer, a similarity matrix (array) can be constructed to show relationships between different organisms. As many as 100 different tests may be run on each organism. Each organism is then given a numerical score indicating the degree of its similarity with every other organism, and the score is shown on a matrix **(a).** In another matrix **(b),** organisms with similar test results are grouped together. Groups I, II, and III represent three species each containing four strains. **(c)** A dendrogram constructed from the similarity values.

mentary to the strand of DNA from which it was transcribed, and will hybridize with that separated strand of DNA. DNA–RNA hybridization can thus be used to determine relatedness between DNA from one organism and RNA from another organism.

It is important to realize, however, that this method, as well as nucleic acid composition and amino acid sequencing tests, is too complex to use for routine identification in the clinical laboratory.

Genetic Recombination

Genetic recombination might provide information useful to the classification and identification of bacteria. Recall from Chapter 8 that genetic recombination (that is, the rearrangement of genes to new combinations) can occur by transformation, transduction, or conjugation. The use of transformation in identification of a microorganism is based on the fact that recombination between bacterial chromosomes is likely to occur only between closely related organisms.

A test to identify *Neisseria gonorrhoeae* (nī-se′rē-ä go-nôr-rē′ä) employs transformation. DNA is extracted from bacteria suspected to be *N. gonorrhoeae.* The DNA is then incubated with an auxotrophic *N. gonorrhoeae* mutant. After incubation, samples are inoculated onto a minimal medium that will not support growth of the mutant (Figure 9–11). Only transformed cells, those that have exchanged genetic material with the DNA of the normal *N. gonorrhoeae,* will grow. A relatively high amount of growth suggests that the unknown bacterium is, indeed, *N. gonorrhoeae.* Use of genetic recombination is desirable for identification because few media are used and a pure culture of the unknown organism is not necessary.

Numerical Taxonomy

Numerical taxonomy (Figure 9–12) has been used to infer evolutionary relationships between organisms. Many microbial characteristics, such as the ability to use a certain carbohydrate, to form a pigment, or to produce a certain acid, are listed, and the presence or absence of each characteristic is scored for each organism. Amino acid sequences, the percentage of G–C pairs, and the percentage of hybridization observed between two nucleic acid strands can be listed.

A computer then matches the characteristics of each organism against those of other organisms. The greater the number of characteristics shared by two organisms, the closer the taxonomic relationship is assumed to be. A match of 90% similarity or higher usually indicates a single taxonomic unit, or species. In this method, the listed characteristics are all given equal weight. The computer can also generate a dendrogram (Figure 9–12c) that illustrates phylogenetic relationships derived from similarities from hybridization and other genetics experiments.

In the remaining chapters of this part, we will look at representative groups of microorganisms.

STUDY OUTLINE

FIVE-KINGDOM SYSTEM OF CLASSIFICATION (p. 255)

1. Taxonomy is the science of the classification of organisms with the goal of showing relationships between organisms.
2. Living organisms are classified into five kingdoms.
3. Procaryotic organisms include bacteria and cyanobacteria and are placed in the kingdom Monera or Procaryotae.
4. Eucaryotic organisms may be classified into the kingdoms Protista, Fungi, Plantae, or Animalia.
5. Fungi are absorptive heterotrophs that develop from spores.
6. Multicellular photoautotrophs that develop from an embryo are placed in the Plantae.
7. Multicellular, ingestive heterotrophs that develop from a blastula (embryo) are classified as Animalia.
8. Protista are mostly unicellular organisms.

NAMING ORGANISMS (pp. 255–257)

1. According to scientific nomenclature, each organism is assigned two names (binomial nomenclature), a genus and a specific epithet.

The Archaeobacteria—Neither Procaryotes nor Eucaryotes?

The division of all living organisms into procaryote and eucaryote has been challenged recently by a number of biologists who claim that there is a third basic category. The organisms belonging to this new category are bacteria, that is, procaryotes. Now it is becoming clear that these bacteria are no more closely related to other bacteria (called *eubacteria* or *true bacteria* by these biologists) than they are to eucaryotic organisms such as animals and plants.

Archaeobacteria

The bacteria belonging to this new category look like typical bacteria and lack a nucleus and other membrane-bounded organelles. For years, however, it has been known that they are unusual in a number of ways. For example, their cell walls never contain peptidoglycan, they live in extreme environments, and they carry out unusual metabolic processes. They include three groups: (1) the methanogens, strict anaerobes that produce methane (CH_4) from carbon dioxide and hydrogen; (2) extreme halophiles, which require high concentrations of salt for survival; and (3) thermoacidophiles, which normally grow in hot, acid environments. Because they are thought to have descended from very ancient organisms, these microbes have been dubbed *archaeobacteria* (*archaios*–meaning ancient).

Evidence

What is the evidence that the archaeobacteria constitute a group separate from the true bacteria and the eucaryotes? The evidence is biochemical, and most of it comes from the sequencing of ribosomal RNA (rRNA) molecules from many kinds of organisms. Ribosomes are essential to all forms of life on Earth, and the sequence of nucleotides in rRNA reflects the evolutionary history of the cell in which the rRNA is found. On the basis of similarities and differences among the rRNAs of many different species, the archaeobacteria form a group distinct from the other bacteria and the eucaryotes. Data on membrane lipids, transfer RNA molecules, and sensitivity to antibiotics also support this distinction.

Evolutionary Relationships

From the information gathered so far, researchers have concluded that the archaeobacteria split off the evolutionary tree to form a separate branch at the same time that the eubacterial and eucaryotic branches diverged (see diagram). The physiology of the methanogenic bacteria certainly seems to be suited to the kind of atmosphere thought to have existed billions of years ago on Earth—one containing carbon dioxide and some hydrogen but little or no oxygen.

CLASSIFICATION OF EUCARYOTIC ORGANISMS (p. 257)

1. Organisms are grouped into taxa according to degrees of relatedness.
2. A eucaryotic species is a group of organisms that interbreeds but does not breed with individuals of another species.
3. Similar species are grouped into a genus; similar genera are grouped into a family; families into an order; orders into a class; classes into a phylum; and phyla into a kingdom.

CLASSIFICATION OF BACTERIA (pp. 257–260)

1. A group of bacteria derived from a single cell is called a strain.
2. Closely related strains constitute a species.
3. Related species are arranged into a genus; genera into a family; and families into an order; orders into classes; and classes into divisions (phyla).

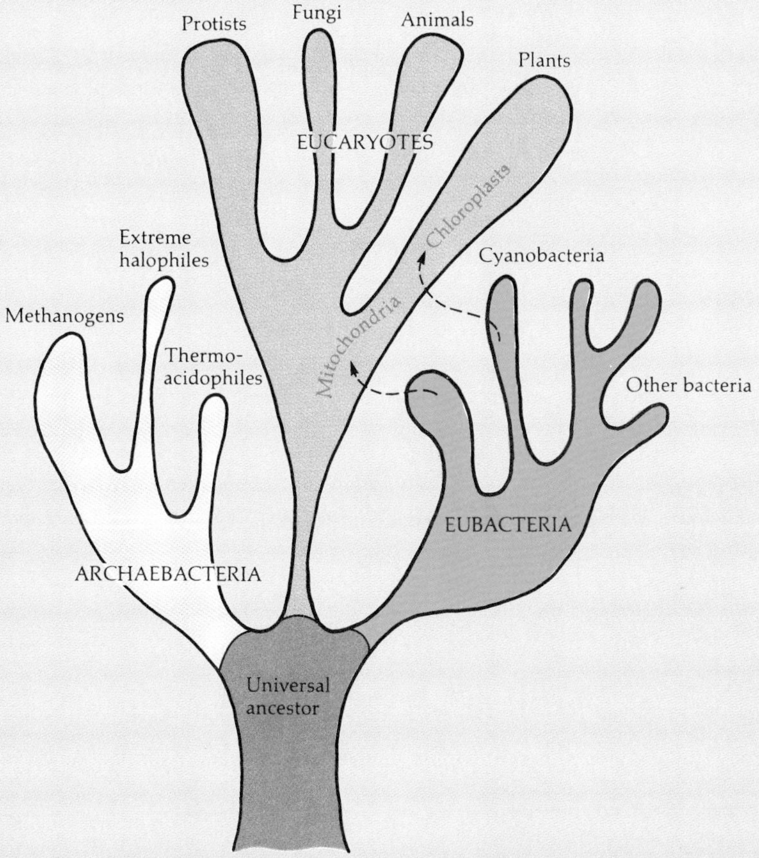

The research on archaeobacteria suggests some changes in the current theory to explain the origin of eucaryotic cells. It has been commonly believed that mitochondria and chloroplasts are descended from bacteria that lived symbiotically within another cell, and it has been assumed that the host cell was an ordinary bacterium. Now, it appears that the original host cell may have been a microorganism that evolved along a line separate from the true bacteria. This concept would help account for many of the basic molecular differences between the true bacteria, the eucaryotes, and the archaeobacteria.

4. *Bergey's Manual of Systematic Bacteriology* (1st edition) is the standard reference on Procaryotae classification.

5. Rules for the assignment of names to bacteria are published in the *Bacteriological Code*.

CLASSIFICATION OF VIRUSES (pp. 260–261)

1. Viruses are not placed in a kingdom. They are not composed of cells and are incapable of growth without a host cell.

CRITERIA FOR THE CLASSIFICATION AND IDENTIFICATION OF MICROORGANISMS (pp. 261–269)

1. *Bergey's Manual of Systematic Bacteriology, Volume 1*, and until the remaining volumes are published, *Bergey's Manual of Determinative Bacteriology*, 8th edition, are references for laboratory identification of bacteria.

2. Morphological characteristics, especially when aided by differential staining techniques, are useful in the identification of microorganisms.

3. Their possession of various enzymes as determined by biochemical tests is used in the identification of microorganisms.
4. Serological tests, involving the reactions of microorganisms with specific antibodies, are useful in determining the identity of strains and species as well as relationships between organisms.
5. Phage typing is the identification of bacterial species and strains by the determination of their susceptibility to various phages.
6. The sequences of amino acids in proteins of related organisms are similar.
7. Related organisms have identical proteins; this characteristic can be ascertained by PAGE "fingerprints."
8. The percentage of G–C pairs in the nucleic acid of cells can be used in the classification of organisms.
9. Single strands of DNA or DNA and RNA from related organisms will hydrogen-bond to form a double-stranded molecule; this bonding is called DNA hybridization.
10. Transformation of an auxotrophic mutant by the DNA of an unknown bacterium can be used in identification of the "unknown."
11. Grouping many different characteristics to show relatedness is called numerical taxonomy.

STUDY QUESTIONS

REVIEW

1. What is taxonomy?
2. List and define the five kingdoms used in the five-kingdom system of classification.
3. What characterizes the kingdom Procaryotae, and what organisms are placed in this kingdom?
4. What is binomial nomenclature?
5. Why is binomial nomenclature preferable to the use of common names?
6. Put the following terms in the correct sequence from the most general to the most specific: order, class, genus, kingdom, species, phylum, family.
7. Define species.
8. List the ten bases discussed in this chapter for the classification of microorganisms. Separate your list into those tests used primarily for taxonomic classification and those used primarily for identification of microorganisms already classified.
9. Higher organisms are arranged into taxonomic groups on the basis of evolutionary relationships. Why is this type of classification only just being developed for bacteria?
10. Using *Bergey's Manual,* identify the organism with the following characteristics: gram-negative rod; facultative anaerobe; motile; oxidase-negative; fermentative; acid and gas from lactose; indole produced; methyl red +; Voges–Proskauer −; citrate −; H_2S not produced; does not grow in KCN broth.
11. Can you tell which of these organisms are most closely related?

Organism	G–C moles %
A	48–52
B	49–53
C	50–52
D	46–48

CHALLENGE

1. Here's some additional information on the organisms in question 11:

Organisms	% DNA hybridization
A and B	70–100
A and C	40–50
A and D	20–30

Which of these organisms are most closely related? Compare this answer to your response to question 11.

2. In *Bergey's Manual,* the gram-positive cocci includes the genera *Micrococcus* and *Staphylococcus.* The reported G–C content of *Micrococcus* is 66 to 75 moles %, and of *Staphylococcus,* 30 to 40 moles %. According to this information, would you conclude that these two genera are closely related?

3. Compare and contrast *Bergey's Manual of Systematic Bacteriology* and *Bergey's Manual of Determinative Bacteriology.*

FURTHER READING

Barghoorn, E. S. May 1971. The oldest fossils. *Scientific American* 224:30–42. Evidence, including photographs of bacteria and algae that lived more than three billion years ago, of early life.

Dickerson, R. E. March 1980. Cytochrome *c* and the evolution of energy metabolism. *Scientific American* 242:137–153. A practical application of amino acid sequencing.

Fox, G. E. et al. 1980. The phylogeny of prokaryotes. *Science* 209:457–463. A discussion of the use of RNA base-sequencing to determine relatedness between microorganisms.

Gerhardt, P., ed. 1981. *Manual of Methods for General Bacteriology.* Washington, D.C.: American Society for Microbiology. Provides useful methods for standard bacteriological tests and identification schemes.

Goodfellow, M. and R. G. Board., eds. 1980. *Microbiological Classification and Identification.* New York: Academic Press. Comprehensive coverage of modern identification techniques including genetics, immunology, and protein electrophoresis.

Holt, J. G., ed. 1984. *Bergey's Manual of Systematic Bacteriology,* 1st ed., vol. 1. Baltimore, MD: Williams and Wilkins. pp 1–36. Discussion of bacterial classification provides a perspective on the relative importance of classification and identification.

Margulis, L. 1982. *Early Life.* Boston, MA: Science Books International. A good book that explains hypotheses and experiments pertaining to the evolution of procaryotes, eucaryotes, fermentation, and respiration.

Woese, C. R. June 1981. Archaebacteria. *Scientific American* 244:98–122. Detailed descriptions of the archaeobacteria and comparisons with eubacteria and eucaryotes.

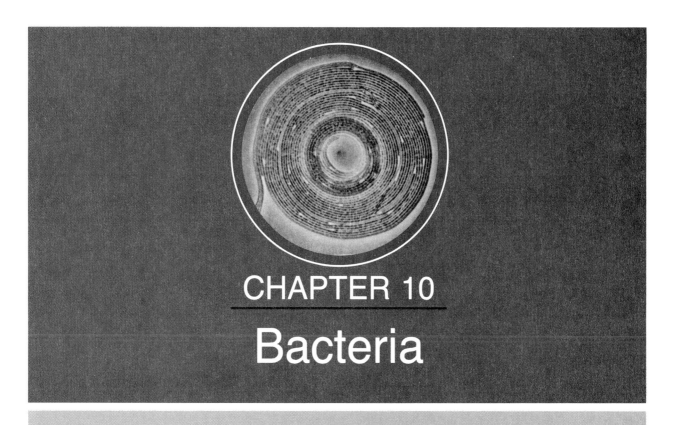

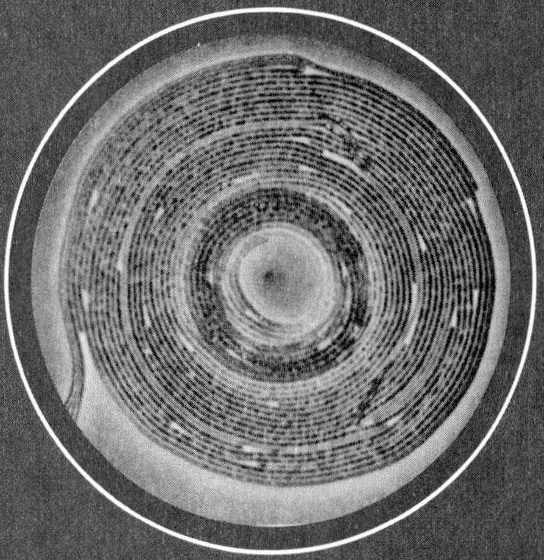

CHAPTER 10
Bacteria

OBJECTIVES

After completing this chapter you should be able to

- List at least six characteristics used to classify and identify bacteria according to *Bergey's Manual of Systematic Bacteriology.*
- Compare and contrast the four divisions of the kingdom Procaryotae (Monera).
- List one major characteristic of each section of *Bergey's Manual of Systematic Bacteriology* described in this chapter.
- Identify the sections that contain species of medical importance.

Bacteria are the most diverse and most important of all microorganisms. Relatively few species of bacteria cause disease in humans or other organisms, and without the activities of the many other kinds of bacteria, all life on Earth would cease. Indeed, eucaryotic organisms probably evolved from bacterialike organisms. Because bacteria are simple in structure and many are readily cultured and controlled in the laboratory, microbiologists have devoted intensive study to their life processes. From such studies we have learned that the basic processes of life are the same for all organisms.

This chapter will introduce you to common bacteria and emphasize those that cause disease in humans. When the bacterium also has an important nonmedical role, that role will also be noted. Many of the bacteria covered in this chapter will be discussed in more detail in the chapters on microbial disease, in Part Four.

BERGEY'S MANUAL: BACTERIAL TAXONOMY

There is no universally accepted evolutionary classification of bacteria. Bacteria exhibit great diversity aside from the shared property of procaryotic cellular organization, and it is difficult to correlate structural and functional characteristics to show evolutionary relationships. The most widely accepted taxonomic classification for bacteria is actually a classification of convenience rather than evolutionary relationships. Formerly called *Bergey's Manual of Determinative Bacteriology,* the most recent version of this reference is called *Bergey's Manual of Systematic Bacteriology,* hereafter called *Bergey's Manual.*

Bergey's Manual, first published in 1923, is the microbiologist's most important taxonomic reference. In the first seven editions, bacteria were classified in a hierarchical way according to kingdom, phylum, class, order, family, genus, and species—similar to schemes used for higher organisms. However, modern molecular evidence from studies of base composition, nucleic acid hybridization, and amino acid sequences began to challenge many aspects of the old hierarchy. The most recent revision of *Bergey's Manual* divides bacteria into four divisions (or phyla) according to the characteristics of cell walls (see Table 9–1). The bacteria are grouped into numbered *sections* according to characteristics such as Gram-stain reaction, cell shape, cell arrangements, oxygen requirements, motility, and nutritional and metabolic properties, among others. Accordingly, there are group names such as spirochetes, gram-negative aerobic rods and cocci, chemoautotrophic bacteria, and endospore-forming rods and cocci. Each section consists of a number of genera. In some sections, genera are grouped into families and orders; in other sections they are not (see Appendix A at the back of the book). *Bergey's Manual of Determinative Bacteriology* is being revised into four volumes, to be collectively called *Bergey's Manual of Systematic Bacteriology.* Volume 1, which presents most of the gram-negative bacteria, was published in 1984. Volume 2 will discuss the gram-positive bacteria other than actinomycetes. Volume 3 will include the archaeobacteria, cyanobacteria, and certain gram-negatives not included in Volume 1. Volume 4 will describe the actinomycetes. The revision is scheduled to be completed in 1986. This chapter will briefly describe these principal bacterial groups according to the basic organization of the revised *Bergey's Manual.*

BACTERIAL GROUPS

Our discussion will emphasize bacteria considered to be of practical importance, particularly in the field of medicine (Table 10–1).

Spirochetes

The spirochetes are found in contaminated water, in sewage, soil, and decaying organic matter, and within the bodies of humans and animals. One of the first microorganisms described by Leeuwenhoek in the 1600s was a large spirochete taken from saliva and tooth scrapings. These bacteria are typically coiled, like a metal spring. Some are tightly coiled, others resemble a spring that is stretched.

All spirochetes are actively motile and achieve their motility by means of two or more axial filaments. The axial filaments are wound around the body of the cell. One end of each axial filament is attached near a pole of the cell (Figure 10–1a and b). By stretching and relaxing its *axial filament,* the cell moves through liquids by rotating as a corkscrew. This mechanism is especially well adapted for movement through liquids that are too viscous to allow motility by flagella, the method used by most other motile bacteria. This adaptation might explain why several species of spirochete parasites are restricted to animal tissue and fluids. An especially interesting group of spirochetes attaches to protozoans that live in certain species of termites. These bacteria move in unison and function as flagella to propel the protozoan.

The spirochetes can be aerobic, facultatively anaerobic, or anaerobic, and they do not have flagella or endospores. All divide by transverse fission.

The spirochetes include a number of important pathogenic bacteria. The best known is the genus *Treponema,* which includes *Treponema pallidum* (tre-pō-nē'mä pal' li-dum), the cause of syphilis (Figure 10–1c). A number of other species of *Treponema* commonly inhabit the mouth. Members of the genus *Borrelia* (bôr-rel'ē-ä) cause relapsing fever, a serious disease that is usually transmitted by either ticks or lice.

Leptospirosis is a disease usually spread to humans by waters contaminated by *Leptospira* (lep-tō-spī'rä) species. Because the bacteria are excreted in

Table 10–1 Summary of Selected Characteristics of Bacterial Groups, from *Bergey's Manual of Systematic Bacteriology*, 1st edition.

Name of Group	Important Genera	Habitat	Special Features
Spirochetes	*Treponema, Borrelia, Leptospira*	Aquatic; animal parasites	Helical morphology; motility by axial filaments; several important pathogens; gram-negative
Aerobic/microaerophilic, motile, helical/vibrioid, gram-negative bacteria	*Spirillum, Campylobacter*	Soil and aquatic environments; human intestinal tract and oral cavity	Helical morphology; motility by flagella, not axial filaments; vibrioids do not have a complete turn
Nonmotile (or rarely motile), gram-negative, curved bacteria	*Spirosoma, Meniscus*	Aquatic and sedimentary environments	Uncommon; mostly aquatic, not pathogenic; form S-shapes, C-shapes, rings
Gram-negative, aerobic rods and cocci	*Pseudomonas, Brucella, Bordetella, Francisella, Legionella*	Soil; water; animal parasites	Contains organisms of medical, industrial, and environmental importance
Facultatively anaerobic, gram-negative rods	*Salmonella, Shigella, Klebsiella, Yersinia, Vibrio, Hemophilus, Streptobacillus, Pasteurella*	Soil; plants; animal intestinal tracts	Contains many important pathogens
Anaerobic, gram-negative straight, curved, or helical rods	*Bacteroides, Fusobacterium*	Animals and insects	Obligate anaerobes, mostly of intestinal tract; some common in mouth and genital tract
Dissimilatory sulfate- or sulfur-reducing bacteria	*Desulfovibrio*	Anaerobic sediments	Reduce oxidized forms of sulfur to H_2S. Gram negative
Anaerobic gram-negative cocci	*Veillonella*	Mostly animal intestinal tracts	Nonmotile anaerobes
Rickettsias and chlamydias	*Rickettsia, Coxiella, Chlamydia*	Parasites of arthropods and animals	Obligate intracellular bacteria; many important pathogens; gram-negative
Mycoplasmas	*Mycoplasma*	Parasites of animals, plants, insects	Pleomorphic; lack cell walls; some pathogens; gram-negative
Endosymbionts	*Holospora, Blattabacterium*	Intracellular symbionts of protozoa, insects, helminths, plants, etc.	Assorted bacteria that live symbiotically in protozoa, insects, and fungi; gram-negative
Gram-positive cocci	*Staphylococcus, Micrococcus, Streptococcus*	Soils; skin and mucous membranes of animals	Some important pathogens
Endospore-forming rods and cocci	*Bacillus; Clostridium*	Soil; animal intestinal tract	*Bacillus*, aerobic or facultative anaerobes; *Clostridium*, anaerobic; both form endospores; gram-positive
Regular, non-sporing, gram-positive rods	*Lactobacillus, Listeria*	Dairy products; genital and oral cavities; animal feces	*Lactobacillus* forms lactic acid from carbohydrates; important industrially; *Listeria*, an animal pathogen

(*continued on next page*)

Table 10–1 Summary of Selected Characteristics of Bacterial Groups, from *Bergey's Manual of Systematic Bacteriology,* 1st edition. (*continued*)

Name of Group	Important Genera	Habitat	Special Features
Mycobacteria	*Mycobacterium*	Soil; plants; animals	*Mycobacterium* is an important pathogen; acid fast; gram-positive
Nocardioforms	*Nocardia*	Soil and animals	Form branched filaments; reproduce by fragmentation; often acid-fast; some pathogens
Gliding, sheathed, and budding/or appendaged bacteria	*Cytophaga, Sphaerotilus, Hyphomicrobium, Caulobacter*	Mostly aquatic, some soil	*Cytophaga* is a cellulose-degrading, gliding soil inhabitant; *Sphaerotilus* is sheathed; *Hyphomicrobium* reproduces by budding; *Caulobacter* is stalked
Gram-negative, chemoautotrophic bacteria	*Nitrosomonas, Nitrobacter, Thiobacillus*	Soil	Nitrifying and sulfur-oxidizing bacteria; agriculturally and environmentally important
Archaeobacteria	*Methanobacterium, Halobacterium, Sulfolobus*	Anaerobic sediments; environments of extreme temperature and osmotic pressure	Not related to other bacterial groups; no peptidoglycan in cell walls; methane producers useful in sewage treatment
Anoxygenic, phototrophic bacteria	*Chromatium, Rhodospirillum, Chlorobium*	Anaerobic sediments	Includes the green and purple sulfur and nonsulfur baçteria; green and purple sulfur bacteria use H_2S as an electron donor, and release sulfur
Cyanobacteria	*Chroococcus, Anabaena*	Aquatic	Produce oxygen during photosynthesis; many species fix atmospheric nitrogen
Actinomycetes	*Streptomyces, Frankia, Micromonospora*	Soil; some aquatic	Common in soil; branching filaments with reproductive conidiospores; *Frankia* involved in nitrogen-fixing symbiosis with plants

the urine of animals such as dogs, rats, and swine, domestic dogs and cats are routinely immunized against it. The distinguishing characteristics of *Leptospira* are that they are aerobic and typically have hooked ends.

Helical/Vibrioid Gram-Negative Bacteria

A distinctly helical shape does not serve as a universal criterion for classification of a bacterium as a spirochete. Some helical bacteria are not included with the spirochetes because they lack an axial filament. They are motile instead by means of flagella. These bacteria possess a single flagellum at one or both poles, or sometimes tufts of flagella at these locations (Figure 10–2). Spiral bacteria, unlike spirochetes, are rigid helices or curved rods. These bacteria are aerobic or microaerophilic.

Most spiral bacteria are harmless aquatic organisms, such as *Spirillum volutans* (spī-ril'um vol'ü-tans). This aquatic bacterium might be specially adapted to the very low concentrations of organic matter found in its habitats.

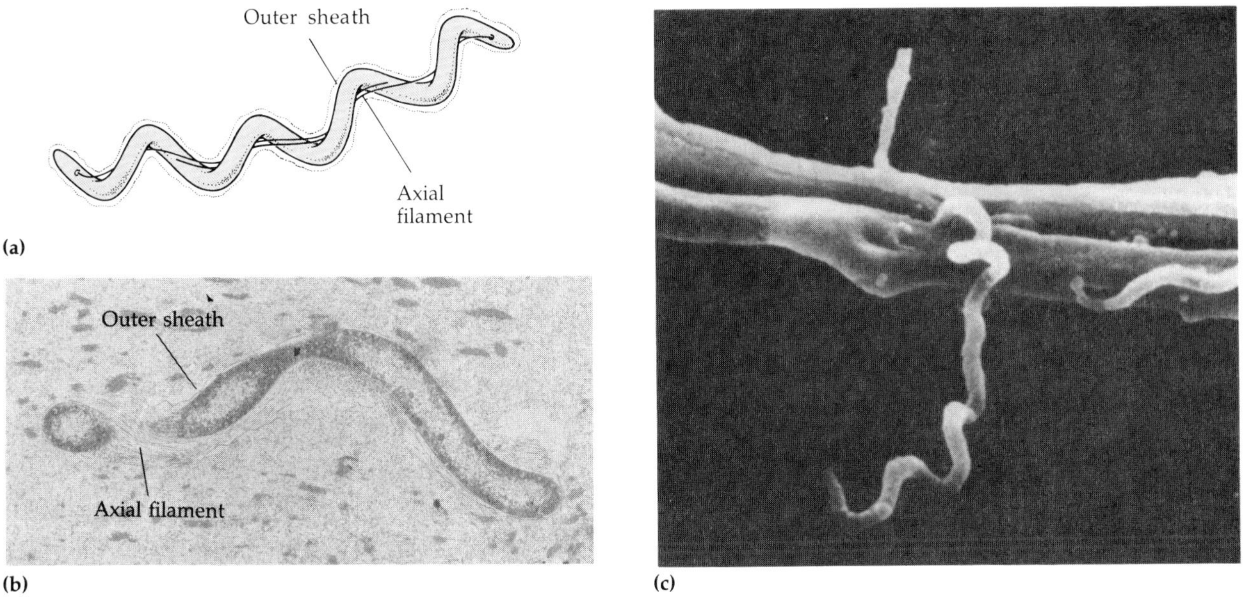

Figure 10–1 Spirochetes are helically shaped bacteria that are motile by means of an axial filament. **(a)** There are at least two axial filaments per cell. Each is anchored to one end of the cell but not the other. By flexing and causing the cell to rotate in a corkscrew fashion, the axial filaments cause locomotion of the cell in fluids. **(b)** A thin section of a large spirochete. Notice the numerous axial filaments beneath the outer sheath (× 15,800). **(c)** The spirochete *Treponema pallidum*, the causitive agent of syphilis, is shown attached to a cell membrane.

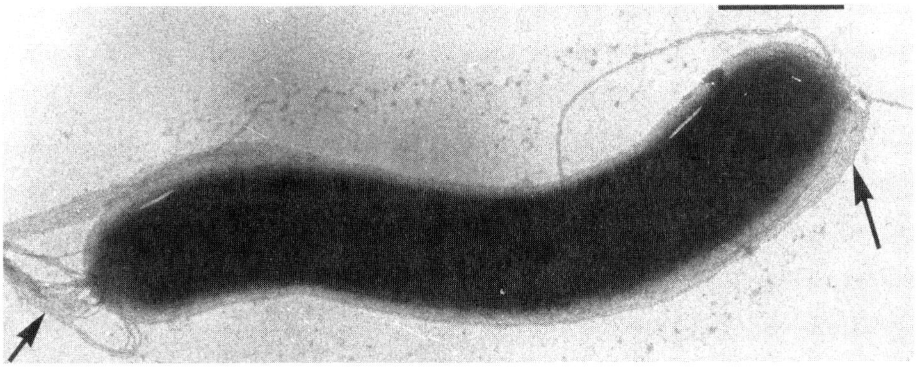

Figure 10–2 Spiral cell of *Aquaspirillum bengal*, which is motile by means of flagella. The arrows point to the tufts of flagella at each end of the cell (bar = 1 μm).

The spiral bacteria also include pathogenic organisms. One species, *Campylobacter fetus* (kam-pī-lō-bak'tėr fē'tus) causes abortion in domestic animals. Another species, *Campylobacter jejuni*, (jē-jū' nē), causes outbreaks of foodborne enteritis (see Chapter 23).

Vibrioid is the term applied to helical bacteria that do not have one complete turn or twist. They look like commas (see Figure 10–3). An interesting vibrioid is the genus *Bdellovibrio*, which attacks other bacteria (see Box in Chapter 3).

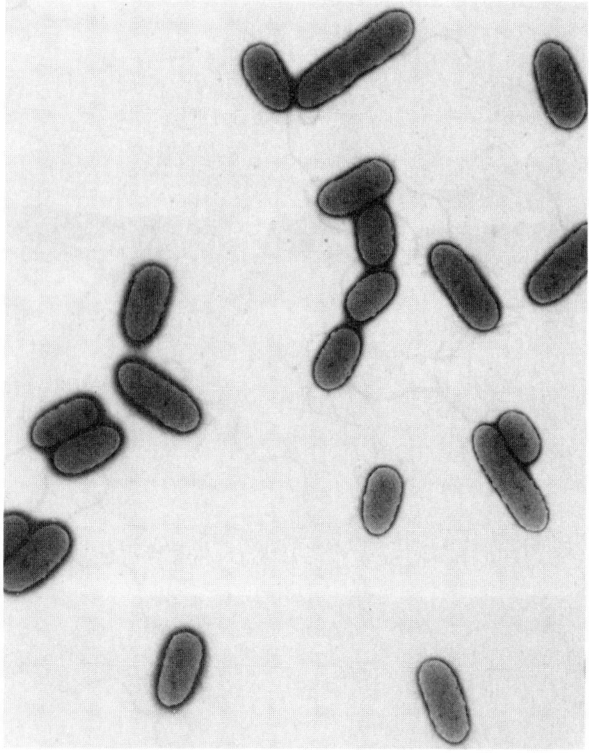

Figure 10–3 *Pseudomonas aeruginosa.* Notice each cell's single polar flagellum characteristic of *Pseudomonas* spp (×7000).

Gram-Negative Aerobic Rods and Cocci

This grouping contains many microorganisms of medical, industrial, and environmental interest. Some of the more interesting are those of the genus *Pseudomonas* (Figure 10–3). The common name for these bacteria is *pseudomonads*. These organisms have polar flagella, and many species excrete extracellular, water-soluble pigments that diffuse into their media. One species, *Pseudomonas aeruginosa* (sū-dō-mō'nas ā-rü-ji-nō'sä), the organism found in blue pus, produces a characteristically soluble, blue-green pigmentation (see Microview 1–2B). Under the right conditions, particularly in weakened hosts, this organism can infect the urinary tract, burns, and wounds, and can cause septicemia, abscesses, and meningitis. Other pseudomonads produce soluble, fluorescent pigments, which glow when illuminated by ultraviolet light.

Pseudomonads are very common in soil and other natural environments and are generally no threat to a healthy individual. These bacteria are less efficient than some other heterotrophic bacteria in utilizing many of the common nutrients, but pseudomonads have compensating characteristics. For example, many are psychrophilic (or psychrotrophic) and grow at refrigerator temperatures; these impart off tastes and colors to foods. Pseudomonads are also capable of synthesizing an unusually large number of enzymes and probably contribute significantly to the decomposition of chemicals, such as pesticides, that are added to soils.

In hospitals, or other places where pharmaceuticals are prepared, the ability of pseudomonads to grow on minute traces of unusual carbon sources has led to their growth in antiseptic solutions, whirlpool baths, and other places where they have been unexpectedly troublesome. And their resistance to most antibiotics has also been a source of medical interest. Some of these considerations will be discussed in more detail later. Many pseudomonad genes for unusual metabolic characteristics, including resistance to antibiotics, are carried on plasmids (see Chapter 8).

Although pseudomonads are classified as aerobic, like many other bacteria they are capable of substituting nitrate for oxygen as a terminal electron acceptor. This process, which is called *anaerobic respiration*, yields almost as much energy as aerobic respiration yields (see Chapter 5). Nitrate is the form of fertilizer nitrogen most easily used by plants. Under anaerobic conditions, as in waterlogged soil, pseudomonads eventually convert nitrate into nitrogen gas (N_2), which is lost to the atmosphere (see Chapter 26). Pseudomonads are one of the bacteria that cause important losses of valuable nitrogen in fertilizer and soil.

Several genera of gram-negative aerobic rods and cocci are of medical importance.

Legionella is a recently discovered genus, which now contains six species. Originally isolated during a search for the cause of a pneumonia now known as Legionnaires' disease, these bacteria do not grow on usual clinical isolation media, nor do they stain with usual histological staining techniques. After intensive effort, they were isolated through tests of guinea pigs inoculated with infected tissue taken from patients. A specially buffered charcoal-yeast extract agar has since been developed for use in their isolation and growth. It has been demonstrated that this is a relatively common organism in streams, and it colonizes such habitats as warm-water supply lines in hospitals and water in cooling towers of air conditioning systems.

Organisms of industrial and environmental importance will be discussed later (Chapters 25 and 26), among them are a number of free-living soil organisms that fix atmospheric nitrogen (not a common characteristic). These include genera such as *Azotobacter* (ä-zō-tō-bak'tėr) and *Azomonas* (a-zō-mōn'as). Especially important to agriculture are the *Rhizobium* (rī-zō'bē-um) bacteria. When these form a symbiotic relationship with legumes such as beans or clover, they fix nitrogen from the air. *Acetobacter* (ä-sē-tō-bak'tėr) and *Gluconobacter* (glü-kon-ō-bak'tér) are industrially important aerobic organisms that convert ethanol into vinegar. *Zoogloea* (zō-ō-glē-ä) will be presented later in a discussion of aerobic sewage treatment processes such as the activated sludge system. As they grow, *Zoogloea* form unusual, flocculant slimy masses that are essential to the proper operation of such systems. The activities of many of these organisms will be discussed in Chapter 25.

Neisseria is a genus of non-endospore-forming diplococci that are aerobic or facultatively anaerobic, parasitic on human mucous membranes, and able to grow well only at temperatures near body temperature. Pathogenic *Neisseria* include the gonococcus bacterium *Neisseria gonorrhoeae,* the causative agent of gonorrhea (Figure 10–4) and *Neisseria meningitidis* (me-nin-ji'ti-dis), the agent of meningococcal meningitis.

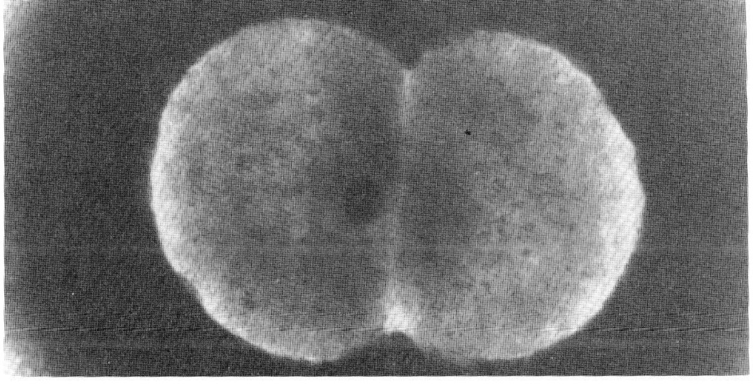

Figure 10–4 Gram-negative cocci. *Neisseria gonorrhoeae,* the organism causing gonorrhea. Note the typical diplococcus (paired) arrangement of this organism.

Members of the genus *Moraxella* are strictly aerobic coccobacilli. Coccobacilli are egg-shaped, a structure between cocci and rods. *Moraxella lacunata* (mô-raks-el'lä la-kü-nä'tä) is implicated in conjunctivitis, an inflammation of the conjuctiva.

Brucella (brü-sel'lä) is a small nonmotile coccobacillus. All species of Brucella are obligate parasites of mammals, and cause brucellosis. Of medical interest is the unusual ability of *Brucella* to survive phagocytosis, an important element of our defense against bacteria (Chapter 15).

Bordetella (bôr-de-tel'lä) is a nonmotile rod. Virulent forms have capsules. Found only in humans, *Bordetella pertussis* is the primary cause of whooping cough.

Francisella is a genus of small, pleomorphic bacteria that grow only on complex media enriched with blood or tissue extracts. *Francisella tularensis* (fran-sis-el'lä tü-lä-ren'sis) causes rabbit fever (tularemia).

Facultatively Anaerobic Gram-Negative Rods

From a medical viewpoint, gram-negative, facultatively anaerobic rods are a very important group of bacteria. Many of them cause diseases of the gastrointestinal tract, as well as other organs. Discussed here are three families: Enterobacteriaceae, Vibrionaceae, and Pasteurellaceae.

Enterobacteriaceae (enterics)

The family *Enterobacteriaceae,* or *enterics,* as they are commonly called, includes a group of bacteria that inhabit the gastrointestinal tracts of humans and other animals. Some species are permanent common residents, others are found in only a fraction of the population, and still others are present only under disease conditions. Most enterics are active fermenters of glucose and other carbohydrates.

Because of the clinical importance of enterics, there are many techniques for their isolation and

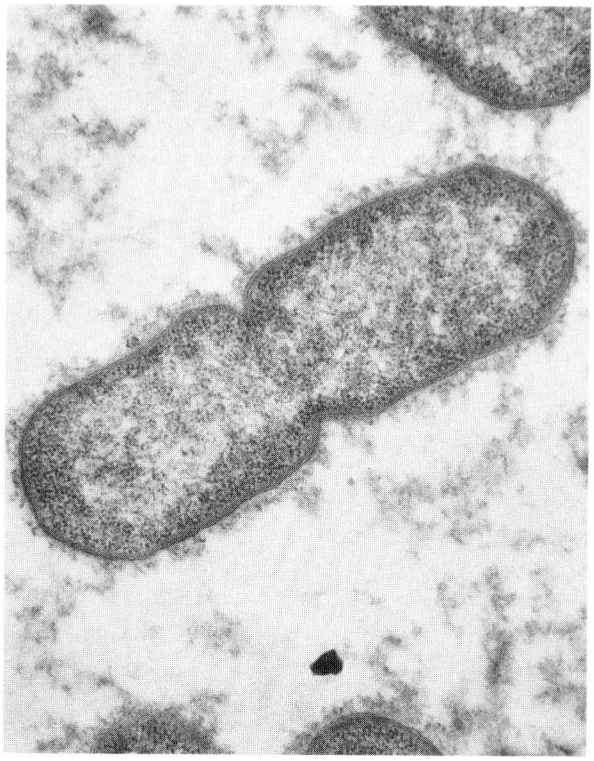

Figure 10–5 Electron micrograph showing dividing *Escherichia coli*, a typical gram-negative rod (×25,000).

identification (see Microview 3–3). A combination of four of the most useful tests is known as the IMViC test (see Chapter 25), which is used to differentiate particular enterics when food or water is being tested for fecal contamination. Biochemical tests are especially important in the detection of foodborne contamination by *Salmonella*. Enterics can also be distinguished from each other according to antigens present on their surfaces.

Enterics include motile as well as nonmotile species, and those that are motile have peritrichous flagella. Many enterics have *pili* (see Figure 4–8), which help them adhere to surfaces of mucous membranes. Specialized *sex pili* exchange genetic information, which often includes antibiotic resistance, between cells (see Figure 8-22 and "Conjugation" in Chapter 8).

Within the intestinal tract, many enterics can produce proteins called *bacteriocins*. These substances cause the lysis of related species of bacteria. Bacteriocins might help maintain the ecological balance of various enterics in the small intestine.

Among the important genera included as enterics are *Escherichia, Salmonella, Shigella, Klebsiella, Serratia, Proteus, Yersinia, Erwinia,* and *Enterobacter*.

Escherichia The facultative anaerobe *Escherichia coli* is one of the most common inhabitants of the intestinal tract and probably the most familiar organism of microbiology (Figure 10–5 and Microview 1–1B). For a lighthearted look at how *E. coli* got its name, see the accompanying box. As you may remember from earlier chapters, a great deal is known about the biochemistry and genetics of *E. coli*, and it continues to be an important tool for basic biological research. Its presence in water or food is also important as an indication of fecal contamination (Chapter 26). *E. coli* is not usually considered to be pathogenic. However, it can be a common cause of urinary tract infections, and certain strains produce enterotoxins that commonly cause traveler's diarrhea (Chapter 23).

Salmonella Almost all members of this group are potentially pathogenic. Because of this, there are extensive biochemical and serological tests to clinically isolate and identify salmonellae. They are common inhabitants of the intestinal tracts of many animals, especially poultry and cattle. Under unsanitary conditions, they can contaminate food.

Typhoid fever, caused by *Salmonella typhi* (sal-mōn-el'lä tī'fē), is the most severe illness caused by a member of the genus *Salmonella*. Caused by other salmonellae, gastrointestinal diseases less severe than typhoid fever are termed salmonelloses. Salmonellosis is one of the most common forms of foodborne illness.

Although many members of the genus *Salmonella* have species-like names such as *Salmonella typhimurium* and *Salmonella dublin*, no individual species are recognized. Instead, this group of bacteria is taxonomically divided into hundreds of **serovars** (serotypes). That is, they are differentiated by serological means. (Serovars can be further differentiated by special biochemical or physiological properties. These subdivisions are called **biovars**, or biotypes.) When the salmonellae are injected into appropriate animals, their flagella, capsules, and cell walls cause the formation of antibodies that are specific for each of these structures. (This topic is fully discussed in Chapter 16.) These specific antibodies, which are available commer-

How *E. coli* Got Its Name

The rise of *Escherichia coli* from obscurity to the role of superstar of modern science must be one of the great success stories of all time. The status of this bacterium now shakes Wall Street, reverberates in the boardrooms of America's great corporations, and furrows the brows of Supreme Court justices. *E. coli* lies at the center of all recombinant DNA research and is thus at the focus of genetic engineering, a technology that is destined to alter the very foundations of our society. The current status of *E. coli* is well known, but what of its past?

Our story goes back to 1885 in Munich, where a young pediatrician, Theodor Escherich, held clinical assistantships at the Children's Polyclinic and Huner's Children's Hospital. Dr. Escherich was trained in the great era of bacteriology that followed the momentous discoveries of Robert Koch and Louis Pasteur. In 1884 he had gone to Naples as a scientific assistant in the study of a cholera epidemic. Not confining himself to narrow clinical interests, Escherich also carried out research, taking a special interest in the intestinal flora of children as a possible clue to epidemics of diarrhea.

In 1885 Escherich was at the height of his working career. His studies of that year culminated in an original research paper entitled, "Die Darmbacterien des Neugeborenen und Säuglings" ("The Intestinal Bacteria of the Newborn and Infant"). On page 518 of Volume 3 of the *Fortschritte* we read for the first time of *Bacterium coli commune*, the original name of the modern-day *E. coli*. Today's wonder bug entered human history in the messy diaper of a Munich infant, a truly modest start for the most widely chronicled organism in modern biology.

Escherich went on to become the leading bacteriologist in the field of pediatrics. He developed into an authority on infant nutrition and was a strong advocate of breastfeeding. His work continued until 1911, when he succumbed to a cerebral hemorrhage.

Leaving the discoverer, we return to his discovery. *Bacterium coli commune* immediately became an object of considerable research interest. In the early days a number of names were applied to our organism. In 1889 it was referred to as *Bacillus escherichii*. By 1895 the strain was called *Bacillus coli*, and in 1900, it was known variously as *Bacterium verus*, *Bacillus coli communis*, and *Aerobacter coli*. The matter was not finally resolved until 1919, when the genus *Escherichia* first proposed by Migula in 1895 was firmly established by Castellani and Chalmers in the third edition of the *Manual of Tropical Medicine*. The 34-year-old species finally had a permanent name. Over the years many species of *Escherichia* have been proposed, but modern techniques for classification have shown that most of these species belong to other genera. At present, there is one other species of *Escherichia*—*E. blattae*, which differs from *E. coli* in that it does not produce gas from lactose and some other sugars.

Even before the name was firmly established, events were occurring that were destined to propel *E. coli* on the path to fame and fortune. Frederick Twort, in 1915, and Felix d'Herelle, in 1917, independently discovered the bacteriophages—viruses that attach to bacteria, enter the cells, reproduce, and ultimately destroy the hosts. Among the species found to be sensitive to bacteriophages was the ever-present *E. coli*. When molecular biology underwent its grand growth period following World War II, there in the center ring was *E. coli* along with its numerous viruses.

The appellation *coli* informs us that the organism so named is the inhabitant of the colon or large intestine. It is a primary component of feces and therefore one of the most ubiquitous bacteria on the face of the earth. Because of this association with feces, the presence of *E. coli* has been used since the early days of bacteriology to determine the

(continued on next page)

sewage contamination of water—particularly drinking water. Since the microbial quality of drinking water has been so central to public health pursuits, testing water for the presence of this fecal bacterium was a skill known to every trained bacteriologist. For medical reasons, *E. coli* was a widely studied and well-known organism.

During the late 1940s there was a resumption of fundamental research, much of which had been delayed for six or more years by the activities of Adolf Hitler. This was a period characterized by the number of individuals earlier trained in the physical sciences shifting their attention to biology. These investigators wanted research materials that were simple to work with and well described. Being impatient, they also wanted a fast-growing bacterium. *E. coli* neatly fit all these requirements and was chosen by many. This activity had positive feedback. Each fact that was discovered made the species a more likely subject for the next experiment. Papers on *E. coli* had a wide audience, and there are few researchers who do not appreciate a sizable public out there interested in their results. Thus, by inherent qualities, by character, and by luck, *E. coli* became the superstar of the microbes.

And so it happened that Dr. Theodor Escherich has been immortalized, his name or the abbreviation *E.* appearing thousands of times each month in the scientific literature. Yet few of today's scientists have probably ever heard of the Austrian pediatrician or his efforts to help the children of the world. He was an idealist whose vigorous career was devoted to improving the conditions under which infants and children might grow and mature. One hopes that in the battle for technologic advance and medical breakthroughs that characterizes modern *E. coli* technology, the generous spirit of Escherich might survive along with the organism that bears his name.

Adapted by permission from an article by Harold J. Morowitz, Professor at Yale University.

cially, can be used to differentiate *Salmonella* serovars by a system known as the Kauffmann–White scheme. The Kauffmann–White scheme represents an organism by numbers and letters that correspond to specific antigens on the organism's capsule, body, and flagella. For example, the antigenic formula of *S. typhimurium* is 1,4,[5],12,i,1,2. Some salmonellae are named only by their antigenic formulas.

Shigella Species of *Shigella* are responsible for a disease called bacillary dysentery or shigellosis. These organisms are second only to *E. coli* as a cause of traveler's diarrhea. Some strains of *Shigella* can cause a life-threatening dysentery.

Klebsiella The organism *Klebsiella pneumoniae* is a major cause of septicemia in pediatric wards and also a cause of one form of pneumonia, which is contracted especially by persons with upper respiratory tract infections or chronic alcoholism (see Microview 1–1D).

Serratia *Serratia marcescens* (ser-rä'tē-ä mär-ses'sens), which is distinguished by its production of red pigment, has been used as an experimental organism for tests of air dispersal of bacteria for biological warfare. *S. marcescens* has become increasingly important in recent years because of its connection with nosocomial infections, that is, infections acquired in hospitals. The organism, which causes urinary and respiratory tract infections, has been found on catheters, in saline irrigation solutions, and in other supposedly sterile solutions.

Proteus *Proteus* (prō'tē-us) bacteria are very actively motile and are implicated in infections of the urinary tract and wounds, and in infant diarrhea (see Microview 1–2C).

Yersinia *Yersinia pestis* (yėr-sin'ē-ä pes'tis) is the causative agent of bubonic plague ("black death"). Urban rats in some parts of the world, and ground squirrels in the American southwest, are animal reservoirs of these organisms. Fleas are usually the vectors that transmit the organisms among animals and to humans, although contact with animals and respiratory droplets from infected persons can be involved with transmission.

Erwinia *Erwinia* (ėr-wi'nē-ä) species are primarily plant pathogens. Some cause plant-rot diseases. These species produce enzymes that dissolve the pectin between individual plant cells and cause the plants to rot.

Enterobacter Two *Enterobacter* species, *Enterobacter cloacae* (klō-ā'kī) and *Enterobacter aerogenes* (ā-rä'jen-ēz), can cause urinary tract infections and nosocomial infections. However, these species are more commonly found in vegetation and soil, and are good examples of nonfecal coliforms (Chapter 25).

Vibrionaceae

It was noted earlier that the gram-negative, facultative, anaerobic rod group includes the family Vibrionaceae.

Vibrio Members of the genus *Vibrio* are slightly curved rods (Figure 10–6). Although most are nonpathogenic, one important pathogen is *Vibrio cholerae* (vib'rē-ō kol'ėr-ī), the causative agent of Asiatic cholera. The disease is characterized by a profuse and watery diarrhea. *Vibrio parahaemolyticus* causes a less serious form of gastroenteritis. *V. parahaemolyticus* usually inhabits coastal salt waters, and is transmitted to humans mostly by shellfish.

Pasteurellaceae

The third medically important family among the gram-negative, facultatively, anaerobic rods includes *Pasteurella* and *Hemophilus.*

Pasteurella This genus is primarily known as a pathogen of domestic animals. It causes septicemia in cattle, fowl cholera in chickens and other fowl, and pneumonias in several types of animals. The best known species is *Pasteurella multocida* (pas-tyėr-el'lä mul-tō'ci-dä), which can be transmitted to humans by dog and cat bites.

Hemophilus *Hemophilus,* also spelled haemophilus, is a very important group of pathogenic bacteria that receives less attention than it should. These organisms commonly inhabit the mucous membranes of the upper respiratory tract, mouth, vagina, and intestinal tract. The best known species that affects humans is *Hemophilus influenzae* (hē-mä'fi-lus in-flü-en'zi), named long ago after the erroneous conclusion that it was responsible for influenza. *Hemophilus* is cultured on complex media enriched with hemoglobin. Species of *Hemophilus* can be identified partly by their requirements for X and V factors. X factor is a precursor needed to synthesize respiratory enzymes. It is supplied by hemin. V factor is nicotinamide adenine dinucleotide (NAD or NADP).

Hemophilus is responsible for several important diseases. It is the most common cause of meningitis in young children, and one of the common causes of otitis media (earaches). Other clinical conditions caused by *H. influenzae* include epiglottitis (a sore throat complication in which the epiglottis becomes infected), septic arthritis in children, bronchitis, and pneumonia. It can also be involved in endocarditis and pericarditis (infections of the heart). If it enters the bloodstream dur-

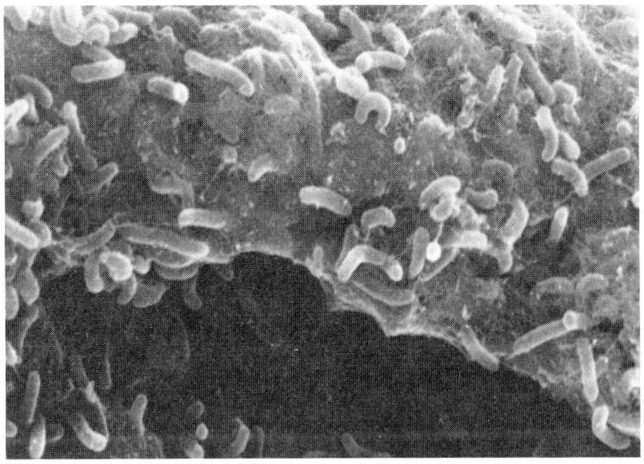

Figure 10–6 Numerous cells of *Vibrio cholerae* are visible here on the surface of a ball of mucus from the small intestine of a mouse. Notice the characteristic slight curve of the rods (×4000).

ing childbirth, it can cause puerperal sepsis, a complication that threatens the mother's life.

Gardnerella

Gardnerella vaginalis (gärd-nė-rel'lä va-jin-al'is), a bacterium that causes one of the most common forms of vaginitis, is not presently assigned to any family. In 1955, researchers named it *Hemophilus vaginalis,* and it was placed with *Hemophilus* in *Bergey's Manual,* 8th edition, because it did not seem to fit anywhere else. *G. vaginalis* varies in its gram-staining reaction but is generally considered to be gram-negative; however, it has a gram-positive type cell wall. Its pleomorphic form and occasional gram-positive reaction have caused it to be also classified with the corynebacteria. Corynebacteria will be discussed later in this chapter.

Anaerobic, Gram-Negative, Straight, Curved, and Helical Rods

Among the anaerobic, gram-negative bacteria, the genus *Bacteroides* (bak-tė-roi'dēz) is a large group of microbes that live in the human intestinal tract (Figure 10–7a). Some of this genus also reside in the oral cavity, genital tract, and upper respiratory tract. *Bacteroides* are non-endospore-forming and nonmotile. Infections due to *Bacteroides* often result from puncture wounds or surgery and are a frequent cause of peritonitis.

Another genus of gram-negative anaerobic bacteria is *Fusobacterium* (fü-sō-bak-ti'rē-um). These microbes are long and slender with pointed rather than blunt ends (Figure 10–7b). In humans, they are found most often in the gums and are responsible for dental abscesses.

Dissimilatory Sulfate- or Sulfur-reducing Bacteria

Although the sulfur-reducing bacteria are of no medical significance, their means of energy production is physiologically interesting and ecologically crucial. They are obligately anaerobic bacteria that use somewhat oxidized forms of sulfur, such as sulfur (S) and sulfates (SO_4^{2-}), as electron acceptors; these forms are reduced to hydrogen sulfide (H_2S). The activity of these bacteria releases millions of tons of H_2S into the atmosphere every year. H_2S is an essential element of the sulfur cycle and is discussed in Chapter 25. These bacteria are found in anaerobic muds and sediments and also in the intestinal tracts of humans and animals. *Desulfovibrio* (dē-sul-fō-vib'rē-ō) is the best-known genus.

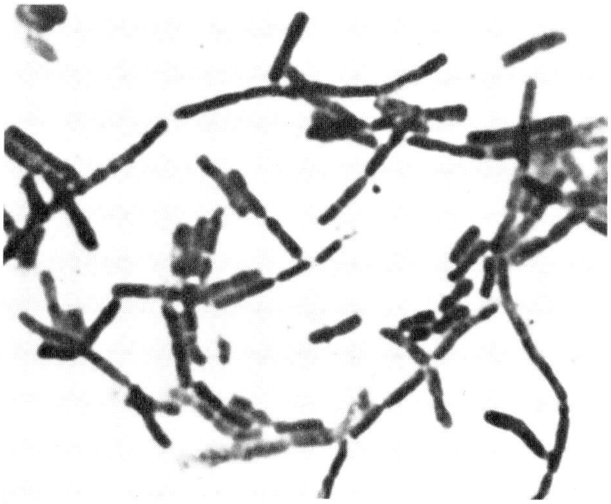

(a)

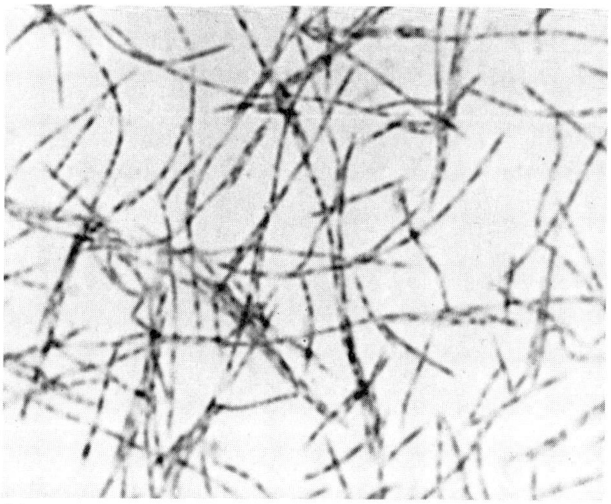

(b)

Figure 10–7 Gram-negative anaerobes. **(a)** Chains of *Bacteroides hypermegas.* **(b)** Chains of slender cells of *Fusobacterium;* note the tapering ends of the cells.

Table 10–2 Comparison of Biochemical Properties of Rickettsias, Chlamydias, and Viruses

Property	Rickettsias	Chlamydias	Viruses
Nucleic acid	RNA and DNA	RNA and DNA	RNA or DNA
Ribosomes	Present	Present	Absent
Structural integrity maintained during multiplication	Yes	Yes	No
Macromolecular synthesis	Carried out	Carried out	Only with use of host cell
ATP-generating system	Present	Absent	Absent
Sensitivity to antibacterial antibiotics	Sensitive	Sensitive	Resistant

Source: Thomas D. Brock, *Biology of Microorganisms,* 3rd ed., © 1979, p. 720. Reprinted by permission of Prentice-Hall, Inc., Englewood Cliffs, NJ.

Anaerobic Gram-Negative Cocci

The cells of gram-negative anaerobic cocci typically occur in pairs, but they can occur singly, in clusters, or in chains. They are nonmotile and nonendospore-forming. Bacteria of the genus *Veillonella* are found as part of the normal flora of the mouth and are components of dental plaque.

Rickettsias and Chlamydias

Both rickettsias and chlamydias are obligate, intracellular parasites, which means that they can reproduce only within a host cell. In this respect, they are similar to viruses. In fact, they are even smaller than some of the largest viruses. However, in morphological and biochemical aspects, they resemble bacteria and are therefore classified as such. A comparison of rickettsias, chlamydias, and viruses is in Table 10–2.

Rickettsias (Figure 10–8a) are rod-shaped bacteria or coccobacilli that have a high degree of pleomorphism. They are gram-negative and nonmotile and divide by transverse binary fission. Rickettsias range in length from 1 to 2μm. One distinguishing feature of most rickettsias is that they are transmitted to humans by insects and ticks. The one exception is *Coxiella burnetii* (käks-ē-el'lä bėr- ne' tē-ē), which causes Q fever; it can be transmitted by airborne or foodborne routes. Recently a sporulation cycle has been reported in *C. burnetii.* This cycle might explain the bacterium's relatively high resistance to pasteurization temperatures and antimicrobial chemicals.

Diseases caused by rickettsias include epidemic typhus caused by *Rickettsia prowazekii* (ri-ket'sē-ä prou-wä-ze'kē-ē) and transmitted by lice; endemic murine typhus caused by *Rickettsia typhi* and transmitted by rat fleas; and Rocky Mountain spotted fever caused by *Rickettsia rickettsii* (ri-ket'sē-ē) and transmitted by ticks (see Microview 1-1H). In humans, rickettsial infections damage the permeability of capillaries; severe infections can cause the circulatory system to collapse. Rickettsias are usually cultivated in the yolk sac of chicken embryos.

Chlamydias (Figure 10–8b) are coccoid bacteria that range in size from 0.2 to 1.5 μm. They are gram-negative and nonmotile and, unlike most rickettsias, do not require insects for transmission. They are transmitted by interpersonal contact or by airborne respiratory routes. The developmental cycle of chlamydias is perhaps their most distinguishing characteristic. First the microbe's infectious form, called the **elementary body,** attaches to a host cell. The host cell phagocytizes the elementary body and houses it in a vacuole. This cycle differs from that of the rickettsias, which usually multiply in the cytoplasm of the host cell. Within the host cell, the chlamydia's elementary body reorganizes and develops into a larger, less infective **reticulate body**. This then divides successively and condenses to produce the smaller infectious elementary bodies. Finally, these are released from the host cell and spread to infect surrounding host cells.

There are only two species of chlamydias. *Chlamydia trachomatis* (kla-mi'dē-ä trä-kō'mä-tis) is the causative agent of trachoma, the most common cause of blindness in humans (see Microview 1–1I). *Chlamydia trachomatis* seems to be the primary causative agent of nongonoccocal urethritis (NGU),

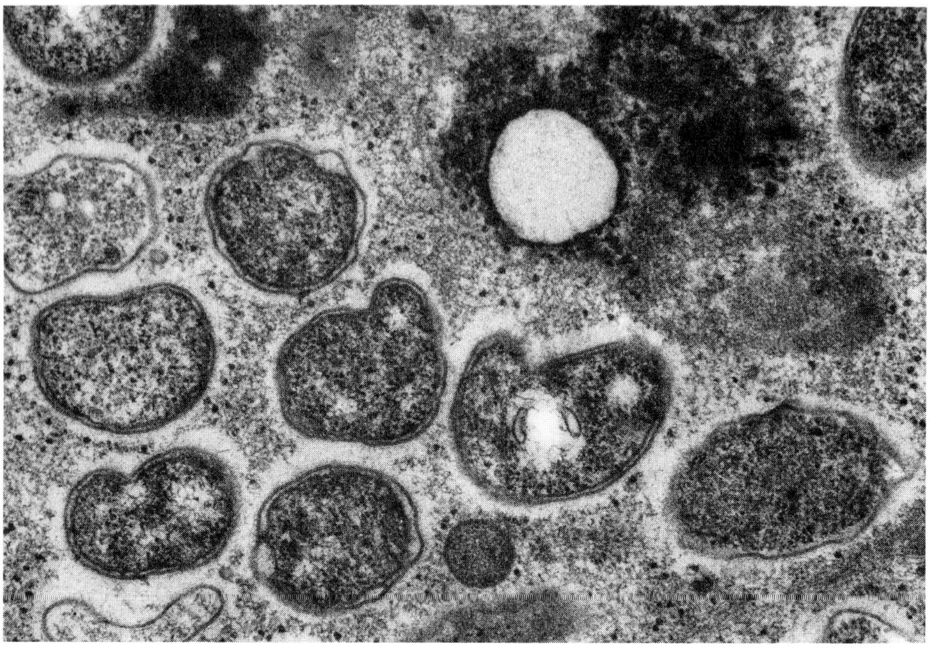

(a)

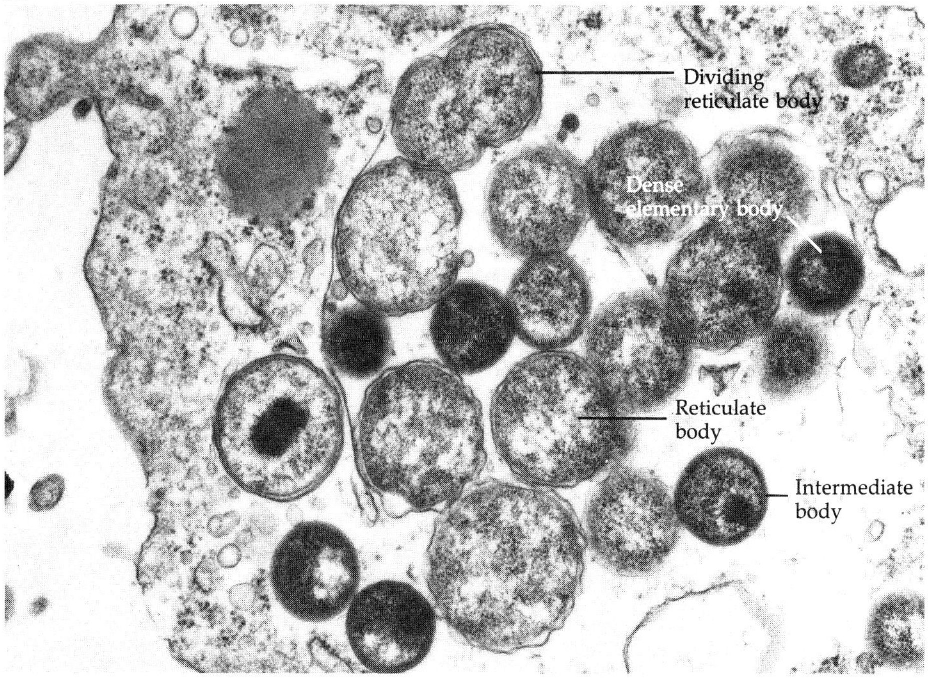

(b)

Figure 10–8 Rickettsias and chlamydias. **(a)** Electron micrograph showing *Rickettsia prowazekii* in experimentally infected tick tissue (×42,500). The inner and outer membranes are clearly visible around the cells in the tissue. **(b)** *Chlamydia psittaci* in the cytoplasm of a host cell (×25,000). The dense, dark, relatively small elementary bodies are the infectious forms. The lighter reticulate bodies with fibrillar internal material have thin walls similar to those of other gram-negative bacteria. One reticulate body is shown dividing; it can divide only in a host cell. The intermediate bodies shown are transitional stages between elementary and reticulate forms.

Table 10–3 Comparison Between Mycoplasmas and L Forms

Property	Mycoplasmas	L Forms
Stability	Cannot revert to cell-wall-containing type	May revert to cell-wall-containing type
Medium	Do not require high salt medium to maintain cellular integrity	Require high salt medium to maintain cellular integrity
Plasma membrane	High sterol content	No sterols
Reaction to penicillin	No adverse effect	Reproduction is inhibited

which might now be the most common venereal disease in the United States, and lymphogranuloma venereum, another venereal disease. *Chlamydia psittaci* (sit'tä-sē) is the causative agent of psittacosis (ornithosis).

Chlamydia can be cultivated in laboratory animals, cell cultures, or the yolk sac of chicken embryos.

Mycoplasmas

Mycoplasmas are bacteria that do not form cell walls. They should not be confused with L forms, which are mutants of wall-forming bacteria that fail to form normal cell walls. Mycoplasmas cannot revert to normal cell-wall-containing types, as L forms can. Moreover, mycoplasmas do not require high concentrations of salt to maintain their cellular integrity, whereas L forms do. And the plasma membranes of mycoplasmas have a high sterol content, while those of L forms do not. Finally, the growth of mycoplasmas is not inhibited by penicillin. These differences between mycoplasmas and L forms are summarized in Table 10–3.

The majority of *Mycoplasma* (mī-kō-plaz'mä) species are aerobes or facultative anaerobes, and because they lack cell walls, they are highly pleomorphic (Figure 10–9). They can produce filaments that resemble fungi, hence their name (*myco* means "fungus"). *Mycoplasma* cells are very small, ranging in size from 100 to 250 nm. The most significant human pathogen among mycoplasmas is *Mycoplasma pneumoniae,* the causative agent of primary atypical pneumonia.

Mycoplasmas can be grown on artifical media that provide them with sterols, if necessary, and other special nutritional or physical requirements. Colonies are less than 1 mm in diameter and have a characteristic "fried egg" appearance through a microscope. Due to the small colony size and poor growth on artificial media, cell culture methods are often more satisfactory.

Gram-Positive Cocci

Most gram-positive cocci of medical importance are members of the genera *Staphylococcus* or *Streptococcus.*

Staphylococci typically occur in grapelike clusters (Figure 10–10 and Microview 1–1A). The most important staphylococcal species is *Staphylococcus aureus* (staf-i-lō-kok'kus ô'rē-us), named for its yellow-pigmented colonies (*aureus* means "golden"). Members of this species are aerobes or facul-

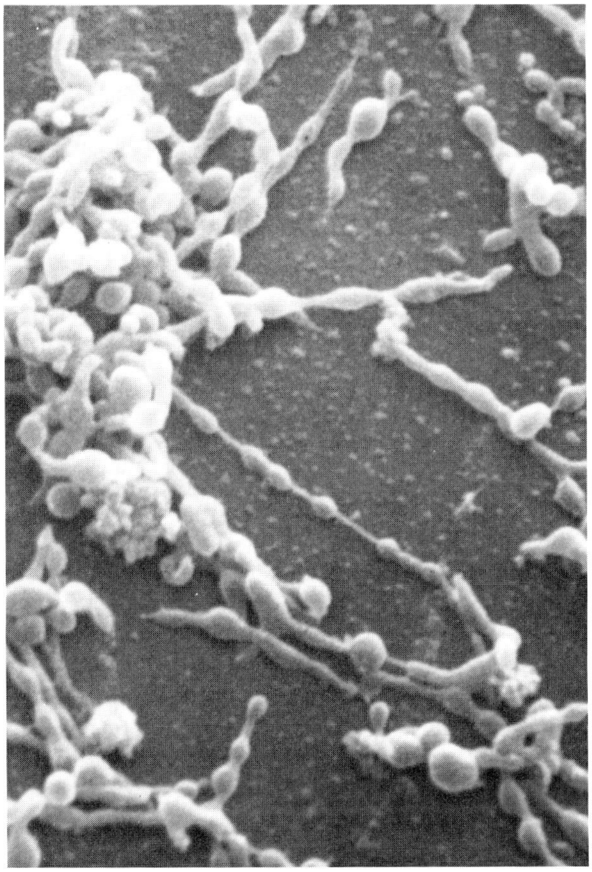

Figure 10–9 Scanning electron micrograph showing the irregular (pleomorphic) form of *Mycoplasma pneumoniae.* The unusual morphology is due to the lack of a cell wall.

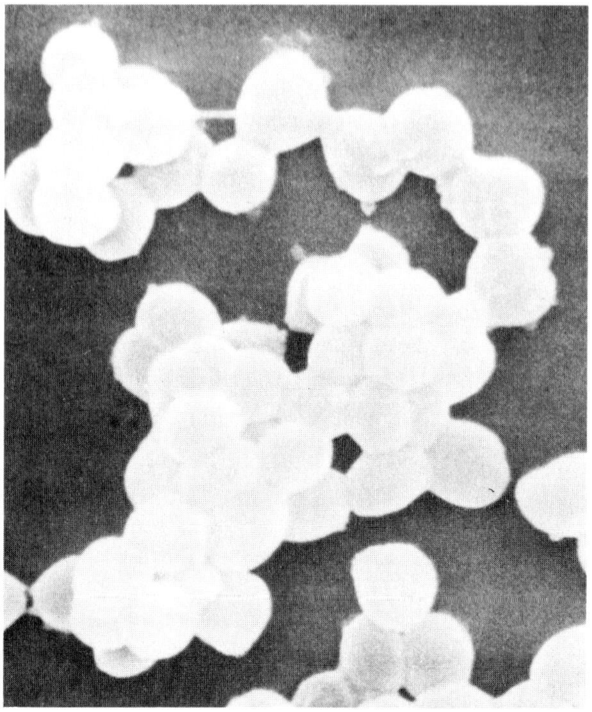

Figure 10–10 Scanning electron micrograph showing the grape-like clusters of cells of *Staphylococcus aureus.*

tative anaerobes. Understanding the characteristics of the staphylococci may help in understanding the reasons for their pathogenicity, which takes many forms. The spherical shape, combined with the strength of the gram-positive cell wall, enables them to survive and grow at high osmotic pressures. This ability partially explains why they can grow and survive in nasal secretions (many of us carry the bacteria there) and on the skin. This ability also explains how *S. aureus* can grow in foods that tend to inhibit the growth of competing organisms. Furthermore, the yellow pigment probably confers some protection from the antimicrobial effects of sunlight.

S. aureus produces many toxins, which contribute to the bacterium's pathogenicity by increasing its ability to invade the body. One type of toxin is *leukocidin;* leukocidin destroys leukocytes and macrophages, which, as discussed in Chapter 15, are essential parts of the body's phagocytic defense system. Tissue damage is caused by a *necrotizing* (*tissue killing*) *exotoxin* that appears in infections of hair follicles (boils) or wounds. The infection of surgical wounds by *S. aureus* is a common problem in hospitals, as we will see later. The bacterium's ability to quickly develop enzymatic resistance to antibiotics such as penicillin contributes to its being a problem in hospitals. The *exfoliative* or *epidermolytic toxin* produced by some strains of *S. aureus* causes the skin to peel off in sheets, a symptom called the scalded skin syndrome. This symptom is almost diagnostic for a systemic staphylococcal infection. *S. aureus* is the agent of toxic shock syndrome, a severe infection causing high fever and vomiting. Finally, *S. aureus* produces an *enterotoxin* that causes vomiting and nausea when ingested; this toxin causes one of the most common types of food poisoning.

Members of the genus *Streptococcus* are spherical, gram-positive bacteria. They are probably responsible for more illnesses and cause a greater variety of diseases than do any other group of bacteria. Streptococci typically appear in chains that can contain as few as four to six cocci or as many as 50 or more (Figure 10–11a). One species, *Streptococcus pneumoniae,* is usually found only in pairs (Figure 10–11b). Streptococci are nonmotile and do not form endospores. They do not use oxygen, although most are aerotolerant (will grow in air). A few are obligately anaerobic. Streptococci do not produce catalase; therefore, the results of a test called the catalase test show them to be *catalase negative.* This simple test is useful in distinguishing streptococci from aerobic or facultatively anaerobic bacteria, for example, staphylococci.

One basis for the classification of streptococci is their action on blood agar. *Alpha-hemolytic* species produce a substance called alpha hemolysin that reduces hemoglobin (red) to methemoglobin (green). This reduction causes a greenish zone to surround the colony. *Beta-hemolytic* species produce a hemolysin that forms a clear zone of hemolysis on blood agar (see Microview 1–2A). Some species have no apparent effect on red blood cells. These are referred to as *nonhemolytic,* or sometimes, less logically, as gamma-hemolytic.

Like staphylococci, streptococci produce a number of extracellular substances that contribute to disease. These include substances that produce erythema (redness) of the skin (*erythrogenic toxin*) and substances that lyse blood cells (*streptolysins*), as well as enzymes (*hyaluronidase* and *streptokinase*) that help spread the infection to other tissues. Because streptokinase has proven to be medically useful in dissolving blood clots, it is used to treat circulatory problems and help prevent heart attacks.

Endospore-Forming Gram-Positive Rods and Cocci

The formation of endospores by bacteria is important to both medicine and the food industry, because of the endospores' resistance to heat and many chemicals. The majority of endospore-forming rods and cocci are gram-positive. With respect to oxygen requirements, they can be strict aerobes, facultative anaerobes, obligate anaerobes, or microaerophiles. The two important genera are *Bacillus* and *Clostridium* (Figure 10–12).

Bacillus anthracis (bä-sil'lus an-thrā'sis) causes anthrax, a disease of cattle, sheep, and horses that can be transmitted to humans. The anthrax bacillus is a nonmotile, facultative anerobe ranging in length from 4 to 8 μm (Figure 10–12a). It is one of the largest bacterial pathogens. The endospores of *Bacillus anthracis* are centrally located, and smears prepared from tissue show a capsule.

Members of the genus *Clostridium* are obligate anaerobes. They vary in length from 3 to 8 μm, and in most species the cells containing endospores appear swollen (Figure 10–12c). Some clostridial endospores can withstand temperatures of 120°C for 15 to 20 minutes. Diseases associated with clostridia include tetanus, or lockjaw, caused by *Clostridium tetani* (te'tan-ē), botulism caused by *Clostridium botulinum*, and gas gangrene caused by *Clostridium perfringens* (pėr-frin'jens).

Regular, Nonsporing, Gram-Positive Rods

A chief representative of nonsporing, gram-positive rods is the genus *Lactobacillus* (lak-tō-bä-sil'lus). Lactobacilli are aerotolerant rods that produce lactic acid from simple carbohydrates and grow well in acidic environments (pH 5). In humans, lactobacilli are located in the vagina, intestinal tract, and oral cavity. Common industrial uses of lactobacilli are in the production of sauerkraut, pickles, buttermilk, and yogurt. A pathogen in this group is *Listeria monocytogenes* (lis-te'rē-ä mo-nō-sī-tô'je-nēz), which is associated with abscess formation, encephalitis, and endocarditis.

Irregular, Nonsporing, Gram-Positive Rods

The organisms in this group are often grouped under the general term of **corynebacteria** (*coryne* means "club shaped"). They tend to be highly irregular in morphology, which often varies with the age of the cells. They may be aerobic, anaerobic, or microaerophilic. The best known and most widely studied species is *Corynebacterium diphtheriae* (kô-rī-nē-bak-ti'-rē-um dif-thi'rē-ä), the causative agent of diphtheria (see Figure 22–3). One related species, *Propionibacterium acnes* (prō-pē-on-ē-bak-ti'rē-um ak'nēz), is commonly found on human skin and has been implicated in acne (Figure 10–13) .

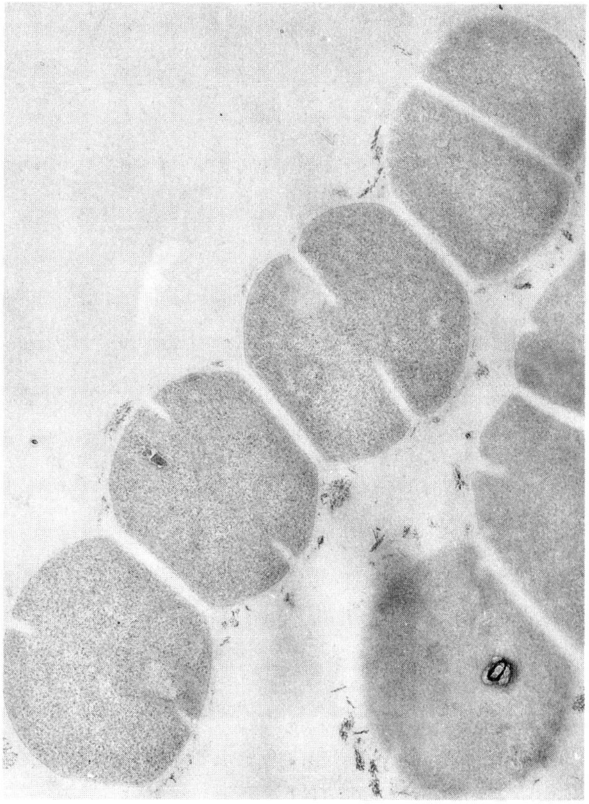

Figure 10–11 *Streptococcus pyogenes* showing dividing cells in a typical chain.

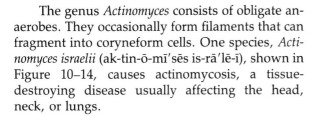

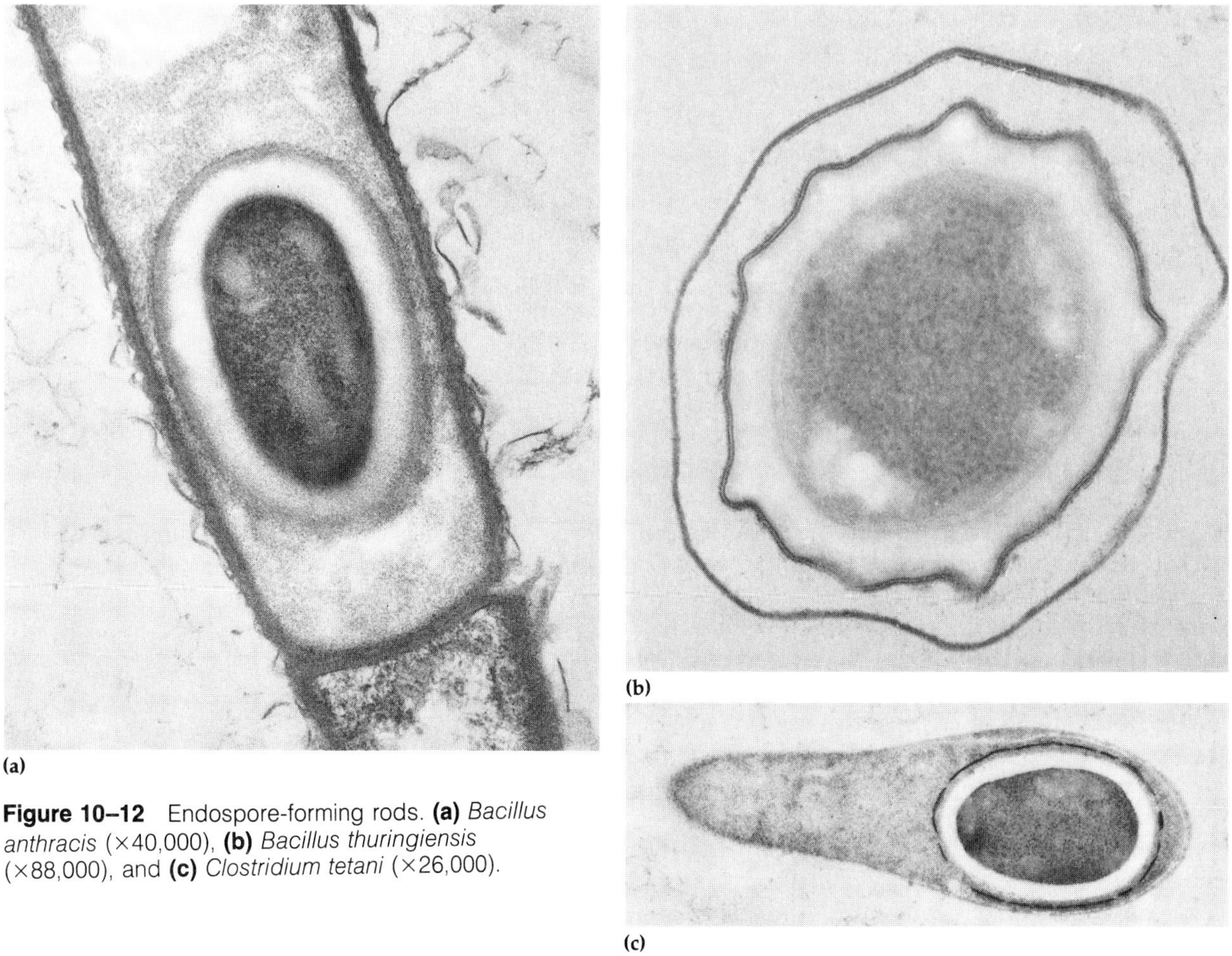

Figure 10–12 Endospore-forming rods. **(a)** *Bacillus anthracis* (×40,000), **(b)** *Bacillus thuringiensis* (×88,000), and **(c)** *Clostridium tetani* (×26,000).

The genus *Actinomyces* consists of obligate anaerobes. They occasionally form filaments that can fragment into coryneform cells. One species, *Actinomyces israelii* (ak-tin-ō-mī'sēs is-rā'lē-ī), shown in Figure 10–14, causes actinomycosis, a tissue-destroying disease usually affecting the head, neck, or lungs.

Mycobacteria

Mycobacteria are aerobic, non-endospore-forming, nonmotile, rod-shaped organisms. Their name, mycobacteria (*myco* means "fungus"), was suggested by their occasional exhibition of filamentous growth. Most of the pathogenic species are acid-fast. A number of species of *Mycobacterium* are found in the soil. Others are important pathogens: *Mycobacterium tuberculosis* (mī-kō-bak-ti'rē-um tü-ber-kū-lō'sis) causes tuberculosis and *Mycobacterium leprae* (lep'rī) causes leprosy (see Microview 1–1E and 1–2E).

Nocardioforms

The best known genus in this group is *Nocardia*. They morphologically resemble Actinomyces; however, *Nocardia* are aerobic. To reproduce, they form rudimentary filaments, which fragment into short rods. The structure of their cell wall resembles that of the mycobacteria; therefore, they are often acid-fast. *Nocardia* are common in soil. Some species, such as *Nocardia asteroides* (nō-kär'dē-ä as-tėr-oi'dēz) (Figure 10–15), occasionally cause chronic, difficult to treat, pulmonary nocardiosis (see Chapter 22). *N. asteroides* is also one of the causative agents of mycetoma, a localized destructive infection of the feet or hands.

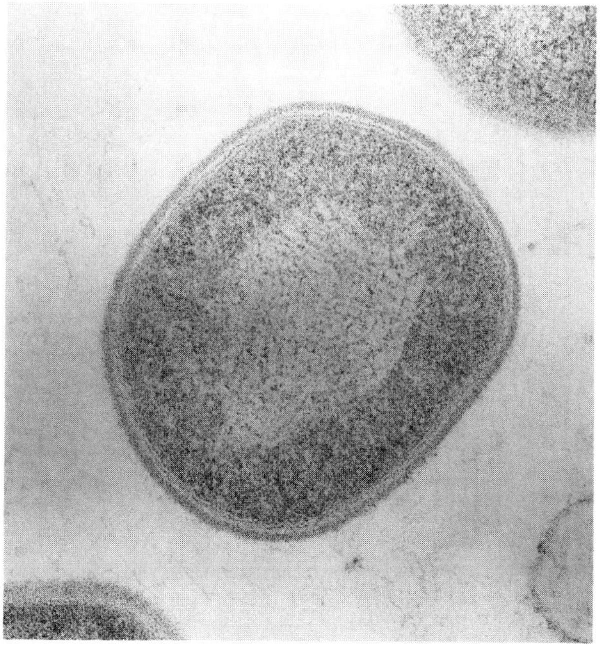

Figure 10–13 Newly divided daughter cell of *Propionibacterium acnes.*

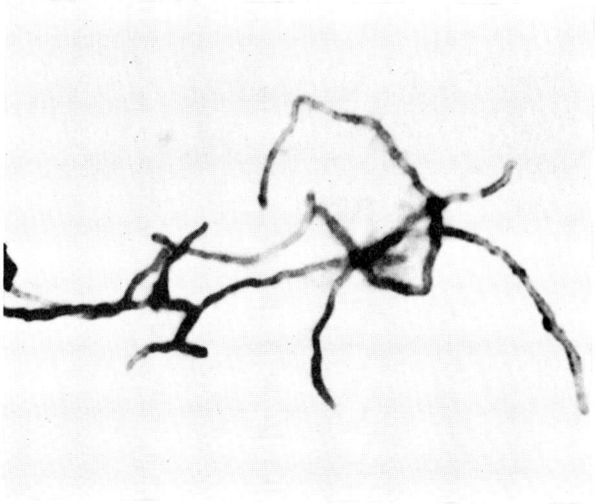

Figure 10–14 *Actinomyces israelii* showing branched filaments.

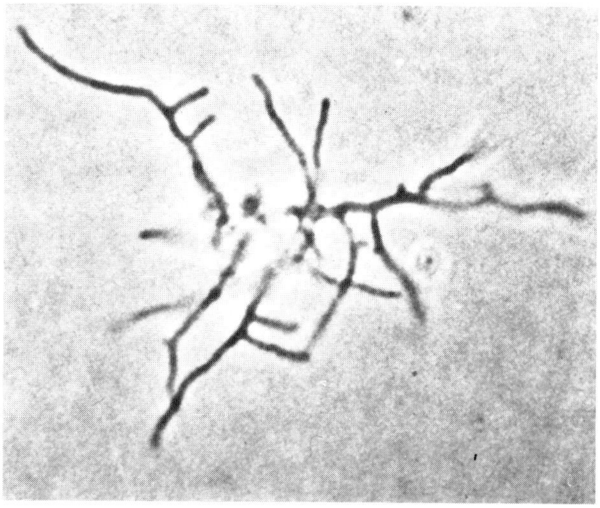

Figure 10–15 The filamentous morphology of a colony of *Nocardia asteroides.*

Gliding, Fruiting, Sheathed, and Budding and/or Appendaged Bacteria

Some morphologically diverse bacteria are included in the group of gliding, fruiting, sheathed, and budding and/or appendaged bacteria. **Gliding** bacteria are gram-negative organisms classified by their method of motility. Embedded in a layer of slime, they glide over surfaces and often leave a visible trail. Members of the genus *Cytophaga* (sī-täf'äg-ä) are important cellulose-degraders in soil.

The **gliding and fruiting** myxobacteria (myxa means "slime") have a remarkable life cycle. Large numbers of cells converge upon a single location. Here they construct a stalked fruiting body (Figure 10–16). Under proper conditions the fruiting body germinates and forms new vegetative gliding cells.

Sheathed bacteria, such as *Sphaerotilus natans* (sfe-rä'ti-lus nā'tans), are found in fresh water and in sewage, and are the cause of an important problem in sewage treatment (see Chapter 25). These gram-negative, polarly-flagellated bacteria form a hollow, filamentous, surrounding sheath (Figure 10–17).

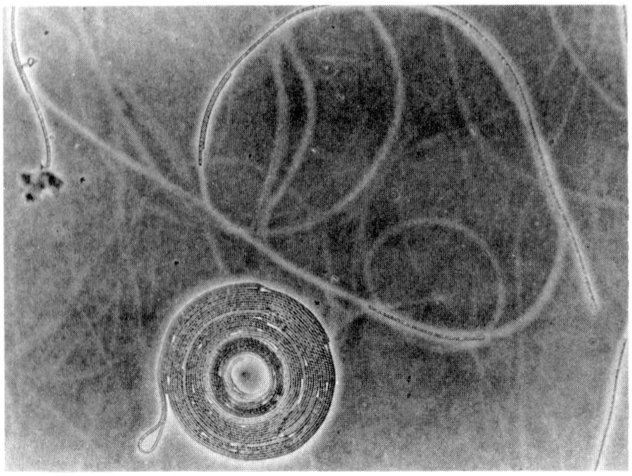

(a)

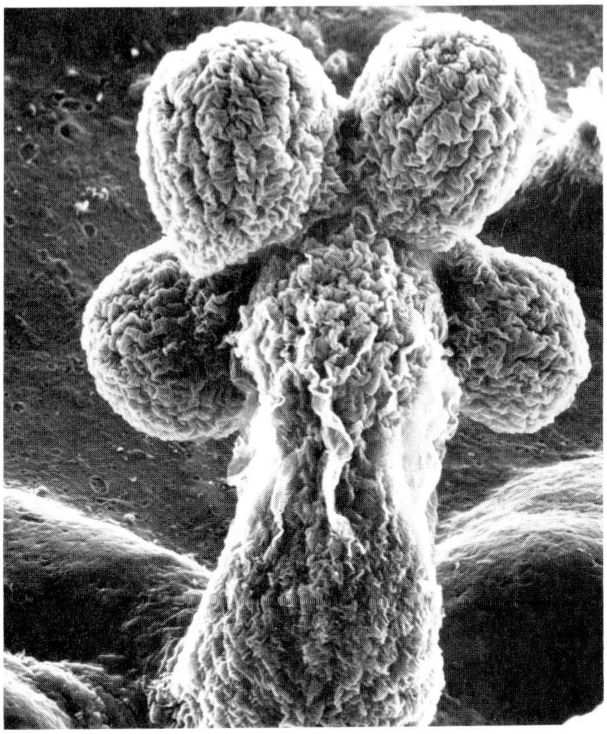

(b)

Figure 10–16 Gliding myxobacteria. **(a)** Slime trails form as the vegetative cells of *Beggiatoa alba* move over a surface. **(b)** Scanning electron micrograph of the stalked fruiting body formed after numerous vegetative cells have aggregated (×350). When suitable conditions occur, the spores shown on this organism, *Stigmatella aurantiaca,* can germinate, forming new vegetative cells.

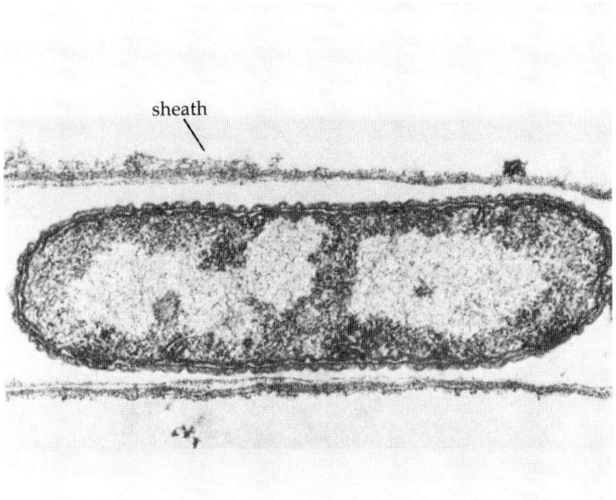

Figure 10–17 *Sphaerotilus natans.* This organism, which is found in sewage and aquatic environments, forms elongated sheaths in which the flagellated bacteria are located.

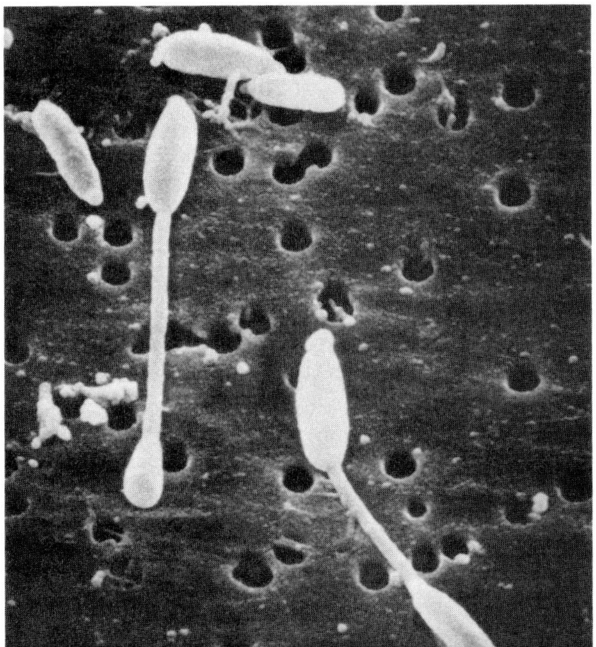

Figure 10–18 The budding bacteria *Hyphomicrobium.*

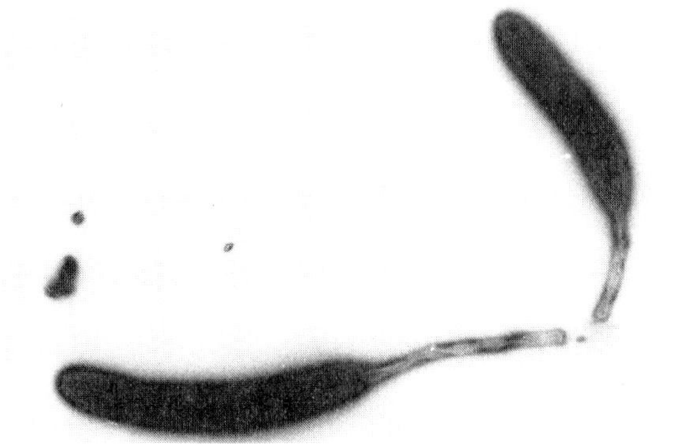

Figure 10–19 Two stalked cells of the stalked bacterium *Caulobacter bacteroides* (×9000).

Budding bacteria do not divide by fission into redundant halves. Members of the genus *Hyphomicrobium* (hī-fō-mī-krō'bē-um) and a few other genera form buds (Figure 10–18). The process, in principle, resembles the asexual reproductive processes of many yeasts. The parent cell retains its identity while the bud increases in size until it separates as a complete, new cell. (See Chapter 11.) Some **appendaged** aquatic microorganisms, mostly of the genus *Caulobacteria,* (kaw-lō-bak-te'rē-um), have stalks by which they anchor themselves to surfaces (Figure 10–19). The stalk is able to absorb nutrients. If the stalk is anchored to the surface of another organism, the bacterium is able to utilize excretions from that host. Appendaged bacteria apparently have adapted these habits to increase their exposure to the nutrients that exist in low levels in their aquatic habitats.

Chemoautotrophic Bacteria

These organisms are of great importance to the environment and to agriculture. They are autotrophs capable of using inorganic chemicals as energy sources and carbon dioxide as the only source of carbon. They possess great synthesizing ability. Especially important for agriculture are the nitrifying bacteria. The energy sources of genera *Nitrobacter* (nī-trō-bak'tėr) and *Nitrosomonas* (nī-trō-sō-mō'näs) are reduced nitrogenous compounds such as ammonium (NH_4^+) and nitrite (NO_2^-); the bacteria convert these compounds into nitrates (NO_3^-). Nitrate is a nitrogen form that is mobile in soil and therefore likely to be encountered and used by plants. *Thiobacillus* (thī-ō-bä-sil'lus) and other sulfur-oxidizing bacteria are parts of the important sulfur cycle. By oxidizing the reduced forms of sulfur such as hydrogen sulfide (H_2S) or elemental sulfur (S) into sulfates (SO_4^{2-}), these bacteria are capable of using the energy stored there. The activities of these organisms are discussed in Chapter 25.

Archaeobacteria

This is an exceptionally interesting group of bacteria. Their evolutionary importance has been outlined in a special box in Chapter 9. They include extreme halophiles, such as *Halobacterium* (hal-ō-bak-te'rē-um) (see Box in Chapter 3 and Microview 1–3a) and *Halococcus* (hal-ō kok'kus), which are not only unusually tolerant of high concentrations of sodium chloride (NaCl), but actually require such environmental conditions for growth. When a microbiological loop is used to transfer these bacteria from their normal habitat, the loop must usually first be dipped into a concentrated NaCl solution, so that the cells do not lyse from exposure to low concentrations of salt. Other archaeobacteria thrive in environments such as acidic, sulfur rich, hot springs. Such an organism is *Sulfolobus,* (sul-fō-lō'bus), which has a pH optimum of about 2, and a temperature optimum of more than 70°C.

Of considerable practical importance are the methane-producing bacteria (methanogens) that are used in sewage treatment processes (Figure 10–20; see also Chapter 26). These archaeobacteria derive energy from combining hydrogen (H_2) with CO_2 to form methane (CH_4). An essential part of the treatment of sewage sludge is encouraging the growth of these organisms in anaerobic digestion tanks—to convert sewage sludge into CH_4.

Phototrophic Bacteria

There are three groups of procaryotes that are **phototrophic;** that is, they use light as an energy source. These are the purple bacteria, the green bacteria, and the blue-green bacteria (cyanobacteria). The energetics of phototrophic bacteria are discussed more completely in Chapter 5.

Purple or green phototrophic bacteria

The purple or green bacteria (which are not necessarily these colors) are generally anaerobic. Their

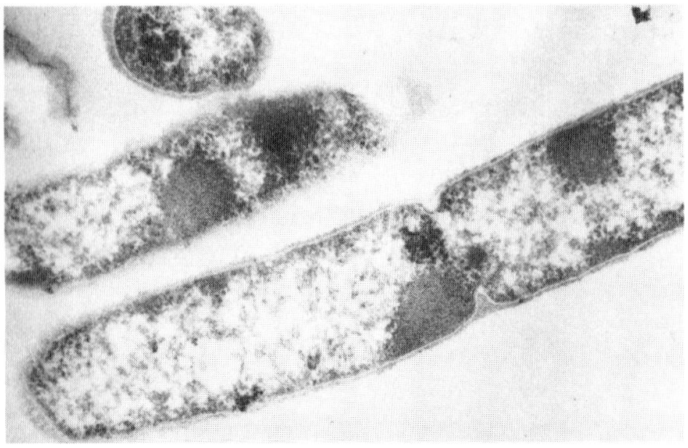

(a)

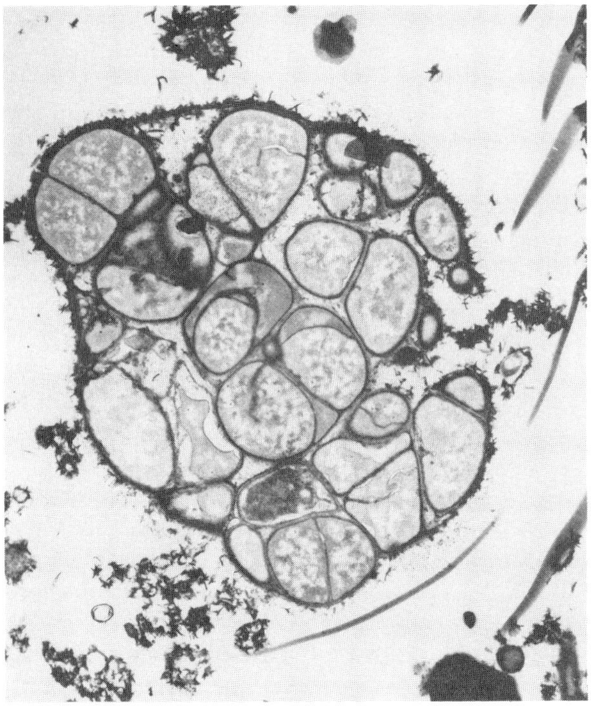

(b)

Figure 10–20 Methanogens. **(a)** *Methanobacterium* MOH and **(b)** Methanosarcina are found in anaerobic sewage digesters and in anaerobic sediments of natural waters.

habitat is usually the deep sediments of lakes and ponds. Unlike plants, algae, and the cyanobacteria, they are **anoxygenic**: their photosynthesis does not produce oxygen. During photosynthesis, water is normally split into hydrogen (H_2) and oxygen (O_2). Hydrogen serves as an electron donor in the reduction of carbon dioxide (CO_2); carbohydrates are produced (see page 125). Some phototrophs, the **purple sulfur** and **green sulfur bacteria** cannot split water for photosynthesis; they instead use reduced sulfur compounds. Hydrogen sulfide (H_2S), for example, is split into hydrogen (H_2) and elemental sulfur (S). This process releases elemental sulfur (S) rather than oxygen (O_2). The sulfur often accumulates in the bacteria, in the form of intracellular granules (Figure 10–21). Other phototrophs, the **purple nonsulfur** and **green nonsulfur bacteria,** use organic compounds such as acids and carbohydrates for the photosynthetic reduction of carbon dioxide. Morphologically, the photosynthetic bacteria are very diverse, and have spirals, rods, cocci, and even budding forms.

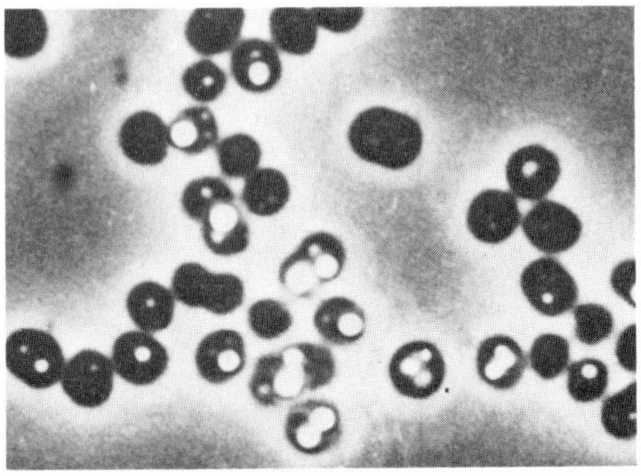

Figure 10–21 Intracellular sulfur granules in *Thiocapsa floridana,* a photosynthetic purple sulfur bacterium.

Cyanobacteria

The **cyanobacteria** are essentially aerobic. They carry out oxygen-producing photosynthesis much as do plants and the eucaryotic algae (Chapter 11). Some are capable of using reduced sulfur compounds for anoxygenic photosynthesis. These produce simple carbohydrates for energy. A considerable number of cyanobacteria are capable of fixing nitrogen. Structures called *heterocysts* fix nitrogen gas (N_2) into ammonium (NH_4^+) for use by the growing cell (Figure 10–22a). Other structures include *gas vacuoles,* found in many species that grow in water. The gas vacuole is a series of chambers or gas vesicles surrounded by a protein wall that is permeable to air but not to water. The gas vacuoles provide buoyancy that helps the cell move to favorable light. Cyanobacteria that are motile move about by gliding.

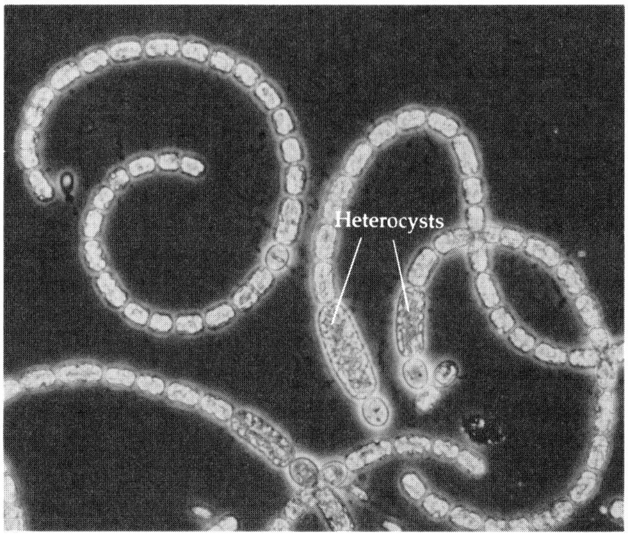

(a)

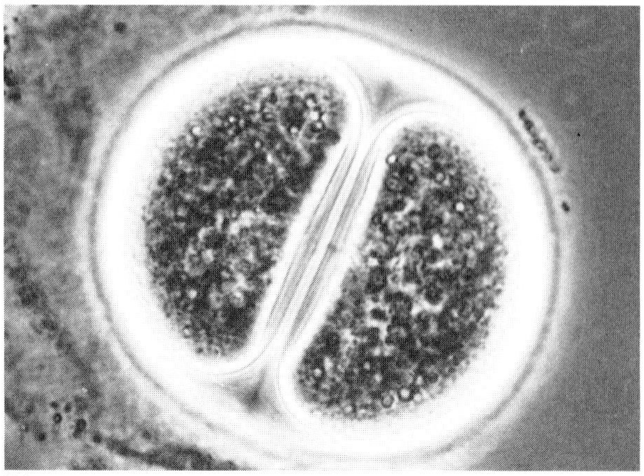

(b)

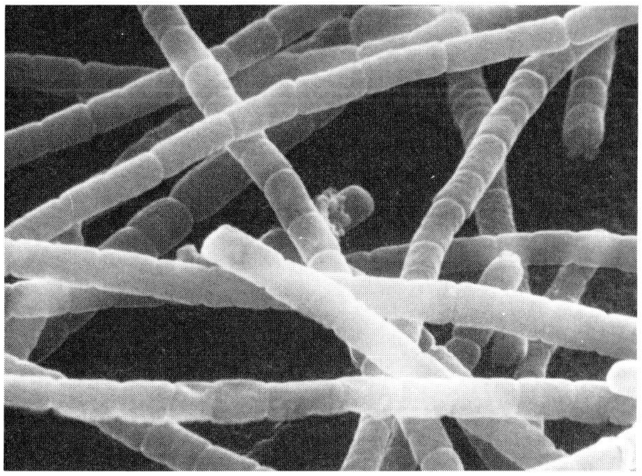

(c)

Figure 10–22 Cyanobacteria. **(a)** Light micrograph of a filamentous cyanobacterium. Note the heterocysts in which nitrogen-fixing activity is located. **(b)** A nonfilamentous cyanobacterium, *Chroococcus turgidus*, as seen with phase-contrast optics (×930). The single cell has just divided. Sometimes two or four cells of this species remain within a single surrounding sheath after cell division. **(c)** Scanning electron micrograph of *Phormidium luridum*, showing long chains of rod-shaped cells (×4400).

The cyanobacteria are morphologically varied. They have unicellular forms that divide by simple fission (Figure 10–22b), colonial forms that divide by multiple fission, and filamentous forms (Figure 10–22c). The filamentous forms usually exhibit some differentiation into cells, which are often bound together within an envelope or sheath. Filamentous cyanobacteria reproduce by fragmentation of the filaments.

The cyanobacteria, especially those that fix nitrogen, are extremely important to the environment. They occupy environmental niches similar to those occupied by the eucaryotic algae described in Chapter 11 (see "The Role of Algae in Nature"), but the ability of cyanobacteria to fix nitrogen makes them even more adaptable (see Microview 1–3b). The environmental role of the cyanobacteria is presented more fully in Chapter 25, in the discussion of eutrophication.

Actinomycetes

The **actinomycetes** are long, branched, filamentous bacteria. They are divided into five sections in *Bergey's Manual*, according to the location of spores and arrangement of filaments. Masses of these filaments are called *mycelia.* A singular mass is called a mycelium. The filaments are of bacterial dimensions, with diameters much smaller than those of molds. Most of the actinomycetes reproduce by forming asexual spores.

Actinomycetes are very common inhabitants of soil. One genus, *Frankia* (frank'ē-ä), causes nitrogen-fixing nodules to form in alder tree roots, much as *Rhizobium* bacteria cause nodules on the roots of legumes (see Chapter 25). The best known genus of actinomycetes is *Streptomyces* (strep-tō-mī'sēs), which is one of the bacteria most commonly isolated from soil (Figure 10–23). The re-

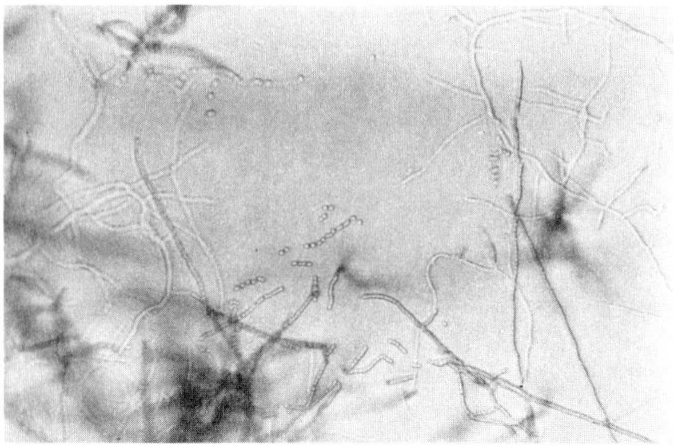

Figure 10–23 *Streptomyces* spp. with branching filaments and chains of conidiospores.

productive asexual spores of *Streptomyces*, conidiospores, are formed at the ends of aerial filaments. If it lands on a suitable substrate, each conidiospore is capable of germinating into a new colony. These organisms are strict aerobes. They often produce extracellular enzymes that enable them to utilize proteins, polysaccharides such as starch or cellulose, and many other organic materials found in soil. *Streptomyces* produce a gaseous compound called *geosmin* that gives fresh soil its typical musty odor. The most important function of various species of *Streptomyces* is the production of most of our commerical antibiotics (Chapter 18).

In the next chapter, we will turn our attention to algae, fungi, protozoans, and helminths.

STUDY OUTLINE

BERGEY'S MANUAL: BACTERIAL TAXONOMY (pp. 275–276)

1. *Bergey's Manual* divides bacteria into sections based on Gram-stain reaction, cellular morphology, oxygen requirements, and nutritional properties.
2. In some cases, the sections include families and orders, and some bacteria are included as genera of uncertain affiliation.

BACTERIAL GROUPS (pp. 276–298)

1. Spirochetes are long, thin, helical cells that move by means of an axial filament.
2. Helical/vibrioid gram-negative bacteria move by means of one or more polar flagella.
3. Gram-negative aerobic rods and cocci have polar flagella, if flagellated, and can utilize a wide variety of organic compounds.
4. Facultatively anaerobic gram-negative rods have peritrichous flagella and include the enterics, Vibrionaceae, Pasteurellaceae, and the genus *Gardnerella*.
5. Members of the anaerobic gram-negative straight, curved, and helical section can be found in humans.
6. Dissimilatory sulfate-reducing bacteria are anaerobes that are important in the sulfur cycle.
7. Anaerobic gram-negative cocci are normal flora of the human mouth.
8. Rickettsias and chlamydias are obligate intracellular parasites.
9. Mycoplasmas are bacteria that lack cell walls.
10. Gram-positive cocci include the catalase-positive *Staphylococcus* and catalase-negative *Streptococcus*.
11. Endospore-forming rods and cocci may be aerobic, facultatively anaerobic, or anaerobic.
12. The diverse group of regular, nonsporing, gram-positive rods includes the important genus, *Lactobacillus*.
13. Irregular, nonsporing, gram-positive rods include the irregular-staining coryneform bacteria.
14. Pathogenic species of mycobacteria are acid-fast.
15. Nocardioform bacteria may be acid-fast; they form short filaments.
16. Bacteria with unusual morphologies are discussed in the section on these bacteria: gliding and nonfruiting; gliding and fruiting; sheathed; and budding and/or appendaged bacteria.

17. The chemoautotrophic bacteria play important roles in the cycles of elements in the environment.

18. Extreme halophiles, acidophiles, thermophiles, and methane-producing bacteria are included in the archaeobacteria division.

19. Photosynthetic purple and green bacteria are included in the group of anoxygenic phototrophic bacteria; they do not produce molecular oxygen.

20. Cyanobacteria produce molecular oxygen during photosynthesis.

21. Actinomycetes produce mycelia and conidiospores.

STUDY QUESTIONS

REVIEW

1. The following is a key that can be used to identify the medically important groups of bacteria. Fill in the name of the group indicated by the key.

	Name of section & representative genus
I. Gram-positive	
A. Endospore-forming	________
B. Nonsporing	
1. Cocci	________
2. Rods	
a. Regular	________
b. Irregular	________
c. Acid-fast	________
C. Mycelium produced	
1. Acid-fast	________
2. Produce chains of conidia	________
II. Gram-negative	
A. Cells are helical or curved	
1. Axial filament	________
2. No axial filament	
a. Aerobic	________
b. Anaerobic	________
B. Cells are rods or cocci	
1. Aerobic, nonfermenting	________
2. Facultatively anaerobic	________
3. Anaerobic	________
C. Intracellular parasites	________
III. Lacking cell wall	________

2. *Bergey's Manual of Systematic Bacteriology* divides the kingdom procaryotae (Monera) into four divisions (phyla). What are they?

3. List six criteria used to classify bacteria according to *Bergey's Manual.*

4. Matching:
   - I. Gram-positive
     - A. Nitrogen-fixing __________
   - II. Gram-negative
     - A. Phototrophic
       1. Anoxygenic __________
       2. Oxygenic __________
     - B. Chemoautotrophic
       1. Oxidize inorganics such as NO_2 __________
       2. Reduce CO_2 to CH_4 __________
     - C. Chemoheterotrophic
       1. Move via a slime layer __________
       2. Form microcysts __________
       3. Reduce sulfate to H_2S
          - a. Anaerobic __________
          - b. Thermophilic __________
       4. Long filaments, found in sewage __________
       5. Form projections from the cell __________
   - III. Unusual cell wall (lacking peptidoglycan) __________

(a) Archaebacteria
(b) Cyanobacteria
(c) *Cytophaga*
(d) *Desulfovibrio*
(e) *Frankia*
(f) *Hyphomicrobium*
(g) Methanogenic bacteria
(h) Myxobacteria
(i) *Nitrobacter*
(j) Purple bacteria
(k) *Sphaerotilus*
(l) *Sulfolobus*

CHALLENGE

1. Does *Bergey's Manual* provide any phylogenetic information regarding the relationships between bacteria?
2. Should *Bergey's Manual* be considered a tool of classification or identification? Briefly explain your answer.

FURTHER READING

Carr, N. G. and B. A. Whitton. 1982. *The Biology of Cyanobacteria.* Botanical Monographs, vol. 19. Berkeley, CA: University of California Press. A comprehensive reference on morphology, metabolism, and reproduction of cyanobacteria.

Finegold, S. M. and W. J. Martin. 1982. *Diagnostic Microbiology*, 6th ed. St. Louis, MO: C. V. Mosby. A good reference for procedures used in the clinical microbiology laboratory.

Goodfellow, M., M. Mordarski, and S. Williams. 1984. *Biology of the Actinomycetes.* New York: Academic Press. A survey of the biology, ecology, and pathogenicity of this diverse group of bacteria.

Holt, J. G., ed. 1984. *Bergey's Manual of Systematic Bacteriology*, 1st ed., vol. 1. Baltimore, MD: Williams and Wilkins. The standard reference of identification and classification of bacteria. Subsequent volumes are planned for release as follows: Vol. 2 (tent. 1985); Vol. 3 (tent. 1986); Vol. 4 (tent. 1986).

Kloos, W. E. 1980. Natural populations of the genus *Staphylococcus. Annual Review of Microbiology* 34:559–592. Illustrates the use of modern technology to determine a species; includes a discussion of staphylococci that live on birds and mammals.

Maniloff, J. 1983. Evolution of wall-less prokaryotes. *Annual Review of Microbiology* 37:477–499. Mycoplasmas and archaeobacteria are used in this discussion of evolution; information comes from current advances in molecular biology.

CHAPTER 11

Fungi, Algae, Protozoans, and Multicellular Parasites

OBJECTIVES

After completing this chapter you should be able to

- Differentiate between asexual and sexual reproduction, and use as examples the processes of a fungus and a protozoan.
- List the defining characteristics of the four phyla of fungi described in this chapter.
- Describe the outstanding characteristics of the six phyla of algae described in this chapter.
- List the distinguishing characteristics of lichens and describe their nutritional needs.
- Compare and contrast cellular slime molds and plasmodial slime molds.
- Describe the outstanding characteristics of the four phyla of protozoans described in this chapter.
- List the characteristics of the three classes of parasitic helminths and give an example of each class.
- Define intermediate host, definitive host, and arthropod vector.

In this chapter, we will examine the eucaryotic microorganisms: fungi, algae, protozoans, and parasitic helminths, or worms (Table 11–1). Although most adult helminths are not microscopic, they do have microscopic stages in their development, and some of these stages cause human diseases; therefore, helminths are studied by microbiologists.

FUNGI

As was noted in Chapter 9, fungi constitute a kingdom of eucaryotes. All fungi are heterotrophs, requiring organic compounds for energy and carbon. Fungi are aerobic or facultatively anaerobic; no strictly anaerobic fungi are known. The majority of fungi are saprophytes in soil and water; there they primarily decompose plant material. Fungi are chemoheterotrophs and, like bacteria, they contribute significantly to the decomposition of matter and recycling of nutrients. By using extracellular enzymes, such as cellulases and pectinases, fungi are the primary decomposers of the hard parts of plants that cannot be digested by most animals. Table 11–2 lists the basic differences between fungi and bacteria. Of the more than 100,000 species of fungi, only about 100 are pathogenic for humans and other animals. On the other hand, thousands of fungi are pathogenic to plants. Virtually every economically important plant is attacked by one or more of the fungi.

The study of fungi is called **mycology**. The fungi include yeasts, molds, and fleshy fungi. Yeasts are unicellular organisms. Molds are multicellular, filamentous organisms such as mildews, rusts, and smuts. And fleshy fungi include multicellular mushrooms, puffballs, and coral fungi.

VEGETATIVE STRUCTURES

A **thallus** (colony) of a mold or fleshy fungus consists of long filaments of cells joined together; these filaments are called **hyphae** (singular, **hypha**). In most molds, the hyphae contain crosswalls, termed **septa** (singular, **septum**), which divide the hyphae into distinct, uninucleate cell-like units. These hyphae are called **septate hyphae** (Figure 11–1a). In a few classes of fungi, however, the hyphae contain no septa and appear as continuous, long cells with many nuclei. These hyphae are referred to as **coenocytic hyphae** (Figure 11–1b). Even in fungi with septate hyphae, there are usually openings in the septa that make the cytoplasm of adjacent "cells" continuous; so these fungi are actually coenocytic organisms too. The hyphae of a thallus grow by elongating at the tips (Figure 11–1c). Each part of a hypha is capable of growth, and when a fragment breaks off, it can elongate to form a new hypha.

When environmental conditions are suitable, the hyphae grow, intertwine, and form a mass called a **mycelium,** which is visible to the naked eye. The portion of the mycelium concerned with obtaining nutrients is called the **vegetative mycelium;** the portion concerned with reproduction is the **reproductive** or **aerial mycelium,** so called be-

Table 11–1 Major Differences Between Fungi, Algae, Protozoans, and Helminths (All are Eucaryotes.)

Property	Fungi	Algae	Protozoans	Helminths
Kingdom	Fungi	Protista and Plantae	Protista	Animalia
Mode of nutrition	Chemoheterotroph	Photoautotroph	Chemoheterotroph	Chemoheterotroph
Multicellular	All, except yeasts	Some	None	All
Cellular arrangement	Unicellular, filamentous, fleshy (i.e., mushrooms)	Unicellular, colonial, filamentous; tissues	Unicellular	Tissues and organs
Food acquisition	Absorptive	Absorptive	Absorptive; cytostome	Mouth (ingestive); absorptive
Characteristic features	Sexual and asexual spores	Pigments	Motility; some form cysts	Many have elaborate life cycles including egg, larva, and adult
Embryo	None	Some	None	All

Table 11–2 Comparison of Selected Features of Fungi and Bacteria

Property	Fungi	Bacteria
Cell type	Eucaryotic with well-defined nuclear membrane	Procaryotic
Cell membrane	Sterols present	Sterols absent except in *Mycoplasma*
Cell wall	Glucans; mannans; chitin (no peptidoglycan)	Peptidoglycan
Spores	Produce a wide variety of sexual and asexual reproductive spores	Endospores (not for reproduction); some asexual reproductive spores
Metabolism	Limited to heterotrophic; aerobic, facultatively anaerobic	Heterotrophic, chemoautotrophic, photoautotrophic; aerobic, facultatively anaerobic, anaerobic

Source: Modified from B. D. Davis, et al., *Microbiology*, 3rd ed., Harper & Row, New York, 1980, p. 829.

cause it projects above the surface of the medium on which the fungus is growing, as shown in Figure 11–2. The aerial mycelium often bears reproductive spores, which we will discuss later.

Yeasts

Yeasts are nonfilamentous, unicellular fungi that are typically spherical or oval in shape. Like molds, yeasts are widely distributed in nature; yeasts are frequently found as a white powdery coating on fruits and leaves. Because most yeasts are colonies of unicellular organisms, they do not reproduce as a unit. Instead, the colony grows as the number of yeast cells increases. This increase usually happens by **budding.** In budding, the parent cell forms a protuberance (bud) on its outer surface. As the bud elongates, the parent cell's nucleus divides, and one nucleus migrates into the bud. Cell-wall material is then laid down between the bud and parent cell, and the bud eventually breaks away (Figure 11–3). One yeast cell can produce up to 24 daughter

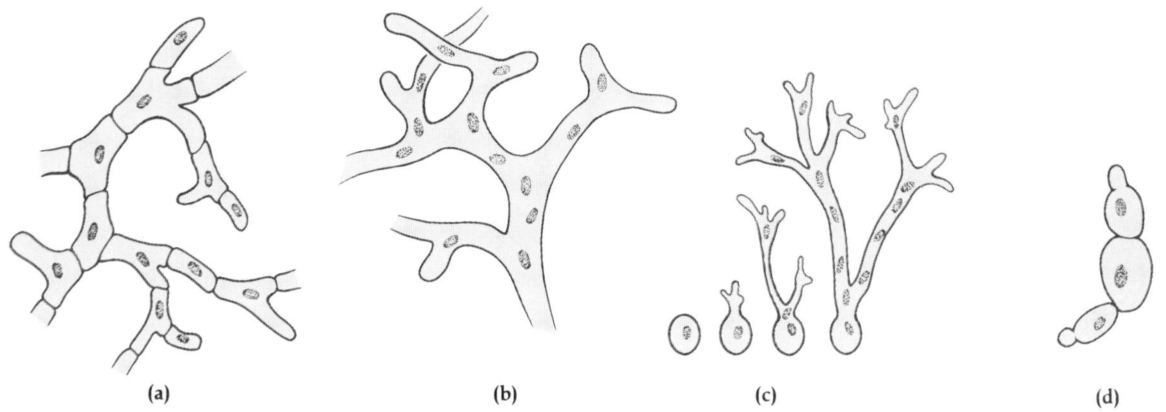

Figure 11–1 Vegetative structures of fungi. **(a)** Septate hyphae have crosswalls dividing the hyphae into cell-like units. **(b)** Coenocytic hyphae lack crosswalls. **(c)** Hyphae grow by elongating at the tips. **(d)** Pseudohyphae are short chains of cells formed by some yeasts.

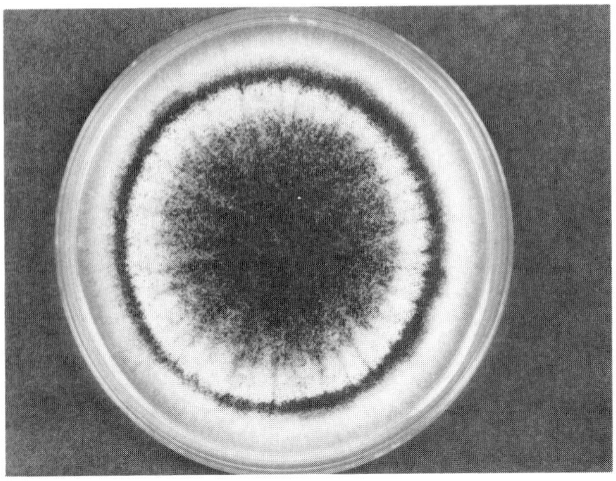

(a)

(b)

Figure 11–2 *Aspergillus niger* grown on agar. **(a)** Vegetative mycelium. **(b)** Aerial mycelia bearing reproductive spores.

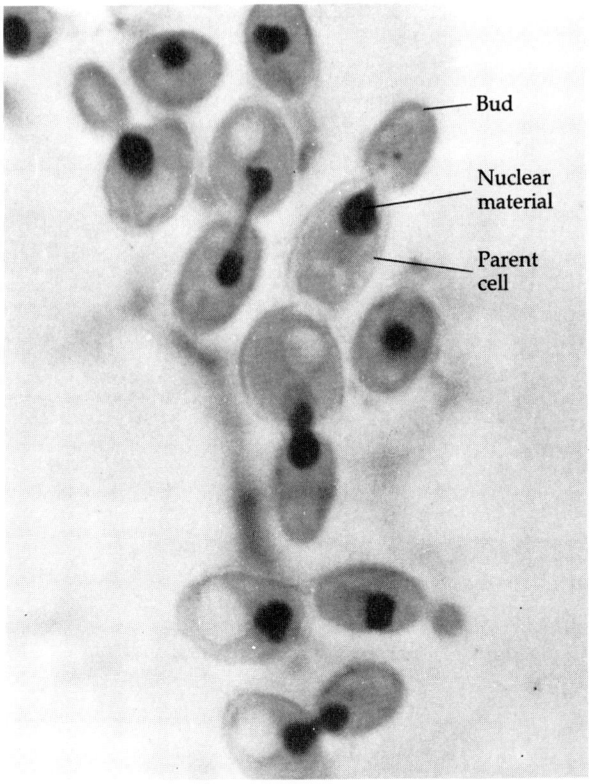

Figure 11–3 Various stages of budding Bakers' yeast, *Saccharomyces cerevisiae* (×1570). Note the distribution of the darkly stained nuclear material from some of the parent cells into the buds.

cells by budding. Some species of yeasts produce buds that fail to detach themselves; these buds form a short chain of cells called a **pseudohypha** (see Figure 11–1d). A few types of yeast grow by fission. During fission, the parent cell itself elongates, its nucleus divides, and two daughter cells are produced. Increases in the number of yeast cells on a solid medium produce a colony similar to a bacterial colony (see Microview 1–2D).

Yeasts are capable of facultative anaerobic growth. As you may recall from Chapter 5, yeasts can use oxygen or an organic compound as the final electron acceptor; this is a valuable attribute. If given access to oxygen, yeasts perform aerobic respiration to metabolize carbohydrates to carbon dioxide and water. Denied oxygen, they ferment carbohydrates and produce ethanol and carbon dioxide. This fermentation is the basis of the brewing, winemaking, and baking industries. Species of *Saccharomyces* (sak-ä-rō-mī'sēs) produce ethanol for brewing beverages and carbon dioxide for raising breads.

Dimorphic Fungi

Some fungi, most notably the pathogenic species, exhibit **dimorphism,** that is, two forms of growth. Such fungi can grow either as a mold or as a yeast. Frequently, dimorphism is temperature dependent: at 37°C the fungus is yeastlike, and at 25°C it is moldlike (see Figure 22–11).

REPRODUCTIVE STRUCTURES

Viewed with a microscope, the hyphae of almost all fungi look alike. When fungi are identified, the reproductive structures, or **spores,** must be examined.

Reproduction in fungi occurs by spore formation. These spores, however, are quite different from bacterial endospores. Bacterial endospores are formed so the vegetative cell will survive adverse environmental conditions. A single, vegetative bacterial cell forms an endospore, which eventually starts to grow (germinates) to produce a single, vegetative bacterial cell. Formation of an endospore by bacteria is not reproduction in the sense that it increases the total number of bacterial cells. But after a mold forms a spore, the spore detaches from the parent and germinates into a new mold (see Figure 11–1c). Unlike the bacterial endospore, this is a true reproductive spore: one cell gives rise to an entire multicellular organism. Spores are formed from the aerial mycelium in a great variety of ways, depending on the species.

Fungal spores can be asexual or sexual. **Asexual spores** are formed by the aerial mycelium of one organism. When these spores germinate, they become organisms that are genetically identical to the parent. **Sexual spores** result from the fusion of nuclei from two opposite mating strains of the same species of fungus. Organisms that grow from sexual spores will have genetic characteristics of both parental strains. Because spores are of key importance in the identification of fungi, we will now look at some of the various types of asexual and sexual spores.

Asexual Spores

Asexual spores are produced by an individual fungus through mitosis and subsequent cell division; there is no fusion of the nuclei of cells. Several types of asexual spores are produced by fungi. One type, an **arthrospore,** is formed by the fragmentation of a septate hypha into single, slightly thickened cells (Figures 11–4a and 22–13). One species that produces such spores is *Coccidioides immitis*

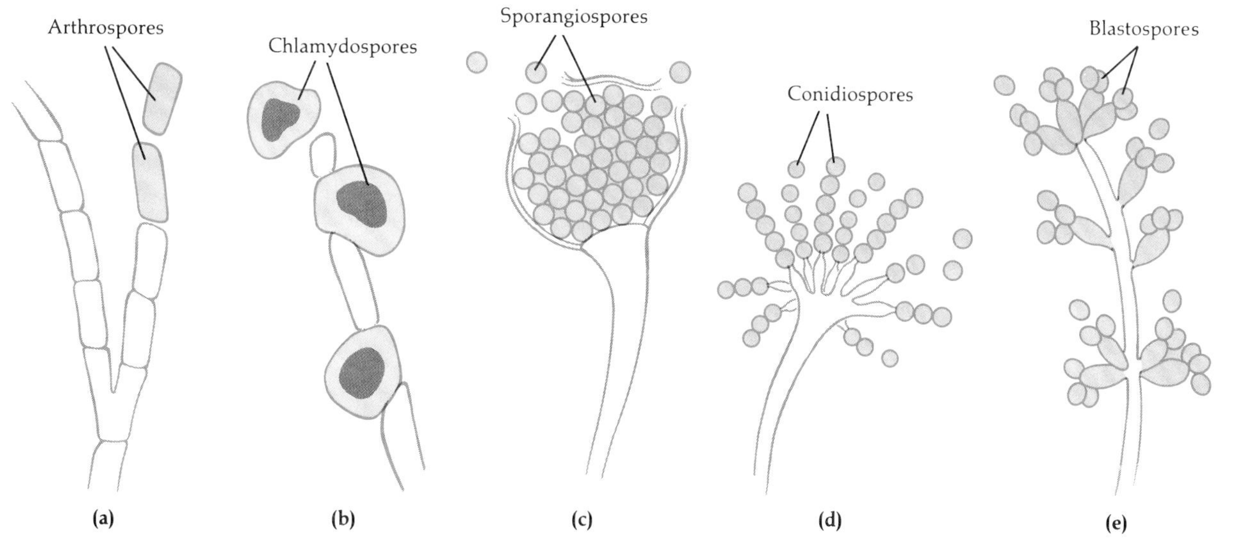

Figure 11–4 Representative asexual spores. **(a)** Fragmentation of hyphae results in formation of arthrospores. **(b)** Chlamydospores are thick-walled cells within the hyphae. **(c)** Sporangiospores are formed within a sporangium (spore sac). **(d)** Conidiospores are arranged in chains at the end of a conidiophore. **(e)** Blastospores are formed from buds of the parent cell.

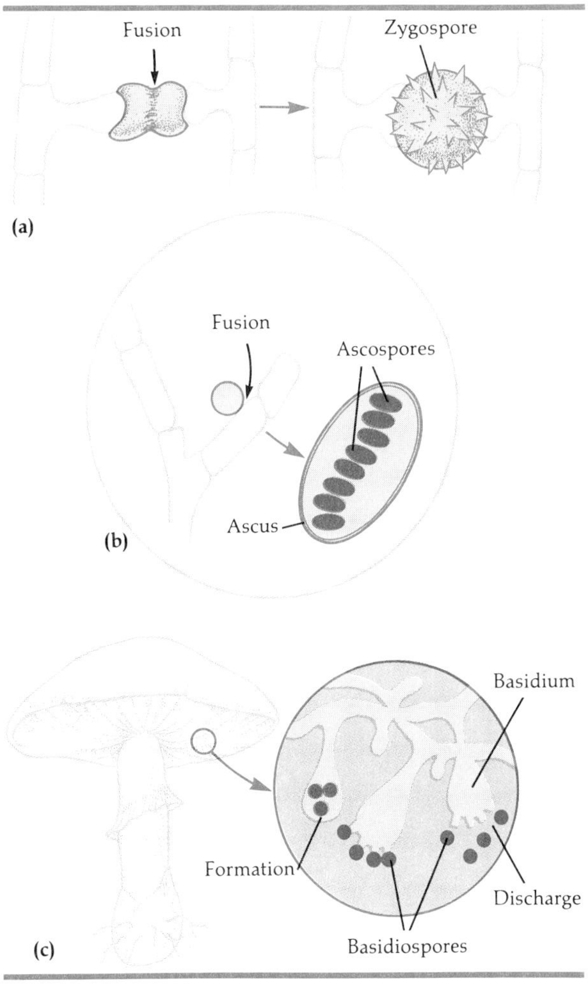

Figure 11–5 Sexual spores. **(a)** Zygospores, characteristic of the phylum Zygomycota, are produced from the fusion of two cells that are morphologically alike. **(b)** Ascospores, produced by the Ascomycota, are formed within an ascus. **(c)** Basidiospores, produced only by Basidiomycota, are formed on the tip of a pedestal called a basidium.

(kok-sid-ē-oi′dēz im′mi-tis). Another type of asexual spore is a **chlamydospore,** a thick-walled spore formed as segments within a hypha (Figure 11–4b). A fungus that produces chlamydospores is *Candida albicans* (kan′did-ä al′bi-kans) (see Microview 1–4C). A **sporangiospore** is an asexual spore formed within a sac **(sporangium)** at the end of an aerial hypha called a **sporangiophore.** The sporangium can contain hundreds of sporangiospores (Figure 11–4c). Such spores are produced by *Rhizopus.* A fourth principal type of asexual spore is a **conidiospore,** which is a unicellular or multicellular spore that is not enclosed in a sac (Figure 11–4d). Conidiospores are produced in a chain at the end of a **conidiophore.** Such spores are produced by *Penicillium.* A final type of asexual spore, a **blastospore,** consists of a bud coming off the parent cell (Figure 11–4e). Such spores are found in some yeasts.

Sexual Spores

A fungal sexual spore results from sexual reproduction consisting of three phases:

1. A haploid nucleus of a donor cell (+) penetrates the cytoplasm of a recipient cell (−).
2. The (+) and (−) nuclei fuse to form a diploid, zygote nucleus.
3. By meiosis, the diploid nucleus gives rise to haploid nuclei (sexual spores), some of which may be genetic recombinants.

Fungi produce sexual spores less frequently than they do asexual spores. Often, the sexual spores are produced only under special circumstances.

One kind of sexual spore is a **zygospore,** a large spore enclosed in a thick wall (Figure 11–5a). This type of spore results when the nuclei of two cells that are morphologically similar to each other fuse. Such spores are produced by fungi of the phylum Zygomycota. A second type of sexual spore is an **ascospore,** a spore resulting from the fusion of the nuclei of two cells that can be either morphologically similar or dissimilar. These spores are produced in a saclike structure called an **ascus** (Figure 11–5b). There are usually two to eight ascospores in an ascus. Such spores are produced by the phylum Ascomycota. A final type of sexual spore is the **basidiospore,** a spore formed externally on a base pedestal called a **basidium** (Figure 11–5c). There are usually four basidiospores per basidium. Basidiospores are produced only by the phylum Basidiomycota.

The sexual spores produced by fungi are a criterion used to group the fungi into several phyla. We will now examine the genera of medical importance contained within these phyla: Zygomycota, Ascomycota, Basidiomycota, and Deuteromycota.

MEDICALLY IMPORTANT PHYLA OF FUNGI

Zygomycota

The Zygomycota, or conjugation fungi, are saprophytic molds that have coenocytic hyphae. A common example is *Rhizopus nigricans,* the common black bread mold (Figure 11–6). The asexual spores of *Rhizopus* are sporangiospores. The dark sporangiospores inside the sporangium give *Rhizopus* its descriptive common name. When the sporangium breaks open, the sporangiospores are dispersed, and if they fall on a suitable medium, they will germinate into a new mold thallus. The sexual spores are zygospores.

Ascomycota

The Ascomycota, or sac fungi, include molds with septate hyphae and yeasts. They are called sac fungi because their sexual spores are ascospores produced in an ascus. Their asexual spores are usually conidiospores produced in long chains from the conidiophore. The arrangements of conidiospores in *Penicillium* (pe-ni-sil'lē-um) and *Aspergillus* (a-spėr-jil'lus) are shown in Figure 11–7 (see also Microview 1–4B). Conidia means "dust," and these spores freely detach from the chain at the slightest disturbance and float in the air like dust.

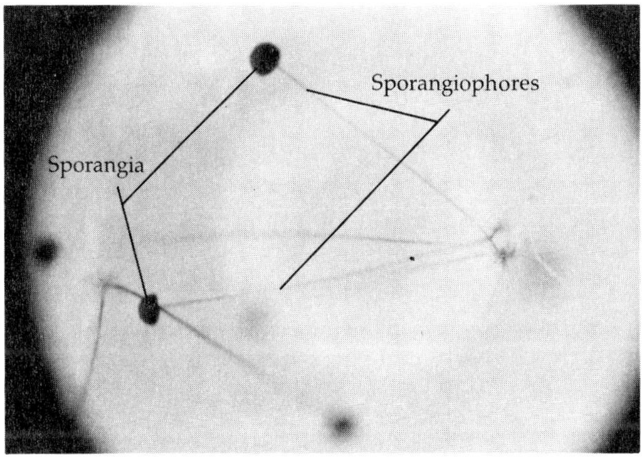

Figure 11–6 *Rhizopus nigricans.* Note the sporangia at the tops of the sporangiophores.

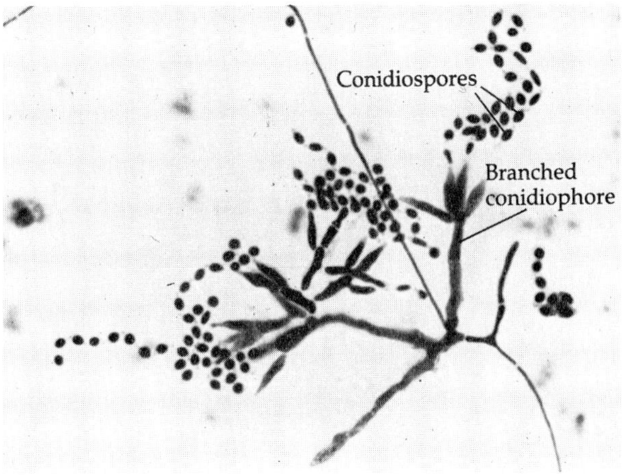

(a)

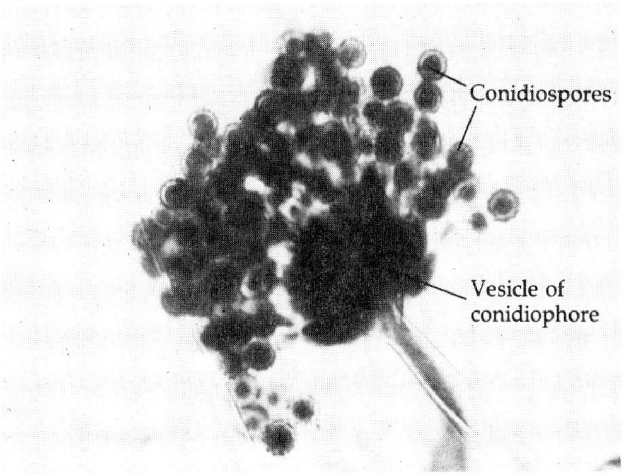

(b)

Figure 11–7 Conidiospores. The arrangement of conidiospores is useful in identification of fungi. **(a)** *Penicillium marneffei* produces conidiospores from a branched conidiophore. **(b)** Conidiospores of *Aspergillus tamarii* are produced from the enlarged terminal end (vesicle) of the conidiophore.

Basidiomycota

The Basidiomycota, or club fungi, also possess septate hyphae. The common name "club fungi" is derived from the shape of the basidium that bears the sexual basidiospores. Some of the basidiomycota produce asexual conidiospores. Representative basidiomycetes are shown in Figure 11–8 (see also Microview 1–3F and G).

Deuteromycota

The Deuteromycota are also known as the Fungi Imperfecti. These fungi are "imperfect" because no sexual spores of them have yet been found. Members of this phylum produce the asexual chlamydospores, arthrospores, and conidiospores; budding also occurs. Deuteromycota have septate hyphae.

Most of the pathogenic fungi are, or once were, classified as Deuteromycota. This phylum might be described as a "holding category" in which fungi are placed until sexual spores are observed and the fungus can be properly classified. Table 11–3 lists some well-known Fungi Imperfecti that have only recently been classified. Note that the generic names are changed with reclassification. When different species within a deuteromycete genus are observed to have morphologically different sexual spores, the species might be reclassified into two or more different genera, as was done with *Penicillium.*

(a) (b) (c)

Figure 11–8 Some representative basidiomycota. **(a)** *Calvulina cristata,* one kind of coral fungus. **(b)** *Agaricus augustus* is a popular produce item. **(c)** Puffballs of the species *Lycoperdon perlatum* (about $\frac{1}{2}$ actual size).

FUNGAL DISEASES

Characteristics of the medically important fungi are summarized in Table 11–4. Any fungal infection is referred to as a **mycosis.** Mycoses are classified as one of five groups, according to the level of infected tissue and mode of entry into the host. Mycoses are classified as systemic, subcutaneous, cutaneous, superficial, or opportunistic.

Systemic mycoses are fungal infections deep within the patient. Not restricted to any region of the body, they can affect a number of tissues and organs. Systemic, or deep, mycoses are usually caused by saprophytic fungi that live in the soil. Inhalation of spores is the route of transmission; these infections typically begin in the lungs and then spread to other body tissues. They are not contagious from another animal to human or from human to human. Two systemic mycoses, histoplasmosis and coccidioidomycosis, are discussed in Chapter 22.

Subcutaneous mycoses are fungal infections beneath the skin, and they are caused by saprophytic fungi that live in soil and on vegetation. Infection occurs by direct implantation of spores or mycelial fragments into a puncture wound in the skin.

Fungi that infect only the epidermis, hair, and nails are known as **dermatophytes,** and their infections are called **dermatomycoses** or **cutaneous mycoses** (see Figure 19–12). Dermatophytes secrete *keratinase,* an enzyme that degrades keratin. Keratin is a protein found in hair, skin, and nails. Infec-

Table 11–3 Several Genera of Reclassified Imperfect Fungi

Imperfect Name	Reclassified to Phylum	Perfect Name
Aspergillus	Ascomycota	*Sartorya, Eurotium, Emericella*
Blastomyces	Ascomycota	*Ajellomyces*
Candida	Ascomycota	*Pichia*
Cryptococcus	Basidiomycota	*Filobasidiella*
Histoplasma	Ascomycota	*Emmonsiella, Gymnoascus*
Microsporum	Ascomycota	*Nannizia*
Penicillium	Ascomycota	*Talaromyces, Carpenteles*
Petriellidium	Ascomycota	*Allescheria*
Trichophyton	Ascomycota	*Arthroderma*

tion is transmitted from person to person or animal to person by direct contact or by contact with infected hairs and epidermal cells (as from barber shop clippers or shower room floors) (see Microview 1–4A).

The fungi that cause **superficial mycoses** are localized along hair shafts and in superficial (surface) epidermal cells. These infections are prevalent in the tropics.

An **opportunistic** pathogen is generally harmless in its normal habitat but can become pathogenic in a host who is seriously debilitated or traumatized, or who is under treatment with broad-spectrum antibiotics or immunosuppressive drugs. A person's normal flora can become opportunistic pathogens under the above conditions.

Mucormycosis is an opportunistic mycosis caused by *Rhizopus* and *Mucor;* the infection occurs mostly in patients with ketoacidosis resulting from diabetes mellitus, leukemia, or treatment with immunosuppressive drugs. **Aspergillosis** is also an opportunistic mycosis; it is caused by *Aspergillus* (see Figure 11–2). This disease occurs in persons who have debilitating lung diseases or cancer and have inhaled *Aspergillus* spores (see Microview 1–4B). The mycosis **candidiasis** (see Figure 19–13) is most frequently caused by *Candida albicans* and may occur as vulvovaginal candidiasis during pregnancy. **Thrush,** a mucocutaneous candidiasis, is an inflammation of the mouth and throat; it frequently occurs in newborns.

UNDESIRABLE ECONOMIC EFFECTS OF FUNGI

Some of the beneficial industrial uses of fungi are discussed in Chapter 26. But now let us discuss the less desirable economic effects.

Fungi have the following nutritional and physiological characteristics that relate them to the human economy:

1. Fungi usually grow better in an acidic pH (5.0), which is too acidic for the growth of most common bacteria.
2. Most molds are aerobic.
3. Most fungi are more resistant to osmotic pressures than bacteria are; most fungi are therefore able to grow in high sugar or salt concentrations.
4. Fungi are capable of growing on substances with a very low moisture content, generally too low to support the growth of bacteria.
5. Fungi require somewhat less nitrogen for growth than do bacteria, and fungi are nutritionally very adaptable and efficient. These characteristics are important factors in fungi being able to grow on such unlikely substrates as painted walls or shoe leather.

As most of us have observed, mold spoilage of fruits, grains, and vegetables is relatively common, but bacterial spoilage of such foods is not. There is little moisture on the unbroken surfaces of such foods, and the interiors of fruits are too acidic for many bacteria to be able to grow there. Jams and jellies also tend to be acidic, and what's more, they have a high osmotic pressure from the sugars they contain. These factors all discourage bacterial growth but readily support the growth of mold. A paraffin layer on top of a jar of homemade jellies will help deter mold growth because molds are aerobic and the paraffin layer keeps out the oxygen. On the other hand, foods like fresh meats are such good substrates for bacterial growth that bacteria not only outgrow molds, but will actively suppress mold growth in such foods.

The ability of fungi to grow at low moisture levels is of particular importance in their role as plant pathogens. Bacterial plant pathogens are far

Table 11–4 Characteristics of Some Parasitic Fungi

Phylum	Growth Characteristics	Asexual Spore Types	Human Pathogens
Zygomycota	Nonseptate hyphae	Sporangiospores	*Rhizopus*, *Mucor*
Ascomycota	Septate hyphae	Conidiospores	*Allescheria boydii*
	Dimorphic	Conidiospores	*Aspergillus** *Blastomyces dermatitidis** *Histoplasma capsulatum**
	Septate hyphae, strong affinity for keratin	Conidiospores, arthrospores	*Microsporum** *Trichophyton**
Basidiomycota	Septate hyphae includes the fleshy fungi (mushroom), rust and smuts, and plant pathogens; yeastlike encapsulated cells	Conidiospores	*Cryptococcus neoformans**
Deuteromycota	Septate hyphae	Conidiospores Chlamydospores	*Epidermophyton* *Cladosporium werneckii*
	Dimorphic	Conidiospores Arthrospores	*Sporothrix schenkii* *Coccidioides immitis*
	Yeastlike, pseudohyphae	Chlamydospores	*Candida albicans**
	Dimorphic	Chlamydospores, arthrospores, conidiospores	*Paracoccidioides brasiliensis*
	Septate hyphae, produce melaninlike pigments	Conidiospores	*Phialophora* *Fonsecaea*
	Yeastlike pseudohyphae	Arthrospores	*Trichosporon*

*Imperfect name. Refer to Table 11–3 for current classification and names.

less common than fungal pathogens are. In fact, the fungus that caused the great potato blight in Ireland during the early 1800s (*Phytophthora infestans*) was one of the first microorganisms to be associated with a disease. Other fungal toxins are discussed in Chapter 14.

The spreading chestnut tree, of which the poet wrote, no longer grows in this country except in a few widely isolated sites. An important fungal blight killed essentially all the trees. First seen in the United States in 1904, the ascomycete *Endothia parasitica* (en-dō′thē-ä par-ä-si′ti-kä) was introduced from China. The tree roots live and put forth shoots regularly, but the shoots are just as regularly killed by the fungus. Another devastating fungal plant disease that was also imported to this country is Dutch elm disease, caused by *Ceratocystis ulmi* (sē-rä-tō′sis-tis ul′mē). Carried from tree to tree by a bark beetle, the fungus blocks the afflicted tree's circulation.

ALGAE

Algae are photosynthetic autotrophs. Using the energy produced in photosynthesis, they convert the carbon dioxide in the atmosphere into carbohydrates. Oxygen is a byproduct of photosynthesis. Chlorophyll *a* and accessory pigments involved in photosynthesis are responsible for the distinctive colors of many algae. Some algae exist as single cells; others exist as colonies of hundreds or thousands of cells. Algae are classified into one of two kingdoms according to their structures, pigments, and other qualities.

Table 11–4 Characteristics of Some Parasitic Fungi *continued*

Clinical Notes	Type of Mycosis	Habitat	Refer to Chapter
Opportunistic pathogen	Systemic	Ubiquitous	22
Opportunistic pathogen	Systemic	Ubiquitous	22
Primary cause of maduromycosis	Subcutaneous	Soil	19
Opportunistic pathogen	Systemic	Ubiquitous	22
Inhalation	Systemic	Unknown	22
Inhalation	Systemic	Soil	22
Tinea capitis	Cutaneous	Soil, animals	19
Tinea pedis	Cutaneous	Soil, animals	19
Inhalation	Systemic	Soil, bird feces	20
Tinea cruris, tinea unguium	Cutaneous	Soil, humans	19
Tinea nigra	Superficial	Ubiquitous	19
Puncture wound	Subcutaneous	Soil	—
Inhalation	Systemic	Soil	22
Opportunistic pathogen	Cutaneous, systemic, mucocutaneous	Human normal flora	19, 24
Inhalation	Systemic	Presumably soil	—
Chromomycosis	Subcutaneous	Soil, plant debris	19
Chromomycosis	Subcutaneous	Soil, plant debris	19
White piedra	Superficial	Soil (?), humans	19

Dinoflagellates (Figure 11–9a) are unicellular, **planktonic** or free-floating, algae. Their cell walls consist of many individual plates made of cellulose and silica. Certain marine dinoflagellates produce neurotoxins that cause paralytic shellfish poisoning. This disease spreads to humans when large numbers of dinoflagellates are eaten by mollusks—such as mussels or clams—which, in turn, are eaten by humans. **Euglenoids** (Figure 11–9b) are unicellular, flagellated algae with a semi-rigid cell membrane. Some euglenoids are facultative chemoheterotrophs and are frequently studied with protozoans.

Diatoms (Figure 11–10a) are unicellular or filamentous algae with complex cell walls, consisting of pectin and a layer of silica. The two parts of the wall fit together as do the halves of a Petri plate. The distinctive patterns of the walls are a useful tool in diatom identification. Diatoms store the energy captured through photosynthesis in the form of oil. Much of the world's petroleum was formed from diatoms that lived over 300 million years ago.

The **brown algae** or **kelp** (see Figure 11–9c) are macroscopic; some reach lengths of 50 meters. Most brown algae are found in coastal waters (see Microview 1–3C). Brown algae have a phenomenal growth rate. Some grow at rates exceeding 20 cm per day and therefore can be harvested regularly. *Algin*, a thickener used in many foods, such as ice cream and cake decorations, is extracted from their cell walls. Algin is also used in the production of a wide variety of non-food goods, including rubber tires and hand lotion. Most **red algae** (see Figure

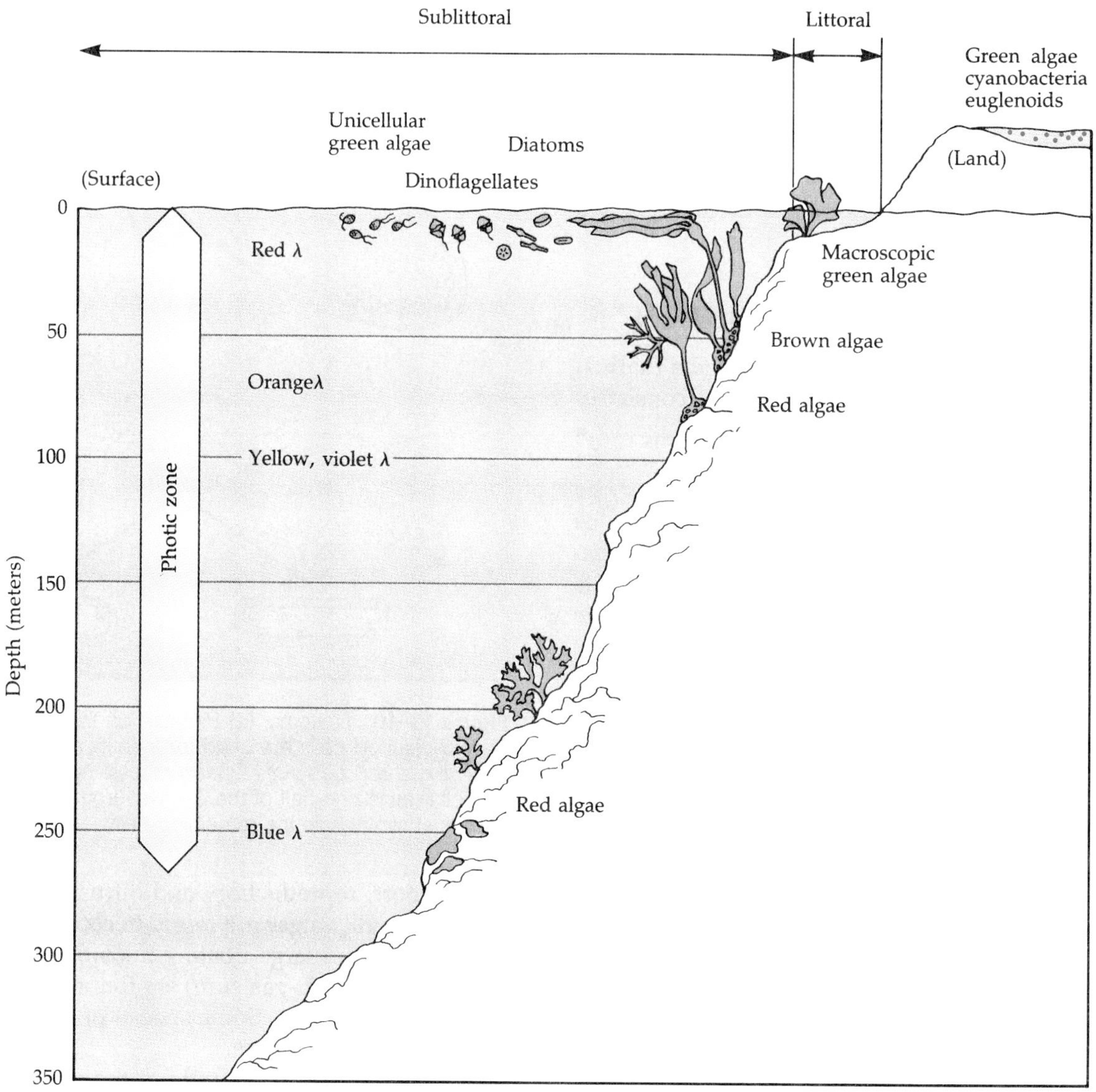

Figure 11–11 Algal habitats. Unicellular and filamentous algae can be found on land; they exist as plankton in aquatic environments. Multicellular green, brown, and red algae require a suitable attachment site, adequate water for support, and light of the appropriate wavelengths.

perature cause fluctuations in algal populations; periodic increases in numbers of planktonic algae are called *blooms*. Blooms of dinoflagellates are responsible for seasonal *red tides*. Blooms of a few species indicate that the water in which they grow is polluted, because these algae thrive in high concentrations of organic materials that exist in sewage or industrial wastes. When algae die, the decomposition of the large numbers of cells associated with an algal bloom depletes the level of dissolved oxygen in the water. (This phenomenon is discussed in Chapter 25.)

LICHENS

A **lichen** is a combination of a green alga (or cyanobacterium) and a fungus. The two organisms exist in a *mutualistic* relationship, in which each partner benefits. The lichen is very different from either the alga or fungus growing alone, and if the partners are separated, the lichen no longer exists. There are approximately 20,000 species of lichens occupying quite diverse habitats. Because lichen can inhabit areas in which neither fungi nor algae could survive alone, lichens are often the first life forms to colonize on newly exposed soil or rock. Lichens secrete organic acids that chemically weather rock, and accumulate nutrients needed for plant growth. Lichens are some of the slowest growing organisms on earth.

Lichens can be grouped into three morphologic categories. **Crustose** lichens (Figure 11–12a and Microview 1–3D) grow flush or encrusting onto the substratum, **foliose** lichens (Figure

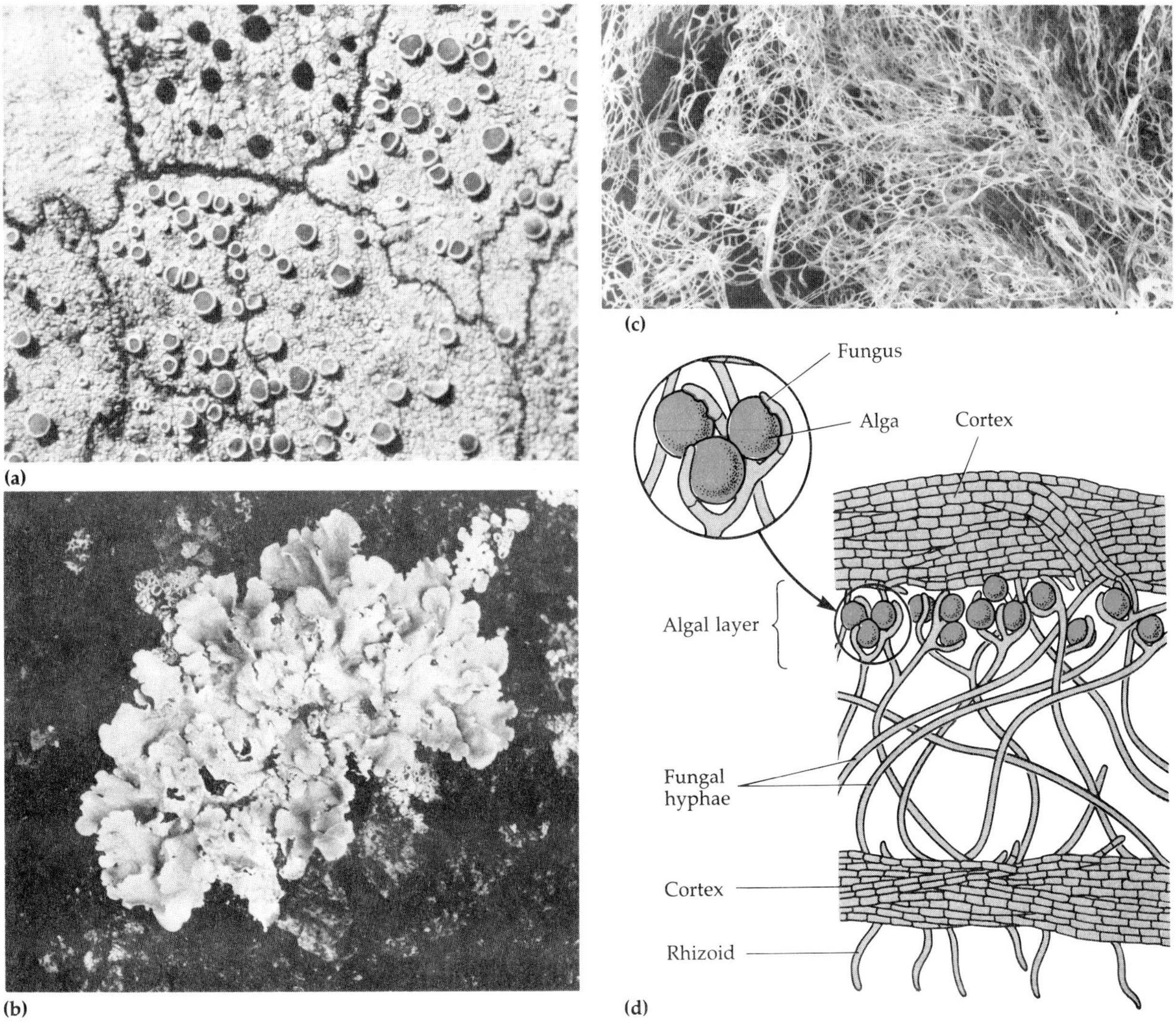

Figure 11–12 Lichens. **(a)** Crustose. **(b)** Foliose. **(c)** Fruticose. **(d)** Cross-section of a lichen thallus. The medulla is composed of fungal hyphae and surrounds the algal layer. The protective cortex is a layer of irregularly organized fungal hyphae that covers the surface and sometimes the bottom of the lichen.

11–12b) are more leaflike, and **fruticose** lichens (Figure 11–12c) have fingerlike projections. The lichen's thallus, or body, forms when fungal hyphae grow around algal cells. After incorporation into a lichen thallus, the alga continues to grow by mitosis in the algal layer of the lichen (Figure 11–12d), and the growing hyphae can incorporate new algal cells. In a lichen, the fungus reproduces sexually.

When the algal partner is cultured separately in vitro, about 1 percent of the carbohydrates produced during photosynthesis are released into the culture medium; however, when the alga is associated with a fungus, the algal plasma membrane is more permeable and up to 60 percent of the products of photosynthesis are released to the fungus or are found as end products of fungus metabolism. It is clear that the fungus benefits from its association. The alga, which appears to be giving up valuable nutrients, is compensated by the protection from desiccation and the holdfast that the fungus provides.

Populations of lichens readily incorporate cations (positively charged ions) into their thalli. The concentrations and types of cations in the atmosphere can be determined by chemical analyses of lichen thalli. The presence or absence of species that are quite sensitive to pollutants can be used to ascertain air quality.

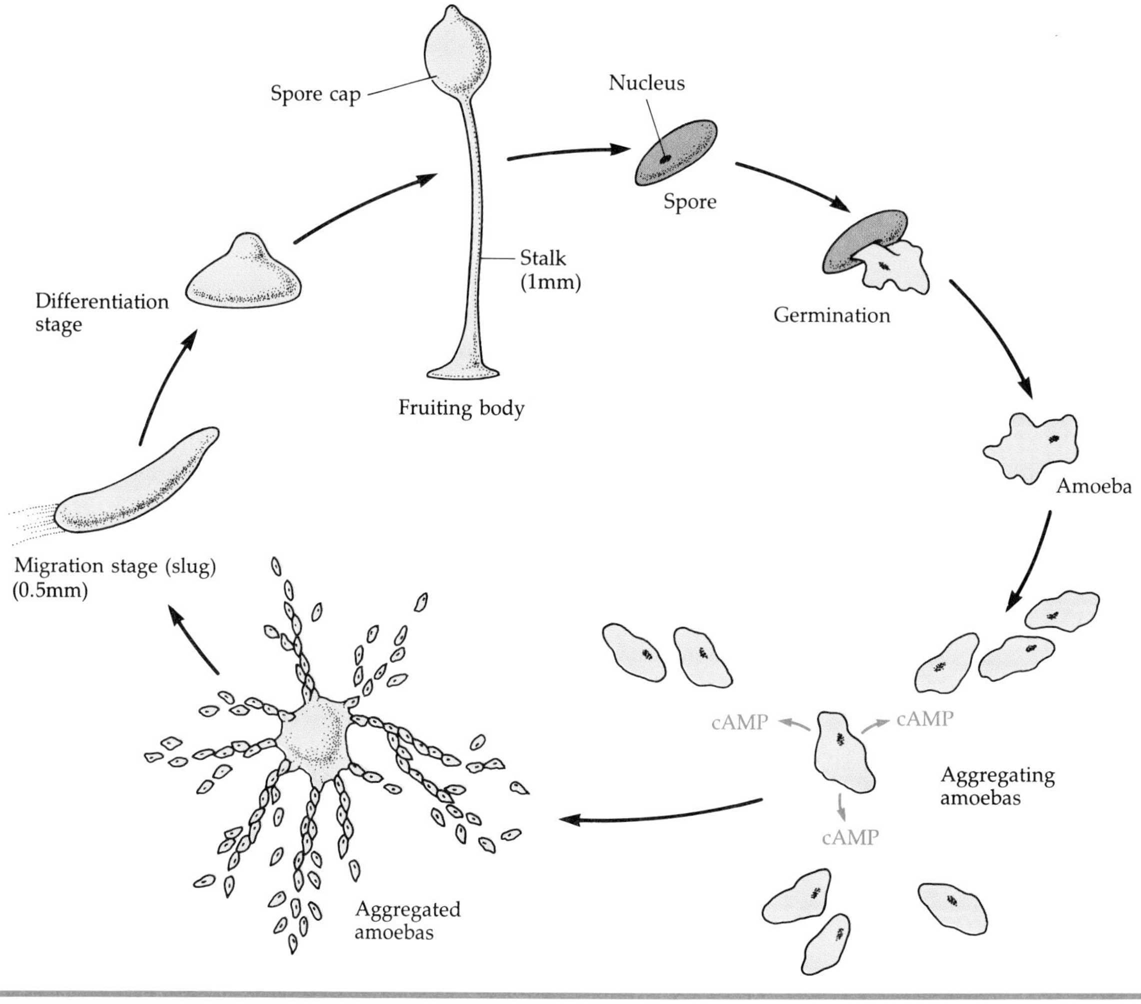

Figure 11–13 Generalized life cycle of a cellular slime mold.

SLIME MOLDS

Slime molds have both fungal and animal characteristics and have membrane-bounded nuclei. Thus they are classified as protists. **Cellular slime molds** (Figure 11–13) are typical eucaryotic cells that resemble amoebas. The amoeboid cells ingest other microorganisms and bacteria by phagocytosis. Cellular slime molds are of interest to biologists who study cellular migration and aggregation. When conditions are unfavorable, large numbers of amoeboid cells aggregate to form a single structure. This aggregation happens because some individual amoebas produce cyclic AMP (cAMP) toward which the other amoebas migrate. The aggregated amoebas are enclosed in a slimy sheath called a *slug*. The slug migrates as a unit toward light. After a period of hours, the slug ceases to migrate and becomes vertically oriented. Some of the amoeboid cells form a stalk; others swarm up the stalk to form a spore cap—most of these differentiate into spores. When spores are released under favorable conditions, they germinate to form single amoebas.

Plasmodial (acellular) slime molds belong to a separate phylum. A plasmodial slime mold exists as a mass of protoplasm with many nuclei (it is multinucleated). This mass of protoplasm is called a **plasmodium** (see Microview 1–3E).The entire plasmodium moves as a giant amoeba (see page 000). It engulfs organic debris and bacteria. In 1973, a pulsating red blob was discovered in a Dallas resident's yard. News media claimed that a "new life form" had been found. Biologists knew it was an unusually large plasmodial slime mold. Plasmodial slime molds were scientifically reported in 1729. Biologists have found that musclelike proteins account for their movement. When plasmodial slime molds are grown in laboratories a phenomenon called **protoplasmic streaming** is observed. During protoplasmic streaming the protoplasm within the plasmodium moves and changes both its speed and direction so that oxygen and nutrients are evenly distributed.

The plasmodium grows as long as there is enough food and moisture. When either is in short supply, the plasmodium separates into many groups of protoplasm; each of these groups forms a stalked sporangium in which spores, a resistant, resting form of the slime mold, develop (Figure 11–14). Nuclei within these spores undergo meiosis and form uninucleate haploid amoebalike cells. When conditions improve, these spores germinate and develop into a multinucleate plasmodium.

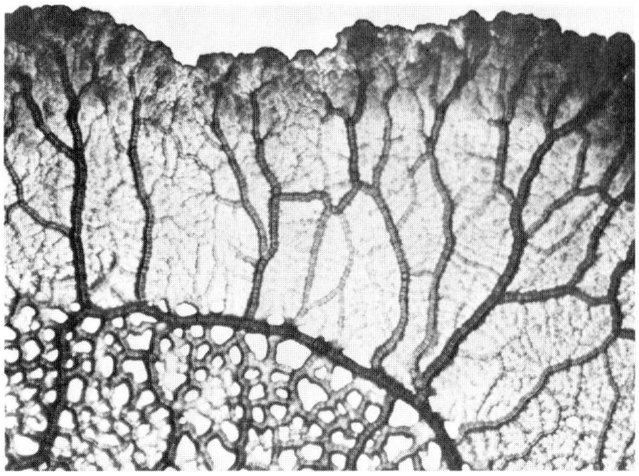

Figure 11–14 Plasmodium of an acellular plasmodial slime mold.

PROTOZOANS

Protozoans are one-celled, eucaryotic organisms that belong to the kingdom Protista. Recall what you have learned about eucaryotic cell structure from Chapter 4. Among the protozoans are many variations on this cell structure, as we shall see. Protozoans inhabit water and soil, and feed upon bacteria and small particulate nutrients. Some protozoans are part of the normal flora of animals. For example, the gastrointestinal tracts of termites and cows contain protozoans that help them digest cellulose. Of the nearly 20,000 species of protozoans, relatively few cause disease.

Protozoans are classified into phyla on the basis of their means of motility. We will restrict our study to the four phyla of protozoans that include disease-causing species. One such phylum, the **Sarcodina,** consists of amoebas (Figure 11–15a). The amoebas move by extending usually blunt, lobelike projections of the cytoplasm called **pseudopods.** Any number of pseudopods can flow from

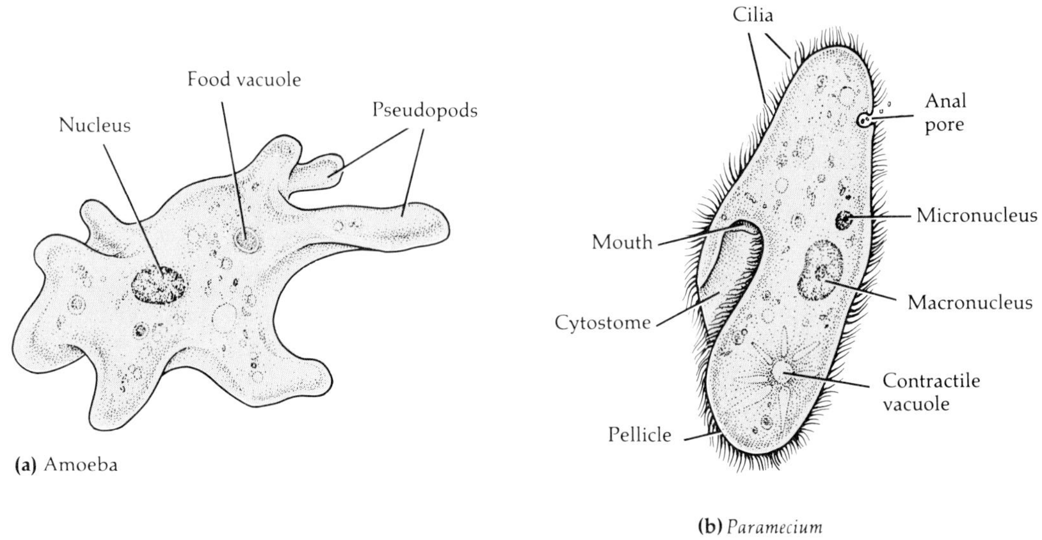

Figure 11–15 Examples of protozoans. **(a)** Amoeba. To move and to engulf food, amoebas extend cytoplasmic portions called pseudopods. Once surrounded by the pseudopods, the food is in a food vacuole. **(b)** *Paramecium*, a ciliate. Rows of cilia cover the cells of ciliates. The cilia are moved in unison for locomotion and to bring food particles to the protozoan. *Paramecium* has specialized structures for ingestion (mouth), elimination of wastes (anal pore), and regulation of osmotic pressure (contractile vacuoles). The macronucleus is involved with protein synthesis and other ongoing cellular activities. The micronucleus functions in sexual reproduction.

one side of the amoeba cell, and the rest of the cell will flow toward the pseudopods. Another phylum, the **Mastigophora,** or flagellates, possess flagella. Flagella are capable of whiplike movements that pull the cell through the medium (see Figure 11–18). Most flagellates have one or two flagella, but some of the parasitic species have as many as eight. Species in the phylum **Ciliata** have projections called cilia that are similar to but shorter than flagella. The cilia are in a precise arrangement over the cell (Figure 11–15b) and move to propel the cell through its medium. The species in the fourth phylum we will study, the **Sporozoa,** are incapable of independent movement.

PROTOZOAN BIOLOGY

Nutrition

Protozoans are anaerobic heterotrophs. One chlorophyll-containing flagellate, *Euglena* (ū-glē′nä) (see Figure 11–9b), can also grow in the dark as a heterotroph and is often included with the protozoans.

All protozoans live in areas with a large supply of water. Some protozoans transport food across the cell membrane. However, some have a protective covering called the **pellicle** and require specialized structures to take in food. The ciliates take in food by waving their cilia toward a mouthlike opening called the **cytostome.** The amoebas engulf food by surrounding it with pseudopods and phagocytizing it. In the protozoans, digestion takes place in membrane-bounded **vacuoles,** and waste might be eliminated through the cell membrane or through a specialized **anal pore.**

Reproduction

Protozoans reproduce asexually by fission, budding, or schizogony. **Schizogony** is *multiple* fission. The nucleus undergoes multiple divisions before the cell divides. After many nuclei are formed, a small portion of cytoplasm concentrates around each nucleus, and separation of the single cell into daughter cells follows.

Sexual reproduction has been observed in some protozoans. The ciliates reproduce sexually by **conjugation** (Figure 11–16), which is very different from the bacterial process of the same name. During protozoan conjugation, two cells fuse and a haploid nucleus (the micronucleus) from each migrates to the other cell. That haploid nucleus fuses with the haploid nucleus within the cell. The parent cells separate, each now a fertilized cell. When the cells later divide, they produce daughter cells with recombined DNA. Some protozoans produce **gametes** or **gametocytes,** haploid sex cells. In reproduction, two gametes fuse to form a diploid zygote.

Encystment

Under certain adverse conditions, some protozoans are capable of producing a protective capsule called a **cyst.** A cyst permits the organism to survive when food, moisture, or oxygen are lacking, when temperatures are not suitable, or when toxic chemicals are present. A cyst also enables a parasitic species to survive outside a host. This is important because parasitic protozoans can have life cycles that involve more than one host.

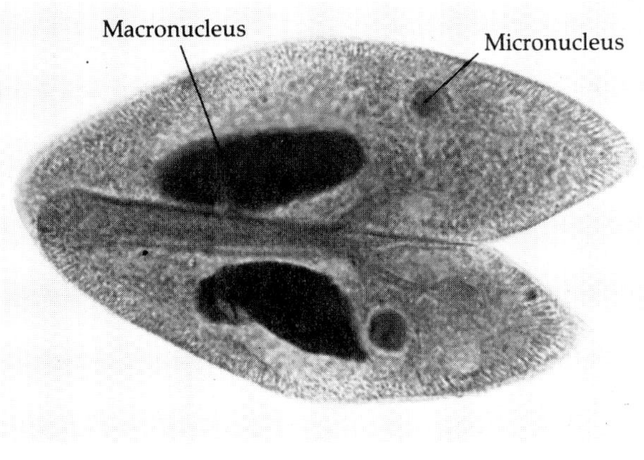

Figure 11–16 *Paramecium* conjugation. Sexual reproduction in ciliates is by conjugation. Each cell has two nuclei, a micronucleus and a macronucleus. The micronucleus is haploid and specialized for conjugation. One micronucleus from each cell will migrate to the other cell during conjugation. (Courtesy of Carolina Biological Supply Company.)

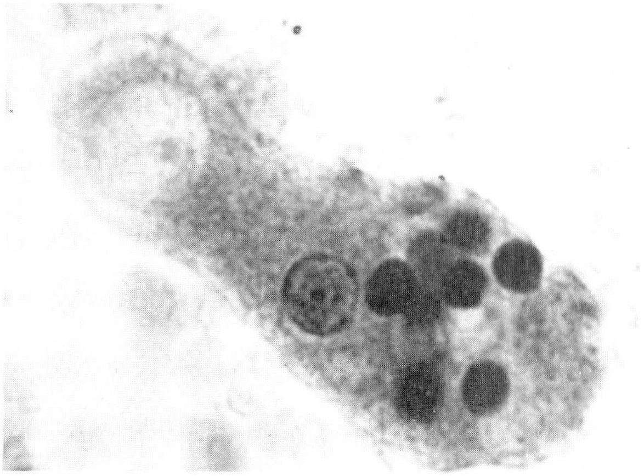

Figure 11–17 *Entamoeba histolytica* in the vegetative form. The dark circles inside the amoeba are engulfed red blood cells from its human host.

We will now examine representatives of the medically important phyla of protozoans.

MEDICALLY IMPORTANT PHYLA OF PROTOZOANS

Sarcodina

A well-known parasitic amoeba is *Entamoeba histolytica* (en-tä-mē'bä his-tō-li'ti-kä), the causative agent of amoebic dysentery (Figure 11–17). *E. histolytica* is transmitted from human to human through cysts passed out with feces and ingested by the next host.

Mastigophora

As previously mentioned, members of the phylum Mastigophora move by means of flagella. Their outer membrane is a tough, flexible pellicle. Flagellates are typically spindle shaped, with flagella projecting from the front end. Food is ingested through a cytostome. Some flagellates have an **undulating membrane,** which appears to consist of highly modified flagella (see Figure 11–18).

An example of a flagellate that is a human parasite is *Giardia lamblia* (jē-är'dē-ä lam'lē-ä). (See Figures 11–19 and 23–15, which shows the vegetative form, or **trophozoite,** of *G. lamblia,* as well as

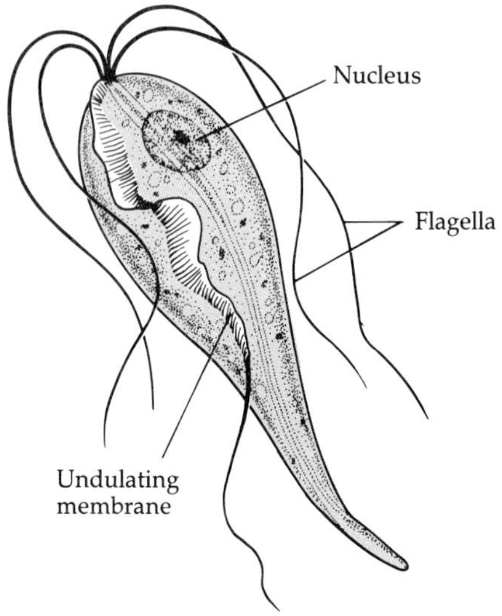

Figure 11–18 *Trichomonas vaginalis.* This flagellate is the cause of urinary and genital tract infections. It has a small undulating membrane. This flagellate does not have a cyst stage.

the cyst stage.) The parasite is found in the small intestine of humans and other mammals. It is passed out of the intestine and survives in a cyst before being ingested by the next host. Diagnosis of giardiasis, the disease caused by *G. lamblia,* is based on identification of cysts in feces.

Another parasitic flagellate is *Trichomonas vaginalis* (trik-ō-mōn'as va-jin'al-is), shown in Figure 11–18. *T. vaginalis* does not have a cyst stage and must be transferred from host to host quickly before desiccation occurs. *T. vaginalis* is found in the vagina and in the male urinary tract. It is usually transmitted by sexual intercourse but can be transmitted by toilet facilities or towels.

The hemoflagellates are transmitted by the bites of bloodsucking insects and are found in the circulatory system of the bitten host. To survive in this viscous fluid, hemoflagellates have long, slender bodies and an undulating membrane. The genus *Trypanosoma* (tri-pan-ō-sō'mä) includes the species that cause African sleeping sickness (e.g. *T. b. gambiense,* which is transmitted by the tsetse fly) (see box and Microview 1–4E), and Chagas' disease (see Box). *T. cruzi* (see Figure 21–16), the causative

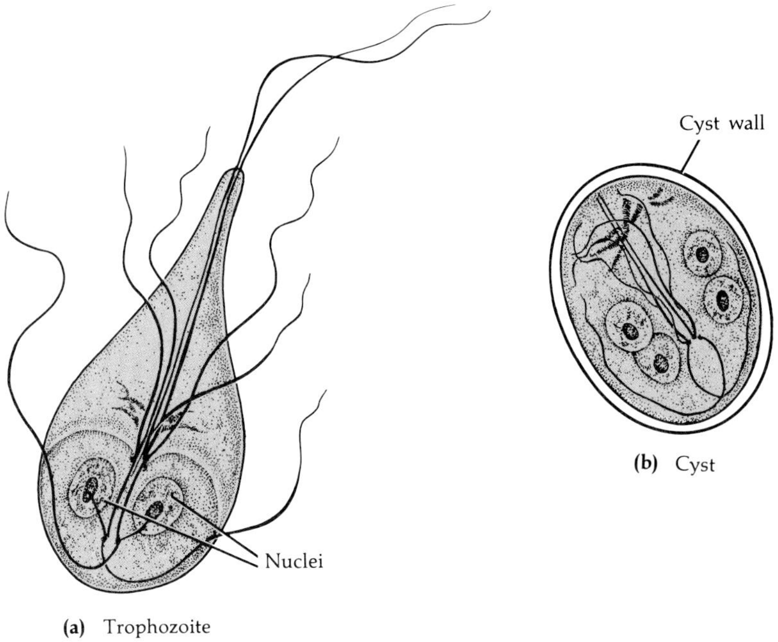

Figure 11–19 *Giardia lamblia.* **(a)** The trophozoite of this intestinal parasite has eight flagella and two prominent nuclei, giving it a distinctive appearance. **(b)** The cyst provides protection in the environment before being ingested by a new host.

agent of Chagas' disease, is transmitted by the "kissing bug." After entering the insect, the trypanosome rapidly multiplies by fission. If the insect then defecates while biting a human, the insect can release trypanosomes that can contaminate the bite wound.

Ciliata

The only ciliate that is a human parasite is *Balantidium coli* (bal-an-tid'ē-um kō'lē), the causative agent of a severe though rare type of dysentery. When cysts are ingested by the host, they enter the colon, into which the trophozoites are released. The trophozoites feed on bacteria and fecal debris as they multiply, and cysts are passed out with feces (see Figure 23–16).

Sporozoa

Sporozoans are not motile in their mature forms and are obligate intracellular parasites. They have complex life cycles that ensure their survival and transmission from host to host. An example of a sporozoan is *Plasmodium* (plaz-mō'dē-um), the causative agent of malaria. A vaccine against malaria is now being investigated.

Plasmodium grows by schizogony in human red blood cells (Figure 11–20), in which it com-

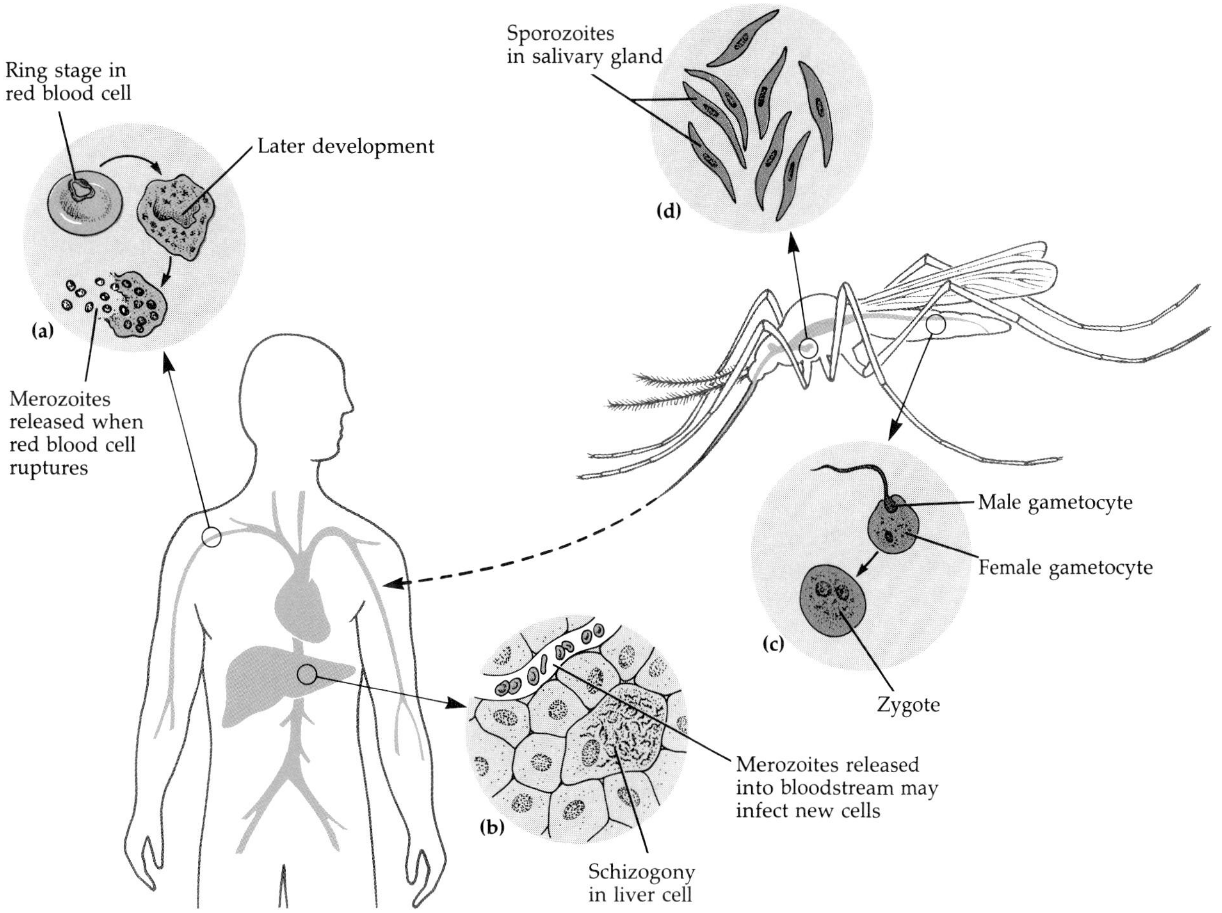

Figure 11–20 Life cycle of *Plasmodium vivax.* Asexual reproduction, schizogony, of the merozoites takes place **(a)** in the red blood cells or **(b)** in the liver of a human host. **(c)** Sexual reproduction of the parasite occurs in the intestine of an *Anopheles* mosquito after the mosquito has ingested gametocytes. **(d)** Sporozoites resulting from sexual reproduction migrate to the salivary gland. From there, they will be injected into the mosquito's next host.

Microbiology in the News

Parasite Outwits the Immune System, Intrigues Scientists

"A leopard," the old saying goes, "can't change its spots." The fearsome hunter of the African forest could never hide its identity by altering its coat. But a tiny yet far more menacing native of Africa can do just that. The trypanosome, a protozoan parasite that causes the devastating disease known as sleeping sickness (see photo), can rapidly change its protein coat to disguise itself from the immune system of its host. In the last few years this strange creature has attracted a great deal of attention from molecular biologists. Though they hope their research will ultimately lead to a vaccine or drug against sleeping sickness, these scientists find the parasite fascinating in its own right. Its remarkable defense mechanisms may shed light on one of the most pressing questions of molecular genetics—how individual genes express themselves.

An Unflagging Invader

Sleeping sickness, or African trypanosomiasis, is spread by the tsetse fly, which hosts the African trypanosome (*Trypanosoma sp.*). When an infected fly bites a human or animal, trypanosomes enter the mammalian host's bloodstream and multiply, eventually invading the central nervous system. The resulting disease is one of the most terrible known. The victim suffers increasing fever, lethargy, and mental deterioration, finally lapsing into a coma. Death is nearly certain without treatment. And while existing drug therapies improve the prognosis somewhat, they are not highly effective and have severe side effects.

In the broad belt across central Africa that is home to the tsetse fly, the impact of sleeping sickness has been enormous. During one epidemic around the turn of the century, two-thirds of the population of the north shore of Lake Victoria died. Today the disease affects about a million people, with 20,000 new cases appearing each year. And no livestock can be kept throughout this vast region, encompassing over a third of the African continent.

What makes the trypanosome so deadly, scientists have found, is its ability to rebound from an attack by the host's immune system. When a foreign organism invades a human or animal, the host's immune system makes antibodies against it. These antibodies are produced in response to specific antigens (large, antibody-provoking molecules) on the invader's surface, which they recognize and attack. In the case of trypanosomes, the antigens are the protein molecules that form its coat.

In the early stages of sleeping sickness, when trypanosomes are circulating in the blood, the host produces enormous numbers of antibodies, destroying over 99 percent of the invaders within a week. With most pathogens this response would end the disease before it became serious—but not with trypanosomes. For the remaining 1 percent have changed their coat proteins so that the antibodies no longer recognize them. These survivors multiply in the blood while the body manufactures new antibodies against them, and eventually they too are destroyed—except the 1 percent that have changed coats again. Each time the host produces new antibodies, the parasite produces new antigens, until finally the parasite penetrates the nervous system and kills its host.

Probing the Mystery

Molecular biologists, intrigued by the trypanosome's quick-change artistry, have been probing the mysteries of its gene expression. The trypanosome's coat is made of a single protein, called *variable surface glycoprotein,* or VSG, so it can

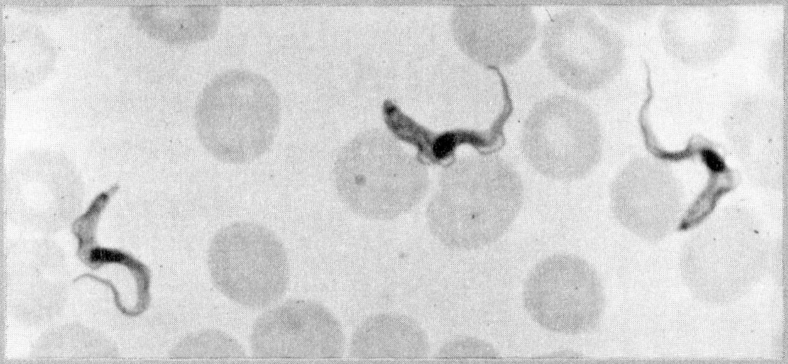

be entirely transformed by the alteration of a single gene. The trypanosome has genes for many variants of VSG, and these genes are switched on and off in quick succession.

An individual trypanosome probably has several hundred to several thousand VSG genes, but only those on the ends of chromosomes are expressed. Most VSG genes are normally located in the interior portions of chromosomes, so when these genes are to be transcribed they are first duplicated and moved to an "expression site" at a chromosome end. No one yet knows what determines which gene is to be transcribed, or how the moving is accomplished. Complicating the picture further, genes normally found at ends of chromosomes are sometimes not expressed, which means that additional on-and-off mechanisms are at work.

Researchers are excited by the possibility that understanding these mechanisms will lead to a better understanding of gene expression in all organisms. The very fact that trypanosomes alter so rapidly makes them excellent subjects for study. Most organisms are so stable at the molecular level that there is little to observe, but trypanosomes turn their genes on and off before the observers' eyes.

A Grim Future?

Will understanding how trypanosomes outwit the immune system lead to development of a vaccine for sleeping sickness? The prospects seem poor. To eliminate an organism that can interchange hundreds or thousands of surface antigens, most of which appear in seemingly random order, vaccination might have to produce hundreds or thousands of kinds of antibodies. The discovery that the same 12 to 14 antigens always appear in the first five days or so of an infection provides a slim ray of hope; perhaps producing antibodies against all of these simultaneously could halt the disease in its earliest stage. But it seems quite possible that some of the trypanosomes would simply produce other antigens and survive the onslaught.

What about new drug therapies? Trypanosomes are remarkable not only for their mechanisms of gene expression, but for a host of unique metabolic pathways. They are very ancient organisms and their relationship to most other living things is very distant, so it is not surprising that they are distinctive in many ways. But just because they are unique, there is enormous interest in studying trypanosomes, and metabolic research could prompt the development of specific antitrypanosomal drugs. Pharmaceutical companies, however, have shown little interest in developing drugs against parasites, since parasitic infections are common only where most people are very poor. The scourge of a continent is proving its worth as a subject of basic research—but the end of the affliction is not in sight.

Table 11–5 Some Representative Parasitic Protozoans

Phylum	Human Pathogens	Distinguishing Features	Disease	Source of Human Infections	See Figure	Refer to Chapter
Sarcodina (amoebas)	*Entamoeba histolytica*	Pseudopods	Amoebic dysentery	Fecal contamination of drinking water	11–17	23
	Naegleria fowleri	Some flagellated forms	Micro-encephalitis	Water in which people swim	—	20
Mastigophora (flagellates)	*Giardia lamblia*	Two nuclei, eight flagella	Giardial dysentery	Fecal contamination of drinking water	11–19	23
	Trichomonas vaginalis	No encysting stage	Urethritis; vaginitis	Contact with vaginal/urethral discharge	11–18	24
	T.b. gambiense, T.b. rhodesiense	Undulating membrane	African trypano-somiasis	Bite of tsetse fly	—	20
	Trypanosoma cruzi		Chagas' disease	Bite of *Triatoma* (kissing bug)	—	21
Ciliata	*Balantidium coli*	Only parasitic ciliate of humans	Balantidial dysentery	Fecal contamination of drinking water	23–16	23
Sporozoa	*Cryptosporidium*	Complex life cycles may require more than one host	Diarrhea	Cows, other animals	—	23
	Plasmodium		Malaria	Bite of *Anopheles* mosquito	11–20	21
	Pneumocystis carinii		Pneumonia	Other animals	—	22
	Toxoplasma gondii		Toxoplas-mosis	Cats, other animals; congenital	21–15	21

pletes its asexual cycle. The young trophozoite looks like a ring in which the nucleus and cytoplasm are visible. This is called a **ring stage** (see Microview 1–4D). The ring stage enlarges and divides repeatedly, and the red blood cells eventually rupture and release the microorganisms, now called **merozoites.** Upon release of the merozoites, their waste products, which cause the fever and chills, are also released. Most of the merozoites infect new red blood cells and perpetuate their cycle of asexual reproduction. However, some develop into male and female sexual forms called **gametocytes.** Even though the gametocytes themselves cause no further damage, they can be picked up by the bite of an *Anopheles* (an-of'el-ēz) mosquito (see Figure 11-27b); they then enter the mosquito's intestine and begin their sexual cycle. Here the male and female gametocytes unite to form a zygote. The zygote forms a cyst in which cell division occurs, and asexual **sporozoites** are formed. When the cyst ruptures, the sporozoites migrate to the salivary glands of the mosquito. They can then be injected into a new human host by the biting mosquito.

When the sporozoites enter a new human host, they are carried by the blood to the liver. Here, the sporozoites undergo schizogony and produce thousands of merozoites. These cells enter the bloodstream and infect red blood cells—the start of another cycle.

Malaria is diagnosed in the laboratory by microscopic observation of thick blood smears for the presence of *Plasmodium* (see Figure 21–18).

Toxoplasma gondii (toks-ō-plaz'mä gon'dē-ī) is another intracellular parasite of humans. The life cycle of this sporozoan is not known but appears to involve domestic cats. The trophozoites reproduce sexually and asexually in an infected cat, and oocysts are passed out with feces. If the oocysts are ingested by humans or other animals, the trophozoites can reproduce in the tissues of the new host (see Figure 21–15). *T. gondii* is dangerous to pregnant women, as it can cause congenital infections in utero. Tissue examination and observation of *T. gondii* is used for diagnosis. Antibodies may be detected by ELISA and indirect fluorescent antibody tests (see Chapter 16).

Table 11–5 lists some typical parasitic protozoans and the diseases they cause.

HELMINTHS

There are a number of parasitic animals that spend part or all of their lives in humans. Most of these animals belong to two phyla: **Platyhelminthes** (flatworms) and **Aschelminthes** (roundworms). These worms are commonly called **helminths.** There are also free-living species in these phyla, but we will limit our discussion to the parasitic species.

HELMINTH BIOLOGY

Helminths are multicellular, eucaryotic animals generally possessing digestive, circulatory, nervous, excretory, and reproductive systems. Parasitic helminths are highly specialized to live inside their hosts. When parasitic helminths are compared to their free-living relatives, the following generalizations can be made about parasitic helminths:

1. They either lack a digestive system or have a greatly simplified one. They can absorb nutrients from the host's food, body fluids, and tissues.
2. Their nervous system is *reduced*. They do not need an extensive nervous system because they do not have to search for food or adapt to their environment. The environment within a host is fairly constant.
3. Their means of locomotion is either *reduced* or *completely lacking*. Because they are transferred from host to host, they do not have to search for a suitable habitat.
4. Their reproductive system is often more complex; it produces more fertilized eggs by which a suitable host is infected.

Reproduction

Adult helminths may be **dioecious:** male reproductive organs are in one individual and female reproductive organs in another. In those species, reproduction occurs only when two adults of opposite sex are in the same host.

Adult helminths may also be **hermaphroditic:** one animal has both male and female reproductive organs. Two hermaphrodites may copulate and simultaneously fertilize each other. A few types of hermaphrodites fertilize themselves.

Life Cycle

The life cycle of parasitic helminths can be extremely complex, involving a succession of hosts. The term **definitive host** is given to the organism that harbors the adult, sexually mature helminth. One or more **intermediate hosts** might be necessary for completion of each **larval,** or developmental, stage of the parasite.

PLATYHELMINTHES

Members of the phylum Platyhelminthes, flatworms, are dorsoventrally flattened. They have what is called an *incomplete* digestive system. This type of digestive system has only one opening (mouth) through which food enters and wastes leave. The classes of parasitic flatworms include the trematodes and cestodes.

Trematodes

Trematodes, or flukes, have flat, leaf-shaped bodies with a ventral sucker and an oral sucker (Figure 11–21). The suckers hold the organism in place and suck fluids from the host. Flukes can also obtain food by absorbing it through their nonliving outer covering, called the cuticle. Flukes are named according to the tissue of the definitive host in which the adults live (for example, lung fluke, liver fluke, blood fluke).

To exemplify a fluke's life cycle, let us look at the lung fluke, *Paragonimus westermanni* (par-ä-gōn'ē-mus we-ster-man'nī) shown in Figure 11–22. The adult lung fluke lives in the bronchioles of humans and other mammals. The adults are ap-

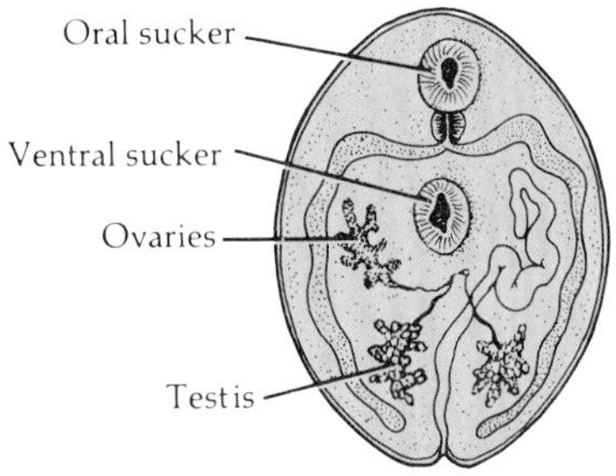

Figure 11–21 This generalized diagram of the anatomy of an adult fluke shows the oral and ventral suckers. The suckers attach the fluke to the host. The mouth is located in the center of the oral sucker. Flukes are hermaphroditic; each animal contains both testes and ovaries.

proximately 6 mm wide by 12 mm long. The hermaphroditic adults liberate eggs into the bronchi. Because sputum containing eggs is frequently swallowed, the eggs are usually excreted in feces from the host.

To continue the life cycle, the eggs must be excreted into a body of water. Inside the egg a **miracidial larva,** or **miracidium,** then develops. When the egg hatches, the larva enters a suitable snail. Only certain species of aquatic snails can be the intermediate host. Inside the snail, the lung fluke undergoes asexual reproduction and produces **rediae.** Each redia develops into a **cercaria** that bores out of the snail and penetrates the cuticle of a crayfish. The parasite encysts as a **metacercaria** in the muscles and other tissues of the crayfish. When the crayfish is eaten by a human, the meta-

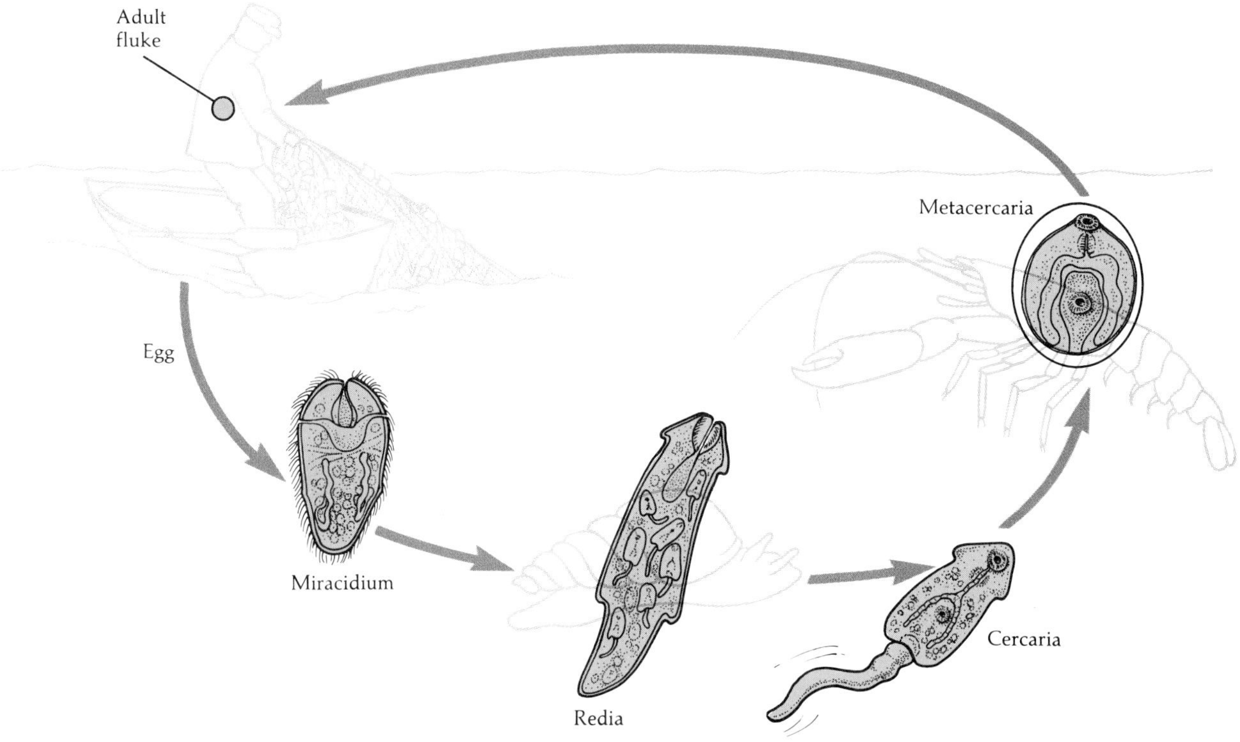

Figure 11–22 Life cycle of *Paragonimus westermanni.* The free-swimming miracidium invades the first intermediate host, a snail. Two generations of rediae develop within the snail, and the rediae give rise to cercariae. The cercariae leave the snail and penetrate the second intermediate host, a crayfish. Here they encyst as metacercariae. When a human eats raw or undercooked crayfish, the metacercariae are released to develop into adults.

cercaria is freed in the human's small intestine. It bores out and wanders around until it penetrates the lungs, enters the bronchioles, and develops into an adult lung fluke.

In a laboratory diagnosis, sputum and feces are examined microscopically for eggs. Infection results from eating undercooked crayfish, and the disease could be prevented by thoroughly cooking crayfish.

The cercariae of the blood fluke *Schistosoma* are not ingested. Instead, they burrow through the skin of the human host and enter the circulatory system. The adults are found in mesenteric and pelvic veins. The disease schistosomiasis is a major world health problem and will be discussed further in Chapter 21.

Cestodes

Cestodes, or tapeworms, are intestinal parasites. Their structure is shown in Figure 11–23. The head, or **scolex,** has suckers and might have small hooks for attaching to the intestinal mucosa of the host (see Microview 1–4F). Tapeworms do not ingest the tissues of their hosts; in fact, they completely lack a digestive system. To obtain nutrients from the small intestine, they absorb food through their cuticle. The body consists of segments called **proglottids** (see Microview 1–4G). Proglottids are continually produced by the neck region of the scolex, as long as the scolex is attached and alive. Each proglottid contains both male and female reproductive organs. The proglottids farthest away from the scolex are the mature ones containing fertilized eggs.

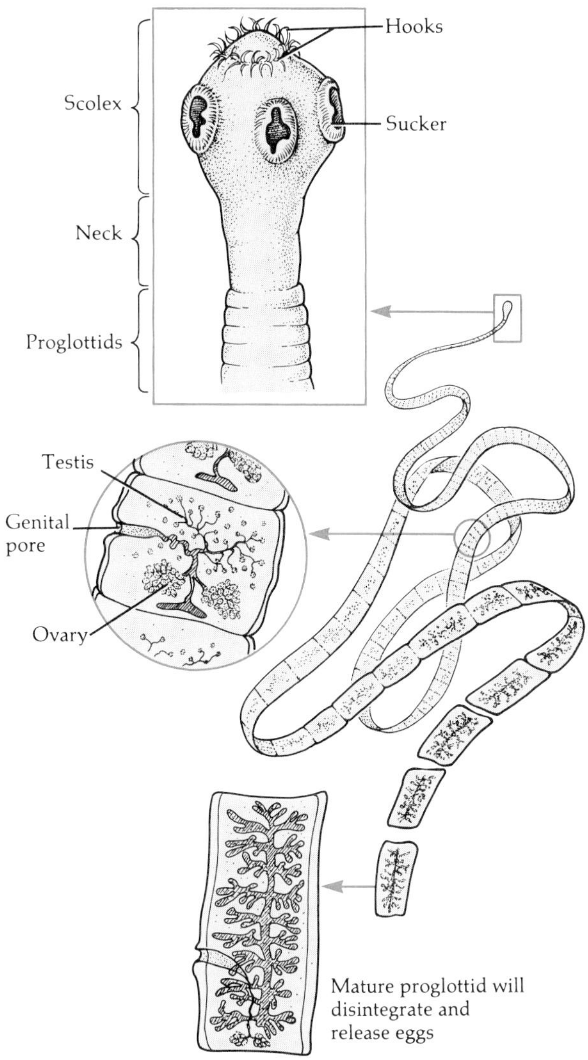

Figure 11–23 General anatomy of an adult tapeworm. The scolex consists of suckers and hooks that attach to the host's tissues. The body lengthens as new proglottids form at the neck. Each proglottid contains both testes and ovaries.

Humans as definitive hosts

The adults of *Taenia saginata,* (te′nē-ä sa-ji-nä′tä) or beef tapeworm, live in humans and can reach a length of 6 meters. The scolex is about 2 mm in length and is followed by a thousand or more proglottids. The life cycle of the beef tapeworm is illustrated in Figure 23–18. The feces of an infected human contain mature proglottids, each proglottid containing thousands of eggs. As the proglottids wriggle away from the fecal material, they increase their chances of being ingested by an animal that is grazing. Upon ingestion by cattle, the larvae hatch from the eggs and bore through the intestine. The larvae migrate to muscle (meat), in which they encyst as **cysticerci.** When the cysticerci are ingested by humans, all but the scolex is digested. The scolex anchors itself in the small intestine and begins producing proglottids.

Diagnosis is based on the presence of mature proglottids and eggs in feces. Cysticerci can be seen macroscopically in meat; their presence is referred to as "measly beef." Beef for human consumption is inspected for "measly" appearance, and such inspection is one way that prevents infec-

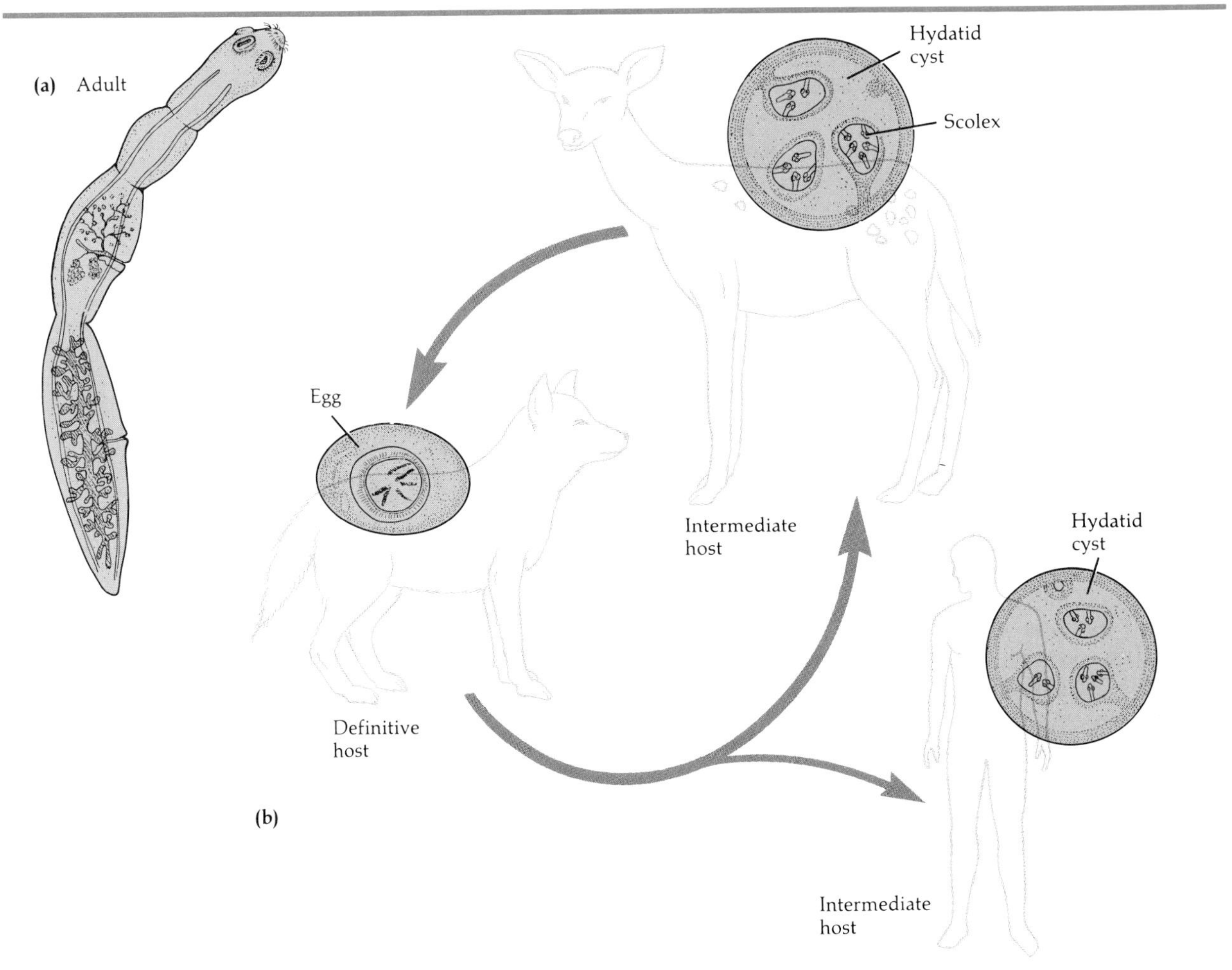

Figure 11–24 *Echinococcus granulosus.* This tiny tapeworm is found in the intestines of dogs, cats, wolves, and foxes. **(a)** Adult. **(b)** Life cycle. Eggs are excreted from the definitive host and ingested by an intermediate host, such as a deer. In the intermediate host, the eggs hatch and the larvae form hydatid cysts in the host's tissues. A hydatid cyst is a fluid-filled sac containing many scoleces. For the cycle to be complete, the cysts must be ingested by a definitive host eating the intermediate host. A human serving as the intermediate host is a dead end for the parasite unless the human is eaten by an animal.

tions by beef tapeworm. Another method of prevention is to avoid the use of untreated human sewage as fertilizer in grazing pastures.

Humans as intermediate hosts

Humans are the intermediate hosts for *Echinococcus granulosus* (e-kīn-ō-kok'kus gra-nū-lō'sus), shown in Figure 11–24. Dogs and cats, both wild and domestic, are the definitive hosts for this minute (2 to 8 mm) tapeworm. Eggs are excreted with feces and ingested by deer or humans. Humans can contaminate their hands with dog feces or can be contaminated by saliva from a dog's tongue. In the human's small intestine, the eggs hatch and the larvae migrate to the liver or lungs. The host forms a cyst around the larvae in these organs. This cyst, called a **hydatid cyst,** is a "brood capsule" in which thousands of scoleces (singular, scolex) might be produced. In the wild, the cysts might be in a deer that is then eaten by a wolf. The scoleces would be able to attach themselves in the wolf's intestine and produce proglottids.

Diagnosis of hydatid cysts is frequently only made on autopsy, although X rays can detect the cysts in lungs.

ASCHELMINTHES

Members of the phylum Aschelminthes, the roundworms, are cylindrical in shape and tapered at each end. Roundworms have a *complete* digestive system, consisting of a mouth, an intestine, and an anus. Most species are dioecious. Parasites of humans belong only to the class consisting of nematodes.

Nematodes

Some species of nematodes are free-living in soil and water, while others are parasites on plants and animals. Parasitic nematodes do not have the succession of larval stages exhibited by flatworms. Some nematodes pass their entire life's cycle, from egg to mature adult, in a single host.

Nematode infections of humans can be divided into two categories: those in which the egg is infective and those in which the larva is infective.

Eggs infective for humans

The pinworm *Enterobius vermicularis* (en-te-rō'bē-us ver-mi-kū-lar'is) spends its entire life in a human host (Figure 11–25). Adult pinworms are found in the large intestine. From there, the female pinworm migrates to the anus to deposit her eggs

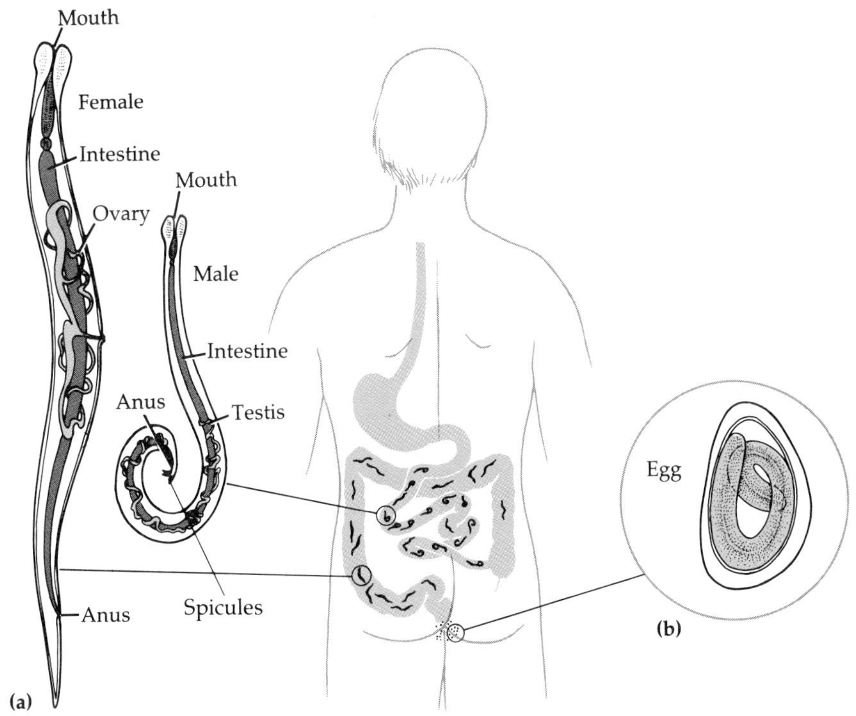

Figure 11–25 *Enterobius vermicularis.* **(a)** Adult pinworms, which live in the intestines of humans. Roundworms have a complete digestive system with a mouth, intestine, and anus. Most roundworms are dioecious, and the female (left) is often distinctly larger than the male (right). **(b)** Eggs, which the female deposits on the perianal skin at night.

in the perianal skin. The eggs can be ingested by the host or by another person exposed through contaminated clothing. Pinworm infections are diagnosed by the Graham sticky-tape method. A piece of tape is placed on the perianal skin in such a way that the sticky side is exposed. The tape is microscopically examined for the presence of eggs adhering to it.

Ascaris lumbricoides (as'kar-is lum-bri-koi'dēz) is a large nematode (30 cm in length). It is dioecious with **sexual dimorphism;** that is, the male and female worms look distinctly different, the male being smaller with a curled tail. The adult *Ascaris* lives in the small intestines of humans and domestic animals (such as pigs and horses); it feeds primarily on semidigested food. Eggs, excreted with feces, can survive in the soil for long periods until accidentally ingested by another host. The eggs hatch and mature into adults in the intestine of the host.

Table 11–6 Representative Parasitic Helminths

Phylum	Class	Human Parasites
Platyhelminthes	Trematodes	*Paragonimus westermanni*
		Schistosoma
	Cestodes	*Taenia saginata*
		Echinococcus granulosus
Aschelminthes	Nematodes	*Ascaris lumbricoides*
		Enterobius vermicularis
		Necator americanus
		Trichinella spiralis

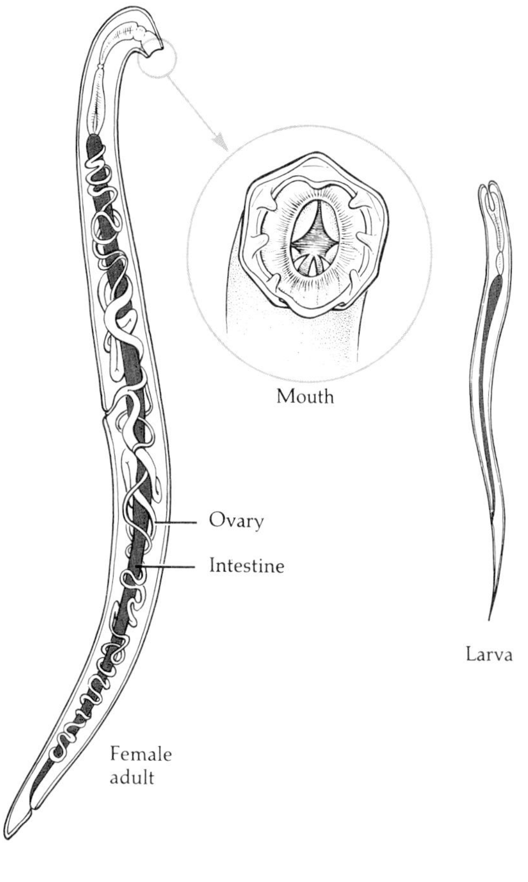

Figure 11–26 The hookworm *Necator americanus*, which is actually a roundworm. Adults are found in the intestine of humans. The hooks located around the mouth are used for attachment and for removing food from the host tissue. The free-living larvae are in the soil and penetrate the skin of the definitive host.

Diagnosis is frequently made when the adult worms are excreted with feces. Prevention of the disease in humans is managed by proper sanitary habits. The life cycle of *Ascaris* in pigs can be interrupted by keeping pigs in areas free of fecal material.

Larvae infective for humans

Adult hookworms, *Necator americanus* (ne-kā'tor ä-me-ri-ka'nus), live in the small intestine of humans (Figure 11–26); the eggs are excreted with feces. The larvae hatch in the soil and feed on bacteria there. A larva enters its host by penetrating the host's skin. The larva then enters a blood or lymph vessel, which carries it to the lungs. The larva is swallowed with sputum and finally carried to the small intestine.

Diagnosis is based on the presence of eggs in feces. People can avoid hookworm infections by wearing shoes.

Table 11–6 Representative Parasitic Helminths *continued*

Intermediate Host	Definitive Host	Stage Passed to Humans	Disease	Location in Humans	See Figure
Freshwater snails and crayfish	Humans, lungs	Metacercaria in crayfish	Paragonimiasis (lung fluke)	Lungs	11–22
Freshwater snails	Humans	Cercariae through skin	Schistosomiasis	Veins	21–19
Cattle	Humans, small intestine	Cysticercus in beef	Tapeworm	Small intestine	23–18
Humans	Dogs and other animals, intestines	Eggs from other animals	Hydatid cyst	Lungs	11–24
—	Humans, small intestine	Ingestion of eggs	Ascariasis	Small intestine	—
—	Humans, large intestine	Ingestion of eggs	Pinworm	Large intestine	11–25
—	Humans, small intestine	Penetration of larvae through skin	Hookworm	Small intestine	11–26
—	Humans, swine, and other mammals, small intestine	Ingestion of larvae in meat, especially pork	Trichinosis	Muscles	23–19

Trichinella spiralis (tri-kin-el'lä spī-ra'lis) infections, called *trichinosis,* are usually acquired by eating encysted larvae in pork or bear meat. In the human digestive tract, the larvae are freed from the cysts. They mature into adults in the intestine and sexually reproduce there. Eggs develop in the female, and she gives birth to live nematodes. The larvae enter lymph and blood vessels in the intestines and migrate from there throughout the body. They encyst in muscles and other tissues, and remain there until ingested by another host (see Figure 23–19).

Diagnosis of trichinosis is made by microscopic examination for larvae in a muscle biopsy. Trichinosis can be prevented by the thorough cooking of meat prior to consumption.

Table 11–6 lists representative parasitic helminths of each phylum and class, and the diseases they cause.

ARTHROPODS

Arthropods are jointed-legged animals. With nearly 1 million species, this is the largest phylum in the animal kingdom. We will briefly describe arthropods here because a few suck the blood of humans and other animals, and while doing so can transmit microbial diseases. Arthropods that carry disease-causing microorganisms are called **vectors.**

Representative classes of arthropods include the following:

1. Arachnida (eight legs): spiders, mites, ticks
2. Crustacea (four antennae): crabs, crayfish
3. Chilopoda (two legs per segment): centipedes
4. Diplopoda (four legs per segment): millipedes
5. Insecta (six legs): bees, flies

Table 11–7 lists those arthropods that are important vectors, and Figures 11–27 and 11–28 illustrate some of them. These insects and ticks reside on an animal only when they are feeding. An exception to this is the louse, which spends its entire life on its host and cannot survive long away from a host. Among insect vectors only the adult female flies (including mosquitoes) bite.

Some vectors are just a mechanical means of transport for a pathogen. For example, houseflies are attracted to decaying organic matter in which to lay their eggs. A housefly can then pick up a pathogen on its feet or body from feces and transport the pathogen to food.

Table 11–7 Important Arthropod Vectors of Human Diseases

Class	Order	Vector	See Figure	Disease	Refer to Chapter
Arachnida	Mites and ticks	*Dermacentor* (tick)	11–28a	Rocky Mountain spotted fever	21
		Ixodes (tick)	—	Lyme disease	21
Insecta	Sucking lice	*Pediculus* (human louse)	11–28b	Epidemic typhus	21
	Fleas	*Xenopsylla* (rat flea)	11–28c	Endemic murine typhus, plague	21
	True flies	*Chrysops* (deer fly)	11–28d	Tularemia	21
		Aedes (mosquito)	11–27a	Dengue fever, yellow fever	21
		Anopheles (mosquito)	11–27b	Malaria	21
		Culex (mosquito)	—	Arboviral encephalitis	20
		Glossina (tsetse fly)	—	African trypanosomiasis	20
	True bugs	*Triatoma* (kissing bug)	11–28e	Chagas' disease	21

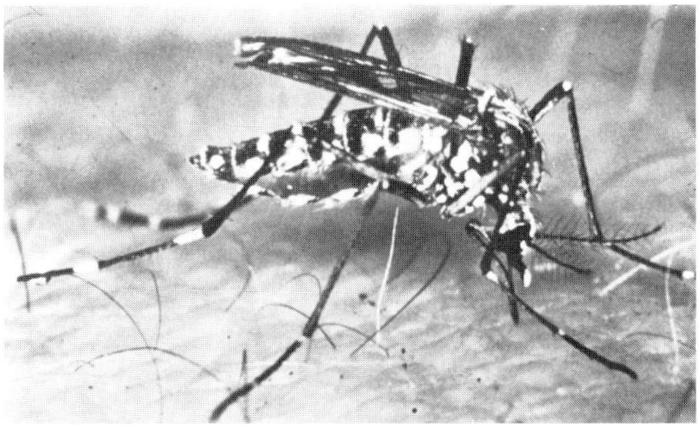

(a)

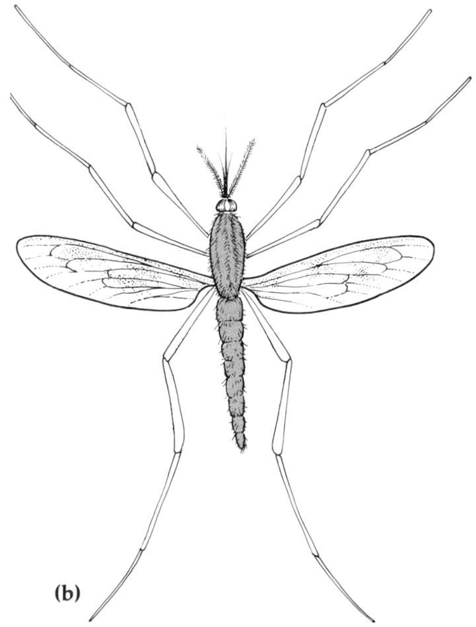

(b)

Figure 11–27 Mosquitoes. **(a)** *Aedes aegypti* (female), sucking blood from human skin. This insect vector transmits several diseases, including the viral diseases yellow fever and dengue fever, from person to person. **(b)** *Anopheles* transmits malaria.

Some parasites multiply in their vectors. When this happens, the parasites can accumulate in the vector's feces or saliva. Large numbers of parasites can be deposited on the skin of the host while the vector is feeding there.

Plasmodium is an example of a parasite that requires that its vector also be its host. *Plasmodium* can sexually reproduce only in the gut of an *Anopheles* mosquito (see Figure 11–27). *Plasmodium* is introduced into a human host with the mosquito's saliva, which acts as an anticoagulant to keep blood flowing.

Elimination of vector-borne diseases is concentrated in efforts to eradicate the vectors.

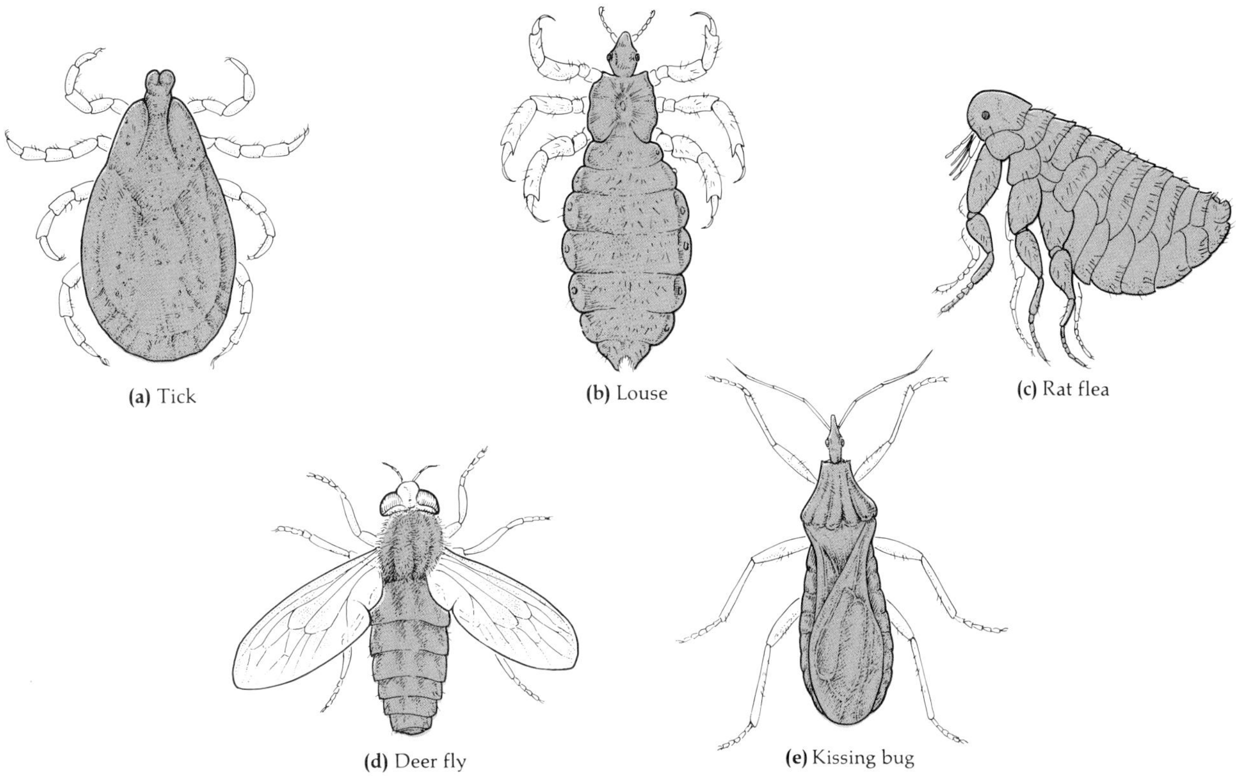

Figure 11–28 Arthropod vectors. **(a)** The tick, *Dermacentor.* **(b)** The human louse, *Pediculus.* **(c)** The rat flea, *Xenopsylla.* **(d)** The deer fly, *Chrysops.* **(e)** The kissing bug, *Triatoma.* See Table 11–7.

STUDY OUTLINE

FUNGI (pp. 302–314)

1. The kingdom Fungi includes yeasts, molds, and fleshy fungi (mushrooms).
2. Fungi are aerobic or facultatively anaerobic chemoheterotrophs.
3. Most fungi are decomposers, and a few are parasites of plants and animals.

VEGETATIVE STRUCTURES (pp. 302–304)

1. A colony of molds and fleshy fungi consists of filaments of cells called hyphae; many hyphae are called a mycelium.
2. Yeasts are unicellular fungi that grow by budding and occasionally by fission.
3. Buds that do not separate form pseudohyphae.
4. Dimorphic fungi are yeastlike at 37°C and moldlike at 25°C.

REPRODUCTIVE STRUCTURES (pp. 305–307)

1. The following spores can be produced asexually: arthrospores, chlamydospores, sporangiospores, conidiospores, and blastospores.
2. Fungi are classified according to the type of sexual spore that they form.

3. Sexual spores include zygospores, ascospores, and basidiospores.

MEDICALLY IMPORTANT PHYLA OF FUNGI (pp. 307–308)

1. The Zygomycota have coenocytic hyphae and produce sporangiospores and zygospores.
2. The Ascomycota have septate hyphae and produce ascospores and frequently conidiospores.
3. Basidiomycetes produce basidiospores and may produce conidiospores.
4. Deuteromycota reproduce asexually by one or more of the following: chlamydospores, arthrospores, conidiospores, or budding.
5. Deuteromycota are called the Fungi Imperfecti because sexual spores have not yet been seen.

FUNGAL DISEASES (pp. 308–309)

1. Systemic mycoses are fungal infections deep within the body and affect many tissues and organs.
2. Subcutaneous mycoses are fungal infections beneath the skin.
3. Cutaneous mycoses affect keratin-containing tissues such as hair, nails, and skin.
4. Superficial mycoses are localized on hair shafts and superficial skin cells.
5. Opportunistic mycoses are caused by normal flora or fungi that are not usually pathogenic.
6. Opportunistic mycoses include mucormycosis, caused by some Zygomycetes; aspergillosis, caused by *Aspergillus;* and candidiasis, caused by *Candida.*
7. Opportunistic mycoses can infect any tissues. However, they are usually systemic.

UNDESIRABLE ECONOMIC EFFECTS OF FUNGI (pp. 309–310)

1. Mold spoilage of fruits, grains, and vegetables is more common than bacterial spoilage of these products.

ALGAE (pp. 310–314)

1. Algae are photoautotrophs that produce oxygen.
2. Algae are classified as plants or protists according to their structures and pigments.
3. Unicellular dinoflagellates can produce a neurotoxin that causes paralytic shellfish poisoning.
4. Euglenoids and diatoms are unicellular plantlike protists.
5. Multicellular algae include the brown algae or kelp, green algae, and red algae.

STRUCTURE AND REPRODUCTION (p. 313)

1. The thallus (or body) of multicellular algae usually consists of a stipe, holdfast, and blades.
2. Algae reproduce asexually by cell division and fragmentation.
3. Many algae reproduce sexually.

ROLES OF ALGAE IN NATURE (pp. 313–314)

1. Algae are the primary producers in aquatic food chains.
2. Planktonic algae produce most of the molecular oxygen in the earth's atmosphere.

LICHENS (pp. 315–316)

1. A lichen is a symbiotic relationship between an alga and a fungus.
2. The alga photosynthesizes providing carbohydrates for the lichen; the fungus provides a holdfast.
3. Lichens colonize habitats that are unsuitable for either the alga or fungus alone.
4. Lichens may be classified on the basis of morphology as crustose, foliose, or fruticose.
5. In the lichen thallus, the alga reproduces by cell division and the fungus forms sexual spores.

SLIME MOLDS (p. 317)

1. Cellular slime molds resemble amoebas and ingest bacteria by phagocytosis.
2. A plasmodial (acellular) slime mold is a multinucleated mass of protoplasm that engulfs organic debris and bacteria as it moves.

PROTOZOANS (pp. 317–325)

1. Protozoans are unicellular eucaryotes in the kingdom Protista.
2. Protozoans are found in soil, water, and as normal flora in animals.
3. Protozoans are classified by their means of locomotion: members of the phylum Sarcodina move by amoeboid motion; the Mastigophora (flagellates) use flagella for motility; the Ciliata possess cilia; and the Sporozoa lack a means of locomotion and are obligate parasites.

PROTOZOAN BIOLOGY (pp. 318–319)

1. Most protozoans are heterotrophs and feed on bacteria and particulate organic material.
2. Protozoans are complex cells that can have a pellicle, a cytostome, and an anal pore.
3. Asexual reproduction is by fission, budding, or schizogony.
4. Sexual reproduction is by conjugation.
5. During protozoan conjugation, two haploid nuclei fuse to produce a zygote.
6. Some protozoans can produce a cyst for protection during adverse environmental conditions.

MEDICALLY IMPORTANT PHYLA OF PROTOZOANS (pp. 319–325)

1. Parasitic sarcodinae that cause amoebic dysentery are found in the genus *Entamoeba.*
2. Parasitic flagellates (Mastigophora) include the following: *Giardia lamblia,* causing an intestinal infection called giardiasis; *Trichomonas vaginalis,* causing genitourinary infections that may be transmitted by coitus; and *Trypanosoma cruzi,* which is found in the blood of humans with Chagas' disease.
3. The only ciliate that is a parasite of humans is *Balantidium coli,* the cause of one form of dysentery.
4. *Plasmodium* is the sporozoan that causes malaria.
5. Asexual reproduction of *Plasmodium* occurs in red blood cells and the liver of humans.
6. Sexual reproduction of *Plasmodium* takes place in the intestine of the female *Anopheles* mosquito.
7. *Toxoplasma gondii* is a sporozoan that infects humans; it can be transmitted to a fetus in utero.

HELMINTHS (pp. 325–331)

1. Some parasitic helminths (worms) belong to the phylum Platyhelminthes.
2. Parasitic roundworms belong to the phylum Aschelminthes (helminths).

HELMINTH BIOLOGY (p. 325)

1. Helminths are multicellular animals; a few are parasites of humans.
2. The anatomy and life cycles of parasitic helminths are modified for parasitism.
3. Helminths can be hermaphroditic or dioecious.
4. The adult stage of a parasitic helminth is found in the definitive host.
5. Each larval stage of a parasitic helminth requires an intermediate host.

PLATYHELMINTHES (pp. 325–329)

1. Flatworms are dorsoventrally flattened animals that do not have a complete digestive system.
2. Adult trematodes or flukes have an oral and ventral sucker with which they attach to and feed on host tissue.
3. Eggs of trematodes hatch into free-swimming miracidia that enter the first intermediate host; two generations of rediae develop in the first intermediate host; the rediae become cercariae that bore out of the first intermediate host and penetrate the second intermediate host; cercariae encyst as metacercariae in the second intermediate host; after they are ingested by the definitive host, the metacercariae develop into adults.
4. A cestode, or tapeworm, consists of a scolex (head) and proglottids.
5. Humans serve as the definitive host for the beef tapeworm and cattle are the intermediate host.
6. Humans serve as the intermediate host for *Echinococcus granulosus;* the definitive hosts are dogs, wolves, and foxes.

ASCHELMINTHES (pp. 329–331)

1. Members of the phylum Aschelminthes are roundworms with a complete digestive system.
2. Parasitic members of Aschelminthes are in the class consisting of nematodes.

3. Aschelminthes having eggs infective for humans are *Ascaris lumbricoides* and *Enterobius vermicularis* (pinworm).
4. Aschelminthes having larvae infective for humans are *Necator americanus* (hookworm) and *Trichinella spiralis*.

ARTHROPODS (pp. 331–333)

1. Jointed-legged animals, including ticks and insects, belong to the phylum Arthropoda.
2. Arthropods that carry diseases are called vectors.
3. Elimination of vector-borne diseases is best done by the control or eradication of the vectors.

STUDY QUESTIONS

REVIEW

1. Contrast the mechanism of conidiospore and ascospore formation by *Penicillium*.
2. Fill in the following table.

Phylum	Spore Type(s)	
	Sexual	Asexual
Zygomycota		
Ascomycota		
Basidiomycota		
Deuteromycota		

3. Fungi are classified into phyla on the basis of ______________.
4. The following is a list of fungi, their methods of entry into the body, and sites of infections they cause. Categorize each type of mycosis (systemic, subcutaneous, cutaneous, superficial).

Genus	Method of Entry	Site of Infection	Mycosis
Blastomyces	Inhalation	Lungs	
Sporotrix	Puncture	Ulcerative lesions	
Microsporum	Contact	Fingernails	
Trichosporon	Contact	Hair shafts	

5. What is the role of the alga in a lichen? Of the fungus?
6. Briefly discuss the importance of lichens in nature. The importance of algae.
7. Complete the following table.

Phylum	Cell Wall Composition	Special Features/Importance
Dinoflagellates		
Euglenoids		
Diatoms		
Red algae	—	
Brown algae	—	
Green algae	—	

Indicate which phyla consist primarily of unicellular forms. Which phylum would you include in the plant kingdom? Why? Into which kingdom would you place the others?

8. Differentiate between cellular and plasmodial slime molds. How does each survive adverse environmental conditions?

9. Why is it significant that *Trichomonas* does not have a cyst stage? Name a protozoan parasite that does have a cyst stage.

10. Protozoans are classified on the basis of ________________.

11. Recall the life cycle of *Plasmodium*. Where does asexual reproduction occur? Where does sexual reproduction occur? Identify the definitive host. Identify the vector.

12. Transmission of helminthic parasites to humans usually occurs by ________________.

13. To what phylum and class does this animal belong? List two characteristics that put it in this phylum. Name the body parts. What is the name of the encysted larva of this animal?

14. Most nematodes are dioecious. What does this term mean? To what phylum do nematodes belong?

15. Vectors can be divided into three major types according to the roles they play for the parasite. List the three types of vectors.

CHALLENGE

1. A generalized life cycle of the liver fluke (*Clonorchis sinensis*) is shown below. Identify the intermediate host(s). Identify the definitive host(s). To what phylum and class does this animal belong?

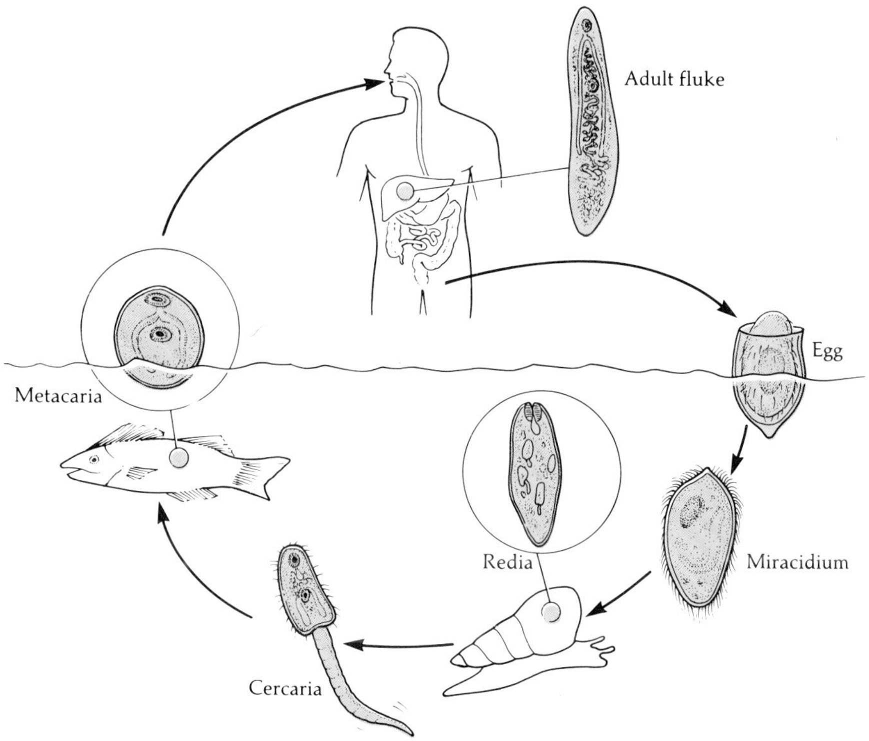

2. *T. b. gambiense* is the causative agent of African sleeping sickness. To what phylum and class does it belong? (Look at part **(a)**.) Part **(b)** shows a generalized life cycle for *T. b. gambiense*. Identify the host and vector of this parasite.

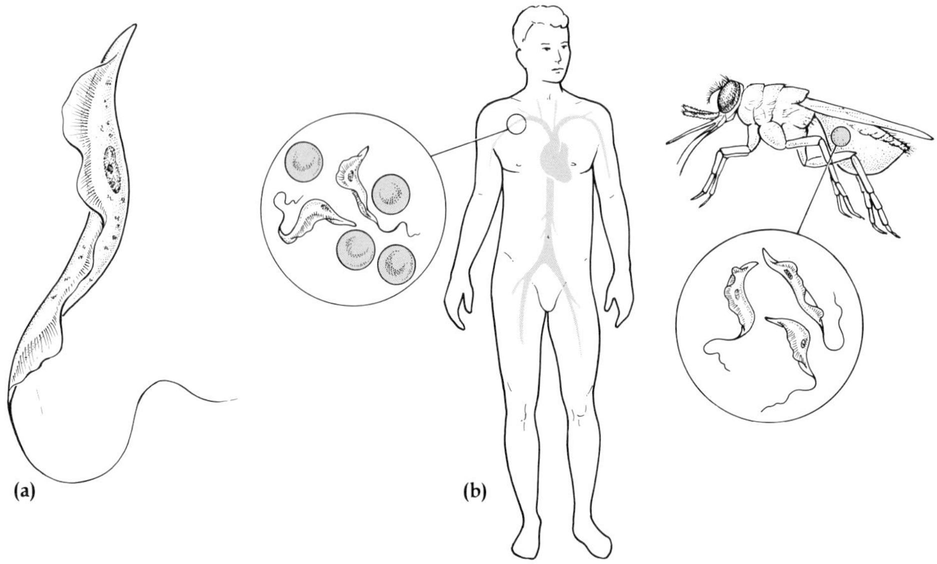

3. The size of a cell is limited by its surface-to-volume ratio; that is, if the volume gets too great, internal heat cannot be dissipated and nutrients and wastes cannot be efficiently transported. How do plasmodial slime molds manage to circumvent the surface-to-volume rule?

FURTHER READING

Barnett, H. L. and B. B. Hunter. 1972. *Illustrated Genera of Imperfect Fungi*. Minneapolis, MN: Burgess Publishing Company. Good line drawings facilitate identification of saprophytic molds.

Barnett, J. A., R. W. Payne, and D. Yarrow. 1984. *Yeasts: Characteristics and Identification*. New York: Cambridge University Press. A complete key to yeasts based on morphologic and biochemical characteristics.

Beaver, P. C. and R. C. Jung. 1985. *Animal Agents and Vectors of Human Disease*, 5th ed. Philadelphia, PA: Lea and Febiger. Includes a discussion of arthropod vectors and culture techniques for protozoans and helminths.

Bonner, J. T. June 1969. Hormones in social amoebae and mammals. *Scientific American* 220:78–91. Description of "slug" formation and function of cAMP in slime molds.

Boyd, H. C. and M. J. Payne. 1985. *Introduction to the Algae*, 2nd ed. Englewood Cliffs, NJ: Prentice-Hall. Covers biology of freshwater and marine unicellular and multicellular algae.

Deans, J. A. and S. Cohen. 1983. Immunology of Malaria. *Annual Review of Microbiology* 37:25–49. Describes the complexities of parasitic diseases including the host response and research on vaccinations.

Emmons, C. W., C. H. Binford, J. P. Utz, and K. J. Kwon-Chung. 1977. *Medical Mycology*, 3rd ed. Philadelphia, PA: Lea and Febiger. Discussions of fungal diseases that affect humans.

Hale, M. E. 1983. *Biology of Lichens*, 3rd ed. Baltimore, MD: Edward Arnold. An authoritative work on the structure, physiology, and ecology of lichens; some discussion of identification and culture techniques.

Stark, A-A. 1980. Mutagenicity and carcinogenicity of mycotoxins. *Annual Review of Microbiology* 34:235–262. A description of fungal toxins and their effects on eucaryotic cells. Possible binding of DNA is discussed as a method of action.

Webster, J. 1980. *Introduction to Fungi*, 2nd ed. New York: Cambridge University Press. A textbook on the biology of fungi.

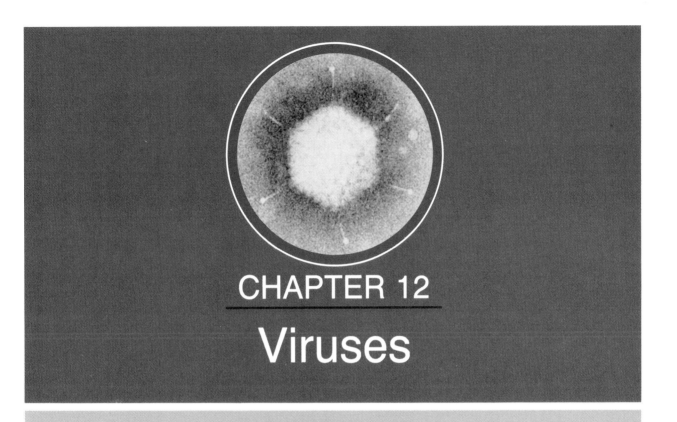

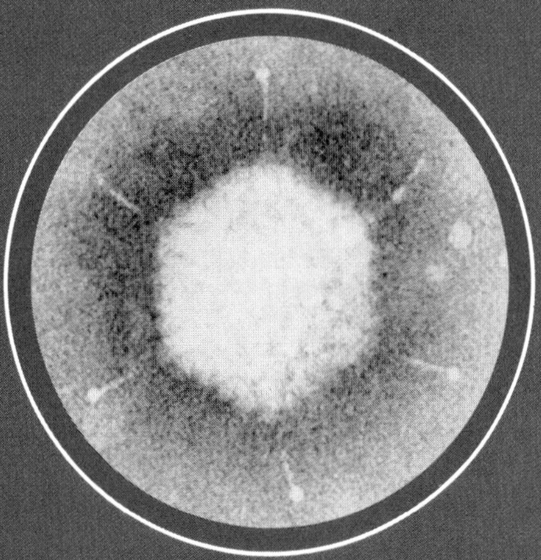

CHAPTER 12

Viruses

OBJECTIVES

After completing this chapter you should be able to

- Define the following terms: virus; viroid; prion; transformed cell; oncogene.
- Describe the chemical composition of a typical virus.
- Classify viruses according to their morphology and list other criteria used to classify viruses.
- Explain how viruses are cultured.
- Describe the lytic cycle of T-even bacteriophages.
- Compare and contrast the multiplication cycle of DNA- and RNA-containing animal viruses.
- Differentiate between slow viral infections and latent viral infections.
- List the effects of animal viral infections on host cells.
- Describe four mechanisms of oncogene activation.
- Explain what a tumor is and distinguish between malignant and benign tumors.
- Discuss the relationship of DNA- and RNA-containing viruses to cancer.

Until the latter part of the nineteenth century, the term **filterable virus** designated all infectious agents capable of passing through filters that retained all known bacteria, fungi, and protozoans. At that time, the term *filterable* was dropped and the word *virus* (meaning "poison" or "venom") was used specifically to refer to submicroscopic, filterable, infectious agents; that is, infectious agents too small to be viewed with a light microscope. During the first decades of the twentieth century, most scientists believed that viruses were a distinct group of infectious agents differing from others only in size. However, it was soon discovered that viruses have their own method of reproduction and a distinctive chemical makeup. With the advent of the electron microscope and advanced analytical procedures, exciting discoveries of the structural and functional characteristics of viruses were made.

Viruses are found as parasites in all types of cells. Viral diseases of humans are fairly well known; some diseases of agriculturally important animals and plants are also known to be caused by viruses. Viruses infect fungi and bacteria, and probably protists as well.

Viruses are of particular interest to microbiologists for several reasons. As you will see, viruses differ fundamentally, in both structure and life cycle, from the microorganisms discussed in previous chapters. Because there are so many drugs available to combat bacterial infections, but so few antiviral drugs, viruses have become the most threatening agents of infectious disease in developed countries such as the United States. Viruses might also be related to certain types of cancer in humans.

GENERAL CHARACTERISTICS OF VIRUSES

For our introduction to viruses, let us first consider their definition, host range, and size.

Definition

In first attempting to define a virus, we should consider whether or not viruses are living organisms. Life can be defined as a complex set of processes resulting from the actions of processes specified by nucleic acids. The nucleic acids of living cells are in action all the time. Because viruses are inert outside of living host cells, in this sense they are not considered living organisms. However, once viruses enter a host cell, the viral nucleic acids become active and viral multiplication results. In this sense, viruses are alive when they multiply in host cells they infect. From a clinical point of view, viruses can be considered alive because they cause infection and disease just as pathogenic bacteria, fungi, and protozoans do. Depending on one's viewpoint, a virus may be regarded as an exceptionally complex aggregation of nonliving chemicals or as an exceptionally simple living microorganism.

How, then, do we define a virus? Viruses were originally distinguished from other infectious agents because they are especially small (filterable viruses) and because they are **obligatory intracellular parasites**—that is, they absolutely require living host cells in order to multiply. However, both of these properties are shared by certain small bacteria, such as some rickettsias. The truly distinctive features of viruses are now known to relate to their simple structural organization and composition, and their mechanism of multiplication. Accordingly, **viruses** are entities that (1) contain a single type of nucleic acid, either DNA or RNA; (2) contain a protein coat (sometimes enclosed by an envelope of lipids, proteins, and carbohydrates) that surrounds the nucleic acid; (3) multiply inside living cells using the synthesizing machinery of the cell; and (4) cause the synthesis of specialized elements that can transfer the viral nucleic acid to other cells.

Since viruses have few or no enzymes of their own for metabolism—for example, enzymes that synthesize proteins and nucleic acids and generate ATP—in order to multiply, viruses must take over the metabolic machinery of the host cell. This fact has considerable medical significance for the development of antiviral drugs, because most drugs that would interfere with viral multiplication would also interfere with the functioning of the host cell and hence are too toxic for clinical use. However, the presence of lipids in the coverings of some viruses makes these viruses sensitive to disinfection or damage by lipid solvents, such as ether, and emulsifying agents, such as bile salts and detergents.

When we describe the structure and multiplication of viruses in more detail, reference will be made to the term virion. A **virion** is a complete, fully developed viral particle composed of nucleic acid surrounded by a coat that protects it from the environment and serves as a vehicle of transmission from one host cell to another.

Host Range

The **host range** of a virus is the spectrum of host cells that the virus can infect. Viruses multiply only in cells of particular species and thus are divided into three main classes: **animal viruses, bacterial viruses (bacteriophages),** and **plant viruses.** (Pro-

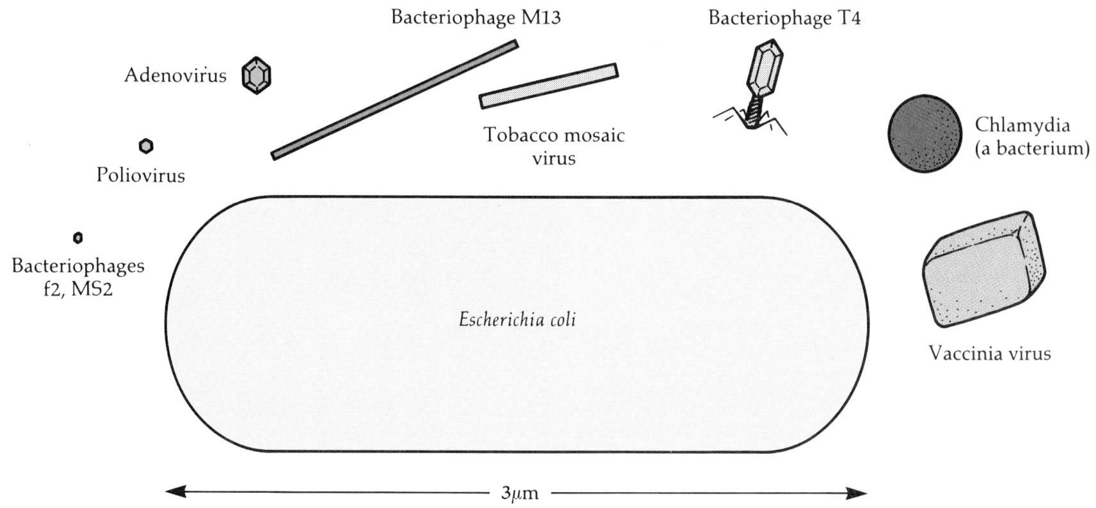

Figure 12–1 Comparative sizes of several viruses and bacteria.

tists and fungi can also be hosts for viruses, but they will not concern us here. In this chapter, we are concerned mainly with animal viruses and bacterial viruses.) Within each class, each virus is usually able to infect only certain species of cells.

The particular host range of a virus is determined by the virus' requirements for its specific attachment to the host cell and the availability within the potential host of cellular factors required for viral multiplication. In order for the virus to infect the host cell, the outer surface of the virus must chemically interact with specific receptor sites on the surface of the cell. The two complementary components are held together by weak bonds, such as hydrogen bonds. For some bacteriophages, the receptor site is a chemical constituent of the cell wall of the host; in other cases, it is a constituent of pili or flagella. For animal viruses, the receptor sites are on the plasma membranes of the host cells.

Size

The sizes of viruses were first estimated by filtration through membranes of known pore diameter. Viral sizes are determined today by ultracentrifugation and electron microscopy. This last technique seems to produce the most accurate results. Viruses vary considerably in size. Although most are quite a bit smaller than bacteria, some of the larger viruses (such as the smallpox virus) are about the same size as some very small bacteria (such as the mycoplasmas, rickettsias, and chlamydias). Viruses range from 20 to 300 nm in diameter. The comparative sizes of several viruses and bacteria are shown in Figure 12–1.

VIRAL STRUCTURE

Nucleic Acid

As noted earlier, the core of a virus contains a single kind of nucleic acid, either DNA or RNA, which is the genetic material. The percentage of nucleic acid in relation to protein is about 1% for the influenza virus and about 50% for certain bacteriophages. The total amount of nucleic acid varies from a few thousand nucleotides (or pairs) to as many as 250,000 nucleotides.

In contrast to procaryotic and eucaryotic cells, in which DNA is always the primary genetic material (and RNA plays an auxiliary role), a virus can have either DNA or RNA, but never both. The nucleic acid of a virus can be single- or double-stranded. Thus, there are viruses with the familiar double-stranded DNA, with single-stranded DNA,

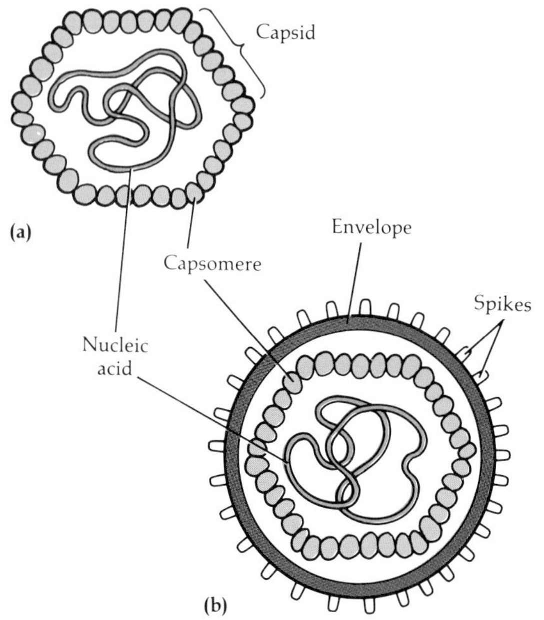

Figure 12–2 General structure of two types of viruses. **(a)** Naked virus. **(b)** Enveloped virus with spikes.

with double-stranded RNA, and with single-stranded RNA. Depending on the virus, the nucleic acid can be linear or circular, and in some viruses (such as the influenza virus), the nucleic acid is in several separate molecules.

Capsid and Envelope

The nucleic acid of a virus is surrounded by a protein coat called the **capsid** (Figure 12–2a). The capsid, whose architecture is ultimately determined by the viral nucleic acid, accounts for most of the mass of a virus, especially of small ones. Each capsid is composed of protein subunits referred to as **capsomeres.** In some viruses, the proteins composing the capsomeres are of a single type; in other viruses, several types of protein may be present. Individual capsomeres are often visible in electron micrographs (see Figure 12–4b, for example). The arrangement of capsomeres is characteristic of a particular virus.

In some viruses, the capsid is covered by an **envelope** (Figure 12–2b), which usually consists of some combination of lipids, proteins, and carbohydrates. The molecular organization of these envelopes is generally unknown. Some animal viruses are released from the host cell by an extrusion process that coats the virus with a layer of the host cell's plasma membrane; that layer becomes the viral envelope. In many cases the envelope contains proteins determined by viral nucleic acid, and it contains materials derived from the normal cell components.

Depending on the virus, envelopes may or may not be covered by **spikes,** which are carbohydrate–protein complexes that project from the surface of the envelope. Some viruses attach to host cells by means of spikes. Spikes are such a reliable characteristic of some viruses that they can be used as a means of identification. The ability of certain viruses, such as the influenza virus, to clump red blood cells is associated with spikes. Such viruses bind to red blood cells and form bridges between them. The resulting clumping is called **hemagglutination** and is the basis for several useful laboratory tests.

Viruses whose capsid is not covered by an envelope are known as **naked viruses** (Figure 12–2a). The capsid of a naked virus protects the nucleic acid from nuclease enzymes in biological fluids and promotes the virus' attachment to susceptible host cells.

General Morphology

According to the architecture of capsids as revealed by electron microscopy, viruses may be classified into several morphological types.

Helical Viruses Helical viruses resemble long rods that may be rigid or flexible. Surrounding the nucleic acid, their capsid is a hollow cylinder with a helical structure. An example of a helical virus that is a rigid rod is the tobacco mosaic virus (Figure 12–3). Another is bacteriophage M13.

Polyhedral Viruses Many animal, plant, and bacterial viruses are polyhedral viruses; that is, they are many-sided. The capsid of most polyhedral viruses is in the shape of an *icosahedron,* a regular polyhedron with 20 triangular faces and 12 corners (Figure 12–4a). The capsomeres of each face form an equilateral triangle. An example of a polyhedral

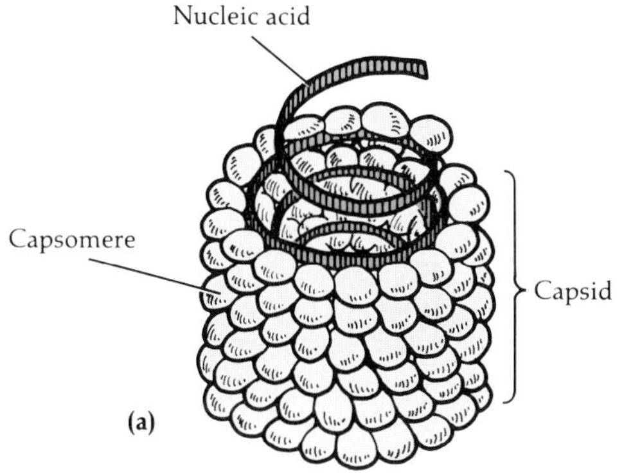

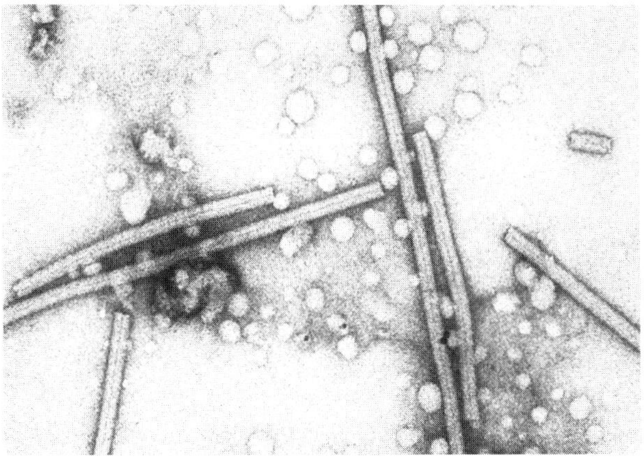
(b)

Figure 12–3 Morphology of a helical virus. **(a)** Diagram of a portion of a tobacco mosaic virus. Several rows of capsomeres have been removed to reveal the nucleic acid. **(b)** Electron micrograph of a tobacco mosaic virus showing helical rods (×147,000).

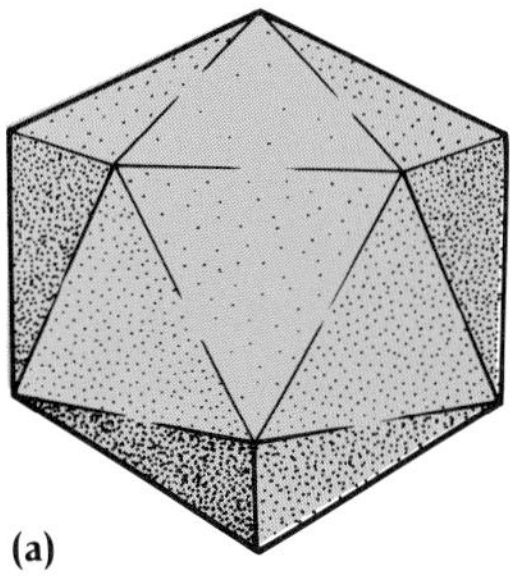
(a)

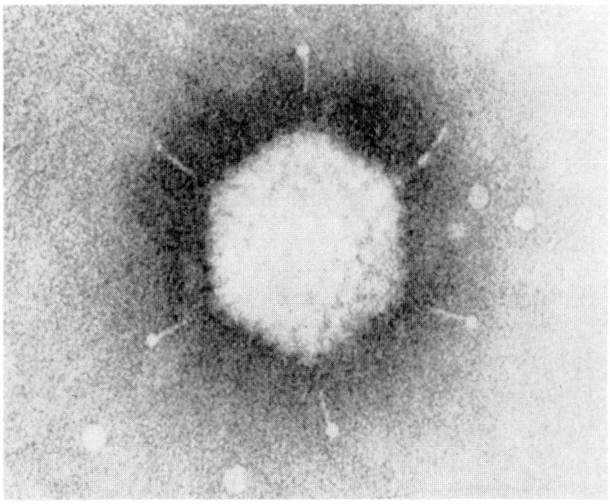
(b)

Figure 12–4 Morphology of a polyhedral virus in the shape of an icosahedron. **(a)** Diagram of an icosahedron. **(b)** Electron micrograph of an adenovirus (×174,000). Individual capsomeres in the protein coat are visible.

virus in the shape of an icosahedron is the adenovirus (Figure 12–4b). Another polyhedral virus is the poliovirus.

Enveloped Viruses As noted earlier, the capsid of some viruses is covered by an envelope, and these viruses are referred to as enveloped viruses. Enveloped viruses are roughly spherical but highly pleomorphic (variable in shape) because the envelope is not rigid. When helical or polyhedral viruses are enclosed by envelopes, they are referred to as **enveloped helical** and **enveloped polyhedral** viruses. An example of an enveloped helical virus is the influenza virus (Figure 12–5). An example of an enveloped polyhedral (icosahedral) virus is the herpes simplex virus (Figure 12–6).

Complex Viruses Some viruses, particularly bacterial viruses, have very complicated structures and are referred to as complex viruses. Examples of complex viruses are poxviruses, which do not contain clearly indentifiable capsids but have several coats around the nucleic acid (Figure 12–7a), and certain bacteriophages that have a capsid to which additional structures are attached (Figure 12–7b and c). If you take a close look at the bacteriophages shown in Figure 12–7b and c, you will note that the capsid (head) is polyhedral and the tail is helical. The head contains the nucleic acid. Later in the chapter we will discuss the functions of the additional structures such as the tail sheath, tail fibers, plate, and pin.

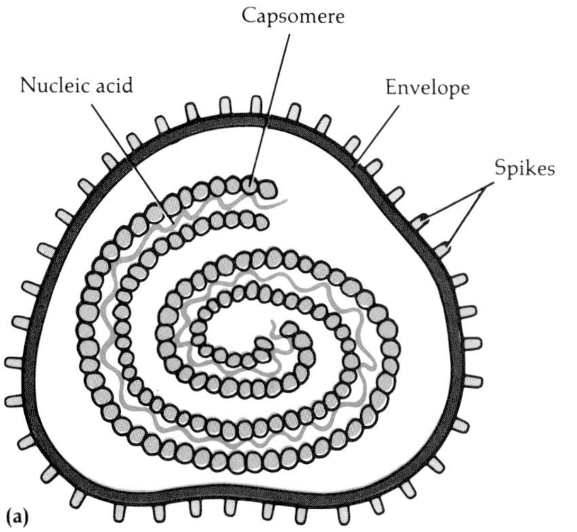

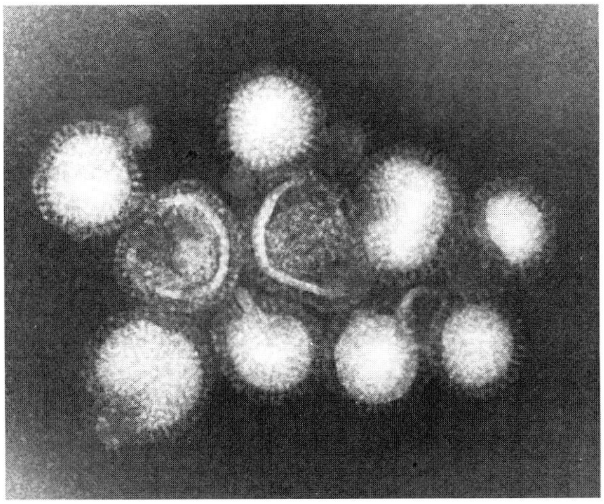

Figure 12–5 Morphology of an enveloped helical virus. **(a)** Diagram of an enveloped helical virus. **(b)** Electron micrograph of influenza viruses (×119,000). Note the halo of spikes projecting from the outer surface of each envelope (see Chapter 22).

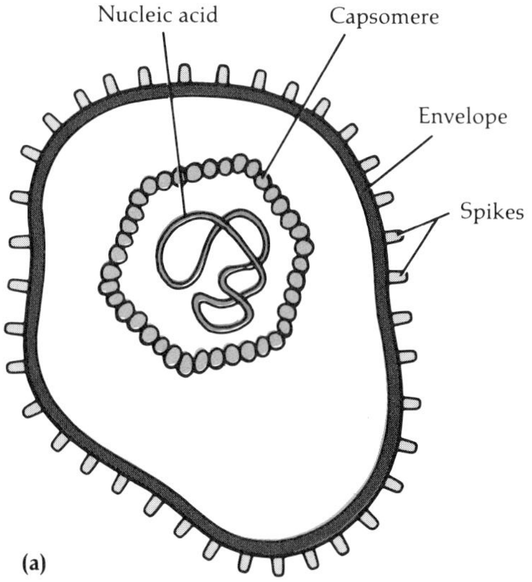

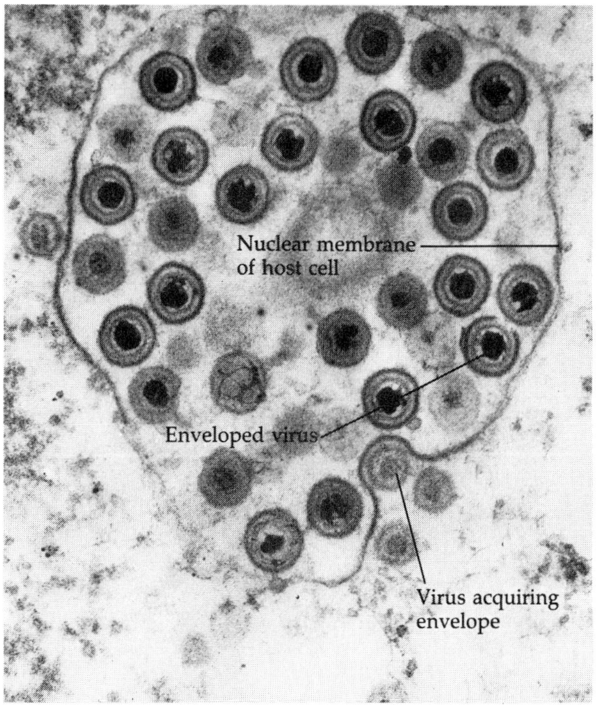

Figure 12–6 Morphology of an enveloped polyhedral (icosahedral) virus. **(a)** Diagram of an enveloped polyhedral virus. **(b)** Electron micrograph of a group of herpes simplex viruses (×48,000). To the lower right, a virus particle is acquiring its envelope as it buds out through a nuclear membrane of the host cell.

CLASSIFICATION OF VIRUSES

It was mentioned earlier that for practical purposes, viruses are classified as animal viruses, bacterial viruses, or plant viruses, according to host range. This classification is convenient, but not scientifically acceptable.

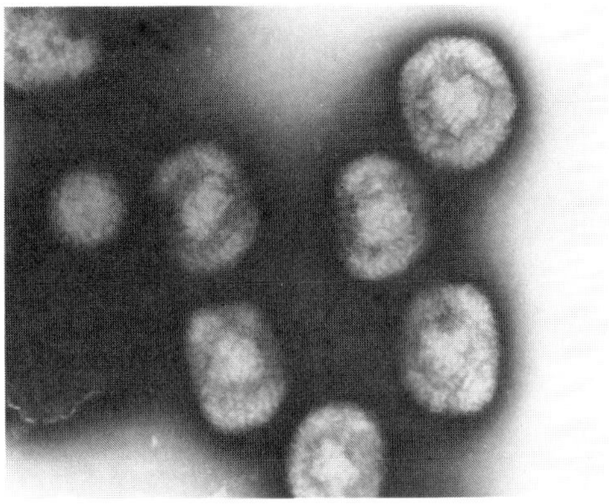

(a)

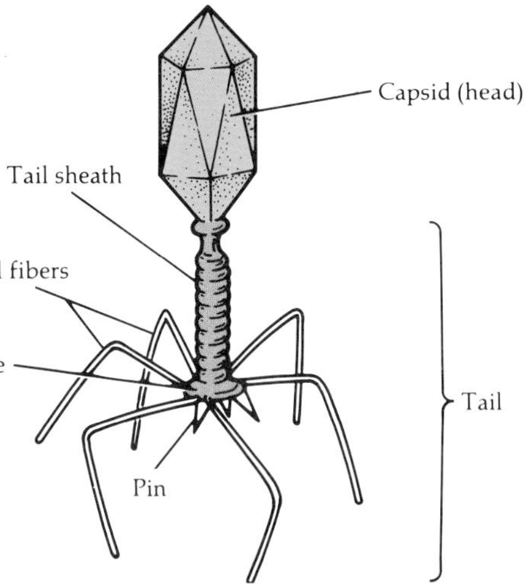

(b)

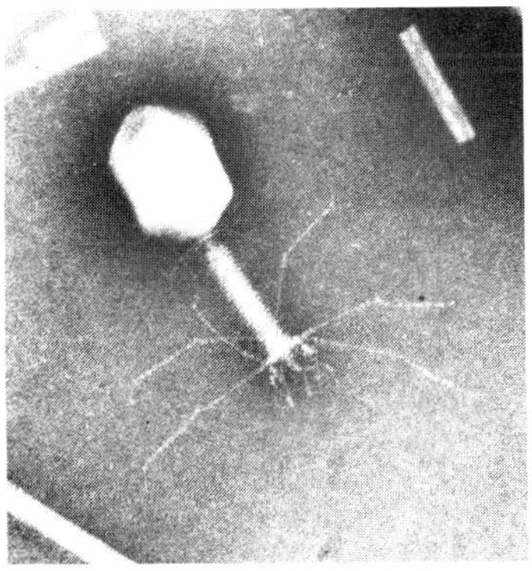

(c)

Figure 12–7 Morphology of complex viruses. **(a)** Electron micrograph of vaccinia virus, a strain of poxvirus used to vaccinate humans against smallpox. In rare cases it causes complications and should not be used on persons with eczema or on pregnant women (×42,000). **(b)** Diagram of a T-even bacteriophage, showing its component parts. **(c)** T4 bacteriophage (×110,000).

Table 12–1 Classification of Viruses by Symptomatology

Classification	Diseases
Generalized Diseases: Diseases in which a virus spreads throughout the body via the blood or lymph, affecting several organs. In some cases, skin rash occurs.	Smallpox, cowpox, measles, German measles, chickenpox, yellow fever, and dengue
Diseases Primarily Affecting Specific Organs:	
Nervous System	Poliomyelitis, aseptic meningitis, rabies, encephalitis
Respiratory system	Influenza, common cold, pharyngitis, respiratory syncytial pneumonia, bronchitis
Skin and mucous membranes	Cold sores, warts, shingles, molluscum contagiosum
Eye	Various types of conjunctivitis
Liver	Infectious hepatitis, serum hepatitis, yellow fever
Salivary glands	Mumps and cytomegalovirus
Gastrointestinal tract	Gastroenteritis A virus and gastroenteritis B virus

Table 12–2 Classification of Animal Viruses According to Morphological, Chemical, and Physical Properties

Viral Group and Specific Examples	Morphological Class	Nucleic Acid*	Dimensions of Capsid (Diameter) (nm)	Clinical or Special Features
Parvoviruses (adenosatellite)	Naked polyhedral	SS DNA	18–26	Very small viruses, most of which depend on coinfection with adenoviruses for growth; probably infect only rats, mice, and hamsters.
Papovaviruses (papilloma, polyoma, simian)	Naked polyhedral	DS circular DNA	40–57	Small viruses that induce tumors; the human wart virus (papilloma) and certain viruses that produce cancer in animals (polyoma and simian) belong to this family. Refer to Chapter 19.
Adenoviruses	Naked polyhedral	DS DNA	70–80	Medium-sized viruses that cause various respiratory infections in humans; some cause tumors in animals.
Herpesviruses (herpes simplex, herpes zoster)	Enveloped polyhedral	DS DNA	150–250	Medium-sized viruses that cause various human diseases, such as fever blisters, chickenpox, shingles, and infectious mononucleosis; implicated in a type of human cancer called Burkitt's lymphoma. Refer to Chapters 19 and 21.
Poxviruses (variola, cowpox, vaccinia)	Enveloped complex	DS DNA	200–350	Very large, complex, brick-shaped viruses that cause diseases such as smallpox (variola), molluscum contagiosum (wartlike skin lesion), cowpox, and vaccinia; vaccinia virus gives immunity to smallpox. Refer to Chapter 19.
Picornaviruses (poliovirus, rhinovirus)	Naked polyhedral	SS RNA Sense strand	18–38	Smallest RNA-containing viruses; at least 70 human enteroviruses are known, including the polio-, coxsackie-, and echoviruses; more than 100 rhinoviruses exist and are the most common cause of colds. Refer to Chapters 20 and 22.

In the oldest system, an animal virus was classified according to the organs that were affected by the diseases it caused. This is a classification by symptomatology (Table 12–1). Because the same virus can cause more than one disease, this scheme is not satisfactory for the microbiologist, but it does offer certain conveniences to the physician.

In the past few decades, hundreds of viruses have been isolated from plants, animals, and humans. As the list grows, the problem of viral classification becomes more and more complex. Present classification systems are based on factors such as type of nucleic acid present, morphological class, presence of an envelope, size of capsid, and number of capsomeres. A summary of the classification of animal viruses based on these morphological, chemical, and physical properties is

Table 12–2 Classification of Animal Viruses According to Morphological, Chemical, and Physical Properties *continued*

Viral Group and Specific Examples	Morphological Class	Nucleic Acid*	Dimensions of Capsid (Diameter) (nm)	Clinical or Special Features
Togaviruses (alpha virus, flavivirus)	Enveloped polyhedral	SS RNA Sense strand	40–60	Included are many viruses transmitted by arthropods; diseases include eastern equine encephalitis (EEE), Venezuelan equine encephalitis (VEE), St. Louis encephalitis (SLE), yellow fever, and dengue. Refer to Chapters 20 and 21.
Orthomyxoviruses (influenza A, B, C)	Enveloped helical	SS segmented RNA Antisense strand	80–200	Medium-sized viruses with a spiked envelope; have the ability to agglutinate red blood cells; cause influenza. Refer to Chapter 22.
Paramyxoviruses (measles, mumps)	Enveloped helical	SS RNA Antisense strand	150–300	Morphologically similar to myxoviruses, but generally larger; cause parainfluenza, measles, mumps. Refer to Chapters 19 and 23.
Coronaviruses	Enveloped helical	SS RNA Antisense strand	80–130	Associated with upper respiratory tract infections and the common cold. Refer to Chapter 22.
Retroviruses	Enveloped helical	DS RNA Two identical sense strands with reverse transcriptase	100–120	Includes all of the RNA tumor viruses; cause leukemia and tumors in animals; cause AIDS.
Rhabdoviruses (rabies)	Enveloped helical	SS RNA Antisense strand	70–180	Bullet-shaped viruses with a spiked envelope; cause rabies and Newcastle disease of chickens. Refer to Chapter 20.
Arenaviruses (lassa)	Enveloped helical	SS segmented RNA Antisense strand	50–300	Viruses contain RNA-containing granules; some members produce "slow" viral infections.
Reoviruses	Naked polyhedral	DS segmented RNA	60–80	Relation to human disease not clear; may be involved in mild respiratory infections and infantile gastroenteritis.

*SS = Single-stranded, DS = Double-stranded.

presented in Table 12–2. Other classification schemes also take into account susceptibility to physical and chemical agents, immunologic properties, site of multiplication (nucleus or cytoplasm), and natural methods of transmission.

Figure 12–8 shows electron micrographs of several DNA-containing animal viruses, and Figure 12–9 shows electron micrographs of several RNA-containing animal viruses.

ISOLATION, CULTIVATION, AND IDENTIFICATION OF VIRUSES

The fact that viruses cannot multiply outside of a living host cell complicates their detection, enumeration, and identification. It is necessary to provide viruses with living cells instead of a fairly simple chemical medium. Living plants and animals are difficult and expensive to maintain, and dis-

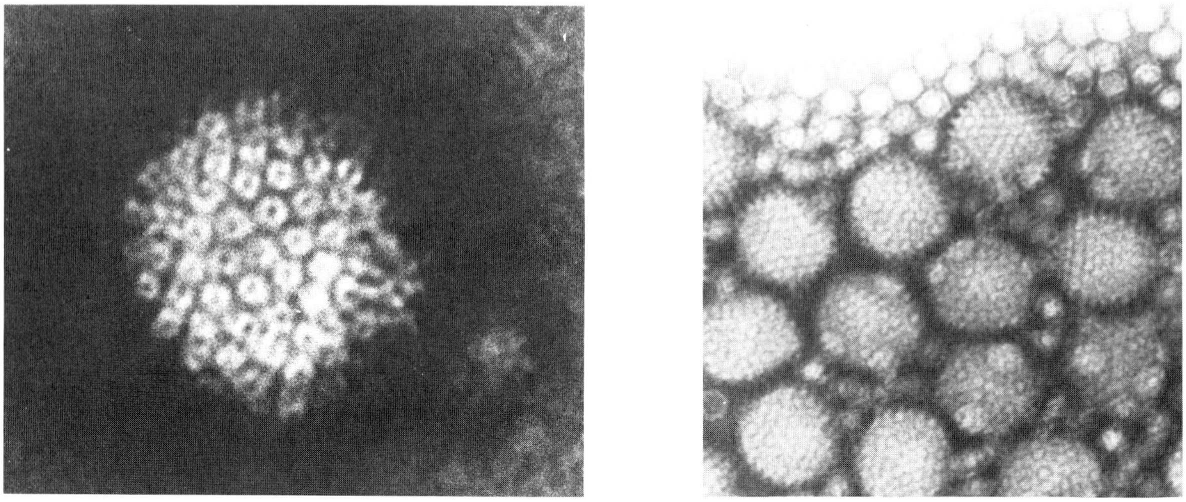

(a) (b)

Figure 12–8 DNA-containing viruses. **(a)** Negatively stained capsid of a herpesvirus. The individual capsomeres are clearly visible. **(b)** Negatively stained viruses that have been concentrated in a centrifuge gradient (×198,000). The larger capsids are an adenovirus and the smaller "adeno-associated particles" are a parvovirus.

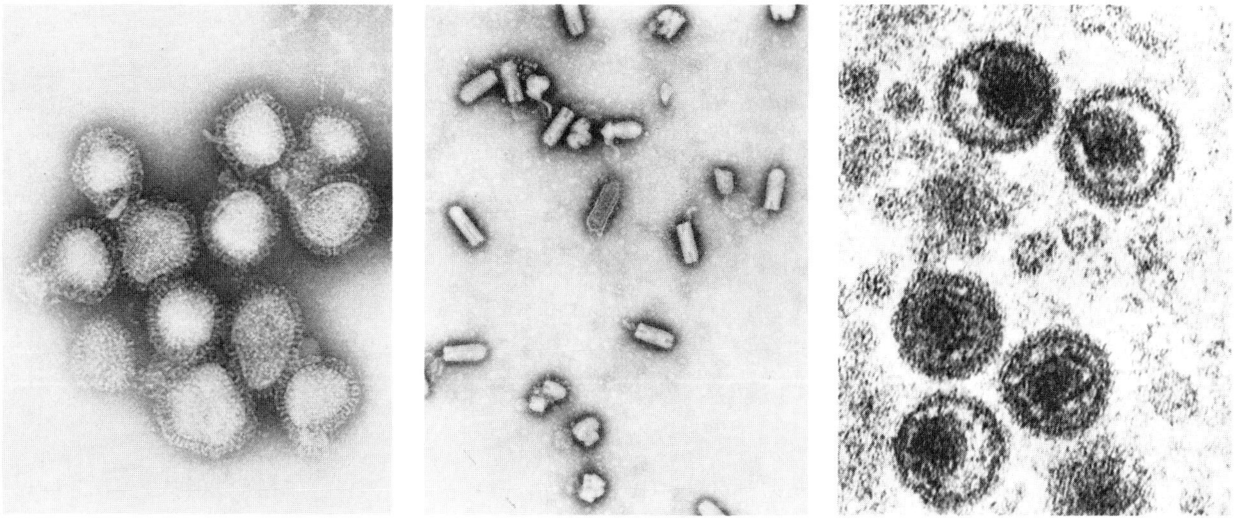

(a) (b) (c)

Figure 12–9 RNA-containing viruses. **(a)** Capsids of A_2 influenza virus (×92,000). Notice the halo of spikes projecting from each capsid. **(b)** Particles of VSV rhabdovirus, a rod-shaped virus similar to rabies (×37,000). **(c)** Mouse mammary tumor virus, a B-type retrovirus (×160,000).

ease-causing viruses that grow only in higher primates and human hosts cause additional complications. On the other hand, viruses that use bacterial cells as a host (bacteriophages) are rather easily grown on bacterial cultures. This condition is one reason why so much of our understanding of viral multiplication has come from bacteriophages.

Growth of Bacteriophages in the Laboratory

Bacteriophages can be grown either in suspensions of bacteria in liquid medium or in bacterial cultures on solid medium. The use of solid medium makes possible the **plaque method** for detecting and counting viruses. Only simple materials and equipment are needed for this procedure, which was first developed for use with bacteriophages. A sample of bacteriophage is mixed with host bacteria and melted agar. The agar plus bacteriophage and host bacteria is then poured into a Petri plate containing a hardened layer of agar growth medium. The virus–bacteria mixture solidifies into a thin top layer that contains a layer of bacteria approximately one cell thick. Each virus infects a bacterium, multiplies, and releases several hundred new viruses. These newly produced viruses infect other bacteria in the immediate vicinity, and more new viruses are produced. Following several viral-multiplication cycles, all the bacteria in the area surrounding the original virus are destroyed. This produces a number of clearings, or **plaques,** visible against a "lawn" of bacterial growth on the surface of the agar (Figure 12–10). While the plaques form, uninfected bacteria elsewhere in the Petri plate multiply rapidly by fission and produce a turbid background.

Each plaque theoretically corresponds to a single virus in the initial suspension. Because a single plaque can arise from more than one virion and because some virions may not be infectious, the concentrations of viral suspensions measured by the number of plaques are usually given in terms of **plaque-forming units (pfu).**

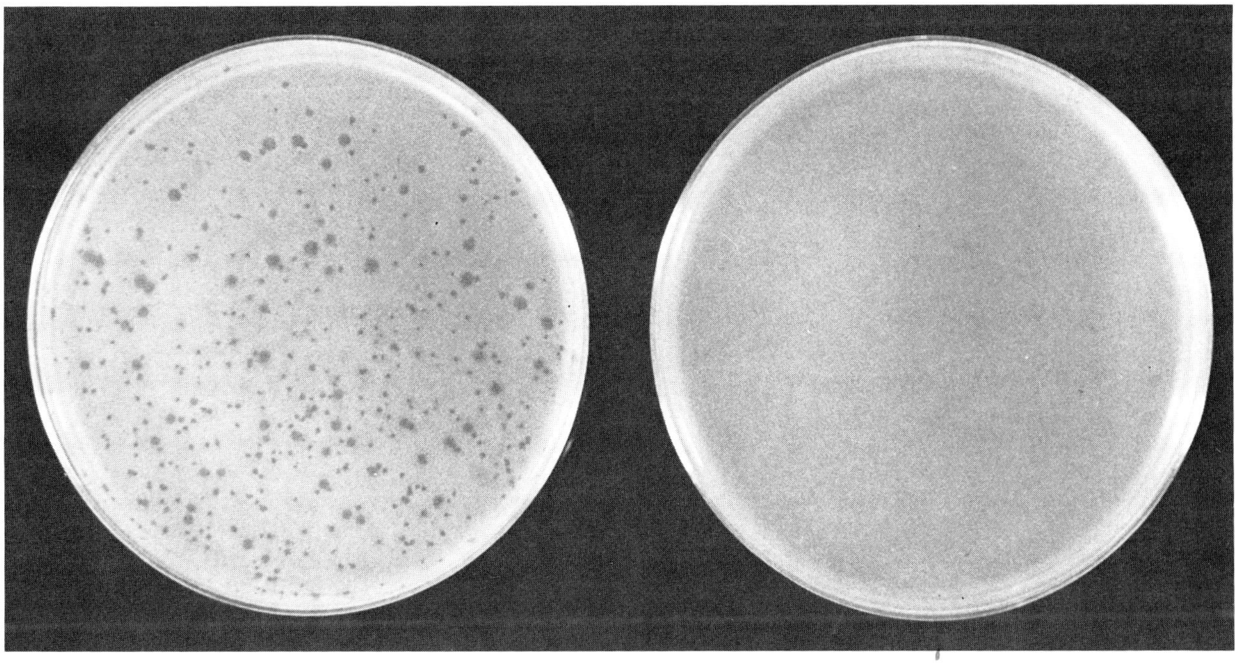

Figure 12–10 In the Petri plate on the left, clear viral plaques of varying sizes have been formed by the bacteriophage lambda on a lawn of *E. coli*. For comparison, the Petri plate on the right contains a bacterial culture without phage.

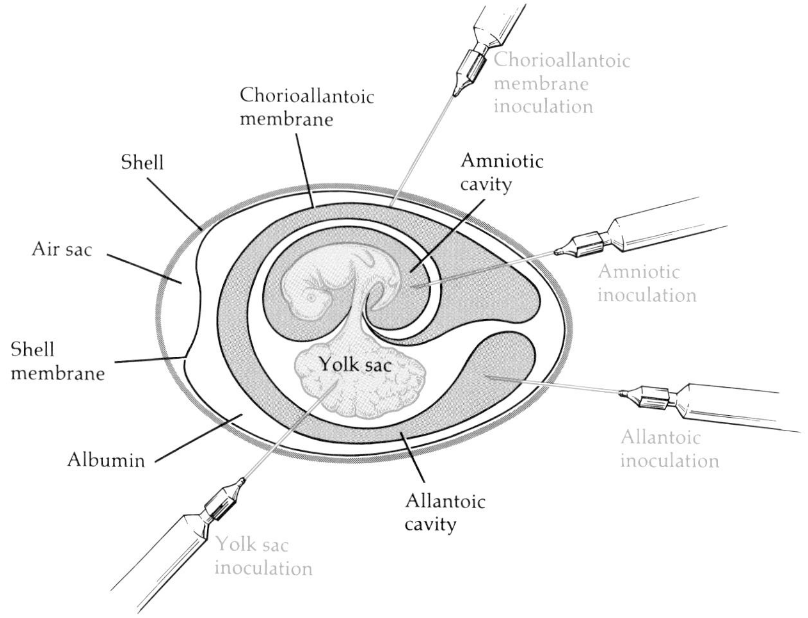

Figure 12–11 Inoculation of embryonated eggs. The injection site determines the membrane on which the viruses will grow.

Growth of Animal Viruses in the Laboratory

In living animals

Some animal viruses can only be cultured in **living animals,** such as mice, rabbits, and guinea pigs. Thus, most experiments to study the immune system's response to viral infections must also be performed in virally infected whole animals. Generally, animal inoculation is used as a diagnostic procedure for identifying a virus from a clinical specimen. After the animal is inoculated with the specimen, the animal is observed for signs of disease or is killed so that infective tissues can be examined and evaluated.

In embryonated eggs

If the virus will grow in it, the **embryonated egg** is a fairly convenient and inexpensive form of animal host for many animal viruses (Figure 12–11). A hole is drilled in the shell of the embryonated egg, and a viral suspension or suspected virus-containing tissue is injected into the fluid of the egg. There are several membranes in the egg on which the virus can be made to grow, and the virus is injected into the proper location in the egg. Viral growth is signaled by the death of the embryo, by embryo cell damage, or by the formation on the membranes of the egg of typical pocks or lesions that result from viral growth. This method was once the most widely used method of viral isolation and growth and is still used to grow viruses for some vaccines. You may be asked if you are allergic to eggs before receiving a vaccination, since egg proteins may be present in the viral vaccine preparations. (Allergy is discussed in Chapter 17.)

In cell culture

Recently, **cell culture** (sometimes called **tissue culture,** although that is not the best term) has replaced embryonated eggs as growth media for many viruses. Cell cultures are animal cells grown in culture media in the laboratory. Because these cultures are generally rather homogeneous collections of cells and can be propagated and handled much like bacterial cultures, they are more convenient to work with than whole animals or embryonated eggs.

Cell culture lines are readily started by treatment of a slice of animal tissue with enzymes that separate the individual cells. These cells are suspended in a solution that provides the osmotic pressure, nutrients, and growth factors needed for the cell to grow. The cells tend to adhere to the glass or plastic container and reproduce to form a monolayer. Viruses infecting such a monolayer sometimes cause the cells of the monolayer to deteriorate as they multiply. This tissue deterioration is called the **cytopathic effect (CPE)** and is illustrated in Figure 12–12. The CPE can be detected and counted in much the same way as are plaques caused by bacteriophages on a lawn of bacteria.

Primary cell lines, derived from tissue slices, tend to die out after only a few generations. Certain cell lines developed from human embryos can be maintained for about a hundred generations and are widely used for diagnostic work concerning human diseases. Cell lines developed from embryonic human cells are used to culture rabiesvirus for a rabies vaccine called human diploid culture vaccine (see Chapter 20).

When viruses are routinely grown in a laboratory, **continuous cell lines** are used. These are "transformed" cells that can be maintained through an indefinite number of generations, and they are sometimes called "immortal" cell lines (see the discussion of transformation at the end of this chapter). One of these, the HeLa cell line, was isolated from the cancer of a woman who died in 1951. After years of laboratory cultivation, many such cell lines have lost almost all the original characteristics of the cell, but these changes have not interfered with the use of the cells for viral propagation. In spite of the success of cell culture in viral isolation and growth, there are still some viruses that have never been successfully cultivated in cell culture.

A major problem with cell culture is that the cell lines must be kept free of microbial contamination. In fact, the idea of cell culture is not new, but dates back to the end of the last century. However, it was not a practical laboratory technique until the development of antibiotics in the years following World War II. The maintenance of cell culture lines still requires considerable experience and trained technicians working on a full-time

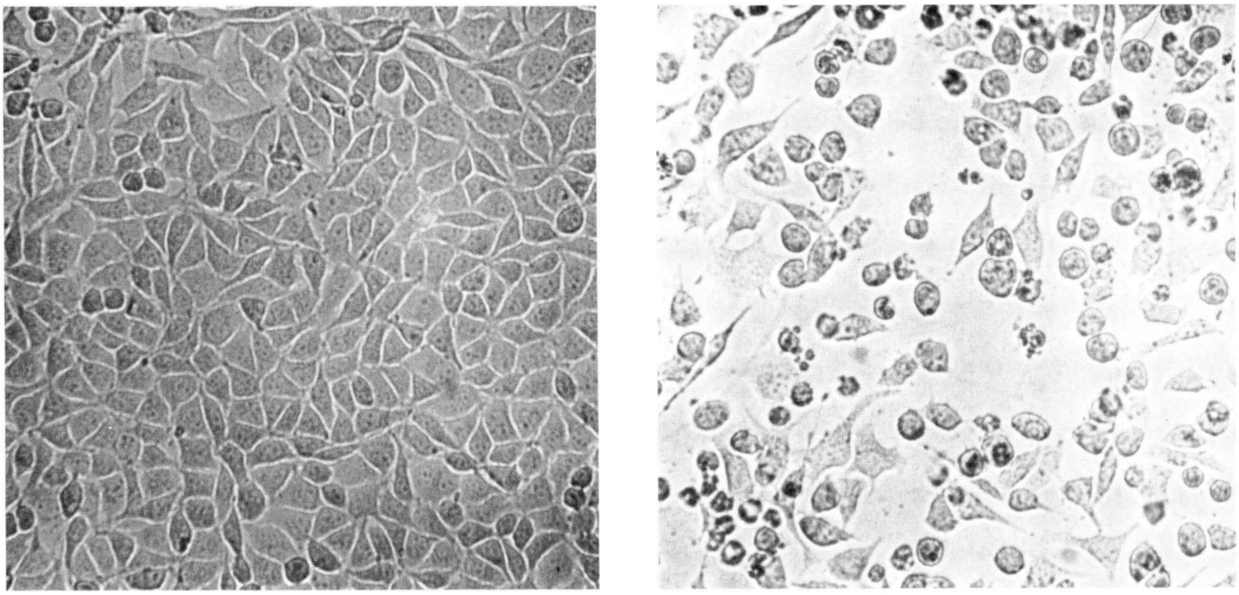

(a) **(b)**

Figure 12–12 Cytopathic effect of viruses on a cell culture. **(a)** A monolayer of uninfected mouse L cells. **(b)** The same cells 24 hours after infection with vesicular stomatitis virus. Note the separation and "rounding up" of the cells.

basis. Because of these difficulties, most hospital laboratories and many state health laboratories do not isolate and identify viruses in clinical work. Instead, the tissue or serum samples are sent to central laboratories that specialize in this work.

The identification of viral isolates is not an easy task. For one thing, viruses cannot be seen at all without the use of an electron microscope. Immunological methods are the most commonly used means of identification. In these tests, the virus is detected and identified by its reaction with antibodies. Antibodies, which are proteins produced by animals in response to exposure to a virus, are highly specific for the virus that caused their formation. We shall discuss antibodies in detail, along with a number of immunological tests for identifying viruses, in Chapter 16. Later in this chapter we will describe how viruses can be identified by observation of their effects on host cells.

Using modern biochemical methods, molecular biologists can identify and characterize purified viruses by isolating the particular nitrogenous base sequences of the virus's nucleic acids. For DNA viruses, restriction enzymes are used to cut DNA into specific nitrogenous base sequences, or fragments (see Chapter 8). Fragments of closely related viruses have similar patterns. Using more sophisticated methods, researchers can even determine detailed nitrogenous-base sequences of entire viral genomes. However, these research-laboratory methods are not practical for the routine identification of clinical samples.

VIRAL MULTIPLICATION

The nucleic acid in a virion contains only a few of the genes needed for the synthesis of new viruses. These include genes for the virion's structural components such as the capsid proteins, and genes for a few of the enzymes used in the viral life cycle. Most **viral enzymes** (or **virus-specific enzymes**) —the enzymes encoded in viral nucleic acid—are not part of the virion. They are synthesized and caused to function only when the virus is within the host cell. Viral enzymes are almost entirely concerned with replicating or processing viral nucleic acid, and almost never with the machinery of protein synthesis or energy production. Although the smallest naked virions do not contain any preformed enzymes, the larger virions may contain one or a few enzymes, which usually function in helping the virus penetrate the host cell or replicate its own nucleic acid.

Thus, for a virus to multiply, it must invade a host cell and take over the host's metabolic machinery. A single virus can give rise to several, or even thousands of, similar viruses in a single host cell. This process drastically changes the host cell and often causes the death of the host cell.

Multiplication of Bacteriophages

Although the means by which a virus enters and exits a host cell may vary, the basic mechanism of viral multiplication is similar for animal viruses, plant viruses, and bacteriophages. The best understood viral life cycles are those of the **bacteriophages,** or simply **phages.** Because the so-called *T-even bacteriophages* (T2, T4, and T6) have been studied most extensively, we will first describe the multiplication of T-even bacteriophages in their host, *E. coli.*

T-even bacteriophages

The T-even bacteriophages have large, complex naked virions, whose characteristic head-and-tail structure was shown in Figure 12–7b and c. The length of DNA contained in these bacteriophages is only about 6% of that contained in *E. coli;* the bacteriophage has enough DNA for over 100 genes. The multiplication cycle of these phages, like those of all viruses, can be divided into several distinct stages: adsorption, penetration, biosynthesis of viral components, maturation, and release.

Adsorption of Phage to Host Cell After a chance collision between phage particles and bacteria, **adsorption** occurs. During this process, an adsorption site on the virus attaches to a complementary receptor site on the bacterial cell. This attachment is a chemical interaction in which weak bonds are formed between the adsorption and receptor sites. T-even bacteriophages use fibers at the end of the tail as adsorption sites. The complementary receptor sites are on the bacteria's cell wall (Figure 12–13a). (Other phages adsorb to flagella or pili.)

Penetration After adsorption, the T-even bacteriophage injects its DNA (nucleic acid) into the bac-

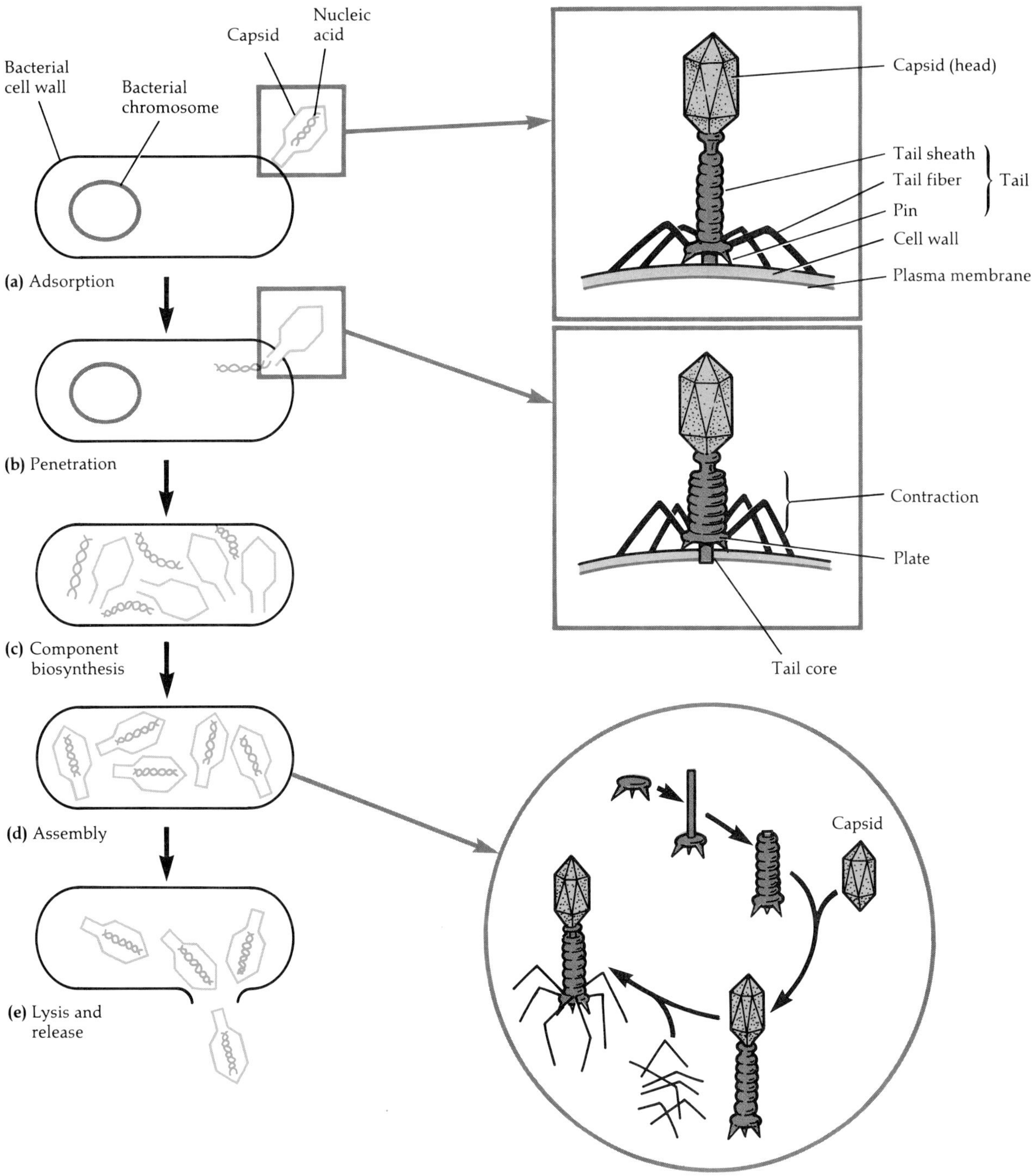

Figure 12–13 Multiplication cycle of a T-even bacteriophage. **(a)** Adsorption. **(b)** Penetration. **(c)** Biosynthesis of viral components (capsids and DNA are shown). **(d)** Maturation. Virions are assembled. **(e)** Lysis of the host cell and release of new virions.

terium. To do this, the bacteriophage's tail releases an enzyme, *phage lysozyme,* which breaks down a portion of the bacterial cell wall. During the process of **penetration,** the tail sheath of the phage contracts, and the tail core is driven through the cell wall. When the tip of the core reaches the plasma membrane, the DNA from the bacteriophage's head passes through the tail core, through the plasma membrane, and enters the bacterial cell. The capsid of most bacteriophages remains outside the bacterial cell (Figure 12–13b).

Biosynthesis of Viral Components Once the bacteriophage DNA has reached the cytoplasm of the host cell, the **biosynthesis** of viral nucleic acid and protein occurs. In this process, the viral DNA takes over the metabolic machinery of the host cell. Transcription of RNA from the host chromosome stops, since the host DNA is broken down. Any RNA subsequently transcribed is mRNA transcribed from phage DNA. Because host enzymes continue to function, they produce energy for the biosynthesis of phage DNA and protein. Along with host enzymes, enzymes encoded in the phage DNA are synthesized and used by the phage.

Initially, the phage uses the host cell's nucleotides and several of its enzymes to synthesize many copies of phage DNA. Soon after, the biosynthesis of viral proteins begins. The host cell's ribosomes, enzymes, and amino acids are used to synthesize viral proteins, including capsid proteins (Figure 12–13c).

Recall that during penetration, the protein coat (capsid) remains outside while the phage DNA is injected into the host cell. This means that the phage DNA must provide the template for the production of all viral components including new phage DNA. mRNA is transcribed from the phage DNA for the translation of phage enzymes and capsid protein.

For several minutes following infection, complete phages cannot be found in the host cell. Only separate components can be detected: DNA and proteins. The time during viral multiplication when complete, infective virions are not yet present is called the **eclipse period.**

Maturation In the next sequence of events, **maturation** occurs. In this process, bacteriophage DNA and capsids are assembled into complete virions. The assembly process is guided by the products of certain viral genes in a step-by-step sequence. The phage heads and tails are separately assembled from protein subunits, the head is packaged with phage DNA, and the tail is attached (Figure 12–13d). (For many simpler viruses, the nucleic acid and capsid proteins assemble spontaneously to form virions, without the intervention of other phage gene products.)

Release The general term for the final stage of viral multiplication is **release,** and it refers to the release of the new virions from the host cell. The term **lysis** is generally used for this stage in the multiplication of T-even phages. Lysozyme, whose code is provided by a phage gene, is synthesized within the cell. This enzyme causes a breakdown of the bacterial cell wall, and the newly produced bacteriophages are released from the host cell (Figure 12–13e). The released bacteriophages infect other susceptible cells in the area, and the viral multiplication cycle is repeated within those cells.

Lysozyme is not synthesized early in the multiplication cycle, because it would cause lysis of the host cell before complete phages could be assembled. Accordingly, there are genetic controls that regulate when different regions of phage DNA are transcribed into mRNA during the multiplication cycle. For example, there are early messages that are translated into early phage proteins, the enzymes used in the synthesis of phage DNA. Also, there are late messages that are translated into late phage proteins for the synthesis of capsid proteins and lysozyme. This control mechanism is mediated by RNA polymerase.

The time elapsed from phage adsorption to release is known as **burst time** and averages between 20 to 40 minutes. The number of newly synthesized phage particles released from a single cell is referred to as **burst size** and usually ranges from about 50 to 200.

The various stages involved in the multiplication of phages can be demonstrated experimentally in what is known as a **one-step growth experiment** (Figure 12–14). In this procedure, a phage suspension is diluted until a sample containing only a few phage particles can be obtained. These particles are then introduced into a culture of host cells. Periodically, samples of phage particles are removed from the culture and inoculated onto

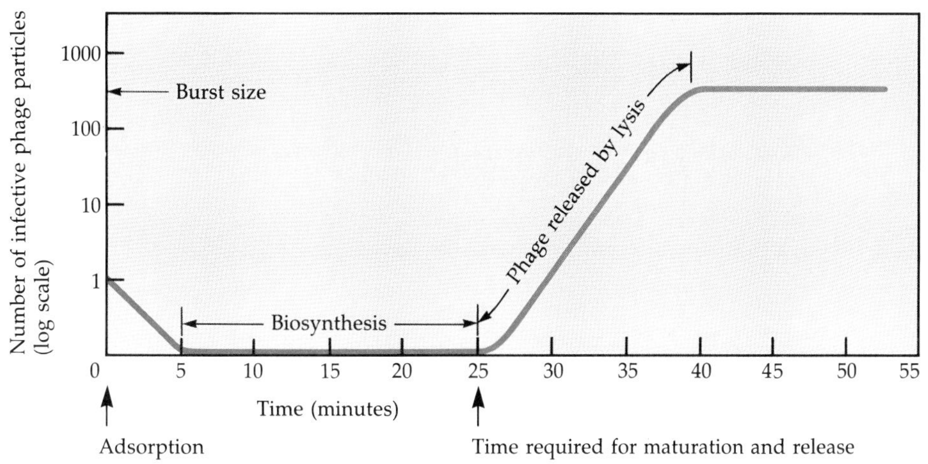

Figure 12–14 A bacteriophage one-step growth curve. No new infective phage particles are found in a culture until after biosynthesis and maturation have taken place.

a plate-culture of susceptible host cells; the plaque method is used to determine the number of infective phage particles on this culture (see Figure 12–10). A few minutes after adsorption, there are no infective particles present. However, phage nucleic acid is found inside the infected cells and capsid proteins are then sythesized. The time required for maturation is the interval between the appearance of phage nucleic acid and the synthesis of mature phages. After a few minutes, the number of infective phage particles found in subcultures begins to rise. The burst size is determined once the number of infective phage particles remains constant, indicating no further phage multiplication will occur.

Lysogeny

In the sequence of events just described for the multiplication of T-even bacteriophages, the release of the phages causes lysis and death of the host cell. Thus, such a sequence is referred to as a **lytic cycle.** Some phages can either proceed through a lytic cycle or incorporate their DNA into the host cell's DNA. In the latter state, the phage remains latent and does not cause lysis of the host cell. Such a state is called **lysogeny,** and such phages are referred to as **lysogenic phages** or **temperate phages.** The participating bacterial host cells are known as **lysogenic cells.**

Let us go through the steps in lysogeny for the bacteriophage λ (lambda), a well-studied temperate phage (Figure 12–15). (Lambda is the name of a Greek letter.) Upon penetration into an *E. coli* cell, the linear phage DNA forms a circle. This circle can multiply and be transcribed, leading to the production of new phage and to cell lysis. Or, the circle can recombine with and become part of the circular bacterial DNA. The inserted phage DNA is now called a **prophage.** Most of the prophage genes are repressed by a repressor protein that is the product of one phage gene. Thus, the phage genes that would otherwise direct the synthesis and release of new virions are turned off in much the same way that the genes of the *E. coli* lac operon are turned off by the lac repressor (see Chapter 8).

Every time the host cell's machinery replicates the bacterial chromosome, it also replicates the prophage DNA. The prophage remains latent within the progeny cells. However, a rare spontaneous event or the action of ultraviolet light or certain chemicals can lead to the excision (popping-out) of the phage DNA—and initiation of the lytic cycle.

There are three important outcomes of lysogeny. First, the lysogenic cells are immune to reinfection by the same phage. (However, the host

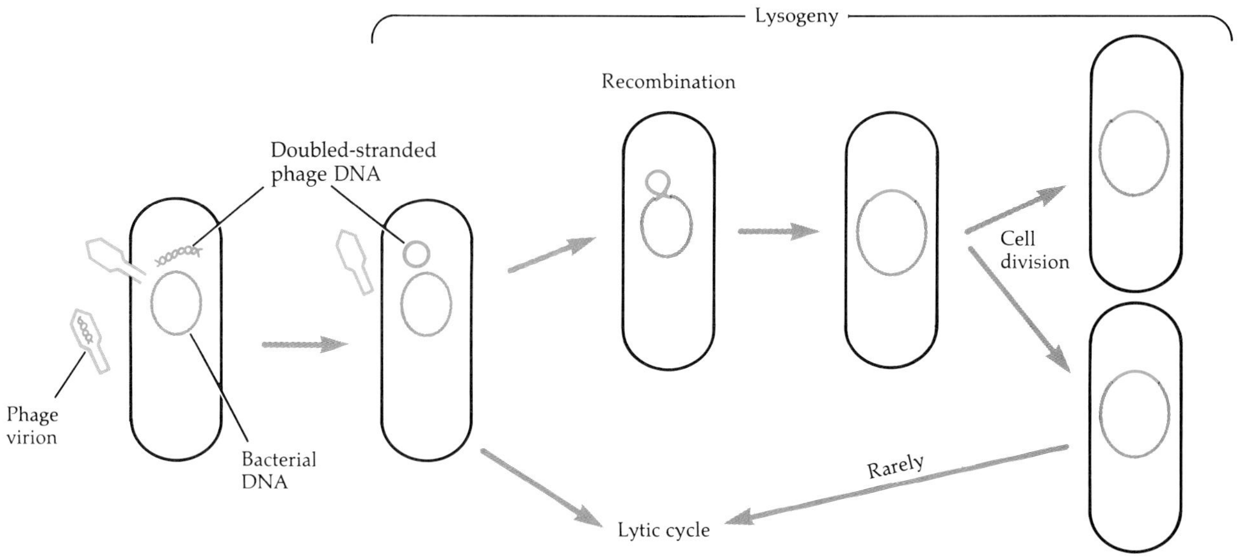

Figure 12–15 Lysogeny. After entering the host cell, the DNA from the temperate phage forms a circle, and by a single recombination event, becomes integrated into the host DNA. The host cell, now called a lysogenic cell, replicates the phage DNA (prophage) every time it divides. Infection by a temperate phage can also lead directly to a lytic cycle, and on rare occasions, a prophage excises from the bacterial chromosome and initiates a lytic cycle. In this and the next figure, all DNA shown is double stranded.

cell is not immune to infection by other phage types.) The second outcome of lysogeny is that the host cell may exhibit new properties. For example, the bacterium *Corynebacterium diphtheriae,* which causes diphtheria, is a pathogen whose disease-producing properties are related to the synthesis of a toxin. The organism is capable of producing toxin only when it carries a temperate phage, because the prophage carries the gene coding for the toxin. As another example, only streptococci carrying a temperate phage are capable of producing the toxin associated with scarlet fever. And the toxin produced by *Clostridium botulinum,* which causes botulism, is coded by a prophage gene.

The third possible outcome of lysogeny is **specialized transduction.** Recall from Chapter 8 that bacterial genes can be picked up in a phage coat and transferred to another bacterium, in a process called **generalized transduction.** Any bacterial genes can be transferred by generalized transduction because the host chromosome is broken down into fragments, any of which can be packaged into a phage coat. In specialized transduction, however, only certain bacterial genes can be transferred.

Specialized transduction is mediated by a lysogenic phage, which packages bacterial DNA *along with* its own DNA in the same capsid. When a prophage is excised from the host chromosome, adjacent genes from either side may remain attached to the phage DNA (Figure 12–16). The genes that remain attached to bacteriophage λ are genes for galactose utilization and biotin synthesis. A phage particle newly synthesized from this DNA may thus contain one or more of the bacterial genes along with phage DNA. Upon lysogenizing a new host cell, the phage can confer new characteristics to the cell; a cell infected by bacteriophage λ, for example, could gain the ability to metabolize galactose or synthesize biotin.

Certain animal viruses may be able to undergo processes very similar to lysogeny. Animal viruses that can remain latent in cells for long periods without multiplying or causing disease may become inserted in a host chromosome or remain separate from host DNA, in a repressed state (as some lysogenic phages). Cancer-causing viruses may also be lysogenic, as will be discussed later.

Multiplication of Animal Viruses

The multiplication of animal viruses follows the basic pattern of bacteriophage multiplication, but has several notable differences. Animal viruses differ from phages in their mechanisms of entering the host cell. And once the virus is inside, the synthesis and assembly of the new viral components are somewhat different, partly because of the differences between procaryotic cells and eucaryotic cells. Animal viruses may have certain types of enzymes not found in phages. Finally, the mechanisms of maturation and release, and the effects on the host cell, differ between animal viruses and phages.

In the following discussion of the multiplication of animal viruses, we will first consider the processes that are shared by both DNA- and RNA-containing animal viruses. These processes are adsorption, penetration, and uncoating. Then we will examine how DNA- and RNA-containing viruses differ with respect to their processes of biosynthesis and release from host cells.

Adsorption

Like bacteriophages, animal viruses have adsorption sites that attach to complementary receptor sites on the host cell's surface. However, the receptor sites of animal cells are proteins and glycoproteins of the plasma membrane (Figure 12–17a). Moreover, animal viruses do not possess appendages like the tail fibers of some bacteriophages. The adsorption sites of animal viruses are distributed over the surface of the virus. The sites themselves vary from one group of viruses to another. In adenoviruses, which are naked icosahedral viruses, the adsorption sites are small fibers at the corners of the icosahedron. In many of the enveloped viruses, such as the myxoviruses, the adsorption sites are spikes located on the surface of the envelope (see Figure 12–5a). As soon as one spike attaches to a host receptor, additional receptor sites on the same cell migrate to the virus. Adsorption is completed when many adsorptior sites are attached. Moreover, adsorption is enhanced by the presence of cations.

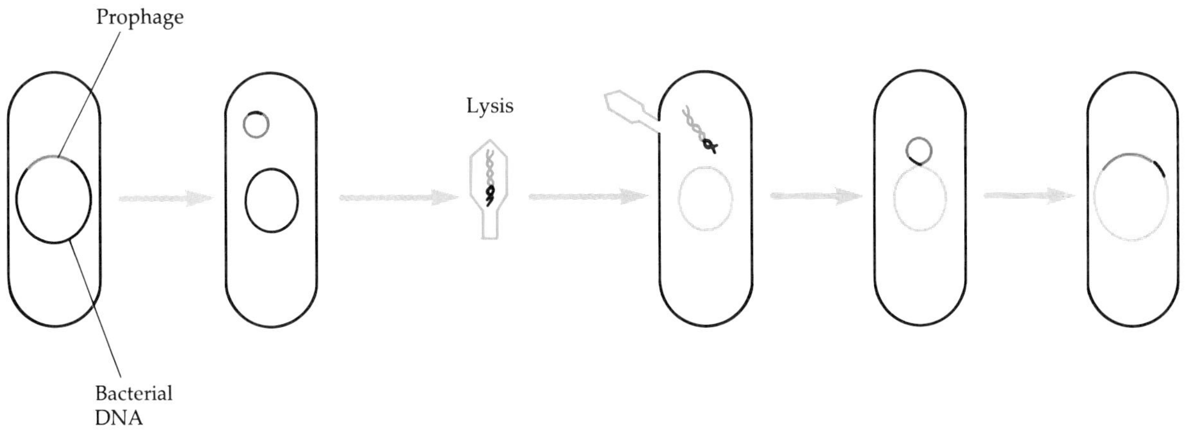

Figure 12–16 Specialized transduction. When a prophage is excised from its host chromosome, it can take with it a bit of the DNA from the bacterial chromosome. Later, when the phage lyses the host cell and enters a new host, it carries with it this bit of the original host's chromosome. Along with the prophage, these genes become integrated into new bacterial DNA.

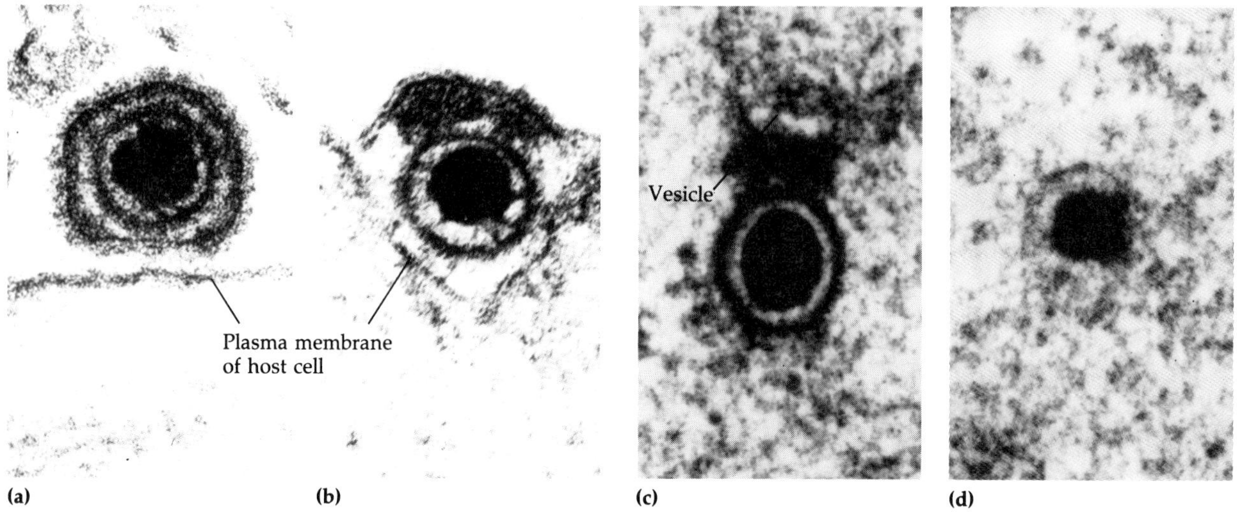

Figure 12–17 Entry of herpes simplex virus into an animal cell. **(a)** Adsorption of the viral envelope to the plasma membrane (×180,000). **(b)** Formation of a vesicle around the virus, which results in loss of the envelope (×180,000). **(c)** Unenveloped capsid entering the cytoplasm of the cell from the vesicle (×180,000). **(d)** Digestion of the capsid leaves only the nucleic acid core (×180,000).

Penetration

Following adsorption, penetration occurs. In enveloped animal viruses, penetration occurs by **endocytosis.** Endocytosis is an active cellular process by which nutrients and other molecules are brought into a cell. A cell's membrane continuously folds inward to form vesicles. These vesicles contain elements that originate outside the cell and are brought into the interior of the cell to be digested. If a virion attaches to a small outfolding (called a microvillus) on the plasma membrane of a potential host cell, the host cell will enfold the virion into its fold of plasma membrane, the vesicle (Figure 12–17b). Once the virion is enclosed within the vesicle, its viral envelope is destroyed. The capsid is digested when the cell attempts to digest the vesicle's contents. The viral nucleic acid is then released into the cytoplasm of the host cell (Figure 12–17c).

An alternate method of penetration has been hypothesized: the viral envelope fuses with the plasma membrane and releases the capsid into the host cell's cytoplasm.

Uncoating

Uncoating is the separation of the viral nucleic acid from its protein coat. It is a poorly understood process, and one that apparently varies with the type of virus. Uncoating occurs only in animal viruses: recall that phage DNA is inserted directly into the host cell. Some viruses accomplish uncoating by the action of lysosomal enzymes contained inside phagocytic vacuoles and coated vesicles. These enzymes degrade the proteins of the viral capsid. The uncoating of poxviruses is completed by a specific enzyme coded by the viral DNA and synthesized soon after infection. For other viruses, uncoating appears to be exclusively caused by enzymes in the host cell cytoplasm (Figure 12–17d). For at least one virus, the poliovirus, uncoating seems to begin while the virus is still attached to the host cell's plasma membrane.

Biosynthesis for DNA-containing viruses

When a poxvirus multiplies, the biosynthesis of viral components and their subsequent assembly occur in the cytoplasm of the host cell. The multiplication of other DNA-containing viruses (parvoviruses, papovaviruses, adenoviruses, and herpesviruses) differs in that their viral DNA is replicated in the nucleus of the host cell, whereas their viral protein is synthesized in the cytoplasm. This stage is followed by the migration of the proteins into the nucleus of the host cell; there they are assembled into complete viruses. To show an example of the multiplication of a DNA virus, we will follow the sequence of events in adenovirus (Figure 12–18).

After adsorption, penetration, and uncoating, the viral DNA is released into the nucleus of the host cell. Next, the transcription of a portion of the viral DNA—the "early" genes—occurs; translation follows. The products of these genes are enzymes required for the multiplication of viral DNA. In most DNA viruses, early transcription is carried out with the host's transcriptase (RNA polymerase); poxviruses, however, contain their own transcriptase. Some time after the initiation of DNA multiplication, transcription and translation of the remaining, "late," viral genes occur. This leads to the synthesis of capsid proteins, which occurs in the cytoplasm of the host cell. After the capsid proteins migrate into the nucleus of the host cell, maturation occurs: the viral DNA and capsid proteins assemble themselves to form complete viruses.

Biosynthesis for RNA-containing viruses

The multiplication of RNA viruses is essentially the same as that of DNA viruses, except that several different mechanisms of mRNA formation occur among different groups of RNA viruses. Although the details of these mechanisms are beyond the scope of this text, for comparative purposes we will trace the multiplication cycles of some representative RNA viruses (Figure 12–19). Multiplication of RNA viruses takes place in the host cell's cytoplasm. The major differences among the multipli-

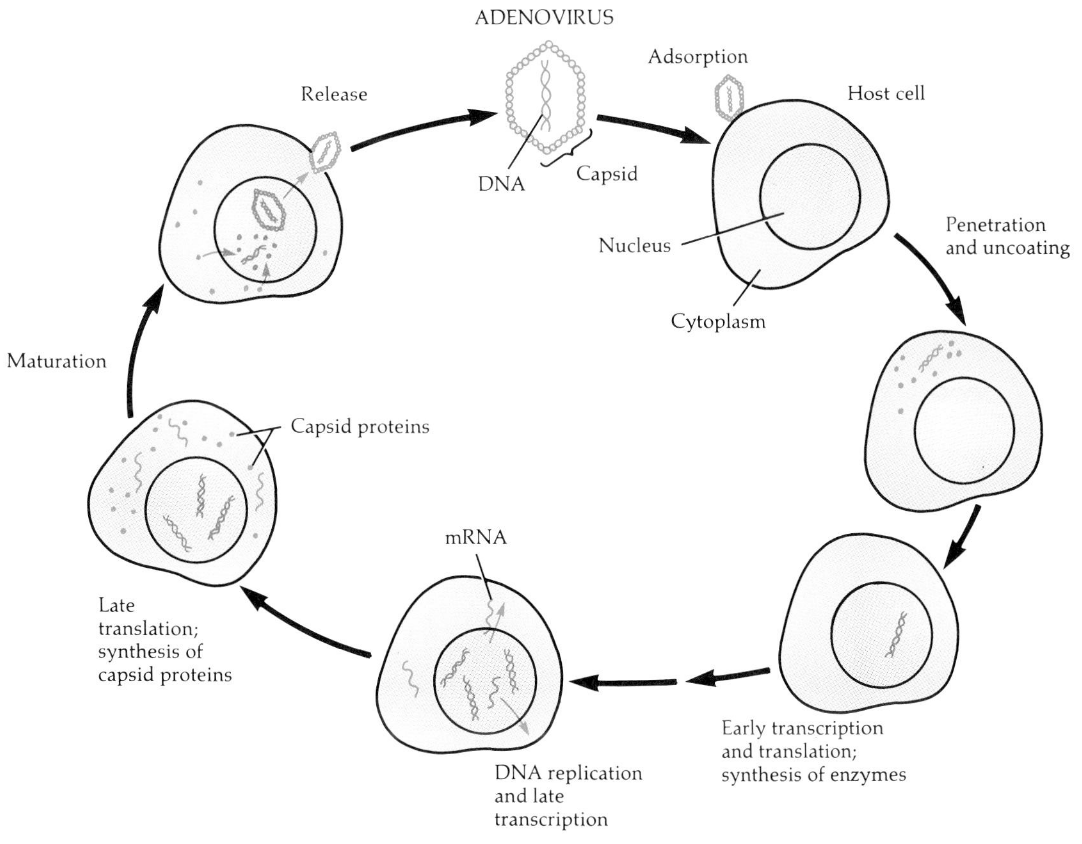

Figure 12–18 Multiplication of adenovirus, a DNA-containing virus.

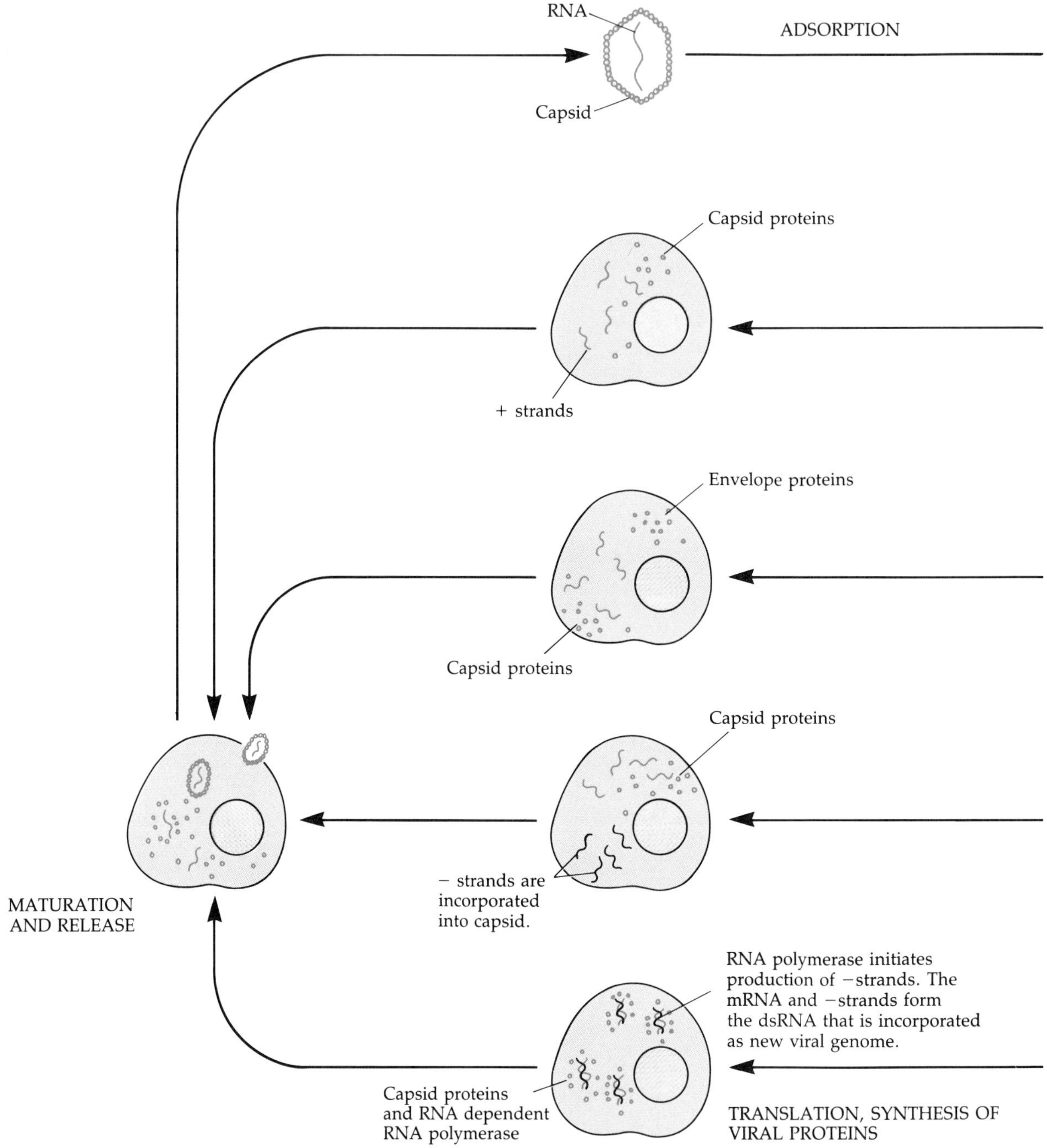

Figure 12–19 Pathways of multiplication used by various RNA-containing viruses. After uncoating, ssRNA viruses with a + strand genome are able to synthesize proteins directly from their + strand. Using the + strand as a template, they transcribe − strands to produce additional + strands to serve as mRNA and be incorporated into capsid protein as viral genome. The ssRNA viruses with a − strand genome must transcribe a + strand to serve as mRNA before they begin synthesizing proteins. The mRNA transcribes additional − strands for incorporation into capsid protein. Both ssRNA and dsRNA must use mRNA (+ strand) to code for proteins, including capsid protein.

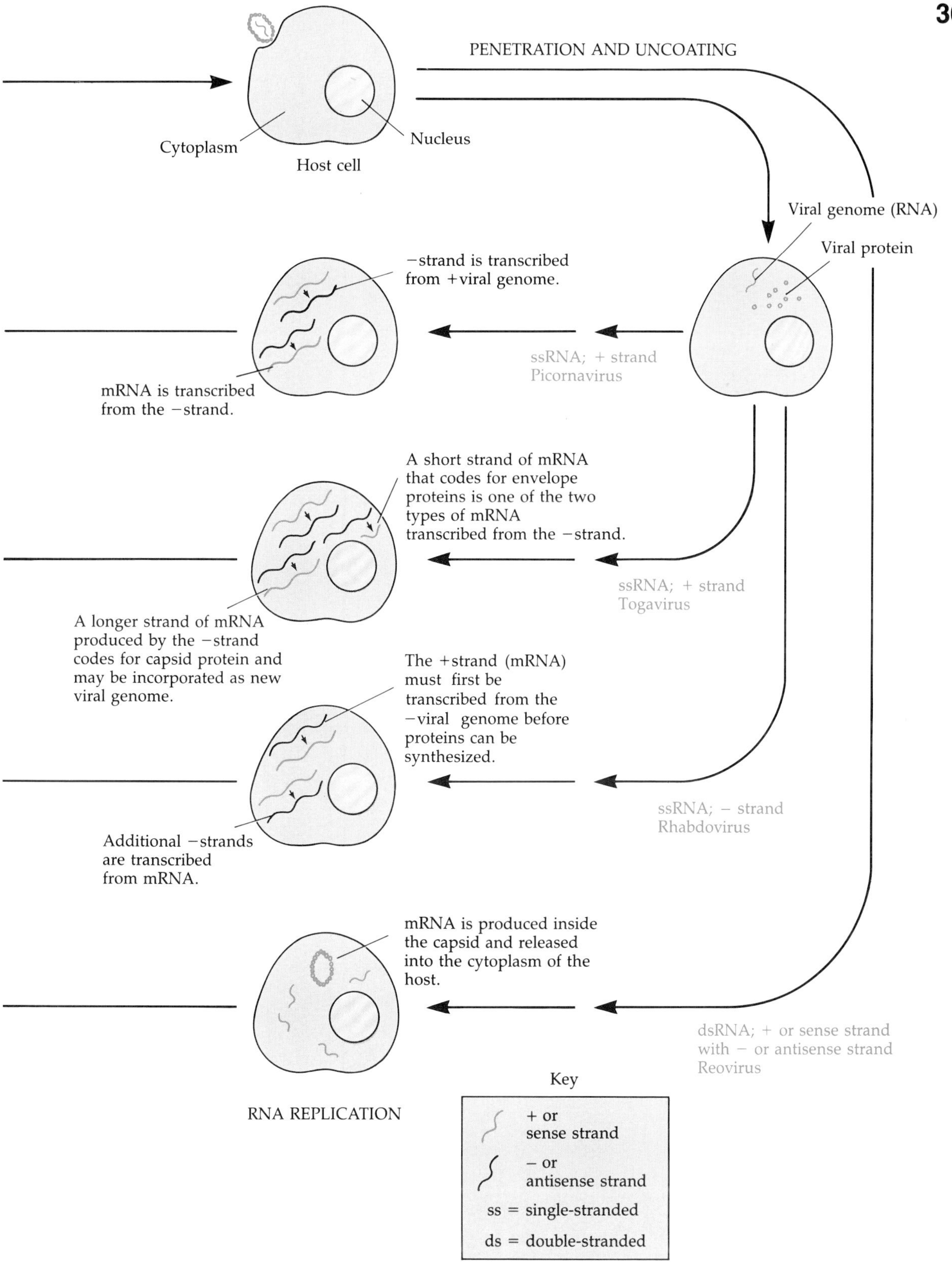
PENETRATION AND UNCOATING
Cytoplasm
Host cell
Nucleus
Viral genome (RNA)
Viral protein
−strand is transcribed from +viral genome.
mRNA is transcribed from the −strand.
ssRNA; + strand
Picornavirus
A short strand of mRNA that codes for envelope proteins is one of the two types of mRNA transcribed from the −strand.
A longer strand of mRNA produced by the −strand codes for capsid protein and may be incorporated as new viral genome.
ssRNA; + strand
Togavirus
The +strand (mRNA) must first be transcribed from the −viral genome before proteins can be synthesized.
Additional −strands are transcribed from mRNA.
ssRNA; − strand
Rhabdovirus
mRNA is produced inside the capsid and released into the cytoplasm of the host.
dsRNA; + or sense strand with − or antisense strand
Reovirus
RNA REPLICATION
Key
+ or sense strand
− or antisense strand
ss = single-stranded
ds = double-stranded

cation processes of these viruses lie in how mRNA and viral RNA are produced. Once viral RNA and viral proteins are synthesized, maturation occurs by similar means.

Picornavirus Picornaviruses, such as poliovirus, are single-stranded RNA viruses. The RNA within the virion is identified as a + or **sense strand** because it can act as mRNA. After adsorption, penetration, and uncoating are completed, the single-stranded viral RNA is translated into two principal proteins, which inhibit the host cell's synthesis of RNA and protein and which form an enzyme called *RNA-dependent RNA polymerase.* This enzyme catalyzes the synthesis of another strand of RNA, which is complementary in base sequence to the original infecting strand. This new strand is called a − or **antisense strand.** It serves as a template to produce additional + strands. The + strands may serve as mRNA for the translation of capsid protein, may become incorporated into capsid protein to form a new virus, or may serve as a template for continued RNA multiplication. Once viral RNA and viral protein are synthesized, maturation occurs.

Togavirus Togaviruses, which include arthropod-borne viruses (Chapter 21), also contain a single + strand of RNA. After a − or antisense strand is made from the + strand, two types of mRNA are transcribed from the − strand. One type of mRNA is a short strand that codes for envelope proteins. The other, longer strand serves as mRNA for capsid proteins and can become incorporated into a capsid; this process forms a new virus.

Rhabdovirus Rhabdoviruses, such as rabiesvirus, contain a single − strand of RNA. They also contain an RNA-dependent RNA polymerase that uses the − strand as a template from which to produce a + strand. The + strand serves as mRNA and as a template for synthesis of new viral RNA.

Reovirus In addition to its structural role for the virus, one of the reovirus capsid proteins serves as an RNA-dependent RNA polymerase. After the capsid containing the double-stranded RNA enters a host cell, mRNA is produced inside the capsid and released into the cytoplasm. There it is used to synthesize more viral proteins. One of the newly synthesized viral proteins acts as RNA-dependent RNA polymerase to produce more − strands of RNA. The mRNA + and − strands form the double-stranded RNA that is then surrounded by capsid proteins.

Retrovirus Retroviruses are widely found infecting vertebrates, but, contrary to their reputation, most do not cause cancer. Some retroviruses do cause cancerous tumors and leukemias in various animals, and one type of retrovirus has been implicated in **acquired immune deficiency syndrome (AIDS)**.

The formation of mRNA and RNA for new retrovirus virions is quite interesting. These viruses carry their own polymerase, which uses the RNA of the virus to synthesize a complementary strand of DNA, which in turn is replicated to form double-stranded DNA. This enzyme also degrades the original viral RNA. The enzyme is an RNA-dependent DNA polymerase called *reverse transcriptase,* so called because it carries out a reaction (RNA → DNA) that is exactly the reverse of the familiar transcription of DNA → RNA (Figure 12–20). The formation of complete viruses requires

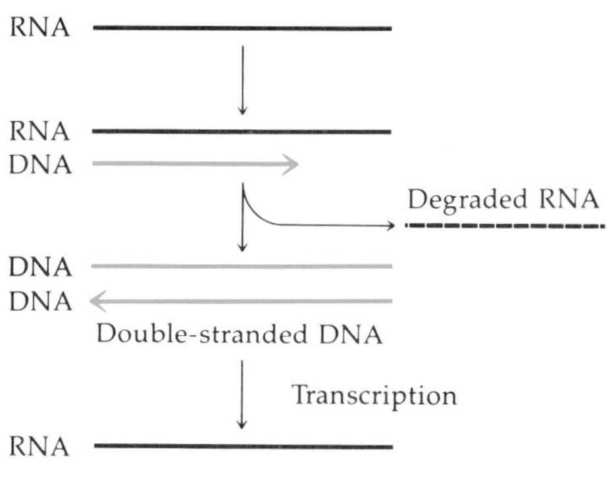

Figure 12–20 Reverse transcriptase. This enzyme catalyzes the synthesis of a strand of DNA on an RNA template. The RNA strand is then degraded, and another DNA strand, complementary to the first, is synthesized. The double-stranded DNA thus produced serves as a template for transcription to form both mRNA and RNA for new virions.

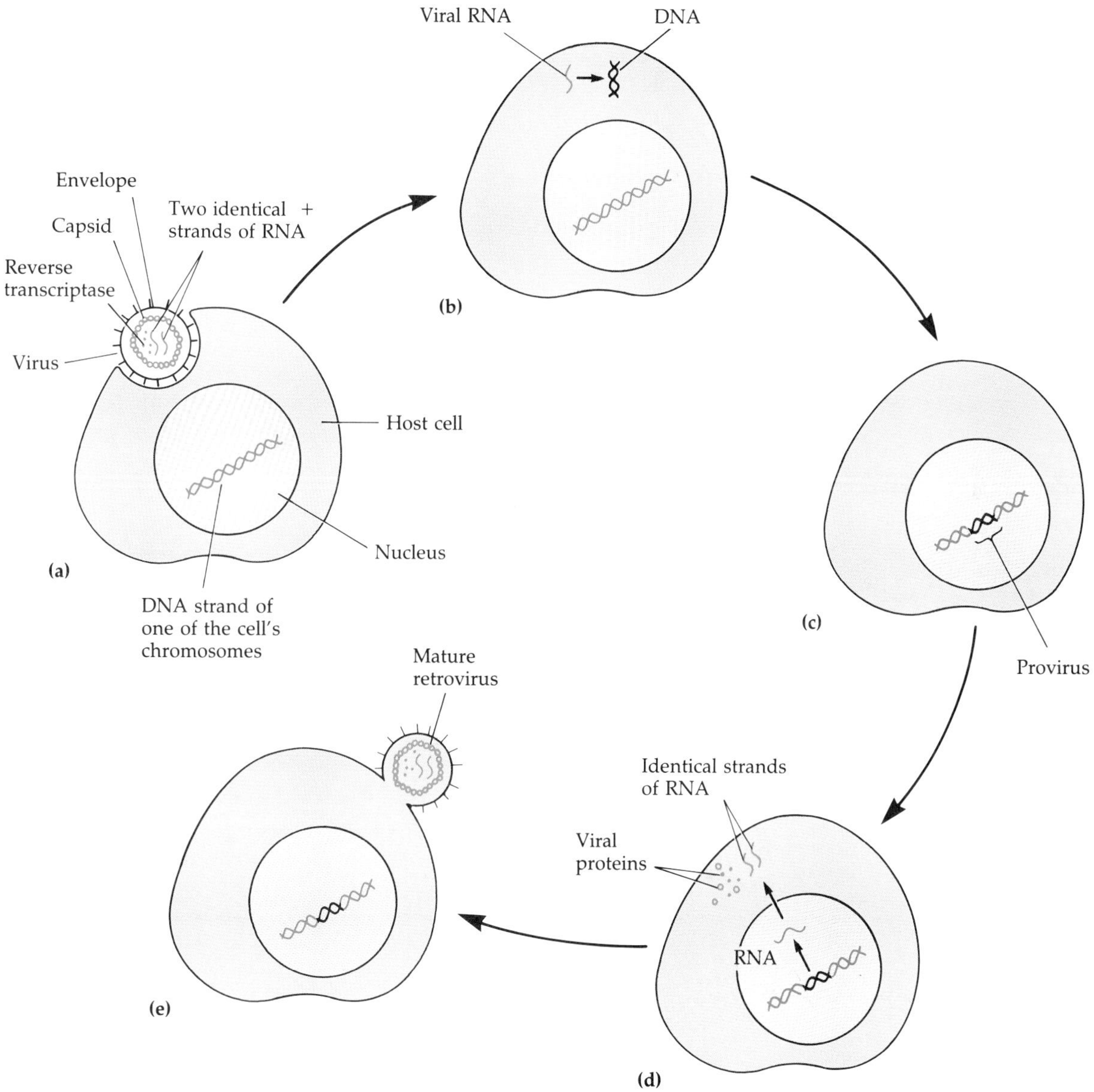

Figure 12–21 Multiplication and inheritance processes of a retrovirus. **(a)** Virus penetrates host cell by endocytosis. **(b)** After uncoating, reverse transcription of the viral RNA genome produces double-stranded DNA (see Figure 12–20). **(c)** The double-stranded DNA is transported into the host cell nucleus and integrated, as a provirus, into the cellular DNA. At this stage, the host cell may divide normally indefinitely. All progeny cells will inherit the viral genes, but the genes remain latent until they are transcribed. **(d)** If transcription of RNA does occur, the new + strands serve both as mRNA to code for capsid protein and as viral genome. **(e)** Two identical + strands of RNA are incorporated into a capsid and released.

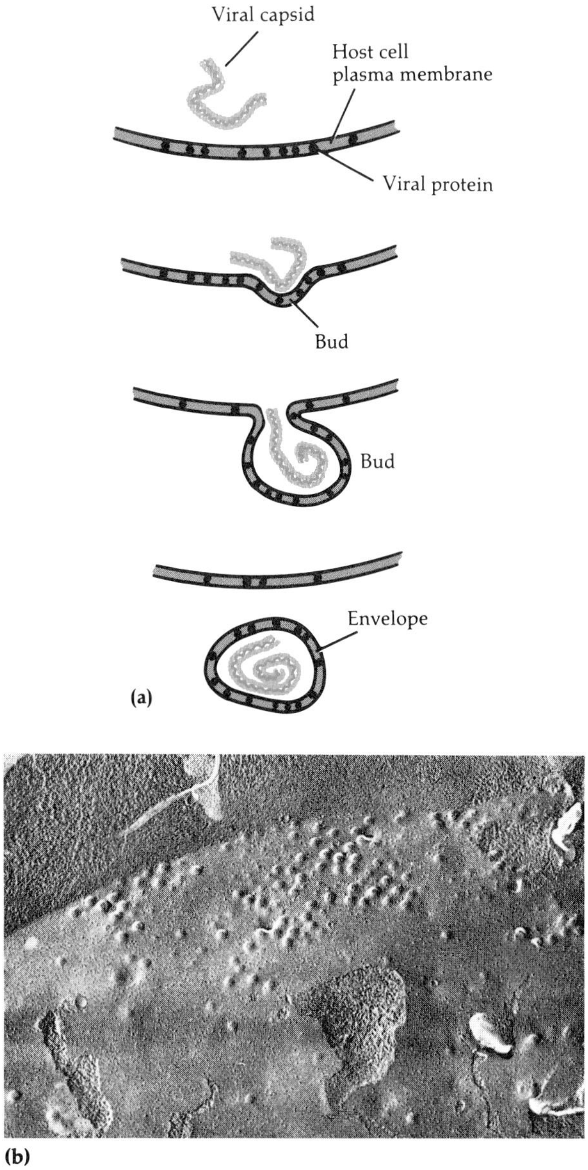

Figure 12–22 Budding of an enveloped virus. **(a)** Diagram of the budding process. **(b)** The small "bumps" seen on this freeze-fractured plasma membrane are Sindbis virus particles caught in the act of budding out from an infected cell (×38,000).

that DNA be transcribed back into the RNA that will serve as mRNA for viral-protein synthesis and be incorporated into new virions. However, before transcription can take place, the viral DNA must be integrated into the DNA of a host cell chromosome. In this integrated state, the viral DNA is called a **provirus.**

Once the provirus is integrated into the host cell's DNA, several things can happen. How a retrovirus infects and multiplies inside a host cell is summarized in Figure 12–21. Sometimes the provirus simply remains in a latent state and duplicates when the DNA of the host cell duplicates. In other cases, the provirus becomes transcribed and produces new viruses, which may infect adjacent cells. The provirus can also convert the host cell into a tumor cell; possible mechanisms will be discussed later.

Maturation and release

The first step in viral maturation is the assembly of the protein capsid; this assembly is usually a spontaneous process. The capsids of many of the RNA-containing animal viruses are enclosed by an envelope consisting of proteins, lipid, and carbohydrate, as noted earlier. Examples of such viruses include myxoviruses and paramyxoviruses. The envelope protein is synthesized by the virus and is incorporated into the plasma membrane of the host cell. The envelope lipid and carbohydrate are synthesized by host cell enzymes and are present in the plasma membrane. The envelope actually develops around the capsid by a process called **budding** (Figure 12–22). After the sequence of adsorption, penetration, uncoating, and biosynthesis of viral nucleic acid and protein, the assembled capsid-containing nucleic acid pushes through the plasma membrane. As a result, a portion of the plasma membrane, now the envelope, adheres to the virus. This extrusion of a virus from a host cell is one method of release. Budding does not immediately kill the host cell, and in some cases the host cell survives.

Naked viruses are released through ruptures in the host cell plasma membrane. In contrast with budding, this type of release usually results in the death of the host cell.

EFFECTS OF ANIMAL VIRAL INFECTION ON HOST CELLS

Infection of a host cell by an animal virus usually kills the host cell. Death can be caused by the accumulation of large numbers of multiplying viruses, by the effects of viral proteins on the permeability of the host cell's plasma membrane, or by inhibition of host DNA, RNA, or protein synthesis. The different abnormalities that can lead to damage or death of a host cell are known as **cytopathic effects (CPE).** These effects are frequently used as tools for the diagnosis of many viral infections.

One type of CPE is an **inclusion body,** some abnormal body of material in a cell. Some inclusion bodies arise as a result of the accumulation of assembled or unassembled viruses in the nucleus and/or cytoplasm of the host cell. Other inclusion bodies arise at sites of earlier viral synthesis but do not contain assembled viruses or their components. Inclusion bodies are important because their presence can help to identify the virus causing an infection. For example, the rabies virus produces inclusion bodies (Negri bodies) in the cytoplasm of nerve cells, and their presence in the brain tissue of animals suspected of being rabid is still sometimes used as a diagnostic tool for rabies. Diagnostic inclusion bodies are also associated with the measles virus, vaccinia virus, smallpox virus, herpesvirus, and adenoviruses (Figure 12–23).

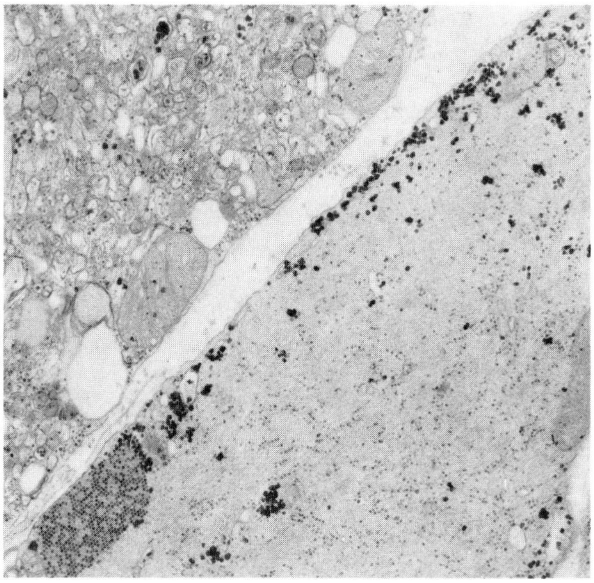

Figure 12–23 Viral inclusion bodies. The dark spots in the sections of two mouse muscle cells shown here are intranuclear inclusion bodies of coxsackie A_4 virus (×5600). Notice the large crystalline array of virus particles in the lower left.

As a result of viral infection, some host cells exhibit another type of CPE. At times, several adjacent infected cells fuse to form giant cells called **polykaryocytes.** These are produced from infections from myxoviruses. Other viral infections result in chromosomal damage to the host cell, most often chromosomal breakage. Some viral infections result in changes in the host cell's functions; there may be no visible changes in the infected cells. For example, if a virus changes the production level of a hormone, then the target cells for that hormone will not function properly.

One very significant response of some virus-infected cells is their production of a substance called **interferon.** Although production of it is induced by viral infection, interferon is coded by the host cell's DNA. Interferon protects neighboring uninfected cells from viral infection. More will be said about interferon in Chapter 15.

VIRUSES AND CANCER

When cells multiply in an uncontrolled way, the excess tissue that develops is called a **tumor.** A cancerous tumor is known as a **malignant tumor,** whereas a noncancerous tumor is referred to as a **benign tumor.** Tumors are generally named by attachment of the suffix *-oma* to the name of the tissue from which the tumor arises. Cancer is not a single disease, but many diseases. The human body contains more than a hundred different kinds of cells, each of which can malfunction in its own distinctive way to cause cancer. Malignant cells multiply in an uncontrolled way, often rapidly. Most persons who die from cancer are not killed by the primary tumor that develops, but by **metastasis,** the spread of the cancer to other parts of the body. Metastatic tumors are harder to detect and eliminate than primary tumors.

Not all cancerous cells form solid tumors. When the malignant cells are those that give rise to white blood cells, the resulting cancer, characterized by an excess of white cells in the circulatory system, is called **leukemia.** The relationship between cancers and viruses was in fact first demonstrated in 1908 when it was shown that certain

Table 12–3 Representative Cellular Oncogenes Isolated in Human Cells

Oncogene	Found in: Cancer cells	Found in: Virus	Location in Human Chromosome (Chromosome #)
abl	Chronic myelogenous leukemia	Abelson murine leukemia virus	9
fes	Acute promyelocytic leukemia	Fujinami (feline) sarcoma virus	15
fos	Burkitt's lymphoma	FBJ osteosarcoma virus	2
mos	Acute myeloblastic leukemia	Moloney murine sarcoma virus	8
myb	Acute lymphocytic leukemia	Avian myeloblastosis virus	6
myc	Burkitt's lymphoma	Avian myelocytomatosis virus	8
ras^{H}-1	Wilms' tumor	Harvey murine sarcoma virus	11
ras^{K}-2	Acute lymphocytic leukemia	Kristen murine sarcoma virus	12
ras^{N}	Myeloid tumors	Not associated with viruses	1
sis	Burkitt's lymphoma; chronic myelogenous leukemia	Simian sarcoma virus	22

Source: From P. D'Eustachio. Gene mapping and oncogenes. 1984. *American Scientist* 72:32–40; and L. Fink. Unraveling the molecular biology of cancer. 1984. *Bioscience* 34:75–77.

chicken leukemias could be transferred to healthy chickens by cell-free filtrates. Three years later it was found that a chicken **sarcoma** (cancer of connective tissue) can be similarly transmitted. Virus-induced **adenocarcinomas** (cancers of glandular tissue) in mice were discovered in 1936. At that time, it was clearly shown that mouse mammary tumors are transmitted from mother to offspring through the mother's milk.

The etiology of cancer can go unrecognized for several reasons. First, most of the particles of some viruses infect cells but do not induce cancer. Second, cancer might not develop until long after infection. And third, cancers do not seem to be contagious.

Transformation of Normal Cells into Tumor Cells

Almost anything that can alter the genetic material of a cell has the potential to make a normal cell cancerous. It is now believed that these cancer-causing alterations to cellular DNA affect parts of the genome called **oncogenes.** All nucleated cells have these oncogenes, which are capable of transforming normal cells into cancerous cells. Under normal conditions these genes probably code for proteins that are necessary to the cell's growth, but mutations can cause oncogenes to bring about cancerous transformations of cells. (Oncogenes that function normally are sometimes referred to as *proto-oncogenes,* to differentiate them from their abnormally activated counterparts.) Twenty-four oncogenes have been isolated from animal cells; those associated with human cancers are listed in Table 12–3. Oncogenes are given three-letter code names. Oncogenes can be activated by a variety of agents, including mutagenic chemicals, high-energy radiation, and viruses. Viruses capable of producing tumors in animals are referred to as **oncogenic viruses.**

Both DNA- and RNA-containing viruses are capable of inducing tumors in animals. When this occurs, the tumor cells are **transformed** in such a way that they acquire properties that are distinct from properties of uninfected cells or infected cells in which tumors are not produced. An outstanding feature of all oncogenic viruses is that their genetic material integrates into the host cell's DNA and replicates along with the host cell's chromosome. This mechanism is similar to the phenomenon of lysogeny in bacteria and can alter the host cell's characteristics.

Transformed cells also lose a property called contact inhibition. Normal animal cells in tissue culture move about randomly by amoeboid movement and divide repeatedly until they come into contact with each other. Then both movement and cell division stops—the phenomenon known as **contact inhibition.** Transformed cells in tissue cul-

Microbiology in the News

Establishing a Link Between EB Virus and Cancer

About 13 years ago, a baby in Texas was born without a functioning immune system. Faced with what was, back then, a fatal situation, doctors placed the boy in a completely sterile environment. They were buying time until they could figure out what to do.

That experiment ended last year when David, the "bubble boy," died. His death provided a clear link between Epstein-Barr virus and cancer.

With no "compatible" relatives to provide him with immune cell-producing bone marrow, David had to wait for the technology to rid donated marrow of the cells that would attack his own. After David spent 12 years in a bubble, doctors hoped the technology was ready, and the boy received treated marrow from his sister.

Eighty days after the transplant and still in a germ-free environment, David developed some of the clinical signs of mononucleosis, a condition caused by Epstein-Barr virus. Doctors brought him out of isolation for easier treatment, hoping his sister's marrow cells had taken hold and would protect him from the microbes the rest of us encounter every day. Her cells hadn't established themselves, and David died about four months after the transplant.

What killed him was not the immediate failure of the transplant but cancer. His own B cells had run amok—a proliferation induced by Epstein-Barr virus. The autopsy revealed small, whitish-pink cancer nodules throughout his body, and closer study showed that these cells all contained Epstein-Barr virus—a virus he could only have gotten through his sister's bone marrow.

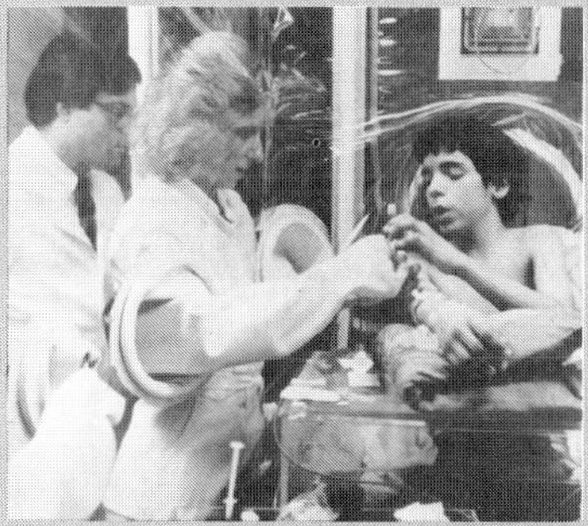

"We're certain (the cancer) came from the transplant itself," says William T. Shearer of Baylor College of Medicine in Houston, who was the lead physician on David's case. David had a B cell cancer of a type similar to Burkitt's lymphoma, and while Shearer can't absolutely say the two types of cancer are kicked off in the same way, "it seems likely that some of these same processes occur."

David's case, says Jeffrey Sklar of Stanford University, "confirms our suspicions about Epstein-Barr virus being an inducer of cancer in immunosuppressed individuals." Moreover, because of the peculiarities of David's situation, the time sequence—how quickly Epstein-Barr virus can induce cell transformation—is now known. "We can now without any doubt describe the very clear progression from infection to the development of cancer," says Shearer.

"This is a very clear demonstration of a virus causing a cancer," says Sklar. "It's also clear as to how the tumors evolved." First, David's B cells were activated by the Epstein-Barr virus, and began dividing. Then a handful of cells took hold, with some of the cancer nodules arising from single cells and others apparently arising from several different cells.

One of the requisites for proving viral transmission of a disease is to infect a test subject with the virus and see if the disease occurs. "Inadvertently this is what happened," says Sklar.

David had been germ-free—there's no way he could have come into contact with Epstein-Barr virus before the transplant. "In a way you have a documented transmission of virus followed by development of tumor," says Sklar.

Says Shearer, "I think this study documents that this common virus produced this cancer." While there's a tragic human story behind the finding, it resulted in an important advance in knowledge, he says. "This will be the beginning of many studies to come."

ture do not exhibit contact inhibition, but instead form tumorlike cell masses. Transformed cells sometimes produce tumors when injected into susceptible animals.

After being transformed by viruses, many tumor cells contain a virus-specific antigen on their cell surface called **tumor-specific transplantation antigen (TSTA)** or an antigen in their nucleus called the **T antigen.** Finally, transformed cells tend to be rounder than normal cells and to exhibit certain chromosomal abnormalities, such as unusual numbers of chromosomes and fragmented chromosomes.

DNA-Containing Oncogenic Viruses

Oncogenic viruses are found within several groups of DNA-containing viruses. These groups include adenoviruses, herpesviruses, poxviruses, and papovaviruses. Among the papovaviruses are the papilloma viruses that cause benign warts in humans and other animals, polyoma viruses that cause several kinds of tumors when injected into newborn mice, and simian virus 40 (SV40), which was originally isolated from cell cultures being used to cultivate polioviruses for vaccine production.

During polyoma and SV40 infection, there is an increase in the host cell's synthesis of DNA. This increase is followed by the appearance of TSTA and T antigen. When the host cell is transformed, viral DNA is integrated into the host cell's DNA as a provirus. This mechanism, similar to lysogeny in bacteria, results in the transformed cells' distinctive properties noted earlier.

Among the herpesviruses that affect humans are the herpes simplex virus, the herpes zoster virus, and the Epstein–Barr (EB) virus. The EB virus has an attraction for lymphocytes (a type of white blood cell) and has the potential for transforming lymphocytes into highly proliferating cells. This EB virus, in addition to being the cause of infectious mononucleosis, has recently been implicated as the causative agent of two human cancers: Burkitt's lymphoma (see page 367) and nasopharyngeal carcinoma (see box on page 367 and see page 591). Burkitt's lymphoma is a rare cancer of the lymphatic system; it affects mostly children in certain areas of Africa. Nasopharyngeal carcinoma, a cancer of the nose and throat, is worldwide in distribution. Some researchers have also suggested that EB virus is involved in Hodgkin's disease, a cancer of the lymphatic system.

It is estimated that about 80% of the U.S. population carry the latent stage of the EB virus in their lymphocytes but have no disease. This latent stage is indicated by the presence of antibodies to EB virus in blood serum. Although the EB virus leads to no apparent symptoms in healthy persons, it can cause infectious mononucleosis, mostly in teenagers. In specimens from persons with either Burkitt's lymphoma or nasopharyngeal cancer, TSTA and T antigen specific for the EB virus have been found. Moreover, DNA from EB virus is always found in malignant cells from these cancers.

One type of herpes simplex virus (HSV 1) produces cold sores. Another type, HSV 2, is associated with more than 90% of genital herpes infections. Women with cervical cancer have more HSV 2 antibodies than do asymptomatic patients, so it has been suggested that HSV 2 might be associated with cervical cancer.

RNA-Containing Oncogenic Viruses

Among the RNA-containing viruses, only the retroviruses seem to be oncogenic. These include the leukemia, lymphoma, and sarcoma viruses of cats, chickens, and mice and the mammary tumor viruses of mice. Retroviruses have also been isolated from humans and monkeys with acquired immune deficiency syndrome (AIDS).

The ability of these viruses to induce tumors is related to their production of a reverse transcriptase by the mechanism described earlier (see Figures 12–20 and 12–21). The provirus, which is the double-stranded DNA molecule synthesized from the viral RNA, becomes integrated into the host cell's DNA; new genetic material is thereby introduced into the host's genome. The key reason why retroviruses can contribute to cancer is that they introduce new genetic material into the host's DNA. Some retroviruses contain oncogenes; others contain promoters that turn on oncogenes or other cancer-causing factors.

Activation of Oncogenes

Cancer appears to be a *multistep* process. At least two oncogenes must be activated to abnormal functioning in order for malignancy to result. At present, four mechanisms for the activation of oncogenes are proposed. Of these, viruses are responsible for transduction and may facilitate translocation and gene amplification; however, other mechanisms might also be responsible.

1. A **single mutation** such as a nitrogenous base-pair substitution resulting from exposure to a chemical mutagen or radiation could alter the amino acid sequence of a gene product. The resulting protein may contribute directly to transformation. Alternatively, the mutation might affect a gene whose product normally regulates an oncogene. If that oncogene codes for a cell growth factor, and is no longer properly regulated, the growth factor might be produced in an uncontrolled way and lead to cancer.
2. **Transduction** of oncogenes by a virus could remove the oncogenes from normal cellular controls and place them under control of viral regulatory proteins. The viral proteins could cause an oncogene product to be produced at an abnormal time or in abnormal amounts.
3. **Translocation** of oncogenes from one chromosomal locus to another can occur during normal cellular activities and could place the oncogenes into a site in which normal controls are not active.
4. **Gene amplification** could cause the cell to produce unusually large amounts of oncogenic products. A gene is **amplified** when an unknown mechanism causes the gene to be replicated several times.

LATENT VIRAL INFECTIONS

A virus can remain in equilibrium with the host and not actually produce disease for a long period, often many years. The classic example of such a **latent viral infection** is the infection of the lip by herpes simplex virus, which produces cold sores. This virus can inhabit the host cells (nerve cells) but cause no damage until it is activated by stimuli such as fever or sunburn—hence the term *fever blister*.

It is interesting that in some individuals, viral production occurs but the symptoms never appear. Even though a large percentage of the human population carries the herpes simplex virus, most people never exhibit the disease. The virus of some latent infections can exist in a lysogenic state within host cells.

The varicella–zoster virus (also referred to as herpes zoster virus) is another virus that can exist in a latent state. Chickenpox (varicella) is a childhood disease in which the viral agent is indistinguishable from the viral agent that causes shingles (herpes zoster), an adult neurological disorder. Shingles occurs in only a small fraction of the population who have had chickenpox. Many researchers believe that shingles may be caused by reactivation of chickenpox virus that has remained latent in certain nerve cells for years.

SLOW VIRAL INFECTIONS

The term **slow viral infection** refers to a disease process that occurs gradually over a long period of time and which is thought to be caused by a virus. (It does not imply that viral multiplication is unusually slow.) Typically, slow viral infections are fatal. Several examples of slow viral infections are listed in Table 12–4.

As the table indicates, a common virus is sometimes the source of a slow viral infection. For example, the measles virus, several years after causing measles, can be responsible for a rare form of encephalitis called *subacute sclerosing panencephalitis*. The mechanism by which a common virus can also cause a slow viral infection is not known.

PRIONS

Several diseases formerly called slow viral infections are now thought to be caused by peculiar agents called **prions.** Like viruses, prions contain protein and reproduce inside cells, but prions seem to have no nucleic acid. Three hypotheses to explain the mechanism of their reproduction have been offered.

1. Prions might be conventional viruses whose nucleic acid is extraordinarily difficult to detect.
2. Prion genomes might be located within the genomes of host animals.
3. Prion proteins might not be encoded by nucleic acids at all; they might be reproduced via reverse translation (protein → mRNA) or pro-

Table 12–4 Examples of Slow Viral Infections

Disease	Host	Organ Primarily Affected	Virus
Subacute sclerosing panencephalitis	Human	Brain	Measles (paramyxovirus)
Progressive encephalitis	Human	Brain	Rubella (togavirus)
Progressive multifocal leukoencephalopathy	Human	Brain	Papovavirus
Progressive pneumonia	Sheep	Lung	Retrovirus
Lymphocytic choriomeningitis	Mouse	Kidney, brain, liver	Arenavirus
Aleutian mink disease	Mink	Reticuloendothelial system	Parvovirus
Canine demyelinating encephalomyelitis	Dog	Brain, spinal cord	Distemper (paramyxovirus)

Source: From B. D. Davis, et al., *Microbiology*, 3rd ed., Harper & Row, New York, 1980, p. 1043.

tein-directed protein synthesis (protein → protein).

All diseases thought to be caused by prions are neurological diseases; these include Creutzfeldt–Jakob disease and scrapie. It is hypothesized that Alzheimer's disease, which is the most common type of senile dementia and a leading cause of death in the U.S., might also be caused by prions (see Chapter 20).

PLANT VIRUSES AND VIROIDS

Plant viruses resemble animal viruses in many respects; plant viruses are morphologically similar to animal viruses, and they have similar types of nucleic acid (see Table 12–5). In fact, some plant viruses can multiply inside insect cells. Plant viruses cause many diseases of economically important crops, including tomatoes (Tomato Spotted Wilt Virus), corn and sugarcane (Wound Tumor Virus), and potatoes (Potato Yellow Dwarf Virus). Viruses can cause color change, deformed growth, wilting, and stunted growth in their plant hosts. Some hosts, however, remain symptomless and only serve as reservoirs of infection.

Plant cells are generally protected from disease by an impermeable cell wall. Viruses must enter through wounds or be assisted by other plant parasites, including nematodes, fungi, and, most often, insects that suck the plant's sap. Once one plant is

Table 12–5 Classification of Some Major Plant Viruses According to Morphological, Chemical, and Physical Properties

Viral Group (Type of Virus Identified)	Morphological Class	Nucleic Acid*	Dimensions of Capsid (nm)	Method of Transmission	Shows Similarities To
Tobacco mosaic virus	Helical	SS RNA	300 (length)	Wounds	Picornavirus
Brome grass mosaic virus	Polyhedral	SS RNA	23 (diameter)	Pollen	Picornavirus
Potato yellow dwarf virus	Bullet-shaped	SS RNA	380 (length)	Leafhoppers and aphids	Rhabdovirus
Wound tumor virus	Polyhedral	DS segmented DNA	70 (diameter)	Leafhoppers	Reovirus
Cauliflower mosaic virus	Polyhedral	DS DNA	50 (diameter)	Aphids	Papilloma

*SS = Single-stranded; DS = Double-stranded.

infected, it can spread infection in its pollen and seeds.

In laboratories, plant viruses are cultured in protoplasts (plant cells with the cell walls removed) and in insect-cell cultures.

Some plant diseases are caused by **viroids.** Viroids are very short pieces of naked RNA; only 300 to 400 nucleotides long, they have no protein coat. The nucleotides are often internally paired, so the molecule has a closed, folded, three-dimensional structure that presumably helps protect it from attack by cellular enzymes. Thus far, virions have been conclusively identified as pathogens only of plants. One of the best-studied viroids is the Potato Spindle Tuber Viroid, or PSTV (Figure 12–24). The PSTV seems capable of infection and multiplication without being incorporated into a capsid. In some hosts, such as potatoes and tomatoes, disease results.

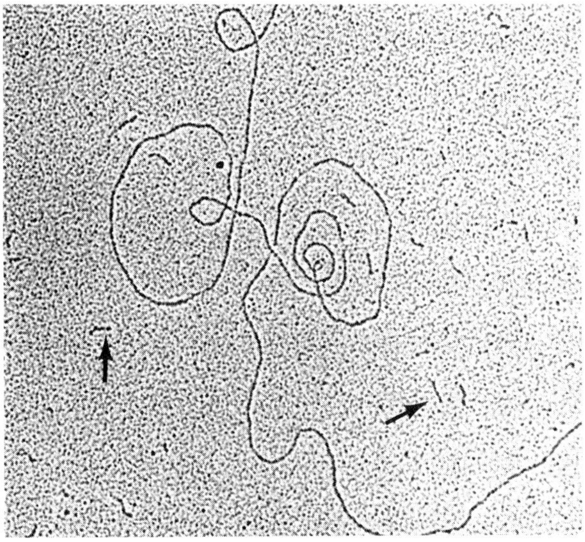

Figure 12-24 Electron micrograph of the potato spindle tuber viroid (PSTV) (×56,300). The arrows point to the viroids, which are adjacent to much longer viral nucleic acid from bacteriophage T7 for comparison.

STUDY OUTLINE

GENERAL CHARACTERISTICS OF VIRUSES (pp. 340–341)

Definition (p. 340)

1. Depending on one's viewpoint, viruses may be regarded as exceptionally complex aggregations of nonliving chemicals or exceptionally simple living microbes.
2. Viruses contain a single type of nucleic acid (DNA or RNA) and a protein coat, sometimes enclosed by an envelope composed of lipids, proteins, and carbohydrates.
3. Viruses are obligate, intracellular parasites. They multiply by using the host cell's synthesizing machinery to cause the synthesis of specialized elements that can transfer the viral nucleic acid to other cells.
4. A virion is a complete, fully developed viral particle composed of nucleic acid surrounded by a coat.

Host Range (pp. 340–341)

1. Host range refers to the spectrum of host cells in which a virus can multiply.
2. Depending on its host range, a virus is generally classified as an animal virus, bacterial virus (bacteriophage), or plant viruses. A virus can infect only certain species within each class.
3. Host range is determined by the specific attachment site on the host cell's surface and the availability of host cellular factors.

Size (p. 341)

1. Viral size is determined by filtration through membrane filters, ultracentrifugation, and electron microscopy.
2. Viruses range from 20 to 300 nm in diameter.

VIRAL STRUCTURE (pp. 341–343)

Nucleic Acid (pp. 341–342)

1. The proportion of nucleic acid in relation to protein in viruses ranges from about 1% to about 50%.
2. Viruses contain either DNA or RNA, never both, and the nucleic acid may be single- or double-stranded, linear or circular, or divided into several separate molecules.

Capsid and Envelope (p. 342)

1. The protein coat surrounding the nucleic acid of a virus is called the capsid.
2. The capsid is composed of subunits, the capsomeres, which can be of a single type of protein or several types.
3. The capsid of some viruses is enclosed by an envelope consisting of lipids, proteins, and carbohydrates.
4. Some envelopes are covered with carbohydrate–protein complexes called spikes.
5. Viruses without envelopes are called naked viruses.

General Morphology (pp. 342–343)

1. Helical viruses (for example, tobacco mosaic virus) resemble long rods, and their capsids are hollow cylinders surrounding the nucleic acid.
2. Polyhedral viruses (for example, adenovirus) are many-sided. Usually the capsid is an icosahedron.
3. Enveloped viruses are covered by an envelope and are roughly spherical but highly pleomorphic. There are also enveloped helical viruses (for example, influenza) and enveloped polyhedral viruses (for example, herpes simplex).
4. Complex viruses have complex structures. For example, many bacteriophages have a polyhedral capsid with a helical tail attached.

CLASSIFICATION OF VIRUSES (pp. 344–347)

1. Classification of viruses is based on morphological class, type of nucleic acid, size of capsid, and number of capsomeres.
2. Other classification schemes take into account the virus' susceptibility to microbial control agents, immunological properties, site of multiplication, and method of transmission.

ISOLATION, CULTIVATION, AND IDENTIFICATION OF VIRUSES (pp. 347–352)

1. Viruses must be grown in living cells.
2. The easiest viruses to grow are bacteriophages.

Growth of Bacteriophages in the Laboratory (p. 349)

1. The plaque method mixes bacteriophages with host bacteria and nutrient agar.
2. After several viral multiplication cycles, the bacteria in the area surrounding the original virus are destroyed, and the area of lysis is called a plaque.
3. Each plaque can originate with a single viral particle or more than one; the number of viral particles required to initiate a plaque is termed a plaque-forming unit.

Growth of Animal Viruses in the Laboratory (pp. 350–352)

1. Cultivation of some animal viruses requires whole animals.
2. Some animal viruses can be cultivated in embryonated eggs.
3. Cell cultures are animal cells that will grow in culture media in the laboratory.
4. Continuous cell lines can be maintained in vitro indefinitely.
5. Viral growth can cause cytopathic effects in the cell culture.
6. Viruses may be identified by serologic tests and nucleic acid base sequencing.

VIRAL MULTIPLICATION (pp. 352–364)

1. Viruses do not contain enzymes for energy production or protein synthesis.
2. For a virus to multiply, it must invade a host cell and direct the host's metabolic machinery to produce viral enzymes and components.

Multiplication of Bacteriophages (pp. 352–355)

1. The T-even bacteriophages that infect *E. coli* have been studied extensively.
2. In adsorption, sites on the phage's tail fibers attach to complementary receptor sites on the bacterial cell.
3. In penetration, phage lysozyme opens a portion of the bacterial cell wall, the tail sheath contracts to force the tail core through the cell wall, and phage DNA enters the bacterial cell. The capsid remains outside.
4. In biosynthesis, the host's DNA is degraded and transcription of phage DNA produces mRNA coding for proteins necessary for phage multiplication. Phage DNA is replicated, and capsid proteins are produced. During the eclipse period, separate phage DNA and protein can be found.

5. During maturation, phage DNA and capsids are assembled into complete viruses.
6. During release, phage lysozyme breaks down the bacterial cell wall, and the multiplied phages are released.
7. The time from phage adsorption to release is called burst time (20 to 40 minutes). Burst size, the number of newly synthesized phages produced from a single infected cell, ranges from 50 to 200.

Lysogeny (pp. 355–357)

1. During a lytic cycle, a phage causes the lysis and death of a host cell.
2. Some viruses can either cause lysis or have their DNA incorporated as a prophage into the DNA of the host cell. The latter situation is called lysogeny.
3. Prophage genes are regulated by a repressor coded for by the prophage. The prophage is replicated each time the cell divides.
4. Exposure to certain mutagens can lead to excision of the prophage and initiation of the lytic cycle.
5. Because of lysogeny, lysogenic cells become immune to reinfection with the same phage, and the host cell can exhibit new properties.
6. A lysogenic phage can transfer bacterial genes from one cell to another through transduction. Any genes can be transferred in generalized transduction, and specific genes in specialized transduction.

Multiplication of Animal Viruses (pp. 357–364)

1. Animal viruses adsorb to the plasma membrane of the host cell.
2. Penetration of enveloped viruses occurs by pinocytosis. In penetration by naked viruses, the complete virus enters the cell.
3. Animal viruses are uncoated by viral or host-cell enzymes.
4. The DNA of most DNA viruses is released into the nucleus of the host cell. Transcription of viral DNA and translation produce viral DNA and, later, capsid protein. Capsid protein is synthesized in the cytoplasm of the host cell.
5. Multiplication of RNA viruses occurs in the cytoplasm of the host cell. RNA-dependent RNA polymerase synthesizes a double-stranded RNA. Viral RNA acts as mRNA for translation.
6. Picornavirus (+) RNA acts as mRNA and directs the synthesis of RNA-dependent RNA polymerase.
7. Togavirus (+) RNA acts as a template for RNA-dependent RNA polymerase, and mRNA is transcribed from a new (−) RNA strand.
8. Rhabdovirus (−) RNA is a template for viral RNA-dependent RNA polymerase, which transcribes mRNA.
9. mRNA is produced inside the capsid of reoviruses.
10. Retroviruses carry reverse transcriptase (RNA-dependent DNA polymerase), which transcribes DNA from RNA.
11. After maturation, viruses are released. One method of release (and envelope formation) is budding. Naked viruses are released through ruptures in the host cell membrane.

EFFECTS OF ANIMAL VIRAL INFECTION ON HOST CELLS (p. 365)

1. Cytopathic effects (CPE) are abnormalities that lead to damage or death of a host cell.
2. Cytopathic effects include inclusion bodies, polykaryocytes, and altered function.
3. Interferon is produced by virus-infected cells and protects neighboring cells from viral infection.

VIRUSES AND CANCER (pp. 365–369)

1. An excess of tissue due to unusually rapid cell multiplication is called a tumor. Tumors are malignant (cancerous) or benign (noncancerous). Metastasis refers to the spread of cancer to other parts of the body.
2. Tumors are usually named by attachment of the suffix *-oma* to the name of the tissue from which the tumor arises.
3. The earliest relationship between cancer and viruses was demonstrated in the early 1900s when chicken leukemia and chicken sarcoma were transferred to healthy animals by cell-free filtrates.

Transformation of Normal Cells into Tumor Cells (pp. 366–368)

1. Eucaryotic cells have proto-oncogenes that code for proteins necessary for the cells' normal growth. When activated to oncogenes, these genes transform normal cells into cancerous cells.
2. Mutagenic chemicals, radiation, and viruses can activate oncogenes.

3. Viruses capable of producing tumors are called oncogenic viruses.
4. Several DNA viruses and retroviruses can be oncogenic.
5. The genetic material of oncogenic viruses becomes integrated into the host cell's DNA.
6. Transformed cells lose contact inhibition, contain virus-specific antigens (TSTA and T antigen), exhibit chromosomal abnormalities, and can produce tumors when injected into susceptible animals.

DNA-Containing Oncogenic Viruses (p. 368)

1. Oncogenic viruses are found among adenoviruses, herpesviruses, poxviruses, and papovaviruses.
2. The EB virus, a herpesvirus, causes infectious mononucleosis and has been implicated in Burkitt's lymphoma and nasopharyngeal carcinoma. One type of herpes simplex (HSV 2), associated with over 90% of genital herpes infections, might be implicated in cervical cancer.

RNA-Containing Oncogenic Viruses (p. 368)

1. Among the RNA viruses, only retroviruses seem to be oncogenic.
2. The virus' ability to produce tumors is related to the production of reverse transcriptase. The DNA synthesized from the viral RNA becomes incorporated as a provirus into the host cell's DNA.
3. A provirus can remain latent, can be multiplied, or can transform the host cell.

Activation of Oncogenes (pp. 368–369)

1. A single mutation can result in the production of a protein required for transformation.
2. Transduction of oncogenes could result in oncogene products being made in abnormal amounts or at the wrong time.
3. Translocation of oncogenes could remove normal controls.
4. Gene amplification causes unusually large amounts of oncogene products.

LATENT VIRAL INFECTIONS (p. 369)

1. A latent viral infection is one in which the virus remains in the host cell for long periods of time without producing an infection.
2. Examples are cold sores and shingles.

SLOW VIRAL INFECTIONS (p. 369)

1. Slow viral infections are disease processes that occur over a long period of time and are generally fatal.

PRIONS (pp. 369–370)

1. Prions are infectious proteins that appear to have no nucleic acid.
2. Prions are thought to cause some neurologic diseases such as Creutzfeldt–Jakob disease and scrapie.

PLANT VIRUSES AND VIROIDS (pp. 370–371)

1. Plant viruses must enter plant hosts through wounds or with invasive parasites such as insects.
2. Some plant viruses also multiply in insect (vector) cells.
3. Viroids are infectious pieces of RNA that cause some plant diseases, e.g. Potato Spindle Tuber Viroid (PSTV) disease.

STUDY QUESTIONS

REVIEW

1. Viruses were first detected because they are filterable. What do we mean by the term *filterable* and how could this property have helped their detection before invention of the electron microscope?
2. Why do we classify viruses as obligate intracellular parasites?
3. List the four properties that define a virus. What is a virion?

4. Describe the four morphological classes of viruses, then diagram and give an example of each.
5. Describe how bacteriophages are detected and enumerated by the plaque method.
6. Explain how animal viruses are cultured in each of the following:
   (a) Whole animals
   (b) Embryonated eggs
   (c) Cell cultures
7. Why are continuous cell lines of more practical use than primary cell lines for culturing viruses? What is unique about continuous cell lines?
8. Describe the multiplication of a T-even bacteriophage. Be sure to include the essential features of adsorption, penetration, biosynthesis, maturation, and release.
9. *Streptococcus pyogenes* produces erythrogenic toxin and is capable of causing scarlet fever only when it is lysogenic. What does this mean?
10. Describe the principal events of adsorption, penetration, uncoating, biosynthesis, maturation, and release of an enveloped DNA-containing virus.
11. Assume that this strand of RNA is the nucleic acid for an RNA-containing animal virus: UAGUCAAGGU.
    (a) Describe the steps of RNA replication for a virus that contains a (+) strand of RNA.
    (b) Describe the steps of RNA replication for a virus that contains a (−) strand of RNA.
    (c) Describe the steps of RNA replication for a virus that contains double-stranded RNA.
    (d) Describe the steps of RNA replication for a virus that contains reverse transcriptase.
12. Provide an explanation for the chronic recurrence of cold sores in some people. (*Note:* The cold sores almost always return at the same site.)
13. Slow viral infections such as ______________ might be caused by ______________ that are ______________.
14. What are cytopathic effects?
15. Recall from Chapter 1 that Koch's postulates are used to determine the etiology of a disease. Why is it difficult to determine the etiology of
    (a) a viral infection like influenza?
    (b) cancer?
16. The DNA of DNA-containing oncogenic viruses can become integrated into the host DNA. When integrated, the DNA is called a ______________. This process results in transformation of the cell. Describe the changes of transformation. How can an RNA-containing virus be oncogenic?
17. Contrast viroids and prions. Name a disease thought to be caused by each.
18. Plant viruses cannot penetrate intact plant cells because ______________; therefore they enter cells by ______________. Plant viruses can be cultured in ______________.

CHALLENGE

1. Discuss the arguments for and against the classification of viruses as living organisms.
2. In some viruses, capsomeres function as enzymes as well as structural supports. Of what advantage is this to the virus?
3. Discuss four mechanisms of oncogene activation and how viruses may be implicated.
4. Prophages and proviruses have been described as being similar to bacterial plasmids. What similar properties do they exhibit? How are they different?

FURTHER READING

Bishop, J. M. March 1982. Oncogenes. *Scientific American* 246:80–92. Includes a discussion of their evolution and initial discovery of them in viruses.

Croce, C. M. and G. Klein. March 1985. Chromosome translocations and human cancer. *Scientific American* 252:54–60. Describes one method of oncogene activation.

Diener, T. O. 1982. Viroids and their interactions with host cells. *Annual Review of Microbiology* 36:239–258. Comprehensive coverage of the structure, pathogenesis, and possible origin of viroids; by their discoverer.

Frankel-Conrat, H. and P. Kimball. 1982. *Virology*. Englewood Cliffs, NJ: Prentice-Hall. A general textbook on the biology of viruses.

Fucillo, D. A., J. E. Kurent, and J. L. Sever. 1974. Slow virus diseases. *Annual Review of Microbiology* 28:231–264. A summary of information about viral diseases caused by conventional and unconventional agents.

Hunter, T. August 1984. The proteins of oncogenes. *Scientific American* 251:70–79. Explains what the products of oncogenes do.

Karpas, A. 1982. Viruses and leukemia. *American Scientist* 70:277–285. Report on cell transformation and transmission of leukemia.

Prusiner, S. B. October 1984. Prions. *Scientific American* 251:50–59. The scientist who discovered and named them poses intriguing and, as yet, unanswered questions regarding the origin and nature of the prion and its genome.

Ptashne, M., A. D. Johnson, and C. O. Pabo. November 1982. A genetic switch in a bacterial virus. *Scientific American* 247:128–140. Discussion of gene regulation in a temperate phage.

Simons, K., J. Garoff, and A. Helenius. February 1982. How an animal virus gets into and out of its host cell. *Scientific American* 246:58–66. The authors describe their work on Semliki forest virus (togavirus).

Weinberg, R. A. November 1983. A molecular basis of cancer. *Scientific American* 249:126–141. Describes two methods of oncogene activation: by point mutation and by viruses.

PART THREE

INTERACTION BETWEEN MICROBE AND HOST

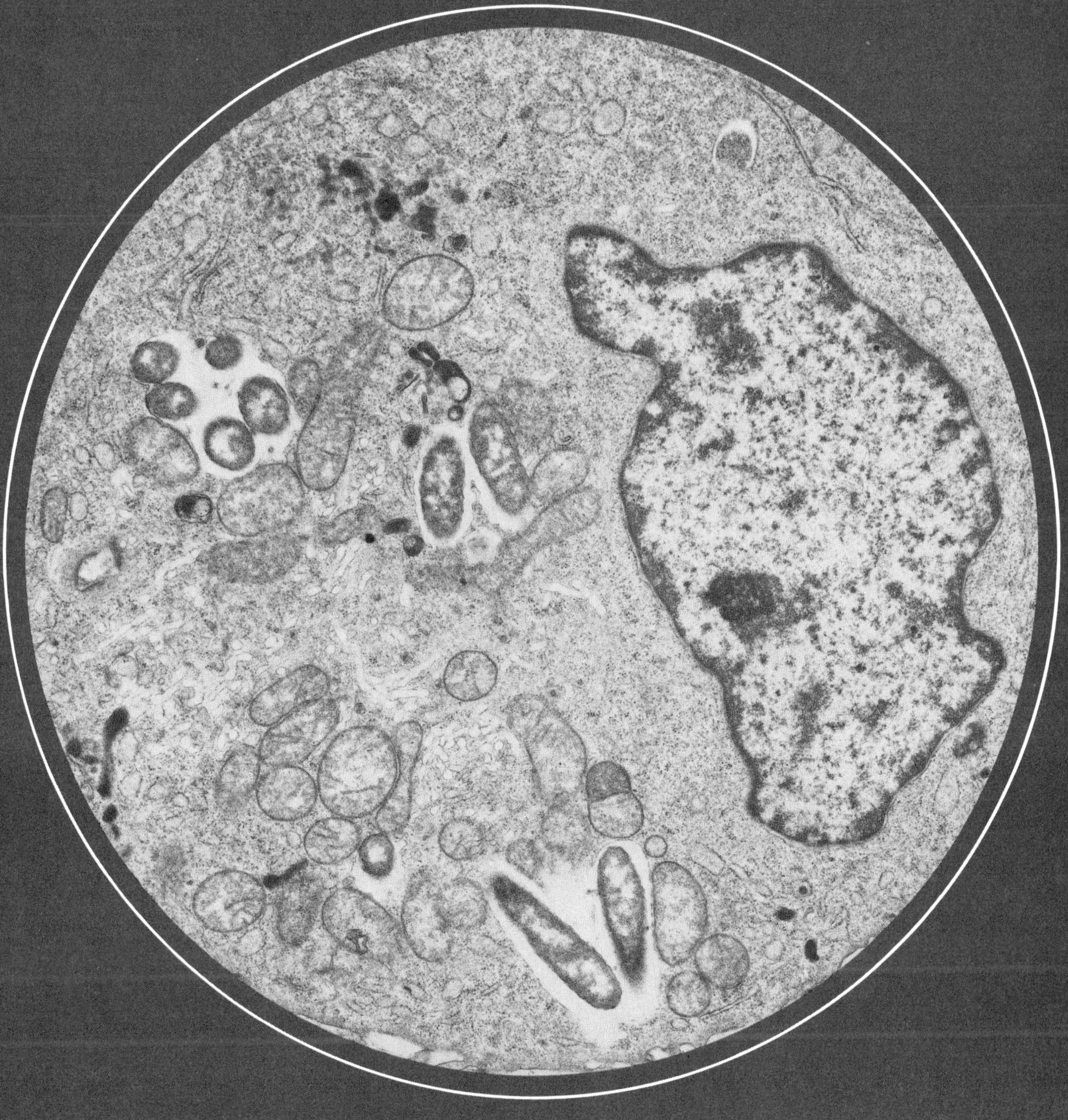

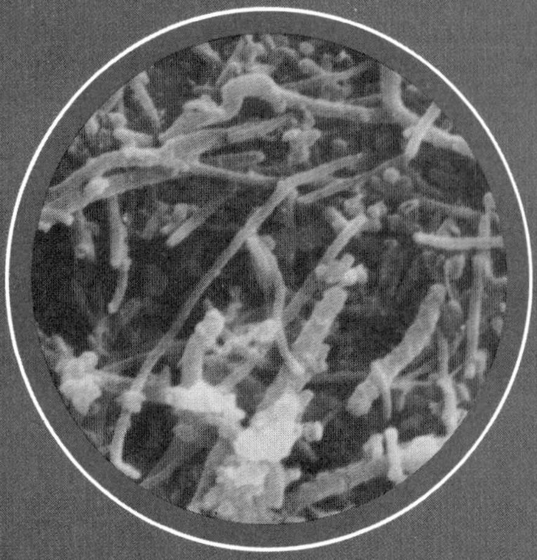

CHAPTER 13

Principles of Disease and Epidemiology

OBJECTIVES

After completing this chapter you should be able to

- Define normal flora and compare commensalism, mutualism, and parasitism and provide one example of each.
- List Koch's postulates.
- Define a reservoir of infection; contrast human, animal, and nonliving reservoirs, and state one example of each.
- Define nosocomial infections and explain their importance.
- Explain four methods of transmission of disease.
- Explain how to categorize diseases according to incidence.
- Define epidemiology and describe three methods of epidemiologic investigation.
- Differentiate between a communicable and noncommunicable disease.
- Define each of the following terms: pathogen; etiology; infection; host; disease; acute; chronic; subacute.

Now that you have a basic understanding of the structural and functional aspects of microorganisms, and of the variety of microorganisms that exist, we can consider the ways in which the human body and microorganisms are related in terms of health and disease.

We all have defense mechanisms that are always operating to keep us healthy. For instance, unbroken skin and mucous membranes are effective barriers against microbial invasion and infection. Tears, saliva, perspiration, and gastric juice also play important defensive roles. The inflammatory response, phagocytosis, and fever are additional attempts to keep us healthy. And under certain conditions, we can produce proteins called antibodies, which combine with particular microorganisms and contribute to their destruction. However, microorganisms have properties that make them **pathogenic,** that is, capable of causing disease. Some bacteria can invade our tissues and resist our defenses by producing capsules or enzymes. Other bacteria release poisonous substances, called toxins, that can seriously affect our health. A rather delicate balance exists between our defenses and the disease-producing mechanisms of microorganisms. When our defenses resist the disease-producing capabilities of the microorganism, we maintain our health. But when the ability of the microorganism to cause disease overcomes our defenses, disease results. After the disease has become established, an infected person can recover completely, suffer permanent damage, or die, depending on many factors.

In Part III, we will examine some of the principles of infection and disease, the mechanisms of pathogenicity, the body's defenses against disease, and the ways in which microbial disease can be prevented by immunization and controlled by drugs. This chapter will discuss the general principles related to disease and start with a discussion of the meaning and scope of pathology.

PATHOLOGY, INFECTION, AND DISEASE

Pathology is the scientific study of disease (*pathos,* meaning "suffering"; *logos,* meaning "science"). Pathology is first involved with the cause, or **etiology,** of disease. Second, it deals with the manner in which a disease develops; this stage is called **pathogenesis.** And third, pathology is concerned with the structural and functional changes brought about by disease and with its final effects on the body.

Although the terms *infection* and *disease* are sometimes used interchangeably, they do differ somewhat in meaning. **Infection** is the invasion or colonization of the body by pathogenic microorganisms. In this relationship, the body is referred to as the **host,** an organism that shelters and supports the growth of a microorganism. We are in contact with microorganisms throughout our lives, for microbial life is all around us. Moreover, microorganisms are spread from one person to another in the course of everyday living by bodily contact, sneezing, coughing, contact with the same objects, and even speaking and breathing. For the most part, the balance between health and disease favors the health of the host. But when the scale is tipped—as when our defenses are penetrated by a pathogen—disease results. **Disease** is any change from a state of health. It is an abnormal state in which part or all of the body is not properly adjusted or is not capable of carrying on its normal functions.

The term "infection" indicates that microorganisms are present, but does not necessarily indicate disease. For example, most throat cultures taken at random do contain large numbers of particular streptococci. Because these bacteria are normally present, they do not necessarily indicate disease. However, if the same bacteria are repeatedly cultured from an individual's blood, which is normally sterile, then it can be concluded that disease is present. Similarly, although large numbers of *Escherichia coli* (esh-ėr-i'kē-ä kō'lē) are normally present in stools, its presence in the urinary tract usually indicates disease.

The presence of microorganisms can even benefit the host. Let us consider this concept by examining the relationship of normal flora to the human body.

NORMAL FLORA

Animals, including humans, are germfree in utero. At birth, however, normal and characteristic microbial populations begin to establish themselves. Just before a woman gives birth, lactobacilli in her vagina multiply rapidly. The newborn's first contact with microorganisms is usually with these lactobacilli, and they are the predominant organisms in the newborn's intestine. After birth, *E. coli* and other bacteria acquired from foods and from contact with other humans begin to inhabit the intestine.

Many other microorganisms grow abundantly inside the normal adult body and on its surface. The microorganisms that establish more or less permanent residence (colonize) but do not produce disease under normal conditions are known as **normal flora** or **normal microbiota** (Figure 13–1). Others, called **transient flora,** may be present for several days, weeks, or months and then disap-

caused the disease, but could instead have been there as a result of the disease, he experimented further. He took a sample of blood from a diseased animal and injected it into another, healthy animal. The second animal became diseased and died. He repeated this procedure many times, always with the same results. (A key criterion in the validity of any scientific proof is that experimental results be repeatable.) Koch also cultivated the microorganism in fluids outside the animal's body, and demonstrated that the bacterium would cause anthrax even after many culture transfers. In short, Koch showed that a specific infectious disease (anthrax) is caused by a specific microorganism (*Bacillus anthracis*) that can be isolated and cultured on artifical media.

Koch later used the same methods to show that the bacterium *Mycobacterium tuberculosis* (mī-kō-bak-ti′rē-um tū-ber-kū-lō′sis) is the causative agent of tuberculosis. As part of this work, he also developed a specific stain for use in microscopic examination of the bacteria. Koch's research provides a framework for the study of the etiology of any infectious disease. Today, we refer to Koch's experimental requirements as **Koch's postulates.** They are summarized as follows:

1. The same pathogen must be present in every case of the disease.
2. The pathogen must be isolated from the diseased host and grown in pure culture.
3. The pathogen from the pure culture must cause the disease when inoculated into a healthy, susceptible laboratory animal.
4. The pathogen must again be isolated from the inoculated animal and must be shown to be the original organism.

It should be noted that although Koch's postulates are useful in determining the causative agent of most bacterial diseases, there are some exceptions. For instance, it is known that the bacterium *Treponema pallidum* (tre-pō-nē′mä pal′li-dum) is the causative agent for syphilis, but virulent strains have never been cultured on artificial media. Moreover, many rickettsial and all viral pathogens cannot be cultured on artifical media because they multiply only within cells.

The discovery of microorganisms that cannot grow on artificial media has made necessary some modifications of Koch's criteria and the use of alternative methods of culturing and detecting particular microbes. For example, when investigators searching for the microbial cause of Legionnaires' disease were unable to isolate the microbe directly from a victim, they took the alternative step of inoculating a victim's lung tissue into guinea pigs. These guinea pigs developed the disease's coldlike symptoms, whereas guinea pigs inoculated with tissue from an unafflicted person did not. Then tissue samples from the diseased guinea pigs were cultured in yolk sacs of chick embryos (a method that reveals the growth of extremely small bacteria). After the embryos were incubated, electron microscopy uncovered rod-shaped bacteria in the chick embryos. Finally, modern immunologic techniques (which will be discussed in Chapter 16) were used to show that the bacteria in the chick embryos were the same bacteria as that in the guinea pigs and in afflicted humans.

There are a number of situations in which a human host exhibits certain signs and symptoms that can be distinguished from all others and which are associated only with a certain pathogen and its disease. Good examples are diphtheria and tetanus. The pathogens responsible for these diseases always give rise to distinguishing signs and symptoms that can be produced by no other microbe. They are unequivocally the only organisms that produce their respective diseases. But some infectious diseases are not quite so clear-cut. For example, nephritis (inflammation of the kidneys) can involve any of several different pathogens, all of which give rise to the same signs and symptoms. Thus, it is often difficult to know which particular microorganism is causing a disease. Other infectious diseases that have similar, poorly defined etiologies are pneumonia, peritonitis, and meningitis.

There also exist pathogens that can cause several pathologies. *Mycobacterium tuberculosis,* for example, is implicated in diseases of the lungs, skin, bones, and internal organs. *Streptococcus pyogenes* (pī-äj′en-ēz) can cause sore throat, scarlet fever, skin infections (erysipelas), puerperal fever, and osteomyelitis, among other diseases. When laboratory methods and clinical signs and symptoms are used, these infections can usually be recognized and distinguished from infections of the same organs caused by other pathogens.

We will now turn our attention to the means by which infections are spread among a population. We will take into account the sources of pathogenic microorganisms and the different ways in which these microbes are transmitted.

SPREAD OF INFECTION

In considering how diseases are spread throughout a population, we will first take a look at the places inhabited by microorganisms prior to their transmission.

Reservoirs

For a disease to perpetuate itself, there must be a continual source of the infection. This source must be either a living organism or a nonliving substance that provides a pathogen with adequate conditions for survival and multiplication and an opportunity for transmission. Such a source is called a **reservoir of infection.** These reservoirs can be classified as human reservoirs, animal reservoirs, or nonliving reservoirs.

Human reservoirs

The principal living reservoir of microorganisms that cause human disease is the human body itself. Many people harbor pathogens and transmit them directly or indirectly to others. People who exhibit signs and symptoms of a disease are obvious transmitters of the disease. However, some people can harbor pathogens and transmit them to others but not exhibit any sign of illness. These people, called **carriers,** are important living reservoirs of infection. Some carriers have infections for which no signs or symptoms are ever exhibited. Other people carry a disease during its symptom-free stages: during the incubation period (before symptoms appear) or during the convalescent period (recovery). Human carriers play an important role in the spread of diseases such as diphtheria, typhoid, hepatitis, gonorrhea, amoebic dysentery, and streptococcal infections.

Animal reservoirs

Animals other than humans are another group of living reservoirs that can transmit diseases to humans. This group includes both sylvatic (wild) and domestic animals. Diseases that occur primarily in wild and domestic animals but which can be transmitted to humans are called **zoonoses.** About 150 zoonoses are known. The transmission of zoonoses to humans can occur via one of many routes: by direct contact with the infected animals; by contamination of food and water; by contact with contaminated hides, fur, or feathers; by consumption of infected animal products; or by insect vectors (insects that transmit pathogens). A few representative zoonoses are presented in Table 13–2.

Nonliving reservoirs

The two major nonliving reservoirs of infectious disease are soil and water. Soil harbors pathogens, such as fungi that cause mycoses and *Clostridium botulinum* (klôs-tri'dē-um bo-tū-lī'num), the agent of botulism. Water that has been contaminated by the feces of humans and other animals is a reservoir for several pathogens, most notably the gastrointestinal pathogens.

Nosocomial (Hospital-Acquired) Infections

The hospital is a unique environment. Bedding, utensils, and instruments can become contaminated with microorganisms, and the hospital personnel are human reservoirs. Meanwhile, hospital patients are usually in a weakened condition from disease or surgery and can be predisposed to infections. A **nosocomial infection** is one that develops during a hospital stay—that is, the patient was not infected when admitted. The word nosocomial is derived from the Greek word for "hospital." The Centers for Disease Control estimate that 5 to 15% of all hospital patients acquire some type of nosocomial infection. Tubs used to bathe patients must be disinfected between uses so that bacteria from the last patient will not contaminate the next. Respirators and humidifiers provide both a suitable growth environment for some bacteria and a method of transmission (aerosols). These sources of nosocomial infections must be kept scrupulously clean and disinfected, and materials used for bandages and intubation should be sterilized prior to use. Moreover, packaging used to maintain sterility should be removed aseptically.

Surgical patients are especially susceptible to infection. In patients who have had certain types of

Table 13–2 Selected Zoonoses That Can Be Transmitted to Humans

Disease	Etiology	Reservoir	Method of Transmission	Refer to Chapter
Viral:				
Influenza (some strains)	Myxovirus	Swine	Direct contact	22
Rabies	Rhabdovirus	Bats, skunks, foxes	Direct contact (bite)	20
Western equine encephalitis	Arbovirus	Horses, birds	*Culex* mosquito bite	20
Yellow fever	Arbovirus	Monkeys	*Aedes* mosquito bite	21
Bacterial:				
Anthrax	*Bacillus anthracis*	Domestic livestock	Direct contact with contaminated hides or animals; air; food	21
Brucellosis	*Brucella*	Domestic livestock	Direct contact with contaminated milk, meat, or animals	21
Bubonic plague	*Yersinia pestis*	Rodents	Flea bites	21
Cat-scratch fever	*Gram-negative bacillus*	Domestic cats	Direct contact	21
Leptospirosis	*Leptospira*	Wild mammals, domestic dogs and cats	Direct contact with urine, soil, water	24
Pneumonic plague	*Yersinia pestis*	Rodents	Direct contact	21
Psittacosis and Ornithosis	*Chlamydia*	Birds, especially parrots	Direct contact	22
Q fever	*Coxiella burnetii*	Domestic livestock	Inhalation, tick bites (between animals)	22
Rocky Mountain spotted fever	*Rickettsia rickettsii*	Rodents	Tick bites	21
Salmonellosis	*Salmonella* spp.	Poultry, rats, turtles	Ingestion of contaminated food, water	23
Tularemia	*Francisella tularensis*	Wild and domestic mammals, especially wild rabbits	Direct contact with infected animals; deer-fly bites	21
Typhus fever	*Rickettsia typhi*	Rodents	Flea bites	21
Fungal:				
Ringworms	*Trichophyton Microsporum Epidermophyton*	Domestic mammals	Direct contact; fomites	19
Protozoan:				
Chagas' disease	*Trypanosoma cruzi*	Wild mammals	"Kissing bug" bite	21
Malaria	*Plasmodium*	Monkeys	*Anopheles* mosquito bite	21
Toxoplasmosis	*Toxoplasma gondii*	Cats and other mammals	Direct contact with infected tissues or fecal material	21
Helminthic:				
Hydatid cyst	*Echinococcus granulosus*	Dogs	Direct contact with fecal material	11
Tapeworm (beef)	*Taenia saginata*	Cattle	Ingestion of contaminated beef	23
Trichinosis	*Trichinella spiralis*	Pigs, bears	Ingestion of contaminated meat	23

operations, such as surgery on the large intestine and amputations, the infection rate approaches 30%. Many other hospital procedures, such as catheterization, intravenous feeding, respiratory aids, and injections, also bypass the normal outer defenses of the body and are therefore sources of infection (see box).

Burn patients are also very susceptible to nosocomial infections because their skin is no longer an effective barrier to microorganisms. A **burn** is tis-

Centers for Disease Control

MMWR

Morbidity and Mortality Weekly Report

Endotoxic Reactions Associated with the Reuse of Cardiac Catheters

During a two-week period in July 1978, three cases of suspected endotoxic reactions (fever, chills, and hypotension) occurred in patients undergoing cardiac catheterization at a Massachusetts hospital. An investigation revealed that reusable intravascular catheters, although sterile, were contaminated with endotoxin and that this contamination was related to procedures employed to clean and disinfect the catheters.

At the time the endotoxic reactions occurred, used catheters were rinsed with hospital distilled water, wiped to remove clotted blood, and then soaked in Detergicide, a quaternary ammonium compound. Next, the bore of the catheter was flushed continuously for 2 hours with distilled water, followed by a second flush with 1 liter of commercial pyrogen-free, sterile, distilled water. Finally, the catheters were wrapped and gas-sterilized. The administration set for the delivery of the pyrogen-free fluid had not been changed regularly.

The hospital-supplied distilled water contained greater than 0.2 nanograms (ng) of endotoxin per ml even when samples were taken directly from the storage tanks. Cultures of samples taken from the cardiac catheterization laboratory's distilled water tap and from the research laboratories contained up to 310 colonies of *Pseudomonas cepacia* per ml, while storage tank samples were sterile.

Five previously used, but subsequently cleaned and sterilized, catheters contained levels of endotoxin ranging form 0.3 ng to 7.4 ng per catheter, as measured by the Limulus Amebocyte assay. (A level of 0.05 ng is considered pyrogenic by the laboratory performing the assay.) All catheters were sterile when cultured. New, sterile catheters were free of endotoxin. After only one use and cleaning, catheter flushes yielded excessive endotoxin.

This investigation resulted in the recommendation that disposable catheters be used. When this was not possible, it was recommended that pyrogen-free, sterile, distilled water be used in the cleaning procedure, and that both the outer and inner surfaces of the catheter be flushed. Detergicide was freshly prepared daily, using pyrogen-free, sterile, distilled water. After the catheters were cleaned and packaged, they were stored at 4°C until sterilized. In the five-month period since this outbreak, there have been no further suspected endotoxin reactions.

Editorial Note: The precise mechanism by which catheters used for cardiac catheterization were contaminated with endotoxin was not proven, but this investigation suggests that contamination was introduced during the cleaning procedure. Hospital supplies of distilled water, used in cleaning, were contaminated. Furthermore, the detergent-disinfectant preparation that was used in the preliminary cleaning, an aqueous quaternary ammonium formulation, has been shown to be ineffective against *P. cepacia* and may even permit the selective growth of this and other microorganisms. Presumably, viable microorganisms were introduced during cleaning from the distilled water, the Detergicide failed to kill the contaminants or permitted their growth, and the final sterilization killed the contaminants but allowed high levels of residual endotoxin to persist. Although a variety of detergent-disinfectants may be used safely for environmental sânitation of floors, surfaces, and the like, these agents must be used with utmost caution with critical medical devices that come in contact with any normally sterile body tissue. An effective sterilizing procedure, such as ethylene oxide, will kill microorganisms but will not remove endotoxin.

This report further highlights the hazards that may result from reuse of some medical devices. Some, but not all, medical devices can be cleaned and safely sterilized. If practical and economical, disposable devices are preferable to reusable devices that are difficult to sterilize.

Source: *MMWR* 28:25 (1/26/79)

sue damage caused by heat, electricity, radiation, or chemicals. The damaging agent causes protein destruction and cell death in affected tissues. Burns are classified as first degree, second degree, or third degree, depending on severity. Criteria of severity include color, presence or absence of sensation, blister formation, and depth of tissue damage. No matter how serious a burn appears on the surface, the systemic effects are far more serious than the local effects. One of the principal systemic effects is bacterial infection.

Another group of patients susceptible to nosocomial infections are cancer patients who have had their immune system suppressed by radiation and immunosuppressive drugs. Organ transplant patients receiving immunosuppressive drugs are also susceptible.

Normal flora from the human body present a particularly strong danger to hospital patients. Most of the bacteria that cause nosocomial infections are not harmful under ordinary conditions, but are opportunists. At one time, most nosocomial infections were caused by gram-positive microbes such as streptococci. At that time, gram-positive *Staphylococcus aureus* (staf-i-lō-kok'kus ô'rē-us) was the primary cause of nosocomial infections. Although antibiotic-resistant strains of that organism are still significant factors, the major causes today are gram-negative bacteria such as *Escherichia coli* and *Pseudomonas aeruginosa,* (sū-dō-mō'nas ā-rü-ji-nō'sa), especially the antibiotic-resistant strains (Table 13–3). *P. aeruginosa* has the ability to cause opportunistic skin infections—especially in surgical and burn patients. *P. aeruginosa,* as well as other gram-negative bacteria, tend to be difficult to control with antibiotics because of their R factors, which carry genes that determine resistance to antibiotics (Chapter 8).

It is estimated that about 20,000 people die of nosocomial infections each year. A summary of the principal kinds of nosocomial infections is presented in Table 13–4.

Accredited hospitals should have an infection control committee, and most hospitals at least have an infection control nurse or epidemiologist. The role of these personnel is to identify problem sources such as antibiotic-resistant strains of bacteria and improper sterilization techniques. The infection control officer should make periodic examinations of equipment to determine the amount of microbial contamination. Samples should be taken from tubing, catheters, respirator reservoirs, and other equipment.

Transmission of Disease

The causative agents of disease can be transmitted from the reservoir of infection to a susceptible host by four routes: (1) contact, (2) a common vehicle, (3) the airborne route, and (4) vectors.

Contact transmission

Contact transmission is the spread of the agent of disease by direct contact, indirect contact, or by droplet.

Direct contact transmission refers to the direct transmission of an agent from its source to a susceptible host; an intermediate object is not involved. This is also known as person-to-person transmission. The most common forms of direct contact transmission are touching, kissing, or sexual intercourse. Among the diseases transmitted from person to person by direct contact are viral

Table 13–3 Six Bacteria That Cause More than 60% of All Nosocomial Infections

Bacterium	Service with Most Frequent Reports	Percentage of Infections in that Service	Percentage of Total Infections
Escherichia coli	Gynecology	32.5	18.6
Staphylococcus aureus	Newborn	34.1	10.8
Enterococci	Gynecology	17.2	10.7
Pseudomonas aeruginosa	Surgery	17.4	10.6
Klebsiella species	Medicine	8.7	7.4
Coagulase–negative *Staphylococci*	Newborn	14.6	6.1

Table 13–4 The Principal Kinds of Nosocomial Infections

Type of Infection	Comment
Urinary tract infection	Most common, usually accounting for about 50% of all cases. Typically related to urinary catheterization.
Surgical wound infection	Ranks second in incidence (about 25%). It is estimated that 5–12% of all surgical patients develop postoperative infections; the percentage can reach 30% for certain surgeries, such as colon surgery and amputations.
Lower respiratory infection	Nosocomial pneumonias rank third in incidence (about 12%) and have high mortality rates. Most of these pneumonias are related to respiratory devices that aid breathing or administer medications.
Bacteremia infection	Bacteremias account for about 6%. Intravenous catheterization is implicated in nosocomial infections of the bloodstream, particularly infections caused by bacteria and fungi.
Cutaneous infection	Among the least common of all nosocomial infections. However, newborns have a high rate of susceptibility to skin and eye infections.

respiratory tract diseases (common cold and influenza), hepatitis A, measles, scarlet fever, smallpox, and sexually transmitted diseases (syphilis, gonorrhea, and genital herpes). Agents of disease can be transmitted by direct contact from animals (or animal products) to humans. An example is rabies.

Indirect contact transmission occurs when the agent of disease is transmitted from its reservoir to a susceptible host by means of a nonliving object. Any nonliving object involved in the spread of an infection is called a **fomite**. Examples of fomites are handkerchiefs, towels, bedding, diapers, drinking cups, eating utensils, and thermometers. Staphylococcal skin infections are spread by indirect contact.

Droplet infection is a third type of contact transmission. Agents of disease spread by droplets are contained in saliva or mucus that is discharged into the air by activities such as coughing, sneezing, laughing, or talking. Such droplets travel a very short distance through the air; that is, less than one meter from the reservoir to the susceptible host. Disease agents that travel such short distances are not regarded as airborne. (Airborne transmission is discussed below.) An example of a disease spread by droplet infection is influenza.

Common vehicle transmission

Common vehicle transmission refers to the transmission of disease agents by a common inanimate reservoir (food, water, drugs, blood) to a large number of individuals. Examples of diseases transmitted in this way are water-borne shigellosis, food-borne salmonellosis, and various kinds of nosocomial infections.

Airborne transmission

Airborne transmission refers to the spread of agents of infection by droplet nuclei or dust; the particles travel more than one meter from the reservoir to the susceptible host. Some microbes are carried in small drops of mucus discharged in a fine spray from the mouth and nose during coughing, sneezing, laughing, and even talking (Figure 13–3). These mucous droplets can be so small that they evaporate, travel more than one meter from the source, and remain airborne for prolonged periods of time. Such droplets are called droplet nuclei. Viruses that cause smallpox and measles and the bacterium that causes tuberculosis are transmitted via droplet nuclei. Infectious agents contained in dust particles are also carried in airborne transmission. Q Fever is one disease transmitted by the airborne route. Spores produced by certain fungi are also transmitted by the airborne route and can cause diseases such as histoplasmosis, coccidioidomycosis, and blastomycosis (see Chapter 23).

Vectors

Arthropods are the most important group of disease **vectors,** animals that carry pathogens from one host to another. (Insects and other arthropod

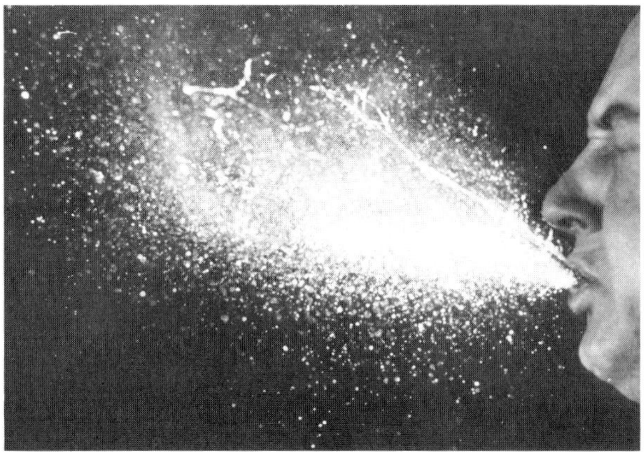

Figure 13–3 This high-speed photograph shows the spray of small droplets that come from the mouth during a sneeze.

vectors were discussed in Chapter 11.) Arthropod vectors transmit disease by two general methods. *Mechanical transmission* is the transport of the pathogens on the insect's feet or other body parts. If the insect makes contact with a host's food, pathogens can be transferred to the food and later swallowed by the host. Houseflies, for instance, are capable of transferring the pathogens of typhoid fever and bacillary dysentery (shigellosis) from the feces of infected persons to food.

Biological transmission is more complex. The arthropod bites an infected person or animal and ingests some of the infected blood. The pathogens then reproduce in the vector, and the increased numbers of pathogens increase the possibility that they will be transmitted to another host. Some parasites reproduce in the gut of the arthropod; these are passed with feces. If the arthropod defecates while biting a potential host for the parasite, the parasite can enter the wound. Other parasites reproduce in the vector's gut and migrate to the salivary gland. These are directly injected into a bite. Some protozoan and helminthic parasites use the vector as a host for a developmental change in their life cycle.

Table 13–5 lists a few important arthropod vectors and the diseases they transmit. Figure 13–4 summarizes all the ways in which pathogenic microorganisms are transmitted from the source of infection to the human host.

KINDS OF DISEASES

So that discussions in this and later chapters can be readily understood, we will define certain key terms that relate to the nature and scope of disease. For the most part, this and subsequent discussions of disease refer to infectious diseases. Any disease that spreads from one host to another, either directly or indirectly, is said to be a **communicable disease.** Typhoid fever and tuberculosis are examples. A **noncommunicable disease** is caused by microorganisms that normally inhabit the body and only occasionally produce disease, or that reside outside the body and produce disease only when introduced into the body. These diseases are not spread from one host to another. An example is tetanus. *Clostridium tetani* (te'tan-ē) produces disease only when it is introduced into the body via

Table 13–5 Representative Arthropod Vectors and the Diseases They Cause

Disease	Causative Agent	Arthropod Vector	Refer to Chapter
Malaria	*Plasmodium* species	*Anopheles* mosquito	21
African trypanosomiasis	*Trypanosoma* sp.	*Glossina* species (tsetse fly)	20
Yellow fever	Arbovirus (yellow fever virus)	*Aedes aegypti* mosquito	21
Dengue	Arbovirus (dengue fever virus)	*Aedes aegypti* mosquito	21
Arthropod-borne encephalitis	Arbovirus (encephalitis virus)	*Culex* mosquito	20
Epidemic typhus	*Rickettsia prowazekii*	*Pediculus vestimenti* (louse)	21
Endemic murine typhus	*Rickettsia typhi*	*Xenopsylla cheopis* (rat flea)	21
Rocky Mountain spotted fever	*Rickettsia rickettsii*	*Dermacentor andersoni* and other species (tick)	21
Plague	*Yersinia pestis*	*Xenopsylla cheopis* (rat flea)	21
Relapsing fever	*Borrelia* species	*Ornithodorus* species (soft ticks)	21

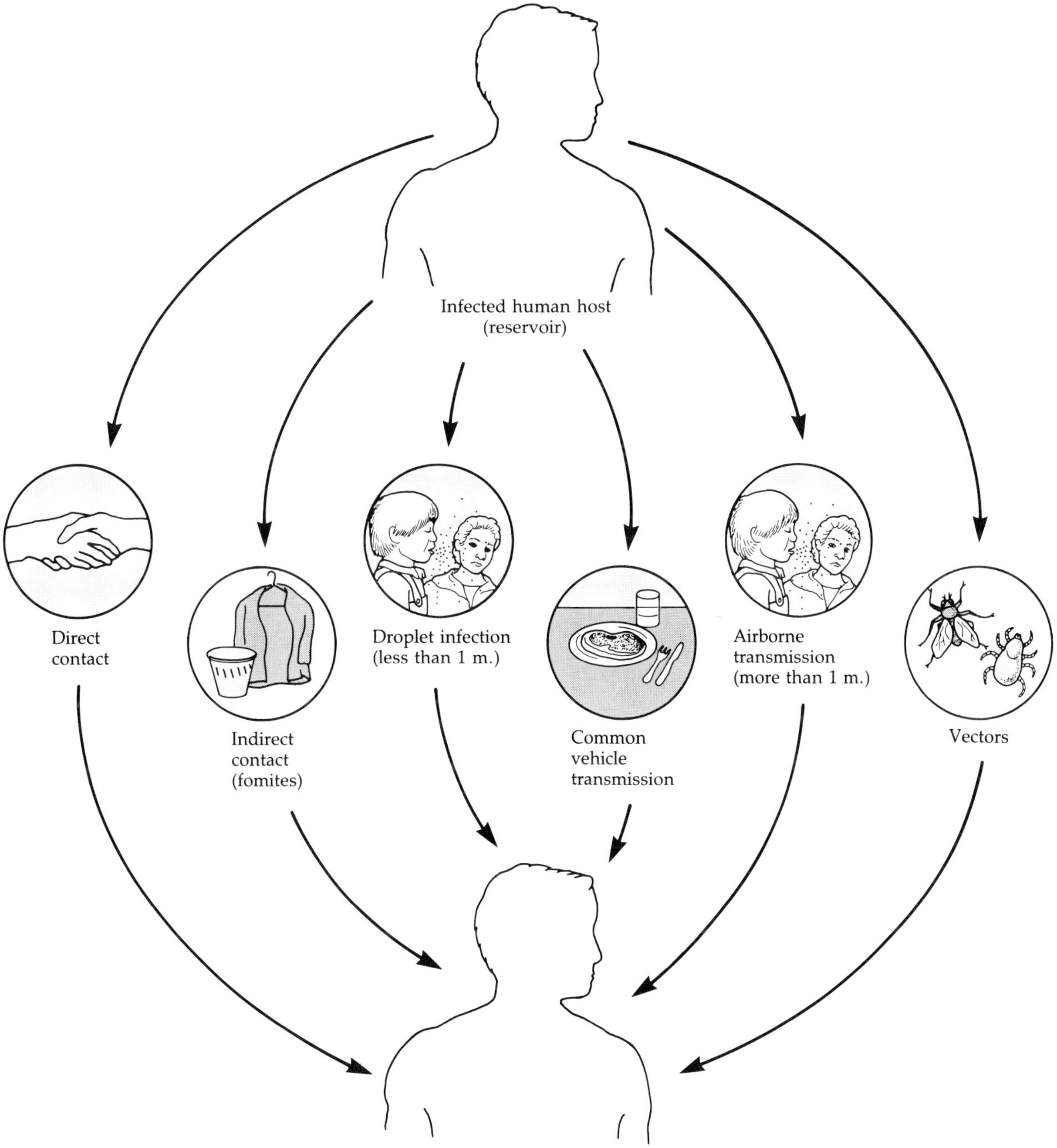

Figure 13–4 Routes by which human diseases are transmitted.

abrasions or wounds. A **contagious disease** is one that is *easily* spread from one person to another.

To understand the scope of a disease, one should know something about its occurrence. The **incidence** of a disease is the fraction of a population that contracts it during a particular length of time. The **prevalence** of a disease is the fraction of the population having the disease at a specified time. These numbers indicate the range of a disease's occurrence and its tendency to affect some groups of people more than others.

One useful way of defining the scope of a disease is in terms of its severity or duration. Accordingly, an **acute disease** is one that develops rapidly but lasts only a short time. A good example is influenza. A **chronic disease** develops more slowly, and the body's reactions are less severe, but the disease is likely to be continuous or recurrent for long periods of time. Tuberculosis, syphilis, and leprosy fall into this category. A disease that is intermediate between acute and chronic is described as a **subacute disease.** A **latent disease** is one in which the causative agent remains inactive for a period of time but then becomes active to produce symptoms of the disease. An example is the cold sore caused by the herpes simplex virus.

Frequency of occurrence is another criterion that is used in the classification of diseases. If a particular disease occurs only occasionally, it is called a **sporadic disease.** Polio might be considered such a disease. A disease constantly present in a population is called an **endemic disease;** an example of such a disease is the common cold. If many people in a given area acquire a certain disease in a relatively short period of time, it is referred to as an **epidemic disease.** Influenza and measles are examples of diseases that often achieve epidemic status. Some authorities consider gonorrhea and certain other sexually transmitted diseases epidemic at this time. Finally, an epidemic disease that occurs worldwide is referred to as a **pandemic disease.** We experienced a pandemic of influenza in the late 1950s.

Infections can also be classified according to the extent to which the host's body is affected. A **local infection** is one in which the invading microorganisms are limited to a relatively small area of the body. Examples of local infections are boils and abscesses. In a **systemic,** or **generalized, infection,** microorganisms or their products are spread throughout the body by the blood or lymphatic system. Typhoid fever is an example. Very frequently, agents of a local infection enter a blood or lymph vessel and spread to other parts of the body. We refer to this condition as a **focal infection.** Focal infections can arise from infections in the teeth, tonsils, or sinuses. The presence of bacteria in the blood is known as **bacteremia,** and if the bacteria actually multiply in the blood, the condition is called **septicemia** (see Microview 3–2).

The health of the body also provides a basis for the classification of infections. A **primary infection** is an acute infection that causes the initial illness. A **secondary infection** is one caused by an opportunist after the primary infection has weakened the body's defenses. Secondary infections of the skin and respiratory tract are common and are sometimes more dangerous than the primary infections. Streptococcal bronchopneumonia following whooping cough, measles, or influenza is an example of a secondary infection. An **inapparent,** or **subclinical, infection** is one that does not cause any noticeable illness. Polio and infectious hepatitis, for example, can be carried by persons who show no illness because they are protected by antibodies.

When a particular disease affects the body, it can manifest itself in many different ways. In other words, it might signal the body that something is wrong. Let us now consider some of the signals of disease.

SIGNALS OF DISEASE

Each disease affecting the body alters body structures and functions in particular ways, and these alterations are usually indicated by several kinds of evidence. For example, the patient may experience certain **symptoms,** changes in body function, such as pain and malaise. These subjective changes are not apparent to the observer. The patient can also exhibit **signs,** which are objective changes that the physician can observe and measure. Signs frequently evaluated include lesions (changes produced in tissues by disease), swelling, fever, and paralysis. Sometimes a specific group of symptoms or signs always accompanies a particular disease. Such a group is called a **syndrome.** The **diagnosis**

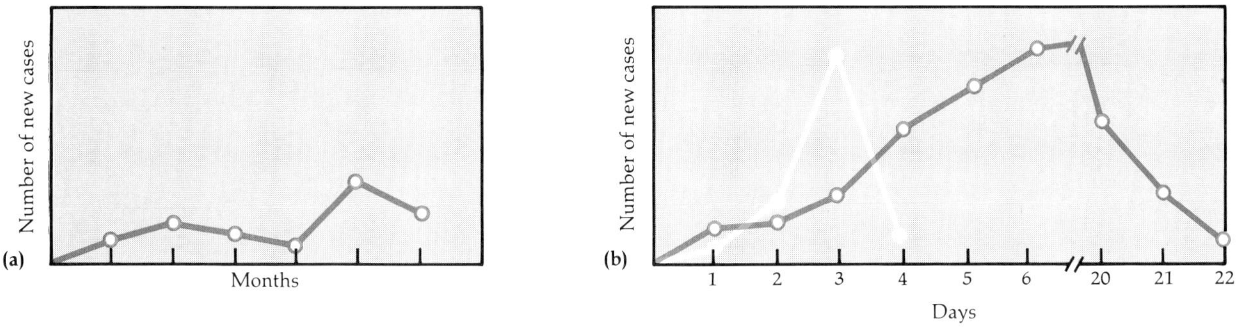

Figure 13–5 Patterns of disease spread. **(a)** Sporadic cases of a disease. **(b)** An epidemic spread from a common source, such as contaminated food (black line), and an epidemic spread from person to person (colored line).

of a disease, that is, its identification, is achieved by evaluation of the signs and symptoms, as well as the results of certain laboratory tests.

EPIDEMIOLOGY

So that a disease can be effectively controlled and treated, it is desirable to identify its causative agent. It is also desirable to understand the mode of transmission and distribution of the disease. The science that deals with when and where diseases occur and how they are transmitted in the human population is called **epidemiology.** In today's crowded, overpopulated world, where frequent travel and mass production and distribution of foods and other goods are a way of life, epidemiology assumes an important and ever-growing role. A contaminated water supply or contaminated food can affect many thousands of people. The occurrence of Legionnaires' disease not so long ago (see Chapter 1) underscored the importance of an epidemiologic investigation in determining the source, means of transmission, and distribution of a disease.

An epidemiologist not only determines the etiology of a disease, but also identifies other possibly important factors, such as geographical distribution, nutrition, gender, and age group of the persons affected. The epidemiologist also evaluates how effectively a disease is being controlled in a community—for example, the effectiveness of a vaccination program. By determining the frequency of a disease in a population and identifying the factors responsible for its transmission, an epidemiologist provides physicians with important information concerning the treatment and prognosis of a disease. (Figure 13–5 shows the graphs obtained from some hypothetical epidemiologic data, namely, the number of new cases of a disease reported each day within a certain time period. Such graphs provide information about whether disease outbreaks were sporadic or epidemic, and, if epidemic, how the disease was probably spread.) Finally, an epidemiologist can provide data to help in the evaluating and planning of overall health care for a community.

Epidemiologists employ three basic types of investigations when analyzing the occurrence of a disease: (1) descriptive, (2) analytical, and (3) experimental.

During the process of **descriptive epidemiology,** all data that describe the occurrence of the disease under study are collected. Such a collection is generally retrospective (after the epidemic has ended). Relevant information usually includes information about the affected persons, and the place and time period in which the disease occurred. In analyzing the data about persons affected by a particular disease, an epidemiologist would take into account factors such as age, sex, occupation, personal habits, socioeconomic status, history of immunization, and the presence of any underlying diseases. The place of infection refers to the site at which a susceptible host *comes* in contact with the agent of infection. It is important to know this in order to prevent further outbreaks. The time period over which the disease occurs can be considered on a seasonal basis (to indicate whether the disease is prevalent during the summer or winter) or on a yearly basis (to indicate the effects of hygiene or immunization).

The study of **analytical epidemiology** analyzes a particular disease to determine its probable cause. This study can be done in two ways. With one method, the epidemiologist looks for factors that might have preceded the disease. A group of persons who have the disease is compared with another group free of the disease. For example, one group with meningitis and one without might be matched by age, sex, socioeconomic status, and location. These statistics are compared so it can be determined which of all the possible factors—genetic, environmental, nutritional, and so forth—might be responsible for the meningitis. Using another method, the epidemiologist studies two populations, one that has had contact with the agent causing a disease and of another that has not. For example, a comparison of two matched groups, one composed of people who have received blood transfusions and one composed of people who have not, could reveal an association between blood transfusions and the hepatitis B virus, the causative agent of serum hepatitis.

Experimental epidemiology begins with a hypothesis about a particular disease; experiments to test the hypothesis are then conducted on a group of people. One such hypothesis could be the assumed effectiveness of a drug. A group of infected individuals is selected and divided randomly so that some receive the drug and others receive a placebo, a substance that has no effect. If all other factors are kept constant between the two groups, and it is determined that those people receiving the drug recover more rapidly than those who receive the placebo, it can be concluded that the drug is the experimental factor (variable) that made the difference.

Epidemiology is a major concern of the state and federal public health departments. The Centers for Disease Control (CDC), a U.S. Public Health Service branch located in Atlanta, Georgia, are a central source of epidemiologic information in the United States.

The CDC issues a publication called *Morbidity and Mortality Weekly Report (MMWR),* which is read by microbiologists, physicians, and other hospital and public health personnel. The *MMWR* contains data on the incidence of specific notifiable diseases **(morbidity)** and the deaths from these diseases **(mortality)**; this data is usually organized by state. (**Notifiable diseases** are those for which physicians must report cases to the Public Health Service. Examples are measles, tetanus, and gonorrhea.) Publication articles include reports of disease outbreaks, case histories of special interest, and summaries of the status of particular diseases over a recent period. These articles often include recommendations for procedures for diagnosis, immunization, and treatment. A number of graphs and other data in this book are from MMWR, and some of the boxes are direct excerpts from this publication.

In the next chapter, we will consider the mechanisms of pathogenicity. There we will discuss in more detail the methods by which microorganisms enter the body and cause disease, the patterns of infection, the effects of disease on the body, and the means by which pathogens leave the body.

STUDY OUTLINE

INTRODUCTION (pp. 378–379)

1. Pathogenic microorganisms have special properties that allow them to invade the human body or produce toxins.
2. When the microorganism overcomes the body's defenses, a state of disease results.

PATHOLOGY, INFECTION, AND DISEASE (p. 379)

1. Pathology is the scientific study of disease.
2. Pathology is concerned with the etiology (cause) of disease, pathogenesis, and effects of disease.

3. Infection is the invasion and growth of pathogens in the body.

4. A host is an organism that shelters and supports the growth of pathogens.

5. Disease is an abnormal state in which part or all of the body is not properly adjusted or is not capable of carrying on normal functions.

NORMAL FLORA (pp. 379–382)

1. Animals, including humans, are germfree in utero.

2. Microorganisms begin colonization in and on the surface of the body soon after birth.

3. Microorganisms that establish permanent colonies inside or on the body without producing disease are called normal flora.

4. Transient flora are present for varying amounts of time and then disappear.

Functions (pp. 380–381)

1. Normal flora can prevent pathogens from causing an infection.

Relationships Between Normal Flora and Host (pp. 380–382)

1. The normal flora and human exist in a symbiosis (living together).

2. The three types of symbiosis are commensalism (one organism benefits and the other is unaffected); mutualism (both organisms benefit); and parasitism (one organism benefits and one is harmed).

3. Opportunists (opportunistic pathogens) do not cause disease under normal conditions, but cause infections under special conditions.

CAUSES OF DISEASE (pp. 382–385)

Main Categories of Disease (pp. 382–383)

1. Diseases that have a known etiology include infectious diseases (e.g., pneumonia), nutritional deficiency diseases (e.g., rickets), congenital diseases (e.g., cleft palate), inherited diseases (e.g., hemophilia), metabolic diseases (e.g., diabetes mellitus), degenerative diseases (e.g., emphysema), neoplastic diseases (e.g., cancer), immunologic diseases (e.g., rheumatoid arthritis), iatrogenic diseases (e.g., Staphylococcus infections), and psychogenic diseases (e.g., ulcers).

2. Diseases that have no known etiology are referred to as idiopathic diseases.

Koch's Postulates (pp. 383–385)

1. Koch's postulates establish that specific microbes cause specific diseases.

2. Koch's postulates state the following requirements: (1) the same pathogen must be present in every case of the disease; (2) the pathogen must be isolated in pure culture; (3) the pathogen isolated from pure culture must cause the same disease in a healthy, susceptible laboratory animal; and (4) the pathogen must be reisolated from the inoculated laboratory animal.

SPREAD OF INFECTION (pp. 385–390)

Reservoirs (pp. 385)

1. A continuous source of infection is called a reservoir of infection.

2. People who have a disease or are carriers of pathogenic microorganisms constitute human reservoirs of infection.

3. Zoonoses (diseases that occur in wild and domestic animals) can be transmitted to humans from animal reservoirs of infection.

4. Some pathogenic microorganisms grow in nonliving reservoirs such as soil and water.

Nosocomial (Hospital-Acquired) Infections (pp. 385–388)

1. Nosocomial infections are acquired during the course of stay in a hospital.

2. About 5 to 15% of all hospitalized patients acquire nosocomial infections.

3. Normal flora are often responsible for nosocomial infections when they are introduced into the body through medical procedures such as surgery and catheterization.

4. Opportunists and saprophytes can cause nosocomial infections in hospitalized patients.

5. Gram-negative bacteria (e.g., *E. coli* or *P. aeruginosa*) are most often involved in nosocomial infections.

6. Patients with large intestinal surgery, burns, or who are immunosuppressed are most susceptible to nosocomial infections.

7. A hospital's infection-control personnel are responsible for overseeing proper storage and handling of equipment and supplies.

Transmission of Disease (pp. 388–390)

1. Transmission by direct contact involves close physical contact between the source of the disease and a susceptible host.
2. Transmission by fomites (inanimate objects), food, or water constitutes indirect contact.
3. Transmission by saliva or mucus is called droplet infection.
4. Transmission to a large number of people from the same inanimate reservoir is called common vehicle transmission.
5. Airborne transmission refers to pathogens carried on water droplets or dust for a distance greater than one meter.
6. Arthropod vectors carry pathogens from one host to another by both mechanical and biological transmission.

KINDS OF DISEASE (pp. 390–392)

1. Communicable diseases are transmitted directly or indirectly from one host to another.
2. Noncommunicable diseases are caused by microorganisms that normally grow outside the human body and are not transmitted from one host to another.
3. A contagious disease is one that is easily spread from one person to another.
4. Disease occurrence is reported by incidence (number of people with the disease) and prevalence (incidence at a particular time).
5. The scope of a disease may be defined as acute, chronic, or subacute.
6. Diseases are classified by frequency of occurrence: sporadic, endemic, epidemic, and pandemic.
7. A local infection affects a small area of the body; a systemic infection is spread throughout the body through the circulatory system.
8. A secondary infection can occur after the host is weakened from a primary infection.
9. An inapparent, or subclinical, infection does not cause any signs of disease in the host.

SIGNALS OF DISEASE (pp. 392–393)

1. A patient may exhibit symptoms (subjective changes in body functions) and signs (measurable changes), which are used by a physician to make a diagnosis (identification of the disease).
2. A specific group of symptoms or signs that always accompanies a specific disease is called a syndrome.

EPIDEMIOLOGY (pp. 393–394)

1. The science of epidemiology is the study of the transmission, incidence, and frequency of disease.
2. The Centers for Disease Control are the main source of epidemiologic information in the United States.
3. The CDC publishes *Morbidity and Mortality Weekly Report* to provide information on morbidity (incidence) and deaths (mortality).

STUDY QUESTIONS

REVIEW

1. Differentiate between the following pairs of terms.
   (a) Etiology and pathogenesis
   (b) Infection and disease
   (c) Communicable disease and noncommunicable disease
2. What is meant by normal flora? How do they differ from transient flora?
3. Define symbiosis. Differentiate between commensalism, mutualism, and parasitism, and give an example of each.
4. Describe how Koch's postulates establish the etiology of many infectious diseases. Why don't Koch's postulates apply to all infectious diseases?

5. What is a reservoir of infection? Match the following diseases with their reservoirs.

______ Influenza	(a) Nonliving
______ Rabies	(b) Human
______ Botulism	(c) Animal

6. Describe the various ways diseases can be transmitted in each of the following categories.
   (a) Transmission by direct contact
   (b) Transmission by indirect contact
   (c) Transmission by arthropod vectors
   (d) Droplet transmission
   (e) Common vehicle transmission
   (f) Airborne transmission
7. Indicate whether each of the conditions described is typical of subacute, chronic, or acute infections.
   (a) Patient experiences rapid onset of malaise; symptoms last five days.
   (b) Patient experiences cough and breathing difficulty for months.
   (c) Patient has no apparent symptoms and is a known carrier.
8. Of all the hospital patients with infections, one-third do not enter the hospital with an infection. How do they acquire these infections? What is the method of transmission of these infections? What is the reservoir of infection?
9. Differentiate between an endemic and epidemic state of infectious disease.
10. What is epidemiology? What is the role of the Centers for Disease Control (CDC)?
11. Distinguish between symptoms and signs as signals of disease.
12. How can a local infection become a systemic infection?
13. Why are some organisms that constitute normal flora described as commensals while others are described as mutualistic?

CHALLENGE

1. Ten years before Koch published his work on anthrax, Anton De Bary showed that potato blight was caused by the fungus *Phytophthora infestans*. Why do you suppose we use Koch's postulates instead of something called "De Bary's postulates"?
2. Florence Nightingale gathered the following data in 1855.

Population sampled	Deaths due to contagious diseases
Englishmen	0.2%
English soldiers (in England)	18.7%
English soldiers (in Crimean War)	42.7%
English soldiers (in Crimean War) after Nightingale's sanitary reforms	2.2%

   Discuss how Nightingale employed the three basic types of epidemiologic investigation. The contagious diseases were primarily cholera and typhus; how are these diseases transmitted? prevented?
3. Name the method of transmission of the following diseases.
   (a) Malaria
   (b) Tuberculosis
   (c) Nosocomial infections
   (d) Salmonellosis
   (e) Streptococcal pharyngitis
   (f) Mononucleosis
   (g) Measles
   (h) Hepatitis A
   (i) Tetanus
   (j) Hepatitis B
   (k) Chlamydial urethritis

4. Mark the graph below to show when this disease occurred sporadically, endemically, and epidemically. What would have to be shown to indicate a pandemic of this disease?

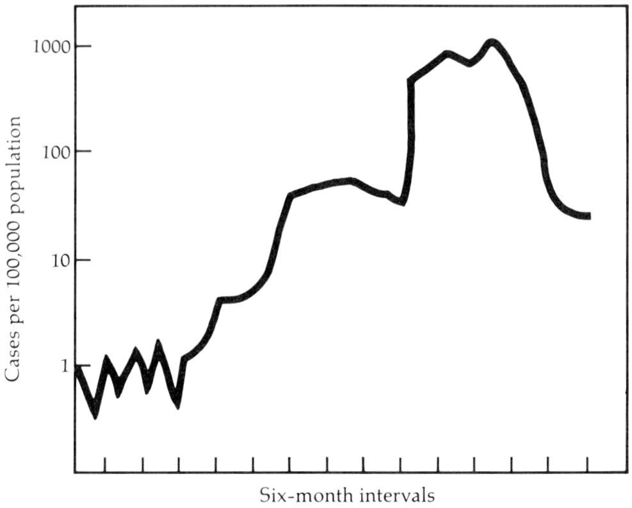

FURTHER READING

Ayliffe, G. A. and L. J. Taylor. 1982. *Hospital-Acquired Infections: Principles and Prevention*. Littleton, MA: John Wright-PSG. Discusses nosocomial infections and how to prevent them.

Busvine, J. R. 1975. *Arthropod Vectors of Disease*. London: Edward Arnold. Describes vector-borne diseases and the different ways vectors can transmit diseases.

Mandell, G. L., R. G. Douglas, and J. E. Bennett. 1979. Nosocomial infections, pp. 2213–2256. In *Principles and Practice of Infectious Disease*, vol. 2. New York: Wiley. Discussions of specific nosocomial infections, such as pneumonia.

Mausner, J. S. and S. Kramer. 1984. *Mausner and Bahn Epidemiology*, 2nd ed. Philadelphia, PA: W. B. Saunders. A comprehensive reference on the collection and interpretation of epidemiological data.

Pike, R. M. 1979. Laboratory-associated infections: incidence, fatalities, causes, and prevention. *Annual Review of Microbiology* 33:41–66. A summary of the particular diseases acquired by laboratory personnel.

Roueche, B. 1984. *The Medical Detectives*, 2 vols. New York: Truman Talley Books. Volume I by Times Books, Volume II by E. P. Dutton, Inc. Narratives describe the scientific investigation of epidemic disease.

Savage, D. C. 1977. Microbial ecology of the gastrointestinal tract. *Annual Review of Microbiology* 31:107-133. An examination of the ecological niches occupied by gastrointestinal flora.

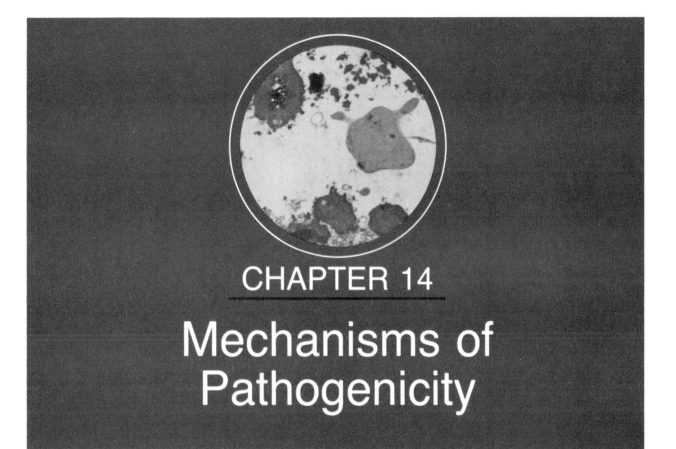

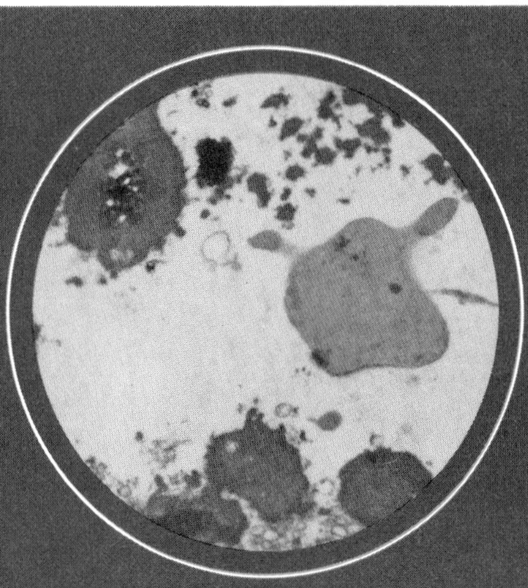

CHAPTER 14

Mechanisms of Pathogenicity

OBJECTIVES

After completing this chapter you should be able to

- Define each of the following terms: portal of entry; pathogenicity; virulence; LD_{50}.
- Explain how adherence, capsules, cell wall components, and enzymes contribute to pathogenicity.
- Compare the effects of hemolysins, leukocidins, coagulase, kinases, and hyaluronidase.
- Contrast the nature and effects of exotoxins and endotoxins.
- List seven cytopathic effects of viral infections.
- Discuss the causes of symptoms in fungal, protozoan, and helminthic diseases.
- Identify four predisposing factors to disease.
- Put the following terms in proper sequence in terms of the pattern of disease: period of decline; period of convalescence; period of illness; crisis; prodromal period; period of incubation.

Now that you have a basic understanding of the principles of disease, we will take a look at some of the specific properties of microorganisms that contribute to **pathogenicity,** the ability to cause disease in a host, and **virulence,** the degree of pathogenicity. Our major goal is to outline some of these key properties. Keep in mind, however, that many of the properties contributing to microbial pathogenicity and virulence are unclear or unknown. If the microbe's attack overpowers the host defenses, disease results. But if the factors contributing to virulence are neutralized or overcome by the host defenses, health is maintained.

In order for a pathogen to cause disease, it must gain access to the host, adhere to host tissues, resist or evade host defenses, and damage the host tissue. The starting point of our discussion will be the routes microorganisms take to gain entrance into the human body and other hosts.

ENTRY OF A MICROORGANISM INTO THE HOST

Pathogens invading the body can use several avenues. We call the avenue by which a microbe gains access to the body a **portal of entry.**

Mucous Membranes

To gain access to the body, many bacteria and viruses can penetrate mucous membranes lining the conjunctiva, respiratory tract, gastrointestinal tract, and genitourinary tract.

The mucous membrane of the respiratory tract offers the easiest and most frequently traveled access route for infectious microorganisms. Microbes are inhaled into the nose or mouth along with the drops of moisture and dust particles that contain them. Some diseases contracted via the respiratory tract are the common cold, pneumonia, tuberculosis, influenza, measles, and smallpox.

Another common portal of entry is the mucous membrane of the gastrointestinal tract. Microorganisms contracted from food, water, milk, and contaminated fingers enter the body this way, although most of these microbes are destroyed by hydrochloric acid and enzymes in the stomach and by bile and enzymes in the small intestine. Those that survive can cause diseases such as poliomyelitis, infectious hepatitis, typhoid fever, amoebic dysentery, bacillary dysentery (shigellosis), and cholera. The pathogens are eliminated with feces and can be transmitted to other hosts via water, food, or contaminated fingers. Most pathogens enter through the mucous membranes of the gastrointestinal and respiratory tracts.

An important pathogen capable of penetrating mucous membranes of the genitourinary tract is *Treponema pallidum,* the causative agent of syphilis.

Skin

Because most microorganisms cannot penetrate unbroken skin, many gain access to the body through openings such as hair follicles and sweat ducts. It has been demonstrated that the hookworm *Necator americanus* (ne-kā'tôr ä-me-ri-ka'nus) (see Figure 11–26) actually bores through intact skin, and some fungi grow on the keratin in skin or infect the skin itself.

Parenteral Route

Other microorganisms gain access to the body when they are deposited directly into the tissues beneath the skin or into mucous membranes; this direct access can occur when these barriers are penetrated or injured (traumatized). This route is referred to as the **parenteral route.** The parenteral route is established by punctures, injections, bites, cuts, wounds, surgery, and splitting due to swelling or drying.

Preferred Portal of Entry

Even after microorganisms have entered the body, they do not necessarily cause disease. The occurrence of disease depends on several factors. One of these factors is the portal of entry. Many pathogens have a preferred portal of entry: the use of that portal is a prerequisite to their being able to cause disease. If they gain access to the body by another route, disease might not occur. For example, the bacterium of typhoid fever, *Salmonella typhi* (sal-mōn-el'lä tī'fē) produces all the signs and symptoms of the disease when it is swallowed (preferred route). But if the same bacteria are rubbed on the skin, no reaction, or only a slight inflammation, occurs. Streptococci that are inhaled (preferred route) can cause pneumonia. Those that are swallowed generally do not produce signs or symptoms. Some pathogens, like the microorganism that causes plague, can initiate disease from more than one portal of entry. The portals of entry for some common pathogens are given in Table 14–1.

Numbers of Invading Microbes

If only a few microbes enter the body, it is likely that they will be overcome by the host's defenses. If, on the other hand, large numbers of microbes

Table 14–1 Causative Agents for Some Common Diseases, Arranged by Portal of Entry

Portal of Entry	Causative Agent*	Disease	Incubation Period
Mucous Membrane			
Respiratory tract	*Corynebacterium diphtheriae*	Diphtheria	2–5 days
	Neisseria meningitidis	Bacterial meningitis	1–7 days
	Streptococcus pneumoniae	Pneumococcal pneumonia	Variable
	*Mycobacterium tuberculosis***	Tuberculosis	Variable
	Bordetella pertussis	Whooping cough (pertussis)	12–20 days
	Myxovirus	Influenza	18–36 hours
	Paramyxovirus	Measles (rubeola)	11–14 days
	Togavirus	German measles (rubella)	2–3 weeks
	Epstein–Barr virus (herpesvirus)	Infectious mononucleosis	2–6 weeks
	Zoster (herpesvirus)	Chickenpox (varicella)	14–16 days
	Poxvirus	Smallpox (variola)	12 days
	Coccidioides immitis (fungus)	Coccidioidomycosis (primary infection)	1–3 weeks
	Histoplasma capsulatum (fungus)	Histoplasmosis	5–18 days
Gastrointestinal tract	*Shigella* species	Bacillary dysentery (shigellosis)	1–2 days
	Brucella melitensis	Brucellosis (undulant fever)	6–14 days
	Vibrio cholerae	Cholera	1–3 days
	Salmonella enteritidis, Salmonella typhimurium, Salmonella cholerae-suis	Salmonellosis	7–22 hours
	Salmonella paratyphi	Paratyphoid fever	7–24 days
	Salmonella typhi	Typhoid fever	5–14 days
	Hepatitis A virus (picornavirus)	Infectious hepatitis	15–50 days
	Paramyxovirus	Mumps	2–3 weeks
	Picornavirus	Poliomyelitis	4–7 days
	Trichinella spiralis (helminth)	Trichinosis	2–28 days
Genitourinary tract	*Neisseria gonorrhoeae*	Gonorrhea	3–8 days
	Treponema pallidum	Syphilis	9–90 days
Skin or parenteral route	*Clostridium perfringens*	Gas gangrene	1–5 days
	Clostridium tetani	Tetanus	3–21 days
	Leptospira interrogans	Leptospirosis	2–20 days
	Yersinia pestis	Plague	2–6 days
	Rickettsia rickettsii	Rocky Mountain spotted fever	3–12 days
	Hepatitis B virus** (picornavirus)	Serum hepatitis	6 weeks–6 months
	Rhabdovirus	Rabies	10 days–1 year
	Togavirus	Yellow fever	3–6 days
	Plasmodium species (protozoan)	Malaria	2 weeks

*All causative agents are bacteria, unless indicated otherwise. For viruses, only the viral group is given, except where the virus has a name different from the disease it causes.

**These pathogens can also cause disease after entering the body via the gastrointestinal tract.

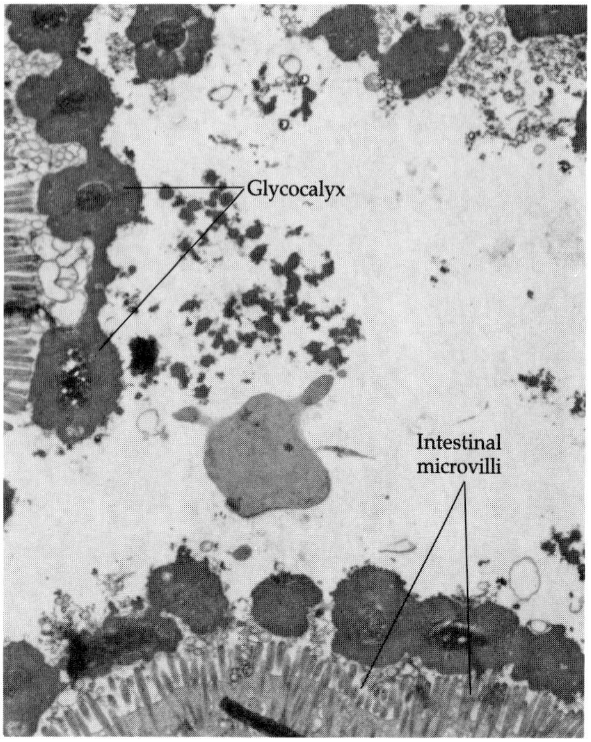

Figure 14–1 Attachment of *E. coli* to intestinal wall. Note the thick glycocalyx surrounding and joining the bacterial cells.

gain entry, then it is probable that the stage is set for disease. Thus, the likelihood of disease increases as the number of pathogens increases and decreases as the resistance of the host increases. The virulence of a microbe or the potency of its toxin (the poisonous substance it produces) is often expressed as the **LD_{50} (lethal dose)**, the number of microbes that will kill 50% of inoculated hosts under standard conditions. The dose required to produce a demonstrable infection in 50% of the hosts is called the **ID_{50} (infectious dose)**.

ADHERENCE

Almost all pathogens have some means of facilitating their attachment to host tissues. For most pathogens **adherence** is a necessary step in pathogenicity. (It should be pointed out, however, that nonpathogens also have structures for attachment.) The attachment between pathogen and host takes place by means of surface projections, on the pathogen, called **ligands** (**adhesins**), and complementary surface **receptors** on the host cells. Whereas the majority of ligands thus far studied are glycoproteins or lipoproteins, the receptors are typically sugars, such as mannose. Ligands on different strains of the same species of pathogen can vary and are frequently associated with other microbial surface structures such as pili. Different cells of the same host can have different receptors.

Some examples will illustrate the diversity of ligands. *E. coli* is usually a commensal in the human intestinal tract. Both nonpathogenic and pathogenic strains exist and both contain ligands on pili (pili are shown in Figure 4–8). With these ligands, *E. coli* attach to intestinal epithelial cells (Figure 14–1). *Streptococcus mutans* (mū'tans) is the microorganism primarily implicated in tooth decay. *S. mutans* adheres to the surface of the tooth enamel by producing an extracellular polysaccharide. *Neisseria gonorrhoeae* (go-nôr-rē'ä), the causative agent of gonorrhea, has pili containing ligands that permit attachment to cells with appropriate receptors in the genitourinary tract, eye, and pharynx. As you saw in Chapter 12, viruses also have specific means of attachment.

HOW PATHOGENS RESIST HOST DEFENSES

Capsules

It was noted in Chapter 4 that some bacteria form capsules around their cell walls (Figure 14–2); this property increases the virulence of the species. The capsule resists the host's defenses by impairing phagocytosis, a process by which certain cells of the body engulf and destroy microbes (discussed in Chapter 15). It appears that the chemical nature of the capsule prevents the phagocytic cell from adhering to the bacterium. However, the human body is capable of producing antibodies against the capsule, and when these antibodies are present, the encapsulated bacteria are easily destroyed by phagocytosis.

One bacterium that owes its virulence to the presence of a polysaccharide capsule is *Streptococcus pneumoniae*, the causative agent of lobar pneumonia (see Figure 4–5). This organism is virulent with its capsule but avirulent and easily susceptible to phagocytosis without it. Avirulent *S. pneumoniae* normally resides in the upper respiratory tract. When the capsule is experimentally removed by enzymes, the bacterium loses its pathogenicity. Other bacteria that produce capsules related to virulence are *Klebsiella pneumoniae* (kleb-sē-el'lä nü-mō'nē-ī), a causative agent of bacterial pneumonia, and *Hemophilus influenzae* (hē-mä'fi-lus in-flü-en'zī), a cause of pneumonia and meningitis in children. Keep in mind that many nonpathogenic bacteria produce capsules and that the virulence of other pathogens is not related to the presence of a capsule.

Components of the Cell Wall

The cell walls of certain bacteria contain chemical substances that contribute to virulence. For example, *Streptococcus pyogenes* contains a heat- and acid-resistant protein called **M protein** (see Figure 19–3). This protein is found on both pili (fimbriae) and the cell surface. The M protein mediates attachment of the bacterium to epithelial cells of the host and helps the bacterium resist phagocytosis by white blood cells. The protein thereby increases the virulence of the microorganism. Immunity to *S. pyogenes* depends on the body's production of an antibody specific to M protein.

Enzymes

The virulence of some bacteria is thought to be aided by the production of extracellular enzymes (exoenzymes) and related substances. These chemicals have the ability to break cells open, dissolve materials between cells, form blood clots, and dissolve blood clots, among other functions. However, the importance of most of these enzymes to bacterial virulence has not been proved conclusively.

Substances produced by some bacteria, **leukocidins** have the ability to destroy neutrophils, white blood cells (leukocytes) that are very active in phagocytosis. Leukocidins are also active against phagocytic cells (macrophages) present in tissue. Among the bacteria that secrete leukocidins are

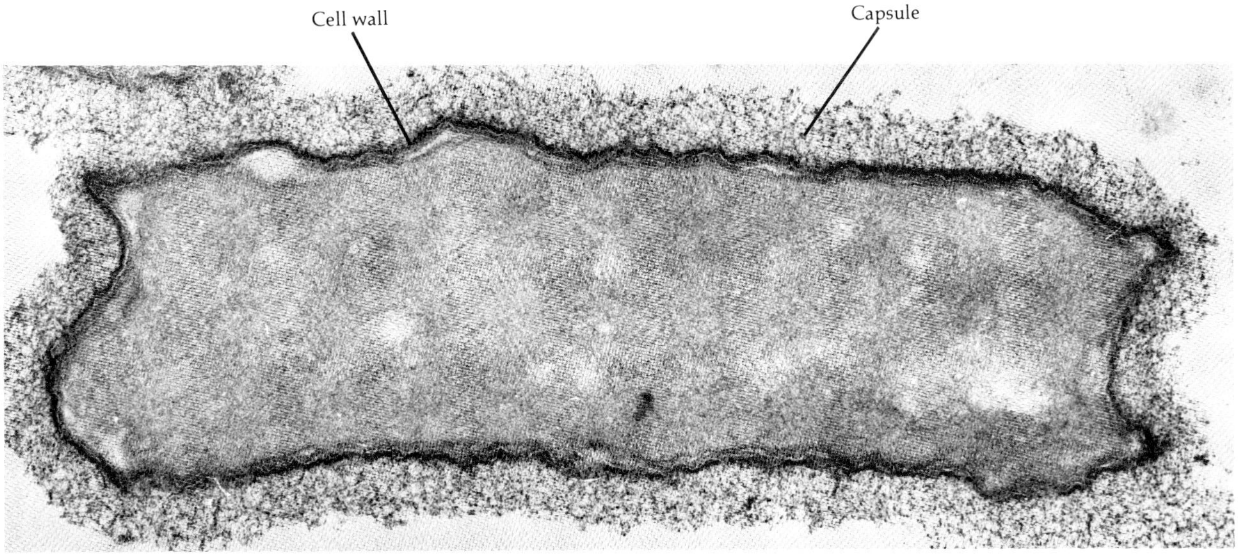

Figure 14–2 Electron micrograph of a thin section through a cell of an encapsulated strain of *Klebsiella pneumoniae* (×53,000). The capsule appears as a thick, gray layer of material outside of the dark cell wall. Capsules play an important role in the resistance of some pathogenic bacteria to phagocytosis.

staphylococci and streptococci. Leukocidins produced by streptococci cause degradation of lysosomes within leukocytes, and thereby cause the death of the white blood cell. Hydrolytic enzymes released from the leukocytic lysosomes can damage other cellular structures and thus intensify streptococcal lesions. This type of damage to white blood cells decreases host resistance.

Hemolysins are another group of enzymes produced by bacteria that might contribute to the virulence of some bacteria. Hemolysins cause the lysis (breakage) of red blood cells. Bacteria produce a number of different hemolysins, and they differ from one another according to the kind of red blood cell they lyse (human, sheep, rabbit, and so on) and the type of lysis they cause. One "type" of lysis is the "hot and cold" lysis exhibited by staphylococcal beta toxin. Beta toxin enzymatically destroys the membrane surrounding a red blood cell. When studied in laboratory culture, the effectiveness of beta toxin can be increased by the incubation of the culture first at 37°C and then at a colder temperature (in a refrigerator, for example). Important producers of hemolysins are staphylococci, streptococci, and *Clostridium perfringens* (klôs-tri'dē-um pėr-frin'jens), the causative agent of gas gangrene.

Coagulases are bacterial enzymes that coagulate (clot) the fibrinogen in the blood. (Fibrinogen, a plasma protein produced by the liver, is converted into fibrin, the threads that form a blood clot.) The fibrin clot caused by the coagulase may protect the bacterium from phagocytosis and isolate it from other defenses of the host. Coagulase is produced by some members of the genus *Staphylococcus* (staf-i-lō-kok'kus); it may, in fact, be involved in the walling-off process in boils produced by staphylococci. It should be noted, however, that some staphylococci that do not produce coagulase are still virulent. A capsule, and not coagulase, may be the more important factor in the virulence of these bacteria.

Bacterial kinases are another group of enzymes that may contribute to bacterial virulence. By breaking down fibrin, the kinases dissolve clots formed by the body to isolate the infection. One of the better-known kinases is **streptokinase (fibrinolysin)**, which is produced by streptococci. Another is **staphylokinase,** produced by staphylococci. (Injected directly into the blood, streptokinase has been used successfully to dissolve some types of blood clots.)

Hyaluronidase is yet another enzyme secreted by certain bacteria and possibly related to microbial virulence. It dissolves hyaluronic acid, a mucopolysaccharide that holds together certain cells of the body, particularly cells in connective tissue. It is believed that this dissolving action helps the microorganism spread from its initial site of infection. Hyaluronidase is produced by streptococci and some clostridia that cause gas gangrene. For therapeutic use, hyaluronidase may be mixed with a drug to promote the spread of the drug through a body tissue. The spread of gas gangrene is also facilitated by another enzyme. **Collagenase,** produced by several species of *Clostridium,* breaks down the protein collagen, which forms the framework of muscles.

Other bacterial substances believed to contribute to virulence are *necrotizing factors,* which cause the death of body cells, *hypothermic factors,* which affect body temperature, and *edema-producing factors,* which contribute to the development of edema (swelling).

Other Factors

In addition to capsules, cell wall components, and enzymes are other factors related to the pathogen's resistance to its host's defenses. Some pathogens are intracellular parasites; that is, they can multiply within phagocytes. These pathogens frequently give rise to chronic diseases, such as tuberculosis or brucellosis. Another factor is flagellation: flagellated pathogens, by virtue of their motility, can escape phagocytosis.

DAMAGE TO HOST CELLS

Pathogenic bacteria cause damage to host cells in three basic ways: (1) by causing direct damage in the *immediate* vicinity of the invasion; (2) by producing toxins, poisonous substances transported by blood and lymph that damage sites *far removed* from the original site of invasion; and (3) by inducing hypersensitivity reactions. This last mechanism is considered in detail in Chapter 17.

Direct Damage

Once a pathogen is attached to host cells, the pathogen is capable of passing through them to invade other tissues. During this invasion, the pathogens metabolize and multiply to kill host cells. Some bacteria, such as *Escherichia coli, Shigella* (shi-gel'lä), *Salmonella,* and *Neisseria gonorrhoeae,* can induce host epithelial cells to engulf them by a process that resembles phagocytosis. These pathogens can also be extruded from the host cells by a reverse phagocytosis process. Some bacteria can also penetrate host cells by excreting enzymes and by means of their own motility both of which can damage cells. Most damage by bacteria, however, is done by toxins.

Toxins

We will now turn our attention to the second main mechanism of pathogenicity, the production of toxins. **Toxins** are poisonous substances that are produced by certain microorganisms. They may be almost entirely responsible for the pathogenic properties of those microbes. The capacity of microorganisms to produce toxins is called **toxigenicity.** Toxins transported by the blood or lymph can cause serious, and sometimes fatal, effects. Some toxins produce fever, circulatory disturbances, diarrhea, and shock. Toxins can also inhibit protein synthesis, destroy blood cells and blood vessels, and disrupt the nervous system by causing spasms. The term **toxemia** refers to symptoms caused by toxins in the blood. Toxins are of two types: exotoxins and endotoxins.

Exotoxins

Some bacteria produce **exotoxins** as part of their growth and metabolism and release the exotoxins into the surrounding medium. Most bacteria that produce exotoxins are gram-positive. Exotoxins are proteins, and the genes for most of them (perhaps all) are carried on bacterial plasmids or phages. Because they are soluble in body fluids, exotoxins easily diffuse into the blood and are rapidly transported throughout the body.

Exotoxins work by destroying particular parts of the host's cells or by inhibiting certain metabolic functions. Exotoxins are among the most lethal substances known, and only 1 mg of botulinum toxin is enough to kill 1 million guinea pigs. Fortunately, only a few bacteria produce exotoxins.

Diseases caused by bacteria that produce exotoxins are often the result of the exotoxins, and not the bacteria themselves. It is the exotoxins that produce the signs and symptoms of the disease. Thus, exotoxins are disease-specific.

The body produces antibodies called **antitoxins** that provide immunity to exotoxins. When exotoxins are inactivated by heat, formaldehyde, iodine, or other chemicals, they no longer cause the disease but are still able to stimulate the body to produce antitoxins. Such altered exotoxins are called **toxoids.** As will be discussed in Chapter 16, when toxoids are injected into the body, they stimulate antitoxin production so that immunity is produced to diseases such as diphtheria and tetanus. Here we will briefly describe a few of the more notable exotoxins.

Botulinum Toxin Eight different types of botulinum toxin are produced by *Clostridium botulinum.* Each toxin possesses a different potency. Botulinum toxin is not a typical exotoxin in that it is made within the clostridial cell and not released into the medium until cell death occurs. Botulinum toxin acts at the neuromuscular junction (the junction between nerve cell and muscle cell), and prevents the transmission of nerve impulses. A toxin that destroys nerve tissue or interferes with nerve impulse transmission is called a **neurotoxin;** it causes flaccid (soft) paralysis.

Tetanus Toxin *Clostridium tetani* produces tetanus neurotoxin. This toxin causes excitation of the central nervous system, and thereby results in the convulsive symptoms (spasmodic contractions) of tetanus, or "lockjaw."

Diphtheria Toxin *Corynebacterium diphtheriae* (kô-rī-nē-bak-ti'rē-um dif-thi'rē ī) produces diphtheria toxin only when it is infected by a lysogenic phage carrying the *tox* gene. Diphtheria toxin inhibits protein synthesis in eucaryotic cells (see box on page 407 for a report on a rare fatal case of diptheria).

Staphylococcal Enterotoxin *Staphylococcus aureus* (ô'rē-us) produces a type of exotoxin called an **en-**

Table 14–2 Diseases Produced by Exotoxins

Disease	Bacterium	Exotoxin's Effect
Botulism	*Clostridium botulinum*	A neurotoxin that prevents transmission of nerve impulses; flaccid paralysis results
Tetanus	*Clostridium tetani*	A neurotoxin that promotes transmissions of nerve impulses; convulsions result
Gas gangrene and food poisoning	*Clostridium perfringens*	One exotoxin causes massive red blood cell destruction (hemolysis); another exotoxin (enterotoxin) is related to food poisoning and causes diarrhea
Diphtheria	*Corynebacterium diphtheriae*	Inhibits protein synthesis, especially in nerve, heart, and kidney cells
Scalded skin syndrome, food poisoning, and toxic shock syndrome (TSS)	*Staphylococcus aureus*	One exotoxin causes skin layers to separate and slough off; another exotoxin (enterotoxin) produces diarrhea and vomiting; still another exotoxin produces symptoms associated with toxic shock syndrome (TSS)
Cholera	*Vibrio cholerae*	Induces diarrhea
Scarlet fever	*Streptococcus pyogenes*	Causes vasodilation that results in the characteristic rash
Anthrax	*Bacillus anthracis*	Causes increased blood vessel permeability that results in hemorrhaging and tissue swelling
Traveler's diarrhea	*Escherichia coli* *Campylobacter jejuni*	Causes excessive secretion of ions and water; diarrhea results
Bacterial dysentery	*Shigella dysenteriae*	Neurotoxin damages blood vessels that supply brain; paralysis and hemorrhaging result

terotoxin because it affects the intestines. The enterotoxin induces vomiting and, by preventing absorption of water from the intestine, diarrhea. A strain of *S. aureus* also produces an exotoxin that results in the symptoms associated with toxic shock syndrome (TSS).

***Vibrio* Enterotoxin** *Vibrio cholerae* (vib'rē-ō kol'ėr-ī), growing in a host's intestines, produces an enterotoxin that alters the water and electrolyte balance, and causes severe diarrhea.

Representative diseases produced by exotoxins are listed in Table 14–2.

Endotoxins

Endotoxins differ from exotoxins in several ways. Endotoxins are part of the outer portion of the cell wall of most gram-negative bacteria. Recall from Chapter 4 that gram-negative bacteria have an outer membrane surrounding the peptidoglycan layer of the cell wall. This outer membrane consists of lipoproteins, phospholipids, and lipopolysaccharides (LPS) (see Figure 4–11). The lipid portion of LPS, called *lipid A,* is the endotoxin. Thus, endotoxins are lipopolysaccharides, whereas exotoxins are proteins.

Endotoxins exert their effects when the gram-negative bacteria die and their cell walls undergo lysis, thus liberating the endotoxin. Antibiotics used to treat diseases caused by gram-negative bacteria can lyse the bacterial cells; this reaction releases endotoxin and may lead to an immediate worsening of the symptoms. All endotoxins produce the same signs and symptoms, regardless of the species of microorganism, although not to the same degree. Responses by the host include fever, weakness, generalized aches, and in some cases, shock. Inability of the blood to clot and miscarriage are also effects caused by endotoxins. Endotoxins do not promote the formation of effective antitoxins. Antibodies are produced, but they tend not to counteract the effect of the toxin; sometimes, in fact, they actually enhance its effect. When medical supplies have been tested for endotoxins and they are not found, the term "pyrogen free" is used (see Box 14–1).

Representative of microorganisms that produce endotoxins are *Salmonella typhi* (the causative agent of typhoid fever), *Proteus* species (the frequent causative agents of urinary tract infections), and *Neisseria meningitidis* (the causative agent of epidemic meningitis). A comparison of exotoxins and endotoxins appears in Table 14–3.

PATHOGENIC PROPERTIES OF OTHER MICROORGANISMS

Viruses

Viruses use the host cell's metabolism to produce substances that can cause observable changes in infected cells. These observable changes are called **cytopathic effects (CPE).** The point in the viral infection cycle at which these cytopathic effects occur varies with the virus. Some viral infections result in early changes in the host cell, while in other infec-

Centers for Disease Control

MMWR

Morbidity and Mortality Weekly Report

Fatal Diphtheria

A fatal case of diphtheria was reported to the Wisconsin State Department of Health and Social Services. A 9-year-old unimmunized female developed listlessness and a sore throat on June 30, 1982, 10 days after arriving at a camp in Colorado operated by a religious group that does not accept immunizations. On July 3, she returned to Wisconsin on a camp bus along with other unimmunized children and adults who had also attended the camp. On July 6, a physician evaluated the patient for her sore throat; a throat culture was taken and oral penicillin prescribed. The throat culture was reported to contain normal flora, group A beta hemolytic streptococci, and large numbers of diphtheroids (non-pathogenic corynebacteria). The patient was hospitalized on July 8.

On admission, she had no fever but had moderate upper airway obstruction, bleeding from the nose and gums, and swelling of the jaw and throat. Examination of the pharynx revealed severe hemorrhagic and necrotic tonsillitis; a membrane was not observed. Treatment with penicillin G, gentamicin, moxalactam, peritoneal dialysis, and platelet transfusions was instituted. The patient died on July 14. A *Corynebacterium* species isolated from a throat culture obtained July 10 was subsequently confirmed to be a toxigenic strain of *C. diphtheriae*.

Editorial Note: Nationally, the number of reported diphtheria cases declined steadily form 435 in 1970 to two in 1982. The mortality rate from all types of diphtheria has declined from 13 deaths in 1971 to 1 in 1980, the last year for which mortality information is available.

The clinical manifestations of diphtheria depend on the anatomic location of infection, the virulence and toxigenicity of the infecting strain, and the host's immunity to diphtheria toxin. In the usual pharyngeal form of diphtheria, an adherent grayish-white membrane covers, to some degree, the pharyngeal and/or tonsillar areas. Infrequently, as in this case, diphtheria may appear as a necrotic tonsillitis. Common complications fall into two groups: 1) the membrane and associated tissue swelling, which may cause airway obstruction; and 2) the bacterial toxin, which may cause myocarditis. Mortality occurs predominantly among noncutaneous cases.

Diphtheria is acquired primarily by contact with the infected respiratory droplets or nasopharyngeal secretions of another patient or a carrier. Disease occurs most frequently and more severely among unimmunized or partially immunized persons. Carrier status may occur among both immunized and unimmunized persons.

Source: *MMWR* 31:553 (10/22/82)

Table 14–3 Comparison of Exotoxins and Endotoxins

Property	Exotoxin	Endotoxin
Bacterial source	Mostly from gram-positive bacteria	Almost exclusively from gram-negative bacteria
Relation to microorganism	Metabolic product of growing cell	Present in lipopolysaccharide (LPS) of outer membrane of cell wall and released only with destruction of cell
Chemistry	Protein or short peptide	Lipid portion (lipid A) of LPS of outer membrane
Heat stability	Unstable; can usually be destroyed at 60–80°C (except staphylococcal enterotoxin)	Stable; can withstand autoclaving (121°C for one hour)
Toxicity (power to cause disease)	High	Low
Immunology (relation to antibodies)	Can be converted to toxoids and neutralized by antitoxin	Cannot be converted to toxoids and are not easily neutralized by antitoxin
Pharmacology (effect on body)	Specific for a particular cell structure or function in the host	General, such as fever, weakness, aches, and shock; all produce the same effects
Lethal dose	Small	Considerably larger
Representative diseases	Gas gangrene, tetanus, botulism, diphtheria, scarlet fever	Salmonellosis, epidemic meningitis, and tularemia

tions, changes are not seen until a much later stage. Moreover, some viruses cause **cytocidal** effects (changes resulting in cell death), while other viruses are **noncytocidal.** A virus can produce one or more of the following cytopathic effects:

1. At some stage in their multiplication, cytocidal viruses cause the macromolecular synthesis within the host cell to stop. Some viruses, such as herpes simplex, irreversibly stop mitosis.
2. Cell death resulting from infection by a cytocidal virus can be due to the release of enzymes from lysosomes. These enzymes cause autolysis of the cell.
3. Inclusion bodies (Figure 14–3a) are granules found in the cytoplasm or nucleus of some infected cells. These granules are viral parts, nucleic acids or proteins in the process of being assembled into virions. The granules vary in size, shape, and staining properties, according to the virus. Some inclusion bodies stain with an acidic stain (acidophilic), while others stain with a basic stain (basophilic).
4. Some viral infections cause host cells to fuse, producing multinucleated "giant" cells (Figure 14–3b).
5. Many viral infections induce antigenic changes on the surface of the infected cells. These antigens elicit a host antibody response against the infected cell, and thus target the infected cell for destruction by the host's immune system.
6. Some viruses induce chromosomal changes in the host cell.
7. Normal cells cease growing when they come close to another cell; this phenomenon is *contact inhibition*. Viruses capable of causing cancer *transform* host cells, as discussed in Chapter 12. Transformation results in an abnormal, spindle-shaped cell that does not recognize contact inhibition (Figure 14–4). Loss of contact inhibition results in unregulated cell growth.

Some representative viruses that cause cytopathic effects are presented in Table 14–4. In subsequent chapters, the pathological properties of viruses will be discussed in more detail.

Fungi, Protozoans, Helminths, and Algae

Although fungi do cause disease, they do not have a well-defined set of virulence factors. *Cryptococcus neoformans* (kryp-tō-kok'kus nē-ō-fôr'manz) is a fungus that produces a capsule that helps this organism resist phagocytosis. Some products of fungal growth are toxic to human hosts. But such a toxin is an indirect cause of a fungal disease, as the fungus is already growing in or on the host. An

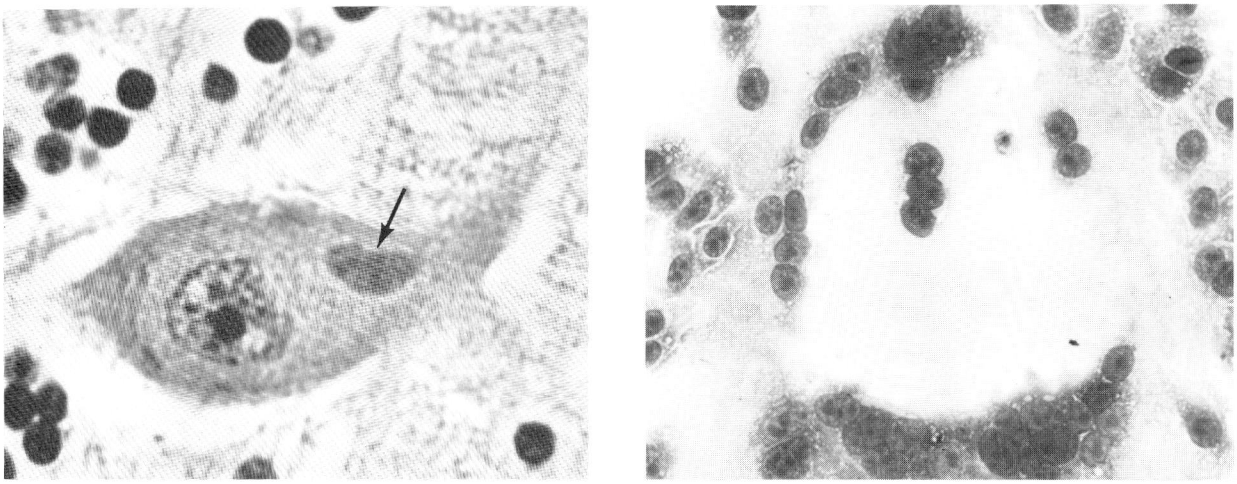

(a) (b)

Figure 14–3 Effects of viruses on cells. **(a)** Cytoplasmic inclusion body (arrow) in a human brain cell from a fatal case of rabies. **(b)** The relatively clear area in the center of this micrograph is a giant cell formed in a culture of Vero cells infected with measles virus. The numerous dark, oval bodies around the cell's edges are its multiple nuclei.

Table 14–4 Cytopathic Effects (CPE) of Selected Viruses

Virus	CPE
Poliovirus	Cell death
Papovavirus	Acidophilic inclusion bodies in nucleus
Adenovirus	Basophilic inclusion bodies in nucleus
Rhabdovirus	Acidophilic inclusion bodies in cytoplasm
Measles, cytomegalovirus	Acidophilic inclusion bodies in nucleus and cytoplasm
Measles	Cell fusion
Polyoma	Transformation

allergic response can also result from a fungal infection, and the symptoms of these fungal infections are typical of allergic reactions.

The presence of protozoans or helminths often produces the symptoms of protozoan or helminthic diseases. Some of these organisms actually use host tissues for their own growth, and the resulting cellular damage evokes the symptoms. Waste products of the metabolism of these parasites can also contribute to the symptoms of a disease. Additionally, a few species of algae produce neurotoxins. Colonization of the skin by the green alga *Prototheca* (prō-tō-thā′ kä) directly damages cells.

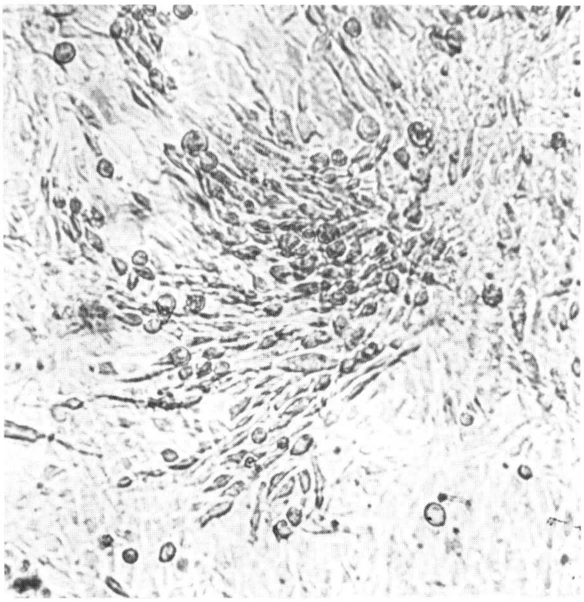

Figure 14–4 Transformed cells in culture. In the center of this micrograph is a "jumble" of chick embryo cells transformed by Rous sarcoma virus. Such a concentration of transformed cells in a cell culture is called a focus; it results from multiplication of a single transforming virus that originally infected a cell at the site. Note how the transformed cells of the focus appear dark in contrast to the monolayer of light, flat, normal cells around them. This appearance is due to their spindle shapes and uninhibited growth on top of one another; there is no contact inhibition.

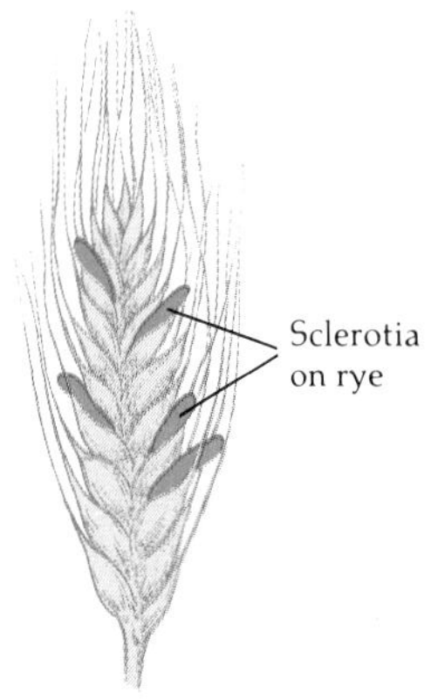

Figure 14–5 The sclerotia of *Claviceps purpurea* are visible on this rye flower. *Claviceps* is the source of the drug LSD.

Specific diseases caused by fungi, protozoans, and helminths will be discussed, along with the pathological properties of these organisms, in later chapters.

The human disease called **ergotism,** which prevailed in Europe during the Middle Ages, is caused by a toxin produced by an ascomycete plant pathogen (*Claviceps purpurea*) that grows on grains. The toxin is contained in **sclerotia** (Figure 14–5), which are highly resistant mycelia. The toxin itself, **ergot,** is an alkaloid that can cause hallucinations resembling those of LSD; in fact, ergot is a natural source of LSD. Ergot also constricts capillaries and can cause, through prevention of proper blood circulation, gangrene of the extremities. Although *Claviceps* still occasionally occurs on grains, modern milling usually removes the sclerotia.

A number of other toxins are produced by fungi that grow on grains. For example, peanut butter is occasionally recalled because of excessive amounts of **aflatoxin,** a toxin that has been found to cause cancer of the liver in laboratory animals. No link to human cancer has been found, but it is believed that aflatoxin might be altered in the animal's body to a mutagenic compound. Aflatoxin is produced by the growth of the mold *Aspergillus flavus* (a-spėr-jil'lus flā'vus), from which the name of the toxin is abridged.

A few mushrooms produce toxins that exert their effects after they have been ingested. Perhaps the most notable mycotoxins are those produced by *Amanita phalloides,* commonly known as the death angel. **Phalloidin** and **amanitin** are potent neurotoxins that affect nerve transmission. Ingestion of the mushroom may result in death.

PATTERN OF DISEASE

A definite sequence of events usually occurs during infection and disease. As you already know, there must be a reservoir of infection as a source of pathogens for an infectious disease to occur. Next, there must be transmission of the pathogen either by direct contact, indirect contact, or vectors to a susceptible host. This event is followed by invasion, in which the microorganism enters the host and multiplies. Entrance is usually made through a preferred portal of entry. Following invasion, the microorganism injures the host through a process called pathogenesis. The extent of injury is dependent on adherence, the effectiveness of the pathogens' resistance to the host's defenses, and the extent to which host cells are damaged either directly or by toxins. Despite the effects of all these factors, the occurrence of disease actually depends on the resistance of the host to the offensive weapons of the pathogen.

Certain predisposing factors also affect the occurrence of disease. A **predisposing factor** is one that makes the body more susceptible to a disease and may alter the course of the disease. Gender is sometimes a predisposing factor. For example, females show a higher incidence of scarlet and typhoid fevers than males. Males, on the other hand, show higher rates of pneumonia and meningitis. Other aspects of genetic background may play a role, too. Individuals with sickle-cell anemia, for instance, are actually more resistant to malaria than others are.

Climate and weather seem to have some effect on the incidence of infectious disease. In temperate regions, the incidence of respiratory diseases increases during the winter. This increase correlates with people staying indoors and having closer contact with each other; these conditions facilitate the spread of respiratory pathogens. Other predisposing factors include inadequate nutrition, fatigue, age, unhealthy environment, habits, life style or occupation, preexisting illness, chemotherapy, and emotional disturbances. It is often difficult to know what the exact relative importance is of different predisposing factors, such as gender and life style.

Once the microorganism does overcome the defenses of the host, a certain pattern of disease evolves. This pattern usually assumes the following sequence, whether the disease is acute or chronic.

Period of Incubation The period of incubation is the time interval between the actual infection and the first appearance of any signs or symptoms. The incubation period in some diseases is constant; in others it is quite variable. The time of incubation depends on the specific microorganism involved, its virulence, the number of infecting microorganisms, and the resistance of the host. Table 14–1 lists the incubation periods for a number of microbial diseases.

Prodromal Period The prodromal period is a relatively short period of time that sometimes follows the period of incubation. The prodromal period is characterized by the first symptoms of disease, such as headache and malaise.

Period of Illness During the period of illness, the disease is most acute. The person exhibits overt signs and symptoms of disease, such as fever, chills, muscle pain (myalgia), sensitivity to light (photophobia), sore throat (pharyngitis), lymph node enlargement (lymphadenomegaly), and gastrointestinal disturbances such as diarrhea. It is during the period of illness that increases or decreases in the number of white blood cells occur.

Period of Decline During the period of decline, the signs and symptoms subside. The fever decreases and the feeling of malaise diminishes. If the period of decline is short, such as less than 24 hours, it is said to occur by *crisis*. If, instead, it takes several days, with the fever decreasing a little each day until it returns to normal, then the period of decline is said to occur by *lysis*.

Period of Convalescence During the period of convalescence, the person regains strength and the body returns to its prediseased state. Recovery has occurred.

We all know that during the period of illness, people serve as reservoirs of disease and can easily spread infections to other persons. However, you should also know that persons can spread infection during incubation and convalescence. This is especially true in cases where the convalescing person carries the pathogenic microorganism for months or even years.

Let us now conclude the chapter by considering how microbes exit the body.

PORTALS OF EXIT

To spread disease throughout a population, a pathogen must exit the body. Just as pathogens have preferred portals of entry, they also have definite routes of exit, called **portals of exit.** In general, portals of exit are related to the part of the body that has been infected.

The most common portals of exit are the respiratory and gastrointestinal tracts. For example, many pathogens living in the respiratory tract exit in discharges from the mouth and nose; these discharges are expelled in coughing or sneezing. These microorganisms are found in droplets formed from mucus. Pathogens that cause tuberculosis, whooping cough, pneumonia, scarlet fever, meningococcal meningitis, measles, mumps, smallpox, and influenza are discharged through the respiratory route. Other pathogens exit from the gastrointestinal tract, in feces or saliva. Feces may be contaminated with pathogens associated with cholera, typhoid fever, paratyphoid fever, bacillary dysentery (shigellosis), amoebic dysentery, and poliomyelitis. And saliva can contain pathogens such as the rabies virus.

Another important route of exit is the genital tract. Microbes responsible for sexually transmitted diseases are found in secretions from the penis or vagina. Urine can also contain the pathogens responsible for typhoid fever and brucellosis, because these pathogens exit via the urinary tract. Skin or wound infections are other portals of exit. Drainage from these wounds can spread infections to another person directly or by contact with a contaminated fomite. Finally, infected blood removed and reinjected by either biting insects or injections

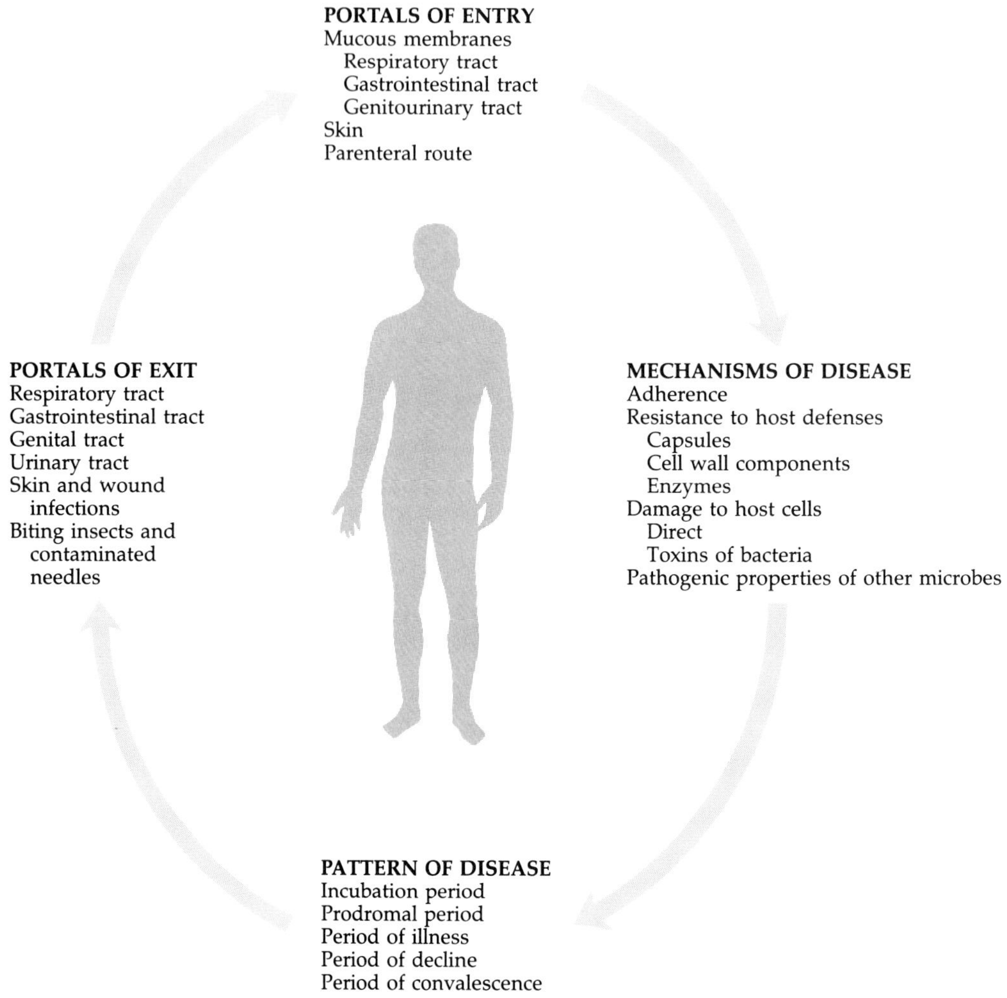

Figure 14–6 How microbes cause disease: a summary of the key concepts of pathogenicity.

with contaminated needles and syringes can spread infection within the population. Examples of diseases so transmitted are yellow fever, Rocky Mountain spotted fever, tularemia, malaria, and serum hepatitis (hepatitis B).

In the next chapter, we will examine a group of nonspecific defenses of the host against disease. But before proceeding, examine Figure 14–6 very carefully. It summarizes some key concepts of the mechanisms of pathogenicity we have discussed.

STUDY OUTLINE

INTRODUCTION (pp. 399–400)

1. Pathogenicity is the ability of a pathogen to produce a disease by overcoming the defenses of the host.
2. Virulence is the degree of pathogenicity.

ENTRY OF MICROORGANISMS INTO THE HOST (pp. 400–402)

1. The specific route by which a particular pathogen gains access to the body is called its portal of entry.
2. Many microorganisms can only cause infections when they gain access through their specific portal of entry.

Mucous Membranes (p. 400)

1. Many microorganisms can penetrate mucous membranes of the conjunctiva and the respiratory, gastrointestinal, and genitourinary tracts.
2. Microorganisms that are inhaled with droplets of moisture and dust particles gain access to the respiratory tract.
3. The respiratory tract is the most frequently used portal of entry.
4. Microorganisms enter the gastrointestinal tract via food, water, and fingers.
5. Most microorganisms are destroyed by the stomach's hydrochloric acid.

Skin (p. 400)

1. Most microorganisms cannot penetrate intact skin; they enter hair follicles and sweat ducts.
2. Some fungi infect the skin itself.

Parenteral Route (p. 400)

1. Some microorganisms can gain access to tissues by inoculation through the skin and mucous membranes in bites, injections, and other wounds.
2. This route of penetration is called the parenteral route.

Numbers of Invading Microbes (pp. 400–402)

1. Virulence can be expressed as LD_{50} (lethal dose for 50% of the inoculated hosts) or ID_{50} (infectious dose for 50% of the inoculated hosts).

ADHERENCE (p. 402)

1. Surface projections on a pathogen called ligands adhere to complementary receptors on the host cells.
2. Ligands can be glycoproteins or lipoproteins and are frequently associated with pili.

HOW PATHOGENS RESIST THE HOST'S DEFENSES (pp. 402–404)

Capsules (pp. 402–403)

1. Some pathogens have capsules that prevent them from being phagocytized.

Components of the cell wall (p. 403)

1. Proteins in the cell wall can facilitate adherence or prevent a pathogen from being phagocytized.

Enzymes (pp. 403–404)

1. Leukocidins destroy neutrophils and macrophages.
2. Hemolysins lyse red blood cells.
3. Local infections can be protected in a fibrin clot caused by the bacterial enzyme coagulase.
4. Bacteria can spread from a focal infection by means of kinases (which destroy blood clots) and hyaluronidase (which destroys a mucopolysaccharide that holds cells together).

Other Factors (p. 404)

1. Some microbes can reproduce inside phagocytes or escape phagocytosis through motility.

DAMAGE TO HOST CELLS (pp. 404–407)

Direct Damage (p. 405)

1. Host cells can be destroyed when pathogens metabolize and multiply inside the host cells.

Toxins (p. 405)

1. Poisonous substances produced by microorganisms are called toxins; toxemia refers to symptoms caused by toxins in the blood.
2. The ability to produce toxins is called toxigenicity.
3. Exotoxins are produced by bacteria and released into the surrounding medium.

4. Exotoxins, not the bacteria, produce the disease symptoms.
5. Antibodies produced against exotoxins are called antitoxins.
6. Endotoxins are lipid A of the lipopolysaccharide component of the cell wall of gram-negative bacteria.
7. Endotoxins cause fever, weakness, generalized aches, and shock.
8. Antibiotics may cause release of endotoxins.
9. Antibodies may enhance the effects of the endotoxins.

PATHOGENIC PROPERTIES OF OTHER MICROORGANISMS (p. 407–410)

Viruses (pp. 407–408)

1. Signs of viral infections are called cytopathic effects (CPE).
2. Some viruses cause cytocidal effects (cell death), and others cause noncytocidal effects.
3. Cytopathic effects include the stopping of mitosis, lysis, formation of inclusion bodies, cell fusion, antigenic changes, chromosomal changes, and transformation.

Fungi, Protozoans, Helminths, and Algae (pp. 408–410)

1. Symptoms of fungal infections can be due to capsules, toxins, and allergic responses.
2. Symptoms of protozoan and helminthic diseases can be due to damage to host tissue or metabolic waste products of the parasite.

PATTERN OF DISEASE (pp. 410–411)

1. A predisposing factor is one that makes the body more susceptible to disease or alters the course of a disease.
2. Examples include climate, age, fatigue, and inadequate nutrition.
3. The period of incubation is the time interval between the actual infection and the first appearance of signs and symptoms.
4. The prodromal period is characterized by the appearance of the first mild signs and symptoms.
5. During the period of illness, the disease is at its height, and all disease signs and symptoms are apparent.
6. During the period of decline, the signs and symptoms subside.
7. During the period of convalescence, the body returns to its prediseased state and health is restored.

PORTALS OF EXIT (pp. 411–412)

1. Just as pathogens have preferred portals of entry, they also have definite portals of exit.
2. Three common portals of exit are: the respiratory tract via coughing or sneezing; the gastrointestinal tract via saliva or feces; the genital tract via secretions from the vagina or penis.

STUDY QUESTIONS

REVIEW

1. List three portals of entry and describe how microorganisms gain access through each.
2. Compare pathogenicity with virulence.
3. How are capsules and cellular wall components related to pathogenicity? Give specific examples.
4. Describe how hemolysins, leukocidins, coagulase, kinases, and hyaluronidase might contribute to pathogenicity.
5. Complete the following table comparing exotoxins with endotoxins.

	Exotoxin	Endotoxin
Bacterial source		
Chemistry		
Toxicity		
Pharmacology		
Example		

6. Define cytopathic effects and give five examples.
7. Write a one-sentence description of factors contributing to the pathogenicity of fungi. Of protozoans and helminths.
8. Put the following in the correct order to describe the pattern of disease: period of convalescence; crisis; prodromal period; period of decline; period of incubation; period of illness.
9. List four predisposing factors to disease.
10. The LD_{50} for botulin toxin is 0.000025 μg. The LD_{50} for *Salmonella* toxin is 200 μg. Which of these is the more potent toxin? How can you tell from the LD_{50} values?

CHALLENGE

1. Food poisoning can be divided into two categories: food infection and food intoxication. On the basis of toxin production by bacteria, explain the difference between these two categories.
2. The following is a case history of a 49-year-old man. Identify each period in the pattern of disease that he experienced.

   On February 7, he handled a parakeet with a respiratory illness. On March 9, he experienced intense pain in the legs, followed by severe chills and headaches. On March 16, he had chest pains, cough, and diarrhea, and his temperature was 40°C. Appropriate antibiotics were administered on March 17, and his fever subsided within 12 hours. He continued taking antibiotics for 14 days. (*Note:* The disease is psittacosis. Can you find the etiology?)
3. Within a three-day period at a large hospital, five patients undergoing hemodialysis developed fever and chills. *Pseudomonas aeruginosa* and *Klebsiella pneumoniae* were isolated from three of the patients. *P. aeruginosa, K. pneumoniae,* and *Enterobacter agglomerans* were isolated from the dialysis system. Why do all three bacteria cause similar symptoms?

FURTHER READING

Bradley, S. G. 1979. Cellular and molecular mechanisms of action of bacterial endotoxins. *Annual Review of Microbiology* 33:67–94. Discusses targets and effects of endotoxin on cellular metabolism, lysosomes, and mitochondria.

Hirschhorn, N. and W. B. Greenough. August 1971. Cholera. *Scientific American* 225:15–21. The scientific method is used to find out how cholera toxin causes loss of body fluids in the host.

Roueche, B. 1967. *Field Guide to Diseases.* Boston: Little, Brown and Company. A concise reference of specific diseases, shows the patterns of diseases.

Smith, I. M. February 1968. Death from staphylococci. *Scientific American* 218:84–94. Effects of the growth of staphylococci in animal hosts.

Van Heyningen, W. E. April 1968. Tetanus. *Scientific American* 218:69-77. An overview of the effects of bacterial exotoxins and specific treatment of the incidence of tetanus.

von Graevenitz, A. 1977. The role of opportunistic bacteria in human disease. *Annual Review of Microbiology* 31:447–471. A good review of bacteria that cause infections in a compromised host.

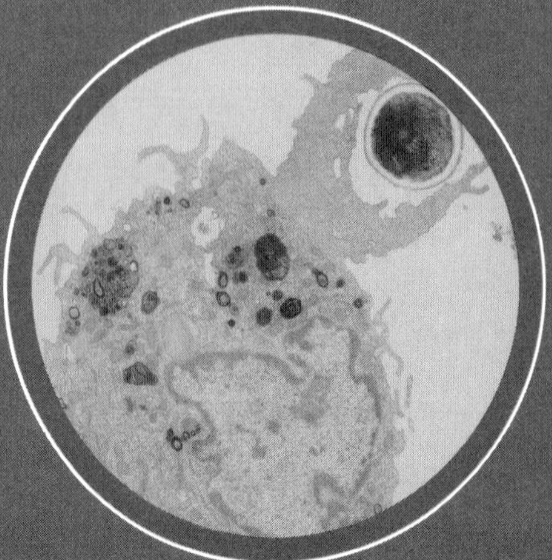

CHAPTER 15

Nonspecific Defenses of the Host

OBJECTIVES

After completing this chapter you should be able to

- Define resistance and susceptibility.
- Define nonspecific resistance.
- Describe the role of the skin and mucous membranes in nonspecific resistance, and differentiate between mechanical and chemical factors.
- Define phagocytosis and include the stages of adherence and ingestion.
- Classify phagocytic cells, and describe the roles of granulocytes and monocytes.
- Describe the stages of inflammation and their relation to nonspecific resistance.
- Discuss the role of fever in nonspecific resistance.
- Discuss the role of interferon.
- Discuss the function of complement.

From what has been said thus far, you can see that pathogenic microorganisms are endowed with special properties that—given the right opportunity—enable them to cause disease. If microorganisms never encountered resistance from the host, we would be constantly ill and would eventually die of various diseases. But in most cases, our body defenses prevent this from happening. Some of our body defenses are designed to keep out microorganisms altogether. Other defenses remove the microorganisms if they do get in, and still others combat them if they remain inside. Our ability to ward off disease through our defenses is called **resistance.** Vulnerability or lack of resistance is known as **susceptibility.**

In discussing resistance, we will divide our body defenses into two general kinds: nonspecific and specific. **Nonspecific resistance** refers to defenses that protect us from *any kind* of pathogen. **Specific resistance,** or **immunity,** is the defense that the body offers against a particular microorganism. Specific defenses are based on the production of specific proteins called antibodies and special cells of the immune system. Chapter 16 deals with specific resistance. In this chapter, we will consider nonspecific resistance.

Contributing to nonspecific resistance are the skin and mucous membranes, phagocytosis, inflammation, fever, and the production of antimicrobial substances other than antibodies. As we discuss these factors, keep in mind that nonspecific resistance protects us from invasion by pathogens in general; it is not a resistance to a specific microorganism.

SKIN AND MUCOUS MEMBRANES

The skin and mucous membranes of the body are commonly regarded as the first line of defense against disease-causing microorganisms. This function results from both mechanical and chemical factors.

Mechanical Factors

The **intact skin** consists of two distinct portions. The *dermis* is the inner portion and is composed of connective tissue. The *epidermis* is the outer portion, the part that is in direct contact with the external environment. It consists of continuous sheets of tightly packed epithelial cells; these sheets are layered with little or no material between the layers. The top layer of epidermal cells contains a waterproofing protein called *keratin.*

If we consider the closely packed cells, continuous layering, and presence of keratin, we can see why the intact skin provides such a formidable physical barrier to the entrance of microorganisms (Figure 15–1). The intact surface of healthy epider-

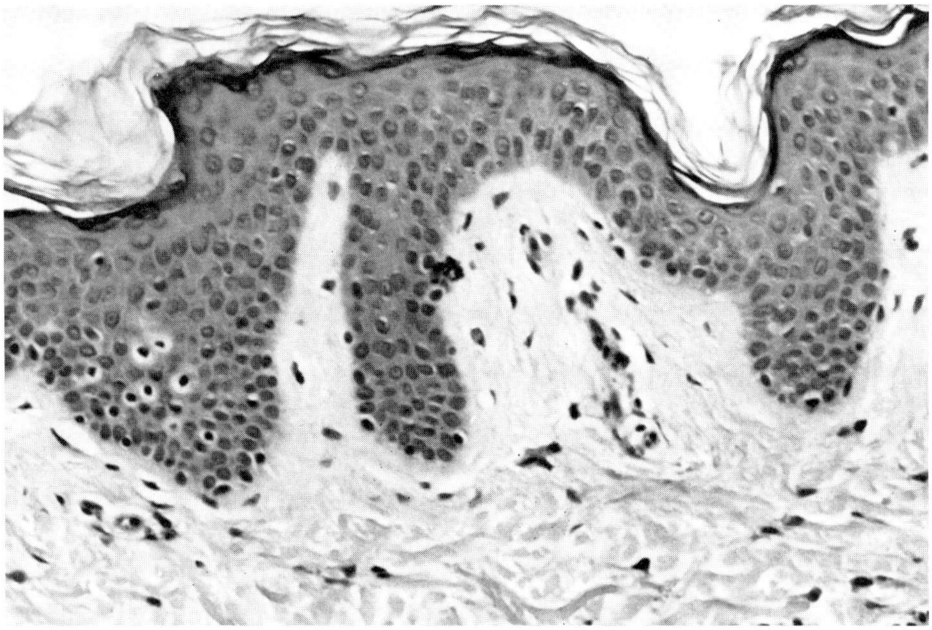

Figure 15–1 Section through human skin (approx. ×200). The thin layer at the top of the photo is keratin. This layer and the darker cells beneath it make up the epidermis. The lighter material in the bottom half of the photo is the dermis.

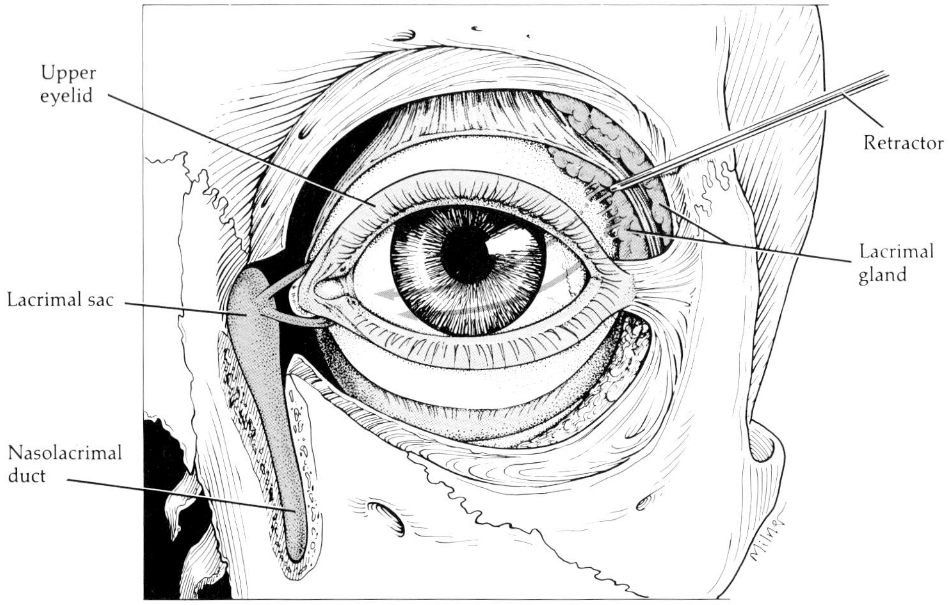

Figure 15–2 The lacrimal apparatus. The washing action of the tears, shown by the arrows, prevents microorganisms from settling on the surface of the eyeball.

mis is rarely, if ever, penetrated by bacteria. But when the epithelial surface is broken, a subcutaneous (below the skin) infection often develops. The bacteria most likely to cause such an infection are staphylococci that normally inhabit the hair follicles and sweat glands of the skin. Infections of the skin and underlying tissues frequently occur as a result of burns or other conditions that break the skin, such as cuts and stab wounds. Moreover, when the skin is moist, as in hot, humid climates, skin infections, especially fungus infections such as athlete's foot, are quite common.

Mucous membranes also consist of an epithelial layer and an underlying connective tissue layer. Mucous membranes line the entire digestive, respiratory, urinary, and reproductive tracts. The epithelial layer of a mucous membrane secretes a fluid called *mucus,* which prevents the tracts from drying out. Some pathogens that can thrive on the moist secretions of a mucous membrane are able to penetrate the membrane if the microorganism is present in sufficient numbers. *Treponema pallidum, Mycobacterium tuberculosis,* and *Streptococcus pneumoniae* are such pathogens. This penetration may be related to toxic substances produced by the microorganism, prior injury by viral infection, or mucosal irritation. Although mucous membranes do inhibit the entrance of many microorganisms, they offer less protection than the skin.

While the skin and mucous membranes serve as physical barriers, there are several other mechanical factors that help protect certain epithelial surfaces. One such mechanism that protects the eyes is the **lacrimal apparatus** (Figure 15–2), a group of structures that manufactures and drains away tears. The lacrimal glands, located toward the upper, outermost portions of each eye socket, produce the tears and pass them under the upper eyelid. From here, tears pass toward the corner of the eye near the nose and into two small holes that lead to the nose. After being secreted by the lacrimal glands, the tears are spread over the surface of the eyeball, by blinking. Normally, the tears evaporate or pass into the nose as fast as they are produced. This continual washing action helps to keep microorganisms from settling on the surface of the eye.

If an irritating substance or large numbers of microorganisms come in contact with the eye, the lacrimal glands start to secrete heavily, and the tears accumulate more rapidly than they can be carried away. This excessive production is a protective mechanism, because the excess tears dilute and wash away the irritating substance or microorganisms.

In a cleansing action very similar to that of tears, **saliva,** produced by the salivary glands, washes microorganisms from both the surface of the teeth and the mucous membranes of the mouth. This mechanism helps prevent colonization by microbes.

The respiratory and digestive tracts have many mechanical forms of defense. Mucus is slightly viscous (thick); it traps many of the microorganisms that enter the respiratory and digestive tracts. The mucous membrane of the nose also has mucus-coated hairs that filter inspired air and trap microorganisms, dust, and pollutants. The cells of the mucous membrane of the lower respiratory tract contain **cilia,** which are microscopic, hairlike projections (Figure 15–3). By moving synchronously, these cilia propel inhaled dust and microorganisms that have become trapped in mucus upward toward the throat. This so-called **ciliary escalator** keeps the mucous blanket moving toward the throat at a rate of 1 to 3 cm per hour, and coughing and sneezing speed up the escalator. You might be interested to know that some substances in cigarette smoke are toxic to cilia and can seriously impair the functioning of the ciliary escalator. Microorganisms are also prevented from entering the lower respiratory tract by a small lip of cartilage called the **epiglottis,** which covers the larynx (voice box) during swallowing.

The cleansing of the urethra by the flow of urine is a mechanical factor that prevents microbial colonization in the urinary system. Vaginal secretions likewise move microorganisms out of the female body.

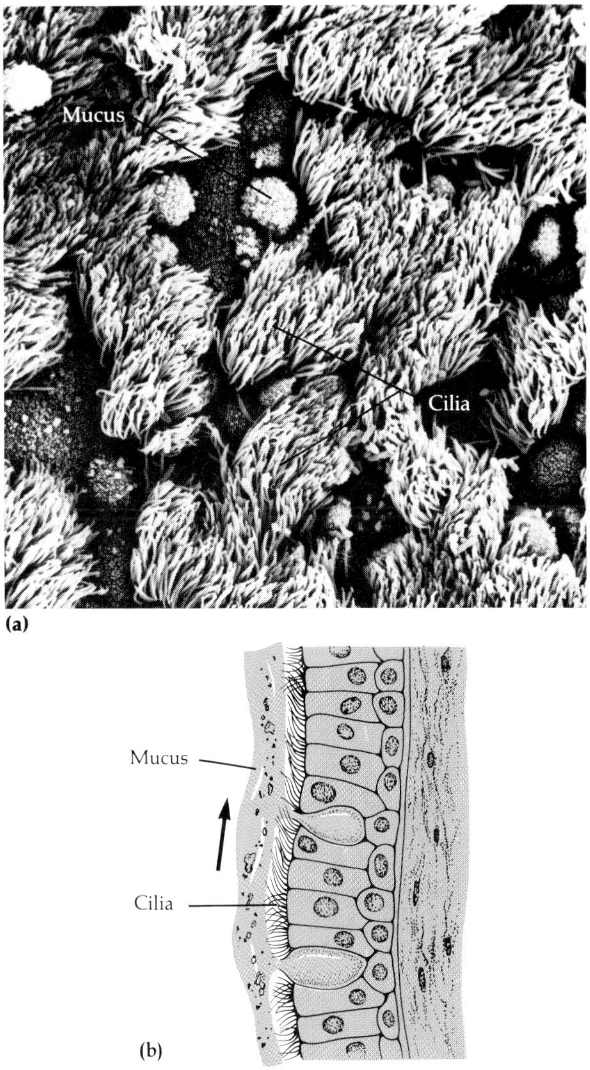

Figure 15–3 Mucous membrane of the trachea (windpipe). **(a)** Micrograph of cilia and mucus-secreting goblet cells, from which blobs of mucus are emerging (×1500). **(b)** Action of the ciliary escalator.

Chemical Factors

Mechanical factors alone do not account for the high degree of resistance of skin and mucous membranes to microbial invasion. Certain chemical factors also play important roles.

Oil (sebaceous) glands of the skin produce an oily substance called **sebum** that prevents hair from drying and becoming brittle. Sebum also forms a protective film over the surface of the skin. One of the components of sebum is unsaturated fatty acids, which inhibit the growth of certain pathogenic bacteria. The low pH of the skin, between pH 3 and 5, is due in part to the secretion of fatty acids and lactic acid. Commensal bacteria on the skin decompose sloughed-off skin cells and these organic molecules, and the end products of this metabolism produce body odor. The skin's acidity probably discourages the growth of many other microorganisms. As you will see in Chapter 19, certain bacteria commonly found on the skin metabolize sebum, and this metabolism forms free fatty acids that cause the inflammatory response associated with acne. A fairly recent treatment for a very severe type of acne, called cystic acne, is isotretinoin (Accutane), a derivative of vitamin A that prevents sebum formation.

The sweat glands of the skin produce perspiration, which helps to maintain body temperature, eliminate certain wastes, and flush microorganisms from the surface of the skin. Perspiration also contains **lysozyme,** an enzyme capable of breaking down cell walls of gram-positive bacteria and a few gram-negative bacteria under certain conditions (see Figure 4–11). Lysozyme is also found in tears, saliva, nasal secretions, and tissue fluids, where it exhibits its antimicrobial activity. Alexander Fleming was actually studying lysozyme in 1929 when he accidentally discovered the antimicrobial effects of penicillin.

Gastric juice is produced by the glands of the stomach. It is a mixture of hydrochloric acid, enzymes, and mucus. The very high acidity of gastric juice (pH 1.2 to 3.0) is sufficient to preserve the usual sterility of the stomach. This acidity destroys bacteria and most bacterial toxins except those of *Clostridium botulinum* and *Staphylococcus aureus.* However, it is probable that many enteric pathogens are protected by food particles and thus are able to enter the intestines via the digestive system.

The resident flora on the skin and mucous membranes usually compete successfully with other microorganisms for available nutrients and may produce metabolic end products that inhibit the growth of other microorganisms. The presence of normal flora thus prevents colonization by other, potentially pathogenic microorganisms.

We will next discuss several aspects of nonspecific resistance commonly regarded as the body's second line of defense against infection. These defense mechanisms all operate within the body. We shall start the discussion with phagocytosis.

PHAGOCYTOSIS

Phagocytosis (from the Greek words for "eat" and "cell") is the ingestion of a microorganism or any particulate matter by a cell. We have previously mentioned phagocytosis as the method of nutrition of certain protozoans. In this chapter, phagocytosis is discussed as a means by which cells in the hu-

Table 15–1 Formed Elements in Blood

Type of Cell	Numbers per Cubic Millimeter (mm^3)	Function
Erythrocytes (Red Blood Cells)	4.8–5.4 million	Transport of O_2 and CO_2
Leukocytes (White Blood Cells)	5000–9000	
A. Granulocytes		
1. Neutrophils (PMNs) (60–70% of leukocytes)		Phagocytosis
2. Basophils (0.5–1%)		Production of heparin and histamine
3. Eosinophils (2–4%)		Phagocytosis
B. Agranulocytes		
1. Lymphocytes (20–25%)		Antibody production
2. Monocytes (3–8%)		Phagocytosis
Thrombocytes (Platelets)	250,000–400,000	Blood clotting

man body counter infection. The human cells that perform this function are collectively called **phagocytes.** All are types of white blood cells or derivatives of blood cells.

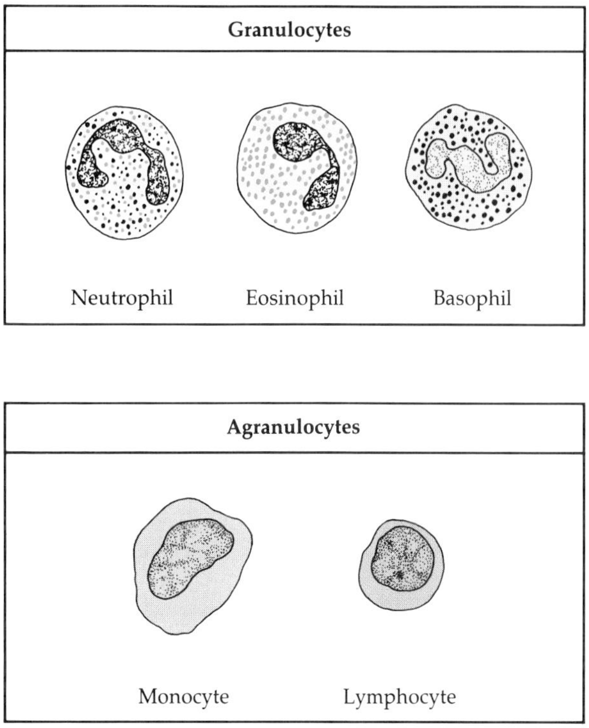

Figure 15–4 Principal kinds of leukocytes.

Formed Elements of the Blood

Blood consists of a fluid called *plasma,* which contains *formed elements;* that is, cells and cell fragments (Table 15–1). Of the cells listed in Table 15–1, those that concern us at present are the **leukocytes,** or white blood cells (Figure 15–4).

During many kinds of infections, there is an increase in the total number of white blood cells (*leukocytosis*). During the active stage of infection, the leukocyte count might double, triple, or quadruple, depending on the severity of the infection. Diseases that might cause such an elevation in the leukocyte count are meningitis, mononucleosis, appendicitis, pneumococcal pneumonia, and gonorrhea. Other diseases, such as typhoid fever, tuberculosis, measles, and influenza, cause a decrease in the leukocyte count (*leukopenia*). The source of leukocyte increase or decrease can be detected by a *differential count,* in which the percentage of each kind of white cell in a sample of white blood cells is calculated. The percentages in a normal differential count of leukocytes are shown in parentheses in the first column of Table 15–1.

Leukocytes are divided into two categories, granulocytes and agranulocytes. *Granulocytes* owe their name to the presence of granules in their cytoplasm. They are differentiated into three types on the basis of how the granules stain. The granules of *neutrophils* stain with a mixture of acidic and basic dyes; those of *basophils* stain with the basic dye methylene blue; and those of *eosinophils* stain with the acid dye eosin.

Neutrophils are also commonly called *polymorphonuclear leukocytes* (or *PMNs*). Neutrophils are highly phagocytic. They have the ability to leave the blood, enter an infected tissue, and destroy microbes and foreign particles. The details of this mechanism will be described shortly. The role of basophils is not clear. However, they release substances such as histamine that are important in inflammation and allergic responses. They also release heparin, an anticoagulant. Eosinophils are somewhat phagocytic and have the ability to leave the blood. They are believed to ingest antigen–antibody complexes, and their number increases greatly during certain parasitic worm infections and hypersensitivity (allergy) reactions.

Agranulocytes are so named because they lack granules in their cytoplasm. The two kinds of agranulocytes are *lymphocytes* and *monocytes.* Lymphocytes occur in lymphoid tissues (tonsils, lymph nodes, spleen, thymus gland, bone marrow, appendix, and Peyer's patches of the small intestine) and circulate in blood (Figure 15–5). As you will see in Chapter 16, lymphocytes assume a key role in antibody production. Monocytes are poorly phagocytic until stimulated to become more active by a progressing infection.

We will now group the phagocytes of the body and describe the details of phagocytosis.

Kinds of Phagocytic Cells

Phagocytic cells fall into two broad categories: granulocytes and macrophages, which include monocytes. Granulocytic phagocytes are sometimes called **microphages.**

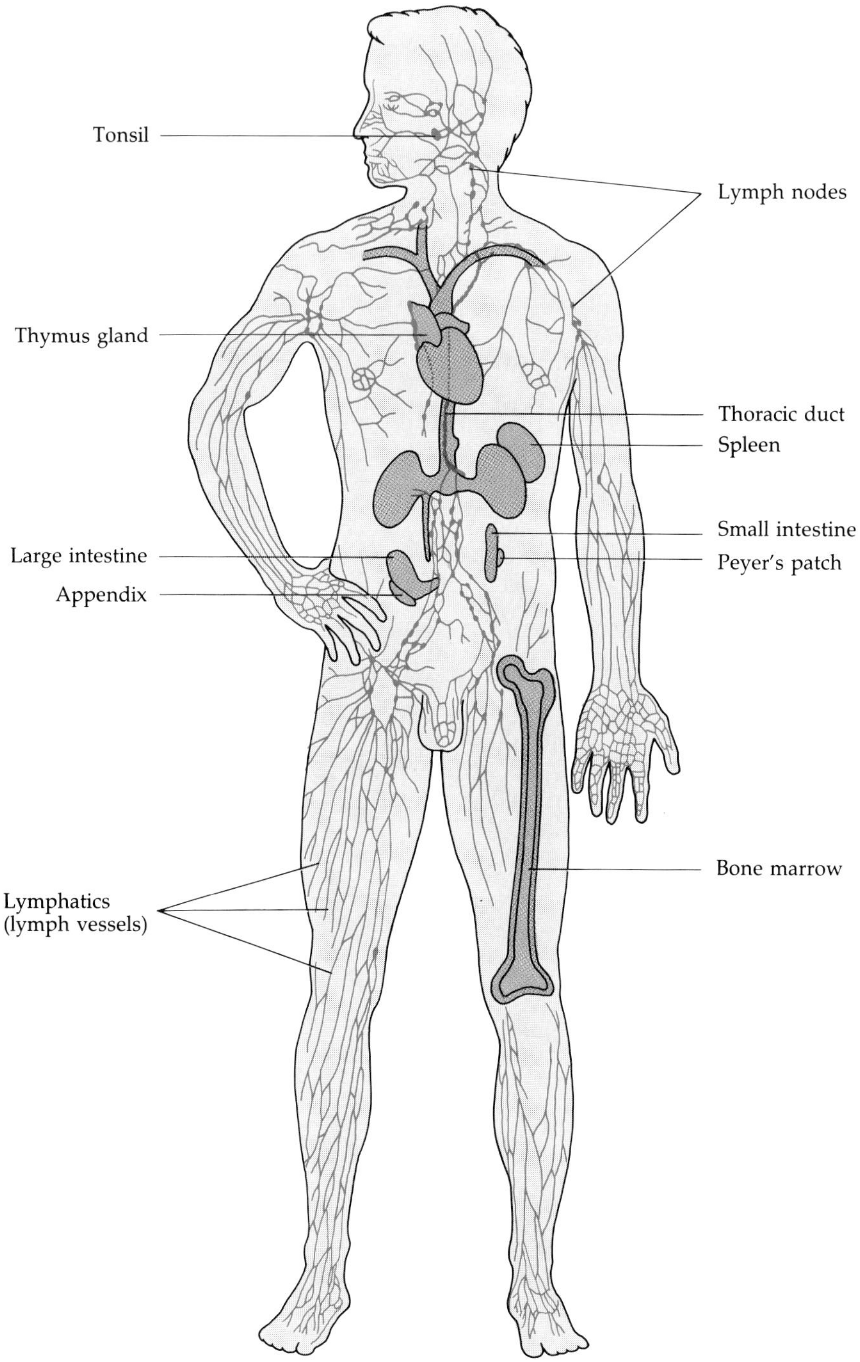

Figure 15–5 Components of the lymphatic system.

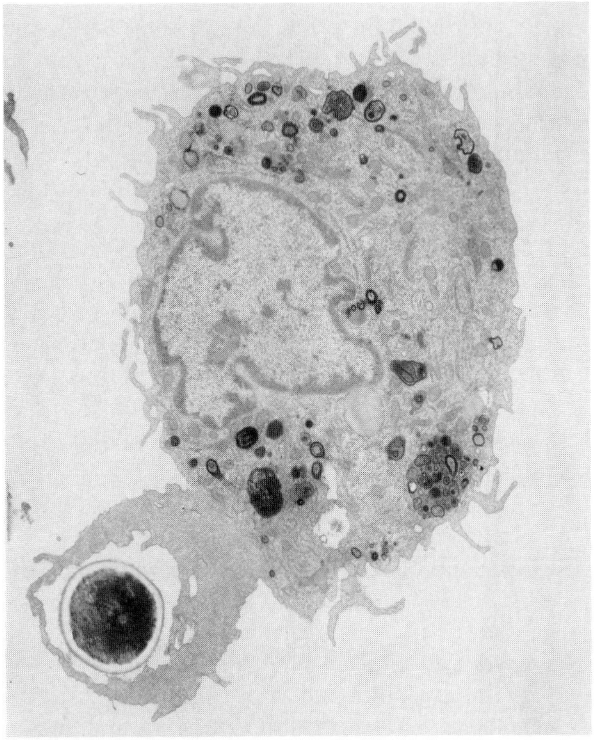

Figure 15–6 Electron micrograph of a macrophage engulfing a yeast particle (×5600). Note the numerous fingerlike extensions of the plasma membrane.

When an infection occurs, both granulocytes (especially neutrophils) and monocytes migrate to the infected area. During this migration, monocytes enlarge and develop into actively phagocytic cells called **macrophages** (Figure 15–6). Because these cells leave the blood and migrate through tissue to infected areas, they are called *wandering macrophages.* Some macrophages, called *fixed macrophages* or *histiocytes,* enter certain tissues and organs of the body and remain there. Fixed macrophages are found in the liver (Kupffer's cells), lungs (alveolar macrophages), nervous system (microglial cells), bronchial tissue, spleen, lymph nodes, bone marrow, and the peritoneal cavity surrounding abdominal organs. They are also part of a system called the **mononuclear phagocytic (reticuloendothelial) system.**

During the course of an infection, there is a shift in the type of white blood cell that predominates. Granulocytes, especially neutrophils, predominate during the inital phase of infection during which time they are actively phagocytic; this dominance is indicated by their number in a differential count. But as the infection begins to subside, the monocytes predominate; they scavenge and phagocytize dead or dying bacteria after the neutrophils have performed phagocytosis. (This in-

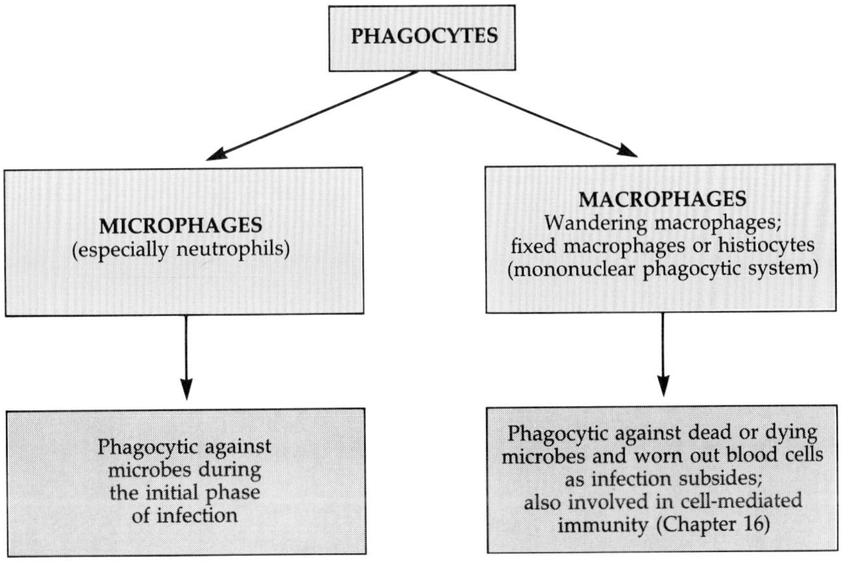

Figure 15–7 Classification of phagocytes.

creased activity of monocytes will be reflected in a differential count.) As blood and lymph containing microorganisms pass through organs with fixed macrophages, cells of the mononuclear phagocytic system remove the microorganisms by phagocytosis. The mononuclear phagocytic system also disposes of worn out blood cells.

Figure 15–7 summarizes the roles of phagocytic cells.

Mechanism of Phagocytosis

How does phagocytosis occur? For convenience of study, we will divide phagocytosis into two phases: adherence and ingestion.

Adherence

Adherence as it pertains to phagocytosis is the *attachment* between the cell membrane of the phagocyte and the surface of the microorganism (or other foreign material). Adherence is facilitated by **chemotaxis,** the attraction of phagocytes to microorganisms by certain chemicals. Among these chemicals are microbial products, components of white blood cells and tissue cells, and peptides derived from complement components (discussed later in the chapter). In some instances, adherence occurs easily and the microorganism is readily phagocytized (Figure 15–8a). Adherence can be more difficult, as it is with *Streptococcus pneumoniae* and *Klebsiella pneumoniae,* both of which have large capsules. Such heavily encapsulated microorganisms can be phagocytized only if the phagocyte traps the microorganism against a rough surface, such as a blood vessel, blood clot, or connective tissue fibers, where the microorganism cannot slide away. This is sometimes called *nonimmune* or *surface phagocytosis.* Microorganisms can be more readily phagocytized if they are first coated with certain plasma proteins that promote the attachment of the microorganism to the phagocyte. This coating process is called **opsonization.** The proteins that act as *opsonins* include, among others, some components of the complement system (described later in this chapter) and antibody molecules (Chapter 16).

Ingestion

Following adherence, ingestion occurs. During the process of **ingestion,** the cell membrane of the phagocyte extends projections, called pseudopods, that engulf the microorganism (Figure 15–8b). Once the microorganism is surrounded, the membrane folds inward and surrounds the microorganism with a sac called a *phagosome* or *phagocytic vacuole* (Figure 15–8c). The phagosome pinches off from the membrane and enters the cytoplasm. Within the cytoplasm, it collides with lysosomes that contain digestive enzymes and bactericidal substances (see Chapter 4). Upon contact, the phagosome and lysosome membranes fuse to form a single, larger structure called a *phagolysosome,* or *digestive vacuole* (Figure 15–8d). The contents of the phagolysosome take only 10 to 30 minutes to kill many types of bacteria (Figure 15–8e). It is assumed that microbial destruction occurs because of the contents of the lysosomes.

Lysosomal enzymes that attack microbial cells directly include lysozyme, which hydrolyzes peptidoglycan in bacterial cell walls, and a variety of other enzymes, which hydrolyze other macromolecular components of microorganisms. The hydrolytic enzymes are most active at around pH 4, which is the phagolysosome's usual pH because of lactic acid produced by the phagocyte. Another kind of lysosomal enzyme, myeloperoxidase, reacts with hydrogen peroxide and chloride ions to bond chlorine atoms to bacteria and viruses—resulting in death of the microorganisms.

Not all phagocytized microorganisms are killed by lysosomal enzymes. Some, such as toxin-producing staphylococci, are not necessarily killed by ingestion. In fact, their toxins can actually kill the phagocytes. Other microorganisms, such as *Mycobacterium tuberculosis,* can multiply within the phagolysosome and eventually destroy the phagocyte. And still others, such as the causative agents of tularemia and brucellosis, can remain dormant in phagocytes for months or years at a time.

Phagocytosis often occurs as part of another nonspecific mechanism of resistance, inflammation.

INFLAMMATION

Damage to the body's tissues triggers a response called **inflammation.** The damage can be caused by microbial infection, physical agents (such as heat, radiant energy, electricity, or sharp objects), or chemical agents (acids, bases, and gases). Inflammation is usually characterized by four funda-

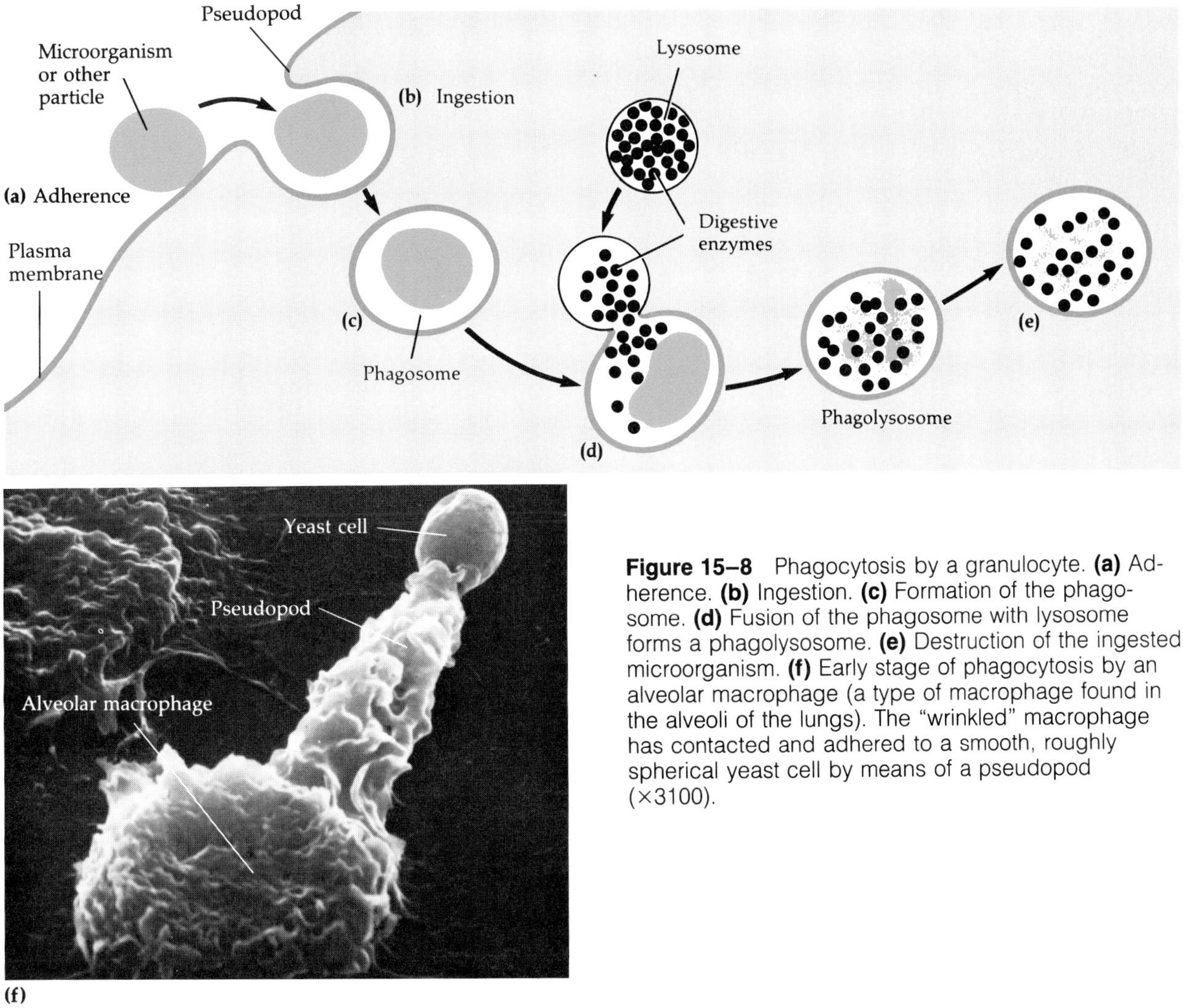

Figure 15–8 Phagocytosis by a granulocyte. **(a)** Adherence. **(b)** Ingestion. **(c)** Formation of the phagosome. **(d)** Fusion of the phagosome with lysosome forms a phagolysosome. **(e)** Destruction of the ingested microorganism. **(f)** Early stage of phagocytosis by an alveolar macrophage (a type of macrophage found in the alveoli of the lungs). The "wrinkled" macrophage has contacted and adhered to a smooth, roughly spherical yeast cell by means of a pseudopod (×3100).

mental symptoms: redness, pain, heat, and swelling. Sometimes a fifth symptom, loss of function, is present; its occurrence depends on the site and extent of damage. In apparent contradiction to the symptoms observed, however, the inflammatory response is actually a beneficial one. Inflammation has the following functions: (1) to destroy the injurious agent, if possible, and to remove it and its by-products from the body; (2) if destruction is not possible, to limit the effects on the body, by confining or walling off the injurious agent and its by-products; and (3) to repair or replace tissue damaged by the injurious agent or its by-products.

For purposes of discussion, we will divide the process of inflammation into three stages: (1) vasodilation and increased permeability of blood vessels; (2) phagocyte migration; and (3) repair.

Vasodilation and Increased Permeability of Blood Vessels

Immediately following tissue damage, blood vessels in the area of damage vasodilate and their permeability increases. **Vasodilation** is an increase in the diameter of blood vessels in the area of the injury. Vasodilation increases blood flow to the damaged area, and is responsible for the redness and heat associated with inflammation.

Increased permeability permits defensive substances normally retained in the blood to pass through the walls of the blood vessels and enter the injured area. The increase in permeability, which permits fluid to move from the blood into tissue spaces, is responsible for the swelling (edema) of inflammation. The pain of inflammation can be caused by nerve damage, irritation by toxins, or the pressure of edema.

What causes vasodilation and the increase in permeability of blood vessels? Damaged cells responding to injury release chemicals. One such substance is **histamine,** a chemical present in many tissues of the body, especially in mast cells in connective tissue, circulating basophils, and blood platelets. Histamine is released in direct response to any injury of cells that contain it; it is also released by certain factors of the complement system (to be discussed later). Phagocytic granulocytes attracted to the site of injury can also produce chemicals that cause the release of histamine.

Kinins are another group of substances that cause vasodilation and increase the permeability of blood vessels. These chemicals are present in blood plasma. Kinins also attract phagocytic granulocytes—chiefly PMNs—to the injured area. However, because they also affect some nerve endings, their presence accounts for much of the pain associated with inflammation. **Prostaglandins** are yet another group of chemicals that cause vasodilation.

Vasodilation and the increase in permeability of blood vessels also deliver clotting elements of blood into the injured area. The blood clots around the site of activity prevent the microorganism (or its toxins) from spreading to other parts of the body. As a result, there is a localized collection of pus in a cavity formed by the breakdown of body tissues. This focus of infection is called an **abscess.** Common abscesses include pimples and boils.

The next stage in inflammation involves the migration of phagocytes to the injured area.

Phagocyte Migration

Generally within an hour after the process of inflammation is initiated, phagocytes appear on the scene. As the flow of blood gradually decreases, phagocytes (both PMNs and monocytes) begin to stick to the inner surface of the endothelium (lining) of blood vessels. This sticking process is called **margination.** Then the collected PMNs begin to squeeze through the wall of the blood vessel to reach the damaged area. This migration, which resembles amoeboid movement, is called **diapedesis;** the migratory process can take as little as two minutes (Figure 15–9).

As mentioned earlier, certain chemicals attract PMNs to the site of injury (chemotaxis). These include chemicals produced by microorganisms and even other PMNs; other chemicals are kinins and components of the complement system. The availability of a steady stream of PMNs is ensured by

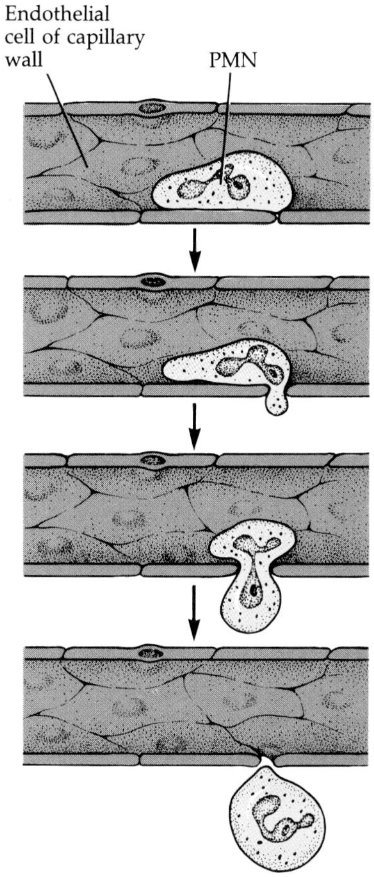

Figure 15–9 Diapedesis of a PMN through the wall of a capillary in inflammation. Monocytes can also squeeze through blood vessel walls in this way.

the production and release of additional granulocytes from bone marrow. This production is brought about by a substance called **leukocytosis-promoting factor,** which is released from inflamed tissues.

As the inflammatory response continues, monocytes follow the granulocytes into the infected area. Once the monocytes are contained in the tissue, they become wandering macrophages. The granulocytes predominate in the early stages of infection but tend to die off rapidly. Macrophages enter the picture during a later stage of the infection once granulocytes have accomplished their function. They are several times more phagocytic than granulocytes and are large enough to phagocytize tissue that has been destroyed, granulocytes that have been destroyed, and invading microorganisms.

After granulocytes or macrophages engulf large numbers of microorganisms and damaged tissue, they themselves eventually die. After a few days, a cavity containing dead phagocytes and damaged tissue forms in the inflamed tissue. This collection of dead cells and various body fluids is called **pus.** Pus formation usually continues until the infection subsides. At times, the pus pushes to the surface of the body or into an internal cavity for dispersal. On other occasions, the pus remains, even after the infection is terminated. In this case, the pus is gradually destroyed over a period of days and is absorbed by the body.

Repair

The process by which tissues replace dead or damaged cells is called **tissue repair,** and is the final stage of inflammation. Repair begins during the active phase of inflammation, but cannot be completed until all harmful substances have been removed or neutralized at the site of injury. The ability of a tissue to repair depends, in part, on the tissue involved. For example, skin has a high capacity for regeneration, whereas nervous tissue in the brain and spinal cord does not regenerate at all.

A tissue is repaired when the stroma or parenchyma of a tissue produce new cells. The *stroma* is the supporting connective tissue, and the *parenchyma* is the functioning part of the tissue. If only parenchymal cells are active in repair, a perfect or near-perfect reconstruction of the tissue occurs. A good example is a minor cut on the epidermis of the skin. However, if repair cells of the stroma are involved, scar tissue is formed.

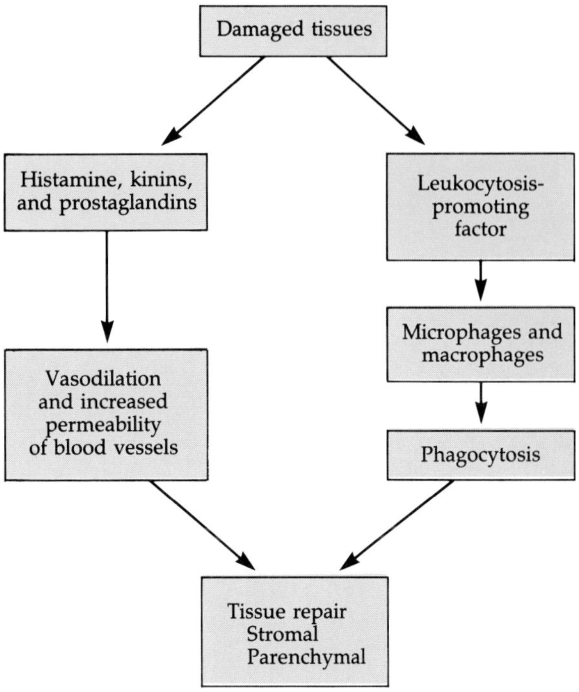

Figure 15–10 Summary of the process of inflammation.

The process of inflammation is summarized in Figure 15–10.

FEVER

We talked in the preceding section about inflammation, the local response of the body to microbial invasion. There are also systemic, or overall, responses, one of the most important being **fever,** an abnormally high body temperature. The most frequent cause of fever is infection from bacteria (and their toxins) or viruses.

Body temperature is controlled by a part of the brain called the hypothalamus. The hypothalamus is sometimes referred to as the body's thermostat, and it is normally set at 37°C (98.6°F). It is believed that certain antigens affect the hypothalamus by setting it at a higher temperature. As little as 1 mg of endotoxin (lipid A) from the bacterium that causes typhoid fever can set the thermostat as high as 43°C, and this temperature will be maintained until the antigen is eliminated. Perhaps more important is the fact that during inflammation, phagocytic granulocytes, especially neutrophils, release a protein called *leukocytic pyrogen* that also has the ability to raise the thermostat.

Assume that the body is invaded by pathogens and the thermostat setting is increased to 39°C. To adjust to the new thermostat setting, the body responds with blood vessel constriction, increased rate of metabolism, and shivering, all of which raise body temperature. Thus, even though body temperature is climbing higher than normal, to say 38°C, the skin remains cold and shivering occurs. This condition, called a **chill,** is a definite sign that temperature is rising. When body temperature reaches the setting of the thermostat, the chills disappear. But the body will continue to regulate its temperature at 39°C until the bacterial lipolysaccharides or pyrogen is eliminated. When elimination is achieved, the thermostat is reset at 37°C. As the infection subsides, heat-losing mechanisms such as vasodilation and sweating go into operation. (The skin becomes warm and the person begins to sweat.) This phase of the fever is called the **crisis** and indicates that body temperature is falling.

Up to a certain point, fever is considered beneficial. The high body temperature is believed to inhibit the growth of some microorganisms. And because the high temperature speeds up the body's reactions, it may help body tissues to repair themselves more quickly. As a rule, however, death results if body temperature rises to about 45°C. But most tissues can withstand marked cooling to less than 7°C, a fact that is useful in some types of surgery.

ANTIMICROBIAL SUBSTANCES

The body produces certain antimicrobial substances in addition to the chemical factors mentioned earlier. Among the most important of these are interferon and the proteins of the complement system.

Interferon (IFN)

Because viruses depend on their host cells to provide many functions of viral multiplication, it is difficult to inhibit viral multiplication without at the same time affecting the host cell itself. One way the infected host counters viral infections is with interferons. **Interferons** are a class of similar antiviral proteins produced by certain animal cells; the function of interferons is to interfere with viral multiplication. One of the most interesting features of interferons is that they are host cell specific but not virus specific. This means that interferon produced by human cells protects human cells but will produce little antiviral activity for cells of other species, such as mice or chicks. However, the interferon of a species is active against a number of different viruses.

Not only do different animal species produce different interferons, but different types of cells in an animal produce different interferons. Human interferons are of three principal types: (1) *alpha interferon* (α-IFN) (formerly called leucocyte IFN); (2) *beta interferon* (β-IFN) (formerly called fibroblast IFN); and (3) *gamma interferon* (γ-IFN) (formerly called immune IFN). There are also various subtypes of interferon within the principal groups. In the human body, interferon is produced by fibroblasts in connective tissue, by lymphocytes, and by other leukocytes. The three types of interferon produced by these cells can each have a slightly different effect on the body.

All interferons are small proteins, with molecular weights between 15,000 and 30,000. They are quite stable at low pH and fairly resistant to heat.

Produced by virus-infected host cells only in very small quantities, interferon diffuses to uninfected neighboring cells and there exerts its antiviral activity. (In this way, interferon protects these neighboring cells.) Interferon molecules bind to the surface of uninfected cells and somehow induce these cells to manufacture another antiviral protein. This newly synthesized protein remains in the cell and acts to prevent the synthesis of viral nucleic acid. In this way the protein inhibits viral multiplication (Figure 15–11).

Low concentrations of interferon are nontoxic to uninfected cells; interferon inhibits only viral multiplication. Because of its beneficial properties, interferon would seem to be an ideal antiviral substance. But certain problems do exist. For one thing, interferon is effective for only short periods of time. It typically plays a major role in infections that are acute and short-term, such as colds and influenza. Another problem is that it has no effect on viral multiplication in cells already infected.

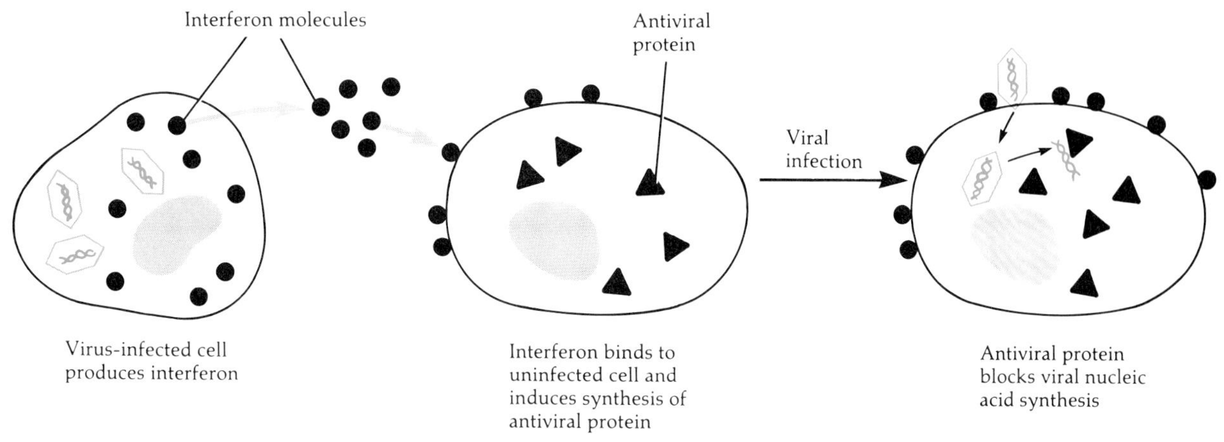
Interferon molecules
Antiviral protein
Viral infection
Virus-infected cell produces interferon
Interferon binds to uninfected cell and induces synthesis of antiviral protein
Antiviral protein blocks viral nucleic acid synthesis

Figure 15–11 Action of interferon.

The importance of interferon in protecting the body against viruses, as well as its potential as an anticancer agent, has made its production in large quantities a top public health priority. Recently, several groups of scientists have successfully applied recombinant DNA technology in inducing certain species of bacteria to produce interferon. (This technique is described in Chapter 8.) The interferons produced with recombinant DNA techniques are called recombinant interferons (rINFs), and are important for two reasons: they are pure, and they are plentiful. Before rINFs were available, investigators had to use more expensive, less pure interferons obtained from human cells. It was difficult to tell whether the effects of interferon were caused by the interferon itself or by contaminants in the preparation. Another advantage of rINF is that its availability in bulk quantities has made it possible to carry out large-scale clinical trials.

Clinical trials to determine the anticancer effects of interferon were begun in 1981. At present it appears that large doses of rINFs have only limited effects against some tumors and no effect on others. For example, rINFs appear to be a somewhat useful treatment for two types of malignant tumors that do not respond well to other therapies: melanoma (a type of skin cancer) and renal (kidney) cell cancer. However, rINFs do not seem to exert anticancer effects against malignant tumors in the breast, colon, or lung. The clinical trials have also revealed that high doses of interferon can have side effects that range from minor to quite serious. Among these are fatigue, malaise, loss of appetite, fever, chills, pains in the joints, mental confusion, seizures, and cardiac complications.

Although glowing predictions about interferon's anticancer properties have not been justified by the results of clinical trials to date, research is continuing for several reasons. First of all, the rINFs currently available are only a few of the possible subtypes yet to be produced by recombinant DNA techniques. Perhaps other subtypes will exhibit definitive anticancer properties. Second, some laboratory studies indicate that interferon might work together with other chemotherapeutic agents so that the effects of interferon are enhanced. Encouraging results have been obtained by use of interferon in combination with doxorubicin or cimetidine, for example. Third, it has been observed that patients who did not respond well to either specific chemotherapy or subsequent treatment with interferon showed improvement when placed back on the original chemotherapy. Some research even suggests that interferon might have a role in the treatment of genital herpes.

Complement and Properdin

Complement is a group of proteins found in normal blood serum. These proteins are very important to both nonspecific and specific defenses against microbial infection. The overall function of the complement system is to attack and destroy invading microorganisms (Figure 15–12). Complement is so named because in its "classical" role, it complements (fills out or completes) certain immune reactions involving antibodies. For this reason, many textbooks do not discuss complement until antigen–antibody reactions have been discussed. We will discuss it here, however, because we regard the complement system as basically an area of *nonspecific* resistance. For one thing, complement has the same general role in all antigen–antibody reactions in which it participates. Also, complement itself is not specific for particular antigens and does not appear in greater amounts after immunization. Furthermore, complement participates in important defense reactions that do not involve antibody. (If you wish, however, you

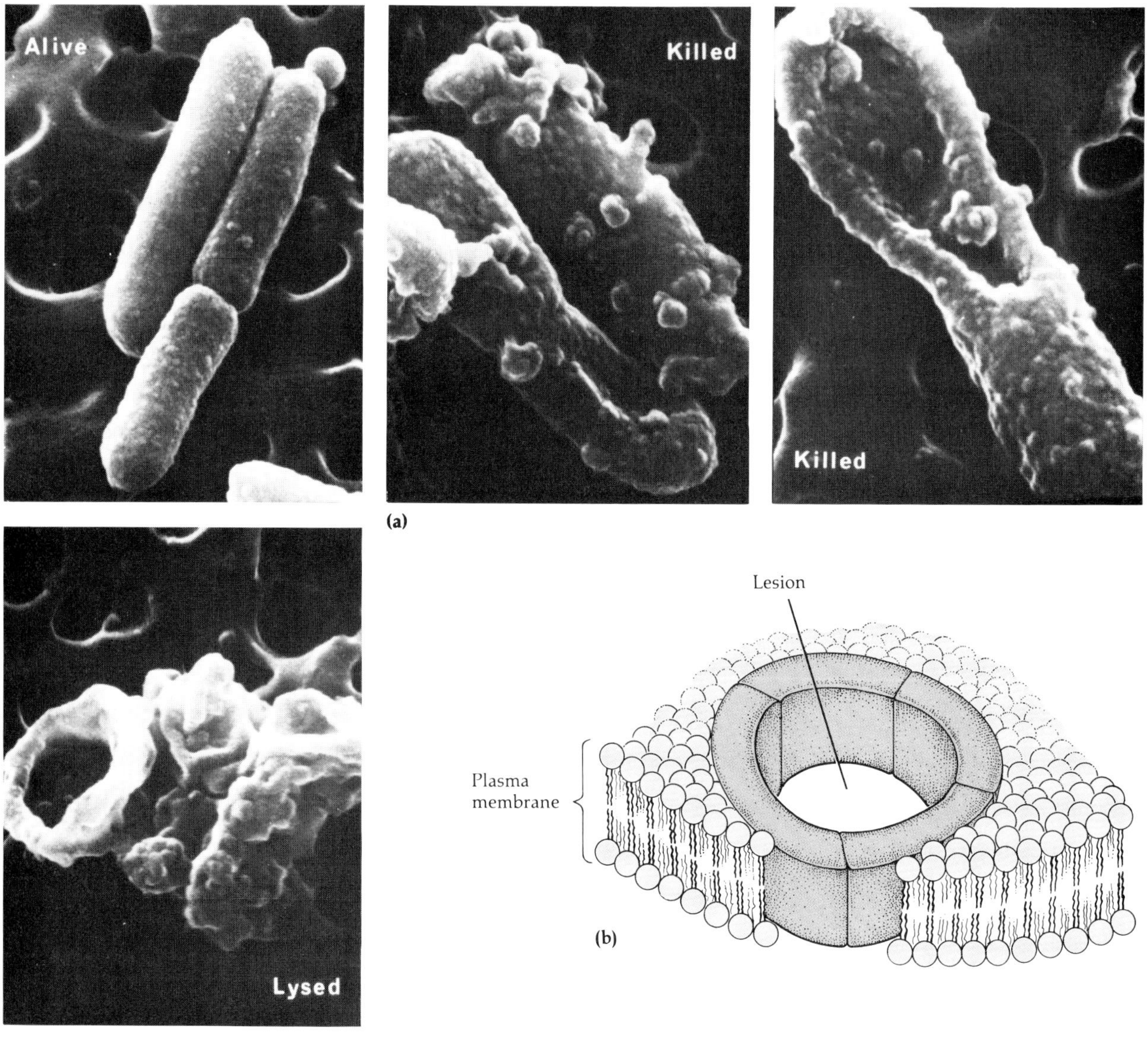

Figure 15–12 Destructive action of complement. **(a)** Scanning electron micrographs of *Escherichia coli* killed and then lysed by the action of complement. **(b)** Diagram of a complement lesion.

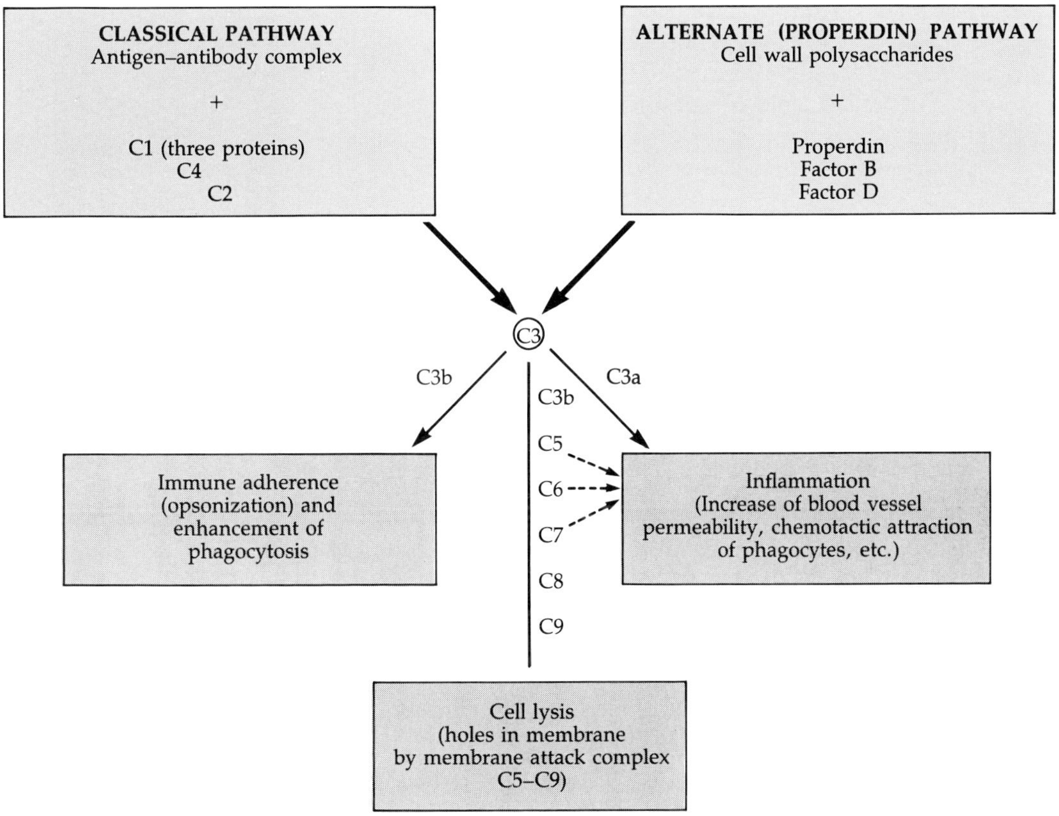

Figure 15–13 Complement activation via classical and alternative (properdin) pathways. Note the cleavage of C3 into C3a and C3b. These fragments induce three kinds of consequences destructive to microorganisms: immune adherence, complement fixation (which causes cell lysis), and inflammation.

may delay discussion of this topic until after Chapter 16.)

Components of the complement and properdin system

The complement system, abbreviated **C** is made up of eleven proteins. These proteins are designated C1 through C9; C1 is actually a complex of three proteins. The molecular weights of these proteins are high, ranging from about 80,000 to over 400,000. The related **properdin** system, used in the alternate pathway (discussed below), is composed of three other serum proteins: properdin itself, factor B, and factor D. Together all these proteins comprise more than 10% of the proteins in serum.

Except for C4, the proteins of the complement and properdin systems act in an ordered sequence, or cascade, that follows the numerical designations. In a series of steps, the proteins *activate* one another, usually by cleaving the next protein in the series. The fragments of a cleaved protein have new enzymatic or physiological functions. For example, a whole protein cannot cleave the next protein in the series, but one of its fragments can.

Pathways of complement activation

C3 plays a central role in the complement system. As you can see in Figure 15–13, there are two pathways to the activation of C3. On the upper left is the *classical pathway,* which is initiated by interaction between an antigen–antibody complex (described in Chapter 16) and the C1 complex. On the upper right is the so-called *alternate* (*properdin*) *pathway*, which is initiated by the interaction between certain polysaccharides and the proteins of the properdin system. Most of these polysaccharides are contained in the cell walls of certain bacteria and fungi (although they also include molecules on the surface of some foreign mammalian red blood cells). The properdin pathway is of particular importance in combating enteric gram-negative bacteria. The outer membrane of the bacteria's cell wall contains a lipopolysaccharide that releases endotoxin (lipid A). This toxin triggers the alternate pathway. Note that this pathway does not require an antibody, and does not involve C1, C2, or C4.

Consequences of complement activation

How does the complement system contribute to microbial destruction? Both the classical and alternative pathways lead to the cleavage of C3 into two fragments, C3a and C3b. These fragments induce three processes that are destructive to microorganisms.

Complement fixation C3b can initiate a sequence of reactions involving C5 through C9, which are known collectively as the **membrane attack complex.** The activated components of these proteins attack the invading cell's membrane and produce circular lesions through which the cell's contents leak out. The lesions are probably formed as a result of insertion of C8 and C9 into the membrane (see Figure 15–12). The utilization of the complement components in this process is called **complement fixation** and is the basis of an important clinical laboratory test (see Chapter 16).

Immune Adherence When C3b is bound to the surface of a microorganism, the C3b can interact with special receptors on phagocytes to promote phagocytosis. This phenomenon is called **immune adherence** or opsonization.

Inflammation C3a and the cleavage products from C5, C6, and C7 can contribute to the development of acute inflammation. They increase the permeability of blood vessels and act as chemotactic agents to attract large numbers of PMNs to the area. (The antimicrobial effects of inflammation were discussed earlier in this chapter.)

Once complement is activated, its destructive capabilities usually cease very quickly in order to minimize destruction of the host's cells. This is accomplished by the spontaneous breakdown of activated complement and by interference from inhibitors and destructive enzymes in the host's body fluids.

There are defense mechanisms other than the nonspecific defenses discussed in this chapter and immunity, which will be discussed in Chapter 16.

FACTORS THAT LOWER RESISTANCE

Despite the best available preventive measures, more than one-half of all compromised hosts develop infections. A **compromised host** is one whose resistance to infection is impaired by disease, therapy, or both. In general, impaired resistance can result from any of the following conditions: (1) reduced numbers of functional phagocytes; (2) decreased numbers of lymphocytes (T cells) that directly destroy antigens; (3) faulty production of antibodies; and (4) injury to mechanical barriers against infection.

A person with diabetes mellitus has both elevated blood glucose levels that impair leukocyte mobilization and acidosis (blood pH below 7.35) that impairs phagocytosis. Persons with acute leukemias can have lowered resistance associated with leukopenia, deficient leukocyte mobilization, and defective phagocytosis. Patients with Hodgkin's disease or AIDS experience a deficiency of lymphocytes (T cells). Persons suffering from multiple myelomas, chronic lymphocytic leukemia, or histiocytic lymphoma experience depression of antibody production. Patients whose mechanical barriers are broken down by burns, trauma, or medical procedures are susceptible to infection. Therapy directed against underlying diseases can further impair resistance. For example, steroids given to persons who have had organ transplants or have other conditions can cause depression of leukocyte mobilization, impairment of phagocytosis, and depression of lymphocytes (T cells).

In addition to diseases and medications there are several other factors that contribute to compromising a host. Nutritional deficiences can interfere with phagocytosis and antibody production, and can cause the skin and mucous membranes to develop defects that permit microbes to penetrate to deeper tissues. Alcoholism depresses the inflammatory response to bacterial infections. Mechanical obstructions of drainage routes, such as the urinary system, bile ducts, and lacrimal apparatus, interfere with mobilization and functioning of phagocytes and prevent removal of infected contents from a structure. Circulatory disturbances, either local (edema) or general (shock), interfere with the mobilization and functioning of phagocytes. Physical and emotional fatigue, overexposure to cold, and smoking apparently increase susceptibility to bacterial infection; these conditions probably impair the functioning of the ciliary escalator. There is increasing evidence that stress also has an immunosuppressive effect. Finally, as age increases there is a general decrease in antibody production and other immune responses.

The next chapter will take a detailed look at the principal factors that contribute to immunity.

STUDY OUTLINE

INTRODUCTION (pp. 416–417)

1. The ability to ward off disease through body defenses is called resistance.
2. Lack of resistance is called susceptibility.
3. Nonspecific resistance refers to all body defenses that protect the body from any kind of pathogen.
4. Specific resistance refers to defenses (antibodies) against specific microorganisms.

SKIN AND MUCOUS MEMBRANES (pp. 417–420)

Mechanical Factors (pp. 417–419)

1. The structure of intact skin and keratin provide resistance to microbial invasion.
2. Some pathogens, if present in large numbers, can penetrate mucous membranes.
3. The lacrimal apparatus protects the eyes from irritating substances and microorganisms.
4. Saliva washes microorganisms from teeth and gums.
5. Mucus traps many microorganisms that enter the respiratory and gastrointestinal tracts; in the lower respiratory tract the ciliary escalator moves mucus up and out.
6. The flow of urine moves microorganisms out of the urinary tract, and vaginal secretions move microorganisms out of the vagina.

Chemical Factors (pp. 419–420)

1. Sebum contains unsaturated fatty acids, which inhibit the growth of pathogenic bacteria. Some bacteria commonly found on the skin can metabolize sebum and cause the inflammatory response associated with acne.
2. Perspiration washes microorganisms off the skin.
3. Lysozyme is found in tears, saliva, nasal secretions, and perspiration.
4. The high acidity (pH 1.2 to 3.0) of gastric juice prevents microbial growth in the stomach.

PHAGOCYTOSIS (pp. 420–424)

1. Phagocytosis is the ingestion of microorganisms or particulate matter by a cell.
2. Phagocytes are certain types of white blood cells or derivatives of them.

Formed Elements of the Blood (p. 421)

1. Blood consists of plasma (fluid) and formed elements (cells and cell fragments).
2. Leukocytes (white blood cells) are divided into two categories: granulocytes (neutrophils, basophils, and eosinophils) and agranulocytes (lymphocytes and monocytes).
3. During many infections, the number of leukocytes increases (leukocytosis); some infections are characterized by leukopenia (decrease in leukocytes).

Kinds of Phagocytic Cells (pp. 421–424)

1. Among the granulocytes, neutrophils are the most important phagocytes.
2. Enlarged monocytes become wandering macrophages and fixed macrophages.

3. Fixed macrophages are located in selected tissues and are part of the mononuclear phagocytic system.
4. Granulocytes predominate during the early stages of infection, whereas monocytes predominate as the infection subsides.

Mechanism of Phagocytosis (p. 424)

1. Phagocytes are attracted to microorganisms by chemotaxis.
2. The phagocyte then adheres to the microbial cell; adherence may be facilitated by opsonization.
3. Pseudopods engulf the microorganism, and enclose it in a phagosome to complete ingestion.
4. Many phagocytized microorganisms are killed by lysosomal enzymes.

INFLAMMATION (pp. 424–427)

1. Inflammation is a bodily response to cell damage, and it is characterized by redness, pain, heat, and swelling.
2. Sometimes loss of function results.

Vasodilation and Increased Permeability of Blood Vessels (pp. 425–426)

1. The release of histamine, kinins, and prostaglandins causes vasodilation and increased permeability.
2. Blood clots can form around an abscess to prevent dissemination of the infection.

Phagocyte Migration (pp. 426–427)

1. Phagocytes have the ability to stick to the lining of the blood vessels (margination).
2. They also have the ability to squeeze through blood vessels (diapedesis).
3. Pus is the accumulation of damaged tissue and dead microbes, granulocytes, and macrophages.

Repair (p. 427)

1. A tissue is repaired when the stroma or parenchyma of a tissue produces new cells.
2. Stromal repair by fibroblasts produces scar tissue.

FEVER (pp. 427–428)

1. Fever is an abnormally high body temperature produced in response to a bacterial or viral infection.
2. Bacterial endotoxins and leukocytic pyrogen can induce fever.
3. A chill indicates a rising body temperature; crisis (sweating) indicates that the body's temperature is falling.

ANTIMICROBIAL SUBSTANCES (pp. 428–432)

Interferon (IFN) (pp. 428–429)

1. Interferon is an antiviral substance produced in response to viral infection.
2. There are three types of human interferon: α-IFN, β-IFN, and γ-IFN. Recombinant interferon has also been produced.
3. Interferon's mode of action is to induce uninfected cells to produce an antiviral protein that prevents viral replication.
4. Interferons are host-cell-specific but not virus-specific.

Complement and Properdin (pp. 429–432)

1. Complement and properdin make up about 10% of the serum proteins.
2. These proteins activate one another to destroy invading microorganisms. Methods of destruction are membrane lysis, immune adherence, or inflammation.

FACTORS THAT LOWER RESISTANCE (pp. 432–433)

1. Compromised hosts are highly susceptible to infection.
2. A compromised host has lowered resistance.
3. Lower resistance may be due to preexisting disease, therapy, nutritional deficiency, fatigue, age, and certain environmental factors.

STUDY QUESTIONS

REVIEW

1. Define the following terms.
   (a) Resistance
   (b) Susceptibility
   (c) Nonspecific resistance
2. Describe the mechanical factors of the skin and mucous membranes that assume a role in nonspecific resistance.
3. Describe the chemical factors of the skin and mucous membranes that assume a role in nonspecific resistance.
4. Define phagocytosis.
5. Compare the structures and functions of granulocytes and monocytes in phagocytosis.
6. How do fixed and wandering macrophages differ?
7. Define inflammation and list its characteristics.
8. Why are redness, heat, and swelling observed during inflammation?
9. Why is inflammation beneficial to the body?
10. How is fever related to nonspecific defense?
11. What is the importance of chill and crisis during fever?
12. What is interferon? Discuss its role in nonspecific resistance.
13. What is complement?
14. Summarize the major outcomes of complement activation.
15. What is properdin? What are its functions in nonspecific resistance?
16. Why is a compromised host more susceptible to infection than is a noncompromised host?

CHALLENGE

1. Diagram the following processes that result in phagocytosis: margination, diapedesis, adherence, and ingestion.
2. A variety of drugs with the ability to reduce inflammation are available. Comment on the danger of misuse of these anti-inflammatory drugs.
3. A hematologist often performs a differential count on a blood sample. A differential count determines the relative numbers of white blood cells. Why are these numbers important? What do you think the hematologist would find in a differential count of a patient with mononucleosis? with neutropenia? with eosinophilia?

FURTHER READING

Collier, H. O. J. August 1962. Kinins. *Scientific American* 207:111–118. Roles of chemical mediators in inflammation.

Edelson, R. L. and J. M. Fink. June 1985. The immunologic function of skin. *Scientific American* 256: 46–53.

Friedman, R. M. 1981. *Interferons: A Primer.* New York: Academic Press. Comprehensive coverage on the status of in vivo and in vitro mechanisms of action, and clinical uses of interferons.

Goren, M. B. 1977. Phagocyte lysosomes: Interactions with infectious agents, phagosomes, and experimental perturbations in function. *Annual Review of Microbiology* 31:507–533. Includes a discussion of how bacteria and protozoans avoid phagocytic destruction.

Mayer, M. M. November 1973. The complement system. *Scientific American* 229:54–66. Detailed description of the functions of the complement enzymes.

McNabb, P. C. and T. B. Tomasi, 1981. Host defense mechanisms at mucosal surfaces. *Annual Review of Microbiology* 33:477–496. Describes various host defenses, including normal microbial flora and IgA antibodies, and how they prevent growth of pathogens.

Trager, W. March 1981. The biochemistry of resistance to malaria. *Scientific American* 244:154–164. Describes the evolution of sickle-cell anemia and how this disease offers protection from malaria.

Wood, W. B. February 1951. White blood cells versus bacteria. *Scientific American* 184:48–52. Electron micrographs illustrate the action of phagocytes.

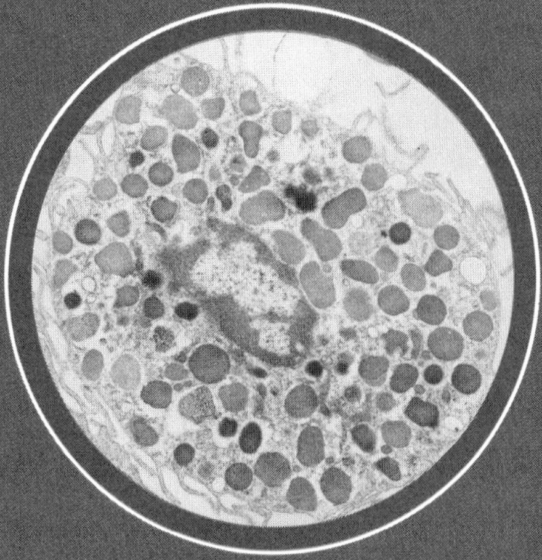

CHAPTER 16

Specific Defenses of the Host: The Immune Response

OBJECTIVES

After completing this chapter you should be able to

- Contrast the kinds of immunity.
- Define the following terms: immunity; antigen; anamnestic response.
- Describe the clonal selection theory.
- Explain what an antibody is and describe the structural and chemical characteristics of antibodies.
- Provide at least one function for each of the five classes of antibodies.
- Compare and contrast humoral and cell-mediated immunity.
- Define lymphokine; differentiate between types of T cells.
- Explain how each of the following can be used in the diagnosis of a disease: precipitation; agglutination; complement fixation; neutralization; immunofluorescence; RIA; ELISA.
- Define the following terms: serology; immunodiffusion; vaccine; toxoid; attenuated; monoclonal antibody.

The last chapter discussed several nonspecific host defenses including the intact skin and mucous membranes, phagocytosis, inflammation, and fever. These are integral components of our resistance to disease. We have, in addition to these nonspecific defenses, an **innate resistance** to certain illnesses. For example, as humans we are resistant to many infectious animal diseases such as canine distemper and chicken and hog cholera. On the other hand, resistance to a human disease such as measles can vary: although the illness is often relatively mild for persons of European ancestry (probably because natural selection has affected the many generations exposed to the measles virus), the disease was devastating and caused many deaths among South Sea islanders first exposed to measles by European explorers. A person's resistance to disease is also dependent on genetic factors, age, nutritional status, and general health.

In this chapter we will study yet another aspect of the body's defenses, called **acquired immunity.** Immunity, in contrast to resistance, involves the production of a *specific* defensive response when the host is invaded by foreign organisms or substances. This immunity is developed (acquired) during a person's life—it is not inherited. The invading organisms might be pathogenic bacteria, viruses, fungi, protozoans, or helminths. They might be foreign materials such as pollen or insect venom, or transplanted tissue from another vertebrate. The body always recognizes these substances as not belonging to itself, and develops an immune response to inactivate or destroy them. We will see in this chapter and the next that the immune system is essential to our survival, but that the system can also cause harm when its efforts are misdirected.

THE DUALITY OF THE IMMUNE SYSTEM

The human immune system, like that of all vertebrates, has two components. One is called the **humoral immune system** because it involves antibodies that are dissolved in the blood plasma and lymph, the body's "humors" (fluids). **Antibodies** are proteins produced by specialized lymphocytes, called **B cells,** after the B cells have been exposed to an antigen. **Antigens** are substances, usually foreign to the host, that cause antibody production and react only with their specific antibody. Antibodies are specifically directed against the antigen that caused their formation.

The second component of the immune system is called the **cell-mediated immune system** because it involves specialized lymphocytes called **T cells.** These cells are located in both the blood and the lymphoid tissues (see Figure 15–5). T cells do not produce antibodies, but they do have antibodylike molecules, called *antigen receptors,* attached to their surfaces. There are several different kinds of T cells, each of which carries out a specific function.

Whereas the humoral immune response defends mostly against bacteria, bacterial toxins, and viruses in the body's fluids, the cell-mediated immune response is most effective against bacteria and viruses located within phagocytic or infected host cells, and against fungi, protozoans, and possibly helminths. The cell-mediated immune system also responds to transplanted tissue (such as a foreign skin graft) by mounting an immune response to reject it, and is thought to be important in our defenses against cancer.

ACQUIRED IMMUNITY

Acquired immunity refers to immunity gained during life as a result of contact with a microbe or another foreign substance. This immunity can be obtained either actively or passively, and by natural or artificial means.

Naturally Acquired Immunity

Naturally acquired active immunity is obtained when a person's immune system comes into contact with microbial antigens in the course of daily life, and the immune system responds by producing antibodies or sensitized T cells. In some cases the immunity may be lifelong, as with measles, chickenpox, and yellow fever. In other cases, as with diphtheria, the immunity usually lasts for only a few years. There are also diseases, such as pneumococcal pneumonia, gonorrhea, and the common cold, for which there appears to be only short-lived or no natural immunity to reinfection. Sometimes subclinical infections (those that produce no evident symptoms) can also confer immunity. Immunity to mumps, hepatitis, and polio, for example, can be acquired from subclinical infections.

Naturally acquired passive immunity involves the natural transfer of antibodies from an immunized donor to a nonimmunized recipient. For example, an expectant mother is able to pass some of her antibodies to the fetus across the placenta; this mechanism is called *placental transfer*. Accordingly, if the mother is immune to diseases such as diphtheria, rubella, or polio, placental transfer will make the newborn infant immune as well. Certain antibodies are also passed from the mother to her nursing infant in breast milk, especially in the first secretions called colostrum. This type of immunity is transmitted naturally but is considered passive, in that the baby does not synthesize the antibodies but has them provided by an outside source. Naturally acquired passive immunity generally lasts only a few weeks or months.

Artificially Acquired Immunity

Artifically acquired active immunity results from vaccination. **Vaccines** consist of inactivated bacterial toxins **(toxoids),** killed microorganisms, or living but attenuated (weakened) microorganisms. The attenuated pathogens are no longer able to cause disease, but they are still able to stimulate an immune response much as naturally acquired pathogens do. Vaccines are discussed in more detail later in this chapter.

Artificially acquired passive immunity is gained by the injection of antibodies from an outside source. For example, a person bitten by a snake might be given snake-venom antibodies produced by a horse. Human antibodies formed in one person can also be transferred to another person. Pooled blood serum from individuals who have either been vaccinated against a certain disease, or have recovered from it, contains antibodies against the disease antigens to which the individuals are immune. These humoral antibodies circulate in the blood and can be isolated from it.

Serum (plural, sera) is the fluid remaining after blood has clotted and the blood cells and clotted matter have been removed. Because most of the antibodies remain in the serum, the word **antiserum** has become a generic term for fluids containing antibodies. When serum is subjected to an electric current in the laboratory in a process called electrophoresis, the proteins within it move at rates determined by their electrical charges, as shown in Figure 16–1. One of the fractions resulting from this procedure contains most of the antibodies that were present in the original sample. Because this fraction is called the gamma fraction and antibodies belong to a group of proteins called **globulins,** the term **gamma globulin** is used to describe the antibody-rich serum component. The term **immune serum globulin** is now preferred.

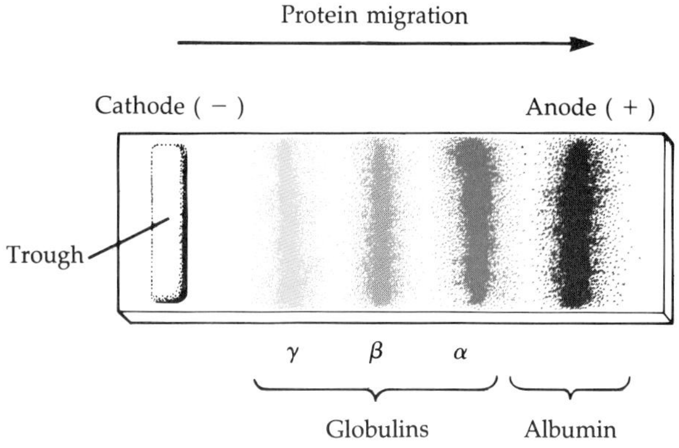

Figure 16–1 Separation of blood serum into protein components. In this procedure, serum to be tested is placed in a trough containing a gel. In response to an electric current, the negatively charged proteins of the serum migrate through the gel from the cathode (−) to the anode (+). Antibodies are concentrated mainly in the gamma (γ) fraction of globulins, hence the familiar term gamma globulin. (The β, α, and albumin fractions include other types of serum proteins.) The separation of proteins on the basis of their movement in an electric field is called *electrophoresis* (see page 453).

Passive immunity can be established after someone who has not been previously immunized against an infectious disease or a toxin is exposed to it. After such exposure there is usually not enough time to build immunity with a single administration of vaccine. However, injections of immune serum globulin can give immediate protection. Sometimes specific immune sera are administered in the hope of preventing a specific disease, such as rabies. In this case the donors of immune serum might be laboratory personnel who must be actively immunized because of their high risk of exposure to the disease.

Immune sera can also be administered as a more general preventive measure. A pooled immune serum from the general population contains antibodies against most common diseases such as measles, rubella, and hepatitis. This type of protective immunization is sometimes recommended to travelers going overseas, for example.

Although artificially acquired passive immunity is immediate, it is short-lived. The body produces no new antibodies, and the half-life of an injected antibody is typically about three weeks.

The various types of immunity are summarized in Table 16–1.

ANTIGENS

As we mentioned earlier, an **antigen** (sometimes called **immunogen**) causes the body to produce a highly specific immune response in the form of antibodies or specially sensitized cells. Normally the immune system recognizes components of the body it protects as "self" and not as foreign matter ("nonself"). This recognition is the reason why people's defense systems do not usually produce antibodies against their own body tissues (though we will see in Chapter 17 that this sometimes occurs).

The vast majority of antigens are proteins, nucleoproteins (nucleic acid + protein), lipoproteins (lipid + protein), glycoproteins (carbohydrate + protein), or large polysaccharides. These proteins and polysaccharides are often components of the invaders: the capsules, cell walls, flagella, pili, and toxins of bacteria; the coats of viruses; and the surfaces of many other types of cells. Nonmicrobial antigens include pollen, egg white, blood cells from other persons or species, and transplanted tissues and organs.

In general, antigens have molecular weights of 10,000 or more. Antibodies are not generally formed against the whole antigen, but only against specific regions of it. On the antigen's surface, the sites at which the antigen and antibody interact are called **antigenic determinant sites** (Figure 16–2). The nature of this interaction depends on the size and shape of the antigenic determinant site in relation to the chemical structure of the antibody. The antigen and antibody fit together much as a lock and key, or an enzyme and its substrate.

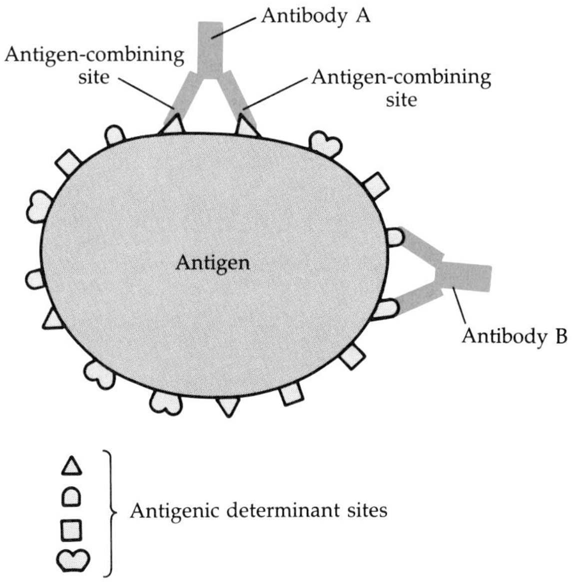

Figure 16–2 Relationship of an antigen to antibodies. Most antigens contain more than one antigenic determinant site, and there can be several different kinds of sites. Each antibody molecule has at least two sites that are specific for a particular kind of determinant site on the antigen and can bind to it. The two antibodies shown here are specific for two different determinant sites on the antigen molecule; note that the antigen-combining sites on any given antibody molecule are always the same. It is possible for the antibody to bind simultaneously with identical antigenic determinant sites on different molecules, which would cause neighboring antigens to aggregate.

Table 16–1 Summary of the Categories of Acquired Immunity

Naturally acquired active immunity: Antigens enter the body → Antibodies and sensitized T cells produced
Naturally acquired passive immunity (placental and milk transfer): Antibodies from an immunized mother → Fetus or baby receives antibodies
Artificially acquired active immunity: Prepared antigens in vaccines → Injected into susceptible individual → Antibodies and sensitized T cells produced
Artificially acquired passive immunity: Immune serum → Exposed individual receives preformed antibodies

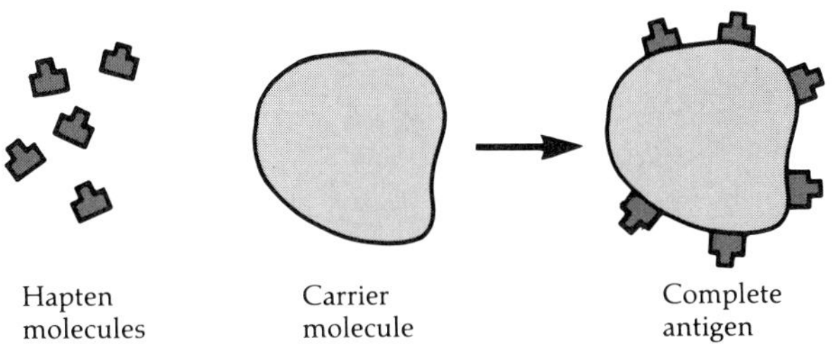

Figure 16–3 Haptens. A hapten is a low-molecular-weight molecule that does not stimulate antibody formation by itself. If the hapten is combined with a larger carrier molecule, such as a serum protein, the hapten and its carrier together function as an antigen and can provoke an immune response. Antibodies can then bind to the hapten whether or not it is attached to the carrier.

These sites have molecular weights of about 200 to 1000, so each site is small in relation to the whole antigen, whose molecular weight is 10,000 or more. An antigenic determinant site cannot function as an antigen (i.e., stimulate antibody production) by itself—but it does react with a compatible antibody once that antibody has been formed.

The number of antigenic determinant sites on the surface of an antigen is that antigen's **valence.** Most antigens are multivalent; that is, they have more than one determinant site.

A foreign substance that has a low molecular weight is often not antigenic unless it is made larger by attachment to a carrier molecule. These low-molecular-weight substances are called **haptens** (haptein, to grasp). Once an antibody against the hapten has been formed, however, the hapten alone will react with antibodies independently of its carrier (Figure 16–3). Penicillin is the best known hapten. This drug is not antigenic by itself, and relatively few people develop allergic reactions to it. However, the penicillin combines with the serum proteins of some persons, and the resulting molecule does initiate an immune response.

ANTIBODIES

An **antibody** is a protein that B lymphocytes produce in response to the presence of an antigen; the antibody is capable of combining specifically with that antigen. Like most antigens, antibodies have more than one antigen–antibody combining site, or valence. Most human antibodies have two combining sites (are bivalent); a bivalent antibody unit is called a **monomer.** Monomers can combine to form multivalent antibodies; some of these have as many as ten combining sites (are multivalent). Antibodies belong to a group of proteins called globulins, so they are known as **immunoglobulins (Ig).** Immunoglobulins are found in the blood serum, mainly in the gamma fraction, and in some body secretions.

Structure of Antibodies

Figure 16–4 shows a typical antibody monomer with its two antigen-combining sites. The Y-shaped molecule is flexible and can assume a T-shape (note the hinge area indicated on the antibody molecule). Figure 16–2 shows an antibody combining with determinant sites on an antigen. In the example illustrated, the antigen might be a bac-

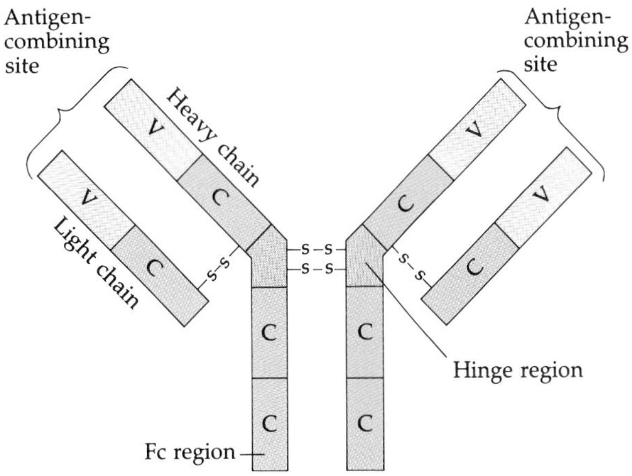

Figure 16–4 Structure of a typical antibody molecule. The molecule is composed of two light and two heavy chains linked together in a flexible Y shape (note hinge region). The constant regions, which are the same for all antibodies of the same class, are indicated as C. The amino acid sequence of the variable V regions differs from molecule to molecule, and can be specific for an almost infinite variety of antigens. The molecule is bivalent; that is, it has two antigen-combining sites, both specific for the same type of antigenic determinant site. The Fc region allows the antibody to bind to the surface of certain host cells. Disulfide bonds are indicated by S—S.

terial cell with several different antigenic determinant sites capable of stimulating an immune response. Accordingly, more than one type of antibody can react with the bacterial cell. If the two antigen-combining sites of an antibody combine with antigenic determinant sites on two different antigens, the antigens can aggregate into clumps. (As we will discuss later in the chapter, this clumping can be an important factor in the diagnosis of some diseases.)

A typical antibody monomer (Figure 16–4) is composed of four protein chains, two identical *light chains* (of low molecular weight) and two identical *heavy chains* (of higher molecular weight). The chains are linked in a characteristic shape by disulfide bridges (Chapter 2) as well as by other bonds. Most of the molecule has a *constant* structure, shown as C in the illustration; this means that its amino acid sequence is the same in all monomers of a given class. The regions at the ends of the Y, however, are *variable* (V). The amino acid sequences of these V regions vary from one monomer to another, and the resulting variations in chemical structure account for the ability of different antibodies to recognize and bind with many different antigens. Any single antibody molecule has only one type of variable region in its protein chains, no matter how many antigen-combining sites it has, and hence will bind to only one type of antigenic determinant site.

Notice that at the base of the molecule, opposite the variable region, is the *Fc region* (standing for crystallizable fragment). The Fc region is the site at which the antibody molecule can attach to a host cell; the two antigen-combining sites are left free to bind to compatible antigenic determinant sites.

The molecule just described is the basic structural unit of all classes of immunoglobulin molecules. Bear in mind, however, that although the basic structural units are similar in appearance, the different classes of immunoglobulins differ chemically in their constant regions, including the Fc regions, and the type of bonding between chains.

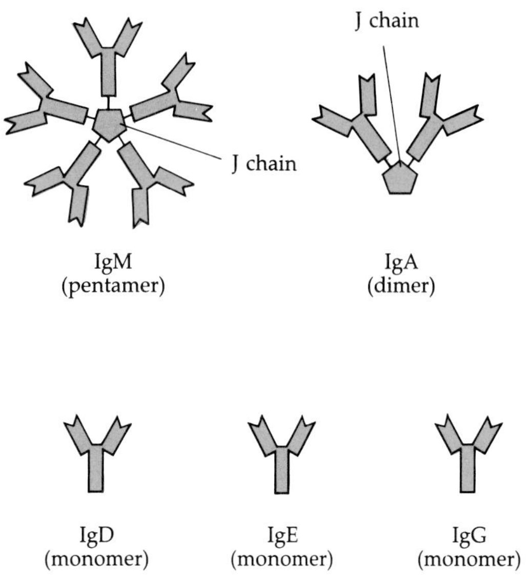

Figure 16–5 Structure of the five principal classes of human immunoglobulins. Note that IgM and IgA are made up of five and two monomers, respectively.

Classes of Immunoglobulins

The five classes of immunoglobulins (Ig) are designated IgG, IgM, IgA, IgD, and IgE according to the properties of their heavy chains. The structures of IgG, IgD, and IgE molecules resemble the structure shown in Figure 16–4. Molecules of IgM and IgA are usually two or more monomers joined together (Figure 16–5). The characteristics of the immunoglobulin classes are summarized in Table 16–2.

IgG

IgG antibodies account for about 80–85% of all antibodies in serum. These antibodies readily cross the walls of blood vessels and enter tissue fluids; maternal IgG antibodies, for example, can cross the placenta and confer passive immunity on the fetus. IgG antibodies have many functions: protection against circulating bacteria and viruses; neutralization of bacterial toxins; and attachment to antigens, to enhance the work of phagocytic cells. This attachment process is called *opsonization.*

IgG also participates in complement reactions (Chapter 15). "Complement" refers to a group of protein components found in the blood. We will have more to say about these reactions in the next chapter.

Table 16–2 Summary of Immunoglobulin Classes

Characteristics	IgG	IgM	IgA	IgD	IgE
% total serum antibody	80–85	5–10	15	0.2	0.002
Location	Blood, lymph, intestinal fluids	Blood, lymph, B cell surface (as monomer)	Secretions (tears, saliva, mucus, intestine, milk), blood, lymph	B cell surface, blood, lymph	Bound to mast cells throughout body, blood
Molecular weight	150,000 (monomer)	900,000 (pentamer)	170,000 (serum, monomer); 400,000 (secretory, dimer)	185,000 (monomer)	190,000 (monomer)
Half-life (days)* in serum	25	5	6	3	2
Complement fixation	+	+	–**	–	–
Placental transfer	+	–	–	–	–
Functions	Enhances phagocytosis, neutralizes toxins and viruses, and protects fetus and newborn	Especially effective against microorganisms and agglutinating antigens; first antibodies produced in response to initial infection	Localized protection on mucosal surfaces	Not known, but presence on B cells may indicate function in initiation of immune response	Allergic reactions; possibly expulsion or lysis of protozoan parasites

*Time required for one-half the antibodies to disappear.
**May be + via alternate pathway.

IgM

Antibodies of the **IgM** class comprise 5–10% of the antibodies in serum. IgM has a pentamer structure, consisting of five Y-shaped monomers similar to the single IgG monomer. The monomers are held in position by attachment to a *J (joining) chain,* as shown in Figure 16–5.

Antibodies of the IgM class are the first ones to appear in response to the initial exposure to an antigen. However, the high IgM concentration in the blood rapidly declines and the IgG concentration increases (Figure 16–6). A later, secondary injection of antigen results mostly in the increased production of IgG.

IgM class antibodies do not generally leave the blood vessels and enter surrounding tissues; nor do they cross the placenta to provide immunity to the fetus. The larger size of the molecule may prevent it from moving about as freely as IgG does. Because of its numerous antigen-combining sites, the IgM molecule is especially effective at cross-linking particulate antigens and causing their aggregation. IgM is the predominant antibody involved in the response to the ABO blood group antigens on the surface of red blood cells (discussed in the next chapter). It is also very effective in reactions involving complement and can enhance the ingestion of target cells by phagocytic cells, as IgG does. IgM is very important in humoral immunity because of its rapid appearance and its high effectiveness as a protective antibody.

IgA

IgA class antibodies account for about 15% of the serum antibodies. Although a molecule of IgA as it occurs in serum is somewhat similar to IgG, more

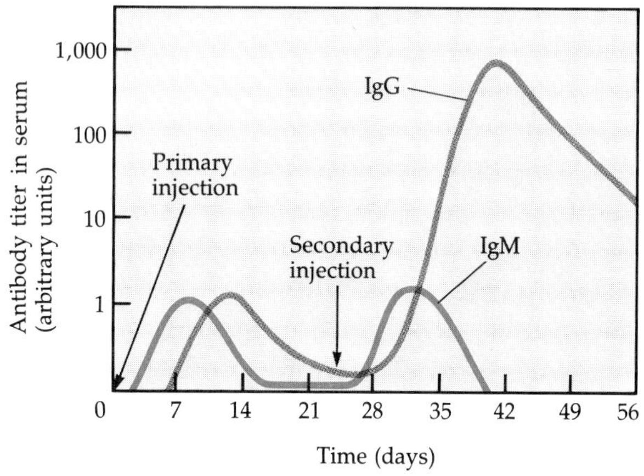

Figure 16–6 The primary and secondary immune response to an antigen. IgM appears first in response to the primary injection, and is followed by IgG, which provides longer-term immunity. The secondary (booster) injection results in a much faster and greater production of antibody (higher titer). The antibody response to a secondary antigen injection is mostly IgG, which appears almost as rapidly as IgM.

often IgA occurs as a dimer of two immunoglobulin subunits connected by a J chain, much like the pentamer of IgM.

A protein bound to IgA, called the *secretory component,* is acquired during the transport of IgA from the blood to the surface of secretory tissues. Secretory IgA, as this molecule is called, is found mostly in secretions of the gastrointestinal tract and other mucous membranes, as well as in saliva, tears, and breast milk. Its presence in colostrum probably helps protect the infant's gastrointestinal tract from infection and assists in the colonization of the tract by normal flora. The main function of secretory IgA is probably to prevent the attachment of pathogens, particularly viruses, to mucosal surfaces.

IgD

IgD antibodies comprise only about 0.2% of the total serum antibodies. Their structure resembles that of the IgG molecules. IgD antibodies are in blood and lymph and on the surfaces of B cells. IgD antibodies do not fix complement and cannot cross the placenta. The fact that IgD antibodies are found in high concentrations on the surface of B cells, especially in newborns, might indicate that these antibodies help to initiate the immune response. However, little is known about their functions.

IgE

Antibodies of the **IgE** class are slightly larger than IgG molecules, but they constitute only 0.002% of the total serum antibodies. The IgE molecules bind tightly to mast cells and basophils, which are specialized cells that participate in allergic reactions (Chapter 17). When an antigen such as pollen reacts with the IgE antibodies attached to a mast cell or basophil, that specialized cell releases histamine and other chemical mediators. The mediators cause dilation of local blood vessels and contraction of smooth muscles such as those in the bronchi; these effects are considered representative of what is known as a type I allergic response. A person who is highly allergic or is a host to a gastrointestinal parasite has an abnormally high serum level of IgE.

MECHANISMS OF THE IMMUNE RESPONSE

As noted at the beginning of the chapter, there are two basic kinds of immune responses. The first, the production of humoral antibodies that circulate in the blood and lymph, is carried out by B cells. The second is the production of specialized cells, T cells, that carry out specific immune reactions. Let us now look in more detail at the mechanisms by which these responses occur.

B cells, T cells, and macrophages all have a common origin, arising from **stem cells** in bone marrow (Figure 16–7). (Other types of cells including red blood cells, mast cells, and leukocytes also develop from stem cells.) In the fetus, the liver appears to give rise to stem cells. Stem cells that become B cells eventually produce humoral antibodies when stimulated by the appropriate antigen. Other stem cells pass through, or are influenced by, the thymus (Chapter 15) and become T cells, which are the effectors of cell-mediated immune reactions. Some of the macrophages produced from stem cells encounter antigens and process them in certain ways; this is part of an essential step in the generation of an immune response by both B and T cells.

B Cells and Humoral Immunity

B cells newly formed from stem cells migrate to lymphoid organs, where some of them mature into cells that circulate in the blood and the lymph. These B cells carry on their surface as many as 100,000 antibody molecules that can function as specific antigen-receptor sites. These sites are usually monomers of IgM or IgD. The total B cell population of each person carries receptors specific for an enormous assortment of antigens, although each B cell carries receptors for only one type of antigen.

When the appropriate antigen contacts the antigen receptors on a B cell, the cell proliferates into a large clone of cells (Figure 16–8). This phenomenon is called **clonal selection.** Some of these cells differentiate into **plasma cells,** each programmed to secrete antibodies specific to the antigen that caused its formation. Each plasma cell lives for only a few days, but is capable of producing about 2000 humoral antibody molecules per second.

Stimulation of a B cell by an antigen also results in the production of a population of **memory B cells.** These cells have a lifetime measured in years, or even decades, and are responsible for the immune system's rapid response to a secondary exposure (booster) to the same antigen.

B cells often work cooperatively with both macrophages and T cells. Before being presented to B cells to stimulate them into activity, most antigens are processed by macrophages. After phagocytic digestion of the antigen, some of its processed components appear on the surface of the macrophage. T cells also interact with the antigen to stimulate an antibody response. Antigens that are "assisted" in this way by T cells are called *T-dependent antigens;* examples are bacteria, foreign red blood cells, proteins, and hapten–carrier combinations that have many different antigenic determinant sites. For a T-dependent antigen to trigger a mature B cell to become an antibody-secreting plasma cell, a helper T cell is required. All immune responses that produce IgG, IgA, and IgE involve helper T cells, discussed in more detail below. *T-independent antigens,* which can provoke an immune response from mature B cells without helper T cell intervention, are presented directly to the B cell. These T-independent antigens are composed of repeating polysaccharide or protein subunits. The B cell then produces antibodies, almost exclusively of the IgM class. Only a few memory cells are capable of producing IgG, which they do in great numbers upon a secondary or subsequent antigen exposure (discussed below).

Antibody Responses to Antigens

Antibodies reacting with antigens form **antigen–antibody complexes.** These complexes can protect a host in several ways. For example, they can neutralize toxins by blocking their active sites, or inactivate viruses by combining with them and preventing their attachment to a host cell. These

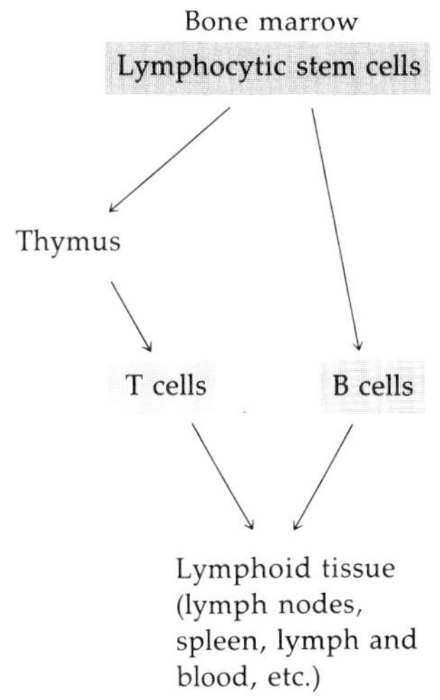

Figure 16–7 Differentiation of T cells and B cells. T cells and B cells, as well as macrophages, originate from stem cells in adult bone marrow. Some of the cells pass through or are influenced by the thymus, where they differentiate into mature T cells. Other cells from the same source become B cells. Both types then migrate to lymphoid tissues such as the lymph nodes or spleen.

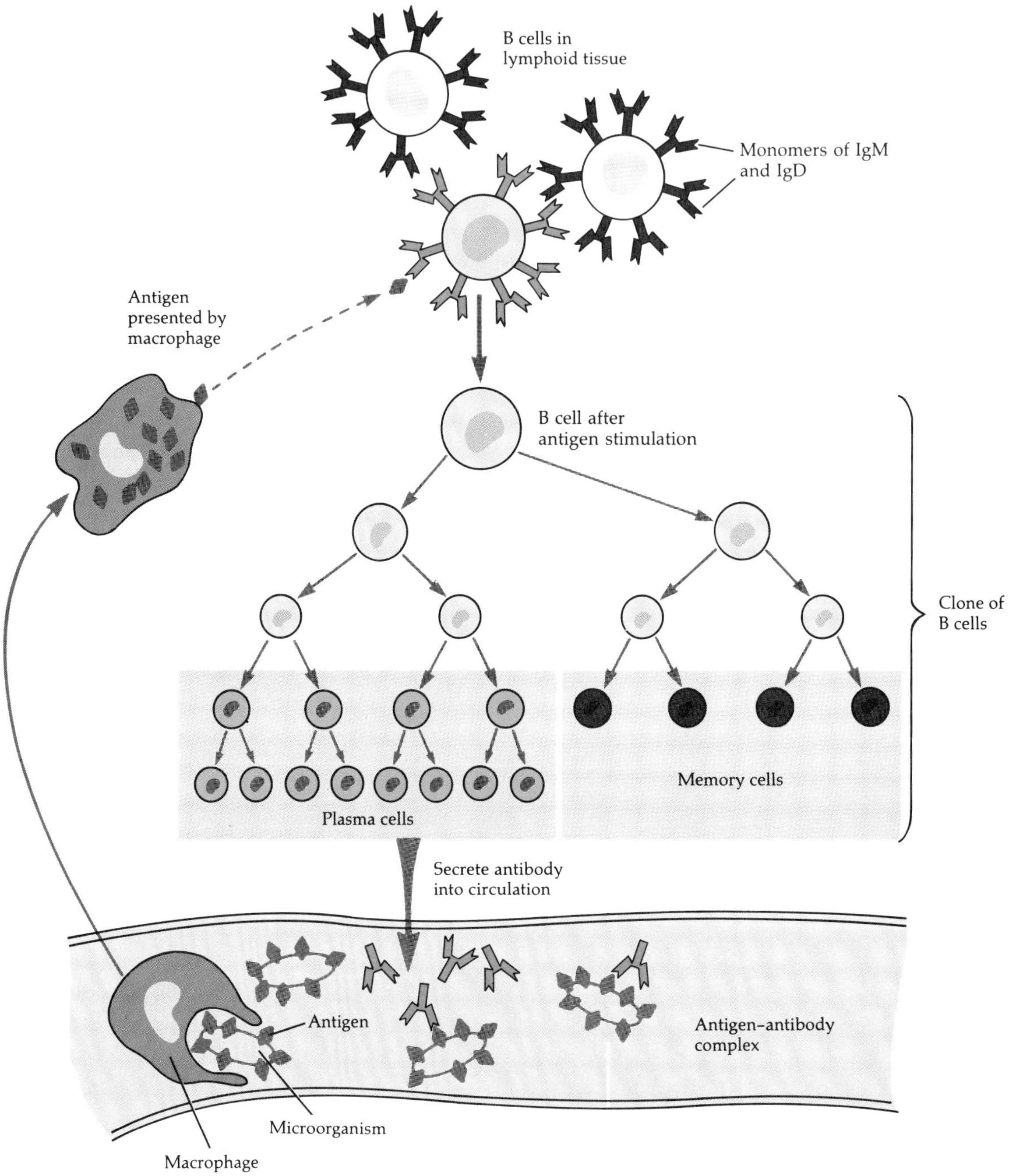

Figure 16–8 Activation of B cells in humoral immunity. Macrophages process antigens; when the antigens are then presented to B cells, some of the B cells proliferate into clones of antibody-producing plasma cells. Other B cells replicate and form memory cells that can later be activated by invasion by the same antigen.

complexes can also fix complement and lead to the lysis of invading cells.

On the other hand, as we will see in the next chapter, the action of antibodies can damage the host. This may occur when immune complexes of antibody, antigen, and complement damage host tissue; when antigens combining with IgE on mast cells initiate allergic reactions; or when antibodies react with host cells and cause autoimmune disorders.

Both the humoral and cell-mediated immune responses of the host intensify after a second exposure to an antigen. This secondary response is called the **memory** or **anamnestic response** (anamne means "recall"). The intensity of the humoral response is reflected by the **antibody titer,** which is the amount of antibody in the serum. After the initial contact with an antigen, the exposed person's serum contains no detectable antibody for several days. Then there is a slow rise in antibody titer, followed by a gradual decline. This pattern is characteristic of the primary response. One component of the primary response is the production of clones of long-lived memory cells. On the second or any subsequent exposure to the same antigen, some of these memory cells differentiate into antibody-producing plasma cells, and are thus responsible for the much faster and higher rise in antibody titer than occurred during the primary response (see Figure 16–6). The primary response is characterized first by the appearance of IgM antibodies, which then decline, and are replaced by IgG antibodies. The secondary response, in contrast, involves mainly the production of IgG antibodies by activated memory cells.

Probably the most familiar clinical example of the anamnestic response is seen after a tetanus toxoid vaccine is injected into someone who has received a serious cut, puncture wound, or bite. Individuals who have an established immunity to tetanus (most acquire this in their childhood vaccination series) display a secondary response to the booster injection of tetanus toxoid. This response renews protection against tetanus quickly enough to be effective against any toxin that might be produced if tetanus bacteria infect the wound. For individuals without previous immunity, the response is too slow to be effective. For these persons, immune serum globulin can provide passive, temporary immunity.

Monoclonal Antibodies

A revolutionary technique that has already had a major impact on medicine is the production of **monoclonal antibodies.** As described in more detail in the box on page 448–449, a single plasma cell from a clone programmed to secrete a specific antibody is combined with a cancer cell; the new hybrid cell is called a **hybridoma.** Hybridomas are capable of producing large amounts of specific antibodies, called monoclonal antibodies; these are beginning to find wide use in the diagnosis of disease, and might someday be used in the treatment of cancer.

T Cells and Cell-Mediated Immunity

In the early days of immunological experimentation, attempts were made to transfer humoral immunity between animals by transferring blood serum from immunized animals to nonimmunized animals. It was found that some forms of immunity are not transferred with blood serum. Much later it was learned that these forms of immunity are transferred only when certain lymphocytes are transferred; the immunity associated with these lymphocytes is now called cell-mediated immunity. Immunity to tuberculosis, for example, was found to be most effectively transferred by these lymphocytes.

T cells are the key component of cell-mediated immunity. Like B cells, they derive from precursor stem cells in bone marrow, but unlike those of B cells, these precursors develop and differentiate into T cells within the thymus. After differentiation, the T cells migrate to lymphoid organs such as the lymph nodes or the spleen. When stimulated by an antigen, T cells do not secrete antibodies, but differentiate into **effector cells** of various types and into some long-lived memory cells (Figure 16–9). In their various roles in the cell-mediated immune response, some T cells function as effector cells that directly attack target cells, while others regulate the immune response.

An important group of effector cells is the **cytotoxic T cells (Tc).** These destroy target cells, such as cancer cells and those of transplanted tissue, upon contact. Cytotoxic T cells are also important in the immune response to some eucaryotic parasites and intracellular bacteria and viruses. For ex-

Microbiology in the News

Nobel Prize-winning Immunology—Theory and Practice

On October 15, 1984, three immunologists were awarded the 1984 Nobel Prize in medicine and physiology. Niels K. Jerne received the prestigious award for his contributions to a basic understanding of the immune system. Georges J.F. Köhler and César Milstein were honored for discovering how to produce monoclonal antibodies.

The 72-year-old Dr. Jerne, a citizen of Britain and Denmark, is professor emeritus of the Basel Institute of Immunology in Switzerland. In announcing the award, the Karolinska Institute in Stockholm described him as the "leading theoretician in immunology during the last 30 years." The first of Jerne's major theories, presented in 1955, explained how the immune system can produce antibodies against an endless variety of foreign substances. The most recent, known as the network theory, describes the elaborate mechanisms by which the body regulates the immune response. The announcement declared that "Jerne's view on the nature of the immune system constitutes the basis for modern immunology." Jerne's work laid the foundation for the production of monoclonal antibodies—antibodies of a single type.

Dr. Milstein, 57, and Dr. Köhler, 38, first produced monoclonal antibodies in 1975, while working together at the British Medical Research Council's Laboratory of Molecular Biology at Cambridge University. Dr. Milstein, a British and Argentinian citizen, is still at Cambridge. Dr. Köhler, a German, is now at the Basel Institute. Milstein and Köhler's reasons for producing pure antibodies had to do with basic immunological research, but they immediately realized the far-reaching implications of their discovery. They never sought to patent their technique, which is now used by laboratories throughout the world and has become the basis for commercial antibody production.

Making Monoclonal Antibodies

Scientists have long known how to produce antibodies by injecting an antigen, such as a virus or a tumor cell, into an animal. The B cells of the animal's immune system respond by releasing antibody molecules into the blood—molecules that combine specifically with that antigen. It is virtually impossible to extract a pure batch of a desired antibody from serum. Laboratory procedures using antibody, such as diagnostic tests for diseases, formerly depended on impure preparations that made it difficult to standardize tests or achieve consistently reproducible results.

Since a single B cell and its descendants manufacture just one kind of antibody, an obvious way to obtain pure antibody preparations would be to grow B cells in the laboratory. But this was not a practical approach, since B cells do not survive long in culture. Köhler and Milstein's historic breakthrough was the discovery that a B cell can be made "immortal" by fusing it with a myeloma cancer cell capable of proliferating endlessly in the laboratory. The resulting hybrid cell is called a *hybridoma.* When the hybridoma cell is grown in culture, its genetically identical descendants continue to produce the type of antibody characteristic of the ancestral B cell. One of a group of genetically identical descendants is called a *clone,* so antibodies produced by a single hybridoma clone are called *monoclonal antibodies.*

To make hybridomas, Köhler and Milstein first injected antigen into a mouse, whose B cells began to produce antibody against it. Then they removed the mouse's spleen, teased the spleen cells apart, and incubated them with myeloma cells from bone marrow tumors. Spleen tissue contains many B cells, and under the right conditions some of the B cells fuse with myeloma cells to produce hybrids. The two scientists screened the hybrid cells for the ones that produced the antibody they wanted and grew these selected hybridomas in culture. The hybridomas proved to have the desirable traits of both parents:

they lived indefinitely while pumping out a never-ending supply of a single antibody. Since they multiplied rapidly like myeloma cells, they bloomed into an extremely rich source of antibody.

In the past few years, the new biotechnology industry has developed methods for growing hybidomas in quantity. Techniques for increasing the density of the cell cultures include growing them in airlift fermentation vats (an approach first developed for growing bacteria on a large scale) or inside capsules that trap the antibody. Nutritional enrichment of the culture medium—with bovine lymph, for example—also increases the yield. In addition, there are new techniques for keeping the parent myeloma cells from producing antibodies of their own and for encouraging the fusion of myeloma cells with B cells. These improved methods allow companies to harvest pure antibody by the gram or kilogram, instead of in the milligram quantities produced by the original laboratory approach. But the fundamental technique of making a hybridoma is the same as in Köhler and Milstein's first experiments.

Uses of Monoclonal Antibodies

Monoclonal antibodies are powerful for three reasons: they are uniform, they are highly specific, and they can be readily produced in large quantities. Like a well-trained battalion, monoclonal antibodies go after one and only one antigen. Armies of antibodies may one day be routinely dispatched on "search-and-destroy" missions to wipe out enemy viruses, bacteria, protozoans, and even cancers that have set up camp in the human body.

In the ten years since Köhler and Milstein developed their method, monoclonal antibodies have assumed enormous importance as tools of biological and medical research and have been used experimentally in the diagnosis and treatment of cancer and other diseases. For example, kits that use monoclonal antibodies to diagnose allergies are available, and diagnostic tests for many infectious diseases are being developed. Monoclonals that react with antigens on certain tumor cells may allow pinpointing of the tumor's location while it is still too small to be detected by conventional means. And monoclonal antibodies helped identify AIDS in 1981 by allowing investigators to determine the type of immune system cell that was being destroyed.

Even more exciting than their diagnostic potential is the prospect of using monoclonal antibodies to treat cancer and infectious diseases. Over 50 years ago. Dr. Paul Ehrlich imagined the possibility of a "magic bullet" that could hunt down and destroy a pathogen without damaging surrounding healthy tissue. Monoclonals used alone or chemically attached to powerful drugs have some qualities of such medically intelligent bullets. Monoclonals linked with toxic agents have been named *immunotoxins*, and the agents used have included radioactive compounds, drugs such as ricin, and diphtheria toxin. In an experimental treatment of human patients with liver cancer, radioactive iodine was attached to monoclonal antibodies against a liver cancer protein; when injected into the patients, the antibodies delivered pinpoint radiation therapy directly to the cancerous tissue. So far the results of such treatments have been encouraging.

Monoclonals may also prove effective in treating disorders caused by the immune system, such as the rejection of transplanted organs. The role of monoclonals in medical research is already well established. They are being used, for example, to investigate infertility, thyroid disorders, juvenile diabetes, inherited brain disorders, and kidney disease. And finally, recombinant DNA firms are starting to use monoclonals to purify their products. Thus, monoclonal antibodies will be useful for extracting almost any molecule produced by recombinant DNA techniques.

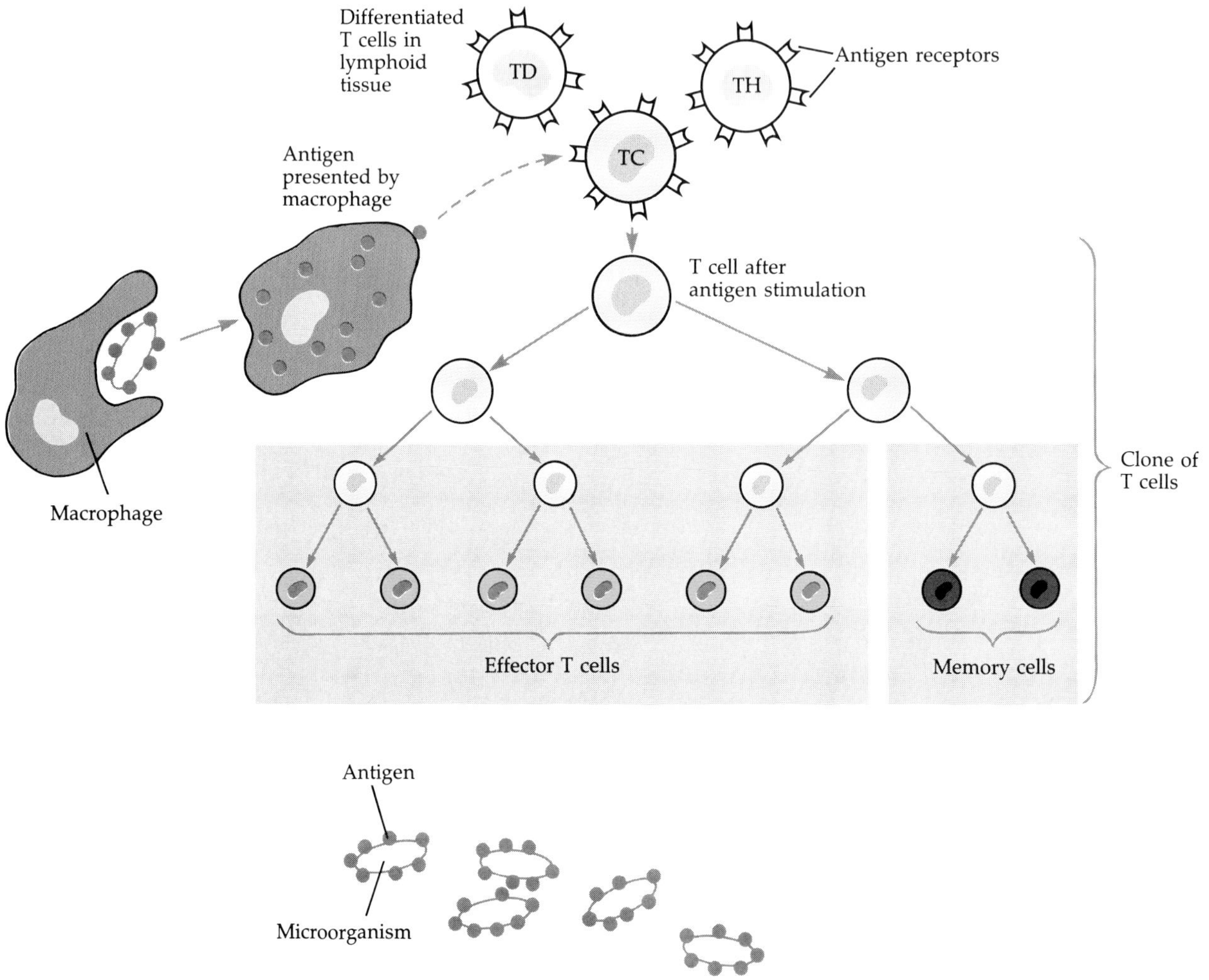

Figure 16–9 Activation of T cells in cellular immunity. The way in which T cells are spurred to proliferate is similar to that of B cells. Unlike B cells, however, T cells do not secrete antibodies. Different types of T cells have different functions: some kill target cells directly; some produce lymphokines; others promote inflammatory responses; and still others help B cells react with antigens or help to regulate the immune response.

ample, many cells infected by a virus are not killed by the virus, but instead produce more viruses. The Tc cells can recognize viral antigens at the surface of the host cell and kill it, and thus discontinue the production of the virus.

Delayed hypersensitivity T cells (Td) are another group of effector cells. Td cells produce substances called *lymphokines*. Lymphokines cause the inflammations characteristic of a positive tuberculin test and associated with the isolation and rejection of transplanted tissue. (Chapter 17 will discuss immunosuppression, the process by which cell-mediated immunity by Td cells is controlled so that tissue can be successfully transplanted.)

We have already seen that certain regulatory T cells, the **helper T cells (Th),** are needed for T-cell dependent antigens to be effectively presented to B

Table 16–3 Comparison of B Cells and T Cells			
Cell	Type of Immunity	Site of Differentiation	Functions
B cell	Humoral	Primarily bone marrow in adults	Differentiates into plasma cells that secrete antibodies
Cytotoxic T cell (Tc)	Cell-mediated	Thymus	Destroys target cells upon contact
Delayed hypersensitivity T cell (Td)	Cell-mediated	Thymus	Provides protection against infectious agents; causes inflammation associated with tissue transplant rejection
Helper T cell (Th)	Cell-mediated	Thymus	Necessary for B cell activation by T-dependent antigens
Suppressor T cell (Ts)	Cell-mediated	Thymus	Regulates immune response and helps maintain tolerance

cells. The functions of another regulatory group, the **suppressor T cells (Ts),** are not completely understood. It is generally accepted that they inhibit the conversion of B cells into plasma cells, which might be a way of turning off the immune response when an antigen is no longer present. Ts cells might also interact with plasma cells directly. They apparently also function in the maintenance of immune tolerance (see Chapter 17), so that immune responses are not developed against host antigens.

Table 16–3 compares B and T cells and summarizes the features of several types of T cells.

Lymphokines

Lymphokines are proteins released by Td cells (and possibly other T cells) that have been stimulated by exposure to antigens at the site of an infection. One lymphokine produced by Td cells is *macrophage chemotactic factor,* which attracts macrophages to the infection site. Others are *macrophage migration inhibition factor,* which inhibits the movement of macrophages away from the infection site, and *macrophage activation factor,* which activates macrophages. Activated macrophages are much more efficient at destroying the cellular antigens they ingest.

Lymphokines are probably not involved in the contact destruction of target cells by Tc cells, although a lymphokine called *lymphotoxin* can destroy nonlymphocytic target cells in vitro (in artificial conditions outside the body).

The functions of these lymphokines are summarized in Table 16–4.

SEROLOGY

Thus far we have discussed antigen–antibody reactions that occur within a host (in vivo). But such reactions can also occur outside the body, under controlled laboratory conditions, and can provide information that is useful in the diagnosis of dis-

Table 16–4 Summary of Some T-Cell Lymphokines	
Lymphokine	Function
Macrophage chemotactic factor	Attracts macrophages to infection site
Macrophage migration inhibition factor	Prevents macrophages from leaving infection site
Macrophage activation factor	Activates macrophages to improve phagocytic activity
Lymphotoxin	Destroys nonlymphocytic target cells in vitro
Interferon	Inhibits viral replication
Transfer factor	Intensifies the effect of sensitized T cells

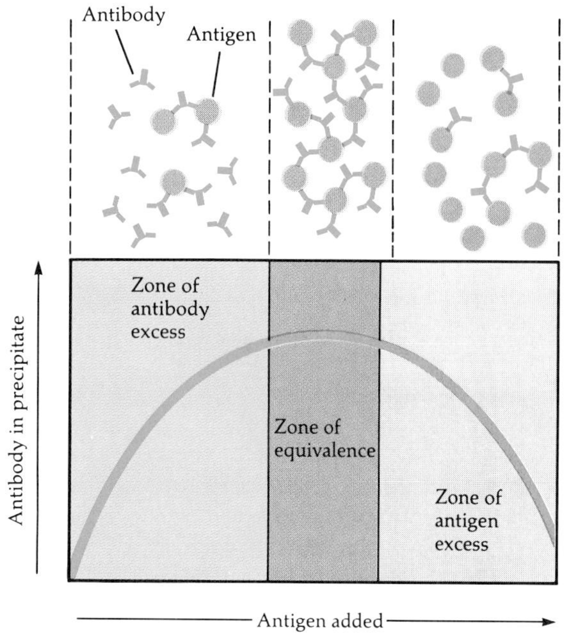

Figure 16–10 Precipitation curve based on the ratio of antigen to antibody. The maximum amount of precipitate forms in the zone of equivalence, in which the ratio is optimal.

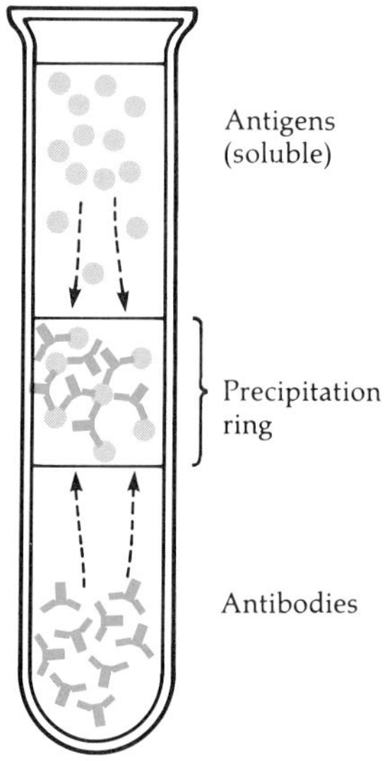

Figure 16–11 Diagrammatic representation of a precipitin ring test. Antigens and antibodies diffuse toward each other in the tube (which usually has a very small diameter) and form a visible line or ring where the zone of equivalence is reached.

ease. The branch of immunology concerned with the study of antigen–antibody reactions in vitro is called **serology.** Serological tests are extremely valuable because they can be performed quickly and offer a high degree of specificity and sensitivity. In this section, we will consider certain tests commonly employed in the diagnosis of disease, examining such immunologic reactions as precipitation, agglutination, complement fixation, and neutralization.

Precipitation Reactions

Precipitation reactions involve soluble antigens. When these react with IgG or IgM antibodies, they form large, interlocking aggregates called *lattices.* The antibodies involved in the formation of precipitates are called **precipitins,** because the lattices they produce precipitate from solution.

Precipitation reactions occur in two distinct stages. First there is the rapid interaction between antigen and antibody to form small antigen–antibody complexes. This interaction occurs within seconds and is followed by a slower reaction, from minutes to hours, in which the antigen–antibody complexes form lattices that precipitate from solution. Precipitation reactions normally occur only when there is an optimal ratio of antigen to antibody. Figure 16–10 shows that no visible precipitate forms when there is an excess of either one. The optimal ratio is produced when separate solutions of antigen and antibody are placed adjacent to each other and allowed to diffuse together. In a **precipitin ring test** (Figure 16–11), a cloudy line of precipitation (ring) appears in the area in which the optimal ratio has been reached (the *zone of equivalence*).

Immunodiffusion tests are precipitation reactions carried out in an agar gel medium. In one such test, the **Ouchterlony test,** wells are cut into a purified agar gel in a Petri plate. A serum containing antibodies is added to one well, usually centrally located, and soluble test antigens are added to each surrounding well. A line of visible precip-

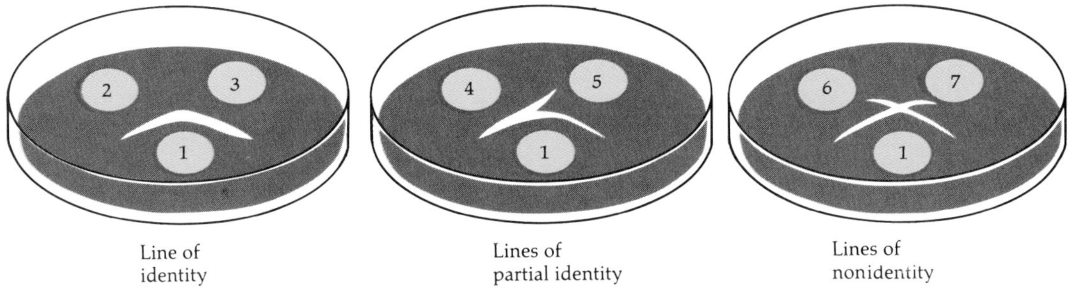

Figure 16–12 Immunodiffusion (Ouchterlony) precipitation test. Well 1 contains antibodies to antigens present in wells 2–7. Wells 2 and 3 contain identical antigens (there is no crossing of the lines); wells 4 and 5 contain closely related antigens (one line crosses, forming a spur); and wells 6 and 7 contain unrelated antigens (both lines cross to form spurs).

itate develops between the wells at the point where the optimal antigen–antibody ratio is reached. The Ouchterlony test can show whether the serum sample contains antibodies against more than one test antigen at a time, and may also indicate whether the antigens are identical, partially identical, or totally different (Figure 16–12).

Other precipitation tests do not depend on the passive diffusion of antigen and antibody in a gel, but use an electric current to speed up their movement. This process is called **electrophoresis.** It is a method by which protein mixtures can be separated rapidly, sometimes in less than an hour. A modification of the precipitation reaction combines the techniques of immunodiffusion with electrophoresis in a procedure called **immunoelectrophoresis.** This procedure is used in research in the separation of proteins in human serum, and is the basis of certain diagnostic tests. The *countercurrent immunoelectrophoresis test* (also called counterimmunoelectrophoresis and abbreviated CIE) can be used in the diagnosis of bacterial meningitis and other diseases. A test serum known to contain certain antibodies is used to identify an unknown antigen in some body fluid. CIE is based on the fact that antigens and antibodies have opposite charges when they are placed at opposite poles in buffers of correct ionic strength and pH (Figure 16–13). When an electrical current is applied, the antigens and antibodies move toward the opposite pole. They pass through each other to do so. A precipitation line appears within an hour if a reaction occurs.

Agglutination Reactions

Precipitation reactions involve soluble antigens; **agglutination reactions** involve particulate antigens. These can be linked together by antibodies to form visible aggregates, a process termed agglutination (Figure 16–14). Antibodies that cause agglutination reactions are called **agglutinins.**

Agglutination reactions have a variety of clinical applications in the diagnosis of disease. For

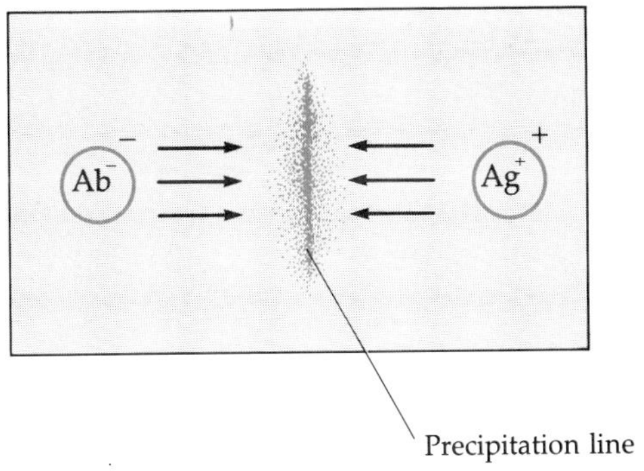

Figure 16–13 Countercurrent immunoelectrophoresis. Antigens and antibodies are placed in opposite wells, and the pH of the surrounding medium is adjusted so that the antigen and antibody have opposite charges. When an electrical current is applied, antigens and antibodies move in opposite directions. A line of precipitate forms where antigen and antibody meet.

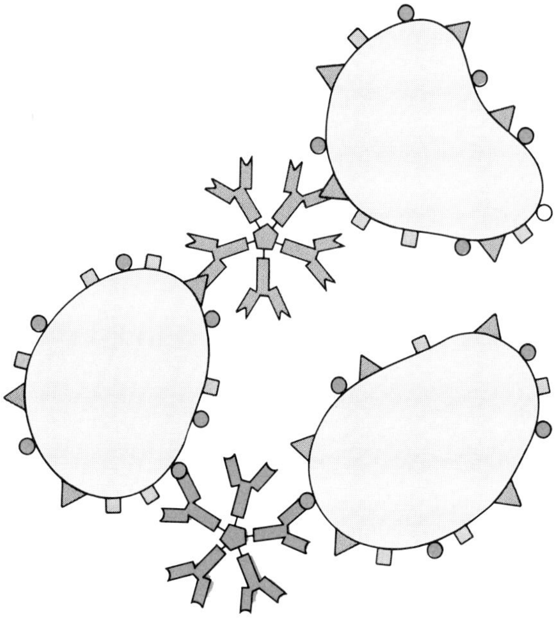

Figure 16–14 Agglutination reaction. When antibodies react with antigenic determinant sites on neighboring antigens such as bacteria or red blood cells, the particulate antigens agglutinate into visible clumps. IgM, the most efficient immunoglobulin for agglutination, is shown here, but IgG can also participate in agglutination reactions.

example, in the **Widal test,** serum from a person with typhoid fever contains antibodies that agglutinate cells of *Salmonella typhi.*

Titer determination

For infectious diseases in general, the higher the serum antibody titer, the greater the immunity to the disease. The titer alone is of limited use in the diagnosis of an existing illness. There is no way of knowing if the measured antibodies were generated in response to the immediate situation or to an earlier illness. For diagnostic purposes a *rise in titer* is significant; that is, the titer is higher later in the disease than at its onset. If it is possible to demonstrate that the person's blood had no antibody titer before the illness but has a significant titer while the disease is progressing, this change is also diagnostic.

The **tube agglutination test,** shown in Figure 16–15, is a simple means by which antibody titer can be determined. For this procedure, a suspension of cellular antigens such as bacterial cells is added to a series of tubes or shallow wells in a plastic plate. The tubes or wells contain serial dilutions of serum (1:5, 1:10, 1:20, 1:40, and so on). The greatest dilution of serum showing an agglutination reaction is determined, and the reciprocal of this dilution is the titer of the original serum. For

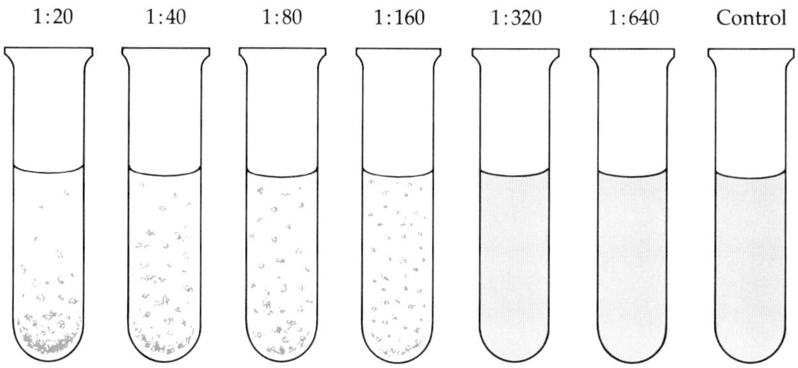

Figure 16–15 Tube agglutination test for determining antibody titer. Here the titer is 160, because there is no agglutination in the next tube in the dilution series (1:320). In practice, these tests are usually performed in microtitration plates containing many small wells.

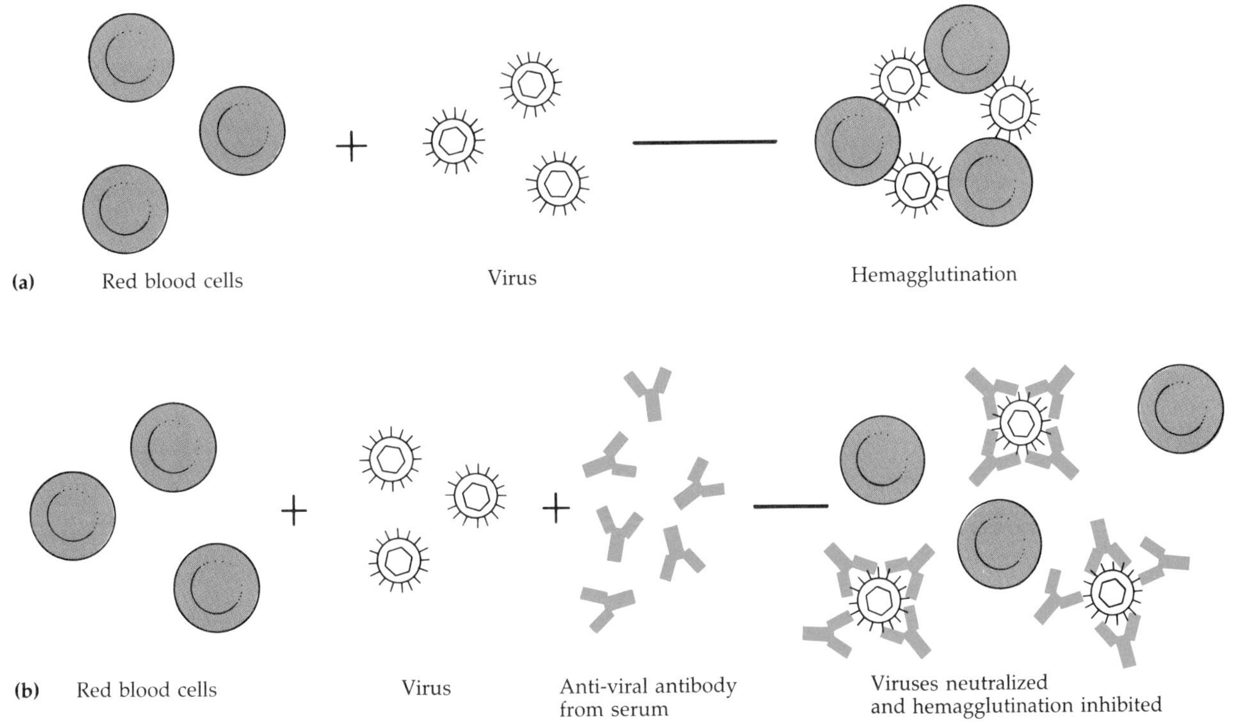

Figure 16–16 Viral hemagglutination. **(a)** Viral hemagglutination is not an antigen–antibody reaction. It simply happens that certain viruses bind to red blood cells and cause their agglutination. **(b)** If serum containing antibodies to the virus is mixed with the red blood cells, the antibodies will neutralize the virus and inhibit hemagglutination.

example, if there is visible agglutination in all tubes through 1:160, but none in dilution 1:320, then the titer is 160.

Hemagglutination

When agglutination reactions involve the clumping of red blood cells, the reaction is termed **hemagglutination.** These reactions are used routinely in blood typing (Chapter 17), and in the diagnosis of infectious mononucleosis.

Certain viruses, such as those causing mumps, measles, and influenza, have the ability to agglutinate red blood cells; this process is called **viral hemagglutination.** If a person's serum contains antibodies against such viruses, these antibodies will react with the viruses and neutralize them (Figure 16–16). For example, if hemagglutination occurs in a mixture of measles virus and red blood cells but does not occur when the patient's serum is added to the mixture, this sequence indicates that the serum contains antibodies that have neutralized the measles virus. This **hemagglutination–inhibition test** is widely used in the diagnosis of influenza, measles, mumps, and a number of other viral infections.

Latex agglutination test for soluble antigens

Soluble antigens can also be detected by an indirect agglutination test. The soluble antigens are attached to latex spheres. Antibodies against the

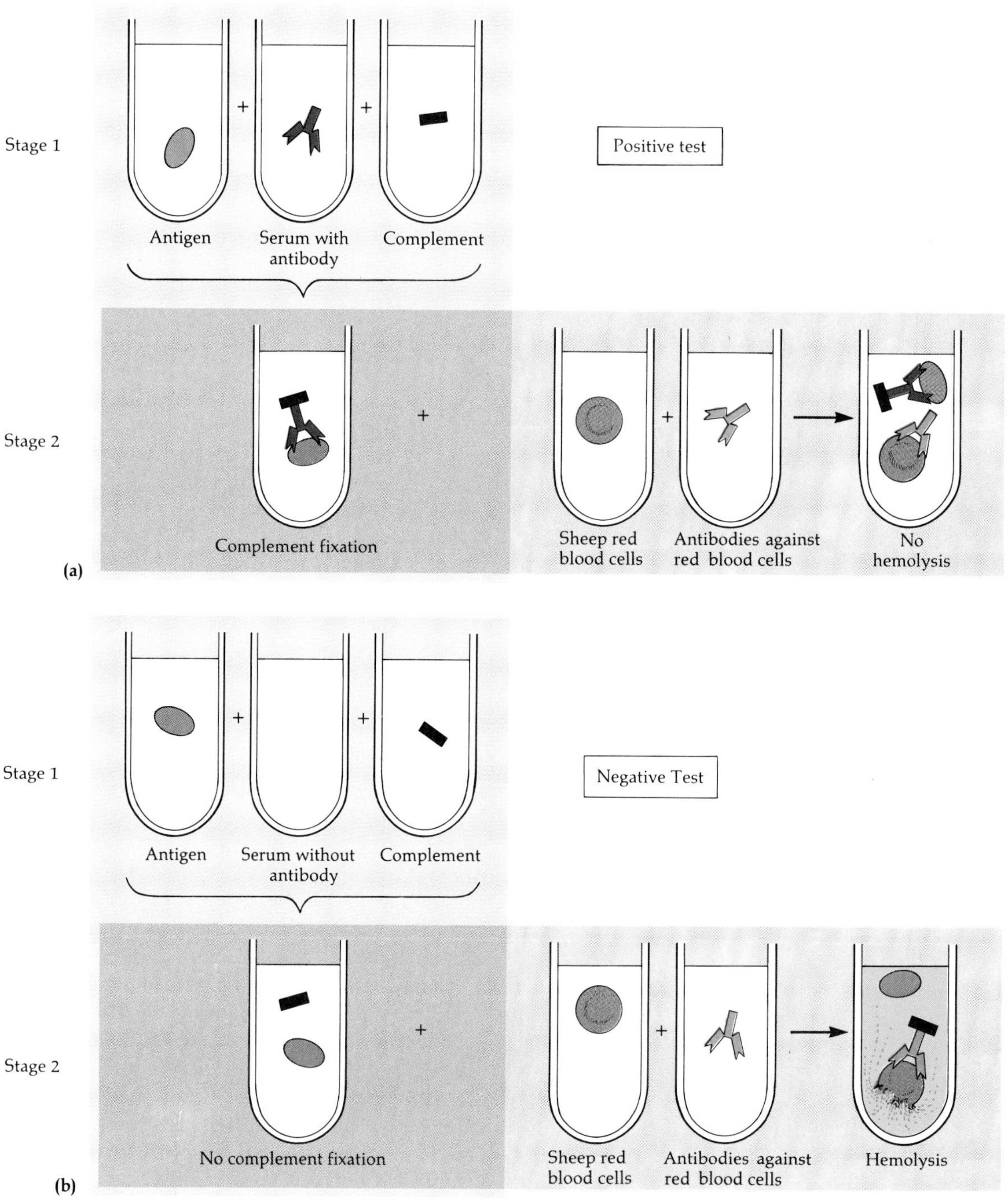

Figure 16–17 Complement fixation test. This test is based on the fact that if complement has been fixed by reacting with antigens and antibodies, it will be used up and there will be no lysis among the indicator red blood cells even though specific antibody against these cells is present. **(a)** Positive test (no hemolysis). **(b)** Negative test (hemolysis).

soluble antigens cause a readily visible agglutination of the latex particles. This procedure is called the **latex agglutination test,** and is used to detect antibodies that develop during certain mycotic and helminthic infections. It is also the basis for some commercial pregnancy testing kits. Recently, several tests using latex agglutination techniques to rapidly identify streptococci have been developed.

Complement Fixation Reactions

In Chapter 15 we discussed a group of serum proteins collectively called complement. During most antigen–antibody reactions, the complement binds to the antigen–antibody complex and is used up, or "fixed." This process of **complement fixation** can be used to detect even very small amounts of antibody. Antibodies that do not produce a visible reaction such as precipitation or agglutination can be demonstrated by the fixing of complement during the antigen–antibody reaction. Complement fixation was once widely used in the diagnosis of syphilis (Wasserman test), and is still used in the diagnosis of certain viral, fungal, rickettsial, and chlamydial diseases.

Execution of the complement fixation test requires great care and good controls. The test is performed in two stages (Figure 16–17). First, the subject's blood serum must be heated at 56°C for 30 minutes to inactivate its complement. The inactivated serum is then diluted and mixed with a specific amount of known antigen and fresh complement. The mixture is then incubated for about 30 minutes. An antigen–antibody reaction cannot be observed at this point. To determine whether the complement present in the mixture is fixed by an antigen–antibody reaction (evidence of the presence of antibodies specific for the antigen used), another stage of the test must be performed.

The second stage uses an indicator system to determine whether complement is free or combined (fixed). The indicator system consists of sheep red blood cells that have specific antibodies attached to their surfaces. The exposure of these "sensitized" cells to complement causes lysis of the red blood cells, and the color of the mixture changes. Therefore, if complement has been fixed by an antigen–antibody reaction during the first stage, it is not available to cause blood cell lysis in the second stage; the test is positive (Figure 16–17a). But if the complement has not been fixed during the first stage, then it is available to cause blood cell lysis in the second stage (Figure 16–17b). This negative test result means that no antibodies specific for the test antigen are present in the patient's serum.

Neutralization Reactions

Toxin neutralization

Neutralization reactions are antigen–antibody reactions in which the harmful effects of a bacterial exotoxin or a virus are eliminated by specific antibodies. These reactions were first described in 1890, when investigators observed that immune serum could neutralize the toxic substances produced by the diphtheria bacillus. An **antitoxin** is a specific antibody produced by a host responding to a bacterial exotoxin or its corresponding toxoid (inactivated toxin) that combines with the exotoxin to neutralize it (Figure 16–18a). Antitoxins produced in an animal can be used in humans to provide passive immunity against a toxin. Antitoxins from horses are routinely used for prevention or treatment of diphtheria and botulism; tetanus antitoxin is usually of human origin.

In vitro neutralization tests are not common in modern clinical laboratories. More frequent are skin tests such as the **Schick test,** which determines the status of a person's immunity to diphtheria. A small amount of diphtheria exotoxin is inoculated into the skin, and if there is sufficient serum antitoxin to neutralize the exotoxin, there is no visible reaction. If antitoxin is insufficient, the exotoxin damages the tissues at the site of entry and produces a swollen, tender, reddish area that turns brown in four to five days. This reaction indicates an unsatisfactory immune response to diphtheria.

Viral neutralization

Neutralization tests are frequently used in the diagnosis of viral infections. The body responds to such infections by producing specific antibodies that bind to receptor sites on the viral surface. The binding of these antibodies prevents the virus from attaching to a host cell, and thus destroys the virus's infectivity (Figure 16–18b). Viruses that exhibit their cytopathic (cell-damaging) effects in tis-

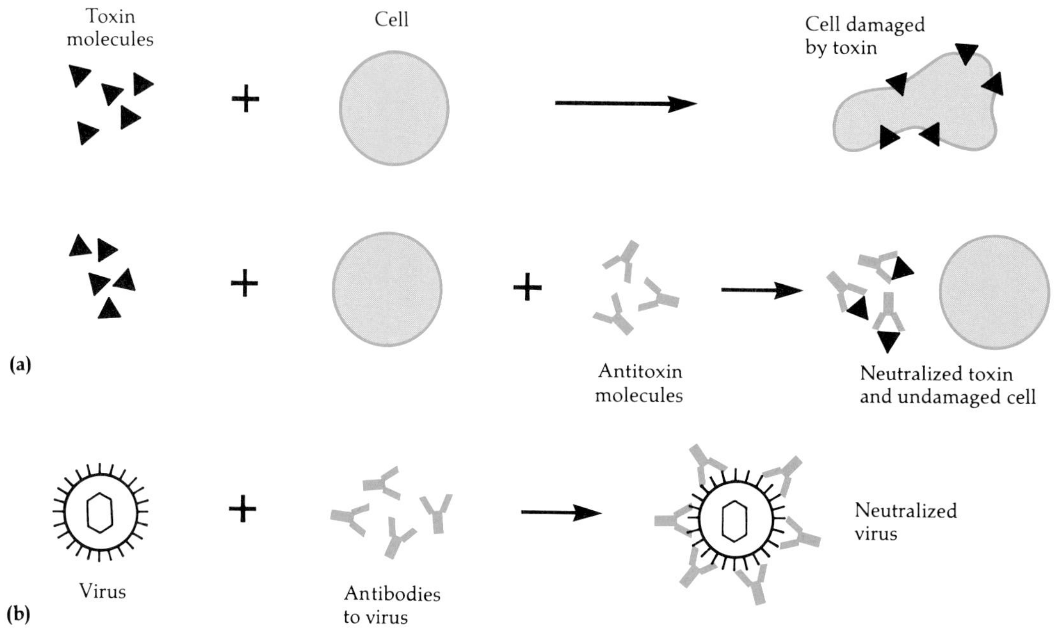

Figure 16–18 Neutralization reactions. **(a)** Effects of a toxin on a susceptible cell and neutralization of the toxin by antitoxin. **(b)** A specific antibody neutralizes a virus (prevents the virus from infecting a cell).

sue culture or embryonic eggs can be used to detect the presence of neutralizing viral antibodies. If the blood serum to be tested contains antibodies against the particular virus, the antibodies will prevent that virus from infecting cells in the tissue culture or eggs, and no cytopathic effects will be seen. The inability of a specific virus to cause cytopathic effects in the presence of immune serum can thus be used to determine the identity of a virus as well as the viral antibody titer.

Immunofluorescence and Fluorescent Antibody Techniques

Fluorescent antibody (F.A.) techniques (Figure 16–19 and Microview 3–4) can identify microorganisms in clinical specimens and detect the presence of a specific antibody in serum. These techniques utilize fluorescent dyes such as *fluorescein isothiocyanate* (FITC), which are combined with antibodies to make them fluoresce when exposed to ultraviolet light. These procedures are quick, sensitive, and very specific; the fluorescent antibody test for rabies can be done in a few hours and has an accuracy rate close to 100%.

Fluorescent antibody tests are of two types, direct and indirect. **Direct F.A. tests** are usually used to identify a microorganism in a clinical specimen. During this procedure, the specimen containing the antigen to be identified is fixed onto a slide. Fluorescein-labeled antibodies are then added and the slide is incubated briefly. The slide is washed to remove any antibody not bound to antigen, and then examined under the ultraviolet (UV) microscope for yellow-green fluorescence (Figure 16–19a).

Indirect F.A. tests (Figure 16–19b) are used to detect the presence of a specific antibody in serum following exposure to a microorganism. During this procedure, a known antigen is fixed onto a slide. The test serum is then added, and if antibody specific to that microbe is present, it reacts with the antigen to form a bound complex. So that the antigen–antibody complex can be seen, fluorescein-labeled anti–human gamma globulin (anti–HGG), an antibody that reacts specifically with human antibody, is added to the slide. After the slide has been incubated and washed (to remove unbound antibody), it is examined under a UV microscope. If the known antigen fixed to the slide appears fluorescent, the antibody specific to the test antigen is present.

Radioimmunoassay

Radioimmunoassay techniques aid in the detection of several compounds, including hormones, drugs, and immunoglobulins. Suppose, for example, we want to determine the amount of insulin in a serum sample. The serum sample is combined with radioactively labeled insulin and with antibodies to insulin. The radioactive insulin and the insulin in the sample will compete for antibody. Antigen–antibody complexes can then be isolated and analyzed for their radioactivity. If the antigen–antibody complexes have a high level of radioactive antigen, the tissue sample did not have much antigen (insulin) to combine with the antibody (Figure 16–20).

Enzyme-Linked Immunosorbent Assay (ELISA)

The **enzyme-linked immunosorbent assay (ELISA)** has become a widely used serological technique. There are two basic methods: the **double**

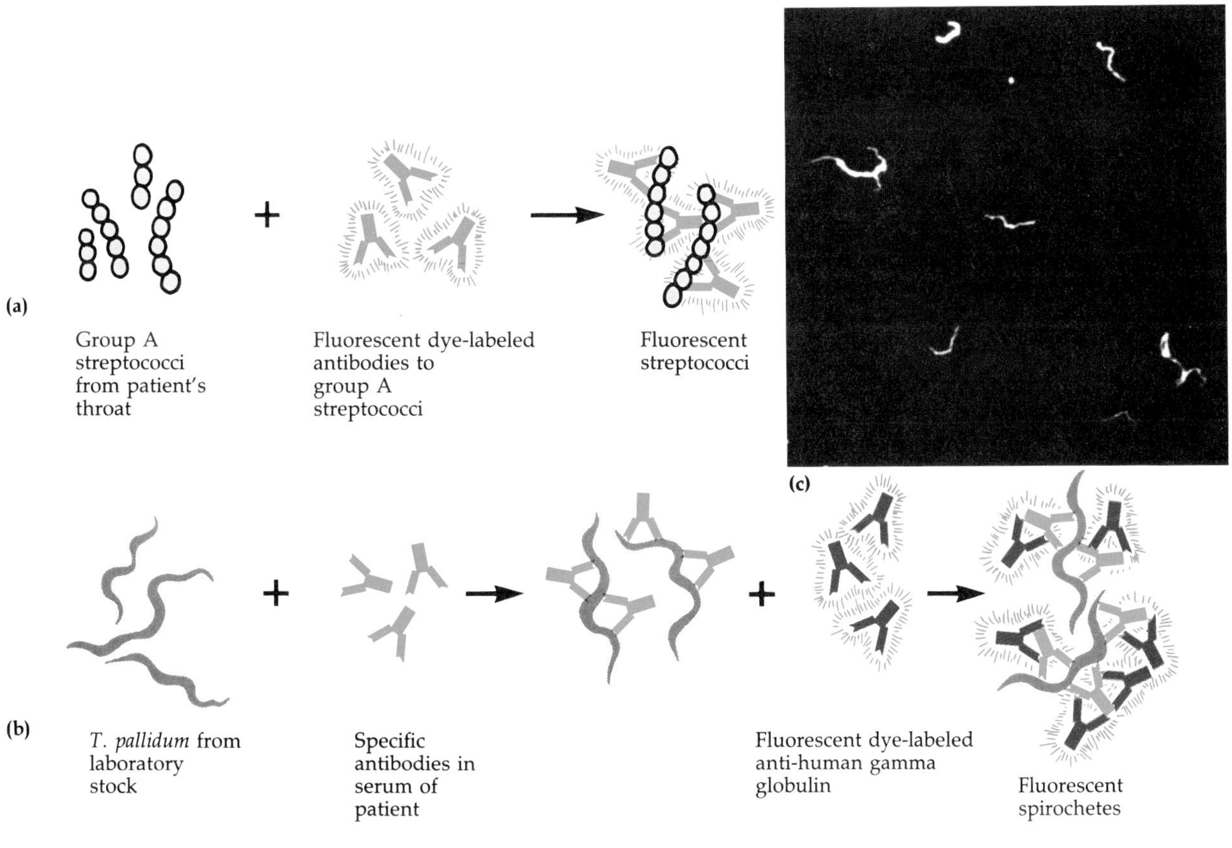

Figure 16–19 Fluorescent antibody techniques. **(a)** A direct F.A. technique identifies group A streptococci. **(b)** An indirect F.A. technique used in the diagnosis of syphilis. The fluorescent dye is attached to anti–human gamma globulin, which reacts with any human immunoglobulin that has previously reacted with antigen. **(c)** Fluorescent cells of *Treponema pallidum* as they appear in a positive indirect F.A. test for syphilis.

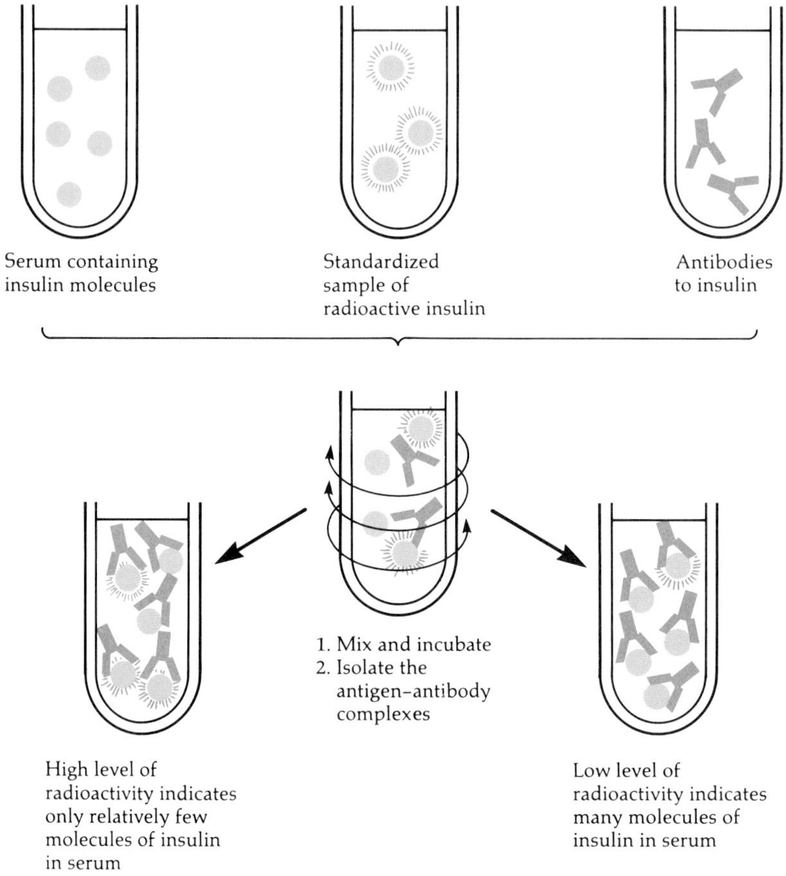
Serum containing insulin molecules
Standardized sample of radioactive insulin
Antibodies to insulin
1. Mix and incubate
2. Isolate the antigen–antibody complexes
High level of radioactivity indicates only relatively few molecules of insulin in serum
Low level of radioactivity indicates many molecules of insulin in serum

Figure 16–20 Above. Radioimmunoassay techniques determine the concentration of material in a clinical sample. Here the procedure is used to detect insulin.

antibody sandwich technique for the detection of antigens, and the enzyme-linked **indirect immunosorbent assay** for the detection of antibodies. A microtiter plate with numerous shallow wells is used in both procedures (see Microview 3–4).

Double antibody sandwich method

The first step of this method (Figure 16–21a) is that antiserum is adsorbed to the surface of the wells on the microtiter plate. A test antigen is then added to each well. If the antigen reacts specifically with the antibodies adsorbed to the well, the antigen will be retained there when the well is washed free of un-

Figure 16–21 Facing page. The ELISA test. **(a)** In the first step of the double-antibody sandwich method for detecting antigens, known antiserum is adsorbed to the wall of the well, and the test antigen is added. If the test antigen is specific for the antibody on the wall, they will bind together. Unbound antigen is rinsed away. An antibody specific for the antigen is then allowed to react with the antigen (assuming that it has bound to antibody adsorbed to the well), and an antibody-antigen-antibody sandwich is formed. The last antibody is linked to an enzyme. When the enzyme's substrate is added there is a visible color reaction, indicating that the antigen is bound to the antibody on the wall of the well. **(b)** In the first step of the indirect ELISA test for antibodies, a known antigen is adsorbed to the wall of the well. Serum under study for the presence of a specific antibody is added. Then anti–human gamma globulin, linked with an enzyme, is allowed to react with the antibodies that have reacted with antigens in the well. When the enzyme's substrate is added, there is a color reaction. This reaction will occur only if the test antiserum contains antibodies specific for the antigen adsorbed to the walls of the wells.

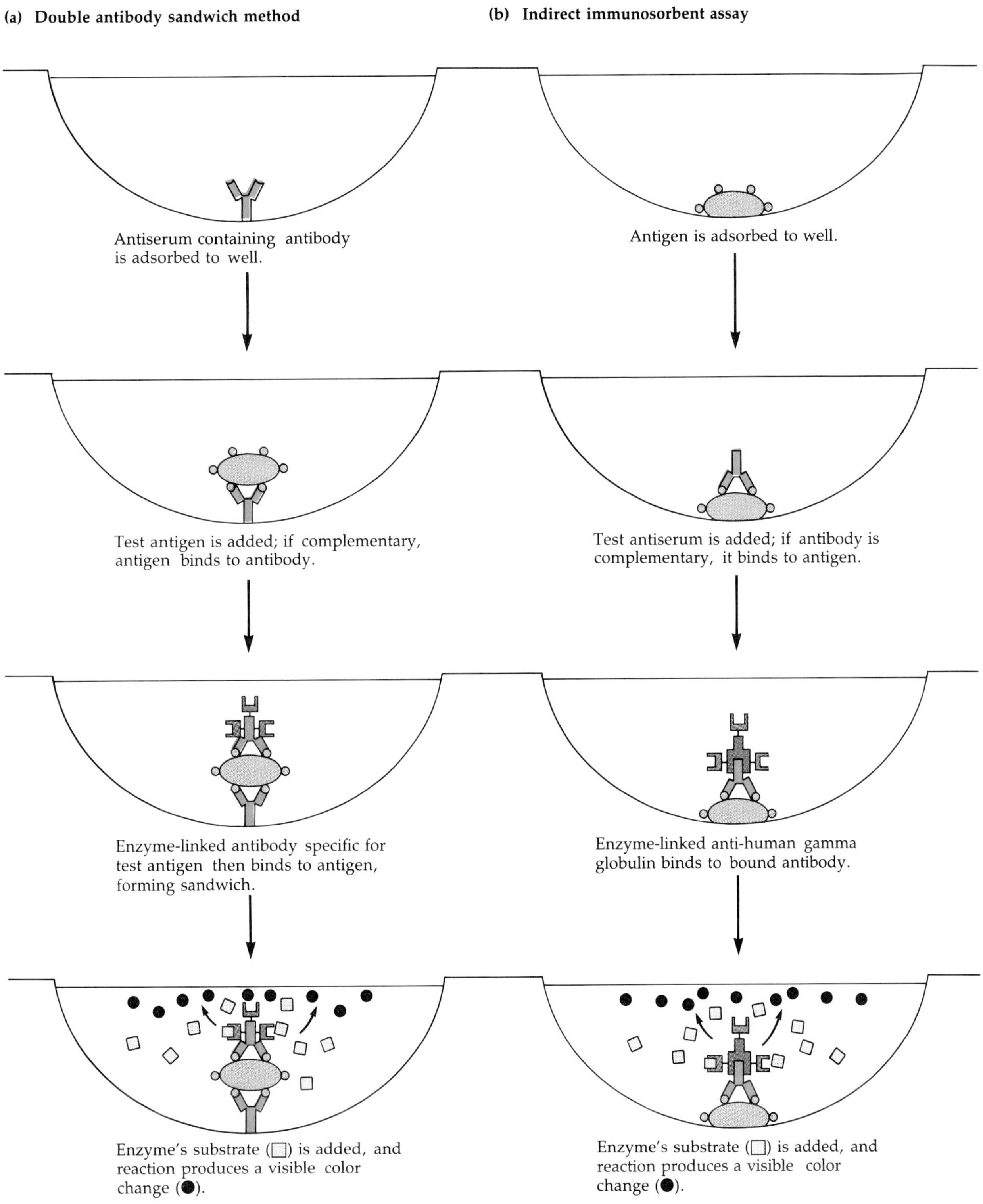
(a) Double antibody sandwich method
Antiserum containing antibody is adsorbed to well.
Test antigen is added; if complementary, antigen binds to antibody.
Enzyme-linked antibody specific for test antigen then binds to antigen, forming sandwich.
Enzyme's substrate (□) is added, and reaction produces a visible color change (●).
(b) Indirect immunosorbent assay
Antigen is adsorbed to well.
Test antiserum is added; if antibody is complementary, it binds to antigen.
Enzyme-linked anti-human gamma globulin binds to bound antibody.
Enzyme's substrate (□) is added, and reaction produces a visible color change (●).

bound antigen. An antibody specific for the antigen is then added. If both the antibody adsorbed to the wall of the well and the antibody known to be specific for the antigen have reacted with the antigen, a "sandwich" with the antigen between two antibody molecules will have formed.

This reaction is visible only because the second added antibody is linked to an enzyme such as horseradish peroxidase or alkaline phosphatase. Unbound enzyme-linked antibody is washed from the well, and then the enzyme's substrate is added to it. Enzymic activity is indicated by a color change that can be seen by eye or read with a special spectrophotometer. The test will be positive if the antigen has reacted with adsorbed antibodies in the first step. If the test antigen was not specific for the antibody adsorbed to the wall of the well, the test will be negative because the unbound antigen will have been washed away.

Indirect immunosorbent assay

In the first step of this method, the antigen, rather than the antibody, is adsorbed to the walls of the shallow wells on the plate (Figure 16–21b). To see whether a serum sample contains antibodies against this antigen, the antiserum is added to the well. If the serum contains antibody specific to the antigen, the antibody will bind to the adsorbed antigen. All unreacted antiserum is washed from the well. Anti-HGG (an antibody that reacts with any human immunoglobulins) is then allowed to react with the antigen–antibody complex. The anti-HGG, which has been linked with an enzyme, reacts with the antibodies that are bound to the antigens in the well. Finally, all unbound, enzyme-linked anti-HGG is rinsed away and the correct substrate for the enzyme is added. A colored enzymatic reaction occurs in the wells in which the bound antigen has reacted with antibody present in the serum sample. This procedure resembles that of the indirect fluorescent antibody test, except that the anti-HGG of the indirect assay is linked with an enzyme rather than a fluorescing dye. This assay is used to test for AIDS.

VACCINATIONS AND PUBLIC HEALTH

The study of immunity has had practical applications beyond the development of the diagnostic laboratory tests just described. The application with the greatest impact on human health has been the development of vaccines. Many pathogens transmitted by food or water are controlled by sanitation, or by antibiotic treatment if prevention fails. However, these methods of control are not possible for all pathogens. Viral diseases are not usually susceptible to treatment once contracted, and transmission of pathogens by air or contact is not easily prevented. In these cases, active immunization may be the only feasible method of control.

Vaccines

A **vaccine** is a suspension of microorganisms or viruses (or some part or product of them) that will produce immunity when it is injected into a host. These microorganisms or viruses may be either inactivated or attenuated. In the latter case, they are still living but are so weakened or altered that they are no longer virulent; yet they will still provoke an immune response. Inactivated bacterial toxins (toxoids) will also induce immunity against their active forms.

Live, attenuated virus vaccines tend to mimic an actual infection and usually provide better immunity than that provided by inactivated viruses. Examples of live vaccines are the Sabin polio vaccines and those used against yellow fever, measles, rubella, and mumps. Most attenuated virus vaccines provide lifelong immunity without booster immunizations, and an effectiveness of 95% is not unusual. This long-term effectiveness probably occurs because the attenuated viruses tend to replicate in the body, and the original dose thereby increases considerably over time. One danger of such vaccines is that the live viruses can mutate to a virulent form, though this very rarely happens.

Viruses for vaccines may be inactivated by treatment with formalin or other chemicals. Heat is not used for this treatment, because it is likely to alter the surface components of the virus and thus interfere with its ability to provoke an effective immune response. Inactivated virus vaccines include those used against rabies (animals sometimes receive a live vaccine considered too hazardous for humans), influenza, polio (the Salk vaccine), and hepatitis B.

Not all vaccines have strong, long-term effects. Some vaccines, especially those containing capsular polysaccharides as the antigens, must be readministered every few years. These are T-independent antigens and memory development against them is poor; there is almost no anamnestic response upon secondary administration. Examples of such vaccines are those for pneumococcal pneumonia and meningococcal meningitis.

A potentially undesirable effect of vaccines is the *interference effect;* that is, one viral vaccine can interfere with the development of effective immunity against an unrelated live vaccine given later. The interference effect is most likely to occur with live viral vaccines. However, there has been no problem with the simultaneous administration of multiple vaccines such as the MMR vaccine for measles, mumps, and rubella.

Harmful side effects are an occasional problem with vaccines. The mass-immunization program of swine-flu vaccine during 1976–77 appears to have been associated with a higher rate of a paralytic disease called the Guillain–Barré syndrome, and there has been concern about the incidence of neurological side effects caused by the pertussis vaccine. Future vaccines will probably contain only the specific portions of the pathogens that produce an immune response. These products of genetic engineering are expected to minimize the current problems.

Vaccines effective against bacteria (including rickettsiae and mycoplasmas) and viruses have been produced, but to date no useful vaccines against chlamydia, fungi, protozoans, or helminthic parasites in humans have been developed. Vaccines against protozoan and helminthic parasites in domestic animals have been developed. However, the application of genetic engineering principles to the problem might help improve this situation. There is special interest in the development of vaccines effective against the protozoan that causes malaria, and against the trypanosome that causes African sleeping sickness (see Box in Chapter 11).

The principal vaccines used to prevent bacterial and viral diseases in the United States are listed in Tables 16–5 and 16–6. Recommendations for

Table 16–5 Principal Vaccines Used to Prevent Bacterial Diseases in Humans

Disease	Vaccine	Recommendation	Booster
Cholera	Crude fraction of *Vibrio cholerae*	For persons who work and live in endemic areas	Every 6 months as needed
Diphtheria	Purified diphtheria toxoid	See Table 16–7	Every 10 years for adults
Meningococcal meningitis	Purified polysaccharide from *Neisseria meningitidis*	For persons with substantial risk of infection	No booster effect with additional doses
Pertussis (whooping cough)	Killed *Bordetella pertussis*	Children prior to school age; see Table 16–7	For high-risk adults
Plague	Crude fraction of *Yersinia pestis*	For persons who come in regular contact with wild rodents in endemic areas	Every 6–12 months as needed
Pneumococcal pneumonia	Purified polysaccharide from *Streptococcus pneumoniae*	For adults over 50 years with chronic systemic diseases	No booster effect with additional doses
Tetanus	Purified tetanus toxoid	See Table 16–7	Every 10 years for adults
Tuberculosis	*Mycobacterium bovis* BCG	For persons who are tuberculin negative and who are exposed for prolonged periods of time to tuberculosis	Every 3–4 years as needed
Typhoid fever and paratyphoid fever	Killed *Salmonella typhi, S. schottmulleri,* and *S. paratyphi*	For persons in endemic areas or areas having outbreak	Every 3 years as needed
Typhus fever	Killed *Rickettsia prowazekii*	For scientists and medical personnel in rural areas endemic for typhus	Every 6–12 months as needed

Table 16–6 Principal Vaccines Used in Prevention of Viral Diseases in Humans

Disease	Vaccine	Recommendation	Booster
Influenza	Inactivated virus	For chronically ill persons, especially with respiratory diseases and over 65 years old	Annual
Measles	Attenuated virus	For infants 15–19 months old	*
Mumps	Attenuated virus	For infants 12–19 months old	*
Poliomyelitis	Attenuated or inactivated virus	For children, see Table 16–7; for adults as risk to exposure warrants (duration unknown)	Adults—as needed
Rabies	Inactivated virus	For field biologists in contact with wildlife in endemic areas; veterinarians	*
Rubella	Attenuated virus	For infants 12–19 months old; for females of childbearing age who are not pregnant	*
Yellow fever	Attenuated virus	For persons traveling to endemic areas; military personnel	Every 10 years

**Note:* The duration of immunity is not known because these vaccines have been in use only a short time.

Table 16-7 Recommended Schedule for Active Immunization and Skin Testing of Children

Recommended Age	Immunizing Agent
2–3 months	DPT vaccine (diphtheria toxoid, pertussis vaccine, and tetanus toxoid) OPV (oral poliomyelitis vaccine), trivalent preparation
4–5 months	DPT vaccine OPV, trivalent preparation
6–7 months	DPT vaccine OPV, trivalent preparation
15–19 months	DPT vaccine OPV, trivalent preparation Mumps vaccine, Measles vaccine, Rubella vaccine } or combined MMR vaccine
4–6 years	DPT vaccine OPV, trivalent preparation Tuberculin skin test (see Chapter 17)
12–14 years	TD vaccine (tetanus and diphtheria toxoid) Tuberculin skin test

their administrations are given in Table 16–7. Travelers who might be exposed to cholera, yellow fever, or other diseases not endemic in this country will find that current inoculation recommendations are available from the U.S. Public Health Service, and should also be available through local public-health agencies.

STUDY OUTLINE

INTRODUCTION (p. 438)

1. An individual's (genetically) predetermined resistance to certain diseases is called innate resistance.
2. Individual immunity is affected by sex, age, nutritional status, and general health.
3. An individual may develop or acquire immunity after birth.
4. Immunity is the ability of the body to specifically counteract foreign organisms or substances.

THE DUALITY OF THE IMMUNE SYSTEM (p. 438)

1. The humoral immune system involves antibodies that are found in blood plasma and lymph.
2. Antibodies are produced by B cells in response to a specific antigen.
3. Antibodies defend against bacteria, viruses, and toxins in body fluids primarily.
4. The cell-mediated immune system depends on T cells and does not involve antibody production.
5. T lymphocytes are coated with antibodylike molecules.
6. Cellular immunity is primarily a response to intracellular viruses, multicellular parasites, transplanted tissue, and cancer cells.

ACQUIRED IMMUNITY (pp. 438–440)

1. Acquired immunity is specific resistance to infection obtained during the life of the individual.
2. Acquired immunity results from the production of humoral antibodies and cellular immune response.
3. Immunity resulting from infection is called naturally acquired active immunity, an immunity that may be long lasting.
4. Humoral antibodies transferred from a mother to a fetus (placental transfer) or to a newborn in colostrum results in naturally acquired passive immunity in the newborn, an immunity that can last up to a few months.
5. Immunity resulting from vaccination is called artificially acquired active immunity, and can be long lasting.
6. Vaccines can be prepared from attenuated, inactivated, or killed microorganisms and toxoids.
7. Artificially acquired passive immunity refers to humoral antibodies acquired by injection, and can last for a few weeks.
8. Humoral antibodies made by a human or other mammal may be injected into a susceptible individual.
9. Serum containing antibodies is often referred to as antiserum.
10. When serum is separated by electrophoresis, antibodies are found in the gamma fraction of the serum, and because antibodies are globulins, antibodies are referred to as gamma globulins or immune serum globulins.

ANTIGENS (pp. 440–441)

1. An antigen is a chemical substance that causes the body to produce specific antibodies or sensitized T cells that the antigen can then combine with.
2. Most antigens are proteins, nucleoproteins, lipoproteins, glycoproteins, or large polysaccharides.
3. As a rule, antigens are foreign substances; they are not part of the body's chemistry.
4. Generally, antigens have a molecular weight greater than 10,000.
5. Antibodies are formed against specific regions on the surface of an antigen called antigenic determinant groups; the number of these groups is the valence.
6. Most antigens are multivalent.
7. A hapten (incomplete antigen) is a molecule that combines with an antibody but cannot cause the formation of antibodies unless combined with another molecule.

ANTIBODIES (pp. 441–444)

1. An antibody is a protein produced by B lymphocytes in response to the presence of an antigen and is capable of combining specifically with the antigen.
2. A single bivalent antibody unit is a monomer; multivalent antibodies are composed of monomers.
3. Since antibodies are globulins, they are called immunoglobulins (Ig).

FURTHER READING

Arnon, R. 1980. Chemically defined antiviral vaccines. *Annual Review of Microbiology* 34:593–618. Immunologic responses to viral vaccines.

Collier, R. J. and D. A. Kaplan. July 1984. Immunotoxins. *Scientific American* 251:56–64. Describes how monoclonal antibodies might target chemicals to treat cancer.

Dead or Alive? September 1977. *Scientific American* 237:96–100. A discussion of the decline in vaccinations against polio in the United States.

Donelson, J. E. and M. J. Turner. February 1985. How the trypanosome changes its coat. *Scientific American* 252:44–51. This parasite evades the immune response by changing its antigenic surface.

Godson, G. N. May 1985. Molecular approaches to malaria vaccines. *Scientific American* 252:52–59. The various techniques of genetic engineering including transformation and monoclonal antibody production are used to identify the parasite's antigens.

Hood, L. E., I. L. Weissman, W. B. Wood, and J. H. Wilson. 1984. *Immunology*, 2nd ed. Menlo Park, CA: Benjamin/Cummings. A concise textbook emphasizing the essential concepts of immunology.

Langer, W. L. January 1976. Immunization against smallpox before Jenner. *Scientific American* 234:112–117. Interesting history of the use of smallpox itself as a vaccination against smallpox.

Leder, P. May 1982. The genetics of antibody diversity. *Scientific American* 246:102–115. Describes the shuffling of DNA and RNA necessary to produce billions of different antibodies.

Lerner, R. A. February 1983. Synthetic vaccines. *Scientific American* 248:66–74. A synthetic viral protein is used to produce antibodies.

Sabin, A. B. 1981. Evaluation of some currently available and prospective vaccines. *Journal of the American Medical Association* 246:236–241. A summary of the currently available vaccines against bacterial pneumonia, poliomyelitis, measles, rubella, and influenza with a discussion of new vaccines under investigation.

Yelton, D. E. and M. D. Scharff. 1980. Monoclonal antibodies. *American Scientist* 68:510–516. Describes production and uses of monoclonal antibodies.

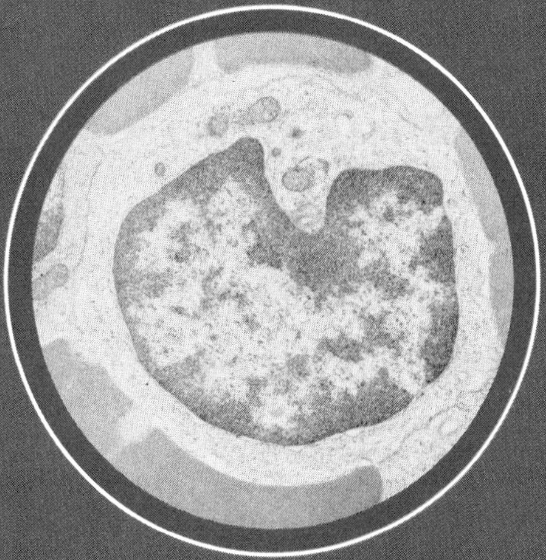

CHAPTER 17

Harmful Aspects of the Immune Response

OBJECTIVES

After completing this chapter you should be able to

- Explain the following terms: hypersensitivity; desensitization; histocompatibility antigens; HLA; immunosuppression; immunologic tolerance.
- Describe the mechanism of anaphylaxis; of contact dermatitis.
- Differentiate between the four types of hypersensitivity reactions.
- Describe the basis of human blood groupings and their relation to blood transfusions and hemolytic disease of the newborn.
- Explain how rejection of a transplant occurs.
- Define and give a possible explanation for autoimmunity; for immune deficiency.
- Discuss one immune-complex disease.
- Describe the immune responses to cancer and immunotherapy.

Not all immune responses against an antigen produce a desirable resistance or immunity. Sometimes the reactions of the immune system are harmful. In this chapter we will discuss several immune responses that do the host more harm than good; we will concentrate on those involving allergy, autoimmunity, immune suppression, and blood group incompatibilities.

HYPERSENSITIVITY

The term **hypersensitivity** implies sensitivity beyond what is considered normal; the term **allergy** is probably more familiar and is essentially synonymous. Hypersensitivity responses occur in people who have been previously "sensitized" by exposure to an antigen, which in this context is sometimes called an **allergen.** Once sensitized, the immune system responds to a subsequent exposure to that antigen by reacting with it in a manner that leads to host tissue damage.

Hypersensitivity reactions are sometimes classified as either *immediate* or *delayed.* While it is true that immediate hypersensitivity reactions appear relatively faster than delayed ones, the main difference between these reactions is in the nature of the immune responses to the antigen.

Hypersensitivity reactions are considered to be of four principal types (Table 17–1). Types I, II, and III involve humoral antibodies; type IV involves effector cells of cell-mediated immunity. Type I (anaphylaxis) reactions are the most common, and are brought about by IgE antibodies bound to certain body cells. Type II (cytotoxic) reactions involve the combination of antibodies with antigens that are normal surface components of an individual's own cells. The action of complement then injures the cells. Type III (immune complex) reactions result when complexes of antibody, antigen, and complement stimulate an inflammatory response that is harmful to body tissues. Type IV (cell-mediated) reactions are brought about by sensitized T cells.

Type I (Anaphylaxis) Reactions

Type I reactions often occur within a few minutes after a person sensitized to an antigen is reexposed to that antigen. **Anaphylaxis** (ana means "against"; phylaxis means "protection") is an inclusive term for the reactions caused by the combining of certain antigens with IgE antibodies. Anaphylactic responses can be divided into *systemic reactions,* which produce shock and breathing difficulties and are sometimes fatal, and *localized reactions,* which include common allergic conditions such as hayfever, asthma, and hives (slightly raised, often itchy and reddened, areas of the skin).

The IgE antibodies produced in response to an antigen such as insect venom or pollen are **cytotrophic;** that is, they bind to the surfaces of mast cells and basophils. This binding is what causes the person to become sensitized, or allergic, to the antigen. *Mast cells* are especially prevalent in the connective tissue of the skin and respiratory tract and in their surrounding blood vessels. *Basophils* are a class of leukocytes thought to be precursors to mast cells; these circulate in the bloodstream.

Table 17–1 Types of Hypersensitivities

Type of Reaction	Immunoglobulin or Effector Cells Involved	Examples
Type I, anaphylaxis	IgE on mast cells or basophils	Anaphylactic shock from drug injections, insect venom; common allergic conditions such as hayfever, asthma
Type II, cytotoxic	IgM, IgG, complement	Transfusion reactions, Rh incompatibility; Graves' disease; myasthenia gravis
Type III, immune complex	IgM, IgG, complement	Systematic lupus erythematosus; rheumatoid arthritis
Type IV, cell-mediated	T cells	Rejection of transplanted tissues; Addison's disease; contact dermatitis

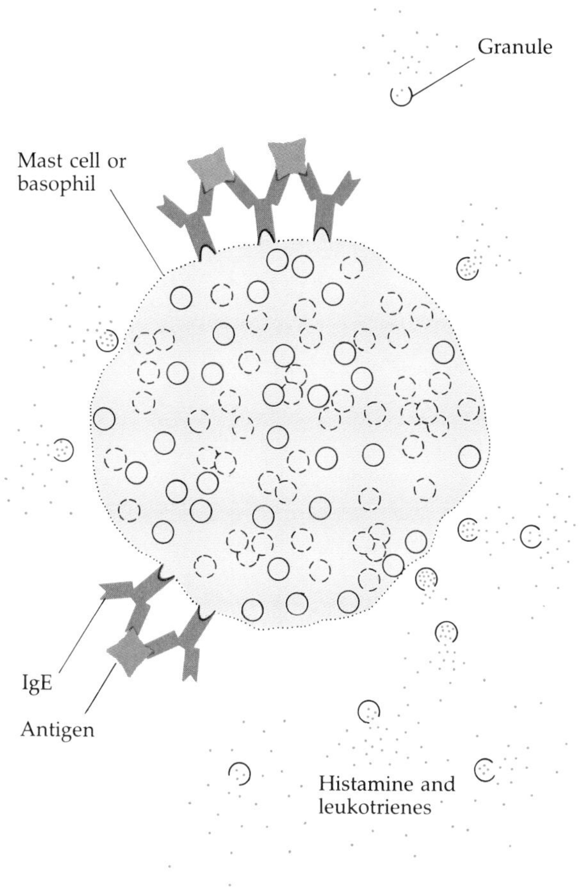

Figure 17–1 Mechanism of anaphylaxis. The mast cell or basophil becomes coated with IgE. When an antigen bridges the gap between two adjacent molecules of the same specificity, the cell undergoes degranulation and releases mediators such as histamine and leukotrienes. The mediators cause allergic effects such as mucus secretion, smooth muscle contraction, and an increase in capillary permeability.

The Fc region of the IgE antibody (discussed in Chapter 16) attaches to a specific receptor site on these cells, leaving its two antigen-combining sites exposed (Figure 17–1). A mast cell can have as many as 500,000 sites for IgE attachment, although not all attached IgE monomers are specific for the same antigen. When one antigen combines with two adjacent IgE antibodies and bridges the space between them, the mast cell or basophil is triggered to release a variety of chemicals called **mediators** from numerous secretory granules (lysosomes) in its cytoplasm. This process is called **degranulation.**

Table 17–2 Mediators of Anaphylaxis

Mediator	Function
Histamine	Increases blood capillary permeability, mucus secretion, and smooth muscle contraction
Leukotrienes	Increase blood capillary permeability and smooth muscle contraction
Prostaglandins	Role not completely defined; probably involved in blood capillary permeability, smooth muscle contraction, mucus secretion

The best known mediator is **histamine.** The pharmacological effects of histamine are to increase the permeability and dilation of blood capillaries, resulting in edema (swelling) and erythema (redness). Other effects include increased mucus secretion (a runny nose, for example) and smooth-muscle contraction, which in the respiratory bronchi results in breathing difficulty. Other mediators include **leukotrienes** of various types. The most widely studied, *SRS-A* (slow-reacting substances of anaphylaxis), is actually a combination of several leukotrienes. Because leukotrienes tend to cause prolonged contractions of certain smooth muscles, their action contributes to the spasms observed in asthmatic attacks. **Prostaglandins** are another group of mediators that affect smooth muscles of the respiratory system, and also cause increases in mucus secretions. A number of other mediators are also released with the degranulation of mast cells and basophils (Table 17–2), although their contributions to the symptoms of anaphylaxis are less direct.

Systemic anaphylaxis

Systemic anaphylaxis (anaphylactic shock) can result when an individual sensitized to an injected antigen receives a subsequent injection of it. The release of mediators causes peripheral blood vessels throughout the body to dilate, resulting in a drop in blood pressure (shock). This reaction can be fatal within a few minutes. The drug epinephrine (adrenaline) counteracts these effects, and kits for epinephrine self-administration are available to persons allergic to insect stings. There is very little time to act once someone develops systemic anaphylaxis.

Penicillin is a case of special interest because many of us are acquainted with people who are sensitive to this drug. In these persons the penicillin, which is a nonimmunogenic hapten, combines with the carrier serum protein albumin and in that form is able to induce antibody formation. Penicillin allergy probably occurs in about 2% of the population. Anyone who has had an adverse reaction to penicillin that included generalized hives, swelling of throat tissues, or chest constriction should not receive the drug again. A childhood rash without hives is probably not a significant reaction. However, there is no completely reliable skin test for sensitivity to penicillin injections.

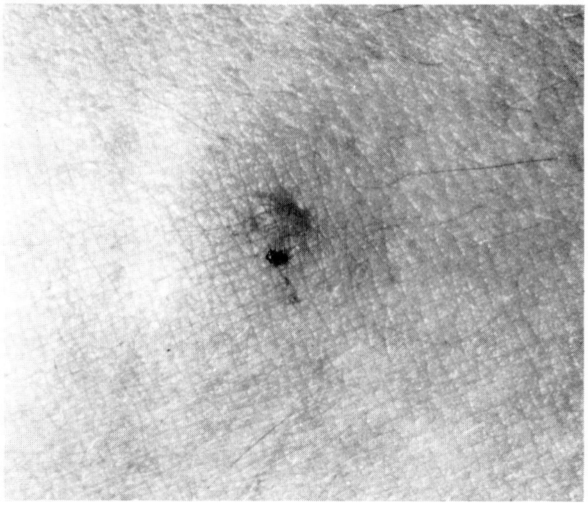

Figure 17–2 Skin test to identify allergens. If there is an allergic response, redness and a raised area, called a wheal, develop where the allergen has been inoculated into the skin.

Localized anaphylaxis

Whereas sensitization to injected antigens is a common cause of systemic anaphylaxis, **localized anaphylaxis** is usually associated with antigens that are ingested (foods) or inhaled (pollen). The symptoms that develop depend primarily on the route by which the antigen enters the body.

In the case of allergies involving the upper respiratory system, such as *hayfever* (allergic rhinitis), sensitization usually involves mast cells localized in the mucous membranes of the upper respiratory tract. Reexposure to the airborne antigen, which might be a common environmental material such as plant pollen, fungal spores, house dust, or animal dander (scaly, dried skin), results in a rapid release of large amounts of histamine, leukotrienes, and other mediators from mast cells in contact with the antigen. The typical symptoms are itchy and tearing eyes, congested nasal passages, coughing, and sneezing. Antihistamine drugs are often used to treat these symptoms.

Asthma is an allergic reaction that affects mainly the lower respiratory system. Symptoms such as wheezing and shortness of breath are caused by the constriction of smooth muscles in the bronchial tubes. Antihistamines are not an effective treatment for asthma, since other mediators such as leukotrienes and prostaglandins are more important than histamine in this reaction. Treatment usually involves administration of epinephrine.

Antigens that enter the body via the digestive tract can also sensitize a host. Many of us may know someone who is allergic to a particular food. Children tend to outgrow these problems, and in fact many food allergies may not be related to hypersensitivity, but are more accurately described as food intolerances. For example, many people are unable to digest the lactose sugar in milk because they lack the necessary enzyme. The diarrhea that results from milk ingestion is a symptom of food intolerance. While it is true that gastrointestinal upset is a common symptom of food allergies, it can also result from many other factors. Hives are more characteristic of a true food allergy, and ingestion of large amounts of the antigen may even result in systemic anaphylaxis. Skin tests are not reliable indicators for the diagnosis of food-related allergies, and completely controlled tests for hypersensitivity to ingested foods are very difficult to perform.

Prevention of anaphylactic reactions

Avoiding contact with the sensitizing antigen is the most obvious way to prevent allergic reactions. Unfortunately, avoidance is not always possible.

Some allergic persons may never know exactly what the antigen is. In other cases, skin tests might be of use in diagnosis (Figure 17–2). These tests involve inoculating small amounts of the suspected antigen just beneath the epidermal skin layer. Sensitivity to the antigen is indicated by a rapid inflammatory reaction that features redness, swelling, and itching at the inoculation site. This small, affected area is called a *wheal.*

Once the responsible antigen(s) has been identified, the person can either try to avoid contact with it or undertake **desensitization.** This procedure consists of a series of dosages of the antigen carefully injected beneath the skin. The object is to cause the production of IgG antibodies rather than those of the IgE class, in the hope that the circulating IgG antibodies will act as *blocking antibodies* to intercept and neutralize the antigens before they can react with cell-bound IgE. Recent evidence indicates that desensitization might also induce the production of suppressor T cells (discussed in Chapter 16). Desensitization is not a routinely successful procedure, but it is effective in 65 to 75% of persons whose allergies are induced by inhaled antigens.

Type II (Cytotoxic) Reactions

Immunologic injury resulting from cytotoxic reactions is due to antibodies that are directed at antigens on the host's tissue cells. These are IgG or IgM antibodies, and their reaction with these antigens usually involves complement. In some cases, complement reacting with the antibody–antigen complex results in the lysis of tissue cells; in other cases, the complex of antigen, antibody, and complement attracts phagocytic cells such as neutrophils or macrophages. The result may be the phagocytosis of circulating cells (as occurs in opsonization), or the accumulation of phagocytic cells that release chemicals toxic to the host's tissue cells.

Several diseases are caused by cytotoxic reactions. *Idiopathic thrombocytopenic purpura (ITP)* (Figure 17–3) is a blood disease in which blood platelets (thrombocytes) are destroyed by antibodies. This condition can occur spontaneously, or in response to drug injections or viral infections. Because platelets are necessary for blood clotting, their loss results in hemorrhages that appear on the skin as purple spots (purpura). *Myasthenia gravis* is a disease in which muscle tone becomes progressively weaker. It is caused by antibodies that coat the receptor sites on the junctions at which nerve impulses reach the muscles. Eventually the muscles controlling the diaphragm and rib cage fail to receive the necessary nerve signals, and respiratory arrest and death result. *Graves' disease* is caused by antibodies called long-acting thyroid stimulators.

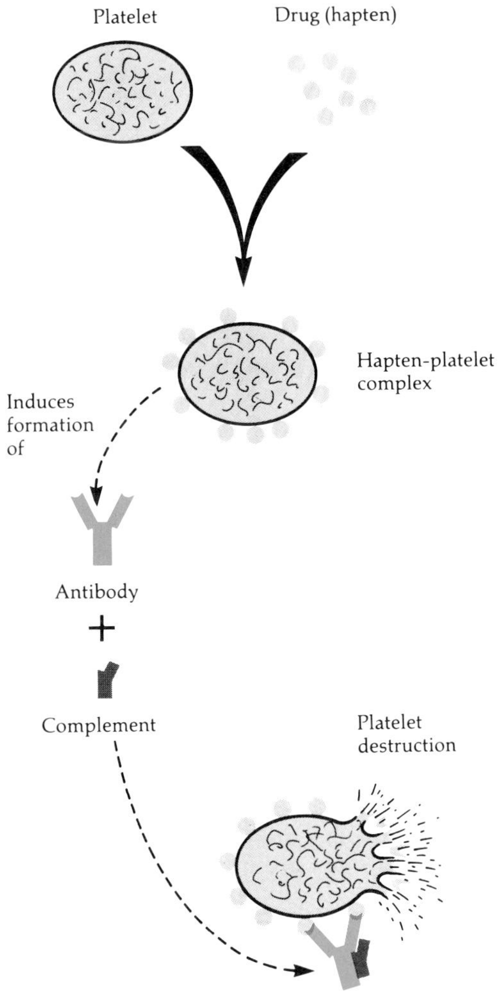

Figure 17–3 Idiopathic thrombocytopenic purpura (ITP), drug-induced in this case. Viral infections are suspected to be the cause of many cases.

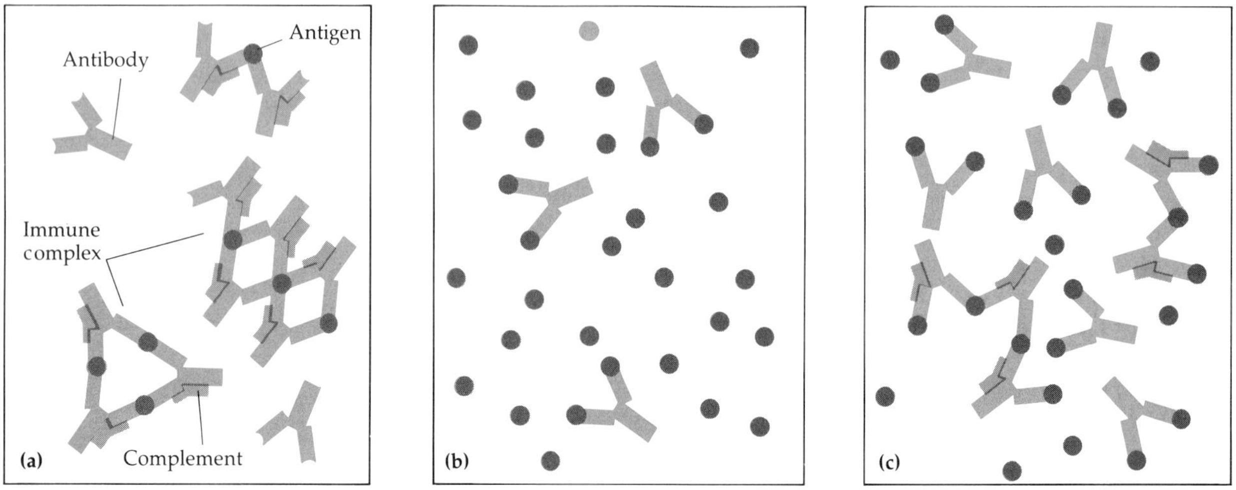

Figure 17–4 Formation of immune complexes. **(a)** Immune complexes formed when antibody is in excess. Most of these are removed from the circulation by phagocytosis. **(b)** Immune complexes formed when antigen is in excess. These generally do not fix complement or cause inflammation. **(c)** Small, soluble complexes formed when there is a certain ratio of antigen to antibody; there is usually a small excess of antigen. These complexes fix complement, but are small enough to escape phagocytosis. They damage tissues when they become trapped in the basement membrane of blood vessels.

These antibodies attach to receptors on the thyroid gland that are intended for the thyroid-stimulating hormone produced by the pituitary gland. The result is that the thyroid gland is stimulated to produce increased amounts of thyroid hormones and becomes greatly enlarged.

Type III (Immune Complex) Reactions

Many damaging reactions can be caused by immune complexes of antigen, antibody, and complement. A primary difference between cytotoxic and immune complex reactions is that the antigens involved in the latter are not part of a host tissue cell.

Immune complexes form when certain ratios of antigen and antibody occur. The antibodies are usually of the IgM or IgG class. A significant excess of antibody (Figure 17–4a) leads to the formation of large complement-fixing complexes that are rapidly removed from the body by phagocytosis. When there is a significant excess of antigen (Figure 17–4b), soluble complexes form that do not fix complement and do not cause inflammation. However, when a certain antigen-antibody ratio exists (usually with a slight excess of antigen; Figure 17–4c), the soluble complexes that form are small and escape phagocytosis. These complexes circulate in the blood, pass between endothelial cells of the blood vessels, and become trapped in the basement membrane beneath the cells. In this location they may activate complement and cause a transient inflammatory reaction. Repeated introduction of the same antigen can lead to more serious inflammatory reactions.

Glomerulonephritis is an immune-complex condition resulting from inflammatory damage to the kidney glomeruli, which are sites of blood filtration. Antibodies generated in response to the M protein of streptococci are believed to be one cause of this disease (see Chapter 14). Another immune complex condition that often involves the kidney glomeruli (as well as other body sites) is *systemic lupus erythematosus*. The cause of this disease is not completely understood, but afflicted persons produce many types of antibodies directed at components of their own cells, primarily nucleic acids. The symptoms of this disease result from deposition of immune complexes in the skin, joints, heart pericardium, and kidney glomeruli.

Crippling *rheumatoid arthritis* is a disease in which immune complexes of IgM, IgG, and complement are deposited in the joints. The chronic inflammation caused by this deposition eventually leads to severe damage to the cartilage and bone of the joint. Cell-mediated immune responses are also thought to be involved in this disease.

Local immune-complex reactions can appear as a region of inflammation or even of necrosis (dead tissue) surrounding the site at which antigen was injected into a person already sensitized to that antigen. This is the *Arthus reaction,* caused by immune complexes deposited in blood vessels of the skin.

Type IV (Cell-Mediated) Reactions

Up to this point, we have discussed humoral types of hypersensitivity involving IgE, IgG, or IgM. Type IV (cell-mediated) reactions involve cell-mediated immune responses, and are effected mainly by T cells, although macrophages may also be involved. Instead of occurring within a few minutes or hours after a sensitized individual is exposed to an antigen, these hypersensitivity reactions are often not apparent for a day or more. A major factor in the delay is the time required for the participating cells to migrate to and accumulate near the foreign antigens.

It is thought that type IV hypersensitivity reactions occur when certain foreign antigens, particularly of a type that bind to tissue cells, are phagocytized by macrophages and then presented to receptors on the T cell surface (Figure 17–5). Contact between the antigenic determinant sites and the appropriate T cell causes the T cell to proliferate. The exact nature of these T-cell receptors is not known, but they seem to at least partially resemble immunoglobulin molecules. If a person sensitized in this way is reexposed to the same antigen, a cell-mediated hypersensitivity reaction might result. A principal factor in this reaction is the release of lymphokines by T cells reacting with the target antigen. Lymphokines may aid directly in destroying target cells or contribute to inflammatory reactions.

Cell-mediated hypersensitivity reactions of the skin

We have seen that the skin is frequently the site on which immune-reaction symptoms are displayed. One cell-mediated hypersensitivity reaction that involves the skin is the familiar skin test for tuberculosis. Because the tubercle bacilli are often located intracellularly in macrophages, this disease can stimulate a cell-mediated immune response. As a screening test, protein components of the tubercle bacilli are injected into the skin. If the recipient has immunity from a prior exposure to tuberculosis, an inflammatory reaction to the injection of these antigens will appear on the skin in one or two days; this interval is typical of delayed hypersensitivity reactions.

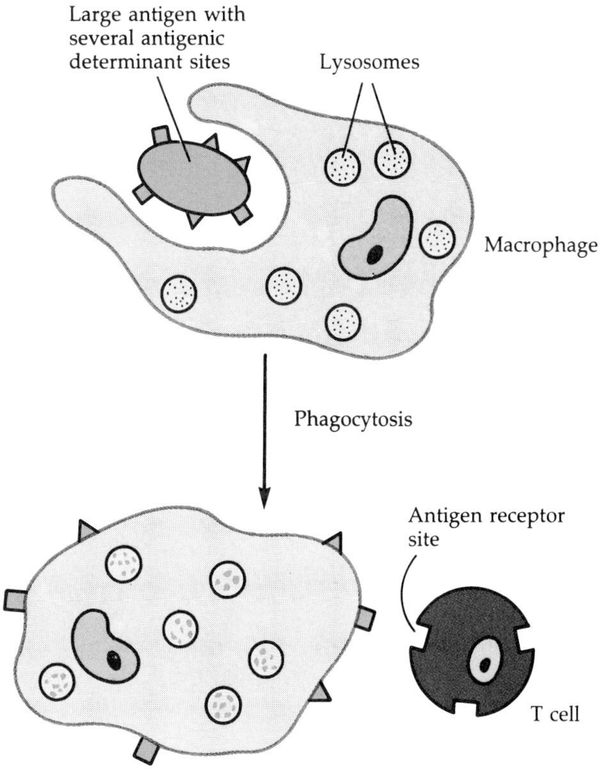

Figure 17–5 Basic mechanism of a cell-mediated hypersensitivity reaction. T cells are usually stimulated by an antigen that has been processed by a macrophage. Some of the antigenic determinant sites appear on the surface of the macrophage after processing, and, upon contact, stimulate a compatible T cell into activity.

Allergic contact dermatitis, another common manifestation of type IV hypersensitivity, is usually caused by haptens that combine with proteins in the skin of some persons to produce an immune response. Reactions to poison ivy (Figure 17–6), cosmetics, and the metals in jewelry (especially nickel) are familiar examples of these allergies.

AUTOIMMUNITY

Loss of Immunologic Tolerance

We have mentioned that the body is normally able to distinguish its own antigens ("self") from all others ("nonself") and does not mount an immunologic attack against them. This phenomenon is called **tolerance,** and apparently arises during fetal development. If a fetus is exposed to an antigen, exposure to the same antigen after birth does not stimulate the production of antibodies or sensitized T cells. It is believed that some clones of lymphocytes *(forbidden clones)* having the potential to respond to self antigens may also be produced during fetal life, but are destroyed or suppressed during embryonic development.

Sometimes the body loses its ability to discriminate between self and nonself. This loss of tolerance leads to **autoimmunity,** a response by antibodies or sensitized T cells against a person's own tissue antigens. We saw in the previous section that cytotoxic reactions are autoimmune. Auto-

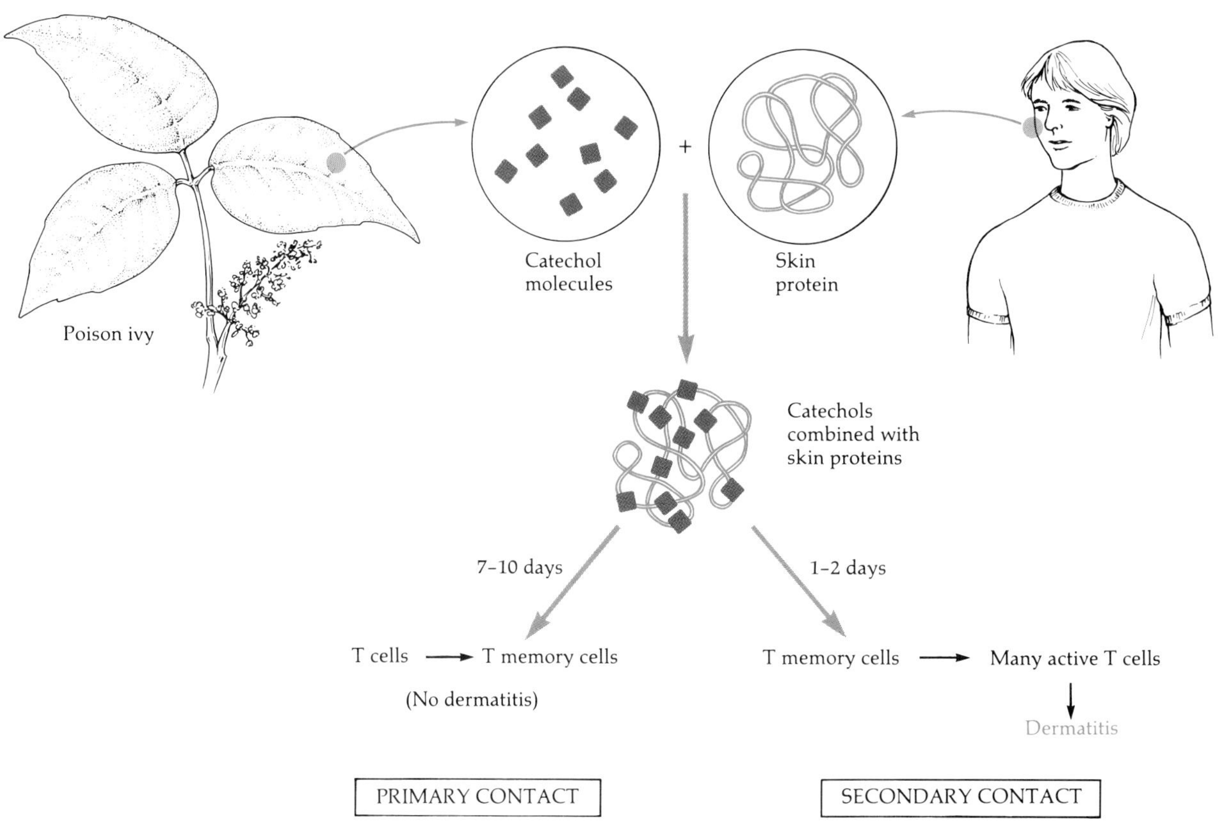

Figure 17–6 The development of allergy (allergic contact dermatitis) to catechols from the poison ivy plant. The catechols are haptens that must combine with skin proteins to provoke an immune response. The first contact with poison ivy sensitizes the susceptible person, and subsequent exposures result in contact dermatitis.

immunity also occurs occasionally when cross-reacting antibodies are produced in the body's response to an infection. For instance, the pathology of rheumatic fever might result from the activity of antibodies formed in response to a streptococcal infection. These antibodies are coincidentally capable of reacting with antigenic determinant sites on heart muscle cells.

Other autoimmune diseases specifically involve the cell-mediated immune system. Like rheumatic fever, some of these diseases might be related to the immune system's response to infection. When the lymphocytic choriomeningitis virus (sometimes spread by contact with pet hamsters) infects the membranes surrounding the brain, the body responds by producing T cells that cause fatal neurological damage. Because animals without a thymus are not affected by the virus, it is believed that cell-mediated immunity is involved in this disease.

There are a number of diseases with less obvious etiologies, in which the cell-mediated immune system destroys the patient's tissues. *Hashimoto's thyroiditis* and *Addison's disease* result from T cells attacking the thyroid and adrenal glands, respectively. *Multiple sclerosis* involves inflammation and destruction of the myelin layer coating the nerve cells; the reason for this attack is unknown, but is thought to be related to a childhood infection.

Many autoimmune diseases have a genetic component. For Addison's disease, Graves' disease, multiple sclerosis, rheumatic fever, and many others, there is a well known association between susceptibility to the disease and a specific determinant in the major histocompatibility complex located on one of the human chromosomes.

Major Histocompatibility Complex

The inherited genetic characteristics of individuals are expressed not only in the color of their eyes or the curl of their hair, but in differences in the antigens on their cell surfaces. These are called **histocompatibility antigens.** The genes controlling the production of these antigens in humans are known as the **major histocompatibility complex (MHC)** or the **human leukocyte antigen (HLA)** complex.

An important medical application of HLA typing is in transplantation surgery: the donor and the recipient must be matched. Family members can be expected to have many of the same histocompatibility antigens, and are usually the best-matched candidates for organ transplantation.

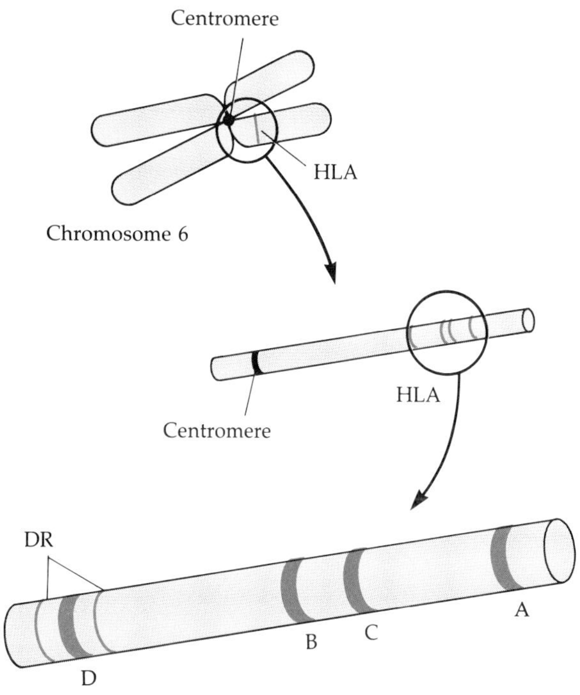

Figure 17–7 The genes that control human leukocyte antigens (HLA) are located on chromosome 6. Note how HLA-A, -B, -C, -D, and -DR are located within a very short length of chromosome.

There are two main classes of HLA antigens that are important in such matches: class I antigens (HLA-A, -B, and -C) and class II antigens (HLA-D and -DR). The genes that control production of these antigens are located close together on human chromosome 6, as shown in Figure 17–7. Matching for class I antigens has long been a standard procedure, but matching for class II antigens might well be more important, especially for tissue from a totally unrelated person (such as a transplant from a cadaver). The donor and recipient must be of the same ABO blood type, and the recipient's serum must be tested for antibodies to the donor's HLA antigens.

As mentioned above, certain HLA antigens are related to susceptibility to specific diseases, and

Table 17–3 Diseases Related to Specific HLA Antigens

Disease	Related HLA Antigen	Increased Risk Compared with General Population	Description
Inflammatory diseases:			
Multiple sclerosis	D(DR2)	5×	Progressive inflammatory disease affecting nervous system
Rheumatic fever	DR(antibody 883)	4–5×	Autoimmune disease caused by cross-reaction with antibodies against streptococcal infection
Endocrine diseases:			
Addison's disease	D(DR3)	4–10×	Deficiency in production of hormones by adrenal gland
Graves' disease	D	10–12×	Antibodies attached to certain receptors in the thyroid gland cause it to enlarge and produce excessive hormones
Malignant diseases:			
Hodgkin's disease	A(A1)	1.5–1.8×	Cancer of lymph nodes

this represents another medical application of HLA typing. A few of these relationships are summarized in Table 17–3.

Transplantation

Human organ transplantation has added productive years to the lives of many individuals. Since the first kidney transplant was performed in 1954, this particular type of transplant has become a nearly routine medical procedure. Other types of transplants now feasible include bone marrow, thymus, heart, liver, and cornea. Tissues and organs for transplantation are usually taken from recently deceased individuals, although nonessential duplicates of an organ such as a kidney might occasionally come from a living donor.

Types of transplants

Some transplantations or grafts do not stimulate an immune response against them. A transplanted cornea, for example, is rarely rejected. Rejection is infrequent mainly because antibodies do not circulate into the anterior chamber of the eye, which is considered an immunologically **privileged site.** It is also possible to transplant **privileged tissue** that does not stimulate an immune rejection. An example is a valve from a pig's heart, which can replace a person's damaged heart valve. However, privileged sites and tissues are more the exception than the rule.

When one's own tissue is grafted to another part of the body, as is done in burn treatment or in plastic surgery, the graft is not rejected. Recent technology has made it possible to use a few cells of a burn patient's uninjured skin to culture extensive sheets of new skin. The graft of this new skin is an example of an **autograft.** Identical twins have the same genetic makeup; therefore, skin or organs such as kidneys may be transplanted between them without provoking an immune response. These are **isografts.**

Most transplants, however, are made between persons who are not identical twins; these transplants do trigger an immune response. Attempts are made to match the HLA antigens of the donor and recipient as closely as possible so that the chances of rejection are reduced. Because they are most likely to match, close relatives, especially siblings, are preferred. Grafts between persons who are not identical twins are called **allografts.**

In terms of organ supply, other considerations aside, it would be helpful if **xenografts,** organs from animals, could be more successfully transplanted to humans. However, the body tends to mount an especially severe immune assault on such transplants. Perhaps the most famous xenograft recipient to date was Baby Fae, a two-week-old baby girl who received the heart of a baboon on October 26, 1984. After her body's attempt to reject the xenograft, the girl died 20 days later from heart and kidney problems.

The various types of grafts are summarized in Table 17–4.

When bone marrow is transplanted to persons with a defective immune system, the transplanted tissue may attack the host. The purpose of the graft is to provide such persons with the B-cell and T-cell manufacturing capability they lack. However, the result can be **graft versus host (GVH) disease.** The transplanted bone marrow contains immunocompetent cells that mount a humoral or cell-mediated immune response against the tissue into which they have been transplanted. Because the recipients lack effective immunity, GVH disease is a serious complication and can even be fatal.

Immunosuppression

To keep the rejection response to transplanted tissue in perspective, it is useful to remember that the immune system is simply doing its job, and has no way of recognizing that its attack against the transplant is not helpful. In an attempt to prevent rejection, the recipient of an allograft usually receives treatment to suppress this normal immune response against the graft. The treatments have involved drugs, antilymphocyte serum directed at T cells, and X-irradiation. Unfortunately, these treatments suppress the immune response to all antigens, by both the humoral and cell-mediated immune systems. The persons so treated then become very susceptible to infectious diseases and cancer.

The drug **cyclosporine** has revolutionized transplantation medicine. Cyclosporine specifically suppresses cell-mediated immunity by acting on T cells. Its mode of action is not completely known at this time. Apparently the presence of cyclosporine blocks the division of T cells, probably by interfering with the synthesis by helper T cells of a T-cell growth factor called interleukin-2. T-cell proliferation resumes when cyclosporine is removed. Yet throughout the term of cyclosporine administration, the patient's humoral immunity against circulating bacteria and viruses remains essentially intact.

As a result of this discovery, the success rate of organ transplantation has risen dramatically. Before cyclosporine came into general use, only about 50 percent of kidneys from donors who were not closely related to the patient were accepted; now as many as 90 percent of these transplants succeed. Liver transplantation became practical only after cyclosporine became available, and heart transplants have achieved previously unheard-of success rates.

The greatest drawback to the use of the drug appears to be an increased risk of cancer, as that disease is combatted by the same cell-mediated immune system that rejects transplanted organs. Certain problems with liver and kidney toxicity have also necessitated careful attention to dose levels.

Natural Immune Deficiencies

A retrovirus that selectively infects the T cells responsible for cell-mediated immunity (HTLV-III) is currently thought to cause a form of naturally acquired immunosuppression, known as **acquired immune deficiency syndrome (AIDS).** There are also a number of inherited immune deficiencies. AIDS is discussed in more detail in the box on pages 482–484.

IMMUNE RESPONSE TO CANCER

Immunologic Surveillance

Persons immunosuppressed by either natural or artificial means are markedly more susceptible to cancers than is the rest of the population. The conventional explanation for this susceptibility is that the immune system normally recognizes and destroys cancer cells before they become established in the body. The tendency of cancer to occur most often in the elderly, whose immune systems are thought to be less efficient, or in the very young, whose immune systems may not have developed

Table 17–4 Types of Grafts

Graft	Description
Autograft (*autos*, "self")	A graft transplanted from one site to another site on the same person (skin from the thigh grafted over a burned area on the arm)
Isograft (*isos*, "equal")	A graft between genetically identical persons—identical twins
Allograft (*allos*, "other")	A graft between genetically different members of the same species
Xenograft (*xenos*, "strange")	A graft between different species (a transplant of a baboon heart to a human)

Acquired Immune Deficiency Syndrome (AIDS) Update: A Medical Detection Report

In 1981, the Centers for Disease Control became aware of an increasing number of cases of *Pneumocystis carinii* pneumonia (PCP) and an atypical and often fatal form of Kaposi's sarcoma (KS). PCP had affected a few hundred adults and children in the United States annually, but typically was seen only in patients receiving drugs to suppress the immune system, such as those used with transplant recipients to prevent rejection. Prior to 1981, KS (a cancer of blood vessel walls) affected elderly men and was seldom fatal. It was obvious that something was depressing the immune response of this new group of PCP and KS patients. The newly identified condition was named *acquired immune deficiency syndrome* (AIDS). Patients may experience fever, appetite loss, weight loss, extreme fatigue, diarrhea, persistent cough, night sweats, and enlarged lymph nodes. Other symptoms depend on the opportunistic infections that develop; these include PCP, KS, yeast infections, cytomegalovirus infections, herpes, and toxoplasmosis.

An Unusual Population

The first cases of AIDS probably occurred in 1979, but the disease was not noticed until an inordinately large number of young men with KS was reported. The most striking factor in the data was that the patients were sexually active homosexual or bisexual males.

Since AIDS was identified in 1981, the number of cases has doubled every six months (see graph). The primary victims are homosexual men, intravenous drug users, and hemophilia patients. Other risk groups are the heterosexual partners of AIDS sufferers and transfusion recipients. Ninety percent of AIDS victims in the United States are between 20 and 49 years of age, and 94 percent are males. AIDS has occurred in 42 states and the disease does not seem to affect any particular ethnic or racial segment of the population. The incubation period is several months to two years. Forty-eight percent of known AIDS cases have died within two years of diagnosis. Though AIDS is underreported worldwide because of lack of recognition and lack of surveillance, the disease pattern reported from 17 European countries and Canada is similar to that in the United States.

Identifying the Means of Transmission

Interviews with AIDS sufferers indicate that AIDS is transmitted from one human to another through intimate sexual contact, sharing of contaminated needles, or, less frequently, through tranfusion of blood or blood products. Pediatric cases have been linked to transfusions (for prematurity) or living in a household with an adult in one of the risk groups; congenital infections also may occur. The approximately 300 infected health-care workers have had some connection to one or more of the high-risk groups. No cases have been due to occupational exposure, but health-care workers should take the same precautions mandated by cases of serum hepatitis.

Recently concern has been expressed about the safety of hepatitis B vaccine, which is made from the plasma of serum hepatitis patients—many of whom are in one or more of the AIDS risk groups. This vaccine is considered safe, however, because the preparation procedure inactivates all known viruses.

Although the AIDS virus has been isolated from both semen and saliva, there is no proof of transmission by saliva. Thousands of family members, co-workers, and friends of AIDS patients do not have AIDS. The virus is transmitted by sexual contact or the parenteral route and not by casual contacts. However, apparently contradictory evidence from Africa leaves the means of transmission of AIDS not unquestionably settled.

The Causative Virus

In 1984, independent research teams in France and the United States isolated and identified the causative agent of AIDS, a retrovirus called HTLV-III (human T cell leukemia virus type 3). Evidence suggests that HTLV-III infects lymphocytes—this leads to an abnormal helper T-to-suppressor T (Th/Ts) ratio. Ultimately the loss of helper T cell functions can halt the production of antibodies. (Helper T cells aid in antibody production and suppressor T cells inhibit the immune response, see Chapter 16.) The normal Th/Ts ratio is 1.0–3.9. AIDS causes this ratio to decrease below 1.0. It is not known for certain whether Th cells are infected and killed or Ts cells are infected and proliferate, but recent DNA hybridization research suggests that HTLV-III causes the death of Th cells and not the proliferation of Ts cells.

HTLV-III has been cultured in T cells. Aided by the ELISA test (see Chapter 16 and Microview III, p. 4), patient serum can be tested for the presence of antibodies against the virus, which would indicate an active case of AIDS, a "pre-AIDS" condition, or a mild or asymptomatic infection. The results of such tests have led the Centers for Disease Control to estimate that at least a quarter of a million people have been infected with the HTLV-III virus. In other words, exposure to the virus may be 30 to 50 times more common than occurrence of AIDS. It appears that the virus does not easily cause an infection. It is possible, in fact, that the milder disease states may result in natural immunization.

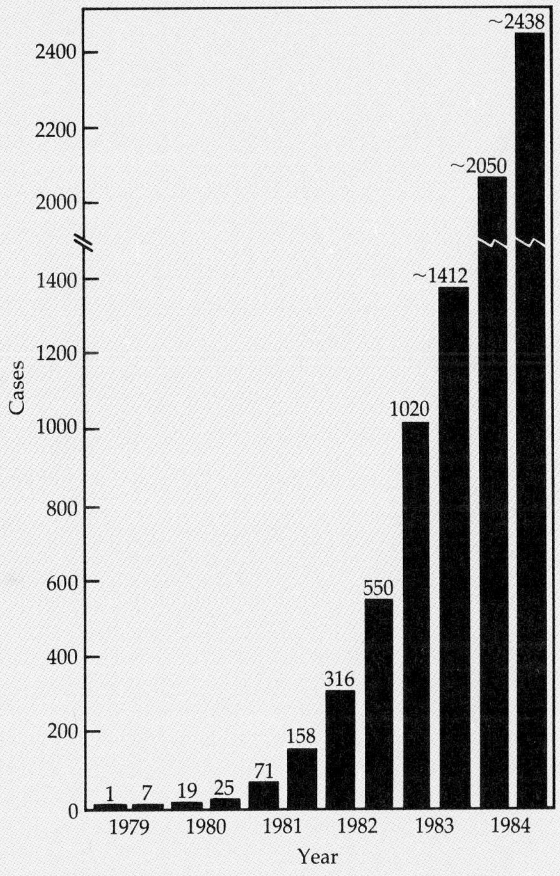

Reported AIDS cases in the United States by half year, 1979–1984.

Treatment and Prevention

At present, the opportunistic infections can be treated, but there is no treatment for AIDS. A trial using interleukin-2, a T cell growth factor, to restore immunity, was not encouraging. Several drugs are currently being tested.

Infectious disease control personnel have been accumulating epidemiologic data, and laboratory researchers have been exploring methods of treatment and vaccination, but the problems in AIDS investigations are numer-

(continued on next page)

ous. There may be many genetic variations of HTLV-III and its genome appears to change in a manner similar to influenza virus (see Chapter 22). There is a 20 percent variation in the gene that codes for the protein coat. This genetic drift will not necessarily interfere with development of a single vaccine against HTLV-III, however. The virus infects cells but does not cause localized symptoms of disease, rather the whole organism is affected. In contrast, poliovirus, for example, infects nerve cells, which kills the cells and results in neurologic symptoms.

Currently, prevention of AIDS in the general public includes screening blood for HTLV-III antibodies and refraining from sexual contact with members of high-risk groups.

Although the virus that causes AIDS has been isolated, many questions concerning the disease remain. It is still not known how HTLV-III weakens the immune system. Nor is it understood how so many individuals have been exposed to HTLV-III without developing AIDS. Complicating the attempt to understand and control AIDS are sociologic factors. So far, instead of being traced to a particular blood type, specific mode of transmission, geographical location, genetic characteristic, or other quantifiable predisposition, AIDS has been linked primarily to a lifestyle. Continuing efforts to control this epidemic must address this issue as well as seek to discover how the virus operates.

properly, is thought to support this view. The immune system's presumed responsibility for patrolling the body for cancer cells is called **immunologic surveillance.**

A cell becomes cancerous when it undergoes transformation (Chapter 12) and loses its normal characteristic of ceasing to grow when it contacts an adjacent cell (this characteristic is called *contact inhibition*). Because of transformation, the surfaces of tumor cells may acquire tumor-specific antigens that mark them as nonself to the immune system (Figure 17–8). It is hoped that monoclonal antibodies will be the tool to help the immune system recognize such tumor-specific antigens and detect cancer in much earlier stages (see box in Chapter 16).

Immunologic Escape

Even in people with presumably normal immune systems, cancer sometimes occurs. Once established, the cancer seems to become resistant to immune rejection, even though it is easy to demonstrate that cell-mediated and humoral immunity are directed against the tumor-specific surface antigens. This resistance to rejection is called **immunologic escape,** and several mechanisms have been proposed to explain it. One is *antigen modulation;* that is, tumor cells are believed to shed their specific antigens, thus evading recognition by the immune system. Another proposed mechanism is that some cancers, such as Hodgkin's disease, release factors that eventually suppress the entire cell-mediated immune system.

An especially interesting experiment illustrating immunologic escape involves the immunization of animals with separated cells of a certain tumor. If the same tumor is later introduced into the animal, the tumor grows readily and kills the animal. However, if the tumor is introduced into a nonimmunized animal, it is rejected in a short time. The growth in the immunized animal is an example of immunologic escape by **immunologic enhancement.** This effect might be due to a form of antigenic modulation. Another possible explanation is that the immunized animal forms humoral antibodies against the tumor-specific surface antigens. These antibodies then cover the tumor surface antigens, so that Tc cells are unable to recognize the tumor as nonself and fail to destroy it. In the nonimmunized animal, the agents of cell-mediated immunity effectively reject the transplanted tumor.

Immunotherapy

Immunotherapy seems likely to be the most effective future approach to cancer treatment. Probably the most promising technique will involve immunotoxins. In Chapter 16 we defined an immunotoxin as a combination of a toxic agent, such as the drug ricin or a radioactive compound, and a monoclonal antibody. The monoclonal antibody is targeted at a particular type of tumor antigen. The antibody selectively locates the cancer cell, and the attached toxic agent destroys the cell but causes little or no damage to healthy tissue. There is great hope that this approach will become analogous to the use of antibiotics in treating infectious diseases.

The principal types of cancers are summarized in Table 17–5.

TRANSFUSION REACTIONS AND Rh INCOMPATIBILITY

It is well known that the immune system will react against the foreign antigens on the surfaces of transfused red blood cells if blood has not been properly typed.

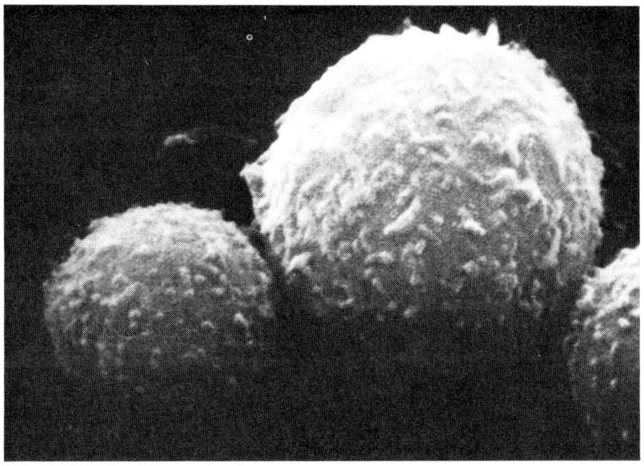

(a)

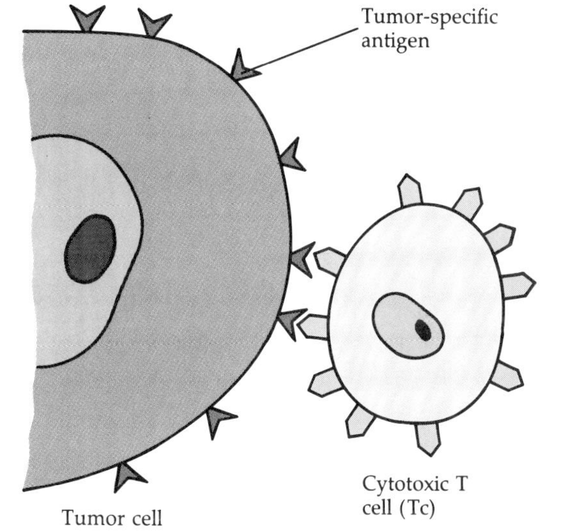

(b)

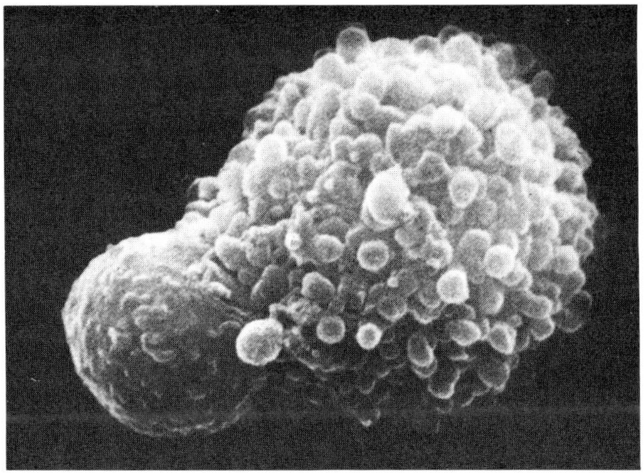

(c)

Figure 17–8 Interaction of a cytotoxic T cell (Tc) and a tumor cell. **(a)** A cytotoxic T cell (smaller sphere at lower left) attaches to the tumor cell after recognizing the tumor-specific antigens (×7250). **(b)** Diagram of a cytotoxic T cell binding to a tumor-specific antigen on the surface of a tumor cell. **(c)** Lysis of the tumor cell is indicated by the deep folds in its surface membrane (×7250).

Table 17–5 Principal Types of Cancers

Type of Cancer	Description
Carcinoma	The most common form of cancer; arises from the cells forming the skin, the glands (such as the breast, uterus, and prostate), and the membranes of the respiratory and gastrointestinal tracts; metastasizes (spreads to other parts of the body) mainly through the lymph vessels
Leukemia	Cancer of the tissues that form blood; characterized by uncontrolled multiplication and accumulation of abnormal white blood cells
Lymphoma	Cancer of the lymph nodes
Sarcoma	Another form of malignant tumor; arises from connective tissues, such as muscle, bone, cartilage, and membranes covering muscles and fat; metastasizes through the blood vessels

Source: Courtesy of Barbara J. Combs, et al., *An Invitation to Health*, second edition. The Benjamin/Cummings Publishing Co., Menlo Park, CA, 1983.

The ABO Blood Group System

In the early 1900s it was discovered that human blood could be grouped into four principal types, which were designated A, B, AB, and O. This method of classification is referred to as the **ABO blood group system.** Since then, at least fifteen other blood group systems have been discovered, but our discussion will be limited to two of the best known: the ABO and the Rh systems.

A person's ABO blood type depends on the presence or absence of two very similar carbohydrate antigens located on the cell membranes of red blood cells, or erythrocytes (Figure 17–9). Persons with type A blood possess a mosaic of antigens designated A on their red blood cells, while persons of blood type B have antigens designated B on theirs. Persons with blood type AB have both A and B antigens on their red blood cells, while persons with blood type O lack both A and B surface antigens.

The serum of persons with type-A antigens on their red blood cells contains antibodies against type-B cells (anti-B antibodies). Persons with type-B blood have antibodies against type-A cells (anti-A antibodies), type-O serum contains antibodies

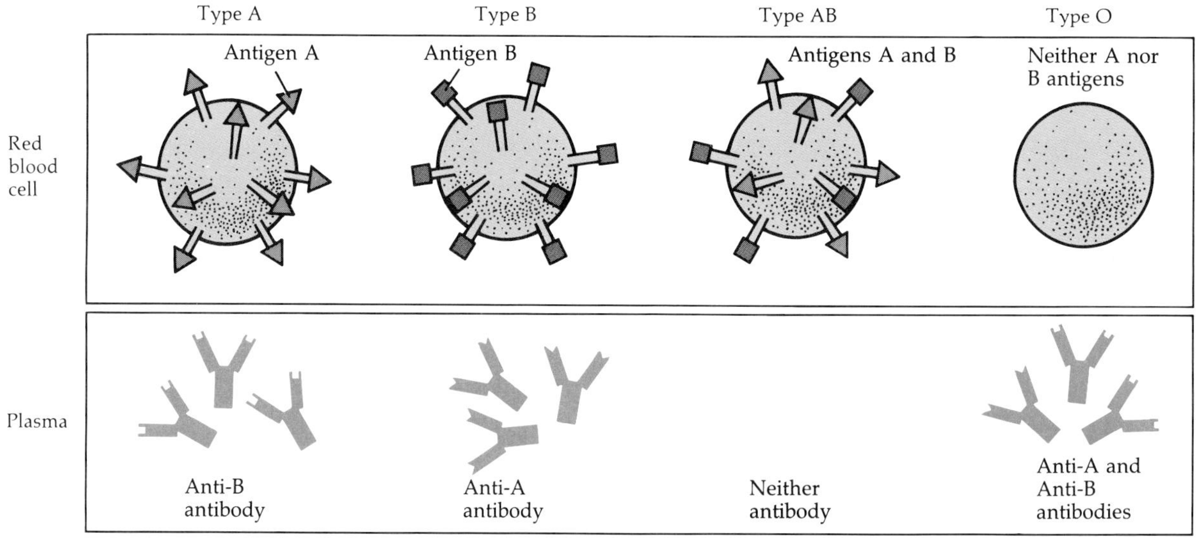

Figure 17–9 Relationships between the antigens on the surface of human red blood cells and the antibodies in the plasma.

against both A and B cells, and type-AB serum contains no anti-A or anti-B antibodies at all. When a transfusion is incompatible, as when type-B blood is transfused into a person with type-A blood, the antigens on the type-B blood cells will react with the anti-B antibodies in the recipient's serum. This antigen–antibody reaction activates complement, which in turn causes lysis of the donor's red blood cells as they enter the recipient's system. The main features of the ABO blood group system are summarized in Table 17–6.

In about 80% of the population (called **secretors)** soluble antigens of the ABO type appear in saliva and other bodily fluids. In criminal investigations it has been possible to type such fluids as saliva residues from a cigarette, or semen in cases of rape.

The Rh Blood Group System

In the 1930s, the presence of a different surface antigen on human red blood cells was discovered. It was found that when rabbits were immunized with red blood cells from Rhesus monkeys, their serum soon contained antibodies, directed against the monkey blood cells, that would also agglutinate 85% of all human red blood cells. This indicated that a common antigen was present on both human and monkey red blood cells. The antigen was named the **Rh factor** (Rh for Rhesus). The roughly 85% of the population whose cells possess this antigen is referred to as Rh^+. Those lacking this antigen (about 15%) are Rh^-. Antibodies that react with the Rh antigen do not occur naturally in the serum of Rh^- individuals, but exposure to this antigen can sensitize them to produce anti-Rh antibodies.

Blood transfusions and Rh incompatibility

If blood from an Rh^+ donor is given to an Rh^- recipient, the donor's red blood cells stimulate the production of anti-Rh antibodies. If the recipient receives Rh^+ red blood cells in a subsequent transfusion, a rapid hemolytic reaction will develop.

Hemolytic disease of the newborn

Blood transfusions are not the only way in which an Rh^- person can become sensitized to Rh^+ blood. When an Rh^- female and an Rh^+ male produce a child, the chances are at least 50% that the child will be Rh^+, because this antigen is inherited as a Mendelian dominant trait. If the child is Rh^+, the Rh^- mother can become sensitized to this antigen during birth, when the placental membranes tear and Rh^+ fetal red blood cells enter the maternal circulation (Figure 17–10). If the fetus in a later pregnancy is Rh^+, the mother will produce anti-Rh antibodies that will cross the placenta and react with fetal red blood cells. The fetus responds to this immune attack by producing large numbers of immature red blood cells called erythroblasts. Hence the name *erythroblastosis fetalis* was once used to describe what is now called **hemolytic disease of the newborn.** Before birth, the maternal circulation removes most of the toxic by-products of fetal red cell destruction. After birth, however, the fetal blood is no longer purified by the mother and the newborn develops jaundice and severe anemia.

Hemolytic disease of the newborn is usually prevented today by passive immunization of the Rh^- mother with anti-Rh antibodies, which are available commercially. These anti-Rh antibodies combine with any fetal Rh^+ red blood cells that

Table 17–6 The ABO Blood Group System

Characteristic	Blood Type A	Blood Type B	Blood Type AB	Blood Type O
Antigen present on the red blood cells	A	B	Both A and B	Neither A nor B
Antibody normally present in the plasma	anti-B	anti-A	Neither anti-A nor anti-B	Both anti-A and anti-B
Plasma causes agglutination of red blood cells of these types	B, AB	A, AB	None	A, B, AB
Percent in a mixed Caucasian population	41	10	4	45
Percent in a mixed black population	27	20	7	46

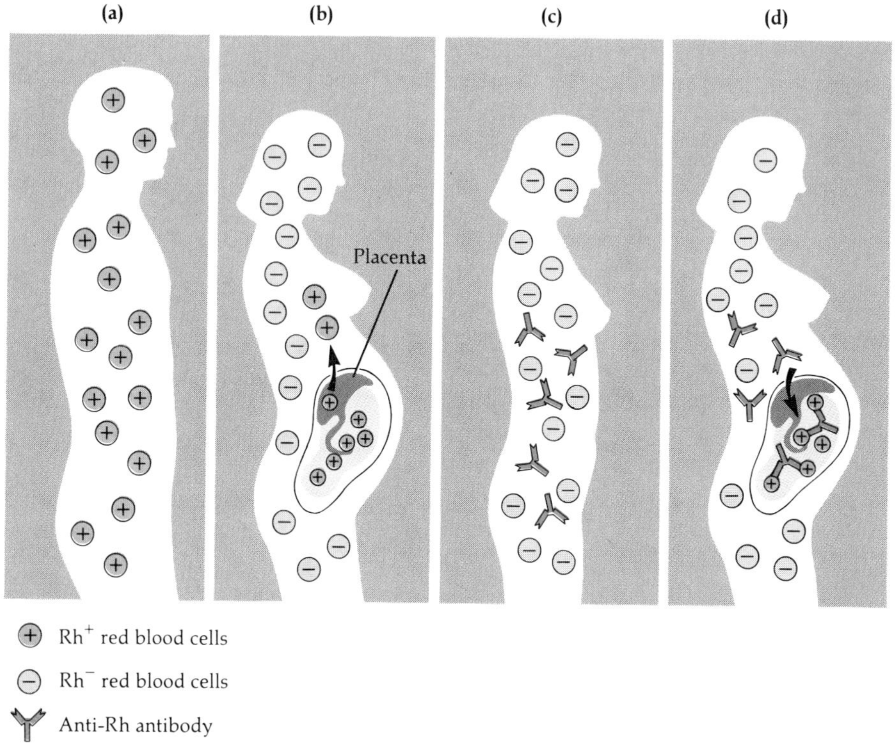

Figure 17–10 Hemolytic disease of the newborn. **(a)** Rh^+ father. **(b)** Rh^- mother carrying her first Rh^+ fetus. Rh antigens from the developing fetus can enter the mother's blood across the placenta during delivery. **(c)** In response to the fetal Rh antigens, the mother will produce anti-Rh antibodies. **(d)** If the woman becomes pregnant again with an Rh^+ fetus, her anti-Rh antibodies will pass through the placenta into the blood of the fetus. The result is damage to fetal red blood cells.

have entered the mother's circulation, so it is much less likely that she will become sensitized to the Rh antigen. If the disease is not prevented, it might be necessary that the newborn's Rh^+ blood, contaminated with maternal antibodies, be replaced with an uncontaminated transfusion.

STUDY OUTLINE

HYPERSENSITIVITY (pp. 472–478)

1. Hypersensitivity reactions represent immunologic responses to an antigen (allergen), which lead to tissue damage rather than immunity.
2. Hypersensitivity reactions occur only when a person has been sensitized to an antigen.
3. Hypersensitivity reactions can be divided into two groups: Types I, II, and III are based on humoral immunity and Type IV is based on cell-mediated immunity.

Type I (Anaphylaxis) Reactions (pp. 472–475)

1. Anaphylactic reactions involve the production of IgE antibodies that bind to mast cells and basophils to sensitize the host.

2. Binding of two adjacent IgE antibodies to an antigen causes the target cell to release chemical mediators, such as histamine, leukotrienes, and prostaglandins, which cause the observed allergic reactions.
3. Systemic anaphylaxis may develop in minutes after injection or ingestion of the antigen; this may result in circulatory collapse and death.
4. Localized anaphylaxis is exemplified by hives, hayfever, and asthma.
5. Skin testing is useful in determining sensitivity to an antigen.
6. Desensitization to an antigen can be achieved by repeated injections of the antigen, which leads to the formation of blocking (IgG) antibodies and Ts cells.

Type II (Cytotoxic) Reactions (pp. 475–476)

1. Type II reactions are mediated by IgG or IgM antibodies and complement.
2. The antibodies are directed toward host cell antigens; complement fixation may result in cell lysis or phagocytosis.
3. Myasthenia gravis is caused by antibodies reacting with neuromuscular synapses.

Type III (Immune Complex) Reactions (pp. 476–477)

1. Immune complex diseases occur when IgM or IgG antibodies and antigen form small complexes that lodge in the basement membranes of cells.
2. Subsequent complement fixation results in inflammation.
3. Glomerulonephritis, systemic lupus erythematosus, and rheumatoid arthritis are immune complex diseases.

Type IV (Cell-Mediated) Reactions (pp. 477–478)

1. Delayed hypersensitivity responses exemplified by the positive tuberculin skin test or contact sensitivities are cell-mediated reactions.
2. Type IV reactions develop slowly (one to three days).
3. Sensitized T cells secrete lymphokines in response to the appropriate antigen.
4. Lymphokines attract and activate macrophages and initiate tissue damage.
5. Allergic contact dermatitis is a reaction to environmental chemicals, which are haptens that combine with skin proteins.

AUTOIMMUNITY (pp. 478–481)

Loss of Immunologic Tolerance (pp. 478–479)

1. Immunologic tolerance represents a state of unresponsiveness to a specific antigen.
2. Tolerance to self-antigens occurs under natural conditions and develops during fetal development.
3. Autoimmunity is a humoral or cell-mediated immune response against self-antigens. Autoimmune responses frequently result in disease.
4. Some antibodies formed against foreign antigens can cross-react with self-antigens (for example, rheumatic fever).
5. Cytotoxic reactions are autoimmune.

Major Histocompatibility Complex (pp. 479–480)

1. Histocompatibility antigens located on cell surfaces express genetic differences between individuals; these antigens are coded for by MHC or HLA gene complexes.
2. To prevent rejection, HLA and ABO blood group antigens are matched as closely as possible.

Transplantation (pp. 480–481)

1. Transplantation to a privileged site (such as the cornea) or of a privileged tissue (such as pig heart valves) does not cause an immune response.
2. Four types of transplants have been defined on the basis of genetic relationships between the donor and the recipient: autografts, isografts, allografts, and xenografts.
3. Rejection of transplanted tissue represents an immune response by the recipient to foreign HLA antigens on the transplanted tissue.
4. Rejection may be suppressed by drugs, X-irradiation, or antilymphocyte serum.
5. Immunosuppression can lead to susceptibility to infectious disease and cancer.

Natural Immune Deficiencies (p. 481)

1. AIDS is caused by a virus that infects T cells.
2. Immune deficiencies may be inherited.

IMMUNE RESPONSE TO CANCER (pp. 481–485)

Immunologic Surveillance (pp. 481–484)

1. Cancer cells have undergone transformation and fail to exhibit contact inhibition and possess tumor-specific antigens.
2. The response of the immune system to cancer is called immunologic surveillance.

Immunologic Escape (pp. 484–485)

1. The ability of tumor cells to escape immune responses mounted against them is called immunologic escape.
2. Antigen modulation and immunologic enhancement represent two ways tumor cells escape detection and destruction by the immune system.

Immunotherapy (p. 485)

1. Immunotoxins are the most promising cancer treatment being researched.
2. Immunotoxins are chemical poisons with a monoclonal antibody; the antibody selectively locates the cancer cell for release of the poison.

TRANSFUSION REACTIONS AND Rh INCOMPATIBILITY (pp. 485–488)

The ABO Blood Group System (pp. 486–487)

1. Human blood may be grouped into four principal types designated A, B, AB, and O.
2. The presence or absence of two carbohydrate antigens designated A and B on the surface of the red blood cell determine a person's blood type.
3. Naturally occurring antibodies are present or absent in serum against the opposite AB antigen.
4. Incompatible blood transfusions lead to the complement-mediated lysis of the donor red blood cells.

The Rh Blood Group System (pp. 487–488)

1. Approximately 85% of the human population possesses another blood group antigen designated the Rh antigen; these individuals are designated Rh^+.
2. The absence of this antigen in certain individuals (Rh^-) can lead to sensitization upon exposure to it.

Blood Transfusions and Rh Incompatibility (p. 487)

1. An Rh^+ person can receive Rh^+ or Rh^- blood transfusions.
2. When an Rh^- person receives Rh^+ blood, that person will produce anti-Rh antibodies.
3. Subsequent exposure to Rh^+ cells will result in a rapid hemolytic reaction.

Hemolytic disease of the newborn (pp. 487–488)

1. An Rh^- mother carrying an Rh^+ fetus will produce anti-Rh antibodies.
2. Subsequent pregnancies involving Rh incompatibility may result in hemolytic disease of the newborn.
3. The disease may be prevented by passive immunization of the mother with anti-Rh antibodies just after giving birth.

STUDY QUESTIONS

REVIEW

1. Define hypersensitivity.
2. Compare and contrast the characteristics of the four types of hypersensitivity reactions. Give an example of each type.
3. List three mediators released in anaphylactic hypersensitivities and explain their effects.
4. Contrast systemic and localized anaphylactic reactions. Which type is more serious? Cite an example of each.

5. Explain what happens when a person develops a contact sensitivity to poison oak.
   (a) What causes the observed symptoms?
   (b) How did the sensitivity develop?
   (c) How might this person be desensitized to poison oak?
6. Discuss the roles of antibodies and antigens in an incompatible tissue transplant.
7. What happens to the recipient of an incompatible blood type?
8. Explain how hemolytic disease of the newborn develops and how this disease might be prevented.
9. Which of the following blood transfusions are compatible? Explain your answers.

	Donor	Recipient
(a)	AB, Rh^-	AB, Rh^+
(b)	B, Rh^+	B, Rh^-
(c)	A, Rh^+	O, Rh^+

10. Which type of graft (autograft, isograft, allograft, or xenograft) is most compatible? Least compatible?
11. Define immunosuppression. Why is it used and what problems does it cause?
12. Define autoimmunity. Present a theory that could explain autoimmune responses. Discuss one autoimmune disease in relation to this theory.
13. In what ways do tumor cells differ antigenically from normal cells? Explain how tumor cells may be destroyed by the immune system.
14. If tumor cells can be destroyed by the immune system, how does cancer develop? What does immunotherapy involve?

CHALLENGE

1. When and how does our immune system discriminate between self and nonself antigens?
2. The first preparations used for artifically acquired passive immunity were antibodies in horse serum. A complication that resulted from the therapeutic use of horse serum was immune-complex disease. Why did this occur?
3. After working in a mushroom farm for several months, a worker develops these symptoms: hives, edema, and swelling of lymph nodes.
   (a) What do these symptoms indicate?
   (b) What mediators cause these symptoms?
   (c) How may sensitivity to a particular antigen be determined?
   (d) Other employees do not appear to have any immunologic reactions. What could explain this?
   (*Note:* The allergen is conidiospores from molds growing in the mushroom farm.)
4. Physicians administering live, attenuated mumps and measles vaccines prepared in chick embryos are instructed to have epinephrine available. Since epinephrine will not treat these viral infections, what is the purpose of keeping this drug on hand?
5. Possible adverse reactions to rabies vaccines are given below. Explain the causes of these reactions. (Refer to pp. 562–564.)

Vaccine	Adverse reaction
DEV	Type I hypersensitivity
HDCV	Type III hypersensitivity

FURTHER READING

Bach, J. F. (ed.). 1978. *Immunology*. New York: Wiley. A collection of papers on immediate and delayed hypersensitivities, autoimmunity, and cancer.

Brodt, P. 1983. Cancer Immunology—Three Decades in Review. *Annual Review of Microbiology*. 37:447–476. A discussion of tumor antigens and cellular and humoral responses.

Buisseret, P. D. August 1982. Allergy. *Scientific American* 247:86–95. Discusses the roles of mast cells, IgE, and antigens in hypersensitivity responses.

Cunningham, B. A. October 1977. The structure and function of histocompatibility antigens. *Scientific American* 237:96–107. The role of cell-surface antigens in transplant rejection and protection against infections and cancer.

Myrvik, Q. N. and R. S. Weiser. 1984. *Fundamentals of Immunology*. Philadelphia: Lea and Febiger. A textbook on immunology that covers basic principles and includes the immune responses to fungi, parasites, and viruses.

Notkins, A. L. and H. Koprowski. January 1973. How the immune response to a virus can cause cancer. *Scientific American* 228:22–31. Allergies, immune-complex diseases, and autoimmunity may be due to immune responses to viruses.

Old, L. J. May 1977. Cancer immunology. *Scientific American* 236:62–79. An explanation of why cancer cells can escape the immune system.

Parker, C. W. (ed.). 1980. *Clinical Immunology*, 2 vols. Philadelphia, PA: W. B. Saunders. A comprehensive textbook and reference for immunology emphasizing topics covered in this chapter.

Samter, M. (ed.). 1978. *Immunological Diseases*, 3rd ed., 2 vols. Boston, MA: Little, Brown and Company. Mechanisms and symptoms of hypersensitivities, atopic diseases, allergies, and other immunologic diseases.

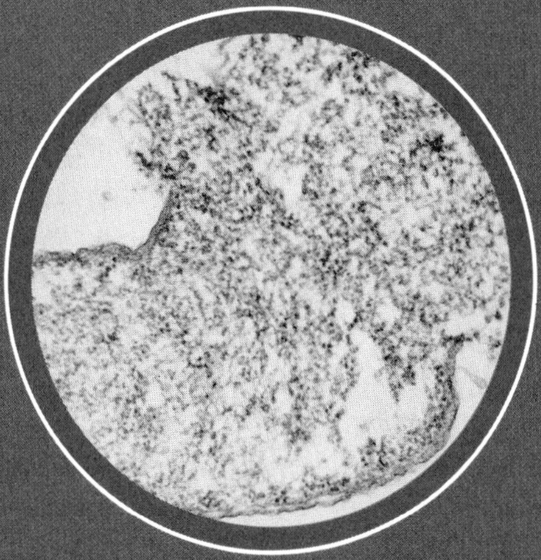

CHAPTER 18

Antimicrobial Drugs

OBJECTIVES

After completing this chapter you should be able to

- Define a chemotherapeutic agent and distinguish between a synthetic drug and an antibiotic.
- Identify the contributions of Ehrlich and Fleming in the field of chemotherapy.
- List the criteria used to evaluate antimicrobial agents.
- Identify five methods of action of antimicrobial agents.
- Describe the methods of action of each of the commonly used antibacterial drugs.
- Describe the problems of chemotherapy for viral, fungal, protozoan, and helminthic infections.
- Explain the actions of currently used antiviral, antifungal, antiprotozoan, and antihelminthic drugs.
- Describe three tests for microbial susceptibility to chemotherapeutic agents.
- Describe the mechanisms of drug resistance.

Sometimes the balance between microorganism and host tilts in the direction of the microbe, and the body's normal defenses cannot prevent or overcome the disease. When this occurs, we might turn to **chemotherapy,** the treatment of disease with chemicals (drugs) taken into the body. The term **chemotherapeutic agent** applies to any drug used for any disease—whether a simple headache, a "strep" throat, malaria, high blood pressure, or cancer. For example, aspirin and penicillin are both chemotherapeutic agents.

In this chapter, our discussion is limited to **antimicrobial drugs;** that is, the class of chemotherapeutic agents used to treat infectious diseases. Like the disinfectants discussed in Chapter 7, these chemicals act by interfering with the growth of infectious microorganisms. Unlike disinfectants, however, they must act *within* the host. Therefore, their effects on cells and tissues of the host are important; the ideal antimicrobial drug kills the harmful microorganism without damaging the host.

The drugs used in the chemotherapy of infectious disease fall into two groups. Drugs that have been synthesized by chemical procedures in the laboratory are called **synthetic drugs.** Others produced by bacteria and fungi are called **antibiotics.**

HISTORICAL DEVELOPMENT

The birth of modern chemotherapy is largely due to the efforts of Dr. Paul Ehrlich in Germany during the early part of this century. While attempting to stain bacteria without staining the surrounding tissue, he speculated about some "magic bullet" that would selectively find and destroy pathogens but not harm the host. This idea provided the basis for chemotherapy, a term he coined. Ehrlich eventually discovered a chemotherapeutic agent called salvarsan, an arsenic derivative that is effective against syphilis (at least by the standards of the day). Prior to Ehrlich's discovery, there had really been only one effective chemotherapeutic agent in the medical arsenal. This was quinine, a drug used for the treatment of malaria.

In the late 1930s, the discovery of miracle sulfa drugs, a group of synthetic drugs, touched off a new interest in chemotherapy. This discovery evolved from a systematic survey of chemicals, including many synthesized derivatives of aniline dyes. The chemical *prontosil,* which had been first synthesized and used as a dye many years before, was found to be an effective antimicrobial agent. Oddly enough, it worked only within living creatures such as rabbits, and was completely ineffective in the test tube against the same bacteria. It turned out that the active ingredient in prontosil is *sulfanilamide,* which forms as the rabbit metabolizes prontosil. This discovery led to the rapid development of a number of related drugs, the *sulfonamides,* or *sulfa drugs,* which are still used today. The sulfa drugs closely resemble a bacterial metabolite known as para-aminobenzoic acid (PABA), which some bacteria require for their synthesis of a vitamin called folic acid. Sulfa drugs competitively inhibit the incorporation of PABA into folic acid in bacteria (this process will be discussed later). Because humans meet their need for folic acid by

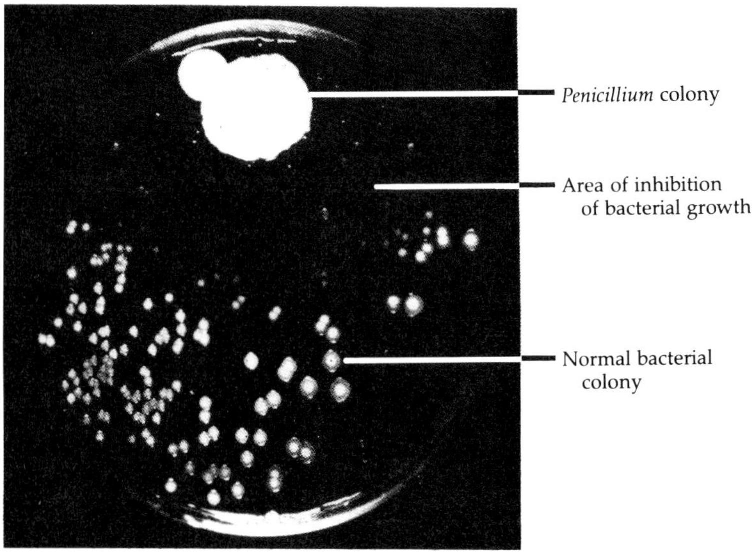

Figure 18–1 Discovery of penicillin. Near the contaminating colony of *Penicillium*, the growth of *Staphylococcus aureus* is inhibited and cells are lysing. This is a photograph of a plate from Fleming's laboratory.

ingesting it in food, they are not adversely affected by sulfonamides.

In London, in 1929, Dr. Alexander Fleming observed that the growth of the bacterium *Staphylococcus aureus* was inhibited in the area surrounding the colony of a mold that had contaminated a Petri plate (Figure 18–1). The mold was identified as *Penicillium notatum* (pe-ni-sil'lē-um nō-tä-tum), and its active compound, which was isolated a short time later, was named *penicillin*. Similar inhibitory reactions between colonies on solid media are commonly observed in microbiology and the mechanism of inhibition is referred to as *antibiosis*. From this term comes the term **antibiotic,** a substance that is produced by microorganisms and that in small amounts will inhibit another microorganism. Therefore, the wholly synthetic sulfa drugs are not technically antibiotics.

Because microorganisms capable of antibiosis are fairly common, it is surprising that antibiotics were not developed much earlier. In fact, there is evidence in the scientific literature that penicillin had been seriously investigated in the late 1800s. But the earlier researchers had the same problem that Fleming later had: the extract of the mold was unstable and production of penicillin by the mold was unreliable, so the drug was not very useful clinically. In 1940, a group of scientists at Oxford University headed by H. Florey and E. Chain succeeded in the first clinical trials of penicillin. Intensive research in the United States then led to the isolation of especially productive *Penicillium* strains for use in mass production of the antibiotic. (The most famous of these high-producing strains was originally isolated from a cantaloupe bought at a market in Peoria, Illinois.) Penicillin is still one of our most effective antibiotics, and its enormous success led to the search for others, a search that has revolutionized medicine. You might be interested to know that penicillin has a singular toxicity for guinea pigs, and it is entirely possible that this drug would never have left the laboratory given the more demanding safety requirements that must be met today in the development of drugs.

Antibiotics are actually rather easy to discover, but few are of medical or commercial value, although some are used commercially other than to treat disease (see box on page 496). Many antibiotics are toxic to humans or lack any advantage over antibiotics already in use.

Table 18–1 Representative Sources of Antibiotics

Microorganism	Antibiotic (Trademarked Name in Parentheses)
Gram-Positive Rods:	
Bacillus subtilis	Bacitracin
Bacillus polymyxa	Polymyxin
Bacillus brevis	Tyrothricin
Actinomycetes:	
Streptomyces parvullus	Actinomycin D (Dactinomysin)
Streptomyces nodosus	Amphotericin B (Fungizone)
Streptomyces venezuelae	Chloramphenicol (Chloromycetin)
Streptomyces aureofaciens	Chlortetracycline (Aureomycin) and Tetracycline
Streptomyces erythraeus	Erythromycin (Ilotycin)
Streptomyces kanamyceticus	Kanamycin (Kantrex)
Streptomyces fradiae	Neomycin
Streptomyces noursei	Nystatin (Mycostatin)
Streptomyces rimosus	Oxytetracycline (Terramycin)
Streptomyces griseus	Streptomycin
Micromonospora purpureae	Gentamicin
Fungi:	
Cephalosporium	Cephalothin (Keflin)
Penicillium griseofulvum	Griseofulvin (Fulvisin)
Penicillium notatum	Penicillin

More than half of our antibiotics are produced by species of *Streptomyces* (strep-tō-mī'sēs), a filamentous bacteria numerous in soil. A few antibiotics are produced by bacteria of the genus *Bacillus*, and others by molds, mostly of the genera *Penicillium* or *Cephalosporium* (sef-ä-lō-spô'rē-um) (Table 18–1).

CRITERIA FOR ANTIMICROBIAL DRUGS

The value of a chemotherapeutic agent can be measured against a number of criteria, including the following:

1. The drug should demonstrate *selective toxicity*. This means that the drug should be toxic for the microorganism but not for the host.

Microbiology in the News

Controversy Surrounds the Use of Antibiotics in Feed

Thirty years ago livestock growers began using antibiotics in the feed of closely penned animals that were being fattened for market. The drugs helped reduce bacterial infections and their spread in such ripe conditions. They also had an unsuspected, and as yet unexplained, effect of accelerating the animals' growth. Today, 70 percent of all cattle and 90 percent of all veal calves and swine are reared on feed containing either penicillin or tetracycline. Nearly half of the antibiotics sold in this country find their way into livestock feed.

Nevertheless, this subtherapeutic use of antibiotics, as it is called by the drug industry, has come under attack. Critics contend that such use might create strains of bacteria immune to drugs that are commonly used to treat human infections. Such resistant bacteria, scientists argue, could theoretically transfer their resistance to bacteria that infect humans in what is called the cross-over effect. Scientists also argue that some resistant bacteria, such as *Salmonella,* could be transferred to humans directly in meat or milk.

In 1978, the Food and Drug Administration proposed cutting back on the use of penicillin and tetracycline in animal feed. Opposition from livestock growers, feed producers, and the drug industry, however, stopped plans to cut down on antibiotic use. However, new evidence cropping up in the past few years confirms the microbiologists' suspicions.

In 1982, Dr. Thomas O'Brien of the Harvard University School of Medicine showed that bacteria that infect humans and other animals freely share genetic information by exchanging plasmids. Another study, by researchers at the Centers for Disease Control, confirmed the suspicion that resistant microorganisms might be transferred directly from meat to humans. Published in August 1984, this study showed that most of the outbreaks of antibiotic-resistant *Salmonella* in the past ten years could be traced to meat from animals that fed on grains laced with antibiotics. These findings were corroborated by a 1984 report on an outbreak of *Salmonella* poisonings caused by meat from South Dakota infected with an antibiotic-resistant strain called *Salmonella newport.*

The scientists at the Centers for Disease Control also noted that 20 to 30 percent of all *Salmonella* poisonings involve antibiotic-resistant strains. In those cases reported to the Center, the death rate was 4.2 percent among victims of drug-resistant strains, compared to only 0.2 percent among victims of drug-sensitive bacteria.

Cattle growers, feed manufacturers, and the drug industry still argue that the link between antibiotics in feed and human disease is sketchy, and claim that the benefits derived from the use of antibiotics far outweigh the potential deleterious effects. Research in progress, however, may eventually cause U.S. farmers to follow the example set by Europe, where antibiotic use in animal feed has been limited since the early 1970s.

With some exceptions, most chemotherapeutic agents are somewhat toxic to host cell metabolism, but usually not toxic enough to interfere with normal body functions.

2. The drug should not produce *hypersensitivity* (allergy) in most hosts (see Chapter 17). Allergic reactions, which are produced by the immune system, are not the same as toxicity. Penicillin, for example, sometimes causes allergic reactions but is one of the antibiotics least toxic for humans.

3. A drug must have *solubility* in body fluids so that it can rapidly penetrate body tissues. The

rates at which the drug is broken down and/or excreted from the body must be low enough that the drug remains in the infected body tissue long enough to exert its effects. Furthermore, a good chemotherapeutic agent has a *long shelf life* at normal refrigeration temperatures.

4. Microorganisms should not readily become *resistant* to the drug. Resistance can be due to a number of mechanisms and will be discussed later in this chapter.

Not all of these properties are displayed by every chemotherapeutic agent. However, the best possible combination is sought.

SPECTRUM OF ACTIVITY

It is a comparatively simple task to find or develop drugs effective against procaryotic cells (bacteria) that do not affect the eucaryotic cells of humans. These two cell types differ substantially in many ways such as in the presence or absence of cell walls, fine structure of their ribosomes, and details of their metabolism. Selective toxicity has numerous targets. It becomes a more difficult problem when the pathogen is another eucaryotic cell such as a fungus, protozoan, or helminth. At the cellular level, these organisms resemble the human cell much more closely than does a bacterial cell. We shall see that our arsenal against these types of pathogens is much more limited than our arsenal of antibacterial drugs. Viral infections are a particularly difficult problem to deal with. Here the pathogen is within the human host's cells and the genetic information of the virus is directing the human cell to make viruses rather than to synthesize normal cellular materials. The more we discover about the reproduction, metabolism, and structure of pathogens, the better equipped we are to discover fresh targets for antimicrobials. No doubt the search will be never ending.

Penicillin has a *narrow spectrum of activity*, affecting gram-positive bacteria and very few gram-negative bacteria. Other drugs have a *broad spectrum of activity*, affecting a large number of gram-positive or gram-negative bacteria and are therefore termed broad spectrum drugs. A broad spectrum drug would seem to have an advantage in the treatment of disease. Since the identity of the pathogen is not always immediately known, valuable time could be saved. The disadvantage is that much of the normal flora of the host is destroyed. The normal flora ordinarily competes with and checks the growth of pathogens or other microbes. If certain organisms in the normal flora are not destroyed by the antibiotic, they can flourish and become *opportunistic pathogens*. An example is the overgrowth by the yeastlike fungus *Candida albicans* (kan'did-ä al'bi-kans), which is not sensitive to antibacterial antibiotics. This overgrowth is called a **superinfection,** a term also applied to growth of a target pathogen that has developed resistance to the antibiotic. Such an antibiotic-resistant strain replaces the original sensitive strain and the infection continues.

ACTION OF ANTIMICROBIAL DRUGS

Antimicrobial drugs either kill microorganisms directly (**bactericidal**) or simply prevent them from growing (**bacteriostatic**). In bacteriostasis, the host's own defenses, such as phagocytosis and antibody production, usually destroy the microorganisms.

We will now examine the various ways in which antibiotics exert their antimicrobial activity. The major modes of action are summarized in Figure 18–2 and Table 18–2.

Inhibition of Cell Wall Synthesis

You recall from Chapter 4 that the cell wall of a bacterium consists of a macromolecular network called *peptidoglycan*. Peptidoglycan is found only in bacterial cell walls. Penicillin and certain other antibiotics prevent the synthesis of intact peptidoglycan; the cell wall is greatly weakened and the cell lyses (Figure 18–3). Because the drug affects only the synthesis process, only actively growing cells are affected by these antibiotics. And, because human cells do not have peptidoglycan cell walls, penicillin has very little toxicity for the host cells. You may also recall that gram-negative bacteria have an outer membrane of lipopolysaccharides and other materials over a comparatively thin layer of peptidoglycan. Because this outer membrane apparently prevents penicillin from reaching the site of peptidoglycan synthesis, gram-negative bacteria are relatively resistant to penicillin.

There are several ways in which cell-wall synthesis can be affected. For example, cycloserine, a toxic and seldom used antibiotic, interferes with the formation of a precursor necessary to pep-

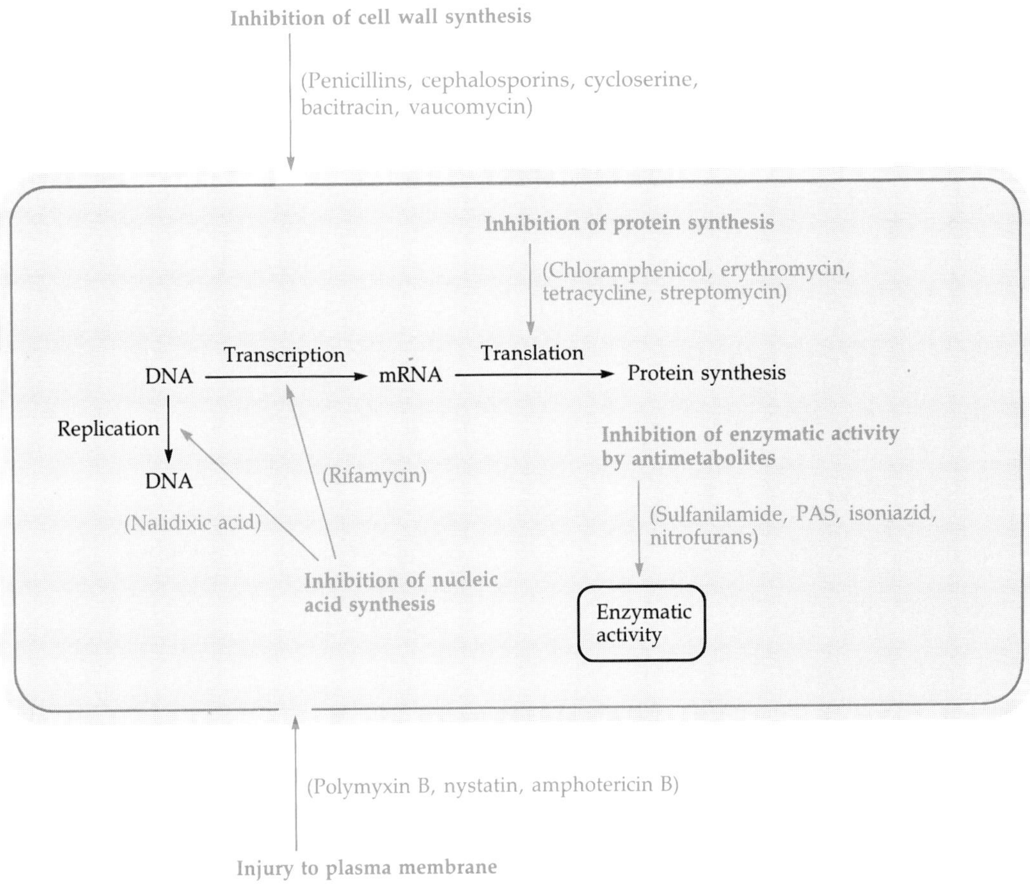

Figure 18–2 Summary of the actions of antimicrobial drugs in a highly diagrammatic, composite microbial cell. All of these drugs act selectively on procaryotic (bacterial) cells, except for nystatin and amphotericin B, which act on eucaryotic (fungal) cells.

Table 18–2 Summary of Actions of Chemotherapeutic Agents

Chemotherapeutic Action	Example
Inhibition of cell wall synthesis	Penicillin, cephalosporins, cycloserine, bacitracin, vancomycin
Inhibition of protein synthesis	Chloramphenicol (Chloromycetin), erythromycin (Ilotycin), tetracycline, streptomycin
Injury to plasma membrane	Polymyxin B, nystatin (Mycostatin), amphotericin B
Inhibition of nucleic acid synthesis	Rifamycin, nalidixic acid, 5-fluorocytosine, trimethoprim
Inhibition of enzymatic activity	Sulfanilamide, para-aminosalicylic acid (PAS), isoniazid, nitrofurans

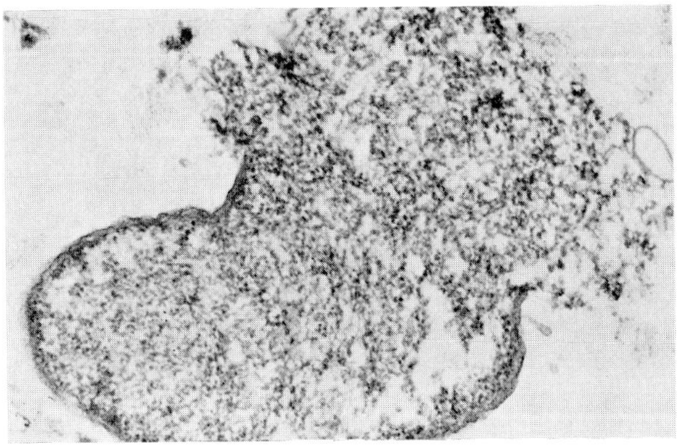

Figure 18–3 Effect of antibiotics on a bacterial cell wall. Treatment with Tobramycin causes lysis of the cell wall of *Pseudomonas aeruginosa*. Treatment with Gentamicin or Polymyxin B (either is recommended) would achieve the same results.

tidoglycan synthesis. Two commonly used antibiotics, bacitracin and vancomycin, interfere with the synthesis of the linear strands of peptidoglycan. Penicillin and cephalosporins prevent the final cross-linking of the peptidoglycans, which interferes with the construction of the macromolecular cell wall.

Inhibition of Protein Synthesis

Because protein synthesis is a common feature of all cells, both procaryotic and eucaryotic, it would seem an unlikely target for selective toxicity. One notable difference between procaryotes and eucaryotes, however, is the structure of their ribosomes. As discussed in Chapter 4, eucaryotic cells have 80S ribosomes; procaryotic cells have 70S ribosomes. (Recall that the 70S ribosome is made up of a 50S and a 30S unit.) The difference in ribosomal structure accounts for the selective toxicity of antibiotics that affect protein synthesis. However, remember that mitochondria, important eucaryotic organelles, also contain 70S ribosomes similar to those of bacteria. Antibiotics targeting the 70S ribosomes can therefore have adverse effects on cells of the host. Among the antibiotics that interfere with protein synthesis are chloramphenicol, erythromycin, streptomycin, and tetracyclines (Figure 18–4).

Reacting with the 50S portion of the 70S procaryotic ribosome, chloramphenicol inhibits the formation of peptide bonds in the growing polypeptide chain. Erythromycin also reacts with the

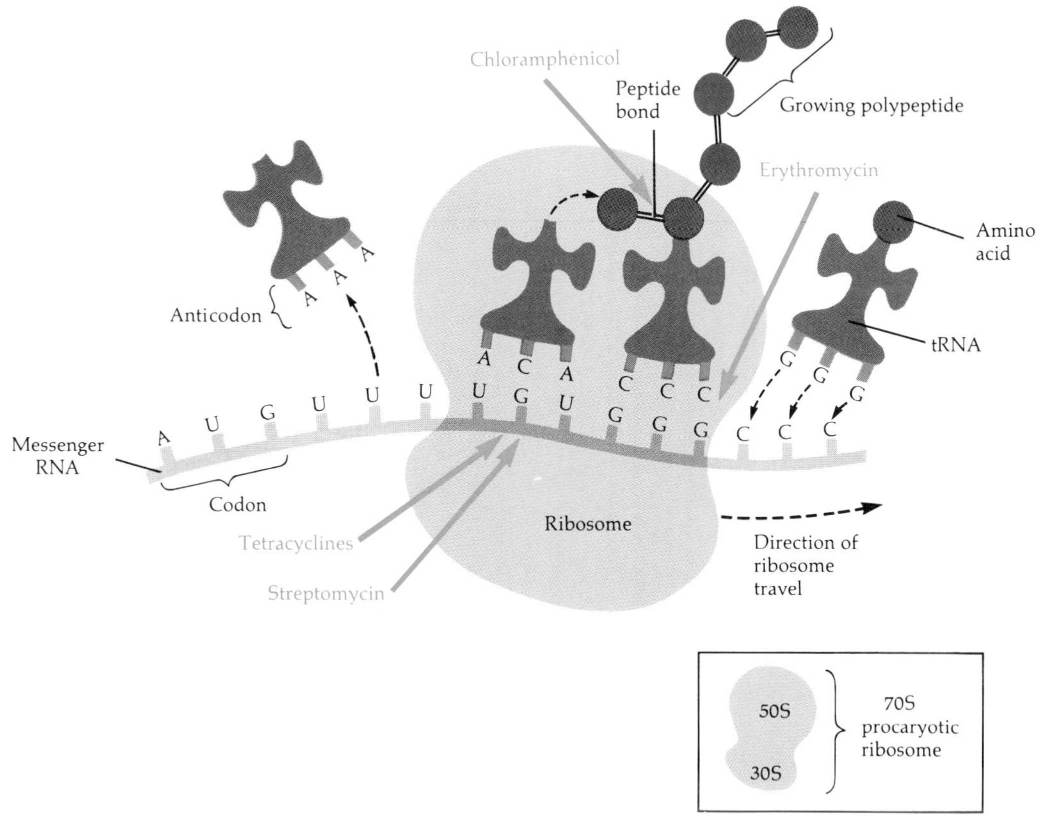

Figure 18–4 Inhibition of protein synthesis by antibiotics. The arrows indicate the specific points at which chloramphenicol, erythromycin, tetracyclines, and streptomycin exert their activity.

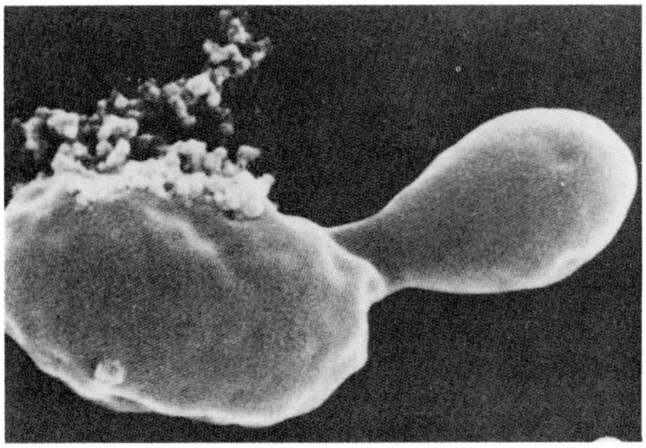

Figure 18–5 Disintegration of a yeast cell, with release of cytoplasmic contents, as the plasma membrane is disrupted by the antifungal drug miconazole.

50S portion of the 70S procaryotic ribosome. Most drugs that inhibit protein synthesis have a broad spectrum of activity; erythromycin is an exception. Because it does not penetrate the gram-negative cell wall, it affects mostly gram-positive bacteria. To inhibit protein synthesis, some other antibiotics react with the 30S portion of the 70S procaryotic ribosome. The tetracyclines interfere with the attachment of the amino acid carrying tRNA to the ribosome and prevent the addition of amino acids to the growing polypeptide chain. Tetracyclines can also interfere with cell-free mammalian ribosomes as well, but apparently this drug does not readily penetrate the intact mammalian cell. Fortunately, the intracellularly located rickettsias and chlamydias are sensitive to tetracyclines. Compared to the host cell, perhaps they are unusually susceptible.

By changing the shape of the 30S portion of the 70S procaryotic ribosome, aminoglycoside antibiotics such as gentamicin and streptomycin interfere with the initial steps of protein synthesis. This interference causes the genetic code on the mRNA to be read incorrectly.

Injury to the Cell Membrane

Certain antibiotics, especially polypeptide antibiotics, bring about changes in the permeability of the plasma membrane; these changes result in the loss of important metabolites from the microbial cell. One such antibiotic is polymyxin B: by attaching to the phospholipids of the plasma membrane, it causes disruption of the membrane.

Some antifungal drugs, such as nystatin, amphotericin B, and the recently introduced ketoconazole, are effective against many systemic fungal diseases (Figure 18–5). They combine with sterols in the fungal plasma membrane to disrupt the membrane. Because bacterial plasma membranes generally lack sterols, these antibiotics do not act on bacteria. However, because the plasma membranes of animal cells do contain sterols, it is not surprising that nystatin and amphotericin B can be toxic to the host. Luckily, animal cell membranes have mostly *cholesterol,* and fungal cells mostly *ergosterol,* against which the drug is most effective, so that the balance of the toxicity is tilted toward the fungus.

Inhibition of Nucleic-Acid Synthesis

A number of antibiotics interfere with nucleic-acid metabolism of microorganisms. As was found to interfere with protein synthesis, some mechanism of action must be found that does not interfere with mammalian DNA-RNA synthesis as well. Some drugs with this mode of action, such as the antiviral idoxuridine, have an extremely limited usefulness because they are not selective enough in their toxicity. Others such as rifamycin, nalidixic acid, and trimethoprim are more widely used in chemotherapy.

Inhibition of Enzymatic Activity

You may recall from Chapter 5 that an enzymatic activity of a microorganism can be *competitively inhibited* by a substance (antimetabolite) that closely resembles the normal substrate for the enzyme. An example of competitive inhibition already considered is the relationship between the antimetabolite sulfanilamide (a sulfa drug) and para-aminobenzoic acid (PABA). In many microorganisms, PABA is the substrate for an enzymatic reaction leading to the synthesis of folic acid, a vitamin that functions as a coenzyme for the synthesis of the purine and pyrimidine bases of nucleic acids. In the presence of sulfanilamide, the enzyme that normally converts PABA to folic acid combines with the drug instead of PABA. This combination prevents folic-acid synthesis and stops the growth of the microorganism. Because humans do not produce folic acid from PABA (they obtain it as a vitamin in ingested foods), sulfanilamide exhibits selective toxicity: it affects microorganisms that synthesize their own folic acid but does not harm the human host. Many bacteria, some fungi, and some protozoans must synthesize folic acid. Other chemotherapeutic agents that act as antimetabolites are the sulfones and trimethoprim.

SURVEY OF COMMONLY USED ANTIMICROBIAL DRUGS

We will now examine the properties and activities of some commonly used antimicrobial drugs. The drugs discussed will be divided into four groups: synthetic antimicrobial drugs, antibiotics, antiviral drugs, and antiprotozoan and antihelminthic drugs.

Synthetic Antimicrobial Drugs

Isoniazid (INH)

Isoniazid (INH) is a very effective, synthetic, antimicrobial drug against *Mycobacterium tuberculosis.* It has little effect on nonmycobacteria. It is a structural analog similar to vitamin B6 (pyridoxine), which serves as a coenzyme in metabolic reactions. The primary effect of isoniazid might be to inhibit synthesis of mycolic acids, components of cell walls only of the mycobacteria, but the specifics of its mode of action are not entirely known. When used to treat tuberculosis, INH is usually administered simultaneously with other drugs such as streptomycin, rifampin, or ethambutol.

Ethambutol

Ethambutol is effective only against mycobacteria. Apparently by inhibiting the synthesis of certain cellular metabolites, the drug eventually kills the microbes. It is a comparatively weak antitubercular drug and its principal use is as the secondary drug to avoid resistance problems; for this purpose it has essentially replaced para-aminosalicylic acid (PAS).

Sulfonamides

As noted earlier, **sulfonamides (sulfa drugs)** were among the first synthetic antimicrobial drugs used to treat microbial diseases. Unfortunately, many pathogens have developed resistance to them. Moreover, sulfonamides cause allergic reactions in many people. Because of these reasons, antibiotics have tended to diminish the importance of sulfonamides in chemotherapy. The sulfonamides continue to be used to treat certain urinary tract infections and have other specialized uses, as in the combination drug *silver sulfadiazine,* used in the control of infections in burn patients. Sulfonamides are bacteriostatic, and as discussed, their action is due to their structural similarity to para-aminobenzoic acid (PABA) (see Figure 18–6). They competitively inhibit the synthesis of folic acid, an important precursor to nucleic acids. Probably the most widely used sulfa today is a combination of *trimethoprim* and *sulfamethoxazole (TMP–SMZ).* This combination is an excellent example of drug *synergism.* When the drugs are used in combination, only about one-tenth of the concentration necessary when each is used alone is needed; the combination also has a broader spectrum of action and the emergence of resistant strains is much reduced. Figure 18–7 illustrates how the two drugs interfere with different steps of the same metabolic sequence. Trimethoprim (see Figure 18–8) is not a sulfa drug, but is a structural analog of a portion of dihydrofolic acid and competitively inhibits an enzyme necessary to make folinic acid. TMP–SMZ is useful against most pathogens of the urinary tract and intestinal tract and is the drug of choice for treatment of *Pneumocystis* pneumonia, a protozoan disease.

Other synthetic antimicrobial drugs

Nitrofurans constitute a family of derivatives that are active against a large number of gram-positive and gram-negative bacteria, protozoans, and fun-

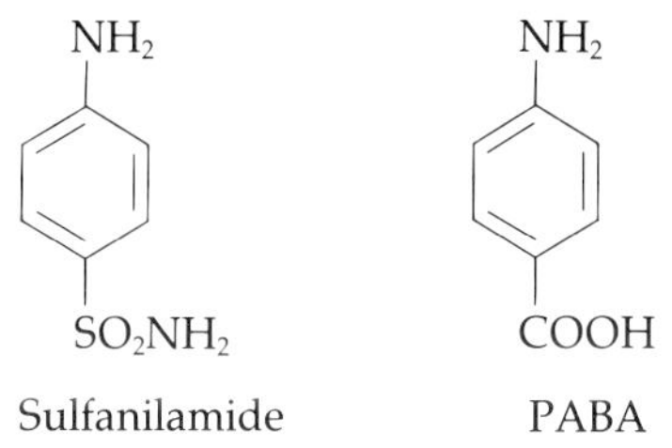

Figure 18–6 Structure of sulfanilamide, a representative sulfa drug. This drug is a competitive inhibitor of an enzyme whose normal substrate is PABA, which is shown for comparison.

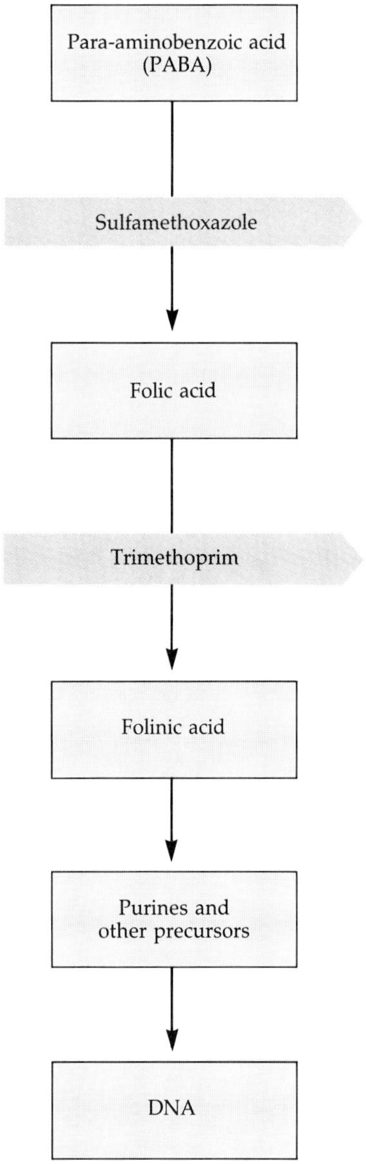

Figure 18–7 Sites of action of sulfamethoxazole and trimethoprim on synthesis of folic acid and folinic acid. Each drug interferes with a step in the sequence and so prevents DNA synthesis. Use of the drugs together is more efficient than use of either drug alone.

gi. *Nitrofurantoin* is widely administered orally for urinary tract infections. Like nalidixic acid (see below) it usually does not reach a high enough concentration elsewhere than in urine to be useful in systemic chemotherapy. Its primary mode of action is not precisely known, but it might cause fractures of DNA polymers. Some other nitrofurans such as *nitrofurazone* and *nifuroxime* are used topically for skin, eye, and vaginal infections.

Nalidixic acid stops bacterial synthesis of DNA and is especially effective against gram-negative enterics. Because it accumulates in the urine, this drug is used to treat urinary tract infections in which *Escherichia* (esh-ėr-i´kē-ä), *Proteus* (prō´tē-us), or *Klebsiella* species are implicated.

The drug *5-fluorocytosine* (Flucytosine) is an antimetabolite of cytosine and was originally synthesized as an anticancer drug; but it has found better use as an antifungal drug. Mammalian cells apparently do not convert the drug enzymatically to 5-fluorouracil, as is required for its activity, but this conversion occurs in fungi sensitive to the drug.

Antibiotics

Penicillin

The term **penicillin** refers to a group of chemically related antibiotics. Penicillin extracted from cultures of the mold *Penicillium* exists in several closely related forms. These are the so-called *natural penicillins* (Figure 18–9a). The prototype compound of all the penicillins is *penicillin G*. It has a narrow but useful spectrum of action and is the drug of choice against pathogenic cocci, gram-positive bacilli, and spirochetes that do not produce penicillinase. When injected intramuscularly, penicillin G is rapidly excreted from the body in 3 to 6 hours (Figure 18–10); when the drug is taken orally, the acidity of the digestive fluids in the stomach diminishes its concentration. *Procaine penicillin,* a combination of the drugs procaine and penicillin G, is retained much longer by the body. Procaine penicillin will be retained at detectable concentrations for up to 24 hours; concentration peaks at about 1 to 4 hours. Still longer retention times can be achieved with *benzathine penicillin,* a combination of benzathine and penicillin G, but at lower concentrations. Although retention times of as long as 4 months can be obtained, the concentration of the

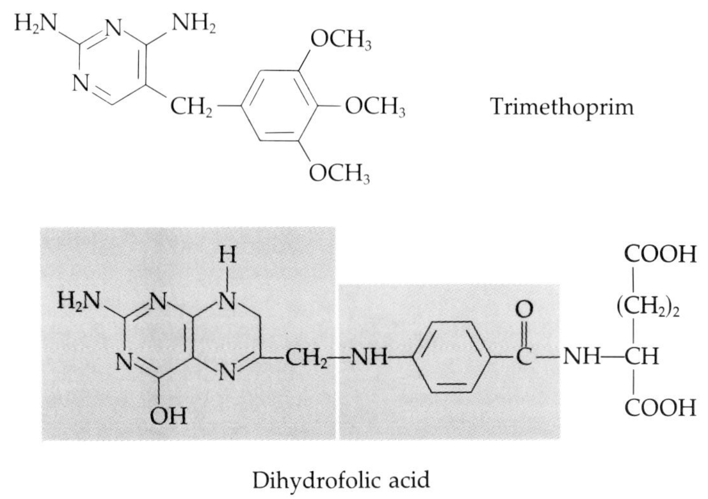

Figure 18–8 Trimethoprim is a structural analog of dihydrofolic acid, a precursor of folinic acid (see Figure 18–7). Trimethoprim is the analog of the left portion of the dihydrofolic acid molecule. Note the PABA moiety in the center portion of this same molecule—for which sulfamethoxazole is the structural analog.

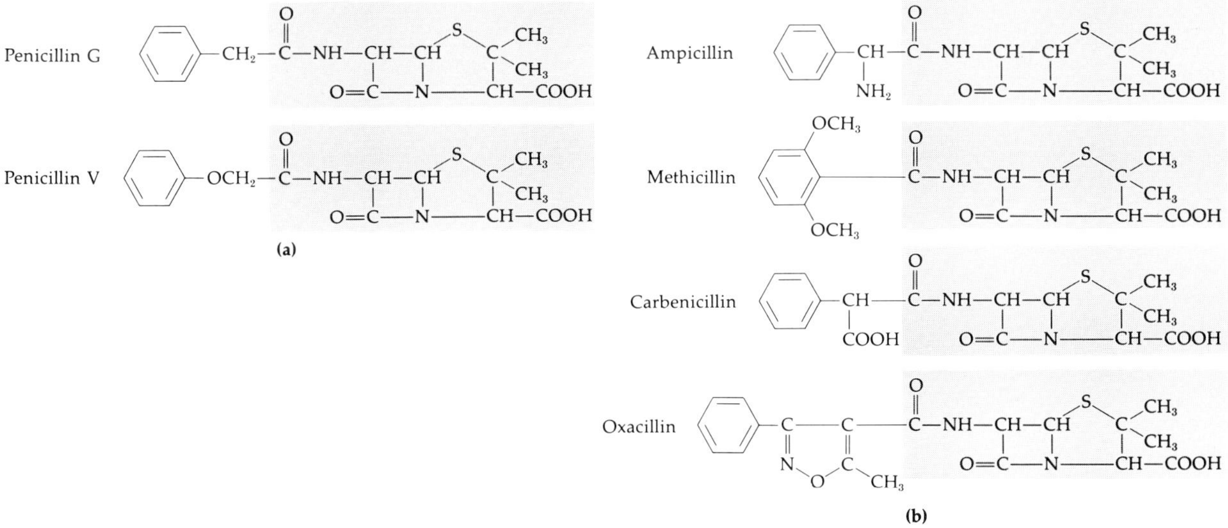

Figure 18–9 Structure of penicillins. The shaded portions of the diagrams represent the common nucleus, which includes the beta-lactam ring. The unshaded portions represent the side chains that distinguish one penicillin from another. **(a)** Natural penicillins **(b)** Semisynthetic penicillins.

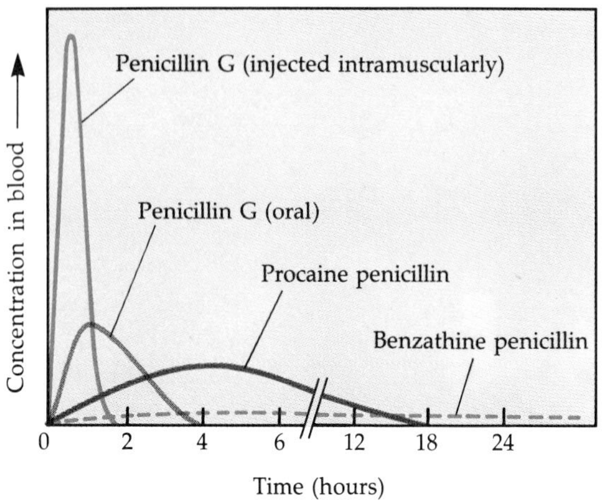

Figure 18–10 Penicillin G reaches different concentrations in the blood when administered by different routes and can be combined with other compounds to prolong its retention in the body.

drug is so low that the organisms must be very sensitive to the drug in order for it to work.

Penicillin G and V are the most often used natural penicillins. A large number of semisynthetic penicillins have been developed in attempts to overcome the natural penicillins' disadvantages, such as their narrow spectrum of activity and destruction by penicillinases. **Penicillinases** are enzymes produced by many bacteria, especially *Staphylococcus* spp., that cleave the beta-lactam ring of the penicillin molecule. Because of this characteristic, penicillinases are sometimes called *beta-lactamases.* The type of a penicillin molecule depends on chemical side chains attached to the beta-lactam ring (see Figure 18–9). In developing semisynthetic penicillins, scientists perfected a technique whereby they could either stop synthesis by *Penicillium* and obtain only the common penicillin nucleus, or remove the side chains from the completed natural molecules, and then chemically add other side chains. Thus the term *semisynthetic* is applied to these penicillins; part of the penicillin is produced by the mold and part is added synthetically.

Semisynthetic penicillins have been found that are more resistant to acid (and therefore more effective when taken orally), are resistant to penicillinases, and have a spectrum of activity that is often greater than that of the natural penicillins (see Figure 18–9b). For example, *ampicillin, amoxicillin,* and *carbenicillin* have a broad spectrum, and are effective against a number of important gram-negative pathogens. And, except for carbenicillin, they are acid-stable enough to be taken orally.

A number of variants resistant to penicillinases have been developed. *Methicillin* has been in use the longest and is administered by injection. For oral administration, *oxacillin* and *dicloxacillin* are useful penicillinase-resistant penicillins.

Some new, broader-spectrum variants, sometimes referred to as fourth generation penicillins, have recently been developed. Some of these variants show antipseudomonas activity and have fewer side effects than those of currently available aminoglycosides used for these infections. These new penicillins are sometimes referred to as *ureidopenicillins.* Examples are *mezlocillin* and *azlocillin.*

Cephalosporins

In structure, the nuclei of **cephalosporins** resemble that of penicillin (Figure 18–11). Cephalosporins are also similar in action to penicillin; they inhibit

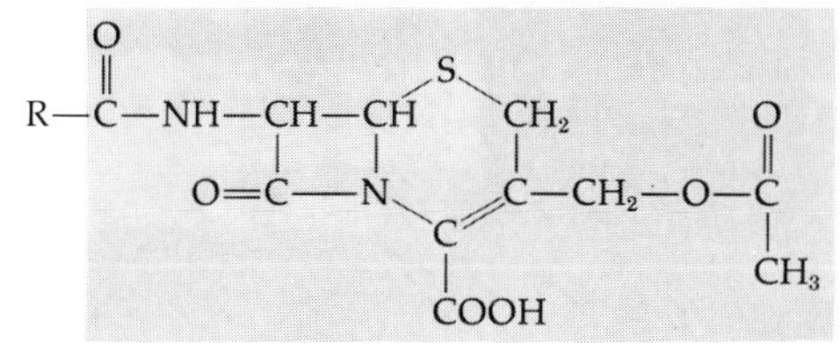

Cephalosporin nucleus

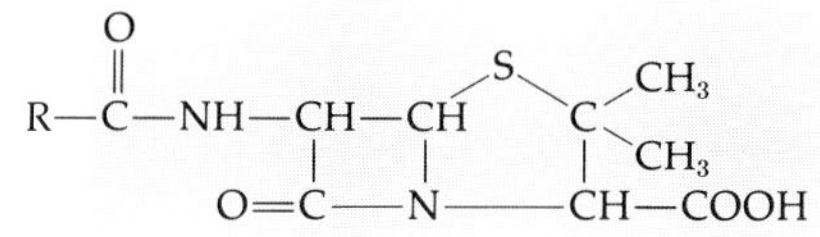

Penicillin nucleus

Figure 18–11 Comparison of the structure of the cephalosporin nucleus with that of penicillin. R is an abbreviation for groups of atoms that make one compound different from another.

the synthesis of cell walls. Although cephalosporins are structurally similar to penicillin, they are sufficiently different so that the cephalosporins are resistant to penicillinases and are effective against more gram-negative organisms than the natural penicillins. However, the cephalosporins are susceptible to their own beta-lactamases.

The number of cephalosporins has proliferated in recent years as what are known as second and third generation cephaloporins have been developed. Some characteristic cephalosporins are *cephalothin, cefamandole,* and *cefotaxime,* but an example from the most recent generation of cephalosporins is *moxalactam.* This has a variation of the beta-lactam ring, an *oxa-beta-lactam* ring. Most cephalosporins are injected, but a few may be taken orally. Cephalosporins generally are more expensive than penicillins.

Aminoglycosides

Aminoglycosides are a group of antibiotics in which amino sugars are linked by glycosidic bonds, hence the name. Probably the best known aminoglycoside is *streptomycin,* which was discovered in 1944 in a culture of *Streptomyces griseus* (gri-sē'us) taken from the throat of a chicken. Historically it is of considerable significance: streptomycin was the first antibiotic effective in the treatment of tuberculosis and against large numbers of gram-negative bacteria. Streptomycin is still used as an alternative drug in the treatment of tuberculosis, but rapid development of resistance and the appearance of serious toxic effects have diminished its usefulness.

Aminoglycosides include many important and widely used antibiotics. All are bactericidal, and their mode of action is that by attaching to the 30S unit of ribosomes, and causing a misreading of the genetic code, they inhibit protein synthesis. Aminoglycosides can affect hearing by causing permanent damage to the auditory nerve, and damage to the kidneys has also been reported. *Neomycin, spectinomycin,* and *gentamicin* (Figure 18–12) are important aminoglycosides. Spectinomycin is frequently used against penicillin-resistant gonorrhea. Neomycin is present in many topical preparations. Gentamicin plays a major role in treatment of many enteric, gram-negative infections and, importantly, in the treatment of *Pseudomonas aeruginosa* (sū-dō-mō'nas ā-rü-ji-nō'sä) infections.

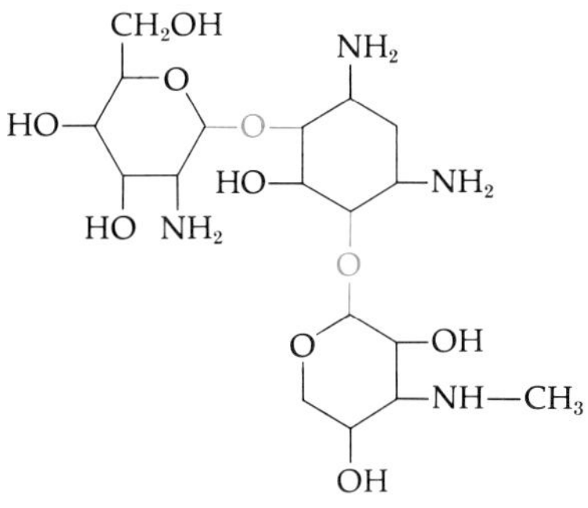

Figure 18–12 Structure of gentamicin, a representative aminoglycoside. Glycosidic bonds (color) are bonds between amino sugars.

When resistance to aminoglycosides is encountered, the semisynthetic aminoglycoside derivative *amikacin* is often prescribed.

Tetracyclines

Tetracyclines are a group of closely related, broad-spectrum antibiotics, produced by *Streptomyces* spp., that inhibit protein synthesis. They probably have the broadest spectrum of antibacterial activity of any known drug. Not only are they effective against gram-positive and gram-negative bacteria, they are also especially valuable against the intracellularly located rickettsias and chlamydias. Three of the more commonly encountered tetracyclines are *oxytetracycline* (*Terramycin*), *chlortetracycline* (*Aureomycin*) and *tetracycline* itself (Figure 18–13).

Some newer semisynthetic tetracylines such as *doxycycline* and *minocycline* are frequently used.

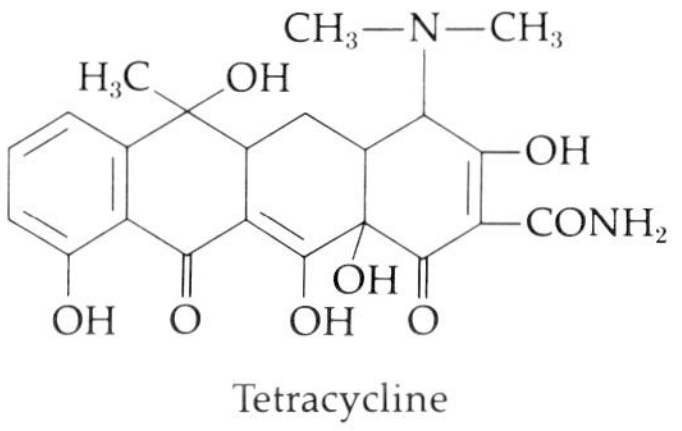

Figure 18–13 Structure of tetracycline. Other tetracycline-type antibiotics closely resemble tetracycline itself.

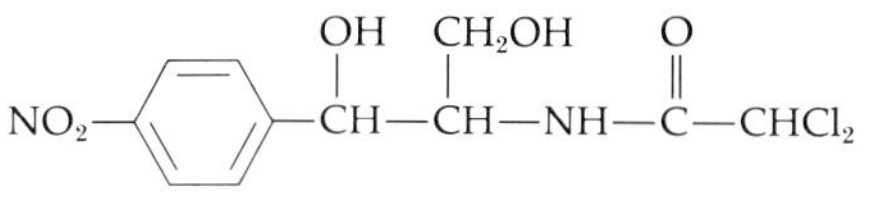

Chloramphenicol

Figure 18–14 Structure of chloramphenicol.

They have the advantage of being retained longer in the body. Tetracyclines are used for many urinary tract infections and are especially useful in treatment of mycoplasmal pneumonia, chlamydial urethritis, and rickettsial infections. They are also frequently used as alternative drugs for diseases such as syphilis and gonorrhea.

When the broad spectrum of the tetracyclines suppresses the normal intestinal flora, they cause gastrointestinal upsets and often lead to superinfections, particularly by *Candida albicans*. They are not advised for administration to children, who might experience a brownish discoloration of the teeth, or to pregnant women, in whom they might cause liver damage.

Chloramphenicol

Chloramphenicol is a broad-spectrum, bacteriostatic antibiotic that interferes with protein synthesis on the 50S ribosome. Because of its relatively simple structure, it is cheaper for the pharmaceutical industry to synthesize it chemically rather than isolate it from *Streptomyces* (Figure 18–14). Its relatively small molecular size promotes its diffusion into portions of the body that are normally inaccessible to many other drugs. However, chloramphenicol has serious side effects; most important is the suppression of bone marrow activity. This suppression affects the formation of blood cells. In about 1 per 40,000 users, the drug appears to cause *aplastic anemia*, a potentially fatal condition; the normal rate for this condition is only about 1 in 500,000 persons. Responsible agencies have advised against use of the drug for trivial conditions or ones in which suitable alternatives are available. It is still the drug of choice in the treatment of typhoid fever, in spite of recent reports of resistant strains, and of certain types of meningitis.

Macrolides

Macrolides are a group of antibiotics named for the presence of a *macrocyclic lactone ring*. Essentially the only macrolide in common clinical use is *erythromycin* (Figure 18–15). Its mode of action is the inhibition of protein synthesis; apparently its site of action is closely related to that of chloramphenicol. However, erythromycin is not able to penetrate the cell wall of most gram-negative bacilli. Its spectrum of activity is similar to that of penicillin G, and it is a frequent alternative drug for penicillin. Erythromycin is, however, effective against the gram-negative agent of Legionnaires' disease and some *Neisseria*—an important gram-negative, pathogenic genus. Because it can be administered orally, an orange-flavored preparation of erythromycin is a frequent penicillin substitute for the treatment of streptococcal and staphylococcal infections in children. Erythromycin is the drug of choice for the treatment of mycoplasmal pneumonia.

Polypeptides

A number of antibiotics are **polypeptides,** chains of amino acids linked by peptide bonds. Almost all of these drugs have been isolated from the genus *Bacillus*. Two well-known examples are *bacitracin* and *polymyxin B* (Figure 18–16). Both drugs are relatively toxic and their use is mostly topical.

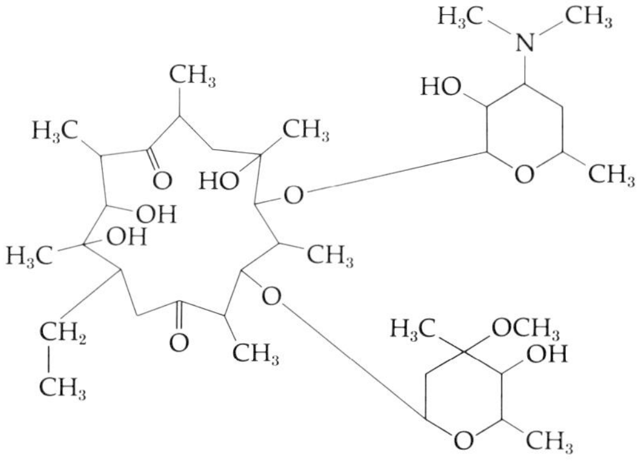

Erythromycin

Figure 18–15 Structure of erythromycin, a representative macrolide.

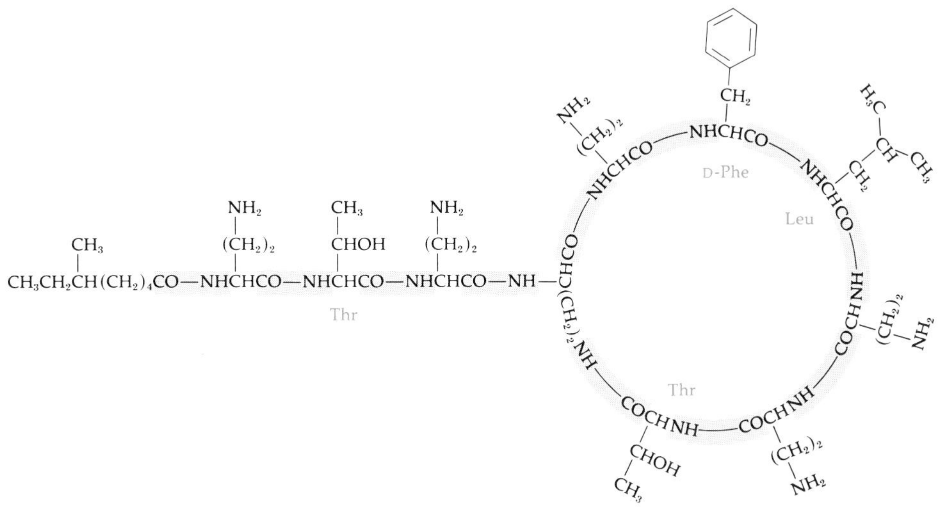

Figure 18–16 Structure of polymyxin B, a representative polypeptide antibiotic; the amino acid subunits (names in color) and peptide bonds (shading) are shown.

Bacitracin is primarily effective against gram-positive bacteria such as staphylococci and streptococci, although it is also used against important gram-negative pathogens such as *Neisseria*. It inhibits the synthesis of cell walls. Its use is restricted to topical application for superficial infections.

Polymyxin B is a bactericidal antibiotic effective against gram-negative bacteria. For many years it was one of very few drugs used against infections by *Pseudomonas* bacteria. The importance of this characteristic has diminished in recent years as additional drugs also effective against these inherently antibiotic-resistant bacteria have been developed. The mode of action of polymyxin B is to injure plasma membranes. Polymyxin B, because of toxicity when it is administered by injection, is seldom used today except in topical treatment of superficial infections.

Both bacitracin and polymyxin B are widely available in nonprescription antiseptic ointments, in which the polypeptides are usually combined with neomycin, a broad-spectrum aminoglycoside.

Vancomycin

Vancomycin is apparently unrelated chemically to any other antibiotic. It is a very toxic drug, is difficult to administer, and has a very narrow spectrum of activity that is based on inhibition of cell wall synthesis (peptidoglycans). The drug is important, nonetheless, because it is probably the most effective drug in clinical use against penicillinase-producing staphylococci. Vancomycin is also used in special situations such as treatment of streptococcal endocarditis and infections by staphylococci of devices such as prosthetic heart valves.

Rifamycins

The best known derivative of the **rifamycin** family of antibiotics is *rifampin*. These drugs inhibit the synthesis of mRNA. By far its most important use is against mycobacteria in the treatment of tuberculosis, but it is effective against a number of gram-positive organisms and *Neisseria*. A valuable characteristic of rifampin is its ability to penetrate tissues and reach therapeutic levels in cerebrospinal fluid and abscesses. This characteristic is probably an important factor in its antitubercular activity because, as we shall see, the tuberculosis pathogen is usually located intracellularly in macrophages. An unusual side effect of rifampin is the appearance of orange-red urine, feces, saliva, sweat, and even tears.

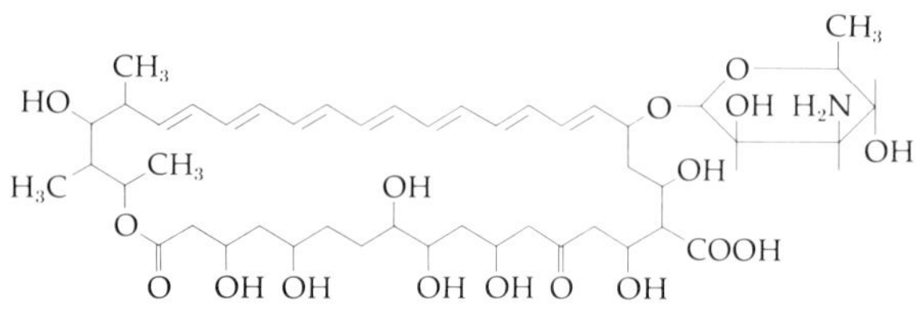

Figure 18–17 Structure of amphotericin B, a representative polyene antibiotic.

Antifungal Drugs

Polyenes

Two of the more commonly used **polyene** antibiotics, both products of *Streptomyces* spp., are nystatin and amphotericin B (Figure 18–17). Both drugs are fungicidal and exert their action by damaging fungal plasma membranes; the drugs combine with the *sterols* in the membranes. Bacterial plasma membranes (except those of mycoplasmas) do not contain sterols, and the drugs therefore do not affect bacteria. *Nystatin* is used to treat local infections of the vagina and skin and can be taken orally for *Candida* infections of the digestive tract. It is so poorly soluble that it does not cross the intestinal lining, so it does not enter the body tissue in toxic amounts. For many years, *amphotericin B* has been a mainstay of clinical treatment for *systemic* fungal diseases such as histoplasmosis, coccidioidomycosis, and blastomycosis. The drug's toxicity, particularly to the kidneys, is a strongly limiting factor in these uses.

Imidazoles

Imidazole antifungals, such as *miconazole* (which has been in use since 1978) and *ketoconazole* (approved for use in 1981), are promising new antifungals. These drugs primarily interfere with sterol synthesis in fungi although they probably have other antimetabolic effects. Miconazole (Figure 18–18) is generally used topically, and is most often used in treatment of cutaneous mycoses such as athlete's foot. When taken orally, ketoconazole is effective and has considerable promise as a less toxic alternative (although occasional liver damage has been reported) to amphotericin B for many systemic fungal infections. Ketoconazole also has the advantage of an unusually wide spectrum of activity.

Griseofulvin is an antibiotic produced by a species of *Penicillium*. It has the interesting property of being active against superficial, dermatophytic, fungal infections of the hair (*tinea capitis*, or ringworm) and nails, even though its route of administration is oral. The drug apparently binds selectively to the keratin found in the skin, hair follicles, and nails. Its mode of action is to interfere with mitosis and thereby inhibit fungal reproduction.

Tolnaftate should be mentioned as a common alternative to miconazole as a topical agent for the treatment of athlete's foot. Its mechanism of action is not known.

Antiviral Drugs

Because viruses use the host cell's own metabolic machinery, it is difficult to damage viruses without damaging the host as well. The first antiviral licensed for systemic use in the United States was *amantadine*. Although it is not known precisely how amantadine and the related drug *rimantadine* work, it has been suggested that these substances interfere with the uncoating of viruses within the host cell or with the transcription of viral RNA. Infection by influenza A viruses can be prevented by amantadine, but the drug has little effect on the course of the disease once it has been contracted—this is a strongly limiting factor for its general use.

Several analogs of purine and pyrimidine nucleosides in DNA have been synthesized and have limited use against herpes viruses. *Idoxuridine* is a pyrimidine analog of thymidine (the thymine-containing nucleoside), and *adenine arabinoside* (also known as vidarabine or Ara-A) is an analog of the purine adenine. Both of these drugs are available as a topical ointment and are useful in treating herpes simplex infections of the eye (herpes keratitis). Adenine arabinoside has also been approved for intravenous use in the treatment of encephalitis caused by herpes viruses.

An antiviral compound more recently licensed in the United States is *acyclovir*. Acyclovir is a nucleoside analog (Figure 18–19). Its specificity is due to the fact that it affects a cell only if acted on by a certain enzyme. This enzyme, thymidine kinase, is present in cells infected by herpesvirus. Thymidine kinase adds a phosphate to acyclovir because acyclovir resembles the nucleoside that the enzyme normally acts upon. Two more phosphates are added to acyclovir by host cell enzymes. This newly formed acyclovir triphosphate inhibits the DNA polymerase activity induced by the herpesvirus; further synthesis of DNA is thereby terminated. Acyclovir's antiviral mechanism has little effect on host cell DNA activity.

Acyclovir has limited usefulness in topical treatment of genital herpes. When administered intravenously, acyclovir is effective in clearing a herpes simplex infection. Intravenous use of acyclovir is recommended for treatment of primary genital herpes infections and herpes simplex infections in immunosuppressed persons. Oral use of acyclovir has been approved for genital herpes (see box in Chapter 24).

A novel approach to viral chemotherapy that may bear good results centers around *interferon* (see Chapter 15), a protein secreted by certain host cells responding to viral infections. Interferon inhibits cell infection by many kinds of viruses. Genetic engineering techniques applied to bacteria have made sufficient amounts of interferon available for clinical studies. These studies seek to determine the possible effectiveness of interferon as a chemotherapeutic agent against viruses and possibly cancer.

Antiprotozoan and Antihelminthic Drugs

For hundreds of years the tropical tree cinchona provided the only drug useful for treatment of parasitic infections—*quinine* for malaria.

Quinine still has some use in the control of malaria but synthetic derivatives such as *chloroquine* have largely replaced it. Chloroquine fits itself between nitrogenous base pairs in DNA *(intercalation)* and prevents DNA from serving as a template for further DNA synthesis. *Quinacrine* functions in a very similar manner and is the drug of choice for treating giardiasis. *Emetine,* which blocks protein synthesis in eucaryotic but not in bacterial cells, has long been a mainstay in the treatment of amoebic dysentery. *Diiodohydroxyquin* is another important amoebicidal drug. Its mode of action is yet unknown. *Metronidazole* is one of the most widely used antiprotozoan drugs. It is unique in having activity not only against parasitic protozoans but also against certain anaerobic bacteria. As an antiprotozoan agent, it is the drug of choice for vaginitis caused by *Trichomonas vaginalis* (trik-ō-mōn'as va-jin-al'is) and *Gardnerella vaginalis* (gärd-né rel'ä). It is also used in treatment of giardiasis and amoebiasis. The most recent view of its mode of action is that it causes the formation of cytotoxic derivatives that disrupt DNA under anaerobic conditions. *Pentamidine* is used against several trypanosomal diseases, but in the United States its most familiar clinical use is probably in the treatment of *Pneumocystis* pneumonia, a common complication in immunocompromised persons and victims of AIDS. The mode of action is unknown but probably related to its effects on DNA.

To estimate the increased incidence of tapeworm infestations due to the ingestion of sushi (raw fish), the Centers for Disease Control recently noted the increase in requests for *niclosamide,* which is the usual first choice in treatment. Niclosamide inhibits oxidative phosphorylation in these helminths. The schistosome parasite causing

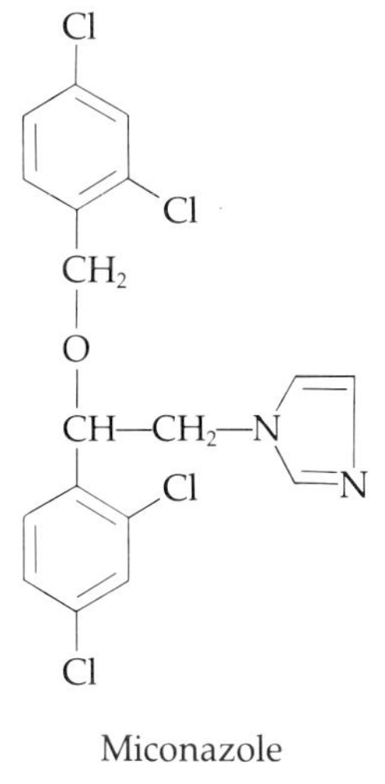

Figure 18–18 Structure of miconazole, a representative imidazole antibiotic.

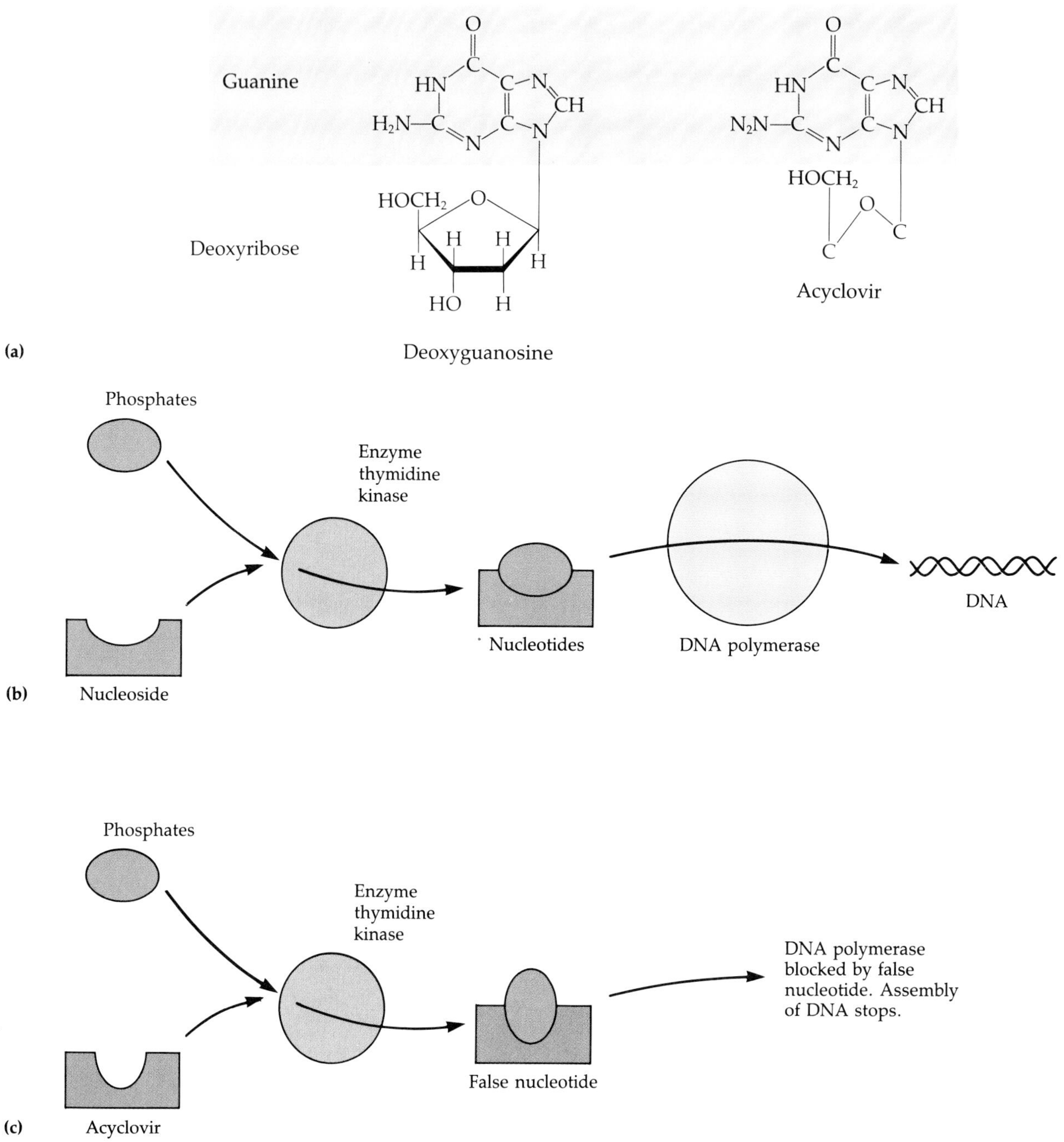

Figure 18–19 Structure and functions of acyclovir. **(a)** Structural resemblance between acyclovir and the nucleoside 2′-deoxyguanosine. **(b)** The enzyme thymidine kinase combines phosphates with nucleosides to form nucleotides. The nucleotides are assembled into DNA. **(c)** Acyclovir is a structural analog of the nucleoside for guanosine (2′-deoxyguanosine). Using acyclovir, thymidine kinase assembles a false nucleotide, which cannot be assembled into DNA; further synthesis of DNA is prevented.

schistosomiasis, which is most commonly found in the U.S. among Puerto Ricans, is usually treated with *antimonial* drugs such as *stibophen.* Antimony interferes with glycolysis of these helminths and is the basic ingredient of a number of other antihelminthic drugs. However, the antimony-based drugs will probably be eventually replaced for this use by single-dose drugs such as *oxamiquine* and *praziquantal,* both of which were recently approved for this use in the United States. *Piperazine* is a drug used in the treatment of pinworm and other intestinal roundworm infestations, among the more common helminthic diseases in this country. The organisms are paralysed and the patient soon passes them out of the body.

A summary of major antimicrobial drugs is presented in Table 18–3.

TESTS FOR MICROBIAL SUSCEPTIBILITY TO CHEMOTHERAPEUTIC AGENTS

We will now turn our attention to the methods researchers use to determine the effectiveness of a chemotherapeutic agent for the treatment of an infectious disease. Different microbial species and strains have different degrees of susceptibility to different chemotherapeutic agents. Moreover, the susceptibility of a microorganism can change with time, even during therapy with a specific chemotherapeutic agent. Because of these factors, a physician must know the sensitivities of the disease-causing microorganism before a particular treatment can be started. Several tests can be employed to indicate to the physician which

Table 18–3 Summary of Commonly Used Antimicrobial Drugs

Antimicrobial Drug	Effect on Microorganisms	Mode of Action	Clinical Use/Toxicity
Synthetic Drugs:			
Sulfonamides	Bacteriostatic	Competitive inhibition	*Neisseria meningitidis* meningitis and some urinary tract infections
Isoniazid (INH)	Bacteriostatic	Competitive inhibition	With streptomycin and rifampin to treat tuberculosis
Ethambutol	Bacteriostatic	Competitive inhibition	With INH to treat tuberculosis
Antibiotics:			
Penicillin, natural (Penicillins G, V)	Bactericidal	Inhibition of cell wall synthesis	Gram-positive bacteria
Penicillin, semisynthetic (Ampicillin, methicillin, oxacillin, and others)	Bactericidal	Inhibition of cell wall synthesis	Broad spectrum; resistant to penicillinase and stomach acid
Aminoglycosides	Bactericidal	Inhibition of protein synthesis	
Streptomycin			Gram-negative bacteria
Neomycin			Topical ointment *Staphylococcus*
Gentamicin			Gram-negative bacteria
Cephalosporins	Bactericidal	Inhibition of cell wall synthesis	Penicillin-resistant *Staphylococci*
Tetracyclines	Bacteriostatic	Inhibition of protein synthesis	Very broad spectrum; some toxic side effects include tooth discoloration and liver and kidney damage
Chloramphenicol	Bacteriostatic	Inhibition of protein synthesis	*Salmonella;* aplastic anemia occasional side effect
Macrolides (Erythromycin)	Bacteriostatic	Inhibition of protein synthesis	Diphtheria; Legionnaires' disease; various infections in persons allergic to penicillin
Polypeptides			
Bacitracin	Bactericidal	Inhibition of cell wall synthesis	Gram-positive bacteria primarily; toxic to kidneys; usually used as topical ointment

(*continued*)

Table 18–3 Summary of Commonly Used Antimicrobial Drugs *continued*

Antimicrobial Drug	Effect on Microorganisms	Mode of Action	Clinical Use/Toxicity
Polymyxin B	Bactericidal	Injury to plasma membranes	Gram-negative bacteria, including *Pseudomonas;* toxic to brain and kidneys
Rifamycins (Rifampin)	Bactericidal	Inhibition of RNA synthesis	Gram-positive bacteria, some gram-negatives, chlamydias, and poxviruses; often used with INH and ethambutol to minimize drug resistance
Polyenes (Nystatin, amphotericin B, and others)	Fungicidal	Injury to plasma membranes	*Candida* infections of skin and vagina (nystatin); histoplasmosis, coccidioidomycosis, and blastomycosis (amphotericin B)
Imidazoles (Ketoconazole, miconazole)	Fungicidal	Injury to plasma membranes	Topical use for fungal infections (miconazole); systematic fungal infections (ketoconazole).
Griseofulvin	Fungistatic	Inhibition of mitotic spindle function	Fungal infections of skin, primarily ringworm
Antiviral Drugs:			
Idoxuridine	Antiviral	Inhibition of DNA synthesis	Herpes eye infections
Adenine arabinoside	Antiviral	Inhibition of viral multiplication	Herpes eye infections, chickenpox, viral encephalitis
Amantadine	Antiviral	Inhibition of nucleic acid release from viruses into host cells	Influenza A prophylaxis and treatment
Antiprotozoan and Antihelminthic Drugs:			
Quinine and chloroquine	Antiprotozoan	Inhibition of DNA and RNA synthesis	Malaria
Emetine	Antiprotozoan	Blocks protein synthesis	Amoebic infections
Metronidazole	Antiprotozoan	Inhibition of certain oxidases	*Trichomonas* infections
Niclosamide	Anithelminthic	Inhibition of oxidative phosphorylation	Tapeworm infections
Piperazine	Antihelminthic	Causes paralysis of helminth	Pinworm infections
Antimony compounds	Antihelminthic	Inhibition of glycolysis	Schistosomiasis and other helminthic infections

chemotherapeutic agent is most likely to combat a specific pathogen. These tests, however, are not always necessary. If the organism has been identified, for example, as *Pseudomonas aeruginosa,* beta-hemolytic streptococci, or gonococci, certain drugs can be selected without specific testing for susceptibility. Only when susceptibility is not as predictable, or when antibiotic resistance problems develop, are these tests necessary.

Disk Diffusion Method

Probably the most widely used, but not necessarily the best, testing method is the **disk diffusion method,** also known as the **Kirby–Bauer test.** A Petri plate containing an agar medium is inoculated ("seeded") uniformly over its entire surface with a standardized amount of a test organism. Next, filter paper disks impregnated with known concentrations of chemotherapeutic agents are placed on the solidified agar surface. During incubation, the chemotherapeutic agents diffuse from the disks into the agar. The farther the agent diffuses from the disk, the lower its concentration. If the chemotherapeutic agent is effective, a *zone of inhibition* forms immediately around the disk (Figure 18–20). The diameter of the zone of inhibition can be measured. Because the size of the zone of inhibition is affected by the diffusion rate of the chemotherapeutic agent, a wider zone does not always indicate greater antimicrobial activity. The results report only that the organism is *sensitive, intermediate,* or *resistant*. This result is often inadequate information for many clinical purposes. However, the test is simple and inexpensive and has considerable value in medical practice.

Tube Dilution and Agar Dilution Tests

When the **tube dilution test** is performed, a series of culture tubes are inoculated with the test organism and exposed (usually in two-fold steps) to an increasing concentration of the chemotherapeutic agent. After incubation for 16 to 20 hours the tubes are examined for turbidity (Figure 18–21). The **minimum inhibitory concentration (MIC)** of the antimicrobial is the lowest concentration that prevented turbidity from appearing in a tube. In clinical practice, the minimum inhibitory concentration is obtained using an MIC plate. See Microview 3–3 for a full discussion.

How does a person know if the inhibition of growth is permanent, that is, if the bacteria have actually been killed? Tubes that show no growth can be subcultured into broth tubes that contain no antibiotic; the tube that contains the lowest concen-

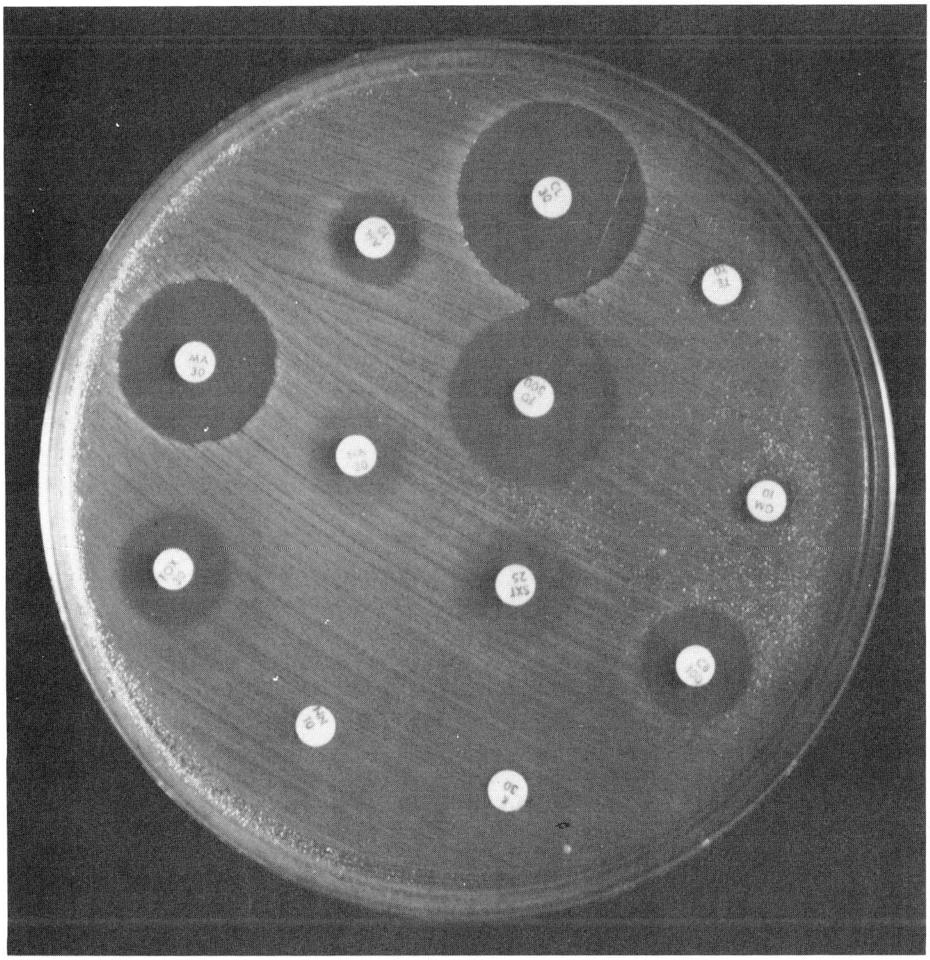

Figure 18–20 Disk diffusion method for determining the activity of chemotherapeutic agents. Each disk contains a different agent, which diffuses into the surrounding agar. Clear zones, indicating that bacterial growth is inhibited occur around disks with agents that are effective against the microorganism being tested.

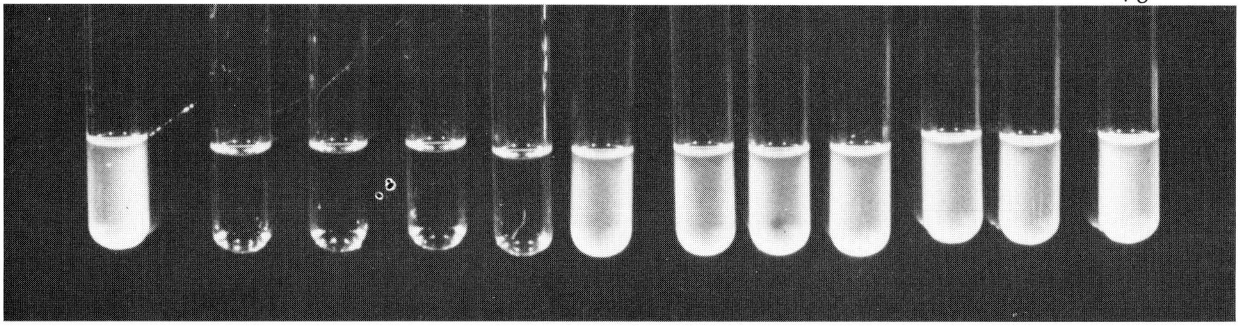

Figure 18–21 Tube dilution test for determining antibiotic activity. This series of tubes shows a test for the sensitivity of *Staphylococcus aureus* to the antibiotic ampicillin. A liquid medium and a different concentration of ampicillin was added to each tube; no ampicillin was added to the control tube. The tubes were then inoculated with *S. aureus* and incubated. Turbidity (cloudiness) in a test tube indicates microbial growth. The tube containing the lowest concentration of the chemotherapeutic agent capable of preventing growth of the test microorganism is said to contain the MIC (minimum inhibitory concentration). This test shows that the minimum concentration of ampicillin needed to inhibit the growth of *S. aureus* is 16 μg/ml, because growth occurred in tubes with lower concentrations of the antibiotic.

tration of the chemotherapeutic agent and contains no growth can be noted. This concentration is referred to as the **minimum bactericidal concentration (MBC)**. The tube dilution test is cumbersome and can fail to detect resistant mutants or contaminated cultures.

The **agar dilution method** is similar in principle to the tube dilution method. Increasing concentrations of the antimicrobial agents are placed on agar plates. With the use of a special replicator device, as many as 36 organisms at a time can be inoculated onto the surface of the plate. The MIC can be determined by measurement of colony growth on the agar's surface. Although the MIC is not usually determined by this method, it is very useful for a laboratory performing large numbers of tests daily.

Automated Tests

Recently, **automated tests** of microbes' sensitivity to drugs have been introduced. To measure the inhibitory effect of drugs in a liquid medium, these tests use light-scattering to determine bacterial growth. These tests are widely employed because of their simplicity and speed; they yield results within a few hours. Automated tests are being used increasingly by laboratories that are called upon to perform large numbers of tests.

DRUG RESISTANCE

Bacteria become resistant to antimicrobials in a number of different ways. Some resistance is due to the ability of a particular mutant to destroy a given antibiotic. As noted earlier, some staphylococci produce the enzyme penicillinase, which destroys natural penicillins. Tetracycline resistance is usually related to cellular changes that decrease the drug's uptake into the cell. Bacteria become resistant to sulfa drugs and antibiotics such as streptomycin and erythromycin when receptor sites for the antibiotic develop less affinity for it. Resistance to trimethoprim can arise when the microbe begins synthesizing very large amounts of the enzyme against which the drug is targeted. Conversely, polyene antibiotics can become less effective when resistant organisms produce small amounts of the sterols upon which the drug produces its effects. Of particular concern is the possibility that such *resistant mutants* will increasingly replace the susceptible normal populations (see box).

Centers for Disease Control
MMWR
Morbidity and Mortality Weekly Report

Penicillin-Resistant Gonococci and Natural Selection

Strains of gonococci resistant to all forms of penicillin have emerged—penicillinase-producing *Neisseria gonorrhoeae* (PPNG). From March 1976, when the first PPNG isolate was identified, through December 1982, some 9100 cases of PPNG in the United States were reported to the Centers for Disease Control (CDC) (see graph). Although the incidence of PPNG remains low, 4457 cases in 1982 comprised approximately 0.5% of all gonorrhea cases reported that year. 3424 cases were reported in the first nine months of 1982, reflecting a 77% increase compared to the same period in 1981.

Importation from Other Countries

A significant proportion of PPNG cases have been related to importation of infection from Southeast Asia and the Philippines. Military personnel or their dependents have been responsible for a majority of the imported cases. Over forty countries have reported cases of PPNG to the World Health Organization. The organism accounts for about 30% of all recent gonococcal isolates in the Philippines and 16% in the Republic of Singapore. (Among the factors thought to contribute to the high prevalence of PPNG in the Philippines and Singapore is the preventive use of oral penicillins, especially by prostitutes.)

Endemic Focus in the United States

In many areas of the United States, the incidence of PPNG infection is increasing. In major metropolitan areas, the cause is sustained endemic transmission. The control of PPNG infections in Los Angeles County and Washington state and the virtual end of endemic transmission in these areas are both attributed to a comprehensive control effort. Unfortunately, control efforts are threatened by the pressure of continued importation

(*continued on next page*)

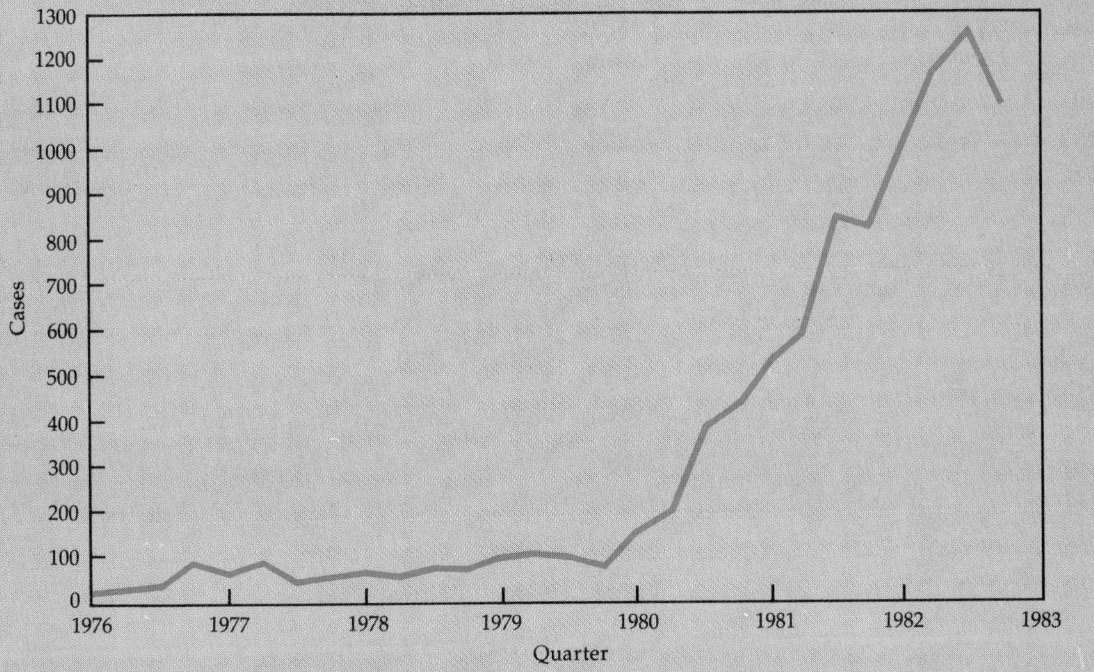

Cases of PPNG by quarter in the United States, January 1976–December 1982 (includes Guam, the Pacific Trust Territories, Puerto Rico, and Virgin Islands). (Source of graph: CDC. *Annual Summary 1982. MMWR* 1983; 31(54), p. 30.)

and delays in identifying early cases of PPNG. All *N. gonorrhoeae* isolates should be tested for penicillinase production. Spectinomycin has been the antibiotic of choice for all PPNG infections. In addition, the Centers for Disease Control specifically recommend spectinomycin for the initial treatment of uncomplicated anogenital gonorrhea in patients who have recently returned from countries with a high prevalence of PPNG infections.

In 1983, 56 cases of penicillin-resistant gonococcal infection occurred in North Carolina. These cases represent the first reported outbreak of gonorrhea caused by bacteria that are resistant to penicillin but do not produce pencillinase. Although this outbreak represents an unusual event because a single resistant strain was transmitted, it is significant because the isolates are negative in a test for PPNG but not treatable with penicillin.

Spectinomycin-Resistant PPNG

In May 1981, a PPNG strain resistant to spectinomycin was reported from Travis Air Force Base, California. The patient had acquired the infection in the Philippines. Although four spectinomycin-resistant isolates of *N. gonorrhoeae* have been reported previously, this is the first to be penicillinase-producing also. The CDC now recommends that all isolates found to be PPNG also be tested for spectinomycin resistance. Patients who have uncomplicated anogenital infections caused by spectinomycin-resistant PPNG should be treated with cefoxitin and probenecid. Spectinomycin was used for 5 years before resistance was seen, which raises the question of how long the lifespan of cefoxitin and other drugs will be.

Natural Selection and Antibiotic Resistance

Through natural selection, organisms best adapted to fit their environment will survive and reproduce. In an antibiotic-laden environment, the "fittest" bacterium is one that is resistant to antibiotics.

Asked to list the reasons for the emergence of drug-resistant bacterial strains, a spokesman for the National Institutes of Health blamed the unnecessary use of prophylaxis, use of antibiotics in animal feeds, the availability of over-the-counter antibiotics in many countries, and misuse by health professionals. It has been pointed out that the indiscriminate use of antibiotics may result in a vast majority of infections being caused by antibiotic-resistant bacteria.

Sources: *MMWR Annual Summary 1982*; *MMWR 29:*241, 381 (1980); *MMWR 30:*221 (1981); *MMWR 31*:1 (1982); *MMWR 32:*181, 273 (1983).

Hereditary drug resistance is often carried by plasmids, extrachromosomal genetic elements. Some plasmids, including those called **resistance (R) factors,** can be transferred between bacterial cells in a population and between different but closely related bacterial populations (see Chapter 8). R factors often contain genes for resistance to several antibiotics at a time.

Strains of bacteria resistant to antibiotics are particularly common among persons who work in hospitals where antibiotics are in constant use. *Staphylococcus aureus,* a common opportunistic pathogen that is carried in the nasal passages, readily develops resistance to antibiotics.

Every day in hospitals, surprisingly large volumes of antibiotics are dispensed as aerosols by the universal practice of clearing syringes of air before an injection is made. As these aerosols are generally inhaled in small amounts, the staphylococci harmlessly residing in humans, as an example, can quickly develop resistance. If these resistant microbes are transferred to surgical wounds and other exposed sites, they can cause *nosocomial* infections (see Chapter 13).

So the chances of survival of resistant mutants are minimized, drugs should not be used indiscriminately, and the concentrations administered should always be of optimum strength. Many hospitals have special monitoring committees who review the use of antibiotics. Only a few years ago, one medical journal reported on a survey of antibiotic use in a hospital. The results indicated that more than 60% of the antibiotics prescribed there were in either the wrong dosage or applications for which they would not be effective. Patients should always finish the prescribed dosage of antibiotic to make sure all microorganisms are killed even after the apparent symptoms may have disappeared. Another way that the development of resistant strains can be reduced is through the administration of two or more drugs simultaneously. If a strain is resistant to one of the drugs, the other might destroy it. The probability that the organism will acquire resistance to both drugs is roughly the product of the two individual probabilities.

It sometimes happens that the chemotherapeutic effect of two drugs given simultaneously is greater than the effect from either alone. As previously discussed, this phenomenon is referred to as **synergism.** For example, in the treatment of bacterial endocarditis, penicillin and streptomycin are much more effective when taken together than when either drug is taken alone. Perhaps damage to bacterial cell walls by penicillin makes it easier for streptomycin to enter. Other combinations of drugs can be **antagonistic.** For example, the simultaneous use of penicillin and tetracycline is often less effective than either drug used alone. By stopping the growth of the bacteria, the bacteriostatic drug tetracycline interferes with the action of penicillin, which requires bacterial growth. Combinations of antimicrobial drugs should be used only for these purposes: (1) to prevent or minimize the emergence of resistant strains; (2) to take advantage of the synergistic effect; (3) to provide optimal therapy in life-threatening situations before a diagnosis can be established with certainty; or (4) to lessen the toxicity of individual drugs by reducing the dosage of each in combination.

Resistance to the drug is not the only problem with antimicrobial therapy. For example, antibiotics have no effect on toxins that might already have been released into the host's body fluids.

Antibiotics are clearly one of the greatest triumphs of medical science, although they are not without their problems and disadvantages. Our most pressing present concern is with the misuse of antibiotics, which increases the likelihood of the emergence of resistant strains of pathogens.

More antiviral drugs are also urgently needed. As research uncovers more aspects of the internal workings of viruses and cells, more chemicals effective against viruses will probaby be developed. There is considerable similarity between the requirements of cancer therapy and the requirements for treatments for viral disease. It will be interesting to see if research for one will benefit the other.

STUDY OUTLINE

INTRODUCTION (pp. 493–494)

1. A chemotherapeutic agent is a chemical that combats disease in the body.
2. An antimicrobial drug is a chemical that destroys disease-causing microorganisms with minimal damage to host tissues.
3. Antimicrobial drugs may be synthetic drugs (prepared in the laboratory) or antibiotics (produced by bacteria or fungi).
4. Antibiotics are produced by species of *Streptomyces, Bacillus, Penicillium,* and *Cephalosporium.*

HISTORICAL DEVELOPMENT (pp. 494–495)

1. The first chemotherapeutic agent, discovered by Paul Ehrlich, was salvarsan, used to treat syphilis.
2. Sulfa drugs came into prominence in the late 1930s.
3. Alexander Fleming discovered the first antibiotic, penicillin, in 1929; its first clinical trials were done in 1940.

CRITERIA FOR ANTIMICROBIAL DRUGS (pp. 495–497)

1. Antimicrobial agents should have selective toxicity for microorganisms.
2. They should not produce hypersensitivity in most patients.
3. They should be stable during storage, be soluble in body fluids, and be retained in the body long enough to be effective.
4. Microbes should not readily become resistant to antimicrobial drugs.

SPECTRUM OF ACTIVITY (p. 497)

1. Antibacterial drugs affect many targets in a procaryotic cell.

2. Fungal, protozoan, and helminthic infections are more difficult to treat because these organisms have eucaryotic cells.

3. Narrow spectrum drugs only affect a select group of microbes, e.g. gram-positive cells.

4. Broad spectrum drugs affect a large number of microbes.

5. Antimicrobial agents should not cause excessive harm to normal flora.

6. Superinfections occur when a pathogen develops resistance to the drug being used or resistant normal flora multiply excessively.

ACTION OF ANTIMICROBIAL DRUGS (pp. 497–501)

1. General action is either by directly killing microorganisms (i.e., bactericidal) or by inhibition of growth (i.e., bacteriostatic).

2. Some agents, such as penicillin, inhibit cell wall synthesis in bacteria.

3. Other agents, such as chloramphenicol, erythromycin, tetracyclines, and streptomycin, inhibit protein synthesis by acting on 70S ribosomes.

4. Agents such as polymyxin B cause injury to plasma membranes.

5. Rifamycin, nalidixic acid, and trimethoprim inhibit nucleic acid synthesis.

6. Agents such as sulfanilamide act as antimetabolites by competitively inhibiting enzyme activity.

SURVEY OF COMMONLY USED ANTIMICROBIAL DRUGS (pp. 501–511)

Synthetic Antimicrobial Drugs (pp. 501–502)

1. Isoniazid (INH) is a structural analog of vitamin B_6 and may inhibit mycolic acid synthesis in mycobacteria. INH is administered with streptomycin, rifampin, and ethambutol to treat tuberculosis.

2. The antimetabolite ethambutol is used with other drugs to treat tuberculosis.

3. Sulfonamides competitively inhibit folic acid synthesis. Sulfonamides are bacteriostatic; they are used to treat urinary tract infections; they can cause allergies and drug resistance.

4. Trimethoprim competitively inhibits folinic acid synthesis. It is used with sulfamethoxazole for treatment of *Pneumocystis* pneumonia and urinary tract and intestinal infections.

5. Nitrofurans may break DNA molecules; they are used to treat urinary tract infections.

6. Nalidixic acid interferes with DNA synthesis; it is used to treat urinary tract infections caused by gram-negative bacteria.

7. The antifungal agent 5-fluorocytosine is an antimetabolite of cytosine.

Antibiotics (pp. 502–507)

1. Penicillins have low toxicity. They inhibit peptidoglycan synthesis; cell death is by osmotic lysis.

2. Penicillin G and V are natural penicillins and are effective against cocci, gram-positive bacilli, and spirochetes.

3. Several semisynthetic penicillins such as ampicillin are resistant to stomach acid and penicillinases and are broad spectrum.

4. Fourth generation penicillins such as mezlocillin are effective against *Pseudomonas* spp.

5. Cephalosporins inhibit cell wall synthesis and are used against penicillin-resistant strains and in treating people allergic to penicillin.

6. Aminoglycosides include streptomycin, neomycin, kanamycin, and gentamicin; all inhibit protein synthesis and are bactericidal.

7. Tetracyclines inhibit protein synthesis and are bacteriostatic toward many bacteria, including rickettsias and chlamydias; they have the broadest spectrum of antibacterial activity.

8. Chloramphenicol inhibits protein synthesis and is bacteriostatic; a side effect of prolonged use is aplastic anemia.

9. Macrolides, such as erythromycin, inhibit protein synthesis and are bacteriostatic; erythromycin is used to treat Legionnaires' disease and mycoplasmal pneumonia.

10. Polypeptides include bacitracin and polymyxin B. They are applied topically to treat superficial infections.

11. Bacitracin inhibits cell wall synthesis primarily in gram-positive bacteria.

12. Polymyxin B damages plasma membranes and is effective against gram-negative bacteria.

13. Vancomycin inhibits cell wall synthesis and may be used to kill penicillinase-producing staphylococci.

14. Rifamycins are bactericidal and are especially effec-

tive against gram-positive bacteria including mycobacteria.

Antifungal Drugs (p. 508)

1. Polyenes, such as nystatin and amphotericin B, combine with plasma membrane sterols and are fungicidal.
2. Imidazoles interfere with sterol synthesis and are used to treat cutaneous and systemic mycoses.
3. Griseofulvin interferes with eucaryotic cell division and is used primarily to treat skin infections caused by fungi.

Antiviral Drugs (pp. 508–509)

1. Antiviral drugs are limited because of toxicity.
2. Amantadine and rimantadine block viral penetration or uncoating.
3. Purine and pyrimidine analogs including idoxuridine, adenine arabinoside, and acyclovir are used to treat herpes infections.

Antiprotozoan and Antihelminthic Drugs (pp. 509–511)

1. Chloroquine, quinacrine, emetine, diiodohydroxyquin, pentamine, and metronidazole are used to treat protozoan infections.
2. Chloroquine and quinacrine stop DNA synthesis by intercalation.
3. Antihelminthic drugs include niclosamide, antimony, oxamiquine, praziquantal, and piperazine.
4. Antimony blocks glycolysis; piperazine paralyzes intestinal roundworms.

TESTS FOR MICROBIAL SUSCEPTIBILITY TO CHEMOTHERAPEUTIC AGENTS (pp. 511–514)

1. These tests are used to determine the degree of susceptibility of different microorganisms to chemotherapeutic agents.
2. They help to determine which chemotherapeutic agent is most likely to combat a specific pathogen.
3. These tests are used when susceptibility cannot be predicted or when drug resistance arises.

Disk Diffusion Method (pp. 512–513)

1. In this test, also known as the Kirby–Bauer test, a bacterial culture is inoculated on an agar medium, and filter paper disks impregnated with chemotherapeutic agents are overlayed on the culture.
2. After incubation, the absence of microbial growth around a disk is called a zone of inhibition.
3. The diameter of the zone of inhibition, compared to a standardized reference table, is used to determine whether the organism is sensitive, intermediate, or resistant to the drug.

Tube Dilution and Agar Dilution Tests (pp. 513–514)

1. In the tube dilution test, the microorganism is grown in tubes of liquid media containing different concentrations of a chemotherapeutic agent.
2. The minimum inhibitory concentration (MIC) is the lowest concentration of chemotherapeutic agent capable of preventing microbial growth.
3. The lowest concentration of chemotherapeutic agent that kills bacteria is called the minimum bactericidal concentration (MBC).
4. In an agar dilution test, up to 36 organisms are replica-plated onto plates containing different concentrations of a drug; colony size is used to determine the MIC.

Automated Tests (p. 514)

1. These tests determine the amount of microbial growth in a liquid medium containing a chemotherapeutic agent by measuring the amount of light that passes through a culture.
2. Such tests are rapid and simple to perform.

DRUG RESISTANCE (pp. 514–517)

1. Drug resistance refers to the ability of a microorganism to resist the antimicrobial effects of a chemotherapeutic agent.
2. Resistance may be due to enzymatic destruction of a drug or cellular or metabolic changes at target areas.
3. Hereditary drug resistance is carried by plasmids called resistance (R) factors.
4. Resistance can be minimized by the discriminate use of drugs in appropriate concentrations and dosages.
5. Combinations of drugs may be used to minimize the development of resistant strains; to employ a synergistic effect; to provide therapy prior to diagnosis; to use small concentrations of each drug to lessen toxicity.

STUDY QUESTIONS

REVIEW

1. Define a chemotherapeutic agent. Distinguish between a synthetic chemotherapeutic agent and an antibiotic.
2. Ehrlich discovered the first ______________. Fleming discovered ______________; it was the first ______________.
3. List and explain five criteria used to identify an effective antimicrobial agent.
4. Fill in the following table.

Antimicrobial Agent	Synthetic or Antibiotic	Method of Action	Principal Use
Isoniazid			
Sulfonamides			
Ethambutol			
Trimethoprim			
Nitrofurans			
Penicillin, natural			
Penicillin, semisynthetic			
Cephalosporins			
Aminoglycosides			
Tetracyclines			
Chloramphenicol			
Macrolides			
Polypeptides			
Vancomycin			
Rifamycins			
Polyenes			
Griseofulvin			
Idoxuridine			
Chloroquine			
Niclosamide			

5. What similar problems are encountered with antiviral, antifungal, antiprotozoan, and antihelminthic drugs?
6. Identify three methods of action of antiviral drugs. Give an example of a currently used antiviral drug for each method of action.
7. Compare and contrast the tube dilution and agar dilution tests. Identify at least one advantage of each.
8. Describe the agar diffusion test for microbial susceptibility. What information can you obtain from this test?
9. Define drug resistance. How is it produced? What measures can be taken to minimize drug resistance?
10. List the advantages of using two chemotherapeutic agents simultaneously to treat a disease. What problem can be encountered using two drugs?

CHALLENGE

1. Why are antiviral drugs like idoxuridine effective if host cells also contain DNA?
2. The following data were obtained from an agar diffusion test.

Antibiotic	Zone of Inhibition
A	15 mm
B	0 mm
C	7 mm
D	15 mm

(a) Which antibiotic was most effective against the bacteria being tested?
(b) Which antibiotic would you recommend for treatment of a disease caused by this bacterium?
(c) Was antibiotic *A* bactericidal or bacteriostatic? How can you tell?

3. The following results were obtained from a tube dilution test for microbial susceptibility.

Tube #	Antibiotic Concentration	Growth	Growth in Subculture
1	200 μg	−	−
2	100 μg	−	−
3	50 μg	−	+
4	25 μg	+	+

(a) The MIC for this antibiotic is ________.
(b) The MBC for this antibiotic is ________.

FURTHER READING

Abraham, E. P. June 1981. The beta-lactam antibiotics. *Scientific American* 244:76–86. A discussion of semisynthetic penicillins and cephalosporins.

Actor, P. (convener). 1975. Laboratory evaluation of antimicrobial agents. *Developments in Industrial Microbiology*. 16:161–214. Washington, D.C.: American Institute of Biological Sciences. Current status of new antibacterial, antifungal, and antiprotozoan drugs.

Baldry, P. 1976. *The Battle Against Bacteria*. Cambridge, England: Cambridge University Press. The history of the fight against diseases and the development of antibacterial drugs.

Bryan, L. E., ed. 1984. *Antimicrobial Drug Resistance*. New York: Academic Press. Discusses mechanisms of resistance to antibacterial, antifungal, antiviral, and antimalarial drugs.

Gauri, K. K. 1981. *Antiviral Chemotherapy: Design of Inhibitors of Viral Functions*. New York: Academic Press. Describes inhibition of viral entry, protein synthesis, and nucleic acid replication as well as toxicology of drugs.

Kobayashi, G. S. and G. Medoff. 1977. Antifungal agents: recent developments. *Annual Review of Microbiology* 31:291–308. Methods of action and uses for antifungal drugs.

Wishnow, R. M. and J. L. Steinfeld. 1978. The conquest of the major infectious diseases in the United States: a bicentennial retrospect. *Annual Review of Microbiology* 30:427–450. A concise summary of control of diseases including tuberculosis, cholera, malaria, and yellow fever.

Inside A Clinical Lab

When a patient is admitted to a hospital for treatment, the patient's illness poses a problem for doctors, nurses, and technicians to solve. The first step toward a solution is to identify the microorganism responsible for the illness. (To the right is a common pathogen, *Staphylococcus aureas*, streaked on a petri plate in a spiral pattern.) The next step is to find a suitable drug and a suitable dosage to combat the pathogen. These two steps are carried out by technicians in special laboratories, utilizing powerful yet simple techniques. To solve the problem, that is, to treat the patient successfully, doctors depend on the results from the hospital laboratories. In the next four pages, we will tour two laboratories in the University of California at San Francisco hospital—the microbiology lab and the immunology lab. To begin, we will visit an area where some samples are prepared for distribution to specialized labs.

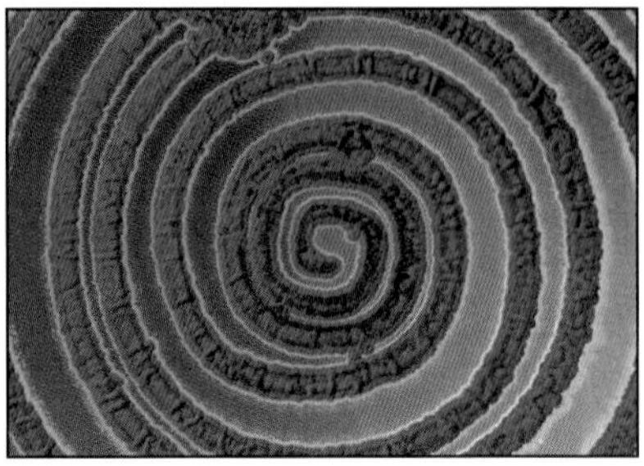

Specimen-Processing Area

The specimen-processing area prepares samples for some of the specialized labs, such as the chemistry, endocrinology, and immunology labs. (The microbiology lab processes its own samples; technicians work with clinical samples that come directly from the patient to the lab—feces, blood, urine, or anything from which bacteria can be cultured.) ■ 1 At UCSF hospital, the history of every patient is filed on a central computer; therefore, the first step in preparing a sample is to enter information about the sample into the patient's file and assign the sample a number. After a technician has labeled each sample with the patient's name and sample number and racked the samples in numerical order, he or she would send them to the aliquot room, part of the specimen-processing area. ■ 2 In the aliquot room, fluids are diluted into portions of known concentration and volume. Because the immunology lab conducts tests using serum, samples destined for the immunology lab must be centrifuged to separate the blood cells from the serum. ■ 3 A technician labels aliquots and sends them to the immunology lab. Fluids destined for other labs, such as spinal fluid and urine, are handled in similar ways.

1

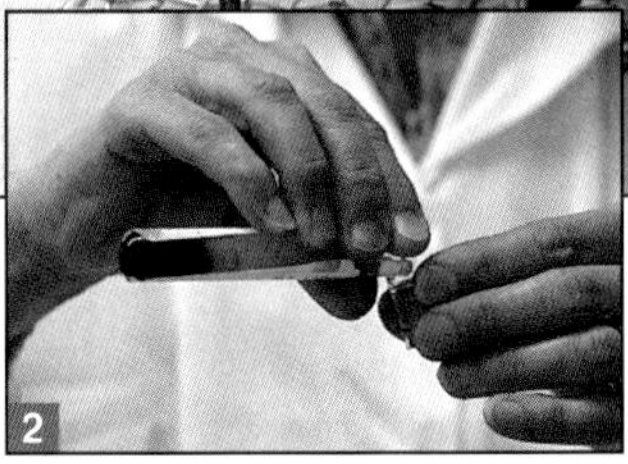

2

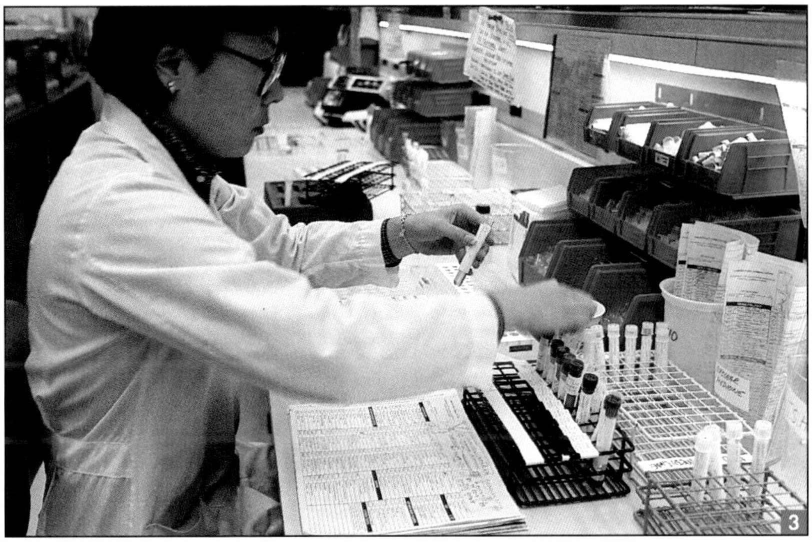

3

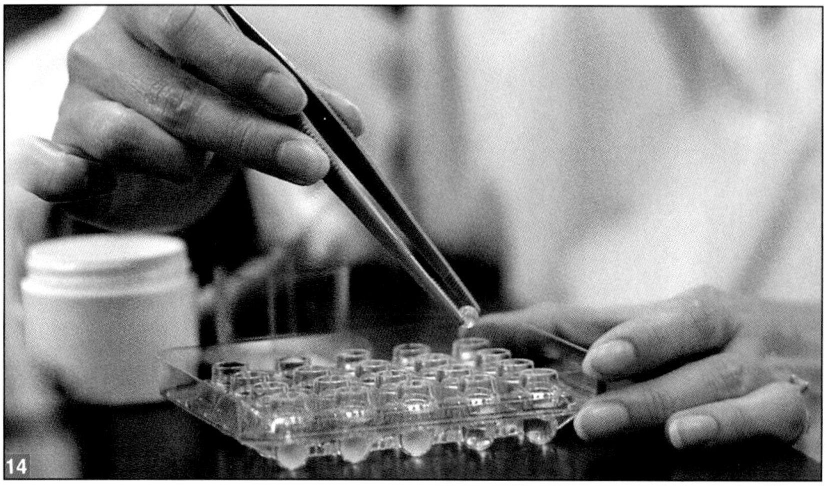
14

15

16

Immunology Lab

The technicians in the immunology lab test clinical samples for antibodies to particular viruses. Immunologic tests are designed to detect reactions between known antigens and unknown antibodies that might be present in the patient's serum. ■ 14 The ELISA test is one method used to test for a specific antibody (see Chapter 16 and Figure 16-21 for a full discussion). The UCSF immunology lab uses the ELISA test to detect rubella and hepatitis viruses in serum. A technician places beads coated with a specific antigen (one that will react to antibodies against the suspected pathogen) in wells. ■ 15 She then adds patient serum. If present, antibodies will adhere to the beads. The number of antibody molecules that adhere determine how many enzyme-antibody-antigen complexes will form on the bead. Each complex corresponds to the presence of an antibody in the serum. ■ 16 When the bead is added to a solution of substrate, the enzyme-antibody-antigen complex turns yellow, and the intensity of color indicates the amount of antibody. A spectrophotometer measures the color in the tube and gives a readout of the corresponding concentration of antibody. ■ 17 A fluorescent antibody test is performed using slides pre-prepared with a specific organism. In this test, the organism acts as the antigen. After serum is added to the wells, the slides are incubated. Antigen-antibody complexes form between the organism and the serum. After incubation, the technician washes off the excess, unbound patient serum. Next, she adds an anti-human fluoresceinated IgG conjugate that binds to the antibody-antigen complex. Again, the slide is incubated and washed. ■ 18 Because the slides are specific for a given organism, the technician need look only for fluorescence under ultraviolet light. The fluorescing organism shown in this micrograph is *Chlamydia*.

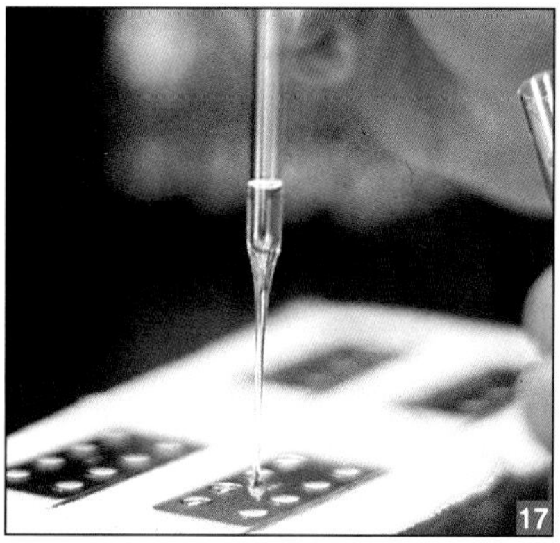
17

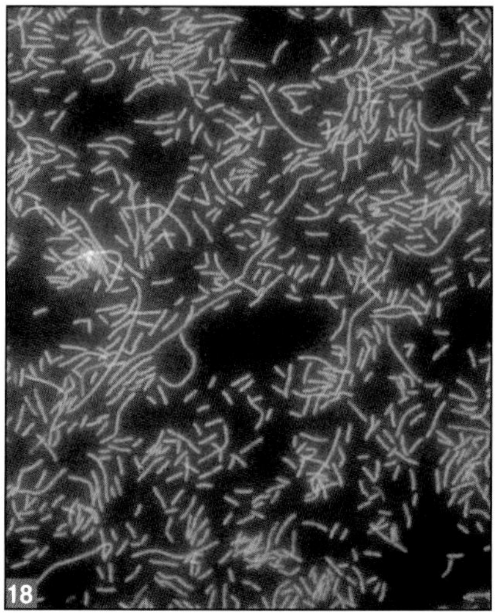
18

PART FOUR

MICROORGANISMS AND HUMAN DISEASE

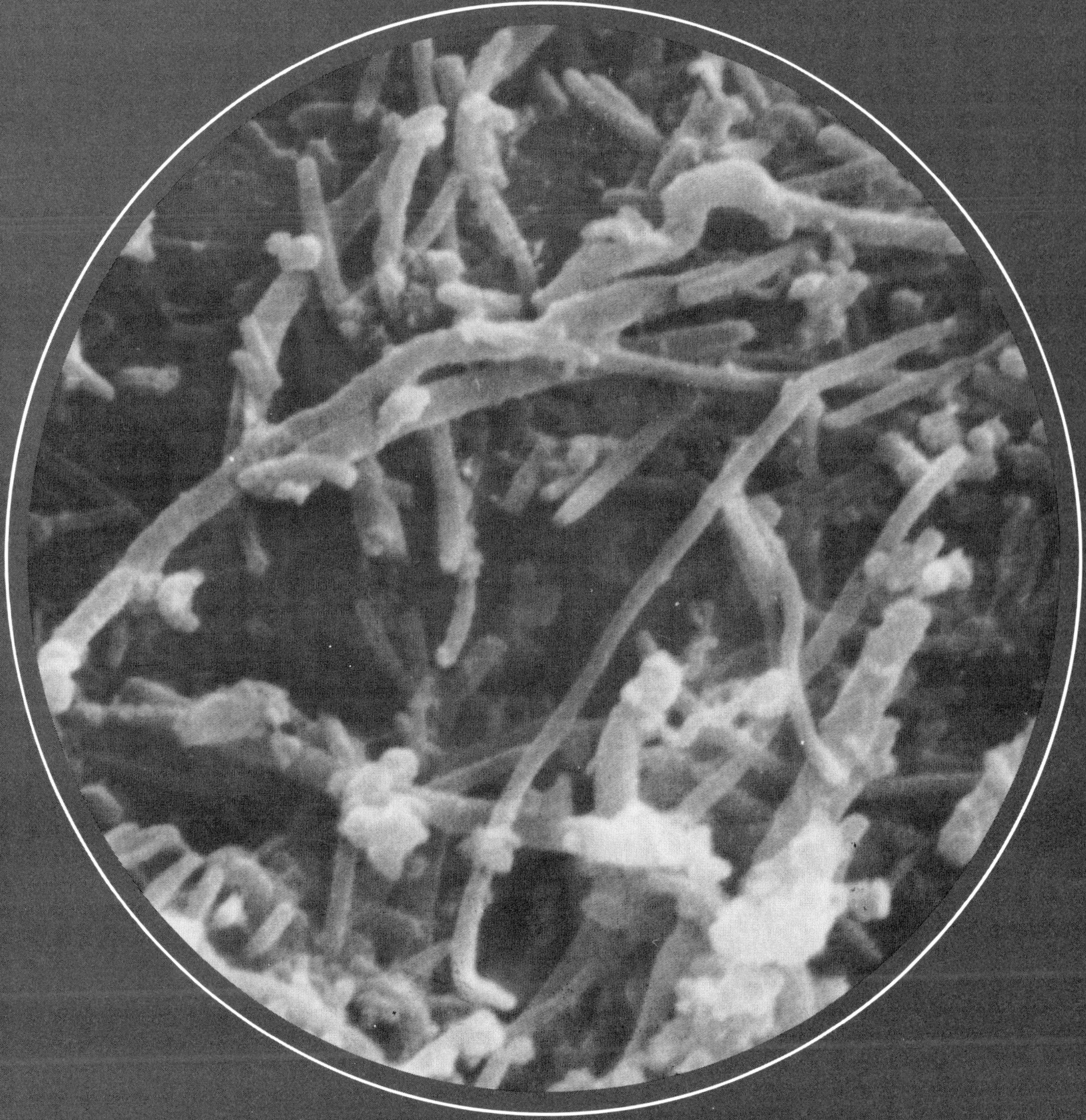

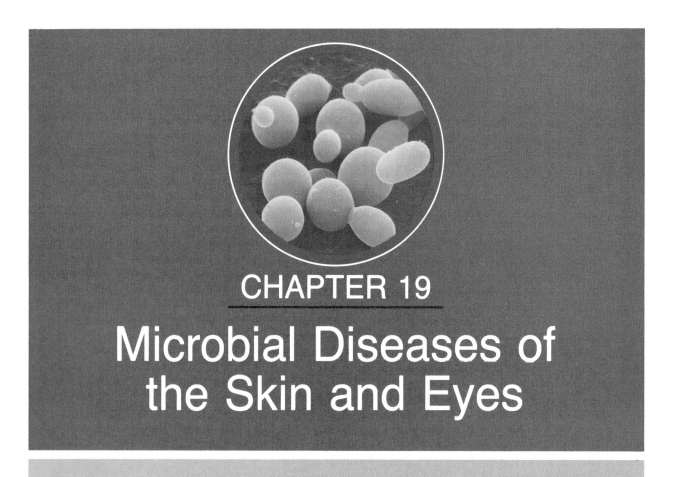

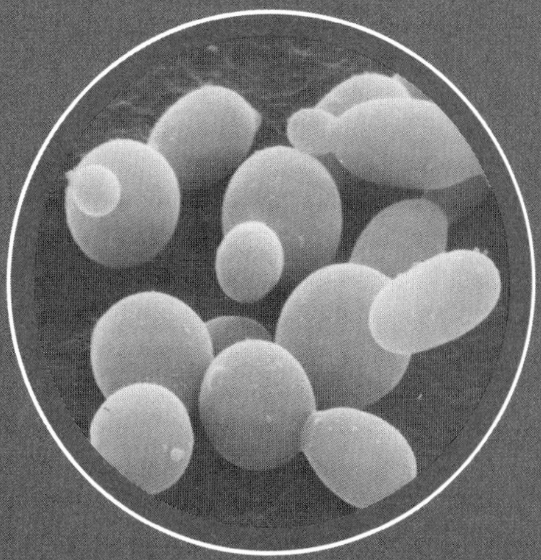

CHAPTER 19

Microbial Diseases of the Skin and Eyes

OBJECTIVES

After completing this chapter you should be able to

- Describe the structure of the skin and the ways in which pathogens can invade the skin.
- Provide examples of normal skin flora, and state their locations and ecological roles.
- Differentiate between staphylococci and streptococci and list skin infections caused by each.
- List the etiologic agent, method of transmission, and clinical symptoms of the following skin infections: acne; warts; smallpox; chickenpox; measles; German measles; cold sores.
- Differentiate between the types of mycoses and provide an example of each.
- Discuss the roles of bacteria, fungi, and viruses in conjunctivitis.
- Describe the epidemiologies of neonatal gonorrheal ophthalmia and trachoma.

In Chapter 15, we saw that the human body possesses a number of defenses that contribute to *nonspecific resistance,* that is, resistance against many types of pathogens. These defenses include the skin and mucous membranes, antimicrobial substances (interferon, complement, and properdin), phagocytosis, the inflammatory response, and fever.

The skin, which covers and protects the body, is the body's first line of defense against pathogens. Basically, the skin is an inhospitable place for most microorganisms because the secretions of the skin are acidic and most of the skin contains little moisture. Moreover, much of the skin is exposed to radiation, which discourages microbial life. Some parts of the body, however, such as the axilla (armpit), have enough moisture to support microbial growth, and the excretions there tend to contain more organic matter than the excretions elsewhere on the body. Thus, the region of the axilla can support relatively large bacterial populations, while regions like the scalp support rather small numbers of microrganisms. The skin is a physical as well as ecological barrier, and it is almost impossible for pathogens to penetrate it, although some can enter through openings not readily apparent to the human eye.

STRUCTURE AND FUNCTION OF THE SKIN

The skin of an average adult occupies a surface area of about 1.9 m^2 and varies in thickness from 0.05 to 3.0 mm. Skin consists of two principal parts: the epidermis and the dermis (Figure 19–1a). The *epidermis* is the thin, outer portion composed of several layers of epithelial cells. The outermost layer of the epidermis, the stratum corneum, consists of dead cells that contain, as hair and nails do, a waterproofing protein called *keratin*. The epidermis, when unbroken, is an effective physical barrier against light, heat, microorganisms, and many chemicals.

The *dermis* is the inner, thick portion of skin that is composed mainly of connective tissue. It contains numerous blood vessels, lymph vessels, nerves, hair follicles, and sweat and oil glands (Figure 19–1a and b). A hair follicle is a small tube in which a hair shaft grows. Sweat glands originate in the dermis and convey their secretion, called perspiration, through ducts that terminate as sweat pores on the surface of the skin. Oil glands contain ducts that feed sebum into the hair follicles. The anatomic structures of hair follicles, sweat-gland ducts, and oil-gland ducts, and their proximity to the surface of the skin provide passageways through which microorganisms can enter the skin and penetrate deeper tissues.

Perspiration is a mixture of water, salt (mostly NaCl), urea, amino acids, glucose, and lactic acid, plus small amounts of other compounds, such as the enzyme lysozyme. This mixture provides moisture and some nutrients for microbial growth. However, the salt creates a hypertonic environment that dehydrates microorganisms, and lysozyme is capable of breaking down the cell walls of certain bacteria. Secreted by oil glands, *sebum* is a mixture of lipids (unsaturated fatty acids), proteins, and salts. It prevents hairs from drying out. Sebum passes upward through the hair follicles and spreads over the surface of the skin, where it prevents excessive evaporation of moisture from the skin. Although the fatty acids inhibit the growth of certain pathogens, sebum, like perspiration, is nutritive for certain microorganisms, most notably the corynebacteria. When oil glands of the face become enlarged because of accumulated sebum, blackheads develop. And because sebum supports the growth of some microorganisms, pimples or boils can then develop. The color of blackheads is due to oxidized oil, not dirt.

In the linings of body cavities such as the mouth, nasal passages, urinary tract, genital tract, and gastrointestinal tract, the outer protective barrier is specialized in various ways. For example, some cells are ciliated; others secrete mucus from glands beneath this outer layer of epithelial cells—hence the name *mucous membrane (mucosa).* In the respiratory system the mucous layer traps particles, including microorganisms, and the ciliary movement sweeps them upward out of the body (see Figure 15–3b). (Coughing up mucus demonstrates the functioning of the cilia.) Mucous membranes are frequently acidic, and this tends to limit their microbial populations. The eyes, meanwhile, are mechanically washed by tears, and the enzyme lysozyme contained in the tears destroys the cell walls of gram-positive bacteria and some gram-negative bacteria.

NORMAL FLORA OF THE SKIN

Although the skin is generally inhospitable to most microorganisms, it supports the growth of particular microbes, which are established as part of the normal flora. For example, some anaerobic bacteria of the genus *Propionibacterium* (prō-pē-on-ē-bak-ti'rē-um) live in hair follicles, where their growth is supported by secretions from oil glands. These same bacteria produce an acid (propionic acid) that helps the pH of the skin remain between 3 and 5. This acid also has a bacteriostatic effect on many

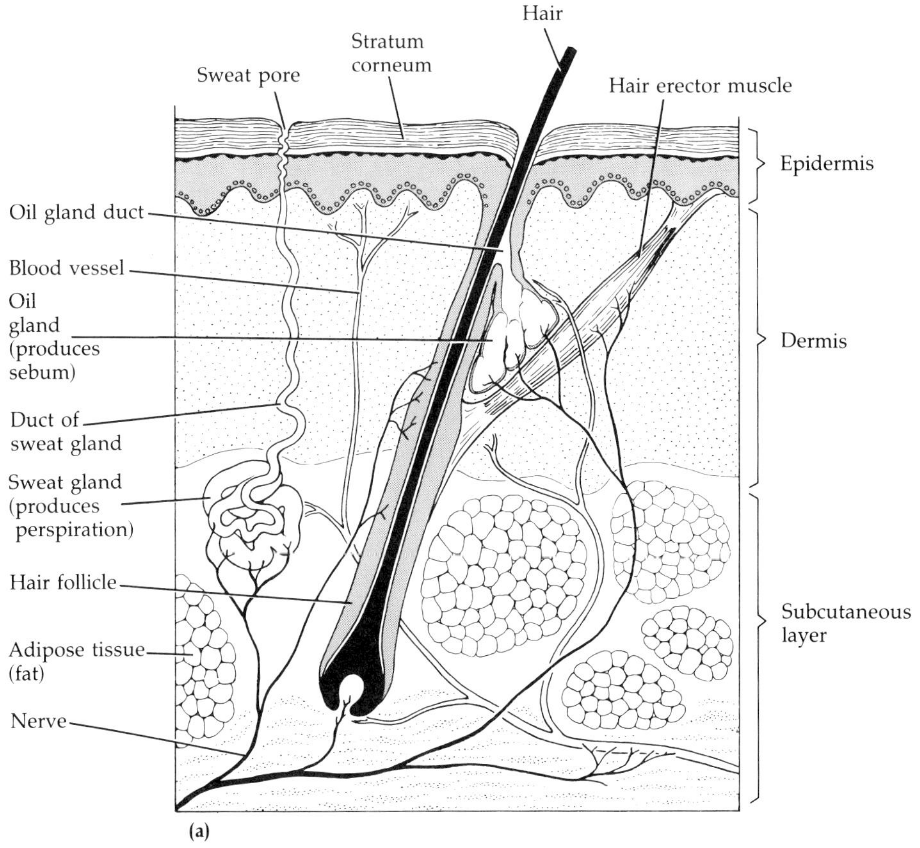

(b)

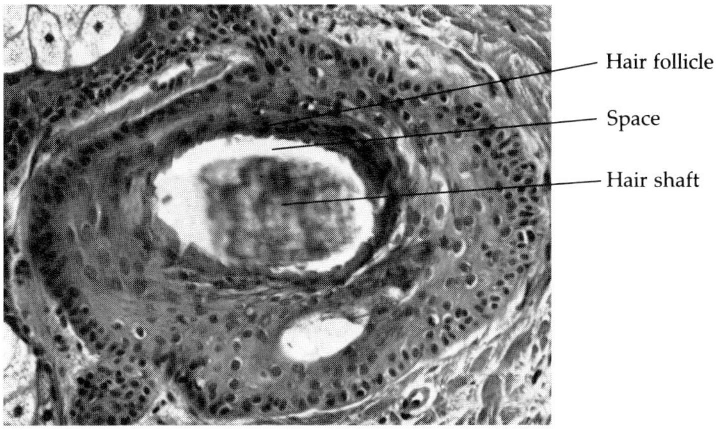

Figure 19–1 Structure of the skin. **(a)** The principal parts of the skin. Note the passageways between the hair follicle and hair shaft through which microbes could penetrate to deeper tissues. Microbes can also enter the skin through sweat pores. **(b)** Cross section through a hair follicle in a human scalp (approx. ×400). Note the space around the hair shaft.

other microorganisms, and prevents their colonization. On superficial skin surfaces, certain aerobic bacteria can, in the presence of oxygen, metabolize fatty acids. This activity contributes to the skin's normal level of fatty acids, which in turn contributes to a pH that is unfavorable to more harmful microbes. The normal flora also contains microorganisms that secrete antimicrobial substances. Because these microorganisms can resist salt and drying, they have the advantage over pathogens that attempt to colonize the skin. Although vigorously scrubbing the skin with soap and water or disinfectants will rid it temporarily of most surface bacteria, microorganisms in hair follicles and sweat glands will soon reestablish the skin's normal floral population.

Even though the skin is considered a single site, its normal flora varies in different regions. For example, the flora of the face reflects that of the oropharynx (portion of the throat behind the mouth), whereas the flora of the anal region is influenced by microorganisms of the lower gastrointestinal tract.

Scanning electron micrographs show that bacteria tend to clump on the skin in small groups. Many of these clumps consist of fairly large, spherical, gram-positive bacteria of the genera *Staphylococcus* (Figure 19–2), pronounced staf-i-lō-kok'kus, and *Micrococcus* (mī-krō-kok'kus). Both genera are capable of producing antimicrobial substances that prevent colonization of the skin by pathogens and help maintain the balance of skin flora. Also part of the skin's normal flora are gram-positive pleomorphic rods referred to as *diphtheroids*. Some diphtheroids, such as *Propionibacterium acnes*, are typically anaerobic and inhabit hair follicles. Others, like *Corynebacterium xerosis* (kô-rī-nē-bak-ti'rē-um ze-rō'sis), are aerobic and inhabit the surface of the skin. A yeast belonging to the genus *Pityrosporum* (pit-i-ros'pô-rum) is capable of growing on oily skin secretions and is a frequent inhabitant of the skin.

BACTERIAL DISEASES OF THE SKIN

Two genera of bacteria, *Staphylococcus* and *Streptococcus* (commonly referred to as staphylococci and streptococci), are frequent causes of skin-related diseases and merit special discussion. We will also discuss these bacteria in later chapters in relation to other organs and conditions. Superficial staphylococcal and streptococcal infections of the skin are very common. The bacteria frequently come into contact with the skin and have adapted fairly well to the physiological conditions there. Both genera also produce invasive enzymes and damaging toxins that contribute to the disease process.

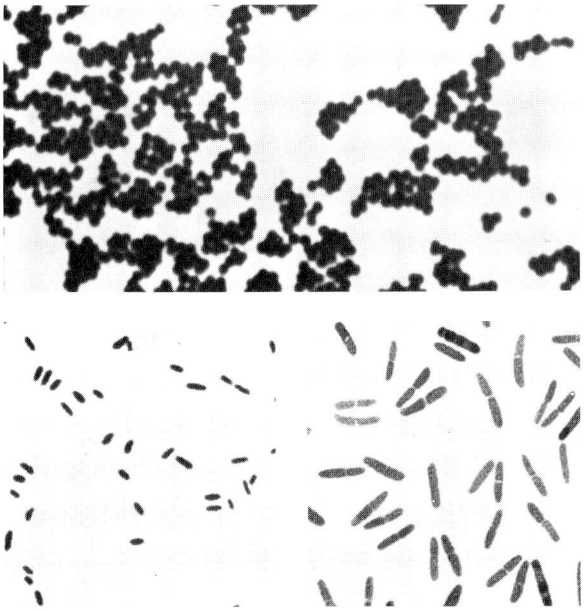

Figure 19–2 Some of the normal bacterial flora on the skin. Staphylococci (top), *Propionibacterium acnes* (lower left), and *Corynebacterium xerosis* (lower right) are among the most common.

Staphylococcal Skin Infections

Staphylococci are spherical, gram-positive bacteria about 0.5 to 1.5 μm in diameter. They tend to form in irregular clusters like those of grapes, because the cells divide at random points about their circumference and the daughter cells do not completely separate from each other.

Staphylococci are facultative anaerobes. As they ferment glucose and other sugars, they produce acid.

Staphylococcus aureus is the most pathogenic of the staphylococci (see Figure 10–11). Typically, it forms golden yellow colonies. Almost all pathogenic strains of *S. aureus* produce **coagulase,** an enzyme that coagulates (clots) the fibrin in blood.

A fibrin clot might protect the microorganisms from phagocytosis and isolate them from other defenses of the host. There is a high correlation between the bacterium's ability to form coagulase and its production of damaging toxins, several of which might injure tissues. Some toxins, called *enterotoxins,* affect the gastrointestinal tract; this topic will be discussed in detail in later chapters. Other toxins, called *leukocidins,* destroy phagocytic leukocytes, aid the staphylococcal cells in invading body tissues, and are responsible for the formation of pus. Because of that last function, *S. aureus* is called a *pyogenic* (pus-producing) coccus.

S. aureus is a very common problem in the hospital environment. Because *S. aureus* is carried on the skin of patients, hospital personnel, and hospital visitors, the danger of infection is very high. Moreover, such infections are difficult to treat because *S. aureus* is exposed to so many antibiotics that it quickly becomes resistant to them. Later we will see how, as opportunists, these organisms seriously infect surgical wounds and other artificial breaches of the skin.

However, as noted earlier, the organism often enters the body through a natural weakness in the skin barrier, the hair follicle's passage through the epidermal layer. Infections of hair follicles are known as folliculitis and often occur as **pimples.** The infected follicle of an eyelash is called a **sty.** A more serious hair follicle infection is the **furuncle (boil),** which is a type of **abscess,** a localized region of pus surrounded by inflamed tissue. When the body fails to wall off the bacterial infection, neighboring tissue can be progressively invaded. The extensive damage is called a **carbuncle,** a hard, round, deep inflammation of tissue under the skin that results in an abscess. At this stage of infection, the patient usually exhibits the symptoms of generalized illness with fever. The condition in which the organisms gain access to the bloodstream and reproduce rapidly is termed **septicemia.** In the days before effective antibiotics were in use, the skin and mucous membranes were the most important barriers to death from septicemia.

Because of the prevalence of drug-resistant staphylococci, especially in hospitals, it is essential that the particular infecting organism's sensitivity to drugs be determined. Many staphylococci produce penicillinase, which degrades natural penicillins. Until the organism's drug sensitivity is determined, patients are usually treated with antibiotics such as methicillin, oxacillin, or cephalosporin (to which penicillinase-producing strains are susceptible). Cephalosporins are drugs related to penicillins in the sense that they both include a beta-lactam ring and have a similar spectrum of action. They are not likely to be affected by penicillinase. Because the bloodstream doesn't reach established abscesses, they do not respond well to antibiotic therapy and should be drained.

Staphylococci are the primary cause of a very troublesome problem in hospital nurseries, **impetigo of the newborn.** Symptoms of this disease are thin-walled vesicles on the skin that rupture and later crust over. To prevent outbreaks, which can reach epidemic proportions, hexachlorophene-containing skin lotions are commonly prescribed. However, if these are used excessively, neurological damage may result (see Chapter 7).

Staphylococcal infections always contain the risk that the underlying tissue will become infected or that the infection will enter the bloodstream. The circulation of toxins, such as those produced by staphylococci, is called toxemia. One such toxin, which is produced by staphylococci lysogenized by certain phage types, causes **scalded skin syndrome.** (See Chapter 12 for a discussion of lysogeny and phage conversion.) This condition is first characterized by a lesion around the nose and mouth. The lesion develops rapidly into a bright red area, and within 48 hours, the skin of the palms and soles peels off in sheets when it is touched. The scalded skin syndrome is frequently observed in children under the age of two, especially in newborns, as a complication of staphylococcal infections. These patients are seriously ill and vigorous antibiotic therapy is required. The scalded skin syndrome is also characteristic of many cases of toxic shock syndrome (TSS). This is a potentially life-threatening condition in which fever, vomiting, and a sunburn-like rash are followed by shock—a sudden drop in blood pressure. The cause of TSS is apparently most often a staphylococcal growth associated with the use of vaginal tampons; the correlation is especially high for cases in which the tampons were retained too long. Staphylococcal toxins enter the bloodstream from the bacterial growth site in and around the tampon, and their circulation causes the symptoms. This syndrome has also been observed with other types of staphylococcal infections.

Streptococcal Skin Infections

Like staphylococci, **streptococci** are gram-positive, spherical bacteria. But unlike the staphylococci, the streptococcal cells grow in chains. Prior to division, the individual cocci elongate on the axis of the chain. Then the cells divide into pairs. When the dividing pairs do not separate, chaining occurs (see Figure 10–12a). The bridges between the individual cocci in the chain consist of cell wall material that has not cleaved. Streptococci are referred to as facultative anaerobes in *Bergey's Manual.* However, because their metabolism is strictly fermentative (they cannot use oxygen), they may better be regarded as aerotolerant anaerobes. Unlike either aerobic or facultative anaerobic bacteria, they do not produce catalase. Although many nonpathogenic streptococci are commonly found inhabiting the mouth, gastrointestinal tract, and upper respiratory system, some streptococci are responsible for important skin infections.

As streptococci grow, they secrete several toxins and enzymes into the growth medium. Among these are **hemolysins,** which damage red blood cells. Depending on the type of destruction caused by the hemolysins, streptococci can be divided into alpha-hemolytic, beta-hemolytic, and gamma-hemolytic streptococci (see Chapter 10). Beta-hemolytic streptococci are further differentiated into a number of immunologic groups, designated A through O, according to the carbohydrates (antigens) in their cell walls.

For their impact on human disease, group A beta-hemolytic streptococci are the most important. The most common species of this group is *Streptococcus pyogenes* (strep-tō-kok'kus pī-äj'en-ēz). Group A beta-hemolytic streptococci can be further subdivided into over 55 immunologic types according to the antigenic properties of a protein, called the *M protein,* distributed on the surface of the pili (Figure 19–3). The M protein has antiphagocytic properties that contribute to the strain's pathogenicity; the M protein also appears to aid the bacteria in adhering to and colonizing on mucous membranes.

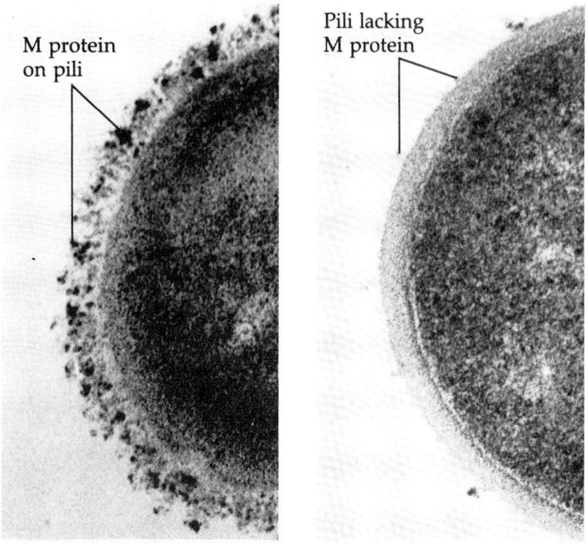

Figure 19–3 Electron micrographs showing portions of group A beta-hemolytic streptococci (approx. ×90,000). The cell on the left is from a strain that has M protein on its pili. The cell on the right lacks M protein.

In addition to hemolysins and M proteins, a number of other substances contribute to the pathogenicity of group A beta-hemolytic streptococci. These substances include *erythrogenic toxin* (responsible for the scarlet fever rash), *deoxyribonucleases* (enzymes that degrade DNA), *NADase* (an enzyme that breaks down NAD), *streptokinases* (enzymes that dissolve blood clots), *hylauronidase* (an enzyme that dissolves hyaluronic acid, the cementing substance of connective tissue), and *leukocidins* (enzymes that kill white blood cells).

Of all pathogens, beta-hemolytic streptococci are among the most susceptible to antimicrobial drugs. Penicillin is the drug of choice, although erythromycin is frequently used. Tetracyclines are no longer recommended because many streptococcal strains are now resistant to them. As described above, group A beta-hemolytic streptococci cause a wide variety of diseases, a number of which will be discussed in later chapters. One species, *Streptococcus pyogenes,* might be implicated in some cases of impetigo.

In discussing staphylococci, we mentioned impetigo of the newborn. **Impetigo** is also common in children of toddler and grade school age and is

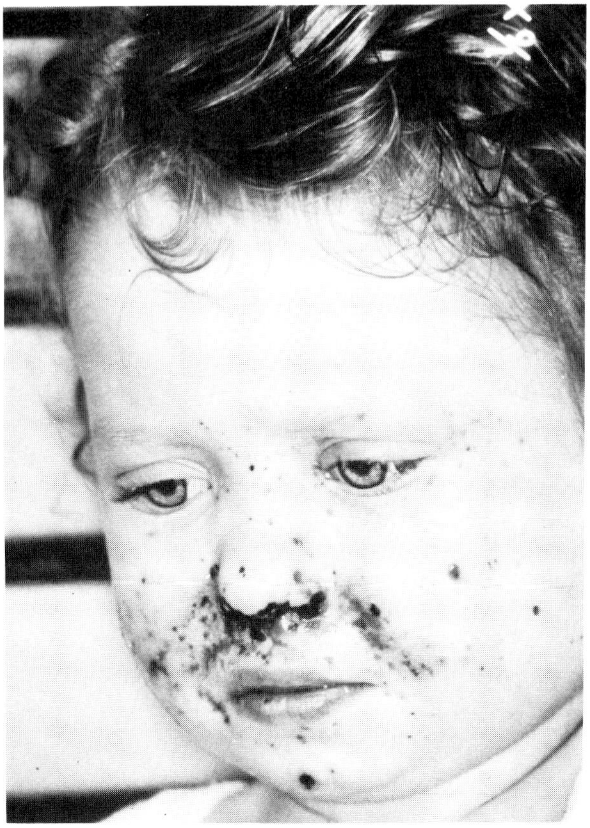

Figure 19–4 Impetigo, caused by streptococci, on the face of a three-year-old child. Note the characteristic pustules—small elevations containing pus.

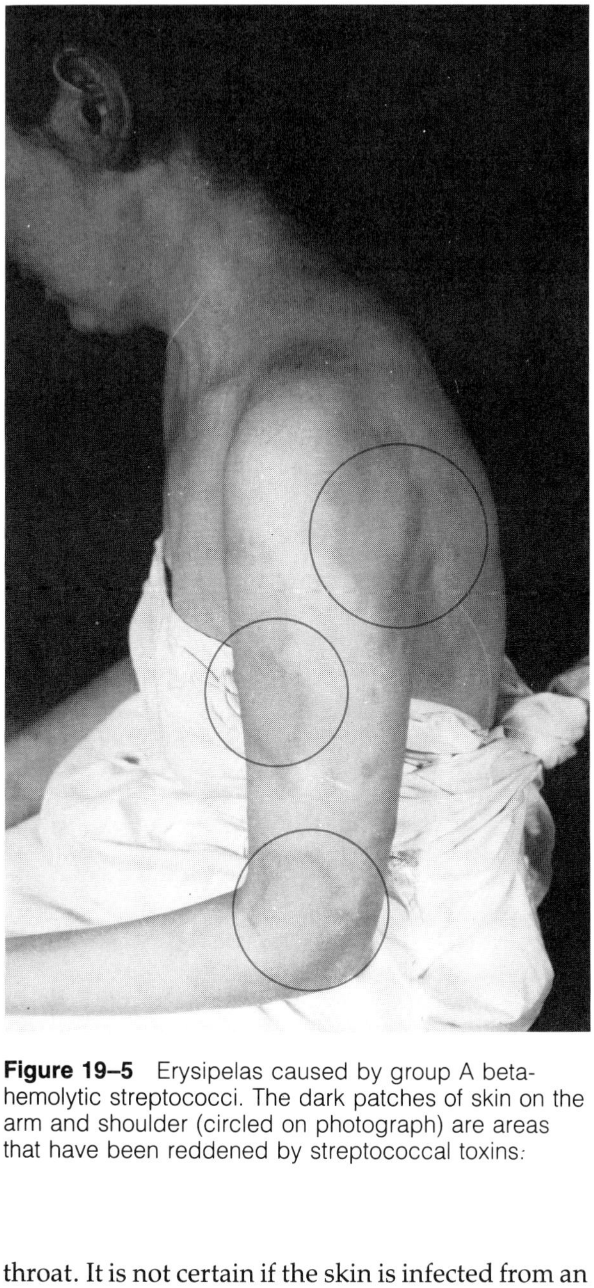

Figure 19–5 Erysipelas caused by group A beta-hemolytic streptococci. The dark patches of skin on the arm and shoulder (circled on photograph) are areas that have been reddened by streptococcal toxins.

likely, in these cases, to be caused by streptococci. A superficial skin infection, impetigo is characterized by isolated pustules (small, round elevations containing pus) that become crusted and rupture (Figure 19–4). The disease is spread largely by contact, the bacteria penetrate the skin through some minor abrasion or wound whose scab has been prematurely removed. Staphylococci can be found in these lesions, but are probably only secondary invaders. Fortunately the condition is seldom serious. The drugs of choice are penicillin and erythromycin.

When streptococcal infections affect the dermis, they cause **erysipelas.** In this disease, the skin erupts into reddish patches that enlarge and thicken and swell at the margins (Figure 19–5). The reddening is due to toxins produced by the streptococci as they invade new areas. Usually, the skin outbreak is preceded elsewhere in the body by a streptococcal infection, such as a streptococcal sore throat. It is not certain if the skin is infected from an external invasion or via some systemic route. Erysipelas is most likely to occur in the very young and the very old. The drugs of choice for the treatment of erysipelas are penicillin and erythromycin.

Infections by Pseudomonads

Pseudomonads frequently cause outbreaks of **folliculitis,** a superficial infection of hair follicles. Often following exposure in hot tubs, sauna-type pools, and even outdoor waterslides, this disease is caused by gram-negative, aerobic organisms that are widespread in soil and water. Surviving in any moist environment, they can grow on traces of even unusual organic matter, and are resistant to comparatively high concentrations of chlorine. The most prominent species is *Pseudomonas aeruginosa* (sū-dō-mō'nas ā-rü-ji-nō'sä). Competition swimmers are often troubled with **otitis externa**, or swimmer's ear, a pseudomonad infection of the ear canal external to the eardrum.

P. aeruginosa produces several exotoxins that account for much of its pathogenicity. It also has an endotoxin. Except for superficial skin infections and otitis externa, infection by *P. aeruginosa* is rare in healthy individuals. However, it often causes respiratory infections in hosts already compromised by immunologic deficiencies (natural or drug-induced) or by chronic pulmonary disease. (Respiratory infections are discussed in Chapter 22.) *P. aeruginosa* is also a very common and serious opportunist in burn patients, particularly those with second- and third-degree burns. In such cases, infection might produce blue-green pus caused by the bacterial pigment *pyocyanin*. Of major concern in many hospitals is the ease with which *P. aeruginosa* is carried by flowers or plants sent by well-wishers. Because the organism can be transmitted from flowers to burn patients or individuals with fresh surgical incisions, many hospitals do not permit such patients to receive flowers.

Pseudomonads are somewhat resistant to many of the commonly used disinfectants (Chapter 7) and antibiotics (Chapter 18). Antibiotic resistance by pseudomonads is still a problem, but in recent years a number of new antibiotics have been developed, so the number of antimicrobials available for chemotherapy of these infections is not so restricted as it once was. Two antibiotics commonly used against pseudomonad infections are gentamicin and carbenicillin; they are often used in combination. Silver sulfadiazine is very useful in the treatment of burn infections by *P. aeruginosa*.

Acne

Acne is probably the most common skin disease of humans. More than 65% of all teenagers have the problem to some degree. In the U.S., about 350,000 persons each year are estimated to suffer from severe acne that produces inflamed cysts and subsequent scarring of the face and upper body. Although its occurrence decreases after the teen years, the scarring from these severe cases often remains.

Acne begins when the sebum produced by the oil glands finds that its channels to the skin surface are blocked. The sebum accumulation leads to the formation of the familiar whiteheads, and later blackheads. In some persons the accumulation of sebum ruptures the lining of the hair follicle. Bacteria, especially *Propionibacterium acnes*, a diphtheroid commonly found on the skin, become involved at this stage. By metabolizing the sebum, *P. acnes* forms free fatty acids that cause an inflammatory response by the body. It is this inflammation that leads to tissue damage and subsequent acne scars. Picking or scratching the lesions—or even having tight collars or other clothing in contact with lesions—increases the incidence of scar formation. Makeup frequently aggravates the condition, but diet—including consumption of chocolate—has been demonstrated to have no significant effect on the disease.

Topical applications of preparations containing benzoyl peroxide are often useful, and antibiotics such as tetracycline have also been used. Severe cases should receive a dermatologist's treatment rather than household remedies. The most important recent development in treatment of severe cystic acne is isotretinoin (Accutane). This drug, taken orally, inhibits sebum formation; then, the numbers of *P. acnes* often reduce and dramatic improvement follows. Isotretinoin is not recommended for use against the usual mild cases of acne.

VIRAL DISEASES OF THE SKIN

Warts

Papovaviruses can cause skin cells to proliferate into uncontrolled growths, called **warts.** Warts are generally benign (noncancerous), and regress spontaneously. If the immune system is operating properly, most warts disappear within about two years. It is possible to spread warts by contact-transfer of the viruses, and sexual contact has been implicated in the transmission of warts on the genitals. After infection, there is an incubation period of several weeks before the warts appear.

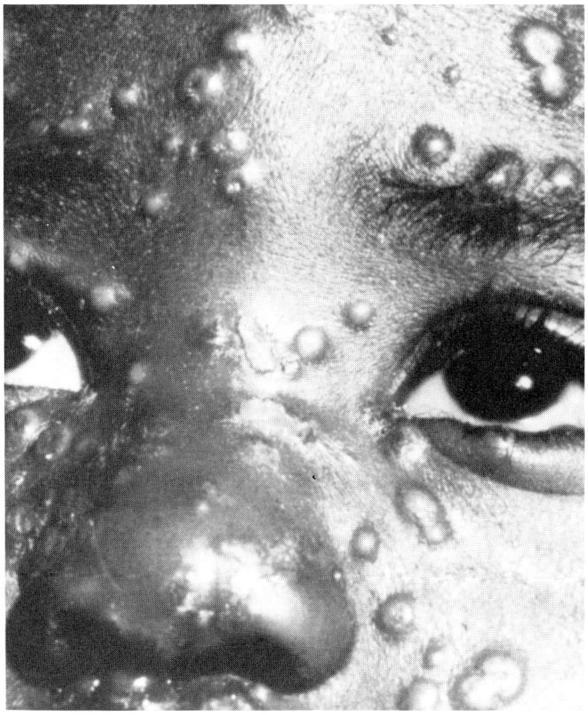

Figure 19–6 Lesions characteristic of smallpox. In some cases, the lesions remain separate from each other; in others, the lesions run together.

The most common medical treatments for warts are applying liquid nitrogen at extremely low temperatures to them (cryotherapy), drying them with electrodesiccation, or burning them with acids. Burning with acids is also the basis of some home remedies.

Smallpox (Variola)

It is estimated that during the Middle Ages, 80% of the population of Europe could expect to contract **smallpox** during their lives. In more recent times, colonial Boston (with a total population of only 10,700) had 855 deaths from 5,889 cases (14% mortality) in about 10 months of 1721–22. Those who recovered from the disease carried disfiguring scars. The disease was even more devastating when contracted by peoples such as the American Indian, who had no previous exposure.

Smallpox is caused by a poxvirus known as the smallpox (variola) virus. There are two basic forms of this disease: **variola major,** which has a mortality rate of 20% or higher, and **variola minor,** which has a mortality rate of less than 1%. Recovery from one form of the disease produces effective immunity against the other.

The transmission of smallpox, and the progression of the disease, roughly parallels that of other viral skin diseases. Transmitted first by the respiratory route, the viruses infect many internal organs before their eventual movement into the bloodstream **(viremia)** leads to infection of the skin and the production of more recognizable symptoms. Symptoms include fever, malaise, headache, severe backache, and, occasionally, abdominal pain. The growth of the virus in the epidermal layers of the skin causes the lesions shown in Figure 19–6. Similar lesions are produced in the mucous membranes of the oral cavity. At any particular time, the lesions throughout the body are at the same stage of development.

It has long been observed that persons who recover from smallpox are immune to further infection. We discussed in Chapters 1 and 16 the early attempts to artificially induce such immunity and the successful use of vaccination by Edward Jenner in the eighteenth century.

Smallpox was the first disease successfully eradicated from the human population. Perhaps this was possible because there are no animal host reservoirs for the disease. It is believed that the last victim was one who recovered from a natural case of variola minor in 1977 in Somalia, Africa. The last outbreak in the United States, which caused the mass inoculation of several million persons, was initiated by a man who succumbed in New York City in 1947. He probably acquired the disease in Mexico. The eradication of smallpox was accomplished by a concerted effort coordinated by the World Health Organization. In nations such as India, rewards were given to persons reporting cases of smallpox, and all persons in contact with the infection were then vaccinated.

At the present time, the smallpox virus collections in laboratories are the most likely sources of new infections. The risk of such infection is not merely a hypothetical concern, as there have already been several such episodes. One death from a laboratory-associated infection occurred in England in 1978. The virus was apparently carried from the laboratory, through the ventilation system, and into a telephone booth on another floor of the building, where the victim, a medical photographer, was infected. In another episode, this one in Germany, the virus escaped from the laboratory in which it was being used and the disease was caught in a part of the hospital considerably distant from the laboratory.

Routine vaccination for smallpox was discontinued in the United States in 1971, a number of years before the recent worldwide eradication of the disease. Until 1971 several persons, usually infants, had died each year from the vaccine. (The vaccination introduces live variola minor virus, which multiplies to produce an ulcerative lesion at the site of vaccination. From there the virus can be spread throughout the body.) This risk had been acceptable so long as smallpox presented a credible threat, particularly when it could be imported by air travelers. But as the disease became less and less common in the world, the risk could no longer be justified.

Chickenpox (Varicella) and Shingles (Herpes Zoster)

Chickenpox (varicella) is a relatively mild childhood disease. After gonorrhea, it is the second most common reportable infectious disease in the United States. It is probably greatly underreported and is estimated to occur in more than two million cases each year in the United States. Disease summaries of the Centers for Disease Control (CDC) show that about 100 deaths per year, usually from encephalitis, are attributed to chickenpox.

Chickenpox is acquired by infection of the respiratory system, and the infection localizes in skin cells after about two weeks. The infected skin is vesicular for 3 to 4 days. During that time, the vesicles fill with pus, rupture, and form a scab before healing (Figure 19–7a). At any particular time, different vesicles may be in different developmental stages. The vesicular rash can also appear in the mouth and throat. When chickenpox occurs in adults, which is not frequent because the high incidence in childhood grants immunity to most persons, it is a more severe disease with a significant mortality rate.

An occasional, severe complication of chickenpox, influenza, and sometimes other viral diseases is **Reye's syndrome.** After the initial infection has receded, the patient persistently vomits and exhibits signs of brain dysfunction. Coma and death can follow. Death, or brain damage in survivors, is due to brain swelling, which prevents blood circulation. At one time, the death rate of reported cases approached 90%, but this has been declining with improved care and is now 10% or lower when the disease is recognized and treated in time. Reye's syndrome affects children and teenagers almost exclusively. It is suspected that the

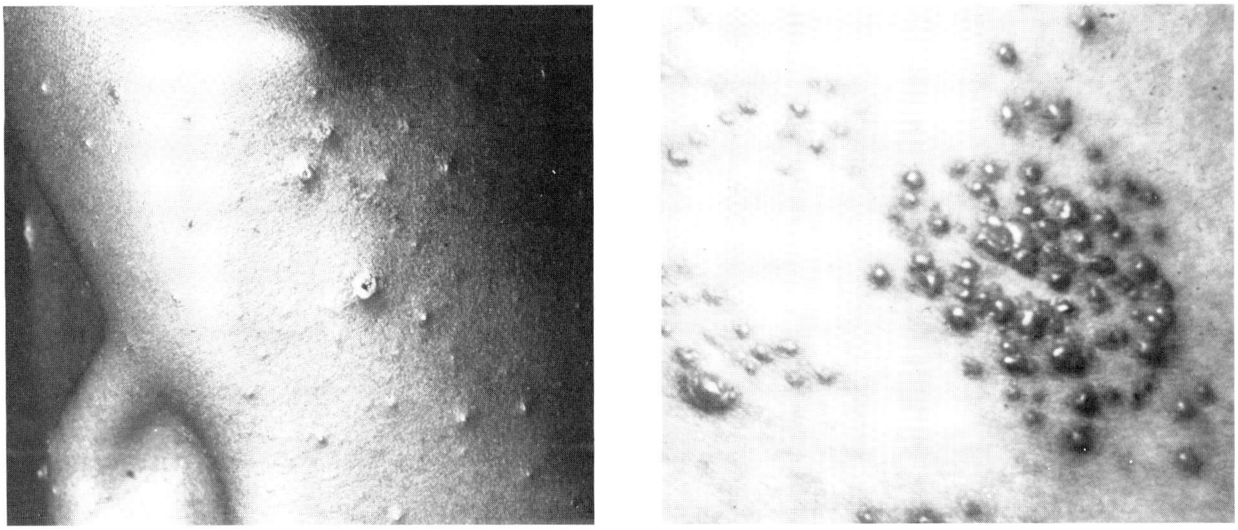

(a) (b)

Figure 19–7 Typical lesions associated with **(a)** chickenpox and **(b)** shingles.

use of aspirin to lower fevers in chickenpox and influenza increases the chances of Reye's syndrome being acquired.

The herpes-type virus that causes chickenpox (herpes zoster virus) appears to remain latent in a patient's nerve cells after he or she has recovered from the childhood infection. (Latency is a characteristic of herpes viruses.) Later in life, some trauma, such as a serious illness or even psychological stress, appears to be able to activate the virus into renewed growth. The result is a disease related to chickenpox, **herpes zoster,** or **shingles.** Vesicles similar to those of chickenpox occur but are localized in distinctive areas (Figure 19–7b). Typically, there is a girdlelike distribution about the waist, although facial shingles and infections of the upper chest also occur. The infection follows the distribution of the affected cutaneous sensory nerves (see Figure 20–1). Because these nerves are unilateral, the infection is usually limited to one side of the body at a time. Occasionally, such nerve infections can result in nerve damage that impairs vision or even causes paralysis. Severe pain is also frequently reported. Shingles is simply a different expression of the virus that caused chickenpox; it expresses differently because the patient, having had chickenpox, now has partial immunity. Exposing children to cases of shingles has led to their contracting chickenpox. Shingles seldom occur in persons under 20, and the elderly population has by far the highest incidence.

Infections with the varicella-zoster virus can be severe in patients with leukemia or in those who are immunosuppressed. Some antiviral drugs appear to have promising results in chemotherapy. Vidarabine is used to treat herpes zoster and is most helpful if administered early in the course of the illness. An attenuated live vaccine for varicella is in the testing stages. If, in a few years, varicella is eliminated, we will probably witness a dramatic decline in the incidence of these diseases.

Measles (Rubeola)

Measles is an extremely contagious disease spread by the respiratory route. Because a person with measles is infectious before symptoms appear, quarantine is not an effective measure of prevention. The causative agent is a paramyxovirus called the measles virus, which resembles the viruses causing influenza or mumps. It is closely related to the virus that causes distemper in dogs. Humans are the only reservoir for measles in most parts of the world, although monkeys are susceptible. Therefore, measles can potentially be eradicated, much as smallpox was (Figure 19–8). A

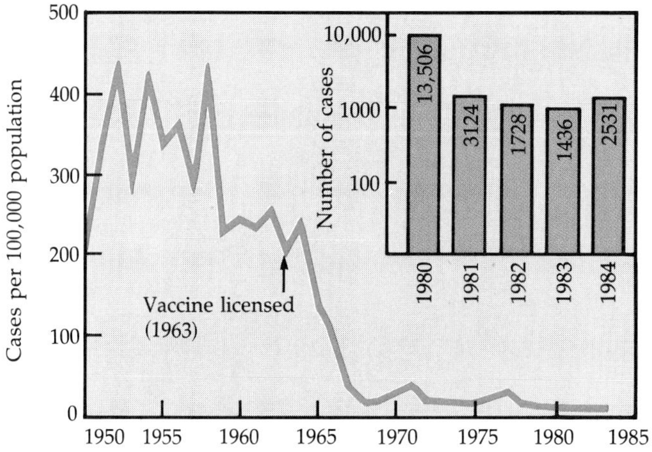

Figure 19–8 Number of reported measles (rubeola) cases, by year, in the United States, 1950–1983. Note the sharp decline in cases after introduction of the vaccine in 1963. Inset shows the number of cases reported in 1980, 1981, 1982, 1983, and the first 39 weeks of 1984.

target to eliminate all indigenous measles in the United States by October 1982 was not met; however, the number of reported cases in 1983 was the lowest since the national reporting of measles cases began in 1912. One-third of the new cases are college students who might not have been vaccinated when the program first began or were vaccinated with the ineffective formalin-inactivated virus preparation initially used. Approximately one-quarter of the new measles cases can be traced to imported measles.

The sequence of the disease is similar to that of smallpox and chickenpox. Infection initiates in the upper respiratory system. After an incubation period of 10 to 12 days there are symptoms resembling those of a common cold—sore throat, headache, and cough. Shortly thereafter, a papular rash appears on the skin (Figure 19–9). Lesions of the oral cavity include the diagnostically useful *Koplik spots* (tiny red patches with central white specks) on the oral mucosa opposite the molars.

Measles is an extremely dangerous disease, especially for the very young and the elderly. It is frequently complicated by middle ear infection or pneumonia caused by the virus itself or by secondary bacterial infection. Encephalitis strikes approximately 1 in 1,000 measles victims; its survivors are often left with permanent brain damage. As many as 1 in 1,000 cases is fatal, usually because of encephalitis or respiratory problems. The virulence of the virus seems to vary with different epidemic outbreaks. Complications such as encephalitis appear about a week after the rash appears.

Measles viruses can remain latent in the recovered patient, as do a number of other viral diseases that give a very solid immunity. The virus might affect only a few cells at a time. In some persons this slow infection can lead to an autoimmune reaction. In fact, one school of thought believes that the measles–encephalitis symptoms are part of an autoimmune reaction. Such latent infections have been suspected as the cause of several autoimmune-type diseases, such as **subacute sclerosing panencephalitis (SSPE).** This disease occurs in patients, usually children or young adults, who have a history of measles. There is a sudden, rapid, and fatal degeneration of the nervous system.

An effective measles vaccine, based on an attenuated live virus, is now recommended for all young children. Immunity seems to be permanent

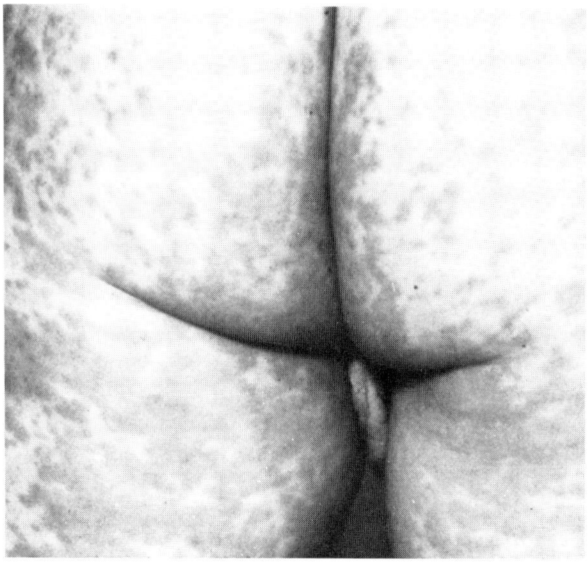

Figure 19–9 The rash of small raised spots typical of measles (rubeola).

and effectiveness is approximately 95%. Early experience with the vaccine showed that children who received it in their first year had a relatively low immunity, and many of these children had to be revaccinated. Now the vaccine is not administered until the child is at least 15 months of age. Frequently, the vaccine is of a trivalent type, consisting of measles, mumps, and rubella (MMR).

German Measles (Rubella)

Rubella is caused by a togavirus (rubella virus). It is a much milder disease than rubeola and often goes undetected (a subclinical case). A rash of small red spots and a light fever are the usual symptoms (Figure 19–10). Complications are rare, especially in children, but encephalitis occurs in about 1 case in 6,000, mostly in adults.

Transmission is airborne, and incubation of two to three weeks is the norm. The seriousness of this disease was not appreciated until 1941, when the association was made between serious birth defects and maternal infection during the first trimester (three months) of pregnancy—the now familiar **congenital rubella syndrome.** When the mother contracts the disease during this time, there is about a 35% incidence of serious fetal damage, including deafness, eye cataracts, heart defects, mental retardation, and death. Fifteen per-

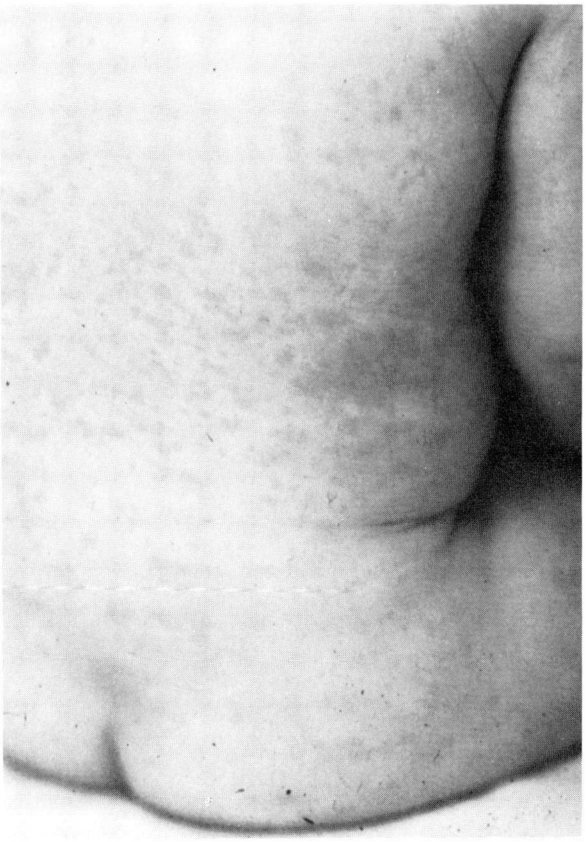

Figure 19–10 The rash of red spots characteristic of German measles (rubella). The spots are not raised above the surrounding skin.

cent of babies with congenital rubella syndrome die in their first year.

The last major epidemic of rubella in the United States was during 1964–65. About 20,000 severely damaged children from this epidemic are still alive.

It is very important to establish whether women who are in early pregnancy or of childbearing age are immune to rubella. Accurate diagnosis always requires laboratory tests; histories alone are very unreliable. Serum antibody against rubella can be assayed by hemagglutination inhibition. ELISA tests (Chapter 16 and Microview 3–4) are commercially available and have comparable results.

Recovery from clinical or subclinical cases of German measles appears to give a firm immunity. In 1969 a rubella vaccine was approved for use. It is important to know if the immunity conferred by the vaccine will be maintained through a woman's childbearing years. Follow-up studies indicate that there has been little decline in antibody levels in children who were the first to receive the vaccine. However, the question of whether vaccination grants permanent immunity will not be answered for a number of years. Nevertheless, vaccinating children prevents them from spreading the disease, even if their own immunity is not permanent. It might prove necessary to cautiously revaccinate women as they enter their childbearing years. In a number of cases, women who did not know they were pregnant received the vaccine. The evidence from these cases suggests that the vaccine itself does not damage the fetus. The vaccine is still not recommended for pregnant women, however.

Cold Sores (Herpes Simplex)

The most striking characterisitic of the herpes simplex virus is its ability to remain latent for long periods of time. During periods of latency, the virus is probably infecting very few cells at a time and propagating slowly. Serologic surveys show that about 90% of the population of the U.S. has been infected with the herpes simplex virus. The initial infection usually occurs in infancy. Frequently this infection is subclinical, but perhaps as many as 15% of the cases develop lesions known as **cold**

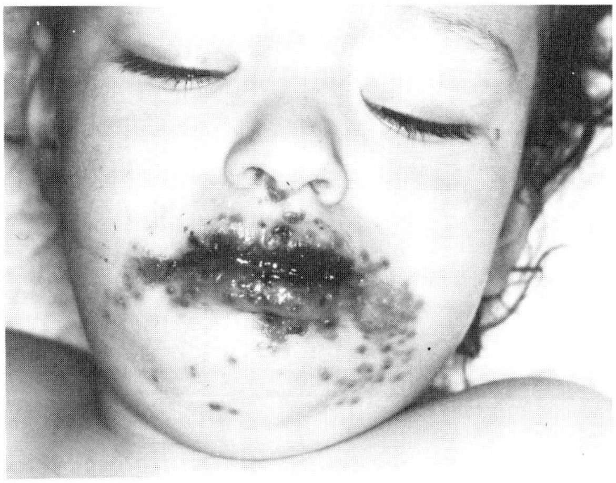

Figure 19–11 Cold sore blisters caused by herpes simplex virus.

sores or **fever blisters** (Figure 19-11). Usually occurring in the oral mucous membrane, these lesions heal as the infection subsides, but recur when the infection strengthens again. These recurrences are usually associated with some trauma; excessive exposure to UV radiation from the sun is a frequent cause, as are the hormonal changes associated with menstrual periods or emotional upset. These infections are due to *herpes simplex type 1 virus,* which is transmitted primarily by oral or respiratory routes. An important complication is herpetic keratitis (discussed later in this chapter), in which the cornea of the eye becomes infected.

A different infection by a very similar virus, *herpes simplex type 2 virus* (differentiated from type 1 by its antigens and by its effect on cells in tissue cultures) is transmitted primarily by sexual contact. We will discuss this infection in a later chapter, in our discussion of diseases of the urinary and genital systems (Chapter 24).

There is no cure for cold sores, genital herpes, or most other herpes virus infections. Acyclovir has been successfully used in treating immunosuppressed patients with systemic herpes infections, and has alleviated the symptoms of genital herpes. Vidarabine has been of limited use in treating herpes simplex viral encephalitis and herpes zoster. Chemotherapy for herpetic keratitis has been fairly successful, largely because the virus is readily accessible on the cornea. Ophthalmic ointments containing idoxuridine, vidarabine, or trifluorothymidine have all proven useful for this application.

FUNGAL DISEASES OF THE SKIN

The skin is most susceptible to microorganisms able to resist high osmotic pressure and low moisture. It is not surprising, therefore, that fungi cause a number of skin disorders. Any fungal infection of the body is referred to as a **mycosis.**

Superficial Mycoses

Superficial mycoses are limited to the external shafts of hair. Such fungal infections of hair are called **piedras.** The fungi form black or white nodules that often give the hair a gritty texture. They are not very common in the United States.

Cutaneous Mycoses

Fungi that colonize the hair, nails, and outer layer (stratum corneum) of the epidermis (see Figure 19–1) are called **dermatophytes,** and their infections are dermatomycoses. These fungi grow on the keratin present in those locations. Dermatophytes cause infections called **tineas,** or **ringworms.** Some of these fungal infections are almost asymptomatic and of interest mainly for cosmetic reasons. An example is **tinea nigra,** caused by *Cladosporium werneckii* (kla-dō-spô'rē-um wėr-ne'kē-i), in which brown or black patches appear on the palms of the hands.

The name ringworm arose from the ancient Greek belief that the infections, which tend to expand circularly, were caused by a worm. The Romans incorrectly associated ringworm with lice, and the term tinea is derived from the Latin for "insect larvae." **Tinea capitis,** or ringworm of the scalp, is fairly common among elementary school children and can result in bald patches (Figure 19–12a). It is usually transmitted by contact with fomites. Dogs and cats are also frequently infected with fungi that cause ringworm in children. Ringworm of the groin, or jock itch, is known as **tinea cruris** and ringworm of the feet, or athlete's foot, is known as **tinea pedis** (Figure 19–12b).

Three genera of fungi are involved in cutaneous mycoses. *Trichophyton* (trik-ō-fī'ton) can infect hair, skin, or nails; *Microsporum* (mī-krō-spô'rum) (see Microview 1–4A) usually involves only the hair or skin; and *Epidermophyton* (ep-i-dėr-mō-fī'ton) affects only skin and nails. The topical drug of choice for tinea infections is usually miconazole or clotrimazole. Other topical agents found in nonprescription remedies are tolnaftate, undecylenic acid, and zinc undecylenate. An oral antibiotic, griseofulvin, is often useful in these infections because it can localize in keratinized tissue.

Subcutaneous Mycoses

Compared to superficial or cutaneous mycoses, subcutaneous mycoses are serious. Cutaneous fungi do not seem to be able to penetrate past the dead cellular material of the stratum corneum, perhaps because they cannot obtain sufficient iron for growth in the living cells of the epidermis and dermis below (see Figure 19–1). Subcutaneous my-

Scabies—An Itchy Infection by a Parasitic Arthropod

Scabies is a disease of human skin caused by the mite *Sarcoptes scabiei,* a parasitic arthropod. The disease is common among school children and is also found in adults. Sometimes it occurs as a nosocomial infection in hospital personnel treating patients with symptoms described as "pruritic dermatitis."

The fingers, wrist, and elbows are the most frequent sites of infection. The mites burrow into the skin and fill the tunnel with their eggs and feces. The eggs hatch and new mites mature, mate, and lay more eggs—perpetuating the life cycle. The symptoms of scabies are the result of hypersensitivity reactions to the mites. Symptoms first occur two to six weeks after the initial infection. The main symptom is itching, especially when the skin is warm (for example, when in bed at night). Red, raised lesions ("erythematous papules") develop, which may become infected with bacteria through scratching. More advanced, chronic cases may result in generalized eczema.

Diagnosis is made by examination of the skin with a 10× hand lens. Burrows can sometimes be seen, and the mites can be picked out with a needle for microscopic examination. Alternatively, scrapings from lesions may be examined microscopically for mites. Scabies is treated by topical application of gamma benzene hexachloride (Kwell). Clothing, bedding, and other personal objects that may contain mites must be thoroughly cleaned. Scabies is transmitted by direct contact with infected persons or with fomites carrying female mites.

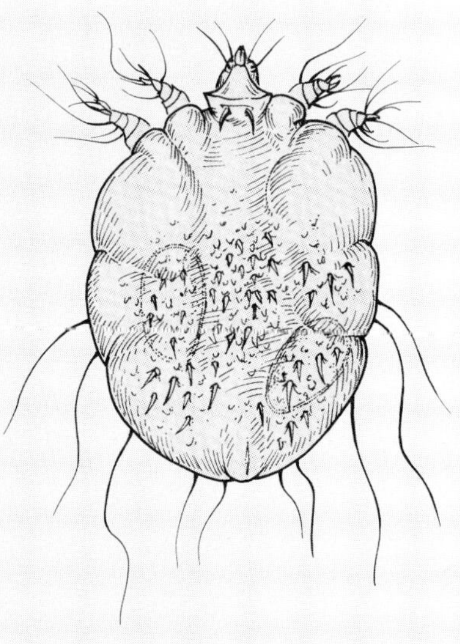

Female *Sarcoptes scabiei.*

coses usually originate in a small wound that allows the fungi, which are rather common in soil, to penetrate into the subcutaneous tissues.

Having penetrated the skin, the fungi that produce subcutaneous mycoses grow and form a nodule just under the skin. The nodule then ulcerates and spreads via the lymphatic system; a series of subcutaneous nodules form along the lymphatic vessels. In one type of subcutaneous mycosis, **chromomycosis,** the lesions are pigmented (chromo means "color") a dark brown. This mycosis is caused by one of several black molds, including *Phialophora verruscosa* (fē-ä-lō'fô-rä ver-rü-sko'sä) and *Fonsecaea pedrosoi* (fon-se'kē-ä pe-drō'sō-ī). Flucytosine, given orally, can be an effective treatment. Another subcutaneous mycosis is **maduromycosis**, which is caused by the fungus *Allescheria boydii* (al-lesh-er'ē-ä boi-dē-ī). Maduromycosis may be referred to as a **mycetoma** or fungal tumor. It destroys subcutaneous tissues and progresses slowly, eventually causing serious deformities. Polyenes, which have not yet been used extensively, may be of value in treating mycetomas; their causative agent is susceptible in vitro. Surgically draining the mycetoma is also important.

Candidiasis

Candida albicans (kan'did-ä al'bi-kans) is a dimorphic fungus (Figure 19–13a and b and Microview 1–2D and 1–4C) that commonly grows on mucous membranes in the genitourinary tract and on the oral mucosa. *Candida albicans* is a very common cause of **vaginitis** (see Chapter 24) and is also common in the newborn, where it appears as an overgrowth of the oral cavity and is referred to as **thrush** (Figure 19–13c). *Candida* is also found in the elderly and those debilitated by diseases such as diabetes and cancer. Because it is not affected by antibacterial drugs, the fungus sometimes overgrows tissue when the normal bacterial flora is suppressed by antibiotics. If **candidiasis,** as an infection caused by this fungus is called, becomes systemic, as can happen in immunosuppressed or immunodeficient individuals, fulminating disease and death can result.

Superficial infections by *C. albicans* are usually treated with topical applications of miconazole, clotrimazole, or nystatin. A summary of diseases associated with the skin is presented in Table 19–1.

INFECTIONS OF THE EYE

The epithelial cells covering the eye can be considered, in one sense, as a continuation of the skin or mucosa. Usually originating from the skin and upper respiratory tract, the organisms most commonly associated with the eye are *Staphylococcus epidermidis* (e-pi-dėr'mi-dis), *S. aureus,* and diphtheroids.

Bacterial Infections

A number of bacteria can infect the eye, largely through the conjunctiva. The role of cosmetics in infection should be mentioned here. Microorganisms can not only survive in eye cosmetics, but can multiply there as well. The cosmetics might be sterile prior to opening, but the initial application can inoculate normal flora into the cosmetics, and subsequent applications can inoculate microorganisms into the eye.

Conjunctivitis is an inflammation of the conjunctiva, the mucous membrane that lines the eyelid and covers the outer surface of the eyeball. One form of conjunctivitis, which occurs primarily in children, is called **contagious conjunctivitis** or **pinkeye.** Contagious conjunctivitis is a common

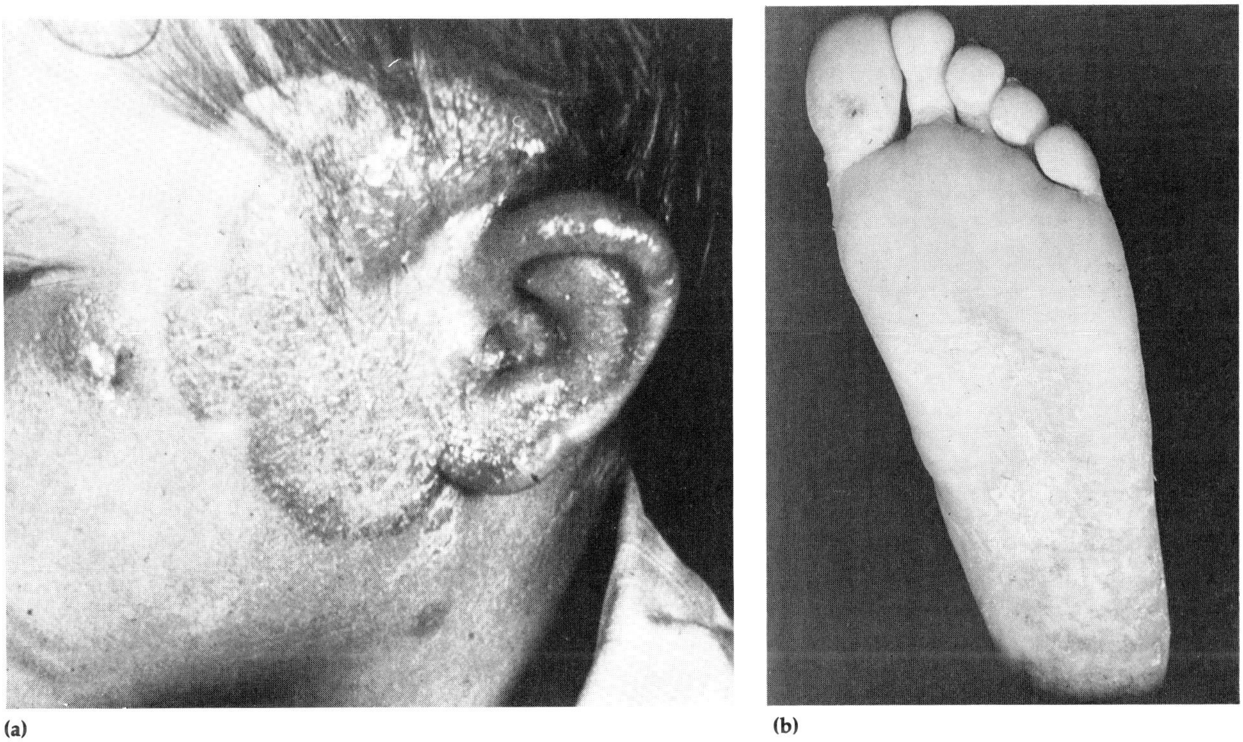

(a) (b)

Figure 19–12 Dermatomycoses. **(a)** A severe case of ringworm on the side of a child's head (tinea capitis). **(b)** Ringworm of the foot, or athlete's foot (tinea pedis).

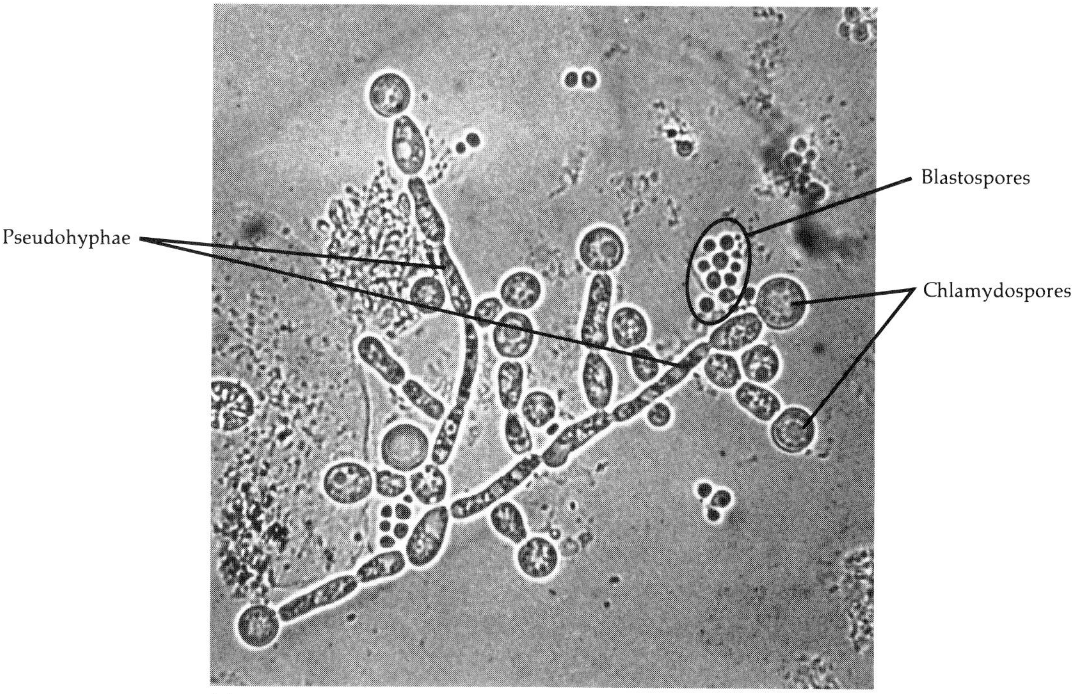

(a)

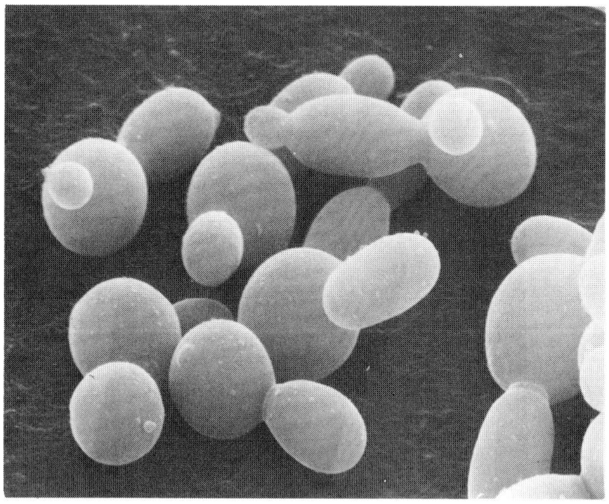

(b)

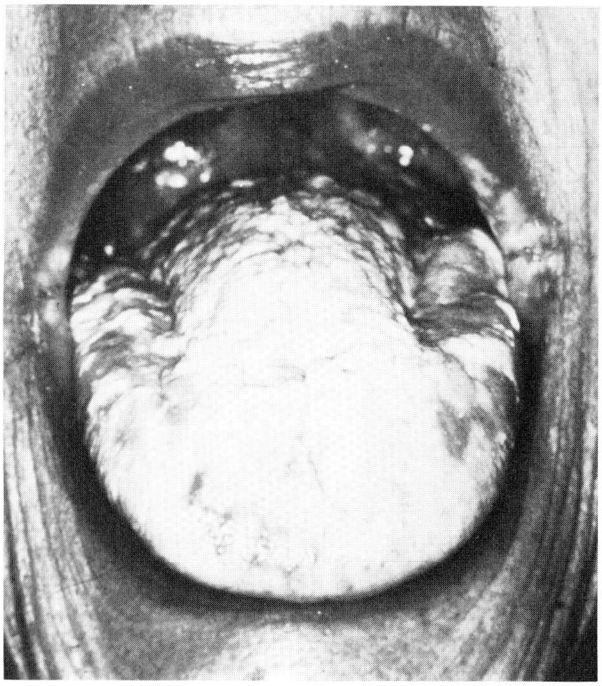

(c)

Figure 19–13 Candidiasis. **(a)** Photomicrograph of *Candida albicans,* an opportunistic fungal pathogen that causes candidiasis. Note the spherical chlamydospores, the smaller blastospores, and the pseudohyphae. **(b)** Scanning electron micrograph of the yeast form of *Candida albicans* (×6720). *Candida albicans* tends to form pseudohyphae in the body tissues but usually grows in the yeast form in cultures. **(c)** Thrush, or oral candidiasis.

Table 19–1 Summary of Diseases Associated with the Skin

Disease	Causative Agent	Treatment
Bacterial Diseases:		
Impetigo	*Staphylococcus aureus;* occasionally, *Streptococcus pyogenes*	Penicillin (for *Streptococcus* infections only), erythromycin
Erysipelas	*Streptococcus pyogenes*	Penicillin, erythromycin
Otitis externa and burn infections	*Pseudomonas aeruginosa*	Polymyxins, gentamicin, carbenicillin, and silver sulfadiazine
Acne	*Propionibacterium acnes*	Benzoyl peroxide, tetracyclines, isotretinoin
Viral Diseases:		
Warts	Papovavirus	May be removed by liquid nitrogen cryotherapy, electrodesiccation, or acids
Smallpox (variola)	Poxvirus (smallpox virus)	None
Chickenpox (varicella)	Herpes virus (herpes varicella-zoster virus)	Vidarabine, for immunocompromised patients
Shingles (zoster)	Herpes virus (varicella-herpes zoster virus)	Vidarabine, for immunocompromised patients
Measles (rubeola)	Paramyxovirus (measles virus)	None
German measles (rubella)	Togavirus (rubella virus)	None
Cold sores (herpes simplex)	Herpesvirus (herpes simplex virus)	Acyclovir may modify symptoms; also see herpetic keratitis
Fungal Diseases:		
Tinea nigra	*Cladosporium werneckii*	Griseofulvin
Ringworm (tinea)	*Microsporum, Trichophyton, Epidermophyton* species	Griseofulvin, miconazole, clotrimazole, tolnaftate
Chromomycosis	*Phialophora verruscosa, Fonsecaea pedrosoi*	Flucytosine
Maduromycosis	*Allescheria boydii*	Polyenes might be useful and surgical drainage
Candidiasis	*Candida albicans*	Topically with miconazole, clotrimazole, nystatin

disease in the Middle East, where the causative organism, *Hemophilus aegyptius,* is transmitted by hand contact or by flies. The disease is usually cured by topical application of antibiotics.

Neonatal gonorrheal ophthalmia (inflammation of the eye in the newborn) is estimated to be responsible for 10% of all cases of blindness. Infection of the infant occurs during its passage through the birth canal. The causative agent, *Neisseria gonorrhoeae* (nī-se'rē-ä go-nôr-rē'ä), is transmitted from a mother infected with gonorrhea. The eyes of all newborns are washed with 1% silver nitrate or a suitable antibiotic. This procedure prevents the growth of *Neisseria* whether or not the mother's infection was diagnosed or treated. Penicillin is used for treatment of neonatal gonorrheal ophthalmia.

Trachoma is the greatest single cause of blindness in the world today. An infection of the epithelial cells of the eye, trachoma is caused by *Chlamydia trachomatis* (kla-mi'dē-ä tra-kō'mä-tis), a species of bacteria that grows only as obligate intracellular parasites. Trachoma causes the formation of scar tissue on the cornea. Millions of people in the world now have the disease and many millions have been blinded by the infection. It is transmitted largely by contact with inanimate objects (fomites) such as towels, or by fingers carrying the organisms from one eye to another. Flies are also suspected of carrying the causative agent. In some parts of the world almost all children are infected early in their lives. Opportunistic secondary bacterial infections frequently contribute to the pathogenicity of the disease. Partial immunity is generated by recovery. Antibiotics, such as sulfonamide and tetracycline, are useful in treatment, particularly in the early stages. However, the control of the disease lies more in sanitation and health education than in treatment.

A disease somewhat similar to trachoma, but

Table 19-2 Summary of Diseases Associated with the Eyes

Disease	Causative Agent	Treatment
Bacterial Diseases:		
Contagious conjunctivitis (pinkeye)	*Hemophilus aegyptii*	Sulfonamides
Neonatal gonorrheal ophthalmia	*Neisseria gonorrhoeae*	Silver nitrate, tetracycline, or erythromycin for prevention; penicillin for treatment
Trachoma	*Chlamydia trachomatis*	Sulfonamide, tetracycline
Inclusion conjunctivitis	*Chlamydia trachomatis*	Sulfonamide, tetracycline
Viral Diseases:		
Epidemic kerato-conjunctivitis	Adenovirus	None
Herpetic keratitis	Herpes virus (herpes simplex type 1)	Idoxuridine, vidarabine, triflurothymidine

much milder, frequently occurs in the more developed countries. **Inclusion conjunctivitis,** which is caused by the same organism as trachoma, differs from trachoma in that it is an infection of the conjunctiva, not the cornea. This disease is apparently acquired by infants as they pass through the birth canal. But it also appears to spread in unchlorinated waters such as swimming pools (swimming pool conjunctivitis). Tetracycline applied as an ophthalmic ointment is an effective treatment.

Viral Infections

Herpetic keratitis is usually caused by herpes simplex type 1 virus. This is a localized infection of the cornea that can recur with an epidemiology similar to that of cold sores. It is characterized by inflammation and corneal ulcers that can be quite deep. Swelling of regional lymph nodes accompanies the infection, and damage to the cornea can lead to blindness. Invasion of the central nervous system and resulting encephalitis occasionally occur in adults.

Idoxuridine and triflurothymidine, analogs of the DNA component thymidine, can be applied topically for treatment of herpetic keratitis. Vidarabine (adenine arabinoside or ara-A), which is also effective against herpetic keratitis, has been approved for use against herpes-caused encephalitis. Herpetic keratitis is one of the few viral diseases for which there is an effective chemotherapeutic treatment.

A summary of diseases associated with the eyes is presented in Table 19–2.

STUDY OUTLINE

INTRODUCTION (pp. 524–525)

1. The skin is a physical and chemical barrier against microorganisms.
2. Moist areas of the skin (such as axilla) support larger populations of bacteria than dry areas (e.g., scalp).

STRUCTURE AND FUNCTION OF THE SKIN (p. 524)

1. The outer portion of the skin, called the epidermis, contains keratin.
2. The inner portion of the skin, or dermis, contains hair follicles, sweat ducts, and oil glands that open to the surface.
3. Sebum and perspiration are secretions of the skin that can remove organisms from openings in the dermis.
4. Sebum and perspiration provide nutrients for some microorganisms.
5. Microorganisms can gain access to the body through the openings in the dermis.

6. Mucous membranes line the body cavities. Mucus traps microorganisms and facilitates their removal from the body.
7. The eye is mechanically washed by tears that contain lysozyme.

NORMAL FLORA OF THE SKIN (pp. 525–527)

1. Members of the genus *Propionibacterium* metabolize oil from the oil glands and colonize hair follicles.
2. Microorganisms that live on human skin are resistant to high concentrations of salt and desiccation.
3. Antimicrobial substances secreted by normal flora can inhibit the growth of potential pathogens.
4. The normal flora of the face is determined by the oropharyngeal organisms, and the normal flora of the anal region is influenced by organisms in the lower gastrointestinal tract.

BACTERIAL DISEASES OF THE SKIN (pp.527–531)

Staphylococcal Skin Infections (pp. 527–529)

1. Staphylococci are gram-positive bacteria, growing often in clusters. They are facultative anaerobes, catalase-positive.
2. Almost all pathogenic strains of *S. aureus* produce coagulase.
3. Pathogenic *S. aureus* are pyogenic and can produce enterotoxins and leukocidins.
4. Localized infections (pimples, abscesses, and carbuncles) result from *S. aureus* entering openings in the skin.
5. Septicemia and toxemia occur when bacteria and toxins, respectively, enter the bloodstream.
6. Many strains of *S. aureus* produce penicillinase.
7. Impetigo of the newborn is a highly contagious superficial skin infection caused by *S. aureus.*

Streptococcal Skin Infections (pp. 529–530)

1. Streptococci are gram-positive cocci, and often reproduce in chains. They are strictly fermentative, catalase-negative.
2. Streptococci are classified according to their hemolytic enzymes and cell wall antigens.
3. Group A beta-hemolytic streptococci (including *S. pyogenes*) are the pathogens most important to humans.
4. Group A beta-hemolytic streptococci produce a number of virulence factors, including erythrogenic toxin, deoxyribonuclease, NADase, streptokinases, and hyaluronidase.
5. Streptococci are susceptible to penicillin.
6. Impetigo (isolated pustules) and erysipelas (reddish patches) are skin infections caused by *S. pyogenes.*

Infections by Pseudomonads (pp. 530–531)

1. Pseudomonads are gram-negative rods. They are strict aerobes.
2. *Pseudomonas aeruginosa* is an opportunistic pathogen found primarily in soil, water, and on plants. It is resistant to many disinfectants and antibiotics.
3. *P. aeruginosa* produces an endotoxin and several exotoxins.
4. Diseases caused by *P. aeruginosa* include otitis externa, post-burn infections, and folliculitis.
5. Infections have a characteristic blue-green pus caused by the pigment pyocyanin.
6. Carbenicillin and gentamicin are useful in treating *P. aeruginosa* infections.

Acne (p. 531)

1. *Propionibacterium acnes* can metabolize sebum trapped in hair follicles.
2. Metabolic end products (fatty acids) cause an inflammatory response known as acne.
3. Treatment with benzoyl peroxide and tetracycline is somewhat effective.

VIRAL DISEASES OF THE SKIN (pp. 531–537)

Warts (pp. 531–532)

1. Papovavirus causes skin cells to proliferate, and produce a benign growth called a wart.
2. Warts are spread by direct contact.
3. Warts may regress spontaneously or be removed chemically or physically.

Smallpox (Variola) (pp. 532–533)

1. Variola virus causes two types of skin infections: variola major and variola minor.
2. Smallpox is transmitted by the respiratory route, and the virus is moved to the skin via the bloodstream.

3. At any particular time, the vesicular or pustular lesions of the skin and mucous membranes will be in the same stage of development.
4. The only host for smallpox is humans. WHO has announced the eradication of this disease.
5. Routine vaccinations with the live virus are no longer made.

Chickenpox (Varicella) and Shingles (Herpes Zoster) (pp. 533–534)

1. Chickenpox is caused by herpes zoster virus.
2. Transmitted by the respiratory route and localized in skin cells, the virus causes a vesicular rash.
3. Complications of chickenpox include encephalitis and Reye's syndrome.
4. After chickenpox, the virus can remain latent in nerve cells, and subsequently activate in shingles.
5. Shingles is characterized by a vesicular rash along the affected cutaneous sensory nerves.
6. The virus can be treated with vidarabine. An attenuated live vaccine is being tested.

Measles (Rubeola) (pp. 534–535)

1. Measles is caused by measles virus (paramyxovirus) and transmitted by the respiratory route.
2. After the virus has incubated in the upper respiratory tract, papular lesions appear on the skin and Koplik spots on the oral mucosa.
3. Complications of measles include middle ear infections, pneumonia, encephalitis, and secondary bacterial infections.
4. Measles virus can remain latent following infection and cause autoimmune reactions such as subacute sclerosing panencephalitis.
5. Vaccination with attenuated live virus provides effective long-term immunity.

German Measles (Rubella) (pp. 535–536)

1. The rubella virus (togavirus) is transmitted by the respiratory route.
2. A red rash and light fever might occur in an infected individual; the disease can be asymptomatic.
3. Congenital rubella syndrome can affect a fetus when a woman contracts rubella during the first trimester of her pregnancy.
4. Damages from congenital rubella syndrome include stillbirth, deafness, eye cataracts, heart defects, and mental retardation.
5. Vaccination with live rubella virus provides immunity of unknown duration.

Cold Sores (Herpes Simplex) (pp. 536–537)

1. Infection of skin and mucosal cells results in cold sores.
2. The virus remains latent in nerve cells, and cold sores can recur when the virus is activated.
3. Herpes simplex type 1 virus is transmitted primarily by oral and respiratory routes.
4. Acyclovir has proven successful in treating immunosuppressed patients with systemic herpes.

FUNGAL DISEASES OF THE SKIN (pp. 537–539)

Superficial Mycoses (p. 537)

1. Piedras are infections of the external hair shaft.

Cutaneous Mycoses (p. 537)

1. Tinea nigra (caused by *Cladosporium wernickii*) is a ringworm characterized by discolored blotches on the palms.
2. *Microsporum, Trichophyton,* and *Epidermophyton* cause dermatomycoses called ringworm, or tinea.
3. These fungi grow on keratin-containing epidermis, such as hair, skin, and nails.
4. Ringworm and athlete's foot are usually treated with topical application of antifungal chemicals.

Subcutaneous Mycoses (pp. 537–538)

1. Chromomycosis and mycetoma result from fungi that penetrate the skin through a wound.
2. The fungi grow and produce subcutaneous nodules along the lymphatic vessels.
3. Polyenes and surgical drainage are used to treat these infections.

Candidiasis (pp. 538–539)

1. *Candida albicans* causes infections of mucous membranes and is a common cause of vaginitis and thrush (in oral mucosa).

2. *C. albicans* is an opportunistic pathogen that may proliferate when normal bacterial flora are suppressed.
3. Topical antifungal chemicals may be used to treat candidiasis.

INFECTIONS OF THE EYE (pp. 539–542)

1. *Staphylococcus epidermidis*, *S. aureus*, and diphtheroids are normal flora of the eye.
2. These microorganisms usually originate from the skin and upper respiratory tract.

Bacterial Infections (pp. 539–542)

1. Conjunctivitis is caused by a number of bacteria.
2. Contagious conjunctivitis, or pinkeye, is caused by *Hemophilus aegyptius* and occurs in children.
3. Neonatal gonorrheal ophthalmia is caused by the transmission of *Neisseria gonorrhoeae* from an infected mother to an infant during its passage through the birth canal.
4. All newborn infants are treated with 1% silver nitrate or an antibiotic to prevent the growth of *Neisseria*.
5. In trachoma, which is caused by *Chlamydia trachomatis*, scar tissue forms on the cornea.
6. Trachoma is transmitted by fingers, fomites, and perhaps flies.
7. Inclusion conjunctivitis is an infection of the conjunctiva caused by *Chlamydia trachomatis*. It is transmitted to infants during birth, and is transmitted in unchlorinated swimming water.

Viral Infections (p. 542)

1. Epidemic keratoconjunctivitis follows trauma to the eye and is caused by an adenovirus.
2. Herpetic keratitis causes corneal ulcers. The etiology is herpes simplex type 1 that occasionally invades the central nervous system and causes encephalitis.
3. Idoxuridine and triflurothymidine are effective treatments against herpes keratitis.

STUDY QUESTIONS

REVIEW

1. Discuss the usual mode of entry of bacteria into the skin. Compare bacterial skin infections with those caused by fungi and viruses with respect to method of entry.
2. Complete the table of epidemiology (below).

Disease	Etiologic Agent	Clinical Symptoms	Method of Transmission
Acne			
Pimples			
Warts			
Smallpox			
Chickenpox			
Measles			
German measles			
Cold sores			

3. How do these infections differ: tinea nigra, mycetoma, athlete's foot? In what ways are they similar?
4. Select a bacterial and viral eye infection and discuss the epidemiology of each.
5. A laboratory test used to determine the identity of *Staphylococcus aureus* is its growth on mannitol salt agar. The medium contains 7.5% sodium chloride (NaCl). Why is it considered a selective medium for *S. aureus*?
6. Why are people immunized against rubella, since the symptoms of the disease are mild or even inapparent?
7. Explain the relationship between shingles and chickenpox.
8. Why are the eyes of all newborn infants washed with an antiseptic or antibiotic?
9. What is the leading cause of blindness in the world?
10. An opportunistic dimorphic fungus that causes skin infections is ______________.
11. Identify the following diseases based on the symptoms (in chart below).

Symptoms	Disease
Koplik spots	
Papular rash	
Vesicular rash	
Small, spotted rash	
Recurrent "blisters" on oral mucosa	
Corneal ulcer and swelling of lymph nodes	
Pustular rash, appearance same over entire body	

12. What complications can occur from herpes simplex type 1 infections?

CHALLENGE

1. You have isolated an organism from what appears to be impetigo. The organisms are gram-positive cocci in singles, pairs, and small groups. What test would help you quickly determine whether your isolate is *Staphylococcus* or *Streptococcus*? What is the result of this test if the organism is *Staphylococcus*?
2. A hospitalized patient recovering from surgery develops an infection that has blue-green pus and a grapelike odor. What is the probable etiology? How might the patient have acquired this infection?
3. Is it necessary to treat a patient for warts? Explain briefly.
4. A 12-year-old diabetic girl using continuous subcutaneous insulin infusion to manage her diabetes developed a fever (39.4°C), low blood pressure, abdominal pain, and erythroderma. She was supposed to change the needle insertion site every 3 days after cleaning the skin with an iodine solution. Frequently she did not change the insertion site more often than every 10 days. Blood culture was negative and abscesses at insertion sites were not cultured. What is the probable cause of her symptoms?

FURTHER READING

Baron, Samuel (editor). 1982. *Medical Microbiology.* Menlo Park, California: Addison-Wesley.

Cliff, A. and P. Haggert. May 1984. Island Epidemics. *Scientific American* 250:138–147. An excellent review of epidemiology using measles as the example.

di Mayorca, G. (chairperson). 1978. Human papovaviruses. In *Microbiology 1978,* pp. 419–442. Washington, D.C.: American Society for Microbiology (Proceedings). Biology and oncogenicity of papovaviruses.

Henderson, D. A. October 1976. The eradication of smallpox. *Scientific American* 235:25–33. A summary of the incidence of smallpox and investigation of the last endemic infection.

Marples, M. J. January 1969. Life on the human skin. *Scientific American* 220:108–115. A study of the microflora of the skin and the relationship between bacterial populations and hair, moisture, and temperature.

Peterson, A.F. (convener). 1976. The microbiology of the eye. In *Developments in Industrial Microbiology* 17:13–47. Washington, D.C.: American Institute of Biological Science. A survey of normal eye flora and transient flora and diseases associated with contact lenses and cosmetics.

Tachibana, D. K. 1976. Microbiology of the foot. *Annual Review of Microbiology* 30:351–375. An interesting report on the normal flora of the foot, foot infections, and foot odor.

See also Further Reading for Part Four at the end of Chapter 24.

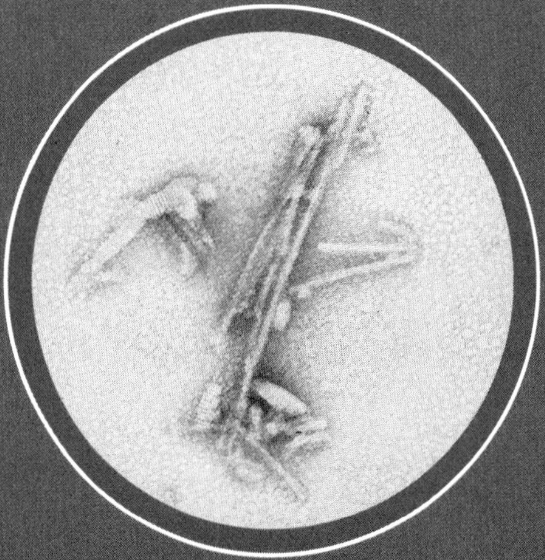

CHAPTER 20

Microbial Diseases of the Nervous System

OBJECTIVES

After completing this chapter you should be able to

- Differentiate between meningitis and encephalitis.
- Discuss the epidemiology of meningococcal meningitis, *H. influenzae* meningitis, and cryptococcosis.
- Define aseptic meningitis.
- Discuss the causative agents, symptoms, and method of transmission of tetanus.
- Provide the etiologic agent, suspect foods, symptoms, and treatment for botulism.
- Discuss the epidemiology of leprosy and poliomyelitis, including method of transmission, etiology, disease symptoms, and preventative measures.
- List the etiology, methods of transmission, and reservoirs for rabies.
- Describe postexposure and preexposure treatments for rabies.
- Identify the causative agent, vector, symptoms, and treatment for arthropod-borne encephalitis and African trypanosomiasis.
- List the characteristics of slow virus diseases.

ORGANIZATION OF THE NERVOUS SYSTEM

The human nervous system consists of billions of nerve cells (neurons) and supporting cells (neuroglia), and is organized into two divisions: the central nervous system and the peripheral nervous system (Figure 20–1). The **central nervous system** consists of the brain and the spinal cord. As the control center for the entire body, it picks up sensory information from the environment, interprets the information, and sends impulses that coordinate the body's activities. The **peripheral nervous system** consists of all the nerves that branch off from the brain and the spinal cord. These nerves are the lines of communication between the central nervous system, the various parts of the body, and the external environment.

The brain is protected from physical injury by skull bones, and the spinal cord is protected by the backbone. Beneath these bones, both the brain and the spinal cord are covered and protected by three continuous membranes called *meninges.* These are the outermost dura mater, the middle arachnoid, and the innermost pia mater. Between the pia mater and arachnoid membranes is a space called *subarachnoid space,* in which *cerebrospinal fluid* circulates (Figure 20–2). This fluid also circulates in spaces within the brain called *ventricles.* A very interesting feature of the brain is a *blood–brain barrier.* This barrier is made of capillaries that permit certain substances to pass from the blood into the brain but restrict others. These capillaries are less permeable than others within the body, and are therefore more selective in passing materials. Oxygen and glucose, for example, can easily cross the barrier, but microorganisms and most antibiotics normally do not.

Even though the central nervous system has considerable protection, it can still be invaded by microorganisms in several ways. For example, microorganisms can gain access through trauma, such as a skull or backbone fracture, or through a medical procedure such as a lumbar puncture (spinal tap), in which a needle is inserted into the subarachnoid space of the spinal meninges and obtains a sample of fluid for diagnosis. Some microorganisms can also move along peripheral nerves. But probably the most common route of central nervous system invasion is by the bloodstream and lymphatic system (Chapter 21). Because cerebrospinal fluid communicates with the lymphatic system, invading microorganisms can enter the cerebrospinal fluid itself. An infection of the meninges is called **meningitis.** An infection of the brain itself is called **encephalitis.**

Just as the central nervous system is resistant to invasion by pathogens, it is also resistant to the entrance of many antibiotics and chemotherapeutic agents used in the treatment of such infections. Although an organism causing meningitis may be sensitive to an antibiotic in a test tube, the antibiotic cannot be clinically effective if it is unable to penetrate the blood–brain barrier.

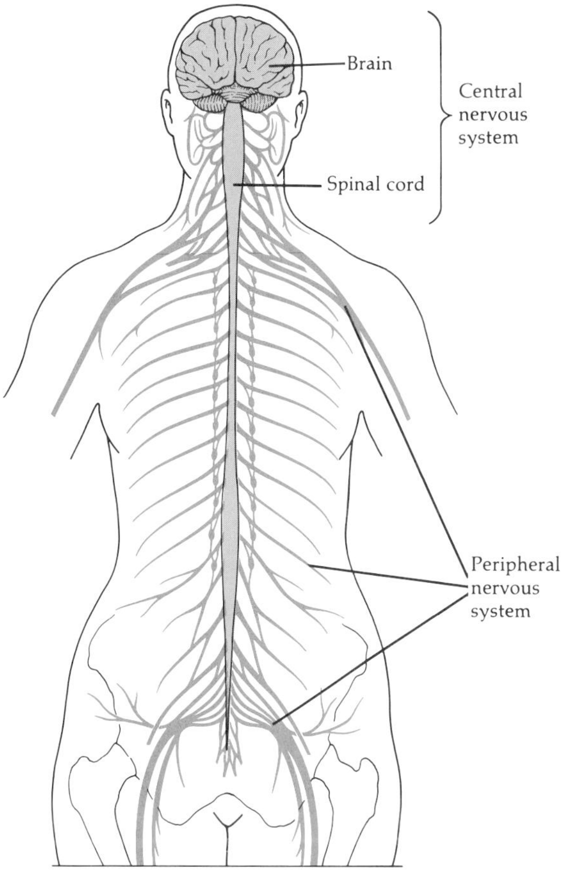

Figure 20–1 Organization of the nervous system.

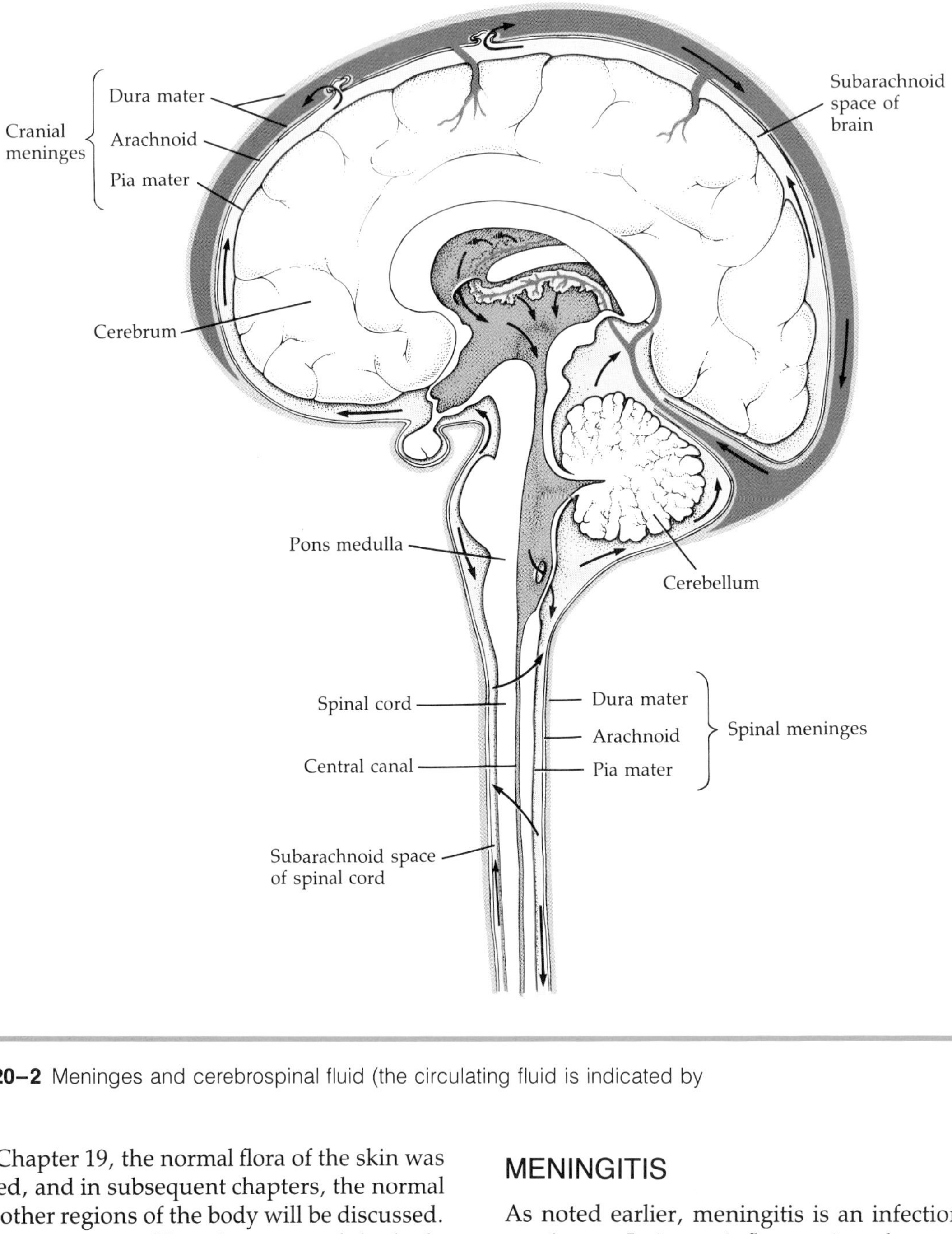

Figure 20–2 Meninges and cerebrospinal fluid (the circulating fluid is indicated by color).

In Chapter 19, the normal flora of the skin was described, and in subsequent chapters, the normal flora of other regions of the body will be discussed. The nervous system, like other parts of the body not in contact with the outside environment, is normally sterile—it has no normal flora.

Now that you have a general understanding of the nervous system, we will examine several diseases that affect it.

MENINGITIS

As noted earlier, meningitis is an infection of the meninges. It is an inflammation that results in swelling and an excess of cerebrospinal fluid. There are three major types of bacterial meningitis: meningococcal meningitis caused by *Neisseria meningitidis*, pneumococcal meningitis caused by *Streptococcus pneumoniae* (strep-tō-kok′ kus nü-mō′nē-ī), and *Hemophilus influenzae* (hē-mä'fi-lus in-flü-en'zī) meningitis. These three comprise the great majority of cases of meningitis. Only meningococcal meningitis is a notifiable (Chapter 13) disease. Table 20–1 presents data about these and some other types of bacterial meningitis. As you can see from the table, meningitis is a very serious disease.

In addition to the aforementioned bacteria, nearly 50 other bacteria have been reported to be opportunistic pathogens that occasionally cause meningitis. These include such common organisms as *Escherichia coli*, *Pseudomonas aeruginosa*, and *Klebsiella pneumoniae* (kleb-sē-el'lä). Later in this chapter we shall discuss cryptococcosis, a meningitis caused by a fungus. Meningitis can also be caused by any of several viruses. Because bacteria or fungi are not isolated, a viral infection is often called **aseptic meningitis.** Meningitis is a frequent complication of protozoan diseases such as toxoplasmosis (Chapter 21) and African trypanosomiasis, which will be discussed later in this chapter.

Neisseria Meningitis (Meningococcal Meningitis)

As we have said, **meningococcal meningitis,** or **cerebrospinal fever,** is caused by the bacterium *Neisseria meningitidis*. The neisseriae are essentially aerobic, nonmotile, gram-negative cocci.

N. meningitidis can inhabit the portion of the throat behind the nose without causing any symptoms. This carrier state can last several days to several months and provides the reservoir for the organism. At the same time, the carrier state enhances the immunity of the carrier. If an individual without adequate immunity acquires the microorganism, usually through contact with a healthy carrier, the resulting throat infection can lead to bacteremia, which is followed by meningitis. Most of the bacteria observed in cerebrospinal fluid are found in leukocytes that have entered the fluid in response to the infection. The ability of the bacteria to leave the blood to make contact with the meninges seems to be related to the concentration of microorganisms in the blood and to the length of time in which they have circulated. Most of the symptoms associated with meningitis are thought to be due to an endotoxin produced by the bacteria.

Meningitis of all types begins with symptoms similar to those of a mild cold. These are followed by a sudden fever, severe headache, and pain and stiffness in the neck and back. The pain is associated with swelling of the meninges and excess fluid. Neurological complications, such as convulsions, deafness, blindness, and minor paralysis, are not uncommon. In cases of meningococcal meningitis, purplish spots might appear on the skin. The *N. meningitidis* endotoxin not only causes extensive damage to blood vessels, but also heightens sensitivity to endotoxin. This increased sensitivity is called the *Shwartzman phenomenon*. In severe cases, death can occur within a few hours.

Meningococcal meningitis is a disease mostly

Table 20–1 Types of Bacterial Meningitis

Organism	Percentage of All Reported Meningitis	Comments
Neisseria meningitidis	27%	Can occur as epidemics. Fatality rate of about 12%
Hemophilus influenzae	43%	Fatality rate of about 6–7%
Streptococcus pneumoniae	11%	Fatality rate of about 27%
Mycobacterium tuberculosis	Uncommon	Complication of tuberculosis. Fatality rate of about 20%
Leptospira interrogans	Uncommon	Meningitis is a common symptom of leptospirosis.
Listeria monocytogenes	Uncommon	Meningitis occurs in about 75% of cases of listeriosis.
Nocardia asteroides	Uncommon	Meningitis is a common complication of pulmonary infections by *Nocardia*

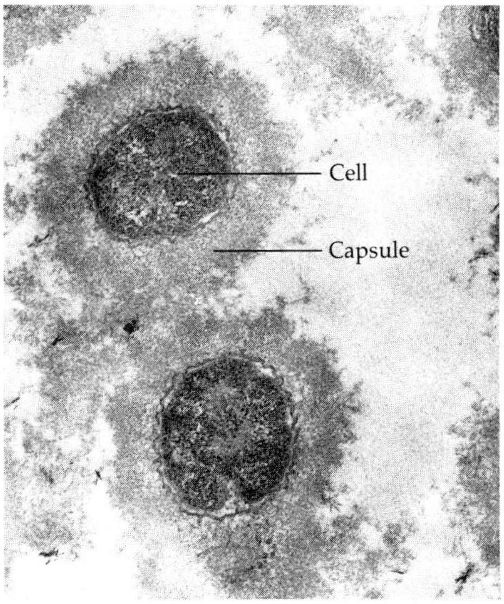

Figure 20–3 Electron micrograph of *Hemophilus influenzae* type b (×36,000). Note the thick layer of capsular material around the cells; in this example, the layer has been accented by treatment with an antibody that reacts with the type b capsule.

of the very young; it usually strikes children younger than four years. The highest incidence is in the first year of life. Children are usually born with maternal immunity and become susceptible as the immunity weakens, at about six months of age. The death rate in untreated cases can be as high as 85%. But if antibiotic therapy is begun soon enough, the mortality rate drops to less than 1%. Explosive epidemics among young adults in military training camps used to be common; such an outbreak occurred in a school in Houston not long ago. Apparently, the close quarters promote the transmission of bacteria from carriers to susceptible individuals. However, a recently developed vaccine, consisting of purified capsular polysaccharide, has proved to be effective in protecting military recruits; this expression of the disease is becoming a thing of the past, although the vaccine is still not in use in the general population.

Preliminary diagnosis is based on gram staining cerebrospinal fluid (CSF) and observing gram-negative diplococci in phagocytes. Countercurrent immuno-electrophoresis (see Chapter 16) of CSF is also useful in differentiating among the several types of bacterial meningitis. The CSF can be inoculated onto blood agar for isolation of the bacteria. The bacteria will not grow on agar devoid of blood, and must be incubated in an environment with 5 to 10% carbon dioxide. This concentration of carbon dioxide can be obtained by the use of a *candle jar* (Chapter 6) or a carbon dioxide incubator. The organism is very susceptible to desiccation and is killed if heated to 55°C for 30 minutes. Because *N. meningitidis* is very sensitive to handling once it is removed from the body, care must be taken to ensure its survival until the cultures can be used in diagnosis.

N. meningitidis, like many other pathogens, is relatively resistant to phagocytosis. Encapsulated strains are the most likely to prove virulent. At one time, sulfonamide was the drug of choice, because it passes readily from the blood into cerebrospinal fluid. However, bacterial resistance to the sulfa drugs has now become common, and other drugs have come into use. Although many antibiotics are unable to pass from the blood into the fluid, a few are capable of passage when the meninges are seriously infected. Examples include penicillin, chloramphenicol, and rifampin.

Hemophilus influenzae Meningitis

Hemophilus influenzae is a nonmotile, aerobic, nonendospore-forming, gram-negative, pleomorphic bacterium (Figure 20–3). A member of the normal throat flora, it can also cause meningitis. The bacterium, which is encapsulated, is further divided into six types on the basis of its antigenic capsular carbohydrates. Only one type has real importance to medical microbiology. It is referred to as type b. The virulence of *H. influenzae* type b is related to its capsule; its endotoxin does not appear to play a major role in pathogenicity, unlike that of *N. meningitidis.*

The name *Hemophilus influenzae* is somewhat misleading. The microorganism was erroneously thought to be the causative agent of the influenza pandemics of 1890 and the first world war—thus the term *influenzae.* The etiologic agent of influenza is now known to be a virus. However, *H. influenzae* was probably a secondary invader during those pandemics. The term *Hemophilus* refers to the fact that the microorganism requires factors in blood for growth (hemo means "blood"). Actually, *Hemophilus influenzae* grows poorly, if at all, on blood agar, but it can be cultured on chocolate agar, which is prepared from lysed red blood cells. Lysis releases the X and V factors required for its growth (see Chapter 10).

H. influenzae type b is by far the most common cause of bacterial meningitis among children under four years of age (Figure 20–4). The incidence of the disease is related to the antibody titer in the blood, whether passively acquired from the mother or actively formed. In children of ages 2 to 36 months, antibody levels are minimal. Thereafter, as the antibody levels increase, the incidence of the disease drops considerably.

Meningitis caused by *H. influenzae* type b follows the same basic pattern as meningitis caused by *N. meningitidis*. The carrier state in children is as high as 30 to 50%. Following a viral infection of the respiratory tract, *H. influenzae* enters the blood from the throat and from there invades the meninges. The fatality rate for untreated cases approaches 100%.

Laboratory diagnosis of *H. influenzae* is based on spinal fluid and blood analysis. If a Gram-stained smear of spinal fluid shows gram-negative pleomorphic rods, the presence of *H. influenzae* can be assumed. A positive quellung test (see Chapter 4) establishes the diagnosis. If the microorganisms obtained from spinal fluid and blood can be cultured on blood agar, the diagnosis is also confirmed.

Virtually all persons can be cured with early treatment. The original drug of choice was ampicillin. However, about 10% of *H. influenzae* type b strains are now resistant to the drug. Therefore, the initial therapy now consists of ampicillin and chloramphenicol in combination. Only a few strains resistant to both ampicillin and chloramphenicol have been found.

The mortality rate of treated persons with *H. influenzae* meningitis is low, but 30% of those who recover have permanent neurological damage. A vaccine has been developed.

Streptococcus pneumoniae Meningitis (Pneumococcal Meningitis)

Children under the age of five are most susceptible to pneumococcal meningitis, caused by *S. pneumoniae*. Like *H. influenzae*, this bacterium is a common inhabitant of the nasopharyngeal region. Pneumococcal meningitis is more common among hospitalized persons than among the general population. Penicillins remain the antibiotics of choice, but the mortality rate of persons with this disease remains very high.

Cryptococcus neoformans Meningitis (Cryptococcosis)

Fungi of the genus *Cryptococcus* are spherical cells resembling yeasts; they reproduce by budding and produce polysaccharide capsules, some much thicker than the cells themselves (Figure 20–5). Only one species, *Cryptococcus neoformans* (kryp-tō-kok'kus nē-ō-fôr'manz), is pathogenic for humans. The organism is widely distributed in soil, especially that enriched by pigeon droppings. It is also found in pigeon roosts and nests on the window ledges of urban buildings. Most cases of cryptococcosis are in urban areas. It is thought to be transmitted by the inhalation of dried, infected pigeon droppings.

The disease caused by *C. neoformans* is called **cryptococcosis.** Inhalation of *C. neoformans* initially causes infection of the lungs, and often the disease does not proceed beyond this stage. However, es-

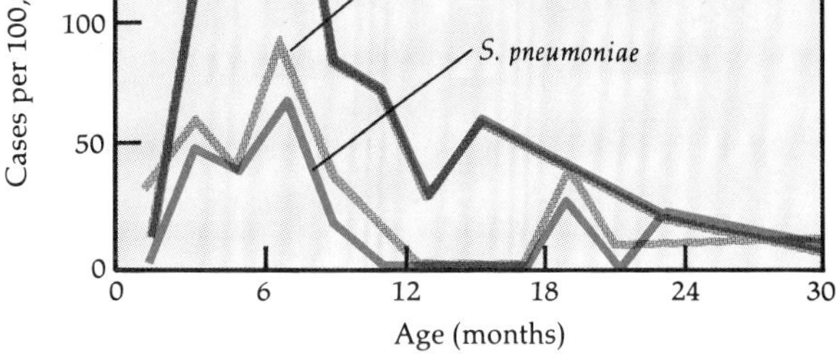

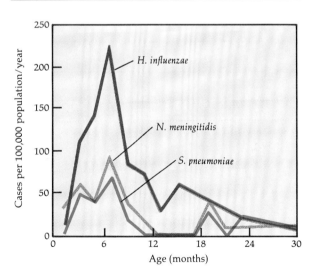

Figure 20–4 Incidence of the most common bacterial meningitis, those caused by *Hemophilus influenzae*, *Neisseria meningitidis*, and *Streptococcus pneumoniae*, in children under 30 months of age.

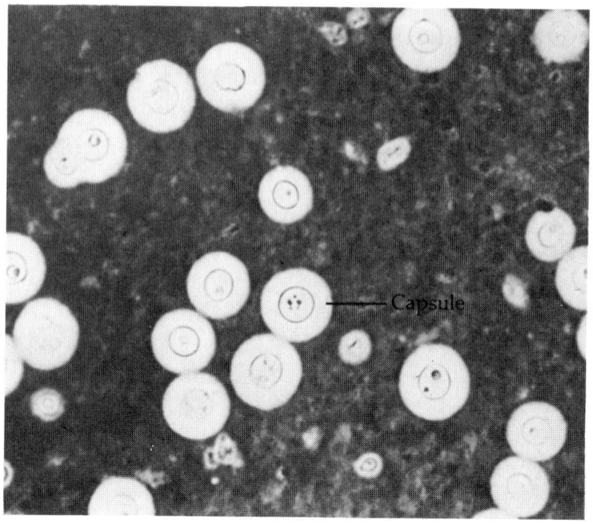

Figure 20–5 *Cryptococcus neoformans.* This is the causative agent of *Cryptococcus neoformans* meningitis (cryptococcosis), a form of chronic meningitis. Note the yeastlike appearance and the prominent capsule. (The capsule is made visible by India ink in the suspension of cells. The carbon particles in the ink darken the background but are excluded from the capsule surrounding the cells.)

pecially in immunosuppressed individuals and persons receiving steroid treaments for major illnesses, it can spread through the bloodstream to other parts of the body, including the brain and meninges. The disease is usually expressed as chronic meningitis.

Laboratory diagnosis of cryptococcosis is by identification of the encapsulated microorganism in an India ink wet mount and by observation of its virulence in mice. Untreated meningitis due to *C. neoformans* is usually fatal. The drugs of choice for treatment are amphotericin B and flucytosine.

TETANUS

The causative organism of **tetanus**, *Clostridium tetani* (klôs-tri'dē-um te' tan-ē), is an obligate anaerobic, endospore-forming, gram-positive rod (see Figure 20–6). It is especially common in soil contaminated with animal fecal wastes. In the American Civil War, it was observed that the cavalry had a higher rate of tetanus than did the infantry, possibly because of animal wastes associated with the cavalry.

The symptoms of tetanus are due to an extremely potent neurotoxin (*tetanospasmin*) that is released upon the death and lysis of the growing bacteria (Chapter 14). The bacteria themselves do not spread from the infection site, however; thus tetanus is sometimes called infectious but noncommunicable. There is often no marked inflammation or observable infection of the wound.

In a muscle's normal operation, a nerve impulse initiates the contraction of the muscle. At the same time, an opposing muscle receives a signal to relax so as not to oppose the contraction. The tetanus neurotoxin blocks the relaxation pathway so that both sets of muscles contract. This results in the characteristic muscle spasms. The muscles of the jaw are affected early in the disease, which prevents opening of the mouth (lockjaw). Gradually other skeletal muscles including those involved in swallowing become affected. Death results from spasms of the respiratory muscles.

Because the organism is an obligate anaerobe, the wound must provide anaerobic growth conditions. Improperly cleaned deep puncture wounds, even those with little or no bleeding, serve very well. Of the approximately 100 tetanus cases occurring each year in the United States (about a third of which are fatal), many arise from trivial but fairly deep injuries.

Most persons in this country have received the DPT immunization, which includes tetanus toxoid that stimulates the formation of antibodies to fight

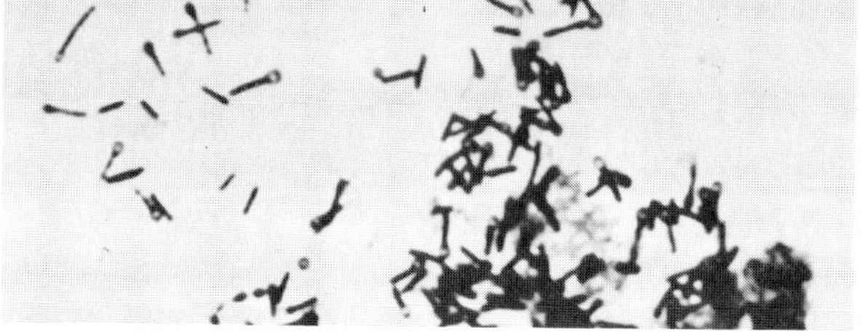

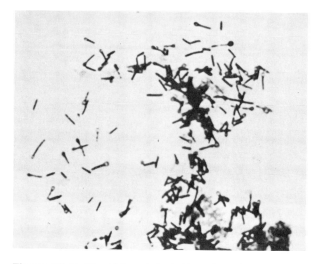

Figure 20–6 *Clostridium tetani.* Note the spherical terminal endospores in many cells.

tetanus toxin. Immunization is nearly 100% effective but, interestingly, recovery from the clinical disease does not confer immunity, because the amount of toxin usually produced is too small to be immunogenic.

When a wound is severe enough to be brought to the attention of a physician, the decision is usually made to provide protection against tetanus, and a booster of toxoid may be given. Even 20 years or so after immunization, the body "remembers" how to manufacture the antibody (the anamnestic response, see Chapter 16). Very rapidly, much more so than after the first exposure to the toxoid, a high level of antibody is attained. If a person has not had previous immunity, however, antibodies resulting from an initial toxoid injection given after the injury may appear too late to prevent the disease. In such a case, a temporary immunity, lasting for about a month, can be given: an injection of immune globulin containing antibodies to tetanus toxoid. The immune globulin is usually taken from the serum of immunized humans. It is also possible to use preformed tetanus antibodies from immunized horses, but this practice increases the chances of there being hypersensitivity reactions to the serum (serum sickness).

Removing the damaged tissue and using antibiotics such as penicillin are also useful in therapy. However, once the toxin has attached to the nerves, such therapy is of little use. Mortality from tetanus exceeds 50% even in the United States; death frequently results from wounds thought to be too minor to bring to the attention of a physician.

In less developed areas of the world, tetanus affecting the severed umbilical cords of infants is a major cause of death. In some cultures, the practice of dressing the cut umbilical cord with materials ranging from soil and clay to cow dung is a major contributor to the development of tetanus. Worldwide, there are probably several hundred thousand cases of tetanus from all causes each year.

BOTULISM

Botulism, often considered to be a form of food poisoning, is caused by *Clostridium botulinum* (klôs-tri'dē-um bo-tū-lī'num), an obligately anaerobic, endospore-forming, gram-positive rod that is found in soil and in the sediments of many freshwater bodies (See Figure 20–7). The microorganism produces an extremely potent exotoxin, a neurotoxin that blocks the transmission of nerve impulses across synapses. The toxin is highly specific for the synaptic end of the nerve, where it blocks the release of acetylcholine, a chemical necessary for nerve activity. Persons suffering from botulism undergo a progressively *flaccid paralysis* for 1 to 10 days, and may die from respiratory and cardiac failure. Compare this with the *spastic paralysis* caused by tetanus toxin. Nausea, but no fever, may precede the neurological symptoms. The initial neurological symptoms vary, but nearly all sufferers have double or blurred vision. Other symptoms include difficulty in swallowing and general weakness. Incubation time varies, but symptoms typically appear within a day or two.

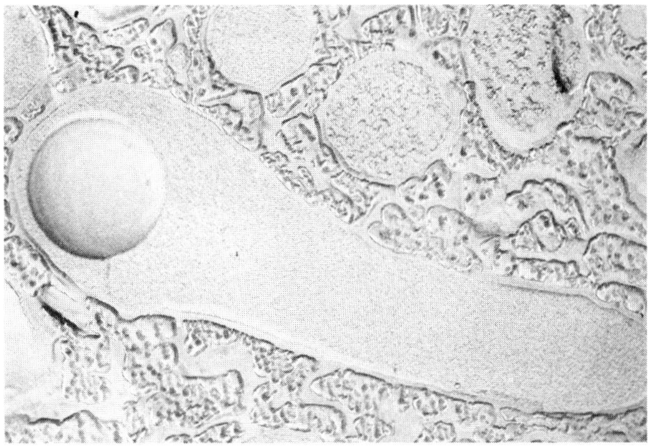

Figure 20–7 Freeze-etch preparation of *Clostridium botulinum* with a terminal endospore (×55,000).

Botulism was first described as a clinical disease in the early 1800s, when it was known as the sausage disease (botulus means "sausage"). Blood sausage was made by filling a pig intestine with blood and ground meats, tying shut all the openings, boiling it for a short time, and smoking it over a wood fire. The sausage was then stored at room temperature. This preparation included most of the requirements for an outbreak of botulism: as an attempt at preservation it killed competing bacteria but allowed the more heat-stable *C. botulinum* endospores to survive, and provided anaerobic conditions and an incubation period for toxin production. Most botulism results from such attempts at preservation that fail to eliminate the *C. botulinum* endospore.

Botulism is not a common disease; only a few cases are reported each year. But, individual outbreaks with sources such as restaurants occasionally involve 20 or 30 cases. As with tetanus, recovery from the disease does not confer immunity, because the toxin is usually not present in amounts large enough to be immunogenic.

There are several serological types of the toxin, produced by different strains of the pathogen. These differ considerably in their virulence and other factors. *Type A* toxin is probably the most virulent. There have been deaths from type A toxin when the food was only tasted but not swallowed. It is even possible to absorb lethal doses through skin breaks while handling laboratory samples. In cases without treatment, there is a 60 to 70% mortality rate. The type A endospore is the most heat-resistant of all botulinum strains. In the United States, it is found mainly in California, Washington, Colorado, Oregon, and New Mexico. Some eastern states have never had a type A outbreak. The type A organism is usually proteolytic (the breakdown of proteins by clostridia releases amines with unpleasant odors), but obvious spoilage odor is not always apparent in low-protein foods, such as corn or beans.

Type B toxin is responsible for most European outbreaks of botulism and is the most common type in the eastern United States. The mortality rate in cases without treatment is perhaps 25%. Type B botulism organisms occur in both proteolytic and nonproteolytic strains. Nonproteolytic strains are less likely to give warning by causing obvious spoilage.

Type E toxin is produced by botulism organisms that are often found in marine or lake sediments. Therefore, outbreaks often involve seafood and are especially common in the Pacific Northwest, Alaska, and the Great Lakes area. The endospore of type E botulism is less heat-resistant than that of other strains, and is usually destroyed by simple boiling. Types A and B will survive much higher temperatures. Type E is nonproteolytic, so the chance of detecting spoilage in high-protein foods such as fish is minimal. It is also capable of producing a toxin at refrigerator temperatures. Type E also requires less strictly anaerobic conditions for growth.

Eskimos and coastal Indians in Alaska probably have the highest rate of botulism (mostly type E) in the world. The problem arises from their attempting to preserve food without using precious fuel for heating or cooking. Cooking the food is important: because all botulinal toxin is very heat labile, it will be destroyed by most ordinary cooking methods that bring the food to a boil. The difficulty in getting prompt treatment for isolated ethnic groups is reflected in the mortality rate of 40% for type E botulism, observed between 1950 and 1973.

Botulinal toxin is not formed in foods with acidity of below pH 4.7. Foods such as tomatoes, therefore, can be safely preserved without the use of a pressure cooker. A concern that newer strains of tomatoes did not contain enough acid for reliable preservation by boiling does not seem to be justified—but some recipes suggest addition of more acid. There have been cases of botulism from acidic foods that normally would not have supported the growth of the botulism organisms. Most of these episodes are related to mold growth, which metabolized enough acid to allow the initiation of growth of the botulism organisms.

Sausage and bacon rarely cause botulism today, largely because of the addition of nitrites to them. It was discovered that nitrites prevent botulinum growth following germination of the endospores. There is considerable research aimed at minimizing or eliminating this use of nitrites, because nitrites are involved in the formation of carcinogenic (cancer-causing) nitrosamines.

The botulism pathogen can also grow in wounds in a manner similar to that of clostridia causing tetanus or gas gangrene. Such episodes of **wound botulism** occur occasionally.

Because the botulism organisms do not seem to be able to compete successfully with the normal intestinal flora, the growth and production of toxin is not generally an important factor in botulism in adults. However, in recent years, the ability of the botulism organisms to grow in the intestinal tracts of infants and cause infant botulism has been confirmed. In some years, more than 40 cases have been reported in the United States. Infants have many opportunities to ingest soil and other materials contaminated with the endospores of the organism, but 30% of recent cases have been associated with honey. Endospores of *C. botulinum* are recovered with some frequency from honey, and the recommendation is to not feed honey to infants under one year of age. There is no problem with older children or adults who have a normally established intestinal flora.

The treatment of botulism relies heavily on supportive care. Antibiotics are of almost no use because the toxin is preformed. Antitoxins aimed at the neutralization of toxins are available and are usually administered as a trivalent ABE type. The antitoxin will not affect the toxin already attached to the nerve ends and is probably more effective on type E than on types A and B. The toxin is quite firmly bound. Recovery requires that the nerve endings regenerate and therefore proceeds slowly. Extended respiratory assistance may be needed, and some neurological impairment may persist for months.

Botulism is diagnosed by the inoculation of mice with samples from patient serum, stools, or vomitus specimens. The toxin in food can also be identified by mouse inoculation. Different sets of mice are immunized with type A, B, or E antitoxins. All the mice are then inoculated with the test toxin; if, for example, those protected with type A antitoxin are the only survivors, the toxin is type A.

LEPROSY

The nonmotile, acid-fast rod that causes leprosy is *Mycobacterium leprae* (mī-kō-bak-ti'rē-um lep' rī) (see Microview 1–1D). It is closely related to the tuberculosis pathogen, *Mycobacterium tuberculosis.* The organism was first isolated and identified about 1870 by G. W. Hansen of Norway; his discovery was one of the first direct links ever made between a specific bacterium and a disease. Leprosy is sometimes called **Hansen's disease** in an effort to avoid using the dreaded name of leprosy.

M. leprae is probably the only bacterium that grows mostly in the peripheral nervous system, although it can also grow in skin cells. In general, the organism shows a preference for the outer, cooler portions of the human body. A very slow generation time of about 12 days has been estimated.

The microorganism has never been grown on artificial media. However, in 1960, *M. leprae* obtained from patients was grown in the footpads of mice. For the next nine years, this procedure was the only means of culturing and testing the organism. Then, in 1969, armadillos were experimentally infected with *M. leprae.* But unlike mice, the armadillos acquired the disease. It is now known that armadillos can contract a leprosylike disease in the wild, although they are not considered a source of human infection. Armadillos are used in studies of the disease itself and in evaluations of the effectiveness of chemotherapeutic agents.

There are two main forms of leprosy, although intermediate forms are also recognized. The *tuberculoid (neural) form* is characterized by regions of the skin that have lost sensation and are surrounded by a border of nodules (Figure 20–8a). In the **lepromin** test, which is analogous to the tuberculin test for tuberculosis, lepromin, an extract of lepromatous tissue, is injected into the skin. This test is positive in most cases of this form of the disease.

In the *lepromatous (progressive) form* of leprosy, skin cells are infected, and disfiguring nodules form all over the body (Figure 20–8b). Mucous membranes of the nose tend to be affected, and a lion-faced appearance is associated with this type of leprosy. Deformation of the hand into a clawed form and considerable necrosis of tissue can also occur. The progression of the disease is unpredictable, and remissions may alternate with rapid deterioration. The lepromin test is negative in the lepromatous form of the disease.

In both forms, leprosy is spread by transfer of the bacteria in exudates (discharges) from lesions on the diseased person to minor abrasions on the skin of another. Inanimate objects, such as clothing, may transmit the disease in a similar manner. Leprosy, however, is not very contagious, and transmission usually occurs only between persons in fairly intimate and prolonged contact. The time from infection to the appearance of symptoms is usually measured in years, although children can have a much shorter incubation period. Death is not usually a result of the leprosy itself, but more often is due to complications caused by other bacteria, such as tuberculosis.

Much of the public's fear of leprosy can probably be attributed to biblical and historical references to the disease. In the Middle Ages, lepers were rigidly excluded from normal European society and sometimes were even given bells to wear so that people could avoid contact with them. This isolation might actually have contributed to the near disappearance of the disease in Europe.

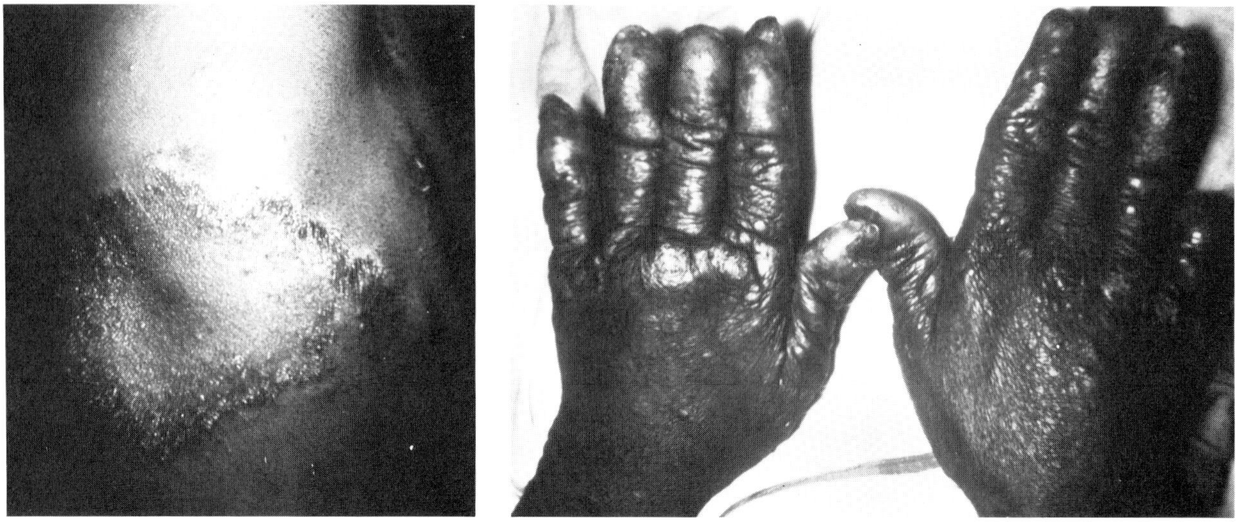

(a) (b)

Figure 20–8 Symptoms of leprosy. **(a)** Tuberculoid leprosy. The lesion below the shoulder on the side of a patient shows a ring of nodules around a relatively clear area in the center. **(b)** Lepromatous leprosy. Tissue necrosis is obvious on the hands of this patient.

But patients with leprosy are no longer kept in isolation. One special treatment center in the United States presently treats about 500 leprosy patients on an ambulatory basis. These outpatients are free to carry out their normal daily routines and return to the hospital for periodic treatment. Patients can be made noncommunicable within four to five days by the administration of sulfone drugs.

The number of leprosy cases in the United States has been gradually increasing. In 1957 there were fewer than 40 cases of leprosy reported. Currently, over 200 cases are reported each year. Many of these cases are imported, for the disease is usually found in tropical climates. Estimates place the total number of cases at about 2600. Most patients are treated at The National Leprosy Hospital in Carville, Louisiana. Millions of persons, most of them in Asia and Africa, suffer from leprosy today. Some researchers believe that unsanitary living conditions, rather than climate, are responsible for the prevalence of disease in these areas. Some individuals are resistant to infection by *M. leprae,* and evidently some individuals have a genetic predisposition to become infected.

Laboratory diagnosis is by identification of acid-fast rods in scrapings from lesions or in fluid from incisions over nonulcerated lesions. (The bacteria can be absent from tuberculoid nodules.) The lepromin test is also useful in some cases.

Although the disease is still difficult to cure, prolonged treatment with sulfone drugs (such as dapsone) has been effective in arresting its progress. Dapsone has been the mainstay of treatment, but resistance is a problem. Rifampin and a fat-soluble dye, clofazimine, are the other main drugs used, often with dapsone. A vaccine for leprosy is undergoing field trials. However, because of the extremely long incubation period for leprosy, it will be many years before the vaccine's effectiveness is established.

POLIOMYELITIS

Poliomyelitis, or **polio,** is best known as a cause of paralysis. Only about 10% of poliomyelitis cases develop identifiable symptoms. The paralytic form of poliomyelitis probably affects less than 1% of those infected with the poliomyelitis virus. The great majority of cases are nonsymptomatic, or cause mild symptoms such as headache, sore throat, fever, and nausea, which are often interpreted as mild meningitis or influenza. Nonsymptomatic or mild cases of polio are most common in the very young, and in parts of the world

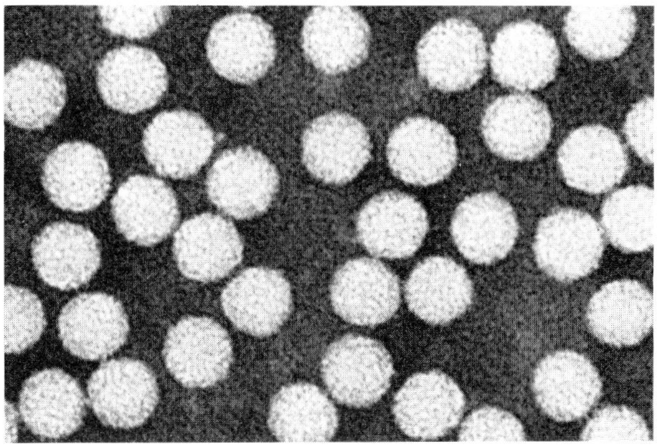

Figure 20–9 Polioviruses (×290,000).

where, due to poor sanitary conditions, most of the population contract nonsymptomatic poliomyelitis as infants and develop immunity. As sanitation improved in more developed regions of the world, the chance of developing the disease in infancy decreased, particularly for those in the upper and middle socioeconomic groups. In those circumstances, when infection did occur in adolescence or early adulthood, the paralytic form of the disease was frequent.

The *poliovirus* (actually, three different serotypes cause the disease) is a picornavirus (Figure 20–9). Humans are the only known natural host for polioviruses. Polioviruses are more stable than most other viruses and can remain infectious for relatively long periods in water and food, so their transmission is facilitated. The primary mode of transmission is ingestion of water that is contaminated with feces containing the virus. Thus polio occurs mainly in the summer months in temperate regions, when people participating in water sports are readily exposed to the virus.

Because the infection is initiated by ingestion of the virus, its primary multiplication is in the throat and small intestine (Figure 20–10). This accounts for the sore throat and nausea. Next, the virus invades the tonsils and the lymph nodes of the neck and ileum (terminal portion of the small intestine). From the lymph nodes, the virus enters the blood and causes viremia. In most cases, the viremia is only transient, the infection does not progress past the lymphatic stage, and clinical disease does not result. If the viremia is persistent, however, the virus penetrates the capillary walls and enters the central nervous system. Once in the central nervous system, the virus displays a high affinity for nerve cells, particularly motor nerve cells, called anterior horn cells, in the upper spinal cord. The virus does not infect the peripheral nerves or the muscles. As the virus multiplies within the cytoplasm of the motor nerve cells, the cells die and paralysis results.

Diagnosis of polio is usually based on isolation of the virus from feces and throat secretions. Cell cultures can be inoculated and cytopathic effects on the cells observed. Serological titrations using viral neutralization tests are also common.

The incidence of polio in the United States has decreased markedly since the availability of the polio vaccines, and now outbreaks usually occur only in parts of the population that lack proper immunization, perhaps because of cultural or religious restrictions. This pattern is found in other developed countries as well. Elsewhere, the disease usually occurs in infancy and is seldom paralytic.

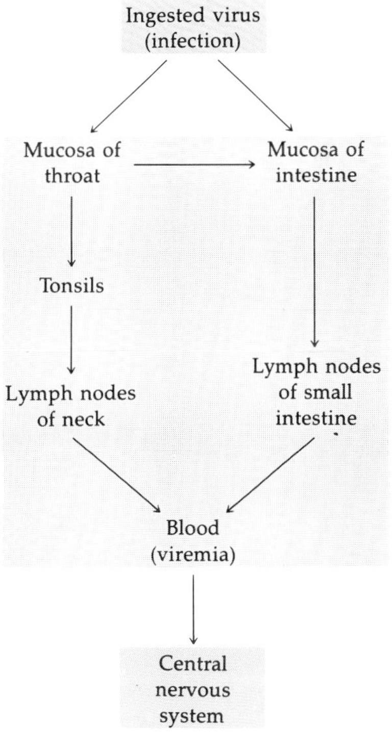

Figure 20–10 Routes taken by poliovirus in the body.

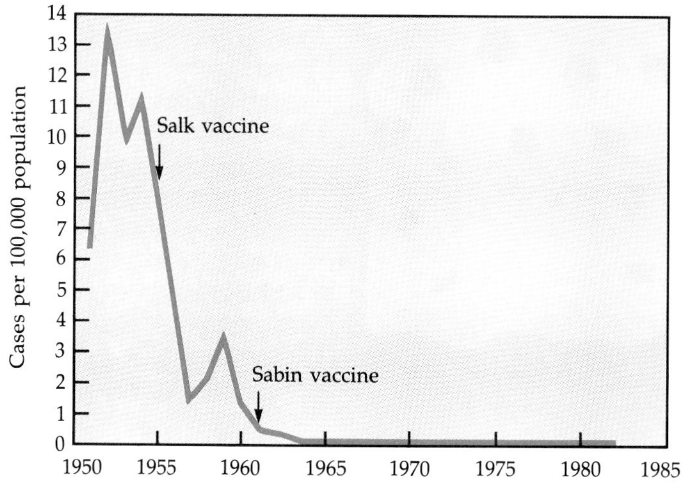

Figure 20–11 Number of cases of poliomyelitis reported in the United States, 1951–1982. Note the decrease after the introduction of the Salk vaccine (1955) and the Sabin vaccine (1961). In 1982, only six new cases were reported: these were probably caused by virulent mutants in the Sabin vaccine.

The development of the first polio vaccine was made possible by the introduction of practical techniques for cell culture, for the virus does not grow in any common laboratory animal. The polio vaccine was the prototype for the vaccines for mumps, measles, and rubella.

Two vaccines are available. The *Salk vaccine,* which was developed in 1954, uses viruses that have been inactivated by treatment with formalin. It requires a series of injections. The effectiveness of this vaccine is high, perhaps 90% against paralytic polio. The antibody levels decline with time and booster shots are needed every few years to maintain full immunity. Nonetheless, using only the Salk vaccine, several European countries have almost eliminated polio from their populations.

The *Sabin vaccine,* developed more recently, contains three living, attenuated strains of the virus and is more popular in the United States than the Salk vaccine. It is less expensive to administer, and most people prefer taking a sip of orange-flavored drink containing the virus to having a series of injections. The immunity achieved with the Sabin vaccine resembles that acquired by natural infection. One disadvantage is that on rare occasions—one in several million cases—one of the attenuated strains of the virus (type 3) seems to revert to virulence and cause the disease. A number of these cases have been secondary contacts, rather than the person who has received the vaccine. This illustrates that recipients of the vaccine can immunize contacts.

Some medical scientists, including Salk himself, have suggested that a return to the Salk vaccine might be desirable despite its disadvantages. As the level of polio in the population declines, the cases that can be linked to the live vaccine become a more and more significant portion of the total cases. There is no reason why polio could not be almost eliminated in this country by an effectively administered program of immunization. Whatever the advantages of either the Salk or Sabin vaccine, both have been used with remarkable effects (Figure 20–11).

RABIES

Rabies is an acute infectious disease that usually results in fatal encephalitis. The causative agent is the *rabies virus,* typically transmitted by the bite of an animal whose saliva contains the virus. The virus is a rhabdovirus that has a characteristic bullet shape (Figure 20–12) and contains single-stranded RNA.

A skin wound or abrasion, usually inflicted by a rabid animal, is the principal portal of entry. It is also possible to contract the disease by inhalation of aerosols of the virus, which can be present in bat caves, for example. It is suspected that aerosol infection can bypass vaccine-based immunity; this route is indicated by cases involving laboratory accidents. The virus can also enter minute skin abrasions that are in contact with virus-containing fluids.

Initially, the virus multiplies in skeletal muscle and connective tissue but remains localized for periods ranging from days to months. Then it travels along the peripheral nerves to the central nervous system, where it produces encephalitis. In the few cases of survival, which are due to the intensive, modern supportive care, neurological damage is often severe.

The initial symptoms of rabies include spasms of the muscles of the mouth and pharynx that occur when liquids are being swallowed. In fact, even the mere sight or thought of water can set off the spasms—thus the common name *hydrophobia* (fear of water). The final stages of the disease result from extensive damage to the nerve cells of the brain and the spinal cord.

Dogs with **furious rabies** are at first restless, then become highly excitable and snap at anything within reach. When paralysis sets in, the flow of saliva increases as swallowing becomes difficult, and nervous control is progressively lost. The disease is almost always fatal within a few days. Some animals suffer from **dumb (paralytic) rabies,** in which there is only minimal excitability. This form is especially common in cats. The animal remains relatively quiet and even unaware of its surroundings, but might snap irritably if handled.

Laboratory diagnosis of rabies in humans and animals is based on several findings. When the patient or animal is alive, a diagnosis can sometimes be confirmed by immunofluorescent studies in which viral antigens are detected in saliva, serum, or cerebrospinal fluid. After death, diagnosis is confirmed with smears from the brain of the infected person or animal. In the past, the diagnostic procedure was to examine the smears for **Negri bodies,** which are masses of viruses or unassembled viral subunits in the cytoplasm of the infected nerve cells (Figure 20–13). Today, however, diagnosis is usually made by a fluorescent-antibody test performed on the smears.

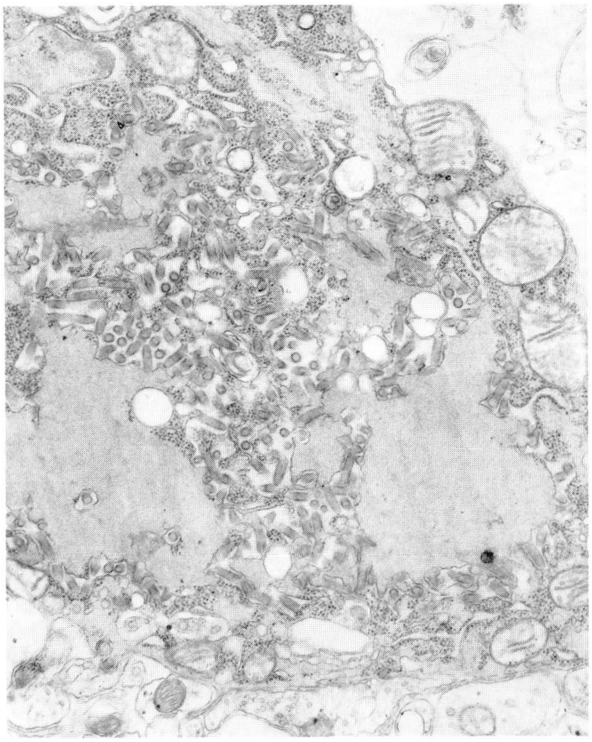

Figure 20–12 A mouse neuron crowded with bullet-shaped rabies virus particles (×30,500).

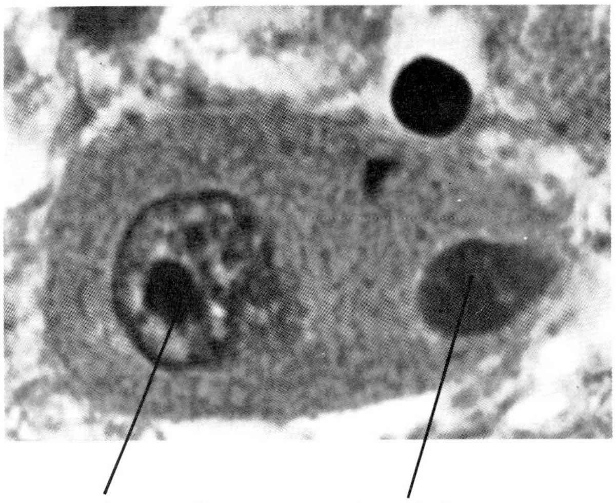

Figure 20–13 A brain cell containing an oval Negri body. First observed in 1905, these viral inclusion bodies probably are accumulations of viral nucleocapsids that failed to be assembled into complete viruses. Now largely replaced by fluorescent antibody tests, detection of these bodies was once the standard criterion for postmortem diagnosis of rabies.

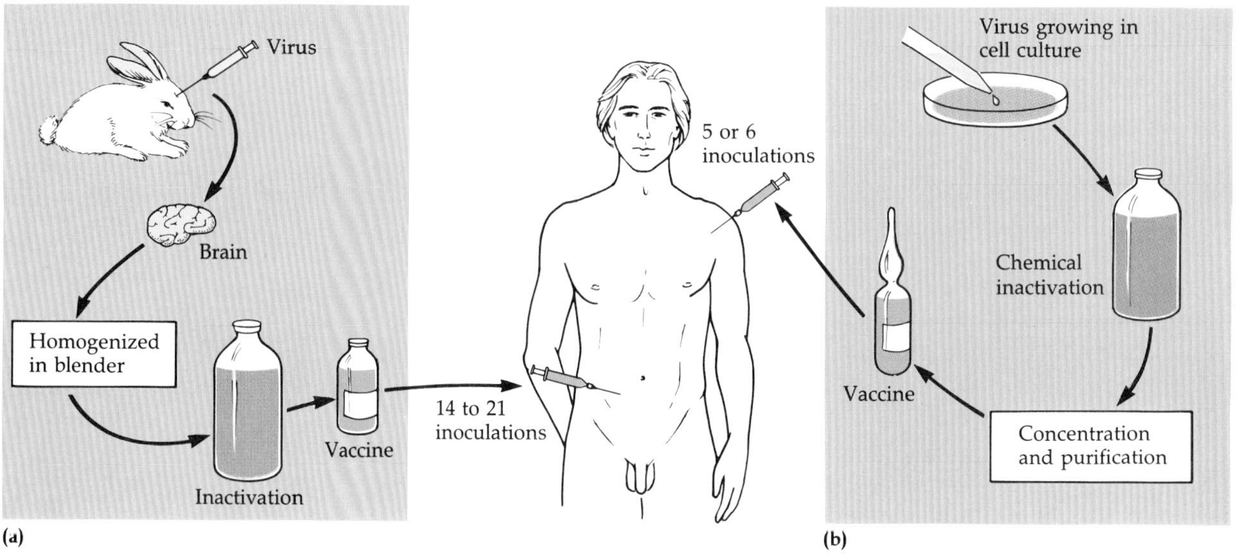

Figure 20–14 Preparation of two rabies vaccines. **(a)** The original Pasteur vaccine as developed in 1885. **(b)** The newly developed Human Diploid Cell Vaccine (HDCV). It is more effective, causes fewer hypersensitivity reactions, and requires fewer injections than previous vaccines, including that grown in duck embryos.

Any person bitten by an animal that is positive for rabies must take antirabies treatment. Another indication for antirabies treatment is any unprovoked bite by a dog, cat, skunk, bat, fox, coyote, bobcat, or raccoon not available for observation and examination. Treatment after a dog or cat bite is determined by the origin of the animal and the prevalence of rabies in the area. (Bites from rodents and rabbits do not require antirabies treatment.)

For many years, the *Pasteur treatment* was the standard means of preventing rabies in exposed persons (Figure 20–14a). To prepare the Pasteur vaccine, a rabbit was first injected with the rabiesvirus, which multiplied in the brain and spinal cord and killed the animal. Then the brain tissue was homogenized in a blender, and the virus inactivated by physical or chemical means. The virus was then incorporated into a vaccine and administered to the infected person. The vaccination procedure consisted of 14 to 21 injections made under the skin of the abdomen over a period of two to three weeks. The patient sometimes developed a hypersensitivity to the rabbit-brain antigens in the vaccine.

Today, postexposure treatment for rabies begins with the administration of a passive immunization called human **rabies immune globulin (RIG),** which is harvested from the serum of human volunteers immunized against rabies. Passive immunization is followed by active immunization with **human diploid cell vaccine (HDCV)**. This vaccine is produced by growing rabies virus in human diploid cell culture and then chemically inactivating it (Figure 20–13b). As a preventative measure, it has better immunogenicity and fewer allergic side effects than the previous vaccine **(DEV)**, which was grown in duck embryos (DEV stands for duck embryo vaccine). In postexposure treatments, five intramuscular injections of HDCV are administered over a four-week period, and a recommended sixth dose can be taken two months later. Passive immunity conferred by RIG is immediate but persists for only a short time. Usually half of the dose is administered into the tissue surrounding the wound to neutralize rabies viruses. The HDVC provides longer term immunity by causing a higher level of antibodies to be formed.

Centers for Disease Control
MMWR
Morbidity and Mortality Weekly Report

Systemic Allergic Reactions Following Immunization with Human Diploid Cell Rabies Vaccine

Human diploid cell rabies vaccine (HDCV) has been licensed for use since June 9, 1980. Approximately 400,000 doses of HDCV have been administered to an estimated 100,000 persons in the United States since that time. The majority of these were for postexposure treatments. Information on possible adverse reactions to HDCV has been collected by CDC from individual physicians and from medical personnel in charge of providing rabies preexposure and postexposure prophylaxis to large cohorts of persons, such as veterinary students and animal-control workers. From June 1980 to April 1984, 108 clinical reports of systemic allergic reactions ranging from hives to anaphylaxis were reported to CDC (11 per 10,000 vaccinees). Few patients required hospitalization, and no deaths secondary to the reactions were reported.

The reports of systemic allergic reactions included nine cases of presumed Type I immediate hypersensitivity, 87 cases of presumed Type III hypersensitivity reactions, and 12 cases of allergic reactions of indeterminate type. These reactions were classified on the basis of clinical observations only. Type I immediate hypersensitivity reactions refer to an immunologic illness occurring within minutes to hours after a dose of HDCV and characterized with either bronchospasm, laryngeal edema, generalized pruritic rash, urticaria, or angioedema. Type III hypersensitivity, a presumed immune complex disease, refers to an immunologic illness occurring 2–21 days after a dose or doses of HDCV and characterized by a generalized pruritic rash or urticaria; the patient may also have arthralgias, arthritis, angioedema, nausea, vomiting, fever, and malaise.

All nine of the presumed immediate hypersensitivity reactions occurred during either primary preexposure immunization or postexposure immunization. However, 81 (93%) of 87 of the presumed Type III hypersensitivity reactions were observed following booster immunization. Although the presumed Type III reactions occurred in six persons during primary immunization series, none were observed following the first dose of the primary series.

Routine boosters of HDCV at 2-year intervals have been recommended for persons with continuing risks of exposure. As increasing numbers of persons received their first routine 2-year boosters, reports of presumed Type III hypersensitivity reactions increased in frequency. Sixty-seven (7%) of 962 persons who received booster immunizations between January 1982 and March 1984 had presumed Type III hypersensitivity reactions.

The table below illustrates the clinical features in three of the cohorts reporting presumed Type III reactions following booster immunization with HDCV. When performed, urinalyses, blood urea nitrogen (BUN), and serum creatinine determinations have been normal. Elevated white blood cell counts ranging from 14,000 to 24,000 were reported in two cases. Serum complement levels (C-3, C-4, and CH-50) were depressed in two patients when serum was drawn at the time of most active clinical symptoms. Serum complement levels were normal in five other patients whose sera were collected at other times. Respiratory distress was infrequently seen. Most patients' symptoms improved within 2-3 days when treated with antihistamines, but a few required systemic corticosteroids and epinephrine.

Preliminary analysis of epidemiologic features of the illness in several cohorts revealed a male/female relative risk of 2.3. No significant associations have been demonstrated between persons who reported presumed Type III hypersensitivity reactions and age, route of primary or booster immunization (intramuscular or intradermal), timing of booster after primary immunization, history of other allergies, or history of previous immunization with rabies vaccines other than HDCV. There is no correlation between HDCV manufacturers and reactions. In two groups for which serologic data were available, no difference was shown in pre-booster antibody titers between those who developed reactions and those who did not, but post-booster titers were significantly higher in those who developed reactions. Most presumed Type III reactions were reported to have occurred following booster doses, but six occurred following two or more doses of HDCV given for primary immunization.

Editorial Note: Primary immunization with HDCV appears to sensitize some recipients to an, as yet, unidentified component of the vaccine. When booster doses of HDCV are then administered, these persons

(*continued on next page*)

Signs and Symptoms In Three Cohorts Reporting Presumed Immune Complex-Type Hypersensitivity Reactions After Booster Immunization with Human Diploid Cell Rabies Vaccine

	Cohort A	Cohort B	Cohort C
No. with reaction/total persons given boosters (%)	23/226 (10%)	22/123 (18%)	6/29 (21%)
Route of booster	Intradermal	Intramuscular	Intramuscular
No. with sign or symptom (%)*:			
Pruritic rash	16 (70%)	5 (18%)	1 (17%)
Urticaria	20 (87%)	20 (91%)	6 (100%)
Edema	10 (43%)	10 (45%)	4 (67%)
Joint pain	4 (17%)	3 (14%)	0
Fever	1 (4%)	0	0
Difficulty breathing	1 (4%)	2 (9%)	0
Mean delay after booster before reaction (range)	9.4 days (3–13)	8.6 days (2–11)	10.5 days (8–11)

*Total in each cohort greater than 100%, because multiple signs and symptoms could be reported in each person.

develop a hypersensitivity reaction clinically consistent with Type III immune complex disease. Until this reaction problem can be resolved, it would be prudent to carefully assess each use of rabies vaccine for routine booster immunization. Persons who have experienced Type III hypersensitivity reactions should receive no further doses of HDCV unless: (1) they are exposed to rabies or (2) they are truly likely to be inapparently and/or unavoidably exposed to rabies virus and have unsatisfactory antibody titers. The routine use of booster immunization in persons without histories of hypersensitivity reactions is clearly indicated only in those subjected to inapparent and/or unavoidable exposures to rabies virus. All available data suggest an anamnestic antibody response will occur in any person who previously received primary preexposure immunization with HDCV, even when the antibody titer at the time of the booster was low or undetectable.

Individuals with histories of presumed Type III hypersensitivity to HDCV may be at higher risk of subsequent hypersensitivity reactions, and vaccine should be administered under appropriate medical supervision.

Sources: *MMWR 33*:185 (4/13/84) and *MMWR 33*:393 (7/20/84)

Veterinarians and laboratory personnel, who have a high risk of exposure to rabies, are immunized on a regular basis, usually every couple of years, and tested for presence of a suitable titer. It is estimated that 30,000 persons each year receive rabies vaccine. When HDCV is available for preexposure treatment, three injections are given at weekly intervals. Alternatively, three or four doses of DEV may be given.

Rabies occurs all over the world. Some islands, such as Australia, New Zealand, and Hawaii, are free of the disease, a condition maintained by rigid quarantine. In North America, rabies is widespread among wildlife (Figure 20–15); the skunk is the wildlife animal most often reported. In a recent year, skunks accounted for 46% of animal rabies cases, bats 19%, foxes 10%, and raccoons 10%. Among domestic animals, cattle accounted for 6% of the reported animal cases, dogs 6%, and cats 4%. Worldwide, however, dogs are the most common carrier of rabies. Rabies is seldom found in squirrels, rabbits, rats, or mice. In fact, the Centers

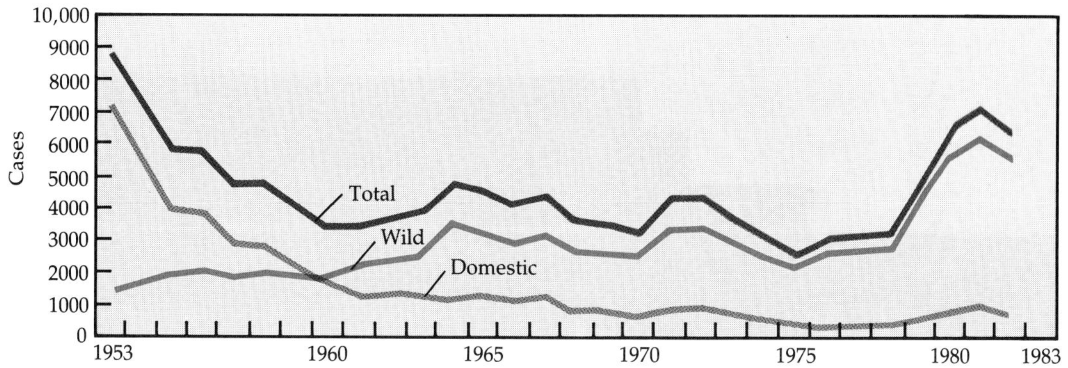

Figure 20–15 Reported cases of rabies in both wild and domestic animals in the United States, 1953–1982. The incidence of disease has declined in domestic animals, probably because of effective vaccination programs. But note that although the number of reported cases is now reduced in dogs, cats, cattle, and oxen (domestic animals), the number of reported cases in skunks, bats, and raccoons (wild animals) has increased.

for Disease Control record only two cases in squirrels. But rabies has long been endemic in vampire bats of South America. And although insect-eating bats in the United States were not known to have rabies until 1953, since that time, bats have been found to be responsible for transmitting rabies to humans and domestic animals in a number of cases. It is not known whether rabies was first introduced into North American bats around 1953, or if it had always existed and was simply not detected until then. Rabies is not always fatal in bats, as it is in other animals.

In the United States, there are up to 6000 cases of rabies diagnosed in animals each year, but in recent years, only one to four cases in humans. Worldwide, there are probably about 1000 human cases each year.

ARTHROPOD-BORNE ENCEPHALITIS

Encephalitis caused by a mosquito-borne arbovirus (encephalitis virus) is rather common in the United States. Figure 20–16 shows the incidence of encephalitis in the United States over a 12-year period. The increase in the summer months is due to the proliferation of adult mosquitoes during these months. A number of clinical types of this disease have been identified; all of these cause inflammation of the brain, meninges, and spinal cord. Active cases of these diseases are characterized by chills, headache, and fever. As the disease progresses, mental confusion and coma are observed. There is evidence that subclinical infections are commonplace.

Outbreaks of encephalitis are reported regularly when the mosquito vectors and reservoirs for the virus coincide. The main viral reservoir is birds, and the mosquito vector is usually a member of the genus *Culex* (kū′ leks).

The Far East also has endemic encephalitis. **Japanese B encephalitis** is the best known; it is a serious public health problem especially in Japan, Korea, and China. Vaccines are used to control the disease in these countries.

Horses are frequently infected by such viruses, so there are terms such as **Eastern equine, Western equine,** and **Venezuelan equine encephalitis (EEE, WEE, and VEE). St. Louis encephalitis (SLE)** virus does not infect horses. **California encephalitis** is another type common in the United States. WEE, VEE, SLE, and California encephalitis tend to cause relatively mild diseases in humans; there can occasionally be complications. Of those prevalent in the United States, EEE is the most

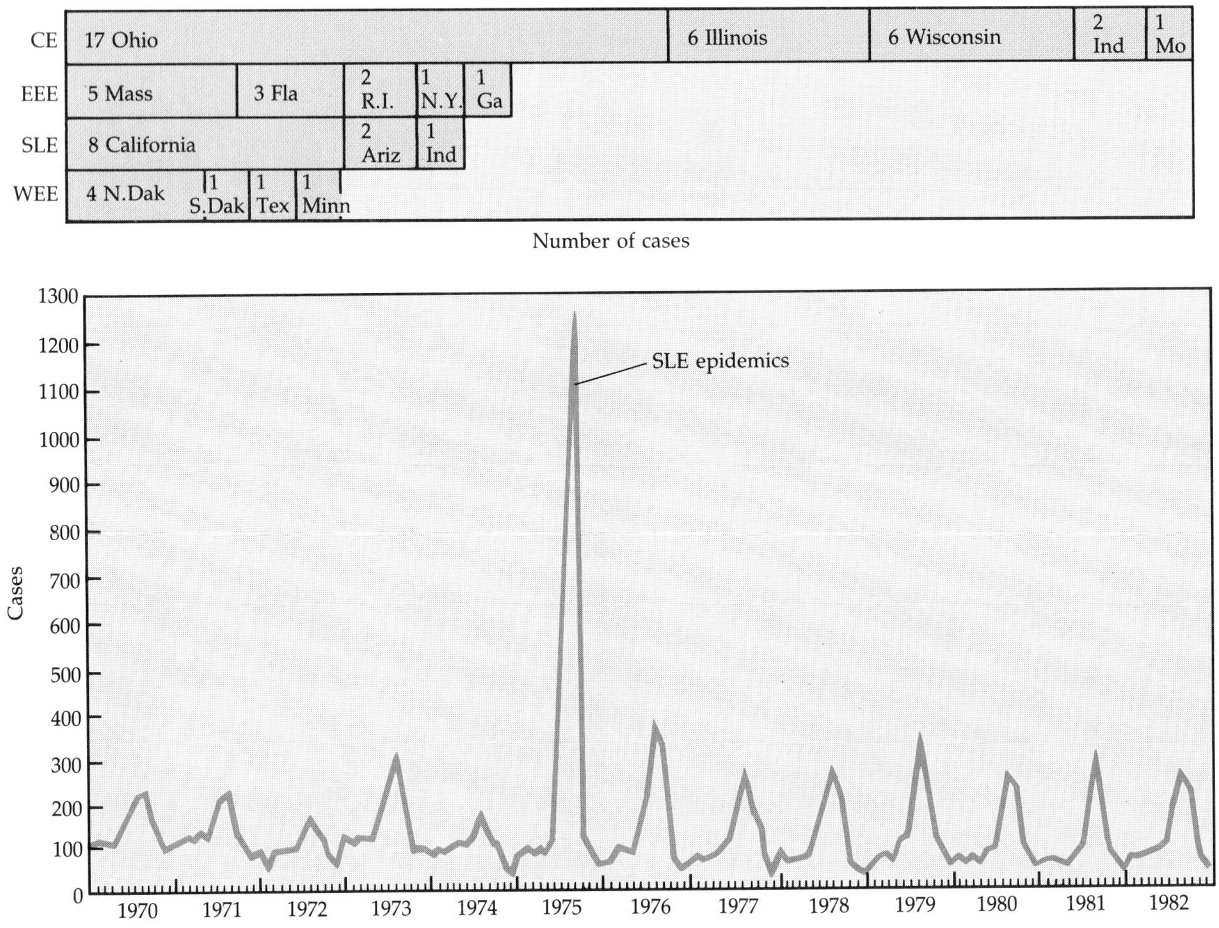

Figure 20–16 Reported cases of arthropod-borne encephalitis by month of onset, United States, 1970–1982. Note the peaks during the summer mosquito season, including one exceptional outbreak of St. Louis encephalitis in 1975. Human cases by etiologic agent and state are shown above.

severe, with a mortality rate of 25% or more and a high incidence of brain damage, deafness, and similar neurological problems in survivors.

Diagnosis of arthropod-borne encephalitis is usually made by serological tests such as complement fixation. The most effective control measure is local elimination of the *Culex* mosquitoes. Persons exposed to the disease can be given immune serum to provide passive immunity, but there is no specific treatment for the disease.

AFRICAN TRYPANOSOMIASIS

African trypanosomiasis is a protozoan disease that affects the nervous system (see box in Chapter 11). The disease is caused by *Trypanosoma brucei gambiense* (tri-pan-ō-sō′mä brü-sē′ī gam-bē-ens′) and *Trypanosoma brucei rhodesiense,* flagellates that are injected by the bite of a tsetse fly *(Glossina)*. Reservoirs for the disease are domestic cattle and wild antelope that frequent the riverbank habitats of the tsetse fly in Africa. During early stages of the stages of the disease, trypanosomes can be found in blood, although they are few. During later stages, the trypanosomes move into the cerebrospinal fluid.

Symptoms of the disease include decreases in physical activity and mental acuity. The name **sleeping sickness** is derived from the lethargy of the host. Untreated, the host enters a coma, and death is almost inevitable.

There are some effective chemotherapeutic agents such as suramin and pentamidine. However, the drugs produce toxic effects and, because they do not cross the blood-brain barrier, they are effective only when the central nervous system (CNS) has not become involved. A toxic arsenical, called melarsoprol, is usually chosen when the central nervous system is involved. A vaccine is being developed, but it might be unsuccessful, because the trypanosome is able to change protein coats at least 100 times and can thus evade antibodies aimed at only one or a few of the proteins (Box in Chapter 11).

NAEGLERIA MICROENCEPHALITIS

Naegleria fowleri (nī-gle'-rē-ä fou'-lė-rī) is a protozoan (amoeba) that is known to cause a neurological disease ***Naegleria* microencephalitis.** Although cases of it are reported from most parts of the world, only a few cases per year are reported in the United States. The most common victims are children who swim in ponds or streams. The organism initially infects the nasal mucosa and later proliferates in the brain. The fatality rate is nearly 100%.

SLOW VIRUS DISEASES

There are a number of fatal diseases of the human central nervous system thought to be caused by what are termed **slow viruses**—however, no one is absolutely sure of the causative agents (Figure 20–17). The name refers to the slow progress of these diseases. In recent years, the study of slow

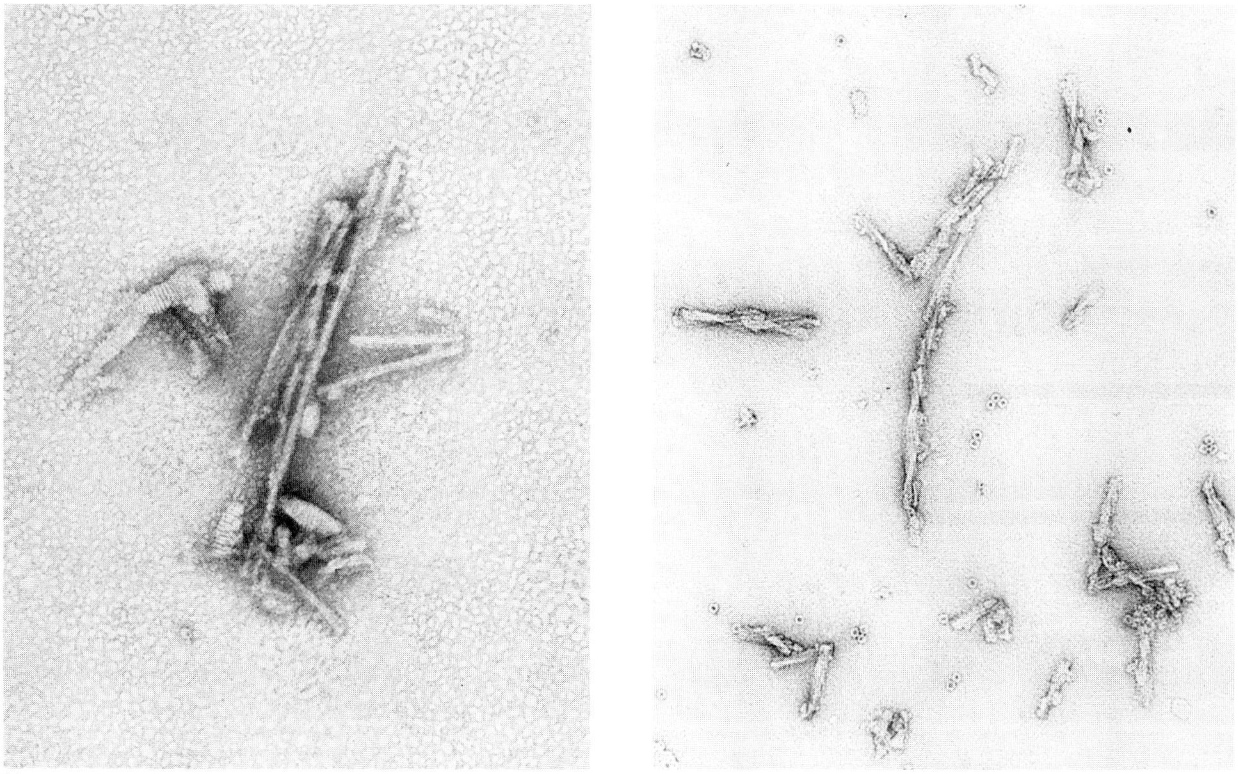

(a) (b)

Figure 20–17 Micrographs of brain tissue infected with slow virus diseases: **(a)** Human Creutzfeldt-Jakob disease; **(b)** scrapie in hamster.

virus diseases has been one of the most interesting areas of microbiology in medical science (see Chapter 12).

A typical slow virus disease is **sheep scrapie.** The infected animal rubs against fences and walls until areas on its body are raw. Over a period of several weeks or months, the animal gradually loses motor control and dies. The agent, whatever it is, can be passed to other animals, such as mice, by the injection of brain tissue from one animal to the next. The scrapie agent can give minks a disease that is indistinguishable from **transmissible mink encephalopathy,** a disease found in minks that are fed mutton at mink farms.

Humans suffer from diseases similar to scrapie. For example, there is **Creutzfeldt–Jakob disease,** which appears at a higher than normal rate in certain Middle Eastern Jewish cultures that eat sheep brain. It has been reported to be transmitted by a corneal transplant and by an accidental self-inflicted wound made by a surgeon during autopsy.

Some tribes in New Guinea have suffered from an apparent slow virus disease called **kuru.** This disease appears most often in women and children, perhaps because one of the funeral rituals involves rubbing the brain of a dead man over the female kin. Cannibalism might also be involved in transmission. Kuru can be transmitted to chimpanzees by their injection with bacteria-free filtrates or the brains of kuru victims.

Many years ago, when rabies vaccine virus was grown in rabbit brains and contained a certain number of rabbit brain cells, a disease known as **experimental allergic encephalitis** was transmitted to some vaccine recipients. The disease, which can still be reproduced in laboratory animals, is similar to multiple sclerosis; a progressive deterioration of the central nervous system was seen with periods of remission. Diseases such as multiple sclerosis might be due to similar etiologic agents. In multiple sclerosis, the myelin sheath (lipid) covering the nerve cells in the central nervous system is destroyed, and the nerve impulses are interrupted.

It has been suggested that slow viruses are the cause of other diseases, including Alzheimer's disease (in which the brain degenerates to premature senility), Parkinson's disease (uncontrolled trembling of the extremities), rheumatoid arthritis, amyotrophic lateral sclerosis (Lou Gehrig's disease), and juvenile-onset diabetes.

Some authorities divide the slow virus dis-

Table 20–2 Summary of Diseases Associated with the Nervous System

Disease	Causative Agent	Mode of Transmission	Treatment
Meningococcal meningitis	*Neisseria meningitidis*	Contact with a healthy carrier via respiratory tract	Penicillin, chloramphenicol, rifampin
Hemophilus influenzae meningitis	*Hemophilus influenzae*	Via respiratory tract in individuals with viral respiratory infections	Ampicillin in combination with chloramphenicol
Cryptococcosis	*Cryptococcus neoformans*	Inhalation of dried infected pigeon droppings	Amphotericin B, flucytosine
Leprosy	*Mycobacterium leprae*	Transfer of exudates from lesions or inanimate objects	Sulfone drugs (dapsone), rifampin, clofazimine
Poliomyelitis	Poliovirus	Ingestion of virus	None
Rabies	Rabies virus	Bite of a rabid animal	RIG and HDCV after exposure
Tetanus	*Clostridium tetani*	Deep, anaerobic wounds	Immune globulin and respiratory support
Botulism	*Clostridium botulinum*	Ingestion of toxin	Antitoxins and respiratory support
Arthropod-borne encephalitis	Arboviruses	*Culex* mosquitoes	Immune globulin
Creutzfeldt-Jakob Disease	Prion	Ingestion or accidental wounds	None
Kuru	Prion has been suggested	Ingestion	None

eases into two groups: those apparently caused by unconventional viruses and those that might be associated with unusual immunologic effects of known viruses such as measles. For example, measles virus can cause the latent expression of subacute sclerosing panencephalitis, which shows many characteristics of a slow virus disease.

The unconventional viruses, which some scientists call "agents" rather than viruses, cannot be inactivated by ultraviolet radiation and are resistant to formalin and heat treatments. Most agree that the agents are a separate class of microorganism; the name **prion** has been suggested for them. They do not seem to stimulate a conventional immune response, although autoimmunity might be part of their pathogenicity. The scrapie agent, for example, appears to contain a major protein required for infectivity but contains no detectable nucleic acids. Yet it is capable of replication.

When it was discovered that certain plant diseases are caused by viroids, which are only short strands of RNA without a protein coat, there was considerable speculation that the agents of slow virus diseases might resemble viroids. This now seems unlikely, although there is evidence that DNA agents resembling viroids might cause other animal diseases. In any event, the slow virus diseases will continue to be the subject of intense research.

A summary of diseases associated with the nervous system is presented in Table 20–2.

STUDY OUTLINE

ORGANIZATION OF THE NERVOUS SYSTEM (pp. 549–550)

1. The central nervous system consists of the brain protected by the skull bones and the spinal cord protected by the backbone.
2. The peripheral nervous system consists of the nerves that branch from the central nervous system.
3. The central nervous system is covered by three layers of membranes called meninges. Cerebrospinal fluid circulates between the inner and middle meninges and in the ventricles of the brain.
4. The blood-brain barrier normally prevents many substances such as antibiotics from entering the brain.
5. Microorganisms can enter the central nervous system through trauma, along peripheral nerves, and through the bloodstream and lymphatic system.
6. An infection of the meninges is called meningitis. An infection of the brain is called encephalitis.

MENINGITIS (pp. 550–554)

1. Aseptic meningitis describes a condition in which bacteria and fungi are not the cause. Viruses are usually suspected.

Neisseria Meningitis (Meningococcal Meningitis) (pp. 551–552)

1. *Neisseria meningitidis* causes meningococcal meningitis. This bacterium is found in the throat of healthy carriers.
2. The bacteria probably gain access to the meninges through the bloodstream. The bacteria may be found in leukocytes in cerebrospinal fluid.
3. Early symptoms resemble a cold but progress to pain and stiffness in the neck and back. Neurological complications may develop. Purplish spots appear on the skin.
4. The disease occurs most often in young children. Military recruits are vaccinated with purified capsular polysaccharide to prevent epidemics in training camps.
5. Encapsulated *N. meningitidis* are resistant to phagocytosis.
6. Diagnosis is based on isolation and identification of the bacteria in blood or cerebrospinal fluid.
7. *N. meningitidis* must be cultured on media containing blood and incubated in an atmosphere containing 5 to 10% carbon dioxide.

Hemophilus influenzae Meningitis (pp. 552–553)

1. *Hemophilus influenzae* is cultured on chocolate agar containing X and V factors.
2. *H. influenzae* is the most common cause of meningitis in children under four years old.

3. The disease often occurs as a secondary infection following a viral infection such as influenza.
4. Diagnosis is based on identification of the organism in spinal fluid by a positive quellung test.

Streptococcus pneumoniae Meningitis (Pneumococcal Meningitis) (p. 553)

1. Hospitalized patients and young children are most susceptible to *S. pneumoniae* meningitis.

Cryptococcus neoformans Meningitis (Cryptococcosis) (pp. 553–554)

1. *Cryptococcus neoformans* is an encapsulated yeastlike fungus that causes *Cryptococcus neoformans* meningitis.
2. The disease may be contracted by inhalation of dried infected pigeon droppings.
3. The disease begins as a lung infection and then spreads to the brain and meninges.
4. Immunosuppressed individuals are most susceptible to *Cryptococcus neoformans* meningitis.
5. Diagnosis is based on observation of the fungus.

TETANUS (pp. 554–555)

1. Tetanus is due to a localized infection of a wound by *Clostridium tetani*.
2. *C. tetani* produces the neurotoxin tetanospasmin, which causes the symptoms of tetanus: spasms, contraction of muscles controlling the jaw, and death resulting from spasms of respiratory muscles.
3. *C. tetani* is an anaerobe that will grow in unclean wounds and wounds with little bleeding.
4. Acquired immunity results from DPT immunization that includes tetanus toxoid.
5. Following an injury, an immunized person may receive a booster of tetanus toxoid. An unimmunized person may receive human or horse immune globulin.
6. Debridement and antibiotics may be used to control the infection.

BOTULISM (pp. 555–557)

1. Botulism is caused by an exotoxin produced by *C. botulinum* growing in foods.
2. Serological types of botulinum toxin vary in virulence, with type A being the most virulent.
3. The toxin is a neurotoxin that inhibits transmission of nerve impulses, resulting in respiratory and cardiac failure.
4. Blurred vision and nausea occur in 1 to 2 days, progressively flaccid paralysis follows for 1 to 10 days. There is as high as 60 to 70% mortality in untreated cases.
5. *C. botulinum* will not grow in acidic foods or in an aerobic environment.
6. Endospores are killed by using a pressure cooker for canning. Addition of nitrites to foods inhibits outgrowth after endospore germination.
7. The toxin is heat labile and destroyed by boiling (100°C) for 5 minutes.
8. Wound botulism occurs when *C. botulinum* grows in anaerobic wounds.
9. Infant botulism results from the growth of *C. botulinum* in an infant's intestines.
10. For diagnosis, mice protected with antitoxin are inoculated with toxin from the patient or foods.

LEPROSY (pp. 557–558)

1. *Mycobacterium leprae* causes leprosy, or Hansen's disease.
2. *M. leprae* cannot be cultured on artificial media. It is grown in the footpads of mice or in armadillos.
3. The tuberculoid form of the disease is characterized by loss of sensation in the skin surrounded by nodules. The lepromin test is positive.
4. In the lepromatous form, disseminated nodules and tissue necrosis occur. The lepromin test is negative.
5. Leprosy is not highly contagious and is spread by prolonged contact with exudates and fomites.
6. Patients with leprosy are made noncommunicable within four or five days with sulfone drugs and then treated as outpatients.
7. Untreated individuals often die of secondary bacterial complications such as tuberculosis.
8. The disease occurs primarily in the tropics. Many people are treated at the National Leprosy Hospital in Carville, LA.

9. Laboratory diagnosis is based on observation of acid-fast rods in lesions or fluids.

POLIOMYELITIS (pp. 558–560)

1. The symptoms of poliomyelitis are usually headache, sore throat, fever, stiffness of the back and neck, and occasionally (1 to 2% of the cases) paralysis.
2. Poliovirus (picornavirus) is found only in humans and is transmitted by ingestion of water contaminated with feces.
3. Poliovirus first invades lymph nodes of the neck and small intestine. Viremia and spinal cord involvement may follow.
4. At present, outbreaks of polio in the United States are uncommon because of the use of vaccines.
5. The Salk vaccination involves injection of formalin-inactivated viruses and boosters every few years. The Sabin vaccine contains three attenuated live strains of poliovirus and is administered orally.
6. Diagnosis is based on isolation of the virus from feces and throat secretions or the presence of virus neutralizing antibodies in the patient's serum.

RABIES (pp. 560–565)

1. Rabiesvirus (a rhabdovirus) causes an acute, usually fatal, encephalitis called rabies.
2. Rabies may be contracted through the bite of a rabid animal, by inhalation of aerosols, or invasion through minute skin abrasions. The virus multiplies in skeletal muscle and connective tissue.
3. Encephalitis occurs when the virus moves along peripheral nerves to the central nervous system.
4. Symptoms of rabies include spasms of mouth and throat muscles followed by extensive brain and spinal cord damage.
5. Laboratory diagnosis may be made by direct immunofluorescent tests of saliva, serum, cerebrospinal fluid, or brain smears.
6. Reservoirs for rabies include skunks, bats, foxes, and raccoons. Domestic cattle, dogs, and cats may get rabies. Rodents and rabbits seldom get rabies.
7. Rabies is fatal in all animals except bats.
8. The Pasteur treatment for rabies involves multiple subcutaneous injections of rabies virus grown in rabbit brain tissue.
9. Current postexposure treatment includes administration of human rabies immune globulin (RIG) followed by multiple intramuscular injections of human diploid cell culture virus (HDCV).
10. Preexposure immunization consists of injections of DEV or HDCV.

ARTHROPOD-BORNE ENCEPHALITIS (pp. 565–566)

1. Symptoms of encephalitis are chills, headache, fever, and eventually coma.
2. Many types of arboviruses (encephalitis viruses) transmitted by *Culex* species mosquitoes cause encephalitis. The reservoir is birds.
3. Horses are frequently infected by EEE, WEE, and VEE viruses.
4. Outbreaks occur primarily during the summer months when adult mosquitoes are present.
5. Diagnosis is based on serological tests.
6. Immune globulin may provide passive immunity, but no treatment is available.
7. Elimination of the vector is the most effective control measure.

AFRICAN TRYPANOSOMIASIS (pp. 566–567)

1. African trypanosomiasis is a disease caused by the protozoans *Trypanosoma brucei gambiense* or *Trypanosoma brucei rhodesiense* and transmitted by the bite of the tsetse fly (*Glossina*).
2. The disease affects the nervous system of the human host, causing lethargy and eventually coma. It is commonly called sleeping sickness.

NAEGLERIA MICROENCEPHALITIS (p. 567)

1. Encephalitis caused by the protozoan *N. fowleri* is almost always fatal.

SLOW VIRUS DISEASES (pp. 567–569)

1. Slow virus diseases are of uncertain etiology. The diseases progress slowly in the host.
2. Sheep scrapie and transmissible mink encephalopathy are examples of slow virus diseases that are transferrable from one animal to another.
3. Creutzfeldt-Jakob disease and kuru are human diseases similar to scrapie that occur in isolated groups of people who eat brains.

4. Multiple sclerosis and Parkinson's disease may be slow virus diseases.
5. Slow virus diseases may result from immunologic complications of conventional viral diseases (e.g., subacute sclerosing panencephalitis following measles).
6. Some slow virus diseases are caused by prions and perhaps viroids.

STUDY QUESTIONS

REVIEW

1. Define the following terms: meningitis; encephalitis.
2. Fill in the following table.

Causative Agent	Disease	Susceptible Population	Method of Transmission	Treatment
N. meningitidis				
H. influenzae				
C. neoformans				

3. Briefly explain the derivation of the name *Hemophilus influenzae.*
4. To what does the term *aseptic meningitis* refer? Look up *aseptic* in the glossary. Why is this word used in a description of meningitis?
5. If *Clostridium tetani* is relatively sensitive to penicillin, then why doesn't penicillin cure tetanus?
6. What treatment is used against tetanus under the following conditions?
   (a) Before a person suffers a deep puncture wound
   (b) After a person suffers a deep puncture wound
7. Why is the following description used for wounds that are susceptible to *C. tetani* infection? ". . . improperly cleaned deep puncture wounds . . . ones with little or no bleeding . . ."
8. List the following information for botulism: etiologic agent, suspect foods, symptoms, treatment, conditions necessary for microbial growth, basis for diagnosis, prevention.
9. Provide the following information on leprosy: etiology, method of transmission, symptoms, treatment, prevention, and susceptible population.
10. Provide the following information on poliomyelitis: etiology, method of transmission, symptoms, prevention. Why aren't Salk and Sabin vaccines considered treatments for poliomyelitis?
11. Compare and contrast the Salk and Sabin vaccines with respect to composition, advantages, and disadvantages.
12. Provide the etiology, method of transmission, reservoirs, and symptoms for rabies.
13. Outline the procedures for treating rabies after exposure. Outline the procedures for preventing rabies prior to exposure. What is the reason for the differences in the procedures?
14. Fill in the following table.

Disease	Etiology	Vector	Symptoms	Treatment
Arthropod-borne encephalitis				
African trypanosomiasis				

15. Why is the incidence of arboviral encephalitis in the United States higher in the summer months?
16. Describe the symptoms and laboratory test results that would lead to a diagnosis of slow virus disease.
17. Provide evidence that slow virus diseases are caused by transmissible agents that resemble viruses.
18. Why are meningitis and encephalitis generally difficult to treat?

CHALLENGE

1. Most of us have been told that a "rusty nail" causes tetanus. What do you suppose is the origin of this adage?
2. A BCG vaccination will result in positive lepromin and tuberculin tests. What is the relationship between leprosy and tuberculosis?
3. Compare the Pasteur treatment with current treatments for rabies. What are the advantages of HDCV over DEV and the Pasteur treatment?
4. Three days before the onset of meningitis symptoms, a 39-year-old nurse assisted in the emergency room evaluation of a patient with *N. meningitidis* in his cerebrospinal fluid. She assisted during the intubation and suctioning of nasopharyngeal secretions. Twenty-four medical personnel had contact with the patient before he was placed in isolation but only this nurse became ill. What two mistakes made by the nurse resulted in her illness?

FURTHER READING

Bingham, R. 1981. Outrageous ardor. *Science 81* 2:54-61. An interesting article on slow viruses and their discoverer, Carleton Gajdusek.

Hall, S. S. 1984. The La Crosse file. *Science 84* 5:54-62. The epidemiology of mosquito-borne encephalitis.

John, D. T. 1982. Primary amebic meningoencephalitis and the biology of *Naegleria fowleri*. *Annual Review of Microbiology* 36:101-123. A good article on the natural history of the disease and the parasite.

Kaplan, M. M., and H. Koprowski. January 1980. Rabies. *Scientific American* 242:120-134. An overview of rabies and a comparison of available vaccines.

Smith, L. 1977. *Botulism: The Organism, Its Toxins, The Disease.* Springfield, IL: C. C. Thomas. A summary of information on botulism.

Spector, D. H., and D. Baltimore. May 1975. The molecular biology of poliovirus. *Scientific American* 232:24-31. Information on viral multiplication is obtained from laboratory-grown poliovirus.

See also Further Reading for Part Four at the end of Chapter 24.

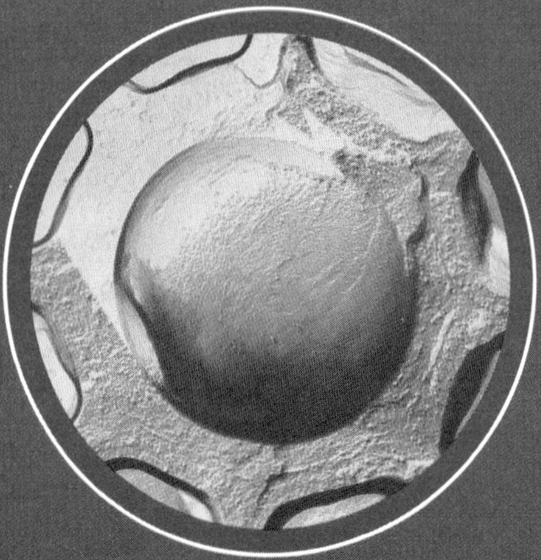

CHAPTER 21

Microbial Diseases of the Cardiovascular and Lymphatic Systems

OBJECTIVES

After completing this chapter you should be able to

- List the symptoms of septicemia and explain the importance of infections that develop into septicemia.
- Discuss the epidemiologies of puerperal sepsis, bacterial endocarditis, and myocarditis.
- Discuss the causes, treatments, and preventive measures of rheumatic fever and infectious mononucleosis.
- Describe the epidemiologies of tularemia, brucellosis, anthrax, listeriosis, and gas gangrene.
- Compare and contrast the causative agents, vectors, reservoirs, symptoms, and treatments of the following diseases: malaria, yellow fever, dengue, relapsing fever, typhus, and Rocky Mountain Spotted Fever.
- List the causative agents and methods of transmission of toxoplasmosis, American trypanosomiasis, schistosomiasis, and viral hemorrhagic fevers. Also describe their world-wide effects on health.
- Identify the best control measure(s) for vector-borne diseases.

The **cardiovascular system** consists of the heart, blood, and blood vessels. The **lymphatic system** consists of the lymph, lymph vessels, lymph nodes, and lymphoid organs (tonsils, appendix, spleen, and thymus gland) (see Figure 15–5). Because both systems circulate various substances throughout the body, they can serve as vehicles for the spread of infection.

STRUCTURE AND FUNCTION OF THE CARDIOVASCULAR SYSTEM

The center of the cardiovascular system is the *heart* (Figure 21–1). The function of the heart is to circulate the blood through the body's tissues so it can deliver certain substances to cells and remove other substances from them. The heart is composed of four chambers, two on the right side and two on the left. The upper chambers, called *atria,* receive blood from different parts of the body, and the lower chambers, called *ventricles,* pump blood to the body. Valves between the chambers control the flow of blood from atrium to ventricle, and prevent the backward flow of blood.

From all parts of the body, deoxygenated blood, which contains more carbon dioxide than oxygen, arrives at the right atrium. From here, it passes into the right ventricle and then through the pulmonary arteries into the lungs. Within the lungs, the carbon dioxide in the blood is exchanged for oxygen. The blood is now oxygenated; that is, it contains more oxygen than carbon dioxide. The oxygenated blood is pumped through pulmonary veins to the left atrium. Passing into the left ventricle, the oxygenated blood is then pumped through the aorta to all parts of the body. The blood circulating to all tissue cells gives oxygen to them and removes their carbon dioxide. This exchange produces deoxygenated blood, which returns to the right atrium to start another cycle.

Blood vessels are the tubes that carry circulating blood throughout the body. An *artery* is a blood vessel that delivers oxygenated blood to different tissues of the body. The *aorta* is the main artery carrying oxygenated blood from the heart. When they reach their destinations, the arteries branch into smaller vessels called *arterioles,* which in turn branch into even smaller vessels called *capillaries.* Capillary walls are only one cell thick, and it is through these walls that blood and tissue cells exchange materials. The capillaries that take the deoxygenated blood away from the tissue cells converge to form small veins called *venules.* Venules unite to form *veins,* and veins return the blood to the right atrium.

The *blood* itself is a mixture of cells and a liquid called plasma. The *plasma* transports dissolved nutrients to body cells and removes wastes from the cells. The cells are known as red blood cells, white blood cells, and platelets. Red blood cells, or erythrocytes, carry oxygen and some carbon dioxide (although most of the carbon dioxide in blood is dissolved in the plasma). White blood cells, or leukocytes, play several roles in defending the body against infection, as discussed in Part Three. Some white cells, in particular the neutrophils, are phagocytes. And the B cells and T cells, two types of lymphocytes, play key roles in immunity. Platelets, or thrombocytes, function in blood clotting.

STRUCTURE AND FUNCTION OF THE LYMPHATIC SYSTEM

As part of the overall pattern of circulation, some plasma filters out of the blood capillaries and into spaces between tissue cells. These spaces are called *interstitial spaces.* The fluid circulating around and between tissue cells is called *interstitial fluid.* Also surrounding tissue cells are microscopic lymphatic vessels called *lymph capillaries,* which are larger and more permeable than blood capillaries. As the interstitial fluid moves around the tissue cells, it is picked up by the lymph capillaries; the fluid is then referred to as *lymph* (Figure 21-1). Because lymph capillaries are very permeable, they readily pick up microorganisms or their products. From lymph capillaries, lymph is transported into larger lymph vessels called *lymphatics,* which contain valves that keep the lymph moving toward the heart. Eventually, all the lymph is returned to the blood just before the blood enters the right atrium. This circulation returns proteins and fluid that have filtered from the plasma back to the blood.

At various points along the lymphatic system are oval or bean-shaped structures called *lymph nodes,* through which lymph flows. Within the lymph nodes are cells capable of phagocytosis (fixed macrophages), which help to clear the lymph of microorganisms. At times, the number of micro-

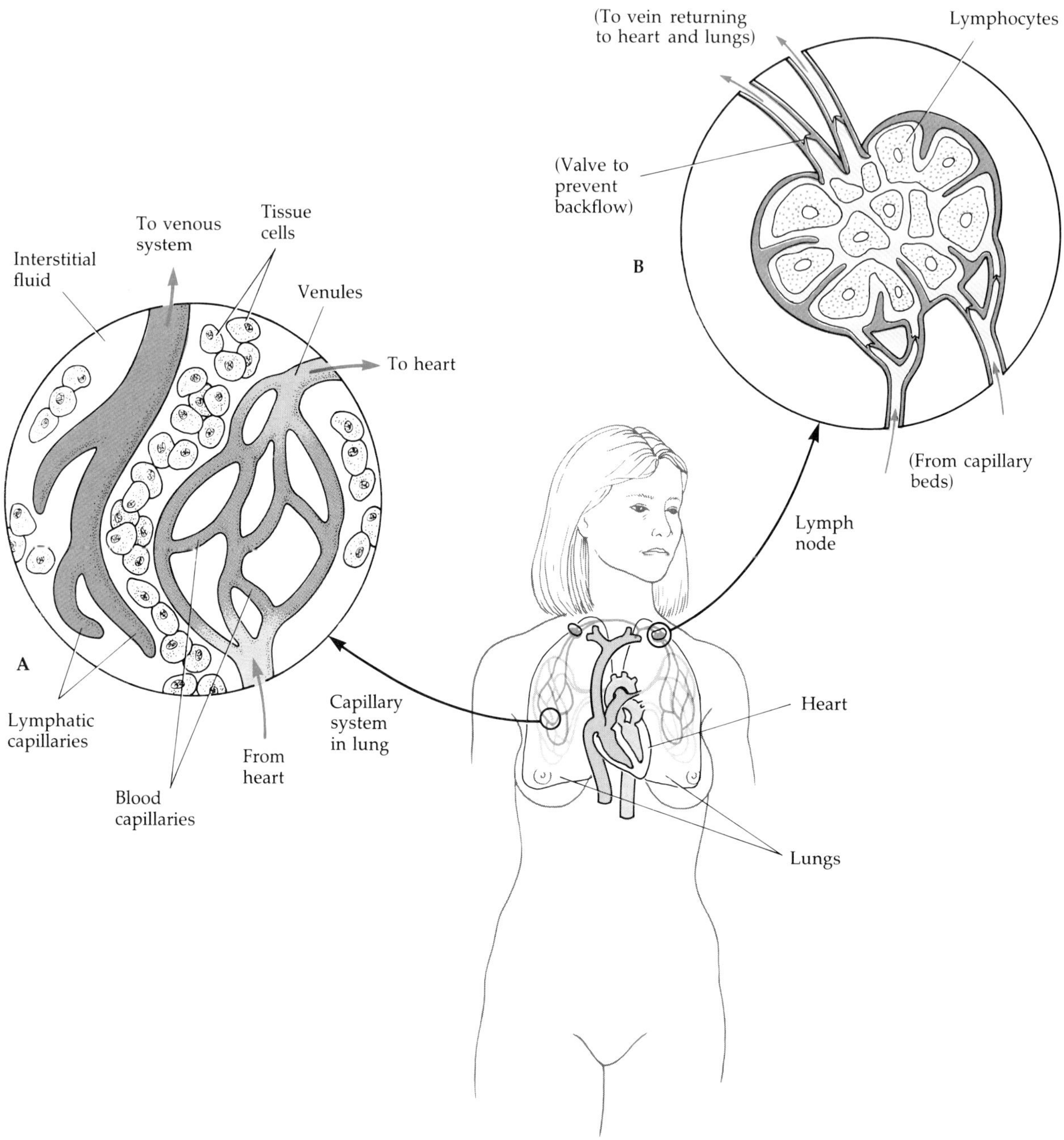

Figure 21–1 The cardiovascular and lymphatic systems (also see Figure 15–5). Details of circulation to head and extremities are not shown in this simplified diagram. The blood circulates from the heart through the arterial system (lighter color) to the blood capillaries in the lungs and other parts of the body. From these capillaries, the blood returns through the venous system (gray color) to the heart. From the blood capillaries, some plasma filters into the surrounding tissue and enters the lymph capillaries. Inset A shows the details of this exchange. This fluid, now called lymph, returns to the heart through the lymphatic circulatory system (darker color). All lymph returning to the heart must pass through at least one lymph node, as shown in inset B.

organisms circulating through nodes is so great that the nodes themselves become infected, and the nodes become enlarged and tender. The lymph nodes are also important to the body's developing an immune reaction against invading microorganisms. Microorganisms entering the lymph nodes and other lymphoid tissues contact B cells that are stimulated to produce humoral antibodies or T cells that differentiate into effector cells. (See Figure 16-9.)

The *lymphoid organs* of the lymphatic system include the tonsils, appendix, spleen, and thymus gland. We have already considered, in Chapter 16, the role of the thymus gland in the maturation of T cells.

BACTERIAL DISEASES OF THE CARDIOVASCULAR AND LYMPHATIC SYSTEMS

Septicemia

Although blood is normally sterile, even moderate numbers of microorganisms are not usually harmful. However, if the defenses of the blood and lymphatic systems fail, the microorganisms can proliferate in the blood without control. That condition is called **septicemia** or **blood poisoning.**

Clinically, septicemia is characterized by fever and a decrease in blood pressure. The decrease in blood pressure can result in ischemia (decreased blood supply) and shock. Another symptom of septicemia is the appearance of **lymphangitis,** inflamed lymph vessels shown as red streaks under the skin, running along the arm or leg from the infection's site. Sometimes the streaks end at a lymph node, where the lymphocytes attempt to stop the invading microorganisms.

The organisms most frequently associated with septicemia are gram-negative rods, although a few gram-positive bacteria and fungi are also implicated. At one time, infections of the bloodstream were more commonly caused by gram-positive bacteria such as staphylococci, which commonly inhabit the skin. Among the gram-negative rods are *Escherichia coli, Serratia marcescens, Proteus mirabilis, Enterobacter aerogenes, Pseudomonas aeruginosa,* and *Bacteroides* species. These bacteria enter the blood from a focus of infection in the body. As you may recall, the cell walls of many gram-negative bacteria contain endotoxins that are released upon the lysis of the cell. It is the endotoxin that actually causes the symptoms. Once released, the endotoxin damages blood vessels; this damage causes the low blood pressure and subsequent shock. In some cases, antibiotics aggravate the condition by causing the lysis of large numbers of bacteria, which in turn release damaging endotoxins. Many of these gram-negative rods are of nosocomial origin, and were introduced into the bloodstream by medical procedures that bypass the normal barriers between the environment and the blood (see box).

Puerperal Sepsis

A nosocomial infection that frequently leads to septicemia is **puerperal sepsis,** also called **puerperal fever** or **childbirth fever.** This begins as an infection of the uterus as a result of childbirth or abortion. *Streptococcus pyogenes,* a group A beta-hemolytic streptococcus, is the most frequent cause, although other organisms might cause infections of this type. Recall from Chapter 19 that group A beta-hemolytic streptococci have an M protein on their pili that increases virulence by improving adherence to the mucous membranes and increasing resistance to phagocytosis.

Puerperal sepsis progresses from an infection of the uterus to an infection of the abdominal cavity (peritonitis), and in many cases to septicemia. At one Paris hospital between 1861 and 1864, of the 9886 women who gave birth, 1226 (12%) died of such infections. At that time, the death rate from puerperal sepsis was frequently twice that. These deaths were largely unnecessary, since some 20 years before, Dr. Oliver Wendell Holmes, in America, and an Austrian physician, Ignaz Semmelweiss, had clearly demonstrated that the disease was transmitted by the hands and instruments of the attending midwives or physicians and that disinfection of hands and instruments could prevent such transmission. Yet Louis Pasteur, in 1879, still thought it necessary to lecture physicians as to the cause of the disease.

Antibiotics and modern hygienic practices have now made puerperal sepsis an uncommon complication of childbirth. In the treatment of the disease, penicillin and erythromycin are effective

Centers for Disease Control
MMWR
Morbidity and Mortality Weekly Report

Bacteremia among Aortic-Valve Surgery Patients

In the period March 19-April 6, 1981, 3 of 5 patients at a Boston hospital who had undergone thoracotomy for aortic-valve replacement, developed bacteremia; the causative agent was a strain of *Enterobacter cloacae* not previously isolated at this hospital. (Bacteremia refers to the presence of microorganisms in the blood. If the microorganisms multiply, the condition is called septicemia.)

Clinical onset of infection ranged from 4 days to 2 weeks after surgery. The infected patients ranged in age from 60 to 83 years. Two of the 3 had also had coronary-artery-bypass grafts (CABG), but 22 other patients who had CABG surgery without valve replacement within the same 19-day period after surgery did not subsequently become infected by *E. cloacae.*

Because 3 of 5 patients who had valve replacements had been infected by the same uncommon bacterial strain, and because 2 different types of valves had been used in surgery, an investigation was initiated to identify procedures or equipment used in this surgery that were not used in other types of open-heart surgery. The only piece of equipment used exclusively for aortic-valve surgery was an anaeroid manometer (see figure) that measured the pressure of cardioplegia solution, used to stop the heart, being injected into the coronary arteries. The manometer, approximately 1 foot of disposable tubing, and a stopcock were connected to sterile tubing extending from the bubble trap (reservoir) of the cardioplegia solution. The manometer was not sterilized or disinfected after each use, and the stopcocks and tubing were not changed on a regular basis. Although reflux of the cardioplegia solution into the manometer tubing had not been observed, it was believed this could have occurred as a result of changes in pressure and fluid levels during surgery.

Use of this system was stopped immediately. The inner surfaces of the manometer and tubing junction were cultured. The cultures were positive for *E. cloacae* of the same biotype that caused infection in the index patients. In addition, 3 subsequent aortic-valve replacements without use of the implicated manometer were observed, and multiple cultures were obtained during surgery; all cultures were negative for *E. cloacae.*

The infusion pressure of cardioplegia solution is now monitored only with a pressure transducer that is sterilized with ethylene oxide after each use and is connected to the bubble trap by sterile tubing and stopcocks.

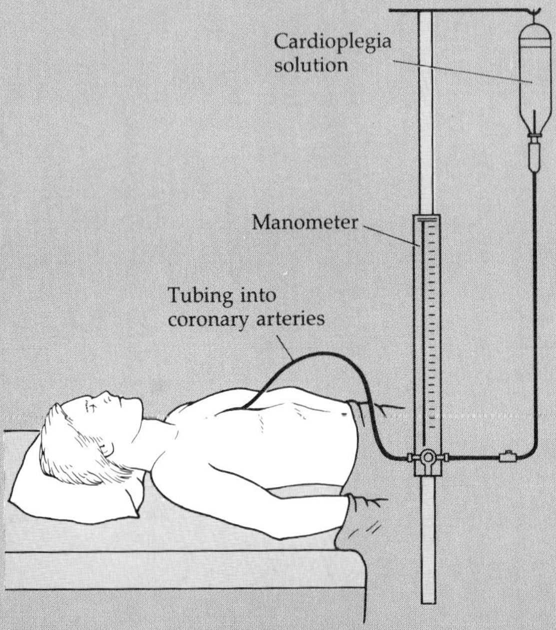

Editorial Note: Bacteremia is not an uncommon complication of cardiac surgery. Clusters of postoperative bacteremia caused by a single strain of bacteria do sometimes occur, although the source of infection for the clusters is not often identified. Over the past years, cases of bacteremia have occurred as a result of intravascular infusion of solutions contaminated by pressure transducers.

Unsterile anaeroid and mercury manometers are used frequently in a variety of medical and surgical settings to calibrate electronic-pressure-monitoring equipment, such as that used in cardiac catheterization studies. Only rarely has contamination of sterile lines or solution by calibration or monitoring procedures been clearly associated with infection. The outbreak investigation reported here clearly demonstrates that a risk exists and reemphasizes the need to keep sterile systems closed and to ensure that the internal surfaces of all equipment directly connected to a sterile system remain sterile.

Source: MMWR 31:88 (2/26/82)

against *S. pyogenes.* Infections related to improperly performed abortions—not strictly puerperal sepsis—are often of mixed bacterial types, many of them caused by anaerobic bacteria of the *Bacteroides* or *Clostridium* genera. Chloramphenicol and penicillin are effective in these cases.

Bacterial Endocarditis

The wall of the heart consists of three layers. The outer layer, called the *pericardium,* is a sac that encloses the heart. The middle layer, known as the *myocardium,* is the thickest layer and consists of cardiac muscle tissue. The inner layer is a lining of epithelium called the *endocardium.* This layer not only lines the heart muscle itself, but also covers the valves in the heart. An inflammation of the endocardium is called **endocarditis.**

One type of bacterial endocarditis is called **subacute bacterial endocarditis.** It is characterized by fever, anemia, general weakness, and heart murmur. It is usually caused by alpha-hemolytic streptococci although beta-hemolytic streptococci or staphylococci (*Staphylococcus epidermidis*) can be involved. About 5 to 10% of the cases are due to enterococci. The condition probably arises from a focus of infection elsewhere in the body, such as the teeth or tonsils. Microorganisms released by tooth extractions or tonsillectomies enter the blood and find their way to the heart. Normally, such bacteria would be quickly cleared from the blood by the body's defensive mechanisms. But in persons whose heart valves are abnormal, because of either congenital heart defects or diseases such as rheumatic fever or syphilis, the bacteria lodge in the preexisting lesions. Within the lesions, the bacteria multiply and become entrapped in blood clots that protect them from phagocytes and antibodies. As multiplication progresses and the clot gets larger, pieces of the clot break off and can occlude blood vessels or lodge in the kidneys. In time, the function of the heart valves is impaired. Left untreated, subacute bacterial endocarditis is invariably fatal.

Another type of bacterial endocarditis is **acute bacterial endocarditis,** which is caused by *Staphylococcus aureus* or *Streptococcus pneumoniae* shown in Figure 21–2. These organisms find their way from the initial site of infection to normal or abnormal heart valves; the rapid destruction of the heart valves is frequently fatal in days or weeks. Strains of *Staphylococcus* that produce penicillinase are usually sensitive to semisynthetic penicillin derivatives such as methicillin or oxacillin. *S. pneumoniae* responds well to penicillin. *Streptococcus* can also cause **pericarditis,** inflammation of the sac around the heart (pericardium).

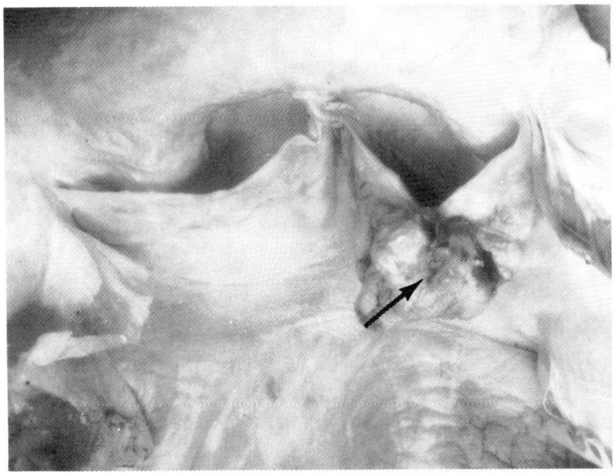

Figure 21–2 Acute bacterial endocarditis. The arrow points to an infected, ulcerated lesion on a heart valve.

In a laboratory diagnosis of bacterial endocarditis, the bacteria isolated from a blood culture. Penicillin is the most commonly used drug in treatment. It is well to use it prophylactically in susceptible persons, such as those suffering from rheumatic fever or undergoing tooth extractions or tonsillectomies. Erythromycin is also effective in many cases.

Rheumatic Fever

Group A beta-hemolytic streptococcal infections, such as those caused by *Streptococcus pyogenes,* sometimes lead to **rheumatic fever,** which is generally considered an autoimmune type of complication (see Chapter 17). The disease is usually expressed as an arthritis, especially in older persons. Another frequent form is an inflammation of the heart, which damages the valves. Other symptoms include fever and malaise. Although rheumatic fever usually does no permanent damage to the joints, there can be permanent heart damage.

The disease is usually precipitated by a streptococcal sore throat. One to five weeks later, after the original infection has probably disappeared, evidence of heart abnormalities appear. Reinfection with streptococci (for example, another sore throat) renews the attack and causes further damage to the heart. Perhaps as many as 3% of children with untreated beta-hemolytic streptococcal infections contract rheumatic fever. However, many of these cases are essentially subclinical.

The exact mechanism by which streptococci produce rheumatic fever is still obscure, although some data point to an immunologic reaction that is localized in the heart and joints. For example, infection with group A beta-hemolytic streptococci may result in streptococcal antigens such as M proteins being deposited in the joints and heart. When the body produces antibodies against the antigens, the antigen–antibody reaction might cause the damage. It is also possible that a streptococcal antigen is cross-reactive with components of heart muscle, and produces antibodies that subsequently damage the heart. Persons with symptoms of damage from rheumatic fever have, in fact, relatively high titers of antibody against streptococcal antigens. Further evidence that rheumatic fever might be the result of an immunologic reaction is the fact that arthritic pain can be induced by injection of sterile filtrates of streptococcal antigens. And finally, postmortem studies of persons who died from rheumatic fever reveal large deposits of antibody and the C3 component of complement in damaged heart tissue. Genetic factors might play a role in this disease. It has been demonstrated that persons of certain HLA types (Chapter 17) are about 15 times more at risk than the population at large.

When beta-hemolytic streptococci are identified, a culture from the patient, usually a throat swab, is streaked onto a blood agar plate. If the colonies are surrounded by clear areas of hemolysis, the microorganisms are Gram-stained. The appearance of gram-positive cocci in chains is a positive diagnosis. However, streptococci might no longer be present at the time of investigation.

The number of cases of rheumatic fever has been declining in this country because of early treatment of streptococcal infections. In 1969, more than 3000 cases were reported; in 1982 only 137 cases. Penicillin administered to rheumatic fever patients is a prophylactic. The penicillin will not alleviate the symptoms (which are treated with antiinflammatory drugs), but it will prevent subsequent streptococcal infections that could cause a recurrence of rheumatic fever.

Tularemia

Tularemia is a disease caused by a small, gram-negative, facultatively anaerobic, pleomorphic, rod-shaped bacterium, *Francisella tularensis* (fran-sis-el'lä tü-lä-ren'sis). The microorganism was named for Tulare County, California, where it was originally observed in ground squirrels in 1911.

Tularemia exhibits one of several forms, depending upon the way the infection is acquired. This might be by inhalation, ingestion, bites, or, most commonly, by contact through minor skin breaks. The first sign of infection is usually local inflammation and a small ulcer at the site of infection. About a week after infection, the regional lymph nodes enlarge; many contain pockets filled with pus. If the disease is not contained by the lymphatic system, the microorganisms can produce septicemia, pneumonia, and abscesses throughout the body. Ingestion of infected, inadequately cooked meat leads to a focus of infection in the mouth and throat. The highest mortality occurs with pneumonic tularemia, which is most often acquired by laboratory workers who have been exposed to aerosols.

Humans most frequently acquire the infection through minor skin abrasions or through the eyes, which are rubbed after small wild mammals have been handled. Probably 90% of the cases in this country are contracted from rabbits; it is estimated that 1% of all American rabbits are infected. The disease can also be spread by the bites of arthropods such as deer flies (see Figure 11–28d), ticks, or rabbit lice, particularly in the western states; contact with small animals is a more likely cause in the central and eastern states (Figure 21–3).

F. tularensis is difficult to grow from serum samples, even on highly specialized media, because they require a greater amount of the amino acid cysteine than is usually present in nutrient media. Colonies may not appear for more than a week, although most cultures will grow within 96 hours. Although no well-defined toxins have been identified to account for its virulence, *F. tularensis* survives for long periods within body cells and phagocytic cells. In fact, resistance to phagocytosis might account for the lowness of the infective dose

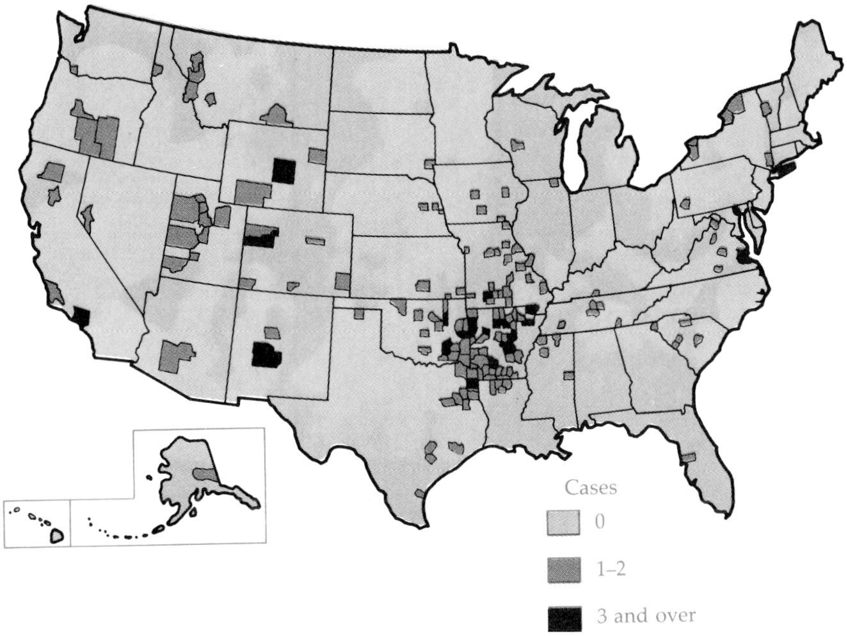

Figure 21–3 The geographical distribution of the 275 tularemia cases reported in the United States during 1982. The disease is not limited to the United States but is found in many countries in the northern hemisphere.

required. Although naturally acquired immunity is usually permanent, recurrences have been reported. An attenuated live vaccine is available for high-risk laboratory workers.

Serological confirmation of a diagnosis of tularemia is routinely and rapidly accomplished by slide agglutination from isolated bacteria. The antisera are available commercially. Fluorescent-labeled antibody and a delayed hypersensitivity skin test are also used for laboratory identification.

Streptomycin is the antibiotic of choice, but prolonged administration is necessary to prevent relapses. Tetracyclines are also effective, but is more likely to allow relapses. The intracellular location of the microorganism is a problem in chemotherapy.

Brucellosis (Undulant Fever)

Like the bacterium that causes tularemia, the bacteria causing **brucellosis** favor intracellular growth, move through the lymphatic system, and travel to organs via the bloodstream. These microorganisms are very small, gram-negative, aerobic rods. They are difficult to grow on artificial media, because they require that not only carbon dioxide be added to the atmosphere of the incubator, but also specially enriched media. There are four main species: *Brucella abortus* (brü-sel'lä ä-bôr'tus), most commonly found in cattle and transmitted by unpasteurized milk or milk products (Figure 21–4); *Brucella melitensis* (me-li-ten'sis), transmitted by goats; *Brucella suis* (sü'is), transmitted by swine; and *Brucella canis* (kā'nis), transmitted by dogs, especially beagles. All species are pathogenic to humans and other mammals. With a fatality rate of 2 or 3%, *B. suis* and *B. melitensis* infections are more severe than *B. abortus* infections. About 200 cases of brucellosis are reported each year in the United States. Most of these cases result from contact with swine carcasses.

The organisms apparently enter the body by passing through minute abrasions either in the skin or mucous membrane of the mouth, throat, or intestinal tract. Once in the body, the microorganisms are ingested by fixed macrophages, in

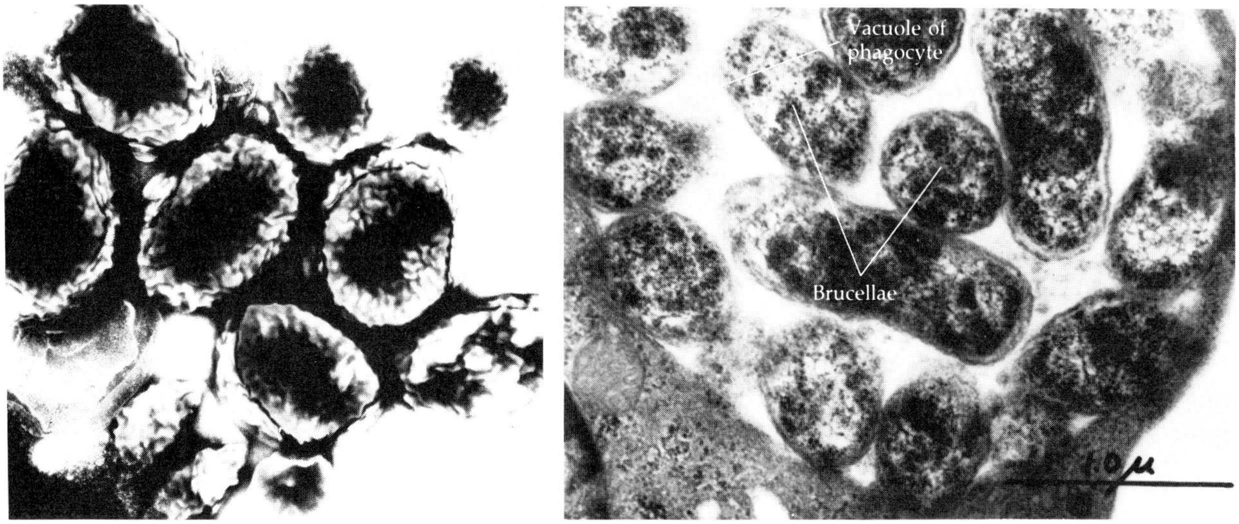

(a) (b)

Figure 21–4 **(a)** *Brucella abortus*, the species transmitted in cow's milk (×34,200). **(b)** Phagocyte containing *B. abortus* in a vacuole. The organisms are able to survive and reproduce in phagocytic cells, which chemical and immunological defenses have difficulty penetrating (×31,000).

which they multiply and travel via lymphatics to the lymph nodes. From here, the microorganisms can be transported to the liver, spleen, or bone marrow. The ability of the microorganisms to grow inside the phagocytes partly accounts for their virulence and resistance to antibiotic therapy and antibodies. If any immunity is achieved, it is not reliable. A vaccine is used for cattle but not for humans.

Most strains of *Brucella* are difficult to treat with antibiotics. Streptomycin alone does not reach bacteria surviving in phagocytes. Tetracycline alone, or in combination with streptomycin in severe cases, is usually effective. However, relapse commonly occurs unless therapy is prolonged for at least three to four weeks.

Brucellosis has an insidious onset. Its symptoms include a general malaise, weakness, and aching. The fever, particularly with *B. melitensis* infections, typically spikes to about 40°C (104°F) each evening; the disease is sometimes called **undulant fever.** Brucellosis is often a chronic disease lasting for many years. Periods of latency and few symptoms can alternate with episodes of renewed chills and fever.

Diagnosis is sometimes difficult and is most accurate when the organism can be isolated. Blood, as well as cerebrospinal fluid, urine, and even bone marrow, can be tested. The bacteria in smears of tissue specimens are seen as small coccobacillary organisms growing intracellularly, and special staining or fluorescent-antibody methods will differentiate them from similar intracellular bacteria. The serological tests for human infections are not always reliable. However, a simple agglutination test can be used on serum or milk to identify brucellosis-positive cattle.

Anthrax

In 1877, Robert Koch isolated *Bacillus anthracis*, the bacterium causing **anthrax** in animals (see Microview 1–1G). His careful work to demonstrate that the rod-shaped, endospore-forming bacteria swarming in the blood of the dead animal was indeed responsible for anthrax, eventually resulted in what we now know as Koch's postulates (Chapter 13). These postulates are the criteria that must be met to prove the microbial etiology of a disease.

The bacillus is a large, aerobic, gram-positive microorganism that is apparently able to grow slowly in soil types having specific moisture conditions. The endospores have survived in soil tests for 60 years. The organism is encapsulated, which

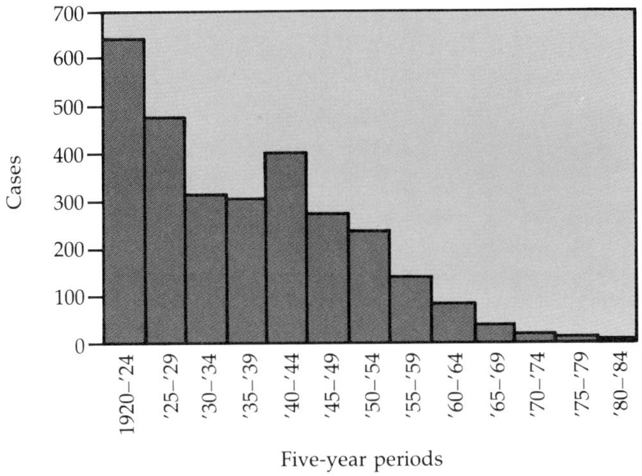

Figure 21–5 Reported cases of anthrax in humans in the United States, by 5-year periods, 1920–1983. In most recent years no cases have been reported.

probably improves its resistance to phagocytosis. The toxins produced also inhibit phagocytosis.

Animal herders in Europe had long associated certain fields with anthrax because animals that grazed there showed high incidence of the disease. The disease strikes primarily grazing animals such as cattle or sheep and is still endemic in some parts of the United States. The *B. anthracis* endospores are ingested with the grass which the endospores inhabit. The endospores pass through the animal's stomach into its intestinal tract. There the endospores germinate, pass through the intestinal mucosa, and invade the blood, causing a fatal, fulminating septicemia. It is possible for animals to be infected directly by infection of small cuts or abrasions on the mouth. Human gastrointestinal anthrax has been reported in the Soviet Union, but never in the United States.

There are only a handful of other types of human anthrax cases in the United States each year (Figure 21–5). Persons at risk are those handling animals that might be infected and those handling hides, wool, and other products from certain foreign countries. Goat hair from the Middle East has been a repeated source of contamination, as have handicrafts containing animal hides. Anthrax in humans differs from that in animals. If contact is made with materials containing anthrax endospores, the organism can enter the skin through a cut or abrasion and cause a malignant pustule there (Figure 21–6). This pustular infection is sometimes kept localized by the defenses of the body, but there is always danger of septicemia. Probably the most dangerous form of anthrax is pulmonary anthrax, contracted when the endospores are inhaled. **Woolsorters disease,** a dangerous form of

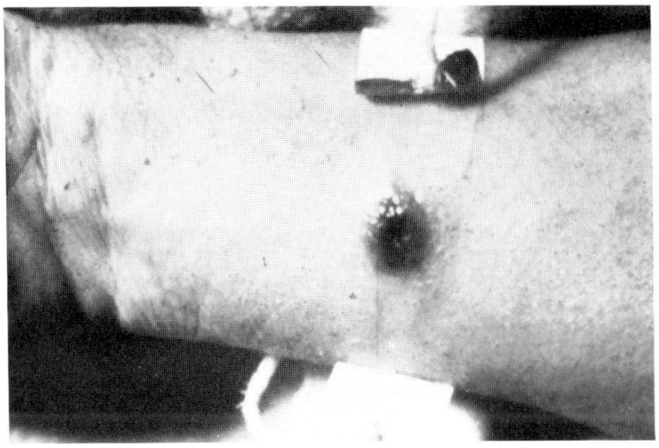

Figure 21–6 A cutaneous pustule of anthrax on a human arm. If this localized infection is not contained by the body's defenses, septicemia might be the next stage.

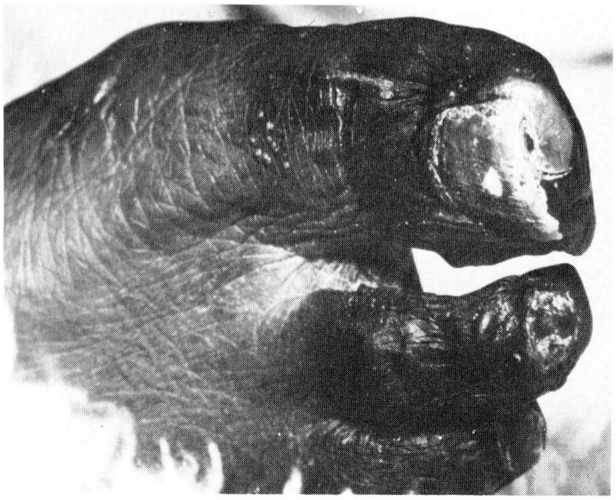

Figure 21–7 Foot of a patient with gangrene, a disease caused by *Clostridium perfringens*. The black, necrotic tissue on the toes furnishes anaerobic growth conditions for the bacterium, which then progressively destroys adjoining tissue.

pneumonia, results. It begins abruptly with high fever, difficulty in breathing, and chest pain. This disease eventually results in septicemia, and the mortality rate is high.

Livestock that have survived naturally acquired anthrax are resistant to reinfection. Second attacks in humans are also extremely rare. Vaccines composed of killed bacilli produce no significant immunity, and the best vaccine for humans appears to be a preparation of the protective antigen of the lethal toxin recovered from culture filtrates. Frequent boosters are necessary. Although a reliable vaccine for the protection of farm animals is available, the sporadic occurrence of bovine anthrax fails to provide ranchers with enough incentive to vaccinate their livestock routinely.

Anthrax is best diagnosed by isolation of the bacterium. A number of morphological and biological tests can be used for precise identification. Animal pathogenicity tests can also be performed, but serological tests are not in common use.

Penicillin is the drug of choice in treatment. Tetracycline, erythromycin, gentamicin, and chloramphenicol are also effective. However, once septicemia is well advanced, antibiotic therapy may prove useless, probably because the exotoxins remain despite the death of the bacteria.

Listeriosis

Listeria monocytogenes (lis-te′rē-ä mo-nō-sī-tô′je-nēz) is a largely intracellular parasite associated with a wide variety of mammals, fish, and birds. **Listeriosis** can be contracted by inhalation, ingestion of contaminated raw milk, or direct contact with infected animals. Meningitis, encephalitis, and monocytosis (an increase in the number of monocytes in the blood) are the most common symptoms of human listeriosis. Bacteremia spreads the microorganisms throughout the body, where they cause granulomatous, necrotic lesions. Untreated infections have a mortality rate of 70%. The bacterium can also be transferred from mother to newborn at birth. Neonatal infections develop within one to four weeks after birth as meningitis, and have a high mortality rate.

Preventative measures for listeriosis include pasteurization of milk. The antibiotics of choice in treatment are penicillin, erythromycin, or tetracycline.

Gangrene

If a wound causes the blood supply to be interrupted (a condition known as *ischemia*), the wound becomes anaerobic. Ischemia leads to *necrosis*, or death of the tissue. These conditions can also occur as a complication of diabetes. The death of soft tissue due to loss of blood supply is called **gangrene** (Figure 21–7). Substances released from dying and dead cells provide nutrients for many bacteria. Various species of the genus *Clostridium*, which are gram-positive, endospore-forming anaerobes widely found in soil and in the intestinal tracts of humans and domesticated animals, grow readily in such conditions. *Clostridium perfringens* (pėr-frin′jens) is the species most commonly involved in gangrene, but other clostridia and an assortment of other bacteria can also grow in such wounds.

Once ischemia and the subsequent necrosis have developed, **gas gangrene** can develop, especially in muscle tissue. As the *C. perfringens* microorganisms grow, they ferment carbohydrates in the tissue and produce gases (carbon dioxide and hydrogen) that swell the tissue. The microorganisms also produce necrotizing exotoxins and enzymes such as collagenase, proteinase, deoxyribonuclease, and hyaluronidase. By further interfering with blood supply, these products favor the spread of the infection. As tissue necrosis spreads, there is opportunity for increased microbial growth, hemolytic anemia, and, ultimately, severe toxemia and death.

One complication of improperly performed abortions is the invasion of the uterine wall by *C. perfringens,* which is resident in the genital tract of about 5% of all women. This infection can lead to gas gangrene and result in a life-threatening invasion of the bloodstream.

The surgical removal of necrotic tissue is called debridement; it and amputation are the most common medical treatments for gas gangrene. When gas gangrene occurs in regions such as the abdominal cavity, the patient can be treated in *hyperbaric chambers,* which contain a pressurized oxygen-rich atmosphere. The oxygen saturates the infected tissues and prevents the growth of the obligately anaerobic clostridia. Prompt cleaning of serious wounds, and precautionary antibiotic treatment, are the most effective steps in the prevention of gas gangrene. Penicillin, clindamycin, and chloramphenicol are antibiotics with good activity against *C. perfringens.*

SYSTEMIC DISEASES CAUSED BY ANIMAL BITES AND SCRATCHES

Animal bites can result in serious infections. Most serious bites are given by domestic animals such as dogs and cats, because they live in close contact with humans. It has been reported that 69% of dog bites become infected unless effectively cleansed. Domestic animals often harbor *Pasteurella multocida* (pas-tyėr-el'lä mul-tō'si-dä), a nonmotile, gram-negative rod similar to that which causes plague. Primarily a pathogen of animals, it causes septicemia, and thus is given the name multicida, meaning "many killing." Humans infected with *P. multocida* may show varied responses. For example, local infections with severe swelling and pain can develop at the site of the wound. Forms of pneumonia and septicemia are possible and life threatening. Penicillin and tetracycline are usually effective in treatment of these infections.

Minor cat scratches sometimes result in **cat scratch fever.** An ulcerated lesion appears, soon followed by fever and painful swelling of the lymph nodes draining the area. Many other organs can become involved and a generalized illness develops. Very recently, a minute gram-negative bacillus suspected of being the causative agent has been observed with use of special staining methods, but it has not been isolated and cultured.

Clostridium species and other anaerobes, such as species of *Bacteroides* and *Fusobacterium* (fū-sō-bak-ti'-rē-um), can also infect deep animal bites. Rat bites are typically infected by organisms of the genus *Streptobacillus* (strep-tō-bä-sil'lus) or *Spirillum* (spī-ril'lum), more commonly in Asiatic countries than in the United States. The disease, often called **rat bite fever,** is characterized by recurring fever, arthritis-like symptoms (inflammation, pain, stiffness), and infections of the lymph vessels.

Plague

Few diseases have affected human history more dramatically than **plague,** known in the Middle Ages as the Black Death. This term comes from one of its characteristics that appear on skin, the blackish areas caused by hemorrhages. In the fourteenth century, this disease destroyed perhaps one-fourth of the total population of Europe. And near the turn of this century, as many as 10 million people are reported to have died of plague in India in a 20-year period. The number of plague cases reported in the United States has been increasing in recent years and there are now about 20 cases annually. A mortality rate of 25% is not unusual (Figure 21–8).

The disease is caused by a gram-negative, rod-shaped bacterium, *Yersinia pestis* (yėr-sin'ē-ä pes'tis). One factor in the virulence of the plague bacterium is its ability to survive and proliferate inside phagocytic cells, rather than being destroyed by them. An increased number of highly virulent organisms eventually emerges. Normally a disease of rats, plague is transmitted from one rat to another by the rat flea (*Xenopsylla cheopis,* pronounced ze-nop-sil'lä chē-ō'pis). See Figure 11–28c. In the United States, particularly the far West and Southwest, the disease is endemic in wild rodents, especially ground squirrels, prairie

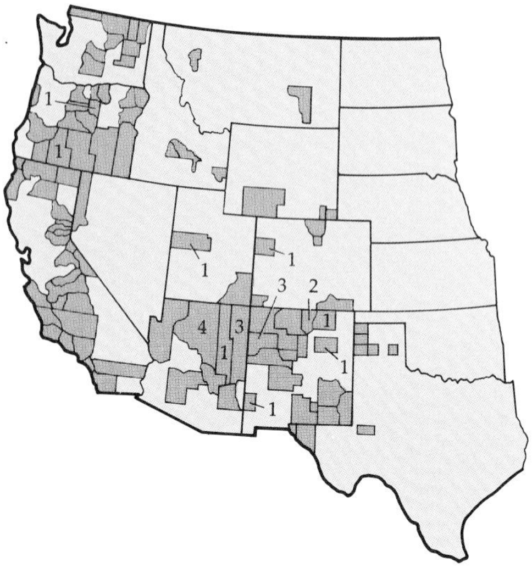

Counties in which animal plague was identified by *Yersinia pestis* isolation and/or presence of antibody in animals.

Figure 21–8 Geographical distribution of human and animal plague in the United States in a typical year. Counties reporting plague in animals are shown in gray. The number of human cases are indicated within each county.

dogs, and chipmunks **(sylvatic plague).** European rats, introduced into the United States many years ago, are the primary reservoirs of plague. If its host dies, the rat flea seeks a replacement, which may be another rodent or a human. A plague-infected flea is hungry for a meal because the growth of the bacteria blocks the digestive tract and the blood the flea ingests is quickly regurgitated. An arthropod vector is not always necessary for plague transmission. Contact from the skinning of infected animals, scratches of domestic cats, and similar contacts have been reported to cause infection.

In parts of the world where human contact with rats is common, the classic form of plague still prevails. From the flea bite, bacteria enter the human's bloodstream and proliferate in the lymph and blood. An overwhelming infection results. The lymph nodes in the groin and armpit become enlarged and fever develops, as the body's defenses react to the infection. Such swellings are called *buboes,* which accounts for the name **bubonic plague** (Figure 21–9). The mortality rate for untreated bubonic plague is 50 to 75%. Death from bubonic plague, if it occurs, is usually within less than a week of the appearance of symptoms.

A particularly dangerous condition arises when the bacteria are carried by the blood to the lungs, and a form of the disease called **pneumonic plague** results. The mortality rate for this type of plague is nearly 100%, and, like influenza, the disease is easily spread by airborne droplets. Even today, this disease can rarely be controlled if it is not recognized within 12 to 15 hours of fever. Great care must be taken to prevent airborne infection of persons in contact with patients. Persons with pneumonic plague usually die within three days.

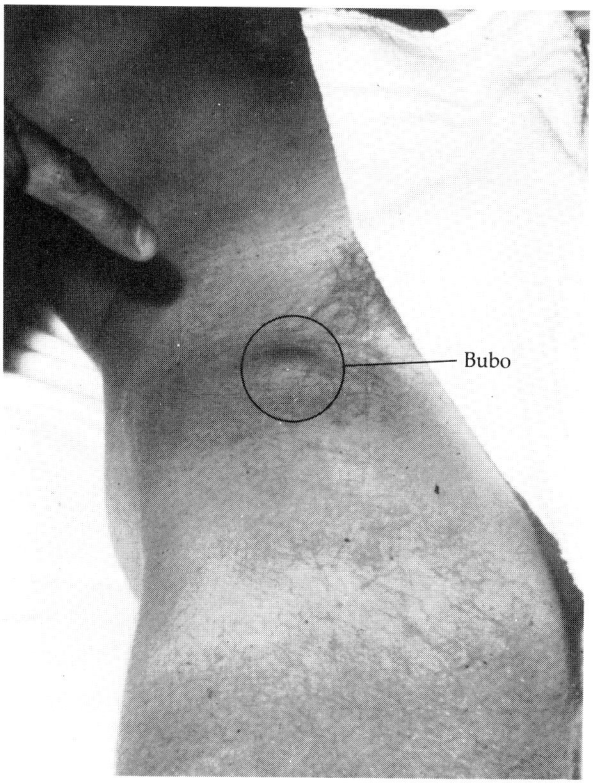

Figure 21–9 Bubonic plague. This photograph shows a bubo (swollen lymph node) in the groin area of the patient. Swollen lymph nodes are a common indication of systemic infection.

Diagnosis is most commonly done by isolation and identification of the bacterium by fluorescent-antibody and phage tests. Animal inoculations are sometimes used. Serological tests must demonstrate at least a fourfold rise in titer to be diagnostic. Persons exposed to infection can be given prophylactic antibiotic protection. A number of antibiotics, including streptomycin and tetracycline, are effective. Recovery from the disease gives a reliable immunity. No vaccines are available except for persons likely to come into contact with infected fleas during field operations or for laboratory personnel exposed to the organism. The long-term effectiveness of the plague vaccine is not high.

Control of the classic rat-based plague is largely due to modern sanitation procedures. Sanitary garbage disposal eliminates a source of food for rats, and ratproofing of buildings denies rats access to hiding places in attics and walls.

Relapsing Fever

Except for that causing Lyme disease (discussed in the next section), all members of the spirochete genus *Borrelia* (bôr-rel'ē-ä) cause **relapsing fever.** The disease is transmitted by soft ticks (*Ornithodorus,* pronounced ôr-nith-ō' dô-rus) that feed on rodents. The incidence of relapsing fever increases during the summer months when the activity of rodents and arthropods increases. The disease is characterized by a fever, sometimes in excess of 40.5°C, jaundice, and rose-colored spots. After 10 days, the fever subsides. Three or four relapses may occur, each shorter and less severe than the initial fever. Diagnosis is made by observation of spirochetes in the patient's blood.

Lyme Disease

A newly identified disease, called **Lyme disease,** was first observed in Old Lyme, Connecticut, in 1975, and has since been reported in 15 states from coast to coast. The disease typically begins with a skin lesion spreading from a single site, muscle aches, fever, and chills. Later complications such as acute arthritis, as well as heart and neurological abnormalities can occur. Once the disease was identified, evidence quickly emerged that it is transmitted by the bite of a tick, *Ixodes dammini* (iks-ō'des dam'-mi-nē). Because penicillin has generally been helpful in treatment, bacteria have been suspected as the cause. Very recently, a previously unknown spirochete organism, classified as a member of the genus *Borrelia,* has been isolated from infected patients.

RICKETTSIAL DISEASES OF THE CARDIOVASCULAR AND LYMPHATIC SYSTEMS

The rickettsia bacteria are obligate intracellular parasites of eucaryotes, and are responsible for a number of diseases spread by arthropod vectors. **Arthropods** are animals with jointed appendages and include mites, ticks and insects such as mosquitoes, fleas and lice (see Figure 11–28). There are several related rickettsial diseases, which differ mainly in their severity and in their arthropod vectors. These include epidemic typhus, endemic murine typhus, and the spotted fevers.

Typhus

Epidemic typhus (louse-born typhus) is caused by *Rickettsia prowazekii* (ri-ket'sē-ä prou-wä-ze'kē-ē) and carried by the human body louse *Pediculus vestimenti* (ped-ik'ū-lus ves-ti-men'tē) (see Figure 11–28b). The pathogen grows in the intestinal tract of the louse and is excreted by it. The pathogen is not transmitted directly by the bite of an infected louse, but rather by the feces of the louse being rubbed into the wound when the bitten host scratches the bite. The disease can flourish only in crowded and unsanitary conditions, when lice can transfer readily from an infected host to a new host. In recent years, a number of cases of epidemic typhus have been traced to contact with a reservoir in flying squirrels in the eastern United States. Although the vector in these cases is unknown, it is probably something other than a louse.

Epidemic typhus disease produces a high and prolonged fever for two or more weeks. Stupor and a rash of small red spots caused by subcutaneous

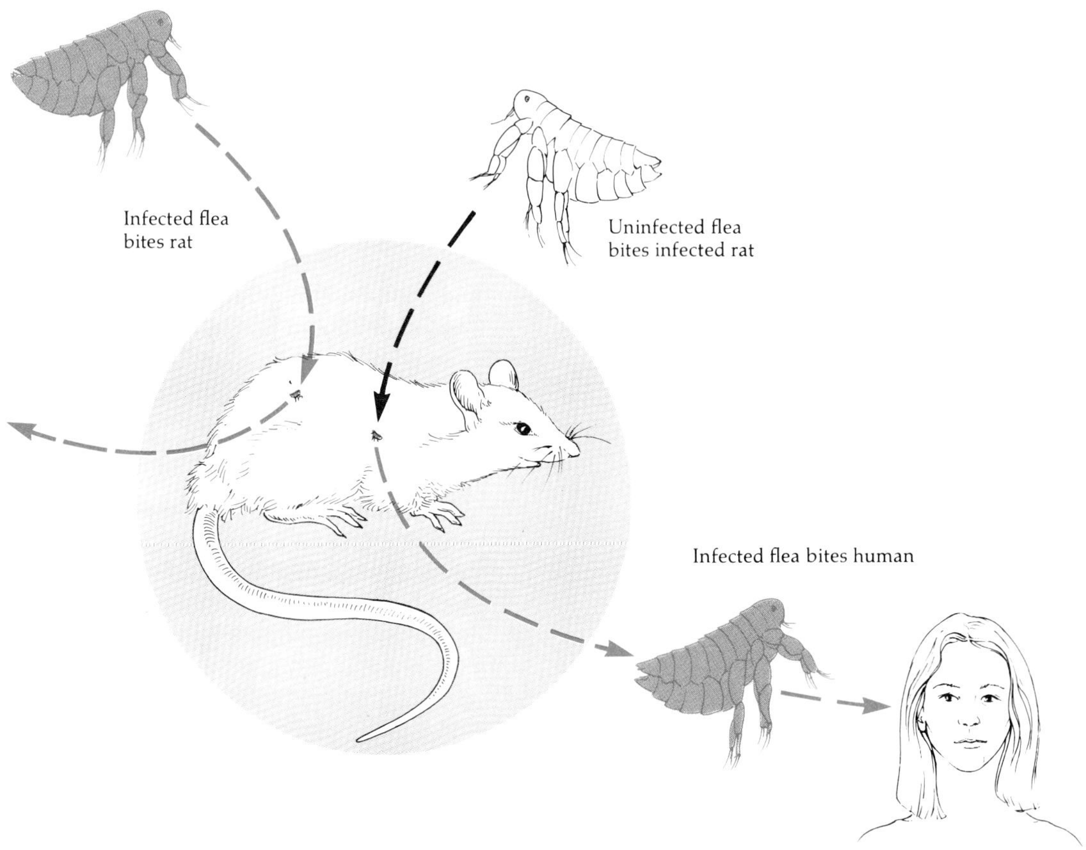

Figure 21–10 The cycle of endemic murine typhus. Normally, the disease is maintained among the rat population. Humans are infected only by accidental contact with the insect vector.

hemorrhaging are characteristic, as the rickettsias invade blood vessel linings. Mortality rates are very high when the disease is untreated.

Tetracycline- and chloramphenicol-type antibiotics are usually effective against epidemic typhus, but the elimination of conditions in which the disease can flourish is more important in its control. Laboratory diagnosis is based on the Weil–Felix test and a complement fixation reaction. Vaccines are available for military populations, which historically have been highly susceptible to the disease. Recovery from the disease gives a solid immunity and also renders a person immune to the related endemic murine typhus.

Endemic murine typhus occurs sporadically rather than in epidemics. The term *murine* refers to the fact that rodents, such as rats and squirrels, are the common hosts for this typhus. Endemic murine typhus is transmitted by rat fleas (*Xenopsylla cheopis*) (see Figure 11–28c), and the pathogen responsible for the disease is *Rickettsia typhi* (tī′fē), a common inhabitant of rats. Humans become involved when the flea finds a human host (Figure 21–10). With a mortality rate of less than 5%, the disease is considerably less severe than the epidemic form. Except for the reduced severity of the disease, endemic murine typhus is clinically indistinguishable from epidemic typhus.

Tetracycline and chloramphenicol are effective treatments for endemic murine typhus, and the diagnostic procedure for the disease is the same as for epidemic typhus. Rat control and avoidance of rats are preventive measures for the disease.

Rocky Mountain Spotted Fever

A number of other rickettsial diseases are grouped as **spotted fevers,** the best known being **Rocky Mountain spotted fever,** which is caused by *Rickettsia rickettsii* (ri-ket'sē-ē) (see Microview 1–1H). Despite its name (it was first recognized in the Rocky Mountain area), it is most common in the southeastern United States and Appalachia (Figure 21–11). The disease is characterized first by fever and headache, which are followed in a few days with a spotty rash that begins at the head and extremities and progresses to the trunk. Death is due to kidney and heart failure. The mortality rate for Rocky Mountain spotted fever is about 5 to 7%, probably because of either incorrect or late diagnosis.

These rickettsias are parasites of ticks, *Dermacentor* (dėr-mä-sen'tôr) species, which are the arthropod vectors of the disease (Figure 21–12; see also Figure 11-28a). The rickettsias can pass from generation to generation of ticks through eggs, a mechanism called *transovarian passage.* In the Eastern states, dog ticks are responsible for most cases of the disease; in the Rocky Mountain area, wood ticks are mainly responsible. No humans or other mammals are needed as reservoirs for the reproduction of the rickettsias.

Serological tests, primarily complement fixation, may be used in diagnosis. Guinea pig inoculation, which results in a characteristic illness, is also used in diagnosis. Although the mortality rate is currently less than 5%, in the days before antibiotic therapy the rate was much higher. Chloramphenicol and tetracyclines are used effectively to treat the disease.

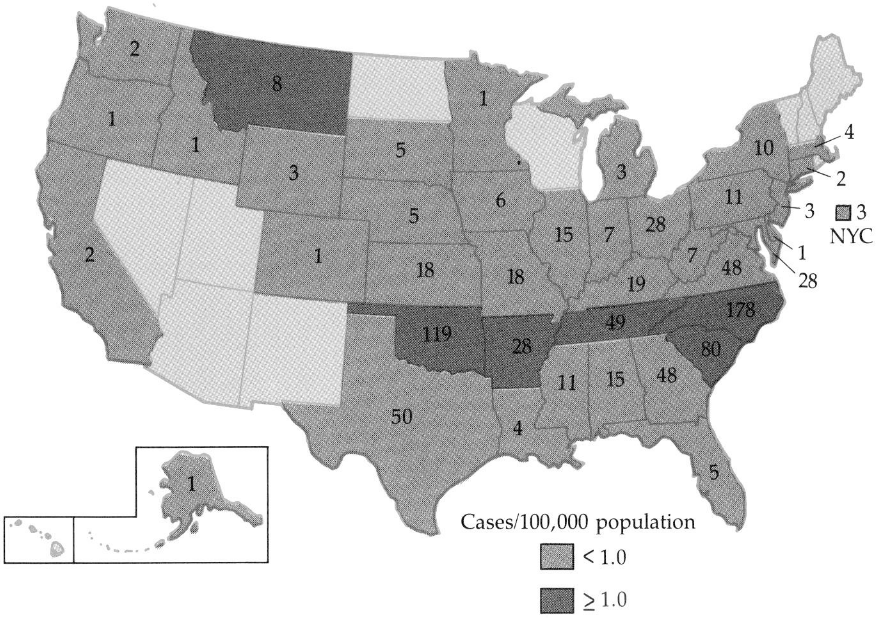

Figure 21–11 Cases of Rocky Mountain spotted fever reported in the United States in 1984. The total number was 847.

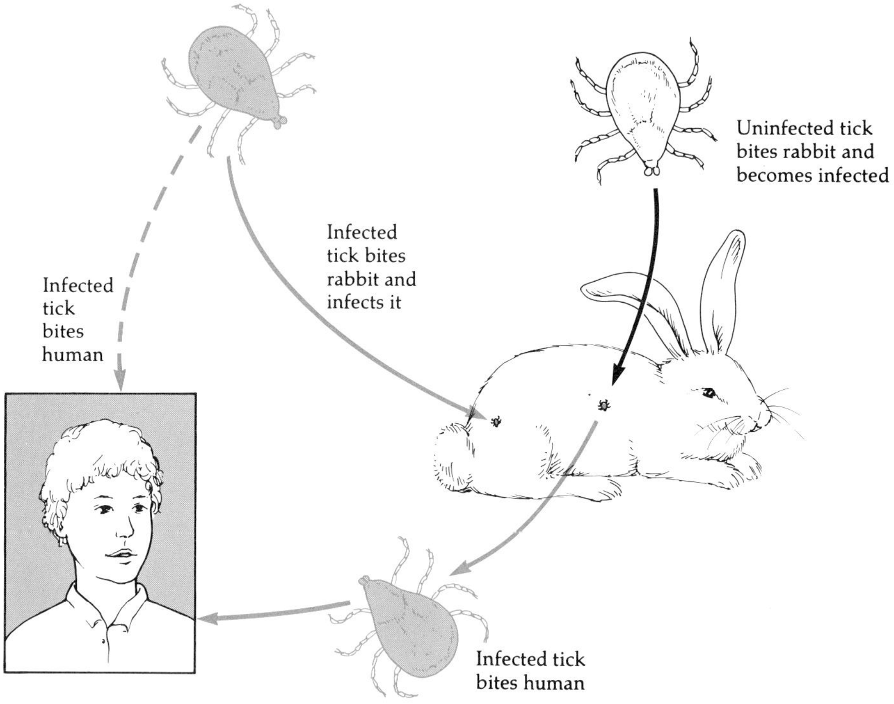
Uninfected tick
bites rabbit and
becomes infected
Infected
tick bites
rabbit and
infects it
Infected
tick
bites
human
Infected tick
bites human

Figure 21–12 Pathways of Rocky Mountain spotted fever. Because the entire reproductive cycle can take place in the tick, the wild mammal is not essential to the tick's life cycle.

It should be mentioned that there are a number of other tick-borne rickettsial diseases in various regions of the world.

VIRAL DISEASES OF THE CARDIOVASCULAR AND LYMPHATIC SYSTEMS

Myocarditis

As noted earlier, the heart muscle, also called the myocardium, is sometimes subject to an inflammatory disease called **myocarditis.** Although usually the result of a viral infection, this disease can also be caused by bacteria, fungi, or protozoans. The most frequently encountered viruses belong to the *coxsackievirus (enterovirus) group.* After infecting the respiratory or gastrointestinal tract, the virus apparently reaches the heart via the blood or lymph. In the heart, the infecting virus damages the myocardium, and leaves scar tissue. Sometimes pericarditis results from myocarditis. **Myocarditis of the newborn,** also caused by a coxsackievirus, is often fatal.

Infectious Mononucleosis

Infectious mononucleosis ("mono") is caused by a herpesvirus, the *Epstein–Barr (EB) virus* (Figure 21–13). Infectious mononucleosis is an acute disease that affects primarily the lymph nodes. It is characterized by enlarged and tender nodes, enlarged spleen, fever, sore throat, headache, nausea, and general weakness. The virus enters the body in oral secretions, multiplies in lymphatic tissue, and infects white blood cells. Infectious mononucleosis causes a proliferation of atypical lymphocytes (mononuclear white blood cells, the origin of the disease's name). In a sense, this result resembles leukemia's proliferation of leukocytes, and infectious mononucleosis has been described as a self-limiting cancer. Fortunately, it is rarely fatal, although one family has been reported to have an apparent genetic anomaly that causes infectious mononucleosis to turn cancerous.

The disease is transmitted by direct or indirect contact, such as kissing and drinking from the same bottle. It has been called the kissing disease, and its association with kissing might indicate that a fairly large inoculum is necessary for transmission. It does not seem to spread in households, so aerosol transmission is unlikely. The normal incubation period is two to six weeks.

The peak incidence of the disease occurs in persons of about 15 to 25 years of age. Collegiate populations, particularly those from the upper socioeconomic strata, have a high incidence of the disease. About 50% of college students have no immunity and about 15% of these can expect to come down with the disease. Persons in low socioeconomic groups tend to acquire the disease in early childhood, in a subclinical case that grants immunity. In some parts of the world, 90% of the children over four years of age have antibodies to the disease. Probably 80% of the general population of the United States eventually acquires the disease in one form or another.

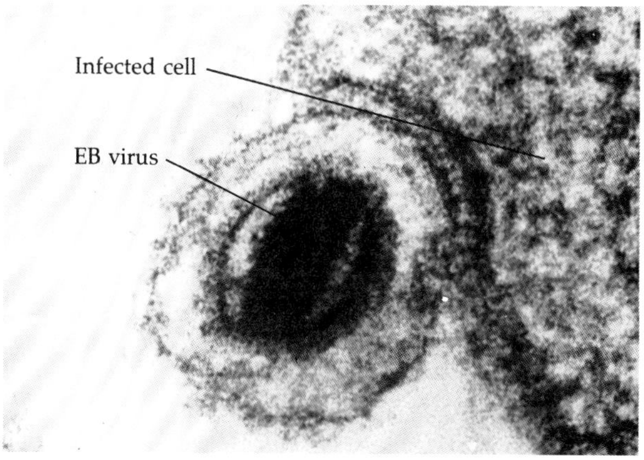

Figure 21–13 An enveloped, polyhedral Epstein–Barr (EB) virus caught in the act of infecting a human cell (right), in culture. This particular virus particle was isolated from cells of Burkitt's lymphoma.

Diagnosis is usually made with a serological test for heterophil antibodies. These antibodies are not specific for the EB virus but cause the nonspecific agglutination of sheep red blood cells. They are developed to high titers and are quite persistent. The antigen that causes the appearance of these antibodies is unknown. Heterophil antibodies are also formed in response to other conditions, and there are simple procedures to differentiate these from the antibodies formed in response to infectious mononucleosis.

Specific antibodies against the EB virus can also be identified with use of immunofluorescent techniques, and it is desirable to use such a test for confirmation of the diagnosis. The antibodies against the virus provide a good immunity to subsequent attacks. Viruses that persist as a latent infection after the apparent infection subsides may help to maintain the persistence of antibodies. The persisting virus is often harbored and shed in the oral secretions as a persistent latent infection.

Burkitt's Lymphoma

As was mentioned in Chapter 12, the EB virus has been linked to two human cancers: **Burkitt's lymphoma** and **nasopharynegeal cancer.** There is a considerable amount of evidence linking the EB virus to these human malignancies. For example, in cell cultures, EB virus will infect only human lymphoid cells. The virus then transforms some of the cells into EB virus-carrying cells that have malignant characteristics. When inoculated into monkeys, it produces fatal lymphomas. There is no doubt that EB virus is potentially oncogenic.

A precondition for Burkitt's lymphoma might be infection by malaria or other tropical diseases. This form of cancer is especially prevalent in a region of Central Africa inhabited by a mosquito that transmits malaria. In peoples of most other areas of the world, the immune system seems to effectively combat tumor cells marked with EB virus antigens and Burkitt's lymphoma is rare, although the EB virus is common. Similarly, the frequency of nasopharyngeal cancer might depend upon the simultaneous presences of the EB virus and other carcinogens, such as the nitrosamines found in salted fish that form a large part of certain regional diets.

Yellow Fever

Arboviruses, or arthropod-borne viruses, can reproduce either in arthropods, such as mosquitoes, or in humans. Although a number of serious human diseases are caused by such viruses, no disease is evident in the arthropods that serve as vectors. Worldwide, there are probably more than 100 arbovirus-caused diseases. **Yellow fever** is caused by an arbovirus (*yellow fever virus*) and is historically important because it was the first such virus discovered and provided first confirmation that an insect could transmit a virus.

The yellow fever virus is injected into the skin by the mosquito. The virus then spreads to local lymph nodes, where it multiplies; from the lymph nodes it advances to the liver, spleen, kidney, and heart, where it can persist for days. In the early stages of the disease, the person experiences fever, chills, headache, and backache, followed by nausea and vomiting. This stage is followed by jaundice, a yellowing of the skin due to the deposition of bile pigments in the skin and mucous membranes. It results from liver damage. (The yellow color of the skin is what gave the disease its name.) In severe cases, the virus produces lesions in the infected organs and hemorrhaging occurs.

Yellow fever is still endemic in many tropical areas such as Central America, tropical South America, and Africa (Figure 21–14). Monkeys are a natural reservoir for the virus. Control of the disease depends on localized control of *Aedes aegypti* (ä-e'dēz ē-jip'tē), the usual mosquito vector (see Figure 11–27a), as was done during the building of the Panama Canal many years ago, or on immunization of the exposed population.

There is no specific treatment for yellow fever. In the early stages of the disease, diagnosis can be made with the following procedure: the serum of patients is used to neutralize known yellow fever virus strains before the virus is inoculated into mice. If the serum contains virus-specific antibodies, the virus is neutralized and the mice do not become ill. The vaccine in use is an attenuated live viral strain and yields a very effective immunity with few side effects.

Dengue

A rather similar, but milder, disease is **dengue,** also a mosquito-borne (*Aedes aegypti*) viral disease endemic in the Caribbean and other tropical environments around the world. It is caused by *dengue fever virus,* an arbovirus, and is characterized by fever, muscle and joint pain, and rash. The muscle and joint pain experienced by sufferers has led to the name **breakbone fever.** Other than the painful symptoms, classic dengue fever is a relatively mild disease and is rarely fatal. The mosquito vector for dengue is common in the Gulf states and there is some concern that the virus will sooner or later be introduced into this region and become endemic. In most years, more than 100 cases are imported into this country, mostly via travelers from the Caribbean and South America.

A more serious form of dengue, **dengue hemorrhagic fever,** is characterized by bleeding from the skin, gums, and gastrointestinal tract, and sometimes circulatory failure and shock.

Serological tests that are useful in diagnosis include complement fixation, hemagglutination, and neutralization. Control measures are directed at eliminating the *Aedes* mosquitoes. There is no specific treatment that is effective against the virus.

Viral Hemorrhagic Fevers

In 1969, three nurses working in missionary hospitals in Nigeria contracted an unknown disease. Two died, and laboratory technicians performing virologic studies of serum from the survivor contracted the disease. The disease became known as **Lassa fever,** one of the **viral hemorrhagic fevers.** In 1967, 31 persons became ill when some African monkeys were imported into Europe. That type of viral hemorrhagic fever is now known as the **Marburg virus outbreak.** Other hemorrhagic fevers of viral origin, such as **Crimean–Congo hemorrhagic fever,** have since been reported. The virus is transmitted very rapidly from person to person, by contact with bodily fluids. Nosocomial transmission via hypodermic needles accounted for one-half of the cases of **Ebola hemorrhagic fever** in Zaire in 1976. The fatality rates of these viral hemorrhagic fevers tend to be very high for an infectious disease. Patients suffer headache, muscle pain, skin or gastrointestinal hemorrhaging, and shock.

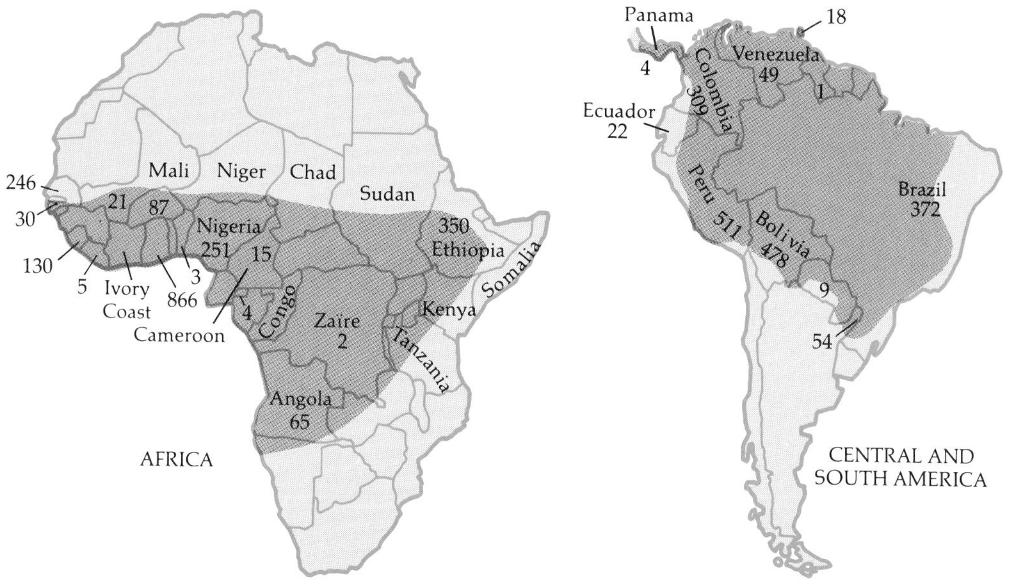

Figure 21–14 Zones of endemic yellow fever are shown in darker color. The numbers of reported cases are also indicated.

PROTOZOAN AND HELMINTHIC DISEASES OF THE CARDIOVASCULAR AND LYMPHATIC SYSTEMS

Toxoplasmosis

Toxoplasmosis, a disease of blood and lymph vessels, is caused by the protozoan *Toxoplasma gondii* (toks-ō-plaz'mä gon'dē-ī), a small, crescent-shaped organism (Figure 21–15a). *T. gondii* is a sporozoan, like the malarial parasite.

Cats seem to be an essential part of the life cycle of *T. gondii* (Figure 21–15b). Random tests on urban cats have shown that a large number are infected with the organism, which causes no apparent illness in the cat. The organism undergoes its only sexual phase in the intestinal tract of the cat. Oocysts are then shed in the cat's feces, and contaminate food or water that can be ingested by other animals. The protozoan reproduces asexually in these new hosts. A human eating the uncooked meat of such an animal, or a cat eating an infected mouse, will be infected with the parasite.

Humans' two main modes of infection appear to be eating undercooked meats and inhaling the dried feces of cats (while cleaning a litter box) or coming into other contact with the feces.

There was a recent outbreak of toxoplasmosis in 34 persons who frequented a riding stable with a large resident population of cats. Some surveys have shown that approximately 30% of the population carries antibodies for this organism; this percentage indicates the high rate of subclinical, unrecognized infections.

The disease is a rather undefined mild illness, and the primary danger is in the congenital infection of a fetus. Thus, pregnant women should avoid cleaning cat's litter boxes or having any unnecessary contact with cats; eating raw meats is also inadvisable. The actions on the fetus are drastic, including convulsions, severe brain damage, blindness, and death. The mother is probably unaware of the disease, which is transmitted across the placenta.

T. gondii can be isolated and grown in cell cultures. The preferred serological tests make use of a fluorescent-antibody method or a form of hemag-

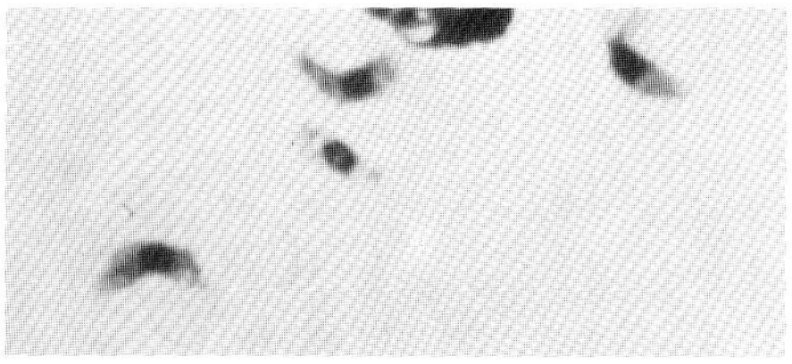

(a)

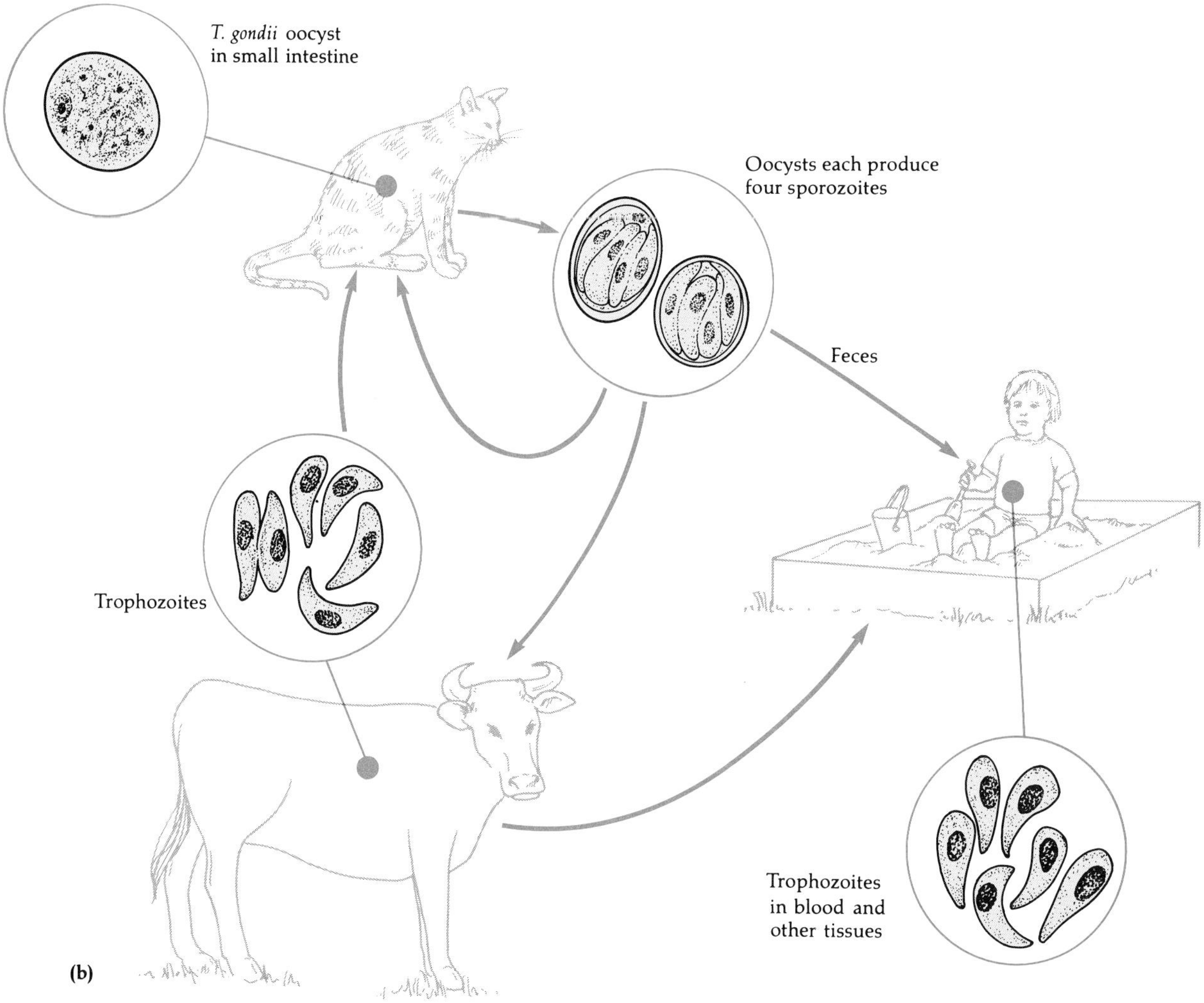

(b)

Figure 21–15 **(a)** Crescent-shaped trophozoites of *Toxoplasma gondii*, the causative agent of toxoplasmosis. **(b)** Life cycle of *T. gondii*. The domestic cat is the definitive host in which the sporozoan reproduces sexually to produce oocysts, which are passed in the feces. Four sporozoites form within each oocyst. The sporozoites are then ingested by one of the intermediate hosts. There, in blood and other tissues, the sporozoites mature into trophozoites.

glutination testing. The latest recommendation is to use the two tests in conjuction. A rising titer in a pregnant woman may be an indication for a therapeutic abortion, but this correlation is very controversial.

Treatment of toxoplasmosis is with pyrimethamine in combination with either trisulfapyrimidines or sulfadiazine.

American Trypanosomiasis (Chagas' Disease)

An example of a protozoan disease of the cardiovascular system is **American trypanosomiasis (Chagas' disease).** The causative agent is *Trypanosoma cruzi* (tri-pan-ō-sō'ma kruz'ē), a flagellated protozoan (Figure 21–16). The disease occurs through the southern portions of the United States, Mexico, Central America, and into South America to Argentina. Although only a few cases have been reported in the United States, the incidence of the disease is 40 to 50% in some rural areas of South America.

The reservoir for *T. cruzi* includes a wide variety of wild animals, including rodents, opossums, and armadillos. The arthropod vector is the reduviid bug, often called the "kissing bug" (see Figure 11–28e) because it often bites persons near the lips. The trypanosomes, which grow in the gut of the bug, are passed on if the bug defecates while feeding. The bitten human or animal often rubs the feces into the bite wound or other skin abrasions by scratching, or into the eye by rubbing. At the site of inoculation, the trypanosomes reproduce and pass through various stages of their life cycle. A swollen lesion develops at the inoculation site (often near the eye). The organisms infect regional lymph nodes and are carried by the blood to other organs of the body, such as the liver and macrophages of the spleen, and frequently the heart. Although central nervous system involvement in adults is rare, it is not uncommon in children. One effect of nervous system involvement is loss of involuntary muscular contractions that move solids through the esophagus and intestinal tract.

Laboratory procedures for the diagnosis of American trypanosomiasis include observing the trypanosomes in blood, growing the microorganisms in blood cultures, or using fluorescent-labeled antibody on blood smears. Treatment of the disease is very difficult when chronic, progressive stages have been reached. The trypanosome multiplies intracellularly and is difficult to reach chemotherapeutically. Cultivation of the parasite in tissue culture has made large-scale screening for antitrypanosomal chemicals possible. Nifurtimox (Lampit) was developed from such screening programs and has proven effective.

Malaria

Malaria is characterized by chills and fever, and often by vomiting and severe headache. These symptoms typically appear for one- to three-day periods, alternating with asymptomatic periods.

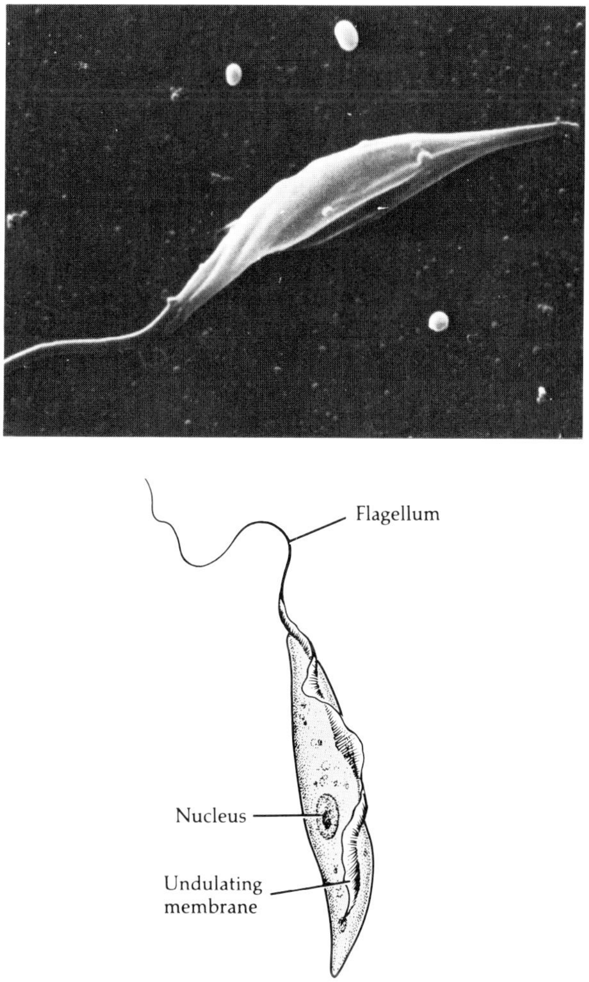

Figure 21–16 Scanning electron micrograph of *Trypanosoma cruzi*, the protozoan that causes American trypanosomiasis (Chagas' disease) (×5600). Trypanosomes have one flagellum that appears as an extension of the undulating membranes.

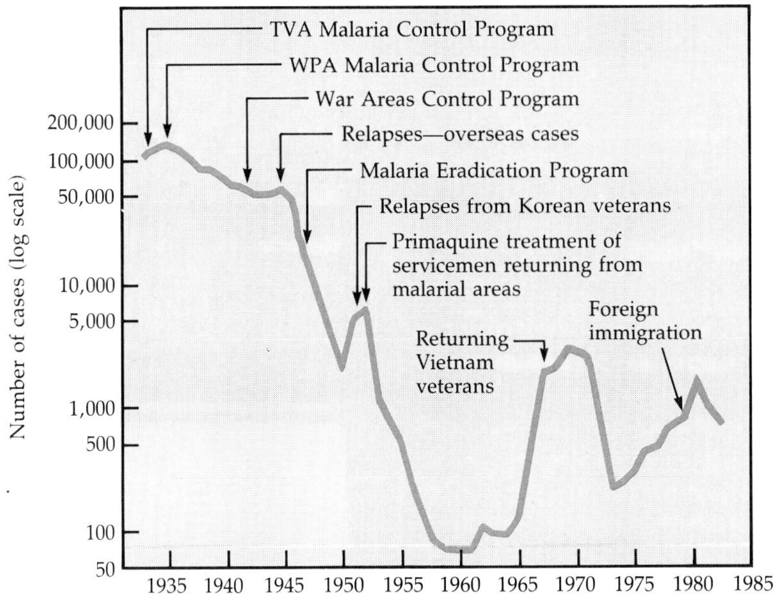

Figure 21–17 Reported cases of malaria in the United States, 1933 to 1983. Malaria was a common disease in this country as recently as 1935. Its incidence declined as government control programs such as the Tennessee Valley Authority (TVA) flood control system got underway, and continued to decline with Work Projects Administration (WPA) programs of Depression years and control programs during World War II. Note the recent increases in incidence as veterans returned from the Vietnam War and foreign immigration increased.

Malaria is found wherever humans host its carrier protozoan parasite and the mosquito vector, certain species of *Anopheles* (an-of'el-ēz), is found (see Figure 11–27b). The disease was once common in the United States, but effective mosquito control and a reduction in the number of human carriers caused the reported cases to drop below 100 by 1960 (Figure 21–17). In recent years, however, an upward trend in the number of U.S. cases, reflects a worldwide resurgence of malaria, increased travel to malarial areas, and the increase in immigrations from malarial areas. Some cases of malaria are due to blood transfusions or unsterilized syringes used by drug addicts. In tropical Asia, Africa, and Central and South America, malaria is still a serious problem.

The causative organisms of malaria are the spore-forming protozoans (sporozoans) of the genus *Plasmodium* (plaz-mō'dē-um) (see Microview I–4D). Four pathogenic species are recognized, each of which causes a distinctive form of the disease. The most dangerous species is probably *Plasmodium falciparum* (fal-sip'är-um), a species that is also widespread geographically. Also widely distributed is *Plasmodium vivax* (vī'vaks). *Plasmodium malariae* (mä-lā'rē-ī) and *Plasmodium ovale* (ō-vä'le) have either a lower infection rate or cause a geographically restricted, relatively milder disease.

When *Plasmodium* first enters a human host, the form of the microorganism called the *sporozoite* enters the liver cells, and there undergoes schizogony (see Figure 11–20). The sporozoites are later released into the bloodstream to infect red blood cells, although some may remain in the liver, only to enter red blood cells months or years later and cause a recurrence of malaria.

The release of the form called the *merozoite* into the bloodstream causes the paroxysms (recurrent intensifications of symptoms) of chills and fever that are the main symptoms of malaria. The fever reaches 40°C and a sweating stage begins as the fever subsides. Between paroxysms the patient feels normal. Anemia results from the loss of red blood cells, and hypertrophy of the liver and spleen are added complications.

Laboratory diagnosis of malaria is made by examination of a blood smear (Figure 21–18). Although there is no effective immunity to malaria, populations in which it has long been endemic have generally developed some resistance to it, and individuals tend to have a less severe disease. In particular, persons who have the genetic sickle-cell trait or sickle-cell anemia—common in many areas where malaria is endemic—are resistant to malaria.

Considerable effort is being expended on the search for an effective vaccine. A two-step approach to the difficult problem of developing a vaccine against malaria is being made. The first step is the identification of the antigens on the surfaces of the parasite's three stages (sporozoite, merozoite, and gametocyte). Once these antigens are positively identified, then the second step—the genetic engineering of antigens for vaccines—will be attempted. The sporozoite stage is of primary interest, because inhibiting its growth would prevent the initial infection from becoming well established. However, some feel that any effective vaccine would have to protect against all three stages.

Malaria has long been fairly effectively treated with quinine. Derivatives of quinine, such as primaquine and especially chloroquine, have been the mainstays for prevention and treatment of malaria for many years. Chloroquine prevents the DNA of the parasite's merozoite stage from replicating, by fitting between base pairs in the DNA. Resistance to these drugs is spreading rapidly and current chemical alternatives are more expensive. Expense is an important factor to the parts of the world most troubled by malaria. One approach to chemotherapy has been to use drug combinations such as Fansidar, a combination of pyrimethamine and sulfadoxine. Both are inhibitors of folic acid synthesis (see the discussion of sulfa drugs in chapter 18); together, they have a synergistic effect. Resistance has been reported against such combinations. Research on anti-malarial chemicals has been much aided in recent years by methods of culturing some stages of the malarial parasite. Culturing makes it possible for large numbers of chemicals to be tested. A U.S. Army research group has screened more than 250,000 such compounds.

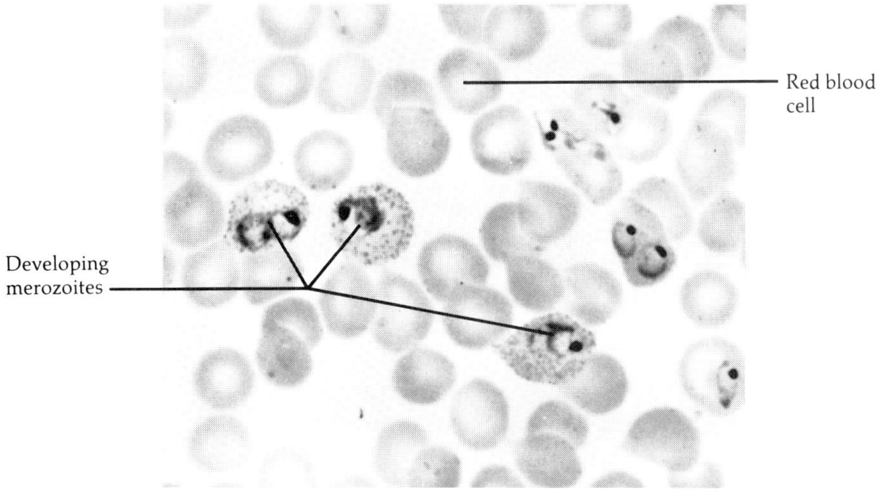

Figure 21–18 Blood smear from a malarial patient. Laboratory diagnosis of malaria is based on the observation of *Plasmodium* in red blood cells. Early in its development, the parasite, called a merozoite at this stage, has a ring form.

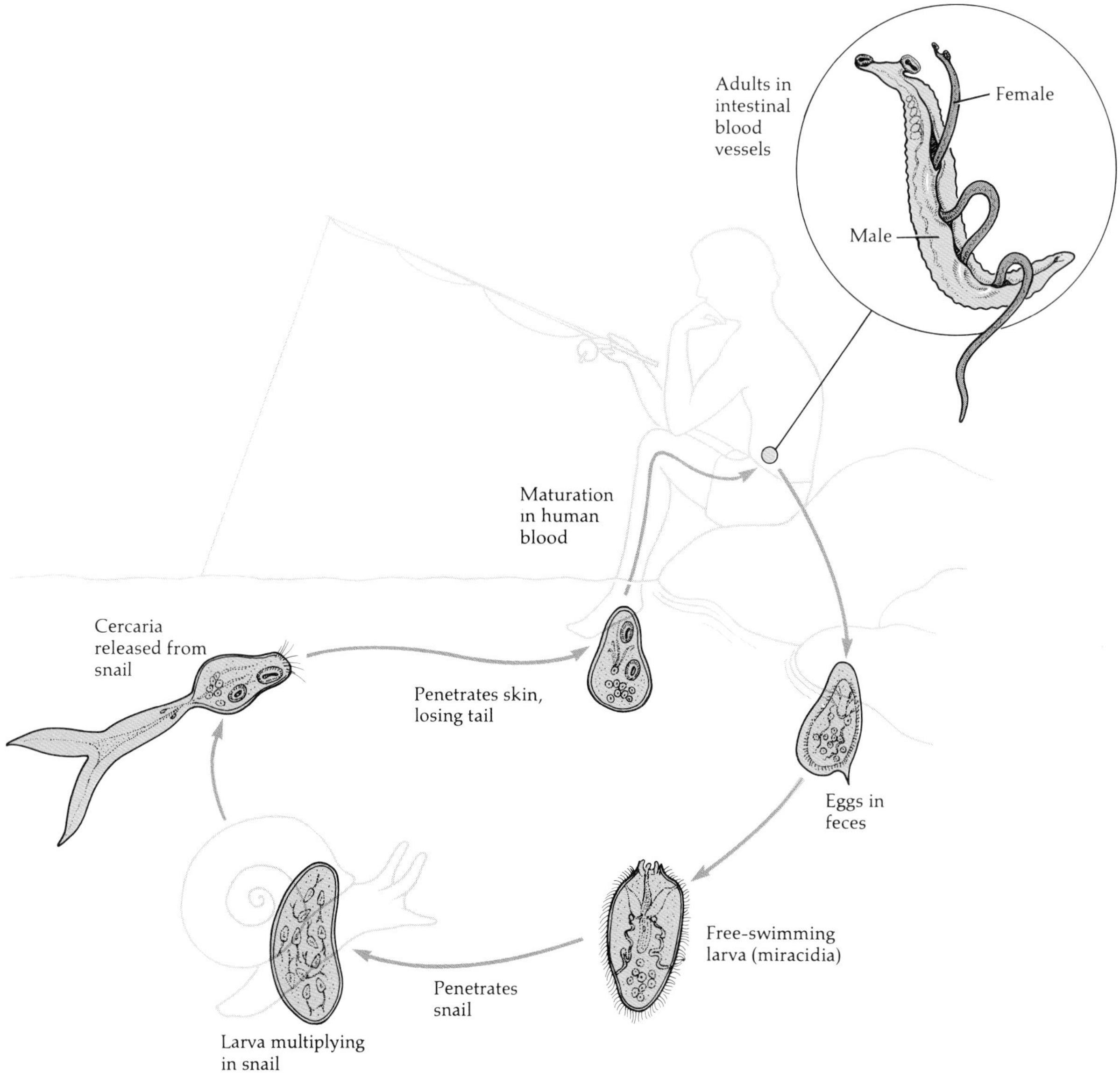

Figure 21–19 Life cycle of *Schistosoma*, the causative agent of schistosomiasis.

Effective control of malaria is not in sight. It will probably require a combination of chemotherapeutic and immunological approaches. The expense and the need for an effective political organization in malarial areas are probably going to be fully as important as advances in medical research.

Schistosomiasis

Schistosomiasis is a disease caused by a flatworm parasite, a fluke. The disease is not contracted in the United States, but more than 400,000 immigrants to the United States have the disease, and millions in Asia, Africa, South America, and the

Caribbean are also affected. Schistosomiasis is a major problem in world health. Species of the genus *Schistosoma* (skis-tō-sō′mä), which causes the disease, vary with geographical location, but the characteristics of the disease are similar.

Waters become contaminated with ova excreted in human wastes (Figure 21–19). A motile, ciliated form of *Schistosoma* called a *larva* is released from the ova, and enters certain species of snails. The lack of a suitable host snail is the primary reason why schistosomiasis cannot be transmitted in the United States. Eventually, the pathogen emerges from the snail in an infective form called the *cercaria.* These cercariae have forked swimming tails, and when they contact the skin of a person wading or swimming in the water, they discard the tail and enzymatically penetrate the skin. They are then carried by the bloodstream to the veins of the liver or urinary bladder. The larvae mature into an adult form in which the slender female lives in a cleft of the male. The union produces a supply of new ova, some of which cause local tissue damage from defensive body reactions. Other ova enter the water to continue the cycle.

The damage caused by the disease is often liver destruction, but sometimes other organs such as the lungs or urinary system become involved. Abscesses and ulcers are formed in response to the infestation, and the body's allergic reactions block various ducts and blood vessels.

Laboratory diagnosis consists of microscopic identification of the flukes or their ova in fecal and urine specimens, intradermal tests, and serological tests such as complement fixation and precipitin tests.

Some new drugs have recently been approved for use against schistosomes and the outlook for treatment has brightened. Drugs approved for use in the United States are praziquantal and oxaminiquine. Current research is underway to identify schistosome enzymes that differ from host enzymes and to develop chemotherapeutic agents that are specific for the parasite. Sanitation and elimination of the host snail are also useful forms of control.

A summary of diseases associated with the cardiovascular and lymphatic systems is presented in Table 21–1.

Table 21–1 Summary of Diseases Associated with the Cardiovascular and Lymphatic Systems

Disease	Causative Agent	Mode of Transmission	Treatment
Bacterial Diseases of the Cardiovascular and Lymphatic System:			
Puerperal sepsis and infections related to abortions	Primarily *Streptococcus pyogenes; Clostridium* and *Bacteroides* species often cause postabortion infections	Unsanitary conditions in childbirth and abortions	Penicillin and erythromycin for *S. pyogenes;* chloramphenicol and penicillin for *Bacteroides* and *Clostridium* species
Endocarditis			
Subacute bacterial	Many organisms, especially alpha-hemolytic streptococci	Bacteremia localizing in the heart	Varies with agent; penicillin and erythromycin for alpha-hemolytic streptococci
Acute bacterial	*Staphylococcus aureus, Streptococcus pneumoniae*	Bacteremia localizing in the heart	Semisynthetic penicillin derivatives such as methicillin or oxacillin for *S. aureus* and penicillin for *S. pneumoniae*
Pericarditis	*Streptococcus pneumoniae*	Bacteremia localizing in the heart	Penicillin
Rheumatic fever	Group A beta-hemolytic streptococci	Inhalation leading to streptococcal sore throat	Penicillin (for sore throat)
Tularemia	*Francisella tularensis*	Animal reservoir (rabbits); skin abrasions, ingestion, inhalation, bites	Streptomycin, tetracycline

(*continued on next page*)

Table 21–1 Summary of Diseases Associated with the Cardiovascular and Lymphatic Systems, *continued*

Disease	Causative Agent	Mode of Transmission	Treatment
Brucellosis	*Brucella* species	Animal reservoir (cows); ingestion in milk, direct contact with skin abrasions	Streptomycin, tetracycline
Anthrax	*Bacillus anthracis*	Reservoir is soil or animals; skin abrasions, inhalation or ingestion of heavy spore concentrations	Penicillin, tetracycline, erythromycin, chloramphenicol
Listeriosis	*Listeria monocytogenes*	Animal reservoir; ingestion	Tetracycline or penicillin
Gangrene	*Clostridium perfringens*	Contamination of open wound by clostridial endospores	Debridement, amputation, hyperbaric chamber, penicillin, clindamycin, chloramphenicol
Cat scratch fever	Unidentified bacillus	Cat scratch or bite	Tetracycline
Rat bite fever	*Spirillum minor* or *Strepto-bacillus moniliformis*	Rat bite	Penicillin or tetracycline
Plague	*Yersinia pestis*	*Xenopsylla cheopis* (rat flea)	Tetracycline and streptomycin
Relapsing fever	*Borrelia* species	*Ornithodorus* species (soft ticks)	Tetracycline
Lyme disease	*Borrelia* species	*Ixodes dammini* (tick)	Penicillin
Epidemic typhus	*Rickettsia prowazekii*	*Pediculus vestimenti* (louse)	Tetracycline and chloramphenicol
Endemic murine typhus	*Rickettsia typhi*	*Xenopsylla cheopis* (rat flea)	Tetracycline and chloramphenicol
Rocky Mountain spotted fever	*Rickettsia rickettsii*	*Dermacentor andersoni* and other species (tick)	Tetracycline and chloramphenicol
Viral Circulatory and Systemic Diseases:			
Myocarditis	Several agents, especially coxsackievirus (enterovirus)	Coxsackievirus; Inhalation and ingestion	None
Pericarditis	Same as above	Complication of viral myocarditis	None
Infectious mononucleosis	Epstein–Barr virus	Oral secretions, kissing	None
Yellow fever	Arbovirus (yellow fever virus)	*Aedes aegypti* mosquito	None
Dengue	Arbovirus (dengue fever virus)	*Aedes aegypti* mosquito	None
Protozoan and Helminthic Circulatory and Systemic Diseases:			
Toxoplasmosis	*Toxoplasma gondii*	Animal reservoir: cats; inhalation and ingestion	Pyrimethamine in combination with either trisulfapyrimidines or sulfadiazine
American trypanosomiasis	*Trypanosoma cruzi*	Bite of reduviid bug	Nifurtimox in early stages
Malaria	*Plasmodium* species	*Anopheles* mosquito	Chloroquine
Schistosomiasis	*Schistosoma* species	Contaminated water; cercariae enter the body through the skin	Praziquantal

STUDY OUTLINE

INTRODUCTION (p. 575)

1. The heart, blood, and blood vessels make up the cardiovascular system.
2. Lymph, lymph vessels, lymph nodes, and lymphoid organs constitute the lymphatic system.

STRUCTURE AND FUNCTION OF THE CARDIOVASCULAR SYSTEM (p. 575)

1. The heart circulates substances to and from tissue cells.
2. Deoxygenated blood enters the right atrium from all parts of the body. It passes to the right ventricle and to the lungs.
3. Oxygenated blood returns to the left atrium. It passes to the left ventricle and to all tissue cells.
4. Arteries and arterioles transport blood away from the heart. Veins and venules bring blood to the heart.
5. Materials are exchanged between blood and tissue cells at capillaries. Capillaries connect arterioles to venules.
6. Blood is a mixture of plasma and cells.
7. Red blood cells carry oxygen. White blood cells are involved in the body's defense against infection. Platelets are involved in blood clotting. Plasma transports dissolved substances.

STRUCTURE AND FUNCTION OF THE LYMPHATIC SYSTEM (p. 575–577)

1. Fluid that filters out of capillaries into spaces between tissue cells is interstitial fluid.
2. Interstitial fluid enters lymph capillaries and is called lymph.
3. Lymphatics return lymph fluid to the blood.
4. Lymph nodes contain fixed macrophages and B cells.
5. Tonsils, appendix, spleen, and thymus gland are lymphoid organs.

BACTERIAL DISEASES OF THE CARDIOVASCULAR AND LYMPHATIC SYSTEMS (pp. 577–590)

Septicemia (p. 577)

1. The growth of microorganisms in blood is called septicemia.
2. Symptoms include fever and decreased blood pressure and lymphangitis (inflamed lymph vessels). Septicemia can lead to ischemia and shock.
3. Septicemia usually results from a focus of infection in the body.
4. Gram-negative rods are usually implicated. Endotoxin causes the symptoms.

Puerperal Sepsis (pp. 577–579)

1. Puerperal sepsis begins as a uterine infection following childbirth or abortion; it can progress to peritonitis or septicemia.
2. *Streptococcus pyogenes* is the most frequent cause.
3. Dr. Oliver Wendell Holmes and Ignaz Semmelweiss demonstrated that puerperal sepsis was transmitted by the hands and instruments of midwives and physicians.
4. Puerperal sepsis is now uncommon because of modern hygienic techniques and antibiotics.
5. Anaerobic bacteria, including *Bacteroides* and *Clostridium,* can cause infections after improperly performed abortions.

Bacterial Endocarditis (p. 579)

1. The outer layer of the heart is the pericardium. The inner layer is the endocardium.
2. Subacute bacterial endocarditis is usually caused by alpha-hemolytic streptococci, although other gram-positive cocci can be involved.
3. The infection arises from a focus of infection such as a tooth extraction.
4. Preexisting heart abnormalities are predisposing factors.
5. Symptoms include fever, anemia, and heart murmur.

6. Acute bacterial endocarditis is usually caused by *Staphylococcus aureus* or *Streptococcus pneumoniae.*
7. The bacteria cause rapid destruction of heart valves.
8. *Streptococcus* can also cause pericarditis.
9. Laboratory diagnosis is based on isolation and identification of the bacteria from blood.

Rheumatic Fever (pp. 579–580)

1. Rheumatic fever is an autoimmune complication of group A beta-hemolytic streptococcal infections.
2. Rheumatic fever is expressed as arthritis or inflammation of the heart, and can result in permanent heart damage.
3. Antibodies against *S. pyogenes* react with streptococcal antigens deposited in joints or heart valves or cross-react with the heart muscle.
4. Rheumatic fever can be a sequel to a streptococcal infection such as streptococcal sore throat. Streptococci might not be present at the time of rheumatic fever.
5. The incidence of rheumatic fever in the United States has declined because of prompt treatment of streptococcal infections.
6. Rheumatic fever is treated with antiinflammatory drugs.

Tularemia (pp. 580–581)

1. Tularemia is caused by *Francisella tularensis*. The reservoir is small wild mammals, especially rabbits.
2. Humans contract tularemia by handling diseased carcasses, eating undercooked meat of diseased animals, and by being bitten by certain vectors (e.g., deer flies).
3. Symptoms include ulceration at the site of entry followed by septicemia and pneumonia.
4. *F. tularensis* is difficult to culture in vitro. In vivo, it is resistant to phagocytosis.
5. Laboratory diagnosis is based on a slide agglutination test on isolated bacteria.

Brucellosis (Undulant Fever) (pp. 581–582)

1. Brucellosis can be caused by *Brucella abortus, B. melitensis, B. suis,* and *B. canis.*
2. Symptoms include malaise and fever that spikes each evening (undulant fever).
3. The bacteria enter through minute breaks in the mucosa or skin, reproduce in macrophages, and spread via lymphatics to liver, spleen, or bone marrow.
4. A vaccine for cattle is available.
5. Diagnosis is based on isolation of the bacteria.

Anthrax (pp. 582–584)

1. *Bacillus anthracis* causes anthrax. In soil, endospores can survive up to 60 years.
2. Grazing animals acquire an infection after ingesting the endospores.
3. Humans contract anthrax by handling hides from infected animals. The bacteria enter through cuts in the skin or through the respiratory tract.
4. Entry through the skin results in a malignant pustule that can progress to septicemia. Entry through the respiratory tract can result in pneumonia.
5. A vaccine for cattle is available.
6. Diagnosis is based on isolation and identification of the bacteria.

Listeriosis (p. 584)

1. *Listeria monocytogenes* causes meningitis, encephalitis, and monocytosis.
2. It is contracted by contact with infected animals and ingestion of contaminated milk.
3. Pasteurization of milk will kill *Listeria.*

Gangrene (pp. 584–585)

1. Soft tissue's dying due to ischemia is called gangrene.
2. Microorganisms grow on nutrients released from gangrenous cells.
3. Gangrene is especially susceptible to the growth of anaerobic bacteria such as *Clostridium perfringens*, the causative agent of gas gangrene.
4. Gas gangrene can be prevented by prompt cleansing of wounds and antibiotic therapy.
5. *C. perfringens* can invade the uterine wall during improperly performed abortions.
6. Debridement, hyperbaric chambers, and amputation are used to treat gas gangrene.

SYSTEMIC DISEASES CAUSED BY ANIMAL BITES AND SCRATCHES (pp. 585–587)

1. Septicemia is caused by *Pasteurella multocida* introduced by the bite of a dog or cat.
2. The causative agent of cat scratch fever is unknown.
3. Anaerobic bacteria such as *Clostridium, Bacteroides,* and *Fusobacterium* infect deep animal bites.
4. Rat bite fever is caused by *Streptobacillus* or *Spirillum.*
5. Arthropods are animals with jointed appendages and include mites, ticks, and insects.
6. Arthropods can act as vectors for pathogenic microorganisms.

Plague (pp. 585–587)

1. Plague is caused by *Yersinia pestis.* The vector is usually the rat flea (*Xenopsylla cheopis*).
2. Reservoirs for plague include European rats and North American rodents.
3. Symptoms of plague include bruises on the skin and enlarged lymph nodes (buboes).
4. The bacteria can enter the lungs and cause pneumonic plague.
5. Laboratory diagnosis is based on isolation and identification of the bacteria.
6. Antibiotics are effective to treat plague, but they must be administered promptly after exposure to the disease.
7. Control of the rat population is an effective deterrent to the spread and incidence of plague.

Relapsing Fever (p. 587)

1. Relapsing fever is caused by *Borrelia* species and transmitted by soft ticks (*Ornithodorus*).
2. The reservoir for the disease is rodents.
3. Symptoms include fever, jaundice, and rose-colored spots. Symptoms recur three or four times after apparent recovery.
4. Laboratory diagnosis is based on the presence of spirochetes in the patient's blood.

Lyme Disease (p. 587)

1. Lyme disease is transmitted by a tick (*Ixodes*) and is probably caused by *Borrelia* species.

RICKETTSIAL DISEASES OF THE CARDIOVASCULAR AND LYMPHATIC SYSTEMS (pp. 587–590)

1. Rickettsias are obligate intracellular parasites of eucaryotic cells.

Typhus (pp. 587–589)

1. The human body louse *Pediculus vestimenti* transmits *Rickettsia prowazekii* in its feces, which are deposited while the louse is feeding.
2. Epidemic typhus is prevalent in crowded and unsanitary living conditions that allow the proliferation of lice.
3. The symptoms of typhus are rash, prolonged high fever, and stupor.
4. Endemic murine typhus is a less severe disease caused by *Rickettsia typhi* and transmitted from rodents to humans by the rat flea.
5. Tetracyclines and chloramphenicol are used to treat typhus.
6. Laboratory diagnosis is based on the Weil–Felix test.

Rocky Mountain Spotted Fever (pp. 589–590)

1. *Rickettsia rickettsii* is a parasite of ticks (*Dermacentor* species) in the southeastern United States, Appalachia, and the Rocky Mountain states.
2. The rickettsia may be transmitted to humans, in whom it causes Rocky Mountain spotted fever.
3. Chloramphenicol and tetracyclines are effective to treat the disease.
4. The complement fixation test is used for laboratory diagnosis.

VIRAL DISEASES OF THE CARDIOVASCULAR AND LYMPHATIC SYSTEMS (pp. 590–592)

Myocarditis (p. 590)

1. Inflammation of the heart muscle (myocardium) is called myocarditis.
2. Coxsackievirus (enterovirus) is the most common cause. It reaches the heart from a respiratory or gastrointestinal infection.

Infectious Mononucleosis (p. 590–591)

1. Infectious mononucleosis is caused by the Epstein–Barr (EB) virus (herpesvirus).
2. The virus multiplies in lymphatic tissue and causes the proliferation of atypical lymphocytes.
3. The disease is transmitted by ingestion of saliva from infected individuals.
4. Diagnosis is made with a serological test for heterophil antibodies and indirect fluorescent-antibody technique.

Burkitt's Lymphoma (p. 591)

1. EB virus is implicated in Burkitt's lymphoma and nasopharyngeal cancer.
2. EB virus is potentially oncogenic.

Yellow Fever (p. 592)

1. Yellow fever is caused by an arbovirus (yellow fever virus). The vector is the mosquito *Aedes aegypti.*
2. The virus multiplies in lymph nodes and then disseminates to liver, spleen, kidney, and heart.
3. Symptoms include fever, chills, headache, nausea, and jaundice.
4. Diagnosis is based on the presence of virus-neutralizing antibodies in the host.
5. No treatment is available. An attenuated, live viral vaccine is available.

Dengue (p. 592)

1. Dengue is caused by an arbovirus (dengue fever virus) and is transmitted by the mosquito *Aedes aegypti.*
2. Symptoms are fever, muscle and joint pain, and rash.
3. Dengue hemorrhagic fever involving bleeding from the skin, gums, and gastrointestinal tract is a more serious form of the disease.
4. Laboratory identification is based on the presence of antibodies.

Viral Hemorrhagic Fevers (p. 592)

1. Viral hemorrhagic fevers are highly contagious, often fatal diseases.

PROTOZOAN AND HELMINTHIC DISEASES OF THE CARDIOVASCULAR AND LYMPHATIC SYSTEMS (pp. 593–600)

Toxoplasmosis (pp. 593–595)

1. Toxoplasmosis is caused by the sporozoan *Toxoplasma gondii.*
2. *T. gondii* undergoes sexual reproduction in the intestinal tract of domestic cats, and oocysts are eliminated in cat feces.
3. Animals that ingest contaminated food or water may contract toxoplasmosis.
4. Humans contract the infection by ingesting undercooked meat from an infected animal or inhaling dried cat feces.
5. Subclinical infections are probably common because the disease symptoms are rather mild.
6. Congenital infections can occur. Symptoms include convulsions, brain damage, blindness, and death.
7. A rising antibody titer in a pregnant woman might indicate the need for a therapeutic abortion.

American Trypanosomiasis (Chagas' Disease) (p. 595)

1. *Trypanosoma cruzi* causes Chagas' disease. The reservoir includes many wild animals. The vector is the "kissing bug."
2. The infection begins in the lymph nodes and disseminates via the bloodstream to other organs.
3. Observation of the trypanosomes in blood confirms diagnosis.

Malaria (pp. 595–598)

1. The symptoms of malaria are chills, fever, vomiting, and headache, and occur in one- to three-day cycles.
2. Malaria is transmitted by *Anopheles* species. The causative agent is any one of four species of *Plasmodium.*
3. Sporozoites reproduce in the liver and release merozoites into the bloodstream.
4. Laboratory diagnosis is based on microscopic observation of merozoites in red blood cells.
5. A vaccine is being developed.
6. New drugs are being developed as the protozoans develop resistance to quinine and chloroquine.

Schistosomiasis (pp. 598–599)

1. Species of the blood fluke *Schistosoma* cause schistosomiasis.
2. Eggs eliminated with feces hatch into larvae that infect the intermediate host. Free-swimming cercariae are released and penetrate the skin of a human.
3. The adult flukes live in the veins of the liver or urinary bladder in humans.
4. Adult flukes reproduce, and eggs are excreted or remain in the host.
5. Symptoms are due to the host's defense to eggs that remain in the body.
6. Swimmer's itch is a cutaneous allergic reaction to cercariae that penetrate the skin. The definitive host for this fluke is wildfowl.
7. Observation of eggs or flukes in feces, skin tests, or indirect serological tests may be used for diagnosis.
8. Chemotherapy is used to treat the disease; sanitation and snail eradication are used to prevent the disease.

STUDY QUESTIONS

REVIEW

1. What are the symptoms of septicemia?
2. How can septicemia result from a single focus of infection such as an abscess?
3. Differentiate between endocarditis, myocarditis, and pericarditis. How are these infections contracted?
4. Complete the following table.

Disease	Frequent Causative Agent	Predisposing Condition(s)
Puerperal sepsis		
Subacute bacterial endocarditis		
Acute bacterial endocarditis		
Myocarditis		

5. Describe the probable cause of rheumatic fever. How is rheumatic fever treated? How is it prevented?
6. Fill in this table.

Disease	Causative Agent	Method of Transmission	Reservoir	Symptoms	Prevention
Tularemia					
Brucellosis					
Anthrax					
Listeriosis					
Rocky Mountain spotted fever					

7. Plot the temperature of a patient with brucellosis for a one-week period.
8. List four bacterial infections that might result from animal bites.
9. Why is *Clostridium perfringens* likely to grow in gangrenous wounds?
10. List the causative agents and methods of transmission of infectious mononucleosis.
11. Cite evidence implicating EB virus in Burkitt's lymphoma. Offer an explanation as to why a large number of people who are infected with EB virus do not get Burkitt's lymphoma or nasopharyngeal cancer.
12. Fill in the following table.

Disease	Causative Agent	Vector	Symptoms	Treatment
Malaria				
Yellow fever				
Dengue				
Relapsing fever				

13. What is the most effective control measure for mosquito-borne diseases?
14. Provide the following information on plague: causative agent; vector; U.S. reservoir; control; treatment; prognosis (probable outcome).
15. Differentiate between the transmission and symptoms of bubonic plague and pneumonic plague.
16. List the causative agent, method of transmission, and reservoir for schistosomiasis, toxoplasmosis, and American trypanosomiasis. Which disease are you most likely to get in the United States? Where are the other diseases endemic?
17. What is infectious mononucleosis? Viral hemorrhagic fever?

CHALLENGE

1. Indirect fluorescent-antibody tests on the serum of three 25-year-old women, each of whom is considering pregnancy, provided the following information.

	Antibody Titer		
Patient	Day 1	Day 5	Day 12
Patient *A*	1:1,024	1:1,024	1:1,024
Patient *B*	1:1,024	1:2,048	1:3,072
Patient *C*	0	0	0

Which of these women may have toxoplasmosis? What advice might be given to each woman with regard to toxoplasmosis?

2. A 19-year-old man went deer hunting. While on the trail, he found a partially dismembered dead rabbit. The hunter picked up the front paws for good luck charms, and gave them to another hunter in the party. The rabbit had been handled with bare hands that were bruised and scratched from the hunter's work as an automobile mechanic. Festering sores on his hands, legs, and knees were noted two days later. What infectious disease do you suspect the hunter has? How would you proceed to prove it?

3. On March 30, a 35-year-old veterinarian experienced fever, chills, and vomiting. On March 31, he was hospitalized with diarrhea, left axillary bubo, and secondary bilateral pneumonia. On March 27, he had treated a cat who had labored respiration; an X-ray revealed pulmonary infiltrates. The cat died on March 28 and was disposed of. Chloramphenicol was administered to the veterinarian. On April 10, his temperature returned to normal and on April 20, he was released from the hospital. Sixty human contacts were given tetracycline. Identify the incubation and prodromal periods for this case. Explain why the 60 contacts were treated. What was the etiologic agent? How would you identify the agent?

FURTHER READING

Freimer, E. H., and M. McCarty. December 1965. Rheumatic fever. *Scientific American* 213:66–74. Discusses the link between streptococcal infection and rheumatic disease.

Gregg, C. T. 1978. *Plague!* New York: Charles Scribner's Sons. An interesting and factual account of plague in the twentieth century.

Harrison, G. 1978. *Mosquitoes, Malaria and Man.* New York: E. P. Dutton. An excellent account of the fight against malaria beginning in 1880.

Henle, W., G. Henle, and E. T. Lennette. July 1979. The Epstein–Barr virus. *Scientific American* 241:48–59. A summary of information on one of the most common viruses infecting humans.

Hubbert, W. T., W. F. McCulloch, and P. R. Schurrenberger. 1975. *Diseases Transmitted from Animals to Man*. 6th ed. Springfield, IL: C. C. Thomas. Thorough coverage of bacterial, fungal, parasitic, and viral diseases transmitted to humans by animal bites and vectors.

Kadis, S., T. C. Montie, and S. J. Ajl. March 1969. Plague toxin. *Scientific American* 220:92–100. The method of action of *Yersinia* toxin.

Weiss, E. 1982. The biology of rickettsiae. *Annual Review of Microbiology* 36:345–370. Ecology of the bacteria and the diseases they cause. Discusses the future role of these diseases in the human population.

See also Further Reading for Part Four at the end of Chapter 24.

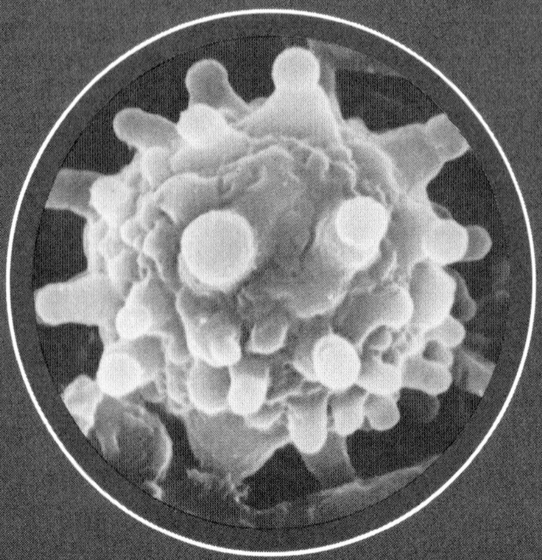

CHAPTER 22

Microbial Diseases of the Respiratory System

OBJECTIVES

After completing this chapter you should be able to

- Describe how microorganisms are prevented from entering the respiratory system.
- List the methods by which infections of the respiratory system are transmitted.
- Characterize the normal flora of the upper and lower respiratory systems.
- Describe the causative agent, symptoms, preferred treatment, and laboratory identification tests for ten bacterial diseases and three viral diseases of the respiratory system.
- Compare these skin tests: Dick test; Schick test; tuberculin skin test.
- Describe vaccines that are available to prevent respiratory system infections.
- List the etiologic agents, methods of transmission, preferred treatments, and laboratory identification tests for several fungal, protozoan, and helminthic diseases of the respiratory system.

The upper respiratory system is in direct contact with the air we breathe—air contaminated with microorganisms—so it is a major portal of entry for pathogens. In fact, respiratory system infections are the most common type of infection—and among the most damaging. Pathogens that enter the body via the respiratory system can sometimes infect other parts of the body, as do the pathogens of measles, mumps, and rubella. Another very serious aspect of respiratory infections is the ease with which they are spread, both by direct contact (such as droplets emitted during sneezing, coughing, or talking) and by fomites. Transmission by droplets is especially common for highly communicable respiratory infections such as pneumonic plague and influenza.

STRUCTURE AND FUNCTION OF THE RESPIRATORY SYSTEM

It is convenient to think of the respiratory system as being composed of two divisions: the upper respiratory system and the lower respiratory system. The **upper respiratory system** consists of the *nose* and *throat* and structures associated with them, including the middle ear and the eustachian tubes (Figure 22–1). Emptying into the nasal cavity are ducts from the sinuses and the nasolacrimal

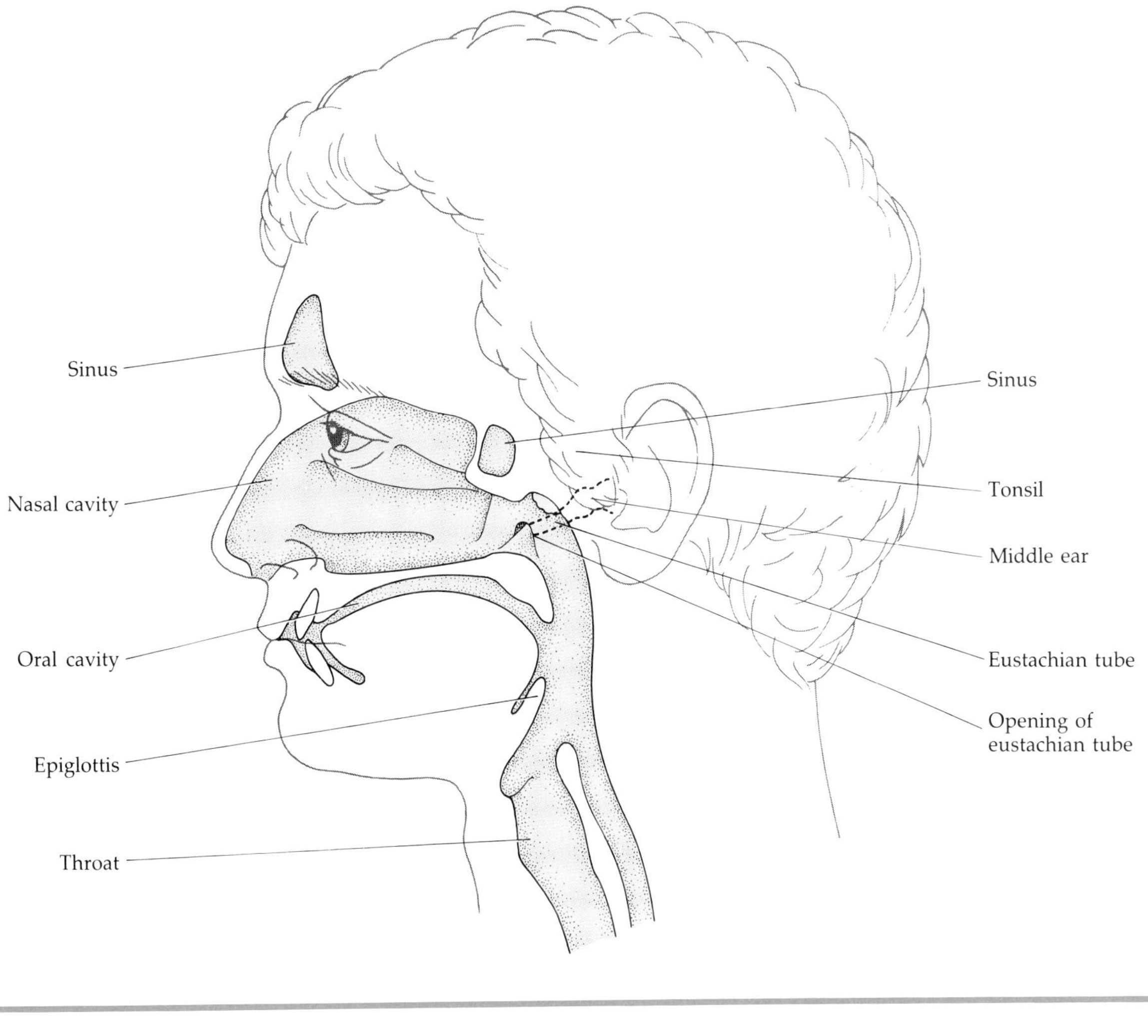

Figure 22–1 Structures of the upper respiratory system.

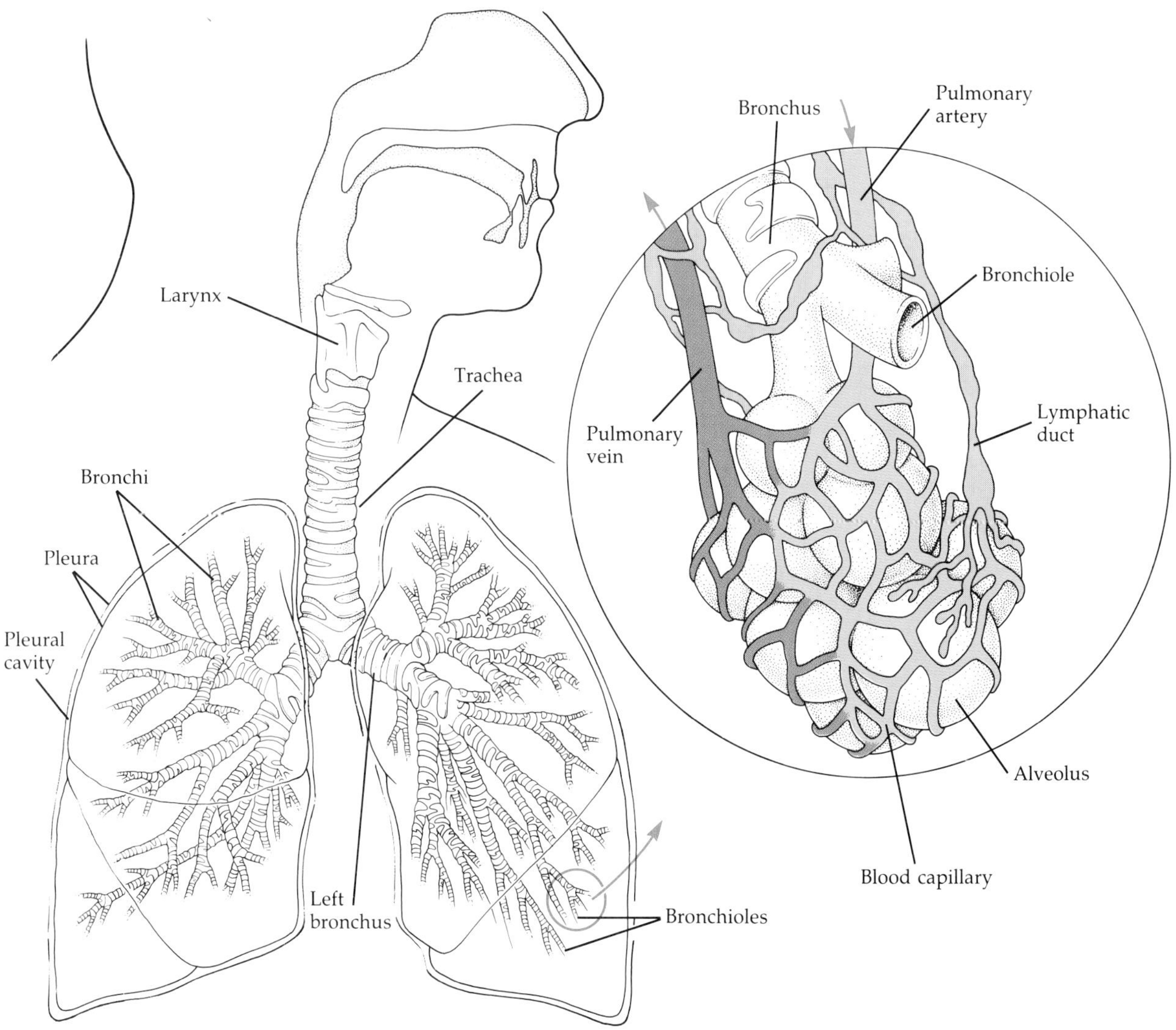

Figure 22–2 Structures of the lower respiratory system.

ducts from the lacrimal apparatus (see Figure 15–2). Emptying into the upper portion of the throat are the eustachian tubes from the middle ear.

The upper respiratory system has several anatomical defenses against airborne pathogens. Coarse hairs in the nose filter large dust particles from the air. And the nose is lined with a mucous membrane that contains numerous mucus-secreting cells and cilia. The upper portion of the throat also contains a ciliated mucous membrane. The mucus moistens inhaled air and traps dust and microorganisms, especially particles larger than 4 to 5 μm. The cilia assist in removing these particles by moving them toward the mouth for elimination. At the junction of the nose and throat are masses of lymphoid tissue, the tonsils and adenoids, that produce immunities to certain infections. Occasionally, however, these tissues become infected and help spread infection to the ears via the eustachian tubes. Because the nose and throat are connected to the sinuses, nasolacrimal apparatus, and middle ear, infections commonly spread from one region to another.

The **lower respiratory system** consists of the *larynx* (voice box), *trachea* (windpipe), *bronchial tubes*, and *alveoli* (Figure 22–2). Alveoli are the air sacs that make up the lung tissue; within them, oxygen and carbon dioxide are exchanged between the lungs and blood. The double-layered membrane around the lungs is the *pleura*. A ciliated mucous membrane lines the lower respiratory system down to the smaller bronchial tubes, and helps prevent microorganisms from reaching the lungs. Particles trapped in the larynx, trachea, and larger bronchial tubes are moved up toward the throat by a ciliary action referred to as the ciliary escalator (see Figure 15–3). The ciliary escalator keeps the trapped particles moving toward the throat. Coughing, sneezing, and clearing the throat speed up the escalator, whereas smoking and air pollution inhibit it. If microorganisms actually reach the lungs, then certain phagocytic cells—the *alveolar macrophages*, or *dust cells*—usually locate, ingest, and destroy most of them. IgA antibodies, in secretions such as respiratory mucus, saliva, and tears, also help protect mucosal surfaces of the respiratory system from many pathogens. Thus, the body has several mechanisms for removing the pathogens that cause airborne infections. But if all these mechanisms fail, then the microorganism wins the host–parasite competition and a respiratory disease results.

NORMAL FLORA OF THE RESPIRATORY SYSTEM

The normal flora of the mouth will be considered in Chapter 23 as part of the gastrointestinal tract. The normal flora of the nasal cavity includes diphtheroids; staphylococci, including *S. aureus;* micrococci; and *Bacillus* species. All are gram-positive. The throat near the nose might contain streptococci, including *Streptococcus pneumoniae,* as well as the gram-negative organisms *Hemophilus influenzae* and *Neisseria meningitidis.* Despite these potentially pathogenic microorganisms' presence as part of the normal flora of the upper respiratory system, epidemics are rare and morbidity is low because of microbial antagonism. Certain microorganisms of the normal flora suppress the growth of other microorganisms by competing with them for nutrients and producing inhibitory substances. For example, alpha-hemolytic streptococci limit the growth of pneumococci in the throat. And although the trachea may contain a few bacteria, the lower respiratory tract is usually sterile because of the normally efficient functioning of the ciliary escalator in the bronchial tubes.

BACTERIAL DISEASES OF THE UPPER RESPIRATORY SYSTEM

As most of us know from personal experience, the respiratory system is the site of many common infections. We will soon discuss **pharyngitis,** or sore throat. When the larynx has such an infection, we suffer from **laryngitis,** which affects our ability to speak. This infection is due to both bacteria such as *H. influenzae* and viruses, often in combination. The organisms that cause pharyngitis also can cause inflamed tonsils, or **tonsillitis.** The tonsils are lymphoid tissue in the rear of the nasal cavity. The nasal sinuses, which are cavities in certain cranial bones that open into the nasal cavity, have a mucous membrane lining that is continuous with that of the nasal cavity. When a sinus becomes infected with organisms such as *S. pneumoniae* or *H. influenzae,* the mucous membranes become inflamed and there is a heavy nasal discharge of mucus. This condition is called **sinusitis.** If the opening by which the mucus leaves the sinus becomes blocked, internal pressure can cause pain or a sinus headache.

Probably the most threatening infectious disease of the upper respiratory system is **epiglottitis,** inflammation of the epiglottis. The epiglottis is a flaplike structure of cartilage that prevents ingested material from entering the larynx (see Figure 22–1). Inflammation of the epiglottis is a rapidly developing disease that results in death within a few hours and is caused by opportunistic pathogens such as *H. influenzae* type b.

The lower respiratory tract can also be infected by many of the same bacteria and viruses that infect the upper respiratory tract. As the bronchi (see Figure 22–2) become involved, **bronchitis** or **bronchiolitis** develops. *Mycoplasma pneumoniae* and a number of common respiratory viruses are suspected causes. In infants the most common agent is probably the respiratory syncytial virus that is discussed later in the chapter. Diseases such as **whooping cough** are also a form of bronchitis. A severe complication of bronchitis is **pneumonia,** in which the pulmonary aveoli become involved. These often interrelated diseases are sometimes lumped under the term **croup.** Except for epiglottitis they are almost always *self-limiting*—meaning that recovery will usually occur even without medical intervention.

Streptococcal Pharyngitis ("Strep" Throat)

Streptococcal pharyngitis ("strep" throat) is an upper respiratory infection caused by beta-hemolytic group A streptococci. They are gram-positive bacteria. The most common of these is *Streptococcus pyogenes,* the same bacterium responsible for many skin and soft tissue infections such as impetigo, erysipelas, and acute bacterial endocarditis. Streptococcal pharyngitis is characterized by inflammation of the mucous membrane of the throat (pharyngitis) and fever. Frequently, the tonsils become inflamed (tonsillitis) and the lymph nodes in the neck become enlarged and tender. Another frequent complication is infection of the middle ear, called otitis media (to be discussed shortly).

In laboratory diagnosis, swabs of bacteria from the inflamed mucous membrane are streaked onto blood agar so that it can be determined which hemolysins are produced (see Microview 1–2A). Without analysis of a throat culture, streptococcal pharyngitis cannot be distinguished from pharyngitis of other causes (mostly viral). Probably no more than half of "strep" throats are actually streptococcal in origin.

Nearly all beta-hemolytic group A streptococci are sensitive to penicillin, the drug of choice. Erythromycin is also effective. Unfortunately, many strains are now resistant to tetracyclines. Treatment of streptococcal pharyngitis is important to prevent complications such as rheumatic fever and glomerulonephritis, both of which are caused by an immune response to streptococcal infections (see Chapter 21).

There are more than 55 serological types of group A streptococci. And because resistance to streptococcal diseases is type-specific, a person who has recovered from infection by one type is not necessarily immune to infection by another type. A general immunity to streptococcal infections does not usually occur.

Strep throat is now most commonly transmitted by respiratory secretions, but epidemics spread by unpasteurized milk were once frequent.

Scarlet Fever

When the *S. pyogenes* strain causing streptococcal pharyngitis produces an *erythrogenic* (reddening) *toxin,* the resulting infection is called **scarlet fever.** When the strains produce this toxin, they have been lysogenized by a bacteriophage. As you may remember, this means that the genetic information of a bacterial virus has been incorporated into the chromosome of the bacterium, so the characteristics of the bacterium have been altered. The toxin causes a pinkish-red skin rash and a high fever; these symptoms are probably the skin's generalized cutaneous hypersensitivity reaction to the circulating toxin. The tongue has a spotted, strawberrylike appearance, and then, as it loses its upper membrane, becomes very red and enlarged. As the disease runs its course, the affected skin frequently peels off as if sunburned. Complications such as deafness can occur.

Scarlet fever seems to increase and decline in severity and frequency over time and in different locations, but the disease generally has been declining in recent years. It is a communicable disease spread mainly by inhalation of infective droplets or dust contaminated by an infected person. Classically, scarlet fever has been considered to be associated with streptococcal pharyngitis, but it might accompany a streptococcal skin infection.

Penicillin is effective in treating the original streptococcal infection. Immunity is developed to the toxin produced by the particular *streptococcus* that was contracted, but there are many types of streptococci, and immunity is specific to each type. Therefore, recovery from scarlet fever does not ensure immunity to streptococcal infections in general. Immunity to scarlet fever can be measured by the *Dick test,* in which erythrogenic toxin is injected into the skin. Local reddening, reaching a maximum at about 24 hours, indicates that the individual is not immune to scarlet fever.

Diphtheria

Another bacterial infection of the upper respiratory system is **diphtheria,** once a major cause of death in children. The disease begins with a sore throat and fever, followed by general malaise and swelling of the neck. The pathogenesis of diphtheria is better understood than that of most other infectious diseases. The organism responsible is *Corynebacterium diphtheriae* (kô-rī-nē-bak-ti'rē-um dif-thi'rē-ä), which is a gram-positive, non-endospore-forming, somewhat pleomorphic rod. The dividing cells are often observed to fold together to form V- and Y-shaped figures resembling Chinese letters (Figure 22–3).

Often found in the throats of symptomless carriers, *C. diphtheriae* has adapted to a generally immunized population. The bacterium is well adapted to airborne transmission and is very resistant to drying. The bacteria are usually spread by respiratory transmission, and once established in the upper respiratory system of a susceptible person, the bacteria rapidly increase in number. Although they do not invade tissues, the bacteria that have been lysogenized by a phage can produce a powerful exotoxin (see Chapter 12). Circulating in the bloodstream, the toxin interferes with protein synthesis. Only 0.01 mg of this highly virulent toxin is enough to kill a 91-kg (200-pound) person. Thus, if antitoxin therapy is to be effective, it must be administered before the toxin enters the tissue cells. When organs such as the heart or kidneys are affected by the toxin, the disease can rapidly be fatal. In other cases, the nerves can be involved, and partial paralysis results. A characteristic of diphtheria is a tough, grayish membrane that forms in the throat. Containing fibrin and dead human and bacterial cells (*diphtheria,* meaning "leather"), this membrane forms in response to the infection, and can totally block the passage of air to the lungs.

Laboratory diagnosis can usually be made if the organisms are isolated from the primary lesion. Typical colonies grow on special media (grayish colonies on a Loeffler slant, black colonies on tellurite agar). Once isolated, the microorganisms are tested for toxigenicity by the guinea pig virulence test or the gel diffusion test. In the *guinea pig test,* the microorganisms are injected into one side of a shaved guinea pig. After four hours, the guinea pig

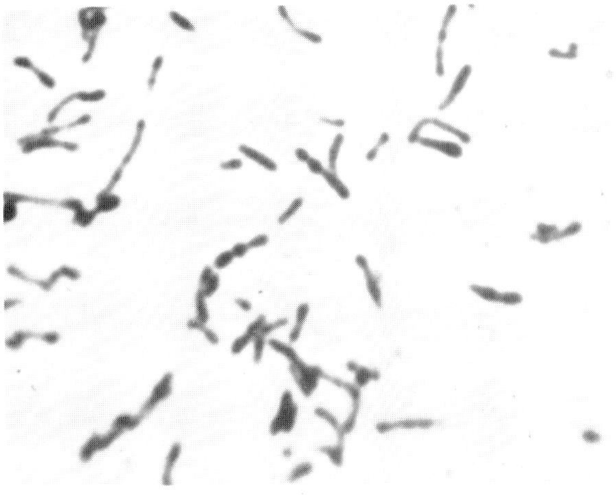

Figure 22–3 The organism causing diphtheria, *Corynebacterium diphtheriae.* Note the typical clubbed shape and V and Y arrangements.

is injected with antitoxin. Thirty minutes later, microorganisms are injected on the opposite side. If the microorganisms are toxigenic, a characteristic lesion will form between 48 and 72 hours at the site injected before the antitoxin was administered. In the *gel diffusion test,* a strip of filter paper containing diphtheria antitoxin is placed on agar containing antitoxin-free calf serum. Then the microorganisms are streaked at a right angle to the filter paper. If the microorganisms are toxigenic, a visible line of antigen–antibody precipitate will form.

Once diagnosis is made, antitoxin therapy must be administered without delay, for once the diphtheria toxin enters susceptible cells, it is no longer neutralized by antitoxin. Even though antibiotics such as penicillin, tetracyclines, and erythromycin control the growth of the bacteria, they do not neutralize the diphtheria toxin. Thus to treat diphtheria, they should only be used in conjunction with antitoxin.

Part of the normal immunization program in the United States is the *DPT vaccine,* which is effective against diphtheria, pertussis (whooping cough), and tetanus. The "D" stands for diphtheria toxoid, an inactivated toxin that causes the body to produce antibodies against the diphtheria toxin. Immunity to diphtheria can be measured by the *Schick test,* in which a standardized amount of toxin is injected into the skin. The absence of a skin reaction shows that circulating antibodies have neutral-

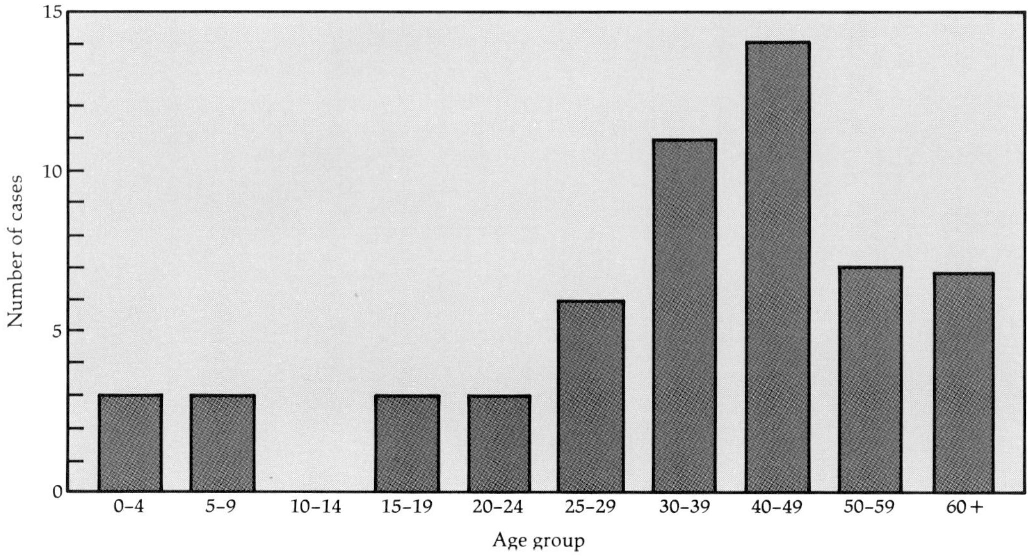

Figure 22–4 Age distribution of diphtheria cases in the United States in a typical year. Note that most cases now occur in the adult population, many of whom were not immunized in childhood. Many of these cases are of cutaneous diphtheria.

ized the toxin, and thus indicates that the subject is immune.

The number of diphtheria cases reported in the United States each year is presently well under 100, but the death rate for respiratory cases is still 5 to 10%, mostly among the elderly or the very young. In young children the disease occurs mainly in groups that, for religious or other reasons, have not been immunized. (See Figure 22–4 and the box in Chapter 14.)

In the past, diphtheria was spread mainly to healthy carriers by droplet infection. Respiratory cases have been known to arise from contact with cutaneous diphtheria (discussed below). With the advent of immunization, however, the carrier rate has decreased dramatically. Because humans are the only important reservoir for the diphtheria microorganism, total eradication is theoretically possible, but only if worldwide immunization becomes a reality.

Cutaneous Diphtheria

Diphtheria is also expressed as what is called **cutaneous diphtheria.** In this form of the disease, there is minimal systemic circulation of the toxin.

In cutaneous infections, *C. diphtheriae* infects the skin, usually at a wound or similar skin lesion and causes slow-healing ulcerations. It is most common among American Indians and among adults in low socioeconomic circumstances.

Cutaneous diphtheria is fairly common in tropical countries. In the United States, this disease is responsible for most of the reported cases of diphtheria in persons over 30 years old (Figure 22–4). When diphtheria was more common, repeated subclinical infections reinforced the immunity, which weakens with time. Many adults now lack immunity because routine immunization was either unavailable or less common during their childhood.

Otitis Media

One of the more uncomfortable complications of the common cold (described shortly) or of any infection of the nose or throat, is infection of the inner ear, leading to earache, or **otitis media.** The inner ear can be infected directly by contaminated water from swimming pools or by some severe trauma such as injury to the eardrum or certain skull fractures. The infecting microorganisms cause the formation of pus, which builds up pressure against the eardrum, and causes it to become inflamed and painful. The condition is more common in early childhood, probably because the eustachian tube connecting the middle ear to the throat is smaller then and more easily blocked by infection (see Figure 22–1). Enlarged adenoids might be a contributing factor for children with an unusually high frequency of infection.

A number of bacteria can be involved. The bacteria most commonly isolated are the gram-positive *Staphylococcus aureus,* alpha-hemolytic *Streptococcus pneumoniae,* assorted beta-hemolytic streptococci, or the gram-negative *Hemophilus influenzae.* Antibiotics such as penicillin and erythromycin are useful in therapy.

VIRAL DISEASE OF THE UPPER RESPIRATORY SYSTEM

Common Cold (Coryza)

There are a number of different viruses involved in the etiology of the **common cold (coryza).** About 30% of all colds are caused by *rhinoviruses. Coronaviruses* probably cause another 15 or 20%. About 10% of all colds are caused by one of an assortment of *adenoviruses, parainfluenza viruses,* and *respiratory syncytial viruses. Coxsackieviruses* and *echoviruses,* considered to be primarily enteric viruses, are known to cause summer colds with fever. No agent has ever been identified as the cause of about 40% of our colds.

Altogether, there are probably more than 200 agents that cause colds. We tend to accumulate immunities against cold viruses during our lifetimes, which may be a reason why older persons tend to get few colds. It has been shown that young children have three or four colds per year, but adults at age 60 have less than one cold per year. Immunity is based on the ratio of IgA antibodies to single serotypes, and has a reasonably high short-term effectiveness. Isolated populations may develop a group immunity, and their colds disappear until a new set of viruses is introduced. Although these immunities can be effective, a vaccine is unlikely. There are at least 113 serotypes of rhinoviruses alone, and a vaccine effective against so many different agents does not seem practical with present technology.

The symptoms of the common cold are familiar to all of us. They include sneezing, excessive nasal secretion, and congestion. (In ancient times one school of medical thought believed that the nasal discharges were waste products from the brain.) The infection can easily spread from the throat to the sinus cavities, the lower respiratory system, and the middle ear, leading to complications of laryngitis and otitis media. The uncomplicated cold usually is not accompanied by fever.

It is popularly thought that colds are spread by coughing and sneezing. However, research indicates that by far the most effective means of spreading colds is by the hands: they contact a contaminated surface then deposit the virus into the nose. A single rhinovirus deposited on the nasal mucosa is sufficient to cause a cold. The better news is that when research required healthy volunteers to kiss cold-sufferers for 60 to 90 seconds only 8% of the volunteers come down with colds.

The incidence of colds definitely increases in frequency with cold weather. The rhinoviruses prefer a temperature slightly below that of normal body temperature—such as might be found in the upper respiratory system, which is open to the outside environment. However, colds will die out in isolated populations even in extremely cold climates as immunity develops in the population. And colds are found in warm climates with little variation in temperature. No one knows exactly why the number of colds seems to increase with colder weather in temperate zones. It is not known if closer indoor contact promotes epidemic-type transmission or if physiologic changes increase susceptibility.

Because colds are caused by viruses, antibiotics are of no use in treatment. The common cold will usually run its course to recovery in about a week's time. Recovery time is not affected by over-the-counter drugs presently available, although such drugs may lessen the severity of certain signs and symptoms. Not yet generally available, a new drug may do more than relieve symptoms (see box).

BACTERIAL DISEASES OF THE LOWER RESPIRATORY SYSTEM

Whooping Cough (Pertussis)

Infection by the bacterium *Bordetella pertussis* (bôr-de-tel'lä pėr-tus'sis) results in **Whooping cough**

Microbiology in the News

A Cure for the Common Cold?

An experimental drug has shown broad antiviral effects on a large number of the picornaviruses against which it has been tested. The tests, say scientists, while still preliminary, could yield the first drug capable of treating this family of disease agents, whose members can cause a wide variety of illnesses ranging in seriousness from polio, hepatitis A, viral meningitis, and neonatal sepsis (a generally fatal disease affecting newborns) to those mild rhinovirus infections responsible for half of all common colds.

Based on the efficacy demonstrated thus far in animals and cell-culture experiments, the drug—known as WIN 512,711— "holds the potential for curing the common cold" and a broad spectrum of other previously untreatable infections, says Guy Diana, group leader for medicinal chemistry at the Sterling-Winthrop Research Institute in Rensselaer, N.Y. Diana reported on relationships between the drug's chemical structure and its antiviral activity in a presentation at the American Chemical Society's 189th national meeting in Miami.

To reproduce, an infectious virus first adheres to a cell's membrane. Then it penetrates the cell, shedding its own outer protein covering and releasing its store of genetic material. If allowed to replicate and reencapsulate, this genetic material would form new viruses that could infect other cells. The new drug halts the reproduction and spread of a virus by preventing the initial uncoating of the protein shell that encapsulates its genetic material.

By halting viral reproduction so early in the virus-uncoating phase, the drug's mechanism of action may be unique, says Mark A. McKinlay, a virologist on the project. Exactly why it selectively binds with viruses and prevents their uncoating, however, is not fully understood.

Used therapeutically, the drug halted development of paralysis in mice that had recently been infected with polio-2 or a paralytic ECHO-9 picornavirus. Used prophylactically, the drug prevented viral infection altogether when given to mice immediately before and again immediately after their exposure to various picornaviruses.

Since there are no animal models other than chimpanzees for studying rhinovirus infections, McKinlay says tests of the drug's efficacy against these cold viruses were performed using several types of

(pertussis). *B. pertussis* is a small, nonmotile, gram-negative coccobacillus that forms a capsule when virulent. It is an obligate aerobe.

Whooping cough, which is primarily a childhood disease, can be quite severe. The initial stage, which is called the *catarrhal stage,* resembles a common cold. Prolonged sieges of coughing characterize the second stage, or *paroxysmal stage.* The disease is transmitted by inhaling pathogens expelled by the coughing of infected patients. The bacterium grows in dense masses in the trachea and larger bronchi and causes the production of thick mucus that impedes ciliary action (Figure 22–5). (As noted earlier, the trachea and larger bronchi are usually kept clear of mucus by ciliary action.) The infected person desperately attempts to cough up these mucus accumulations. The violence of the coughing can actually result in broken ribs. Gasping for air between coughs causes a whoop sound, the reason for the name of the disease. Coughing episodes occur several times a day over an interval of one to six weeks. The *convalescence stage,* the third stage, may last for months. The disease is of unusually long duration for a respiratory infection.

B. pertussis produces an endotoxin as part of its cell wall; the endotoxin resembles that of other pathogenic gram-negative bacteria. In the cytoplasm, the bacteria produce an exotoxin that is apparently released along with the endotoxin when the cell autolyses upon death. The precise role of

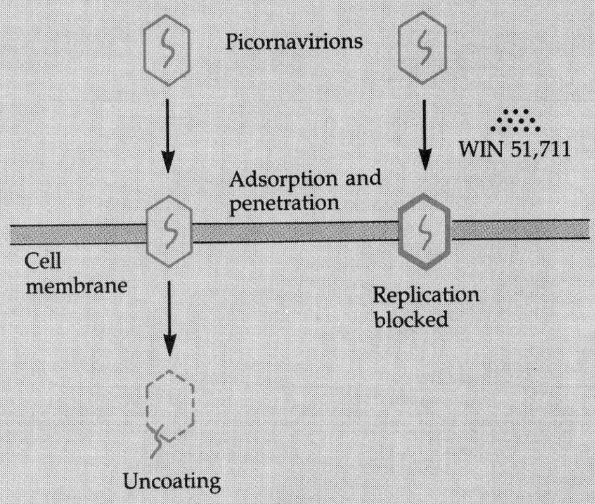

cultured human cells. To date, WIN 512,711 appears effective against 34 of the 40 rhinoviruses tested. Testing isn't over, however; there are more than 120 rhinoviruses capable of causing colds.

So far, McKinlay says, WIN 512,711 is somewhat more effective against enteroviruses—the picornavirus family's major class of nonrhinoviruses causing human disease—than against cold viruses. Last week Diana reported data on a close analog of the drug that showed the opposite effect—slightly better action against rhinoviruses. This leads to speculation that the end product of the research may not be a single antipicornavirus drug but instead a group of related compounds, each of whose activity has been optimized for a target class of the viruses.

Initial toxicity tests suggest the new drug is safe. The next step is clinical trials. A critical question to be answered there will be whether, by the time symptoms appear, these viral diseases are already too advanced to treat.

the toxins in pathogenesis has yet to be determined. The organism does not invade tissues. Very high white blood cell counts (sometimes six to eight times normal levels) are frequently seen in persons with pertussis.

For diagnosis, the organism is cultured from a throat swab inserted through the nose on a thin wire and held in the throat while the patient coughs. *B. pertussis* forms characteristically small, glistening, white colonies on Bordet–Gengou medium. A fluorescent-antibody test or agglutination by antiserum is used for confirmation. Mild cases of whooping cough require no specific treatment. Severe cases, especially in infants, are usually treated with erythromycin. Tetracyclines and chloramphenicol can be used. Although antibiotics might not result in rapid improvement, the drugs do render the patient noninfectious. If secondary pneumonia develops as a complication, antibiotic therapy must be continued.

After recovery, immunity is good. A vaccine prepared from heat-inactivated whole bacteria (the "P" in DPT vaccine stands for *pertussis*) is a part of the regular immunization schedule for children. Vaccination is usually recommended at about two months of age, when the titer of maternal antibodies in the baby's circulation is dropping and the child's own antibodies have not yet replaced them.

There has been considerable concern about the safety of the pertussis vaccine. It has a higher rate

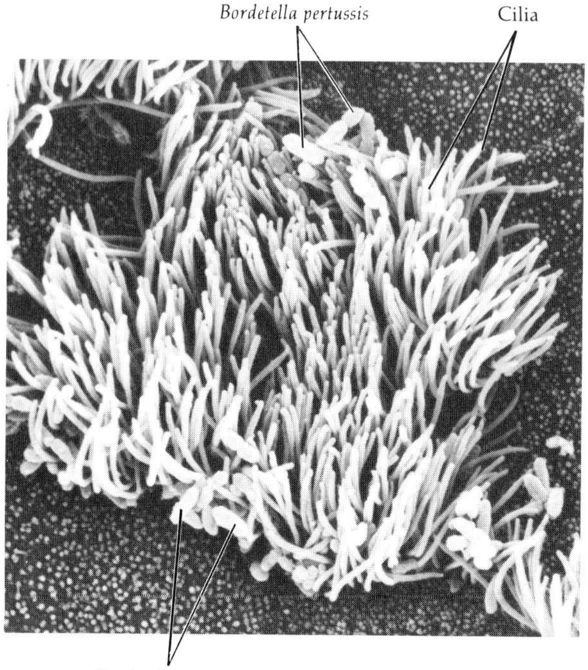

Figure 22–5 Scanning electron micrograph of the whooping cough pathogen, *Bordetella pertussis*, growing among the cilia of respiratory epithelial cells and blocking their action (×3000).

of adverse effects than most vaccines, although the risk of permanent neurological damage or death is not great. However, it lowers the annual number of cases in the United States from more than 250,000 per year to less than 2,000. In Japan and England, adverse publicity has led to widespread avoidance of pertussis vaccination. The result in Japan was 35,000 cases among children, of whom 118 died; and in England 65,000 cases with 14 deaths. Authorities point to these figures as indications that vaccination should continue while research on better vaccines is intensified. Newer vaccines, at least one of which is being tested on a broad scale in Japan, are not whole-cell vaccines but are directed at purified fractions of cells and toxins.

There are presently fewer than 2000 cases of whooping cough each year in the United States, and the number of deaths is usually fewer than 10 per year. Because infants are less capable of coping with the effort of coughing to maintain an airway, they have a relatively high mortality rate.

Tuberculosis

Tuberculosis is an infectious disease caused by the bacterium *Mycobacterium tuberculosis* (mī-kō-bak-ti′rē-um tü-ber-kū-lō′sis). The microorganism is a slender rod and an obligate aerobe. The rods are slow growing (20-hour generation time), sometimes form filaments, and tend to grow in clumps. On the surface of liquid media, their growth appears moldlike, which suggested the genus name *Mycobacterium* (from the Greek mýkēs, meaning "fungus") (see Microview 1–2E). These bacteria are relatively resistant to normal staining procedures. When stained by the Ziehl–Neelson technique of steaming in carbolfuchsin dye, they cannot be decolorized with a mixture of acid and alcohol and are therefore classified as *acid-fast.* This characteristic reflects the unusual composition of the cell wall, which contains large amounts of lipid materials. These lipids might also be responsible for the mycobacteria's resistance to environmental stresses such as drying. In fact, these bacteria can survive for weeks in dried sputum and are very resistant to sunlight and to chemical antimicrobials used as antiseptics and disinfectants. Therefore, particular care must be taken in handling possibly infected material.

Tuberculosis is characterized by the formation of nodules in the infected tissue. The lungs are most commonly infected; the pathogen is usually acquired by inhalation. The disease can also be transmitted via the gastrointestinal tract or by direct contact with infected material. Once the bacteria reach the lungs, the body responds to their presence by walling off the organisms; a small nodule called a *tubercle* forms in two to four weeks. Bacteria can survive in tubercles indefinitely. The tubercle, in time, might undergo necrosis, in which it changes to cheeselike consistency called a *caseous lesion.* Caseous lesions heal by scar formation and calcification and are then referred to as *Ghon complexes,* which can be recognized in chest X rays for the remainder of the person's life (Figure 22–6).

The protective mechanisms that isolate the pathogen from surrounding tissue are sometimes defeated. Rather than healing, the caseous lesion's center may liquefy and eventually form an air-filled *tuberculous cavity.* The tuberculosis bacilli proliferate with little interference in these cavities and are carried to new foci of infections in the lungs or in other organs. This condition of spreading infection is called **miliary tuberculosis** (the name is derived from the numerous millet seed-sized tubercles formed in the infected tissues). This condition leads to a progressive disease characterized by loss of weight, coughing, and general loss of vigor. (Before Koch's discovery of the causative agent, tuberculosis was commonly called *consumption.*) The tissue damage causing caseation and subsequent tissue destruction is probably caused by delayed hypersensitivity reactions. Lymphocytes in the area of inflammation become sensitized to the tubercle bacillus antigens and release toxic substances. The tubercle bacillus itself is not known to produce any injurious toxins.

Among persons infected with *M. tuberculosis,* only a small number develop active cases of the disease. Variations in host resistance levels are related to genetic differences, the presence of other illness, and probably physiological and environmental factors such as malnutrition, overcrowding, and stress. Tuberculosis is a good demonstration of the importance of ecological balance between host and parasite in infectious disease, because infection with *M. tuberculosis* is clearly a necessary but not a sufficient cause of clinical disease.

The bacteria entering the lungs can be ingested by wandering macrophages. Unfortuantely, these bacteria resist lysis by the macrophages, and when the macrophages die they release the pathogen, which has generally increased in numbers. This intracellular growth, which shields the bacteria from drugs, plus the very slow reproductive rate of the organisms (many antibiotics work well only on growing cells) make it difficult to treat the microorganisms with chemotherapy. Intracellular growth also protects the cells from attack from circulating humoral antibodies, so cell-mediated immunity is therefore the most important immune response to tuberculosis.

The first effective antibiotic in the treatment of tuberculosis was streptomycin, although rifampin is currently preferred. Probably the drug most commonly used today is the synthetic chemical isoniazid (INH). A combination of INH and rifampin is particularly effective against phagocytized bacteria. Ethambutol and streptomycin are used in cases of intolerance or resistance to the primary drugs. To be effective, chemotherapy for tuberculosis must be continued over several months. Antibiotics and synthetics are usually given in combinations that minimize the development of resistance. Such resistance is likely in diseases such as tuberculosis that require prolonged chemotherapy, and thus give spontaneous mutants that are resistant to the drug time to arise.

Persons infected with tuberculosis develop sensitized T cells against the bacterium, the basis for the *tuberculin skin test.* In this test, a purified protein derivative (PPD) of the tuberculosis bacterium is injected cutaneously. If the injected person has been exposed to tuberculosis, sensitized T cells react with these proteins, and a delayed hypersensitivity reaction appears in about 48 hours. This reaction appears as an induration (hardening) and reddening of the area about the injection site. Probably the most accurate tuberculin test is the *Mantoux test,* in which dilutions of 0.1 ml of antigen are injected and the reacting area of the skin is measured. However, a number of similar tests are also in common use. A positive tuberculin test in

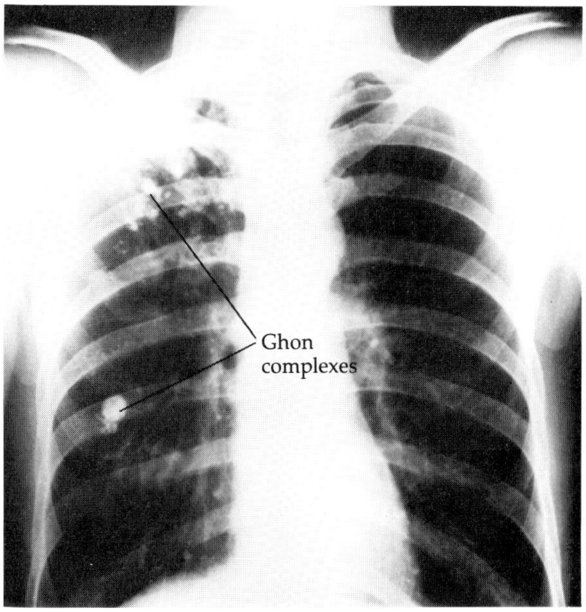

Figure 22–6 X ray showing Ghon complexes in a case of tuberculosis. The lower area of the right lung (left side of photo) has a large complex, and numerous smaller lesions toward the head are visible. The left lung (right side of photo) is clear.

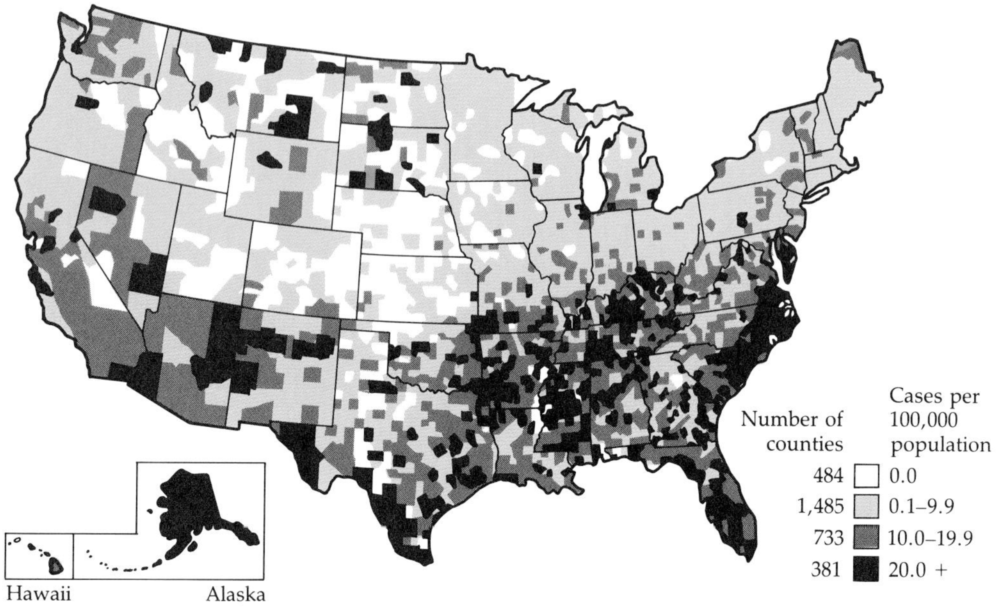

Figure 22–7 The geographical distribution of reported tuberculosis in the United States in a recent year. The map shows the number of reported cases per 100,000 persons, by county. The darker the area in the figure, the greater the incidence of tuberculosis.

the very young is a probable indication of an active case of tuberculosis. In older persons it might indicate only a previous exposure to the disease and immunity to it, not an active case. Nonetheless, it is an indication that further examination, such as a chest X ray for the detection of lung lesions, is needed, as well as attempts to isolate the bacterium.

Confirmation of tuberculosis by isolation of the bacterium is complicated by the very slow growth of the pathogen. Eight weeks might be needed for the growth of an isolate, and at least a week is required for the appearance of an exceedingly minute colony. Special media, such as the Lowenstein–Jensen medium, with a high lipid content in the form of glycerol or eggs, are recommended.

Another species, *Mycobacterium bovis* (bō′vis), is a pathogen mainly of cattle. It is the cause of **bovine tuberculosis,** which is transmitted to humans by contaminated milk or food. Bovine tuberculosis seldom spreads from human to human. But before the days of pasteurized milk and the development of control methods such as tuberculin testing of cattle herds, this disease was a common form of tuberculosis in humans. *M. bovis* infections cause tuberculosis that primarily affects the bones or lymphatic system. At one time, a common manifestation of this type of tuberculosis was hunchbacked deformation of the bone structure.

The *BCG vaccine* is a live culture of *M. bovis* that has been made avirulent by long cultivation on artificial media. (The name stands for "the bacillus of Calmette and Guerin," the persons who originally isolated the strain.) The vaccine is fairly effective (about 76%) in preventing tuberculosis and has been in use since the 1920s. In the United States it is used among high-risk groups only. Apparently, the vaccine enhances cell-mediated immunity, and persons who have received the vaccine show a positive reaction to tuberculin skin tests.

There are still more than 20,000 new cases of

tuberculosis and more than 12,000 deaths reported each year in the United States. In recent years, the mortality rate has been decreasing faster than the case rate. With the new treatments, patients are no longer isolated once chemotherapy has rendered their disease noncontagious. Tuberculosis in the United States is most prevalent among the poor and among Eskimos and American Indians (Figure 22–7). This incidence is probably due to environmental factors, although genetic factors probably also play a role. Tuberculosis cases in the developing countries of the world are estimated to number in the tens of millions.

Bacterial Pneumonias

The term **pneumonia** is applied to many pulmonary infections. In the U.S., the annual number of cases of the various forms of pneumonia probably exceeds 2,000,000, and there are about 50,000 deaths. Most pneumonias are bacterial in origin. The more common pathogens are *Streptococcus pneumoniae, Hemophilus influenzae, Staphylococcus aureus, Legionella pneumophila,* and *Mycoplasma pneumoniae.*

Pneumonias caused by gram-negative organisms are relatively uncommon. These bacteria are most likely to cause disease when the patient's defenses are naturally lowered from conditions such as diabetes or from alcoholism or drug abuse. Among the gram-negative organisms that cause pneumonias are *Klebsiella pneumoniae, Escherichia coli, Pseudomonas aeruginosa,* and several species of *Enterobacter.*

Most of the organisms that cause pneumonia are normal inhabitants of the mouth and throat and therefore cause opportunistic infections. Viruses, fungi, and protozoans are also responsible for different pneumonias. Examples of these will be discussed later in the chapter. In practice, the causes of many pneumonia cases are never diagnosed. The physician usually makes an initial "best-guess" diagnosis based on symptoms, history, and microscopic examination of a sputum specimen. An antibiotic treatment is then directed at the suspected organisms.

Pneumococcal pneumonia

The most common cause of pneumonia in adults is *Streptococcus pneumoniae* (the pneumococcus); the disease it causes is **pneumococcal pneumonia.** *S. pneumoniae* is a gram-positive, catalase-negative, ovoid bacterium. Because it usually forms cell pairs, it was formerly called *Diplococcus pneumoniae.* The organism produces a heavy capsule that makes the pathogen resistant to phagocytosis (Figure 22–8a). These capsules are also the basis of serological differentiation of pneumococci into some 83 serotypes. Before antibiotic therapy was so effective, antisera directed at these capsular antigens were used to treat the disease.

Pneumococcal pneumonia involves both the bronchi and the alveoli. Symptoms are fever, hard breathing, and chest pain. The lungs have a reddish appearance because blood vessels are dilated. In response to the infection, alveoli fill with some red blood cells, polymorphonuclear lymphocytes (PMNs), and edema fluid. The sputum is often rust colored from blood coughed up from the lungs. Pneumococci can invade the bloodstream, the pleural cavity surrounding the lung, and occasionally the meninges. No bacterial toxin has been clearly related to pathogenicity.

An initial diagnosis can be made from a chest X ray, because the fluid in the lungs produces recognizable shadows on the X ray. The diagnosis can be confirmed by isolation of the pneumococci from the throat, sputum, and other fluids. Pneumococci can be distinguished from other gram-positive, alpha-hemolytic streptococci by observing the inhibition of growth next to a disk of optochin (ethylhydrocuprein hydrochloride) or by the solubility of bile in pneumococcal broth cultures. Or they can be serologically typed; the bacteria are observed under a microscope, and when they react with a positive antiserum, their capsules swell (Figure 22–8b). This swelling is called the *quellung reaction* (from the German "quellung" meaning "swelling").

It should be noted that there are many healthy carriers of the pneumococcus. Virulence of the bacteria seems to be based mainly on the carrier's resistance, which can be lowered by stress. Many illnesses of the elderly terminate in pneumococcal pneumonia. Viral infections of the respiratory system are a frequent precursor of pneumonia.

Recurrence of pneumonia is not uncommon, but the second serological type is usually different. Before chemotherapy was available, the fatality

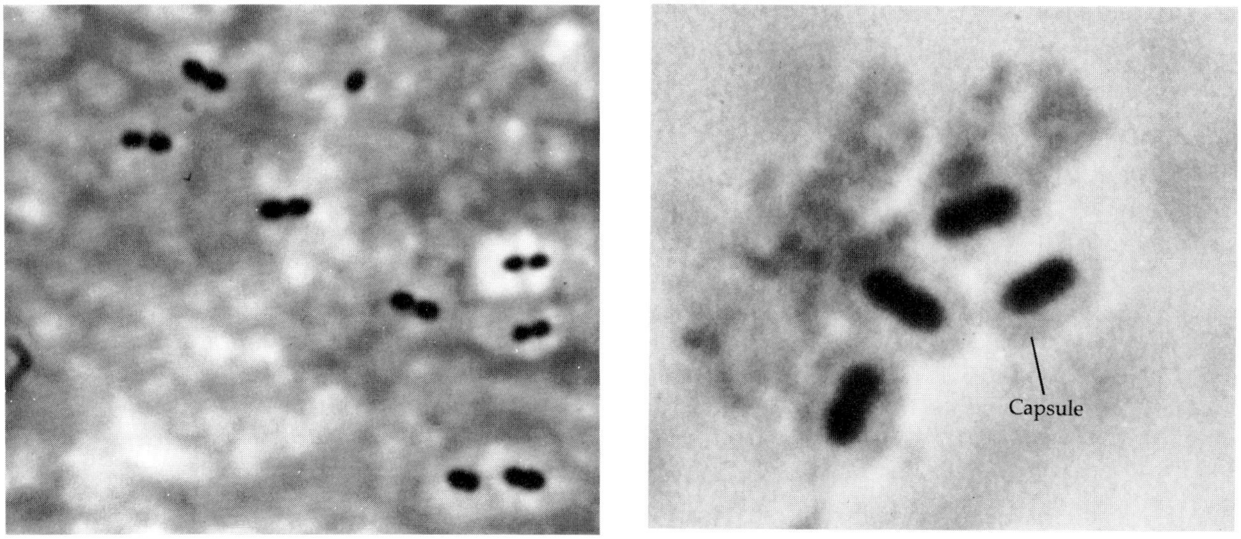

(a) (b)

Figure 22–8 The pneumococcus, *Streptococcus pneumoniae.* **(a)** Note the paired arrangement, the reason the organism was previously called *Diplococcus pneumoniae.* **(b)** A quellung reaction after the bacteria have been exposed to specific pneumococcus antiserum (×12,000). The capsular material, easily visible here, appears to swell in response to the antiserum.

rate was as high as 25%. This has now been lowered to less than 1% for younger patients treated early in the course of their disease. Penicillin is the drug of choice, but a few strains of penicillin-resistant pneumococci have been reported. A vaccine called Pneumovax has been developed from the purified capsular material of the 23 types of pneumococci that cause at least 80% of the pneumococcal pneumonias in the United States. This vaccine is used for the most susceptible group, the elderly, and debilitated individuals.

Klebsiella pneumonia

Klebsiella pneumoniae, the causative agent of **Klebsiella pneumonia,** is a member of the enteric group of bacteria, but is not uncommon in the throat and mouth of healthy persons. The organism is an encapsulated gram-negative rod, and its virulence is strongly related to the presence of the capsule (see Microview 1–1E). This pneumonia is most commonly found in persons who are chronically debilitated, and malnutrition is often a contributing factor. Male alcoholics past 40 years of age are probably the most susceptible. Probably 1–3% of bacterial pneumonias are of this type. Symptoms resemble those of pneumococcal pneumonia, but a prime difference is the formation of lung abscesses and permanent lung damage with *Klebsiella* pneumonia.

The fatality rate in untreated cases can be very high, perhaps exceeding 85%. Early treatment is vital, but considering the nature of the most susceptible population, is usually not received. *K. pneumoniae,* like most gram-negatives, is not sensitive to penicillin. Therefore, the normal penicillin treatment for pneumococcal pneumonia would be harmful, because it would suppress the throat's normal gram-positive flora that affords some protection against invading pathogens. Quick identification is essential, because delay allows the disease to progress without impedance, which is very dangerous. For *Klebsiella* pneumonia, the antibiotics of choice are cephalosporins or gentamicin, but the bacteria are commonly resistant to these drugs.

Mycoplasmal pneumonia (primary atypical pneumonia)

Typical pneumonia has a bacterial origin. If a bacterial agent is not isolated, the pneumonia is considered atypical, and viral agents are usually suspected. However, there could be another source of infection—mycoplasmas. These are bacteria that do not form cell walls. The mycoplasmas do not grow under the conditions normally used to recover most bacterial pathogens. Because of this characteristic, pneumonias caused by mycoplasmas are often confused with viral pneumonias.

The bacterium *Mycoplasma pneumoniae* (mī-kō-plas'mä) is the causative agent of **mycoplasmal pneumonia (primary atypical pneumonia).** This type of pneumonia was first discovered when such atypical infections responded to tetracyclines. That response indicated that the agent was nonviral. The disease is endemic and is a fairly common cause of pneumonia in young adults and children. The pathogens are transmitted in airborne droplets. Mycoplasmas usually infect the upper respiratory tract, and relatively few cases develop into pneumonia. The mortality rate is less than 1%.

When isolates from throat swabs and sputum grow on a medium containing horse serum and yeast extract, they form distinctive colonies of a "fried egg" appearance. The colonies are usually so small that they have to be observed with a hand lens or microscope. *M. pneumoniae* is beta-hemolytic for guinea pig red blood cells; this trait differentiates it from other mycoplasmas. Conclusive identification is made by fluorescent-antibody methods. The mycoplasmas are highly varied in appearance because of the lack of cell wall (see Figure 10–9). Their flexibility allows them to pass through filters with pores as small as 0.2 μm in diameter, which is small enough to strain out most other bacteria.

Diagnosis based on the recovery of organisms might not be useful in treatment, because as much time as two weeks is required for the slow-growing organisms to develop. A complement fixation test, using antigens from *M. pneumoniae* cells, can be used to test for a significantly rising titer of circulating antibodies, although this procedure also takes two to three weeks. Early treatment with tetracycline and erythromycin shortens the illness.

Legionnaires' disease

A type of pneumonia identified only recently is **Legionnaires' disease,** or **legionellosis,** which has already been discussed in Chapters 1 and 13. As you may recall, this disease first received public attention in 1976 when a series of deaths occurred among members of the American Legion who had attended a meeting in Philadelphia. A total of 182 persons, 29 of whom died, contracted a pulmonary disease, apparently at this meeting. Because no obvious bacterial cause could be found for what came to be known as Legionnaires' disease, the deaths were attributed to viral pneumonia. Close investigation, mostly with techniques directed at locating a suspected rickettsial agent, eventually identified the organism. A hitherto unknown bacterium, it was an aerobic gram-negative rod, now known as *Legionella pneumophila* (lē-jä-nel'lä nü-mō'fi-lä). See Figure 1–1b and Microview 1–1C.

The disease is characterized by a high temperature, cough, and general symptoms of pneumonia. No person-to-person transmission seems to be involved. Recent studies have shown that the bacterium can be isolated with some frequency from natural waters. Of particular interest is the fact that the organisms can grow well in the water of air-conditioning cooling towers. This ability might explain some epidemics associated with hotels, urban business districts, and a few hospitals that were suspected of having airborne transmission. The organism has also been found to inhabit the water lines of many hospitals. Most hospitals keep the temperature of hot water lines relatively low (43° to 55°C), as a safety measure, and thus inadvertently maintain the optimum growth temperature for this organism. The bacterium is considerably more resistant to chlorine than are most bacteria and can survive in water with a low level of chlorine for long periods of time.

Although the disease has only recently been identified, it seems now to have been fairly common. The *L. pneumophila* organism is not often enough suspected of causing disease. One hospital that had never recorded a case of nosocomial Legionnaires' disease found upon investigation that more than 14% of their nosocomial pneumonias had this cause, and that most of the sites in their water-distribution system contained *L. pneumophila*. Males over 50 years of age most commonly contract the disease, especially if they are heavy smokers or alcohol abusers, or have a chronic illness.

The best diagnostic method is by culture of the organism on a selective charcoal–yeast extract medium that has recently been developed. Examination of respiratory specimens by fluorescent-antibody methods (Chapter 16) is also used.

Erythromycin and rifampin are the drugs of choice.

Psittacosis (ornithosis)

The term **psittacosis (ornithosis)** was derived from the disease's association with psittacine birds such as parakeets and parrots. It was later found that the disease can also be contracted from many other birds, such as pigeons, chickens, ducks, and turkeys. Therefore, the more general term ornithosis has come into use.

The causative organism is *Chlamydia psittaci* (kla-mi'dē-ä sit'tä-sē), a gram-negative, obligately intracellular bacterium (see Figure 10–9b). One way chlamydias differ from the rickettsias is that chlamydias form tiny **elementary bodies** as one part of their life cycle. Unlike most rickettsias, elementary bodies are resistant to environmental stress; therefore, they can be transmitted through air and do not require an infective bite to transfer the infective agent directly from one host to another. The elementary bodies attach to epithelial cells of the mucous membrane of the respiratory system. They enter the cell by phagocytosis and develop into larger **reticulate bodies** that reproduce by repeated binary fission. Reticulate bodies eventually change back into infectious elementary bodies that leave the cell to infect other cells.

A form of pneumonia, psittacosis usually causes fever, headache, and chills. Subclinical infections are very common, and stress appears to enhance susceptibility to the disease. Disorientation and even delirium in some cases indicate that the nervous system can be involved.

The disease is not usually transmitted from one human to another, but is spread by contact with the droppings and other exudates of fowl. One of the most common modes of transmission is inhalation of dried particles from droppings. The birds themselves usually have symptoms of diarrhea, ruffled feathers, respiratory illness, and a generally droopy appearance. The parakeets and other parrots sold commercially are usually, but not always, free of the disease. However, pet store employees and persons involved in the commercial rearing of turkeys are at greatest risk of contracting the disease.

Diagnosis is made by isolation of the bacterium in embryonated eggs or mice, or by cell culture. With use of a fluorescent-antibody staining technique, the organism can then be identified by the presence of its specific antigens. If organisms are not isolated, a complement fixation test that detects rising serum antibody titer in the patient can be used. No vaccine is available, but tetracyclines are effective antibiotics in treating humans and animals. Effective immunity does not result from recovery, even when high titers of antibody are present in the person's serum.

Most years, fewer than 100 cases and very few deaths are reported in the United States. The main danger is in late diagnosis. Before antibiotic therapy was available, mortality was about 20%.

Q Fever

In Australia in the mid-1930s, a reported disease was characterized by a fever lasting one or two weeks, chills, chest pain, severe headache, and other evidence of a pneumonia-type infection. The disease was rarely fatal. In the absence of an obvious cause, the affliction was labeled **Q** (for query) **fever**—much as one might say X fever. The causative agent was subsequently identified as the obligately parasitic, intracellular bacterium *Coxiella burnetii* (käks-ē-el'lä bėr-ne'tē-ē), a rickettsia (Figure 22–9). Most rickettsias are not resistant enough to survive airborne transmission, but this organism is an exception.

The most serious complication of Q fever is endocarditis, which occurs in about 10% of the cases. Five to ten years might elapse between the initial infection and appearance of endocarditis. The organisms apparently reside in the liver during this interval.

See Table 22–1 for a summary of diseases associated with the respiratory system.

The organism is a parasite of a number of arthropods, and is transmitted among animals by direct bite. Arthropod bites are rarely, if ever, involved in transmission to humans. Cattle ticks are most commonly involved in transmission to humans. The cattle ticks spread the disease among dairy herds, and the organisms are shed in the feces, milk, and urine of infected cattle. In humans, the disease is thus most commonly associated with ingestion of unpasteurized milk and inhalation of aerosols generated in dairy barns. The infection in animals is usually subclinical. Many dairy workers have acquired at least subclinical infections. Workers in meat- and hide-processing plants are also at risk. The pasteurization temperature of milk, which was originally aimed at eliminating tuberculosis organisms, was raised slightly in 1956 to ensure the killing of *C. burnetii.*

Most cases of Q Fever in the United States are reported from the western states. The disease is endemic to California, Arizona, Oregon, and Washington. A vaccine for laboratory workers and similar high-risk personnel is available. Tetracyclines are very effective in treatment.

Identification of the organism may be made by isolation and growth of it in egg embryos or in cell culture. Handling of the organism by laboratory personnel testing for *Coxiella*-specific antibodies in the patient's serum can be avoided by their using serological tests, mostly complement fixation and agglutination types.

VIRAL DISEASES OF THE LOWER RESPIRATORY SYSTEM

Viral Pneumonia

Viral pneumonia can occur as a complication of influenza or even chickenpox. A number of enteric and other viruses have been shown to cause viral pneumonia, but viruses are isolated and identified in fewer than 1% of pneumonia-type infections, because few laboratories are equipped to properly test clincial samples for viruses. In those cases of pneumonia for which no cause is determined, viral etiology can be assumed if mycoplasmal pneumonia has been ruled out. *Respiratory syncytial virus* is probably the most common cause of viral respiratory disease, especially in the very young.

Influenza ("Flu")

The developed countries of the world are probably more aware of **influenza ("flu")** than of any other disease, next to the common cold. The flu is characterized by chills, fever, headache, and general muscular aches. Recovery normally occurs in a few days, and coldlike symptoms appear as fever subsides. Incidentally, diarrhea is not a normal symptom of the disease, and the intestinal discomforts attributed to "stomach flu" are probably due to some other cause.

The *influenza virus* consists of eight distinct RNA fragments of differing lengths, enclosed by an inner layer of protein and an outer lipid bilayer. Embedded in the lipid bilayer are numerous projections that characterize the virus (Figure 22–10). There are two types of projections: *hemagglutinin (H) spikes* and *neuraminidase (N) spikes.* The H spikes, of which there are about 500 on each virus, allow the virus to recognize and attach to body cells before infecting them. Antibodies against the influenza virus are directed mainly at these spikes. The term *hemagglutinin* refers to the agglutination (clumping) of red blood cells that occurs when the viruses are mixed with them. This reaction is important in serological tests such as the hemagglutination-inhibition test used to identify influenza viruses. The N spikes, of which there are about 100 per virus, differ from the H spikes in appearance and function. Apparently, they help the virus to separate from the infected cell as the virus exits after intracellular reproduction. N spikes also stimulate formation of antibodies, but these are less important in the body's resistance to the disease than those produced in response to the H spikes.

Viral strains are identified by variations in their H and N antigens. The different forms of the

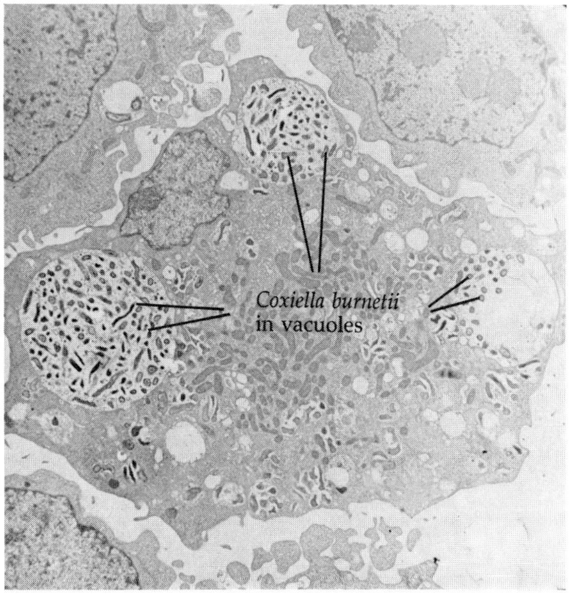

Figure 22–9 The rickettsial organism, *Coxiella burnettii*, that causes Q fever. The organisms are inside vacuoles within the cytoplasm of the cell.

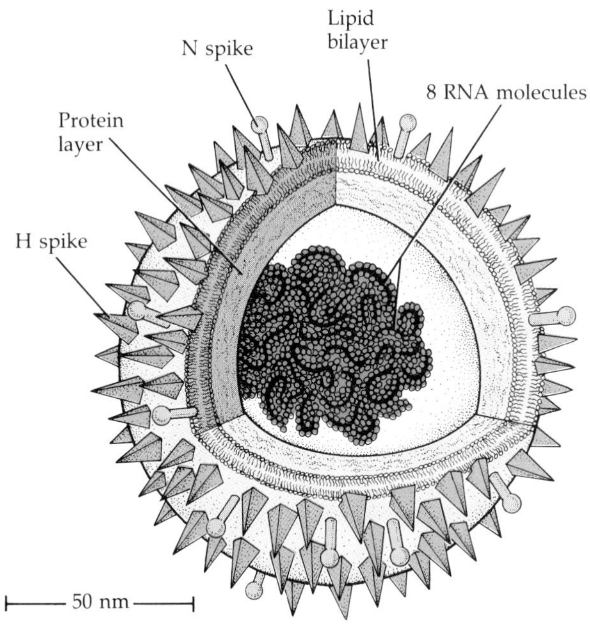

Figure 22–10 Detailed structure of the influenza virus. The envelope is composed of a protein layer, lipid layer, and two types of spikes. Because the eight separate RNA molecules can be rearranged readily, the virus can evade previous immunity by changing its genetic makeup. The outer spikes also change their antigenic character frequently.

antigens are assigned numbers—for example, H_0, H_1, H_2, H_3, and N_1, N_2. Each number reflects a substantial alteration in the protein makeup of the spike. The first antigenic type was isolated in 1933 and named H_0N_1. The Asiatic flu pandemic of 1957 was of type H_2N_2, and the Hong Kong virus of 1968 was designated H_3N_2. The influenza viruses are also classified into major groups according to the antigens of their protein coats. These groups are A, B, and (rarely) C. The A-type viruses are responsible for pandemics, and their H and N antigens vary. Historically, such changes, termed *antigenic shifts,* occur every 10 or 12 years and evade most of the immunity developed in the human population. The B-type virus also circulates, but is usually responsible for infections that are more geographically limited and milder. These antigenic shifts are probably due to a major genetic recombination. Because influenza viral RNA occurs as eight molecules, recombination is likely in infections caused by more than one strain. Recombination between the RNA of animal viral strains (found in swine, horses, and birds, for example) and the RNA of human strains might be involved. Recently, ducks in southern Chinese farming communities have come under suspicion as the animals most likely to be involved in genetic shifts. Wild ducks and other migratory birds then become carriers that spread the virus over large geographical areas.

Between episodes of such major antigenic shifts there are minor year-to-year variations in the antigenic makeup. The virus might still be designated as H_3N_2, for example, but viral strains reflecting minor antigenic changes within the antigenic group arise. These strains are sometimes assigned names related to the locality in which they were first identified, for example, Victoria or Russia. These minor variations are called *antigenic drift.* They usually reflect an alteration of only a single amino acid in the protein makeup of the H or N spike. Such a minor, one-step mutation is probably a response to selective pressure by antibodies (usually IgA in the mucous membranes) that neutralize all viruses except for the new mutations. Such mutations can be expected in about one in each million multiplications of the virus.

The usual result of antigenic drift is that a vaccine effective against H_3, for example, will be less effective against H_3 isolates circulating 10 years after the event. There will have been enough drift in that time that the virus can largely evade the antibodies stimulated originally by the earlier strain. In 1977, the Russian strain of H_1N_1 circulated in the United States. It essentially affected only persons under the age of about 25, apparently because it closely resembled a strain that had circulated in the 1950s. This early reappearance is somewhat out of the normal character of the disease and has led to speculation that an escape from a laboratory collection might have been involved in the later outbreak.

Development of a vaccine for influenza has not been practical. Although it is not difficult to make a vaccine for a particular strain of the virus, the problems are the identification of a new strain of a circulating virus and the development and distribution of the new vaccine for it in time. Unless the strain appears to be unusually virulent, this costly process is also considered to be of questionable value. The usual practice is to limit vaccine administration to the elderly, to hospital personnel, and to similar high-risk groups. The vaccines are often *multivalent,* that is, directed at several strains in circulation at the time. At present, influenza viruses for manufacturing vaccines are grown in egg embryo cultures. The vaccines are usually 70 to 90% effective, but the duration of protection is probably no more than three years.

One new approach to the problem of antigenic variation might be to utilize genetic engineering techniques. A laboratory bacterium can theoretically be made to produce, very quickly, a large number of antigens to the influenza virus, when genes for this purpose have been inserted into the bacterial DNA.

Almost every year, epidemics of the "flu" spread rapidly through large populations. The disease is so readily transmissible that epidemics are quickly propagated through populations susceptible to the newly changed strain of virus. The mortality rate from the disease is not high, usually less than 1%, and these deaths are mainly among the very young and the very old. However, so many persons are infected in a major epidemic that the total number of deaths is often high. Very often the cause of death is not the influenza virus, but secondary bacterial infections. The bacterium *Hemophilus influenzae* was named under the mistaken belief that it was the principal pathogen causing influenza rather than a secondary invader. *Staphylococcus aureus* and *Streptococcus pneumoniae* are other prominent secondary bacterial invaders.

In any discussion of influenza, the great pandemic of 1918–19 must be mentioned. Worldwide, more than 20 million people died. No one is sure why it was so unusually lethal. Usually, the very young and very old are the principal victims, but in 1918–19, young adults had the highest mortality rate. Considerable effort has been expended in examining this unusual pattern of virulence. Recent exhumation of bodies of victims of the epidemic buried in the permafrost above the Arctic Circle failed to recover viable isolates of the virus. It is possible that the 1918 viral strain became endemic in swine in the United States.

In 1976, a recruit at Fort Dix, New Jersey, died of influenza caused by a swine flu strain with many of the characteristics of the 1918 pandemic strain. This case precipitated a national campaign of preventative inoculations for swine influenza. Many feel that this response was ill advised, especially because there proved to be no widespread transmission from the initial focus of infection. Another factor against such widespread vaccination is that the incidence of *Guillain–Barré syndrome (GBS)* was five to six times higher in persons who received flu vaccine that season (1976–77) than it was in unvaccinated persons. (GBS is characterized by a paralysis that is usually self-limited and reversible, although approximately 5% of the cases are fatal.) The reasons for the association between flu vaccination and GBS that year are still a mystery, and the Centers for Disease Control found no association between flu vaccination and GBS for either the 1978–79 or 1979–80 flu season.

The questions raised by the 1976–77 flu vaccination program highlight the trade-offs that must be considered when a large-scale inoculation program is used against a new strain of flu virus. In order to be effective, the program must be started as soon after initial infection as possible, but this urgency does not allow enough time for the new vaccine to be thoroughly tested.

While little can be done to alleviate a viral disease such as influenza except to treat symptoms, the bacterial complications are amenable to treatment with antibiotics. A pandemic such as that of 1918–19 would be unlikely to have the same mortality rate today, because many of these deaths were probably due to secondary bacterial infections.

An antiviral drug, *amantadine,* has been found to significantly reduce the symptoms of influenza if administered promptly. In prophylactic use, it apparently reduces the rate of infection and illness by perhaps as much as 70%. Its mode of action is to inhibit the uncoating of type A viruses in the cell (it has no effect on type B viruses). The inhibition of uncoating interferes with the reproductive cycle of the virus.

FUNGAL DISEASES OF THE LOWER RESPIRATORY SYSTEM

Histoplasmosis

Histoplasmosis superficially resembles tuberculosis. In fact, it was first recognized as a disease in the United States when X ray surveys showed lung

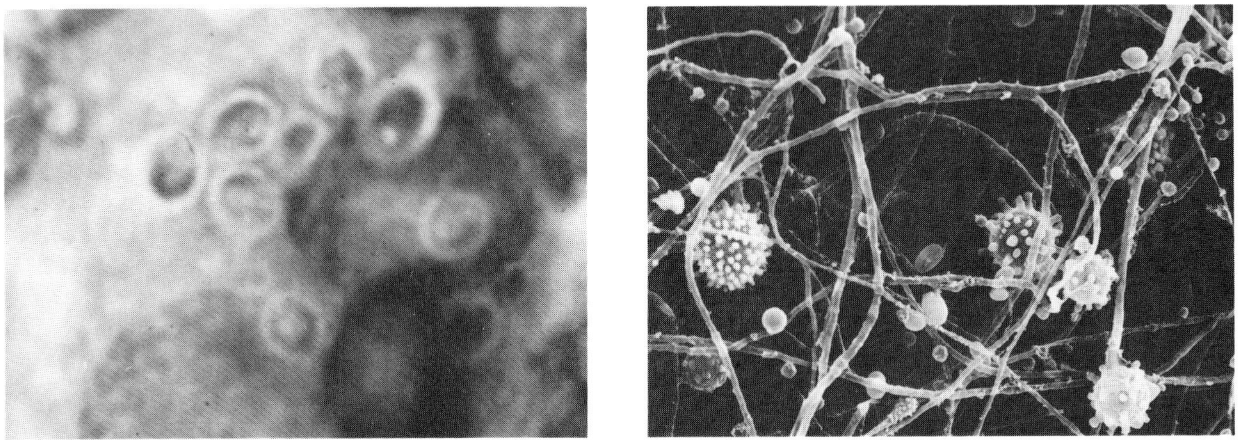

(a) (b)

Figure 22–11 *Histoplasma capsulatum*, a dimorphic fungus that causes histoplasmosis. **(a)** Yeastlike form typical of growth in tissue. **(b)** Spores of the filamentous, spore-forming phase found in soil. These are the infectious particles (×1100).

lesions in many persons who were tuberculin-test negative. Although the lungs are the organs most likely to be initially infected, the organisms cause lesions in almost all organs of the body. Normally, symptoms of the disease are rather ill-defined and mostly subclinical and the disease passes for a minor respiratory infection. In a few cases, histoplasmosis becomes progressive and is a severe, generalized disease. This occurs in only a small number of infections, perhaps fewer than 0.1%

The causative organism, *Histoplasma capsulatum* (hiss-tō-plaz'mä kap-su-lä'tum), is a dimorphic fungus; that is, it has a yeastlike morphology in tissue growth, and in soil or artificial media, it forms a filamentous mycelium carrying reproductive conidia (Figure 22–11). In the body, the yeastlike form is found intracellularly in phagocytic cells.

Although the disease is rather widespread throughout the world, it has a limited geographical range in the United States (Figure 22–12). For example, of 513 cases reported in 1981, 403 were from the states of Minnesota, Iowa, Missouri, and Alabama. In general, the disease is found in the states adjoining the Mississippi and Ohio rivers. More than 75% of the population in some of these states have antibodies against the infection. In other states, Maine, for example, a positive test is a rather rare event. Approximately 50 deaths are reported in the United States each year from histoplasmosis. The total number of infected persons may run into the millions, so the death rate is low.

Humans acquire the disease from airborne conidia produced under conditions of particular moisture and pH levels. These conditions occur particularly where droppings from birds have accumulated. Birds themselves do not carry the disease, but their droppings provide nutrients, particularly a source of nitrogen, for the fungus.

Clinical signs and history, complement fixation tests, and, most importantly, isolation of the organism or its identification in tissue specimens are necessary for proper diagnosis. The organism is found in the spleen and bone marrow, as well as in phagocytic cells. Currently, the most effective chemotherapy is with amphotericin B or ketoconazole and other orally active imidazoles. Less toxic than the polyenes like amphotericin B, the imidazoles have recently received FDA approval to be prescribed for systemic mycoses in the United States.

Coccidioidomycosis

Another fungal pulmonary disease, also rather restricted geographically, is **coccidioidomycosis.** The causative organism is *Coccidioides immitis* (kok-sid-ē-oi'dēz im̄'mi-tis), a dimorphic fungus. The spores are found in dry, highly alkaline soils of the American Southwest and in similar soils of South America and northern Mexico. In tissue, the organism forms a thick-walled body filled with spores (Figure 22–13a). In soil, it forms filaments that reproduce by formation of arthrospores (Figure 22–13b). The wind carries the arthrospores to transmit the infection. Arthrospores are often so abundant that simply driving through an endemic area, especially during a dust storm, can result in infection.

The symptoms of coccidioidomycosis include chest pain and perhaps fever, coughing, and loss of weight. Most infections are unapparent, and almost all cases resolve themselves by recovery in a few weeks, and sound immunity remains. However, in fewer than 1% of the infections, a progressive disease resembling tuberculosis disseminates throughout the body. The resemblance to tuberuculosis is so close that isolation of the organism is necessary to properly diagnose coccidioidomycosis. A tuberculinlike skin test is used in screening for cases. Of those that were tested in California, more than half had positive skin tests, although most had not been aware of the infection.

Farm workers are most likely to be infected, and predisposing factors such as fatigue and poor nutrition can lead to more severe disease.

Approximately 50 to 100 deaths each year are reported from this disease in the United States. Most cases are reported from southern California and Arizona desert regions, with only a scattering of cases reported from other states (Figure 22–14).

A vaccine is currently being tested. If successful, it is expected to be widely used in the areas affected by coccidioidomycosis. Amphotericin B has been used to treat serious cases, but might be replaced by less toxic imidazole drugs such as ketoconazole.

Blastomycosis (North American Blastomycosis)

Blastomycosis is usually called **North American blastomycosis** to differentiate it from a similar South American blastomycosis. It is caused by the fungus *Blastomyces dermatitidis* (blas-tō-mī'sēs dėr-mä-tit'i-dis), a dimorphic fungus found most often in the Mississippi valley, where it probably grows in soil. Approximately 30 to 60 deaths are reported each year, although most infections are asymptomatic. The infection begins in the lungs and can spread rapidly. Cutaneous ulcers commonly appear and there is extensive abscess formation and tissue destruction. The organism can be isolated from pus and biopsy sections. Amphotericin B is usually an effective treatment.

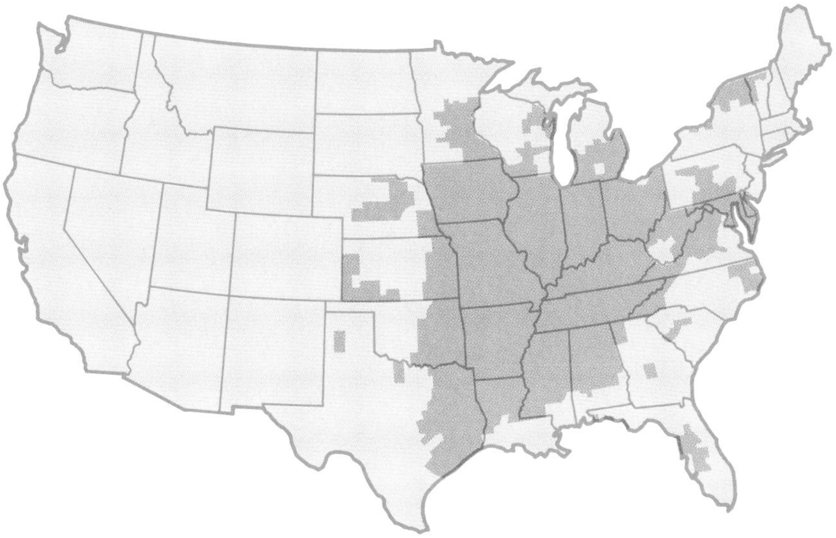

Figure 22–12 Geographical distribution of histoplasmosis (dark area) in the United States.

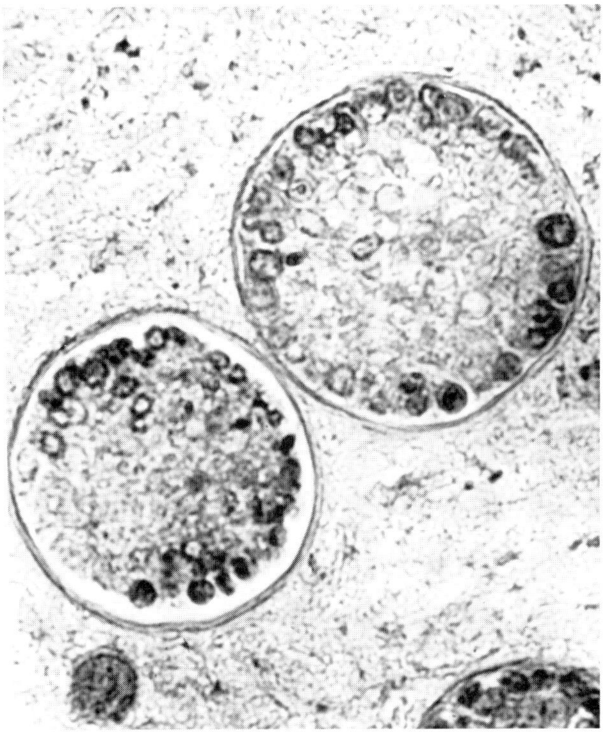

(a)

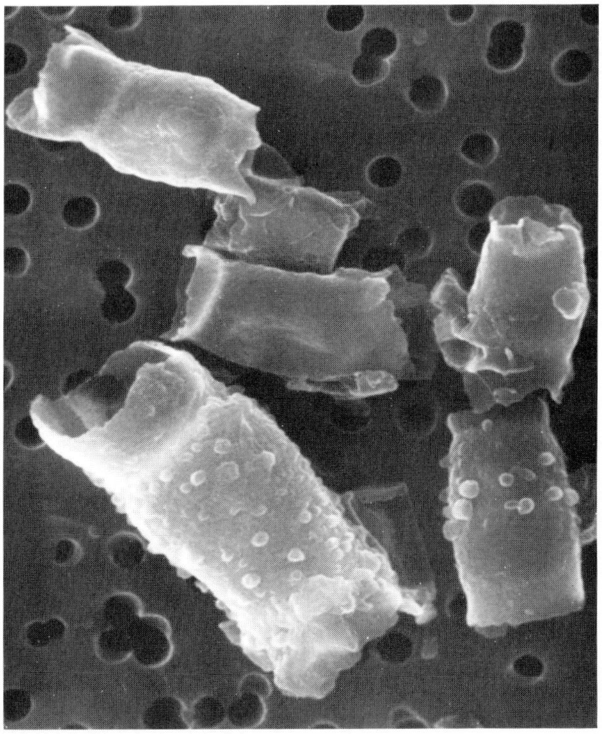

(b)

Figure 22–13 *Coccidioides immitis.* **(a)** Spore-filled spherical bodies growing in tissue. **(b)** Scanning electron micrograph showing arthrospores (also see Figure 11–4a). These are the airborne infectious particles that cause coccidioidomycosis (×4500).

Other Fungi Involved in Respiratory Disease

There are many other opportunistic fungi that may cause respiratory disease, particularly in immunosuppressed hosts or if there is exposure to massive numbers of spores. **Aspergillosis** is caused by the spores of *Aspergillus fumigatus* (a-spėr-jil'lus fü-mi-gä'tus) (see Microview 1–4B) and other species of *Aspergillus,* which are fairly widespread in decaying vegetation. Compost piles are ideal sites for growth, and farmers and gardeners are the ones most often exposed to infective amounts of such spores. Similar pulmonary infections result when some individuals are exposed to spores of mold genera such as *Rhizopus* or *Mucor.* Such diseases, particularly invasive infections of pulmonary aspergillosis, can be very dangerous. Predisposing factors include an impaired immune system, cancer, and diabetes. As with most systemic fungal infections, there is only a limited arsenal of antibiotics available; amphotericin B has proved the most useful.

A similar pulmonary disease, caused by a bacterium, is **nocardiosis,** caused by the soil actinomycete *Nocardia asteroides* (nō-kär'dē-ä as-tėr-oi'dēz). Usually mild, this can be a very serious disease if the organism disseminates throughout the body and causes secondary infections.

PROTOZOAN AND HELMINTHIC DISEASES OF THE LOWER RESPIRATORY SYSTEM

***Pneumocystis* pneumonia** is caused by the sporozoan *Pneumocystis carinii* (nü-mō-sis'tis kär-i'nē-ī). The disease occurs throughout the world and can be endemic in hospitals. It gained importance when it was revealed to be an increasing cause of mortality among immunosuppressed patients.

This group includes persons receiving immunosuppressive drugs to minimize rejection of transplanted tissue and those whose immunity is depressed because of cancer. Persons with acquired immune deficiency syndrome (AIDS) are also very susceptible to this organism.

The life cycle of this protozoan is not well known. However, outbreaks in hospitals suggest that the parasite is transmitted by direct contact between humans. Persons receiving immunosuppressive drugs or who have leukemia are especially susceptible to infection. *Pneumocystis* causes the alveoli to fill with a frothy exudate. Diagnosis is based on the recovery of *Pneumocystis* from the respiratory system. Untreated infections are usually fatal. Pentamidine is the drug of choice.

Humans are occasional intermediate hosts for the tapeworm *Echinococcus granulosus* (ē-kīn-ō kok'kus gra-nū-lō'sus.) (See Figure 11–24). The adult worm is a few millimiters in length and is an inhabitant of the intestinal tract of animals such as dogs. From these definitive hosts, the eggs are excreted and later ingested by intermediate hosts such as sheep or deer. The eggs become larvae in the gut of the new host and journey in the bloodstream to organs such as the lungs or liver. There they slowly form large cysts **(hydatidosis** or **cystic hydatid disease).** The cyst, because of its size, may impair organ function Predators eating the intermediate host acquire the infection. Humans, especially children, may acquire the disease when playing with infected dogs, who harbor eggs around the muzzle. However, the disease can also be transmitted by food or water contaminated by eggs shed in animal feces. The disease is most common in sheep-rearing cultures that feed domestic dogs sheep flesh. In humans the most common symptom is organ dysfunction caused by growth of the cyst—which may take decades. Eventually, in humans the cyst may contain several liters of material. Serious anaphylactic shock (Chapter 17) can result if the cyst ruptures and releases its contents (scolexes, etc.). For humans the only therapy is surgical removal.

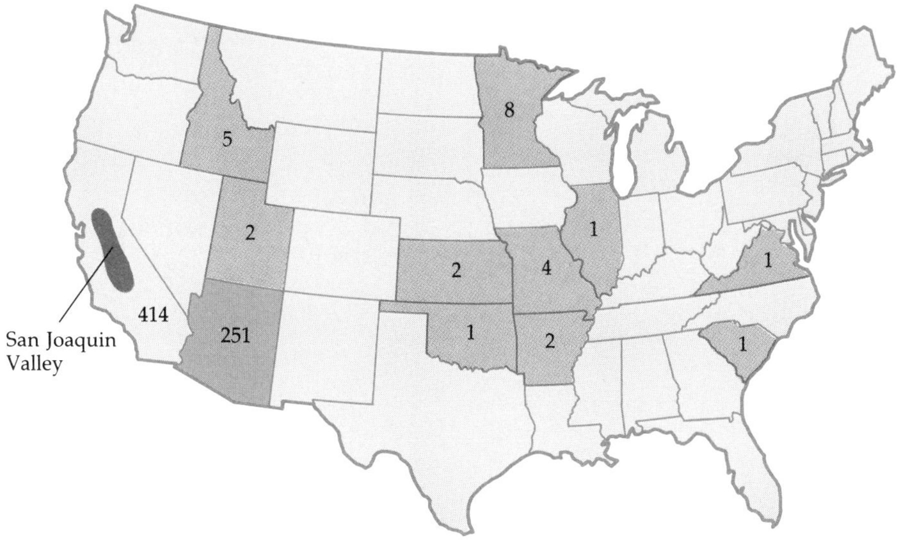

Figure 22–14 Cases of coccidioidomycosis in the United States in 1982. Because most of the cases occur in the San Joaquin Valley in California, this disease is sometimes called San Joaquin Valley fever. A number of the cases shown in states outside the southwest might have been acquired during travel to endemic areas.

Table 22–1 Summary of Diseases Associated with the Respiratory System

Disease	Causative Agent	Mode of Transmission	Treatment
Bacterial Diseases of the Upper Respiratory System:			
Streptococcal pharyngitis ("strep" throat)	Streptococci, especially *Streptococcus pyogenes*	Respiratory secretions	Penicillin and erythromycin
Scarlet fever	Erythrogenic toxin-producing strains of *Streptococcus pyogenes*	Respiratory secretions	Penicillin and erythromycin
Diphtheria	*Corynebacterium diphtheriae*	Respiratory secretions, healthy carriers	Antitoxin and penicillin, tetracyclines, erythromycin
Cutaneous diphtheria	*Corynebacterium diphtheriae*	Respiratory secretions, healthy carriers	Antitoxin and penicillin, tetracyclines, erythromycin
Otitis media	Several agents, especially *Staphylococcus aureus, Streptococcus pneumoniae,* beta-hemolytic streptococci, and *Hemophilus influenzae*	Complication of cold, or nose or throat infection	Penicillin or erythromycin
Viral Disease of the Upper Respiratory System:			
Common cold	Coronaviruses, rhinoviruses	Respiratory secretions	None
Bacterial Diseases of the Lower Respiratory System:			
Whooping cough	*Bordetella pertussis*	Respiratory secretions	Erythromycin, tetracyclines, chloramphenicol in severe cases; none in mild cases
Tuberculosis	*Mycobacterium tuberculosis*	Respiratory secretions; infrequently by food, especially milk	Isoniazid and rifampin
Pneumococcal pneumonia	*Streptococcus pneumoniae*	Healthy carriers; primarily a disease following viral respiratory infection or other stress	Penicillin
Klebsiella pneumonia	*Klebsiella pneumoniae*	Primarily a disease in debilitated hosts, e.g., alcoholics	Cephalosporins or gentamicin
Mycoplasmal pneumonia	*Mycoplasma pneumoniae*	Respiratory secretions probably	Tetracycline, erythromycin
Legionnaires' disease	*Legionella pneumophila*	Thought to be aerosols from contaminated water	Erythromycin, rifampin
Psittacosis (Ornithosis)	*Chlamydia psittaci*	Animal reservoir: aerosols of dried droppings and other exudates of birds; person-to-person transmission is rare	Tetracyclines
Q Fever	*Coxiella burnetii*	Animal reservoir; aerosols in dairy barns and similar places; unpasteurized milk	Tetracyclines
Other bacterial pneumonias	*Pseudomonas* species; *Escherichia coli* and other bacteria	Bacteria are opportunistic	Varies with agent
Viral Diseases of the Lower Respiratory System:			
Viral pneumonia	Several viruses	Aerosols; complication of other viral diseases	None
Influenza	Influenza viral strains; many serotypes	Respiratory secretions	Amantadine

Table 22–1 Summary of Diseases Associated with the Respiratory System, *continued*

Disease	Causative Agent	Mode of Transmission	Treatment
Fungal Diseases of the Lower Respiratory System:			
Histoplasmosis	*Histoplasma capsulatum*	Animal reservoir: aerosols of dried bird droppings; person-to-person transmission is rare	Amphotericin B, ketoconazole
Coccidioidomycosis	*Coccidioides immitis*	Soil organism: aerosols of dust	Amphotericin B, ketoconazole
Blastomycosis	*Blastomyces dermatitidis*	Soil organism: aerosols of dust	Amphotericin B
Other fungal pneumonias	Species of *Aspergillus*, *Rhizopus*, *Mucor*, and other genera	Aerosols of dust containing opportunistic fungi	Amphotericin B
Protozoan and Helminthic Diseases of the Lower Respiratory System:			
Pneumocystis pneumonia	*Pneumocystis carinii*	Direct contact (?)	Pentamidine
Hydatidosis	*Echinococcus granulosus*	Direct contact with dogs	Surgical removal of cysts
Paragonimiasis	*Paragonimus westermanni*	Ingestion of crayfish	Chloroquine

STUDY OUTLINE

INTRODUCTION (p. 609)

1. Infections of the upper respiratory system are the most common type of infection.
2. Pathogens that enter the respiratory system can infect other parts of the body.
3. Respiratory infections are transmitted by direct contact, droplets, and fomites.

STRUCTURE AND FUNCTION OF THE RESPIRATORY SYSTEM (pp. 609–611)

1. The upper respiratory system consists of the nose, throat, and associated structures such as the middle ear and eustachian tubes.
2. Coarse hairs in the nose filter large particles from air entering the respiratory tract.
3. The ciliated mucous membranes of the nose and throat trap airborne particles and remove them from the body.
4. Lymphoid tissue, tonsils, and adenoids provide immunity to certain infections.
5. The lower respiratory system consists of the larynx, trachea, bronchial tubes, and alveoli.
6. The ciliary escalator of the lower respiratory system helps prevent microorganisms from reaching the lungs.
7. Microorganisms in the lungs can be phagocytized by alveolar macrophages.
8. Respiratory mucus contains IgA antibodies.

NORMAL FLORA OF THE RESPIRATORY SYSTEM (p. 611)

1. Normal flora of the nasal cavity are usually gram-positive bacteria, including diphtheroids, staphylococci, micrococci, and *Bacillus* species.
2. The throat can contain *Streptococcus pneumoniae*, *Hemophilus influenzae*, and *Neisseria meningitidis*.
3. The lower respiratory system is usually sterile because of the action of the ciliary escalator.

BACTERIAL DISEASES OF THE UPPER RESPIRATORY SYSTEM (pp. 611–615)

1. *H. influenzae* and respiratory viruses can cause laryngitis.
2. *H. influenzae* type b can cause epiglottitis.

3. Bronchitis is often caused by *Mycoplasma pneumoniae*, respiratory viruses, or (in infants) respiratory syncytial virus.
4. Infections of the lower respiratory tract are often referred to as croup.
5. Most respiratory tract infections are self-limiting.

Streptococcal Pharyngitis ("Strep" Throat) (p. 612)

1. This infection is caused by beta-hemolytic group A streptococci, the group to which *Streptococcus pyogenes* belongs.
2. Symptoms of this infection are inflammation of the mucous membrane and fever; tonsillitis and otitis media may also occur.
3. Diagnosis requires isolation and identification of the bacteria.
4. Penicillin is used to treat streptococcal pharyngitis.
5. Immunity to streptococcal infections does not usually develop, in part because there are many serological types.
6. Streptococcal sore throat is usually transmitted by droplets, but has been associated with unpasteurized milk.

Scarlet Fever (p. 612)

1. Streptococcal sore throat, caused by an erythrogenic toxin-producing *S. pyogenes*, results in scarlet fever.
2. *S. pyogenes* produces erythrogenic toxin when lysogenized by a phage.
3. Symptoms include a pink rash, high fever, and a red, enlarged tongue.
4. Immunity to scarlet fever is measured by the Dick test, a skin test with erythrogenic toxin.

Diphtheria (pp. 613–614)

1. Diphtheria is caused by exotoxin-producing *Corynebacterium diphtheriae*.
2. Exotoxin is produced when the bacteria are lysogenized by a phage.
3. The exotoxin inhibits protein synthesis, and heart, kidney, or nerve damage can result.
4. A membrane, containing fibrin and dead human and bacterial cells, forms in the throat and can block the passage of air.
5. Laboratory diagnosis is based on isolation of the bacteria and appearance of growth on differential media.
6. Toxigenicity is determined with the guinea pig test or gel diffusion test.
7. Antitoxin must be administered to neutralize the toxin, and antibiotics can stop growth of the bacteria.
8. Routine immunization in the United States includes diphtheria toxoid in the DPT vaccine.
9. The Schick test is a skin test used to determine immunity to diphtheria.

Cutaneous Diphtheria (p. 614)

1. Slow-healing skin ulcerations are characteristic of cutaneous diphtheria.
2. There is minimal dissemination of the exotoxin in the bloodstream.

Otitis Media (pp. 614–615)

1. Earache, or otitis media, can occur as a complication of nose and throat infections or through direct inoculation from an external source.
2. Pus accumulation causes pressure on the eardrum.
3. Bacterial causes include *Staphylococcus aureus*, *Streptococcus pneumoniae*, beta-hemolytic streptococci, and *Hemophilus influenzae*.

VIRAL DISEASES OF THE UPPER RESPIRATORY SYSTEM (p. 615)

Common Cold (Coryza) (p. 615)

1. Any one of approximately 200 different viruses can cause the common cold.
2. Symptoms include sneezing, nasal secretions, and congestion.
3. Sinus infections, lower respiratory tract infections, laryngitis, and otitis media can occur as complications of a cold.
4. Colds are most often transmitted by indirect contact.
5. Rhinoviruses prefer temperatures slightly lower than body temperature.
6. The incidence of colds probably increases during cold weather because of increased interpersonal indoor contact or physiological changes.
7. Antibodies are produced against the specific viruses.

BACTERIAL DISEASES OF THE LOWER RESPIRATORY SYSTEM (pp. 615–625)

Whooping Cough (Pertussis) (pp. 615–618)

1. Whooping cough is caused by *Bordetella pertussis.*
2. The initial stage of whooping cough resembles a cold and is called the catarrhal stage.
3. The proliferation of bacteria blocks the trachea and bronchi, and causes deep coughs characteristic of the paroxysmal (second) stage.
4. The convalescence (third) stage can last for months.
5. Laboratory diagnosis is based on isolation of the bacteria on Bordet–Gengou medium, followed by serological tests.
6. Regular immunization for children includes dead *B. pertussis* cells as part of the DPT vaccine.

Tuberculosis (pp. 618–621)

1. Tuberculosis is caused by *Mycobacterium tuberculosis.*
2. Large amounts of lipids in the cell wall account for the bacteria's acid-fast characteristic as well as its resistance to drying and disinfectants.
3. Lesions formed by *M. tuberculosis* are called tubercles; necrosis results in a caseous lesion that might calcify and appear in an X ray as a Ghon complex.
4. New foci of infection can develop when a caseous lesion ruptures and releases bacteria into blood or lymph vessels; this is called miliary tuberculosis.
5. Miliary tuberculosis is characterized by weight loss, coughing, and loss of vigor.
6. *M. tuberculosis* may be ingested by wandering macrophages; the bacteria reproduce in the macrophages and are liberated when the macrophages die.
7. The tuberculin skin test is used to determine whether a person has been exposed to tuberculosis.
8. A positive tuberculin skin test can indicate either an active case of tuberculosis or prior exposure and immunity to the disease.
9. Laboratory diagnosis is based on isolation of the bacteria and requires incubation of up to eight weeks.
10. *Mycobacterium bovis* causes bovine tuberculosis and can be transmitted to humans by unpasteurized milk.
11. *M. bovis* infections usually affect the bones or lymphatic system.
12. BCG vaccine for tuberculosis consists of a live, avirulent culture of *M. bovis.*

Bacterial Pneumonias (pp. 621–625)

1. Most cases of pneumonia are caused by normal flora of the mouth and throat.
2. The most common etiologic agents are *S. pneumoniae, H. influenzae,* and *S. aureus.*

Pneumococcal pneumonia (pp. 621–622)

1. Pneumococcal pneumonia is caused by encapsulated *Streptococcus pneumoniae.*
2. Symptoms of this disease are fever, difficult breathing, chest pain, and rust-colored sputum.
3. Alveoli fill with red blood cells and edema fluid; the appearance of shadows in the X ray of lungs can be used for diagnosis.
4. The bacteria can be identified by the production of alpha hemolysins, inhibition by optochin, and bile solubility.
5. A vaccine called Pneumovax consists of purified capsular material from 14 serotypes of *S pneumoniae.*

Klebsiella pneumonia (p. 622)

1. *Klebsiella pneumoniae* causes *Klebsiella* pneumonia.
2. *Klebsiella* pneumonia results in lung abscesses and permanent lung damage.

Mycoplasmal pneumonia (primary atypical pneumonia) (pp. 622–623)

1. *Mycoplasma pneumoniae* causes primary atypical pneumonia, or mycoplasmal pneumonia.
2. *M. pneumoniae* produces small "fried egg" colonies after two weeks' incubation on enriched media containing horse serum and yeast extract.
3. A complement fixation test, used to diagnose the disease, is based on the rising of antibody titer.

Legionnaires' disease (pp. 623–624)

1. This disease is caused by the aerobic gram-negative rod *Legionella pneumophila.*
2. The bacterium can grow in water such as air-

conditioning reservoirs and then be disseminated in the air.

3. This pneumonia does not appear to be transmitted from person to person.
4. Bacterial culture and fluorescent-antibody tests are used for laboratory diagnosis.

Psittacosis (ornithosis) (p. 624)

1. *Chlamydia psittaci* is transmitted by contact with contaminated droppings and exudates of fowl.
2. Elementary bodies allow the bacteria to survive outside a host.
3. The bacteria are isolated in embryonated eggs, mice, or cell culture; identification is based on fluorescent-antibody staining.
4. Commerical bird handlers are most susceptible to this disease.

Q Fever (p. 624–625)

1. Obligately parasitic, intracellular *Coxiella burnetii* causes Q fever.
2. The disease is usually transmitted to humans through unpasteurized milk or inhalation of aerosols in dairy barns.
3. Laboratory diagnosis is made with the culture of bacteria in embryonated eggs or cell culture.

VIRAL DISEASES OF THE LOWER RESPIRATORY SYSTEM (pp. 625–627)

Viral Pneumonia (p. 625)

1. Many pneumonia cases are thought to be due to viral etiology.
2. A number of viruses can cause pneumonia as a complication of infections such as influenza.
3. The etiologies are not usually identified in a clinical laboratory because of the difficulty in isolating and identifying viruses.

Influenza ("Flu") (pp. 625–627)

1. Influenza is caused by influenza virus and is characterized by chills, fever, headache, and general muscular aches.
2. Hemagglutinin (H) and neuraminidase (N) spikes project from the outer lipid bilayer of the virus.
3. Viral strains are identified by antigenic differences in the H and N spikes; they are also divided by antigenic differences in their protein coats (A, B, and C).
4. Antigenic shifts that alter the antigenic nature of the H and N spikes make natural immunity and vaccination of questionable value.
5. Deaths during an influenza epidemic are usually due to secondary bacterial infections.
6. Multivalent vaccines are available for the elderly and other high-risk groups.
7. Amantadine is an effective prophylactic and curative drug.

FUNGAL DISEASES OF THE LOWER RESPIRATORY SYSTEM (pp. 627–630)

1. The following mycoses can be treated with amphotericin B.

Histoplasmosis (pp. 627–628)

1. *Histoplasma capsulatum* causes a subclinical respiratory infection that only occasionally progresses to a severe, generalized disease.
2. The disease is acquired by inhalation of airborne conidia.
3. Isolation of the fungus or identification of the fungus in tissue samples is necessary for diagnosis.

Coccidioidomycosis (pp. 628–629)

1. Inhalation of the airborne arthrospores of *Coccidioides immitis* can result in coccidioidomycosis.
2. Most cases are subclinical, but when there are predisposing factors such as fatigue and poor nutrition, a progressive disease resembling tuberculosis can result.

Blastomycosis (North American Blastomycosis) (p. 629)

1. *Blastomyces dermatitidis* is the causative agent of blastomycosis.
2. The infection begins in the lungs and can spread to cause extensive abscesses.

Other Fungi Involved in Respiratory Disease (p. 630)

1. Opportunistic fungi can cause respiratory disease in immunosuppressed hosts, especially when large numbers of spores are inhaled.

2. Among these fungi are *Aspergillus, Rhizopus,* and *Mucor.*

PROTOZOAN AND HELMINTHIC DISEASES OF THE LOWER RESPIRATORY SYSTEM

(pp. 630–631)

1. Immunosuppressed patients or patients with certain forms of cancer are susceptible to an endemic hospital sporozoan, *Pneumocystis carinii.*
2. Humans infected with the tapeworm *Echinococcus granulosus* might have hydatid cysts in their lungs.

STUDY QUESTIONS

REVIEW

1. Describe how microorganisms are prevented from entering the upper respiratory system. How are they prevented from causing infections in the lower respiratory system?
2. Respiratory diseases are usually transmitted by ______________. List other methods by which they can be transmitted.
3. How does the normal flora of the respiratory system illustrate microbial antagonism?
4. Complete the following table.

Disease	Causative Agent	Symptoms	Treatment
Streptococcal sore throat			
Scarlet fever			
Diphtheria			
Whooping cough			
Tuberculosis			
Pneumococcal pneumonia			
Klebsiella pneumonia			
Legionnaires' disease			
Psittacosis			
Q fever			
Epiglottitis			
Nocardiosis			

5. How is otitis media contracted? What causes otitis media? Why was otitis media included in a chapter on diseases of the respiratory system?
6. List the causative agent, symptoms, and treatment for three viral diseases of the respiratory system. Separate the diseases according to whether they infect the upper or lower respiratory system.
7. Compare and contrast primary atypical pneumonia and viral pneumonia.

8. A patient has been diagnosed as having pneumonia. Is this sufficient information to begin treatment with antimicrobial agents? Briefly discuss why or why not.

9. List the causative agent, method of transmission, and endemic area for the following fungal diseases: histoplasmosis, coccidioidomycosis, blastomycosis.

10. Under what conditions can the saprophytes *Aspergillus* and *Rhizopus* cause infections?

11. Why is the DPT vaccine routinely given to children in the United States? Of what does it consist?

12. Briefly describe the procedures and positive results of the following skin tests, and indicate what is revealed by a positive test: Dick test; Schick test; tuberculin test.

13. Match the bacteria in question 4 to the following laboratory test results.
    Gram-positive cocci:
      beta-hemolytic, bacitracin inhibition ______________
      alpha-hemolytic, optochin inhibition ______________
    Gram-positive rods:
      not acid-fast ______________
      acid-fast: ______________
        mycelia formed ______________
        mycelia not formed ______________
    Gram-negative rods:
      facultative anaerobes: ______________
      aerobes:
        rods ______________
        coccobacillus ______________

14. Discuss reasons for the increased incidences of colds and pneumonias during cold weather.

CHALLENGE

1. Differentiate between the following.
   (a) *S. pyogenes* causing "strep" throat and *S. pyogenes* causing scarlet fever.
   (b) Diphtheroids and *C. diphtheriae.*

2. Why is vaccination against influenza felt to be of questionable value?

3. Provide reasons for each of the following statements.
   (a) Penicillin should not be used indiscriminately.
   (b) A combination of antimicrobial drugs is used to treat tuberculosis.

4. In a 2-week period, 8 infants in an intensive care nursery (ICN) developed pneumonia caused by respiratory syncytial virus (RSV). FA screening and viral cultures were used to diagnose infections, and positive patients were placed in a separate room. A 2-week-old girl from the newborn nursery, adjacent to ICN, also developed an RSV infection. Toward the end of this outbreak, FA tests and viral cultures were made of 15 ICN personnel. (All 15 reported having had upper respiratory illness during the previous week.) Viral cultures were negative but the RSV–FA test was strongly positive in one nurse and weakly positive in 5 others. Comment on the probable source of this outbreak. Explain the apparent discrepancy between the FA test and viral culture results. How can RSV infections in nurseries be prevented?

FURTHER READING

Al-Doory, Y. 1975. *The Epidemiology of Human Mycotic Diseases.* Springfield, IL: C. C. Thomas. Includes histoplasmosis, coccidioidomycosis, and blastomycosis.

Bordetella pertussis: pathogenesis and prevention of whooping cough. *In* L. Leive and D. Schlessinger, eds., *Microbiology—1984.* Washington, D. C.: American Society for Microbiology, pp. 157–183. A collection of articles on the epidemiology, toxins, and immunity of whooping cough.

Skinner, F. A. and L. B. Quesnel. 1978. *Streptococci.* New York: Academic Press. Pathogenicity, immunology, and taxonomy of the genus *Streptococcus.*

Stuart-Harris, C. 1981. The epidemiology and prevention of influenza. *American Scientist* 69:166–172. The relationship between antigenic shift in the virus and immunity to influenza.

Thomas, G. and M. Morgan-Witts. 1982. *Anatomy of an Epidemic.* New York: Doubleday. A medical detection story about the mysterious epidemic that led to the discovery of a new bacterium *(Legionella).*

See also Further Reading for Part Four at the end of Chapter 24.

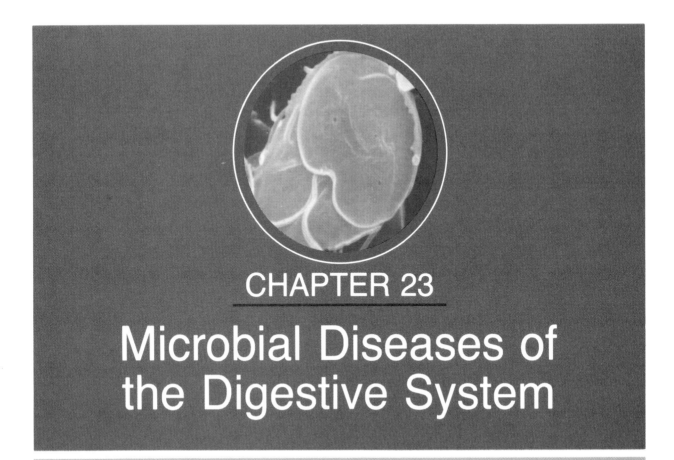

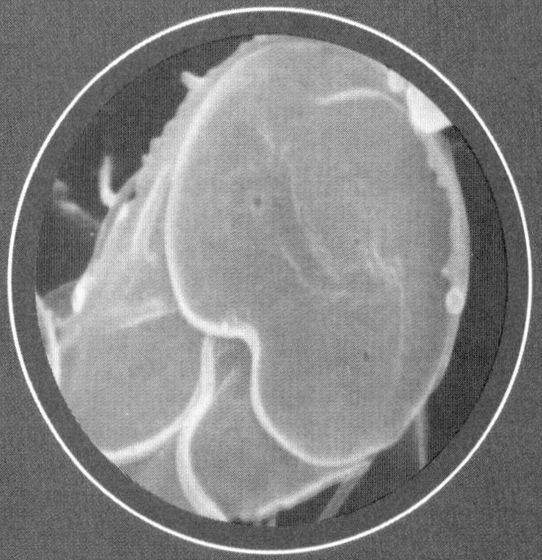

CHAPTER 23

Microbial Diseases of the Digestive System

OBJECTIVES

After completing this chapter you should be able to

- Describe the antimicrobial features of the digestive system.
- Describe the events that lead to the formation of dental caries and periodontal disease.
- List the etiologic agent, suspect foods, symptoms, and treatment for each of the following: staphylococcal food poisoning; salmonellosis; typhoid fever; bacillary dysentery (shigellosis); Asiatic cholera; gastroenteritis.
- List the etiologic agent, method of transmission, site of infection, and symptoms for mumps, cytomegalovirus inclusion disease, infectious hepatitis, and serum hepatitis.
- Compare and contrast giardiasis, balantidiasis, and amoebic dysentery.
- Compare and contrast tapeworm infestations and trichinosis.
- Discuss preventative measures for infections of the gastrointestinal system.

The body's major sites of invasion are the skin and the mucous membranes of the respiratory and digestive systems. The most common routes by which pathogens exit the body are also the respiratory and digestive systems. Having infected the digestive system or other organs that are related by the blood to the digestive system, many pathogens are excreted with feces. Although the various respiratory diseases are the leading cause of illness in the United States, diseases of the digestive system rank second.

Most diseases of the digestive system result from the ingestion of food or water that contain microorganisms and their toxins. When food handlers practice good sanitation methods, including proper handwashing and fly control, contamination of food is prevented. And modern methods of sewage treament and disinfection of drinking water help break the fecal–oral cycle of disease. Preservation of foods by heat and refrigeration also minimizes contamination by disease-causing microorganisms.

STRUCTURE AND FUNCTION OF THE DIGESTIVE SYSTEM

The **digestive system** may be divided into two principal groups of organs (Figure 23–1). One group is the *gastrointestinal (GI) tract* or *alimentary canal,* essentially a tubelike structure that includes the mouth, pharynx (throat), esophagus (food tube), stomach, small intestine, and large intestine. The other group of organs, the *accessory structures,* consists of the teeth, tongue, salivary glands, liver, gallbladder, and pancreas. Except for the teeth and tongue, the accessory structures lie outside the tract and produce secretions conveyed by ducts into the tract.

The purpose of the digestive system is to digest foods, that is, to break them down into small molecules that can be taken up and used by body cells. The combined actions of enzymes in saliva, stomach enzymes, pancreatic enzymes, intestinal enzymes, and bile from the gallbladder convert ingested nutrients into the end products of digestion. Thus, carbohydrates are reduced to simple sugars,

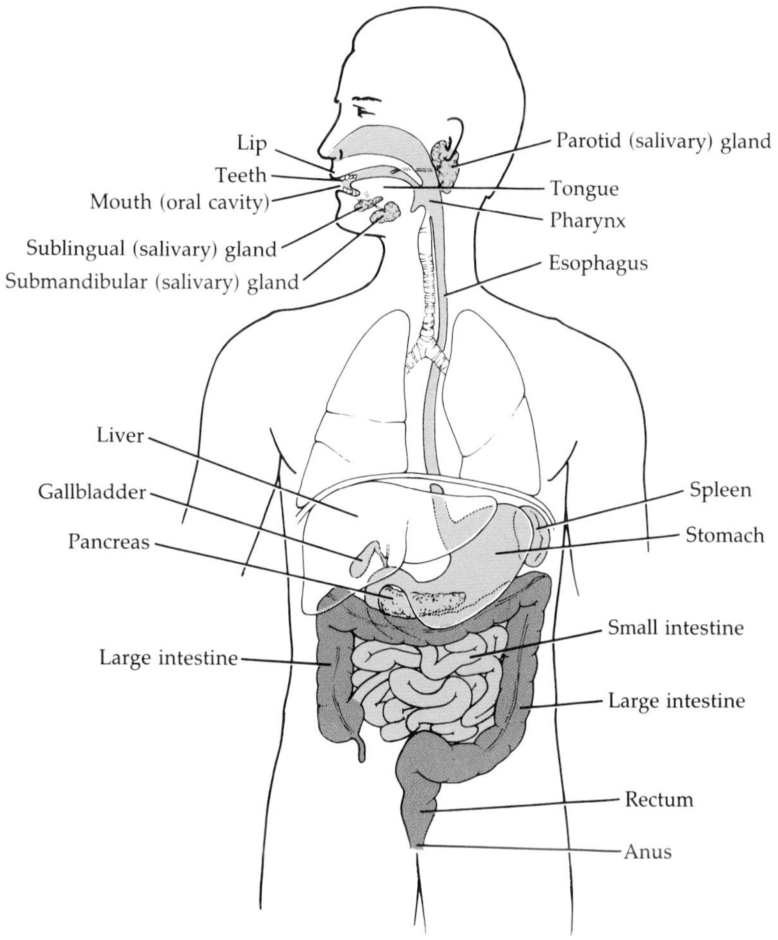

Figure 23–1 Anatomy of the human digestive system.

fats to fatty acids and glycerol, and proteins to amino acids. In a process called *absorption,* these end products of digestion then pass from the small intestine into the blood or lymph, for distribution to body cells.

By the time the remaining food leaves the small intestine, digestion and absorption are almost complete. As the food moves through the large intestine, water, vitamins, and nutrients are absorbed from it and microbial cells are added to it. The resulting undigested solids, called *feces,* are stored in the rectum until they are eliminated from the body through the anus, a process called *defecation.*

ANTIMICROBIAL FEATURES OF THE DIGESTIVE SYSTEM

To prevent being infected from its intimate contact with food, the digestive system has several defenses. The digestive system, like the skin and respiratory system, is in contact with the external environment. And, like the skin and respiratory system, it has certain adaptations that protect it from microorganisms. First of all, saliva contains lysozyme, mucins, and IgA antibodies. Lysozyme exerts antimicrobial activity by bringing about the lysis of gram-positive bacteria. Both mucins and IgA antibodies can coat bacteria and prevent them from attaching to the surfaces of the teeth, gums, and mucous membrane of the mouth. Some of the stomach cells produce a second level of defense: hydrochloric acid (HCl), which destroys many types of microorganisms. Third, the walls of the small and large intestine contain patches of lymphoid tissue that provide immunity against certain microorganisms.

Finally, the normal flora of the intestines inhibits the growth of certain other microorganisms. Like normal floras of other parts of the body, those of the intestines have a number of methods for outgrowing pathogens and other invaders. By secreting acid as a by-product of metabolism, some members of the normal flora create an inhospitably acidic environment for pathogens. Others are more efficient than the invaders in using the available nutrients. Those flora that have shorter generation times crowd out any newcomers.

Highly specific growth inhibitors might also assume a role in maintaining the flora and checking the growth of pathogens. For example, certain strains of *E. coli* can produce an inhibitor called *colicin* that kills several strains of *E. coli* but has no effect on other strains of the same species. There are 22 different kinds of colicin labeled A to V. Similar substances are produced by bacteria other than enterics. Strains of *Pseudomonas aeruginosa* produce *pyocins,* and strains of *Bacillus megaterium* (bä-sil'lus meg-ä-tėr' ē-um) produce *megacins.* Together, colicins, pyocins, and megacins are known as *bacteriocins.* The formation of a particular bacteriocin is due to a corresponding plasmid called a *bacteriocinogen.* Bacteriocins are proteins that attach to specific receptors on the outer membranes of certain microorganisms. Some bacteriocins act by altering membrane permeability, while others inhibit protein synthesis after entering the cell. It is worth noting that if the delicate balance of normal flora is changed by antibiotic therapy, normally harmless microorganisms can become opportunistic pathogens.

NORMAL FLORA OF THE DIGESTIVE SYSTEM

Bacteria heavily populate most of the digestive system. In the mouth, each milliliter of saliva can contain millions of bacteria with various bacteria distributed in specific areas (Figure 23–2). For example, different types of streptococci, which utilize sugars to form lactic acid, are found in different regions of the mouth. *Streptococcus salivarius* (sa-li-vā' rē-us) colonizes the surface of the tongue, *S. mitis* (mī'tis) inhabits mostly the mucosa of the cheek, and *S. sanguis* (san'gwis) colonizes the teeth. Strict anaerobes, such as *Bacteroides* and *Fusobacterium,* and spirochetes inhabit the spaces where the gums and teeth merge.

The stomach and small intestine have relatively few microorganisms, because of the hydrochloric acid produced by the stomach and the rapid movement of food through the small intestine. By contrast, the large intestine has enormous microbial populations exceeding 100 billion bacteria per gram of feces. (Up to 40% of fecal mass is microbial cell material.) The population of the large intestine is mostly anaerobes of the genera *Lactobacillus* (lak-tō-bä-sil'lus) and *Bacteroides* (bak-tė-roi'dēz), and facultative anaerobes such as *E. coli, Enterobacter* (en-te-rō-bak'tėr), *Citrobacter* (sit' rō-bak-tėr), and *Proteus* (prō'tē-us). Most of these bacteria assist in the enyzmatic breakdown of foods, and some of

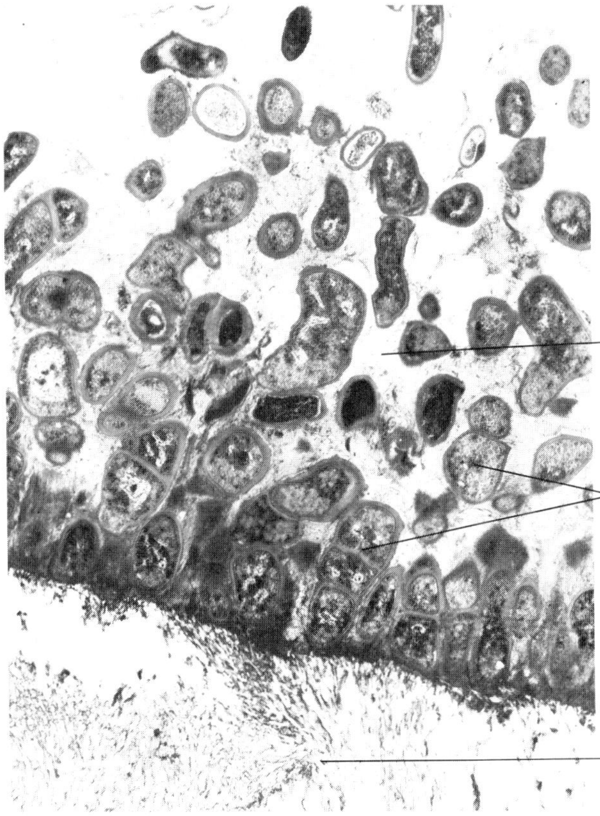

Figure 23–2 Normal flora of teeth (×11,000). This transmission electron micrograph shows a section cut through the enamel (white area at bottom), and the plaque on the surface of the enamel. Plaque is made of dextran and bacteria—normal flora of the mouth. Compare this with Figure 23–4.

them synthesize useful vitamins, such as niacin, vitamins B_1, B_2, B_6, and B_{12}, folic acid, biotin, and vitamin K.

BACTERIAL DISEASES OF THE DIGESTIVE SYSTEM

In discussing bacterial diseases of the digestive system, we will divide the diseases into those associated with the mouth and those associated with the lower digestive system.

Diseases of the Mouth

Dental caries (tooth decay)

Dental caries (tooth decay) involve a gradual softening of the enamel and dentin of a tooth (Figure 23–3). If the condition remains untreated, various bacteria can invade the pulp and cause death (necrosis) of the pulp and abscess of the bone surrounding the tooth. (Such a tooth must be treated by root canal therapy.) You might be surprised to learn that dental caries were not common in the Western world until about the seventeenth century. In human remains from older times, only about 10% or fewer of teeth contain caries. The introduction of table sugar, or sucrose, into the diet correlates highly with our present level of caries in the Western world. Oral bacteria convert sucrose and other carbohydrates into acid, which in turn attacks the tooth enamel.

Bacteria that convert sugar into lactic acid are commonplace in the oral cavity, but most of them are not important contributors to tooth decay. For caries to form, there must be a localization of the acid production. If a single bacterial species can be

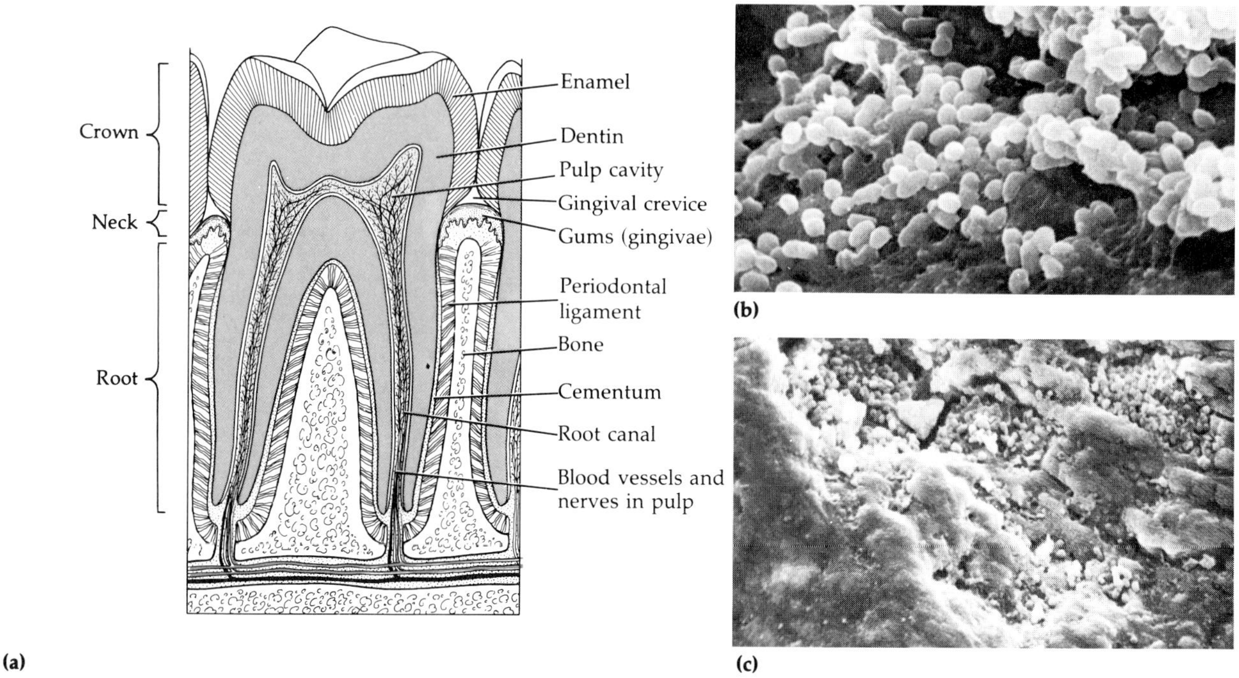

Figure 23–3 Dental caries. **(a)** Structure of a normal tooth. **(b)** Spherical cells of *Streptococcus mutans* adhering to tooth enamel (×3700). **(c)** Scanning electron micrograph showing a carious lesion of tooth enamel, with numerous cells of *S. mutans* in the eroded area (×1300).

said to be cariogenic (causing caries), that species is *Streptococcus mutans* (mū´tans), a gram-positive coccus (Figure 23–3b and c). The organisms (some believe that it is a group of related organisms) become established in the mouth when the teeth erupt from the gums. They adhere to the teeth, especially in crevices and contact points between teeth, not to the epithelial surfaces of the mouth.

S. mutans produces a sticky polysaccharide of glucose molecules called **dextran** (sometimes called glucan), with which the bacteria adhere firmly to the teeth. In the production of dextran, sucrose is first hydrolysed into its component monosaccharides, fructose and glucose. The enzyme glucosyl transferase then assembles the glucose molecules into dextran. The masses of bacterial cells and dextran adhering to the teeth are what we know as **dental plaque** (Figure 23–4). The bacteria in plaque ferment lactic acid from sucrose (Figure 23–5). The primary sugar used in acid production is the fructose that was released by the hydrolysis of sucrose. Because plaque is not very permeable to saliva, the lactic acid is not diluted or neutralized, and breaks down the enamel of the teeth to which the plaque adheres. It is the localized acid-producing bacteria in the plaque that initiate dental caries.

Sucrose is the dominant cause of dental caries. Because the bacteria use both glucose and fructose, people living on high-starch diets (starch is a polysaccharide of glucose) have a low incidence of caries unless sucrose is also part of their diet. Studies have shown that glucose or fructose individually are cariogenic, but are much more so when combined as sucrose. Because the sugars mannitol, sorbitol, and xylitol are not at all cariogenic, they are used to sweeten "sugarless" chewing gums.

No drugs are now available to prevent tooth decay. A vaccine, based on IgA antibodies in the saliva, is theoretically possible, and tests conducted on animals have given encouraging results. However, at present the main strategies in the prevention of caries are minimal ingestion of sucrose;

brushing, flossing, and professional cleaning to remove plaque; and the use of fluorides to provide the teeth with resistance against acidic action.

Periodontal disease

Even persons who avoid tooth decay might, in later years, lose their teeth to **periodontal disease.** Periodontal disease is a collective term for a number of conditions characterized by inflammation and degeneration of the gums, supporting bone, periodontal ligament, and cementum (Figure 23–6). The disease is probably initiated by bacterial growth in the gingival crevice, although the exact mechanism is unclear. In any event, plaque deposited in the gingival crevice causes an initial inflammatory reaction. Species of *Actinomyces*, *Nocardia*, and *Corynebacterium*, among other genera, are involved in plaque formation. The formation of plaque can eventually lead to formation of abscesses, bone destruction, and general tissue necrosis.

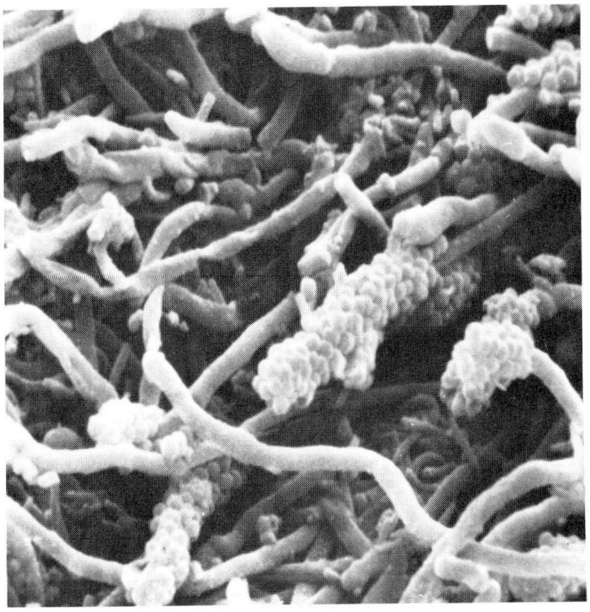

Figure 23–4 Dental plaque. After three days of not being brushed, the surface of a tooth has accumulated these organisms. Also see Figure 24–3b.

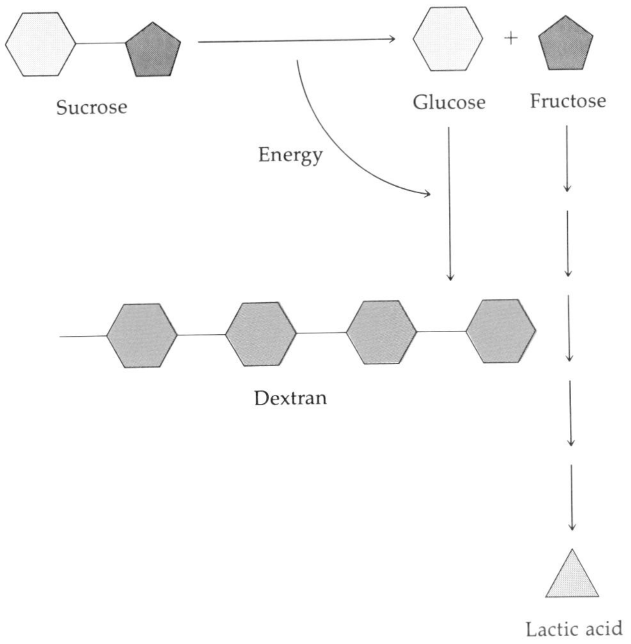

Figure 23–5 Conversion of sucrose to dextran (dental plaque) and lactic acid by *Streptococcus mutans*. Dextran is a polysaccharide formed from the glucose molecule; fructose is the primary sugar fermented to lactic acid.

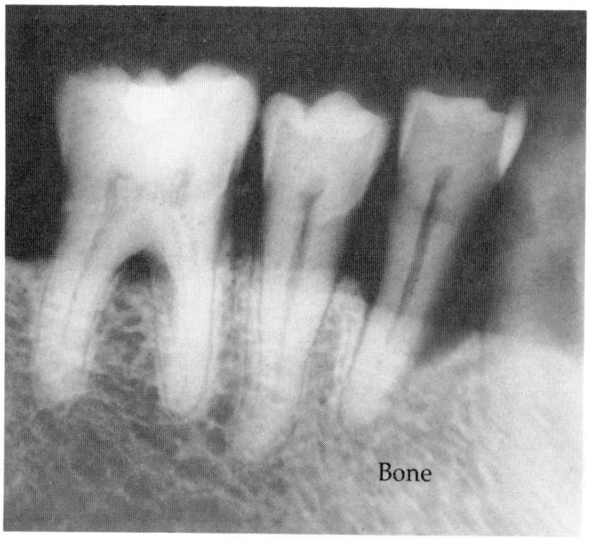

Figure 23–6 X ray of a severe case of periodontal disease. The bone has degenerated away from the upper portions of the two left teeth and from all but the tip of the right tooth.

Some periodontal disease can be controlled by frequent removal of plaque by brushing, flossing, and professional cleaning. Other forms of the disease require antibiotics such as penicillin and tetracycline.

Diseases of the Lower Digestive System

Most diseases of the digestive system are contracted when contaminated food or water is ingested. These diseases are essentially of two types: infections and intoxications. An **infection** occurs when a pathogen enters the gastrointestinal (GI) tract and multiplies. Such a microorganism can penetrate into the intestinal mucosa and grow there or can pass through to other systemic organs. Other intestinal pathogens cause disease by **intoxication,** that is, by elaborating toxins that affect the GI tract. These toxins can be excreted by the pathogens as they grow in the intestines, or they can be released by lysis of the cells upon their death. General characteristics of an infection include a delay in the appearance of symptoms; meanwhile, the pathogen increases in numbers or affects invaded tissue. There is also usually a fever, one of the body's responses to an infective organism.

An intoxication is due to the ingestion of a preformed toxin. Most intoxications, such as that caused by *Staphylococcus aureus,* are characterized by a very sudden appearance (usually in only a few hours) of symptoms of a GI disturbance. Fever is less often one of the symptoms. Both infections and intoxications often cause *diarrhea,* which most of us have experienced. Severe diarrhea, accompanied by blood or mucus, is termed *dysentery.* Both types of digestive system diseases are also frequently accompanied by *abdominal cramps, nausea,* and *vomiting.* The general term *gastroenteritis* is applied to diseases causing inflammation of the stomach and intestinal mucosa. Botulism (Chapter 20) is a special case of intoxication because the ingestion of the preformed toxin affects the nervous system rather than the GI tract.

Food poisoning is a term commonly used for gastroenteritis that results from the ingestion of food. In a purely technical sense, this is not a very satisfactory term but it is still commonly used even in the medical literature. The term *ptomaine poisoning* is a term from days before the association was recognized between illness and microbial contamination of food; the association between illness and obviously spoiled food had been made. At that time, illness was believed to be caused by *ptomaines,* the term then used for partially decomposed proteins. The theory was wrong, but the name has remained in common use. It is not an acceptable medical term.

Staphylococcal food poisoning (staphylococcal enterotoxicosis)

A very common cause of gastroenteritis is **staphylococcal food poisoning.** This is an *intoxication* caused by ingestion of an *enterotoxin* produced by *Staphylococcus aureus.* Staphylococci are comparatively resistant to environmental stresses, as we discussed in a previous chapter. Vegetative cells can tolerate 60°C for half an hour; this is a fairly high resistance to heat. Staphylococci are also resistant to drying, radiation, and high osmotic pressures; these resistances help them survive on skin surfaces and grow in foods in which high osmotic pressure or salts inhibit growth of competitors.

S. aureus (Figure 23–7) is a common inhabitant of the nasal passages, from which it often contaminates the hands. It is also a frequent cause of skin lesions on the hands. From these sources it can readily enter food. If the organisms are allowed to incubate in the food (often termed *temperature abuse*), they reproduce and release enterotoxin into the food. The large numbers of organisms produce the enterotoxin in the food. Generally, a population of about one million bacteria per gram will produce enough enterotoxin to cause illness. The growth of the organism is facilitated if the competing organisms in the food have been eliminated, by cooking, for example. The organism is also more likely to increase to large numbers if competing bacteria are inhibited by a higher-than-normal osmotic pressure or by a relatively low moisture level. *S. aureus* tends to outgrow most competing bacteria under these conditions. Certain foods have an unusually high incidence of association with staphylococcal food poisoning. Custard and cream pies, for example, have a relatively high osmotic pressure because of their sugar content. They also have regions of low moisture that are still adequate for staphylococcal growth. If the pie is cooked during preparation, most competing organisms are eliminated. These foods can be readily inoculated with *S. aureus* by subsequent handling, and the bacteria multiply during storage.

Another high-risk food is ham, which contains many salts as preservatives or flavorings. These curing agents inhibit competing organisms more than they inhibit staphylococci. Poultry products can also harbor staphylococci if they are handled and allowed to stand at room temperatures—but in fact, almost every food has been incriminated in staphylococcal food poisonings at one time or another. On the other hand, foods such as hamburger are rarely associated with any type of food poisoning because they have normal flora that are not competitive with staphylococci, and are cooked immediately before consumption. Picnic foods that are prepared early in the day, are not kept chilled, and are not eaten until later are a potential source of staphylococcal food poisoning.

The toxin itself is heat stable, and can survive up to 30 minutes of boiling. So, once the toxin is formed, it is not destroyed when the food is reheated, although the bacteria will be killed.

Staphylococcal food poisoning is characterized by nausea, vomiting, and diarrhea one to six hours after the contaminated food is ingested. The symptoms usually disappear in 24 hours. The disease has been identified in 25% of the outbreaks of food poisoning of known etiology in recent years.

The mortality rate of staphylococcal food poisoning is almost nil among otherwise healthy individuals, but can be significant in weakened individuals such as residents of nursing homes. No reliable immunity results from recovery. However, there is a great deal of variation in individual susceptibility to the toxin, and it is suspected that an immunological basis due to prior exposure might account for some of this variation.

S. aureus produces several toxins that damage tissues or increase the organism's virulence. The enterotoxins causing food poisoning are classified

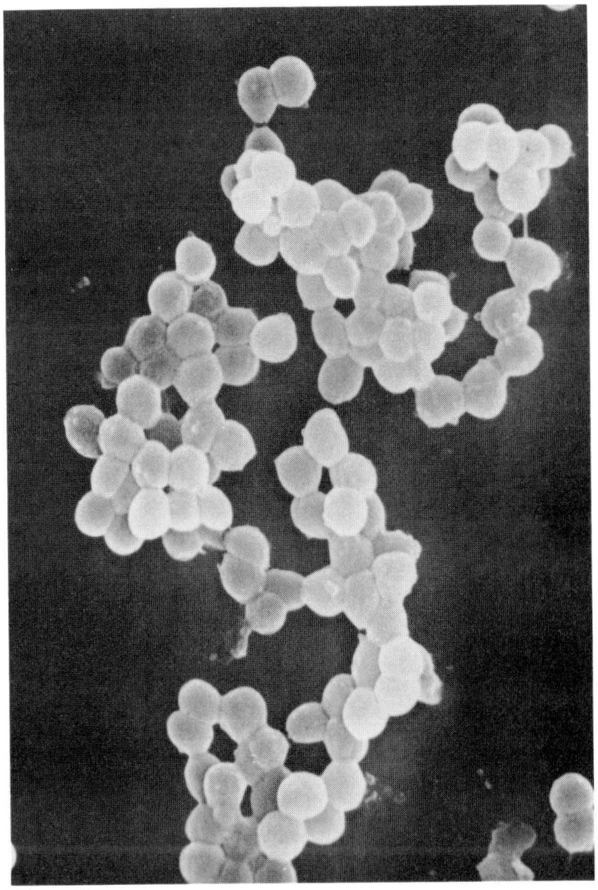

Figure 23–7 Grapelike clusters of *Staphylococcus aureus*.

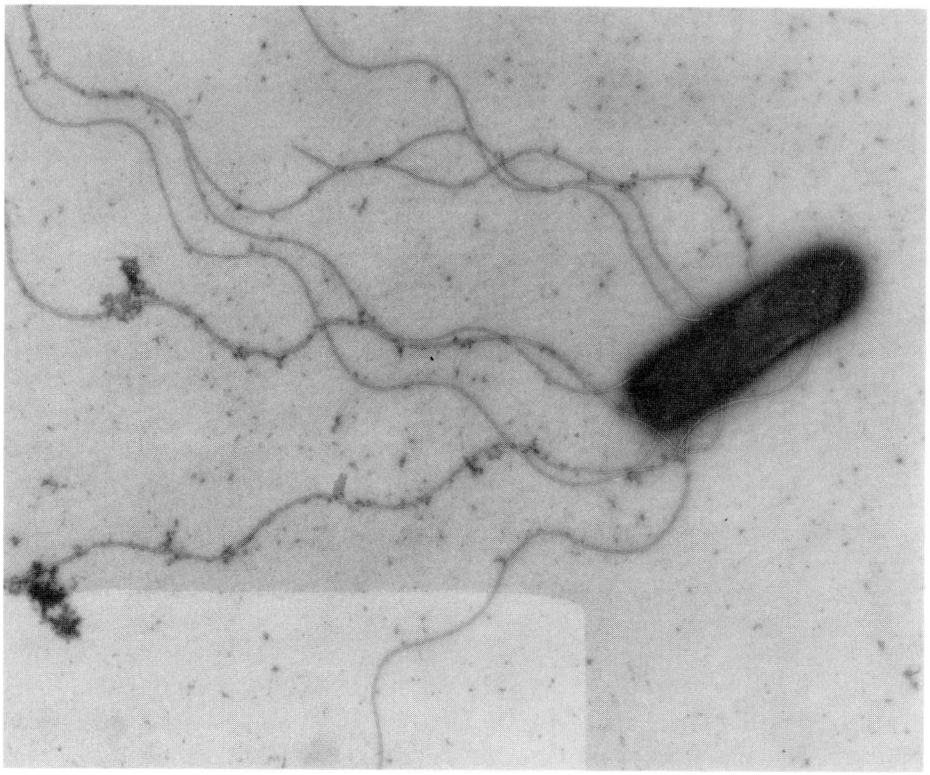

Figure 23–8 *Salmonella* species.

as serological types A (which is responsible for most cases) through D. Production of this toxin is usually correlated with production of an enzyme that coagulates blood plasma. Such bacteria are described as *coagulase-positive* (see Chapter 10). No direct pathogenic effect can be attributed to the enzyme, but it is useful in the tentative identification of types likely to be virulent. Coagulase-positive strains are considered potentially pathogenic.

The diagnosis of staphylococcal food poisoning is usually based on the symptoms, particularly the short incubation time characteristic of an intoxication. If the food has not been reheated and the bacteria killed, the organism can be recovered and grown. *S. aureus* isolates can be tested by the *phage typing* method, which is useful in tracing the source of the contamination (Chapter 9). Because the bacteria will grow well in 10% sodium chloride, this concentration is often used in selective media used for their identification. Pathogenic staphylococci usually ferment mannitol, produce blood hemolysins and coagulase, and form yellow colonies. They cause no obvious spoilage when growing in foods.

No simple procedure will identify the toxin in foods. Identification requires a complex extraction and a method of serological testing against antisera; these procedures are not commercially available. Food can be easily tested, however, for the presence of *thermostable nuclease*. Produced by *S. aureus*, this enzyme survives heating that will kill the cells themselves. It can be detected by a simple test in which an extract of the food is allowed to react with DNA. Presence of detectable amounts of the enzyme is usually related to a microbial population large enough to have produced sufficient toxin to cause illness.

Because contamination of foods cannot be avoided completely, the most reliable method of preventing staphylococcal food poisoning is adequate refrigeration during storage to prevent toxin formation.

Because considerable amounts of water and

electrolytes are lost during episodes of vomiting and diarrhea, treatment for the disease consists of replacing the lost water and electrolytes.

Salmonellosis (*Salmonella* gastroenteritis)

As noted earlier, in bacterial infections the disease results from microbial growth in body tissues rather than the ingestion of food and drink already contaminated by microbial growth. Compared with bacterial intoxications, bacterial infections such as salmonellosis usually have longer incubation periods (from 12 hours to 2 weeks), needed for the organism to grow in the tissues of the host, and by some fever indicative of the host's response to the infection.

The *Salmonella* (sal-mōn-el'lä) bacteria (named for a Dr. Salmon) are gram-negative, facultatively anaerobic, non-endospore-forming, usually motile rods that ferment glucose to produce acid and gas (Figure 23–8). Their normal habitat is the intestinal tracts of humans and many animals. All salmonellae are considered pathogenic to some degree. The relatively mild infection that is known as **salmonellosis, *Salmonella* gastroenteritis,** or ***Salmonella* food infection,** is caused by organisms whose infective dose (often in the millions) varies (see box).

Centers for Disease Control

MMWR

Morbidity and Mortality Weekly Report

Milk-borne Salmonellosis

The number of culture-confirmed cases of salmonellosis reported to the Illinois Department of Public Health during the outbreak of milk-borne salmonellosis, which began March 22, reached 5,770 on April 16. Fifty-eight percent of the first 765 cases occurred among persons under 10 years of age. *Salmonella typhimurium*, resistant to ampicillin and tetracycline, has been isolated from patients and from milk in unopened cartons of two lots of 2% milk, dated March 29 and April 8. Both lots were from the same dairy plant in Illinois, which stopped producing milk April 9. The milk was sold in Jewel, Eisner, and Magna supermarkets in Illinois, Indiana, Iowa, and Michigan. All milk produced by the plant has been removed from sale. Although the plant produces milk with several different concentrations of butterfat, thus far only 2% milk has been strongly implicated. Investigations of the plant by state, federal, and industry officials are continuing to determine the precise cause of the contamination of the milk.

By April 16, the Illinois Department of Public Health had received reports of milk-associated, culture-confirmed cases of salmonellosis from the three other states where the milk was distributed—Indiana (289 cases), Michigan (43), and Iowa (28). In addition, three state health departments (Minnesota, Wisconsin, and Florida) have reported a total of 19 cases among persons returning to their states.

Editorial Note: This is the largest number of culture-confirmed cases ever associated with a single outbreak of salmonellosis in the United States. Although *Salmonella* is sometimes found in dairy cattle and in raw milk, pasteurization kills *Salmonella*. The implicated milk underwent the pasteurization process, suggesting that it was either inadequately pasteurized or contaminated after pasteurization. Pasteurized milk constitutes approximately 99% of all (cow) milk consumed in the United States, but milk-borne outbreaks of *Salmonella* investigated by CDC in the past have almost always involved raw milk because effective pasteurization kills *Salmonella*. The large number of affected persons in this outbreak illustrates how a widely consumed product, once contaminated, can result in many cases. Similar widespread transmissions of *Salmonella* occurred in a waterborne outbreak involving an estimated 16,000 people (100 reported cases) in Riverside, California, in 1965 and in an estimated 3,400 affected Navajo Indians (105 investigated cases) at a barbecue on a reservation in 1974.

Abridged from: *MMWR 34*:200 (4/12/85) and *MMWR 34*:215 (4/19/85).

The nomenclature of the *Salmonella* organisms is confusing. According to *Bergey's Manual,* none of the present methods for naming salmonellae is satisfactory from a scientific viewpoint. Rather than recognized species, there are more than 2,000 *serovars* (closely related organisms differentiated by serological testing). Only about 50 such serovars are isolated with any frequency in the United States. Some of these are named much like species: for example, *Salmonella dublin* and *Salmonella eastbourne* (after the sites where they were first isolated), and *Salmonella typhimurium* (which causes typhoid-like disease in mice). Others are represented by the Kauffmann–White scheme (see discussion of *Salmonella* in Chapter 10), which is generally used in clinical laboratories. This method assigns numbers and letters to different antigens: O (somatic or body), Vi (capsular), and H (flagellar).

Salmonellosis has an incubation time of about 12 to 36 hours. The salmonellae first invade the intestinal mucosa and multiply there. Sometimes they manage to pass through the intestinal mucosa and enter the cardiovascular system, in which they spread to eventually affect many organs. The fever associated with salmonellae infections might be due to endotoxins released by lysed cells, but this relationship is not certain. There is usually a moderate fever accompanied by nausea, abdominal pain and cramps, and diarrhea. As many as one billion salmonellae per gram can be found in the infected person's feces during the acute phase of the illness.

The mortality rate is overall very low, probably lower than 1%. However, the death rate is high in infants and among the very old; death is usually due to septicemia. Individual responses to the infection vary considerably. The severity and incubation time can depend on the number of *Salmonella* ingested. Normally, recovery will be complete in a few days, but many patients will continue to shed the organisms in their feces for up to six months.

Salmonellosis is probably greatly underreported. The 20,000 to 30,000 cases reported each year are probably only 1 to 10% of the actual total.

Meat products are particularly susceptible to contamination by *Salmonella,* and if mishandled, can grow to infective levels rather quickly. The sources of the bacteria are the intestinal tracts of many animals, and meats can be contaminated readily in processing plants. Poultry, eggs, and egg products are often contaminated by *Salmonella.* The organisms are generally destroyed by normal cooking, which heats the food to an internal temperature of at least 68°C (145°F). However, foods can be contaminated after cooking by mishandling.

Prevention of salmonellosis depends on good sanitation practices to deter contamination and on proper refrigeration to prevent increases in bacterial numbers. Bear in mind that contaminated food can contaminate a surface such as a cutting board. Although the food first prepared on the board might later be cooked and its bacteria killed, a food such as salad subsequently prepared on the board might not be cooked.

Diagnosis is generally made on isolation of the organisms from leftover foods or stools of patients. Antibiotic therapy is not usually useful in treating salmonellosis, so treatment consists of replacing water and electrolytes lost through diarrhea.

Typhoid fever

A few serovars of *Salmonella* are much more virulent than others and produce an intestinal disease in which they cross the intestinal wall and become invasive. The most virulent species, *Salmonella typhi* (tī′fē), causes the bacterial infection **typhoid fever.** The incubation period, much longer than that of salmonellosis, is normally about two weeks. Diarrhea is usually absent, but fever and malaise lasting for two or three weeks are typical. The organism becomes disseminated in the body and can be isolated from the blood, urine, and feces. In severe cases, there can be perforation of the intestinal wall. The mortality rate is now about 1 to 2%; at one time it was about 10%. Before the days of proper sewage disposal, water treatment, and food sanitation, typhoid was an extremely common disease. It is still a frequent cause of death in parts of the world with poor sanitation.

A substantial number of recovered patients become carriers. They harbor the pathogen in the gallbladder and continue to shed the pathogen for several months. A certain number of such carriers continue to shed the organism indefinitely. Most of us are familar with the term "Typhoid Mary." This was Mary Mallon, who worked as a cook in New York state in the early part of the century and was responsible for several outbreaks of typhoid and three deaths. Her case became well known through the attempts of the state to restrain her from working at her chosen trade. In each recent year there have been about 500 cases. The Centers for Disease Control usually lists 60 to 70 carriers each year. More than half of the cases in recent years were acquired during foreign travel. Many of those who acquire typhoid in the United States are persons such as migrant workers who use unsanitary facilities. Normally, there are fewer than 10 deaths each year.

Despite its known toxicity, chloramphenicol is often the drug of choice for the treatment of typhoid fever. However, drug-sensitivity testing is required because of the continuing appearance of resistant strains. Ampicillin and trimethoprim-sulfamethoxazole are usually effective alternatives. Recovery from typhoid confers a lifelong immunity.

Immunization for typhoid is not normally done except for high-risk laboratory and military personnel. The vaccine presently used is a killed-organism type and must be injected; infection poses a problem in endemic areas of the world. Field trials with a newly developed, orally ingested, live, attenuated vaccine have been promising.

Bacillary dysentery (shigellosis)

Bacillary dysentery (shigellosis) is a severe form of diarrhea that is characterized by mucus and blood in the stools. Symptoms include abdominal cramps and fever. Bacillary dysentery is a bacterial infection caused by a group of facultatively anaerobic gram-negative rods of the genus *Shigella* (shi-gel'lä). These intestinal bacteria of humans or higher primates are not so invasive as the salmonellae, and *Shigella* infections are usually limited to the large intestine (see Microview 3–3).

There are four species of pathogenic *Shigella: S. sonnei* (sōn'nē-ī), *S. dysenteriae* (dis-en-te'rē-ī), *S. flexneri* (fleks'nėr-ī), and *S. boydii* (boi'dē-ī). *S. sonnei,* the most common in the United States, causes a relatively mild dysentery. At the other extreme, *S. dysenteriae* infection results in a severe dysentery and prostration. Ulcerations are formed in the intestinal mucosa, which eventually heal but form scar tissue. Although some strains of *S. dysenteriae* produce an endotoxin (neurotoxin), its role in pathogenesis is still not clear. It is known, however, that toxin production alone does not cause virulence; invasiveness seems to be the more important factor. Tests with human volunteers indicate that only a few hundred cells will cause shigellosis.

Diagnosis is usually based on recovery of the organisms from rectal swabs. Differentiation from amoebic (protozoan) dysentery, which is discussed later, is usually made from examination of the feces. A large number of leukocytes is likely to be present in the blood of a person with bacillary dysentery.

In recent years, the number of cases reported in the United States has been about 15,000 to 20,000, with 20 to 35 (around 0.2%) deaths. However, the death rate in tropical areas can be much higher, perhaps 20%. The disease is probably more common than the reported numbers indicate. Some cases of so-called tourist diarrhea might be mild forms of bacillary dysentery. This dysentery is especially common in institutionalized patients and in persons living on Indian reservations. Some immunity seems to result from recovery, but no satisfactory vaccine has yet been developed.

In severe cases of bacillary dysentery, antibiotic therapy and fluid and electrolyte replacement are indicated. At present, ampicillin is the drug of choice, but some resistance has appeared. Trimethoprim-sulfamethoxazole is a frequently used alternative.

Asiatic cholera

During the 1800s, the bacterial infection called **Asiatic cholera** crossed Europe and North America in repeated epidemics. Today, it is endemic in Asia, particularly India, and has only occasional outbreaks in Western countries. These outbreaks are due to temporary lapses in sanitation, and the number of cases is limited. In recent years, there have been a number of outbreaks in the Gulf coast region of the United States. Most of these cases have been traced to seafood contaminated by the causative organism *Vibrio cholerae* (vib'rē-ō kol'ėr-i), which can grow in shallow coastal waters. The bacterium is a slightly curved, gram-negative rod, with a single polar flagellum (Figure 23–9).

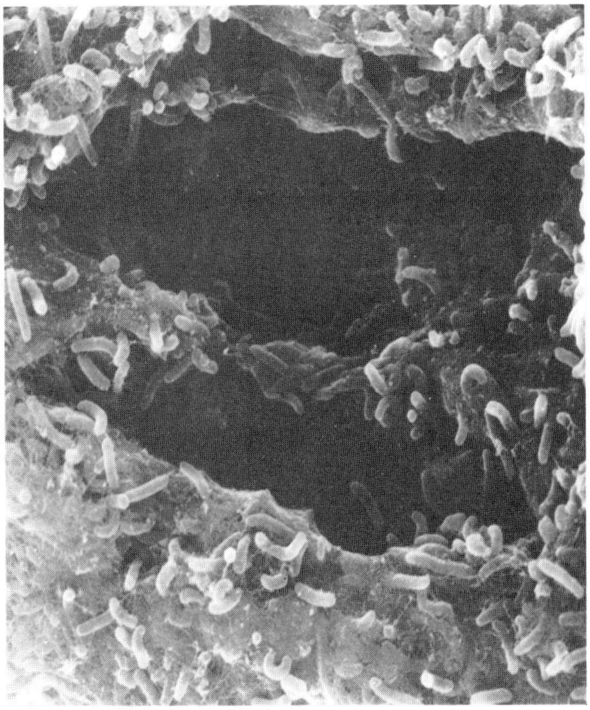

(a)

(b)

(c)

Figure 23–9 *Vibrio cholerae*, the causative agent of Asiatic cholera. **(a)** Note the slightly curved morphology in the scanning electron micrograph. **(b)** Normal small intestine of an infant rabbit. **(c)** Intestine of an infant rabbit after infection with *V. cholerae*.

The serogroup-01 of *V. cholerae* causes the classically recognized epidemic form of the disease. *V. cholerae* that are not of serogroup-01 (non-01) cause a very similar disease, but not the rapidly spreading epidemic characteristic of cholera. The disease, like that of other enteric bacteria, is spread by the fecal-oral route, frequently via contaminated water. The incubation period is usually less than three days.

The organisms remain in the intestinal tract and are not invasive. They produce an exotoxin, called *enterotoxin,* which causes the intestinal wall to become extraordinarily permeable. Body fluids and mineral electrolytes move through the wall and are excreted from the body. While semisolid stools might appear early in the disease, the excretions soon take on the typical appearance of "rice water stools." This appearance is due to the masses of intestinal mucus, epithelial cells, and bacteria. The sudden loss of these fluids and electrolytes (12 liters of fluid might be lost in one day) causes shock, collapse, and often death. Because of the loss of fluid, the blood becomes so viscous that vital organs are unable to function properly. Violent vomiting is also a characteristic of the disease. The severity of the disease varies considerably, usually because different strains of the pathogen can be the cause. The number of subclinical cases might be several times those that are recognized.

The bacteria can be readily isolated from the feces, partly because the organisms can grow in media alkaline enough to suppress the growth of many other organisms.

The most reliable control methods are based on principles of sanitation, particularly of water supplies. The development of several different oral vaccines is underway. These vaccines would be a considerable improvement over currently used, injected vaccines, because the new vaccines would stimulate the production of antibodies in parts of the intestinal tract where the organism is infective.

Recovery from the disease results in an effective immunity based on the antigenic activity of both the cells and the enterotoxin. However, because of antigenic differences between bacterial strains, the same person can have cholera more than once. Most victims in endemic areas are children.

Chemotherapy (tetracycline or chloramphenicol) is usually employed but is not very effective. Treatment essentially replaces the lost fluids and electrolytes. Untreated cases of cholera have a 50% mortality rate, whereas the rate for cases having proper supportive care can be as low as 1%.

Vibrio parahaemolyticus gastroenteritis

Vibrio parahaemolyticus (pa-rä-hē-mō-lī'ti-kus) was first recognized as a pathogen in Japan in 1950. It is morphologically similar to *V. cholerae,* but differs in that it is halophilic, and requires 2 to 7% sodium chloride for growth. It is the most common cause of **gastroenteritis** in Japan, and causes thousands of reported cases annually. It is also a common cause of gastroenteritis in many nations of southeast Asia. The organism is present in coastal waters of the continental United States and Hawaii. Crustaceans such as shrimp and crabs have been associated with several outbreaks caused by *V. parahaemolyticus* in the United States in recent years. Symptoms include a burning sensation in the stomach and abdominal pain. Vomiting and watery stools not unlike those of Asiatic cholera are also characteristic. The organism has also been known to cause cutaneous infections of cuts that have come in contact with contaminated clams and oysters. Incubation time is normally less than 24 hours. Recovery usually follows in a few days and the fatality rate is low. Antibiotics such as tetracycline, chloramphenicol, and penicillin are used only in severe cases.

Because *V. parahaemolyticus* has a requirement for sodium, usually supplied as sodium choride, isolating media containing 2 to 7% sodium chloride are used in diagnosis of the disease.

Escherichia coli gastroenteritis (traveler's diarrhea)

The bacterium most familiar to microbiologists is probably *Escherichia coli,* one of the most prolific microorganisms in the human intestinal tract. *E. coli* is normally harmless, but certain strains acquire the ability to cause gastroenteritis by one of three mechanisms. Some produce an enterotoxin that is plasmid coded. These are referred to as *enterotoxigenic E. coli (ETEC).* Others that invade the epithelial lining of the large intestine are called *enteroinvasive E. coli (EIEC).* Still others have a distinctly different mechanism that is not understood; these are known as *enteropathogenic E. coli* and are the primary cause of **epidemic diarrhea** in nurseries, and probably the major cause of **traveler's diarrhea** acquired in foreign countries. In practice, it is difficult to differentiate isolates of pathogenic and nonpathogenic *E. coli.* In adults, the disease is usually self-limiting, and chemotherapy is not attempted. In any event, antibiotic resistance is widespread in pathogenic strains.

Traveler's diarrhea is treated by replacement of water and electrolytes. Infant diarrhea (epidemic diarrhea in nurseries) has been successfully treated with antibiotics. However, because the bacteria quickly develop resistance, nurseries must maintain constant surveillance to administer antibiotics that are still effective.

Probably 50 to 65% of traveler's diarrhea is caused by *enteropathogenic E. coli.* Of the remaining cases, 10 to 20% are probably caused by *Shigella. Salmonella, Campylobacter,* and an assortment of unidentified bacterial pathogens; viruses, and protozoan parasites are occasionally involved.

Traveler's diarrhea is almost impossible to avoid in parts of the world with poor practices of sanitation. But recently, the tetracycline-type antibiotic doxycycline has been found to greatly lower the incidence among persons who take the drug prophylactically a couple of days before departure and continue with daily doses during their stays. Trimethoprim-sulfamethoxazole has proved similarly useful, is less likely to encounter resistance problems, and is safer for young children.

Diarrhea is not only a nuisance to travelers to underdeveloped parts of the world, it is a serious health problem to the resident population. It is estimated that diarrhea kills at least five million children per year. The main danger of diarrhea in children is dehydration. It has been demonstrated that an oral solution of sodium chloride, potassium chloride, and sodium bicarbonate would save many lives. Even a solution containing only a bit of salt and a handful of table sugar would often be an adequate treatment.

Gastroenteritis caused by other gram-negative bacteria

In addition to the gram-negative organisms discussed so far, a number of others are known to cause gastrointestinal disorders in humans, and more are being discovered all the time. *Yersinia enterocolitica* (yėr-sin'ē-ä en-tėr-ō-kōl-it'ik-ä) and *Campylobacter jejuni* (kam-pī-lō-bak'tėr jē-jū'nē) are two gram-negative rods whose importance as human enteric pathogens has become apparent only in recent years. Both of these bacteria are found in the intestinal tract of several domestic and wild animals. They are transmitted to humans by contaminated food or water. They cause a gastroenteritis that is characterized by diarrhea, severe inflammation, and even intestinal ulceration.

Y. enterocolitica is noteworthy because it is capable of growing at refrigerator temperatures. Although the organism in milk is killed by standard pasteurization, there is evidence that if very large numbers are originally in the milk, a few bacteria can survive pasteurization and subsequently grow under refrigeration. *Y. enterocolitica* can cause an acute intestinal upset with fever, vomiting, and diarrhea. Abdominal pain has been severe enough to have occasionally caused a diagnosis of appendicitis.

C. jejuni was once best known as a pathogen of animals. Only in recent years has it been recognized as a significant cause of intestinal illness in humans. Some cases involve only a brief episode of diarrhea; persons with severe cases might suffer a prolonged illness with high fever and bloody stools. One billion organisms per gram of stool have been isolated at the peak of an infection.

Clostridium perfringens gastroenteritis

Probably one of the more common, if underrecognized, forms of food poisoning in the United States is caused by *Clostridium perfringens*, a large gram-positive, endospore-forming, obligately anaerobic rod. This is the organism responsible for human gas gangrene. It produces a wide array of toxins that damage tissues and cause gastrointestinal disturbances.

Most outbreaks are associated with meats or stews containing meats. The organism's nutritional requirement for amino acids is met by such foods, and when the meats are cooked, the oxygen level is lowered enough for clostridial growth. The endospores survive most routine heatings, and the generation time of the vegetative bacterium is less than 20 minutes under its ideal conditions. Large populations can therefore build up rapidly when foods are being held for serving, or when inadequate refrigeration leads to too-slow cooling.

The organism grows in the intestinal tract and produces an exotoxin that causes the typical symptoms of abdominal pain and diarrhea. The toxin alters the permeability of the intestinal wall, and the resulting loss of water and electrolytes produces the diarrheal symptoms. Most cases are mild and self-limiting and probably are never clinically diagnosed. The symptoms usually appear about 8 to 12 hours after ingestion.

Diagnosis is usually accomplished by isolation and identification of the organism in stool samples. Treatment consists of replacing lost water and electrolytes.

Bacillus cereus gastroenteritis

Bacillus cereus (se'rē-us) is a large, gram-positive, spore-forming organism that is very common in the environment and is usually considered harmless. It has, however, been identified as the cause of outbreaks of food-borne illness. Some cases resemble *C. perfringens* infections and are almost entirely diarrheal in nature. Other episodes involve nausea and vomiting. It is suspected that different toxins are involved in the differing symptoms. Both forms of the disease are self-limiting.

VIRAL DISEASES OF THE DIGESTIVE SYSTEM

Mumps

The parotid glands, the target of the mumps virus, are one of the three pairs of the salivary glands of the digestive system. Hence, the inclusion of

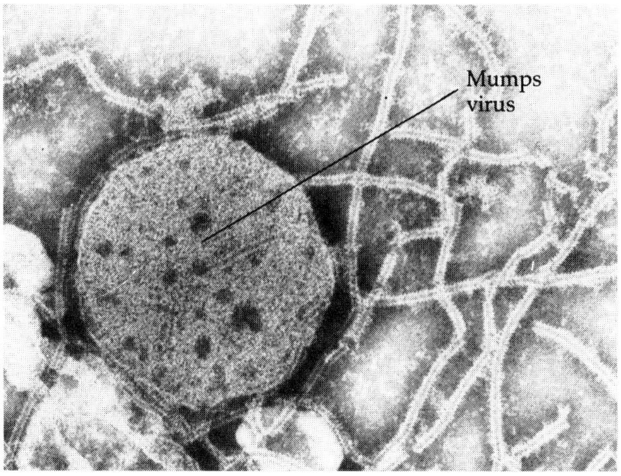

Figure 23–10 Transmission electron micrograph of a mumps virus; a single virus and released helical nucleocapsids are shown (×45,600).

mumps in this chapter. The parotid glands are located just below and in front of the ear (see Figure 23–1). The virus itself is a *paramyxovirus,* the group to which the measles virus belongs. The mumps virus is an enveloped helical virus containing single-stranded, fragmented RNA (Figure 23–10).

Mumps typically begins with painful swelling of one or both parotid glands 16 to 18 days after exposure to the virus. The virus is transmitted in saliva and respiratory secretions, and its portal of entry is the respiratory tract. Once the viruses have begun to multiply in the respiratory tract and local lymph nodes in the neck, they reach the salivary glands via the blood (Figure 23–11). Viremia (viral infection of the blood) begins several days before the onset of mumps symptoms and before the virus appears in saliva. The virus is present in the blood and saliva for 3 to 5 days after the onset of the disease, and in the urine after 10 days or so. Mumps is characterized by inflammation and swelling of the parotid glands, fever, and extreme pain during swallowing. About 4 to 7 days after the onset of symptoms, the testes can become inflamed (orchitis). This happens in about 20 to 35% of males past puberty. Sterility is a possible consequence, but rarely occurs. Other complications of mumps include meningitis, inflammation of the ovaries, and pancreatitis. In children, mumps is less common than chickenpox or measles because the

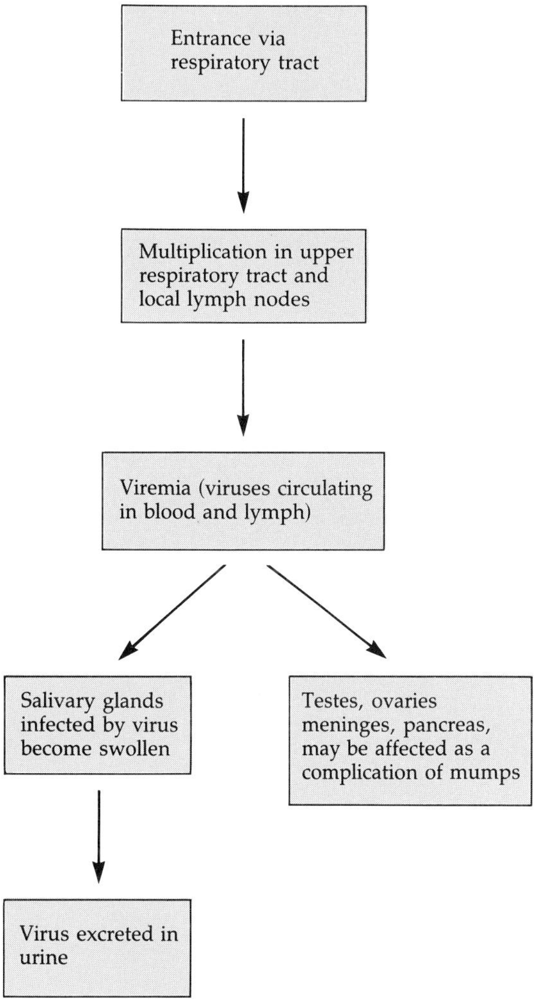

Figure 23–11 Pathogenesis of mumps.

disease is less infectious, and so many children escape infection.

An effective vaccine is available, and is often administered as part of the trivalent measles–mumps–rubella (MMR) vaccine. The number of cases of mumps has dropped sharply since the introduction of the vaccine in 1968. For example, in 1971 there were 125,000 cases of mumps reported, whereas in 1983 there were 3,285 cases. Second attacks are rare, and cases involving only one parotid gland, or subclinical cases, are as effective as bilateral mumps in conferring immunity.

Serological diagnosis is not usually necessary. But if laboratory confirmation of a diagnosis based

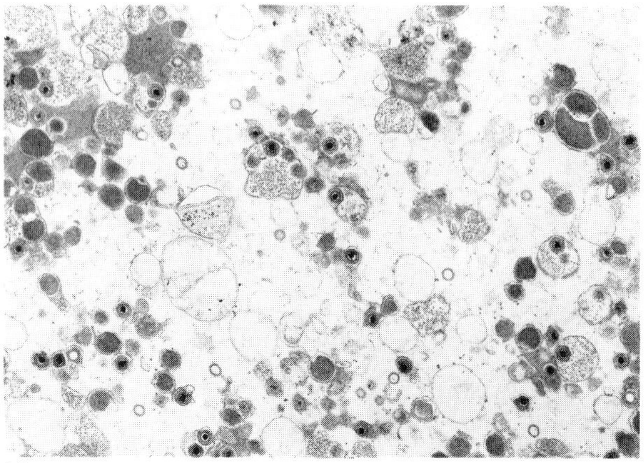

(a)

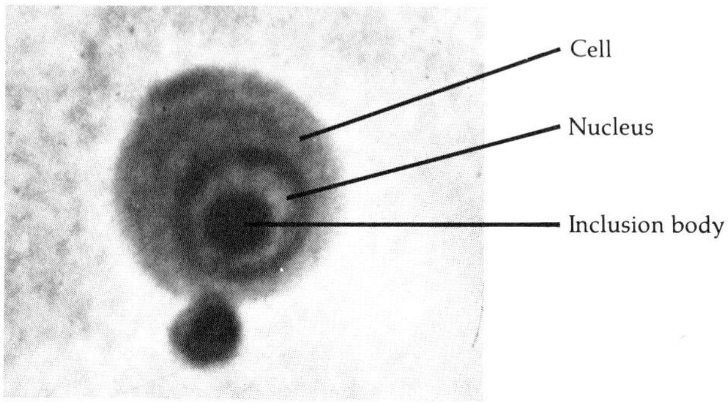

(b)

Figure 23–12 Cytomegalovirus (CMV). **(a)** Electron micrograph showing individual particles in tissue culture (approx. ×13,300). **(b)** Light micrograph of a single animal cell infected with cytomegalovirus. The dark, intranuclear inclusion body has the typical "owl-eyed" appearance.

only on symptoms is desired, the virus can be isolated by embryonated egg or cell culture techniques. The virus can be identified with hemagglutination-inhibition tests.

Cytomegalovirus (CMV) Inclusion Disease

The *cytomegalovirus (CMV)* is a herpes virus that induces a cellular swelling characterized as cytomegaly (large size), and the unique appearance of intranuclear inclusion bodies (Figure 23–12). CMV causes mostly asymptomatic infections, but it sometimes produces a mild disease, **cytomegalovirus inclusion disease,** that resembles infectious mononucleosis. The virus is shed in saliva, urine, and other secretions of the body and can be spread by kissing and other personal contacts. It is also transmitted via transfused blood or transplanted organs. Immunosuppressed patients, including victims of AIDS, are frequently affected by CMV infections. In some cases, the disease in these patients is progressive and fatal, and in others it is mild and self-limiting.

The most serious manifestation of CMV is **congenital infection of the fetus.** In the United States, CMV infections are implicated in up to 7,600 cases of birth defects each year. These infections are most common in teenage mothers from lower socioeconomic groups. The virus can be transmitted across the placental barrier to infect the fetus. It appears that the virus can also be transmitted in breast milk. The mother usually does not show any symptoms of the infection and the infant does not appear affected at the time of birth. Some children acquire only a benign infection; others have serious symptoms involving many organs, such as the liver and kidneys. Neurological involvement can result in brain damage with severe mental retardation or mild speech or hearing defects. Sometimes defects become apparent several years after birth.

Laboratory diagnostic procedures include the demonstration of intranuclear inclusion bodies and cytomegalic cells in specimens, isolation of CMV in tissue culture, and serological tests such as immunofluorescence, hemagglutination, and complement fixation.

Despite the relative rarity of active CMV inclusion disease, latent infection seems to be worldwide and common. From 10 to 18% of all stillborns show the characteristic CMV lesions, and antibodies appear in the blood of 53% of the population in the United States between 18 to 25 years of age and 81% of the population above 35 years of age. A vaccine for CMV inclusion disease is being studied.

Hepatitis

An infection of the liver is called **hepatitis.** Viral hepatitis is now the second most frequently reported infectious disease in the United States. At least three closely related viruses commonly cause hepatitis. **Hepatitis A (infectious hepatitis)** is usually transmitted by fecal-oral contamination. **Hepatitis B (serum hepatitis)** and **non-A non-B (NANB)** hepatitis are usually transmitted by contact with virus-containing blood.

Hepatitis A (infectious hepatitis)

The *hepatitis A virus (HAV)* is the etiological agent of **hepatitis A (infectious hepatitis).** The HAV contains single-stranded RNA and lacks an envelope (Figure 23–13). After a typical entrance via the oral route, HAV multiplies in the epithelial lining of the intestinal tract. Viremia eventually occurs, and the virus spreads to the liver, kidney, and spleen. The virus is shed in the feces and can also be detected in the blood and urine.

Recent outbreaks of HAV disease indicate that the virus might also be carried in oral secretions and can be spread when fingers put in the mouth are then used to handle food. The amount of virus excreted is greatest before symptoms appear and then declines rapidly. Therefore, the food handler responsible for spreading the virus might not appear to be ill. The virus is probably able to survive several days on surfaces such as cutting boards. Contamination of food or drink by feces is aided by the resistance of HAV to disinfectants such as chlorine at concentrations ordinarily used in water. Moreover, mollusks, such as oysters, that live in contaminated waters are also a source of infection.

The initial symptoms of hepatitis A are anorexia (loss of appetite), general malaise, nausea, diarrhea, abdominal discomfort, fever, and chills. These symptoms last from 2 to 21 days. Eventually, jaundice, with the skin typically yellowing due to liver infection, appears. At this point, the liver becomes tender and enlarged.

The mortality rate is low (below 1%), and infections, especially those in children, are often unrecognized. Something like 45 to 75% of tested adults are found to have serum antibodies, yet few of these have ever been aware of a diagnosable illness.

The annual reported incidence of hepatitis A in the United States is about 30,000 cases, but the actual incidence has been estimated at one-half million. Hepatitis A is generally a milder disease than hepatitis B. Most cases of hepatitis A resolve in four to six weeks, and recovery yields good immunity.

The hepatitis A virus or antibodies can be identified with serological tests, including complement fixation, immune adherence, and radioimmunoassay. No specific treatment for the disease exists. However, the incidence of hepatitis A might be decreased by passive immunization; that is, to avoid their developing the disease, persons exposed to the disease can be given immune globulin from many persons. The incubation period of two to six weeks for hepatitis A is much shorter than that for hepatitis B, but is still long enough to make epidemiological studies for the source of the infections difficult.

Serum hepatitis: hepatitis B and non-A non-B hepatitis

Hepatitis B, or **serum hepatitis,** is caused by the *hepatitis B virus (HBV).* Because HBV has often been transmitted by blood transfusions, there has been great interest in studying this virus. The serum from patients with hepatitis B contains three distinct particles.

The largest is called the *Dane particle* and is probably the complete virion; it is infectious and

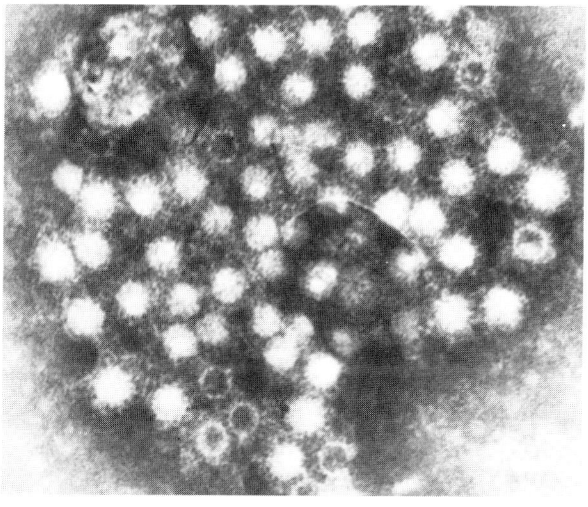

Figure 23–13 Hepatitis A viruses (×163,000).

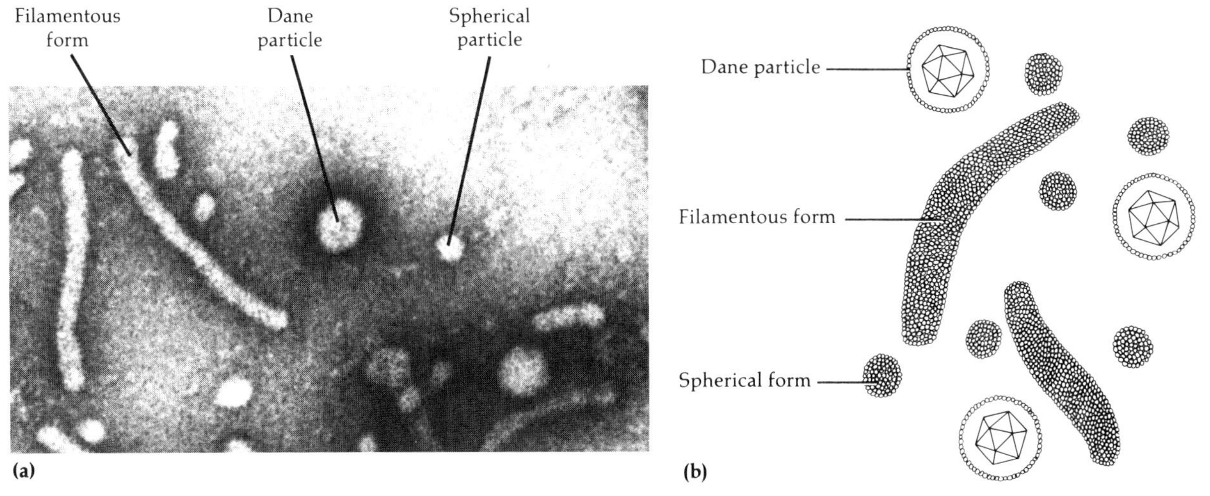

Figure 23–14 **(a)** Electron micrograph showing the three distinct types of hepatitis B particles discussed in the text. **(b)** Diagrammatic representation.

capable of replicating. There are also smaller *spherical particles* about half the size of a Dane particle, and *filamentous particles,* which are tubular particles with a diameter about that of the spherical particles, but a length about 10 times as long (Figure 23–14). The spherical and filamentous particles are probably unassembled components of Dane particles; assembly is evidently not very efficient and large numbers of these unassembled components accumulate. There are about one million times more unassembled than assembled Dane particles in the blood of a hepatitis B patient. These numerous unassembled particles contain *hepatitis B surface antigen (HB_sAg).* The HB_sAg can be detected with antibodies specific against them. Although it would be difficult to detect the HB_sAg on the relatively few Dane particles in serum, relatively insensitive tests are able to detect the numerous other particles that also carry the HB_sAg. Such tests make possible the large-scale screening of blood for HBV and thus minimize the chances of transmitting hepatitis B in transfused blood.

The normal route of transmission of hepatitis B is by any transfer of virus-carrying blood from one person to another. Blood transfusions and contaminated equipment such as syringes have been common modes of transfer. Persons, such as doctors, nurses, dentists, and medical technologists, who are in daily contact with blood, have a considerably higher incidence of this disease than does the population at large. Doctors, for example, have five times, and dentists two to three times, the normal rate of infection. Recent evidence indicates that hepatitis B can be transmitted by any secretion of bodily fluid, such as saliva, sweat, breast milk, and semen. Sexual transmission is sometimes suspected among heterosexuals, and is common among male homosexuals. The incubation period averages about three months, but ranges from one to six months. Because this incubation period is of uneven duration, determination of the origin of an infection can be difficult.

Clinically, hepatitis B varies widely. Probably about one-half of the cases are entirely asymptomatic. Symptoms are highly variable and include in the early stages, loss of appetite, low-grade fever, and joint pains. Later, jaundice usually appears. It is difficult to distinguish between hepatitis A and B solely on clinical grounds. There is no effective treatment for viral hepatitis.

The mortality rate from hepatitis B is significantly higher than that for hepatitis A, but is probably fewer than 1% for the population at large. It might be as high as 2% among hospitalized patients—many of whom are elderly.

Researchers have been unable to cultivate the HBV in cell culture. This type of culture has been the means by which vaccines for polio, mumps, measles, and rubella have been developed. Current research indicates that the genetic coding for the HB_sAg, the major surface antigen, can be inserted into the genome of the vaccinia virus that has been used for many years for smallpox vaccination. Another recent approach to vaccination involves genetically engineered yeast. These yeast are modified in the laboratory to carry a gene encoding the production of HB_sAg. The yeasts then produce HB_sAg, which is extracted and purified for use as a vaccine. The vaccine has been found to induce antibodies in humans, and presumably gives immunity.

In the meantime, a successful vaccine with HB_sAg harvested from the serum of volunteer human carriers has been developed. There are hundreds of thousands of such carriers currently in the population of the United States. Adults are immunized with two doses, spaced one month apart; a booster follows six months later. Immunosuppressed patients receive larger doses and children, smaller doses. Vaccination is recommended for high-risk groups; these include medical personnel exposed to blood and blood products, persons undergoing hemodialysis, patients and staff at institutions for the mentally retarded, and homosexually active males. The incidence of hepatitis B, apparently sexually transmitted, is very high in this latter group. An interesting aspect of the vaccine's development was a test involving more than one thousand male homosexuals. Previously uninfected men who received the vaccine subsequently showed a very low incidence of hepatitis B, whereas approximately 15% would normally have acquired the disease within a year. *Immune globulin (HBIG)* can be used to confer temporary immunity for exposed persons in special situations.

The CDC estimates that 200,000 or more persons, mostly young adults, are infected with hepatitis B each year. About one-half of the cases are asymptomatic. Almost all patients recover completely, but 5 to 10% develop a chronic infection of one sort or another. Such chronic infections can cause drastic and often fatal damage to the liver. A special concern is the strong correlation between the occurrence of liver cancer and the incidence of chronic hepatitis B infections. Liver cancer is the most common form of cancer in sub-Saharan Africa and the Far East, areas where hepatitis B is extremely common. Although it has a population of only 17 million, Taiwan, for example, has twice the number of cases of liver cancer as the United States. Almost all of these cases of liver cancer in

Table 23–1 Characteristics of Viral Hepatitis A, B, and Non-A Non-B

Characteristic	Hepatitis A	Hepatitis B	Non-A Non-B
Transmission	Fecal-oral	Virus-contaminated blood and intimate contact	Virus-contaminated blood and intimate contact
Incubation period	2 to 6 weeks	4 weeks to 6 months	Usually 6 weeks to 6 months
Severity of acute illness	Usually mild to moderate; often no jaundice	Can be severe; jaundice frequent	Moderate; often no jaundice
Chronic liver disease	No	Yes	Yes
Carrier state	No	Yes	Yes
Immune globulin (IG) prophylaxis	IG very effective	HBIG for specific situations; new HBV vaccine recommended for high-risk groups	Unknown
Vaccines	Under development	Available	None

Source: Modified from J.L. Dienstag, MD. Treatment strategy for acute viral hepatitis. *Modern Medicine*, Aug. 1982.

Taiwan occur in patients with chronic hepatitis B infections.

Screening of donated blood for HB_sAg has become so effective that transmission by blood transfusion has declined dramatically. However, it is now apparent that a previously unsuspected form of hepatitis also causes post-transfusion hepatitis infections. This form was termed, by elimination, **non-A non-B hepatitis (NANB)**. It has been estimated that 90% of transfusion-transmitted cases are now of the NANB type. The disease itself is somewhat milder than hepatitis B, and is clinically similar to hepatitis A. Diagnosis is made with clinical symptoms and a negative test for hepatitis A and B. There is no serological test at present.

A recently isolated hepatitis virus is *Delta hepatitis.* The delta virus is transmitted as an infectious agent, but cannot cause disease unless the host is also infected with hepatitis B. To be infectious, the delta virus must acquire a coat of HB_sAg, and it can do that only by coinfection with hepatitis B virus. This type of infection, which is often severe, is widespread in much of the world, but relatively rare in the United States.

The characteristics of the three common forms of viral hepatitis are summarized in Table 23–1.

Viral Gastroenteritis

A number of viruses such as the polioviruses, echoviruses, and coxsackieviruses are transmitted by the fecal–oral route. However, despite the name of enteroviruses, these generally do not directly affect the digestive system. Gastroenteritis related to viral infections is usually caused by reoviruses or parvoviruses. Illness is almost entirely restricted to children under two years, among whom it is probably a common cause of diarrhea.

Major epidemics have been caused by the parvovirus known as the *Norwalk agent* (after an outbreak in Norwalk, Ohio in 1968). All age groups are affected by this virus; infected persons typically suffer nausea, abdominal cramps, diarrhea, and vomiting for one to three days. About one-half of middle-aged Americans show evidence, by serum antibodies, that they have been exposed. Susceptibility to the virus varies considerably; some persons are not affected and do not even respond with an immune reactions. Others develop gastroenteritis and an immunity that might be only short term.

MYCOTOXINS

Some fungi produce toxins called **mycotoxins,** which cause blood diseases, nervous system disorders, kidney damage, and liver damage. Some mycotoxins are associated with poisonous mushrooms, such as *Amanita* (am-an-ī'ta) or *Claviceps purpurea* (kla'vi-seps pür-pū-rē'ä). *Claviceps* causes **ergot poisoning,** which is caused by the ingestion of rye contaminated with the mycotoxin. Ergot poisoning was very widespread during the Middle Ages. The toxin can restrict blood flow in the extremities, and gangrene results. More recently used as an hallucinogenic drug, it can produce bizarre behavior similar to that caused by LSD. A mycotoxin of current interest is *aflatoxin.* It is produced by the fungus *Aspergillus flavus* (a-spėr-jil'lus flā'vus), a common mold. Aflatoxin is highly toxic and can cause serious damage to livestock when their feed is contaminated with *A. flavus.* Although the risk to humans is unknown, there is strong circumstantial evidence that aflatoxin contributes to cirrhosis of the liver and cancer of the liver in parts of the world, such as India and Africa, where food is subject to aflatoxin contamination.

PROTOZOAN DISEASES OF THE DIGESTIVE SYSTEM

Giardiasis

Giardia lamblia (jē-är'dē-ä lam'lē-ä), a flagellated protozoan (Figure 23–15) that is able to attach firmly to a human's intestinal wall, is the cause of a prolonged diarrheal disease in humans called **giardiasis.** The disease, which sometimes persists for weeks, is characterized by malaise, nausea, weakness, weight loss, and abdominal cramps. There are frequent outbreaks of giardiasis in the United States; it is probably the most common cause of epidemic waterborne diarrheal disease. About 7% of the population are healthy carriers and shed the cysts in their feces.

The organism is also shed by a number of wild mammals, especially beavers, and the disease is not uncommon in backpackers drinking from wilderness waters. Most outbreaks in the United States are transmitted by contaminated water supplies. And because the cyst stage of the protozoan is relatively insensitive to chlorine, filtration of wa-

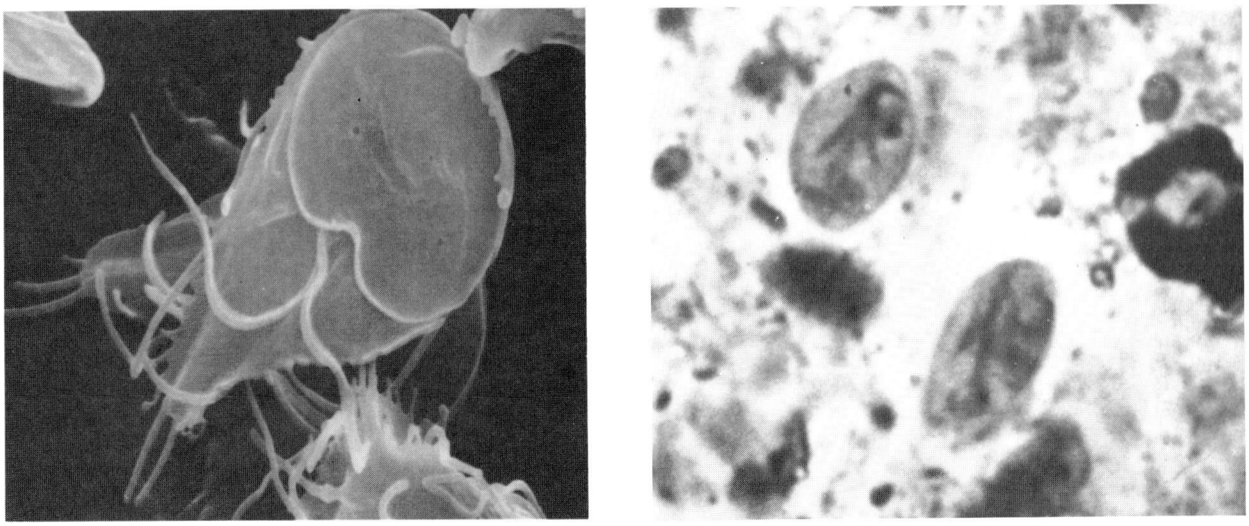

(a) (b)

Figure 23–15 *Giardia lamblia*, a flagellated protozoan causing giardiasis. **(a)** Scanning electron micrograph of the vegetative form (trophozoite). **(b)** Cyst form. See also Figure 11–19.

ter supplies is usually necessary to eliminate the cysts from water.

Giardia are not reliably found in the stools and the *string test* is recommended for diagnosis. This is interesting enough to describe. A gelatin capsule packed with about 140 cm of fine string is swallowed by the patient. One end of the string is taped to the cheek. The gelatin capsule dissolves in the stomach and an enclosed weighted rubber bag attached to the other end of the string enters the upper bowel. After a few hours the string is drawn up through the mouth and examined for parasites.

Treatment by metronidazole and quinacrine hydrochloride have been effective.

Balantidiasis (Balantidial Dysentery)

Balantidium coli (bal-an-tid'ē-um kō'lē) is a ciliated protozoan that causes **balantidiasis,** or **balantidial dysentery.** The organism is the only ciliate known to be pathogenic for humans and is the largest intestinal protozoan of humans. Like *Giardia lamblia*, *B. coli* exists in a vegetative form (trophozoite) and a cyst form (Figure 23–16). *B. coli* lives in the large intestine; in rare cases, it invades the epithelial lining and causes ulceration and fatal dysentery. Typically, the disease is mild, and consists of abdominal pain, nausea, vomiting, diarrhea, and weight loss.

Humans acquire *B. coli* by ingesting cysts in food or water that has been contaminated by feces containing cysts. Following ingestion, the cysts descend to the colon, where the cyst walls dissolve and the released vegetative cells feed on bacteria, fecal debris, and the host's tissue cells. As feces passing through the colon are dehydrated, encystment occurs. Some cysts form after the feces are discharged. Subsequent food and water contamination initiates another cycle.

Laboratory confirmation of balantidiasis is based on the demonstration of vegetative cells in feces. Treatment consists of chlortetracycline or oxytetracycline, followed by diiodohydroxyquin, if necessary.

Amoebic Dysentery (Amoebiasis)

Amoebic dysentery (amoebiasis) is found worldwide, and is spread mostly by food or water contaminated by cysts of the protozoan amoeba *Entamoeba histolytica* (en-tä-mē'bä his-tō-li'ti-kä) (Figure 23–17). Although stomach acid (HCl) can destroy vegetative cells, it does not affect the cysts. In the intestinal tract, the cyst wall is digested away, and the vegetative forms are released. The

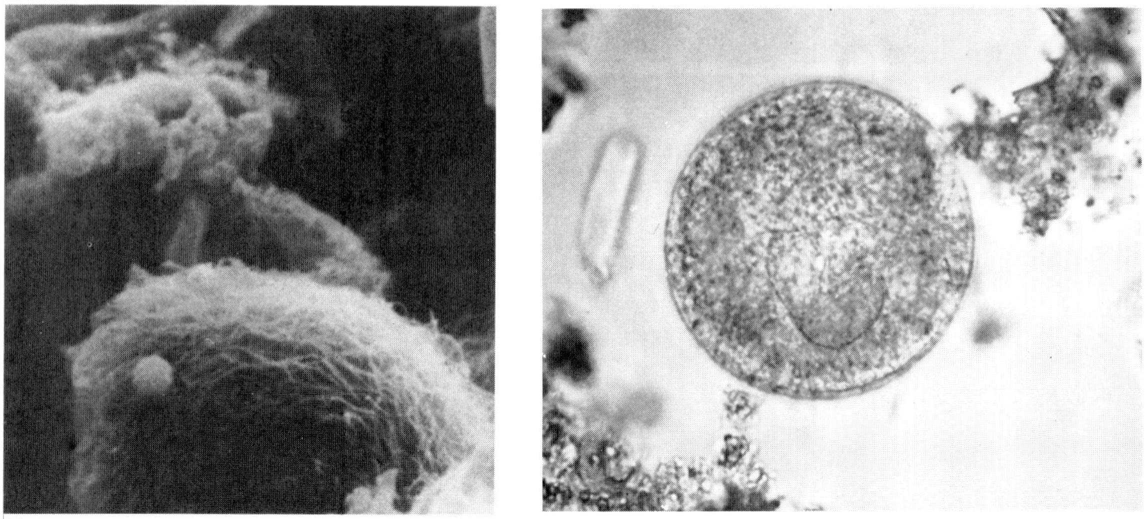

(a) (b)

Figure 23–16 *Balantidium coli.* **(a)** Ciliated trophozoite nestled in the tissue of a primate intestine. **(b)** Spherical cyst.

vegetative forms then multiply in the epithelial cells of the large intestinal wall. A severe dysentery results, and the feces characteristically contain blood and mucus. The vegetative forms feed on red blood cells and destroy tissue in the gastrointestinal tract. Severe infections result if the intestinal wall is perforated. Abscesses might have to be treated surgically, and invasion of other organs, particularly the liver, is not uncommon. Perhaps 5% of the population in the United States are asymptomatic carriers of *E. histolytica.* In recent years, more than 3000 cases per year have been reported in the United States, and the mortality rate approaches 1%.

Diagnosis is largely dependent upon recovery and identification of the organisms (red blood cells observed within the trophozoite stage of an amoeba are considered an indication of *E. histolytica*). There are several serological tests that can also be used for diagnosis, including immunodiffusion and fluorescent-antibody tests.

Metronidazole and diiodohydroxyquin are the drugs of choice in treatment.

Cryptosporidiosis

A protozoan disease called **cryptosporidiosis** is characterized by severe, prolonged (sometimes for several months) diarrhea. This disease has been observed in increasing number of immunosuppressed patients, including those suffering from AIDS. Diagnosis is based on microscopic examination of feces for *Cryptosporidium* oocysts. The organism causing the disease, *Cryptosporidium* (krip-tō-spô-ri'dē-um) had long been recognized as a pathogen in calves, but until 1976 was not known to infect humans. Since then, the number of reported cases has been increasing. Some infections have responded to spiramycin, a macrolide antibiotic similar to erythromycin, but at present there is no reliable treatment.

HELMINTHIC DISEASES OF THE DIGESTIVE SYSTEM

Tapeworm Infestation

Most **tapeworm infestations** of humans result from consumption of undercooked beef, pork, or fish, which are intermediate hosts for the tapeworm. When forms of the tapeworm called cysticerci, encysted in muscles of the intermediate host, are ingested by a human, they develop into adult tapeworms. The adults attach to the intestinal wall of the human host and shed eggs with the host's feces. Where human excrement contaminates the habitat (such as animal pastures) and food of in-

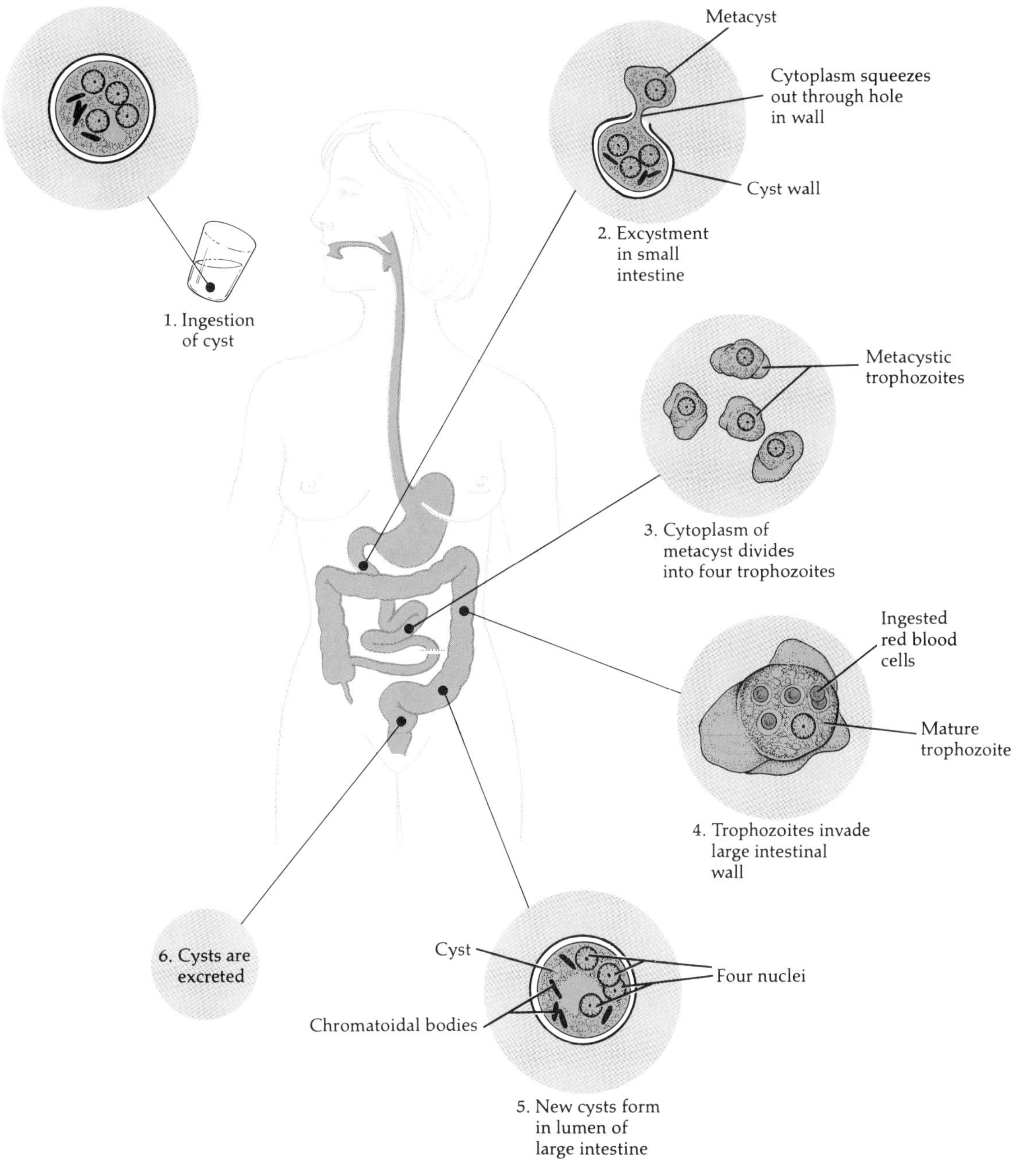
Metacyst
Cytoplasm squeezes out through hole in wall
Cyst wall
2. Excystment in small intestine
1. Ingestion of cyst
Metacystic trophozoites
3. Cytoplasm of metacyst divides into four trophozoites
Ingested red blood cells
Mature trophozoite
4. Trophozoites invade large intestinal wall
6. Cysts are excreted
Cyst
Four nuclei
Chromatoidal bodies
5. New cysts form in lumen of large intestine

Figure 23–17 Life cycle of *Entamoeba histolytica*. Transmission is usually by ingestion of food or water containing cysts. Living in the intestinal epithelial layer, the trophozoite feeds on red blood cells. New cysts are formed by trophozoites in the lumen of the large intestine.

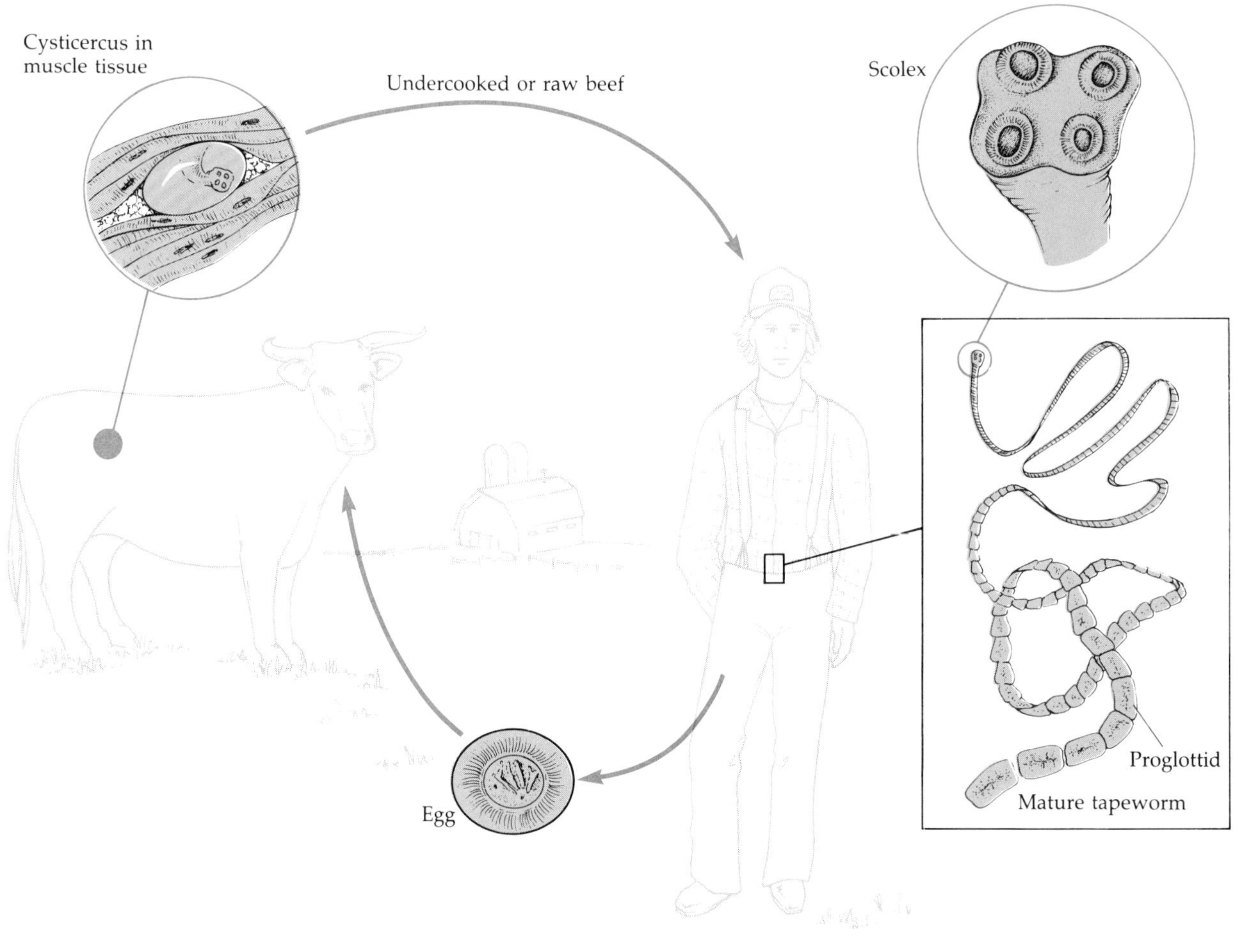

Figure 23–18 Life cycle of the beef tapeworm, *Taenia saginata.* The adult tapeworm lives in the intestine of the human, the definitive host. Tapeworm proglottids and eggs are eliminated with feces and ingested by intermediate hosts such as grazing cattle. The tapeworm eggs hatch, and cysticerci form in the animal's muscles, to be consumed later by humans. The pork tapeworm, *Taenia solium,* has a similar life cycle.

termediate hosts, the life cycle continues (Figure 23–18).

Usually the symptoms of an infestation are so mild that the host is unaware of the parasite. But anemia can result in severe cases, because some tapeworms can absorb vitamin B_{12}, a substance required for red blood cells to form. Tapeworms rarely invade tissue.

Taenia saginata (te'nē-ä sa-ji-nä'tä), the beef tapeworm, can reach a length of 6 m or more. The pork tapeworm, *Taenia solium* (sō'lē-um), is normally 2 to 7 m in length. *Diphyllobothrium latum* (dī-fil-lo-bo'thrē-um lā'tum), the fish tapeworm, is found in pike, trout, perch, and salmon. Fully developed, they can be 3 to 6 m in length. Recently the CDC issued a warning about the risks of fish-tapeworm infestation from the increasingly popular sushi (a Japanese dish prepared from raw fish). About 10 days after eating sushi, one person developed symptoms of abdominal distention, flatulence (intestinal gas), belching, intermittent abdominal cramping, and diarrhea. Eight days later the patient passed a four-foot long tapeworm identified as a species of *Diphyllobothrium.*

Today, fewer than 0.1% of the cattle slaughtered in the United States are infected with tapeworm, and only a few swine are infested. At one time, however, tapeworm infestations were very common. Laboratory diagnosis of human infection consists of tapeworm identification in feces. The drug of choice for eliminating tapeworms is niclosamide.

Trichinosis

Most infestations with the small roundworm *Trichinella spiralis* (tri-kin-el'lä spī-ra'lis), called **trichinosis,** are insignificant. The larvae, in encysted form, are located in muscles of the host. Routine autopsies of human diaphragm muscles in 1970 showed that about 4% of those tested carried this parasite. Severe cases of trichinosis can be fatal—sometimes in only a few days. The severity of the disease is generally proportional to the amount of infestation. Ingestion of undercooked pork is the most common vehicle of infection, but the flesh of other animals that feed on garbage (bears, for example) are also sources of outbreaks.

Any ground meat can be contaminated from machinery previously used to grind contaminated meats. Eating raw sausage or hamburger is a poor habit. One person acquired trichinosis by chewing the fingernails after handling infected pork. Pro-

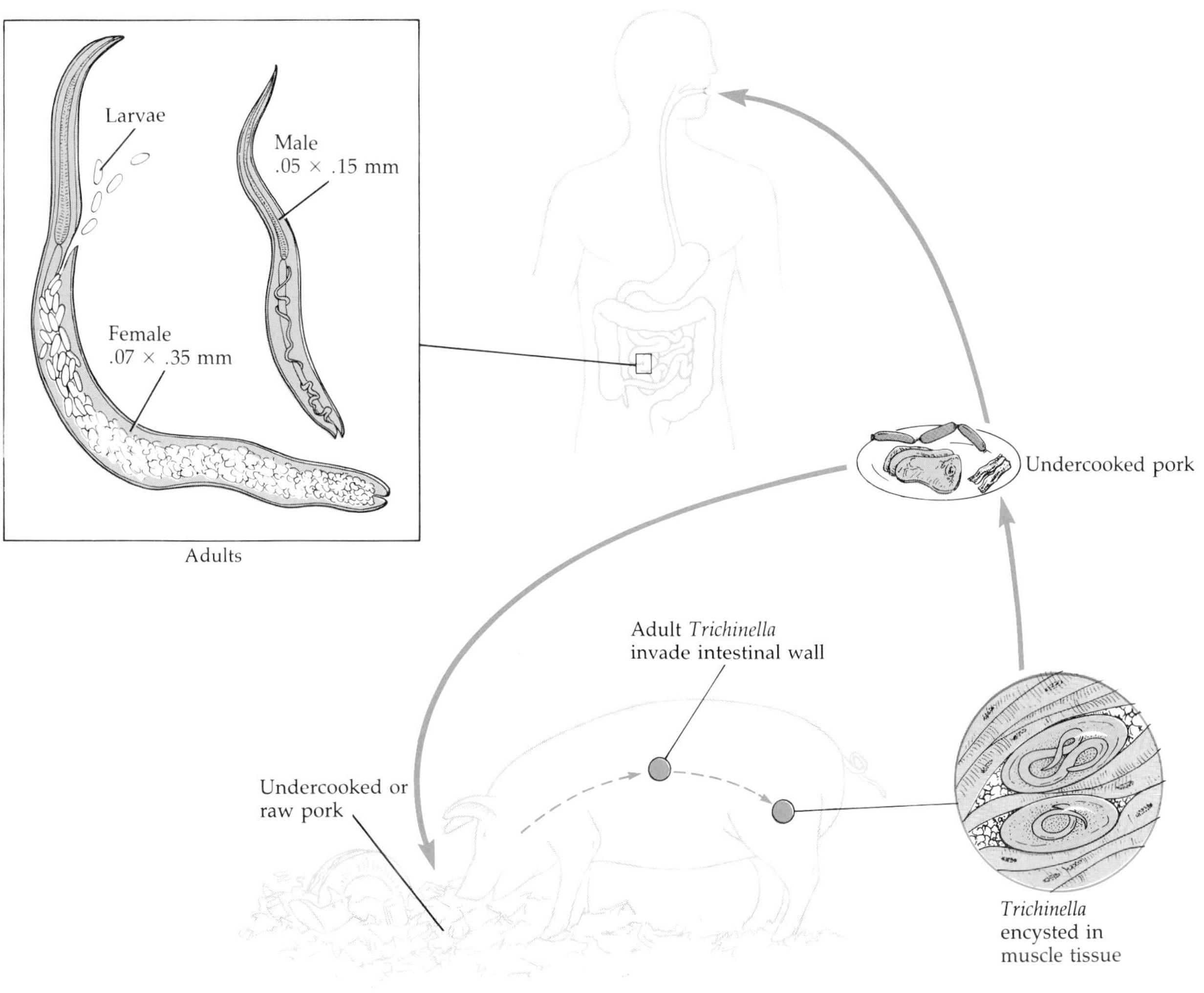

Figure 23–19 Life cycle of *Trichinella spiralis*, the causative agent of trichinosis.

Table 23–2 Summary of Diseases Associated with the Mouth and Gastrointestinal System

Disease	Causative Agent	Mode of Transmission	Prevention or Treatment
Bacterial Diseases of the Mouth:			
Dental caries	*Streptococcus mutans*	Bacteria use sucrose to form plaque	Restrict ingestion of sucrose; brushing, flossing, and professional cleaning to remove plaque; fluoridation
Periodontal disease	*Actinomyces, Nocardia, Corynebacterium*	Plaque initiates an inflammatory response	Same as above plus antibiotics such as penicillin and tetracycline
Bacterial Diseases of the Lower Digestive System:			
Staphylococcal food poisoning (enterotoxicosis)	*Staphylococcus aureus*	Ingestion of exotoxin in food, usually improperly refrigerated	Replace lost water and electrolytes
Salmonellosis	*Salmonella* species	Ingestion of contaminated food and drink	Replace lost water and electrolytes
Typhoid fever	*Salmonella typhi*	Ingestion of contaminated food and drink	Ampicillin and trimethoprim-sulfamethoxazole
Bacillary dysentery (shigellosis)	*Shigella* species	Ingestion of contaminated food and drink	Replace lost water and electrolytes; ampicillin, chloramphenicol, or trimethoprim-sulfamethoxazole
Asiatic cholera	*Vibrio cholerae*	Ingestion of contaminated food and drink	Replace lost water and electrolytes
Vibrio parahaemolyticus gastroenteritis	*Vibrio parahaemolyticus*	Often, ingestion of contaminated shellfish	Tetracycline, chloramphenicol, or penicillin in severe cases
Escherichia coli gastroenteritis	Enterotoxigenic, enteroinvasive, and enterophathogenic strains of *Escherichia coli*	Ingestion of contaminated food and drink	Doxycycline and trimethoprim-sulfamethoxazole may help in prevention
Yersinia enterocolitica gastroenteritis	*Yersinia enterocolitica*	Ingestion of contaminated food and drink	Replace lost water and electrolytes
Campylobacter jejuni gastroenteritis	*Campylobacter jejuni*	Ingestion of contaminated food and drink	Replace lost water and electrolytes
Clostridium perfringens gastroenteritis	*Clostridium perfringens*	Ingestion of contaminated food (usually meat) and drink	Replace lost water and electrolytes
Bacillus cereus gastroenteritis	*Bacillus cereus*	Ingestion of contaminated food	None, self-limiting

longed freezing of meats containing *Trichinella* tends to eliminate them but should not be considered a substitute for thorough cooking.

In the muscles of intermediate hosts such as pigs, the *T. spiralis* larvae are encysted in the form of short worms 1 to 2 mm in length (Figure 23–19). When the flesh of an infected animal is ingested by humans, the cyst wall is removed by digestive action in the intestine. The organism then matures into the adult form. The adult worms spend only about one week in the intestinal mucosa and produce larvae that invade tissue. Eventually, the encysted larvae localize in muscle (common sites include the diaphragm and eye muscles), where they

Table 23–2 Summary of Diseases Associated with the Mouth and Gastrointestinal System *continued*

Disease	Causative Agent	Mode of Transmission	Prevention or Treatment
Viral Diseases of the Digestive System:			
Mumps	Mumps virus (a paramyxovirus)	Saliva and respiratory secretions	None, but a vaccine is available (MMR)
Cytomegalovirus (CMV) inclusion disease	Cytomegalovirus (CMV)	Placental transfer and blood transfusion	None
Hepatitis A	Hepatitis A virus	Ingestion of contaminated food or drink	None, but passive immunization is available
Hepatitis B	Hepatitis B virus	Blood or blood contamination by needles, transfusions, etc.; sexual contact, especially between male homosexuals	Vaccine available
Non-A non-B hepatitis	Non-A non-B hepatitis	Blood or blood contamination by needles, transfusions, etc.	None
Viral gastroenteritis	Reoviruses, parvoviruses (such as Norwalk agent)	Ingestion of contaminated food and water	None
Ergot poisoning	*Claviceps purpurea*	Ingestion of food, usually cereal grain	Antispasmodic drugs
Protozoan Diseases of the Digestive System:			
Giardiasis	*Giardia lamblia*	Ingestion of contaminated water	Metronidazole and quinacrine hydrochloride
Balantidiasis (balantidial dysentery)	*Balantidium coli*	Ingestion of contaminated food and water	Antibiotics (chlortetracycline, oxytetracycline) and diiodohydroxyquin
Amoebic dysentery (amoebiasis)	*Entamoeba histolytica*	Ingestion of contaminated food and water	Metronidazole and diiodohydroxyquin
Cryptosporidiosis	*Cryptosporidium*	Uncertain, probably oral-fecal	Possibly spiramycin
Helminthic Diseases of the Digestive System:			
Tapeworm infestation	*Taenia saginata* (beef tapeworm), *T. solium* (pork tapeworm), *Diphyllobothrium latum* (fish tapeworm)	Ingestion of contaminated food and water	Niclosamide
Trichinosis	*Trichinella spiralis*	Ingestion of contaminated food, especially improperly cooked pork	Thiabendazole and corticosteroids

are barely visible in biopsied specimens.

Symptoms of trichinosis include fever, swelling about the eyes, and gastrointestinal upset. Small hemorrhages under fingernails are often observed. Biopsy specimens, as well as a number of serological tests, can be used in diagnosis. Treatment consists of the administration of thiabendazole to kill intestinal worms and corticosteroids to reduce inflammation.

In recent years, the number of cases reported annually in the United States has varied from 67 to 252. Deaths are rare, and in most years none occur.

See Table 23–2 for a summary of diseases associated with the mouth and gastrointestinal system.

STUDY OUTLINE

INTRODUCTION (pp. 640–641)

1. Diseases of the digestive system are the second most common illnesses in the United States.
2. Diseases of the digestive system usually result from ingestion of microorganisms and their toxins in food and water.
3. The fecal-oral cycle of transmission can be broken by proper disposal of sewage and proper preparation and storage of foods.

STRUCTURE AND FUNCTION OF THE DIGESTIVE SYSTEM (pp. 641–642)

1. The gastrointestinal (GI) tract, or alimentary canal, consists of the mouth, pharynx, esophagus, stomach, small intestine, and large intestine.
2. The teeth, tongue, salivary glands, liver, gallbladder, and pancreas are accessory structures.
3. In the GI tract, with mechanical and chemical help from the accessory structures, carbohydrates are broken down to simple sugars, fats to fatty acids and glycerol, and proteins to amino acids.
4. Feces, resulting from digestion, are stored in the rectum and eliminated by defecation.

ANTIMICROBIAL FEATURES OF THE DIGESTIVE SYSTEM (p. 642)

1. Saliva contains lysozyme, mucins, and IgA antibodies.
2. Hydrochloric acid in the stomach destroys many microorganisms.
3. Lymphoid tissue in the small and large intestines provides immunity against some microorganisms.
4. Normal flora of the intestines inhibits the growth of some microorganisms by producing acids and growing rapidly.
5. Some of the normal flora produce bacteriocins that bind to outer membrane receptors of susceptible bacteria and kill them.

NORMAL FLORA OF THE DIGESTIVE SYSTEM (p. 642)

1. *S. salivarius* is found on the tongue, *S. mitis* on the cheek mucosa, and *S. sanguis* on the teeth.
2. *Bacteroides, Fusobacterium,* and spirochetes are found in anaerobic niches between the teeth and gums.
3. The stomach and small intestine have few resident microorganisms.
4. The large intestine is the habitat for *Lactobacillus, Bacteroides, E. coli, Enterobacter, Citrobacter,* and *Proteus.*
5. Bacteria in the large intestine assist in degrading food and synthesizing vitamins.
6. Up to 40% of fecal mass is microbial cells.

BACTERIAL DISEASES OF THE DIGESTIVE SYSTEM (pp. 643–654)

Diseases of the Mouth (pp. 643–646)

1. Dental caries begin when tooth enamel and dentin are eroded, and the pulp is exposed to bacterial infection.
2. *S. mutans,* found in the mouth, utilizes sucrose to form dextran from glucose and lactic acid from fructose.
3. Bacteria adhere to teeth with a sticky dextran capsule, forming dental plaque.
4. Acid produced during carbohydrate fermentation destroys tooth enamel at the site of the plaque.
5. Carbohydrates such as starch, mannitol, and sorbitol are not used by cariogenic bacteria to produce dextran, and do not promote tooth decay.
6. Caries are prevented by restricting ingestion of sucrose and by physical removal of plaque; a vaccine against *S. mutans* is theoretically possible.
7. Inflammation of the gums, bone destruction, and necrosis of tissues around the teeth can be caused by *Actinomyces, Nocardia,* and *Corynebacterium;* this is called periodontal diseases.

Diseases of the Lower Digestive System (p. 646)

1. A gastrointestinal infection is caused by growth of a pathogen in the intestines.
2. Incubation times, the times required for bacterial cells to grow and their products to produce symptoms, range from 12 hours to 2 weeks. Symptoms generally include a fever.
3. Antibiotic therapy is useful in treatment of bacterial infections.

4. A bacterial intoxication results from ingestion of preformed bacterial toxins.
5. Symptoms appear 1 to 48 hours after ingestion of the toxin.
6. Antibiotics are of no use because symptoms are due to an exotoxin.
7. Gastroenteritis refers to infections or intoxications that result in abdominal pain, diarrhea, or dysentery.

Staphylococcal food poisoning (staphylococcal enterotoxicosis) (pp. 646–649)

1. Staphylococcal food poisoning is due to ingestion of an enterotoxin produced in improperly stored foods.
2. *S. aureus* is inoculated into foods during preparation. The bacteria grow and produce enterotoxin in food stored at room temperature.
3. The polypeptide exotoxin is not denatured by boiling for 30 minutes.
4. Foods with high osmotic pressure and those not cooked immediately before consumption are most often the source of staphylococcal enterotoxicosis.
5. Nausea, vomiting, and diarrhea occur 1 to 6 hours after eating, and the symptoms last approximately 24 hours.
6. Laboratory identification of *S. aureus* isolated from foods or the presence of thermostable nuclease in foods can confirm diagnosis.

Salmonellosis (*Salmonella* gastroenteritis) (pp. 649–650)

1. Salmonellosis, or *Salmonella* gastroenteritis, is caused by *Salmonella* endotoxin.
2. Symptoms include nausea, abdominal pain, and diarrhea and occur 12 to 36 hours after ingestion of large numbers of *Salmonella*. Septicemia can occur in infants and in the elderly.
3. Mortality is lower than 1% and recovery can result in a carrier state.
4. Heating food to 68°C will usually kill *Salmonella*.
5. Laboratory diagnosis is based on isolation and identification of *Salmonella* from feces.

Typhoid fever (pp. 650–651)

1. A few serovars of *Salmonella* cause typhoid fever.
2. Fever and malaise occur after a two-week incubation. Symptoms last two to three weeks.
3. *Salmonella* are harbored in the gallbladder of carriers.
4. A killed-bacteria vaccine is available for high-risk persons.

Bacillary dysentery (shigellosis) (p. 651)

1. Bacillary dysentery is caused by four species of *Shigella*.
2. Symptoms include blood and mucus in stools, abdominal cramps, and fever. Infections of *S. dysenteriae* result in ulceration of the intestinal mucosa.
3. Isolation and identification of the bacteria from rectal swabs is used for diagnosis.

Asiatic cholera (pp. 651–653)

1. *V. cholerae* produces an exotoxin that alters membrane permeability of the intestinal mucosa; vomiting, diarrhea, and loss of body fluids result.
2. The incubation period is approximately three days. The symptoms last for a few days. Untreated cholera has a 50% mortality rate.
3. Diagnosis is based on isolation of *Vibrio* from feces.

Vibrio parahaemolyticus gastroenteritis (p. 653)

1. Gastroenteritis can be caused by the halophile *V. parahaemolyticus*.
2. Onset of symptoms begins within 24 hours after ingestion of contaminated foods. Recovery occurs within a few days.
3. The disease is contracted by ingestion of contaminated crustaceans or handling of contaminated mollusks.

Escherichia coli gastroenteritis (traveler's diarrhea) (pp. 653–654)

1. Gastroenteritis may be caused by enterotoxigenic, enteroinvasive, or enteropathogenic strains of *E. coli*.
2. The disease occurs as epidemic diarrhea in nurseries, as traveler's diarrhea, and as endemic diarrhea in underdeveloped countries.
3. In adults the disease is usually self-limiting and does not require chemotherapy. Children are treated with fluids and electrolytes.

Gastroenteritis caused by other gram-negative bacteria (p. 654)

1. *Yersinia enterocolitica* and *Campylobacter jejuni* cause gastroenteritis.
2. They are transmitted by poultry, dogs, raw milk, and contaminated water.

Clostridium perfringens gastroenteritis (p. 654)

1. A self-limiting gastroenteritis is caused by *C. perfringens.*
2. Endospores survive heating and germinate when foods (usually meats) are stored at room temperature.
3. Exotoxin produced when the bacteria grow in the intestines is responsible for the symptoms.
4. Diagnosis is based on isolation and identification of the bacteria in stool samples.

Bacillus cereus gastroenteritis (p. 654)

1. Ingesting food contaminated with the soil saprophyte, *Bacillus cereus,* can result in diarrhea, nausea, and vomiting.

VIRAL DISEASES OF THE DIGESTIVE SYSTEM (pp. 654–660)

Mumps (pp. 654–656)

1. Mumps virus (paramyxovirus) enters and exits the body through the respiratory tract.
2. About 16 to 18 days after exposure, the virus causes inflammation of the parotid glands, fever, and pain during swallowing. About 4 to 7 days later, orchitis may occur.
3. After onset of the symptoms, the virus is found in the blood, saliva, and urine.
4. A measles–mumps–rubella (MMR) vaccine is available.
5. Diagnosis is based on symptoms or hemagglutination-inhibition tests of viruses cultured in embryonated eggs or cell culture.

Cytomegalovirus (CMV) Inclusion Disease (p. 656)

1. CMV (herpesvirus) causes intranuclear inclusion bodies and cytomegaly of host cells.
2. CMV is transmitted by saliva, urine, blood, and transplanted organs.
3. Cytomegalovirus inclusion disease can be asymptomatic, a mild disease, or progressive and fatal.
4. If the virus crosses the placenta, it can cause congenital infection of the fetus and cause malfunction of the liver and kidneys, neurological damage, and stillbirth.
5. Diagnosis is based on appearance of the host cells and serological tests.

Hepatitis (pp. 657–660)

1. Inflammation of the liver is called hepatitis. Symptoms include loss of appetite, malaise, fever, and jaundice. Many cases are asymptomatic.
2. Viral causes of hepatitis are hepatitis A virus (infectious hepatitis), hepatitis B virus (HBV), and non-A non-B hepatitis virus (NANB). HBV and NANB are types of serum hepatitis.

Hepatitis A (infectious hepatitis) (p. 657)

1. Hepatitis A virus causes infectious hepatitis.
2. HAV is ingested in contaminated food or water, grows in the cells of the intestinal mucosa, and spreads to the liver, kidney, and spleen in the blood.
3. The virus is eliminated with feces.
4. The incubation period is 2 to 6 weeks; the period of disease, 2 to 21 days; and recovery is complete in 4 to 6 weeks.
5. Diagnosis is based on serological tests for the virus or antibodies.

Serum hepatitis: hepatitis B and non-A non-B hepatitis (pp. 657–660)

1. Hepatitis B virus causes serum hepatitis.
2. HBV is transmitted by blood transfusions, contaminated syringes, saliva, sweat, breast milk, and semen.
3. The average incubation period is three months; recovery is usually complete; some patients develop a chronic infection.
4. Dane particles (complete virion), spherical particles (Hb_sAg), and filamentous particles (containing Hb_sAg) are found in serum of patients with hepatitis B.
5. The virus has never been cultured. HBV from serum is used to immunize high-risk individuals. Genetic recombination techniques are being used to develop a vaccine against Hb_sAg.

6. NANB virus presently causes about 90% of post-transfusion hepatitis. Screening methods for NANB are not yet available.
7. Immune globulin is used to confer passive immunity. There is no specific treatment.

Viral Gastroenteritis (p. 660)

1. Gastroenteritis may be caused by reoviruses and parvoviruses (in children) and the Norwalk agent (a parvovirus).

MYCOTOXINS (p. 660)

1. Mycotoxins are toxins produced by some fungi.
2. They affect the blood, nervous system, kidney, or liver.

PROTOZOAN DISEASES OF THE DIGESTIVE SYSTEM (pp. 660–662)

Giardiasis (pp. 660–661)

1. *G. lamblia* grows in the intestines of humans and wild animals and is transmitted in contaminated water.
2. Symptoms of giardiasis are malaise, nausea, weakness, and abdominal cramps that persist for weeks.
3. Diagnosis is based on identification of the protozoan in feces or small intestine.

Balantidiasis (Balantidial Dysentery) (p. 661)

1. *B. coli* causes balantidial dysentery when growing in the large intestine.
2. Infections are acquired by ingesting cysts in contaminated food and water.
3. *B. coli* can cause ulceration of the intestinal wall and fatal dysentery.
4. Diagnosis is based on observation of trophozoites in feces.

Amoebic Dysentery (Amoebiasis) (pp. 661–662)

1. Amoebic dysentery is caused by *E. histolytica* growing in the large intestine.
2. The amoeba feeds on red blood cells and GI tract tissues. Severe infections result in abscesses.
3. Diagnosis is confirmed by observation of trophozoites in feces.

Cryptosporidiosis (p. 662)

1. *Cryptosporidium* causes prolonged diarrhea in immunosuppressed patients.
2. Presence of oocysts in feces confirms diagnosis.

HELMINTHIC DISEASES OF THE DIGESTIVE SYSTEM (pp. 662–667)

Tapeworm Infestation (pp. 662–665)

1. Tapeworms are contracted by the consumption of undercooked beef, pork, or fish containing encysted larvae (cysticerci).
2. The scolex attaches to the intestinal mucosa of humans (the definitive host) and matures into an adult tapeworm.
3. Eggs are shed in the feces and must be ingested by an intermediate host.
4. Adult tapeworms can be undiagnosed in a human. Severe infestations can result in anemia, as the parasite competes with the host for vitamin B_{12}.
5. Diagnosis is based on observation of proglottids and eggs in feces.

Trichinosis (pp. 665–667)

1. *T. spiralis* larvae encyst in muscles of humans, swine, and other mammals to cause trichinosis.
2. The roundworm is contracted by ingesting undercooked meat containing larvae.
3. Adults mature in the intestine and lay eggs. The new larvae migrate to invade muscles.
4. Symptoms include fever, swelling around the eyes, and gastrointestinal upset.
5. Biopsy specimens and serological tests are used for diagnosis.

STUDY QUESTIONS

REVIEW

1. List at least five antimicrobial features of the digestive system and describe their activity.
2. What are bacteriocins? How are they produced and what do they do?
3. State examples of representative normal flora, if any, in these parts of the GI tract: mouth; stomach; small intestine; large intestine; rectum.
4. What properties of *S. mutans* implicates this bacterium in the formation of dental caries? Why is sucrose, more than any other carbohydrate, responsible for the formation of dental caries?
5. List the general symptoms of gastroenteritis. Since there are many etiologies, on what is the laboratory diagnosis usually based?
6. Differentiate between these factors of bacterial intoxication and bacterial infection: prerequisite conditions, etiologic agents, onset, duration of symptoms, and treatment.
7. Complete the following table.

Disease	Etiologic Agent	Suspect Foods	Symptoms	Treatment
Staphylococcal food poisoning				
Salmonellosis				
Bacillary dysentery				
Asiatic cholera				
Gastroenteritis				
Traveler's diarrhea				

8. Differentiate between salmonellosis and typhoid fever.
9. You probably listed *E. coli* in answer to questions 2 and 6. Explain why this one bacterial species is both beneficial and harmful.
10. Provide the information asked for in this table.

Disease	Etiologic Agent	Method of Transmission	Site of Infection	Symptoms	Prevention
Mumps					
CMV inclusion disease					
Infectious hepatitis					
Serum hepatitis					
Viral gastroenteritis					

11. What treatments are currently used for infectious hepatitis? for Hepatitis B? for Non-A Non-B hepatitis?
12. How is blood to be used for transfusions tested for HBV? For NANB?
13. Define mycotoxin.
14. Explain how the following diseases differ and how they are similar: giardiasis; balantidiasis; amoebic dysentery; cryptosporidiosis.
15. Differentiate between amoebic dysentery and bacillary dysentery.
16. Diagram the life cycle for a human tapeworm.
17. Diagram the life cycle for trichinosis, and include humans in the cycle.
18. How can bacterial and protozoan infections of the GI tract be prevented?

CHALLENGE

1. Look at your diagrams for questions 16 and 17. Indicate sequences in the life cycles that could be easily broken to prevent these diseases.
2. Why is a human infection of trichinosis considered a "dead end" for the parasite?
3. Complete the following table.

Disease	Conditions Necessary for Microbial Growth	Basis for Diagnosis	Prevention
Staphylococcal food poisoning			
Salmonellosis			

4. Twenty-eight kindergarten children and seven adults visited a certified raw milk (CRM) bottling plant where they were given ice cream and CRM. Three to six days later, nine children and three adults developed gastroenteritis. The only foods eaten by all these children (ill and non-ill) were in the school-provided lunches. No one else in the school became sick. What was the source of this gastroenteritis outbreak? Stool cultures showed one bacterium common to nine of the ill children and not present in samples from nine non-ill children. This bacterium is a curved gram-negative rod; it neither ferments nor oxidizes glucose. What is the bacterium?

FURTHER READING

Centers for Disease Control. *Salmonella Surveillance Report*. Published throughout each year. Tabulation of incidence and reports on current investigations.

Centers for Disease Control. *Shigella Surveillance Report*. Published throughout each year. Tabulation of data and background on recent food-borne or water-borne outbreaks.

Marsh, P. 1980. *Oral Microbiology*. Washington, D.C.: American Society for Microbiology. Discussion of the relationship of bacteria to dental caries and periodontal disease.

Melnick, J. L., G. R. Dreesman, and F. B. Hollinger. July 1977. Viral hepatitis. *Scientific American* 237:44–62. Epidemiology and immunology of hepatitis.

Prince, A. M. 1983. Non-A, Non-B hepatitis viruses. *Annual Review of Microbiology* 37:217–232. Describes ongoing research seeking the causative agent(s) of NANB hepatitis.

Reimann, H. and F. L. Bryan. 1979. *Food-borne infections and intoxications*, 2nd ed. New York: Academic Press. An authoritative discussion of nearly all food-borne diseases.

See also Further Reading for Part Four at the end of Chapter 24.

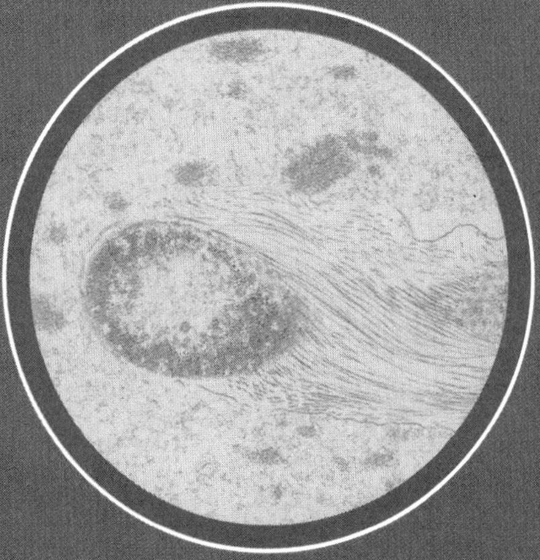

CHAPTER 24

Microbial Diseases of the Urinary and Genital Systems

OBJECTIVES

After completing this chapter you should be able to

- List the normal flora of the urinary and genital systems and their habitats.
- Describe methods of transmission for urinary and genital system infections.
- List microorganisms that cause cystitis and pyelonephritis and name the predisposing factors for these diseases.
- Describe the cause and treatment of glomerulonephritis.
- List the etiologic agent, symptoms, method for diagnosis, and treatment for each of the following: leptospirosis; gonorrhea; syphilis; NGU; vaginitis; LGV; chancroid; granuloma inguinale; candidiasis; trichomoniasis.
- Discuss the epidemiology of genital herpes.
- List genital diseases that can cause congenital and neonatal infections and explain how these infections can be prevented.

The **urinary system** consists of organs that regulate the chemical composition of the blood and excrete waste products of metabolism. The **genital,** or **reproductive, system** consists of organs that produce gametes to propagate the species and that, in the female, support and nourish the developing embryo. Because the urinary and genital systems are closely related anatomically, they will be discussed in the same chapter. In fact, some diseases that affect one system also affect the other, especially in the male. Microbial diseases of both systems tend to cause only minor discomfort at first. But if left untreated, these diseases can spread beyond their original locations and cause serious complications.

Both systems open to the external environment and thus have portals of entry for microorganisms that can cause disease. Normal flora of these and other body systems also cause opportunistic infections of the urinary and genital systems.

STRUCTURE AND FUNCTION OF THE URINARY SYSTEM

The **urinary system** consists of two *kidneys,* two *ureters,* a single *urinary bladder,* and a single *urethra* (Figure 24–1). The kidneys contain microscopic functional units called *nephrons.* As blood circulates through the kidneys, the nephrons control the concentration and volume of blood by removing and adding selected amounts of water and solutes and excreting wastes. The wastes, which include urea, uric acid, creatinine, and various salts, together with water, are referred to as *urine.* The urine passes down the ureters into the urinary bladder, in which it is stored prior to elimination from the body. Elimination occurs through the urethra. In the female, the urethra conveys only urine to the exterior. In the male, the urethra is a common tube for both urine and seminal fluid.

The urinary tract has certain characteristics that help prevent infection. Where the ureters enter the urinary bladder, there are valves that normally prevent the backflow of urine to the kidneys. This mechanism helps to shield the kidneys from lower urinary tract infections. In addition, the acidity of normal urine has some antimicrobial properties. And the flushing action of urine to the exterior prevents microorganisms from setting up foci of infection. Finally, it appears that antibody-forming cells cluster in regions of the urinary system where infection does occur.

STRUCTURE AND FUNCTION OF THE GENITAL SYSTEM

The **female reproductive system** consists of two *ovaries,* two *fallopian tubes,* the *uterus,* the *vagina,* and *external genitals* (Figure 24–2). The ovaries produce female sex hormones and ova (eggs). When an ovum is released, in the process called ovulation, it enters a fallopian tube; if viable sperm are present, fertilization occurs. The fertilized ovum (zygote) descends the fallopian tube and enters the uterus. There it implants in the inner wall, and remains while it develops into an embryo and later a fetus. At birth, the infant is expelled from the uterus through the vagina. The vagina also serves

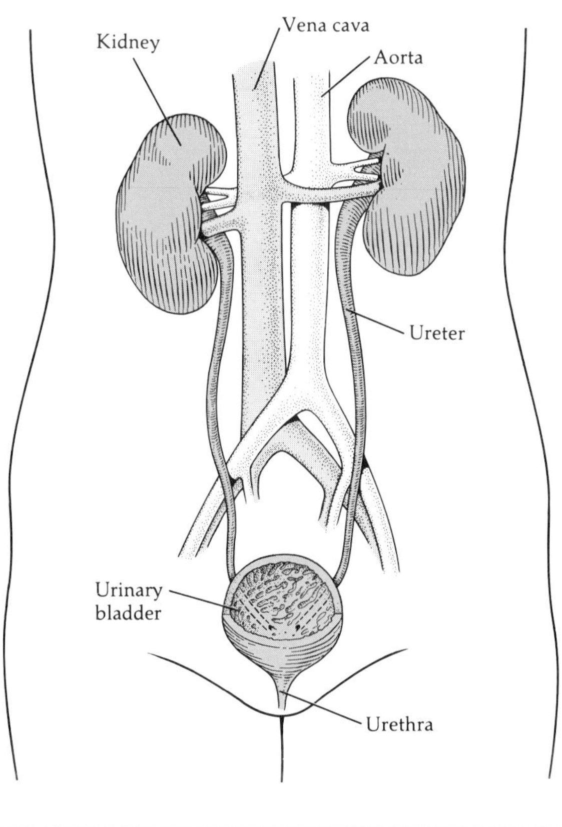

Figure 24–1 Organs of the urinary system.

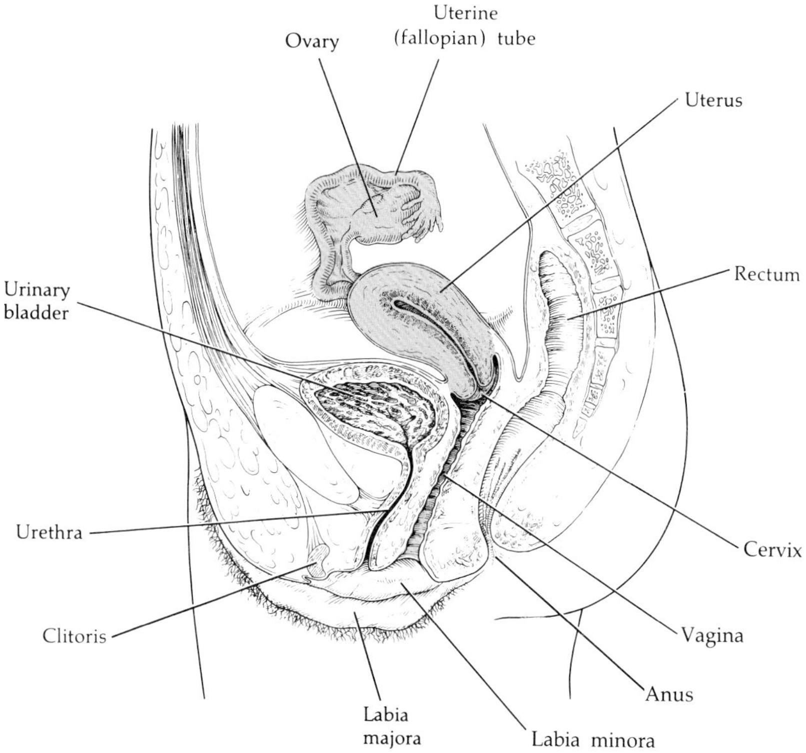

Figure 24–2 Female organs of reproduction.

as a copulatory canal. The external genitals (vulva) consist of the clitoris, labia, and glands that produce a lubricating secretion during copulation.

The **male reproductive system** consists of two *testes,* a system of *ducts, accessory glands,* and the *penis* (Figure 24–3). The testes produce male sex hormones and sperm. Newly produced sperm are moved into the epididymis, in which they are stored until ejaculation. To exit from the body, the sperm cells pass through a series of ducts (epididymis, ductus (vas) deferens, ejaculatory duct, and urethra). During ejaculation, contractions of the ductus (vas) deferens and ejaculatory duct propel the sperm toward the urethra. Along the route, the seminal vesicles, prostate gland, and bulbourethral glands secrete an alkaline fluid into seminal fluid. This alkaline fluid buffers the acidic environment of the vagina; this buffering is an important mechanism, because sperm cells are killed in an acidic environment. On ejaculation, the seminal fluid leaves the body through the urethra. A valve system prevents the seminal fluid from entering the bladder.

NORMAL FLORA OF THE URINARY AND GENITAL SYSTEMS

Normal urine in the urinary bladder and the organs of the upper urinary tract is sterile. The urethra, however, does contain a normal resident flora that includes *Streptococcus, Bacteroides, Mycobacterium, Neisseria,* and a few enterobacteria, and urine becomes contaminated with skin flora during passage.

In the female genital system, the normal flora of the vagina is greatly influenced by sex hormones. For example, within a few weeks after birth, the female infant's vagina is populated by lactobacilli. This population has grown because estrogens, which were transferred from maternal to fetal blood, have caused glycogen to accumulate in the cells lining the vagina. Lactobacilli convert the glycogen to lactic acid, and the pH of the vagina becomes acidic. This glycogen–lactic acid sequence provides the conditions under which the acid-tolerant normal flora in the vagina grows. As the effects of the estrogens diminish several weeks after birth, other bacteria—including corynebacteria and a variety of cocci and bacilli—establish and dominate the flora, and the pH becomes more neutral until puberty. At puberty, estrogen levels increase, lactobacilli again dominate, and the vagina again becomes acidic. During the reproductive years, small numbers of other bacteria and yeasts become part of the flora. In the adult, a disturbance of this ecosystem by a decrease in glycogen (caused by oral contraceptives or pregnancy, for example), or elimination of the normal flora by antibiotics, can lead to **vaginitis,** an infection of the vagina. When the female reaches menopause, estrogen levels again decrease, the flora returns to that of childhood, and the pH again becomes neutral.

BACTERIAL DISEASES OF THE URINARY SYSTEM

Many infections of the urinary system appear to be opportunistic and related to a number of predisposing factors. These factors include disorders of the nervous system, toxemia associated with pregnancy, diabetes mellitus, and obstructions, such as tumors and kidney stones, to the flow of urine. Because the anus, from which feces are excreted, is located close to the urethra, it is not unusual, especially in females, for the urinary tract to become contaminated with intestinal bacteria. These are

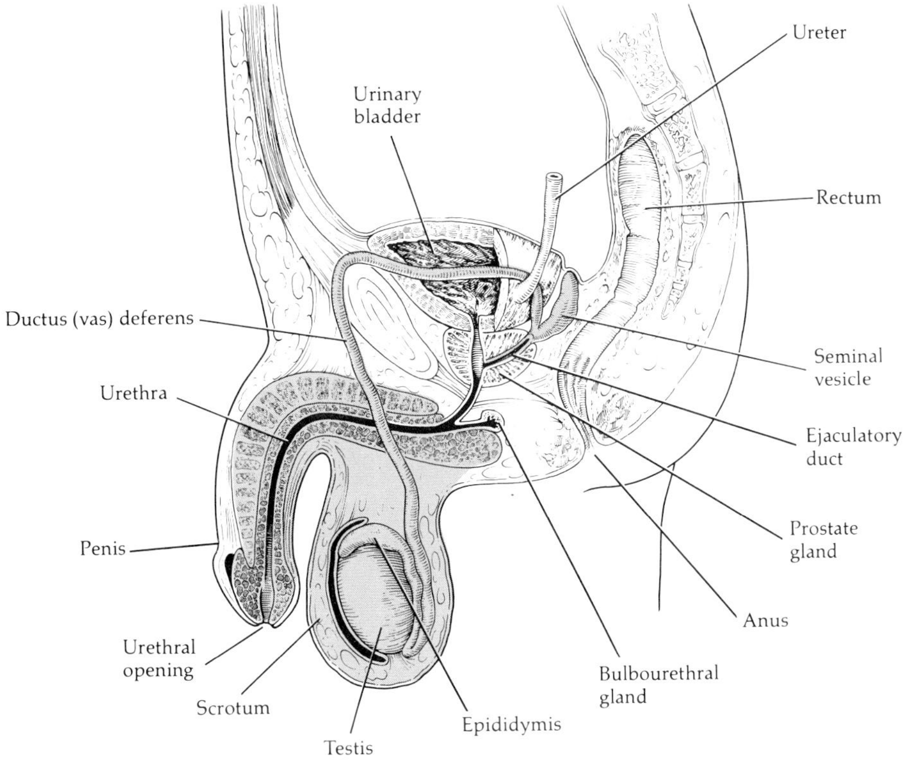

Figure 24–3 Male organs of reproduction.

usually *Escherichia coli, Proteus* species, and other gram-negative enteric bacteria, although pseudomonads, fecal streptococci, and staphylococci are also common in such infections. These *Pseudomonas* infections are, as those elsewhere, unusually troublesome to treat. The fungus *Candida albicans* is also an opportunistic agent of urinary tract infections. Moreover, a number of sexually acquired diseases can cause inflammation in the urinary tract.

Infections usually cause inflammation of the affected tissue. Inflammation of the urethra is called *urethritis;* of the urinary bladder, *cystitis;* and of the ureters, *ureteritis.* The most significant danger from lower urinary tract infections is that they can affect the kidneys (pyelonephritis) and impair their function. The kidneys are also sometimes affected by systemic bacterial diseases such as typhoid fever or leptospirosis; when the kidneys are thus infected, the pathogens causing these diseases can be found in excreted urine.

Many infections of the urinary tract are of nosocomial origin. (In fact, about 35% of all nosocomial infections occur in the urinary tract.) Operations on the urinary bladder or prostate gland and catheterization for draining the urinary bladder are procedures that introduce bacteria into the bladder and ureters. *E. coli* causes more than half of the nosocomial infections of the urinary tract, although fecal streptococci, *Proteus, Klebsiella,* and *Pseudomonas* also commonly cause such infections.

Treatment of diseases of the urinary tract depends on diagnostic isolation of the causative organism and determination of its antibiotic sensitivity. Normal urine contains fewer than 10,000 bacteria per milliliter. When more than 100,000 bacteria per milliliter are found, there is usually an infection.

Cystitis

Cystitis is an inflammation of the urinary bladder and is very common, especially among females. The female urethra has many microorganisms in the area around its opening, and it is shorter than the male urethra, so that microorganisms can traverse it more readily between voidings. Sexual intercourse or careless personal hygiene facilitates such transfer. Contributing factors in females include gastrointestinal system infections and preexisting infections of the vagina, uterus, or urethra. In males, cystitis might be associated with infections of the gastrointestinal system, kidneys, or urethra. Bacteria frequently associated with cystitis in both sexes are the gram-negative rods *Escherichia coli, Proteus vulgaris* (prō'tē-us vul-ga'ris), and *Pseudomonas aeruginosa.*

Treatment of cystitis depends on the bacterium responsible and involves administration of sulfonamides, chloramphenicol, kanamycin, penicillin G, or polymyxin.

Pyelonephritis

Pyelonephritis, an inflammation of one or both kidneys, involves the nephrons and the renal pelvis (the opening into the ureter). The disease is generally a complication of infection elsewhere in the body. In females it is often a complication of lower urinary tract infections. The causative agent in about 75% of the cases is *Escherichia coli.* Other bacteria associated with pyelonephritis are *Enterobacter aerogenes, Proteus* species, *Pseudomonas aeruginosa, Streptococcus pyogenes,* and staphylococci. Should pyelonephritis become chronic, scar tissue forms in the kidneys and severely impairs their function. Depending on the etiological agent, a number of antibiotics can be used to treat pyelonephritis; these include chloramphenicol, gentamicin, kanamycin, polymyxin, methicillin, penicillin, vancomycin, and tetracycline.

Leptospirosis

The causative organism is the spirochete *Leptospira interrogans* (lep-tō-spī'rä in-tėr'rä-ganz), shown in Figure 24–4. Leptospiras have a characteristic shape—an exceedingly fine spiral wound so tightly that it is barely discernible under the darkfield microscope. *L. interrogans,* like other spirochetes, is hard to see under a normal light microscope and is therefore rarely Gram stained. It is an obligate aerobe that can be readily grown in a variety of artificial media supplemented with rabbit serum.

Leptospirosis is a disease of wild mammals, and the microorganism is excreted in the urine of infected animals. Humans and animals in contact with water contaminated by such urine are most likely to contract the disease, which can be transmitted through mucous membranes. Domestic dogs are commonly immunized against the disease.

Leptospirosis is characterized by chills, fever, headache, and muscular pain. Leptospiral organisms can be found during the course of the disease in the blood, liver, cerebral spinal fluid, kidneys, and urine. Jaundice due to liver damage occurs frequently. The human mortality rate is less than 10%, and kidney failure is the most common cause of death. Recovery results in solid immunity. There are about 100 cases per year in the United States.

Diagnosis by serology is complicated by an immunological overlap between the various pathogenic serotypes of *L. interrogans*. Serum antibodies usually appear during the second week of illness and reach a maximum titer during the third or fourth week. However, they are not easily identified because there are so many antigenic types. Agglutination tests, however, can be done with suspensions of killed leptospiras that have been pooled to contain the most common antigenic strains. The treatment of leptospirosis is generally unsatisfactory, partly because the disease is frequently recognized late. The drug of choice is penicillin.

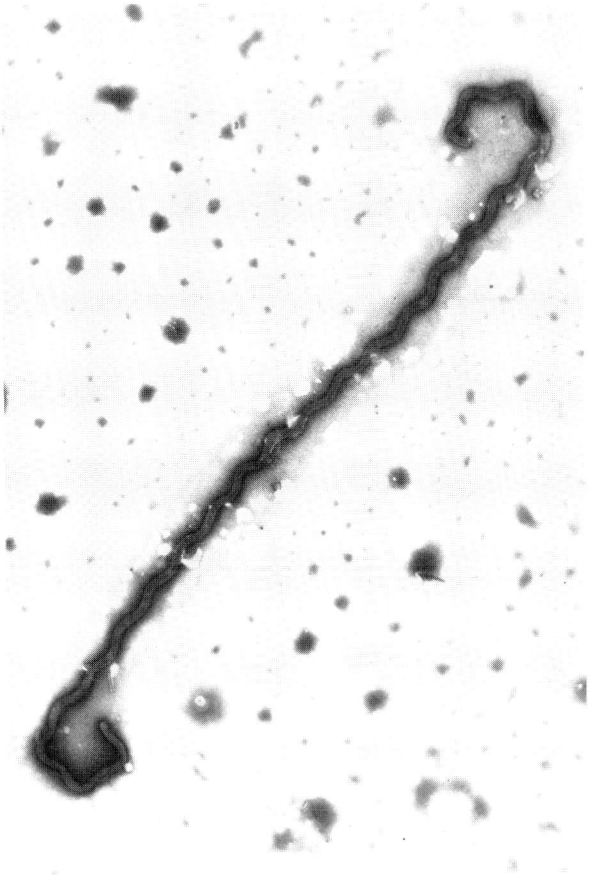

Figure 24–4 Negatively stained cell of the spirochete *Leptospira interrogans*, the causative agent of leptospirosis (×12,000). The "hooked" ends are commonly seen in preparations of this organism, and remnants of the threadlike axial filament are barely visible on the hooks.

Glomerulonephritis

Glomerulonephritis, or **Bright's disease,** is an inflammation of the glomeruli in the kidneys. The glomeruli are blood capillaries that assist in filtering blood as it passes through the kidneys. Most cases of glomerulonephritis are a sequel to infection with beta-hemolytic streptococci *(Streptococcus pyogenes)*, especially type 12. The disease is characterized by fever, high blood pressure, and the presence of protein and red blood cells in the urine. The red blood cells and protein are in the urine because the permeability of the glomeruli increases, a result of inflammation. Glomerulonephritis is an immune-complex disease. Soluble streptococcal antigens combine with specific antibodies to form antigen–antibody complexes that interact with complement (see Chapter 17). The complexes are deposited in the glomeruli, and there cause inflammation and kidney damage. Although a few patients die from glomerulonephritis and some develop chronic conditions, most people recover completely. Treatment of the initial infection consists of administration of chloramphenicol, erythromycin, lincomycin, penicillin, or vancomycin.

BACTERIAL DISEASES OF THE GENITAL SYSTEM

Most diseases of the genital system are transmitted by sexual activity and are therefore called **sexually transmitted diseases (STDs).** These diseases are also called **venereal diseases (VDs),** an older term derived from the name Venus, the Roman goddess

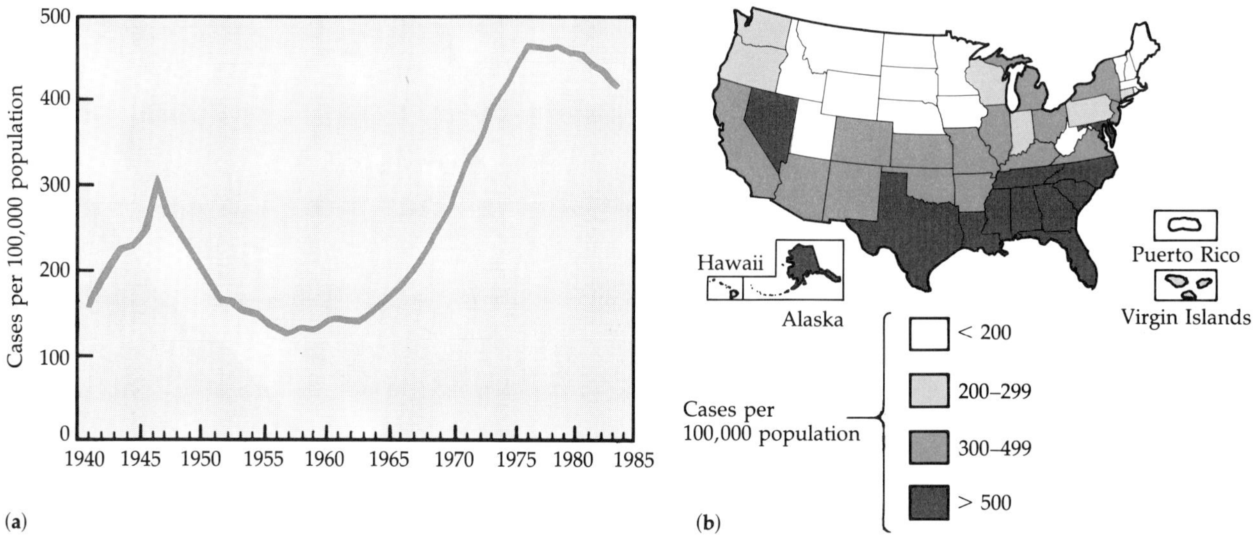

Figure 24-5 Incidence and distribution of gonorrhea. **(a)** Incidence of gonorrhea in the United States since 1941. **(b)** Geographical frequency of cases reported in 1982.

of love. Most of these diseases can be readily cured with antibiotics if treated early and can largely be prevented by the use of condoms. Nevertheless, STDs are a major U.S. public health problem.

Gonorrhea

By far the most common reportable (must be reported to local, state, or federal health agencies) communicable disease in the United States is **gonorrhea,** an STD caused by the gram-negative diplococcus *Neisseria gonorrhoeae* (see Figure 10–9a and Microview 1–1F). An ancient disease, gonorrhea was described and given its present name by the Greek physician Galen in A.D. 150. At the present, the number of cases seems to have reached a plateau after steadily and steeply increasing for many years (Figure 24–5). About 1 million cases in the United States are reported to the Centers for Disease Control each year, and the true number of cases is probably 3 to 4 million. More than 60% of the cases are in the 15- to 24-year-old age group.

To infect, the gonococcus must attach, by means of pili, to the mucosal cells of the epithelial wall. One experimental vaccine is aimed at preventing these pili from attaching. The organism does not infect the layered squamous cells characteristic of the external skin, but invades the spaces separating mucosal cells. Mucosal cells are found in the oral-pharyngeal area, the eyes, joints, and rectum, male and female genitalia, and external genitalia of prepubertal females. The invasion sets up an inflammation, and when leukocytes move into the inflamed area, the characteristic pus formation results.

Males are made aware of a gonorrheal infection by painful urination and discharge of pus-containing material from the urethra (Figure 24–6). About 80% of infected males show these obvious symptoms after an incubation period of only a few days, and most others in less than a week. In the days before antibiotic therapy these symptoms persisted for weeks. In untreated cases, recovery may eventually occur without complication, but when complications do occur, they can be serious. In some cases, the urethra is scarred and partially blocked. And sterility can result when the testes become infected or when the ductus (vas) deferens, the tube carrying sperm from the testes, becomes blocked by scar tissue.

In females, the disease is more insidious. Very few women are aware of the early stages of the infection. Later, however, there might be abdominal pain due to *pelvic inflammatory disease (PID)*. PID is a collective term for any extensive bacterial infection of the pelvic organs, particularly the uterus, cervix, fallopian tubes, and ovaries. The most serious of these is the infection of the fallopian tubes (*salpingitis*). This is a fairly common occurrence if the disease is not treated in the early stage, which is mild and often undiagnosed. The results of salpingitis can be scarring that blocks the passage of ova from the ovaries to the uterus. Only about 20% of such cases of sterility can be reversed surgically. This infection can also lead to pregnancy taking place in the fallopian tubes rather than in the uterus; this life-threatening condition is called *ectopic pregnancy*. Tens of thousands of women have been rendered sterile by such infections in recent years.

In both sexes, untreated gonorrhea can become a serious, systemic infection. Complications of gonorrhea can involve the joints, heart **(gonorrheal endocarditis)**, meninges **(gonorrheal meningitis)**, eyes, pharynx, or other parts of the body. **Gonorrheal arthritis**, which is caused by the growth of the gonococcus in fluids in joints, occurs in about 1% of gonorrhea cases. Joints commonly affected include the wrist, knee, and ankle.

Gonorrheal eye infections occur most often in newborns. If the mother is infected, then the eyes of the infant can become infected as it passes through the birth canal. This condition, **ophthalmia neonatorum**, can result in blindness. Because of the seriousness of this condition, and the difficulty of being certain that the mother is free of gonorrhea, erythromycin, or silver nitrate in dilute solution, is placed in the eyes of all newborn infants. If the mother is known to be infected, an intramuscular injection of penicillin is also administered to the infant. This is required by law in most states. Gonorrheal infections can also be transferred by hand contact from infected sites to the eyes of adults.

Gonorrheal infections can be acquired at any point of sexual contact; pharyngeal and anal gonorrhea are not uncommon. The symptoms of **pharyngeal gonorrhea** often resemble those of the usual septic sore throat; **anal gonorrhea** can be rather painful and accompanied by discharges of pus.

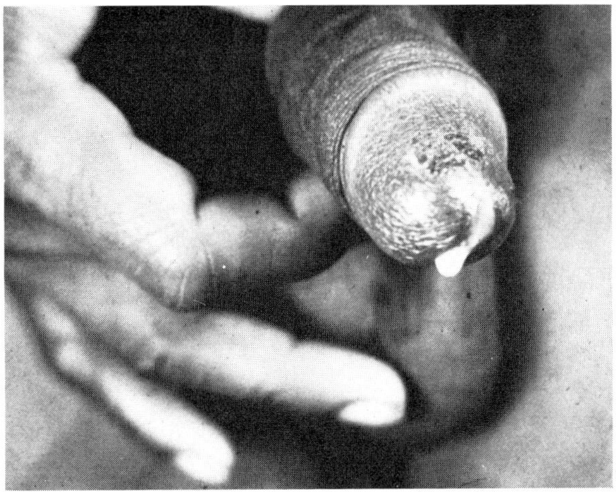

Figure 24-6 Pus-containing discharge from urethra of a male with an acute case of gonorrhea.

The increase of sexual activity with a series of partners and the fact that the disease in the female may go unrecognized have contributed considerably to the increased incidence of gonorrhea and other STDs. The widespread use of oral contraceptives has also contributed to the increase. Oral contraceptives tend to increase the moisture content and raise the pH of the vagina, while increasing the susceptibility of mucosal cells. They often replaced condoms and spermicides that helped prevent disease transmission. All these factors predispose the female to infection, and therefore increase the overall incidence of disease.

Immunity to reinfection does not result from recovery from gonorrhea. Penicillin has been an effective treatment for gonorrhea over the years, although the dosages have had to be much increased because of the appearance of penicillin-resistant bacteria. Penicillin resistance is usually due to the presence of a gene for penicillinase (an enzyme that degrades penicillin). The penicillinase gene might be carried by the bacterial chromosome or by a plasmid (see Chapter 8). Gonococci having such plasmids first appeared in 1976 (see box on pp. 515–516). The plasmid is transmissible among gonococci and might have been derived from *Hemophilus influenzae* or other gram-negative bacteria. Spectinomycin is the favored alternative antibiotic for strains carrying penicillinase plasmids.

Chemotherapy for gonorrhea can vary greatly according to the infection's site, which might be

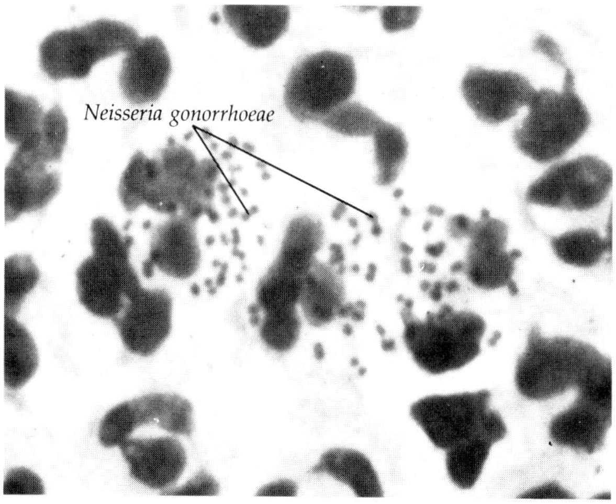

Figure 24–7 A smear from a patient with gonorrhea. Although a finding of *Neisseria gonorrhoeae* in phagocytic leukocytes is a probable indication of gonorrhea, the bacterium should be isolated and identified.

the eyes, throat, anus, joints, heart (endocarditis), or meninges. Another consideration in treatment is that the patient might have concurrent infections by other sexually transmitted pathogens such as *Chlamydia trachomatis* (kla-mi'dē-ä trä-kō'mä-tis). An estimated 20 to 25% of gonorrhea patients have these concurrent infections. Because tetracycline is effective against chlamydiae (as well as the gonococcus) some physicians routinely administer tetracycline, in addition to the penicillin or other antigonococcal antibiotic, to treat possible undiagnosed chlamydial infections.

Gonorrhea in men is diagnosed with a stained smear of pus from the urethra. The typical gram-negative diplococci within the phagocytic leukocytes are readily identified (Figure 24–7 and Microview 1–1F). Gram staining of exudates is not as reliable a diagnostic procedure with women. Usually a culture is taken from within the cervix and grown on special media such as Thayer–Martin. Cultivation of the nutritionally fastidious bacterium requires an atmosphere containing carbon dioxide. The gonococcus is very sensitive to adverse environmental influences (desiccation and temperature) and survives poorly outside the body. It even requires special transporting media to keep it viable for short intervals before the cultivation is underway. After 12 to 16 hours, smears of the early growth can be stained with specific fluorescent antibody for rapid identification.

Syphilis

The earliest reports of **syphilis** date back to the end of the fifteenth century in Europe. This was coincidental with the return of Columbus from the New World, and there has always been a hypothesis that syphilis was introduced to Europe by his men. Although there is no way to know for sure, it appears from descriptions that the disease in earlier times was rather more severe than it is at present. Whether this decline is due to a diminished virulence of the bacterium or an increased resistance in the population is unknown.

In contast to the precipitous increase in the number of gonorrhea cases, the number of new syphilis cases in the United States has recently remained fairly stable at about 30,000 (Figure 24–8). This relative stability compared to gonorrhea is remarkable, because the epidemiology of the two diseases is quite similar and concurrent infections by both diseases are not uncommon. An important difference, however, is that syphilis has a longer incubation time, which facilitates the tracing of cases in time for effective treatment before the new contacts also become infective.

The highest incidence of syphilis is in the 20- to 39-year-old age group, which is the most sexually active. The incidence among male homosexuals is about 25 to 30 times that of the general population. In the heterosexual population, syphilis has actually become an uncommon disease. Many states have discontinued the requirements for premarital syphilis tests because so few cases are detected.

The causative organism of syphilis is a gram-negative spirochete, *Treponema pallidum* (tre-pō-nē'mä pal'li-dum). The spirochete is a thin, tightly coiled helix no more than 15 μm in length. See Figure 24–9. The pathogenic strains of the spirochetes have never been successfully cultured in vitro. Instead, they are propagated in rabbits.

Syphilis is transmitted by sexual contact of all kinds, via syphilitic infections of the genitals or other body parts. The organism is apparently able to penetrate intact mucous membranes, but can penetrate skin only through minor breaks or abrasions. The incubation period averages about three

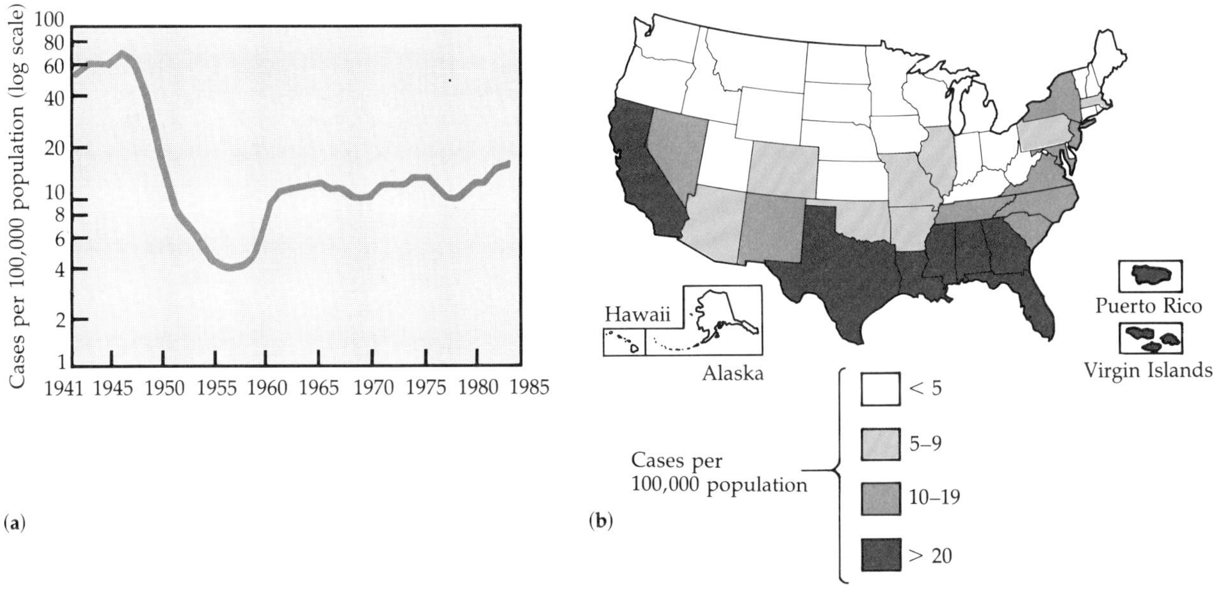

Figure 24–8 Incidence and distribution of syphilis. **(a)** Number of cases in the United States since 1941. **(b)** Geographical frequency of cases reported in 1982.

weeks, but can range from two weeks to several months. The disease progresses through several recognized stages.

In the **primary stage** of the disease, the initial symptom is a small, hard-based **chancre,** or sore, which usually appears at the site of infection (Figure 24–10a). The chancre is painless, and a serous exudate is formed in the center. This fluid is highly infectious, and examination with the darkfield microscope shows many spirochetes. In a few weeks, this lesion disappears, possibly because of local immunity having developed. None of these symptoms cause any distress. In fact, many women are entirely unaware of the chancre, which is commonly on the cervix of the uterus. And in males, the chancre sometimes forms in the urethra and is not visible. Serological diagnostic tests are positive in about 80% of patients in the primary stage. During this stage, bacteria enter the bloodstream and lymphatic system, which widely distribute them in the body.

Several weeks after the primary stage (the exact amount of time varies), the disease enters the **secondary stage,** characterized mainly by skin rashes of varying appearance (Figure 24–10b). Other symptoms often observed are the loss of patches of hair, malaise, and mild fever. The rash is widely distributed on the skin and is also found in the mucous membranes of the mouth, throat, and cervix. At this stage, the lesions of the rash contain many spirochetes and are very infectious. Dentists or other medical workers coming into con-

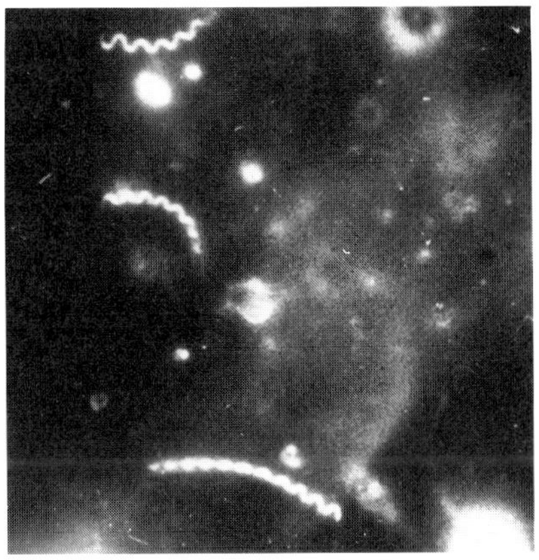

Figure 24-9 Syphilis spirochetes as seen through a darkfield microscope.

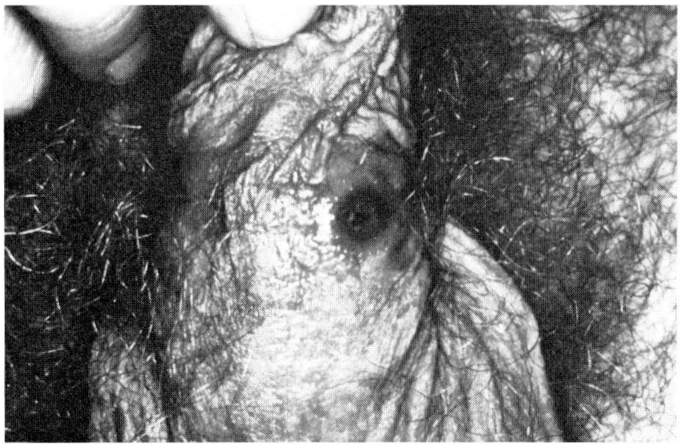
(a)

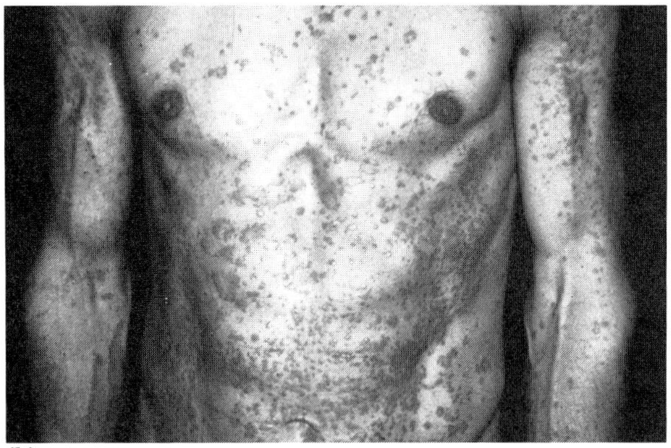
(b)

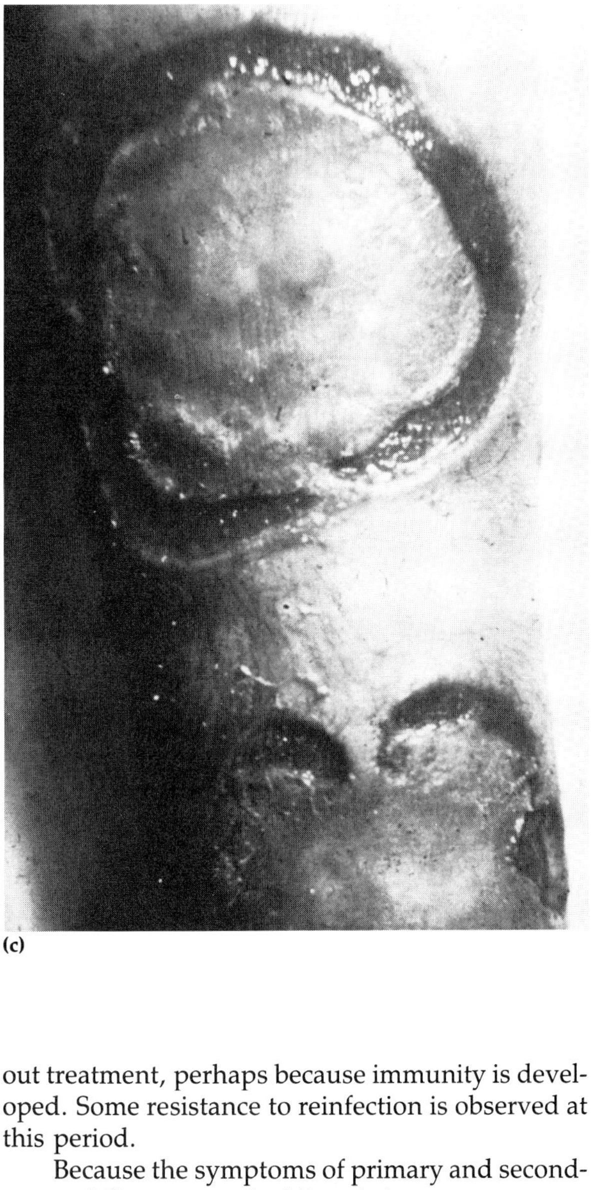
(c)

Figure 24–10 Characteristic lesions associated with syphilis at various stages of development. **(a)** Chancre of primary stage on a male. **(b)** Skin rash of secondary stage. **(c)** Gummas of tertiary stage on the back of an arm.

tact with fluid from these lesions can easily become infected by the spirochete entering through minute breaks in the skin. Such nonsexual transmission is possible, but because the organisms do not survive long on environmental surfaces, they are very unlikely to be transmitted via such objects as toilet seats. Serological tests for syphilis become almost uniformly positive at this stage.

The symptoms of secondary syphilis usually subside after a few weeks, and the disease enters a **latent period.** During this period there are no symptoms of the disease although serological tests are positive. After 2 to 4 years of latency, the disease is not normally infectious, except for transmission from mother to fetus. The majority of cases do not progress beyond the latent stage, even without treatment, perhaps because immunity is developed. Some resistance to reinfection is observed at this period.

Because the symptoms of primary and secondary syphilis are not disabling, it is not uncommon for persons to enter the latent period without having received medical attention. In less than half of these cases, the disease reappears as a **tertiary stage.** This stage occurs only after an interval of many years after the latent phase is entered—usually at least 10. It has been suggested that most of the symptoms of tertiary syphilis are due to the body's hyperimmune reactions to surviving spirochetes. Many of the lesions are not unlike those formed by the body in response to infections by *Mycobacterium tuberculosis.* These lesions, called *gummas*, are rubbery masses of tissue that appear in many organs and sometimes on the external skin (Figure 24–10c). Although many of these lesions are not very harmful, some can cause extensive tissue damage, such as deafness or blindness from lesions in the central nervous system or perforation of the palate (roof of the mouth), which interferes with speech. Few, if any, organisms are found in the lesions of the tertiary stage, and they are not considered very infectious. Today, cases of syphilis allowed to progress to this stage have become uncommon.

One of the most distressing and dangerous forms of syphilis, called **congenital syphilis,** is transmitted across the placenta to the unborn fetus. Damage to mental development and other neurological symptoms are among the more serious consequences. This type of infection is most common among pregnancies occurring during the latent stage of syphilis. A pregnancy during the primary or secondary stage is likely to produce a stillbirth.

To detect primary syphilis, serological testing is not completely reliable. Often the best method for diagnosis is an examination of exudates removed from punctured lymph nodes in the affected area or from lesions. A darkfield microscope is necessary because the organisms do not stain well and are only about 0.2 μm in diameter. This approaches the limits of a brightfield microscope. Figure 24–9 shows the spirochete in such a preparation. The **Venereal Disease Research Laboratory (VDRL)** slide flocculation test is probably the most widely used serological screening test. The **rapid plasma reagin** (***RPR***) card test, which is similar, is also in common use. Both of these tests are nonspecific in that they do not detect antibodies produced against the spirochete itself, but reagin-type antibodies (IgE). Reagin-type antibodies are apparently a response to lipid materials that the body forms in indirect response to infection by the spirochete. The antigen used in such precipitation-type slide tests is thus not the spirochete, but an extract of beef heart (cardiolipin) that seems to contain lipids similar to those that stimulated the reagin-type antibody production. Because these antibodies are formed 3 to 5 weeks after infection, the usefulness of VDRL and RPR tests are limited for primary syphilis. Each of these serological tests will detect only about 70 to 80% of primary syphilis cases, but 99% of secondary syphilis cases.

Most slide tests are likely to produce a percentage of false positive reactions for syphilis. Therefore, positive reactions should be confirmed by a more exacting test based on spirochete-type antigens. These tests often have less than 1% false positive reactions. One such test is the *fluorescent treponemal antibody absorption (FTA–ABS) test.* This is an indirect immunofluorescence test in which a preparation of an avirulent, cultivated strain of *T. pallidum* is allowed to react with a sample of a patient's serum on a microscope slide (see Figure 16–19c). Antibodies in the serum will combine with the spirochetes, but this reaction is not visible. To make it visible, an antibody that will combine with any human antibody (anti–human gamma globulin or anti–HGG) is tagged with a dye that will be visible under a fluorescent microscope. The tagged anti–HGG is added to the slide and will form a combination of spirochete–antibody and tagged anti–HGG. The slide is again washed and examined with a special fluorescent-type microscope. The spirochetes are detected by their fluorescent glow when struck by ultraviolet light. A disadvantage of the FTA–ABS test is that it remains positive even after syphilis has been successfully treated. The non-specific reagin-based tests tend to become negative upon successful treatment, and are therefore more useful for determining the effect of therapy.

Benzathine penicillin, a long-acting formulation that remains effective in the body for about two weeks, is the usual antibiotic treatment of syphilis. The serum concentrations achieved by this formulation are low, but the spirochete has remained very sensitive and shows no significant resistance. Penicillin therapy is especially effective during the primary stage. But antibiotics are generally not effective in tertiary syphilis, probably because the organisms are usually not present.

Antibiotic treatment might be prolonged. One reason for this prolonged exposure is that the spirochete grows slowly and penicillin is effective only against growing, not dormant, organisms. For penicillin-sensitive persons, a number of other antibiotics such as erythromycin and the tetracyclines have also proved effective. Antibiotic therapy aimed at gonorrhea and other infections is not likely to eliminate syphilis as well, because the therapy for those other diseases is usually administered over too short a term. Similarly, penicillin directed at syphilis does not reach a high enough concentration to successfully treat gonorrheal infections.

Nongonococcal Urethritis (NGU)

The term **nongonococcal urethritis (NGU)**, also known as **nonspecific urethritis (NSU)**, can refer to any inflammation of the urethra not caused by *Neisseria gonorrhoeae*. Symptoms include pain during urination and a watery discharge.

Although nonmicrobial factors such as trauma (passage of a catheter) or chemical agents (alcohol and certain chemotherapeutic agents) can cause this condition, probably at least 40% of the cases of NGU are acquired sexually. In fact, NGU might be the most common sexually transmitted disease in the United States today. Although it is not a notifiable disease and exact data are lacking, the Centers for Disease Control estimate that 4 to 9 million Americans have NGU. Because the symptoms are often mild in males, and females are usually asymptomatic, many cases go untreated. Physicians often treat NGU as a male urological disease rather than as an STD. Complications are not common, but can be serious. Males may develop inflammation of the epididymis. And in females, inflammation may cause sterility by blocking the fallopian tubes.

Probably the most common pathogen associated with NGU is *Chlamydia trachomatis*. A substantial number of persons suffering from gonorrhea are co-infected with *C. trachomatis*. *C. trachomatis* is the same organism responsible for trachoma (an eye infection) and the sexually transmitted disease, lymphogranuloma venereum (see next section). Chlamydias are small, gram-negative bacteria that are obligate intracellular parasites (see Microview 1–1I). Although the symptoms caused by chlamydias are relatively mild and often asymptomatic, especially in women, they apparently have an alarming capacity to cause *salpingitis* and subsequent infertility.

It is estimated that as many as 60% of salpingitis cases are due to chlamydial infection. A culture of chlamydia requires cell culture lines not available in all laboratories. However, new serological tests that can be done quickly with common laboratory equipment promise to improve diagnostic methods for chlamydia.

Like gonorrhea, *C. trachomatis* can infect the eyes of the infant during birth. Bacteria other than *N. gonorrhoeae* and *C. trachomatis* can also be implicated in NGU. The next most common cause of urethritis and infertility is probably *Ureaplasma urealyticum* (ū-rē-ä-plas'mä ū-rē-ä-lit'i-kum). This organism is a member of the mycoplasma group, bacteria without a cell wall. Another mycoplasma, *Mycoplasma hominis* (ho'min-nis), commonly inhabits the normal vagina, but can opportunistically cause salpingitis.

For diagnosis, mycoplasma can be grown on specialized media, but two to three weeks of incubation time are required. Both chlamydia and mycoplasmas are sensitive to tetracycline-type antibiotics such as tetracycline and doxycycline. Erythromycin is an acceptable alternative drug.

Gardnerella Vaginitis

Vaginitis is most commonly due to one of three organisms, the fungus *Candida albicans* (kan'did-ä al'bi-kans), the protozoan *Trichomonas vaginalis* (trik-ō-mōn'as va-jin-al'is), or the bacterium *Gardnerella vaginalis*. Infections by *C. albicans* (see discussion starting on page 678) are usually opportunistic infections from the normal vaginal flora, but infections by *T. vaginalis* and *G. vaginalis* are likely to be sexually transmitted.

Vaginal infections by *G. vaginalis* are sometimes considered a nonspecific vaginitis because the disease is an interaction between *G. vaginalis* and anaerobic bacteria in the vagina, neither of which alone will produce the disease. Probably one-third of all vaginitis cases are of this type, which occurs when the vaginal pH is 5 to 6. The condition is characterized by a pronounced fishy odor of a frothy vaginal discharge, which is usually light in volume. Diagnosis is based on the fishy odor, the level of vaginal pH, and the microscopic observation of *"clue cells"* in the discharge. These "clue cells" are sloughed-off epithelial cells from the vagina. Treatment is primarily by metronidazole, a drug that eradicates the anaerobes essential to continuation of the disease, but allows the normal lactobacilli to repopulate the vagina.

Lymphogranuloma Venereum (LGV)

There are a number of STDs that are uncommon in the U.S., but more common in the tropical areas of the world. For example, *Chlamydia trachomatis,* the cause of trachoma (see Chapter 19) and a major cause of NGU, is responsible for **lymphogranuloma venereum (LGV),** a disease found in much of the tropical or near-tropical world. In the United States, there are usually about 250 cases per year, most in the southeastern states.

After a latent period of 7 to 12 days, a small lesion appears at the site of infection, usually on the genitals. The lesion ruptures and heals without scarring. One week to two months later, the microorganisms invade the lymphatic system, and the regional lymph nodes become enlarged and tender. Suppuration may also occur. The inflammation of the lymph nodes results in scarring that can occasionally obstruct the lymph vessels. This scarring leads to edema of the genital skin and massive enlargement of the external genitalia in males, and rectal narrowing in females. In females, rectal narrowing results from involvement of the lymph nodes in the rectal region. These conditions can eventually require surgery.

For diagnosis, pus can be aspirated from infected lymph nodes. When infected cells are properly stained with an iodine preparation, the chlamydias can be seen as inclusions. The isolated organisms can also be grown in cell culture or in embryonated eggs. The drug of choice for treatment is tetracycline, and alternatives are drugs such as doxycycline or erythromycin. Enlarged nodes may be slow to subside, even after successful antibiotic therapy. Several serological tests are available, but all are troubled with frequent, false positive reactions.

Chancroid (Soft Chancre) and Granuloma Inguinale

The STDs known as chancroid and granuloma inguinale are also rather more common in tropical areas. In the United States, there are about 800 cases of chancroid reported per year and fewer than 100 cases of granuloma inguinale. However, these diseases are seen with some frequency in sailors and overseas military personnel.

In **chancroid (soft chancre),** a swollen, painful ulcer that forms on the genitalia involves an infection of the adjacent lymph nodes. About a week elapses between infection and appearance of the symptoms. Infected lymph nodes in the groin area sometimes even break through and discharge pus to the surface. These ulcers are highly infective, as are the other genital lesions that occur. Lesions might also occur on such diverse areas as the tongue and lips. Because it is so seldom seen by some physicians, they might confuse it with primary syphilis or genital herpes infections. The causative organism is *Hemophilus ducreyi* (dü-krā'ē), a small gram-negative rod that can be isolated from exudates of lesions. The recommended antibiotics are erythromycin and trimethoprim-sulfamethoxazole.

Granuloma inguinale is characterized by an initial chancre that ulcerates on or about the genitals. It can easily be confused with chancroid and primary syphilis. Without treatment, however, the ulcer spreads, and large areas of genital tissue and other tissues can be destroyed in time. The disease is most common in India, the Caribbean, and Southeast Asia. Epidemiology of the disease is not completely understood, but the disease is assumed to be sexually transmitted, although it is not highly communicable. Male homosexuals are the most common sufferers of granuloma inguinale, although it also comprises about 2 to 3% of the venereal disease found in newly inducted black military recruits in the United States. There is probably a great deal of individual resistance to the disease.

The causative organism is *Calymmatobacterium granulomatis* (kal-im-mä-tō-bak-ti'rē-um gran-ū-lō'mä-tis), a small, gram-negative, encapsulated bacillus, very similar in many respects to *Klebsiella pneumoniae.* In smears of scrapings from lesions, the encapsulated *C. granulomatis* can often be seen with large mononuclear cells called *Donovan bodies* (Figure 24–11). The bacteria can be grown in embryonated eggs and laboratory media. Diagnosis depends on demonstration of the presence of Donovan bodies in infected lesions and tissues. Gentamicin and chloramphenicol are the drugs of

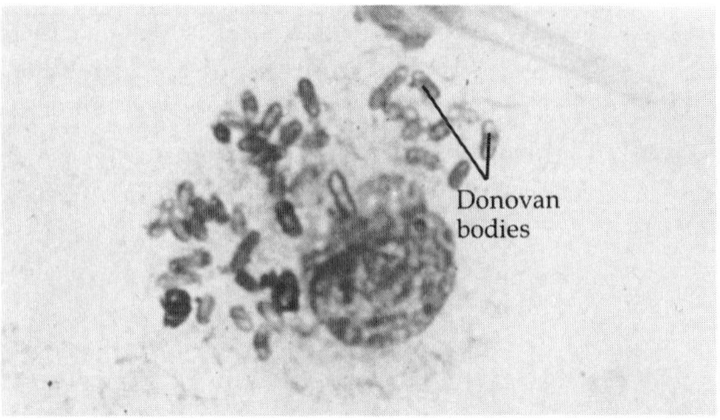

Figure 24–11 Numerous rodlike Donovan bodies are seen in the cytoplasm of this cell from a case of granuloma inguinale. (The darkly stained nucleus is clear but the plasma membrane around the cell is not.)

choice, although tetracyclines and streptomycin are also effective.

VIRAL DISEASE OF THE GENITAL SYSTEM

Genital Herpes

Probably the most publicized STD is genital herpes, caused by *herpes simplex virus* type 2. The herpes simplex virus (see Figures 12–6b and 12–8a) occurs as type 1 or type 2. Herpes simplex type 1 is the virus primarily responsible for the common cold sore or fever blister (see Chapter 19). Herpes simplex type 2, and sometimes type 1, causes a disease of sexual transmission that affects the genitalia; this disease is called **genital herpes.** The lesions appear after an incubation period of about a week, and cause a burning sensation. After this, vesicles appear. In both sexes, urination can be painful and walking quite uncomfortable; the patient is even irritated by clothing. The vesicles contain fluid that is infectious. Usually, the vesicles heal in a couple of weeks.

One of the most distressing characteristics of genital herpes is the possibility of recurrences. As in other herpes infections such as cold sores or chickenpox–shingles, the virus enters a latent state in nerve cells. There is apparently considerable individual variation in latency. Some persons have frequent attacks and for others recurrence is rare. Reactivation appears to be triggered by a number of factors, including menstruation, emotional stress or illness (especially if accompanied by fever—a factor also involved in appearance of cold sores), and perhaps just scratching the affected area.

In the United States, the incidence of genital herpes has increased so much during the past decade that it is now a very common sexually transmitted disease. Females are less likely to be aware that they have the disease. This lack of knowledge can have serious effects, because infants can be infected at birth. Such *neonatal herpes* might be a mild asymptomatic infection or a rapidly fatal one. And an infant who survives can be left with a seriously damaged central nervous system. Therefore, for females known to have genital herpes infections, delivery by Caesarian section might be advised. Occasionally, the infection takes place earlier, in the uterus; this infection also causes birth defects.

Genital herpes infection seems to be associated with a higher-than-normal rate of cervical cancer. It is also associated with a high rate of miscarriages.

At the present time, there is no completely effective treatment for genital herpes, although research on its prevention and cure is intensive. Herpes simplex is difficult to treat for several reasons. First, the virus reproduces very quickly after entering the host cell, new virus particles being released within about eight hours after infection. Newly produced viruses often infect neighboring cells before an effective immune response can be made. They also avoid exposure to extracellular circulating antibodies by migrating directly from the host cell to the neighboring cell at points where the cell membranes make contact. Researchers are hopeful, however, that the new drug acyclovir (see box), which inhibits viral replication by interfering with DNA synthesis, may be able to stop initial outbreaks, prevent recurrences, or decrease the frequency of subsequent attacks if it is administered soon after the individual is infected, while viral replication is still in the early stages. Clinically, however, this would be difficult to accomplish since patients are usually unaware of the infection until lesions appear, at which point replication is well under way.

Considerable effort is being spent on the development of a vaccine for genital herpes. The fact that the disease itself does not prevent latency or recurrence seems to indicate that this search is not likely to produce a reliable vaccine. The approach theoretically most likely to succeed is identification of the genes responsible for latency and then the targeting of these genes with a vaccine, which would prevent latency from developing.

FUNGAL DISEASE OF THE GENITAL SYSTEM

Candidiasis

Candida albicans is a yeastlike fungus that commonly grows on mucous membranes of the mouth, intestinal tract, and genitourinary tract (see Figure 19–13a). Infections are usually a result of opportunistic overgrowth when the competing microflora are suppressed by antibiotics or other reasons. As already discussed in Chapter 19, *C. albicans* is the cause of thrush (oral candidiasis). It is also responsible for occasional cases of NGU in males and an infection of the mucous membrane of the vagina (vaginitis) called **vulvovaginal candidiasis,** which is the most common cause of vaginitis.

The lesions of vulvovaginal candidiasis resemble those of thrush but produce more irritation, severe itching, and a thick, yellow, cheesy discharge. *C. albicans* is an opportunistic pathogen. Predisposing conditions include oral contraceptives and pregnancy. These cause a decrease of glycogen in the vagina (see previous discussion,

Microbiology in the News

New Hope For Treating Herpes

Significant relief for millions of people who suffer from frequent, severe outbreaks of genital herpes may at last be on the way. The Food and Drug Administration recently approved oral use of the antiviral drug acyclovir, the first truly specific antiherpes agent to be licensed for clinical use. Although acyclovir has been used intravenously and topically with moderate success in treating a variety of herpesvirus infections, patients taking the drug orally to treat recurring outbreaks of genital herpes have shown dramatic improvement.

In studies conducted over a period of four months on patients who experience 12 to 16 flare-ups a year (as compared to the average rate of 3 to 4 flare-ups a year), only 25–33 percent of those receiving acyclovir experienced flare-ups, in contrast to 94–100 percent of those receiving placebos. In the cases in which acyclovir did not entirely prevent recurrences, outbreaks did not appear for an average of 120 days after taking the drug; patients taking placebos usually experienced outbreaks in less than 30 days. Also, blistering in those who took acyclovir was very mild, sometimes even difficult to document.

Besides significantly decreasing the frequency and severity of flare-ups, acyclovir decreased the time it took for lesions to heal and decreased virus shedding. Decreased shedding of the virus results in fewer days in which the individual is able to infect someone else; therefore, use of the drug may help prevent the disease from spreading. It is not yet known if acyclovir can prevent the spread of herpes from a pregnant woman to her unborn child.

Although oral acyclovir holds great promise for treating people with severe, frequent outbreaks, it seems to be of little benefit to those who experience only mild attacks. Researchers stress that it is not a "cure" and offers no long-term benefit once treatment is discontinued. Patients usually experience recurrences within one month after they stop taking the drug and continue to suffer outbreaks as frequently as they did before treatment. Oral acyclovir seems to have no serious side effects accompanying short-term use, but studies on its long-term use have yet to be completed. It is still not recommended for use in pregnant or nursing women. There is also some concern that

(continued on next page)

long-term use may lead to the emergence of drug-resistant strains by selectively blocking drug-sensitive strains. However, research completed to date indicates that this is not a serious hazard.

The most striking characteristic of herpesviruses, aptly described as a form of guerrilla warfare, is their capacity for latency. Latency is the ability of the virus to lie dormant in an infected cell for long periods of time, out of the reach of the body's immune system, waiting for an opportune time to launch a new attack and produce new viruses. How is this latent state established? First, the virus enters the body through the mucous membranes of the mouth or genitals or through a cut or wound in the skin. The virus then begins to replicate, producing lesions and infecting nearby cells. Some of the viruses also enter the endings of sensory nerves. The virus then travels up the long axon that projects from a neural cell until it reaches the central ganglion. In the case of genital herpes, the central ganglia are in the sacral region, at the base of the spine. Here the virus "hides out" in its dormant condition.

No one really knows the status of the virus during latency. Its DNA can be isolated (at least parts of it), but it is not known if the DNA is incorporated into the nerve cell's chromosome or if it is free in the cell. In any case, the infectious, intact virus cannot be detected in the cell. The virus is not replicating, but the potential to once again produce infectious viruses seems to remain at all times. From its refuge in the central ganglion, the virus can be triggered, by seemingly trivial stimulation such as sunlight or a slight fever, to emerge and attack. One suggested mechanism for reactivation is that minor irritations might cause the synthesis of prostaglandins (hormonal mediators) that can decrease cell-mediated immunity in the local area, giving the virus a chance to reemerge. When reactivated, the virus begins to multiply in the nerve cells and move outward along the axon to renew the infection.

Antiviral drugs such as acyclovir inhibit the virus when it is reproducing but are ineffective during latency. The virus is vulnerable to these agents, however, each time it reactivates, even if no outward symptoms are expressed. This explains why acyclovir can suppress the virus but not cure the individual.

Table 24–1 Summary of Diseases Associated with the Urinary and Genital Systems

Disease	Causative Agent	Mode of Transmission	Treatment
Bacterial Diseases of the Urinary System:			
Cystitis (bladder infection)	*Escherichia coli*, *Proteus vulgaris*, *Pseudomonas aeruginosa*, and others	Opportunistic infections	Sulfonamides, chloramphenicol, kanamycin, penicillin G, or polymyxin
Pyelonephritis (kidney infection)	*E. coli* is most frequently implicated; other agents include *Enterobacter aerogenes*, *Proteus* species, *Pseudomonas aeruginosa*, *Streptococcus pyogenes*, and staphylococci	From systemic bacterial infections or infections of the lower urinary tract	Depending on agent, antibiotics include chloramphenicol, gentamicin, kanamycin, polymyxin, methicillin, penicillin, vancomycin, and tetracycline
Leptospirosis (kidney infection)	*Leptospira interrogans*	Direct contact with infected animals, urine of infected animals, or contaminated water	Penicillin

Table 24–1 Summary of Diseases Associated with the Urinary and Genital Systems *continued*

Disease	Causative Agent	Mode of Transmission	Treatment
Bacterial Diseases of the Urinary System:			
Glomerulonephritis (kidney infection)	Autoimmune reaction to certain strains of *Streptococcus pyogenes*	Sequel to infection by particular strains of *S. pyogenes* in another part of the body	Chloramphenicol, erythromycin, lincomycin, penicillin, or vancomycin for the initial infection
Bacterial Diseases of the Genital System:			
Gonorrhea	*Neisseria gonorrhoeae*	Direct contact, especially sexual contact	Penicillin for nonresistant strains; spectinomycin for resistant strains; tetracycline for concurrent *Chlamydia* infections
Syphilis	*Treponema pallidum*	Direct contact, especially sexual contact	Penicillin, erythromycin, tetracycline
Nongonococcal urethritis (NGU)	*Chlamydias* or other bacteria, including *Mycoplasma hominis, Ureaplasma urealyticum.*	Sexual contact or opportunistic infections	Tetracycline, erythromycin, or other antibiotics
Gardnerella vaginalis (vaginitis)	*Gardnerella vaginalis*	Opportunistic pathogen, vaginal pH 5 to 6	Metronidazole
Lymphogranuloma venereum (LVG)	*Chlamydia trachomatis*	Direct contact, especially sexual contact	Tetracycline, doxycycline, erythromycin
Chancroid (soft chancre)	*Hemophilus ducreyi*	Direct contact, especially sexual contact	Erythromycin and trimethoprim-sulfamethoxazole
Granuloma inguinale	*Calymmatobacterium granulomatis*	Probably sexual contact	Gentamicin, chloramphenicol, tetracycline, streptomycin
Viral Disease of the Genital System:			
Genital herpes	Herpes simplex virus type 2, and occasionally type 1	Direct contact, especially sexual contact	Acyclovir may help symptoms
Fungal Disease of the Genital System:			
Vulvovaginal candidiasis (occasionally urethritis)	*Candida albicans*	Opportunistic pathogen, may be transmitted by sexual contact	Clotrimazole, miconazole
Protozoan Disease of the Genital System:			
Trichomoniasis (usually vaginal infection)	*Trichomonas vaginalis*	Usually sexually transmitted	Metronidazole

page 676, on normal vaginal flora). Diabetes and treatment with broad-spectrum antibiotics are also commonly associated with the occurrence of *C. albicans* vaginitis.

Vulvovaginal candidiasis is diagnosed by microscopic identification of the fungus in scrapings of lesions and by isolation of the fungus in culture (Figure 24–12). Treatment consists of topical application of clotrimazole, miconazole, or a number of other antifungal drugs.

PROTOZOAN DISEASE OF THE GENITAL SYSTEM

Trichomoniasis

The protozoan *Trichomonas vaginalis* is a fairly normal inhabitant of the vagina in females and of the urethra in many males (Figures 24–13 and 11–18). If the usual acidity of the vagina is disturbed, the protozoan may overgrow the normal microbial population of the genital mucosa and cause **trichomoniasis.** (Males rarely have any symptoms as a result of the presence of the organism.) In response to the protozoan infection, the body accumulates leukocytes at the infection site; the resulting purulent discharge is profuse, yellowish or light cream in color, and characterized by a disagreeable odor. This discharge is accompanied by irritation and itching. The infection is usually sexually transmitted. Together with infections by *C. albicans* and *G. vaginalis,* these are the three most common types of vaginitis.

Diagnosis is easily made by microscopic examination and identification of the organisms in the discharge. Male carriers show the organism in semen or urine. Treatment is by oral metronidazole, which readily clears the infection. The organism can also be isolated and grown on laboratory media.

The major microbial diseases of the urinary and genital systems are summarized in Table 24–1.

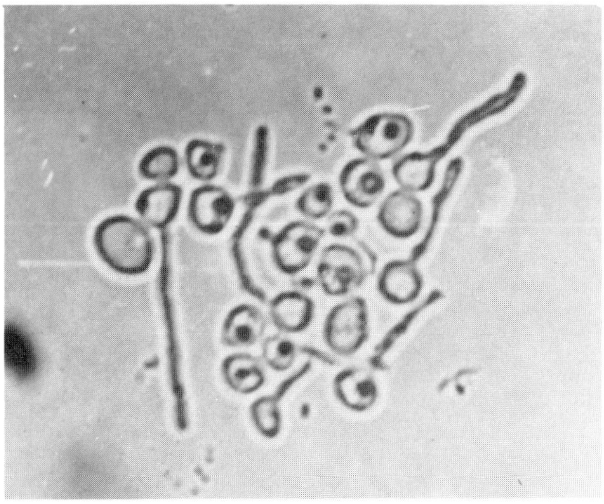

Figure 24–12 *Candida albicans* in a smear prepared from vaginal discharge.

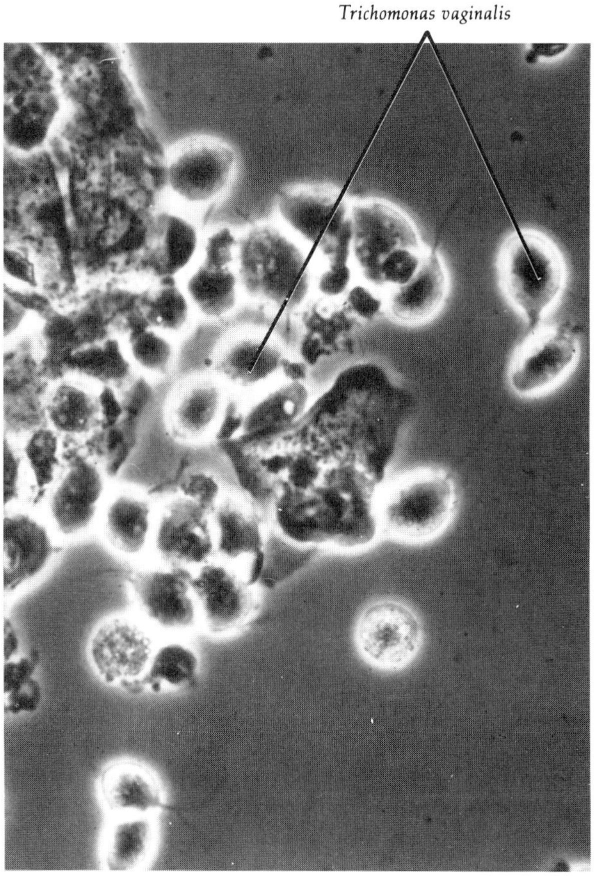

Figure 24–13 *Trichomonas vaginalis*, the protozoan causing trichomoniasis.

STUDY OUTLINE

INTRODUCTION (p. 675)

1. The urinary system regulates the chemical composition of the blood and excretes waste products of metabolism.
2. The genital system produces gametes for reproduction and, in the female, supports the growing embryo.
3. The urinary and genital systems are closely related anatomically, and both open to the external environment.
4. Microbial disease of these systems can result from infection from an outside source or opportunistic infection by normal body flora.

STRUCTURE AND FUNCTION OF THE URINARY SYSTEM (p. 675)

1. The nephrons of the kidneys remove urea, uric acid, creatinine, salts, and water from the blood and form urine.
2. Urine is transported from the kidneys through ureters to the urinary bladder, and is eliminated through the urethra.
3. Valves prevent urine from flowing back to the urinary bladder and kidneys.
4. The flushing action of urine and the acidity of normal urine have some antimicrobial value.

STRUCTURE AND FUNCTION OF THE GENITAL SYSTEM (pp. 675–676)

1. The female genital system consists of two ovaries, two fallopian tubes, the uterus, the vagina, and external genitalia.
2. The vagina functions as the birth canal and copulatory canal.
3. The male genital system consists of two testes, ducts, accessory glands, and the penis.
4. Seminal fluid leaves the male body through the urethra.

NORMAL FLORA OF THE URINARY AND GENITAL SYSTEMS (pp. 676–677)

1. *Streptococcus, Bacteroides, Mycobacterium, Neisseria,* and enterobacteria are resident flora of the urethra.
2. The urinary bladder and upper urinary tract are sterile under normal conditions.
3. Lactobacilli dominate vaginal flora during reproductive years.

BACTERIAL DISEASES OF THE URINARY SYSTEM (pp. 677–679)

1. Opportunistic gram-negative bacteria from the intestines often cause urinary tract infections, especially in females.
2. Predisposing factors of urinary tract infections include nervous system disorders, toxemia, diabetes mellitus, and obstructions to the flow of urine.
3. Urethritis, cystitis, and ureteritis are terms describing inflammations of tissues of the lower urinary tract.
4. Pyelonephritis can result from lower urinary tract infections or systemic bacterial infections.
5. Diagnosis and treatment of urinary tract infections depends on the isolation and antibiotic-sensitivity testing of the etiological agents.
6. More than 100,000 bacteria per milliliter of urine indicates an infection.
7. About 35% of all nosocomial infections occur in the urinary system. *E. coli* causes more than half of these infections.

Cystitis (p. 678)

1. Cystitis, inflammation of the urinary bladder, is common in females.
2. Microorganisms at the opening of the urethra and along the length of the urethra, careless personal hygiene, and sexual intercourse contribute to the high incidence of cystitis in females.

Pyelonephritis (p. 678)

1. Inflammation of the kidneys, or pyelonephritis, is usually a complication of lower urinary tract infections.
2. About 75% of pyelonephritis cases are caused by *E. coli*.

Leptospirosis (pp. 678–679)

1. The spirochete *L. interrogans* causes leptospirosis.
2. Leptospirosis is characterized by chills, fever, headache, and jaundice.
3. Serological identification is difficult because there are many serotypes of *L. interrogans*.

Glomerulonephritis (p. 679)

1. Glomerulonephritis, or Bright's disease, is an immune-complex disease occurring as a sequel to a beta-hemolytic streptococcal infection.
2. Antigen–antibody complexes are deposited in the glomeruli, and cause inflammation and kidney damage there.

BACTERIAL DISEASES OF THE GENITAL SYSTEM (pp. 679–688)

1. Most diseases of the genital system are sexually transmitted diseases (STDs).
2. Most STDs can be prevented by the use of condoms and are treated with antibiotics.

Gonorrhea (pp. 679–682)

1. *N. gonorrhoeae* causes gonorrhea.
2. Gonorrhea is the most common reportable communicable disease in the United States.
3. *N. gonorrhoeae* attaches to mucosal cells of the oralpharyngeal area, genitalia, eyes, joints, and rectum by means of pili.
4. Symptoms in males are painful urination and pus discharge. Blockage of the urethra and sterility are complications of untreated cases.
5. Females might be asymptomatic until the infection spreads to the uterus and fallopian tubes. Blockage of the fallopian tubes, sterility, and pelvic inflammatory disease are complications of untreated cases.
6. Gonorrheal endocarditis, gonorrheal meningitis, and gonorrheal arthritis are complications that can affect both sexes if gonorrheal infections are untreated.
7. Ophthalmia neonatorum is an eye infection acquired by infants during passage through the birth canal of an infected mother.
8. Penicillin has been used effectively to treat gonorrhea. In 1976, penicillinase-producing strains of *N. gonorrhoeae* appeared.
9. Direct fluorescent-antibody tests can be used to confirm identification of gram-negative diplococci isolated from patients.

Syphilis (pp. 682–686)

1. Syphilis is caused by *T. pallidum*, a spirochete that has not been cultured in vitro. Laboratory cultures are grown in rabbits.
2. *T. pallidum* is transmitted by direct contact and can invade intact mucous membranes or penetrate through breaks in the skin.
3. The primary lesion is a small, hard-based chancre at the site of infection. The bacteria then invade the blood and lymphatic system, and the chancre spontaneously heals.
4. The appearance of a widely disseminated rash on the skin and mucous membranes marks the secondary stage. Spirochetes are present in the lesions of the rash.
5. The patient enters a latent period after the secondary lesions spontaneously heal.
6. At least 10 years after the secondary lesion, tertiary lesions called gummas can appear on many organs.
7. Congenital syphilis, resulting from *T. pallidum* crossing the placenta during the latent period, can cause neurological damage in the newborn.
8. *T. pallidum* is identifiable through darkfield microscopy of fluid from primary and secondary lesions.
9. Many serological tests, such as VDRL and FTA–ABS, can be used to detect the presence of antibodies against *T. pallidum* during any stage of the disease.

Nongonococcal Urethritis (NGU) (p. 686)

1. NGU, or nonspecific urethritis, is any inflammation of the urethra not caused by *N. gonorrhoeae*.
2. About 40% of NGU cases are caused by *Chlamydia trachomatis* and are usually transmitted sexually.
3. Symptoms of NGU are often mild or lacking although salpingitis and sterility may occur.
4. *C. trachomatis* can be transmitted to infants' eyes at birth.
5. *Ureaplasma urealyticum* and *Mycoplasma hominis* also cause NGU.

Gardnerella Vaginitis (pp. 686–687)

1. Vaginitis can be caused by *Candida albicans*, *Trichomonas vaginalis*, or *G. vaginalis*.

2. Diagnosis of *G. vaginalis* is based on increased vaginal pH, fishy odor, and the presence of clue cells.

Lymphogranuloma Venereum (LGV) (p. 687)

1. *C. trachomatis* causes LGV, primarily a disease of tropical and subtropical regions.
2. The initial lesion appears on the genitalia and heals without scarring.
3. The bacteria are spread in the lymph system and cause enlargement of lymph nodes, obstruction of lymph vessels, and edema of the genital skin.
4. The bacteria are isolated and identified from pus taken from infected lymph nodes.

Chancroid (Soft Chancre) and Granuloma Inguinale (pp. 687–688)

1. Chancroid, a swollen, painful ulcer on the mucous membranes of the genitalia or mouth, is caused by *H. ducreyi*.
2. *C. granulomatis* causes granuloma inguinale.
3. Granuloma inguinale begins as an ulcer on the genitalia that ulcerates and spreads to other tissues.
4. Diagnosis is based on observation of mononuclear cells called Donovan bodies.

VIRAL DISEASE OF THE GENITAL SYSTEM (pp. 688–689)

Genital Herpes (pp. 688–689)

1. Herpes simplex type 2 causes an STD called genital herpes.
2. Symptoms of the infection are painful urination, genital irritation, and fluid-filled vesicles.
3. Neonatal herpes is contracted during birth. It can be asymptomatic or it can result in neurological damage or infant fatalities.
4. The virus might enter a latent stage in nerve cells. Vesicles reappear following trauma and hormonal changes.
5. The disease is not contagious during the latent periods.
6. Genital herpes is associated with cervical cancer and miscarriages.
7. The drug acyclovir has proven effective in treating the symptoms of genital herpes; however, it does not cure the disease.

FUNGAL DISEASE OF THE GENITAL SYSTEM (pp. 689–692)

Candidiasis (pp. 689–692)

1. *C. albicans* cause NGU in males and vulvovaginal candidiasis in females.
2. Vulvovaginal candidiasis is characterized by lesions that produce itching and irritation.
3. Predisposing factors for candidiasis are pregnancy, diabetes, tumors, and broad-spectrum antibacterial chemotherapy.
4. Diagnosis is based on observation of the fungus and its isolation from lesions.

PROTOZOAN DISEASE OF THE GENITAL SYSTEM (p. 692)

Trichomoniasis (p. 692)

1. *T. vaginalis* causes trichomoniasis when the pH of the vagina increases.
2. Diagnosis is based on observation of the protozoan in purulent discharges from the site of infection.

STUDY QUESTIONS

REVIEW

1. List the normal flora of the urinary system and show their habitats in Figure 24–1.
2. List the normal flora of the genital system and show their habitats in Figures 24–2 and 24–3.
3. How are urinary tract infections transmitted?
4. Explain why *E. coli* is frequently implicated in cystitis in females. List some predisposing factors for cystitis.
5. Name one organism that causes pyelonephritis. What are the portals of entry for organisms that cause pyelonephritis?

6. Complete the following table.

Disease	Causative Agent	Symptoms	Method of Diagnosis	Treatment
Gardnerella vaginitis				
Gonorrhea				
Syphilis				
NGU				
LGV				
Chancroid				
Granuloma inguinale				

7. Name one fungus and one protozoan that can cause genital system infections. What symptoms would lead you to suspect these infections?

8. Leptospirosis is a kidney infection of humans and other animals. How is this disease transmitted? What types of activities would increase one's exposure to this disease? What is the etiology?

9. Describe the symptoms of genital herpes. What is the etiological agent? When is this infection least likely to be transmitted?

10. What is glomerulonephritis? How is it transmitted? How is it treated?

11. List the genital infections that cause congenital and neonatal infections. How can transmission to a fetus or newborn be prevented?

CHALLENGE

1. The tropical skin disease called yaws is transmitted by direct contact. Its causative agent, *Treponema pallidum pertenue*, is indistinguishable from *T. pallidum*. The appearance of syphilis in Europe coincides with the first importation of slaves. How might *T. pallidum pertenue* have evolved into *T. pallidum* in the temperate climate of Europe?

2. Why can frequent douching be a predisposing factor to bacterial vaginitis, vulvovaginal candidiasis, or trichomoniasis?

3. The chart below is a key to selected microorganisms that cause genitourinary infections. Complete this key by listing genera covered in this chapter in the blanks corresponding to their respective characteristics.

Gram-negative bacteria
  Spirochete
    Aerobic ______________
    Anaerobic ______________
  Coccus
    Oxidase-positive ______________
  Bacillus, non-motile
    Requires X factor ______________
    X factor not required ______________
  Obligate intracellular parasite ______________
  Lacking cell wall
    Urease-positive ______________
    Urease-negative ______________
Fungus
  Pseudohyphae ______________
Protozoan
  Flagella ______________
No organism observed/cultured from patient ______________

4. A previously well, 19-year-old female was admitted to a hospital after two days of nausea, vomiting, headache, and neck stiffness. CSF and cervical cultures showed gram-negative diplococci in leukocytes; a blood culture was negative. What disease did she have? How was it probably acquired?

FURTHER READING

Chesney, P. J., M. S. Bergdoll, J. P. Davis, and J. M. Vergeront. 1984. The disease spectrum, epidemiology, and etiology of toxic shock syndrome. *Annual Review of Microbiology* 38:315–338. History and recent developments related to this disease.

Gunby, P. 1983. Genital herpes research: many aim to tame maverick virus. *Journal of the American Medical Association* 250:2417–2427. A summary of prospects for control of genital herpes infections.

Holmes, K. K. 1981. The *Chlamydia* epidemic. *Journal of the American Medical Association* 245:1718–1723. A readable summary of recent information on chlamydial bacteria and their sexually transmitted diseases.

Rapp, F. 1978. Herpesviruses, venereal disease, and cancer. *American Scientist* 66:670–674. Differentiates between the association of a virus in cancer and the cause of the cancer.

Rosebury, T. 1971. *Microbes and Morals.* New York: Viking Press. A history of venereal diseases from Fracastor through modern antibiotics, including laboratory diagnosis, treatment, and the prospects for vaccines.

Schachter, J. 1980. Chlamydiae. *Annual Review of Microbiology* 34:285–309. A good review of the clinical features and epidemiology of chlamydial infections.

FURTHER READING FOR PART FOUR

The following medical microbiology textbooks include principles of microbiology with an emphasis on bacterial pathogens. Some texts also include viral, fungal, and parasitic diseases.

Davis, B. D., R. Dulbecco, H. N. Eisen, and H. S. Ginsberg. 1980. *Microbiology,* 3rd ed. New York: Harper and Row.

Freeman, B. A. 1985. *Burrow's Textbook of Microbiology,* 22nd ed. Philadelphia, PA: W. B. Saunders.

Hoeprich, P. D. (ed.). 1983. *Infectious Diseases.* 3rd ed. New York: Harper and Row.

Joklik, W. K., H. P. Willett, and D. B. Amos (eds.). 1984. *Zinsser Microbiology,* 18th ed. Norwalk, Ct: Appleton-Century-Crofts.

Wilson, G., A. Miles, and M. T. Parker. 1984. *Topley and Wilson's Priniciples of Bacteriology, Virology and Immunity,* 7th ed., 4 vols. Baltimore: Williams and Wilkins.

PART FIVE

MICROBIOLOGY, THE ENVIRONMENT, AND HUMAN AFFAIRS

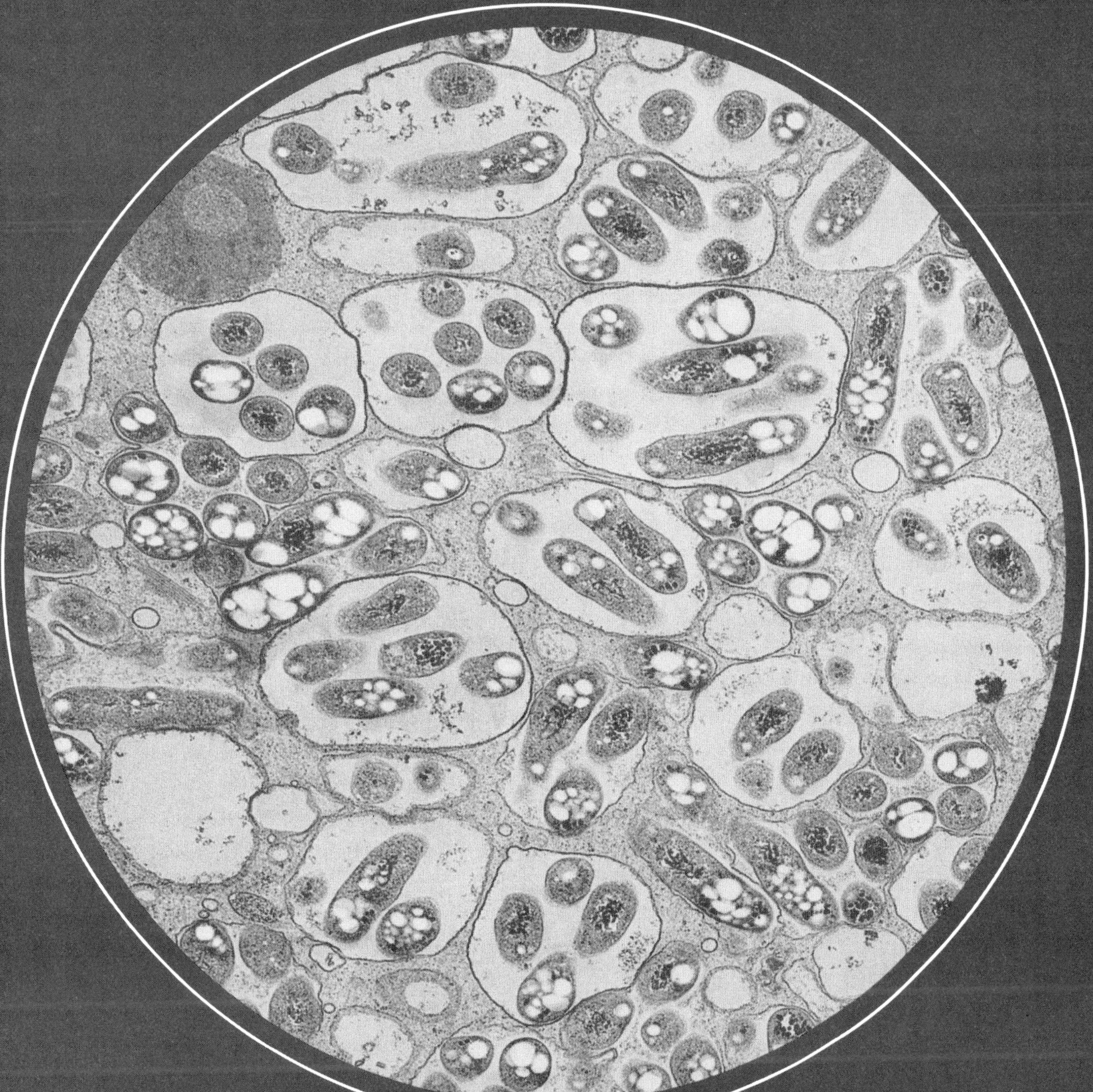

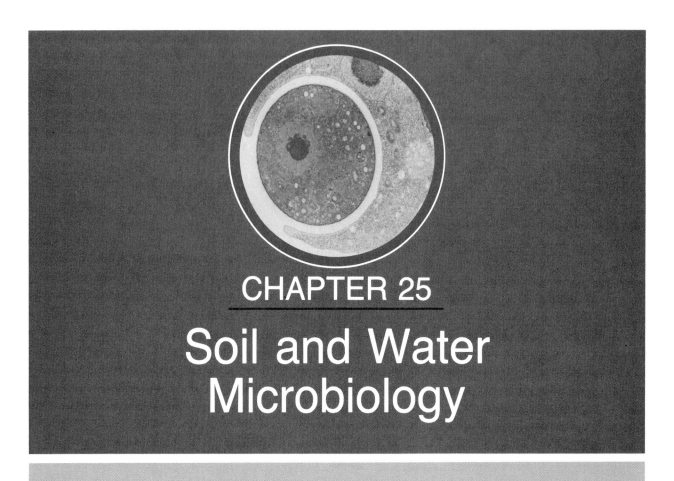

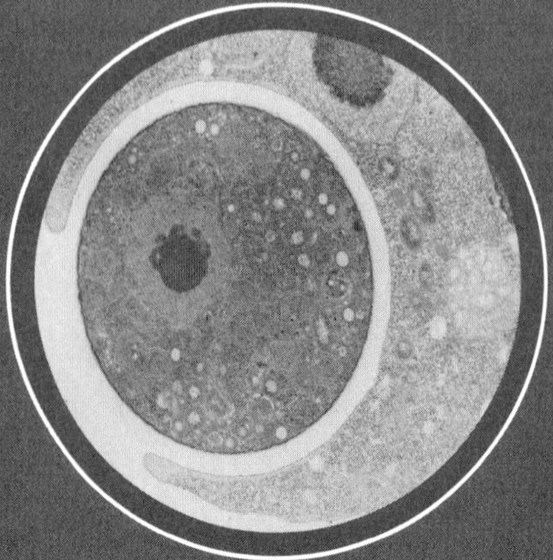

CHAPTER 25

Soil and Water Microbiology

OBJECTIVES

After completing this chapter you should be able to

- Explain how the components of soil affect soil microflora.
- Outline the carbon and nitrogen cycles and explain the roles of microorganisms in these cycles.
- Describe the freshwater and seawater habitats of microorganisms.
- Explain how water is tested for bacteriologic quality.
- Compare primary, secondary, and tertiary sewage treatments.
- List some of the biochemical activities that take place in an anaerobic sludge digester.
- Define each of the following terms: BOD; septic tank; oxidation pond; activated sludge; trickling filter.
- Discuss the causes and effects of eutrophication.

Many people immediately associate bacteria with disease or food spoilage, and think of microorganisms as things to be avoided or controlled, as things that can hurt them. Actually, harmful microorganisms are only a very small fraction of the total. In fact, bacteria and other microorganisms are intimately involved in the very maintenance of life on earth and perform many essential functions. In this chapter, we will consider the importance of microorganisms in soil and water.

SOIL MICROBIOLOGY AND CYCLES OF THE ELEMENTS

THE COMPONENTS OF SOIL

Soil is a somewhat complex mixture of solid inorganic matter (rocks and minerals), water, air, and living organisms and the products of their decay. Numerous physical and chemical changes take place in this mixture. The uppermost layer of soil, called the topsoil, is the most important to living organisms because they directly contact this layer. Topsoils vary considerably in physical texture, chemical composition, origin, depth, and fertility.

Rock Particles and Minerals

Most agricultural soil is a mixture of rock and mineral fragments that have been formed by the weathering of preexisting rock. The weathering processes lead to physical and chemical breakdown and include precipitation, wind, and temperature fluctuations (especially the freezing and thawing of water). Because of weathering, elements such as silicon, aluminum, and iron are added to the soil. Calcium, magnesium, potassium, sodium, phosphorus, and small amounts of other elements are also present in the mineral component of soil.

Soil Water

Soil also contains water, in an amount that depends on the amount of precipitation as well as on the climate and drainage. Found between the soil particles, the water adheres to the soil particles. Through the solvating property of water, various inorganic and organic constituents of soil are dissolved in the soil water and made available to the living inhabitants of the soil.

Soil Gases

The gases in the soil are essentially the same as those found above the soil, and they include carbon dioxide, oxygen, and nitrogen. However, the proportions of soil gases can be different from those of atmospheric gases because of the biological processes occurring in soil. For example, because of respiration, soil contains a high proportion of carbon dioxide and a low proportion of oxygen. Soil gases are mainly between soil particles or dissolved in the soil water.

Organic Matter

The organic matter in soil consists of carbohydrates, proteins, lipids, and other materials. Organic matter comprises 2 to 10% of most agriculturally important soils. Swamps and bogs, by contrast, contain a higher content of organic matter, up to 95% in some peat bog soils. The waterlogged soil of swamps and bogs is an anaerobic environment, and the microbial decomposition of organic matter there occurs only slowly.

All organic matter in soil is derived from the remains of microorganisms, plants, and animals, their waste products, and the biochemical activities of various microorganisms. A great portion of the organic matter is of plant origin, mostly dead roots, wood and bark, and fallen leaves. A second source is the vast numbers of bacteria, fungi, algae, protozoans, small animals, and viruses, which can total billions per gram of fertile soil. Through the actions of these organisms the breakdown of organic substances produces and maintains a continuous supply of inorganic substances that plants and other organisms require for growth. Much of organic matter is ultimately decomposed to inorganic substances such as ammonia, water, carbon dioxide, and various compounds of nitrate, phosphate, and calcium.

A considerable part of the organic matter in soil occurs as **humus,** a dark material composed chiefly of organic materials that are relatively resistant to decay. In this regard, humus is partially decomposed organic matter.

The addition of organic matter, either completely or partially decomposed, is essential to the continuation of soil's fertility. Moreover, because of their spongy nature, organic materials loosen the soil, and so prevent the formation of heavy crusts and increase the pore spaces in the soil. This addition of pore spaces in turn increases aeration and water retention.

Organisms

Fertile soil contains a great many animals, ranging from microscopic forms, including numerous nematodes, to larger forms, including insects, millipedes, centipedes, spiders, slugs, snails, earthworms, mice, moles, gophers, and reptiles. Most of these animals are beneficial in that they promote some mechanical movement of the soil, and thereby help to keep the soil loose and open. All soil organisms also contribute to the organic matter of soils in the forms of their waste products and eventual remains.

Soil also contains the root systems of higher plants and enormous numbers of microorganisms. Without the microorganisms, especially bacteria, the soil would soon become unfit to support life. Bacteria affect the soil in many ways. Most bacteria decompose organic matter into simple products. In these reactions, nutrients are made available for reuse. Other soil bacteria are associated with the transformation of nitrogen and sulfur compounds, so that usable supplies of nitrogen and sulfur are continually provided.

The soil microflora

The soil is one of the main reservoirs of microbial life. A good agricultural soil the size of a football field usually contains a microbial population of size approaching the weight of a cow eating grass on that field. But the metabolic capabilities of these vast numbers of microorganisms is probably about 100,000 times that of the grazing cow. However, measurement of the carbon dioxide evolved from soil and other evidence indicate that these organisms are existing in near starvation and at low reproductive rates. When usable nutrients are added to soil, the microbial populations and their activity rapidly increase until the nutrients are depleted, and then the microbial activity returns to the lower levels.

The most numerous organisms in soil are bacteria (Table 25–1). A typical garden soil will have millions of bacteria in each gram. The population is highest in the top few centimeters of the soil, and declines rapidly with depth. The populations are usually estimated with plate counts on nutrient media, and the actual numbers are probably greatly underestimated in this method. No single nutrient medium or growth condition can possibly meet all the myriad nutrient and other requirements of soil microorganisms.

Actinomycetes are bacteria, but are usually considered separately in enumerations of soil populations. Found in large numbers, the actinomycetes produce a gaseous substance called *geosmin*, which gives fresh soil its characteristic musty odor. The numbers reported might reflect only the formation of asexual spores and fragmentation of the mycelium of these filamentous organisms. The actual biomass (total mass of living organisms in a given volume) of the actinomycetes is probably about that of the conventional bacteria. Interest in these bacteria was greatly stimulated by the discovery that some genera, particularly *Streptomyces* (strep-tō-mī'sēs), produce valuable antibiotics.

Fungi are found in soil in much smaller numbers than bacteria and actinomycetes. Because many of the counted fungal colonies arise from the germination on media of asexual spores, the relationship between the count and the actual fungal population is questionable. Estimates of the biomass of fungi are that it probably equals that of the bacteria and actinomycetes combined. This large biomass is because the dimensions of the fungal mycelium are many times greater than that of bacterial cells. Molds greatly outnumber yeasts in soil.

Algae and cyanobacteria sometimes form vis-

Table 25–1 Distribution of Microorganisms in Numbers Per Gram of Typical Garden Soil at Various Depths

Depth (cm)	Bacteria	Actinomycetes	Fungi	Algae
3–8	9,750,000	2,080,000	119,000	25,000
20–25	2,179,000	245,000	50,000	5,000
35–40	570,000	49,000	14,000	500
65–75	11,000	5,000	6,000	100
135–145	1,400	—	3,000	—

Source: Adapted from M. Alexander, *Introduction to Soil Microbiology*, 2nd ed., Wiley, New York, 1977.

ible **blooms** (abundant growths) on the surface of moist soils; these are also found in dry desert soils. As might be expected of photosynthetic organisms, they are located mainly on the surface layer, where sunlight, water, and carbon dioxide are most abundant. Nonetheless, significant numbers of algae and cyanobacteria are found more than 50 cm below the surface. The environmental tribution of these microorganisms is significant in only special cases. For example, fixation of atmospheric nitrogen (discussed shortly) by some species of cyanobacteria in grasslands, tundra regions, and after rainfall in deserts is significant in the fertility of soils in those regions.

Pathogens in soil

Human pathogens, adapted to life as parasites, find the soil an alien, hostile environment. Even relatively resistant enteric pathogens such as *Salmonella* (sal-mōn-el'lä) species, when introduced into soil, have been observed to survive for only a few weeks or months. Most human pathogens that can survive in soil are endospore-forming bacteria. Endospores of *Bacillus anthracis* (bä-sil'lus an-thrā'sis), which causes anthrax in animals, can survive in certain soils for decades and finally germinate when ingested by grazing animals. Disposing of the body of an animal infected with anthrax requires considerable care, so that the soil is not seeded with the endospores from the dead animal. *Clostridium tetani* (klôs-tri'dē-um te'tan-ē), the causative agent of tetanus, *Clostridium botulinum* (bo-tū-lī'num), the causative agent of botulism, and *Clostridium perfringens* (per-frin'jens), the causative agent of gas gangrene, are also endospore-forming pathogens whose normal habitat is the soil. From soil they are introduced into foods or wounds, where they grow and elaborate toxins.

Pathogens of plants are much more likely to be normal soil inhabitants. Most plant pathogens are fungi, mainly because fungi are capable of growing in the low moisture typical of plant surfaces. (Bacterial diseases of plants are uncommon.) Many of the rusts, smuts, blights, and wilts affecting plants are caused by fungi that pass part of their life cycle in soil.

Some of the microorganisms found in soil are insect pathogens, which are potentially useful for pest control. *Bacillus thuringiensis* (thůr-in-jē-en'sis), for example, is a soil bacterium pathogenic to the larvae of many insects; it is now used as an aid in their control (see box). When ingested by a larva, the endospore germinates, and the growing bacillus forms a toxic protein crystal that eventually kills the insect. A number of other promising insect pathogens, including viruses and fungi, are under investigation as biological pesticides.

MICROORGANISMS AND BIOGEOCHEMICAL CYCLES

Perhaps the most important role of soil microorganisms is their participation in **biogeochemical cycles,** that is, the recycling of certain chemical elements so that they can be used over and over again. Among these elements are carbon, nitrogen, sulfur, and phosphorus. Were it not for the activities of microorganisms in the various biogeochemical cycles, essential elements would become depleted and all life would cease. The first biogeochemical cycle we will consider is the carbon cycle.

Carbon Cycle

As you already know, all organic compounds contain carbon. Most of the inorganic carbon used in the synthesis of organic compounds comes from the carbon dioxide (CO_2) in the atmosphere (Figure 25–1). Some CO_2 is also dissolved in water.

In photosynthesis, the first step in the carbon cycle, carbon dioxide is incorporated into organic compounds by photoautotrophs such as cyanobacteria, green plants, algae, and green and purple sulfur bacteria. In the next step in the cycle, chemoheterotrophs consume the organic compounds—animals eat photoautotrophs, especially green plants, as well as other animals. Thus, the organic compounds of the photoautotrophs are digested and resynthesized. In this way, the carbon atoms of carbon dioxide are transferred from organism to organism.

Some of the organic molecules are used by chemoheterotrophs, including animals, to satisfy their energy requirements. When such energy is released through the process of respiration, carbon dioxide is released into the atmosphere. This carbon dioxide becomes immediately available to start the cycle over again. However, much of the carbon

Microbiology in the News

Waging a Bacterial War on Pests

In the United States, despite enormous expenditures on pest control, one out of every three tons of food goes to the hungry mouths of pests. In the tropics and subtropics where three crops a year are grown on the same plot, and where pests are more abundant, the loss is greater.

Chemists have concocted thousands of chemical pesticides to kill virtually every known form of pest. Experience has shown, however, that many organisms—especially insects—develop genetic resistance to pesticides. To combat this fact of biology, farmers periodically switch to a new chemical preparation or simply increase the amount of chemical pesticides they apply to their fields or the frequency of application. In tropical regions, thirty sprayings may be needed for a single year where a decade ago only six were needed. New pesticides and the constant escalation of pesticide use have brought with them widespread environmental contamination.

A new way of controlling pests utilizes biological agents. One of the most promising methods involves the bacterium *Bacillus thuringiensis*, or BT, see photo. BT occurs naturally and is now successfully used to control leaf-eating caterpillars that do billions of dollars of damage to crops and forests throughout the world. Cultivated in the lab, and sold commercially as a dry powder, BT is either applied as a powder or mixed with water and then sprayed on plants. Once in the caterpillar's stomach, BT releases a toxic substance that kills the pest.

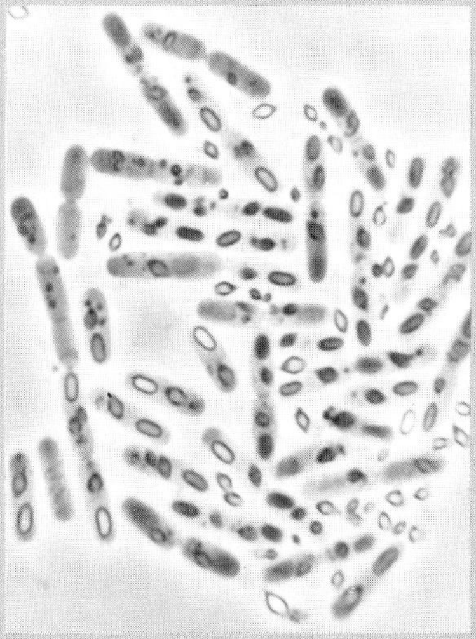

The success of BT has been very encouraging. Farmers in California have used the bacterium to control caterpillars for several decades. Forest managers have used it to help control the devastating gypsy moth, and it has even been successful in mosquito control.

BT carries the added benefit of reducing pesticide use. In California alone, pesticide use has dropped from 600,000 pounds per year to 50,000 pounds. For insect-eating birds, for beneficial insects such as honey bees, for fish, and for humans, the benefits of reduced pesticide use are immeasurable.

Robert Kaufman of the Monsanto Company recently announced a new genetic engineering approach that may revolutionize pest control and keep it safe for farmers and consumers alike. Kaufman and his colleagues have isolated the gene that gives BT its pesticidal action. According to Kaufman, that gene controls the synthesis of a protein that kills insects like the black cutworm, one of the more destructive corn pests. Using recombinant DNA techniques, they have transferred the gene to a naturally occurring organism, *Pseudomonas fluorescens*, which lives on the roots of corn and some other plants.

Currently, one of every three farmers in the corn belt treats his fields with chemical insecticides to cut down on cutworm damage. In the future, farmers may simply plant seeds that have been pretreated with *P. fluorescens* that carry the toxic gene. Greenhouse studies show that the bacterium will colonize the roots of corn plants that have been grown from pretreated seeds and, with its new gene, actually protect the plants from hungry black cutworms.

Kaufman hopes that more insecticidal genes can be added to *P. fluorescens* in the years to come, giving corn a wider range of protection, reducing chemical pesticide use, and protecting the soil and water from contamination.

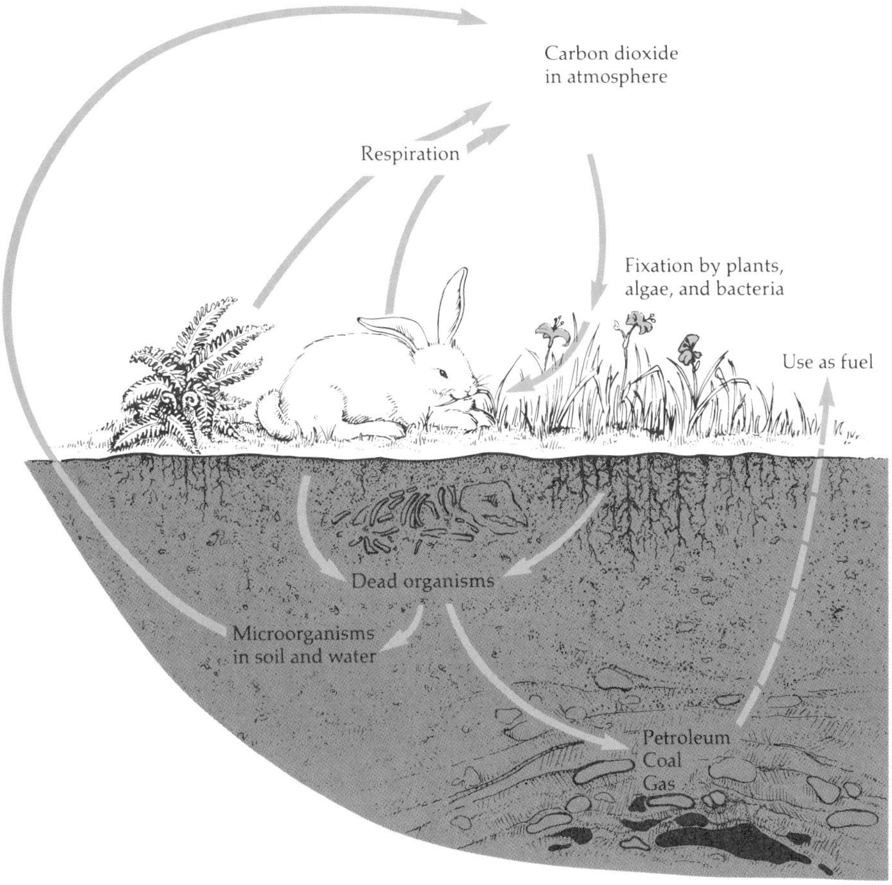

Figure 25–1 The carbon cycle. See also Figure 1–8.

remains with the organisms until they excrete wastes or die. When the organisms die, the organic compounds are deposited in the soil and decomposed by microorganisms, principally bacteria and fungi. During this decomposition, organic compounds are degraded and carbon dioxide is returned to the atmosphere. Although the carbon dioxide of the atmosphere comprises only about 0.03% of the atmospheric gases, it is essential for the synthesis of new living matter.

Carbon is stored in rocks such as limestone ($CaCO_3$) and dissolved in oceans (as CO_3^- ions); it is also stored in organic forms such as coal and petroleum. Burning such fossil fuels releases carbon dioxide into the atmosphere. There is evidence that carbon dioxide in the atmosphere has been increasing because of the burning of fossil fuels. Some feel that this is a potential problem because the increase of atmospheric carbon dioxide could cause a warming of the earth. This warming has been called the greenhouse effect.

Nitrogen Cycle

Nitrogen is needed by all organisms for their synthesis of protein, nucleic acid, and other nitrogen-containing compounds. Molecular nitrogen (N_2) comprises almost 80% of the earth's atmosphere; the atmosphere over every acre of fertile soil contains more than 30,000 tons of nitrogen. Despite

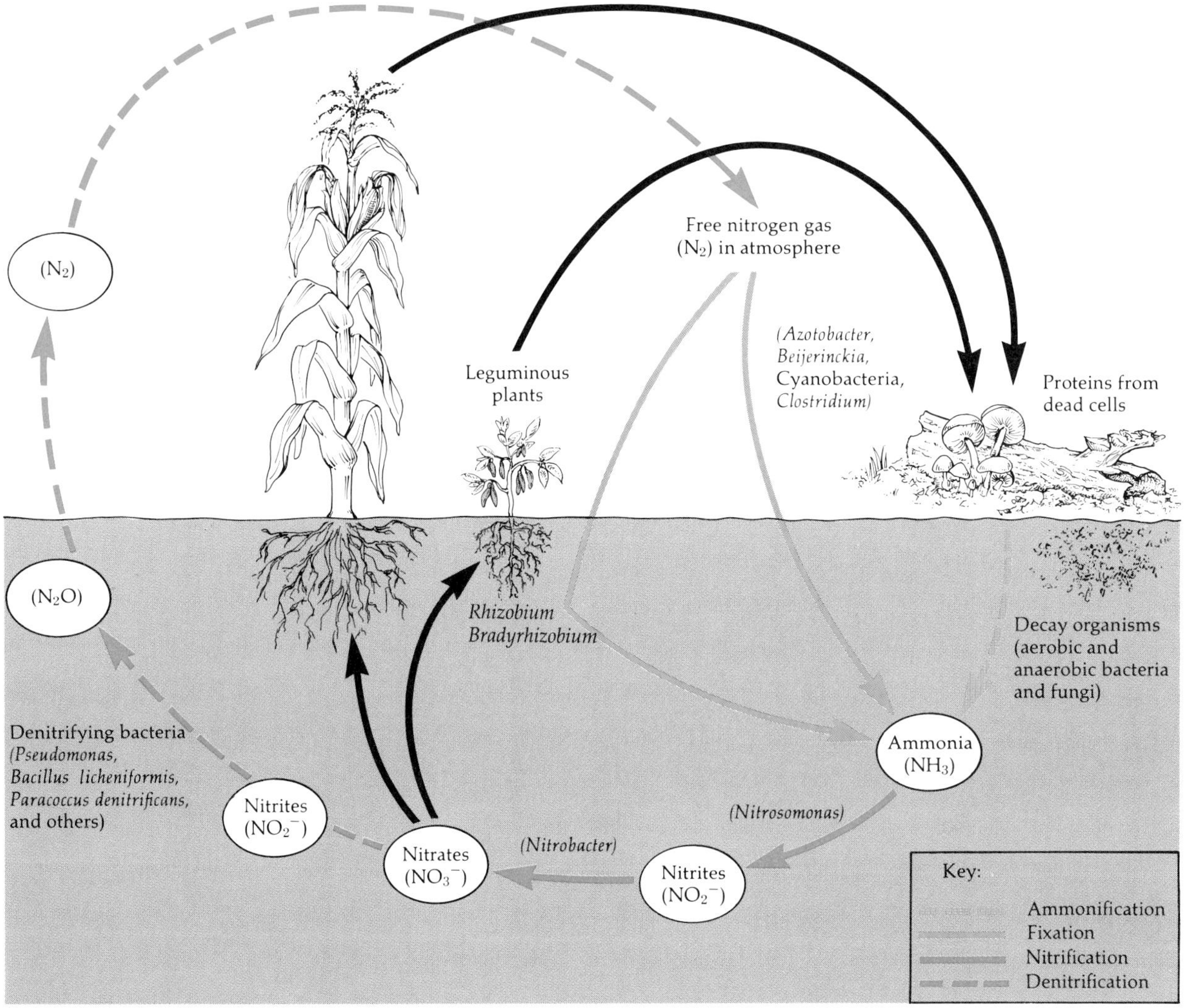

Figure 25–2 The nitrogen cycle.

the abundance of nitrogen, a molecular gas, however, no eucaryote is able to make direct use of it. Instead, the nitrogen must be fixed (combined) with other elements such as oxygen and hydrogen. The resulting compounds, such as nitrate ion (NO_3^-) and ammonium ion (NH_4^+), are then used by autotrophic organisms. The chemical and physical forces operating in the soil, water, and air, together with the activities of specific microorganisms, are important factors in the conversion of nitrogen to usable forms.

Almost all of the nitrogen in the soil exists in organic molecules, and is primarily in proteins. When an organism dies, the process of decomposition results in the hydrolytic breakdown of proteins into amino acids. The amino groups of amino acids are removed by deamination (see Chapter 5) and ammonia is formed (Figure 25–2). This release of ammonia is called **ammonification.** Ammonification is brought about by aerobic and anaerobic bacteria and fungi, and can be represented as follows:

$$\text{Proteins from dead cells} \xrightarrow{\text{Microbial decomposition}} \text{Amino acids}$$

$$\text{Amino acids} \xrightarrow{\text{Microbial ammonification}} \text{Ammonia } (NH_3)$$

Microbial growth releases extracellular proteolytic enzymes that accomplish this simplification of chemicals. The fate of the ammonia produced by ammonification depends on soil conditions. Because ammonia is a gas, it might rapidly disappear from dry soil. But in moist soil, it becomes solubilized in water, and ammonium ions (NH_4^+) are formed.

$$NH_3 + H_2O \rightarrow NH_4OH \rightarrow NH_4^+ + OH^-$$

Ammonium ions are used by bacteria and plants for amino acid synthesis.

The next sequence of reactions in the nitrogen cycle involves the oxidation of the ammonium ion to nitrate, a process called **nitrification.** Living in the soil are autotrophic nitrifying bacteria such as those of the genus *Nitrosomonas* (nī′-trō-sō-mō′näs) and *Nitrobacter* (nī-trō-bak′tėr). These organisms obtain energy by oxidizing ammonia or nitrite. In the first stage, *Nitrosomonas* oxidizes ammonia to nitrites:

$$\underset{\text{Ammonium ion}}{NH_4^+} \xrightarrow{\textit{Nitrosomonas}} \underset{\text{Nitrites}}{NO_2^-}$$

In the second stage, *Nitrobacter* oxidizes nitrites to nitrates:

$$\underset{\text{Nitrites}}{NO_2^-} \xrightarrow{\textit{Nitrobacter}} \underset{\text{Nitrates}}{NO_3^-}$$

Nitrate is the form of nitrogen most commonly used by plants, which use it mostly for protein synthesis.

At various points in the cycle, atmospheric nitrogen is either added or removed. The loss of nitrogen from the cycle involves a process called **denitrification,** the conversion of nitrates to nitrogen gas, and can be represented as follows:

$$\underset{\text{Nitrates}}{NO_3^-} \rightarrow \underset{\text{Nitrites}}{NO_2^-} \rightarrow \underset{\text{Nitrous oxide}}{N_2O} \rightarrow \underset{\text{Atmospheric nitrogen gas}}{N_2}$$

Pseudomonas (sū-dō-mō′nas) species appear to be the most important group of bacteria in denitrification in soils. A number of other genera, including *Paracoccus* (pa-rä-kok′kus), *Thiobacillus* (thī-ō-bä-sil′lus), and *Bacillus,* also have species capable of carrying out denitrification reactions. Denitrifying bacteria are aerobic, but under anaerobic conditions, they can use nitrate in place of oxygen as a final electron acceptor. (This process is called anaerobic respiration; see Chapter 5.) Thus, denitrification occurs in waterlogged soils depleted of oxygen. Because denitrifying bacteria deliver nitrogen to the atmosphere at the expense of removing nitrates from the soil, denitrification is an unfavorable process from the standpoint of soil fertility.

During the final phase of the nitrogen cycle, nitrogen is converted into ammonia in a process called **nitrogen fixation.** Only a few species of bacteria and cyanobacteria are capable of enacting this process. Because the nitrogenase enzyme responsible for nitrogen fixation is anaerobic, it probably arose early in the history of the planet, before the atmosphere contained oxygen and before nitrogen-containing compounds were available from decaying organic matter. Nitrogen fixation is brought about by two types of organisms: nonsymbiotic and symbiotic.

Nonsymbiotic (free-living) *nitrogen-fixing bacteria* are found in particularly high concentrations in the *rhizosphere,* the region where the soil and roots make contact, such as grasslands. Among the nonsymbiotic bacteria that can fix nitrogen are aerobic species such as *Azotobacter* (ä-zō-tō-bak′tėr). These aerobic organisms apparently shield the nitrogenase enzyme from oxygen by, among other things, having a very high rate of oxygen utilization that minimizes the diffusion of oxygen into the cell where the enzyme is located. Another nonsymbiotic obligate aerobe that fixes nitrogen is *Beijerinckia* (bī-jė-rink′ē-ä). Some anaerobic bacteria, such as certain species of *Clostridium,* also fix nitrogen. A dominant, obligate, anaerobic, nitrogen-fixing microorganism is the bacterium *Clostridium pasteurianum* (pas-tyėr-ē-ā′num). Other nonsymbiotic nitrogen-fixing bacteria include certain species of the facultatively anaerobic *Klebsiella* (kleb-sē-el′lä), *Enterobacter* (en-te-rō-bak′tėr), and *Bacillus* and the photoautotrophic *Rhodospirillum* (rō-dō-spī-ril′lum) and *Chlorobium* (klô-rō′bē-um).

Most of the nonsymbiotic nitrogen-fixing organisms are capable of fixing large amounts of nitrogen under laboratory conditions. But in the soil, there is usually a shortage of usable carbohydrates, which supply the energy needed for the reduction of nitrogen to ammonia, which is then incorporated into protein. Nevertheless, the bacteria make important contributions to the nitrogen economy of areas such as grasslands, forests, and the arctic tundra.

There are many species of aerobic, photosynthesizing cyanobacteria that fix nitrogen. Because their energy supply is independent of carbohydrates in soil or water, they are especially useful suppliers of nitrogen to the environment. Cyanobacteria usually carry their nitrogenase en-

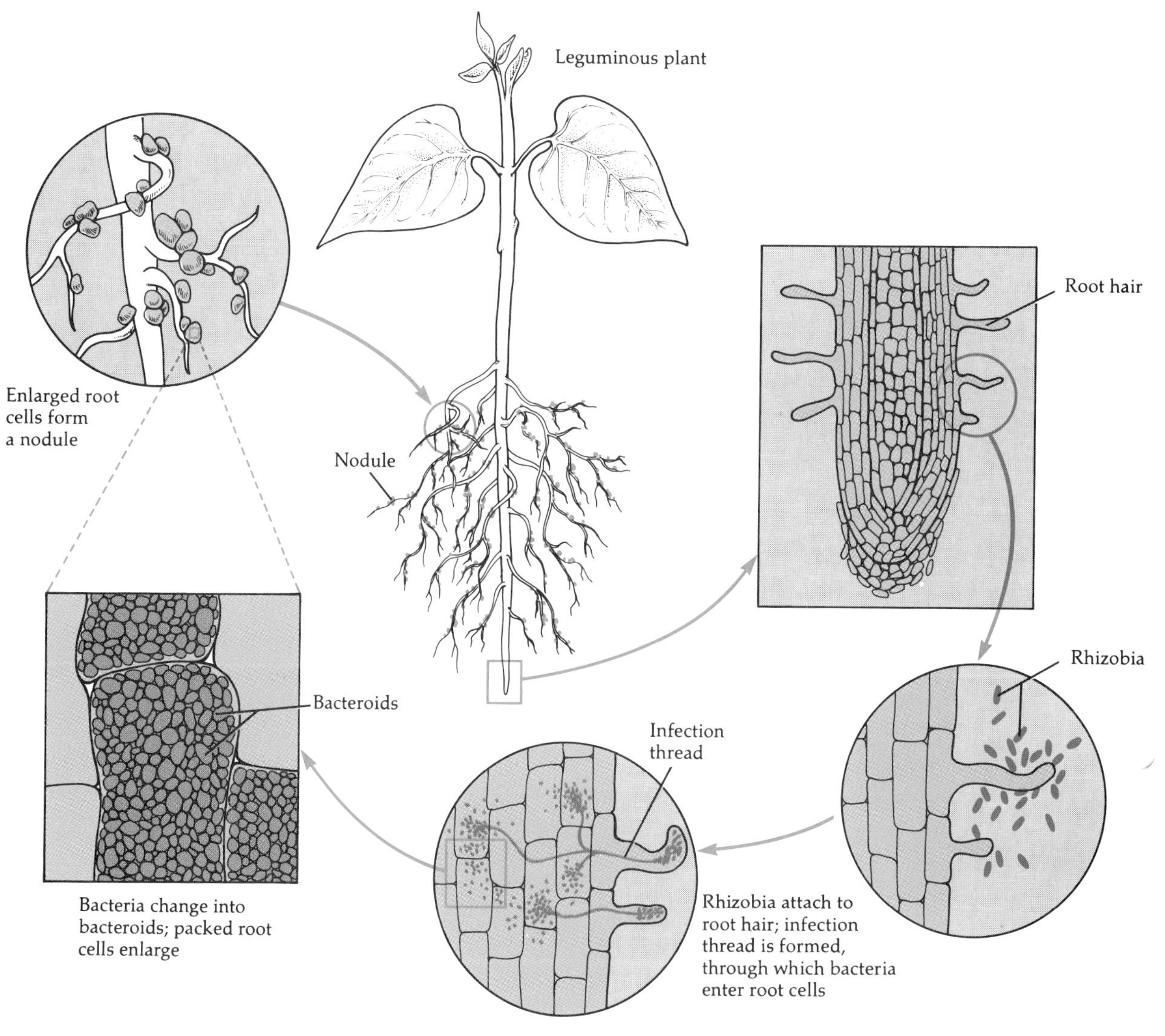
Leguminous plant
Root hair
Enlarged root cells form a nodule
Nodule
Rhizobia
Bacteroids
Infection thread
Bacteria change into bacteroids; packed root cells enlarge
Rhizobia attach to root hair; infection thread is formed, through which bacteria enter root cells

Figure 25–3 Stages in the formation of a root nodule.

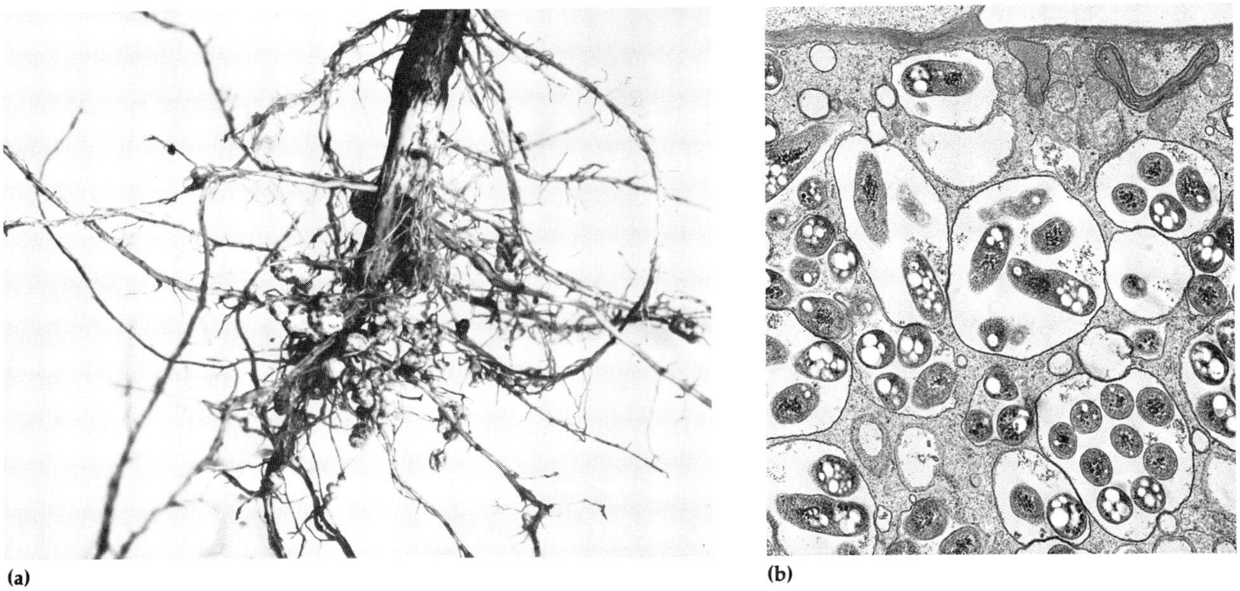

Figure 25–4 **(a)** Soybean root nodules. **(b)** Bacteroids of the nitrogen-fixing bacterium *Rhizobium japonicum* in vacuoles in an infected cell of a soybean root nodule (×12,000).

zymes in specialized structures called *heterocysts* that provide anaerobic conditions for fixation.

Symbiotic (mutualistic) nitrogen-fixing bacteria serve an even more important task in plant growth and crop production. In a symbiotic relationship, two organisms of different species live together, each benefiting from the relationship. Such a relationship is illustrated by members of the genus *Rhizobium* (rī-zō′bē-um), a symbiotic nitrogen-fixing bacterium, and the roots of leguminous plants, such as soybeans, beans, peas, peanuts, alfalfa, and clover. These agriculturally important plants are only a few of the thousands of known leguminous species, many of which are bushy plants or small trees found in poor soils in many parts of the world.

The rhizobia bacteria are specific for a particular leguminous species. The bacteria attach to the root of the host legume, usually at a root hair (Figure 25–3). In response to the bacterial infection, an indentation forms in the root hair, and an *infection thread* that passes down the root hair into the root itself forms. The bacteria follow this infection thread and enter the cells in the root. Inside these cells, the bacteria alter in morphology, and change into larger forms called *bacteroids* that eventually pack the plant cell. The root cells are stimulated by this infection to form a tumorlike *nodule* of bacteroid-packed cells (Figure 25–4). Nitrogen is then fixed by a symbiotic process of the plant and the bacteria. The plant furnishes anaerobic conditions and growth nutrients for the bacteria, and the bacteria fix nitrogen to be incorporated into plant protein.

Millions of tons of such nitrogen are fixed each year in the world. There are similar examples of symbiotic nitrogen fixation in nonleguminous plants such as alder trees. These trees are among the first to appear in forests after fires or glaciation. The alder tree is symbiotically infected with an actinomycete (*Frankia*) and forms nitrogen-fixing root nodules. About 50 kg of nitrogen can be fixed each year by the growth of one acre of alder trees; the trees thus make a valuable addition to the forest economy.

Another important contribution to the nitrogen economy of forests is made by **lichens**—a symbiosis between a fungus and an alga or cyanobacterium. When one symbiont is a nitrogen-fixing cyanobacterium, the product is fixed nitrogen that eventually enriches the forest soil. These cyanobacteria can alone fix significant amounts of ni-

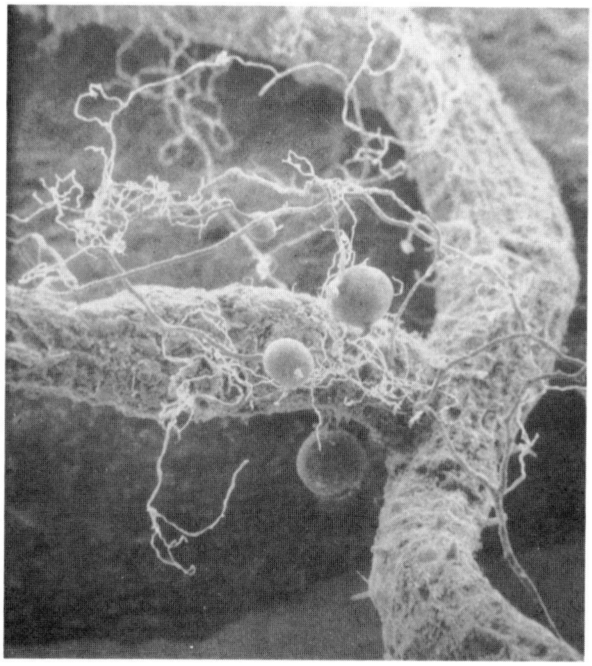

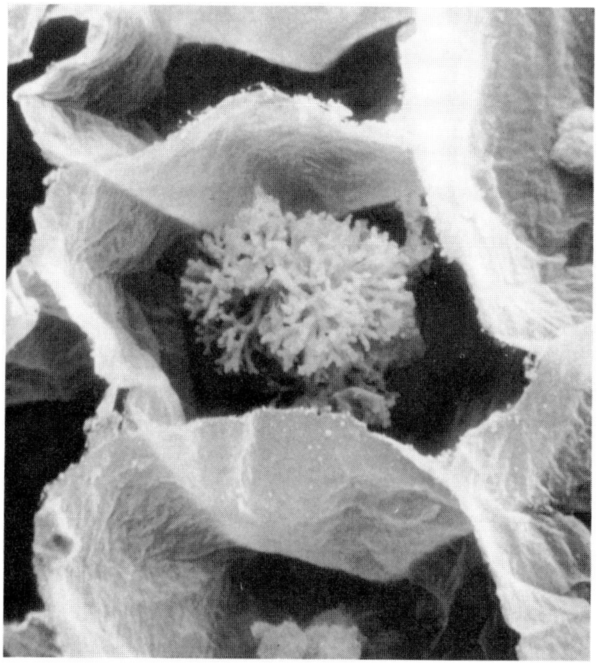

(b)

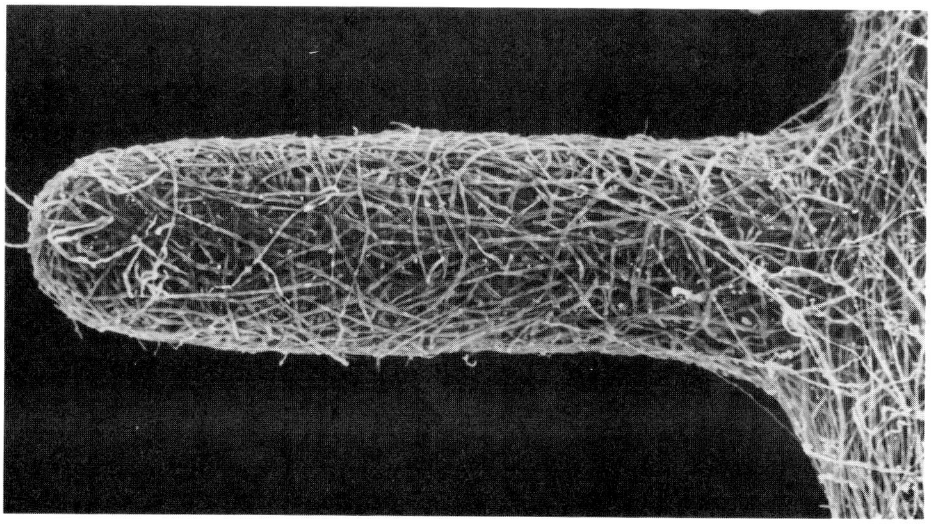

(c)

Figure 25–5 Mycorrhiza. **(a)** Endomycorrhiza on a soybean root. Note the spores and hyphae of *Glomus fasciculatus*. Also note the finely branched lateral hyphae surrounding the root (×140). These hyphae increase the plant's absorption of nutrients. **(b)** A fully developed endomycorrhiza arbuscule in a plant cell (×900). As the arbuscule decomposes it will release accumulated nutrients for the plant. **(c)** Ectomycorrhiza. The mycelial mantle of a typical ectomycorrhizal fungus surrounding a eucalyptus tree root.

trogen on desert soils directly after infrequent rains, and on the surface of arctic tundra soils. In the Orient, rice paddies can accumulate heavy blooms of such nitrogen-fixing organisms. The cyanobacteria also form a symbiosis with a small, floating fern plant, *Azolla*, which grow thickly in rice paddy waters. So much nitrogen is fixed by these organisms that other nitrogenous fertilizers are often unnecessary for rice cultivation.

Mycorrhizae

A very important contribution to plant growth is made by mycorrhizal fungi. There are two types: **endomycorrhiza,** also known as **vesicular-arbuscular mycorrhiza;** and **ectomycorrhiza.** Both types function as root hairs on plants, that is, they extend the surface area through which the plant can absorb nutrients, especially phosphorus, which is not very mobile in soil. Vesicular-arbuscular mycorrhizae (Figure 25–5a) form large spores that sieving can isolate easily from soil. The hyphae from these germinating spores penetrate into the plant root and form two types of structures: vesicles and arbuscules. **Vesicles** are smooth oval bodies that are most likely storage structures. **Arbuscules** (Figure 25–5b) are formed inside plant cells. Nutrients travel from the soil through the fungal hyphae to these arbuscules, which gradually break down and release the nutrients to the plants. Most grasses and plants are surprisingly dependent upon these fungi for proper growth, and their presence is nearly universal in the plant world.

Ectomycorrhizae mainly infect trees such as the pine. The fungus forms a mycelial *mantle* over the smaller roots of the tree (Figure 25–5c). Ectomycorrhizae do not form vesicles or arbuscules. Operators of commercial tree-farms take care to see that seedlings are inoculated with soil containing effective mycorrhizae.

Other Biogeochemical Cycles

Microorganisms are also important in cycles involving other elements, such as sulfur. In addition, there are important microbial transformations of potassium, iron, manganese, mercury, selenium, zinc, and other minerals. The various chemical reactions in these cycles are often essential in making the minerals available to plants in soluble form for their metabolism.

DEGRADATION OF PESTICIDES AND OTHER SYNTHETIC CHEMICALS

We seem to take it for granted that soil microorganisms will degrade materials entering the soil. Natural organic matter such as falling leaves or animal residues are, in fact, readily degraded. However, in this industrial age, there are many chemicals, such as agricultural pesticides and plastics—chemicals that do not occur in nature—that enter the soil in large amounts. Many of these synthetic chemicals are highly resistant to degradation by microbial attack. A well-known example is the insecticide DDT. When first introduced, the property of *recalcitrance* (resistance to degradation) was considered quite beneficial because one application remained effective in the soil for an extended time. But it was soon found that because of their fat solubility, the chemical tended to accumulate and concentrate in parts of the food chain. Eagles and other predatory birds accumulated DDT from contaminated food and suffered impaired reproductive ability. (For example, eggs had soft shells and failed to hatch.)

Not all synthetic chemicals are as recalcitrant as DDT. Some are made up of chemical bonds and subunits that are subject to attack by bacterial enzymes. But small differences in chemical structure can make large differences in biodegradability. The classic example is that of two herbicides— 2,4-D, the common chemical used to kill lawn weeds, and 2,4,5-T, which is used to kill shrubs. The addition of a single chlorine atom to the structure of 2,4-D extends its life in soil from a few days to an indefinite period (Figure 25–6).

AQUATIC MICROBIOLOGY AND SEWAGE TREATMENT

Aquatic microbiology refers to the study of microorganisms and their activities in natural waters such as lakes, ponds, streams, rivers, estuaries, and the sea. Here we will consider the microbial content of fresh water and seawater.

High nutrient levels in water are generally reflected by high microbial numbers. Water con-

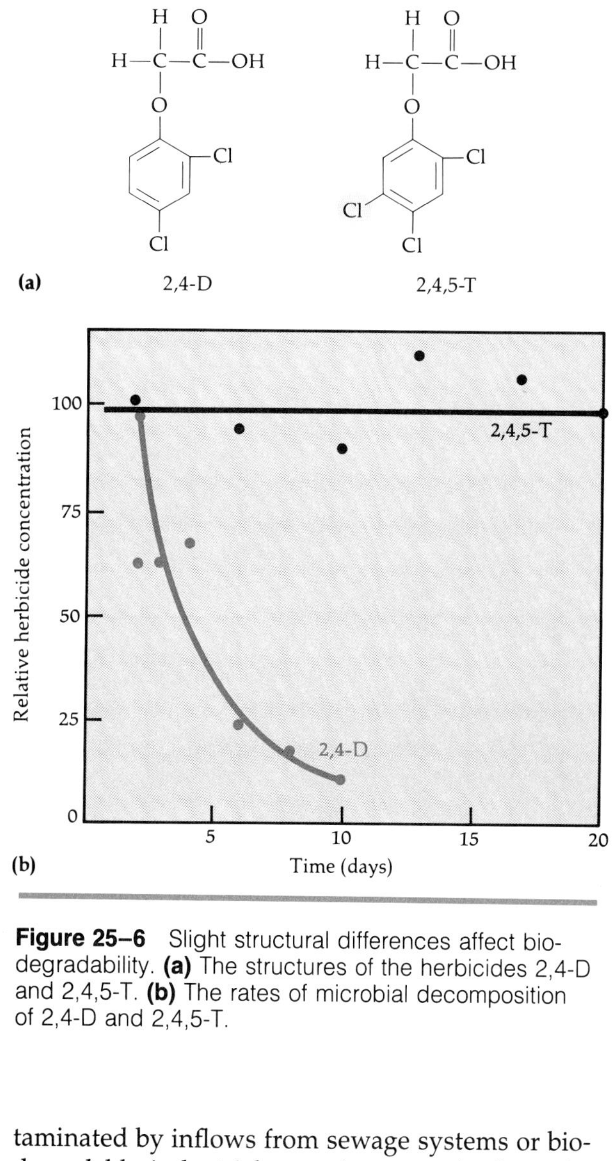

Figure 25–6 Slight structural differences affect biodegradability. **(a)** The structures of the herbicides 2,4-D and 2,4,5-T. **(b)** The rates of microbial decomposition of 2,4-D and 2,4,5-T.

taminated by inflows from sewage systems or biodegradable industrial organic wastes is also relatively high in bacterial counts. Similarly, ocean estuaries (fed by rivers) have higher nutrient levels and therefore higher microbial counts than other shoreline waters. In water, particularly water with low nutrient concentrations, microorganisms tend to grow on stationary surfaces and on particulate matter, rather than being randomly suspended in the water. In this way, the microorganism has contact with more nutrients than if it were freely floating with the current. Many bacteria whose main habitat is water have appendages, such as holdfasts, that attach to various surfaces. One example is *Caulobacter* (kô-lō-bak′tėr). (See Figure 10–20.) Some bacteria also have gas vesicles that they can fill and empty to adjust buoyancy.

FRESHWATER MICROBIAL FLORA

A typical lake or pond can represent the various zones and the kinds of microbial flora in a body of fresh water (Figure 25–7). The **littoral zone** along the shore has considerable rooted vegetation and light penetrates throughout it. The **limnetic zone** consists of the surface of the open water area away from the shore. The **profundal zone** is the deeper water under the limnetic zone. And the **benthic zone** contains the sediment at the bottom.

Microbial populations of freshwater bodies tend to be differentiated mainly by the availability of oxygen and light. Light is in many ways the more important of the two because the photosynthetic algae are the main source of organic matter, and hence energy, for the lake. These organisms are the primary producers of a lake that supports a population of bacteria, protozoans, fish, and other aquatic life. Photosynthetic algae are located in the limnetic zone.

Areas of the limnetic zone with sufficient oxygen contain pseudomonads and species of *Cytophaga* (sī-täf′äg-ä), *Caulobacter*, and *Hyphomicrobium* (hī-fō-mī-krō′bē-um). Oxygen does not diffuse into water very well, as any aquarium owner knows. Microorganisms growing on nutrients in stagnant water quickly use up the dissolved oxygen in the water. In the oxygenless water, fish die and odors (from hydrogen sulfide and organic acids, for example) are produced from anaerobic activity. Wave action in shallow layers, or water movement as in rivers, tends to increase the amount of oxygen throughout the water and aid in the growth of aerobic populations of bacteria. Movement thus improves the quality of the water and aids in degradation of polluting nutrients.

Deeper waters of the profundal zone have low oxygen concentrations and less light. Algal growth near the surface often filters the light, and it is not unusual for photosynthetic microorganisms deeper down to utilize wavelengths of light differing from those used by surface-layer photosynthesizers. The purple and green sulfur bacteria are found in these deeper waters. These bacteria are anaerobic photosynthetic organisms that me-

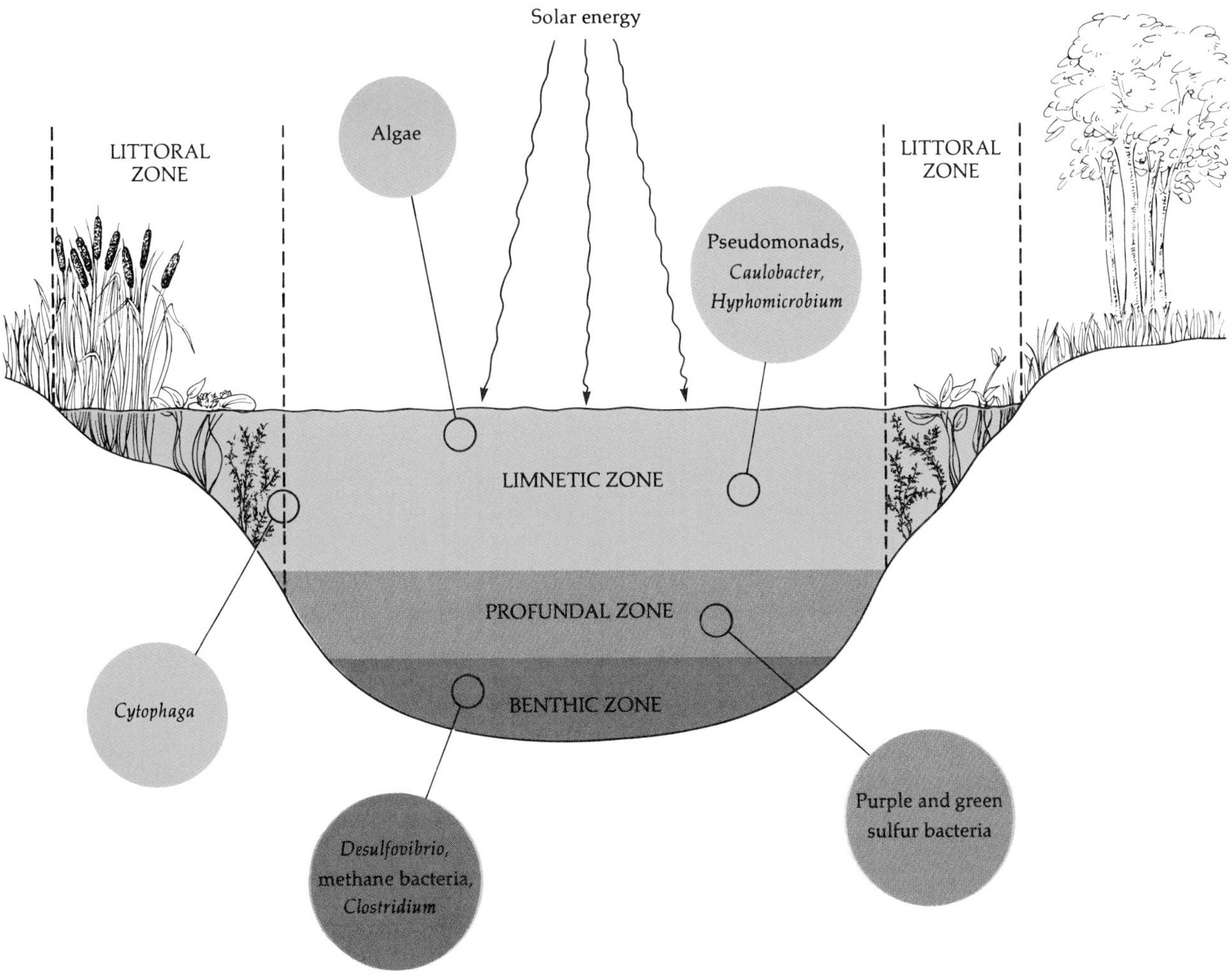

Figure 25–7 The zones in a lake or pond, and some representative microorganisms of each zone.

tabolize hydrogen sulfides to sulfur and sulfates, in the bottom sediments of the benthic zone. The sediment also includes bacteria such as *Desulfovibrio* (dē-sul-fō-vib′rē-ō), which use sulfate (SO_4^{2-}) as an electron acceptor, and reduce it to hydrogen sulfide (H_2S), which is responsible for the rotten-egg odor of many lake muds. Methane-producing bacteria are also part of these anaerobic, benthic populations. In swamps, marshes, or bottom sediments, they produce methane gas. *Clostridium* species are common in bottom sediments and may include botulism organisms, particularly those causing outbreaks of botulism in water fowl.

In freshwater systems, then, the primary producers are the photosynthetic algae and cyanobacteria. These are eventually consumed by other aquatic life and are degraded to more elemental nutrients by bacteria of the limnetic and benthic zones of the lake.

SEAWATER MICROBIAL FLORA

The open ocean is relatively high in osmotic pressure and low in nutrients. The pH also tends to be higher than is optimal for most microorganisms. Bacterial populations in such waters tend to be much lower, therefore, than in estuaries and most small freshwater bodies, which are fed by rivers and streams and have higher microbial counts. Much of the microscopic life of the ocean is composed of photosynthetic diatoms and other algae. Largely independent of preformed organic nutrient sources, these microbes use energy from photosynthesis and atmospheric carbon dioxide for carbon. These organisms constitute the marine **phytoplankton** community, the basis of the oceanic food chain. Oceanic bacteria benefit from the eventual death and decomposition of phytoplankton and also attach to their living bodies. Protozoans in turn feed on bacteria and the smaller phytoplankton. Krill, shrimplike crustaceans, feed on the phytoplankton and in turn are an important food supply of larger sea life. Many fish and whales are also able to feed directly on the phytoplankton.

Microbial luminescence is an interesting minor aspect of deep-sea life. Certain algae produce light flashes when they are agitated by wave action or a boat wake. Many bacteria are also luminescent, and some have established symbiotic relationships with benthic-dwelling fish. These fish sometimes use the glow of their residing bacteria as an aid in attracting and capturing prey in the complete darkness of the ocean depths. These bioluminescent organisms have an enzyme called luciferase that picks up electrons from flavoproteins in the electron transport chain and then emits some of the electron's energy as a photon of light.

EFFECTS OF POLLUTION

Water that moves below the ground's surface undergoes a filtering that tends to remove microorganisms. For this reason, water from springs and deep wells is generally of good quality.

Transmission of Infectious Diseases

Contamination of water supplies by pathogenic microorganisms is an important factor in the spread of many diseases. Sometimes the bacterial pathogens are ingested, as in typhoid fever or Asiatic cholera. Protozoan diseases, such as amoebic dysentery or giardial dysentery, are spread by cysts carried in water. Relatively large parasites, such as the free-swimming trematode cercaria, the causative agent of schistosomiasis, are also dependent upon water for their spread to human hosts (see Chapters 11 and 21). In developed parts of the world, such contaminations are minimized by education of the population and by sophisticated water treatment systems.

Chemical Pollution

Preventing chemical contamination of water is a more difficult problem. Industrial and agricultural chemicals leached from the land enter water in great amounts and in forms resistant to biodegradation. Many of these chemicals become biologically concentrated in some of the organisms in the food chain.

A striking example of industrial water pollution involved mercury used in the manufacture of paper. The metallic mercury was allowed to flow into waterways as waste. It was assumed that the mercury was inert and would remain segregated in the sediments. However, bacteria in the sediments incorporated the mercury into a soluble chemical compound, which was then taken up by fish and invertebrates in the waters. When such seafood is a substantial part of the human diet, the mercury concentrations can accumulate with devastating effects on the nervous system.

Another example is the synthetic detergents developed immediately after World War II. These rapidly replaced many of the soaps then in use. Because these new detergents were not biodegradable, they rapidly accumulated in the waterways. In some rivers, large rafts of detergent suds could be seen traveling downstream; and in some cities, a small head of bubbles might appear on the surface of a glass of water. These detergents were replaced in 1964 by new biodegradable formulations. These, however, brought new problems. Substantial amounts of phosphates were added to many of these detergents to improve effectiveness. Unfortunately, phosphates pass nearly intact through most sewage treatment systems and can cause **eutrophication** of lakes. Eutrophication, meaning "well nourished," is the process in which the addition of large amounts of nutrients to water results in massive growth of algae and eventual loss of oxygen. In the short run, these algae produce oxygen. However, the algae eventually die and are degraded by bacteria; during the degrading process, the oxygen in the water is used up. Undigested remnants settle to the bottom and hasten the filling of the lake.

Eutrophication can also result from the addition of raw sewage, agricultural runoff, or industrial wastes to a lake (Figure 25–8). However, eutrophication is not always a result of chemical pollution, but can occur as a natural process. Algae and cyanobacteria are able to grow by using photosynthesis for energy and atmospheric carbon dioxide dissolved in water to construct their organic compounds. Therefore, the major component of the carbon structures of algae and cyanobacteria requires little or no addition of nutrients to water. Given light and relatively small amounts of nutrients such as nitrogen and phosphorus compounds, lakes and streams can undergo sudden blooms of heavy algal growth.

As mentioned earlier, many species of cyanobacteria also have the capacity to fix nitrogen from the atmosphere. Thus, even in waters in which the polluting nitrogen is minimal, blooms of these cyanobacteria can thrive. This is where the phosphates from detergents become involved. Because these procaryotic organisms get their energy from light and their carbon and nitrogen essentially from the atmosphere, they find most of their requirements for growth even in very clean waters. About the only major growth nutrient not found in water in sufficient amounts is phosphate, which is rather insoluble and tends to be retained in soil. When a few parts per million of phosphate are added to a lake, the result can be a bloom of cyanobacteria. Eventually, these organisms die and the bacteria metabolizing their remains use up the

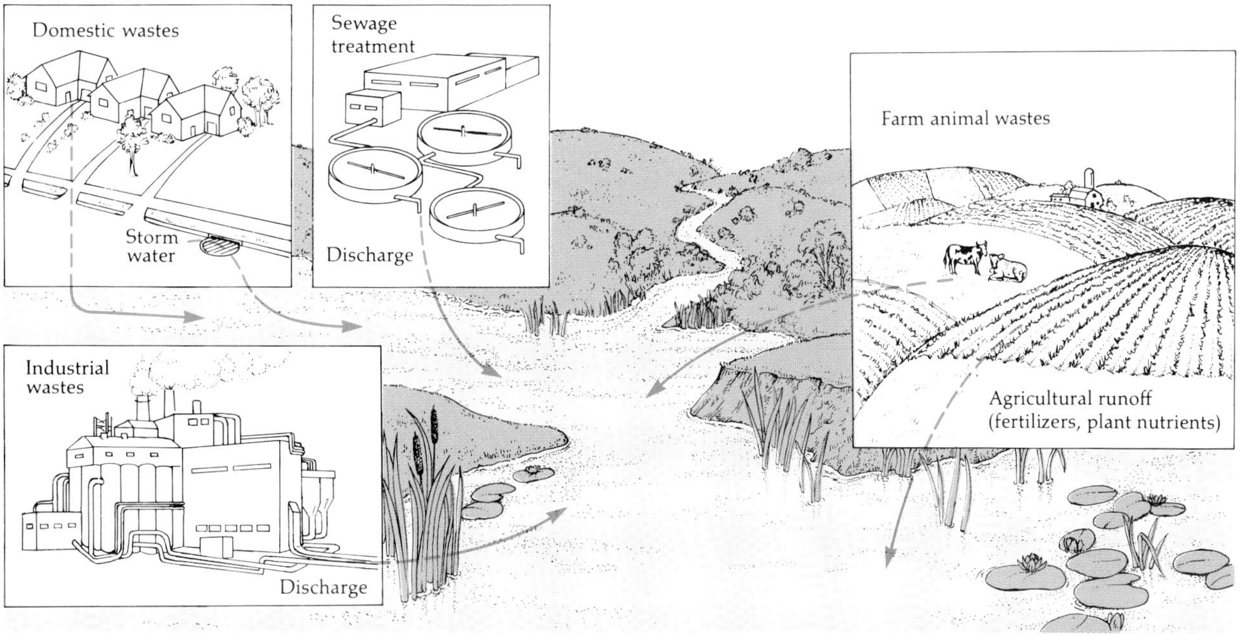

Figure 25–8 Sources of nutrients, mainly phosphorus and nitrogen compounds, that contribute to eutrophication.

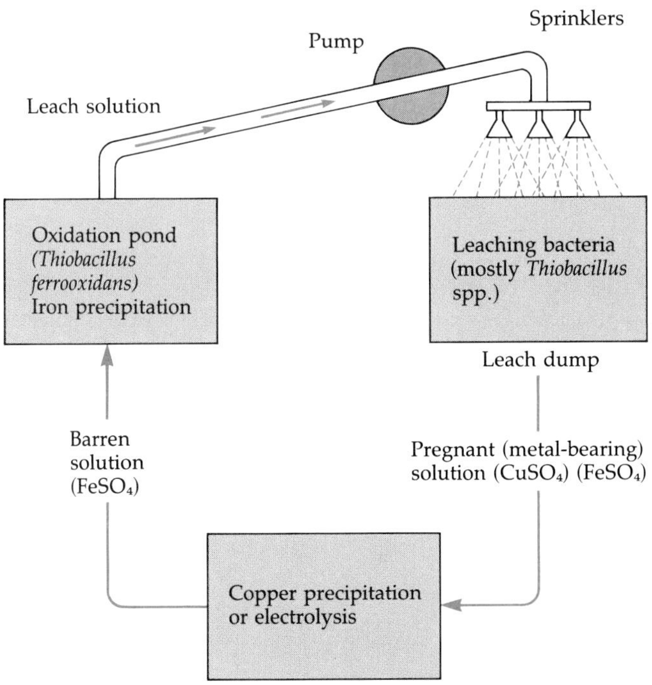

Figure 25–9 The use of bacteria in mineral extraction from ore. As the water flows through the dump of copper-bearing ore, bacteria produce sulfuric acid and copper sulfate (soluble) from the copper sulfides (insoluble) in the ore. The movement of soluble compounds along with moving liquids is called leaching. The copper sulfate leached from the ore is then precipitated or removed by electrolysis. However, the remaining solution, now barren of copper, still contains iron that would interfere with the activity of leaching bacteria. The iron is precipitated in an oxidation pond by the activity of *Thiobacillus ferrooxidans*. The leaching solution is then recycled.

oxygen in the water in a manner similar to that resulting from the addition of excess nutrients such as sewage. Because even small amounts of phosphate lead to these results, phosphate-containing detergents are illegal in many localities.

Coal-mining wastes, particularly in the eastern United States, are very high in sulfur content, mostly iron sulfides (FeS_2). In the process of obtaining energy from the oxidation of the ferrous ion (Fe^{2+}), bacteria such as *Thiobacillus ferrooxidans* (fer-rō-oks'i-danz) convert the sulfide into sulfates. The sulfates enter streams as sulfuric acid, which lowers the pH of the water and damages aquatic life. The low pH also promotes the formation of insoluble iron hydroxides, which form the yellow precipitate often seen clouding such polluted waters. On the other hand, *T. ferrooxidans* is used in the recovery of otherwise unprofitable grades of uranium and copper ores (Figure 25–9). For example, when solutions containing ferric ion (Fe^{3+}) are washed through deposits of insoluble uranium compounds, the uranium is oxidized to soluble compounds. In this process, Fe^{3+} is reduced to Fe^{2+}. The Fe^{2+} can be reoxidized by *T. ferrooxidans.* The soluble uranium then moves out of the ores and is reclaimed.

TESTS FOR WATER PURITY

Historically, most of our concern about water purity has been related to the transmission of disease. Tests to determine the safety of water (many of these tests are also applicable to foods) have therefore been developed.

It is not practical, however, to look only for pathogens in water supplies. For one thing, if we were to find the pathogen causing typhoid or cholera in the water system, the discovery would already be too late to prevent an outbreak of the disease. Moreover, such pathogens would probably be present only in small numbers and might not be included in tested samples.

The tests for water safety in use today are aimed at detecting particular **indicator organisms.** There are several criteria for an indicator organism. The most important criterion is that the organism is consistently present in human intestinal wastes in substantial numbers, so that its detection is a good indication that human wastes are entering the water. The indicator organisms should also survive in the water at least as well as the more likely pathogenic organisms would. The indicator organisms must also be detectable by simple tests that can be carried out by persons with relatively little technical training, for example, operators of water treatment plants in small communities.

In the United States, the usual indicator organisms are the *coliform* bacteria. Coliforms are defined as aerobic or facultative anaerobic, gram-negative, non-endospore-forming, rod-shaped bacteria that ferment lactose to form gas within 48 hours of being placed in a medium at 35°C. Because some coliforms are not solely enteric bacteria, but are more commonly found in plant and soil samples, many standards for food and water specify the determination of *fecal coliforms.* The predominant fecal coliform is *Escherichia coli,* which constitutes a large proportion of the human intestinal population. There are specialized tests to distinguish between fecal coliforms and nonfecal coliforms. It is important to note that coliforms are not themselves pathogenic under normal conditions, although they can cause diarrhea, and also opportunistic urinary tract infections.

The United States Environmental Protection Agency Drinking Water Standards specify the minimum number of water samples to be examined each month and the maximum number of coliform organisms permitted in each 100 ml of water.

For detection and enumeration of coliforms, selective and differential media are required and one of two specialized methods is used. The first method is the **multiple-tube fermentation technique.** In this technique, coliforms are detected in three stages: presumptive, confirmed, and completed (Figure 25–10). In the *presumptive test,* dilutions from the water sample are added to tubes of lactose broth medium containing a pH indicator and incubated for 24 to 48 hours. Fermentation of lactose to gas (trapped in an inverted small vial) is considered a positive reaction. In a common version of the presumptive test, 15 tubes of lactose broth are inoculated with samples of the water being tested; 5 tubes receive water samples of 10 ml each, 5 tubes samples of 1 ml each, and 5 tubes samples of 0.1 ml each. If no gas forms in any of the tubes, the water is usually considered satisfactory. Statistically, this result indicates that there are less than 2.2 coliforms per 100 ml of water. (Most Probable Number tables are consulted; see Chapter 6 and Appendix D.)

If any of the tubes in the presumptive test are positive, additional tests are performed, because organisms other than coliforms are capable of lactose fermentation and gas production. In the *confirmed test,* samples from the positive presumptive tubes at the highest dilutions are streaked onto plates of differential media, which are eosin methylene blue (EMB) agars that contain lactose. Coliforms produce acid from lactose, and the eosin and methylene blue dyes are absorbed under acidic conditions, so if there are coliforms, they form dark-centered colonies with or without metallic sheens. These colonies indicate a positive confirmed test.

In the *completed test,* separate colonies from the confirmed test are inoculated in lactose broth and on a nutrient agar slant for 24 hours at 35°C. If gas is produced in lactose broth, and the isolated microorganism is a gram-negative non-endospore-forming rod, the completed test is positive.

Coliforms may also be detected with the **membrane filter method** (Figure 25–10b). A water sample of 100 ml is drawn through a paper- thin membrane filter with pores of about 0.45 μm, which retain the bacteria on the membrane surface. The membrane is then placed on a pad of nutrient medium similar to EMB. Coliform bacteria growing on the nutrient medium soaking through the membrane form colonies with a distinctive appearance. If only one colony is detected per 100 ml of water, the water is usually considered of satisfactory qual-

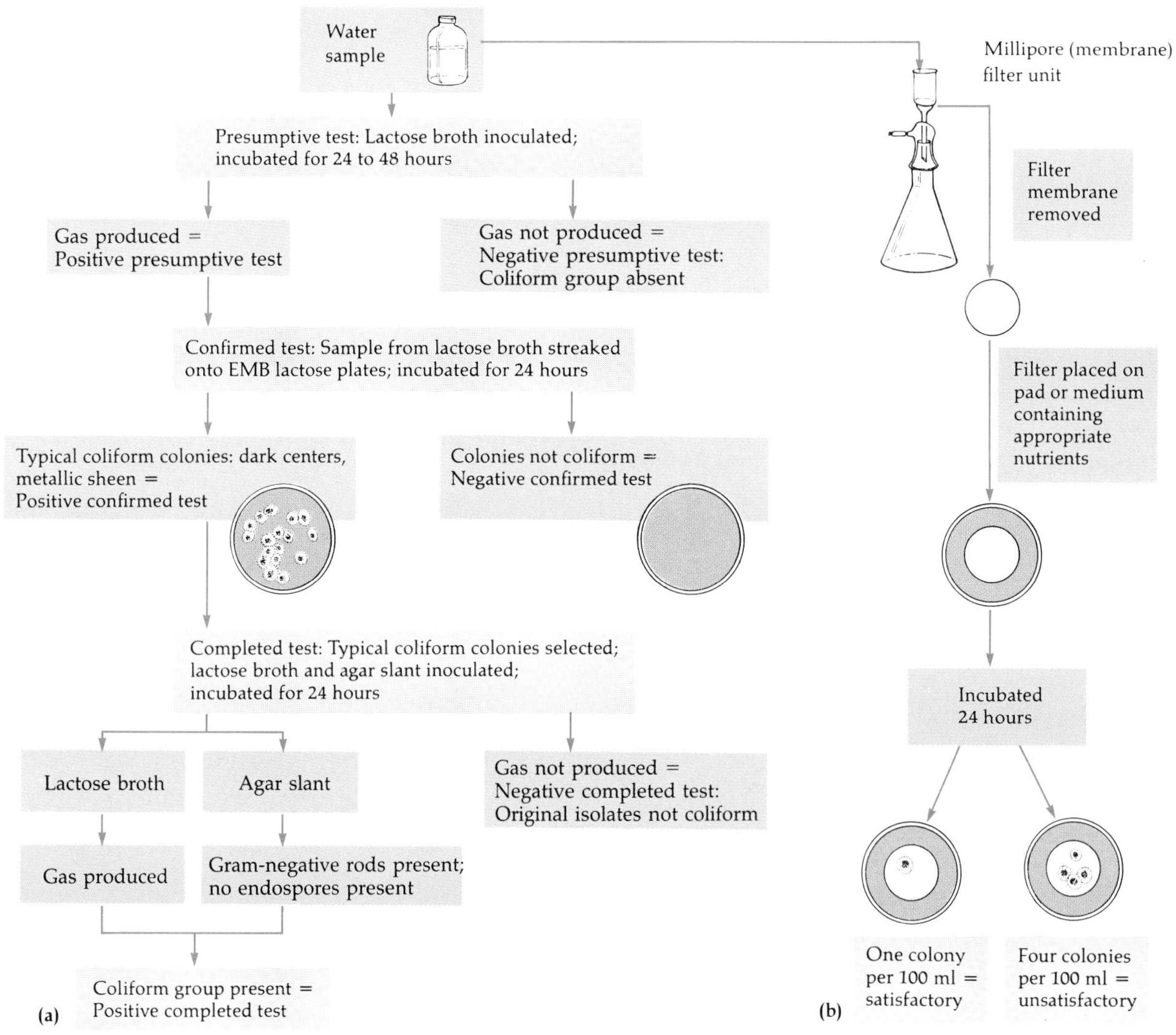

Figure 25–10 Analysis of drinking water for coliforms. **(a)** Multiple-tube fermentation technique. **(b)** Membrane filter method.

ity for drinking. For uses other than drinking, the number of coliforms allowable in a water sample varies with the intended human use.

WATER TREATMENT

When water is obtained from uncontaminated reservoirs fed by clear mountain streams or from deep wells, it requires minimal treatment to make it safe to drink. Many cities, however, obtain their water from badly polluted sources such as rivers that have received municipal and industrial wastes, as well as leaked nuclear cooling water wastes, upstream. In order to allow as much particulate suspended matter as possible to settle out, very turbid (cloudy) water is allowed to stand in a holding reservoir for a period of time (Figure 25-11). The water then undergoes **flocculation treatment**, that is, removal of colloidal materials such as clay, which would remain in suspension indefinitely.

Among the more common flocculant chemicals is aluminum potassium sulfate (alum). This chemical forms a floc, which slowly settles out and carries colloidal material entrapped in it to the bottom. Large numbers of viruses and bacteria are also removed by this treatment. You might be interested to know that alum was used to clear muddy river water during the first half of the nineteenth century in the military forts of the American West, long before the germ theory of disease was developed. Similarly, in the early 1800s, many cities in Germany filtered muddy river water through sand beds to clarify it. There were observations at the time that people who used such filtered water did not have the same incidence of cholera during outbreaks.

After flocculation treatment, water is passed through beds of sand to accomplish **sand filtration.** This treatment removes about 99% of the bacteria and viruses remaining from flocculation. Some protozoan cysts, such as those of *Giardia lamblia* (jē-är'dē-ä lam'lē-ä), appear to be removed from water only by such a filtration treatment. The microorganisms are trapped mostly by surface absorption in the sand beds. Bacteria do not penetrate the torturous routing of the sand beds, even though the openings might be larger than the organisms that are filtered out. These sand filters are periodically backflushed to clear them of accumulations. Water systems of the future might replace or supplement sand filtration with filters of activated charcoal (carbon). Charcoal has the advantage of removing not only particulate matter, but also some organic chemical pollutants dissolved in solution.

Before entering the municipal distribution system, the water is **chlorinated.** Chlorination is very effective in killing pathogenic bacteria but less effective in destroying protozoan cysts and viruses. Because organic matter neutralizes chlorine, the plant operators must pay constant attention to maintain effective levels of chlorine. There has been some concern that chlorine itself might be a health hazard, that it might react with organic contaminants of the water to form carcinogenic compounds. For the moment, this possibility is considered minor compared to the proven usefulness of chlorination of water. As noted in Chapter 7, one substitute for chlorination is ozone treatment. Ozone (O_3) is a highly reactive form of oxygen that is formed by electrical spark discharges and ultraviolet light. The fresh odor of air following an elec-

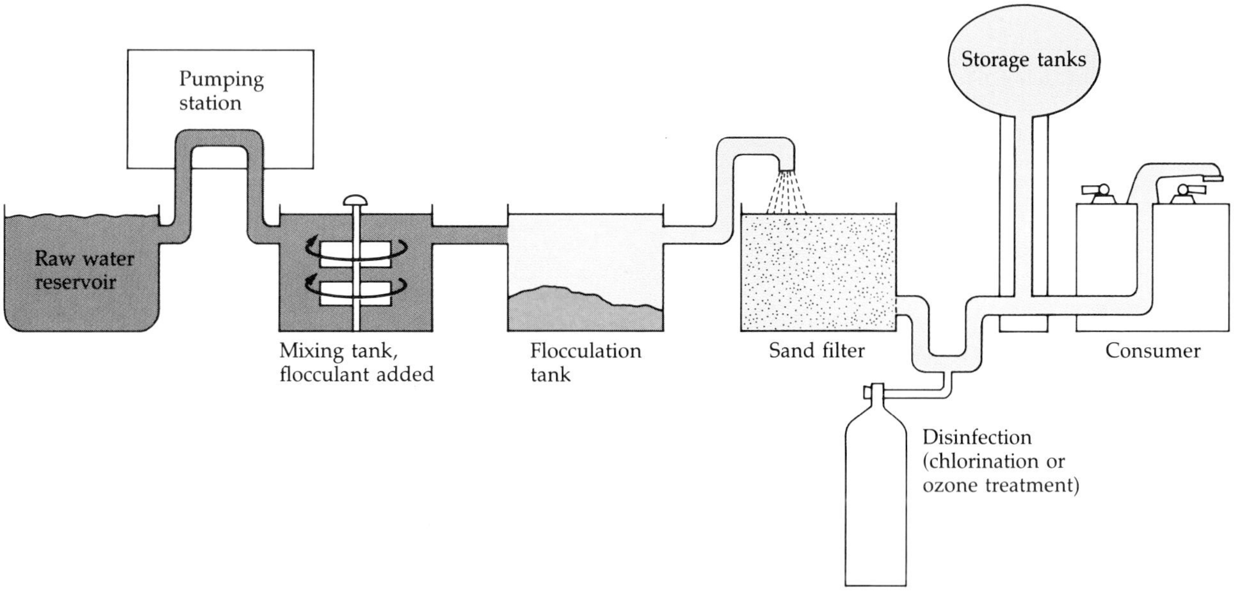

Figure 25–11 Steps of water treatment in a typical municipal water purification plant.

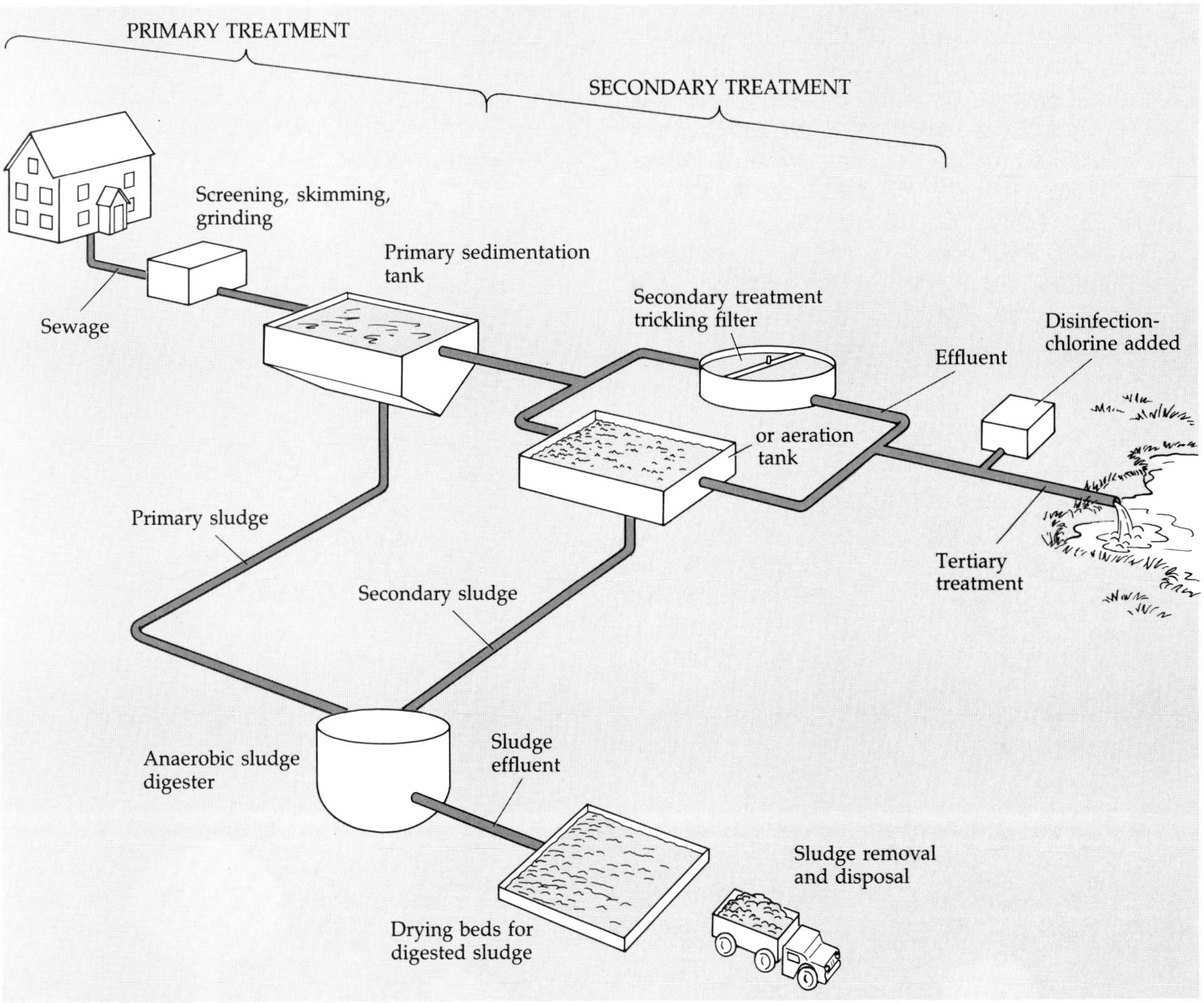

Figure 25–12 Steps involved in sewage waste treatment: primary treatment, secondary treatment, and tertiary treatment.

trical storm or around an ultraviolet light bulb is due to ozone. Ozone for water treatment is generated electrically at the site of treatment. Unlike chlorine, however, ozone does not remain to provide continued protection in the water system.

SEWAGE TREATMENT

After water is used it becomes **sewage.** Sewage includes all the dishwashing, handwashing, and bath water from a household, as well as toilet wastes. Rainwater flowing into street drains enters the sewage system in some cities, as do some industrial wastes. Sewage is mostly water and contains little particulate matter, perhaps only about 0.03%. Even so, in large cities, this solid portion of sewage can daily total more than 1000 metric tons of solid material.

Until environmental awareness intensified in recent years, a surprising number of large cities in this country had only rudimentary sewage treatment systems—or no system at all. Raw sewage, untreated or nearly so, was simply discharged into rivers or oceans. A flowing, well-aerated stream is capable of considerable self-purification. Therefore, until increases in populations and their wastes exceeded this capability, this casual treatment of municipal wastes caused little complaint. Most methods of simple discharge in the United States are now being improved. However, at this time, 90% of the 125 largest cities on the Mediterranean Sea release all sewage into this body of water with no treatment at all. This statistic is surprising, considering that many of these cities are supposed to be part of the well-developed world.

Primary Treatment

The usual first step in sewage treatment is called **primary treatment** (Figure 25–12). In this process, incoming sewage receives preliminary treatment: large floating materials are screened out; the sewage is allowed to flow through settling chambers so that sand and similarly gritty material can be removed; skimmers remove floating oil and grease; and floating debris is shredded and ground. After this step, the sewage passes through sedimentation tanks, where solid matter settles out. (The design of these primary settling tanks varies.) Sewage solids collecting on the bottom are called **sludge;** sludge at this stage is called primary sludge. From 40 to 60% of suspended solids are removed from sewage by this settling treatment, and flocculating chemicals that increase the removal of solids are sometimes added. Biological activity is not particularly important in primary treatment, although some digestion of sludge and dissolved organic matter can occur during long holding times. The sludge is removed on either a continuous or intermittent basis, and the effluent (the liquid flowing out) then undergoes secondary treatment.

Biochemical Oxygen Demand

Primary treatment removes approximately 25 to 35% of the **biochemical oxygen demand (BOD)** of the sewage. BOD is an important concept in sewage treatment and in the general ecology of waste treatment. A measure of the biologically degradable organic matter in water, BOD is determined by the amount of oxygen required by bacteria to metabolize the matter. The classic method of measurement is to use special bottles with airtight stoppers (Figure 25–13). Each bottle is first filled with the test water or dilutions of the test water. The water is initially aerated to provide a relatively high level of dissolved oxygen and is, if necessary, seeded with bacteria. The filled bottles are incubated for five days at 20°C, and the decrease in dissolved oxygen is determined by use of a chemical or electronic testing method. The more oxygen that is used up as the bacteria utilize the organic matter in the sample, the greater the BOD—which is usually expressed in terms of miligrams per liter of oxygen used. The amount of oxygen that normally can be dissolved in water is only about

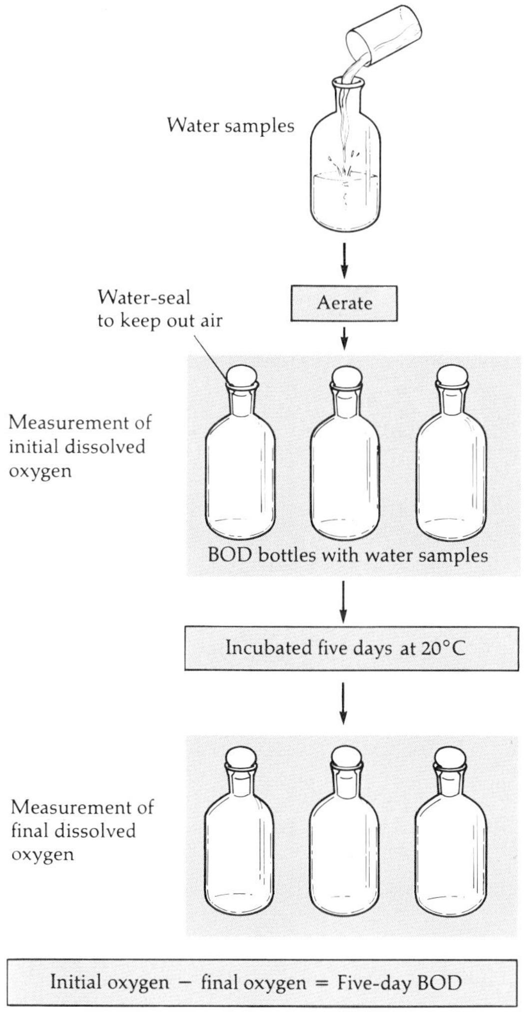

Figure 25–13 Determination of biochemical oxygen demand (BOD).

10 mg/l. A typical BOD value may be 200 mg/l; therefore, as bacteria begin to consume this organic matter (BOD), they rapidly deplete the oxygen in the water.

Secondary Treatment

After primary treatment, the greater part of the BOD remaining in the sewage is in the form of dissolved organic matter. **Secondary treatment,** which is primarily biological, is designed to remove most of this organic matter and reduce the BOD. In this process, the sewage undergoes strong aeration to encourage the growth of aerobic bacteria and other microorganisms that oxidize the dissolved organic matter to carbon dioxide and water. Two commonly used methods of secondary treatment are activated sludge systems and trickling filters.

In the aeration tanks of the **activated sludge system**, air or pure oxygen is added to the effluent from primary treatment (Figure 25–14). The sludge in the effluent contains large numbers of metabolizing bacteria together with yeasts, molds, and protozoans. An especially important ingredient of the sludge are species of *Zooglea* (zō-ō-glē'ä) bacteria, which form flocculant masses (floc) in the aeration tanks (Figure 25–15). The activity of these aerobic microorganisms oxidizes much of the effluent's organic matter into carbon dioxide and water. When the aeration phase is completed, the floc (secondary sludge) is allowed to settle to the bottom as the nonsoluble solids settle in primary treatment.

Soluble organic matter in the sewage is absorbed onto the floc and is incorporated into microorganisms in the floc. As it settles out, this organic matter is removed with the floc and is subsequently treated in an anaerobic sludge digester. More organic matter is probably removed by this process than by the relatively short-term aerobic oxidation.

Most of the settled sludge is removed for treatment in an anaerobic sludge digester; about 20% of the sludge is recycled to the activated sludge tanks as a "starter" for the next sewage batch. The effluent is sent on for final treatment. Occasionally, when aeration is stopped, the sludge will float

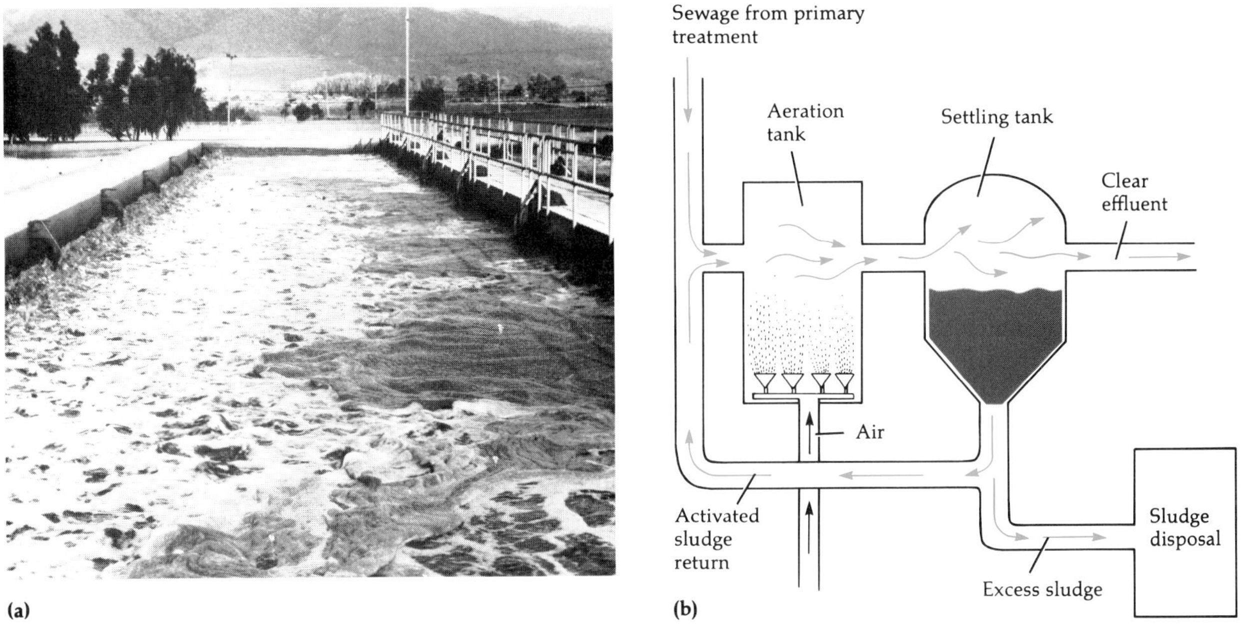

Figure 25–14 Activated sludge system of secondary treatment. **(a)** An aeration tank at the San Jose–Santa Clara Water Pollution Control Plant in San Jose, California. At this facility, sewage passes through four such tanks during secondary treatment. Note that the surface is frothing from aeration. **(b)** Diagram of an activated sludge system.

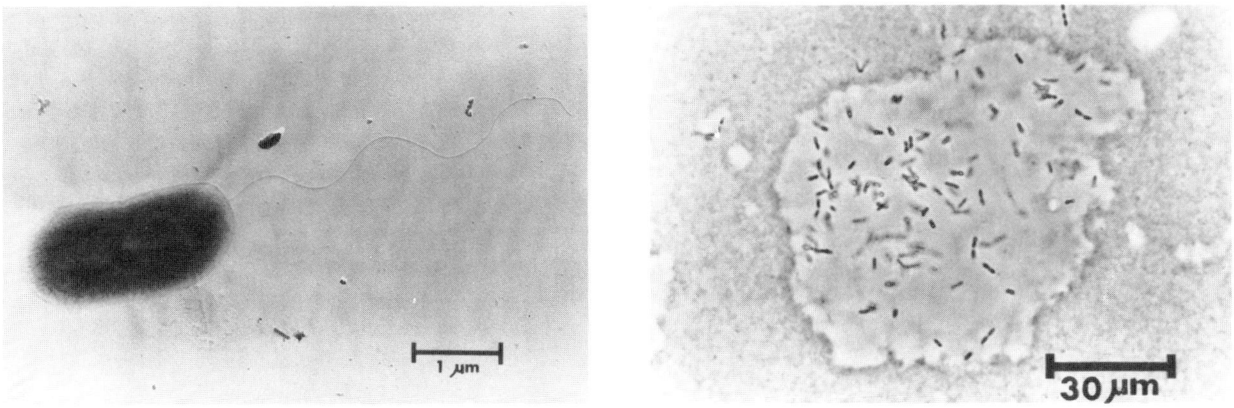

(b)

Figure 25–15 *Zoogloea*, the organism that contributes to floc formation. **(a)** *Zoogloea ramigera* cell with single, polar flagellum. **(b)** Amorphous floc formed by *Z. ramigera*. Note the cells embedded in gelatinous matter synthesized by the bacteria.

rather than settle out; this phenomenon is called *bulking*. When this happens, the organic matter in the floc flows out with the discharged effluent and often causes serious problems of local pollution. A considerable amount of research has been devoted to the causes of bulking and its possible prevention. It is apparently caused by the growth of filamentous bacteria of various types, although the sheathed bacterium *Sphaerotilus natans* (sfe-rä'ti-lus nā'tans) is often mentioned as the primary offender. Activated sludge systems are quite efficient: they remove 75 to 95% of the BOD from sewage.

Trickling filters are the other commonly used method of secondary treatment. In this method, the sewage is sprayed over a bed of rocks or molded plastic (Figure 25–16). The rocks or other components of the bed must be large enough that air penetrates to the bottom, but small enough to maximize the surface area available for microbial activity. Actually, there is no "filtering" action. A slimy, gelatinous film of aerobic microorganisms grows on the rocks' surfaces and is in many ways functionally similar to the organisms found in activated sludge systems. Because air circulates throughout the rock bed, these aerobic microorganisms in the slime layer are able to oxidize much of the organic matter trickling over the surfaces into carbon dioxide and water. Trickling filters remove 80 to 85% of the BOD, so are less efficient than activated sludge systems. On the other hand, trickling filters are usually less troublesome to operate and less subject to problems from overloads or toxic sewage.

Sludge Digestion

Primary sludge accumulates in primary sedimentation tanks; sludge also accumulates in activated sludge and trickling filter secondary treatments. For further treatment, these sludges are often pumped to **anaerobic sludge digesters** (Figure 25–17). The process of sludge digestion is carried out in large tanks from which oxygen is almost completely excluded. In secondary treatment, emphasis is placed on the maintenance of aerobic conditions so that organic matter is converted to carbon dioxide, water, and solids that can settle out. By comparison, anaerobic fermentations are metabolically inefficient, and the microorganisms leave large amounts of only partially digested organic materials in the form of fatty acids, alcohols, and similar products that retain much of their original BOD. An anaerobic sludge digester is designed to encourage the growth of anaerobic bacteria, especially methane-producing bacteria, that decrease organic solids by degrading them to soluble substances and gases, mostly methane (60 to 70%) and carbon dioxide (20 to 30%). The gases are routinely used as a fuel for heating the digester and are frequently also used to power equipment in the plant.

There are essentially three stages in the activity of an anaerobic sludge digester. The first stage is the production of carbon dioxide and organic

(a)

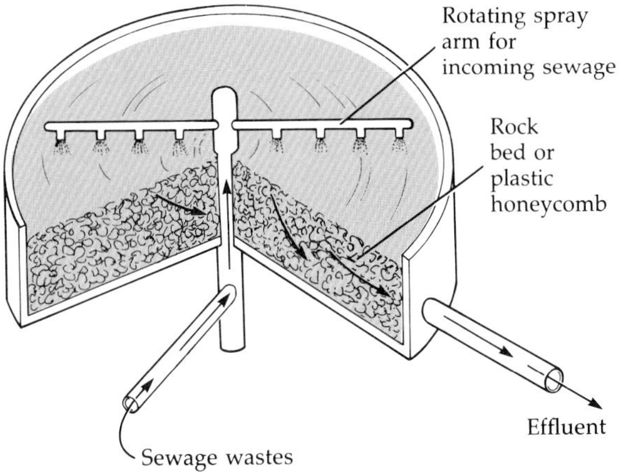

(b)

Figure 25–16 Trickling filter method of secondary treatment. **(a)** A modern trickling filter at the Sunnyvale, California, Water Pollution Control Plant. This type of tank, which is more than 18 feet high, is filled with a plastic honeycomb designed to have a maximum surface area and to allow oxygen to penetrate deeply into the bed. Microorganisms grow on the enormous surface area of the plastic filling (some systems use rocks) and aerobically metabolize the organic matter in the sewage flowing down through the bed. **(b)** Diagram of a trickling filter system.

(a)

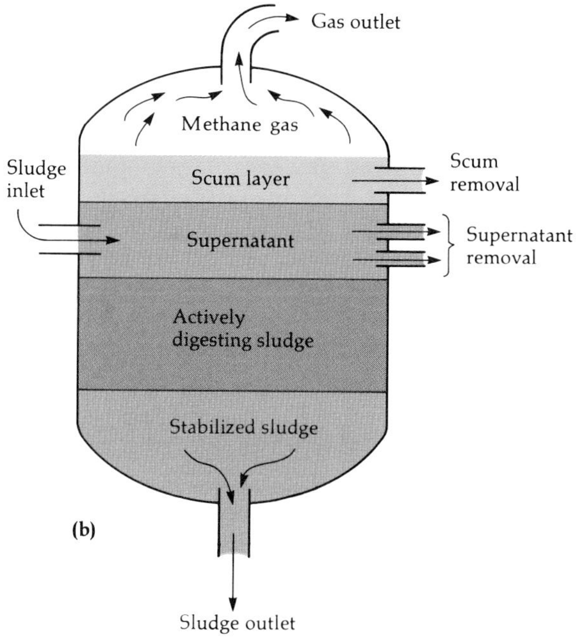

(b)

Figure 25–17 Sludge digestion. **(a)** An anaerobic sludge digester at the San Jose–Santa Clara Water Pollution Control plant. Much of a typical digester is below the ground's surface. In regions with cold climates, the above-ground part of the tank is insulated by mounded earth. Methane from this sludge digester is used to generate electricity for the plant. **(b)** Diagram of a sludge digester.

acids from anaerobic fermentation of the sludge by assorted anaerobic and facultatively anaerobic microorganisms. In the second stage, the organic acids are metabolized to form considerable hydrogen and carbon dioxide as well as organic acids such as acetate. These products are the raw materials for a third stage in which the methane-producing bacteria produce methane (CH_4). Most of the methane is derived from the energy-yielding reduction of carbon dioxide by hydrogen gas.

$$CO_2 + 4H_2 \rightarrow CH_4 + 2H_2O$$

Other methane-producers split acetate to yield methane and carbon dioxide.

$$CH_3COOH \rightarrow CH_4 + CO_2$$

Methane and carbon dioxide are relatively innocuous end products, comparable to the carbon dioxide and water from aerobic treatment. However, considerable amounts of undigested sludge still remain, although it is relatively stable and inert. To reduce its volume, this sludge is pumped to shallow drying beds or filters. It is then carried away for disposal. Some seacoast plants barge sludge out to sea; it can be used for landfill or as a soil conditioner, or it can be incinerated. Sludge has about one-fifth of the value of normal commercial lawn fertilizers but has desirable soil-conditioning qualities much as soil humus does.

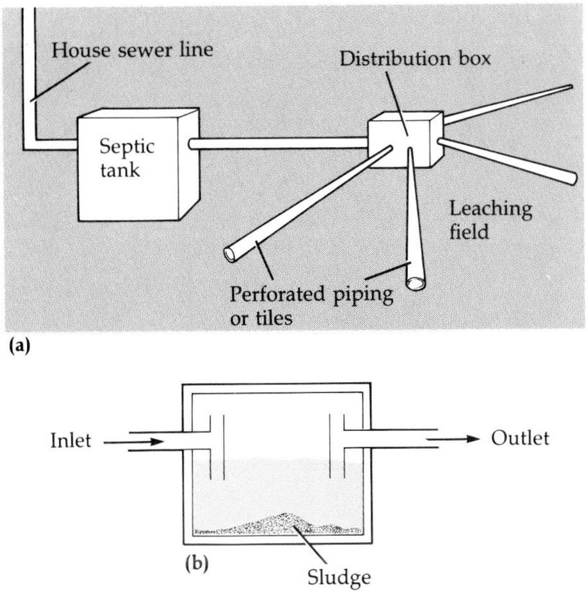

Figure 25–18 Septic tank system. **(a)** Overall plan. **(b)** The inside of a septic tank. The sludge is pumped out at periodic intervals.

Septic Tanks

Homes and businesses not connected to municipal sewage systems often use a **septic tank** (Figure 25–18), a device whose operation is similar in principle to primary treatment. Sewage enters a holding tank and suspended solids settle out. The sludge in the tank must be pumped out periodically and disposed of. The effluent flows through a system of perforated piping into a leaching (soil drainage) field. The effluent entering the soil is decomposed by soil microorganisms. These systems work well when not overloaded and when the drainage system is properly sized to the load and soil type. Heavy clay soils require extensive drainage systems because of the soil's poor permeability. Sandy soils can allow chemical or bacterial pollution of nearby water supplies.

Oxidation Ponds

Many small communities and many industries use **oxidation ponds,** also called **lagoons** or **stabilization ponds,** for water treatment. These are inexpensive to build and operate but require considerable land. Designs vary, but most incorporate two stages. The first stage is analogous to primary treatment—the sewage pond is deep enough that conditions are almost entirely anaerobic. Sludge settles out in this stage. In the second stage, which corresponds to secondary treatment, effluent is pumped into an adjoining pond or system of ponds that are shallow enough to be aerated by wave action. Because it is difficult to maintain aerobic conditions for bacterial growth in ponds with so much organic matter, the growth of algae is encouraged to produce oxygen. Bacterial action in decomposing the organic matter in the wastes generates carbon dioxide. Algae, which uses carbon dioxide in its photosynthetic metabolism, grows and produces oxygen, which in turn encourages activity of aerobic microorganisms in the sewage. Considerable amounts of organic matter in the form of algae accumulate, but this is not a problem because the oxidation pond, unlike a lake, already has a heavy nutrient load.

Some small sewage-producing operations, such as isolated campgrounds or highway rest stop areas, use an **oxidation ditch** for sewage treatment. In this method, a small oval channel in the shape of a race track is filled with sewage water. A paddle wheel similar to that on a Mississippi steamboat propels the water in a self-contained flowing stream aerated enough to oxidize the wastes.

Tertiary Treatment

As we have seen, primary and secondary treatments of sewage do not remove all of the biologically degradable organic matter. Amounts that are not excessive can be added to a flowing stream without causing a serious problem. Eventually, however, the pressures of increased populations might change this picture, and additional treatments might be required. Even now, primary and secondary treatments are inadequate in certain situations, such as when the effluent is discharged into small streams or recreational lakes. Thus, some communities have tertiary treatment plants. Lake Tahoe in the Sierra Nevada mountains, surrounded by extensive development, is the site of one of the best known tertiary systems designed to deal with such problems.

The effluent from secondary treatment plants contains some residual BOD. It also contains about 50% of the original nitrogen and 70% of the original phosphorus—which can have a great impact on a lake's ecosystem. Tertiary treatment is designed to remove essentially all of the BOD, nitrogen, and phosphorus. Tertiary treatment depends less on biological treatment than on physical and chemical treatments (Figure 25–12). Nitrogen is converted to ammonia and evaporated into the air in stripping towers. Some systems encourage denitrifying bacteria to form volatile nitrogen gas. Phosphorus is precipitated out by combining with chemicals such as lime, alum, and ferric chloride. Filters of fine sands and activated charcoal remove small particulate matter and dissolved chemicals. Finally, chlorine is added to the purified water to kill or inhibit any remaining microorganisms and to oxidize any remaining odor-producing substances.

Tertiary treatment provides water that is suitable for drinking, but the process is extremely costly. Secondary treatment is less costly, but water that has undergone only secondary treatment contains many nutrients that pollute water. Much work is being done at present to design secondary treatment plants in which the effluent can be used for irrigation. This design would eliminate a source of water pollution, provide nutrients for plant growth, and reduce the demand on already scarce water supplies. The soil would act as a trickling filter to remove chemicals and microorganisms before the water reaches ground and surface water supplies. Even now, wastewater with coliform counts below 2.2/100 ml are being used to irrigate food crops, orchards, and pastures; water with coliform counts below 23/100 ml are being used to irrigate landscaping and recreational areas.

STUDY OUTLINE

SOIL MICROBIOLOGY AND CYCLES OF THE ELEMENTS (pp. 701–703)

THE COMPONENTS OF SOIL (pp. 701–703)

1. Soil consists of solid inorganic matter, water, air, and living organisms and products of their decay.
2. Rocks and minerals are fragmented by weathering of preexisting rock.
3. Organic and inorganic materials are dissolved in soil water.
4. Soil gases are the same as atmospheric gases. However, the proportions vary due to biological processes in the soil.
5. Organic matter called humus comes from plants, microorganisms and animals, their waste products, and the biochemical activities of microorganisms.
6. Organic matter provides nutrients for growth and affects the soil water and gases.

Organisms (pp. 701–703)

1. Microorganisms in the soil decompose organic matter and transform nitrogen- and sulfur-containing compounds into usable forms.

2. Bacteria are the most numerous organisms in the soil.
3. The characteristic musty odor of soil is due to the geosmin produced by actinomycetes.
4. Soil is not a reservoir for human pathogens except for some endospore-forming bacteria.
5. Insect and plant pathogens are found in the soil.

MICROORGANISMS AND BIOGEOCHEMICAL CYCLES (pp. 703–711)

1. In biogeochemical cycles, certain chemical elements are recycled.
2. Microorganisms are essential to the continuation of biogeochemical cycles.

Carbon Cycle (pp. 703–705)

1. CO_2 is fixed into organic compounds by photoautotrophs.
2. These organic compounds provide nutrients for chemoheterotrophs.
3. Chemoheterotrophs release CO_2 that is then used by photoautotrophs.

Nitrogen Cycle (pp. 705–711)

1. Microorganisms decompose proteins from dead cells and release amino acids.
2. Ammonia is liberated by microbial ammonification of the amino acids.
3. Ammonia is oxidized to nitrates by nitrifying bacteria.
4. Denitrifying bacteria reduce nitrates to molecular nitrogen (N_2).
5. N_2 is converted into ammonia by nitrogen-fixing bacteria.
6. Nitrogen-fixing bacteria include free-living genera such as *Azotobacter;* cyanobacteria; and the symbiotic bacteria *Rhizobium* and *Frankia.*
7. Ammonium and nitrate are used by bacteria and plants to synthesize amino acids.

Mycorrhizae (p. 711)

1. Symbiotic fungi called mycorrhizae live in and on plant roots.
2. Increase the surface area and nutrient absorption of the plant.

Other Biogeochemical Cycles (p. 711)

1. Microorganisms are also involved in the transformation of other elements, including sulfur, potassium, iron, manganese, mercury, and zinc.
2. These reactions make minerals available in soluble form to plants for their metabolism.

DEGRADATION OF PESTICIDES AND OTHER SYNTHETIC CHEMICALS (p. 711)

1. Many synthetic chemicals such as pesticides and plastics are recalcitrant (resistant to degradation).
2. Recalcitrance is based on the nature of the chemical bonding.

AQUATIC MICROBIOLOGY AND SEWAGE TREATMENT (p. 711)

1. The study of microorganisms and their activities in natural waters is called aquatic microbiology.
2. Natural waters include lakes, ponds, streams, rivers, estuaries, and the sea.
3. The concentration of bacteria in water is proportional to the amount of organic material in the water.
4. Most aquatic bacteria tend to grow on surfaces rather than in a free-floating state.

FRESHWATER MICROBIAL FLORA (p. 712)

1. Numbers and location of microbial flora depend on the availability of oxygen and light.
2. Photosynthetic algae are the primary producers of a lake. They are found in the limnetic zone.
3. Pseudomonads, *Cytophaga, Caulobacter,* and *Hyphomicrobium* are found in the limnetic zone, where oxygen is abundant.
4. Microbial growth in stagnant water uses available oxygen and can cause odors and the death of fish.
5. The amount of dissolved oxygen is increased by wave action.
6. Purple and green sulfur bacteria are found in the benthic zone, which contains light and H_2S, but no oxygen.
7. *Desulfovibrio* reduces SO_4 to H_2S in benthic mud.

8. Methane-producing bacteria are also found in the benthic zone.

SEAWATER MICROBIAL FLORA (pp. 713–714)

1. The open ocean is not a favorable environment for most microorganisms because of its high osmotic pressure, low nutrients, and high pH.
2. Phytoplankton, consisting mainly of diatoms, are the primary producers of the open ocean.
3. Some algae and bacteria are bioluminescent. They possess the enzyme luciferase, which can emit light.

EFFECTS OF POLLUTION (pp. 714–716)

1. Microorganisms are filtered from water that percolates into ground water supplies.

Transmission of Infectious Diseases (p. 714)

1. Some pathogenic microorganisms are transmitted to humans in water supplies.

Chemical Pollution (pp. 714–716)

1. Recalcitrant chemical pollutants may be concentrated in animals in an aquatic food chain.
2. Mercury is metabolized by certain bacteria into a soluble compound that is concentrated in animals.
3. Nutrients such as phosphates cause eutrophication of aqautic ecosystems.
4. Eutrophication means "well nourished." It is the result of the addition of pollutants or natural nutrients.
5. Coal-mining wastes contain sulfides that are converted into sulfuric acid by *Thiobacillus ferrooxidans.*
6. The metabolic activities of *T. ferrooxidans* can be used to recover uranium ore.

TESTS FOR WATER PURITY (pp. 716–718)

1. Tests for the bacteriologic quality of water are based on the presence of indicator organisms.
2. The most common indicator organisms are coliforms.
3. Coliforms are aerobic or facultatively anaerobic, gram-negative, non-endospore-forming rods that ferment lactose with the production of acid and gas within 48 hours of being placed in a medium at 35°C.
4. Fecal coliforms, predominately *E. coli,* are used to indicate the presence of human feces.
5. The multiple-tube fermentation technique and membrane filter method are used to detect the presence of coliforms.
6. The most probable number (MPN) of coliforms per 100 ml of water is determined statistically from the results of the multiple-tube fermentation test and directly from the membrane filter method.
7. The allowable number of coliforms in water varies with the intended human use of the water.

WATER TREATMENT (pp. 718–720)

1. Drinking water is held long enough that suspended matter settles.
2. Flocculation treatment uses a chemical such as alum to coalesce and then settle colloidal material.
3. Sand filtration removes bacteria, viruses, and protozoan cysts.
4. Finally, drinking water is disinfected with chlorine to kill remaining pathogenic bacteria.

SEWAGE TREATMENT (pp. 720–726)

1. Domestic wastewater is called sewage.
2. It includes dishwashing, handwashing, bath water, and toilet wastes.

Primary Treatment (p. 721)

1. Primary sewage treatment is the removal of solid matter called sludge.
2. Biological activity is not very important in primary treatment.

Biochemical Oxygen Demand (pp. 721–722)

1. Primary treatment removes approximately 25 to 35% of the biochemical oxygen demand (BOD) of the sewage.
2. BOD is a measure of the biologically degradable organic matter in water.
3. It is determined by the amount of oxygen bacteria required to degrade the organic matter.

Secondary Treatment (pp. 722–723)

1. Secondary treatment is the biological degradation of organic matter in sewage after primary treatment.
2. Activated sludge and trickling filters are methods of secondary treatment.

3. Microorganisms degrade the organic matter aerobically.
4. Secondary treatment removes most of the BOD, about 50% of the nitrogen, and 30% of the phosphorus.

Sludge Digestion (pp. 723–725)

1. Sludge is placed in an anaerobic sludge digester; bacteria degrade organic matter and produce simpler organic compounds, methane, and CO_2.
2. The methane produced in the digester is used to heat the digester and to operate other equipment.
3. Excess sludge is periodically removed from the digester, dried, and disposed of at sea, as landfill, or as soil conditioner.

Septic Tanks (p. 725)

1. Septic tanks can be used in rural areas to provide primary treatment of sewage.
2. They require a large leaching field for the effluent.

Oxidation Ponds (pp. 725–726)

1. Small communities can use oxidation ponds for secondary treatment.
2. These require a large area in which to build an artificial lake.

Tertiary Treatment (p. 726)

1. Tertiary treatment employs physical filtration and chemical precipitation to remove all of the BOD, nitrogen, and phosphorus from water.
2. Tertiary treatment provides drinkable water, whereas secondary treatment provides water usable only for irrigation.

STUDY QUESTIONS

REVIEW

1. Write a one-sentence description of each of the following soil constituents: solid inorganic matter; water; gases; organic matter.
2. The precursor to coal is peat found in bogs. Why does peat accumulate in bogs?
3. The metabolic activities of microorganisms often produce acids. Why would microbial growth increase the solid inorganic matter in soil?
4. Compare and contrast humus and recalcitrant chemicals.
5. Diagram the carbon cycle in the presence and absence of oxygen. Name at least one microorganism that is involved at each step.
6. Fill in the following table with the information provided below.

Process	Chemical Reactions	Microorganisms
Ammonification		
Nitrification		
Denitrification		
Nitrogen fixation		

Choices:

$NO_3^- \rightarrow N_2$
$N_2 \rightarrow NH_3$
$-NH_2 \rightarrow NH_3$
$NH_3 \rightarrow NO_2^-$
$NO_2^- \rightarrow NO_3^-$

Bacillus
Rhizobium
Nitrosomonas
Azotobacter
Nitrobacter
Proteolytic bacteria

7. The listed organisms have important roles as symbionts with plants and fungi. Describe the symbiotic relationship of each organism with its host. Cyanobacteria, mycorrhizae, *Rhizobium, Frankia.*

8. Compare and contrast the physical conditions of the ocean with those of fresh water.

9. Matching.

___ $CO_2 + H_2S \xrightarrow{\text{light}} C_6H_{12}O_6 + SO$	(a) Methane-producing bacteria
___ $SO_4^{-} + H^+ \longrightarrow H_2S$	(b) *Desulfovibrio*
___ $CO_2 + H_2 \longrightarrow CH_4$	(c) Photosynthetic bacteria

10. Indicate which reactions in question 9 require atmospheric oxygen (O_2).

11. Outline the treatment process for drinking water.

12. What is the purpose of a coliform count on water?

13. If coliforms are not normally pathogenic, why are they used as indicator organisms for bacteriologic quality?

14. The following processes are employed in wastewater treatment. Match the type of treatment with the processes. Each choice can be used once, more than once, or not at all.

Processes:
___ Leaching field
___ Removal of solids
___ Biological degradation
___ Activated sludge
___ Chemical precipitation of phosphorus
___ Trickling filter
___ Results in drinking water
___ Effluent can be used for irrigation
___ Produces methane

Types of treatment:
(a) Primary
(b) Secondary
(c) Tertiary

15. Define BOD.

16. Why is activated sludge a more efficient means of removing BOD than a sludge digester?

17. Why are septic tanks and oxidation ponds not feasible for large municipalities?

18. Explain the effect of dumping untreated sewage into a pond on the eutrophication of the pond. The effect of sewage that has primary treatment? The effect of sewage that has secondary treatment? Contrast your previous answers with the effect of each type of sewage on a fast-moving river.

CHALLENGE

1. A sewage treatment plant in Santee, CA, is a tertiary treatment plant. Outline the flow of water and solids through this plant. What is the final quality of the water with respect to BOD, nitrogen, and phosphorus?

2. Here are the formulas of two detergents that have been manufactured:

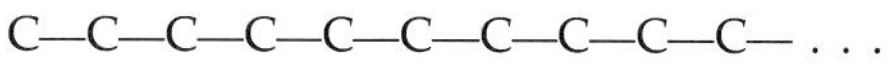

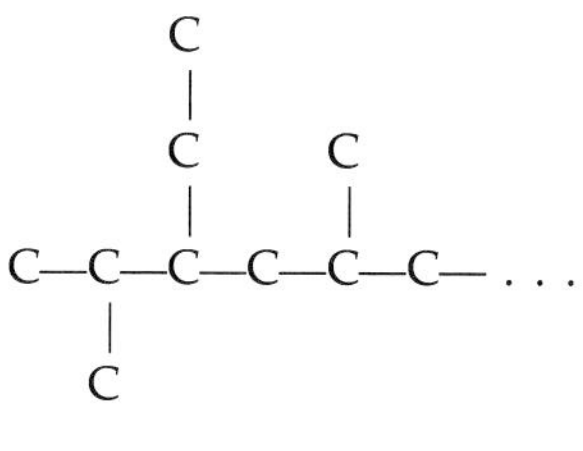

Which of these would be recalcitrant, and which would be readily degraded by microorganisms? (*Hint:* Refer to the degradation of fatty acids in Chapter 5.)

3. What is the MPN of a water sample with these multiple-tube fermentation test results: 10 ml portions: 4 positive for acid and gas; 1 ml portions: 3 positive for acid and gas; 0.1 ml portions: 1 positive. (See MPN table, Appendix D.)

4. Flooding after two weeks of heavy rainfall in Tooele, Utah, preceded the high rate of diarrheal illness shown below. *G. lamblia* was isolated from 26 percent of the patients. A comparison study of a town 65 miles away revealed that there was diarrheal illness in 2.9% of the 103 persons interviewed. Tooele has a municipal water system and a municipal sewage treatment plant. Explain the probable cause of this epidemic and method(s) of stopping it.

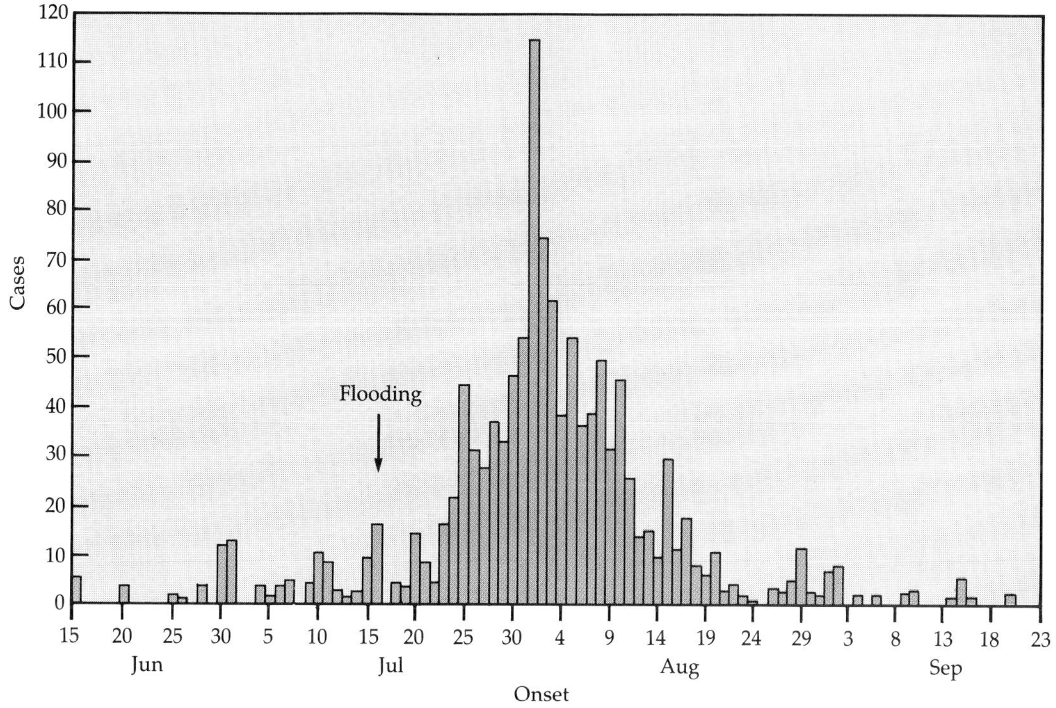

Distribution of cases of diarrheal illness.

FURTHER READING

Alexander, M. 1977. *Introduction to Soil Microbiology*. New York: Wiley. A text on microbial ecology and the role of microorganisms in the transformation of carbon, nitrogen, sulfur, phosphorus, and iron.

Brock, T. D. 1978. *Thermophilic Microorganisms and Life at High Temperatures*. New York: Springer-Verlag. The microorganisms, their metabolism, and growth requirements of the unique environment of the hot springs in Yellowstone National Park.

Cloud, P. September 1983. The biosphere. *Scientific American* 249:176–189. A discussion of how life has been shaped and sustained on earth, in soil, water, and the atmosphere.

Gutnick, D. L. and E. Rosenberg. 1977. Oil tankers and pollution: a microbiological problem. *Annual Review of Microbiology* 31:378–396. The role of microorganisms in the removal of petroleum from an ecosystem.

Imhoff, K., W. J. Muller, and D. K. B. Thistlewayte. 1971. *Disposal of Sewage and Other Water-Borne Wastes*. Ann Arbor, MI: Ann Arbor Science. An excellent reference on wastewater treatment plant designs, methods of disposal, and biological treatment of sewage.

Ourisson, G., P. Albrecht, and M. Rohmer. August 1984. The microbial origin of fossil fuels. *Scientific American* 251:44–51. An introduction to the new field of molecular paleontology.

Padwa, D. (convener). 1983. Benefits to agriculturists from the studies of microorganisms. *Developments in Industrial Microbiology* 24:19–67. Papers on technologic uses of nitrogen-fixing bacteria, mycorrhizae, yeasts, and agrobacteria.

Sieburth, J. M. 1979. *Sea Microbes*. New York: Oxford University Press. Microorganisms and their adaptation for living in ocean waters.

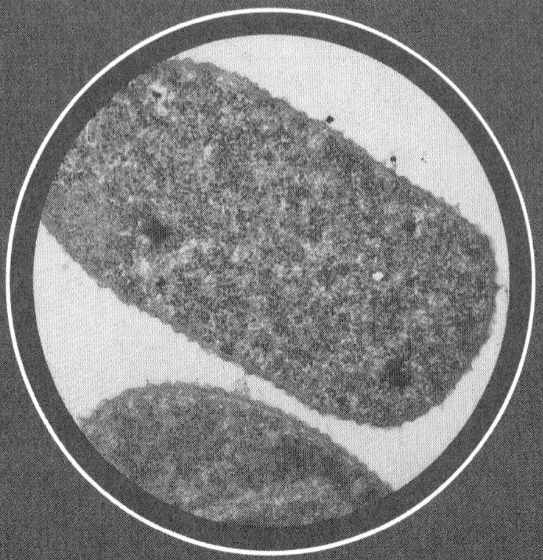

CHAPTER 26

Food and Industrial Microbiology

OBJECTIVES

After completing this chapter you should be able to

- Provide a brief history of the development of food preservation.
- Explain why sterilization of canned foods is important.
- Describe thermophilic anaerobic spoilage, flat sour spoilage, and spoilage by mesophilic bacteria.
- Describe low-temperature preservation and aseptic packaging.
- Define pasteurization and explain why dairy products are pasteurized.
- Provide examples of chemical food preservatives and explain why they are used.
- Outline at least four beneficial activities of microorganisms in food production.
- Describe the role of microorganisms in the production of single-cell protein (SCP), alternative energy sources, industrial chemicals, and chemotherapeutic agents.

We will now turn our attention from soil and aquatic microbiology to food and industrial microbiology. This chapter will discuss food spoilage and preservation, food-borne infections and food poisoning, and food production by microorganisms. We will also look at some important aspects of industrial microbiology and the impact of genetic engineering on industrial microbiology.

FOOD SPOILAGE AND PRESERVATION

Modern civilization and its large populations could not be supported were it not for effective methods of food preservation. In fact, civilization arose only after agriculture produced a year-round stable food supply in a single site, and people were able to give up the nomadic hunting cultures. Many of the methods of food preservation used today were probably discovered by chance in centuries past. Primitive people observed that dried meat or salted fish resisted decay. And the nomads no doubt observed that soured animal milk resisted further decomposition and was still palatable. Moreover, if the curd of the soured milk was pressed to remove moisture and allowed to ripen (in effect, cheesemaking), it was even more effectively preserved and could also develop desirable characteristics. And farmers learned that if grains were kept dry, they did not become moldy.

All such phenomena are readily understood by anyone familiar with the physical methods used to control the growth of microorganisms. Bacteria and fungi require a minimum amount of available moisture for growth. Drying food or adding salt or sugar lowers the available moisture and thus prevents spoilage. And the acidity created by the natural fermentation of milk or vegetable juices, as in sauerkraut, also prevents the growth of many spoilage bacteria. These methods of preservation are still in use today. However, a tour of any supermarket will demonstrate that heat sterilization, pasteurization, refrigeration, and freezing are now much more popular methods for the control of food spoilage. These modern methods preserve the food at much nearer its natural state and palatability than methods available to people in older times.

Preservation by heat (canning) requires considerable sophistication, and refrigeration of foods is an even more technical and therefore recent development. In old times in some parts of the world, ice was used for food preservation and could be kept for warm seasons in specially insulated structures, but it was difficult to transport and its use was only a curiosity in much of the world.

Preservation by heat originated in the early 1800s, before microorganisms were known to cause human disease and food spoilage. The technology of canning was promoted by military necessity of the day. When armies were small professional organizations, they could be supported off the land, with supplements of portable rations of dried grains, cheese, and wine. But with the rise of large armies of citizen soldiers, a development of the French revolution, this diet was no longer sufficient. Thus, the French government offered a prize to anyone who could devise a method of preserving food, particularly meat. The prize was won by a confectioner, Nicholas Appert, who around 1810 showed that food could be preserved if it was sealed in tightly stoppered containers and boiled for specified periods of time. The key to his method was detailed tables of times required for different foods and container sizes. As we know today, there are many endospore-forming bacteria that will survive hours of such boiling. Failures were manageably few, however, and usually attributed to faulty sealing.

The concept of heat preservation was followed quickly by the invention of the metal can (Figure 26–1). Temperatures above the boiling point of water were introduced into processing and were obtained by oil or salt baths. The use of pressurized steam in closed containers, as used today, was introduced only late in the nineteenth century, when reliable pressure controls and safety valves were developed. By then, of course, there was more of a theoretical basis to microbe control.

Actually, the preservation of foods by heating of a properly sealed container is not difficult. Commercially, the problem is determining and using the minimum amount of heat to kill spoilage organisms and dangerous microorganisms, such as the endospore-forming botulism bacterium. Because heating degrades the quality of food, much research was needed to determine the exact heat treatment that would succeed in sterilizing but minimize these quality changes.

Canned goods today undergo what is called **commercial sterilization** (Figure 26–2). The minimum preservation process is aimed at destroying the endospore-forming pathogen *Clostridium botulinum.* If this is accomplished, then any other significant spoilage or pathogenic bacterium will also be destroyed. To ensure complete sterilization, the 12D treatment is applied, by which a theoretical population of botulism bacteria would be decreased by 12 logarithmic cycles. In other words, if there were 10^{12} (1,000,000,000,000) botu-

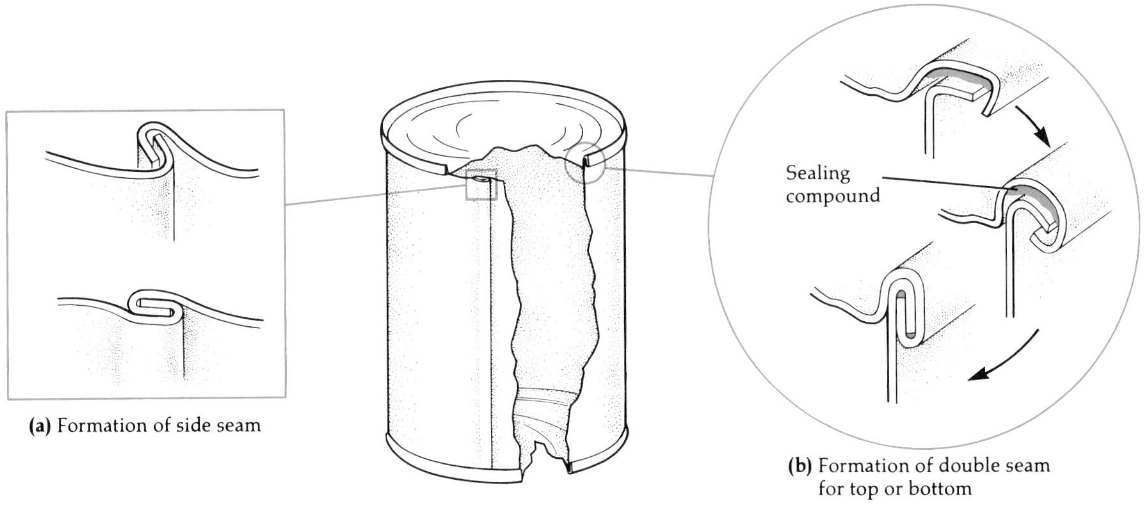

Figure 26–1 Sealing of a metal can. **(a)** Side seam. **(b)** Top and bottom seams.

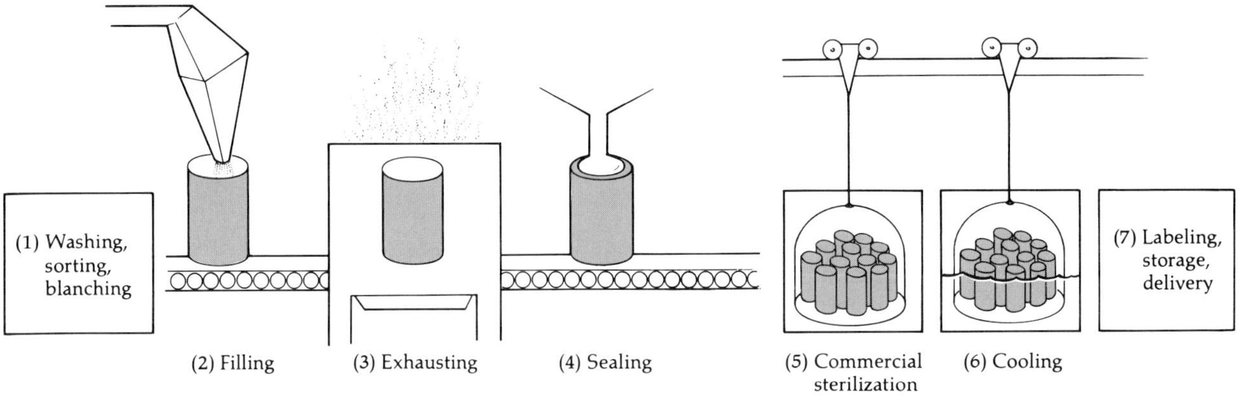

Figure 26–2 Industrial canning. **(1)** Blanching, a treatment with hot water or live steam, softens the product so that the can will fill properly. Blanching also destroys enzymes that might alter the flavor, texture, or color of the product, and lowers the microbial count. **(2)** After the cans are cleaned, they are filled with the product, as full as possible, so that there is no dead space. **(3)** In exhausting, air is removed so that corrosion and spoilage are reduced. **(4)** Sealing is accomplished by machine and is designed to make tight seals (see Figure 26–1). After being sealed, the cans are marked with code numbers indicating product, grade, and packing date in case of recall. **(5)** Commercial sterilization is done in a retort with steam under pressure and is designed to destroy *C. botulinum.* **(6)** After sterilization, the cans are cooled by water sprays or submergence. **(7)** The cans are then labeled, stored, and delivered.

lism organisms in a can, there would be only one survivor. Because 10^{12} is a very improbably large population, this treatment is considered quite safe. Certain thermophilic endospore-forming bacteria have endospores that are more resistant to heat treatment than are those of *C. botulinum*. However, these bacteria are obligate thermophiles and will not grow at normal room temperatures, but generally remain dormant at temperatures lower than about 45°C. Therefore, they are not a problem at normal storage temperatures.

SPOILAGE OF CANNED FOOD

If canned foods are incubated at high temperatures, such as exist in a truck in the hot sun or next to a steam radiator, the thermophilic anaerobic bacteria that often survive commercial sterilization (which is not as rigorous as true sterilization) can germinate and grow. This **thermophilic anaerobic spoilage** is a fairly common cause of spoilage in low-acid canned foods. The can usually swells from gas and the contents have a lowered pH and a sour odor (Figure 26–3). A number of thermophilic species of *Clostridium* can cause this type of spoilage. When thermophilic spoilage occurs but the can is not swollen by gas production, the spoilage is termed **flat sour spoilage.** This type of spoilage is caused by thermophilic organisms such as *Bacillus stearothermophilus* (ste-rō-thėr-mä´fil-us), which is found in the starch or sugars used in food preparation. Many industries have standards for the numbers of such thermophilic organisms permitted in raw materials. Both types of spoilage occur only when the cans are stored at higher than normal temperatures, which permits the growth of bacteria whose endospores are not destroyed by normal processing.

Mesophilic bacteria can spoil canned foods if the food is underprocessed or if the can leaks. Normally, the bacteria are killed by proper processing. Underprocessing is more likely to result in spoilage by endospore-formers; the presence of non-endospore-formers strongly suggests that the can leaks. Leaking cans are often contaminated during the cooling down of cans after processing by heat. The hot cans are sprayed with cooling water or drawn on wheeled carts through a trough filled with water. As the can cools, a vacuum is formed inside and external water can be sucked through a leak, past the heat-softened sealant in the crimped lid. Contaminating bacteria in the cooling water are drawn with the water into the can. Spoilage due to underprocessing or can leakage is likely to produce odors of putrefaction, at least in high-protein foods, and occurs at normal storage temperatures. In such types of spoilage, there is always the potential of botulism bacteria being present.

Figure 26–3 A can with severe swelling due to gas, which is produced in thermophilic anaerobic spoilage or mesophilic spoilage due to underprocessing or can leakage.

A summary of the types of canned food spoilage is presented in Table 26–1.

Home canning is an important use of heat as a preservative. Because of the possibility of botulism food poisoning resulting from improper canning methods, persons involved in home canning should obtain reliable directions and follow them exactly.

Some acidic foods, such as tomatoes or preserved fruits, are preserved by heats of 100°C or below. As a rule, the only important spoilage organisms in acidic foods are molds, yeasts, and occasional species of acid-tolerant, non-endospore-forming bacteria. These organisms are the only

Table 26–1 Types of Canned Food Spoilage

Type of Spoilage	Indications of Spoilage: Appearance of Can	Indications of Spoilage: Contents of Can
Low- and Medium-Acid Foods (pH above 4.5):		
Flat sour *(Bacillus stearothermophilus)*	Possible loss of vacuum on storage	Appearance not usually altered; pH markedly lowered; sour; may have slightly abnormal odor; sometimes cloudy liquid
Thermophilic anaerobic *(Clostridium thermosaccharolyticum)*	Can swells, may burst	Fermented, sour, cheesy, or butyric odor
Sulfide spoilage *(Desulfotomaculum nigrificans)*	Can flat	Usually blackened; "rotten egg" odor
Putrefactive anaerobic *(Cl. sporogenes)*	Can swells, may burst	May be partially digested; pH slightly above normal; typical putrid odor
Aerobic endospore-formers *(Bacillus* species)	Usually no swelling, except in cured meats when nitrate and sugar are present	Coagulated evaporated milk, black beets
High-Acid foods (pH below 4.5):		
Flat sour *(B. coagulans)*	Can flat, little change in vacuum	Slight pH change; off odor and flavor
Butyric anaerobic *(Cl. butyricum)*	Can swells, may burst	Fermented, butyric odor
Non-endospore-formers (mostly lactic acid bacteria)	Can swells, usually bursts, but swelling may be arrested	Acid odor
Yeasts	Can swells, may burst	Fermented; yeasty odor
Molds	Can flat	Surface growth; musty odor

Source: Data from the National Canners Association, 1950 6th Street, Berkeley, CA 94710.

ones capable of growing at the pH of these foods and are easily killed by temperatures less than 100°C. One problem with such acidic foods can be the *sclerotia* of certain species of molds. Sclerotia are specialized resistant bodies that can survive temperatures of 80°C for a few minutes.

ASEPTIC PACKAGING

A recent development in food preservation is the increasing use of **aseptic packaging.** The package is usually made of some material such as laminated paper or plastic that cannot tolerate conventional heat treatment. The packaging materials come in continuous rolls that are fed into a machine that sterilizes the material with a hot hydrogen peroxide solution, sometimes aided by ultraviolet light that enhances its activity. If the containers are metal, sterilization can be done with superheated steam or other high-temperature methods. Still in the sterile environment, the material is then formed into packages, which are filled with liquid foods that have been conventionally sterilized by heat. The filled package is not sterilized after it is sealed.

LOW-TEMPERATURE PRESERVATION

Low temperatures increase the reproduction time of microorganisms. Even so, some molds and bacteria can grow at a significant rate at temperatures below the freezing point of water, 0°C. Food typically does not freeze solid at temperatures several degrees below this point. Growth in foods a few degrees below freezing is common, and growth at even −18°C has been reported. A properly set refrigerator maintains a temperature of 0° to 7°C. Many microorganisms will grow slowly at these temperatures and will alter the taste and appearance of foods stored for too long a time. Pathogenic bacteria, with a few exceptions, will not grow at these temperatures. Among the exceptions are the clostridia that cause type E botulism and the intestinal pathogen, *Yersinia enterocolitica* (yėr-sin´ē-ä en-tėr-ō-kōl-it´ik-ä). Freezing will not immediately kill significant numbers of bacteria, but frozen bacterial populations are dormant and decline slowly with time. Some parasites such as the roundworms that cause trichinosis are killed by several days' freezing. Some important temperatures associated with microorganisms and food spoilage are shown in Figure 26–4.

Anyone responsible for preparing food in large amounts should be aware of the spoilage that can arise from the slow cooling of hot foods in large containers. Hot food placed in a refrigerator cools more slowly than most persons realize, and can spend a considerable time at the incubation temperatures of pathogenic or spoilage bacteria before the temperature of the food drops to that of the refrigerator (Figure 26–5).

RADIATION AND FOOD PRESERVATION

There has been considerable research, especially for military applications, into the use of ionizing radiation for food preservation. It is possible to sterilize food by radiation, but it is unlikely that this procedure will replace conventional heat sterilization for most purposes. Taste and appearance of irradiated food has been much improved in recent years. This improvement has been accomplished mainly by freezing the food in liquid nitrogen and exhausting oxygen from the package before it is irradiated. However, a major factor affecting the adoption of this type of preservation is that the public would probably be apprehensive of foods exposed to radiation, even though there is no evidence that harmful products result from the amount and types of radiation used.

Pasteurization-type radiation treatments, short of total sterilization, might prove commercially practical. Especially promising is exposure to low-energy radiation that kills insects in stored grains; this technique might replace gas fumigation. Irradiation treatment for spices has recently been approved by the FDA. Cobalt 60 is the usual source of gamma rays for sterilization. High-energy electron accelerators are much faster, but have limited penetrating power.

Microwaves, such as used for cooking, kill bacteria only by heating foods, and have little or no direct killing effect on bacteria. Bacteria can survive

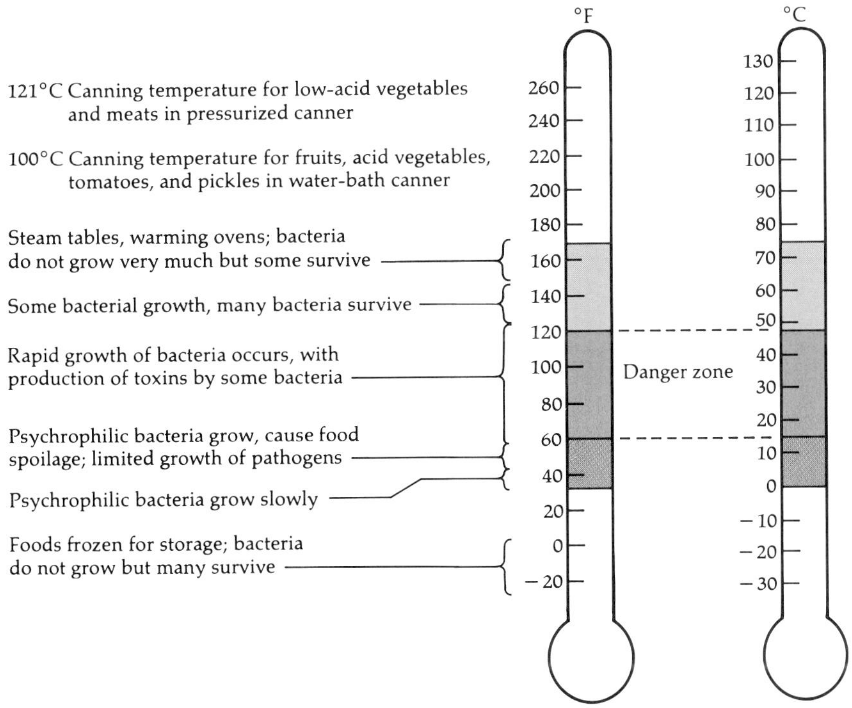

Figure 26–4 Some temperatures important to microorganisms that cause food spoilage.

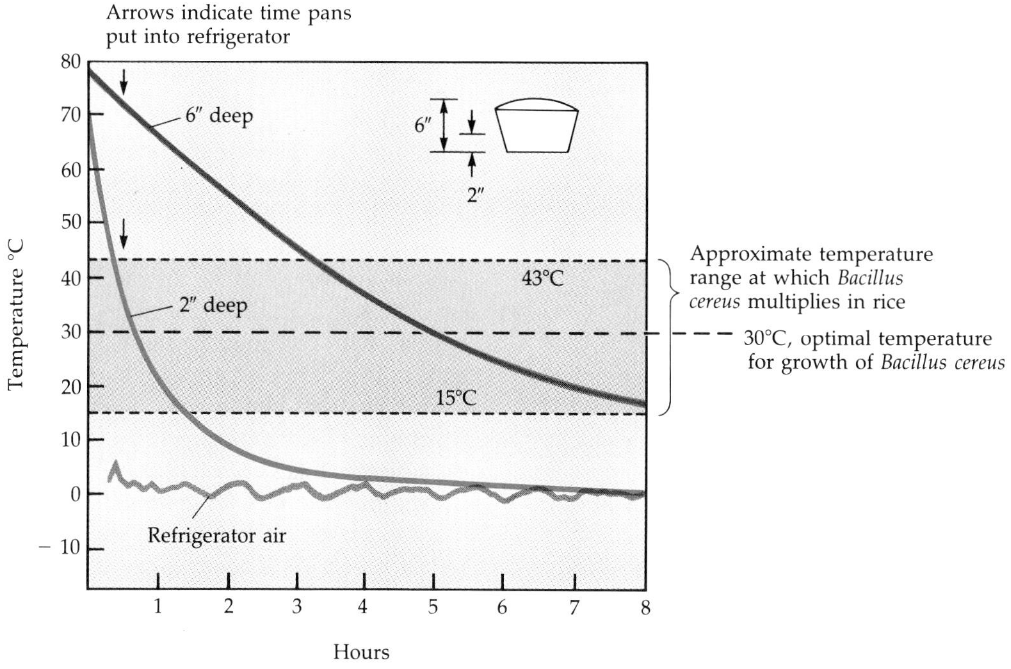

Figure 26–5 The effect of food's mass on its cooling rate. The boiled rice was placed in aluminum pans to a depth of six inches or two inches and then placed in a walk-in refrigerator. Note that the two-inch-deep rice cooled out of the incubation range of *Bacillus cereus* in about an hour, and that the six-inch-deep rice remained at the incubation temperature of this poisoning organism for more than five hours.

(and are easily isolated) on the interior walls of microwave ovens.

PASTEURIZATION

Louis Pasteur began his career as a microbiologist when the French brewing industry commissioned him in 1871 to investigate the causes of spoilage in beer and wine. He concluded that spoilage was due to the growth of microorganisms, and he worked out a method of eliminating them with mild heating. Too much heat would alter the beverage's characteristics to an unacceptable degree, but killing all the microorganisms was found to be unnecessary. The principle of the selective killing of microorganisms, now called **pasteurization,** was later applied to other foods, milk in particular (see Chapter 7). Other products, such as ice cream, yogurt, and beer, all have individual pasteurization times and temperatures, which often differ considerably.

CHEMICAL PRESERVATIVES

Chemical preservatives are frequently added to foods to retard spoilage. Among the more common additives are sodium benzoate, sorbic acid, and calcium propionate. These chemicals are simple organic acids, or salts of organic acids, which the body readily metabolizes and which are generally judged to be safe in foods. Sorbic acid, or its more soluble salt potassium sorbate, and sodium benzoate prevent mold from growing in certain acidic foods such as cheese and soft drinks. Such foods, usually with a pH of 5.5 or less, are most susceptible to mold-type spoilage. Calcium propionate is an effective fungistat used in bread. It prevents the growth of surface molds and the *Bacillus* species of bacteria that causes ropy bread. These organic acids inhibit mold growth not by affecting the pH, but by interfering with mold's metabolism or the integrity of the cell membrane.

Sodium nitrate and sodium nitrite are found in

many meat products, such as ham, bacon, wieners, and sausage. The active ingredient is sodium nitrite, which certain bacteria in the meats can also produce from sodium nitrate. These bacteria use nitrate as a substitute for oxygen under anaerobic conditions, much as it is used in denitrification in soil. The nitrite has two main functions: to preserve the pleasing red color of the meat by reacting with blood components in the meat, and to prevent the germination and growth of any botulism endospores that might be present. There has been some concern that the reaction of nitrites with amino acids can form certain carcinogenic products known as **nitrosamines,** and the amount of nitrites added to foods has generally been reduced lately for this reason. However, nitrites have not been eliminated, partly because of their established value in preventing botulism, and because nitrosamines are formed in the body from other sources, so the added risk posed by limited use of nitrates and nitrites in meats might be lower than was once thought.

FOOD-BORNE INFECTIONS AND MICROBIAL INTOXICATIONS

We have seen in Chapter 23 that illness resulting from microbial growth in food is associated with two principal mechanisms. In the mechanism known as food-borne infection, the contaminating microorganism infects the person who ingests contaminated food. As the pathogen grows in this host, it produces damaging toxins. Diseases caused by this mechanism include gastroenteritis, typhoid fever, and dysentery. In the other mechanism, called microbial intoxication, the toxin is formed in the food by microbial growth and then ingested with the food. Diseases associated with this mechanism include botulism, staphylococcal food poisoning, and mycotoxicoses (intoxications caused by fungal toxins such as ergot or aflatoxin). Several microbial intoxication diseases are described in Table 26–2.

Dairy products, which are often consumed without having been cooked, are particularly likely

Table 26–2 Food-Borne Microbial Intoxications

Disease	Foods Involved	Prevention	Clinical Features	Duration
Aflatoxin	Moldy grains	Avoid contaminated grains	Low doses may induce liver cancer; high doses cause general liver damage; no human cases	
Bacillus cereus intoxication	Custard, cereal, starchy foods	Refrigeration of foods	Cramps, diarrhea, nausea, vomiting	Onset: 8–16 hours Duration: Less than 1 day
Botulism	Canned foods	Proper canning procedures; boiling food prior to consumption	Nausea, vomiting, headache, vertigo, respiratory paralysis	Onset: 2 hours–6 days Duration: Weeks
Ergotism	Moldy grains	Avoid contaminated grains	Burning abdominal pain, hallucinations	Onset: 1–2 hours Duration: Months
Methyl mercury poisoning	Freshwater or ocean fish	Stop dumping mercury into waters	Blurred vision, numbness, apathy, coma	Onset: 1 week Duration: May be chronic
Mushroom poisoning	*Amanita* species	Don't eat poisonous mushrooms	Vomiting, hepatic necrosis, neurotoxic effects	Onset: Less than 1 day Duration: Less than 10 days
Paralytic shellfish poisoning	Bivalve mollusks during red tide (dinoflagellate blooms)	Avoid eating mollusks during red tide	Tingling, rash, fever, respiratory paralysis	Onset: Less than 1 hour Duration: Less than 12 hours
Scombroid poisoning	Histaminelike substance produced by *Proteus* growing on ocean fish	Refrigeration of fish	Headache, cramps, hives, shock (rare)	Onset: Minutes–hour
Staphylococcal intoxication	High osmotic pressure foods not cooked before eating	Refrigeration of foods	Nausea, vomiting, diarrhea	Onset: Less than 1 day Duration: Less than 3 days

to transmit food-related diseases. Standards for sanitation in the dairy industry are therefore very stringent. Because most dairy milk is drawn mechanically and promptly put into cooled holding tanks, most of the bacteria in milk today are gram-negative psychrophiles. Grade A pasteurized cultured products, such as buttermilk and cultured sour cream, have high counts of lactic acid bacteria as a natural result of their method of manufacture. Therefore, the standards require only that they contain fewer than 10 coliforms per milliliter. Pasteurized Grade A milk is required to have a standard plate count of fewer than 20,000 bacteria and not more than 10 coliforms per milliliter. Grade A dry milk products are required to have a standard plate count of fewer than 30,000 bacteria per gram or a coliform count of fewer than 90 per gram.

Some people prefer raw milk because of the lack of additives or because they feel it has not been altered chemically. Pasteurization can denature enzymes found in milk. Unpasteurized (raw) dairy products are available commercially in about half the states in the country. Three states have boards that certify raw milk from dairies meeting the required standards for cleanliness; the dairy herds are also inspected for disease. The standard for certified raw milk is a maximum bacterial count of 10,000 per milliliter and 10 coliforms per milliliter. The microbial standards guaranteed by certification do not prevent transmission of disease in raw milk; in fact, certified raw milk regularly causes outbreaks of salmonellosis.

Figure 26–6 In this Swiss plant, curd is drawn from whey in a porous cheesecloth.

ROLE OF MICROORGANISMS IN FOOD PRODUCTION

Cheese

Leading the world in the manufacture of cheese, the United States produces more than 1.5 million tons each year. Although there are many types of cheeses, all require the formation of a **curd,** which can then be separated from the main liquid fraction, or **whey** (Figure 26–6). The curd is made up of a protein, **casein,** and is usually formed by the action of an enzyme, **rennin,** which is aided by acidic conditions provided by certain lactic acid-producing bacteria. These inoculated lactic acid bacteria also provide the characteristic flavors and aromas of fermented dairy products. Except for a few unripened cheeses, such as Ricotta or cottage cheese, the curd undergoes a microbial ripening process.

Cheeses are generally classified by their hardness, which is produced in the ripening process. The more moisture lost from the curd and the harder it is compressed, the harder the cheese. Romano and Parmesan cheeses, for example, are classified as very hard cheeses; Cheddar and Swiss are hard cheeses. Limburger, blue, or Roquefort cheeses are classified as semisoft; Camembert is an example of a soft cheese. The hard Cheddar and Swiss cheeses are ripened by lactic acid bacteria growing anaerobically in the interior. A *Propionibacterium* (prō-pē-on-ē-bak-ti´rē-um) species of bacteria in Swiss cheese produces carbon dioxide that forms the holes. The longer the incubation time, the higher the acidity and the sharper the taste of the cheese. Such hard interior-ripened cheeses can be quite large. Semisoft cheeses, such as Limburger, are ripened by bacteria and other contaminating organisms growing on the surface. Blue and Roquefort cheeses are ripened by *Penicillium* (pe-ni-sil´lē-um) molds inoculated into the cheese. The texture of the cheese is loose enough that adequate oxygen can reach the aerobic molds. The growth of the *Penicillium* molds is visible as blue-green clumps in the cheese. Camembert cheese is ripened in small packets so that the enzymes of *Penicillium* mold growing aerobically on the surface will diffuse into the cheese for ripening.

Other Dairy Products

Butter is made by the churning of cream until the fatty globules of butter separate from the liquid buttermilk fraction. Lactic acid bacteria produce the *diacetyls* that give butter and buttermilk their typical flavor and aroma, so these bacteria are allowed to grow until the desired acidity is reached. Today, buttermilk is not usually a by-product of butter making, but is skim milk that has been inoculated with bacteria that form lactic acid and the diacetyls. The inoculum is allowed to grow for 12 or more hours before the buttermilk is cooled and packaged. Sour cream is made from cream inoculated with organisms similar to those used to make buttermilk.

A wide variety of slightly acidic dairy products, probably a heritage of a nomadic past, are found around the world. Many of them are part of the daily diet in the Balkans, Eastern Europe, and Russia. One such product is yogurt, which is also popular in the United States. Commercial yogurt is made from low-fat milk from which much of the water has been evaporated in a vacuum pan. (The resulting milk is inoculated with a species of lactic-acid-producing *Streptococcus* (strep-tō-kok´kus) that grows at elevated temperatures.) Incubation at about 45°C for several hours gives the acidity to the product, and added stabilizers aid in the formation of the thick texture. A second bacterial inoculum provides the characteristic flavor of the yogurt. Maintaining the proper balance between the flavor-producing and the acid-producing organisms is the secret of good yogurt. Kefir and kumiss are popular beverages in Eastern Europe. The usual lactic acid-producing bacteria are supplemented with a lactose-fermenting yeast to give these drinks an alcoholic content of 1 or 2%.

Nondairy Fermentations

Microorganisms are also used in baking. The sugars in bread dough are fermented to alcohol and carbon dioxide by yeasts, just as is done in the fermentation of alcoholic beverages (discussed in the next section). The carbon dioxide makes the typical bubbles of leavened bread, and the alcohol evaporates during baking. In some breads such as rye or sourdough, the growth of lactic acid bacteria gives the typical tart flavoring.

Fermentation is also used in the production of foods such as sauerkraut, pickles, and olives. In Asia, extremely large amounts of soy sauce are produced with molds that form starch-degrading enzymes to produce fermentable sugars. This principle is used in making other Asian fermented foods, including sake, the Japanese rice wine. In soy sauce production, molds such as *Aspergillus oryzae* (a-spėr-jil´lus ô-rī´zē) are grown on wheat bran and then allowed to act, along with lactic acid-type bacteria, on cooked soybean and crushed wheat mixtures. After this process has produced fermentable carbohydrates, a prolonged fermentation results in soy sauce.

Alcoholic Beverages

Microorganisms are involved in the production of almost all alcoholic beverages. Beer and ale are grain liquids fermented by yeast, as are a number of other alcoholic beverages (Table 26–3). Because yeasts are unable to use starch directly, the starch from grain must be converted to glucose and maltose, which the yeasts can ferment into ethanol and carbon dioxide. In this conversion, called **malting,** starch-containing grains such as malting barley are allowed to sprout, then dried and ground. This product, called **malt,** contains starch-degrading enzymes used in the conversion of cereal starches into fermentable carbohydrates that can be utilized by yeasts. For distilled spirits such as whiskey, vodka, or rum, carbohydrates from such sources as cereal grains, potatoes, and molasses are fermented to alcohol. The alcohol is then distilled to make a concentrated alcoholic beverage.

Wines are typically made from grapes, which contain sugars that can be used directly by yeasts for fermentation; the malting needed to make a substrate for yeasts in beers and whiskies is unnecessary for wine (Figure 26–7). Grapes usually need no additional sugars, but other fruits might be supplemented with sugars to ensure enough alcohol production.

Lactic acid bacteria are also important in the making of wine from grapes that are especially acidic from malic acid. These bacteria convert the malic acid to the less acidic lactic acid in a process called **malolactic fermentation.** The result is a less acidic, better-tasting wine than would otherwise be produced.

Winemakers who allowed wine to be exposed to air found that it soured from acetic-acid-forming bacteria growing aerobically on the alcohol. The result was bad wine, or vinegar. This process is now used deliberately to make vinegar. Ethanol is first made by anaerobic yeast fermentation. The ethanol is then aerobically oxidized to acetic acid by acetic-acid-producing bacteria such as *Acetobacter* (ä-sē-tō-bak´ tėr) and *Gluconobacter* (glū-kō-nō-bak′ tėr). The basic process of vinegar production is shown in Figure 26–8.

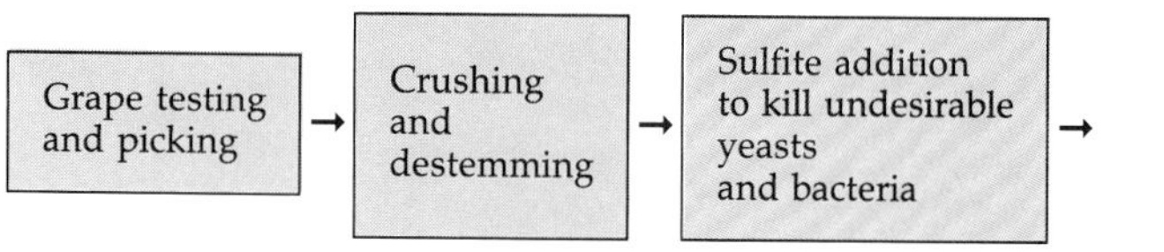

FOOD FROM MICROORGANISMS

In the future, as human populations expand and arable land does not, microorgansims may become more important to the food requirements of the world. Protein is in particularly short supply. Microorganisms, which can double their weight in a few hours, and often in less than an hour, may help ease the problem. Used as a food source, microorganisms are referred to as **single-cell protein (SCP)** (see box). Microorganisms can use as sub-

Table 26–3 Production of Alcoholic Beverages by Yeasts

Beverage	Yeast	Method of Preparation	Function of Yeast
Beer	*Saccharomyces cerevisiae* or *S. carlsbergensis*	Barley malt and starch adjuncts mixed with warm water; after enzymatic starch conversion, wort is filtered, then boiled with hops, and finally fermented with yeast	Converts sugar into alcohol and carbon dioxide; produces changes in proteins and other minor constituents that modify flavor
Rum	*S. cerevisiae* or other yeasts	Blackstrap molasses containing 12 to 14% fermentable sugar; ammonium sulfate and occasionally phosphates may be added as nutrients; distilled after fermentation	Sugar converted to alcohol, which is then removed by distillation
Scotch	*S. cerevisiae* (generally a top yeast)	Grain mash cooked with peated malt and fermented; distillate aged in oak casks at least three years; then blended with grain whiskey	Produces alcohol and congeneric substances (acids, esters, various alcohols), which, with the peated malt, give characteristic Scotch flavor
Bourbon	*S. cerevisiae*	Grain mash consisting of corn (at least 51%); generally with rye cooked with malt and fermented; matured in charred oak barrels	Same as for Scotch whiskey, but flavor is characteristic of bourbon whiskey
Wine	*S. ellipsoideus*, various strains	Grapes with sugar concentration up to 22%; allowed to ferment with special strain of yeast, or with yeast naturally present on the grape; primary fermentation succeeded by a period of storage for maturation	Converts sugar into alcohol and also produces changes in minor constituents that modify flavor and bouquet; amount of alcohol varies according to type of wine

Source: W. S. Spector (ed), *Handbook of Biological Data*, Saunders, Philadelphia, PA, 1956.

Fermentation → Pressing to separate solids from wine → Clarification and aging → Filtration → Bottling

Figure 26–7 Basic steps employed in making red wine. The crushed and destemmed grapes are called must. For red wines, the alcohol formed during fermentation helps extract the red color from the grape skins. The solids are then separated from the liquid wine by a pressing of the fermenting must, usually through cloth. For white wines, the pressing precedes fermentation so that the color is not extracted from the solid matter.

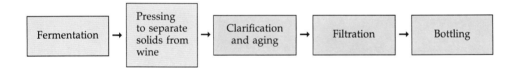

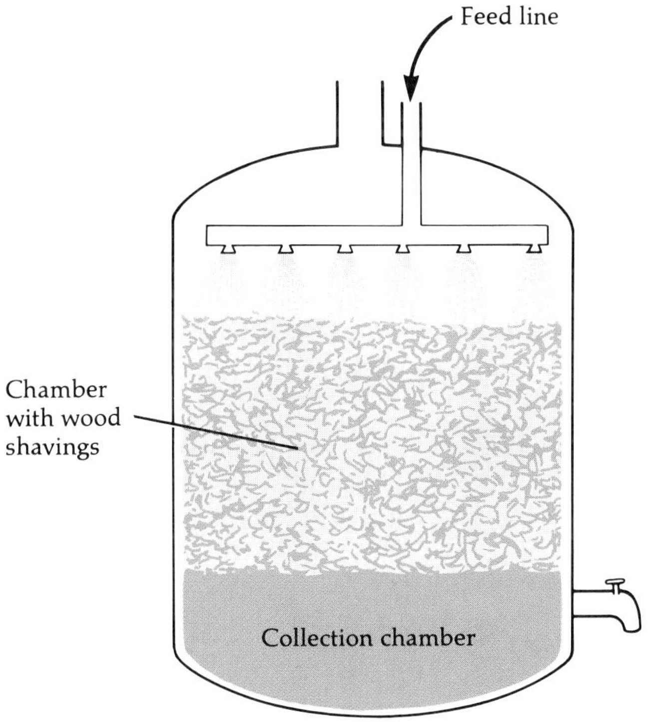

Figure 26–8 Vinegar production. In this vinegar generator, ethanol trickles through wood shavings that are coated with *Acetobacter* or *Gluconobacter*. The bacteria aerobically oxidize the ethanol to acetic acid, which is removed periodically.

Microbiology in the News

Single-Cell Protein May Help Ease World Food Problem

Sheaves of plump bacteria? Cornucopias brimming over with yeast? These may never become popular symbols of plenty, but in the not too distant future bacteria and yeast may make an important new contribution to the world food supply. Bacteria have been extensively investigated as a source of single-cell protein, or SCP, because almost any organic substance can serve as a substrate for some bacterial species. Some microorganisms may even be able to grow on agricultural or industrial byproducts that would otherwise go to waste, turning them into usable protein.

Growing Bacteria for Protein

SCP producers grow bacteria in airlift fermenters, tanks in which the microorganisms and their liquid substrate are continuously stirred by injected air. The diagram shows the production process in simplified form; other airlift fermenters operate in similar fashion. Sterilized air, ammonia (the nitrogen source), and methanol enter the fermenter through pipes; carbon dioxide produced by microbial metabolism is vented from the top of the tank. A jacket of water surrounds the tank, to carry away the heat generated by fermentation. The fermenting mixture is continuously removed from the bottom of the tank and steam treated to kill and coagulate the bacteria. The bacteria are then separated in a centrifuge and dried, forming granules.

In the early 1980s, an English chemical company began full-scale production of Pru-

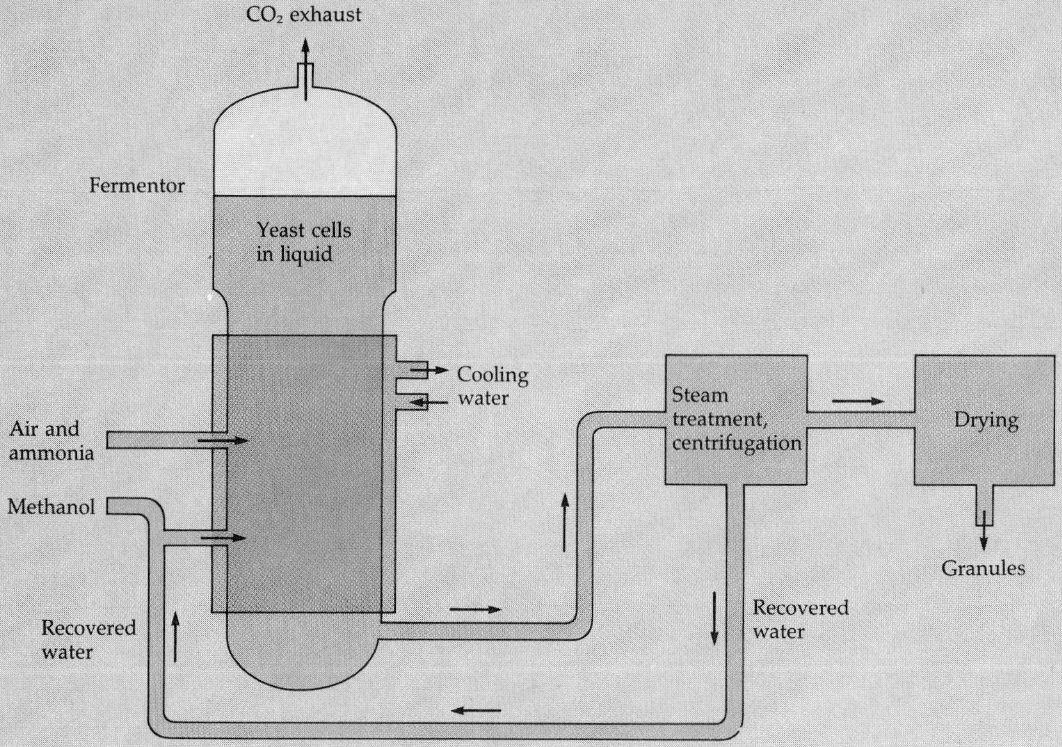

teen, an SCP livestock feed made from bacteria grown on methanol. The Pruteen fermenter, the largest in the world, had a capacity of 1000 cubic meters, or about 260,000 gallons. In 1981, it produced 50,000 tons of high-quality feed.

SCP from Yeast

Some American companies are producing SCP from yeast. An advantage of yeast is that several strains have long been recognized as edible, so they do not require the extensive testing for toxicity that other microorganisms do. The original methods for harvesting yeast on a large scale were similar to those used for bacteria, but biotechnology companies have been developing more productive yeast strains and advanced methods of production that make the process much more efficient. An SCP powder called Provesteen, now on the market, is produced from a yeast that grows at densities similar to those achieved by centrifugation with ordinary strains. This means not only that centrifugation can be omitted, but also that there is no excess water to be recycled; the yeast goes right from the fermenter to a pasteurizer and a drier. Provesteen is 60 percent protein. It is deficient in the essential amino acid methionine, but methionine can be added cheaply to animal feed. When tested on pigs, methionine-supplemented Provesteen proved a satisfactory substitute for soybean meal, the major protein source for young commercially raised pigs.

Genetic engineering may further enhance the productivity of yeast. Harvesting SCP from the growth medium would be more efficient than grinding up batches of yeast cells, and it soon may be possible to do this. Yeast cells secrete some enzymes into the medium, and gene fusion can be used to attach SCP to these enzymes inside the cell. In addition, enzyme secretion can be increased by inserting multiple copies of the appropriate genes. The engineered yeast would secrete large quantities of SCP molecules attached to enzymes, and these could be isolated from the medium by means of monoclonal antibodies. Genetic engineering is also being used to develop a yeast that can utilize liquid starch as a substrate, eliminating the need to convert starch into sugar.

Overcoming Economic Barriers

The question now is whether SCP can succeed economically. Several years after its auspicious beginning, the Pruteen plant stopped production because Pruteen could not compete in price with soybean meal and fish meal. It remains to be seen whether other SCP products can compete in the marketplace. The price of SCP is affected by the cost of the substrate and the cost of converting the substrate to usable protein. Technological improvements and genetic engineering may reduce the cost of fermentation and harvesting, and in some cases cogeneration of products may provide a substrate at minimal cost. One company, for example, plans to produce both ethanol and SCP from wood chips. The cellulose would be metabolized to ethanol by yeast, which would produce proteins in the process. The manufacturers of Pruteen conducted an extensive search for a bacterium that would thrive on methanol; perhaps other microorganisms will be found or created that can utilize various waste products. An enormous amount of research has gone into the biological and technical problems of producing SCP. If the economic problems can be overcome as well, microbial protein may indeed increase the human food supply.

Figure 26–9 Gasohol can be dispensed in the same way as gasoline.

strates materials that are otherwise considered waste, such as cellulose, methanol, or petroleum hydrocarbons. Petroleum yields the most food for its weight, but because it is needed as a fuel, it is becoming expensive. Wastes from meat packing, paper production, agriculture, and forestry might also serve. Cellulose products from wood or other plants are the most obvious candidates, but relatively few microorganisms can use cellulose, or the lignin often associated with it. Considerable effort is being directed at technology to economically convert cellulose into products that can be readily metabolized, such as glucose. Among all the microorganisms from which single-cell protein can be made, bacteria are economically the most attractive. They have shorter generation times compared to yeasts and molds, but even more importantly, their protein content is as high as 80%, as compared to 45% in soybeans and 50% in yeasts. Photosynthetic organisms are of special interest because of their inexhaustible energy source. However, algae and photosynthetic bacteria tend to make relatively unpalatable end products. An exception is the alga *Spirulina* (spī-rü-lī′na), which for centuries has been used in dried form from a lake in Chad, Africa. *Spirulina* was also eaten by ancient Aztecs in Mexico. Experiments on eating *Spirulina* have been conducted in the United States, France, and Mexico, and as a result, the United States and Mexico permit the sale of *Spirulina*.

Yeasts such as *Saccharomyces cerevisiae* or *Candida utilis* (Torula) can be grown in large quantities, but they lead to adverse gastrointestinal reactions in humans. Also, their high nucleic acid content exceeds the human body's capacity to metabolize them and may precipitate ailments such as gout. Yeasts are therefore most likely to be used as an animal feed supplement.

In England the bacterium *Methylophilus methylotrophus* (meth-i-lō-fi′lus meth-i-lō-trōf′us) is being used to produce protein supplements for animal feeds. As much as half a gram of cells, which contain about 70% crude protein, can be produced from a gram of methanol. Methanol, which might seem to be an unlikely substrate, has the advantage of being free of impurities, soluble in water, and readily separated from the end product.

ALTERNATIVE ENERGY SOURCES FROM MICROORGANISMS

Each year, the United States produces many hundreds of metric tons of organic waste from crops, forests, and municipalities. This waste is called **biomass,** and so far it has been disposed of by burning or placing in a dump. Conversion of a biomass into alternative fuel sources, called **bioconversion,** would help meet our energy needs as well as help dispose of troublesome accumulations of solid waste. The bioconversion of organic materials into alternative fuels by microorganisms is a developing industry.

Probably one of the most convenient energy sources produced from biomass conversion is methane. Methane was discussed in Chapter 25 as a product of the anaerobic treatment of sewage sludge. New York City recently opened a plant that will produce 10 million cubic feet of methane per day from wastes at one of its landfill sites. Large cattle-feeding lots must dispose of immense amounts of animal manure and considerable effort has been expended in devising practical methods for producing methane from these wastes. An unexpected problem in using animal wastes has been inhibition of methanogens by antibiotics that have been fed to the animals specifically to prevent methane formation in their rumen. In the animal, fatty acids are a desired fermentation product because they are readily metabolized into meat and milk. By comparison, methane is a wasteful by-product because much of it is usually lost from the animal as gas. A major problem with any scheme for large-scale methane production is the need to economically concentrate the widespread biomass material. If it could be concentrated, the animal and human wastes in the United States could supply much of our energy now supplied by fossil fuels and natural gas.

The agricultural industry has encouraged the production of ethanol from agricultural products. This *gasohol* (90% gasoline + 10% ethanol) is available in many parts of the United States and is used in automobiles worldwide (See Figure 26–9). Corn is presently the most frequently used substrate, but eventually agricultural waste products may be used. A by-product of ethanol production is large quantities of ethanol-producing yeasts. The Food and Drug Administration recently approved use of such yeasts as a food additive, although they are most likely to be used as an animal-feed supplement.

In coming years we will see more and more efforts to recover the energy now lost in industrial, municipal, and rural wastes.

GENETIC ENGINEERING AND INDUSTRIAL MICROBIOLOGY

In the coming years, there will no doubt be a revolution in the application of microorganisms in industry—what is known as **biotechnology.**

As we have seen, the use of microbes in making foods has a history almost as long as that of mankind. Their application in brewing, baking, and cheesemaking have origins lost in antiquity. In the latter part of the nineteenth century, such microbes were grown in pure culture. This step quickly led to improved understanding of the relationships between specific microbes and certain microbial activities. For example, once it was understood that a certain yeast grown under certain conditions produced beer, and that certain bacteria could spoil the beer, the age of industrial microbiology began in earnest. Industry became active in microbiological research once microbes began to be selected for special qualities; for example, the brewing industry extensively investigated the isolation and identification of yeasts able to produce more alcohol. Microbes also proved, under the stress of wartime conditions during World War I, to be useful sources of chemical compounds such as glycerol and acetone. Modern industrial microbiology came into being after World War II, with the rise of the pharmaceutical industry to produce antibiotics. To increase the number of random mutations, antibiotic-producing organisms were subjected to ultraviolet radiation in the hope that some mutations would be useful. These and other techniques soon led to strains that produce antibiotics thousands of times faster than the original isolates did.

Genetic Engineering

Tremendous advances in the understanding of molecular biology have produced techniques that now replace the previous, relatively hit-or-miss methods of mutation and selection. The application of these new techniques is known by the now-familiar term **genetic engineering.** As we saw in Chapter 8, the basic technique of genetic engineering is the recombination of DNA molecules. In this technique, DNA is enzymatically cut up and the resulting fragments (hopefully containing useful genetic characteristics) are spliced into other strands of DNA. These strands of DNA, which then contain new genetic information, are part of a vector molecule, such as a plasmid (usually) or a bacterial virus, that can replicate the DNA now containing the new genetic information. It is then necessary to identify those pieces of recombinant DNA that contain useful genes. Once the recombinant DNA has been identified as useful and has been inserted into a plasmid, it can be transferred to a host cell with any of various methods. Once in the cell, the plasmid is capable of replicating independent of the chromosome. Bacteria normally carry only a few plasmids, but by a pro-

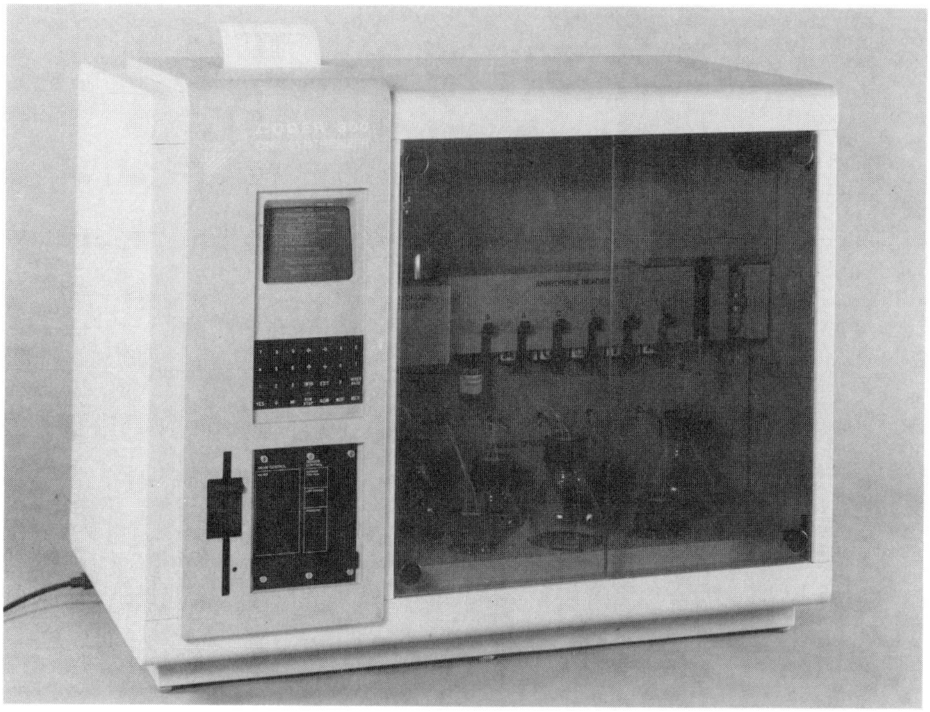

Figure 26–10 Instruments that synthesize short sequences of DNA. The desired sequences of nucleotide bases are entered on the panel to the left. The chemical components necessary are placed in the flasks inside the machine.

cess of *gene amplification* the plasmids can be encouraged to replicate, and a single cell can be made to contain hundreds or even a few thousand plasmids.

Eucaryotic cells such as yeasts also carry plasmids, and can be more useful in some ways than bacteria. For example, eucaryotic cells are more likely to continuously excrete the desired product. A bacterium such as *Escherichia coli* usually must be disrupted to recover the product. The ability of an organism to continuously excrete the product is a very valuable industrial characteristic.

There are available special techniques to produce the DNA segments to be used in genetic engineering. DNA sequences containing a particular gene can be synthesized, for example, with the messenger RNA isolated from the cell of a microorganism or animal. From the messenger RNA, an enzyme, reverse transcriptase, can synthesize a complementary section of DNA that contains the gene. Short sequences of DNA can even be synthesized by machine (Figure 26–10). A keyboard on the machine is used to enter the sequence of bases desired, much as one enters the letters to compose a sentence. A microprocessor controls the synthesis of the DNA from stored supplies of nucleotides and the necessary other reagents. A chain of about 40 nucleotides can by synthesized by this method.

Protoplast fusion is a technique that increases the rate of recombination; it can be used to recombine DNA of bacterial species that normally do not exchange genetic information. The cell walls are enzymatically removed, and spherical *protoplasts* form. Two protoplasts are caused to fuse together and then synthesize the normal cell wall; a hybrid organism with new combinations of genes is the result. This technique is often used on antibiotic-producing streptomycetes to increase the frequency of isolation of useful mutants. Another important use is the transfer of genetic information between plant species (Figure 26–11).

A different approach constructs new antibiotics by isolating a mutant microorganism that synthesizes only a portion of the complete antibiotic. The incomplete antibiotic lacks an essential chemical constituent. Adding the missing chemical constituent to the medium allows the mutant to form the normal antibiotic. However, adding substitute chemicals can result in the synthesis of new, more useful antibiotics.

This is only an outline of some techniques used in an increasingly complex area of microbiology. These methods are beginning to see wide use in industrial microbiology.

Industrial Microbiology

Microbial products have traditionally been termed either primary or secondary metabolites. Primary metabolites, such as amino acids, vitamins, and citric acid, are essential for the growth of the cell. The most important secondary metabolites, not essential for the growth of the cell, are antibiotics and enzymes.

Certain **amino acids,** such as lysine, are termed *essential amino acids* because animals cannot synthesize them. These amino acids are present only at low levels in vegetable proteins and many diets are deficient in them. Therefore, the commercial synthesis of lysine and some other essential amino acids is an important industry. In nature, feedback inhibition (see Chapter 5, page 121) prevents wasteful, excessive production of primary metabolites. When the product accumulates, feedback inhibition shuts down further production. Without feedback inhibition, the amount of wasted energy would rapidly cause the organism to be eliminated from its highly competitive natural environment. Lysine is produced by a bacterium, *Corynebacterium glutamicum* (kô-rī-nē-bak-ti′rē-um glü-tam′i-kum), which normally produces both threonine and lysine. If both accumulate, feedback inhibition prevents further production of both amino acids, which share a common enzyme in their synthetic pathway. A mutant bacterium that lacks the enzymes necessary to make threonine is used to produce lysine in large amounts. Enough threonine is added to the medium to allow this mutant organism to grow, but not enough is added to trigger feedback inhibition. The organism then continues to produce lysine, because lysine alone does not cause feedback inhibition.

Glutamic acid is an amino acid widely used to make the popular flavor enhancer, monosodium glutamate. Hundreds of thousands of tons per year of glutamic acid are synthesized from glucose by microbes. The organism, the same bacterial species used to produce lysine, is induced to excrete glutamic acid when it is provided only a minimal amount of the vitamin biotin. The plasma membrane is weakened by the vitamin deficiency and the glutamic acid leaks into the medium. This technique of inducing the cell to form a leaky membrane is also sometimes used in antibiotic production.

Citric acid constitutes an estimated 60% of all food additives produced in the United States. It is widely used as an antioxidant and provides tartness to food. It is produced by a mold, *Aspergillus niger* (a-spėr-jil′lus nī′jer); molasses is used as a substrate. The mold excretes citric acid when provided only a very limited supply of iron and manganese. Vitamins such as riboflavin and cobalamin are also produced by nutritionally and genetically manipulated organisms.

Even the ancient technology of using yeasts to produce **ethanol** might be improved upon by the use of bacteria such as *Zymomonas* (zī-mō-mō′näs), which might be able to use cellulose as a substrate rather than the traditional carbohydrates. An improved tolerance of ethanol-producing microorganisms for ethanol might also allow them to produce it in higher concentrations. The field of

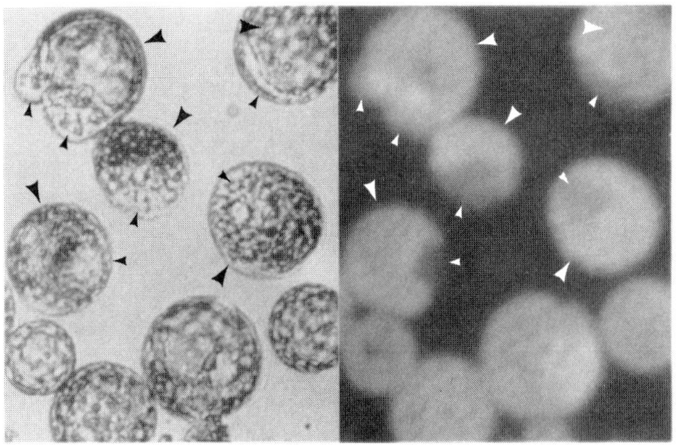

Figure 26–11 Fusion of two plant protoplasts, in this case leaf cell protoplasts from the tobacco species *Nicotiana tabacum* (large arrows) and *N. sylvestris* (small arrows). On the left is a brightfield micrograph and on the right is a fluorescently stained preparation of the same field. Various stages of fusion are seen, from early contact to almost complete.

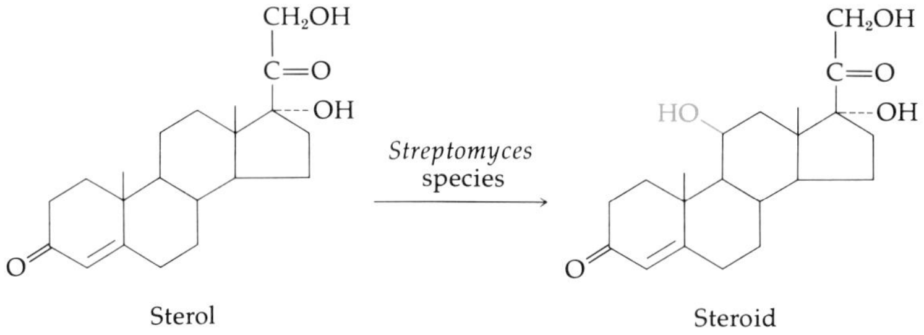

Figure 26–12 Conversion of a precursor compound, such as a sterol, into a steroid by a *Streptomyces.* These changes are extremely difficult for synthetic chemistry to make, so microorganisms are essential to economic production of steroids.

medicine should benefit from the increased production of useful drugs such as *interferon, insulin,* and *growth hormone,* which used to be laboriously and expensively extracted from sources such as transfused blood or slaughtered animals.

Human growth hormone was the first such product to be genetically engineered and commercially produced. A synthetic gene was constructed, even though the gene itself had never been isolated, and was spliced into a plasmid and inserted into *E. coli* bacteria. Similar methods led to the commercial production of insulin (for the treatment of diabetes) and interferon (which may prove of use in treatment of viral diseases or cancer).

Most current **vaccines** are composed of the intact but killed or attenuated (weakened) pathogen. The vaccines can be made safer and more effective by use of the essential protein-antigenic fraction of the antigen, rather than the entire cell. That essential part can now be synthesized.

Antibiotics, another important product of this industry, are produced in surprisingly large tonnages in the United States. Nearly half of these antibiotics are fed to animals to increase meat yields. The search for new and modified antibiotics is increasing as worldwide microbial resistance to antibiotics continues to rise alarmingly.

In Chapter 16 we discussed **monoclonal antibodies.** These are antibodies produced under industrial conditions. Instead of the limited amounts of antibody that can be made by an individual animal, an isolated antibody-producing plasma cell is fused to an "immortal" cancer cell. The plasma cell, which has been stimulated to form a specific antibody, does this continuously in large quantities. Monoclonal antibodies are likely to be another important medical product of industrial microbiology. The most promising use for monoclonal antibodies is in the diagnosis of disease when culturing the organism is difficult, or serological tests are inadequate or absent. There are high hopes, also, that monoclonal antibodies will make earlier diagnoses of cancer possible and will be used to locate cancerous growths with great specificity. The monoclonal antibodies might be combined with *immunotoxins,* plant or bacterial toxins attached to the antibodies, to seek out and attack tumors.

Steroids (Figure 26–12) are a very important group of chemicals that include cortisone, an antiinflammatory, and estrogens and progesterone, both used in birth control pills. Even though it is difficult to recover steroids from animal sources, microorganisms can synthesize steroids from sterols or related compounds derived from animals and plants. Certain strains of molds and bacteria (particularly streptomycetes) can produce steroids by converting chemical groups on sterols and similar molecules.

Enzymes are widely used in different industries. For example, amylases are used in the production of syrups from corn starch, of paper sizing (a coating for smoothness), and of glucose from starch. Glucose isomerase is an important enzyme: it converts the glucose that amylases form from starch into fructose, which is used in place of sucrose as a sweetener in many foods. Probably half of the bread baked in this country is made with proteases, which adjust the amounts of glutens (protein) in wheat so that baked goods are improved or made uniform. Other proteolytic enzymes are used as meat tenderizers, or, in detergents, as an additive to remove proteinaceous stains. There is considerable interest in the development of thermophilic enzymes. These usually operate at high speeds and cooling the reaction vessels would require small amounts of energy.

Proteins other than enzymes also promise to be a very important aspect of industrial microbiology. One type of protein that might be profitably synthesized is immunoglobulin (see Chapter 16).

Several enzymes produced commercially by microorganisms are listed in Table 26–4.

As petroleum products decrease in availability and increase in price, the classic microbial fermentations once used industrially to produce such products as butanol, acetone, and the like might be revived. Currently, petroleum derivatives are used as the starting material for the synthesis of these chemicals.

Agriculture could also benefit from improved species of the nitrogen-fixing bacterium *Rhizobium*. Some day it might be possible to engineer nitrogen-fixing bacteria to make a symbiosis with roots of cereal crops much like the legume–*Rhizobium* symbiosis.

Table 26–4 Some Microbial Enzymes Produced Commercially

Enzyme	Microorganism	Use of Enzyme
Amylase	*Aspergillus niger* *Aspergillus oryzae* *Bacillus subtilus*	Baking: flour supplement Brewing: mashing Food: precooked foods, syrup Pharmaceuticals: digestive aids Starch: cold-water laundry Textile: desizing agent
Cellulase	*Aspergillus niger*	Food: liquid coffee concentrate
Dextransucrase	*Leuconostoc mesenteroides*	Pharmaceuticals: dextran
Glucose oxidase	*A. niger*	Food: glucose removal from egg solids Pharmaceuticals: test papers
Invertase	*Saccharomyces cerevisiae*	Candy: prevents granulation in soft center Food: artificial honey
Lactase	*Saccharomyces fragilis*	Dairy: prevents crystallization of lactose in ice cream and concentrated milk
Lipase	*A. niger*	Dairy: flavor production in cheese
Pectinase	*A. niger*	Wine and juice: clarification
Penicillinase	*B. subtilis*	Medicine: diagnostic agent
Protease	*A. oryzae*	Brewing: beer stabilizer Baking: bread making Food: meat tenderizer Pharmaceutical: digestive aid Textile: desizing
Streptodornase	*Streptococcus pyogenes*	Medicine: reagent, wound debridement

Source: J. R. Porter, Microbiology and the food and energy crisis, *ASM News* 40(11):822, November 1974.

STUDY OUTLINE

FOOD SPOILAGE AND PRESERVATION (pp. 733–735)

1. The earliest methods of preserving foods were drying, the addition of salt or sugar, and fermentation.
2. In 1810, Appert devised a canning process that involved keeping air out of sealed cans and heating food in the cans.
3. In the nineteenth century, steam under pressure (pressure cooking) was used in canning.
4. Today, commercial sterilization heats canned foods to the minimum temperature necessary to destroy *C. botulinum* endospores while minimizing alteration of the food.
5. The commercial sterilization process uses sufficient heat to reduce a population of *C. botulinum* by 12 logarithmic cycles (12D treatment).
6. Endospores of thermophiles can survive commercial sterilization.

SPOILAGE OF CANNED FOOD (pp. 735–736)

1. Canned foods stored above 45°C can be spoiled by thermophilic anaerobes.
2. Thermophilic anaerobic spoilage might be accompanied by gas production; if no gas is formed, the spoilage is called flat sour spoilage.
3. Spoilage by mesophilic bacteria is usually due to improper heating procedures or leakage.
4. Acidic foods can be preserved by heat of 100°C because microorganisms that survive are not capable of growth in a low pH.

ASEPTIC PACKAGING (p. 736)

1. Presterilized materials are assembled into packages and aseptically filled with heat-sterilized liquid foods.

LOW-TEMPERATURE PRESERVATION (pp. 736–737)

1. Many microorganisms, although few pathogens, will grow slowly at refrigeration temperatures (0 to 7°C).
2. Freezing might not kill bacteria, but it will prevent their growth.

RADIATION AND FOOD PRESERVATION (pp. 737–738)

1. Research is being done to determine the utility of ionizing radiation to sterilize or pasteurize foods.
2. Microwaves kill bacteria by heating the food.

PASTEURIZATION (p. 738)

1. Pasteur determined that because spoilage was due to only certain microorganisms, it was not necessary to kill all microorganisms to prevent spoilage.
2. Pasteurization is the application of heat that selectively kills microorganisms without greatly altering the quality of the food.

CHEMICAL PRESERVATIVES (pp. 738–739)

1. Sodium benzoate, sorbic acid or potassium sorbate, or calcium propionate are added to foods to inhibit the growth of molds.
2. These chemicals are metabolized by humans.
3. Sodium nitrate and sodium nitrite are added to meats to preserve the red color and to prevent the germination of *C. botulinum* endospores.
4. Nitrates are converted to nitrites by some bacteria using anaerobic respiration; nitrites are the active ingredient.

FOOD-BORNE INFECTIONS AND MICROBIAL INTOXICATIONS (pp. 739–740)

1. In a food-borne infection, the pathogen grows in the infected host.
2. Gastroenteritis, typhoid fever, and dysentery are food-borne infections.
3. Microbial intoxication results from ingestion of a toxin produced by microbes growing in food.
4. Botulism and staphylococcal food poisoning are examples of microbial intoxications.
5. Pasteurized dairy milk might contain gram-negative psychrophiles.
6. Pasteurized products of cultured milk (i.e., buttermilk and sour cream) contain large numbers of lactic acid bacteria.
7. Public health service standards for Grade A pasteurized milk are a standard plate count of fewer

than 20,000 bacteria per milliliter and/or fewer than 10 coliforms per milliliter.

8. Dry milk products must have a standard plate count of fewer than 30,000 bacteria per gram or a coliform count fewer than 90 per gram.
9. The species of bacteria in unpasteurized (raw) milk are not controlled, and raw milk can transmit disease.

ROLE OF MICROORGANISMS IN FOOD PRODUCTION (pp. 740–742)

Cheese (pp. 740–741)

1. The milk protein casein curdles because of the action by lactic acid bacteria and the enzyme rennin.
2. Cheese is the curd separated from the liquid portion of milk, called whey.
3. Hard cheeses are produced by lactic acid bacteria growing in the interior of the curd.
4. The growth of microorganisms in cheeses is called ripening.
5. Semisoft cheeses are ripened by bacteria growing on the surface.
6. Soft cheeses are ripened by *Penicillium* growing on the surface.

Other Dairy Products (p. 741)

1. Old-fashioned buttermilk was produced by lactic acid bacteria growing during the butter-making process.
2. Commercial buttermilk is made by letting lactic acid bacteria grow in skim milk for 12 hours.
3. Sour cream, yogurt, kefir, and kumiss are produced by lactobacilli, streptococci, and/or yeast growing in low-fat milk.

Nondairy Fermentations (p. 741)

1. Sugars in bread dough are fermented by yeast to ethyl alcohol and CO_2; the CO_2 causes the bread to rise.
2. Sauerkraut, pickles, olives, and soy sauce are the result of microbial fermentations.

Alcoholic Beverages (pp. 741–742)

1. Carbohydrates obtained from grains, potatoes, or molasses are fermented by yeast to produce ethanol in the production of beer, ale, and distilled spirits.
2. The sugars in fruits such as grapes are fermented by yeast to produce wines.
3. *Acetobacter* oxidizes ethanol in wine to acetic acid (vinegar).

FOOD FROM MICROORGANISMS (pp. 742–746)

1. Microorganisms can grow and produce proteinaceous cell material from otherwise unusable substrates such as cellulose.
2. The microbial cells can be made into a food called single-cell protein (SCP).
3. *Spirulina* and yeasts are cultured and used as SCP.

ALTERNATIVE ENERGY FROM MICROORGANISMS (pp. 746–747)

1. Organic waste, called biomass, can be converted by microorganisms into alternative fuels, a process called bioconversion.
2. Fuels produced by microbial fermentation are methane, methanol, ethanol, and hydrogen.

GENETIC ENGINEERING AND INDUSTRIAL MICROBIOLOGY (pp. 747–751)

1. Microorganisms produce alcohol, lactic acid, glycerol, and antibiotics that are used in food manufacturing and other industrial processes.

Genetic Engineering (pp. 747–749)

1. Selected genes can be inserted into bacteria and yeasts by genetic engineering so these cells will produce large quantities of desired hormones, vaccines, and antibiotics.

Industrial Microbiology (pp. 749–751)

1. Amino acids, citric acid, monoclonal antibodies, and enzymes are only some of the products produced industrially by bacteria.

STUDY QUESTIONS

REVIEW

1. Why are attempts made to preserve foods?
2. List five methods used to preserve foods.
3. Matching.

_____ Sorbic acid	(a) Sterilization
_____ 5°C	(b) Pasteurization
_____ 72°C for 15 seconds	(c) Low-temperature preservation
_____ 121°C for 15 minutes	(d) Chemical preservation
_____ 12D treatment	

4. Define pasteurization and aseptic packaging.
5. Pasteurization does not kill all microorganisms, so food can still spoil. Why then are dairy products pasteurized?
6. State one advantage and one disadvantage of the addition of nitrites to foods.
7. Outline the production of cheese, and compare the productions of hard and soft cheeses.
8. Outline the process of wine production.
9. List at least two advantages and two disadvantages of SCP.
10. How can microorganisms provide energy sources? What metabolic processes can result in fuels?
11. Appert's invention was the forerunner of hermetically sealed cans. Why do hermetically sealed cans prevent food spoilage?
12. Why is a can of blackberries preserved by commercial sterilization typically heated to 100°C instead of at least 116°C?
13. Under what conditions would you expect to find thermophiles spoiling foods? Psychrophiles?
14. Discuss the importance of industrial microbiology.
15. How will advances in genetic engineering improve production of inexpensive chemicals? Outline the steps taken to engineer a cell to produce a selected protein.

CHALLENGE

1. What bacteria seem to be most frequently used in the production of food? Can you guess why?
2. Why do the following processes preserve food?
   (a) Fermentation
   (b) Salting
   (c) Drying
3. Explain the processes involved in the production of sourdough bread.
4. Three to five days after eating Thanksgiving dinner at a restaurant, 112 people developed fever and gastroenteritis. All of the food had been consumed except for five "doggie" bags: bacterial analysis of the mixed contents of the bags (containing roasted turkey, giblet gravy, and mashed potatoes) showed the same bacterium as was isolated from the patients. The gravy had been prepared from giblets of 43 turkeys that had been refrigerated for three days prior to preparation. The uncooked giblets were ground in a blender and added to a thickened, hot stock mixture. The gravy was not reboiled and was stored at room temperature throughout Thanksgiving Day. What was the source of the illness? What was the most likely etiologic agent? Was this an infection or an intoxication? (*Hint:* Refer to Chapter 23.)

FURTHER READING

Ayres, J. C., J. O. Mundt, and W. E. Sandine. 1980. *Microbiology of Foods*. San Francisco: W. H. Freeman. Contains chapters on the necessary conditions for microbial growth in food, methods of preserving food, fermentations, and food-borne illnesses.

Biotechnology. 1983. *Science* 219:611ff. A special issue on biotechnology includes discussions of monoclonal antibodies and applications of genetic engineering to industrial microbiology.

Developments in Industrial Microbiology. Proceedings of the general meeting of the Society for Industrial Microbiology are published annually. Each volume contains discussions of current advances in food, dairy, and pharmaceutical microbiology. Arlington, VA: Society for Industrial Microbiology.

Industrial microbiology. September 1981. *Scientific American* 245. The September issue is devoted entirely to the topic of industrial microbiology and genetic engineering.

Kosikowski, F. V. May 1985. Cheese. *Scientific American*. 252:88–99. Describes how types of milk, species of microbes, and ripening times are used to produce 2,000 cheese varieties.

Rose, A. H., ed. 1983. *Economic Microbiology Series*, 8 vols. Volume 7 covers the history and science of food fermentations. New York: Academic Press.

Strobel, G. A. and G. N. Lanier. August 1981. Dutch elm disease. *Scientific American* 245:56–66. An excellent paper on the development of biologic controls for the fungal parasite and insect vector.

Zeikus, J. G. 1980. Chemical and fuel production by anaerobic bacteria. *Annual Review of Microbiology* 34:423–464. A summary of the current technology and practicality of producing hydrocarbon fuels and organic solvents from agricultural and municipal wastes.

APPENDIX A

Classification of Bacteria According to *Bergey's Manual of Systematic Bacteriology*

KINGDOM PROCARYOTAE
divided into four divisions:

DIVISION I. GRACILICUTES;
procaryotes with thin cell walls, implying a gram-negative type of cell wall.

DIVISION II. FIRMICUTES;
procaryotes with thick and strong skin, indicating a gram-positive type of cell wall.

DIVISION III. TENERICUTES;
procaryotes of a pliable and soft nature, indicating the lack of a rigid cell wall.

DIVISION IV. MENDOSICUTES;
procaryotes with faulty cell walls, suggesting the lack of conventional peptidoglycan.

VOLUME 1

SECTION 1
THE SPIROCHETES

Order I. *Spirochaetales*
Family I. *Spirochaetaceae*
Genus I. *Spirochaeta*
Genus II. *Cristispira*
Genus III. *Treponema*
Genus IV. *Borrelia*
Family II. *Leptospiraceae*
Genus I. *Leptospira*
Other organisms: Hindgut spirochetes of termites and *Cryptocercus punctulatus* (wood-eating cockroach).

SECTION 2
AEROBIC/MICROAEROPHILIC, MOTILE, HELICAL/VIBRIOID GRAM-NEGATIVE BACTERIA

Genus I. *Aquaspirillum*
Genus II. *Spirillum*
Genus III. *Azospirillum*
Genus IV. *Oceanospirillum*
Genus V. *Campylobacter*
Genus VI. *Bdellovibrio*
Genus VII. *Vampirovibrio*

SECTION 3
NONMOTILE (OR RARELY MOTILE), GRAM-NEGATIVE CURVED BACTERIA

Family I. *Spirosomaceae*
Genus I. *Spirosoma*
Genus II. *Runella*
Genus III. *Flectobacillus*
Other genera:
Genus *Microcyclus*
Genus *Meniscus*
Genus *Brachyarcus*
Genus *Pelosigma*

SECTION 4
GRAM-NEGATIVE AEROBIC RODS AND COCCI

Family I. *Pseudomonadaceae*
Genus I. *Pseudomonas*
Genus II. *Xanthomonas*
Genus III. *Frateuria*
Genus IV. *Zoogloea*
Family II. *Azotobacteraceae*
Genus I. *Azotobacter*
Genus II. *Azomonas*
Family III. *Rhizobiaceae*
Genus I. *Rhizobium*
Genus II. *Bradyrhizobium*
Genus III. *Agrobacterium*
Genus IV. *Phyllobacterium*
Family IV. *Methylococcaceae*
Genus I. *Methylococcus*
Genus II. *Methylomonas*
Family V. *Halobacteriaceae*
Genus I. *Halobacterium**
Genus II. *Halococcus**
Family VI. *Acetobacteraceae*
Genus I. *Acetobacter*
Genus II. *Gluconobacter*
Family VII. *Legionellaceae*
Genus I. *Legionella*
Family VIII. *Neisseriaceae*
Genus I. *Neisseria*
Genus II. *Moraxella*
Genus III. *Acinetobacter*
Genus IV. *Kingella*
Other genera:
Genus *Beijerinckia*
Genus *Derxia*
Genus *Xanthobacter*
Genus *Thermus*
Genus *Thermomicrobium*
Genus *Halomonas*
Genus *Alteromonas*
Genus *Flavobacterium*
Genus *Alcaligenes*

Genus *Serpens*
Genus *Janthinobacterium*
Genus *Brucella*
Genus *Bordetella*
Genus *Francisella*
Genus *Paracoccus*
Genus *Lampropedia*

SECTION 5 FACULTATIVELY ANAEROBIC GRAM-NEGATIVE RODS

Family I. *Enterobacteriaceae*
Genus I. *Escherichia*
Genus II. *Shigella*
Genus III. *Salmonella*
Genus IV. *Citrobacter*
Genus V. *Klebsiella*
Genus VI. *Enterobacter*
Genus VII. *Erwinia*
Genus VIII. *Serratia*
Genus IX. *Hafnia*
Genus X. *Edwardsiella*
Genus XI. *Proteus*
Genus XII. *Providencia*
Genus XIII. *Morganella*
Genus XIX. *Yersinia*
Other genera of the family *Enterobacteriaceae:*
Genus *Obesumbacterium*
Genus *Xenorhabdus*
Genus *Kluyvera*
Genus *Rahnella*
Genus *Cedecea*
Genus *Tatumella*
Family II. *Vibrionaceae*
Genus I. *Vibrio*
Genus II. *Photobacterium*
Genus III. *Aeromonas*
Genus IV. *Plesiomonas*
Family III. *Pasteurellaceae*
Genus I. *Pasteurella*
Genus II. *Haemophilus*
Genus III. *Actinobacillus*
Other genera:
Genus *Zymomonas*
Genus *Chromobacterium*
Genus *Cardiobacterium*
Genus *Calymmatobacterium*
Genus *Gardnerella**
Genus *Eikenella*
Genus *Streptobacillus*

SECTION 6 ANAEROBIC GRAM-NEGATIVE STRAIGHT, CURVED, AND HELICAL RODS

Family I. *Bacteroidaceae*
Genus I. *Bacteroides*
Genus II. *Fusobacterium*
Genus III. *Leptotrichia*
Genus IV. *Butyrivibrio**
Genus V. *Succinimonas*
Genus VI. *Succinivibrio*
Genus VII. *Anaerobiospirillum*
Genus VIII. *Wolinella*
Genus IX. *Selenomonas*
Genus X. *Anaerovibrio*
Genus XI. *Pectinatus*
Genus XII. *Acetivibrio*
Genus XIII. *Lachnospira**

SECTION 7 DISSIMILATORY SULFATE- OR SULFUR-REDUCING BACTERIA

Genus *Desulfuromonas*
Genus *Desulfovibrio*
Genus *Desulfomonas*
Genus *Desulfococcus*
Genus *Desulfobacter*
Genus *Desulfobulbus*
Genus *Desulfosarcina*

SECTION 8 ANAEROBIC GRAM-NEGATIVE COCCI

Family I. *Veillonellaceae*
Genus I. *Veillonella*
Genus II. *Acidaminococcus*
Genus III. *Megasphaera*

SECTION 9 THE RICKETTSIAS AND CHLAMYDIAS

Order I. *Rickettsiales*
Family I. *Rickettsiaceae*
Tribe I. *Rickettsieae*
Genus I. *Rickettsia*
Genus II. *Rochalimaea*
Genus III. *Coxiella*
Tribe II. *Ehrlichieae*
Genus IV. *Ehrlichia*
Genus V. *Cowdria*
Genus VI. *Neorickettsia*
Tribe III. *Wolbachieae*
Genus VII. *Wolbachia*
Genus VIII. *Rickettsiella*
Family II. *Bartonellaceae*
Genus I. *Bartonella*
Genus II. *Grahamella*
Family III. *Anaplasmataceae*
Genus I. *Anaplasma*
Genus II. *Aegyptianella*
Genus III. *Haemobartonella*
Genus IV. *Eperythrozoon*
Order II. *Chlamydiales*
Family I. *Chlamydiaceae*
Genus I. *Chlamydia*

SECTION 10 THE MYCOPLASMAS

Order I. *Mycoplasmatales*
Family I. *Mycoplasmataceae*
Genus I. *Mycoplasma*
Genus II. *Ureaplasma*
Family II. *Acholeplasmataceae*
Genus I. *Acholeplasma*
Family III. *Spiroplasmataceae*
Genus I. *Spiroplasma*
Other genera:
Genus *Anaeroplasma*
Genus *Thermoplasma**
Mycoplasma-like organisms of plants and invertebrates

SECTION 11 ENDOSYMBIONTS

A. Endosymbionts of Protozoa
Genus I. *Holospora*
Genus II. *Caedibacter*
Genus III. *Pseudocaedibacter*
Genus IV. *Lyticum*
Genus V. *Tectibacter*
B. Endosymbionts of Insects
Genus I. *Blattabacterium*
C. Endosymbionts of Fungi and Invertebrates other than Arthropods

VOLUME 2

SECTION 12 GRAM-POSITIVE COCCI

Family I. *Micrococcaceae*
Genus I. *Micrococcus*
Genus II. *Stomatococcus*
Genus II. *Planococcus*
Genus IV. *Staphylococcus*
Family II. *Deinococcaceae*
Genus I. *Deinococcus*
Other organisms: "Pyogenic" streptococci, "Oral" streptococci, "Lactic" streptococci and enterococci
Genus *Leuconostoc*
Genus *Pediococcus*
Genus *Aerococcus*
Genus *Gemella*
Genus *Peptococcus*
Genus *Peptostreptococcus*
Genus *Ruminococcus*
Genus *Coprococcus*
Genus *Sarcina*

SECTION 13
ENDOSPORE-FORMING GRAM-POSITIVE RODS AND COCCI

Genus *Bacillus*
Genus *Sporolactobacillus*
Genus *Clostridium*
Genus *Desulfotomaculum*
Genus *Sporosarcina*
Genus *Oscillospira*

SECTION 14
REGULAR, NON-SPORING, GRAM-POSITIVE RODS

Genus *Lactobacillus*
Genus *Listeria*
Genus *Erysipelothrix*
Genus *Brochothrix*
Genus *Renibacterium*
Genus *Kurthia*
Genus *Caryophanon*

SECTION 15
IRREGULAR, NON-SPORING, GRAM-POSITIVE RODS

Animal and saprophytic corynebacteria *(Corynebacterium)*
Plant corynebacteria
Genus *Gardnerella**
Genus *Arcanobacterium*
Genus *Arthrobacter*
Genus *Brevibacterium*
Genus *Curtobacterium*
Genus *Caseobacter*
Genus *Microbacterium*
Genus *Aureobacterium*
Genus *Cellulomonas*
Genus *Agromyces*
Genus *Arachnia*
Genus *Rothia*
Genus *Propionibacterium*
Genus *Eubacterium*
Genus *Acetobacterium*
Genus *Lachnospira**
Genus *Butyrivibrio**
Genus *Thermoanaerobacter*
Genus *Actinomyces*
Genus *Bifidobacterium*

SECTION 16
MYCOBACTERIA

Family I. *Mycobacteriaceae*
Genus I. *Mycobacteria*

SECTION 17
NOCARDIOFORMS

Genus *Nocardia*
Genus *Rhodococcus*
Genus *Nocardioides*
Genus *Pseudonocardia*
Genus *Oerskovia*
Genus *Saccharopolyspora*
Genus *Micropolyspora*
Genus *Promicromonospora*
Genus *Intrasporangium*

VOLUME 3

SECTION 18
GLIDING, NON-FRUITING BACTERIA

Order I. *Cytophagales*
Family I. *Cytophagaceae*
Genus I. *Cytophaga*
Genus II. *Sporocytophaga*
Genus III. *Capnocytophaga*
Genus IV. *Flexithrix*
Genus V. *Flexibacter*
Genus VI. *Microscilla*
Genus VII. *Saprospira*
Genus VIII. *Herpetosiphon*
Order II. *Lysobacterales*
Family I. *Lysobacteraceae*
Genus *Lysobacter*
Order III. *Beggiatoales*
Family I. *Beggiatoaceae*
Genus I. *Beggiatoa*
Genus II. *Thioploca*
Genus III. *Thiospirillopsis*
Genus IV. *Thiothrix*
Genus V. *Achromatium*
Family II. *Simonsiellaceae*
Genus I. *Simonsiella*
Genus II. *Alysiella*
Family III. *Leucotrichaceae*
Genus *Leucothrix*
Families and genera *incertae sedis* (status of species questionable):
Genus *Toxothrix*
Genus *Vitreoscilla*
Genus *Chitinophagen*
Genus *Desulfonema*
Family IV. *Pelonemataceae*
Genus *Pelonema*
Genus *Achroonema*
Genus *Peloploca*
Genus *Desmanthus*

SECTION 19
ANOXYGENIC PHOTOTROPHIC BACTERIA
PURPLE BACTERIA

Family I. *Chromatiaceae*
Genus I. *Chromatium*
Genus II. *Thiocystis*
Genus III. *Thiospirillum*
Genus IV. *Thiocapsa*
Genus V. *Amoebobacter*
Genus VI. *Lamprobacter*
Genus VII. *Lamprocystis*
Genus VIII. *Thiodictyon*
Genus IX. *Thiopedia*
Family II. *Ectothiorhodospiraceae*
Genus *Ectothiorhodospira*
Purple nonsulfur bacteria
Genus *Rhodospirillum*
Genus *Rhodopseudomonas*
Genus *Rhodobacter*
Genus *Rhodomicrobium*
Genus *Rhodopila*
Genus *Rhodocyclus*
GREEN BACTERIA
Green sulfur bacteria
Genus *Chlorobium*
Genus *Prosthecochloris*
Genus *Anacalochloris*
Genus *Pelodictyon*
Genus *Chloroherpeton*
Multicellular filamentous green bacteria
Genus *Chloroflexus*
Genus *Heliothrix*
Genus *Oscillochloris*
Genus *Chloronema*
Genera *incertae sedis*
Genus *Heliobacterium*
Genus *Erythrobacter*

SECTION 20
BUDDING AND/OR APPENDAGED BACTERIA

PROSTHECATE BACTERIA
Budding bacteria
Genus *Hyphomicrobium*
Genus *Hyphomonas*
Genus *Pedomicrobium*
Genus *"Filomicrobium"*
Genus *"Dicotomicrobium"*
Genus *"Tetramicrobium"*
Genus *Stella*
Genus *Ancalomicrobium*
Genus *Prosthecomicrobium*
Non-budding bacteria
Genus *Caulobacter*
Genus *Asticcacaulis*
Genus *Prosthecobacter*
Genus *Thiodendron*
NON-PROSTHECATE BACTERIA
Budding bacteria
Genus *Planctomyces*
Genus *Pasteuria*
Genus *Blastobacter*
Genus *Angulomicrobium*
Genus *Gemmiger*
Genus *Ensifer*
Genus *Isophaera*
Non-budding stalked bacteria

Genus *Gallionella*
Genus *Nevskia*
Morphologically unusual budding bacteria involved in iron and manganese deposition
Genus *Seliberia*
Genus *Metallogenium*
Genus *Caulococcus*
Genus *Kuznezovia*
Others: Spinate bacteria

SECTION 21 ARCHAEOBACTERIA

METHANOGENIC BACTERIA
Genus *Methanobacterium*
Genus *Methanobrevibacter*
Genus *Methanococcus*
Genus *Methanomicrobium*
Genus *Methanospirillum*
Genus *Methanosarcina*
Genus *Methanococcoides*
Genus *Methanothermus*
Genus *Methanolobus*
Genus *Methanoplanus*
Genus *Methanogenium*
Genus *Methanothrix*
EXTREME HALOPHILIC BACTERIA
Genus *Halobacterium**
Genus *Halococcus**
EXTREME THERMOPHILIC BACTERIA
Genus *Thermoplasma**
Genus *Sulfolobus*
Genus *Thermoproteus*
Genus *Thermofilum*
Genus *Thermococcus*
Genus *Desulfurococcus*
Genus *Thermodiscus*
Genus *Pyrodictium*

SECTION 22 SHEATHED BACTERIA

Genus *Sphaerotilus*
Genus *Leptothrix*
Genus *Haliscominobacter*
Genus *Lieskeella*
Genus *Phragmidiothrix*
Genus *Crenothrix*
Genus *Clonothrix*

SECTION 23 GLIDING, FRUITING BACTERIA

Order I. *Myxobacterales*
Family I. *Myxococcaceae*
Genus *Myxococcus*
Family II. *Archangiaceae*
Genus *Archangium*
Family III. *Cystobacteraceae*
Genus I. *Cystobacter*
Genus II. *Melittangium*
Genus III. *Stigmatella*
Family IV. *Polyangiaceae*
Genus I. *Polyangium*
Genus II. *Nannocystis*
Genus III. *Chondromyces*
Genus *incerta sedis*
Genus *Angiococcus*

SECTION 24 CHEMOLITHOTROPHIC BACTERIA

NITRIFIERS
Family I. *Nitrobacteraceae*
Genus I. *Nitrobacter*
Genus II. *Nitrospina*
Genus III. *Nitrococcus*
Genus IV. *Nitrosomonas*
Genus V. *Nitrospira*
Genus VI. *Nitrosococcus*
Genus VII. *Nitrosolobus*
SULFUR OXIDIZERS
Genus *Thiobacillus*
Genus *Thiomicrospira*
Genus *Thiobacterium*
Genus *Thiospira*
Genus *Macromonas*
OBLIGATE HYDROGEN OXIDIZERS
Genus *Hydrogenbacter*
METAL OXIDIZERS AND DEPOSITORS
Family I. *Siderocapsaceae*
Genus I. *Siderocapsa*
Genus II. *Naumaniella*
Genus III. *Ochrobium*
Genus IV. *Siderococcus*
OTHER MAGNETOTACTIC BACTERIA

SECTION 25 CYANOBACTERIA

OTHERS
Order I. Prochlorales
Family I. *Prochloraceae*
Genus *Prochloron*

VOLUME 4

SECTION 26 *ACTINOMYCETES* THAT DIVIDE IN MORE THAN ONE PLANE

Genus *Geodermatophilus*
Genus *Dermatophilus*
Genus *Frankia*
Genus *Tonsilophilus*

SECTION 27 SPORANGIATE *ACTINOMYCETES*

Genus *Actinoplanes* (including *Amorphosporangium*)
Genus *Streptosporangium*
Genus *Ampullariella*
Genus *Spirillospora*
Genus *Pilimelia*
Genus *Dactylosporangium*
Genus *Planomonospora*
Genus *Planobispora*

SECTION 28 *STREPTOMYCETES* AND THEIR ALLIES

Genus *Streptomyces*
Genus *Streptovertillium*
Genus *Actinopycnidium*
Genus *Actinosporangium*
Genus *Chainia*
Genus *Elytrosporangium*
Genus *Microellobosporia*
(The last five genera may be merged with *Streptomyces*)

SECTION 29 OTHER CONIDIATE GENERA

Genus *Actinopolyspora*
Genus *Actinosynnema*
Genus *Kineospora*
Genus *Kitasatosporia*
Genus *Microbispora*
Genus *Micromonospora*
Genus *Microtetrospora*
Genus *Saccharomonospora*
Genus *Sporichthya*
Genus *Streptoalloteichus*
Genus *Thermomonospora*
Genus *Actinomadura*
Genus *Nocardiopsis*
Genus *Excellospora*
Genus *Thermoactinomyces*

---

*Genera which have been placed in more than one volume: *Gardnerella*, *Lachnospira*, and *Butyrivibrio* have thin, gram-positive walls but stain as gram-negatives. *Halobacterium*, *Halococcus*, and *Thermoplasma* are ARCHAEOBACTERIA that are also included among the GRAM-NEGATIVE RODS and COCCI or the MYCOPLASMAS.

APPENDIX B

Pronunciation of Scientific Names

RULES OF PRONUNCIATION

Vowels. Pronounce all the vowels in scientific names.

Diphthongs. Two vowels written together are a diphthong. Pronounce diphthongs as one vowel.

Accent. The accented syllable is either the next to last or third to last syllable.

The accent is on the next to last syllable:

1. When the name contains only two syllables, for example, péstis.
2. When the next to last syllable is a diphthong, for example, *Amoéba.*
3. When the vowel of the next to last syllable is long, for example, *Treponéma.*

The vowel in the next to last syllable is long in words ending in these suffixes:

Suffix	Example
-ales	orders such as Eubacteriáles
-ina	*Sarcína*
-anus, -anum	*pasteuriánum*
-uta	*diminúta*

The vowel in the next to last syllable is short in words ending in these suffixes:

Suffix	Example
-atus, -atum	*caudátum*
-ella	*Salmonélla*

The accent is on the third to last syllable in family names. Families end in -aceae and the final -ae is pronounced as ē as in Micrococcáceae.

Consonants. When *c* or *g* is followed by ae, e, oe, i, or y it has a soft sound. When *c* or *g* is followed by a, o, oi, or u it has a hard sound. When a double *c* is followed by e, i, or y it is pronounced as ks (i.e., cocci).

PRONUNCIATION OF ORGANISMS IN THIS TEXT

Pronunciation key:

a	hat	ē	see	o	hot	th	thin	â	care	g	go	ô	order	u̇	put
ā	age	ė	term	ō	go	u	cup	ä	father	i	sit	oi	oil	ü	rule
								ch	child	ī	ice	ou	out	ū	use
								e	let	ng	long	sh	she	zh	seizure

Acetobacter ä-sē-tō-bak'tėr
Acinetobacter ā-si-ne'tō-bak-tėr
Actinomyces israelii ak-tin-ō-mī'sēs is-rā'lē-ī
Aedes aegypti ä-e'dēz ē-jip'tē
Agaricus augustus ä-gär'i-kus au'gus-tus
Ajellomyces ä-jel-lō-mī'sēs
Allescheria boydii al-lesh-er'ē-ä boi'dē-ī
Amanita phalloides am-an-ī'ta fal-loi'dēz
Anabaena an-ä-bē'nä
Anopheles an-of'el-ēz
Aquaspirillum bengal a-kwä-spī-ril'lum ben'gal
A. magnetotacticum mäg-nē-to-tak'ti-kum
Arthroderma är-thrō-dėr'mä
Ascaris lumbricoides as'kar-is lum-bri-koi'dēz
Aspergillus flavus a-spėr-jil'lus flā'vus
A. fumigatus fü-mi-gä'tus
A. niger nī'jėr
A. oryzae ô-rī'zē
A. tamarii ta-mär'ē-ī
Azolla ā-zō'lä
Azomonas ā-zō-mō'nas
Azotobacter ä-zō-tō-bak'tėr
Bacillus anthracis bä-sil'lus an-thrā'sis
B. brevis bre'vis
B. cereus se'rē-us
B. coagulans kō-äg'ū-lanz
B. licheniformis lī-ken-i-fôr'mis
B. megaterium meg-ä-tėr'ē-um
B. polymyxa po-lē-miks'ä
B. sphaericus sfe'ri-kus
B. stearothermophilus ste-rō-thėr-mä'fil-us
B. subtilis su'til-us
B. thuringiensis thu̇r-in-jē-en'sis
Bacteroides fragilis bak-tė-roi'dēz fra'jil-is
B. hypermegas hī-pėr-meg'äs
Balantidium coli bal-an-tid'ē-um kō'lē
Bdellovibrio del-lō-vib'rē-ō
Beggiatoa bej-jē-ä-tō'ä
Beijerinckia bī-jė-rink'ē-ä
Bifidobacterium globosum bī-fi-dō-bak-ti'rē-um glob-ō'sum
B. pseudolongum sū-dō-lông'um
Blastomyces dermatitidis blas-tō-mī'sēs dėr-mä-tit'i-dis
Blattabacterium blät-tä-bak-ti'rē-um
Boletus edulis bō-lē'tus e-dū'lis

Bordetella pertussis bôr-de-tel'lä pėr-tus'sis
Borrelia bôr-rel'ē-ä
Brucella abortus brü-sel'lä ä-bôr'tus
B. canis kā'nis
B. melitensis me-li-ten'sis
B. suis sü'is
Calvulina cristata kal-vū-lī'nä kris'tä-tä
Calymmatobacterium granulomatis kal-im-mä-tō-bak-ti'rē-um gran-ū-lō'mä-tis
Campylobacter fetus kam-pī-lō-bak'tėr fē'tus
C. jejuni jē-jū'nē
Canis familiaris kā'nis fa-mil'yār-is
Candida albicans kan'did-ä al'bi-kans
C. utilis ū'til-is
Carpenteles kär-pen'tel-ēz
Caulobacter bacteroides kô-lō-bak'tėr bak-tėr-oi'dēz
Cephalosporium sef-ä-lō-spô'rē-um
Ceratocystis ulmi sē-rä-tō'sis-tis ul'mē
Chlamydia psittaci kla-mi'dē-ä sit'tä-sē
C. trachomatis trä-kō'mä-tis
Chlamydomonas klam-i-dō-mō'näs
Chlorobium klô-rō'bē-um
Chromatium vinosum krō-mā'shum vi-nō'sum
Chroococcus turgidus krō-ō-kok'kus tėr'gi-dus
Chrysops krī'sops
Citrobacter sit'rō-bak-tėr
Cladosporium werneckii kla-dō-spô'rē-um wėr-ne'kē-ī
Claviceps purpurea kla'vi-seps pur-pu-rē'ä
Clonorchis sinensis klo-nôr'kis si-nen'sis
Clostridium acetobutylicum klô-tri'dē-um a-sē-tō-bū-ti'li-kum
C. bifermentens bī-fėr-men'tans
C. botulinum bo-tū-lī'num
C. butyricum bü-ti'ri-kum
C. pasteurianum pas-tyėr-ē-ā'num
C. perfringens pėr-frin'jens
C. sporogenes spô-rä'jen-ēz
C. subterminale sub-tėr-mi-na'lē
C. tetani te'tan-ē
C. thermosaccharolyticum thėr-mō-sak-kär-ō-li'ti-kum
Coccidioides immitis kok-sid-ē-oi'dēz im'mi-tis
Corynebacterium diphtheriae kô-rī-nē-bak-ti'rē-um dif-thi'rē-ī
C. glutamicum glü-tam'i-kum
C. xerosis ze-rō'sis
Coxiella burnetii käks-ē-el'lä bėr-ne'tē-ē
Cristispira pectinis kris-tē-spī'rä pek'tin-is
Cryptococcus neoformans kryp-tō-kok'kus nē-ō-fôr'manz
Cryptosporidium krip-tō-spô-ri'dē-um
Culex kū'leks
Cytophaga sī-täf'äg-ä
Dermacentor andersoni dėr-mä-sen'tôr an-dėr-sō'nī
Desulfotomaculatum nigrificans dē-sul-fo-to-mak'ū-lä-tum nī-gri'fi-kans
Desulfovibrio dē-sul-fō-vib'rē-ō
Didinium nasutum dī-di'nē-um nä-süt'um
Diphyllobothrium latum dī-fil-lo-bo'thrē-um lā'tum
Dryas octopetala drī'äs ok-tō-pet'äl-ä
Echinococcus granulosus ē-kīn-ō-kok'kus gra-nū-lō'sus
Emericella em-ėr-ē-sel'lä
Emmonsiella em-mon-sē-el'lä
Endothia parasitica en-dō'thē-ä par-ä-si'ti-kä
Entamoeba histolytica en-tä-mē'bä his-tō-li'ti-kä
Enterobacter aerogenes en-te-rō-bak'tėr ā-rä'jen-ēz
E. agglomerans ag-glom'ėr-anz
E. cloacae klō-ā'kä
Enterobius vermicularis en-te-rō'bē-us ver-mi-kū-lar'is
Epidermophyton ep-i-dėr-mō-fī'ton
Erwinia carotovora ėr-wi'nē-ä ka-ro-tov'ô-rä
E. herbicola hėr-bik'ō-lä
Escherichia coli esh-ėr-i'kē-ä kō'lē
Euglena ū-glē'nä
Eurotium yėr-ō'shum
Filobasidiella fi-lō-ba-si-dē-el'lä
Fonsecaea pedrosoi fon-se'kē-ä pe-drō'sō-ī
Francisella tularensis fran-sis-el'lä tü-lä-ren'sis
Frankia frank'ē-ä
Fusobacterium fü-sō-bak-ti'rē-um
Gardnerella vaginalis gärd-nė'rel-lä va-jin'al-is
Gelidium jel-id'ē-um
Giardia lamblia jē-är'dē-ä lam'lē-ä
Gloeocapsa glē-ō'kap-sä
Glossina gläs-sē-nä
Gluconobacter glü-kon-ō-bak'tėr
Gymnoascus jim-nō-as'kus
Halobacterium halobium hal-ō-bak-ti'rē-um hal-ō'bē-um
Halococcus hal-ō-kok'kus
Hebeloma alpinum hē-bel-ō'mä al'pin-um
Hemophilus aegyptius hē-mä'fi-lus ē-jip'ti-us
H. ducreyi dü-krā'ē
H. influenzae in-flü-en'zī
H. parainfluenzae pa-ra-in-flü-en'zī
Histoplasma capsulatum hiss-tō-plaz'mä kap-su-lä'tum
Holospora hō-lo'spô-rä
Homo sapiens hō'mō sā'pē-ens
Hydrogenomonas hī-drō-je-nō-mō'näs
Hyphomicrobium hī-fō-mī-krō'bē-um
Ixodes dammini iks-ō'dēs dam'mi-nē
Klebsiella pneumoniae kleb-sē-el'lä nü-mō'nē-ī
Lactobacillus acidophilus lak-tō-bä-sil'lus a-si-dä'fi-lus
L. bulgaricus bul-ga'ri-kus
L. sanfrancisco sän-fran-sis'kō
Laminaria lam-i-nā'rē-ä
Legionella pneumophila lē-jä-nel'lä nü-mō'fi-lä
Leptospira interrogans icterohaemorrhagiae lep-tō-spī'rä in-tėr'rä-ganz ik-tėr-ō-hem-ôr-raj'ē-ī
Leuconostoc mesenteroides lü-kō-nos'tok mes-en-ter-oi'dēz
Listeria monocytogenes lis-te'rē-ä mo-nō-sī-tô'je-nēz
Lycoperdon perlatum lī-kō-pėr'don pėr-lā'tum
Meniscus mē-nis'kus
Methanobacterium meth-a-nō-bak-ti'rē-um
Methylophilus methylotrophus meth-i-lo'fi-lus meth-i-lō-trōf-us
Microcladia mī-krō-klād'ē-ä
Micrococcus luteus mī-krō-kok'kus lū'tē-us
Micromonospora purpureae mī-krō-mo-nä'spô-rä pur-pu-rē'ī
Microsporum gypseum mī-krō-spô'rum jip'sē-um

Moraxella lacunata mô-raks-el'lä la-kü-nä'tä
Mucor mū-kôr
Mycobacterium bovis mī-kō-bak-ti'rē-um bō'vis
M. leprae lep'rī
M. smegmatis smeg-ma'tis
M. tuberculosis tü-ber-kū-lō'sis
Mycoplasma hominis mī-kō-plaz'mä ho'min-nis
M. pneumoniae nu-mō'nē-ī
Naegleria fowleri nī-gle'rē-ä fou'lėr-ī
Nannizia nan-nī'zē-ä
Necator americanus ne-kā'tôr ä-me-ri-ka'nus
Neisseria gonorrhoeae nī-se'rē-ä go-nôr-rē'ä
N. meningitidis me-nin-ji'ti-dis
Nereocystis nē-rē-ō-sis'tis
Nitrobacter nī-trō-bak'tėr
Nitrosomonas nī-trō-sō-mō'näs
Nocardia asteroides nō-kär'dē-ä as-tėr-oi'dēz
Oocystis ō-ō-sis'tis
Ornithodorus ôr-nith-ō'dô-rus
Paracoccidioides brasiliensis par-ä-kok-sid'ē-oi-dēz bra-sil-ē-en'sis
Paracoccus denitrificans pa-rä-kok'kus dē-nī-tri'fi-kanz
Paragonimus westermanni par-ä-gōn'e-mus we-ster-man'nī
Paramecium multimicronucleatum par-ä-mē'sē-um multē-mī-krō-nü-klē'ä-tum
Pasteurella multocida pas-tyėr-el'lä mul-tō'si-dä
Pediculus vestimenti ped-ik'ū-lus ves-ti-men'tē
Pediococcus pe-dē-ō-kok'kus
Penicillium griseofulvum pe-ni-sil'lē-um gri-sē-ō-ful'vum
P. marneffei mar-nif'fē-ī
P. notatum nō-tä'tum
Peridinium per-i-din'ē-um
Petriellidium pet-rē-el-li'dē-um
Phialophora verruscosa fē-ä-lo'fô-rä ver-rü-skō'sä
Phormidium luridum fôr-mi'dē-um lė-rid'um
Phytophthora infestans fī-tof'thô-rä in'fes-tans
Pichia pik'ē-ä
Pityrosporum pit-i-ros'pô-rum
Plasmodium falciparum plaz-mō'dē-um fal-sip'är-um
P. malariae mä-lā'rē-ī
P. ovale ō-vä'lē
P. vivax vī'vaks
Pneumocystis carinii nü-mō-sis'tis kär-i'nē-ī
Propionibacterium acnes prō-pē-on-ē-bak-ti'rē-um ak'nēz
P. freudenreichii froi-den-rīk'ē-ē
Proteus mirabilis prō'tē-us mi-ra'bi-lis
P. vulgaris vul-ga'ris
Prototheca prō-tō-thā'kä
Pseudomonas aeruginosa sū-dō-mō'nas ā-rü-ji-nō'sä
P. cepacia se-pā'sē-ä
P. diminuta di-mi-nü'tä
Quercus kwer'kus
Rhizobium japonicum rī-zō'bē-um jap-on'i-kum
Rhizopus nigricans rī-zō'pus nī'gri-kans
Rhodomicrobium vannielii rō-dō-mī-krō'bē-um van-yel'ē-ē
Rhodospirillum rō-dō-spī-ril'lum
Rickettsia prowazekii ri-ket'sē-ä prou-wä-ze'kē-ē
R. rickettsii ri-ket'sē-ē
R. typhi tī'fē
Riftia rift'ē-ä
Rosa multiflora rō'sä mul-ti-flō'rä
Saccharomyces carlsbergensis sak-ä-rō-mī'sēs kärls-bėr'gen-sis
S. cerevisiae se-ri-vis'ē-ī
S. ellipsoideus ē-lip-soi-dē'us
S. exiguus egz-ij'ū-us
Salmonella cholerae-suis sal-mōn-el'lä kol-ėr-ä-sü'is
S. dublin dub'lin
S. eastbourne ēst'bôrn
S. enteritidis en-tėr-it'id-is
S. paratyphi pa-rä-tī'fē
S. typhi tī'fē
S. typhimurium tī-fi-mūr'ē-um
Sargassum sär-gas'sum
Sartorya sär-tô'rē-ä
Schistosoma skis-tō-sō'mä
Serratia marcescens ser-rä'tē-ä mär-ses'sens
Shigella boydii shi-gel'lä boi'dē-ī
S. dysenteriae dis-en-te'rē-ī
S. flexneri fleks'nėr-ī
S. sonnei sōn'nē-ī
Sphaerotilus natans sfe-rē'ti-lus nā'tans
Spirillum volutans spī-ril'lum vō-lū'tans
Spirosoma spī-rō-sō'mä
Spirulina spī-rü-lī'nä
Sporosarcina ureae spô-rō-sär-sī'nä yė'rē-ī
Sporothrix schenkii spô-rō'thriks shen'kē-ī
Staphylococcus aureus staf-i-lō-kok'kus ô'rē-us
S. epidermidis e-pi-dėr'mi-dis
Stigmatella aurantiaca stig-mä'tel-lä ô-ran-tē-ä-kä
Streptobacillus strep-tō-bä-sil'lus
Streptococcus mitis strep-tō-kok'kus mī'tis
S. mutans mū'tans
S. pneumoniae nü-mō'nē-ī
S. pyogenes pī-äj'en-ēz
S. salivarius sa-li-vā'rē-us
S. sanguis san'gwis
Streptomyces aureofaciens strep-tō-mī'sēs ô-rē-ō-fa'si-ens
S. erythraeus ā-rith'rē-us
S. fradiae frā'dē-ī
S. griseus gri-sē'us
S. kanamyceticus kan-ä-mī-sē'ti-kus
S. nodosus nō-dō'sus
S. noursei nôr'sē-ī
S. parvullus pär-vū'lus
S. rimosus ri-mō'sus
S. venezuelae ve-ne-zü-e'lē
Sulfolobus sul-fō-lō'bus
Taenia saginata te'nē-ä sa-ji-nä'tä
T. solium sō'lē-um
Talaromyces ta-lä-rō-mī'sēs
Thiobacillus ferrooxidans thī-ō-bä-sil'lus fer-rō-oks'i-danz
T. thiooxidans thī-ō-oks'i-danz
Toxoplasma gondii toks-ō-plaz'mä gon'dē-ī

Trachelomonas trä-kel-ō-mōn′as
Treponema pallidum tre-pō-nē′mä pal′li-dum
Triatoma trī-ä-tō′ma
Trichinella spiralis tri-kin-el′lä spī-ra′lis
Trichomonas vaginalis trik-ō-mōn′as va-jin′al-is
Trichophyton trik-ō-fī′ton
Trichosporon trik-ō-spôr′on
Trypanosoma brucei gambiense tri-pan-ō-sō′mä brü-sē′ī gam-bē-ens′
Trypanosoma brucei rhodesiense T. b. rō-dē-sē-ens′
T. cruzi kruz′ē
Ureaplasma urealyticum ū-rē-ä-plas′mä ū-rē-ä-lit′i-kum
Veillonella vi-lo-nel′lä
Vibrio cholerae vib′rē-ō kol′ėr-ī
V. fischeri fish′ėr-ī
V. parahaemolyticus pa-rä-hē-mō-li′ti-kus
Xenopsylla cheopis ze-nop-sil′lä chē-ō′pis
Yersinia enterocolitica yėr-sin′ē-ä en-tėr-ō-kōl-it′ik-ä
Y. pestis pes′tis
Zoogloea zō-ō-glē′ä
Zymomonas zī-mō-mō′näs

APPENDIX C

Word Roots Used in Microbiology

Since scientific names are latinized, Latin rules of grammar pertaining to singular and plural forms of words are used.

	Gender		
	Feminine	Masculine	Neuter
Singular	-a	-us	-um
Plural	-ae	-i	-a
Examples	Alga, algae	Fungus, fungi	Bacterium, bacteria

a-, an- absence, lack. Examples: abiotic, in the absence of life; anaerobic, in the absence of air.

-able able to, capable of. Example: viable, ability to live or exist.

aer- air. Examples: aerobic, in the presence of air; aerate, to add air.

actino- ray. Example: actinomycetes, bacteria that form star-shaped (with rays) colonies.

albo- white. Example: *Streptomyces albus,* produces white colonies.

ameb- change. Example: ameboid, movement involving changing shapes.

amphi- around. Example: amphitrichous, tufts of flagella at both ends of a cell.

amyl- starch. Example: amylase, enzyme that degrades starch.

ana- up. Example: anabolism, building up.

ant-, anti- opposed to, preventing. Example: antimicrobial, a substance that prevents microbial growth.

archae- ancient. Example: archaeobacteria, "ancient" bacteria, thought to be like the first form of life.

asco- bag. Example: ascus, a baglike structure holding spores.

aur- gold. Example: *Staphylococcus aureus,* gold pigmented colonies.

aut-, auto- self. Example: autotroph, self-feeder.

bacillo- a little stick. Example: bacillus, rod-shaped.

basid- base, pedestal. Example: basidium, a cell that bears spores.

bio- life. Example: biology, the study of life and living organisms.

blast- bud. Example: blastospore, spores formed by budding.

bovi- cattle. Example: *Mycobacterium bovis,* a bacterium found in cattle.

brevi- short. Example: *Lactobacillus brevis,* a bacterium with short cells.

butyr- butter. Example: butyric acid, formed in butter, responsible for rancid odor.

campylo- curved. Example: *Campylobacter,* curved rod.

carcin- cancer. Example: carcinogen, a cancer-causing agent.

-caryo, -karyo a nut. Example: eucaryote, a cell with a membrane-enclosed nucleus.

caseo- cheese. Example: caseous, cheeselike.

caul- a stalk. Example: *Caulobacter,* appendaged or stalked bacteria.

cerato- horn. Example: keratin, the horny substance making up skin and nails.

chlamydo- covering. Example: chlamydospores, spores formed inside hypha.

chloro- green. Example: chlorophyll, green-pigmented molecule.

chrom- color. Examples: chromosome, readily stained structure; metachromatic, intracellular colored granules.

chryso- golden. Example: *Streptomyces chryseus,* golden colonies.

-cide killing. Example: bactericide, an agent that kills bacteria.

cili- eyelash. Example: cilia, a hairlike organelle.

cleisto- closed. Example: cleistothecium, completely closed ascus.

co-, con- together. Example: concentric, common center, together in the center.

cocci- a berry. Example: coccus, a spherical-shaped cell.

coeno- shared. Example: coenocyte, cell with many nuclei not separated by septa.

col-, colo- colon. Examples: colon, large intestine; *Escherichia coli,* bacterium found in large intestine.

conidio- dust. Example: conidia, spores developed at end of aerial hypha, never enclosed.

coryne- club. Example: *Corynebacterium,* club-shaped cells.

-cul small form. Example: particle, a small part.

cyano- blue. Example: cyanobacteria, blue-green pigmented organisms.

cyst- bladder. Example: cystitis, inflammation of the urinary bladder.

cyt- cell. Example: cytology, the study of cells.

de- undoing, reversal, loss, removal. Example: deactivation, becoming inactive.

di-, diplo- twice, double. Example: diplococci, pairs of cocci.

dia- through, between. Example: diaphragm, the wall through or between two areas.

dys- difficult, faulty, painful. Example: dysfunction, disturbed function.

ec-, ex-, ecto- out, outside, away from. Example: excrete, to remove materials from the body.

en-, em- in, inside. Example: encysted, enclosed in a cyst.

entero- intestine. Example: *Enterobacter,* bacterium found in the intestine.

epi- upon, over. Example: epidemic, disease over all the people.

erythro- red. Example: erythma, redness of the skin.

eu- well, proper. Example: eucaryote, a proper cell.

exo- outside, outer layer. Example: exogenous, from outside the body.

extra- outside, beyond. Example: extracellular, outside the cells of an organism.

flagell- a whip. Example: flagellum, a projection from a cell; in eucaryotic cells, it pulls cells in a whiplike fashion.

flav- yellow. Example: *Flavobacterium,* cells produce yellow pigment.

fruct- fruit. Example: fructose, fruit sugar.

-fy to make. Example: magnify, to make larger.

galacto- milk. Example: galactose, monosaccharide from milk sugar.

gamet- to marry. Example: gamete, reproductive cell.

gastr- stomach. Example: gastritis, inflammation of the stomach.

gel- to stiffen. Example: gel, solidified colloid.

-gen an agent that initiates. Example: pathogen, any agent that produces disease.

-genesis formation. Example: pathogenesis, production of disease.

germ, germin- bud. Example: germ, part of an organism capable of developing.

-gony reproduction. Example: schizogony, multiple fission producing many new cells.

halo- salt. Example: halophile, an organism that can live in high salt concentrations.

haplo- one, single. Example: haploid, half the number of chromosomes or one set.

hema-, hemato-, hemo- blood. Example: *Hemophilus,* bacterium that requires nutrients from red blood cells.

hepat- liver. Example: hepatitis, inflammation of the liver.

herpes creeping. Example: herpes, or shingles, lesions appear to creep along the skin.

hetero- different, other. Example: heterotroph, obtains organic nutrients from other organisms; other feeder.

hist- tissue. Example: histology, the study of tissues.

hom-, homo- same. Example: homofermenter, an organism that produces only lactic acid from fermentation of a carbohydrate.

hydr-, hydro- water. Example: dehydration, loss of body water.

hyper- excess. Example: hypertonic, having a greater osmotic pressure in comparison to another.

hypo- below, deficient. Example: hypotonic, having a lesser osmotic pressure in comparison to another.

im- not, in. Example: impermeable, not permitting passage.

inter- between. Example: intercellular, between the cells.

intra- within, inside. Example: intracellular, inside the cell.

io- violet. Example: iodine, a chemical element that produces a violet vapor.

iso- equal, same. Example: isotonic, having the same osmotic pressure when compared to another.

-itis inflammation of. Example: colitis, inflammation of the large intestine.

kin- movement. Example: streptokinase, an enzyme that lyses or moves fibrin.

lacti- milk. Example: lactose, the sugar in milk.

leuko- whiteness. Example: leukocyte, white blood cell.

lip-, lipo- fat, lipid. Example: lipase, an enzyme that breaks down fats.

-logy the study of. Example: pathology, the study of changes in structure and function brought on by disease.

lopho- tuft. Example: lophotrichous, having a group of flagella on one side of a cell.

luc-, luci- light. Example: luciferin, substance in certain organisms that emits light when acted upon by the enzyme luciferase.

lute-, luteo- yellow. Example: *Micrococcus luteus,* yellow colonies.

-lysis loosening, to break down. Example: hydrolysis, chemical decomposition of a compound into other compounds as a result of taking up water.

macro- largeness. Example: macromolecules, large molecules.

meningo- membrane. Example: meningitis, inflammation of the membranes of the brain.

meso- middle. Example: mesophile, an organism whose optimum temperature is in the middle range.

meta- beyond, between, transition. Example: metabolism, chemical changes occurring within a living organism.

micro- smallness. Example: microscope, instrument used to make small objects appear larger.

-mnesia memory. Examples: amnesia, loss of memory; anamnesia, return of memory.

-monas a unit. Example: *Methylomonas,* a unit (bacterium) that utilizes methane as its carbon source.

mono- singleness. Example: monotrichous, having one flagellum.

morpho- form. Example: morphology, the study of form and structure of organisms.

multi- many. Example: multinuclear, having several nuclei.

mur- wall. Example: murein, component of bacterial cell walls.

mus-, muri- mouse. Example: murine typhus, a form of typhus endemic in mice.

mut- to change. Example: mutation, a sudden change in characteristics.

myco-, -mycetoma, -myces a fungus. Example: *Saccharomyces,* sugar fungus, a genus of yeast.

myxo- slime, mucus. Example: Myxobacteriales, an order of slime-producing bacteria.

necro- a corpse. Example: necrosis, cell death or death of a portion of tissue.

nigr- black. Example: *Aspergillus niger,* fungus that produces black conidia.

ob- before, against. Example: obstruction, impeding or blocking up.

oculo- eye. Example: monocular, pertaining to one eye.

-oecium, -ecium a house. Examples: perithecium, ascus with an opening that encloses spores; ecology, the study of the relationships between organisms and between an organism and its environment (household).

-oid like, resembling. Example: coccoid, resembling a coccus.

-oma tumor. Example: lymphoma, a tumor of the lymphatic tissues.

-ont being, existing. Example: schizont, a cell existing as a result of schizogony.

ortho- straight, direct. Example: orthomyxovirus, a virus with a straight, tubular capsid.

-osis, -sis condition of. Examples: lysis, the condition of loosening; symbiosis, the condition of living together.

pan- all, universal. Example: pandemic, an epidemic affecting a large region.

para- beside, near. Example: parasite, an organism that "feeds beside" another.

peri- around. Example: peritrichous, projections from all sides.

phaeo- brown. Example: Phaeophyta, brown algae.

phago- eat. Example: phagocyte, a cell that engulfs and digests particles or cells.

philo-, -phil liking, preferring. Example: thermophile, an organism that prefers high temperatures.

-phore bears, carries. Example: conidiophore, a hypha that bears conidia.

-phyll leaf. Example: chlorophyll, the green pigment in leaves.

-phyte plant. Example: saprophyte, a plant that obtains nutrients from decomposing organic matter.

pil- a hair. Example: pilus, hairlike projection from a cell.

plano- wandering, roaming. Example: plankton, organisms drifting or wandering in water.

plast- formed. Example: plastid, formed body within a cell.

-pnoea breathing. Example: dyspnea, difficulty in breathing.

pod- foot. Example: pseudopod, footlike structure.

poly- many. Example: polymorphism, many forms.

post- after, behind. Example: posterior, places behind (a specific) part.

pre-, pro- before, ahead of. Examples: procaryote, cell with the first nucleus evolutionarily; pregnant, before birth.

pseudo- false. Example: pseudopod, false foot.

psychro- cold. Example: psychrophile, an organism that grows best at cold temperatures.

-ptera wing. Example: Diptera, order of true flies, insects with two wings.

pyo- pus. Example: pyogenic, pus-forming.

rhabdo- stick, rod. Example: Rhabdovirus, an elongated, bullet-shaped virus.

rhin- nose. Example: rhinitis, inflammation of mucous membranes in the nose.

rhizo- root. Examples: *Rhizobium,* bacterium that grows in plant roots; mycorrhiza, mutualism between a fungus and the roots of a plant.

rhodo- red. Example: *Rhodospirillum,* red-pigmented spiral-shaped bacterium.

rod- gnaw. Example: rodents, class of mammals with gnawing teeth.

rubri- red. Example: *Clostridium rubrum,* red-pigmented colonies.

rumin- throat. Example: *Ruminococcus,* bacterium associated with a rumen.

saccharo- sugar. Example: disaccharide, a sugar consisting of two simple sugars.

sapr- rotten. Example: *Saprolegnia,* fungus that lives on dead animals.

sarco- flesh. Example: sarcoma, a tumor of muscle or connective tissues.

schizo- split. Example: schizomycetes, organisms that reproduce by splitting, an early name for bacteria.

scolec- worm. Example: scolex, the head of a tapeworm.

-scope, -scopic watcher. Example: microscope, an instrument used to watch small things.

semi- half. Example: semicircular, having the form of half a circle.

sept- rotting. Example: aseptic, free from bacteria that could cause decomposition.

septo- partition. Example: septum, a cross-wall in a fungal hypha.

serr- notched. Example: serrate, with a notched edge.

sidero- iron. Example: *Siderococcus,* a bacterium capable of oxidizing iron.

siphon- tube. Example: Siphonaptera, order of fleas, insects with tubular mouths.

soma- body. Example: somatic cells, cells of the body other than gametes.

speci- particular thing. Examples: species, the smallest group of organisms with similar properties; specify, to indicate exactly.

spiro- coil. Example: spirochete, bacterium with a coiled cell.

sporo- spore. Example: sporangium, a structure that holds spores.

staphylo- grapelike cluster. Example: *Staphylococcus,* bacterium that forms clusters of cells.

-stasis arrest, fixation. Example: bacteriostasis, cessation of bacterial growth.

strepto- twisted. Example: *Streptococcus,* bacterium that forms twisted chains of cells.

sub- beneath, under. Example: subcutaneous, just under the skin.

super- above, upon. Example: superior, quality or state of being above others.

sym-, syn- together, with. Examples: synapse, the region of communication between two neurons; synthesis, putting together.

-taxi to touch. Example: chemotaxis, response to the presence (touch) of chemicals.

taxis- orderly arrangement. Example: taxonomy, the science dealing with arranging organisms into groups.

thallo- plant body. Example: thallus, an entire macroscopic fungus.

therm- heat. Example: thermometer, an instrument used to measure heat.

thio- sulfur. Example: *Thiobacillus,* a bacterium capable of oxidizing sulfur-containing compounds.

-tome, -tomy to cut. Example: appendectomy, surgical removal of the appendix.

-tone, -tonic strength. Example: hypotonic, having less strength (osmotic pressure).

tox- poison. Example: antitoxic, effective against poison.

trans- across, through. Example: transport, movement of substances.

tri- three. Example: trimester, three-month period.

trich- a hair. Example: peritrichous, hairlike projections from cells.

-trope turning. Example: geotropic, turning towards the earth (pull of gravity).

-troph food, nourishment. Example: trophic, pertaining to nutrition.

-ty condition of, state. Example: immunity, condition of being resistant to disease or infection.

undul- wavy. Example: undulating, rising and falling, presenting a wavy appearance.

uni- one. Example: unicellular, pertaining to one cell.

vaccin- cow. Example: vaccination, injection of a vaccine (originally pertained to cows).

vacu- empty. Example: vacuoles, an intracellular space that appears empty.

vesic- bladder. Example: vesicle, a bubble.

vitr- glass. Example: in vitro, in culture media in a glass (or plastic) container.

-vorous eat. Example: carnivore, an animal that eats other animals.

xantho- yellow. Example: *Xanthomonas,* produces yellow colonies.

xeno- strange. Example: axenic, sterile, free of strange organisms.

xero- dry. Example: xerophyte, plant that tolerates dry conditions.

xylo- wood. Example: xylose, a sugar obtained from wood.

zoo- animal. Example: zoology, the study of animals.

zygo- yoke, joining. Example: zygospore, spore formed from the fusion of two cells.

-zyme ferment. Example: enzyme, protein in living cells that catalyzes chemical reactions.

APPENDIX D

Most Probable Numbers (MPN) Table

MPN Index for Various Combinations of Positive and Negative Results When Five 10-ml Portions, Five 1-ml Portions, and Five 0.1-ml Portions Are Used (see pp. 175, 176)

No. of Tubes Giving Positive Reaction out of			MPN Index per 100 ml
5 of 10 ml Each	5 of 1 ml Each	5 of 0.1 ml Each	
0	0	0	<2
0	0	1	2
0	1	0	2
0	2	0	4
1	0	0	2
1	0	1	4
1	1	0	4
1	1	1	6
1	2	0	6
2	0	0	5
2	0	1	7
2	1	0	7
2	1	1	9
2	2	0	9
2	3	0	12
3	0	0	8
3	0	1	11
3	1	0	11
3	1	1	14
3	2	0	14
3	2	1	17
3	3	0	17
4	0	0	13
4	0	1	17
4	1	0	17
4	1	1	21
4	1	2	26
4	2	0	22
4	2	1	26
4	3	0	27

MPN Index, continued

No. of Tubes Giving Positive Reaction out of			MPN Index per 100 ml
5 of 10 ml Each	5 of 1 ml Each	5 of 0.1 ml Each	
4	3	1	33
4	4	0	34
5	0	0	23
5	0	1	31
5	0	2	43
5	1	0	33
5	1	1	46
5	1	2	63
5	2	0	49
5	2	1	70
5	2	2	94
5	3	0	79
5	3	1	110
5	3	2	140
5	3	3	180
5	4	0	130
5	4	1	170
5	4	2	220
5	4	3	280
5	4	4	350
5	5	0	240
5	5	1	350
5	5	2	540
5	5	3	920
5	5	4	1600
5	5	5	≧2400

Source: *Standard Methods for the Examination of Water and Wastewater*, 13th ed., American Public Health Association, New York, 1971.

APPENDIX E

Methods for Taking Clinical Samples

To diagnose a disease, it is often necessary to obtain a sample of material that may contain the disease-causing organism. Samples must be taken aseptically. The sample container should be labeled with the patient's name, room number (if hospitalized), date, time, and medications being taken. Samples must be transported to the laboratory immediately for culture. Delay in transport may result in growth of some organisms and their toxic products may kill other organisms. Pathogens tend to be fastidious and die without their optimum environmental conditions.

In the laboratory, samples from infected tissues are cultured on differential and selective media in an attempt to isolate and identify any pathogens or organisms that are not normally found in association with that tissue.

Wound or Abscess Culture

1. Cleanse the area with a sterile swab moistened in sterile saline.
2. Disinfect the area with 70% ethyl alcohol or iodine solution.
3. If the abscess has not ruptured spontaneously, a physician will open it with a sterile scalpel.
4. Wipe the first pus away.
5. Touch a sterile swab to the pus taking care not to contaminate the surrounding tissue.
6. Replace the swab in its container and properly label the container.

Ear Culture

1. Clean the skin and auditory canal with 1% tincture of iodine.
2. Touch the infected area with a sterile cotton swab.
3. Replace the swab in its container.

Eye Culture

This procedure is usually performed by an ophthalmologist.

1. Anesthetize the eye with topical application of a sterile anesthetic solution.
2. Wash the eye with sterile saline solution.
3. Collect material from the infected area with a sterile cotton swab. Return the swab to its container.

Blood Culture

1. Close room windows to avoid contamination.
2. Clean skin around selected vein with 2% tincture of iodine on a cotton swab.
3. Remove dried iodine with gauze moistened with 80% isopropyl alcohol.
4. Draw a few milliliters of venous blood.
5. Aseptically bandage puncture.

Urine Culture

1. Provide the patient with a sterile container.
2. Instruct the patient to collect a mid-stream sample. This is obtained by voiding a small volume from the bladder before collection. This washes away extraneous skin flora.
3. A urine sample may be stored under refrigeration (4–6°C) for up to 24 hours.

Fecal Culture

For bacteriological examination, only a small sample is needed. This may be obtained by inserting a sterile swab into the rectum or feces. The swab is then placed in a tube of sterile enrichment broth for transport to the laboratory. For examination for parasites, a small sample may be taken from a morning stool. The sample is placed in a preservative (polyvinyl alcohol, buffered glycerol, saline, or formalin) for microscopic examination for eggs and adult parasites.

Sputum Culture

1. A morning sample is best as microorganisms will have accumulated while the patient is sleeping.
2. Patient should rinse his/her mouth thoroughly to remove food and normal flora.
3. Patient should cough deeply from the lungs and expectorate into a sterile glass wide-mouth jar.
4. Care should be taken to avoid contamination of personnel.
5. In cases such as tuberculosis where there is little sputum, stomach aspiration may be necessary.
6. Infants and children tend to swallow sputum. A fecal sample may be of some value in these cases.

Glossary

abscess A localized accumulation of pus.

acetyl coenzyme A (acetyl-CoA) A substance composed of an acetyl group and a carrier molecule called coenzyme A (CoA); provides the means for pyruvic acid to enter the Krebs cycle.

acetyl group

$$CH_3-\overset{\overset{\Large O}{\|}}{C}-$$

acid A substance that dissociates into one or more hydrogen ions and one or more negative ions.

acidic dyes A salt in which the color is in the negative ion; used for negative staining.

acid-fast stain A differential stain used to identify bacteria that are not decolorized by acid-alcohol.

acquired immunity The ability, obtained during the life of the individual, to produce specific antibodies.

actinomycetes Gram-positive bacteria that tend to form branching filaments; may form true mycelia; may produce conidiospores; see ***Bergey's Manual*** vols. 2 and 4.

activated sludge Aerobic digestion used in secondary sewage treatment.

activation energy The minimum collision energy required for a chemical reaction to occur.

actively acquired immunity Production of antibodies by an individual in response to an antigen.

active site Place on an enzyme that interacts with the substrate.

active transport Net movement of a substance across a membrane against a concentration gradient; requires energy.

acute disease A disease in which symptoms develop rapidly but last for only a short time.

adenine A purine nucleic acid base that pairs with thymine in DNA and uracil in RNA.

adenocarcinoma Cancer of glandular tissue.

adenosine diphosphate (ADP) The substance formed when ATP is split and energy released.

adenosine diphosphoglucose (ADPG) Glucose activated by ATP; precursor for glycogen synthesis.

adenosine triphosphatase The enzyme that catalyzes the following reactions: $ADP + P_i \longrightarrow ATP$ and $ATP \longrightarrow ADP + P_i$.

adenosine triphosphate (ATP) An important intracellular energy source.

adherence Attachment of a microbe to a host cell or other surface.

adhesin *See* ligand.

aerial mycelium A mycelium composed of fungal hyphae that project above the surface of the growth medium and produce asexual spores.

aerobic Requiring oxygen (O_2) to grow.

aerobic respiration Respiration in which the final electron acceptor in the electron transport chain is oxygen (O_2).

aerotolerant anaerobe An organism that does not use oxygen (O_2) but is not affected by its presence.

aflatoxin $C_{17}H_{10}O_6$, a carcinogenic toxin produced by ***Aspergillus flavus.***

agar A complex polysaccharide derived from a marine alga and used as a solidifying agent in culture media.

agglutination A joining together or clumping of cells.

agglutinin An antibody that causes an agglutination reaction.

agranulocyte A leukocyte without granules in the cytoplasm; includes lymphocytes and monocytes.

airborne transmission Spread of pathogens farther than 1 meter in air from reservoir to susceptible host.

alcohol An organic molecule with the functional group $-OH$.

aldehyde An organic molecule with the functional group $-\underset{\underset{\Large O}{\|}}{C}-H$

algae A group of photosynthetic eucaryotes; some are included in the kingdom Protista and some in the kingdom Plantae.

alkaline Having more OH^- ions than H^+ ions; pH is greater than 7.

alkaline phosphatase An enzyme that removes PO_4 from an organic molecule; optimum pH 8.6.

allergen An antigen that evokes a hypersensitivity response.

allergy *See* hypersensitivity.

allograft A graft between persons who aren't identical twins.

allosteric site The site on an enzyme at which an inhibitor binds.

allosteric transition The process in which an enzyme's activity is changed because of binding on the allosteric site.

alpha-amino acid An amino acid with $-COOH$ and $-NH_2$ attached to the same carbon atom.

alum Aluminum sulfate.

amanitin Polypeptide mushroom toxin that causes liver and nerve damage.

Ames test A procedure using bacteria to identify potential carcinogens.

amination The addition of an amino group.

amino acid An organic acid containing an amino group and a carboxyl group.

aminoglycoside An antibiotic consisting of amino sugars and an aminocyclitol ring; for example, streptomycin.

amino group $-NH_2$

ammonification Removal of amino groups from amino acids to form ammonia.

amoeba An organism belonging to the Protista kingdom that moves by means of pseudopods.

amphibolic pathway A pathway that is anabolic and catabolic.

amphitrichous Having tufts of flagella at both ends of a cell.

anabolism All synthesis reactions in a living organism.

anaerobic Not requiring oxygen (O_2) for growth.

anaerobic respiration Respiration in which the final electron acceptor in the electron transport chain is an inorganic molecule other than oxygen (O_2); for example, a nitrate or sulfate.

anaerobic sludge digester Anaerobic digestion used in secondary sewage treatment.

anal pore A site in certain protozoans for elimination of waste.

analytic epidemiology Comparison of a diseased group and a healthy group to determine the cause of the disease.

anamnestic response A rapid rise in antibody titer following exposure to an antigen after the primary response to that antigen.

anaphylaxis A hypersensitivity reaction involving IgE antibodies, mast cells, and basophils.

angstrom (Å) A unit of measurement equal to 10^{-10}m, $10^{-4}\mu$m, and 10^{-1}nm; no longer an official unit.

animalia A kingdom composed of multicellular eucaryotes lacking cell walls.

animal virus A virus that multiplies in animal tissues.

anion An ion with a negative charge.

anoxygenic Not producing molecular oxygen; typically of bacterial photosynthesis.

antagonism Active opposition; for example, between two drugs or two microbes.

antibiotic An antimicrobial agent produced naturally by a bacterium or fungus.

antibody A protein produced by the body in response to an antigen and capable of combining specifically with that antigen.

antibody titer The amount of antibody in serum.

anticodon The three nucleotides by which a transfer RNA recognizes an RNA codon.

antigen Any substance that, when introduced into the body, causes antibody formation and reacts only with its specific antibody.

antigenic determinant site A specific region on the surface of an antigen against which antibodies are formed.

antigenic drift Minor variations in the antigenic makeup of a virus that occur with time.

antigenic shift Major genetic changes in influenza viruses causing changes in H and N antigens.

antigen modulation The process whereby tumor cells shed their specific antigens, thus avoiding the host's immune response.

antigen receptors Antibodylike molecules on T cells.

antihuman gamma globulin Antibodies that react specifically with human antibodies.

antimetabolite Any substance that interferes with metabolism by competitive inhibition of an enzyme.

antimicrobial agent A chemical that destroys pathogens without damaging body tissues.

antisense strand (−strand) Viral RNA that cannot act as mRNA.

antiseptic A chemical for disinfection of the skin, mucous membranes, or other living tissues.

antiseptic surgery Employing antiseptics, disinfection, and asepsis during operations.

antiserum A solution containing antibodies.

antitoxin A specific antibody produced by the body in response to a bacterial exotoxin or its toxoid.

apoenzyme The protein portion of an enzyme, which requires activation by a coenzyme.

A protein Staphylococcal cell wall protein that prevents phagocytosis.

arbuscule Fungal mycelia in plant root cells.

arthropod An animal phylum characterized by an exoskeleton and jointed legs, includes insects and ticks.

arthrospore An asexual fungal spore formed by fragmentation of a septate hypha.

artificially acquired active immunity The production of antibodies by the body in response to a vaccination.

artificially acquired passive immunity The transfer of humoral antibodies formed by one individual to a susceptible individual, accomplished by injection of antiserum.

ascospore A sexual fungal spore produced in an ascus, formed by the Ascomycetes.

ascus A saclike structure containing ascospores.

asepsis The absence of contamination by unwanted organisms.

asexual reproduction Reproduction without opposite mating strains.

asthma An allergic response characterized by bronchial spasms and difficult breathing.

atom The smallest unit of matter that can enter into a chemical reaction.

atomic number The number of protons in the nucleus of an atom.

atomic weight The total number of protons and neutrons in the nucleus of an atom.

attenuation Lessening of virulence of a microorganism.

autoclave Equipment for sterilization by steam under pressure, usually operated at 15 psi and 121°C.

autograph A tissue graft from one's self.

autoimmunity An immunologic response against a person's own tissue antigens.

autotroph An organism that uses carbon dioxide (CO_2) as its principal carbon source.

auxotroph A mutant microorganism with a nutritional requirement not possessed by the parent.

axial filament The structure for motility found in spirochetes.

bacillus Any rod-shaped bacterium; when written as a genus, refers to rod-shaped, endospore-forming, facultatively anaerobic, gram-positive bacteria.

bacteremia A condition of bacteria in the blood.

bacteria All living organisms with procaryotic cells.

bacterial growth curve A graph indicating the growth of a bacterial population over time.

bactericidal Capable of killing bacteria.

bacteriochlorophyll The light-absorbing pigment found in green sulfur and purple sulfur bacteria.

bacteriorhodopsin The light-absorbing purple pigment in Halobacterium's cell membrane.

bacteriocinogenic plasmid Plasmid containing genes for the synthesis of bacteriocins.

bacteriocins Toxic proteins produced by bacteria that kill other bacteria.

bacteriophage A virus that multiplies in bacterial cells.

bacteriostatic Capable of inhibiting bacterial growth.

bacteroid Enlarged *Rhizobium* cells found in root nodules.

basal body A structure that anchors flagella to the cell wall and plasma membrane.

base A substance that accepts hydrogen ions and is capable of uniting with water to form an acid.

base analog A chemical that is structurally similar to the normal nitrogenous bases in nucleic acids but with altered base-pairing properties.

base pairs The arrangement of nitrogenous bases in nucleic acids based on hydrogen bonding; in DNA, base pairs are A–T and G–C; in RNA, base pairs are A–U and G–C.

base substitution The replacement of a single base in DNA by another base, causing a mutation.

basic dye A salt in which the color is in the positive ion; used for bacterial stains.

basidiospore A sexual fungal spore produced in a basidium, characteristic of the Basidiomycetes.

basidium A pedestal that produces basidiospores; found in basidiomycetes.

basophil A granulocyte that readily takes up basic dye.

B cells Stem cells differentiate into plasma cells that secrete antibodies.

BCG vaccine A live, attenuated strain of *Mycobacterium bovis* used to provide immunity to tuberculosis.

benign tumor A noncancerous tumor.

benthic zone The sediment at the bottom of a body of water.

Bergey's Manual The taxonomic reference on bacteria.

beta-oxidation The removal of two carbon units of a long chain of fatty acid to form acetyl-CoA.

binary fission Bacterial reproduction by division into two daughter cells.

binomial nomenclature The system of having two names (genus and specific epithet) for each organism.

biochemical oxygen demand (BOD) A measure of the biologically degradable organic matter in water.

biochemical pathway A sequence of enzymatically catalyzed reactions occurring in a cell.

biochemistry The science of chemical processes in living organisms.

bioconversion Changes in organic matter brought about by the growth of microorganisms.

biogenesis The concept that living cells can only arise from preexisting cells.

biogeochemical cycles The recycling of chemical elements by microorganisms for use by other organisms.

biological transmission Transmission of a pathogen from one host to another when the pathogen reproduces in the vector.

bioluminescence The emission of light from the electron transport chain of certain living organisms.

biomass Organic matter produced by living organisms and measured by weight.

biotechnology The use of microbes in industry.

bladder A bag. In brown algae, an air sac.

blastospore An asexual fungal spore produced by budding from the parent cell.

blocking antibody An IgG antibody that reacts with an allergen to prevent a hypersensitivity reaction.

blood-brain barrier Cell membranes that allow some substances to pass from the blood to the brain but restrict others.

blooms (algal) Abundant growth of microscopic algae, producing visible colonies in nature.

booster dose The administration of antigens to elicit an anamnestic response.

brightfield microscope A microscope that uses visible light for illumination; the specimens are viewed against a white background.

broad-spectrum antimicrobial agent A chemical that has antimicrobial activity against many infectious microorganisms.

Brownian movement The movement of particles including microorganisms in a suspension due to bombardment by the moving molecules in the suspension.

bubo An enlarged lymph node caused by inflammation.

budding Asexual reproduction beginning as a protuberance from the parent cell that grows to become a daughter cell; also, release of an enveloped virus through the plasma membrane of an animal cell.

buffer A substance that tends to stabilize the pH of a solution.

burst size The number of newly synthesized bacteriophage particles released from a single cell.

burst time The time required from bacteriophage adsorption to release.

cancer A malignant, invasive cellular tumor that has the capability of spreading throughout the body or body parts.

candle jar Sealed jar containing a lighted candle, used to incubate bacterial cultures in high CO_2 atmosphere.

capsid The protein coat of a virus that surrounds the nucleic acid.

capsomere A protein subunit of a capsid.

capsule An outer, viscous covering on some bacteria composed of a polysaccharide or polypeptide.

carbohydrates Organic compounds composed of carbon, hydrogen, and oxygen, with the hydrogen and oxygen present in a 2:1 ratio; includes starches, sugars, and cellulose.

carbon cycle The series of processes that converts carbon dioxide (CO_2) to organic substances and back to carbon dioxide in nature.

carbon skeleton The basic chain ring of carbon atoms in a molecule; —C—C—C—, for example.

carboxyl group —COOH.

carboxysome Procaryotic inclusion containing ribulose 1,5-diphosphate carboxylase.

carbuncle An inflammation of the skin and subcutaneous tissue due to dissemination of a furuncle.

carcinogen Any cancer-producing substance.

cardiolipin A beef heart extract used in the venereal disease research laboratory (VDRL) slide test to detect antibodies against syphilis.

carrier An individual who harbors a pathogen but exhibits no signs of illness.

casein Milk protein.

catabolism All decomposition reactions in a living organism.

catalase An enzyme that catalyzes the breakdown of hydrogen peroxide to water and oxygen.

catalyst A substance that affects the rate of a chemical reaction, usually increasing the rate, but isn't changed in the reaction.

cation A positively charged ion.

cDNA (copy DNA) DNA made from mRNA in vitro.

cell The basic microscopic unit of structure and function of all living organisms.

cell culture Animal cells grown in vitro.

cell-mediated immunity An immune response that involves the binding and elimination of antigens by T cell lymphocytes.

cell theory The principle that all living things are composed of cells.

cellulose A polysaccharide that is the main component of plant cell walls.

centrioles Paired, cylindrical structures found in the centrosome of eucaryotic cells.

centrosome A dense area of cytoplasm near the nucleus of eucaryotic cells; involved in mitosis.

cephalosporin An antibiotic produced by the fungus *Cephalosporium* that inhibits the synthesis of gram-positive bacterial cell walls.

cercaria A free-swimming larva of trematodes.

chancre A hard sore, the center of which ulcerates.

chemical bond Attractive force between atoms forming a molecule.

chemical element A fundamental substance composed of atoms that have the same atomic number and behave the same way chemically.

chemical energy The energy of a chemical reaction.

chemically defined medium A culture medium in which the exact chemical composition is known.

chemical reaction The process of making or breaking bonds between atoms.

chemiosmosis The movement of protons across a cytoplasmic membrane; can be used to generate ATP.

chemistry The science of the interactions of atoms and molecules.

chemoautotroph An organism that uses an inorganic chemical as an energy source, and carbon dioxide (CO_2) as a carbon source.

chemoheterotroph An organism that uses organic molecules as a source of carbon and energy.

chemostat An apparatus to keep a culture in log phase indefinitely.

chemotaxis A response to the presence of a chemical.

chemotherapy Treatment of a disease with chemical substances.

chemotroph An organism that uses oxidation-reduction reactions as its primary energy source.

chitin A glucosamine polysaccharide that is the main component of fungal cell walls and arthropod skeletons.

chlamydospore An asexual fungal spore formed within a hypha.

chloramphenicol A broad-spectrum bacteriostatic chemical.

chlorobium vesicle *See* chlorosome.

chlorophyll *a* The light-absorbing pigment in cyanobacteria, algae, and plants.

chloroplast The organelle that performs photosynthesis in photoautotrophic eucaryotes.

chlorosome Plasma membrane folds in green sulfur bacteria containing bacteriochlorophylls.

chromatin Threadlike, uncondensed DNA in an interphase eucaryotic cell.

chromatophore An infolding in the plasma membrane where bacteriochlorophyll is located in photoautotrophic bacteria.

chromosome The structure that carries hereditary information.

chronic disease An illness that develops slowly and is likely to continue or recur for long periods of time.

cilia Relatively short cellular projections that move in a wavelike manner.

ciliate A member of the protozoan phylum Ciliata that uses cilia for locomotion.

cisternae Stacked elements of the Golgi complex.

class A taxonomic ranking between phylum and order.

clone A population of cells that are identical to the parent cell.

coagulase A bacterial enzyme that causes blood plasma to clot.

coccobacillus A bacterium that is an oval-shaped rod.

coccus A spherical or ovoid bacterium.

codon A group of three nucleotides in DNA or mRNA that specifies the insertion of an amino acid into a protein.

coenocytic hyphae Fungal filaments that are not divided into uninucleate cell-like units because they lack septa.

coenzyme A nonprotein substance that is associated with and that activates an enzyme.

coenzyme A A coenzyme that functions in decarboxylation.

cofactor The nonprotein component of an enzyme.

colicins Bacteriocins produced by *Escherichia coli.*

coliforms Aerobic or facultatively anaerobic, gram-negative, non-spore-forming, rod-shaped bacteria that ferment lactose with acid and gas formation within 48 hours at 35°C.

collagen The main structural protein of muscles.

collision theory The principle that chemical reactions are due to energy gained as particles collide.

colony A clone of bacterial cells on a solid medium that is visible to the naked eye.

colony-forming units Visible units counted in a plate count, which may be formed from a group of cells rather than from one cell.

commensalism A system of interaction in which two organisms live in association and one is benefited while the other is neither benefited nor harmed.

commercial sterilization A process of treating canned goods aimed at destroying the endospores of *Clostridium botulinum*.

common vehicle transmission Transmission of a pathogen to a large number of people by an inanimate reservoir.

communicable disease Any disease that can be spread from one host to another.

competence The physiological state in which a recipient cell can take up and incorporate a large piece of donor DNA.

competitive inhibition The process by which a chemical competes with the normal substrate for the active site of an enzyme.

complement (C) A group of 11 serum proteins involved in phagocytosis and lysis of bacteria.

complement fixation The process in which complement combines with an antigen-antibody complex.

complete digestive system A digestive system with a mouth and an anus.

completed test The final test for detection of coliforms in the multiple-tube fermentation test.

complex medium A culture medium in which the exact chemical composition is not known.

complex virus A virus with a complicated structure, such as a bacteriophage.

compound A substance composed of two or more different chemical elements.

compound light microscope An instrument with two sets of lenses that uses visible light as the source of illumination.

compromised host A person whose resistance to infection is impaired.

condensation reaction A chemical reaction in which a molecule of water is released.

condenser A lens system located below the microscope stage that directs light rays through the specimen.

confirmed test The second stage of the multiple-tube fermentation test, used to identify coliforms on solid, differential media.

congenital disease A disease present at birth as a result of some condition that occurred in utero.

conidiophore An aerial hypha bearing conidiospores.

conidiospore An asexual spore produced in a chain from a conidiophore.

conjugation The transfer of genetic material from one cell to another involving cell-to-cell contact.

conjugative plasmid A plasmid with genes for carrying out conjugation.

contact transmission Spread of disease by direct contact, indirect contact, or by droplet.

constitutive enzyme An enzyme that is produced regardless of how much substrate is present.

contact inhibition The cessation of animal cell movement and division due to contact.

contagious disease A disease that is easily spread from one person to another.

continuous cell line Animal cells that can be maintained through an indefinite number of generations in vitro.

convalescent period The period of recovery from a disease.

corepressor The molecule (end product) that brings about repression of a repressible enzyme.

corticosteroids Steroid hormones released by the adrenal gland, derivatives of which are used to treat inflammatory diseases.

countercurrent immunoelectrophoresis (CIE) The movement of antigen and antibody towards each other through an electric field. See also Electrophoresis.

counterstain A stain used to give contrast in a differential stain.

coupled reaction Two chemical reactions that must occur simultaneously.

covalent bond A chemical bond in which the electrons of one atom are shared with another atom.

cresols A mixture of isomers from petroleum.

crisis The phase of a fever characterized by vasodilation and sweating.

cristae Foldings of the inner membrane of a mitochondrion.

cross-over Process by which a portion of a chromosome is exchanged with a portion of another chromosome.

culture Microorganisms that grow and multiply in a container of culture medium.

culture medium The nutrient material prepared for growth of microorganisms in a laboratory.

curd The solid part of milk that separates from the liquid when the milk is fermented.

cutaneous mycosis A fungal infection of the epidermis, nails, and hair.

cuticle Nonliving outer covering of helminths.

cyanobacteria Members of the kingdom Monera, formerly called blue-green algae.

cyclic photophosphorylation Movement of an electron from chlorophyll through a series of electron acceptors, and back to chlorophyll; purple and green bacterial photophosphorylation.

cyst A sac with a distinct wall containing fluid or other material; also, a protective capsule of some protozoans.

cysticercus Encysted tapeworm larva.

cystitis Inflammation of the urinary bladder.

cytochrome oxidase An enzyme that oxidizes cytochrome *c*.

cytochromes Proteins that function as electron carriers in respiration and photosynthesis.

cytocidal Resulting in cell death.

cytomegaly Enlarged cells.

cytopathic effects (CPE) Tissue deterioration caused by viruses.

cytoplasm In a procaryote, everything inside the plasma membrane; in a eucaryote, everything inside the plasma membrane and external to the nucleus.

cytoplasmic streaming The flowing of cytoplasm in a eucaryotic cell.

cytosine A pyrimidine nucleic acid base that pairs with guanine.

cytoskeleton Microfilaments and microtubules which provide support and movement for eucaryotic cytoplasm.

cytostome The mouthlike opening in some protozoans.

cytotoxic T cells (Tc) Cells that destroy antigens.

cytotrophic Binding to cells; for example, IgE antibodies bind to target cells.

D- Prefix describing a stereoisomer.

Dane particle A structure in the serum of a patient with hepatitis B having hepatitis B virus (HBV) surface antigens.

darkfield microscope A microscope that has a device to scatter light from the illuminator so that the specimen appears white against a black background.

deamination Removal of an amino group.

death phase Period of logarithmic decrease in a bacterial population.

debridement Surgical removal of necrotic tissue.

decarboxylation Removal of carbon dioxide (CO_2) from an amino acid.

decimal reduction time The time (in minutes) required to kill 90% of a bacterial population at a given temperature.

decolorization The process of removing a stain.

decomposition reaction A chemical reaction in which bonds are broken to produce smaller parts from a large molecule.

definitive host An organism that harbors the adult, sexually mature form of a helminthic parasite.

degeneracy (of the genetic code) Most amino acids are signaled by several codons.

degenerative disease Loss of function due to wearing down of parts.

degranulation Release of contents of secretory granules.

dehydration The removal of water.

dehydrogenation The loss of hydrogen atoms from a substrate.

delayed hypersensitivity Cell-mediated hypersensitivity.

delayed hypersensitivity T cells (Td) Cells that produce lymphokines.

denaturation A change in the molecular structure of a protein.

denitrification The reduction of nitrates to nitrites or nitrogen gas.

dental plaque A combination of bacterial cells, dextran, and debris adhering to the teeth.

deoxyribonucleic acid (DNA) The nucleic acid of genetic material.

deoxyribose A five-carbon sugar contained in DNA nucleotides.

dermatophyte A fungus that causes a cutaneous mycosis.

dermis The inner portion of the skin.

descriptive epidemiology Analysis of all data regarding the occurrence of a disease to determine the cause of the disease.

desensitization The prevention of allergic inflammatory responses.

desiccation The absence of water.

detergents Any substance that reduces the surface tension of water.

dextran A polymer of glucose.

diacetyl $CH_3COCOCH_3$ produced from carbohydrate fermentation.

diagnosis Identification of a disease.

diapedesis The process by which phagocytes move out of blood vessels.

Dick test A skin test to determine immunity to scarlet fever.

differential count The number of each kind of leukocyte in a sample of 100 leukocytes.

differential interference contrast (DIC) microscope A microscope that provides a three-dimensional image.

differential medium A solid culture medium that makes it easier to distinguish colonies of the desired organism.

differential stain A stain that distinguishes objects on the basis of reactions to the staining procedure.

diffusion The net movement of molecules or ions from an area of higher concentration to an area of lower concentration.

digestion The process of breaking down substances physically and chemically.

digestive vacuole An organelle in which substrates are broken down enzymatically.

dimorphism The property of having two growth forms.

dioecious Referring to organisms in which organs of different sexes are located in different individuals.

diphtheroid Gram-positive pleomorphic rod.

diploid Having two sets of chromosomes; normal state of a eucaryotic cell.

dipocolinic acid Chemical substance found in bacterial endospores and not in vegetative cells.

diplobacilli Rods that divide and remain attached in pairs.

diplococci Cocci that divide and remain attached in pairs.

diploid A cell or organism with two sets of chromosomes.

direct contact A method of spreading infection from one host to another through some kind of close association of the hosts.

direct count Enumeration of cells by observation through a microscope.

direct F.A. test A fluorescent-antibody test to detect the presence of an antigen.

disaccharide A sugar consisting of two monosaccharides.

disease Any change from a state of health.

disinfectant Any substance used on inanimate objects to kill or inhibit the growth of microorganisms.

disk-diffusion test An agar-diffusion test to determine microbial susceptibility to chemotherapeutic agents.

dissimilation plasmids Plasmids containing genes coding for the production of enzymes that catalyze the catabolism of certain unusual sugars and hydrocarbons.

dissociation Transformation of a compound into positive and negative ions in solution.

disulfide bond Two atoms of sulfur held together by a covalent bond (S—S).

division A phylum; used in botany and microbiology.

donor cell A cell that gives DNA to a recipient cell in recombination.

Donovan bodies Inclusion bodies in large mononuclear cells infected with *Calymmatobacterium granulomatis*.

DPT vaccine A combined vaccine used to provide active immunity, containing diphtheria and tetanus toxoids and killed *Bordetella pertussis* cells.

droplet infection The transmission of infection by small liquid droplets carrying microorganisms.

dysentery A disease characterized by frequent, watery stools.

eclipse period The time during viral multiplication when complete, infective virions are not present.

edema An abnormal accumulation of fluid in body parts or tissues, causing swelling.

effector cells *See* cytotoxic T, helper T, suppressor T, delayed hypersensitivity T, and memory cells.

electrical energy Energy from the flow of electrons.

electron A negatively charged particle in motion around the nucleus of an atom.

electron acceptor An ion that picks up an electron which has been lost from another atom.

electron donor An ion that gives up an electron to another atom.

electronic configuration The arrangement of electrons in shells or energy levels in an atom.

electron microscope A microscope that uses a flow of electrons instead of light to produce an image.

electron shells Regions of an atom corresponding to different energy levels.

electron transport chain A series of compounds that transfer electrons from one compound to another, generating ATP by oxidative phosphorylation.

electrophoresis The separation of substances (for example, serum proteins) by their rate of movement through an electric field.

elementary body An infectious form of *Chlamydia*.

ELISA (enzyme-linked immunosorbent assay) A group of serologic tests that use enzyme reactions as indicators.

Embden-Meyerhof pathway *See* glycolysis.

emulsify To mix two liquids that do not dissolve in each other.

encephalitis Inflammation of the brain.

encystment Formation of a cyst.

endemic disease A disease that is constantly present in a certain population.

endergonic reaction A chemical reaction which requires energy.

endocarditis Inflammation of the lining of the heart (endocardium).

endocytosis The process of moving material into a eucaryotic cell.

endoplasmic reticulum A membrane network in eucaryotic cells connecting the plasma membrane with the nuclear membrane.

endospore A resting structure formed inside some bacteria.

endotoxin Part of the outer portion of the cell wall of most gram-negative bacteria; lipid A.

energy The capacity to do work.

energy level The energy of an electron and its position relative to the atomic nucleus.

enrichment culture A culture medium used for preliminary isolation that favors the growth of a particular microorganism.

enterics The common name for bacteria in the family Enterobacteriaceae.

enterotoxin An exotoxin that causes diarrhea; produced by *Staphylococcus*, *Vibrio*, and *Escherichia*.

Entner-Doudoroff pathway An alternate pathway for the oxidation of glucose to pyruvic acid.

envelope An outer covering surrounding the capsid of some viruses.

enzyme A protein that catalyzes chemical reactions in a living organism.

enzyme induction The process by which a substance can cause the synthesis of an enzyme.

enzyme repression The process by which a substance can stop the synthesis of an enzyme.

enzyme-substrate complex A temporary union of an enzyme and its substrate.

eosinophil A granulocyte whose granules take up the stain eosin.

epidemic disease A disease acquired by many people in a given area in a short time.

epidemiology The science dealing with when and where diseases occur and how they are transmitted.

epidermis The outer portion of the skin.

epidermolytic toxin A staphylococcal toxin that causes skin to peel off.

equilibrium The point of even distribution.

ergot A substance produced in sclerotia by the fungus *Claviceps purpurea* that causes contraction of arteries and uterine muscle.

erythema Redness of the skin.

erythrogenic toxin A substance produced by some streptococci that causes erythema.

ester linkage The bonding between two organic molecules (R) as $R—\underset{\underset{O}{\|}}{C}—O—R$.

ethambutol A synthetic antimicrobial agent that interferes with the synthesis of RNA.

etiology The study of the cause of a disease.

eucaryote A cell with DNA enclosed within a distinct membrane-bounded nucleus.

eutrophication The addition of organic matter and subsequent removal of oxygen from a body of water.

exchange reaction A chemical reaction that has both synthesis and decomposition components.

exergonic reaction A chemical reaction which releases energy.

exocytosis The process of exporting material from a eucaryotic cell.

exon A region of a eucaryotic chromosome that codes for a protein.

exotoxins Protein toxins released from bacterial cells into the surrounding medium.

experimental epidemiology The study of a disease using controlled experiments.

extracellular enzyme An enzyme released from a cell to break down large molecules.

extracellular polymeric substance (EPS) *See* glycocalyx.

extreme halophile An organism that requires a high salt concentration for growth.

facilitated diffusion The transfer of a substance across a plasma membrane from an area of higher concentration to an area of lower concentration mediated by carrier proteins (permeases).

facultative anaerobe An organism that can grow with or without oxygen (O_2).

facultative halophile An organism capable of growth in, but not requiring, 1 to 2% salt.

family A taxonomic group between order and genus.

fat An organic compound consisting of glycerol and fatty acids.

fatty acids Long hydrocarbon chains ending in a carboxyl group.

fecal coliforms Coliform organisms found in the human intestine, capable of fermenting lactose at 44.5°C.

feedback inhibition Inhibition of an enzyme in a particular pathway by the accumulation of end product from the pathway.

fermentation The enzymatic degradation of carbohydrates in which the final electron acceptor is an organic molecule, ATP is synthesized by substrate-level phosphorylation, and oxygen (O_2) is not required.

fever An abnormally high body temperature.

F factor Fertility factor; a plasmid found in the donor cell in bacterial conjugation.

fibrinolysin A kinase produced by streptococci.

filamentous form A structure in the serum of a patient with hepatitis B having hepatitis B virus (HBV) surface antigens.

filtration Passage of a liquid or gas through a screenlike material.

fimbria *See* pilus.

fixed macrophage Macrophage that is located in a certain organ or tissue, for example, in liver, lungs, spleen, or lymph nodes.

fixing (in slide preparation) The process of attaching the specimen to the slide.

flagella Thin appendages that arise from one or more locations on the surface of a cell and are used for cellular locomotion.

flagellate A member of the protozoan phylum Mastigophora that uses flagella for locomotion.

flaming The process of sterilizing an inoculating loop by holding it in an open flame.

flat sour Thermophilic spoilage of canned goods not accompanied by gas production.

flatworm An animal belonging to the phylum Platyhelminthes.

flavoprotein Protein with flavin coenzyme that functions as an electron carrier in respiration.

flocculation Removal of colloidal material by addition of a chemical that causes the colloidal particles to coalesce.

flora The microbial population of an area, such as of human skin.

fluid mosaic model A way of describing the dynamic arrangement of phospholipids and proteins comprising the plasma membrane.

fluke A flatworm belonging to the class Trematoda.

fluorescence The ability to give off light of one color when exposed to light of another color.

fluorescent-antibody technique A diagnostic tool using antibodies labeled with fluorochromes and viewed through a fluorescent microscope.

fluorescent microscope A microscope that uses an ultraviolet light source to illuminate specimens that will fluoresce.

fluorochromes Dyes used to stain bacteria that fluoresce when illuminated with ultraviolet light.

focal infection A systemic infection that began as an infection in one place.

folliculitis Infection of hair follicles.

fomite A nonliving object that can spread infection.

forespore Structure consisting of chromosome, cytoplasm, and endospore membrane inside a bacterial cell.

frameshift mutation A mutation due to the addition or deletion of one or more bases in DNA.

free radical A highly reactive particle with an unpaired electron; designated X·.

freeze drying *See* lyophilization.

FTA-ABS test An indirect fluorescent-antibody test used to detect syphilis.

functional groups Arrangement of elements in organic molecules that are responsible for most of the chemical properties of those molecules.

fungi Organisms that belong to the kingdom Fungi; eucaryotic chemoheterotrophs.

furuncle An infection of a hair follicle.

gamete A male or female reproductive cell.

gametocyte A male or female protozoan cell.

gamma globulin *See* immune serum globulin.

gangrene Tissue death due to loss of blood supply.

gastroenteritis Inflammation of the stomach and intestine.

gas vacuole A procaryotic inclusion; for buoyancy compensation.

gene A segment of DNA or a sequence of nucleotides in DNA that codes for a functional product.

gene amplification A mechanism that causes a gene to be replicated many times.

generalized transduction Transfer of bacterial chromosome fragments from one cell to another by a bacteriophage.

generation time The time required for a cell or population to double in number.

genetic engineering Manufacturing and manipulating genetic material in vitro.

genetics The science of heredity.

genotype The genetic makeup of an organism.

genus The first name of the scientific name (binomial); taxon between family and species.

geosmin An alcohol produced by actinomycetes that has an earthy odor.

germ Part of an organism capable of developing.

germicidal Capable of killing microorganisms.

germicidal lamp An ultraviolet light (wavelength = 260 nm) capable of killing bacteria.

germination The process of starting to grow from a spore.

germ theory The principle that microorganisms cause disease.

globulin Any antibody.

glomerulonephritis Inflammation of the glomeruli of the kidneys, but not a result of kidney infection.

glucan A polysaccharide component of yeast cell walls.

glycerol An alcohol; $C_3H_5(OH)_3$.

glycogen A polysaccharide stored by some cells.

glycolysis The main pathway for the oxidation of glucose to pyruvic acid.

glycocalyx A gelatinous polymer surrounding a procaryotic cell wall.

glycoprotein Carbohydrate and protein complex.

Golgi complex An organelle involved in the secretion of certain proteins.

graft-versus-host (GVH) disease A condition that occurs when a transplanted tissue has an immune response to the tissue recipient.

gram-negative cell wall A peptidoglycan layer surrounded by a lipopolysaccharide outer membrane.

gram-positive cell wall Composed of peptidoglycan.

Gram stain A differential stain that divides bacteria into two groups, gram-positive and gram-negative.

granulocyte A leukocyte with granules in the cytoplasm; includes neutrophils, basophils, and eosinophils.

griseofulvin A fungistatic antibiotic.

guanine A purine nucleic acid base that pairs with cytosine.

Guillain-Barré syndrome A neurological syndrome, of unknown etiology, marked by muscular weakness and paralysis.

gumma A rubbery mass of tissue characteristic of tertiary syphilis and tuberculosis.

halogen One of the following elements: fluorine, chlorine, bromine, iodine, or astadine.

halophile An organism which grows in high concentrations of salt.

hanging-drop preparation A slide prepared so that microorganisms can be viewed live in a liquid suspension.

H antigen A flagellar antigen of enterics.

haploid A eucaryotic cell or organism with one of each type of chromosome.

hapten An antigen that has reactivity and not immunogenicity.

heavy metals Certain elements with specific gravity greater than 4 that are used as antimicrobial agents; for example, silver (Ag), copper (Cu), and mercury (Hg).

helminth A parasitic roundworm or flatworm.

helper T cells (Th) Cells that interact with an antigen before B cells interact with the antigen.

hemagglutination Clumping of red blood cells.

hemagglutination-inhibition Process whereby an antibody inhibits viral hemagglutination.

hemoflagellate A parasitic flagellate found in the circulatory system of its host.

hemolysins Enzymes that lyse red blood cells.

heparin A substance that prevents clotting.

hermaphroditic Having both male and female reproductive capacities.

heterocyst A large cell in certain cyanobacteria; site for nitrogen-fixation.

heterophilic antibody An antibody that reacts with distantly related antigens.

heterotroph An organism that requires an organic carbon source.

hexachlorophene A chlorinated phenol used as an antiseptic.

hexose monophosphate shunt *See* pentose phosphate pathway.

hexylresorcinol A benzene derivative used as an antihelminthic.

Hfr A bacterial cell in which the F factor has become integrated into the chromosome.

H (hemagglutination) spikes Antigenic projections from the outer lipid bilayer of influenza virus.

High-temperature short-time (HTST) pasteurization 72°C for 15 seconds.

histamine A substance released by tissue cells that causes vasodilation.

histiocyte *See* fixed macrophage.

histocompatibility antigens Antigens on the surface of human cells.

histones Proteins associated with DNA in eucaryotic chromosomes.

HLA complex *See* major histocompatibility complex (MHC).

holdfast The branched base of an algal stipe.

holoenzyme An enzyme consisting of an apoenzyme and a cofactor.

homologous chromosome A chromosome that has the same base sequence as another. In a diploid cell, one of a pair of chromosomes.

host An organism infected by a pathogen.

hot-air sterilization The use of an oven at 170°C for approximately 2 hours.

humoral immunity Immunity produced by antibodies dissolved in body fluids, mediated by B cells.

humus Organic matter in soil.

hyaluronic acid A mucopolysaccharide that holds together certain cells of the body.

hybridoma A cell made by fusing an antibody-producing B cell with a cancer cell.

hydrogen bond A bond between a hydrogen atom covalently bonded to oxygen or nitrogen and another covalently bonded oxygen or nitrogen.

hydrogen ion (H^+) A proton.

hydrolysis To split by using water.

hydroxy group $-OH$.

hydroxyl ion OH^-.

hyperbaric chamber An apparatus to hold gases at pressures greater than one atmosphere.

hypersensitivity Altered, enhanced immune reactions leading to pathologic changes.

hypertonic Describing a solution that has a higher concentration of solutes than an isotonic solution.

hyphae Long filaments of cells in fungi or actinomycetes.

hypothermic factor A bacterial substance that affects the host's body temperature.

hypotonic Describing a solution that has a lower concentration of solutes than an isotonic solution.

iatrogenic disease A disease caused by health professionals while administering health care.

icosahedron A polyhedron with 20 triangular faces and 12 corners.

ID_{50} The bacterial concentration required to produce a demonstrable infection in 50% of the test host population.

idiopathic disease A disease of undetermined cause.

IgA The class of antibodies found in secretions.

IgD Antibodies found on B cells.

IgE The class of antibodies involved in hypersensitivities.

IgG The most abundant antibodies found in serum.

IgM The first antibodies to appear after exposure to an antigen.

illuminator A light source.

imidazoles Antimetabolites that inhibit the action of histamine.

immediate hypersensitivity Allergic reactions involving humoral antibodies.

immune adherence The attachment of the phagocyte to an antigen.

immune complex A circulating antigen-antibody aggregate capable of fixing complement.

immune-complex disease A condition in which antibodies are formed against the self.

immune serum globulin Serum fraction containing immunoglobulins (antibodies).

immunity The body's defense against a particular microorganism.

immunization A process that produces immunity.

immunoassay Detection of an antigen or antibody by serologic methods.

immunodiffusion test A test consisting of precipitation reactions carried out in an agar-gel medium.

immunoelectrophoresis Identification of proteins by electrophoretic separation and then serologic testing.

immunofluorescence Procedures using the fluorescent-antibody technique.

immunogen *See* antigen.

immunoglobulin (Ig) An antibody.

immunologic disease A disease caused by the immune system attacking the self (autoimmunity) or overreacting (hypersensitivity).

immunologic enhancement The binding of antibodies to tumor antigens.

immunologic escape Resistance of cancer cells to immune rejection.

immunologic surveillance The body's immune response to cancer.

immunologic tolerance A state of specific unresponsiveness to an antigen following initial exposure to that antigen.

immunosuppression Inhibition of the immune response.

immunotherapy Treatment using antibodies.

immunotoxin Immunotherapeutic agent consisting of a poison bound to a monoclonal antibody.

IMViC Biochemical tests used to identify enterics, including indole, methyl red, Voges-Proskauer, and citrate.

inapparent infection *See* subclinical infection.

incidence The fraction of the population that contracts a disease during a particular length of time.

inclusion Material inside a cell.

incomplete digestive system A digestive system with one opening (mouth) for intake of food and elimination of waste.

incubation period The time interval between the actual infection and first appearance of any signs or symptoms of disease.

indirect contact Transmission of pathogens by agents such as food and water.

indirect F.A. test A fluorescent-antibody test to detect the presence of specific antibodies.

inducer A substrate that brings about an increased amount of an enzyme.

inducible enzyme *See* enzyme induction.

inert Inactive.

infection Growth of microorganisms in the body.

infection thread An invagination in a root hair that allows *Rhizobium* to infect the root.

infectious disease A disease caused by pathogens.

inflammation A host response to tissue damage characterized by reddening, pain, heat, and swelling.

inherited disease A disease passed from parent to child through gametes.

innate resistance Resistance of an individual to diseases that affect other species and other individuals of the same species.

inoculate To introduce microorganisms into a culture medium or host.

inoculating loop (needle) Instrument used to transfer bacteria from one culture medium to another.

inorganic compounds Small molecules not usually containing carbon.

interferon An antiviral protein produced by certain animal cells in response to a viral infection.

intermediate host An organism that harbors the larval stage of a helminth.

intoxication Poisoning.

intron A region of a eucaryotic chromosome that does not code for a protein.

invasiveness The ability of microorganisms to establish residence in a host.

in vitro "In glass"; not in a living organism.

in vivo Within a living organism.

iodophor A complex of iodine and a detergent.

ion A negatively or positively charged atom or group of atoms.

ionic bond A chemical bond formed when atoms gain or lose an electron in the outer energy levels.

ionization Separation of a molecule into groups of atoms with electrical charges.

ionizing radiation High-energy radiation that causes ionization; for example, X rays and gamma rays.

ischemia Local loss of blood supply.

isograft Tissue graft from an identical twin.

isomers Two molecules with the same chemical formula but different structures.

isoniazid (INH) A bacteriostatic agent used to treat tuberculosis.

isotonic Referring to a solution in which osmotic pressure is equal across a membrane.

isotope A form of a chemical element in which the number of neutrons in the nucleus is different from the other forms of that element.

keratin A protein found in epidermis, hair, and nails.

kinases Bacterial enzymes that break down fibrin (blood clots).

kinetic energy The energy of motion.

kingdom The highest category in the taxonomic hierarchy of classification.

kinins Substances released from tissue cells that cause vasodilation.

Koch's postulates Criteria used to determine the causative agent of infectious diseases.

Krebs cycle A pathway that converts two-carbon compounds to carbon dioxide (CO_2), transferring electrons to NAD^+ and other carriers.

L- Prefix describing a stereoisomer; L-amino acids are more commonly found in proteins.

lag phase The time interval in a bacterial growth curve with no growth.

lagging strand During DNA replication, the daughter strand synthesized discontinuously.

larva The sexually immature stage of a helminth or arthropod.

latent disease A disease characterized by a period of no symptoms when the pathogen is inactive.

latent viral infection A condition in which a virus remains in the host without producing disease for long periods of time.

latex agglutination test Indirect agglutination test using soluble antigens attached to latex spheres.

LD_{50} The lethal dose for 50% of the inoculated hosts within a given period of time.

leading strand During DNA replication, the daughter strand synthesized continuously.

lepromin test A skin test to determine the presence of antibodies to *Mycobacterium leprae.*

leukemias Cancers characterized by abnormally high numbers of leukocytes.

leukocidins Substances produced by some bacteria that can destroy neutrophils and macrophages.

leukocyte White blood cell.

leukocytic pyrogen A protein produced by PMNs that causes an increase in body temperature.

leukocytosis-promoting factor A substance released by inflamed tissues that increases the production of granulocytes.

leukopenia A condition in which the number of leukocytes is less than normal.

leukotrienes Mediators of anaphylaxis.

L form A mutual bacterium with a defective cell wall.

lichen A symbiosis between a fungus and an alga or cyanobacterium.

ligands Projections on procaryotic cells for adherence.

ligase An enzyme that joins together pieces of DNA.

limnetic zone The surface zone of a body of water away from the shore.

lipase An exoenzyme that breaks down fats into their component fatty acids and glycerol.

lipid A molecule composed of glycerol and fatty acids; fat.

lipid A Component of gram-negative outer membrane; endotoxin.

lipopolysaccharide (LPS) A molecule consisting of a lipid and a polysaccharide, forming the outer layer of gram-negative cell walls.

lipoprotein A molecule consisting of a lipid and protein.

liposome A fatty globule that may be used to administer chemotherapeutic agents.

lithotroph *See* autotroph.

littoral zone The region along the shore of an inland body of water where there is considerable vegetation and where light penetrates to the bottom.

local infection An infection in which pathogens are limited to a small area of the body.

localized reaction An anaphylaxis-type reaction such as hayfever, asthma, and hives.

log phase Period of bacterial growth or logarithmic increase in a cell number.

lophotrichous Having two or more flagella at one end of a cell.

lymphangitis Inflammation of the lymph vessels.

lymphocyte An agranulocyte involved in antibody production.

lymphokines Proteins released by T cells.

lymphotoxin A lymphokine which can destroy target cells in vitro.

lymphoma Cancer of lymphoid tissue.

lyophilization Freeze-drying; freezing a substance and evaporating the ice in a vacuum.

lysis Disruption of the plasma membrane. In disease, a gradual period of decline.

lysogeny A state in which phage DNA is incorporated into the host cell without lysis.

lysosome An organelle containing digestive enzymes.

lysozyme An enzyme capable of lysing bacterial cell walls.

lytic cycle A sequence for replication of phages that results in host cell lysis.

macrolides Antibiotics that inhibit protein synthesis; for example, erythromycin.

macromolecules Large organic molecules.

macrophage A phagocytic cell; an enlarged monocyte.

macrophage activation factor Increases macrophages' efficiency at destroying ingested cells.

macrophage chemotactic factor Attracts macrophages to infection site.

macrophage migration-inhibiting factor Prevents macrophages from leaving infection site.

macular rash Small red spots appearing on the skin.

major histocompatibility complex (MHC) The genes that code for the histocompatibility antigens; also known as human leukocyte antigens (HLA).

malaise A feeling of general discomfort.

malignant Cancerous.

malolactic fermentation Conversion of malic acid to lactic acid by lactic acid bacteria.

malt Barley grains containing maltose and amylase.

malting Germination of starchy grains resulting in glucose and maltose production.

mannan A polysaccharide component of yeast cell walls.

Mantoux test A tuberculin skin test.

margination The process by which phagocytes stick to the lining of blood vessels.

mast cell Type of cell found throughout the body that contains histamine and other substances that stimulate vasodilation.

maximum growth temperature The highest temperature at which a species can grow.

mechanical energy The energy involved in movement.

mechanical transmission The process by which arthropods transmit infections by carrying pathogens on their feet and other body parts.

megacin A bacteriocin produced by *Bacillus megaterium.*

meiosis The process that leads to the formation of haploid gametes in a diploid organism.

membrane filter A screen-like material with pores small enough to retain microorganisms.

memory cell Long-lived B or T cell responsible for anamnestic response.

meningitis Inflammation of the meninges covering the central nervous system.

merozoite A trophozoite of *Plasmodium* found in red blood cells.

mesophile An organism that grows between 25° and 40°C.

mesosome An irregular fold in the plasma membrane of a procaryotic cell.

messenger RNA (mRNA) The type of RNA molecule that directs the incorporation of amino acids into proteins.

metabolic disease A disease resulting from an abnormality in the biochemistry of bodily functions.

metabolism The sum of all the chemical reactions that occur in a living cell.

metacercaria The encysted stage of a fluke in its final intermediate host.

metachromatic granule The intracellular volutin stored by some bacteria.

metalloprotein A protein conjugated with a metal atom.

metastasis The spread of cancer from a primary tumor to other parts of the body.

meter The standard unit of length in the metric system; one ten-millionth of the distance from the equator to the pole.

microaerophile An organism that grows best in an environment with less oxygen (O_2) than is found in air.

microbiota *See* flora.

micrometer (μm) A unit of measure equal ato 10^{-6} m.

microorganism A living organism too small to be seen with the naked eye; includes bacteria, fungi, protozoans, microscopic algae, and viruses.

microphage A granulocytic phagocyte.

microtubule The structure of the proteins comprising eucaryotic flagella and cilia.

microwave An electromagnetic wave; wavelength between 10^{-1} and 10^{-3} m.

minimal bactericidal concentration (MBC) The lowest concentration of a chemotherapeutic agent that will result in no growth.

minimal inhibitory concentration (MIC) The lowest concentration of a chemotherapeutic agent that will prevent growth of the test microorganism.

minimum growth temperature The lowest temperature at which a species will grow.

miracidium The free-swimming, ciliated larva of a fluke that hatches from the egg.

missense mutation A mutation that results in substitution of an amino acid in a protein.

mitochondria Organelles containing the respiratory ATP-synthesizing enzymes.

mitosis The division of the cell nucleus, often followed by division of the cytoplasm of the cell.

mixed culture A culture containing more than one kind of microorganism.

MMWR A weekly publication of the Centers for Disease Control containing data on notifiable diseases and topics of special interest.

molds Fungi that form mycelia and appear as cottony tufts.

mole An amount of a chemical equal to the atomic weights of all the atoms in a molecule of the chemical.

molecular biology The science dealing with proteins of living organisms.

molecular weight The sum of the atomic weights of all atoms making up a molecule.

molecule A combination of atoms forming a specific chemical compound.

Monera The kingdom to which all procaryotic organisms belong.

monoclonal antibodies Specific antibodies produced by in vitro clones of B cells hybridized with cancerous cells.

monocyte A phagocytic agranulocyte.

monolayer A single layer of cells due to the cessation of cell division by contact inhibition.

monomers The units that combine to form polymers.

mononuclear phagocytic system A system of fixed macrophages located in the spleen, liver, lymph nodes, and bone marrow.

monotrichous Having a single flagellum.

morbidity The incidence of a specific notifiable disease.

mordant A substance added to a staining solution to make it stain more intensely.

morphology The external appearance.

mortality The deaths from a specific notifiable disease.

most probable number (MPN) A statistical determination of the number of coliforms per 100 ml of water or food.

motility The ability of an organism to move by itself.

M protein A heat- and acid-resistant protein of streptococcal cell walls.

multiple-tube fermentation test A method of detecting the presence of coliforms.

murein *See* peptidoglycan.

mutagen An agent in the environment that brings about mutations.

mutation Any change in the base sequence of DNA.

mutation rate The probability of a gene mutating each time a cell divides.

mutualism A symbiosis in which both organisms are benefited.

mycelium A mass of long filaments of cells that branch and intertwine, typically found in molds.

mycetoma A chronic infection caused by certain fungi and *Nocardia* characterized by a tumor-like appearance.

mycology The science dealing with fungi.

mycorrhiza A fungus growing in symbiosis with plant root hairs.

mycosis A fungal infection.

mycotoxin A toxin produced by a fungus.

myocarditis Inflammation of the heart muscle.

naked virus A virus without an envelope.

nanometer (nm) A unit of measurement equal to 10^{-9} m, 10^{-3} μm, and 10 Å.

naturally acquired active immunity Antibody production in response to an infectious disease.

naturally acquired passive immunity The natural transfer of humoral antibodies; transplacental transfer.

necrosis Tissue death.

negative (indirect) selection *See* replica plating.

negative stain A procedure that results in colorless bacteria against a stained background.

neoplastic disease A disease resulting in new growth of cells; tumors.

neurotoxin A chemical that is poisonous to the nervous system.

neutralizing antibody An antibody that inactivates a bacterial exotoxin or virus.

neutron An uncharged particle in the nucleus of an atom.

neutrophil Also called polymorphonuclear leukocyte; a highly phagocytic granulocyte.

nicotinamide adenine dinucleotide (NAD) A coenzyme that functions in the removal and transfer of H^+ and electrons from substrate molecules.

nicotinamide adenine dinucleotide phosphate (NADP) A coenzyme similar to NAD.

nitrification The oxidation of nitrogen from ammonia to nitrites and nitrates.

nitrofuran Synthetic antimicrobial drug.

nitrogen cycle The series of processes that converts nitrogen (N_2) to organic substances and back to nitrogen in nature.

nitrogen fixation The conversion of nitrogen (N_2) into ammonia.

nitrosamine A carcinogen formed by the combination of nitrite and amino acids; nitroso-: —N=O.

N (neuraminidase) spikes Antigenic projections from the outer lipid bilayer of influenza virus.

nomenclature The system of naming things.

noncommunicable disease A disease that is not transmitted from one person to another.

noncyclic photophosphorylation Movement of an electron from chlorophyll to NAD; plant and cyanobacterial photophosphorylation.

nonionizing radiation Radiation that does not cause ionization; for example, ultraviolet radiation.

nonsense codon A special terminator codon that does not code for any amino acid.

nonsense mutation A base substitution in DNA that results in a nonsense codon.

nonspecific resistance Host defenses that tend to afford protection from any kind of pathogen.

normal flora Microorganisms that colonize an animal without causing disease.

nosocomial infection An infection that develops during the course of a hospital stay and was not present at the time the patient was admitted.

notifiable disease A disease that physicians must report to the public health service.

nuclear envelope The double membrane that separates the nucleus from the cytoplasm in a eucaryotic cell.

nucleic acid A macromolecule consisting of nucleotides; for example, RNA and DNA.

nucleic acid hybridization The process of combining single complementary strands of DNA.

nucleoid The region in a bacterial cell containing the chromosome.

nucleoli Areas in a eucaryotic nucleus where rRNA is synthesized.

nucleoplasm The gell-like fluid within the nuclear envelope.

nucleoprotein A macromolecule consisting of protein and nucleic acid.

nucleoside A compound consisting of a purine or pyrimidine and pentose sugar.

nucleotide A compound consisting of a purine or pyrimidine base, a five-carbon sugar, and a phosphate.

nucleus The part of a eucaryotic cell that contains the genetic material; also, the part of an atom consisting of the protons and neutrons.

numerical taxonomy A method of comparing organisms on the basis of many characteristics.

nutrient broth (agar) A complex medium made of meat extracts that may contain sugar.

nutritional deficiency disease A disease caused by lack of nutrients.

O antigen A cell antigen of enterics.

objective lens In a compound light microscope, the lens closest to the specimen.

obligate anaerobe An organism that is unable to use oxygen.

ocular lens In a compound light microscope, the lens closest to the viewer.

oligodynamic action The ability of small amounts of a heavy metal compound to exert antimicrobial activity.

oncogene A gene that can bring about malignant transformation.

oncogenic virus A virus that is capable of producing tumors.

operator The region of DNA adjacent to structural genes that controls their transcription.

operon The operator site and structural genes it controls.

opportunistic pathogen An organism that does not ordinarily cause a disease but can become pathogenic under certain circumstances.

opsonization The enhancement of phagocytosis by coating microorganisms with certain serum proteins (opsonins).

optical density A measure of the amount of light that is absorbed by (or does not pass through) a culture.

optimum growth temperature The temperature at which a species grows best.

oral groove On some protozoans, the site at which nutrients are taken in.

order A taxonomic classification between class and family.

organelles Membrane-bounded structures within eucaryotic cells.

organic compounds Molecules that contain carbon and hydrogen.

organic growth factor An essential organic compound that an organism is unable to synthesize.

organotroph *See* heterotroph.

osmosis The net movement of solvent molecules across a selectively permeable membrane from an area of higher concentration to an area of lower concentration.

osmotic pressure The force with which a solvent moves from a solution of lower solute concentration to a solution of higher solute concentration.

Ouchterlony test An immunodiffusion test.

outer membrane Outer layer of a gram-negative cell wall consisting of lipoproteins, lipopolysaccharides, and phospholipids.

oxidase test A diagnostic test for the presence of cytochrome c; using *p*-aminodimethylaniline.

oxidation The removal of electrons from a molecule or the addition of oxygen to a molecule.

oxidation pond A method of secondary sewage treatment.

oxidation-reduction (redox) reaction A coupled reaction in which one substance is oxidized and one is reduced.

oxidative phosphorylation The synthesis of ATP coupled with electron transport.

oxygen cycle The processes that convert molecular oxygen (O_2) to oxides, water, organic compounds, and back to O_2.

PABA Para-aminobenzoic acid; a precursor for folic acid synthesis.

PAGE Polyacrylamide gel electrophoresis; *see* electrophoresis.

pandemic disease An epidemic that occurs worldwide.

papular rash A skin rash characterized by raised spots.

parasite An organism that derives nutrients from a living host.

parasitism A symbiosis in which one organism (the parasite) exploits another (the host) without providing any benefit in return.

parenteral route Deposition directly into tissues beneath the skin and mucous membranes.

passively acquired immunity Immunity acquired when antibodies produced by another source are transferred to the individual who needs them.

pasteurization The process of mild heating to kill particular spoilage organisms or pathogens.

pathogen A disease-causing organism.

pathology Study of the causes and development of disease.

pellicle The flexible covering of some protozoans.

penicillins A group of antibiotics produced either by *Penicillium* (natural penicillins) or by adding side chains to the beta-lactam ring (semisynthetic penicillins).

pentose phosphate pathway A metabolic pathway that can occur simultaneously with glycolysis to produce pentoses and $NADH_2$ without ATP production.

peptide A chain of two (di-), three (tri-), or more (poly-) amino acids.

peptide bond A bond joining the amino group of one amino acid to the carboxyl group of a second amino acid with the loss of a water molecule.

peptidoglycan The structural molecule of bacterial cell walls consisting of the molecules N-acetylglucosamine, N-acetylmuramic acid, tetra-peptide side chain, and peptide side chain.

peptone Short chains of amino acids produced by the action of acids or enzymes on proteins.

pericarditis Inflammation of the sac around the heart (pericardium).

periplasmic space The region between the outer membrane and the cytoplasmic membrane.

peritrichous Having flagella distributed over the entire cell.

permease A carrier protein in the plasma membrane.

peroxidase An enzyme that breaks down hydrogen peroxide; $H_2O_2 + NADH_2 \longrightarrow 2H_2O + NAD$.

peroxide An oxygen oxide consisting of two atoms of oxygen.

pH The symbol for hydrogen ion concentration; a measure of the relative acidity of a solution.

phage *See* bacteriophage.

phage typing A method of identifying bacteria using specific strains of bacteriophages.

phagocyte A cell capable of engulfing and digesting particles that are harmful to the body.

phagocytosis The ingestion of solids by cells.

phagolysosome A digestive vacuole.

phalloidin *See* amanitin.

phase-contrast microscope A compound light microscope that allows examination of structures inside cells through the use of a special condenser.

phenol C_6H_5OH; carbolic acid.

phenol coefficient A standard of comparison for the effectiveness of disinfectants; the disinfecting action of a chemical is compared to phenol for the same length of time on the same organism under identical conditions.

phenolic A synthetic derivative of phenol.

phenotype The external manifestations of the genetic makeup of an organism.

phosphate group A portion of a phosphoric acid molecule attached to some other molecule;

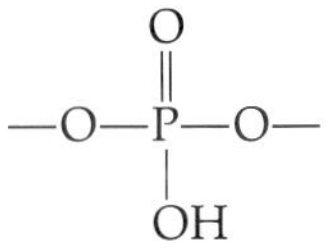

phospholipid A complex lipid composed of glycerol, two fatty acids, and a phosphate group.

phosphorylation The addition of a phosphate group to an organic molecule.

photoautotroph An organism that uses light for its energy source and carbon dioxide (CO_2) as its carbon source.

photoheterotroph An organism that uses light for its energy source and an organic carbon source.

photophosphorylation The production of ATP by photosynthesis.

photosynthesis The light-driven synthesis of carbohydrate from carbon dioxide (CO_2).

phototroph An organism that uses light as its primary energy source.

phylogeny The evolutionary history of a group of organisms.

phylum A taxonomic classification between kingdom and class.

phytoplankton Free-floating photoautotrophs.

pilus An appendage on a bacterial cell for attachment and conjugation.

pinocytosis The engulfing of liquid by a cell.

planktonic Free-floating.

plankton Free-floating.

plantae A kingdom composed of multicellular eucaryotes with cellulose cell walls.

plant virus A virus that multiplies in plant tissues.

plaque A clearing in a confluent growth of bacteria due to lysis by phages. Also, see dental plaque.

plaque-forming units Visible plaques counted, perhaps due to more than one phage.

plasma The liquid portion of blood in which the formed elements are suspended.

plasma cells Cells produced by B lymphocytes and which manufacture specific antibodies.

plasma membrane The selectively permeable membrane enclosing the cytoplasm of a cell; the outer layer in animal cells, internal to the cell wall in other organisms.

plasmid A small cyclic DNA molecule in bacteria in addition to the chromosome.

plasmodium A multinucleated mass of protoplasm; written as a genus the etiology of malaria.

plasmolysis Loss of water in a hypertonic environment.

plate count A method of determining the number of bacteria in a sample by counting the number of colony-forming units on a solid culture medium.

pleomorphic Having many shapes.

pneumonia Inflammation of the lungs.

point mutation *See* base substitution.

polar molecule A molecule with an unequal distribution of charges.

poly-β-hydroxybutyric acid A fatty acid storage material unique to bacteria.

polyene An antimicrobial agent that alters sterols in eucaryotic plasma membranes and contains more than four carbon atoms and at least two double bonds.

polyhedron A many-sided solid.

polykaryocyte A multinucleated giant cell.

polymer A molecule consisting of a sequence of similar units or monomers.

polymerase An enzyme that synthesizes specific polymers.

polymorphonuclear (PMN) leukocyte A neutrophil.

polypeptide A chain of amino acids. Also, a group of antibiotics that causes disintegration of phospholipids.

polyribosome An mRNA strand with several ribosomes attached to it.

portal of entry The avenue by which a pathogen gains access to the body.

portal of exit The route by which a pathogen leaves the body.

positive (direct) selection A procedure for picking out mutant cells by growing them.

potential energy Energy that is stored.

pour plate method A method of inoculating a solid nutrient medium by mixing bacteria in the melted medium and pouring the medium into a Petri plate to solidify.

PPD (Purified Protein Derivative) Proteins extracted from a boiled culture of *M. tuberculosis;* the antigen used in the tuberculin skin test.

precipitation reaction A reaction between soluble antigens and multivalent antibodies to form aggregates.

precipitin An antibody that causes a precipitation reaction.

precipitin ring test A precipitation test performed in a capillary tube.

predisposing factor Anything that makes the body more susceptible to a disease or alters the course of a disease.

presumptive test The first step in the detection of coliforms; used to determine MPN values.

prevalence The fraction of a population having a specific disease at a given time.

primary cell line Human tissue cells that grow only a few generations in vitro.

primary infection An acute infection that causes the initial illness.

primary response Antibody production to the first contact with an antigen.

primary treatment Physical removal of solid matter from wastewater.

prion Infectious protein.

procaryote A cell whose genetic material is not enclosed in a nuclear membrane.

prodromal period The time following incubation when the first symptoms of illness appear.

product The substance formed in a chemical reaction.

profundal zone The deeper water under the limnetic zone in an inland body of water.

proglottid A body segment of a tapeworm containing male and female organs.

promoter site The starting point on DNA for transcription of RNA by RNA polymerase.

properdin system Three serum proteins that function with complement to kill bacteria.

prophage Phage DNA inserted into the host cell's DNA.

prostaglandins Hormonelike substances that are synthesized in many tissues and circulate in the blood, and the exact function of which is unknown.

prosthetic group A coenzyme that is bound tightly to its apoenzyme.

protein A large molecule containing carbon, hydrogen, oxygen, and nitrogen (and sulfur); some have a globular structure and others are pleated sheets.

proteolytic An enzyme capable of hydrolyzing a protein.

Protista The kingdom to which protozoans belong; unicellular, eucaryotic organisms.

proton A positively charged particle in the nucleus of an atom.

protooncogene An oncogene that is functioning normally.

protoplasmic streaming Movement of protoplasm in a plasmodial slime mold.

protoplast A gram-positive bacterium without a cell wall.

protozoa A unicellular eucaryotic organism belonging to the kingdom Protista.

provirus Viral DNA that is integrated into the host cell's DNA.

pseudohypha A short chain of cells that results from the lack of separation of daughter cells after budding.

pseudopods Extensions of a cell that aid in locomotion and feeding.

psychogenic disease A condition in which emotional factors contribute to disease.

psychrophile An organism that grows best at 15°C and does not grow above 20°C.

psychrotroph An organism that is capable of growth at 4°C and above 20°C.

pure culture A culture with only one kind of microorganism.

purines The class of nucleic acid bases that includes adenine and guanine.

pus An accumulation of dead phagocytes, dead bacterial cells, and fluid.

pustules Small elevations containing pus.

pyocin A bacteriocin produced by *Pseudomonas aeruginosa.*

pyocyanin Blue-green pigment produced by ***Pseudomonas aeruginosa.***

pyogenic Pus-forming.

pyrimidines The class of nucleic acid bases that includes uracil, thymine, and cytosine.

quaternary ammonium compound A cationic detergent with four organic groups attached to a central nitrogen atom, used as a disinfectant.

quellung reaction Swelling of a bacterial capsule in the presence of a specific antibody.

quinine An antimalarial drug derived from the cinchona tree and effective against sporozoites in red blood cells.

quinones Low molecular weight, nonprotein carriers in the electron transport chain.

radiant energy Energy traveling through space; radiation.

radioimmunoassay A method of measuring the amount of a compound using radioactive antigens.

rapid plasma reagin A serologic test for syphilis.

reactants Substances that are combined in a chemical reaction.

reactivity The ability of an antigen to combine with an antibody.

reagin IgE antibodies made in response to a treponemal infection characterized by their ability to combine with lipids; reagin–lipid complex will fix complement.

recalcitrance Being resistant to degradation.

receptor An attachment for a pathogen on a host cell.

recipient cell A cell that receives DNA from a donor cell in recombination.

recombinant DNA A DNA molecule produced by recombination.

recombinant DNA techniques *See* genetic engineering.

recombinant RNA technology Techniques used to make RNA molecules.

recombination The process of joining pieces of DNA from different sources.

redia A trematode larval stage which may reproduce asexually one or two times before developing into a cercaria.

red tide A bloom of planktonic dinoflagellates.

reducing medium A culture medium containing ingredients that will remove dissolved oxygen from the medium to allow the growth of anaerobes.

reduction The addition of electrons to a molecule; the gain of hydrogen atoms.

refractive index The relative velocity with which light passes through a substance.

regulator gene The gene that codes for a repressor protein.

rennin A proteolytic enzyme obtained from a calf's stomach.

replica plating A method of inoculating a number of solid minimal culture media from an original plate of complete medium, mutant colonies that don't grow on the minimal media can be selected from the original plate.

replication fork The point where DNA separates and new strands of DNA will be synthesized.

replicative form A double-stranded RNA molecule produced during the multiplication of certain RNA viruses.

repressible enzyme *See* enzyme repression.

repressor A protein that binds to the operator site to prevent transcription.

reservoir of infection A continual source of infection.

resistance The ability to ward off diseases through nonspecific and specific defenses.

resistance (R) factor A bacterial plasmid carrying genes that determine resistance to antibiotics.

resistance transfer factor (RTF) A group of genes for replication and conjugation on the R factor.

resolution The ability to distinguish fine detail with a magnifying instrument.

respiration An ATP-generation process in which chemical compounds are oxidized and the final electron acceptor is usually an inorganic molecule; also, the process by which living organisms produce carbon dioxide (CO_2).

restriction enzyme An enzyme that cuts DNA.

reticulate body An intracellular stage of *Chlamydia.*

reverse transcriptase RNA-dependent DNA polymerase; an enzyme that synthesizes a complementary DNA from an RNA template.

reversible reaction A chemical reaction in which the end products can readily revert to the original molecules.

R genes Genes carried on the R factor that code for enzymes that inactivate certain drugs.

rhizosphere The region in soil where the soil and roots make contact.

riboflavin A B vitamin which functions as a flavoprotein.

ribonucleic acid (RNA) The class of nucleic acids that comprises messenger RNA, ribosomal RNA, and transfer RNA.

ribose A five-carbon sugar that is part of ribonucleotide molecules and RNA.

ribosomal RNA (rRNA) The RNA molecules that form the ribosomes.

ribosomes The site of protein synthesis in a cell, composed of RNA and protein.

rifamycin An antibiotic that inhibits bacterial RNA synthesis.

ring stage A young *Plasmodium* trophozoite which looks like a ring.

RNA-dependent RNA polymerase An enzyme that synthesizes a complementary RNA from an RNA template.

RNA primer A short strand of RNA used to start synthesis of the lagging strand of DNA.

root nodule A tumorlike growth on the roots of certain plants containing a symbiotic nitrogen-fixing bacterium.

roundworms Animals belonging to the phylum Aschelminthes.

S (Svedberg unit) Notes the relative rate of sedimentation during ultra-high speed centrifugation.

Sabin vaccine A preparation containing three attenuated strains of polio virus administered orally.

saccharide Sugar; general formula $(CH_2O)_n$.

Salk vaccine A preparation of a formalin-inactivated polio virus that is injected.

salt A substance that dissolves in water to cations and anions, neither of which is H^+ or OH^-.

saphrophyte An organism that obtains its nutrients from dead organic matter.

sarcina A group of eight bacteria that remain in a packet after dividing.

sarcoma A cancer of fleshy, nonepithelial tissue or connective tissue.

saturation The condition in which the active site on an enzyme is occupied by the substrate or product at all times.

scanning electron microscope An electron microscope that provides three-dimensional views of the specimen magnified about 10,000 times.

Schaeffer-Fulton stain An endospore stain that uses malachite green to stain the endospores and safranin as a counterstain.

Schick test A skin test to detect the presence of antibodies to diphtheria.

schizogony The process of multiple fission, where one organism divides to produce many daughter cells.

sclerotia The reddish hardened ovaries of a grain that are filled with mycelia of the fungus *Claviceps purpurea.*

scolex The head of a tapeworm, containing suckers and possibly hooks.

secondary infection An infection caused by an opportunistic pathogen after a primary infection has weakened the host's defenses.

secondary treatment Biological degradation of the organic matter in wastewater, following primary treatment.

secretion The production and release of fluid containing a variety of substances from a cell.

secretory granule A vesicle containing protein produced from the Golgi complex.

selective medium A culture medium designed to suppress the growth of unwanted bacteria and encourage the growth of desired microorganisms.

selective permeability The property of a plasma membrane to allow certain molecules and ions to move through the membrane while restricting others.

selective toxicity The property of some antimicrobial agents to be toxic for a microorganism and nontoxic for the host.

semiconservative replication The process of DNA replication in which each double-stranded molecule of DNA contains one original strand and one new strand.

sense codon A codon that codes for an amino acid.

sense strand (+strand) Viral RNA that can act as mRNA.

sepsis The presence of unwanted bacteria.

septate hypha A hypha consisting of uninucleate cell-like units.

septicemia A condition characterized by the multiplication of bacteria in the blood.

septic tank A tank, built into the ground, in which wastewater is treated by primary treatment.

septum A crosswall dividing two parts.

serial dilution The process of diluting a sample several times.

serology The branch of immunology concerned with the study of antigen–antibody reactions in vitro.

serum The liquid remaining after blood plasma is clotted, and which contains immunoglobulins.

sewage Domestic wastewater.

sex pilus A pilus used for the transfer of genetic material during bacterial conjugation.

sexual dimorphism The distinctly different appearance of adult male and female organisms.

sexual reproduction Reproduction that requires two opposite mating strains, usually designated male and female.

signs Changes due to a disease that a physician can observe and measure.

simple stain A method of staining microorganisms with a single basic dye.

single-cell protein (SCP) A food substitute consisting of microbial cells.

singlet oxygen Highly reactive O_2.

skin test The intradermal injection of an antigen or antibody to determine susceptibility to an antigen.

slide agglutination test A method of identifying an antigen by combining it with a specific antibody in a slide.

slime layer *See* capsule.

slime mold Fungallike protists.

slow-reacting substance of anaphylaxis (SRS-A) Leukotrienes released by target cells after being bound by IgE antibodies.

slow virus infection A disease process that occurs gradually over a long period of time, caused by a virus.

sludge Solid matter obtained from sewage.

smear A thin film of material on a slide.

soap A surface-active agent made from animal fats and lye (NaOH).

soil A mixture of solid inorganic matter, water, air, organic matter, and living organisms.

solubility The ability to be dissolved, usually in water.

solute A substance dissolved in another substance.

solvent A dissolving medium.

somatic Relating to the cell itself.

specialized transduction The process of transferring a piece of cell DNA adjacent to a prophage to another cell.

species The most specific level in the taxonomic hierarchy. **Bacterial species:** A population of cells with similar characteristics.

specific epithet The second name in a scientific binomial.

specific resistance *See* immunity.

spherical particle A particle in the serum of a patient with hepatitis B having hepatitis B virus (HBV) surface antigens.

spheroplast A gram-negative bacterium lacking a complete cell wall.

spike A carbohydrate-protein complex that projects from the surface of certain enveloped viruses.

spirillum A spiral or corkscrew-shaped bacterium.

spirochete A corkscrew-shaped bacterium with an axial filament.

spontaneous generation The idea that life could arise spontaneously from nonliving matter.

spontaneous mutation A mutation that occurs without a mutagen.

sporadic disease A disease that occurs occasionally in a population.

sporangiophore Aerial hypha supporting a sporangium.

sporangiospore An asexual fungal spore formed within a sporangium.

sporangium A sac containing one or more spores.

spore A reproductive structure formed by fungi and actinomycetes.

sporogenesis The process of spore and endospore formation.

sporozoite A trophozoite of *Plasmodium* found in mosquitoes, infective for humans.

stability The condition of not deteriorating with time.

staphylococcus Broad sheet of cells.

stationary phase The period in a bacterial growth curve when the number of cells dividing equals the number dying.

STD Sexually transmitted disease.

stem cells Fetal cells that give rise to bone marrow, blood cells, and B and T cells.

stereoisomers Two molecules consisting of the same atoms, arranged in the same manner but differing in their relative positions; mirror images.

sterile Free of microorganisms.

steroids A specific group of chemical substances, including cholesterol and hormones.

sterol A lipid–alcohol found in the plasma membranes of fungi and *Mycoplasma.*

stipe Stemlike supporting structure of multicellular algae and basidiomycetes.

strain A group of cells all derived from a single cell.

streak plate method A method of inoculating a single culture medium by spreading microorganisms over the surface of the medium.

streptobacilli Rods that remain attached in chains after cell division.

streptococci Cocci that remain attached in chains after cell division.

structural gene A gene that codes for an enzyme.

sty An infection of an eyelash follicle.

subacute disease A disease with symptoms between acute and chronic.

subclinical infection An infection that does not cause a noticeable illness.

subcutaneous mycosis A fungal infection of tissue beneath the skin.

substrate Any compound with which an enzyme reacts.

substrate-level phosphorylation The synthesis of ATP by direct transfer of a high-energy phosphate group from an intermediate metabolic compound to ADP.

sulfa drugs Any synthetic chemotherapeutic agent containing sulfur and nitrogen; *see* sulfonamides.

sulfhydral group —SH.

sulfonamides Bacteriostatic compounds that interfere with folic acid synthesis by competitive inhibition.

superficial mycosis A fungal infection localized in surface epidermal cells and along hair shafts.

superinfection Growth of the target pathogen that has developed resistance to the antimicrobial drug being used.

superoxide dismutase Enzyme that destroys O_2^-; $O_2^- + O_2^- + 2H^+ \rightarrow H_2O_2 + O_2$.

superoxide free radical O_2^-.

suppressor T cells (Ts) Cells that inhibit an immune response.

surface-active agent (surfactant) Any compound that decreases the tension between molecules lying on the surface of a liquid.

susceptibility The lack of resistance to a disease.

sylvatic Belonging in the woods; a wild animal.

symbiosis The living together of two different organisms.

symptom A change in body function that is felt by the patient due to a disease.

syndrome A specific group of signs or symptoms accompanying a particular disease.

synergistic effect The principle whereby the effectiveness of two drugs used simultaneously is greater than either drug used alone.

synthesis reaction A chemical reaction in which two or more atoms combine to form a new, larger molecule.

synthetic chemotherapeutic agent An antimicrobial agent that is prepared in a laboratory.

synthetic drug A chemotherapeutic agent that is prepared from chemicals in a laboratory.

systemic anaphylaxis Hypersensitivity reaction causing vasodilation and resulting in shock.

systemic (generalized) infection An infection throughout the body.

systemic mycosis A fungal infection in deep tissues.

T antigen An antigen in the nucleus of a tumor cell.

tapeworm A flatworm belonging to the class Cestoda.

target cell A basophil or mast cell to which IgE antibodies bind.

taxon A taxonomic category.

taxonomy The science of classification.

T cells Stem cells processed in the thymus gland responsible for cellular immunity.

teichoic acid A polysaccharide found in gram-positive cell walls.

temperate phage A bacteriophage existing in lysogeny with a host cell.

terminator site The site on DNA at which transcription ends.

tertiary treatment Physical and chemical treatment of wastewater to remove all BOD, nitrogen, and phosphorus, following secondary treatment.

tetracyclines Broad-spectrum antibiotics that interfere with protein synthesis.

tetrad A group of four cocci.

tetrahedron A four-sided solid structure.

T-even bacteriophage A complex virus with double-stranded DNA that infects ***E. coli;*** for example, T2, T4, T6.

thallus The entire vegetative structure or body of a fungus, lichen, or alga.

thermal death point (TDP) The temperature required to kill all the bacteria in a liquid culture in 10 minutes at pH 7.

thermal death time (TDT) The length of time required to kill all bacteria in a liquid culture at a given temperature.

thermoduric Heat resistant.

thermophile An organism whose optimum growth temperature is between 50 and 60°C.

thermophilic anaerobic spoilage Spoilage of canned foods due to the growth of thermophilic bacteria.

thermostable nuclease A heat-stable enzyme produced by ***Staphylococcus aureus.***

thylakoid Chlorophyll-containing membrane in chloroplast.

thymine A pyrimidine nucleic acid base in DNA that pairs with adenine.

tincture An alcoholic or aqueous solution.

tinea A cutaneous fungal infection; ringworm.

tissue culture *See* cell culture.

titer Reciprocal of a dilution.

total magnification Magnification of a specimen determined by multiplying the ocular lens magnification by the objective lens magnification.

toxemia Symptoms due to toxins in the blood.

toxigenicity The capacity of a microorganism to produce a toxin.

toxin Any poisonous substance produced by a microorganism.

toxoid An inactivated toxin.

trace element A chemical element required in small amounts for growth.

transamination The transfer of an amino group from an amino acid to an organic acid.

transcription The process of synthesizing RNA from a DNA template.

transfer RNA (tRNA) The class of molecules that brings the amino acids to the site where they are incorporated into proteins.

transformation The process in which genes are transferred from one bacterium to another as "naked" DNA in solution; also, the changing of a normal cell into a cancerous cell.

transient flora Microorganisms that are present on an animal for a short time without causing a disease.

translation The use of RNA as a template in the synthesis of protein.

translocation Movement of a gene from one chromosomal locus to another.

transmission electron microscope An electron microscope that provides high magnifications of thin sections of a specimen.

transverse fission *See* binary fission.

transverse septum A crosswall that separates genetic material into two daughter cells in binary fission.

trickling filter A method of secondary sewage treatment.

trophozoite The vegetative form of a protozoan.

tube agglutination test A method for determining antibody titer using serial dilutions of serum mixed with antigen.

tube dilution test A method of determining the MIC using serial dilutions of an antimicrobial drug.

tuberculin test A skin test used to detect the presence of antibodies to *Mycobacterium tuberculosis.*

tumor Excessive tissue due to uncontrolled cell growth.

tumor-specific transplantation antigen (TSTA) A viral antigen on the surface of a transformed cell.

turbidity The cloudiness of a suspension.

turnover number The number of substrate molecules metabolized per enzyme molecule per second.

12D treatment A sterilization process that results in a decrease of the bacterial population by 12 logarithmic cycles.

UDP-N-acetyl glucosamine (UDPNAG) A compound necessary for the biosynthesis of peptidoglycan.

ultrastructure Fine detail not seen with a compound light microscope.

ultraviolet (UV) radiation Radiation from 10 to 390 nm.

uncoating The separation of viral nucleic acid from its protein coat.

undulating membrane A highly modified flagellum on some protozoans.

uracil A pyrimidine nucleic acid base in RNA that pairs with adenine.

uridine diphosphoglucose (UDPG) Precursor for synthesis of glycogen.

use-dilution test A method of determining effectiveness of a disinfectant using serial dilutions.

vaccination The process of conferring immunity using a vaccine.

vaccine A preparation of killed, inactivated, or attenuated microorganisms or toxoids to induce artificially acquired active immunity.

vacuole An intracellular inclusion, in eucaryotic cells, surrounded by a plasma membrane containing raw food; in procaryotic cells, surrounded by proteinaceous membrane containing gas.

valence The combining capacity of an atom or molecule.

vancomycin Antibiotic that inhibits cell wall synthesis.

vasodilation Dilation or enlargement of blood vessels.

VDRL test A rapid screening test to detect the presence of antibodies against *Treponema pallidum.*

vector An arthropod that carries disease-causing organisms from one host to another.

vegetative cells Cells involved with obtaining nutrients, as opposed to reproduction or resting.

venereal disease A sexually transmitted disease.

vesicle The expanded, terminal area of the Golgi complex; also, a fluid-filled blister. In a procaryote, protein-covered hollow cylinder in a gas vacuole.

V factor NAD or NADP.

vibrio A curved or comma-shaped bacterium.

viral hemagglutination The ability of certain viruses to cause agglutination of red blood cells.

viremia The presence of viruses in the blood.

virion A fully developed complete viral particle.

viroid An infectious piece of "naked" RNA.

virulence The degree of pathogenicity of a microorganism.

virus A submicroscopic, parasitic filterable agent consisting of a nucleic acid surrounded by a protein coat.

visible light Radiation from 400 to 700 nm, which the human eye can see.

volutin Stored phosphate in a procaryotic cell.

wandering macrophage A macrophage that leaves the blood and migrates to infected tissue.

Wassermann test A complement fixation test used to diagnose syphilis.

wheal An area of edema of the skin resulting from a skin test.

whey The fluid portion of milk that separates from the curd.

Widal test An agglutination test used to detect typhoid fever.

X factor A precursor necessary to synthesize cytochromes.

xenograft A tissue graft from another species.

yeast A unicellular fungus belonging to the phylum Ascomycetes.

zone of inhibition The area of no bacterial growth around an antimicrobial agent in the agar diffusion test.

zoonosis A disease that occurs primarily in wild and domestic animals but can be transmitted to humans.

zygospore A sexual fungal spore characteristic of the Zygomycetes.

zygote A fertilized ovum produced by the fusion of two gametes.

zymogen An enzyme storage inclusion in eucaryotic cells.

Acknowledgments

Note: CDC = Centers for Disease Control, Public Health Service, U.S. Department of Health and Human Services, Atlanta, Georgia.

BPS = Biological Photo Service, Moss Beach, California.

Chapter 1 Title: CDC, Div. of Viral Diseases, E. M. Lab. Figures: 1–1a: CDC. 1–1b: D. L. Smalley, Univ. of Tennessee Center for the Health Sciences and D. D. Ourth, Memphis State Univ. 1–2a and 1–5: C. L. Case, Skyline College. 1–6a: H. S. Pankratz, Michigan State Univ./BPS. 1–6b: G. T. Cole, Univ. of Texas-Austin/BPS. 1–6c: H. S. Wessenberg and G. A. Antipa. *J. Protozool.* 17:250–270 (1970). 1–6d: J. R. Waaland, Univ. of Washington/BPS. 1–6e: S. C. Holt, Univ. of Texas Health Science Center, San Antonio/BPS. 1–9: Z. Skobe/BPS.

Chapter 2 Title: N. L. Max, Univ. of California/BPS.

Chapter 3 Title: S. C. Holt, Univ. of Texas Health Science Center, San Antonio/BPS. Figures: 3–1a: Reichert Scientific Instruments, Buffalo, N. Y. 3–2: J. R. Waaland, Univ. of Washington/BPS. 3–4b: CDC. 3–6a: R. Rodewald, Univ. of Virginia/BPS. 3–4b: E. Golub, Purdue Univ. 3–10a: S. C. Holt, Univ. of Texas Health Science Center, San Antonis/BPS. 3–10b and c: CDC. Box: p. 71: J. C. Burhnam and S. F. Conti. *Journal of Bacteriology* 976:1374 (1968).

Chapter 4 Title: Z. Skobe/BPS. Figures: 4–1a: CDC. 4–1b: Z. Skobe/BPS. 4–1c: T. J. Beveridge, Univ. of Guelph/BPS. 4–1d: A. E. McKee, Naval Medical Research Institute, Bethesda, MD. 4–2a and b: Z. Skobe/BPS. 4–2c: J. J. Cardamone, Jr., Univ. of Pittsburgh/BPS. 4–3a: G. T. Cole, Univ. of Texas-Austin/BPS. 4–3b: J. P. Burans, Naval Medical Research Institute, Bethesda, MD. 4–3c: N. S. Hayes et al., Univ. of North Carolina, Chapel Hill. 4–5 and 4–6c upper left: CDC. 4–6c lower left: T. J. Beveridge, Univ. of Guelph/BPS. 4–6c middle: P. W. Johnson & J. McN. Sieburth, Univ. of Rhode Island/BPS. 4–6c right: L. E. Simon, Rutgers Univ., New Brunswick, NJ. 4–7: S. C. Holt, Univ. of Texas Health Science Center, San Antonio/BPS. 4–8: S. Abraham and E. H. Beachey, V. A. Medical Center, Memphis, TN. 4–12: M. E. Bayer and T. W. Starkey. *Virology* 49:236–256 (1972). 4–13c: D. Branton, Harvard Univ. 4–14: H. S. Pankratz, Michigan State Univ./BPS. 4–15: T. J. Beveridge, Univ. of Guelph/BPS. 4–19b: S. C. Holt, Univ. of Texas Health Science Center, San Antonio/BPS. 4–19c left: H. S. Pankratz, Michigan State Univ./BPS. 4–19c middle and right: T. J. Beveridge, Univ. of Guelph/BPS. 4–21a: E. B. Small, Univ. of Maryland, courtesy of G. A. Antipa. 4–21b: H. S. Wessenberg and G. A. Antipa. *J. Protozool.* 17:250–270 (1970). 4–22b: C. L. Sanders, Battelle Pacific Northwest Labs./BPS. 4–23a: D. Branton, Harvard Univ. 4–24a and 4–25a: G. E. Palade, Yale Univ. Medical School. 4–26a: K. R. Porter, Univ. of Colorado. 4–27: G. E. Palade, Yale Univ. Medical School. 4–28: W. P. Wergin and E. H. Newcomb, Univ of Wisconsin-Madison/BPS. Box: p. 97: D. Balkavill and D. Maratea.

Chapter 5 Title: T. J. Beveridge, Univ. of Guelph/BPS.

Chapter 6 Title: CDC. Figures: 6–7: L. J. Le Beau, Univ. of Illinois Hospital/BPS. 6–9: CDC. 6–10a, b, and d: CDC. 6–10c: R. J. Hawley, Georgetown Univ. School of Dentistry. 6–11b: L. E. Simon, Rutgers Univ. 6–16a: Nalge Company, Div. of Sybron Corp., Rochester, NY. 6–16b: H. W. Jannasch, Woods Hole Oceanographic Institution. Box: p. 155: K. Crane, Woods Hole Oceanographic Institution.

Chapter 7 Title: C. L. Case, Skyline College. Figures: 7–2: R. Humbert, Stanford Univ./BPS. 7–5: B. R. Funke, North Dakota State Univ. 7–8: R. Humbert, Stanford Univ./BPS.

Chapter 8 Title: L. Caro and R. Curtiss. Figures: 8–2a: J. Griffith, Univ. of North Carolina, Chapel Hill. 8–2b: G. F. Bahr, Armed Forces Institute of Pathology. 8–6a: J. Cairns, Imperial Cancer Research Fund Lab., Mill Hill, London. 8–7b: O. L. Miller and B. R. Beatty, Oak Ridge National Lab. 8–26a: R. Welch, Medical School, Univ. of Wisconsin.

Chapter 9 Title: S. M. Awramik, Univ. of California/BPS. Figures: 9–3: S. M. Awramik, Univ. of California/BPS. 9–6: L. J. Le Beau, Univ. of Illinois Hospital/BPS. 9–7: B. D. Davis, et al., *Microbiology*, 3rd ed. Hagerstown, MD: Harper & Row, 1980. 9–9: Biavatti, et al., *Int'l. J. of Systematic Bact.*, 32:358–373 (1982).

Chapter 10 Title: R. P. Burchard, Univ. of Maryland Baltimore County. Figures: 10–1b: P. W. Johnson & J. McN. Sieburth, Univ. of Rhode Island/BPS. 10–1c: N. S. Hayes et al., Univ. of North Carolina, Chapel Hill. 10–2: T. J. Beveridge, Univ. of Guelph/BPS. 10–3: J. J. Cardamone, Jr., Univ. of Pittsburgh/BPS. 10–4: CDC. 10–5:

T. J. Beveridge, Univ. of Guelph/BPS. 10–6: G. T. Cole, Univ. of Texas-Austin/BPS. 10–7: CDC. 10–8a: W. Burgdorfer, Rocky Mountain Lab., Hamilton, MT. 10–8b: R. C. Cutlip, National Animal Disease Center, Ames, IA. 10–9: M. G. Gabridge, Bionique Labs. 10–10: B. R. Funke, North Dakota State Univ. 10–11, 10–12, and 10–13: T. J. Beveridge, Univ. of Guelph/BPS. 10–14: CDC. 10–15: R. J. Hawley, Georgetown Univ. School of Dentistry. 10–16a: R. P. Burchard, Univ. of Maryland Baltimore County. 10–16b: K. Stephens, Stanford Univ./BPS. 10–17: J. F. M. Hoeniger, Univ. of Toronto/BPS. 10–18: R. L. Moore, BioTechniques Labs./BPS. 10–19 and 10–20: T. J. Beveridge, Univ. of Guelph/BPS. 10–21: S. C. Holt, Univ. of Texas Health Science Center, San Antonio/BPS. 10–22a: P. W. Johnson & J. McN. Sieburth, Univ. of Rhode Island/BPS. 10–22b: J. R. Waaland, Univ. of Washington/BPS. 10–22c: J. C. Burnham, Medical College of Ohio. 10–23: CDC.

Chapter 11 Title: G. T. Cole, Univ. of Texas-Austin/BPS. Figures: 11–2: C. L. Case, Skyline College. 11–3: C. Robinow, Univ. of Western Ontario. 11–6: C. L. Case, Skyline College. 11–7: CDC. 11–8a: R. Humbert/BPS. 11–8b and c: J. R. Waaland, Univ. of Washington/BPS. 11–9d and 11–12: C. L. Case, Skyline College. 11–14 and 11–16: Photograph by Carolina Biological Supply Company. 11–17 and 11–27: CDC. Box: p. 323: CDC.

Chapter 12 Title: R. C. Valentine and H. G. Pereira. *J. Mol. Biol.* 13:13–20 (1965). Figures: 12–3b: J. Griffith, Univ. of North Carolina, Chapel Hill. 12–4b: R. C. Valentine and H. G. Pereira. *J. Mol. Biol.* 13:13–20 (1965). 12–5b: F. A. Murphy, Viral Pathology Branch, CDC. 12–6b: A. K. Harrison, Viral Pathology Branch, CDC. 12–7a and c: R. C. Williams, Univ. of California, Berkeley. 12–8a: B. Roizman, Univ. of Chicago. 12–8b: C. Garon and J. Rose, National Institute of Allergy and Infectious Diseases. 12–9a: A. K. Harrison, Div. of Viral Diseases, CDC. 12–9b: J. J. Cardamone, Jr., Univ. of Pittsburgh/BPS. 12–9c: G. H. Smith, National Cancer Institute. 12–10: R. Humbert, Stanford Univ./BPS. 12–12: G. Wertz, School of Medicine, Univ. of North Carolina, Chapel Hill. 12–17: C. Morgan, H. M. Rose, and B. Mednis. *Journal of Virology* 2:507–516 (1968). 12–22b: D. R. Brown, et al. *Journal of Virology* 10:524–536 (1972). 12–23: A. K. Harrison, Div. of Viral Diseases, CDC. 12–24: T. O. Diener, U.S. Dept. of Agriculture. Box: p. 367: Edmondson/Gamma-Liaison.

Chapter 13 Title: Z. Skobe/BPS. Figures: 13–1a: D. C. Savage, Univ. of Illinois at Urbana-Champaign. 13–1b: Z. Skobe/BPS. 13–1c: D. C. Savage, Univ. of Illinois at Urbana-Champaign. 13–3: M. W. Jennison and the Dept. of Biology, Syracuse Univ.

Chapter 14 Title: J. W. Costerton, Univ. of Calgary. Figures: 14–1: J. W. Costerton, Univ. of Calgary. 14–2: S. C. Holt, Univ. of Texas Health Science Center, San Antonio/BPS. 14–3: F. A. Murphy, Viral Pathology Branch, CDC. 14–4: J. P. Bader, National Cancer Institute.

Chapter 15 Title: C. L. Sanders/BPS. Figures: 15–1: L. Winograd, Stanford Univ./BPS. 15–3a: K. E. Muse, Duke Univ. Medical Center. 15–6: C. L. Sanders/BPS. 15–8f: J. G. Hadley, Battelle-Pacific Northwest Labs./BPS. 15–12a: Schreiber, et al. *J. Exp. Med.* 149:870–882 (1979).

Chapter 16 Title: J. J. Cardamone, Jr., and S. Salvin, Univ. of Pittsburgh/BPS. Figures: 16–19c: CDC.

Chapter 17 Title: J. J. Cardamone, Jr., and S. Salvin, Univ. of Pittsburgh/BPS. Figures: 17–2: L. J. Le Beau, Univ. of Illinois Hospital/BPS. 17–8a and c: A. Liepins, Sloan-Kettering Institute for Cancer Research.

Chapter 18 Title: T. J. Beveridge, Univ. of Guelph/BPS. Figures: 18–1: A. Fleming. *British Journal of Experimental Pathology* 10:226–236 (1929). 18–3: T. J. Beveridge, Univ. of Guelph/BPS. 18–5: M. Bastide, S. Jouvert, and J.-M. Bastide. *Canadian J. of Microbiol.* 28:1119–1126 (1982). 18–20 and 18–21: L. J. Le Beau, Univ. of Illinois Hospital/BPS.

Chapter 19 Title: G. T. Cole, Univ. of Texas-Austin/BPS. Figures: 19–1b: L. Winograd, Stanford Univ./BPS. 19–2 top: L. J. Le Beau, Univ. of Illinois Hospital/BPS. 19–2 bottom: CDC. 19–3: P. P. Cleary, Univ. of Minnesota School of Medicine/BPS. 19–4: L. J. Le Beau, Univ. of Illinois Hospital/BPS. 19–5: Armed Forces Institute of Pathology, Neg. No. 58-6180-6. 19–6 and 19–7a: World Health Organization, Geneva, Switzerland. 19–7b, 19–9, and 19–10: CDC. 19–11: P. Weary, Univ. of Virginia School of Medicine. 19–12 and 19–13a: CDC. 19–13b: G. T. Cole, Univ. of Texas-Austin/BPS. 19–13c: N. Goodman, Univ. of Kentucky Medical College.

Chapter 20 Title: P. A. Merz, New York State Institute for Basic Research. Figures: 20–3: F. L. A. Buckmire, The Medical College of Wisconsin. 20–5: N. Goodman, Univ. of Kentucky Medical College. 20–6: CDC. 20–7: T. J. Beveridge, Univ. of Guelph/BPS. 20–8: CDC. 20–9: B. A. Phillips, Univ. of Pittsburgh. 20–12: A. K. Harrison, Division of Viral Diseases, CDC. 20–13: CDC. 20–17: P. A. Merz, New York State Institute for Basic Research.

Chapter 21 Title: S. C. Holt, Univ. of Texas Health Science Center, San Antonio/BPS. Figures: 21–2: Armed Forces Institute of Pathology, Neg. No. N-64686. 21–4a: J. J. Cardamone, Jr., Univ. of Pittsburgh/BPS. 21–4b: M. J. Tufte, Univ. of Wisconsin-Platteville. 21–6: CDC. 21–7: Armed Forces Institute of Pathology, Slide No. 79-18280-2. 21–9: CDC. 21–13: J. Griffith, Univ. of North Carolina, Chapel Hill. 21–15a: CDC. 21–16a: S. G. Baum, Albert Einstein College of Medicine/BPS. 21–18: CDC.

Chapter 22 Title: G. T. Cole, Univ. of Texas-Austin/BPS. Figures: 22–3: CDC. 22–5: K. E. Muse, Duke Univ. Medical Center/BPS. 22–6: R. B. Morrison, Austin, TX.

22–8a: L. J. Le Beau, Univ. of Illinois Hospital/BPS. 22–8b: G. L. Goodhart, V.A. Hospital, Philadelphia. 22–9: J. A. Stuekemann and D. Paretsky, Univ. of Kansas. 22–11a: G. D. Roberts, Mayo Clinic, Rochester, MN. 22–11b: G. T. Cole, Univ. of Texas-Austin/BPS. 22–13a: CDC. 22–13b: G. T. Cole, Univ. of Texas-Austin/BPS.

Chapter 23 Title: J. J. Paulin, Univ. of Georgia/BPS. Figures: 23–2: M. A. Listgarten, Univ. of Pennsylvania/BPS. 23–3b, 23–3c, and 23–4: Z. Skobe/BPS. 23–6: M. A. Listgarten, Univ. of Pennsylvania/BPS. 23–7: A. E. McKee, Naval Medical Research Institute, Bethesda, MD. 23–8: W. L. Dentler, Univ. of Kansas/BPS. 23–9a: G. T. Cole, Univ. of Texas-Austin/BPS. 23–9b and c: J. M. Madden, B. A. McCardell, and R. B. Read. *Food Technology* March 1982:93–96. 23–10: F. A. Murphy, Viral Pathology Branch, CDC. 23–12a: A. K. Harrison, Div. of Viral Diseases, CDC. 23–12b: CDC. 23–13: A. Z. Kapikian, National Institutes of Health, Bethesda, MD. 23–14a: J. Griffith, Univ. of North Carolina, Chapel Hill. 23–15a: J. J. Paulin, Univ. of Georgia/BPS. 23–15b: I. Armstrong, Washington Hospital Center, Washington, DC. 23–16a: A. E. McKee, Naval Medical Research Institute, Bethesda, MD. 23–16b: CDC.

Chapter 24 Title: P. W. Johnson & J. McN. Sieburth, Univ. of Rhode Island/BPS. Figures: 24–4: D. Bromley, West Virginia Univ. Medical Center. 24–6, 24–7, 24–9, 24–10, 24–11, 24–12, and 24–13: CDC.

Chapter 25 Title: R. N. Band and H. S. Pankratz, Michigan State Univ./BPS. Figures: 25–4a: R. Toja, Charles R. Kettering Laboratory. 25–4b: E. H. Newcomb and S. R. Tandon, Univ. of Wisconsin-Madison/BPS. 25–5a and b: M. F. Brown, Univ. of Missouri-Columbia. 25–5c: R. L. Peterson, Univ. of Guelph. 25–13a: C. W. May/BPS. 25–14: R. F. Unz. *International Journal of Systematic Bacteriology* 21:91–99 (1971). 25–15a and 25–16a: C. W. May/BPS. Box: p. 704: P. C. Fitz-James, Univ. of Western Ontario.

Chapter 26 Title: T. Beveridge, Univ. of Guelph, and J. Fein, Weston Research Centre, Toronto/BPS. Figures: 26–3: C. W. May/BPS. 26–6: Switzerland Cheese Association, New York, NY. 26–9: D. M. Munnecke/BPS. 26–10: Courtesy of Vega Biotechnologies, Inc., Tucson, AZ. 26–11: D. W. Galbraith, Univ. of Nebraska-Lincoln.

Part Openings: One: Z. Skobe/BPS. Two: S. Dales, Univ. of Western Ontario. Three: J. J. Cardamone, Jr., and P. W. Dowling, Univ. of Pittsburgh/BPS. Four: Z. Skobe/BPS. Five: E. H. Newcomb and S. R. Tandon, Univ. of Wisconsin-Madison/BPS.

Microview 1 Figures by page number: 1-1, A,B,D: Leon J. Le Beau, University of Illinois Hospital at the Medical Center, Chicago/BPS. All others CDC. 1-2, A,B,C,D: Christine L. Case. E: CDC. F,G: Christine L. Case. 1-3, A,B,C,D: Christine L. Case. E: © Richard Humbert. F,G: Christine L. Case. 1-4, All from CDC.

Microview 2 2-1, Left: Damon Biotech, Inc.; four at right: Genentech, Inc. 2-2, 1,2,3,4: © Nita Winter, 1985/Cetus Corporation and art: Carla Simmons. 2-3, 5: Cetus Corporation. 6: © Nita Winter, 1985/Cetus Corporation. 7: © Chuck O'Rear, 1985/Cetus Corporation. A,B,C,D: © Nita Winter, 1985/Cetus Corporation. 2-4, Top Left: Genentech, Inc. Center: Genentech, Inc. Top Right: Genentech, Inc. Bottom left: © Chuck O'Rear, 1985. Bottom right: University of California, Davis

Microview 3 3-1, Top: Squibb Institute For Medical Research. 1,2,3: © David Powers, 1985/University of California, Davis. 3-2, 4,5,6: © David Powers, 1985/University of California, Davis. 7: Medical Center, Chicago/BPS. 3-3, 8,9,10,11,12,13: © David Powers, 1985/University of California, Davis. 3-4 14,15,16,17: © David Powers, 1985/University of California, Davis 18: Genentech, Inc.

Index

Note: Italic type refers to figures or tables.

Abdominal cramps, gastrointestinal intoxications and, 646
ABO blood group system, 486–487, *487*
Abortion, 279, 383, 577
Abscesses, 426, 528
Absorbance, 176
Absorption, nutrient, 642
Accessory glands, 676, *677*
Accessory structures, 641
Accutane; *see* 13-cis Retinoic acid, Isotretinoin
Acetobacter spp., 281
Acetyl coenzyme A, 131–132
  formation from pyruvic acid, *131*
N-Acetylglucosamine, 86–87
  chemical structure of, *87*
N-Acetylmuramic acid, 86–87
  chemical structure of, *87*
Acid-base balance, 37–38
Acid-fast stains, 73
Acidic dyes, 69
Acidity, 37–38
  microbial growth and, 153
Acids, 36–37, *37*
  enzyme denaturation and, 118
Acinetobacter spp., 236
Acne, 291, 420, 531, *541*
Acquired immune deficiency syndrome, 3–4, 481, 482–484
  cryptosporidiosis and, 662
  genetic engineering and, 244
  monoclonal antibodies and, 449
  pentamidine and, 509
  Pneumocystis pneumonia and, 631
  resistance and, 432
  retroviruses and, 362, 368
Acquired immunity, 438–440, *440*
Acridine dyes, mutagenesis and, 230
Actinomyces israelii, 292, *293*
Actinomyces spp., 291
  periodontal disease and, 645
Actinomycetes, 297–298
  in soil, 702, *702*
  symbiosis and, 709
Actinomycosis, 292
Activated sludge systems, 722–723, *722*
Activation energy, 33, 115
Active processes, 93
Active sites, enzymic, 34
Active transport, 94–95, *95*, 103
Acyclovir, 509
  genital herpes and, 688, 689–690
  herpesvirus infections and, 537
  structure and function of, *510*
Addison's disease, 479
  human leukocyte antigen and, *480*
Adenine, 51, 52
Adenine arabinoside, 508, *512*
Adenine nucleotide, 51
Adenocarcinomas, 366
Adenoids, 610
Adenosine diphosphate, 52–53
Adenosine diphosphoglucose, 140
Adenosine triphosphatase, 124
Adenosine triphosphate, 52–53
  active transport and, 94
  aerobic respiration and, *135*
  anabolism, catabolism and, *114*
    chemiosmosis and, 123–124
  electron transport chain and, 132–134
  glycolysis and, 130–131
  phosphorylating enzymes and, 117
    phosphorylation and, 122–123
  production of, 129, *129*
  respiration and, 131–135
  structure of, *53*
  synthesis of, nitrogen and, 156
Adenoviruses, 343, *343*, 381
  common cold and, 615
  multiplication of, 358–359, *359*
  oncogenic, 368
  properties of, *346*
  type 3, inclusion bodies of, *365*
  virions of, *365*
Adherence
  of glycocalyx, 83
  pathogenic, 402
  phagocytosis and, 424
Adhesins, 402
ADP; *see* Adenosine diphosphate
ADPF; *see* Adenosine diphosphoglucose
Adsorption
  of animal viruses, 357
  of phages to host cells, 352
Aedes aegypti, *332*, 592
Aerial mycelium, 302–303
Aerobes, 157
Aerobic respiration, 131–135
  ATP and, *135*
  fermentation and, *139*
Aerotolerant anaerobes, 159
Aflatoxin, 660
  liver cancer and, 410
  mutagenesis and, 230
African sleeping sickness, 320, 322–323, 463
African trypanosomiasis, 566–567
  arthropod vectors and, *390*
  meningitis and, 551
Agar, 159–160
  algae and, 313
  bismuth sulfite, 163
  blood, 164, 290
  brilliant green, 164
  MacConkey, 164
  nutrient, 161
  Sabouraud's glucose, 164
  selective and differential, *165*
Agar dilution method, 514
Agaricus augustus, *308*
Agglutination, 453–457, *454*
  C. burnetii and, 625
  leptospirosis and, 679
  whooping cough and, 617
Agglutinins, 453
Agranulocytes, 421, *421*
Agriculture, biotechnology and, Microview 2–4
AIDS; *see* Acquired immune deficiency syndrome
Airborne transmission, 389
Alcoholic beverages
  fermentation and, 6, 138, *138*
  yeasts and, 741–742, *742*
Alcoholism
  pneumonia and, 284, 621
  resistance and, 433
Alcohols
  antimicrobial activity of, 198 *203*
  enzyme denaturation and, 118
  hydroxy groups in, 40–41
  S. pyogenes and, *198*
Aldehydes, antimicrobial activity of, 200–201, *204*
Algae, 15, 310–314
  brown, 311, *312*
  carbon cycle and, 17
  cell walls of, 102
  characteristics of, *302*
  chloroplasts of, 107, 126
  copper sulfate and, 199
  eutrophication and, 715
  flagella of, 101
  green, *312*, 313
  habitats of, *314*
  metachromatic granules in, 96
  oxidation ponds and, 725
  pathogenicity of, 408–410
  photosynthesis in, 125–126
  red, 311–313, *312*
  roles in nature, 313–314
  in seawater, 714
  in soil, 702–703, *702*
  structure and reproduction of, 313
  use as food, 746
Algin, 311
Alimentary canal, 641
Alkalinity, 37–38
  microbial growth and, 153
Allergens, 472
Allergic contact dermatitis, 478
Allergy, 383, 472
  antimicrobial agents and, 496
  development of, *478*
Allescheria boydii, 538
Allografts, 480, *481*
Allosteric sites, 120
Allosteric transitions, 120
Alveolar macrophages, 611
Alveoli, 611
Alzheimer's disease, 370
  prions and, 15
  slow viruses and, 568
Amanita phalloides, toxins of, 410
Amanita spp., mycotoxins of, 660
Amanitin, 410
Amantadine, 508, *512*
  influenza and, 627
American trypanosomiasis, 595, *600*
Ames test, 233–234, *233*
Amikacin, 505
Amination, 143

Amino acid analyzer, Microview 2–2
Amino acids, 46–49
  biosynthesis of, 142–143, *142*
  commercial production of, 749
  essential, 749
  exchange reactions and, 33
  functional groups in, 41–42
  isomers of, 46, *46*
  peptidoglycan and, 87
  in proteins, *47*
  sequences of, *50*
    microorganism classification and, 264
  structural formula for, *46*
para-Aminobenzoic acid, 494
  enzyme inhibition and, 500–501
  inhibition of, 120
Aminoglycosides, 505, *511*
  protein synthesis and, 500
2-Aminopurine, 229, *229*
Ammonification, 706–707
Ammonium ion, structure of, *200*
Amniocentesis, 383
Amoebas, 317–318, *318*
  locomotion of, 15
Amoebiasis, 661–662
  metronidazole and, 509
Amoebic dysentery, 319, 400, 411, 661–662
  emetine and, 509
  human carriers and, 385
  water supply contamination and, 712
Amoxicillin, spectrum of activity of, 504
Amphibolic pathways, 145
Amphitrichous flagella, 84, *85*
Amphotericin B, 508, *512*
  aspergillosis and, 630
  blastomycosis and, 629
  coccidioidomycosis and, 628
  cryptococcosis and, 554
  histoplasmosis and, 628
  plasma membrane and, 500
  structure of, *508*
Ampicillin, *511*
  bacillary dysentery and, 651
  H. influenzae type b and, 553
  spectrum of activity of, 504
  typhoid fever and, 651
Amyotrophic lateral sclerosis, 568
Anabolic reactions, 113
Anabolism, 32, 113–114
  catabolism, ATP and, *114*
Anaerobes; *see also specific type*
  aerotolerant, 159
  facultative, 157
  obligate, 157
    oxygen sensitivity of, 159
Anaerobic chamber, *163*
Anaerobic container, *162*
Anaerobic respiration, 135, 280
  fermentation and, *139*
Anaerobic sludge digesters, 723–725, *724*
Anal gonorrhea, 681
Anal pore, protozoan, 318, *318*
Analytical epidemiology, 394
Anamnestic response, 447, 555
Anaphylactic shock, 473–474
  hydatidosis and, 631
Anaphylaxis, 472–475
  IgE antibodies and, 472
  localized, 474
  mechanism of, *473*
  mediators of, *473*
  prevention of, 474–475
  systemic, 473–474
Angstrom, 60
Animal bites, 585–587
Animalcules, 5, 6
Animal feed, antibiotics and, 496
Animalia, 13, 255
Animal viruses, 340–341
  classification of, *346*
  growth in laboratory, 350–352
  host cells and, 365
  multiplication of, 357–364
Anions, 29
Anopheles mosquito, 332, *332*, 596
  sporozoan reproduction in, 321, 324
Anoxygenicity, 296
Antagonism, drug, 517
Anthrax, 10, 67, 291, *386*, 582–584, *600*, 703
  exotoxins and, *406*
  pustule of, *583*
  reported cases of, *583*
Antibiosis, 495
Antibiotics, 11–12, 16, 494, 495, 502–507, *511*
  animal feed and, 496
  aspergillosis and, 630
  bacillary dysentery and, 651
  bacterial cell wall and, 90
  botulism and, 557
  brucellosis and, 584
  C. perfringens and, 585
  chancroid and, 687
  commercial production of, 750
  contagious conjunctivitis and, 541
  epidemic typhus and, 588
  gastroenteritis and, 653
  gonorrhea and, 681–682
  listeriosis and, 584
  nongonococcal urethritis and, 686
  nucleic acid synthesis and, 500
  otitis media and, 615
  periodontal disease and, 646
  plague and, 587
  plasma membrane and, 500
  polyene, 508, *512*, 538
  protein synthesis and, 499–500, *499*
  pseudomonads and, 531
  puerperal sepsis and, 577–579
  pyelonephritis and, 678
  resistance to, 514–517
  ribosomal protein synthesis and, 96
  salmonellosis and, 650
  sources of, *495*
  staphylococci and, 528
  syphilis and, 684–685
  trachoma and, 541
  tuberculosis and, 619
  whooping cough and, 617
Antibodies, 438, 441–444
  anaphylaxis and, 472–475
  antigens and, *440*, 445–447
  blocking, 475
  cell-mediated reactions and, 477–478
  in colostrum, 439
  cytotoxic reactions and, 475–476
  cytotrophic, 472
  Epstein-Barr virus and, 591
  fluorescent, 67
  humoral, 439
    hypersensitivity reactions and, 472
  IgA, in saliva, 642
  IgE, cytotrophic, 472
  structure of, 441–442, *441*
Antibody titer, 447
Antifungal agents, 508
  vulvovaginal candidiasis and, 692
Antigen-antibody complexes
  complement activation and, 431
  immune complex reactions and, 472
  precipitation reactions and, 452
  radioimmunoassay and, 459
Antigenic determinant sites, 440, *440*, 442
Antigenic drifts, 626
Antigenic shifts, 626
Antigen modulation, 484
Antigen receptors, 438
Antigens, 67, 438, 440–441
  antibodies and, *440*, 445–447
  histocompatibility, 479–480
  human leukocyte, 479–480
  latex agglutination test for, 455–457
  localized anaphylaxis and, 474
  red blood cell, 486–487, *486*
  T-dependent, 445
  T-independent, 445
  valence of, 441
Antihelminthic agents, 509–511, *512*
Antihistaminic agents, hayfever and, 474
Antimetabolite agents, enzyme inhibition and, 500–501
Antimicrobial agents, 493–517; *see also specific type*
  actions of, 185, 497–501, *498*
  antibiotics, 502–507
  antifungal, 508
  antihelminthic, 509–511
  antiprotozoan, 509–511
  antiviral, 508–509
  criteria for, 495–497
  plasma membrane and, 92
  resistance to, 514–517
  spectrum of activity of, 497
  synthetic, 501–502
Antimonial agents, *512*
  schistosomiasis and, 511
Antiprotozoan agents, 509–511, *512*
Antisense strands, 218
  picornavirus, 362
Antisepsis, *184*
Antiseptic surgery, 183
Antisera, 263, 439
Antitoxins, 405, 457
Antiviral agents, 508–509, *512*
  herpesvirus infections and, 537
  herpes zoster and, 534

herpetic keratitis and, 542
Aorta, 575
Aortic-valve surgery, bacteremia and, 578
Aplastic anemia, chloramphenicol and, 506
Apoenzymes, 115
Appendicitis, leukocyte count in, 421
Aquaspirillum bengal, spiral cell of, *279*
Aquaspirillum graniferum, flagella of, *85*
Aquaspirillum magnetotacticum, magnetosomes of, 97
Aquatic microbiology, 711–726
Ara-A, 508
Arachnids, 331
Arachnoid membranes, 549
Arboviruses
dengue and, 592
encephalitis and, 565
yellow fever and, 592
Arbuscules, *710,* 711
Archaeobacteria, 270, 295
Arenaviruses, properties of, *347*
Arsenic, enzyme inhibition and, 119
Arteries, 575
Arterioles, 575
Arthritis, gonorrheal, 681
Arthropod-borne encephalitis, 565–566, *566, 568*
Arthropods, 331–332
Arthropod vectors, *332, 333,* 389–390, *390*
Arthrospores, 305, *305*
Arthus reaction, 477
Ascaris lumbricoides, 330
Aschelminthes, 325, 329–331
Ascomycota, 307
characteristics of, *310*
spores of, 306
Ascospores, 306, *306*
Ascus, 306, *306*
Asepsis, *184*
Aseptic meningitis, 551
Aseptic packaging, 736
Aseptic techniques, 8
Asexual spores, of fungi, 305–306, *305*
Asiatic cholera, 285, 651–653
water supply and, 712
Asiatic influenza, 626
Aspartic acid
purine biosynthesis and, 143
pyrimidine biosynthesis and, 143
Aspergillosis, 309, 630
Aspergillus flavus, aflatoxin and, 230, 410, 660
Aspergillus fumigatus
aspergillosis and, 630
hyphae of, Microview 1–4
Aspergillus niger, 304
citric acid and, 749
Aspergillus oryzae, soy sauce and, 741
Aspergillus spp., 309
spores of, 307
Aspergillus tamarii, conidiospores of, *307*
Aspirin, Reye's syndrome and, 534
Asthma, 383, 472, 474
Athlete's foot, 537, *539*
tolnaftate and, 508
zinc undecylenate and, 199
Atomic number, 25
Atomic weight, 25
Atoms, 25
chemical bonding of, 28–32
electronic configurations of, 26–28, *27*
structure of, 25–28, *25*
ATP; *see* Adenosine triphosphate
ATPase; *see* Adenosine triphosphatase
Atria, of heart, 575
Auramine O, 65–67
Aureomycin; *see* Chlortetracycline
Autoclave, 186, *187*
Autoclaving, 186–188
Autografts, 480, *481*
Autoimmunity, 383, 447, 478–481
prions and, 569
rheumatic fever and, 579
Automated tests, 514
Autotrophs, 125
Avery, O. T., 236
Axial filaments, 84, *86,* 276, *279*
Axilla, microbial growth of, 525
Azlocillin, 504
Azomonas spp., 281
Azotobacter spp., 281
lipid inclusions in, 96
nitrogen fixation and, 707

Bacillary dysentery, 284, 400, 411, 651
transmission of, 390
Bacilli, 13, 81, *82*
in vagina, 677
Bacillus anthracis, 291, 383–384, 582–583
colonies of, *169*
endospore of, *99*
fluorescein isothiocyanate and, 67
quellung reaction and, 83–84
rods of, Microview 1–1
in soil, 703
Bacillus cereus, 14
gastroenteritis and, 654
Bacillus licheniformis, division of, *170*
Bacillus megaterium
endospore of, *99*
megacins and, 642
Bacillus sphaericus, endospore of, *99*
Bacillus spp., 291
anaerobic respiration and, 135
antibiotics and, 495
denitrification and, 707
endospores of, 96
genetic transformation in, 236
lipid inclusions in, 96
of nasal cavity, 611
nitrogen fixation and, 707
Bacillus stearothermophilus, 735
Bacillus subtilis, 131
Bacillus thuringiensis
pest control and, 18, 704
in soil, 703
Bacitracin, 506–507, *511*
peptidoglycan synthesis and, 499
Bacteremia, 392
aortic-valve surgery and, 578
listeriosis and, 584
Bacteria, 13, 275–298; *see also specific type,* Procaryotes
budding, *294,* 295
cell walls of, 497–499
chemoautotrophic, 295
classification of, 257–260
coagulase-positive, 648
coliform, 717
conjugation in, 236–237, *237*
division of, 168–169, *170*
enteric, 263, 281–285
features of, *303*
fruiting, 293
gene expression in, 223–226
genetic transformation in, 234–236
gliding, 293
groups of, 276–298
characteristics of, *277*
identification of, *267*
Leeuwenhoek's drawings of, *4*
metachromatic granules in, 96
mineral extraction and, *716*
nitrifying, 295
nitrogen fixation and, 707
nonsymbiotic, 707–709
pest control and, 704
phototrophic, 295–297
phylogenetic relationships of, *260*
sheathed, 293
in soil, 702, *702*
symbiotic, 709
taxonomy of, 275–276
transduction in, 237–239, *239*
vaccines for, *463*
Bacterial chromosomes, 95
Bacterial diseases
of cardiovascular system, 577–590
of digestive system, 643–654, *666*
of genital system, 679–688, *691*
of lower respiratory system, 615–625, *632*
of lymphatic system, 577–590
of mouth, 643–646, *666*
of skin, 527–531, *542*
of upper respiratory system, 611–615, *632*
of urinary system, 677–679, *690*
Bacterial dysentery, exotoxins and, *406*
Bacterial endocarditis, 579
acute, 579, *579, 599*
subacute, 579, *599*
Bacterial growth curve, 171–172, *171*
Bacterial infections, of eye, 539–542, *542*
Bacterial kinases, 404
Bacterial meningitis, 550–555
incidence of, *553*
types of, *551*
Bacterial pneumonias, 403
Bacteriochlorophylls, 126
Bacteriocinogenic plasmids, 241
Bacteriocins, 241, 282, 642
Bacteriophages, *14,* 237–239, 340–341
growth in laboratory, 349
lambda
lysogeny and, 355–356
viral plaques formed by, *349*
lysogenic, 355

M13, 342
multiplication of, 352–357, *353*
T-even, 352–355
T4, *345*
transduction by, *239*
Bacteriorhodopsin, 127–128
Bacteriostasis, *184*, 497
Bacteroides fragilis, plasmid of, *240*
Bacteroides spp., 286
animal bites and, 585
chains of, *286*
in large intestine, 642
in mouth, 642
puerperal sepsis and, 579
septicemia and, 577
in urethra, 676
Baker's yeast, *304*
Balantidial dysentery, 661
Balantidiasis, 661
Balantidium coli, 321, *662*
balantidiasis and, 661
Basal body, flagellar, 84
Base analogs, 229, *229*
Base pair mutagens, 229
Bases, 36–37, *37*
enzyme denaturation and, 118
Base substitution, 226–228, *227*
Basic dyes, 69
Basicity, 37–38
Basidiomycota, 307, *308*, *310*
Basidiospores, Microview 1–3, 306, *306*
Basidium, 306, *306*
Basophils, 421, *421*
anaphylaxis and, 472–473
B cells, 438, 575, 577
activation of, *446*
differentiation of, *445*
humoral immunity and, 445
memory, 445
stem cells and, 444
T cells compared, *451*
BCG vaccine, 620
Bdellovibrio spp., 279
behavior of, 71
Beef tapeworm, 327–328
life cycle of, *648*, *664*
Beggiatoa spp., energy sources for, 126
Beijerinckia spp., nitrogen fixation and, 707
Benign tumors, 365, 383
Benthic zone, 713, *714*
Benzalkonium chloride
antimicrobial activity of, 198, 200
structure of, *200*
Benzathine penicillin, 502
syphilis and, 685
Benzethonium chloride, 200
Benzoic acid, antifungal action of, 200
Benzoyl peroxide
acne and, 531
antimicrobial activity of, 202
Benzypyrene, mutagenesis and, 230
Betadine, 196
Beta-lactamases, 504
Beta oxidation, 139
Beta-propiolactone, gaseous sterilization and, 202
Binary fission, 13
Binomial nomenclature, 255
Biochemical assay devices, Microview 2–3
Biochemical oxygen demand, 721–722
determination of, *721*
Biochemical testing, microorganism classification and, 263
Bioconversion, 746
Biodegradability, *712*
Biogenesis, 8
Biogeochemical cycles, microorganisms and, 703–711
Biological transmission, 390
Biomass, 746
Biosynthesis
of DNA-containing viruses, 358–359
of RNA-containing viruses, 359–362
of viral components, 354
Biotechnology, Microview 2–2, 747–751
benefits of, Microview 2–4
Biovars, 282
Birth defects, 383
Black Death, 585
Blackheads, 525, 531
Blades, algal, 313
Blastoconidia, *C. albicans*, Microview 1–4
Blastomyces dermatitidis, 629
Blastomycosis, 629, *633*
airborne transmission of, 389
amphotericin B and, 508
Blastospores, *305*, 306
Blindness
neonatal gonorrheal ophthalmia and, 541
trachoma and, 287, 541
Blocking antibodies, 475
Blood, 575
formed elements in, *420*, 421
Blood agar, 290
Blood-brain barrier, 549
Blood fluke, 327
Blood groups
ABO, 486–487, *487*
Rh, 487–488
Blood poisoning, 577
Blood transfusion
hepatitis B and, 658
Rh incompatibility and, 487
Blood vessels, permeability of, host defense and, 425–426
Blooms, algal, 314
BOD; *see* Biochemical oxygen demand
Body temperature, regulation of, 427
Boiling, microbial control and, 186
Boils, 426, 525, 528
Boletus edulis, Microview 1–3
Bordetella pertussis, 281, *618*
portal of entry of, *401*
whooping cough and, 615–617
Bordetella spp., 281
Borrelia spp., 276
relapsing fever and, 587
Botulinum toxin, 405, 556
Botulism, 291, 356, 385, 555–557, *568*, 646, 703, 739
exotoxins and, *406*
food preservation and, 736
home canning and, 735
wound, 556
Bovine tuberculosis, 620
Breakbone fever, 592
Breast milk, secretory IgA in, 444
Bright's disease, 679
Brightfield illumination, 61, *63*
5-Bromouracil, 229, *229*
Bronchial tubes, 611
Bronchiolitis, 611
Bronchitis, 285, 611–612
Bronchopneumonia, secondary, 392
Brown algae, Microview 1–3, 311, *312*
Brownian movement, 68
Brucella abortus, *582*
brucellosis and, 581
Brucella canis, brucellosis and, 581
Brucella melitensis
brucellosis and, 581
portal of entry of, *401*
Brucella spp., 281
Brucella suis, brucellosis and, 581
Brucellosis, 163, 281, *386*, 404, 411, 581–582, *600*
Buboes, 586, *586*
Bubonic plague, 284, *386*, 586, *586*
Budding, 168–169
of bacteria, *294*, 295
of yeasts, 303–304
Buffers, 153
Burkitt's lymphoma, 368
Epstein-Barr virus and, 591
Burns
nosocomial infection and, 387–388
P. aeruginosa and, 531
resistance and, 432
Burst size, 354
Burst time, 354
Butter, lactic acid bacteria and, 741
Buttermilk, lactic acid bacteria and, 741

Calcium, ion of, 29
Calcium hypochlorite, disinfection and, 198
Calcium ions, enzyme-substrate complexes and, 117
Calcium propionate
food preservation and, 738
mold growth and, 200
California encephalitis, 565
Calvulina cristata, *308*
Calymmatobacterium granulomatis, 687
Campylobacter fetus, 279
Campylobacter jejuni, 279
gastroenteritis and, 654
Campylobacter spp., traveler's diarrhea and, 653
Cancer
acquired immune deficiency syndrome and, 4
aspergillosis and, 630
causes of, 383
cell-mediated immune system and, 438
immune response to, 481–485
interferon and, 429
monoclonal antibodies and, 447
nosocomial infections and, 388

types of, *486*
viruses and, 365–369
Candida albicans, Microview 1–4, 309, *540, 692*
candidiasis and, 538–539, 689–692
colonies of, Microview 1–2
spores of, 306
superinfection and, 497, 506
urinary tract infections and, 678
vaginitis and, 380–381, 686
Candida spp., nystatin and, 508
Candida utilis, use as food, 746
Candidiasis, 309, 538–539, *540, 541,* 689–692
oral, 539, *540,* 689
vulvovaginal, 689–692, *691*
Candle jar technique, 162, *163,* 552
Canned food, spoilage of, 735–736
Canning, 733, *734,* 735
Capillaries, 575
Capsids, 342, *342, 348*
Capsomeres, 342, *342*
Capsules, 83, *83*
host defenses and, 402–403
staining of, 74, *74*
Carbenicillin
pseudomonads and, 531
spectrum of activity of, 504
Carbohydrates, 42–43
biosynthesis of, *141*
catabolism of, 130–138, *136*
synthesis of, Golgi complex and, 105
Carbolfuchsin, 69, 73, 618
Carbolic acid; *see* Phenol
Carbon
atomic structure of, *25*
bonding patterns of, *40*
in carbohydrates, 42
in complex lipids, 45
covalent bonding of, 31
in lipids, 43
microbial growth and, 154–156
in proteins, 45
tetrahedral compounds of, 40, *40*
valence of, 28
Carbon cycle, 17, *17,* 703–705, *705*
Carbon dioxide
aerobic respiration and, 132
anaerobic sludge digestion and, 723–725
carbon cycle and, 17
Carbon dioxide incubators, 163
Carbon skeleton, 40
Carboxysomes, 96
Carbuncles, 528
Carcinogens, identification of, 233–234
Carcinomas, *486*
Cardiac catheters, endotoxic reactions and, 387
Cardiovascular system, *576*
bacterial diseases of, 577–590
helminthic diseases of, 593–599
protozoan diseases of, 593–599
rickettsial diseases of, 587–590
structure and function of, 575
viral diseases of, 590–592
Carrageen, 313
Carrier proteins, 94
Carriers, 385
Casein, 740
Caseous lesions, 618
Catabolic reactions, 113
Catabolism, 33, 114
anabolism, ATP and, *114*
carbohydrate, 130–138, *136*
lipid, 138–139, *140*
protein, 139–140
Catalase, 158
Catalase test, streptococci and, 290
Catalysts, 34, 115
Cations, 29
Cat scratch fever, *386,* 585, *600*
Caulobacter bacteroides, 295
Caulobacter spp., 295
fresh water and, 713
holdfasts of, 712
cDNA, 242
reverse transcription of, *243*
Ceepryn; *see* Cetylpyridinium chloride
Cefamandole, 505
Cefotaxime, 505
Cell cultures, virus cultivation in, 350–352
Cell lines, 351
Cell-mediated immunity, 438
hypersensitivity reactions and, 472
T cells and, 447–451
Cell-mediated reactions, 477–478
mechanism of, *477*
Cell morphology, microorganism classification and, 262
Cell walls, bacterial, antimicrobial agents and, 497–499
Cellular slime molds, *316,* 317
Cellulose, 43
in eucaryotic cell walls, 102
use as food, 746
Central nervous system, 549
Centrioles, *100,* 107
Centrosomes, *100,* 107
Cephalosporins, 504–505, *511*
peptidoglycans and, 499
eucaryotic cells and, 103
Klebsiella pneumonia and, 622
staphylococci and, 528
structure of, *504*
Cephalosporium spp., 495
Cephalothin, 505
Ceratocystis ulmi, 310
Cercaria, 326
schistosomal, 599
Cerebrospinal fever, 551–552
Cerebrospinal fluid, 549, *550*
Cestodes, 327–329
Cetus Corporation, Microview 2–2
Cetylpyridinium chloride, 200
Chagas' disease, 320, *386,* 595
Chancres, syphilitic, 683, *684*
Chancroid, 687, *691*
Cheesemaking, 740–741, *740*
Chemical bonds, 28–32
Chemical carcinogens, 233–234
Chemical elements, 25–26, *26*
Chemical energy, 35
Chemical factors, host defense and, 419–420
Chemically defined media, 161
Chemical mutagens, 230
Chemical pollution, 712–713
Chemical preservatives, 738–739
Chemical reactions, 32–35
energy requirements of, *34*
water and, 36
Chemiosmosis, 123–124, *124*
Chemistry, 25
Chemoautotrophs, 125, 126–128, 155–156, 295
chemically defined media for, *160*
Chemoheterotrophs, 125, 128, 302, 311, 703
Chemosterilizers, 201–202, *203*
Chemosynthesis, 155–156
Chemotaxis, phagocytosis and, 424
Chemotherapeutic agents, 493, 511–514
Chemotherapy, 11–12, 183, 493
Asiatic cholera and, 653
gonorrhea and, 681–682
historical development of, 494–495
targeting and, 12
Chemotrophs, 125
Chickenpox, 369, 533–534, 535, *541,* 655
immunity and, 438
lesions of, *533*
pneumonia and, 625
Childbirth fever, 577
Chills, 428
Chilopoda, 331
Chitin, 13
in cell wall of fungi, 102
Chlamydia psittaci, 289
psittacosis and, 624
Chlamydia trachomatis, 287
gonorrhea and, 682
inclusion bodies of, Microview 1–1
lymphogranuloma venereum and, 687
nongonococcal urethritis and, 686
trachoma and, 541
Chlamydia spp., 287–289, *288*
artificial media and, 163
biochemical properties of, *287*
Chlamydospores, *305,* 306
Chloramines, disinfection and, 198
Chloramphenicol, 506
Asiatic cholera and, 653
brucellosis and, 584
C. perfringens and, 585
cystitis and, 678
endemic murine typhus and, 588
epidemic typhus and, 588
gastroenteritis and, 653
glomerulonephritis and, 679
granuloma inguinale and, 687
H. influenzae type b and, 553
N. meningitidis and, 552
protein synthesis and, 499
puerperal sepsis and, 579
pyelonephritis and, 678
Rocky Mountain spotted fever and, 589
structure of, *506*
typhoid fever and, 651

whooping cough and, 617
Chloride ions, 29
Chlorination, 719–720
Chlorines
antimicrobial activity of, 196–198
ionic bonding of, 29
phenolics and, 195
pseudomonads and, 531
Chlorobium spp.
nitrogen fixation and, 707
vesicles of, 126
Chlorophyll
chloroplasts and, 107
cytochromes and, 117
photophosphorylation and, 123
Chlorophyll a, 126
Chloroplasts, 107, *107*, 126
DNA in, 103
ribosomes of, 105
Chloroquine, *512*
malaria and, 509, 597
Chlorosomes, 126
Chlortetracycline, 505
balantidiasis and, 661
Cholera, 400, 411
Asiatic, 285, 651–653
chicken and hog, 438
exotoxins and, *406*
water supply and, 712
Cholesterol, 45, 500
structure of, *45*
Chrococcus turgidus, 297
Chromatin, 104
Chromatium vinosum, 296
Chromatophores, 91, *91*
Chromomycosis, 538, *541*
Chromophores, 69
Chromosome number, 239
Chromosomes, 210
bacterial, 95
DNA and, 212–213
eucaryotic, 104, 212, *213*
homologous, 240
procaryotic, 212, *213*
Chronic disease, 392
Chronic lymphocytic leukemia, 432
Chrysops spp., *333*
Cidex, 201
CIE; *see* Countercurrent immunoelectrophoresis
Cilia
eucaryotic, 101–102, *101*
protozoan, 15, 318
respiratory, 610
host defense and, 419
Ciliary escalator, 611
host defense and, 419
Ciliata, 318, 321
Cimetidine, interferon therapy and, 429
Cirrhosis, 383
aflatoxin and, 660
Cisternae, of Golgi complex, 105, *105*
Citric acid, production of, 749
Citric acid cycle; *see* Krebs cycle
Citrobacter spp., in large intestine, 642
Cladosporium werneckii, 537
Class, 13, 257
Claviceps purpurea
mycotoxins of, 660
sclerotia of, *410*
Cleft palate, 383
Clindamycin, *C. perfringens* and, 585
Clinical laboratory, Microview 3–1
Clofazimine, leprosy and, 558
Clonal selection, 445
Clones, 448
forbidden, 478
Clorox, 198
Clostridium botulinum, 291, 385
botulism and, 555
endospores of, 184
gastric juice and, 420
in soil, 703
toxins of, 356, 405
Clostridium pasteurianum, 707
Clostridium perfringens, 291
gangrene and, 584–585
gastroenteritis and, 654
hemolysins of, 404
phospholipase C and, 51
portal of entry of, *401*
in soil, 703
Clostridium spp., 291
animal bites and, 585
collagenase of, 404
endospores of, 96
food preservation and, 736
nitrogen fixation and, 707
oxygen sensitivity of, 159
puerperal sepsis and, 579
superoxide free radical formation in, 158
thermophilic anaerobic spoilage and, 735
Clostridium tetani, 291, 390, *554*
neurotoxin of, 405
portal of entry of, *401*
in soil, 703
tetanus and, 554
Clotrimazole
C. albicans and, 539
tinea infections and, 537
vulvovaginal candidiasis and, 692
Clubfoot, 383
Clue cells, 686
CoA; *see* Coenzyme A
Coagulase, 404, 527–528
Coagulase-positive bacteria, 648
Coal tar, phenolics derived from, 195
Cobalamin, production of, 749
Cobalt 60, food preservation and, 737
Cobalt ions, enzyme-substrate complexes and, 117
Cocci, 13, 81
arrangements of, *80*
gram-negative, 280–281, 287
gram-positive, endospore-forming, 291
pyogenic, 528
vaginal, 677
Coccidioides immitis, 630
coccidioidomycosis and, 628
portal of entry of, *401*
spores of, 305–306
Coccidioidomycosis, 308, 628–629, *633*
airborne transmission of, 389
amphotericin B and, 508
geographic distribution of, *631*
Coccobacilli, 81, *82*, 281
Codons
nonsense, 221–222
sense, 222
Coefficients, 39
Coenocytic hyphae, 302, *303*
Coenzyme A, 116; *see also* Acetyl coenzyme A
decarboxylation and, 132–132
Krebs cycle and, 132
Coenzymes, 115–116
oxidation-reduction and, 122
Cofactors, 115, 117
Cold sores, 368, 369, 392, 536–537, *536, 541*
Colicin, 642
Coliform bacteria, water purity and, 717
Colitis, 383
Collagenase, 404, 585
Collision theory, 33
Colonies, 168, *169*
Colony-forming units, 175
Colostrum, antibodies in, 439
Commensalism, 381
Commercial sterilization, 733–735, *734*
Common cold, 389, 392, 400, 615, 616–617, *632*
immunity and, 438
Common vehicle transmission, 389
Communicable disease, 390
Competence, genetic, 236
Complement
action of, *430*
activation of, 431–432, *431*
host defense and, 429–432
immune complex reactions and, 472
Complement fixation, 432, *456*, 457
C. burnetii and, 625
dengue and, 592
hepatitis A virus and, 657
Complex lipids, 44–45
Complex media, 161
Complex viruses, 343, *345*
Compound light microscope, 61, *62*
images produced by, *63*
Compound light microscopy, 61
Compounds, 28
inorganic, 35–38
organic, 38–53
Compromised host, 432
Concentration gradients, 92–93
Condensation reactions, 42
Condenser, in compound light microscope, 61
Condiophores, 306
Congenital disease, 383
Congenital heart disease, 383
Congenital infection of fetus, 656
Congenital rubella syndrome, 535–536
Congenital syphilis, 685
Conidia, histoplasmosis and, 628
Conidiospores, 298, *298*, *305*, 306, *307*
Conjugated proteins, 51

Conjugation
  bacterial, 236–237, *237*
  mechanism for, *238*
  protozoan, 319, *319*
Conjugative plasmids, 240
Conjunctivitis, 281, 539–541
  contagious, 539–541, *542*
  inclusion, 542, *542*
Constitutive enzymes, 226
Constitutive genes, 226
Contact dermatitis, 478
Contact inhibition, 366–368, 408, 484
Contact transmission, 388–389
Contagious conjunctivitis, 539–541, *542*
Continuous cell lines, 351
Convalescence period, 411
Copper, microbial growth and, 157
Copper ions, enzyme-substrate complexes and, 117
Copper sulfate, algicidal activity of, 199
Corepressors, 224
Coronaviruses
  capsids of, *348*
  common cold and, 615
  properties of, *347*
Corticosteroids, trichinosis and, 667
Corynebacteria, 291–292
  sebum and, 525
  vaginal, 677
Corynebacterium diphtheriae, 291, *293, 613*
  diphtheria and, 613
  metachromatic granules in, 96
  portal of entry of, *401*
  toxin of, 356, 405
Corynebacterium glutamicum, 749
Corynebacterium spp., periodontal disease and, 645
Corynebacterium xerosis, 527, *527*
Coryza, 615
Countercurrent immunoelectrophoresis, 453, *453*
  bacterial meningitis and, 552
Counterstains, 72
Covalent bonds, 31
  formation of, *30*
Cowpox, 10
Coxiella burnetii, 287, *625*
  Q fever and, 624–625
Coxsackieviruses
  common cold and, 615
  gastroenteritis and, 660
  myocarditis and, 590
CPEs; *see* Cytopathic effects
Cresols, disinfection and, 195
Cretinism, 383
Creutzfeldt-Jakob disease, 370, 568, *568*
  prions and, 15
Crimean-Congo hemorrhagic fever, 592
Cristae, of mitochondrion, 106, *106*
Crossing over, 234, *234,* 240
Crustacea, 331
Crustose lichens, Microview 1–3, 315, *315*
Cryotherapy, warts and, 532
Cryptococcosis, 553–554, *568*
Cryptococcus neoformans, 554
  meningitis and, 553–554
  virulence of, 408
Cryptosporidiosis, 662
Cryptosporidium spp., 662
Culex mosquito, encephalitis and, 566
Culture media, 159–168
  anaerobic, 161–162
  chemically defined, 160–161
  complex, 161
  reducing, 161
  selective and differential, 163–165,
Cultures, 159
  enrichment, 168
  growth of, 168–177
  microbial, Microview 1–2
  mixed, 167
  preservation of, 168
  pure, isolation of, *166,* 167–168
Cutaneous diphtheria, 614, *632*
Cutaneous mycoses, 308–309, 537
Cyanide, enzyme inhibition and, 119
Cyanobacteria, 296–297, *297; see also* Procaryotes
  alkaline environments and, 38
  carboxysomes in, 96
  eutrophication and, 715
  gas vacuoles in, 96
  microscopic images of, *63*
  nitrogen cycle and, 17
  nitrogen fixation and, 157, 707
  photosynthesis in, 125–126
  in soil, 702–703, *702*
  thermophilic, Microview 1–3
Cyclic AMP, amoebas and, 317
Cyclic photophosphorylation, 123, *123*
Cycloserine, peptidoglycan synthesis and, 497–499
Cyclosporine, organ transplantation and, 481
Cystic hydatid disease, 631
Cysticerci, tapeworm, 327
Cystitis, 678, *690*
Cystostomes, protozoan, 318, *318*
Cysts, protozoan, 319
Cytochrome c, 116
  amino acid sequencing and, 264, *265*
Cytochrome oxidase, 118
Cytochromes, 116–117, 133–134
Cytocidal effects, 407
Cytomegalovirus, 656, *656*
  acquired immune deficiency syndrome and, 482
Cytomegalovirus inclusion disease, 656
Cytopathic effects, 351, *351,* 365, 407, *409*
Cytophaga spp., 293
  fresh water and, 713
Cytoplasm
  eucaryotic, *100,* 103
  procaryotic, *83, 92,* 95–96
Cytoplasmic inclusions, *100,* 107
Cytoplasmic membrane; *see* Plasma membrane
Cytoplasmic streaming, 103
Cytosine, in DNA, 51
Cytoskeleton, 103
Cytotoxic reactions, 475–476
Cytotoxic T cells, 447–450
  tumor cells and, 485, *485*
Cytotrophic antibodies, 472

Dairy products, food-borne infections and, 739–740
Dane particle, 657–658, *658*
Dapsone, leprosy and, 558
Darkfield microscope, 61–64, *64*
Darkfield microscopy, 61–64
DDT, 18, 711
Deamination, 140, 706
Death angel, 410
Death phase, of bacterial growth, *171, 172*
Decarboxylation, 131–132
  coenzyme A and, 116
Decimal reduction time, 186
Decomposition reactions, 32–33
Deep-freezing, bacterial cultures and, 168
Deer fly, *333*
Defecation, 642
Definitive hosts
  helminthic, 325
  humans as, 327–328
Degeneracy, of genetic code, 222
Degenerative disease, 383
Degermation, *184*
Degranulation, 473
Dehydration synthesis, 42, *43*
  peptide bond formation and, *49*
Dehydrogenases, 116, 118
Dehydrogenation, 116, 122
Delayed hypersensitivity T cells, 450–451
Delta hepatitis, 660
Denaturation, 118, *119*
  alcohols and, 198
Dengue, 592, *600*
  arthropod vectors and, *390*
Dengue hemorrhagic fever, 592
Denitrification, 707
Dental abscess, 286
Dental caries, 402, 643–645, *644*
Dental plaque, 644, *645*
Deoxyribonucleases, 529, 585
Deoxyribonucleic acid; *see* DNA
Deoxyribose, in DNA, 42, 51
Dermacentor spp., *333*
  Rocky Mountain spotted fever and, 589
Dermatitis, 383, 478
Dermatomycoses, 308–309, 537, *539*
Dermatophytes, 308, 537
Dermis, 417, 525
Descriptive epidemiology, 393–394
Desensitization, 475
Dessication, microbial control and, 189–190
Desulfovibrio spp., 286
  anaerobic respiration and, 135
  fresh water and, 713
Detergents
  antimicrobial activity of, 199–200, *203*
  water supply and, 712–713

Deuteromycota, 307–308
  characteristics of, *310*
DEV; *see* Duck embryo vaccine
Dextran, 43, 644
  sucrose conversion to, *645*
Diabetes mellitus, 383
  aspergillosis and, 630
  juvenile-onset, slow viruses and, 568
  mucormycosis and, 309
  pneumonia and, 621
  recombinant DNA and, 19
  resistance and, 432
  urinary tract infections and, 677
Diagnosis, 392
Diapedesis, 426, *426*
Diarrhea; *see also* Traveler's diarrhea
  enterotoxins and, 406
  epidemic, 653
  infant, 653
  intoxications and, 646
Diatoms, 311, *312, 313*
Dick test, 612
Dicloxacillin, penicillinases and, 504
Didinium nasutum, 14, 101
Diethylstilbesterol, 383
Differential count, 421
Differential interference contrast microscopy, 65
Differential media, 163–164, *164, 165*
Differential staining, microorganism classification and, 262
Diffusion
  facilitated, 94, *95,* 103
  simple, 93, *93,* 103
Digestion, 43
  catabolism and, 33
Digestive system
  accessory structures of, 641
  anatomy of, *641*
  antimicrobial features of, 642
  bacterial diseases of, 643–654, *666*
  helminthic diseases of, 662–667, *667*
  normal flora of, 642
  protozoan diseases of, 660–662, *667*
  structure and function of, 641–642
  viral diseases of, 654–660, *667*
Digestive vacuoles, *102,* 103, 424, *425*
Diglycerides, 44
Dihydrofolic acid, 501, *503*
Dihydroxyacetone phosphate, 139
Diiodohydroxyquin
  amoebic dysentery and, 662
  balantidiasis and, 661
Dimorphic fungi, 304
Dimorphism, 304, 330
Dinoflagellates, 311, *312*
Dioecious, 325, 329
Dipeptides, 49
Diphtheria, 291, 356, 384, 457, 613–614, *632*
  age distribution of, *614*
  cutaneous, 614, *632*
  exotoxins and, *406*
  fatal, 407
  human carriers and, 385
  immunity and, 438
  newborn and, 439
Diphtheria toxin, 405
Diphtheroids
  eye infections and, 539
  of nasal cavity, 611
  on skin, 527
Diphyllobthrium latum, 664
Dipicolinic acid, 98
Diplobacilli, 81, *82*
Diplococci, 81
  arrangements of, *80*
Diplococcus pneumoniae, 621
Diploid cells, 240
Diploid number, 240
Diplopoda, 331
Direct flaming, sterilization and, 188
Direct microscopic count, microbial growth and, 172–173, *173*
Direct selection, 231
Disaccharides, 42–43
Disease, 379, 382–385, 388–394, 410–411; *see also specific type*
  germ theory of, 8–10
Disinfectants
  effective, qualities of, 192
  evaluation of, 194, *195*
  selection of, 184
  temperature and, 183
  types of, 194–202
Disinfection, 183, *184*
  effective, principles of, 192–194
Disk diffusion method, 512–513, *513*
Dissimilation plasmids, 241
Dissociation, 36–37
Distemper, canine, 438
Division, 257
DNA, 51–52, 211
  base substitution in, 226–228, *227*
  chromosomes and, 212–213
  complementary strands of, 211
  deoxyribose in, 42, 51
  double helix of, 221
  in endospore core, 98, *99*
  eucaryotic, 103–104, 241–243
  gene expression and, 225
  genetic engineering and, 747–749
  hybridization, *266,* 267–269
  instruments synthesizing, *748*
  mutation of, 211
  nitrogen in, 16
  plasmid, preparation of, Microview 2-2
  recombination of, 234, *234*
  replication fork of, 213–218, *216*
  replication of, 213–218, *214, 217*
    semiconservative, 213
    thymine dimers and, 192
  strands of
    antisense, 218
    sense, 218
  structure of, *52*
  sugar-phosphate bonds in, 213–215, *215*
  synthesis of, 215–218
    nitrogen and, 156
  viral, 15
DNA ligase, 216
DNA polymerase, 213–218
  RNA-dependent, viral multiplication of, 362
DNA viruses, 341–342
  isolation of, 352
  multiplication of, 358–359
Donor cells, genetic transfer and, 234
Donovan bodies, 687, *688*
Double antibody sandwich method, 460–462
Double helix, 52
Doubling time, 169–171
Down's syndrome, 383
Doxorubicin, interferon therapy and, 429
Doxycycline, 505
  lymphogranuloma venereum and, 687
  nongonococcal urethritis and, 686
  traveler's diarrhea and, 653
DPT immunization, 554–555
DPT vaccine, 613, 617
Droplet infection, 389
Drug abuse, pneumonia and, 621
Drug resistance, 514–517
Drugs
  sulfa, 11
  synthetic, 11
Dry heat sterilization, 188
Dry weight, bacterial growth and, 176–177
Duck embryo vaccine, 562
Dumb rabies, 561
Dura mater, 549
Dust cells, 611
Dutch elm disease, 310
Dyes, 69
Dysentery, 646, 739; *see also* Amoebic dysentery, Bacillary dysentery

Eastern equine encephalitis, 565
Ebola hemorrhagic fever, 592
Echinococcus granulosus, 328
  hydatidosis and, 631
  life cycle of, *328*
Echoviruses, 381
  common cold and, 615
  gastroenteritis and, 660
Eclipse period, 354
Ectomycorrhizae, *710,* 711
Ectopic pregnancy, gonorrhea and, 681
Edema-producing factors, 404
Effector cells, 447
  hypersensitivity reactions and, 472
Electrical energy, 35
Electrodessication, warts and, 532
Electron acceptors, 29
Electron donors, 28
Electronic configurations, atomic, 26–28, *27*
Electron microscopes, *66,* 67
Electron microscopy, 67
Electrons, 25
  shells, 26–28
Electron transport chain, 123, 132–134, *134*
Electrophoresis, 264–265, 439
Elementary bodies, 287, 624
ELISA; *see* Enzyme-linked immunosorbent assay
Embden-Meyerhof pathway, 130

Embryonated eggs
inoculation of, *350*
virus cultivation in, 350
Emetine, *512*
amoebic dysentery and, 509
Emphysema, 383
Emulsification, 199
Encephalitis, 291, 549
adenine arabinoside and, 508
arthropod-borne, *390*, 565–566, *566*, *568*
California, 565
Eastern equine, 565
experimental allergic, 568
herpes simplex virus and, 537
herpetic keratitis and, 542
Japanese B, 565
listeriosis and, 584
measles and, 535
rabiesvirus and, 561
rubella and, 535
St. Louis, 565
Venezuelan equine, 565
Western equine, *386*, 565
Encephalopathy, transmissible mink, 568
Endemic disease, 392
Endemic murine typhus, 287, 588–589, *600*
arthropod vectors and, *390*
cycle of, *588*
Endergonic reactions, 35
Endocarditis, 285, 291
bacterial, 579, *579*, *599*
drug synergism and, 517
gonorrheal, 681
Q fever and, 624
streptococcal, vancomycin and, 507
Endocardium, 579
Endocytosis, 103
Endomycorrhizae, *710*, 711
Endoplasmic reticulum, *100*, *104*, 105
Endospores, 96–100, *99*
B. anthracis, 583
boiling and, 186
chemical resistance of, 184
clostridial, 555–556
dessication and, 190
in soil, 703
staining of, 74–75, *74*
Endosymbiosis, 156
Endothia parasitica, 310
Endotoxins, 88, 406–407, 531, 551, 577, 616, 651
exotoxins compared, *408*
End product inhibition, 121
Energy
forms of, 35
production of, 129, *129*
Enrichment culture, 168
Entamoeba histolytica, 319, *319*
amoebic dysentery and, 661–662
life cycle of, *647*, *663*
Enteric bacteria, 263, 281–285
Enteric cytopathogenic human orphan viruses, 381
Enteritis, foodborne, 279
Enterobacter aerogenes, 285
pyelonephritis and, 678
septicemia and, 577
Enterobacter cloacae, 285
bacteremia and, 578
Enterobacter spp., 285
identification of, 263
in large intestine, 642
nitrogen fixation and, 707
pneumonia and, 621
urethral, 676
Enterobacteriaceae, 281–285
Enterobius vermicularis, 329, *329*
Enterococci, nosocomial infection and, *388*
Enterotoxicosis, staphylococcal, 646–649
Enterotoxins, 290, 405–406, 528, 646–647, 652, 653
Enteroviruses, myocarditis and, 590
Entner-Doudoroff pathway, 131
amino acid synthesis and, 143
nucleotides and, 143
Enveloped viruses, 342, *342*, 343, *344*
budding of, 364, *364*
Environment, microbial growth and, 184–185
Environmental microbiology, Microview 1–3
Enzyme-linked immunosorbent assay, 459–462, *460*
HTLV-III and, 483
indirect, 462
rubella and, 536
Enzymes, 114–121
actions of
factors influencing, 118–120, *119*
mechanism of, 117, *117*
bacterial, host defenses and, 403–404
biologic function of, 34–35
classification of, *118*
constitutive, 226
endospore, 98
extracellular, 94
feedback inhibition of, 121
inducible, 224
induction of, 224
inhibition of, antimicrobial agents and, 500–501
lysosomal, 424
naming of, 118
oxygen-related, *158*
parts of, 115–117
production of, 750–751, *751*
repressible, 224
repression of, 224
restriction, 241
DNA viruses and, 352
viral, 352
Enzyme-substrate complexes, 34, 117
Eosinophils, 421, *421*
Epidemic diarrhea, 653
Epidemic disease, 392
Epidemic keratoconjunctivitis, *542*
Epidemic meningitis, endotoxins and, 407
Epidemic typhus, 287, 587–588, *600*
arthropod vectors and, *390*
Epidemiology, 393–394
analytical, 394
descriptive, 393–394
experimental, 394
Epidermis, 417, 525
Epidermolytic toxin, 290
Epidermophyton spp., 537
Epiglottis, host defense and, 419
Epiglottitis, 285, 611
Epinephrine
anaphylactic shock and, 473–474
asthma and, 474
EPS; *see* Glycocalyx
Epstein-Barr virus, 368, *591*
antibodies and, 591
Burkitt's lymphoma and, 591
infectious mononucleosis and, 590–591
nasopharyngeal cancer and, 591
portal of entry of, *401*
Equilibrium, 93
Equivalent treatments, 189
ER; *see* Endoplasmic reticulum
Ergosterol, 500
Ergot, 410
Ergotism, 410
Ergot poisoning, 660
Erwinia spp., 285
Erysipelas, 384, 530, *530*, *541*
Erysiphe grammis, *14*
Erythroblastosis fetalis, 487–488
Erythrocytes, 575
Erythrogenic toxin, 290, 529, 612
Erythromycin, *511*
bacterial endocarditis and, 579
beta-hemolytic streptococci and, 529
brucellosis and, 584
chancroid and, 687
erysipelas and, 530
glomerulonephritis and, 679
impetigo and, 530
legionellosis and, 3, 624
listeriosis and, 584
lymphogranuloma venereum and, 687
neonatal ophthalmia neonatorum and, 681
nongonococcal urethritis and, 686
otitis media and, 615
primary atypical pneumonia and, 623
protein synthesis and, 499–500
puerperal sepsis and, 577
resistance to, 514
streptococcal pharyngitis and, 612
structure of, *506*
syphilis and, 685
whooping cough and, 617
Escherichia coli, 71, 282, *282*
amino acid synthesis in, 142
attachment to intestinal cells, *402*
chemically defined media for, 161, *161*
colicin and, 642
complement and, *430*
conjugation in, 236–237, *237*
cystitis and, 678
DNA of, 212–213, 221–222
DNA synthesis in, 216–218

doubling time of, 170–171
enteroinvasive, 653
enterotoxigenic, 653
feedback inhibition in, 121, *121*
gastroenteritis and, 653–654
genetic engineering and, 748
genetic transformation in, 236
host cells and, 405
lactose metabolism in, 224
in large intestine, 642
meningitis and, 551
mutualism and, 381
name derivation of, 12
naming of, 283–284
newborns and, 379
nosocomial infection and, 388, *388*
operon model and, 224
operons of, 355
opportunistic, 381
pentose phosphate pathway and, 131
pili of, *86*
plasmid of, *240*
pneumonia and, 621
protoplast of, *89*
pyelonephritis and, 678
rods of, Microview 1–1
septicemia and, 577
transducing phage of, 237–239
urinary tract infections and, 678
Escherichia spp., 282
identification of, 263
nalidixic acid and, 502
Essential amino acids, 749
Ester linkage, 44
Ethambutol, 501, *511*
isoniazid and, 501
tuberculosis and, 619
Ethanol
commercial production of, 749
disinfection and, 198
hydroxy group in, 40
S. pyogenes and, *198*
Ethyl alcohol; *see* Ethanol
Ethylene oxide, antimicrobial activity of, 201
Etiology, 379
Eucaryotes, Microview 1–3, 13–15
chromosomes of, 212, *213*
classification of, 255, 257
genetic recombination in, 239–240
origins of, 80–81
oxidative phosphorylation in, 123
photosynthesis in, 125–126, *126*
sexual reproduction in, *240*
Eucaryotic cells, 100–107
cell wall of, 102–103
characteristics of, 81
composite, *100*
cytoplasm of, 103
cytoplasmic inclusions of, 107
DNA of, 241–243
flagella and cilia of, 101–102, *101*
genetic engineering and, 748
lysozyme in, 89
organelles of, 100, 103–107
plasma membrane of, 103
procaryotic cells compared, *108*
Eucaryotic pathogens, Microview 1–4
Euglena spp., *312*, 318
Euglenoids, 311, *312*
Eustachian tubes, 609–610
Eutrophication, 714, 715
Evolution, genes and, 244
Exchange reactions, 33
Exergonic reactions, 35
Exfoliative toxin, 290
Exocytosis, 103
Exons, 241–242
Exotoxins, 405–406, 531, 555, 584–585, 616, 654
diseases caused by, *406*
endotoxins compared, *408*
necrotizing, 290
Experimental allergic encephalitis, 568
Experimental epidemiology, 394
Exponential growth phase, of bacteria, 171–172, *171*
Exponential notation, 39
Exponents, 39
Extracellular enzymes, 94
Extracellular polymeric substance; *see* Glycocalyx
Extreme halophiles, 154
Eyepiece, in compound light microscope, 61
Eyes
bacterial infections of, 539–542
diseases associated with, *542*
viral infections of, 542

Facilitated diffusion, 94, *95*, 103
Factor VIII, recombinant DNA and, 19
Facultative anaerobes, 157
Facultative halophiles, 154
FAD; *see* Flavin adenine dinucleotide
Fallopian tubes, 675, *676*
Family, 12, 257
Fansidar, malaria and, 597
Fat inclusions, 107
Fats, 44
Fatty acids, 44
F+ cells, 236–237
F− cells, 236–237
Fc region, 442
Feces, 642
pathogens in, 411
Feedback inhibition, 121, *121*, 749
Fermentation, 135–138
aerobic respiration and, *139*
alcoholic, 138, *138*
anaerobic respiration and, *139*
end products of, *137*
industrial, *139*
lactic acid, 137–138, *138*
malolactic, 742
microorganisms and, 5–6
nondairy, 741
Fever, host defense and, 427–428
Fever blisters, 369, 536–537, *536*
F factors, 236–237, 240
Fibrinogen, coagulases and, 404
Filamentous particles, 658, *658*
Filaments, flagellar, 84
Filterable viruses, 339
Filter paper method, 194, *195*
Filters, membrane, 189, *190*
Filtration, 175, *175*
microbial control and, 189
Fimbriae, procaryotic, 84–86
Five-kingdom system, 255, *256*
Fixed macrophages, 423
Flaccid paralysis, 555
Flagella
bacterial, 13
eucaryotic, 101–102, *101*
procaryotic, 84, *85*
protozoan, 15, 318
staining of, *74*, 75
Flagellates, 318–320
Flat sour spoilage, 735
Flatworms, 15; *see also* Platyhelminthes
Flavin adenine dinucleotide, 116
Flavin mononucleotide, 116
Flavoproteins, 133
Flocculation, 718–719
Flora, 19
of fresh water, 712
of human mouth, *19*
normal, 379–382, *381*, *382*, *383*
of respiratory system, 611
of seawater, 713–714
of skin, 525–527, *527*
transient, 379–380
Flucytosine
chromomycosis and, 538
cryptococcosis and, 554
Fluid mosaic model, 91
Flukes, 325–327, 598–599
anatomy of, *326*
Fluorescein isothiocyanate, 458
fluorescence microscopy and, *67*
Fluroescence microscopy, 65–67
Fluorescent-antibody stains, *L. pneumophila*, Microview 1–1
Fluorescent-antibody technique, *65*, 67
Fluorescent-antibody test, 458–459, *459*
direct, 458
indirect, 458–459
legionellosis and, 624
primary atypical pneumonia and, 623
whooping cough and, 617
Fluorescent treponemal antibody absorption test, syphilis and, 685
Fluorochromes, 65–67
5-Fluorocytosine, 502
FMN; *see* Flavin mononucleotide
Focal infections, 392
Foliose lichens, 315, *315*
Fomites, 389, 609
Fonsecaea pedrosoi, 538
Food, from microorganisms, 746–747
Food-borne enteritis, 279
Food-borne infections, intoxications and, 739–740, *739*
Food chain, microorganisms and, 5
Food intolerance, 474
Food poisoning, 646
exotoxin and, *406*
S. aureus, 290
staphylococcal, 646–649, 739
Food preservation, 733–735
chemical preservatives and, 738–739

low-temperature, 736–737
pasteurization and, 738
radiation and, 191, 737–738
Food production, microorganisms and, 740–742
Food spoilage, 733–735
canned, types of, *736*
flat sour, 735
thermophilic anaerobic, 735, *735*
Forbidden clones, 478
Forespores, 98, *99*
Formaldehyde, 200–201
Formalin, 200–201
Formed elements, *420*, 421
Fossilized cells, *259*
Fowl cholera, 285
Frameshift mutagens, 230
Frameshift mutations, 228, *228*
Francisella spp., 281
Francisella tularensis, 281, 580
Frankia spp., 297
Freeze-drying, bacterial cultures and, 168
Fresh water
flora of, 713
zones of, 713, *714*
Fructose, sucrose formation and, 42
Fruiting bacteria, 293
Fruticose lichens, *315*, 316
Functional groups, 40–42, *41*
Fungal diseases, 308–309
of genital system, 689–692, *691*
of lower respiratory system, 627–630, *633*
of skin, 537–539
Fungi, 13–15, 255, 302–310
acidic environments and, 38
cell walls of, 102
characteristics of, *302*
dimorphic, 304
diseases caused by, 308–309
economic effects of, 309–310
features of, *303*
imperfect, *309*
isolation of, 164
medically important, 307–308
metachromatic granules in, 96
mycotoxins of, 660
parasitic, characteristics of, *310*
pathogenicity of, 408–410
pneumonia and, 621
reproductive structures of, 305–307
speticemia and, 577
in soil, 702, *702*
vegetative structures of, 302–303, *303*
Fungi Imperfecti, 307–308
Furious rabies, 561
Furuncles, 528
Fusobacterium spp., 286
animal bites and, 585
chains of, *286*
in mouth, 642

Galactose, lactose formation and, 43
beta-Galactosidase, genes coding for, 224
Galactoside permease
active transport and, 95
lactose transport and, 224
Gametes, 239
protozoan, 319
Gametocytes
protozoan, 319
sporozoan, *321*, 324
Gamma globulin, 439
Gamma rays, 191
food preservation and, 737
mutagenesis and, 230
Gangrene, 584–585, *584*, *600*; *see also* Gas gangrene
Gardnerella vaginalis, 286, *691*
metronidazole and, 509
vaginitis and, 686–687
Gaseous chemosterilizers, antimicrobial activity of, 201–202, *203*
Gases, in soil, 701
Gas gangrene, 291, 404, 584–585, 703
exotoxins and, *406*
phospholipase C and, 51
Gas vacuoles, 96, 296
Gasohol, *746*, 747
Gastric juice, host defense and, 420
Gastroenteritis, 285, 739
B. cereus, 654
C. perfringens, 654
E. coli, 653–654
gram-negative, 654
Salmonella, 649–650
V. parahaemolyticus, 653
viral, 660
Gastrointestinal tract, 641
microbial invasion through, 400, *401*
Gel diffusion test, 613
Gelidium spp., 313
Gene amplification, 747–748
oncogenic, 369
Generalized infections, 392
Generalized transduction, 239, 356
Generation time, 169–171
Genes
constitutive, 226
evolution and, 244
expression of, regulation of, 223–226
R, 241
regulator, 225
structural, 224
Genetic code, 222–223, *223*
degeneracy of, 222
Genetic engineering, 18–19, 241–244, 747–749
promise and peril of, 244–245
Genetic material, structure and function of, 211–223
Genetic recombination, 234, *234*
eucaryotic, 239–240
microorganism classification and, 269
Genetic transfer, 234
Genetic transformation, *235*
bacterial, 234–236
mechanism of, *236*
Genetics, 210
Genital herpes, 537, 688–689, *691*
acyclovir and, 509
interferon and, 429
transmission of, 389
Genital system
bacterial diseases of, 679–688, *691*
female, 675–676, *676*
fungal disease of, 689–692, *691*
male, 676, *677*
normal flora of, 676–677
protozoan disease of, *691*, 692
structure and function of, 675–676
viral disease of, 688–689, *691*
Genitourinary tract, microbial invasion through, 400, *401*
Genotype, 211–212
Gentamicin, 505, *511*
brucellosis and, 584
granuloma inguinale and, 687
Klebsiella pneumonia and, 622
protein synthesis and, 500
pseudomonads and, 531
pyelonephritis and, 678
structure of, *505*
Gentian violet, 69
Genus, 12, 255, 257
Geosmin, 298
German measles, 535–536
rash of, *536*
Germicide, *184*
Germination, endospore, 98
Germ theory of disease, 8–10
Gohn complexes, 618, *619*
Giardia lamblia, 319–320
giardiasis and, 660–661
trophozoites of, 319, *320*, *661*
water treatment and, 719
Giardiasis, 320, 660–661
metronidazole and, 509
quinacrine and, 509
water supply and, 712
Gliding bacteria, 293
Gliding myxobacteria, 293, *294*
Globulins, 439; *see also specific type*
Gloeocapsa spp., microscopic images of, *63*
Glomerulonephritis, 476, 679, *691*
streptococcal pharyngitis and, 612
Glossina spp., African trypanosomiasis and, 566
Glucan, 644
in cell wall of yeasts, 102
Gluconobacter spp., 281
Glucose
formula for, 42
lactose formation and, 43
sucrose formation and, 42
Glucose-6-phosphate, 117
Glutamic acid, commercial production of, 749
Glutamine
purine biosynthesis and, 143
pyrimidine biosynthesis and, 143
Glutaraldehyde, antimicrobial activity of, 201
Glycerol, 44
structural formula for, *44*
Glycine
purine biosynthesis and, 143
pyrimidine biosynthesis and, 143

Glycocalyx, 82–84
Glycogen, 43, 107
Glycolipids, bacterial cell walls and, 45
Glycolysis, 130–131
  principal reactions of, *130*
Glycoproteins, 51, 105
Goiter, 382
Golgi complex, *100*, 105, *105*
Gonococci, penicillin-resistant, 515–516
Gonorrhea, 86, 163, 382, 392, 394, 533, 541, 680–682, *691*
  anal, 681
  discharge of, *681*
  human carriers and, 385
  immunity and, 438
  incidence and distribution of, *680*
  leukocyte count in, 421
  pharyngeal, 681
  spectinomycin and, 505, 516
  tetracyclines and, 506
  transmission of, 389
Gonorrheal ophthalmia neonatorum, 199, 541, *542*
Grafts, 480, *481*
Graft versus host (GVH) disease, 481
Gram-negative anerobes, *286*
Gram-negative bacteria, 72–73
  cell wall of, structure of, *88*
  characteristics of, *73*
  cystitis and, 678
  enteric, 263
  flagella of, 84, *85*
  gastroenteritis and, 654
  helical/vibrioid, 278–279
  lysozyme and, 525
  nosocomial infections and, 388
  opportunistic, 183
  penicillin and, 90, 497
  peptidoglycan of, 88
  pili of, 86
  plasma membranes of, 92
  pneumonia and, 621
  polymyxin B and, 507
  septicemia and, 577
  of throat, 611
  urinary tract infections and, 678
Gram-negative cocci, *281*
  aerobic, 280–281
  anaerobic, 287
Gram-negative rods
  aerobic, 280–281
  facultatively anaerobic, 281–296
Gram-positive bacteria, 72–73
  bacitracin and, 507
  cell wall of, structure of, *88*
  characteristics of, *73*
  detergents and, 199
  exotoxins of, 405
  flagella of, 84
  hexachlorophene and, 196
  lysozyme and, 525
  of nasal cavity, 611
  penicillin and, 90
  peptidoglycan of, 87–88
  septicemia and, 577
  on skin, 527
Gram-positive cocci, 289–291
  endospore-forming, 291
Gram-positive rods
  endospore-forming, 291, *292*
  irregular, non-sporing, 291–292
Gram stain, 70–73, 87
  procedure for, *72*
Granulocytes, 421, *421*
  phagocytosis by, *425*
Granuloma inguinale, 687–688
Graves' disease, 475–476, 479
  human leukocyte antigen and, *480*
Green algae, *312*, 313
Green nonsulfur bacteria, 296
Green plants
  carbon cycle and, 17
  photosynthesis in, 125–126
Green sulfur bacteria, 296
  photosynthesis in, 125–126
Griseofulvin, 508, *512*
  tinea infections and, 537
Growth hormone; *see* Human growth hormone
Guanine, in DNA, 51
Guillain-Barre syndrome, swine flu vaccine and, 463, 627

Hair, fungal infections of, 537
Hair follicles, 525, *526*
  microbial invasion through, 400
Halazone, 198
Halobacterium halobium, 127–128
Halobacterium spp., Microview 1–3, 295
  gas vacuoles in, 96
  photosynthetic system in, 127–128
Halococcus spp., 295
Halogens, antimicrobial activity of, 196–198, *203*
Halophiles, 296
  extreme, 154
  facultative, 154
Hanging-drop preparation, 68, *69*
Hansen's disease, 557–558
Haploid cells, 239
Haploid number, 239
Haptens, 441, *441*, 474
  allergic contact dermatitis and, 478
Hashimoto's thyroiditis, 479
Hayfever, 472, 474
HDCV; *see* Human diploid cell vaccine
Heat
  food preservation and, 733
  microbial control and, 186–189
Heavy chains, 442
Heavy metals, antimicrobial activity of, 199, *203*
HeLa cells, 351, *365*
Helical viruses, 342, *343*
Helium, electron shells of, 28
Helminthic diseases
  of cardiovascular system, 593–599
  of lower respiratory system, 630–631, *633*
  of lymphatic system, 593–599
Helminths, 15, 325–331
  characteristics of, *302*
  life cycle of, 325
  parasitic, *330*
  pathogenicity of, 408–410
  portal of entry of, *401*
  reproduction of, 325
Helper T cells, 450–451
Hemagglutination, 342, 455, *455*
  dengue and, 592
Hemagglutination inhibition, 455
  rubella and, 536
Hemagglutinin, 625
Hemagglutinin (H) spikes, influenza virus and, 625
Hemodialysis, hepatitis B and, 659
Hemoflagellates, 320
Hemoglobin, cytochromes and, 117
Hemolysins, 404, 529
Hemolytic disease of newborn, 487–488, *488*
Hemophilia, 383
Hemophilus aegyptius, contagious conjunctivitis and, 541
Hemophilus ducreyi, chancroid and, 687
Hemophilus influenzae, 285
  epiglottitis and, 611
  influenza and, 627
  meningitis and, 551, *551*, 552–553, *568*
  otitis media and, 615
  pneumonia and, 621
  sinusitus and, 611
  of throat, 611
  type b, 552–553, *552*
  upper respiratory system and, 611
  virulence of, 403
Hemophilus spp., 285–286
  genetic transformation in, 236
Hemophilus vaginalis, 286
Hemorrhagic fevers, 592
Hepatitis, 392, 657–660
  characteristics of, *659*
  Delta, 660
  genetic engineering and, 244
  human carriers and, 385
  immune sera and, 439–440
  immunity and, 438
  infectious, 400, 657
  non-A, non-B, 657–660
  serum, 394, 657–660
    transmission of, 412
Hepatitis A, 657
  characteristics of, *659*
  transmission of, 389
  WIN 51, 711 and, 616
Hepatitis A virus, 657, *657*
  portal of entry of, *401*
Hepatitis B, 657–660
  characteristics of, *659*
  recombinant DNA and, 19
  transmission of, 412
  vaccine for, 11, 462, 482
Hepatitis B surface antigen, 658–660
Hepatitis B virus, 394, 657
  portal of entry of, *401*
Hepatitis viruses, boiling and, 186
Heptoses, 42
Hermaphroditic, 325
Herpes; *see also* Genital herpes
  acquired immune deficiency syndrome and, 482
  genetic engineering and, 244
  recombinant DNA and, 19

  vaccines and, 11
Herpes simplex virus, 343, *344*, 368, 369, 392, 536–537, *541*
  entry into animal cell, *358*
  latency of, 369, 536
  type 1, 537
  type 2, 537
    genital herpes and, 688
Herpesviruses
  adenine arabinoside and, 508
  capsids of, *348*
  inclusion bodies and, 365
  infectious mononucleosis and, 590
  latency of, 534, 690
  multiplication of, 358
  oncogenic, 368
  portal of entry of, *401*
  properties of, *346*
Herpes zoster, 533–534
  vidarabine and, 537
Herpes zoster virus, 368, 534
Herpetic keratitis, 542, *542*
  adenine arabinoside and, 508
  chemotherapy for, 537
Heterocysts, 296
Heterofermentation, 138
Heterophil antibodies, infectious mononucleosis and, 591
Heterotrophs, 125, 302
  chemically defined media for, *160*, *161*
Hexachlorophene, antimicrobial activity of, 195–196
Hexose monophosphate shunt, 130–131
Hexoses, 42
Hfr cells, 237
High-energy electron beams, microbial control and, 191
High-temperature, short-time pasteurization, 189
Histamine
  anaphylaxis and, *473*
  blood vessel permeability and, 426
  pharmacologic effects of, 473
Histiocytes, 423
Histiocytic lymphoma, 432
Histocompatibility antigens, 479–480
Histones, 104
Histoplasma capsulatum, *628*
  histoplasmosis and, 628
  portal of entry of, *401*
Histoplasmosis, 308, 627–628, *633*
  airborne transmission of, 389
  amphotericin B and 508
  geographic distribution of, *629*
Hives, 472, 474
HLA; *see* Human leukocyte antigen
Hodgkin's disease, 368, 484
  human leukocyte antigen and, *480*
  resistance and, 432
Holdfasts, algal, 313
Holoenzymes, 115
  components of, *116*
Homofermentation, 138
Homologous chromosomes, 240
Hong Kong influenza, 626
Hook, flagellar, 84
Hookworms, 330
Host, 379
  compromised, 432
  entry of microorganisms, 400–402
  helminthic, 325
  normal flora and, 381–382
Host cells
  animal viral infection and, 365
  pathogenic bacteria and, 404–407
Host defense
  antimicrobial substances and, 428–432
  fever and, 427–428
  inflammation and, 424–427
  mucous membranes and, 417–420
  nonspecific, 416–433
  pathogenic resistance to, 402–404
  phagocytosis and, 420–424
  skin and, 417–420
  specific, 437–464
Hot air sterilization, 188
Houseflies, disease transmission and, 390
HTLV-III, acquired immune deficiency syndrome and, 4, 481, 483–484
Human chromosome 6, 479, *479*
Human diploid cell vaccine, 351, 562, 563
Human growth hormone
  bacterial synthesis of, 243
  commercial production of, 750
  recombinant DNA and, 19
Human leukocyte antigen, 479–480
  diseases related to, *480*
Humidifiers, nosocomial infections and, 385
Humoral antibodies, hypersensitivity reactions and, 472
Humoral immunity, 438
  B cells and, 445
Humus, 701
Hyaluronic acid, 404
Hyaluronidase, 290, 404, 529, 585
Hybridomas, 447
  production of, 448–449
Hydatid cyst, 328–329, *386*, 631, *633*
Hydatidosis, 328–329, *386*, 631, *633*
Hydrochloric acid
  dissociation of, *37*
  formation of, 31
Hydrogen
  in carbohydrates, 42
  in complex lipids, 45
  covalent bonding of, 31
  in lipids, 43
  in proteins, 45
  valence of, 28
Hydrogen bonds, 31–32
  formation in water, *32*
Hydrogen ions, 31
  acidity and, 37–38
Hydrogen peroxide, antimicrobial activity of, 202
Hydrogenomonas spp., 126
Hydrolysis, 43, 44
Hydrophobia, 561
Hydrothermal vents, ecosystem of, 155–156
Hydroxy groups, 40, 41
  in amino acids, 46
  in glycerol, 44
poly-beta-Hydroxybutyric acid, 96
Hydroxyl free radicals, 159
Hydroxyl ions, 41
  alkalinity and, 37–38
Hypersensitivity, 383, 472–478
  antimicrobial agents and, 496
  human diploid cell rabies vaccine and, 563
  types of, *472*
Hypertension, 383
Hypertonic solutions, 94
Hyphae, Microview 1–4, 302, *303*
  coenocytic, 302, *303*
  septate, 302, *303*
Hyphomicrobium spp., 295
  fresh water and, 713
Hypochlorous acid, killing power of, 196–198
Hypothalamus, body temperature and, 427
Hypothermic factors, 404
Hypotonic solutions, 94

Iatrogenic disease, 383
Icosahedron, 342–343
ID_{50}, 402
Idiopathic disease, 383
Idiopathic thrombocytopenic purpura, 475, *475*
Idoxuridine, 508, *512*
  herpetic keratitis and, 537, 542
  nucleic acid synthesis and, 500
IgA, *442*, 443–444
  antibodies, in saliva, 642
  secretory, 444
IgD, *442*, 444
IgE, *442*, 444
  antibodies
    anaphylaxis and, 472
    cytotrophic, 472
IgG, 442, *442*
IgM, *442*, 443
Illuminator, in compound light microscope, 61
Imidazoles, 508, *512*
  coccidioidomycosis and, 628
  histoplasmosis and, 628
Immune adherence, 432
  hepatitis A virus and, 657
Immune complexes, 476–477
  formation of, *476*
Immune complex reactions, 476–477
Immune deficiency, natural, 481
Immune globulin, hepatitis B and, 659
Immune response, 471–488
  autoimmunity and, 478–481
  cancer and, 481–485
  hypersensitivity and, 472–478
  mechanisms of, 444–451
  Rh incompatibility and, 485–488
  transfusion reactions and, 485–488
Immune serum globulin, 439
Immune system
  cell-mediated, 438
  humoral, 438

Immunity, 10, 417, 575
  acquired, 438–440, *440*
  cell-mediated
    hypersensitivity reactions and, 472
    T cells and, 447–451
  humoral, B cells and, 445
  leptospirosis and, 679
  meningococcal meningitis and, 551–552
  passive, 439
  poliomyelitis and, 560
  proteins and, 46
  psittacosis and, 624
  rubella and, 536
  scarlet fever and, 612
  whooping cough and, 617
Immunization
  DPT, 554–555
  rabies, 562–564
  schedule for children, *464*
  typhoid fever and, 651
Immunodiffusion tests, 452–453, *453*
Immunoelectrophoresis, 453
  countercurrent, *453*, 552
Immunofluorescence, 67, 458–459
  principle of, *65*
Immunogens, 440–441
Immunoglobulins, 441
  classes of, 442–444, *443*
Immunologic disease, 383
Immunologic enhancement, 484
Immunologic escape, 484–485
Immunologic surveillance, 481–485
Immunologic tolerance, 478–479
Immunology laboratory, Microview 3–4
Immunosuppression, 450, 481
  candidiasis and, 539
  cryptosporidiosis and, 662
  Pneumocystis pneumonia and, 630–631
Immunosuppressive agents, 309
Immunotherapy, 485
Immunotoxins, 12, 449
  commercial production of, 750
Impetigo, 529–530, *530*, *541*
  of newborn, 528
IMViC test, 282
Inapparent infection, 392
Incineration, sterilization and, 188
Inclusion bodies, 408
  C. trachomatis, Microview 1–1
  viral, 365, *365*
Inclusion conjunctivitis, 542, *542*
Inclusions, procaryotic, 96
Incubation period, 411
Indirect selection, 231
Inducible enzymes, 224
Induction, enzyme, 224
Industrial microbiology, 749–751
Infant diarrhea, 284, 653
Infection, 379; *see also specific type*
  focal, 392
  gastrointestinal, 646
  generalized, 392
  inapparent, 392
  local, 392
  nosocomial, 385–388
  primary, 392
  reservoirs of, 385
  secondary, 392
  spread of, 385–390
  subclinical, 392
  systemic, 392
Infection threads, 709
Infectious diseases, 382
  water supply contamination and, 714
Infectious dose, 402
Infectious hepatitis, 400, 657
Infectious mononucleosis, 368, 590–591, *600*, 656
Inflammation, 424–427
  complement activation and, 432
  process of, *427*
Influenza, 285, *386*, 392, 400, 411, 534, 609, 625–627, *632*
  Asiatic, 626
  droplet infection and, 389
  genetic engineering and, 244
  Hong Kong, 626
  leukocyte count in, 421
  pneumonia and, 625
  recombinant DNA and, 19
  transmission of, 389
  vaccine for, 11, 462, *464*
Influenza virus, 343, *344*, 625–626
  structure of, *626*
Ingestion, phagocytosis and, 424
INH; *see* Isoniazid
Inherited diseases, 383
Inhibitors
  competitive, 120
  enzyme, 119–120, *120*
  noncompetitive, 120
Innate resistance, 438
Inoculating loop, 168
Inoculating needle, 168
Inoculation, 168
Inorganic compounds, 35–38
Insect control, microorganisms and,
Insecta, 331
Insulin
  bacterial synthesis of, 243
  commercial production of, 750
  recombinant DNA and, 19
Intact skin, host defense and, 417
Interference effect, 463
Interferon
  action of, *429*
  bacterial synthesis of, 243
  cancer and, 429
  commercial production of, 750
  host defense and, 428–429
  recombinant DNA and, 19
  viral chemotherapy and, 509
  viral infection and, 365
Interleukin-2, acquired immune deficiency syndrome and, 483
Intermediate hosts
  helminthic, 325
  humans as, 328–329
Interstitial fluid, 575
Interstitial spaces, 575
Intoxications
  food-borne infections and, 739–740, *739*
  intestinal pathogens and, 646
Introns, 242
Ioclide, 196
Iodide ions, 29
Iodine
  activity of, mechanism for, *196*
  antimicrobial activity of, 196
  deficiency, 382
Iodophors
  antimicrobial activity of, 196
  contaminated, peritonitis and, 197
Ionic bonds, 28–31
  formation of, *29*
Ionization, 36–37
Ionizing radiation
  microbial control and, 191–192
  mutagenesis and, 230
Ions, 28–31; *see also specific type*
Iron, microbial growth and, 157
Iron ions, enzyme-substrate complexes and, 117
Isocitric acid, decarboxylation of, 132
Isodine, 196
Isografts, 480, *481*
Isoleucine, synthesis of, inhibition of, 121
Isomers, 43
  in amino acids, 46, *46*
Isoniazid, 501, *511*
  tuberculosis and, 619
Isopropanol
  disinfection and, 198
  hydroxy group in, 40
Isopropyl alcohol; *see* Isopropanol
Isotonic solutions, 94
Isotopes, 25–26
Isotretinoin, acne and, 420, 531
Ixodes dammini, 587

Japanese B encephalitis, 565
J chain, 443

Kanamycin
  cystitis and, 678
  pyelonephritis and, 678
Kaposi's sarcoma, 482
Kauffmann-White scheme, 282–284
Kefir, lactic acid bacteria and, 741
Kelp, 311, *312*
Keratin, 308–309, 417, 525
Keratinase, 308
Keratoconjunctivitis, epidemic, *542*
Ketoacidosis, mucormycosis and, 309
Ketoconazole, 508, *512*
  coccidioidomycosis and, 628
  histoplasmosis and, 628
  plasma membrane and, 500
alpha-Ketoglutaric acid, decarboxylation of, 132
Kidneys, 675, *675*
Kinases, bacterial, 404
Kinetic energy, 35
Kingdom, 13, 257
Kinins, blood vessel permeability and, 426

Kirby-Bauer test, 512–513, *513*
Kissing bug, 321, *333*
Klebsiella pneumoniae, 257, 284
  capsule stain of, Microview 1–1
  colonies of, *169*
  encapsulated, *403*
  meningitis and, 551
  phagocytosis and, 424
  pneumonia and, 621, 622, *632*
  quellung reaction and, 83–84
  virulence of, 403
Klebsiella spp., 284
  nalidixic acid and, 502
  nitrogen fixation and, 707
  nosocomial infections and, *388*
  urinary tract infections and, 678
Koch's postulates, 10, 383–385, 582
Koplik spots, 535
Krebs cycle, 116, 132, *133*
  amino acid synthesis and, 142
Krill, 714
Kumiss, lactic acid bacteria and, 741
Kuru, 568, *568*

Lacrimal apparatus, host defense and, 418–419, *418*
D-Lactate dehydrogenase, 118
Lactic acid, dehydrogenation of, 122
Lactic acid bacteria, 138
  dairy products and, 741
Lactic acid fermentation, 137–138, *138*
Lactobacillus sanfrancisco, 7
Lactobacillus spp., 291
  chemically defined media for, *160*
  homofermentative, 138
  in large intestine, 642
  newborns and, 379
  vaginal, 676–677
Lactose
  formation of, 43
  metabolism in *E. coli,* 224
Lactose operon, *225*
Lag phase, of bacterial growth, 171, *171*
Laminaria spp., Microview 1–3
Lampit; *see* Nifurtimox
Large intestine, flora of, 642
Larvae
  helminthic, 325
  miracidial, 326
  schistosomal, 599
Laryngitis, 611
Larynx, 611
Lassa fever, 592
Latency
  herpes simplex virus and, 536
  herpesviruses and, 534, 690
  syphilis and, 684
Latent diseases, 392
Latent viral infections, 369
Latex agglutination test, 455–457
Lauric acid, structural formula for, *44*
LD_{50}, 402
Legionella pneumophila
  fluroescent-antibody stain of, Microview 1–1
  legionellosis and, 3, *3*, 623
  name derivation of, 12
  pneumonia and, 621
Legionella spp., 280–281
Legionellosis, 3, 623–624
  erythromycin and, 3
  susceptibility to, 19–20
Legionnaires' disease, 280, 384, 623–624, *632; see also* Legionellosis
  erythromycin and, 506
  initial outbreak of, 2–3
Lepromin test, 557
Leprosy, 162, 292, 392, 557–558, *568*
  lepromatous, 557, *558*
  symptoms of, *558*
  tuberculoid, 557, *558*
Leptospira interrogans, 679
  leptospirosis and, 678–679
  meningitis and, *551*
  portal of entry of, *401*
Leptospira spp., 276–278
  axial filaments of, *86*
Leptospirosis, 276, *386*, 678–679, *690*
Lesch-Nyhan syndrome, genetic engineering and, 244
Lethal dose, 402
Leuconostoc mesenteroides, 131
Leuconostoc spp., homofermentative, 138
Leukemia, *Pneumocystis* pneumonia and, 631
Leukemias, 365–366, *486*
  mucormycosis and, 309
  resistance and, 432
Leukocidins, 290, 403–404, 528, 529
Leukocytes, 421, *421*, 575
Leukocytic pyrogen, 428
Leukocytosis, 421
Leukocytosis-promoting factor, 427
Leukopenia, 421
Leukotrienes, 473, *473*
LGV; *see* Lymphogranuloma venereum
Lice, *333*, 537
  epidemic typhus and, 587
Lichens, 315–316, *315*
  crustose, Microview 1–3, 315, *315*
  foliose, 315, *315*
  fruticose, *315*, 316
Ligands, pathogenic adherence and, 402
Light chains, 442
Light microscopy, specimen preparation for, 68–75
Limnetic zone, 713, *714*
Lincomycin, glomerulonephritis and, 679
Lipases, 138–139
Lipid A, 88
  body temperature and, 427–428
Lipid inclusions, 96
Lipids, 43–45
  biosynthesis of, 140–142, *141*
  catabolism of, 138–139, *140*
  in plasma membrane, 91
Lipopolysaccharides, in outer membrane, 88, *88*
Lipoproteins, 51
  in outer membrane, 88, *88*
Liposomes, 12
Liquid solutions, sterilization times for, *188*
Lister Institute, 89
Listeria monocytogenes, 291
  listeriosis and, 584
  meningitis and, *551*
Listeriosis, 584, *600*
Littoral zone, 713, *714*
Liver cancer
  aflatoxin and, 410, 660
  hepatitis B and, 659
  monoclonal antibodies and, 19
Living animals, virus cultivation in, 350
Lobar pneumonia, 402
Local infection, 392
Localized anaphylaxis, 474
Lockjaw; *see* Tetanus
Log phase, of bacterial growth, 171–172, *171*
Logarithmic decline phase, of bacterial growth, *171*, 172
Logarithms, 39
Long-acting thyroid stimulators, Graves' disease and, 475
Lophotrichous flagella, 84, *85*
Lou Gehrig's disease, slow viruses and, 568
Louse-borne typhus, 587–588
Low temperatures, microbial control and, 189
Lower respiratory system, 611
  bacterial diseases of, 615–625, *632*
  fungal diseases of, 627–630, *633*
  helminthic diseases of, 630–631, *633*
  protozoan diseases of, 630–631, *633*
  structures of, *610*
  viral diseases of, 625–627, *632*
LSD, 410
Lung fluke, 352–327
Lycoperdon periatum, 308
Lyme disease, 587, *600*
Lymph, 575
Lymphangitis, 577
Lymphatic system, *576*
  bacterial diseases of, 577–590
  components of, *422*
  helminthic diseases of, 593–599
  protozoan diseases of, 593–599
  rickettsial diseases of, 587–590
  structure and function of, 575–577
  viral diseases of, 590–592
Lymph capillaries, 575
Lymph nodes, 575–577
Lymphocytes, 421, *421*, 575
  forbidden clones of, 478
  micrographs of, *68*
Lymphocytic choriomeningitis virus, autoimmunity and, 479
Lymphogranuloma venereum, 289, 687, *691*
Lymphoid organs, 577
Lymphokines, 450–451
  cell-mediated hypersensitivity and, 477
  functions of, *451*
Lymphomas, *486*
  resistance and, 432

Lymphotoxin, 451
Lyophilization, 190
bacterial cultures and, 168
Lysine
C. glutamicum and, 749
commercial production of, 749
Lysis
osmotic, 90
viral multiplication and, 354
Lysogenic cells, 355
Lysogenic phages, 355
Lysogeny, 528
viral multiplication and, 355–357, *356*
Lysol, 195
Lysosomes, *100*, 106–107, *107*
leukocidins and, 404
Lysozyme, 525
host defense and, 420
phage, 354
salivary, 642
viral multiplication and, 354

Macroconidia, Microview 1–4
Macrocyclic lactone ring, 506
Macrolides, 506, *511*
cryptosporidiosis and, 662
Macrophage activation factor, 451
Macrophage chemotactic factor, 451
Macrophage migration inhibition factor, 451
Macrophages, 423, *423*, 575
alveolar, 611
fixed, 423
leukocidins and, 403
stem cells and, 444
wandering, 423
Maduromycosis, 538, *541*
Magnesium, valence of, 28
Magnesium ions, enzyme-substrate complexes and, 117
Magnetosomes, 97–98
Magnetotactic bacteria, 97–98
Magnetotaxis, 97
Major histocompatibility complex, 479–480
Malachite green, 75
Malaria, 324, *386*, 595–598, *600*
arthropod vectors and, *390*
chloroquine and, 509
laboratory diagnosis of, *597*
predisposing factors in, 410
quinine and, 11, 494, 509, 597
recombinant DNA and, 19
reported cases of, *596*
transmission of, 412
vaccine for, 463
Malignant tumors, 365, 383
Malnutrition, 382–383
Malolactic fermentation, 742
Malt, 741
Malting, 741
Manganese ions, enzyme-substrate complexes and, 117
Mannan, in cell wall of yeasts, 102
Mantoux test, 619
Marburg virus, 592
Margination, 426
Mast cells, anaphylaxis and, 472–473
Mastigophora, 318–321
Matrix, of mitochondrion, 106, *106*
Maturation, viral multiplication and, 354
Maximum growth temperature, 152
MBC; *see* Minimum bactericidal concentration
Measles, 382, 392, 394, 400, 411, 534–535, *541*, 609, 655
immune sera and, 439–440
immunity and, 438
leukocyte count in, 421
rash of, *535*
reported cases of, *534*
resistance to, 438
transmission of, 389
vaccine for, 10, 462–463, *464*
Measles virus, 534–535
inclusion bodies and, 365
subacute sclerosing panencephalitis and, 369
Measurement, units of, 60, *60*
Mechanical energy, 35
Mechanical factors, host defense and, 417–419
Mechanical transmission, 390
Mediators, anaphylactic, 473, *473*
Medicine, biotechnology and, Microview 2–4
Megacins, 642
Meiosis, 234, 240
Membrane attack complex, 432
Membrane filter method, water purity and, 717–718, *718*
Membrane filters, 189, *190*
Memory B cells, 445
Memory response, 447
Meninges, 549, *550*
Meningitis, 285, 381, 382, 384, 394, 403, 549, 550–555
aseptic, 551
bacterial
incidence of, *553*
types of, *551*
C. neoformans, 553–554
chloramphenicol and, 506
epidemic, endotoxins and, 407
gonorrheal, 681
H. influenzae, 551, *551*, 552–553, *586*
leukocyte count in, 421
listeriosis and, 584
meningococcal, 163, 281, 411, 550, 551–552, *551*, *568*
Neisseria, 550, 551–552, *551*
pneumococcal, 550, *551*, 553
predisposing factors in, 410
S. pneumoniae, 550, *551*, 553
vaccine for, 463
WIN 51, 711 and, 616
Meningococcal meningitis, 281, 411, 550, 551–552, *551*, *568*
vaccine for, 463
Menopause, 677
Menstruation, genital herpes and, 688
Mercuric chloride, antimicrobial activity of, 199
Mercurochrome, 199
Mercury
enzyme inhibition and, 119
water supply contamination and, 712
Merozoites
plasmodial, 596–597
sporozoan, *321*, 324
Merthiolate, 199
Mesophilic bacteria, 152–153
flat sour spoilage and, 735
Mesosomes, *83*, 92
Messenger RNA, 52
rifamycins and, 507
viral multiplication and, 359–364
Metabolic activity, bacterial growth and, 176
Metabolic diseases, 383
Metabolism, 113
integration of, 143–145, *144*
microbial, 113–145
microorganism classification and, *262*
Metacercaria, 326
Metachromatic granules, 96
Metal cans, 733–735
sealing of, *734*
Metal ions
enzyme denaturation and, 118
enzyme-substrate complexes and, 117
Metalloproteins, 51
Metastasis, 365
Meter, 60
Methane
anaerobic sludge digestion and, 723–725
bioconversion and, 746–747
covalent bonding in, 31
Methanol, hydroxy group in, 40
Methicillin, *511*
bacterial endocarditis and, 579
penicillinases and, 504
pyelonephritis and, 678
staphylococci and, 528
Methylene blue, 69, 176
metachromatic granules and, 96
Methylophilus methylotrophus, 746
Methylparaben, mold growth and, 200
Metronidazole, 509, *512*
amoebic dysentery and, 662
giardiasis and, 661
trichomoniasis and, 692
vaginitis and, 509, 686
Mezlocillin, 504
MHC; *see* Major histocompatibility complex
MIC; *see* Minimum inhibitory concentration
Miconazole, 508, *512*
C. albicans and, 539
structure of, *509*
tinea infections and, 537
vulvovaginal candidiasis and, 692
yeast cell disintegration and, *500*
Microaerophilicity, 159
Microbes
death of, patterns of, 185, *185*
genetics of, 210–244

growth of, 151–177
metabolism of, 113–145
Microbial antagonism, 380–381
Microbial cultures, Microview 1–2
Microbiology
aquatic, 711–726
environmental, Microview 1–3
history of, 5–12
industrial, 18–19, 71, 749–751
medical, 261
soil, 701–711
Microbiology laboratory, Microview 3–2
Microbiota, normal, 379, *380*
Microcladia spp., *312*
Micrococcus luteus, colonies of, *169*
Micrococcus spp.
of nasal cavity, 611
on skin, 527
Microencephalitis, *Naegleria*, 567
Microfilaments, in eucaryotic cytoplasm, 103
Micrometer, 60
Microorganisms, 4–5
biogeochemical cycles and, 703–711
classification of, 12–13, 254–269
criteria for, 261–269
diversity of, 13–15
energy sources from, 746–747
entry into host, 400–402
fermentation and, 5–6
food from, 742–746
food production and, 740–742
human disease and, 19–20
human welfare and, 15–19
pathogenic, 11, 379
portals of exit of, 411–412
in soil, 701–703, *702*
types of, *14*
Microphages, 421
Microscopes
compound light, 61, *62*
images produced by, *63*
darkfield, 61–64, *64*
electron, *66*, 67
Leeuwenhoek's, *4*
phase-contrast, 64
scanning electron, *66*, 67
transmission electron, *66*, 67
Microscopy, 60–67
compound light, 61
darkfield, 61–64
differential interference contrast, 65
electron, 67
fluorescence, 65–67
light, specimen preparation for, 68–75
phase-contrast, 64–65
scanning electron, 67
transmission electron, 67
Microsporum gypseum, Microview 1–4
Microsporum spp., 537
Microtubules, 103
Microwaves
food preservation and, 737–738
microbial control and, 192
Miliary tuberculosis, 619
Milk
food-borne infections and, 739–740
pasteurization of, 189, 738
Minerals, in soil, 701
Minimum bactericidal concentration, 513–514
Minimum growth temperature, 152
Minimum inhibitory concentration, 513
Mining, bacteria and, 716
Minocycline, 505
Miracidial larvae, 326
Miracidium, 326
Missense mutations, 228, *228*
Mitochondria, *100*, 106, *106*
DNA in, 103
ribosomes of, 105
MMR vaccine, 463, 535, 655
Molds, 13, 302
growth of, inhibition of, 200
osmotic pressure and, 190
Molecular weight, 32
Molecules, 25, 28
polar, 35–36
Moles, 32
Molybdenum, microbial growth and, 157
Monera, 13, 255
divisions and classes in, *259*
Monoclonal antibodies, 12, 447
commercial production of, 750
liver cancer and, 19
production of, 448–449
uses of, 449
Monocytes, 421, *421*
Monocytosis, listeriosis and, 584
Monoglycerides, 44
Monomers, 42, 441–442, *441*
Mononuclear phagocytic system, 423
Mononucleosis, leukocyte count in, 421
Monosaccharides, 42
Monosodium glutamate, 749
Monotrichous flagella, 84, *85*
Moraxella lacunata, 281
Moraxella spp., 281
Morbidity, 394
Mordants, 69
Mortality, 394
Most probable number, 175–176
Motility, 68
flagella and, 84
Mouth
bacterial diseases of, 643–646, *666*
flora of, 642
Moxalactam, 505
MPN; *see* Most probable number
M protein, 403, 529, 577, 580
Mucins, 642
Mucor spp., 309
Mucormycosis, 309
Mucous membranes, 525
host defense and, 417–420
microbial invasion through, 400, *401*
of respiratory system, 610
of trachea, *419*
Mucus, host defense and, 418, 419
Multiple myeloma, resistance and, 432
Multiple sclerosis, 479
human leukocyte antigen and, *480*
Multiple-tube fermentation technique, water purity and, 717, *718*
Mumps, 411, 534, 609, 654–656
immunity and, 438
pathogenesis of, *655*
vaccine for, 10, 462–463, *464*
Mumps virus, 655–656, *655*
Mushrooms, mycotoxins of, 410, 660
Mutagenesis, 228–230
by nitrous acid, *229*
radiation and, 230
Mutagens, 228–230
base pair, 229
chemical, 230
frameshift, 230
Mutations, 211, 226–234
drug-resistant, 514
frameshift, 228, *228*
frequency of, 230
identification of, 230–233
missense, 228, *228*
nonsense, 228, *228*
oncogenic, 369
point, 226–228, *227*
rate, 230
spontaneous, 228
types of, 226–228, *228*
Mutualism, 381
Myasthenia gravis, 475
Mycelia, 13, *14*, 297, 302–303
aerial, 302–303
vegetative, 302
Mycetoma, 292, 538
Mycobacteria, 292
Mycobacterium bovis, 620
Mycobacterium leprae, Microview 1–1, 73, 292
culturing of, 162–163
leprosy and, 557
Mycobacterium smegmatis, *169*
Mycobacterium spp.
acid-fast staining and, 73
lipid inclusions in, 96
urethral, 676
Mycobacterium tuberculosis, 73, 292, 384, 557
auramine O and, 65–67
cell wall of, 45
chemical resistance of, 183–184
colonies of, Microview 1–2
isoniazid and, 501
meningitis and, *551*
mucous membranes and, 418
phagolysosomes and, 424
portal of entry of, *401*
tuberculosis and, 618–619
Mycolic acids, isoniazid and, 501
Mycology, 302
Mycoplasma hominis, 686
Mycoplasma pneumoniae, 289, *289*
bronchitis and, 611–612
pneumonia and, 621, 623
Mycoplasma spp., 289
antibiotics and, 90
L forms and, *289*

plasma membrane of, 88
pneumonia and, 622–623, *632*
Mycorrhizae, *710*, 711
Mycoses, 308–309, 537
cutaneous, 308–309, 537
subcutaneous, 308, 537–538
superficial, 309, 537
systemic, 308
Mycotoxicoses, 739
Mycotoxins, 660
Myocarditis, *600*
of newborn, 590
viral, 590
Myocardium, 579
Myxobacteria, 293, *294*
Myxoviruses
enveloped capsids of, 364
portal of entry of, *401*
prokaryocytes and, 365

NAD; *see* Nicotinamide adenine dinucleotide
NADase, 529
NADP; *see* Nicotinamide adenine dinucleotide phosphate
Naegleria fowleri, 567
NAG; *see* N-Acetylglucosamine
Nalidixic acid, 502
nucleic acid synthesis and, 500
NAM; *see* N-Acetylmuramic acid
Nanometer, 60
Nasal cavity, normal flora of, 611
Nasopharyngeal cancer, 368
Epstein-Bass virus and, 591
Natural selection, penicillin-resistant gonococci and, 515–516
Nausea, gastrointestinal intoxications and, 646
Necator americanus, 330, *330*, 400
Necrotizing factors, 404
Negative selection, 231
Negative staining, 69
Negri bodies, 365, 561, *561*
Neisseria gonorrhoeae, *80*, 281, *281*, 680, *682*
adherence of, 402
diplococci of, Microview 1–1
host cells and, 405
identification of, 269
neonatal gonorrheal ophthalmia and, 541
pili of, 86
portal of entry of, *401*
spectinomycin-resistant, 516
Neisseria meningitidis, 281, 381
endotoxins of, 407
meningitis and, 550, 551–552, *551*
portal of entry of, *401*
of throat, 611
Neisseria spp., 281
bacitracin and, 507
genetic transformation in, 236
rifamycins and, 507
urethral, 676
Nematodes, 329–331
Neomycin, 505, *511*
ribosomal protein synthesis and, 96
Neonatal gonorrheal ophthalmia; *see* Gonorrheal neonatal ophthalmia
Neonatal herpes, 688
Neonatal sepsis, WIN 51, 711 and, 616
Neoplastic disease, 383
Nephritis, 384
Nereocystis spp., *312*
Nerve gas, enzyme inhibition and, 119
Nervous system
central, 549
diseases associated with, *568*
organization of, 549–550, *549*
peripheral, 549
Neuraminidase (N) spikes, influenza virus and, 625
Neurotoxins, 405, 554
Neutralization reactions, 457–458, *458*
Neutrons, 25
Neutrophils, 421, *421*, 575
leukocidins and, 403
NGU; *see* Nongonococcal urethritis
Niacin; *see* Nicotinic acid
Niclosamide, *512*
tapeworm and, 509
Nicotinamide adenine dinucleotide, 115–116, 285
redox reactions and, 122
Nicotinamide adenine dinucleotide phosphate, 115–116
redox reactions and, 122
Nicotinic acid, 115
Nifuroxime, 502
Nifurtimox, Chagas' disease and, 595
Nitrification, 707
Nitrifying bacteria, 295
carboxysomes in, 96
Nitrobacter spp., 295
energy sources for, 126
nitrification and, 707
Nitrofurans, 501–502
Nitrofurantoin, 502
Nitrofurazone, 502
Nitrogen
hydrogen bonding of, 31
in complex lipids, 44
microbial growth and, 156–157
in nucleic acids, 16
in proteins, 45
Nitrogen cycle, 16–17, *16*, 705–711, *706*
Nitrogen fixation, 157, 295, 707–711
Nitrosamines, 739
Nitrosomonas spp., 295
energy sources for, 126
nitrification and, 707
Nitrous acid, mutagenesis by, *229*
Nocardia asteroides, 292, *293*
meningitis and, *551*
nocardiosis and, 630
Nocardia spp., 292
acid-fast staining and, 73
periodontal disease and, 645
Nocardioforms, 292
Nocardiosis, 292, 630
Noncommunicable diseases, 390
Noncyclic photophosphorylation, 123, *123*
Nongonococcal urethritis, 287, 686, *691*
Nonhistones, 104
Nonimmune phagocytosis, 424
Nonionizing radiation, microbial control and, 192
Nonsense codons, 221–222
Nonsense mutations, 228, *228*
Nonspecific resistance, 417, 430, 524
Nonspecific urethritis, 686
Nonsymbiotic bacteria, 707–709
North American blastomycosis, 629
Norwalk agent, 660
Nose, 609
Nosocomial infections, 284, 285, 383, 385–388, *389*
Notifiable diseases, 394
NSU; *see* Nonspecific urethritis
Nuclear area, procaryotic, *92*, 95–96
Nuclear envelope, 103, *104*
Nuclease, thermostable, 648
Nucleic acids, 51–52
antimicrobial agents and, 185
hybridization, *266*, 267–269
microorganism classification and, 266–269
nitrogen in, 16
synthesis of, antibiotics and, 500
viral, 341–342
Nucleoid, procaryotic, *92*, 95–96
Nucleolus, of eucaryotic organelle, *100*, 104, *104*
Nucleoplasm, of eucaryotic organelle, 104, *104*
Nucleoproteins, 51
Nucleosides, 51
Nucleotides, 51, 143, 211
RNA, 218
Nucleus
atomic, 25
of eucaryotic organelle, *100*, 103–104, *104*
Numerical taxonomy, *268*
microorganism classification and, 269
Nutrient agar, 161
Nutrient broth, 161
Nutritional deficiency, 432–433
Nutritional deficiency disease, 382–383
Nystatin, 508, *512*
C. albicans and, 539
plasma membrane and, 500

Objective lens, in compound light microscope, 61
Obligate aerobes, 157
Obligate anaerobes, 157
oxygen sensitivity of, 159
Ocular lens, in compound light microscope, 61
Oil glands, 525
Oligodynamic action, 199, *199*
Olives, fermentation and, 741
Oncogenes, 368
activation of, 368–369
in human cells, *366*
Oncogenic viruses, 368
DNA-containing, 368
RNA-containing, 368

One-step growth experiment, 354–355, *355*
Oocystis spp., *14*
Operon model, 224–226
Operons, 226
  E. coli, 355
  lactose, *225*
Opportunistic pathogens, 183, 309, 497
Opportunists, 381
Opsonins, 424
Opsonization, 424, 432, 442
Optimum growth temperature, 152
Optochin, 621
Oral candidiasis, 689
Oral contraceptives, gonorrhea and, 681
Order, 13, 257
Organ transplantation, 480–481
  nosocomial infections and, 388
  resistance and, 432
Organelles, eucaryotic, 100, 103–107
Organic acids, antimicrobial activity of, 200, *203*
Organic compounds, 38–53
  ATP, 52–53
  carbohydrates, 42–43
  functional groups of, 40–42, *41*
  lipids, 43–45
  nucleic acids, 51–52
  proteins, 45–51
Organic growth factors, 159
Organic matter, in soil, 701
Organisms
  classification of, *258*
  nutritional classification of, *125*
  in soil, 701–703
Ornithodorus spp., 587
Ornithosis, *386*, 624, *632*
Orthomyxoviruses, properties of, *347*
Osmosis, 93–94, *93*, 103
Osmotic lysis, 90
Osmotic pressure, 94
  microbial control and, 190
  microbial growth and, 153–154
Osteoarthritis, 383
Osteomyelitis, 384
Otitis externa, 531, *541*
Otitis media, 285, 614–615, *632*
Ouchterlony test, 452–453, *453*
Ovaries, 675, *676*
Oxacillin, *511*
  bacterial endocarditis and, 579
  penicillinases and, 504
  staphylococci and, 528
Oxamiquine, schistosomiasis and, 511
Oxidases, 118
Oxidase test, 134
Oxidation ditches, 726
Oxidation ponds, 725–726
Oxidation reactions, 122
Oxidation-reduction reactions, 122
Oxidative phosphorylation, 123, 127
Oxidizing agents, antimicrobial activity of, 202, *204*
Oxidoreductases, 118
Oxygen
  carbohydrates and, 42
  in complex lipids, 45
  covalent bonding of, 31
  hydrogen bonding of, 31
  in lipids, 43
  microbial growth and, 157–159
  in proteins, 45
  toxic forms of, 157–159
  valence of, 28
Oxygen cycle, 17
Oxytetracycline, 505
  balantidiasis and, 661
Ozone
  antimicrobial activity of, 202
  water treatment and, 719–720

PABA; *see* para-Aminobenzoic acid
PAGE; *see* Polyacrylamide gel electrophoresis
Pandemic diseases, 392
Panencephalitis, subacute sclerosing, 369, 535, 569
Pantothenic acid, 116
Papovaviruses
  multiplication of, 358
  oncogenic, 368
  properties of, *346*
  warts and, 531
Parabens, mold growth and, 200
Paracoccus spp., denitrification and, 707
Paragonimiasis, *633*
Paragonimus westermanni, 325
  life cycle of, *326*
Parainfluenza viruses, 615
Paralysis
  flaccid, 555
  spastic, 555
Paralytic rabies, 561
Paramecium multimicronucleatum, *14*
  cilia of, *101*
Paramecium spp.
  cilia of, 101, 318
  conjugation in, *319*
  locomotion of, 15
Paramyxoviruses
  enveloped capsids of, 364
  measles and, 534
  mumps and, 655
  portal of entry of, *401*
  properties of, *347*
Parasites, 15, 128
  artificial media and, 163
  cytotoxic T cells and, 447
  obligately intracellular, 340
Parasitism, 381
Paratyphoid fever, 411
  vaccine for, *463*
Parenchyma, 427
Parenteral route, microbial invasion through, 400, *401*
Parkinson's disease, slow viruses and, 568
Parotid glands, mumps and, 654–655
Parvoviruses
  capsid of, *348*
  gastroenteritis and, 660
  multiplication of, 358
  properties of, *346*
Passive immunity, 439
Passive processes, 93
Pasteur's experiment, spontaneous generation and, *9*
Pasteur treatment, 562
Pasteurella multocida, 285, 585
Pasteurellaceae, 285–286
Pasteurization, 6, 188–189, 738
  high-temperature, short-time, 189
  Salmonella and, 649
Pathogenicity, 379
  capsules and, 74
  mechanisms of, 399–412
Pathogens
  eucaryotic Microview 1–4
  opportunistic, 183, 309, 497
  in soil, 703
Pathology, 379
Pedicularis spp., *333*
Pediculus vestimenti, 587
Pellicles, protozoan, 15, 102, 318, *318*
Pelvic inflammatory disease, 681
Penetration
  of animal viruses, 358
  of phages to host cells, 352–354
Penicillin, 11–12, 502–504, *511*
  allergic reactions and, 496
  anaphylactic shock and, 474
  animal feed and, 496
  bacterial cell wall and, 90, *498*
  bacterial endocarditis and, 517, 579
  beta-hemolytic streptococci and, 529
  brucellosis and, 584
  C. perfringens and, 585
  discovery of, 11, *494*, 495
  erysipelas and, 530
  eucaryotic cells and, 103
  glomerulonephritis and, 679
  gonococcal resistance to, 515–516
  gonorrhea and, 681
  gram-negative bacteria and, 497
  impetigo and, 530
  Klebsiella pneumonia and, 622
  leptospirosis and, 679
  listeriosis and, 584
  N. meningitidis and, 552
  natural, 502
  neonatal gonorrheal ophthalmia and, 541
  otitis media and, 615
  outer membrane and, 88
  P. multocida and, 585
  peptidogylcans and, 499
  pneumococcal meningitis and, 553
  pneumococcal pneumonia and, 622
  puerperal sepsis and, 577–579
  pyelonephritis and, 678
  S. aureus and, 290
  scarlet fever and, 612
  semisynthetic, 504
  spectrum of activity of, 497
  staphylococci and, 528
  streptococcal pharyngitis and, 612
  streptomycin and, 517
  structure of, *503*
  syphilis and, 684–685

  tetanus and, 555
  tetracyclines and, 517
  V. parahaemolyticus gastroenteritis and, 653
Penicillinase, 502–504, 514, 528, 579, 681
Penicillin G, 502–504, *511*
  cystitis and, 678
  retention of, *504*
Penicillin V, 504, *511*
Penicillium chrysogenum, *11*
Penicillium marneffei, *307*
Penicillium notatum, 11, 495
Penicillium spp.
  antibiotics and, 495
  cheese and, 741
  spores of, 306–307
Penis, 676, *677*
Pentamidine
  African trypanosomiasis and, 567
  Pneumocystis pneumonia and, 509, 631
  trypanosomal disease and, 509
Pentose phosphate pathway, 130–131
  amino acid synthesis and, 143
  nucleotides and, 143
Pentoses, 42, 51
Peptidases, 139
Peptide bonds, 49
  dehydration synthesis and, *49*
Peptide cross bridges, 87
Peptidoglycan, 13, 86–87
  antimicrobial agents and, 497
  chemical structure of, *87*
  eucaryotes and, 103
Pericarditis, 285, 579, *599*, *600*
Pericardium, 579
Peridinium spp., *312*
Periodontal disease, 645–646, *646*
Peripheral nervous system, 549
Periplasmic space, 88
Peritonitis, 286, 384
  contaminated iodophor solution causing, 197
Peritrichous flagella, 84, *85*
Permeability, of plasma membrane, 91
Peroxide, 158
Perspiration, 525
  host defense and, 420
Pertussis, 613, 615–618
  vaccine for, *463*
Pest control, bacteria and, 704
Pesticides, degradation of, 711
Petri plates, 173, *174*
Petroff-Hausser counter, 172–173, *173*
Petroleum, use as food, 746
pH
  buffers, 38
  enzymatic activity and, 118
  microbial growth and, 153
  scale, 37–38, *38*
Phage lysozyme, 354
Phages; *see* Bacteriophages
Phage typing
  microorganism classification and, 264
  S. aureus, *264*, 648
Phagocytes, 421, 575
  classification of, *423*
  migration of, 426–427
Phagocytic cells, 421–424
Phagocytic vacuoles, 424, *425*
Phagocytosis, *102*, 103, 402, 404, 405, 420–424, 575
  capsules and, 83
  granulocytic, *425*
  immune complexes and, 476
  mechanism of, 424
  N. meningitidis and, 552
  nonimmune, 424
  surface, 424
Phagolysosomes, 424, *425*
Phagosomes, 424, *425*
Phalloidin, 410
Pharyngeal gonorrhea, 681
Pharyngitis, 611, 612, *632*
Phase-contrast microscope, 64
Phase-contrast microscopy, 64–65
Phemerol; *see* Benzethonium chloride
Phenol
  antimicrobial activity of, 183, 194–196, *202*
  infection control and, 9
  O-phenyl-, 195
  structure of, *195*
Phenol coefficient test, 194
Phenolics
  antimicrobial activity of, 194–196, *202*
  chlorines and, 195
  structure of, *195*
Phenotype, 211–212
Phenylketonuria, 383
Phialophora verruscosa, 538
pHisoHex, 196
Phormidium luridum, *297*
Phosphates, water supply contamination and, 713
Phospholipase C, 51
Phospholipids, 45
  bilayers, in plasma membrane, *90*, 91
  in outer membrane, 88, *88*
  in plasma membrane, 91
  structure of, *45*
Phosphoproteins, 51
Phosphorus
  in complex lipids, 44
  microbial growth and, 156–157
Phosphorylase a, quaternary structure of, 51
Phosphorylation, 122–123, *123*
  oxidative, 123, 127
  substrate-level, 123
Photoautotrophs, 125–126, 310, 703
Photoheterotrophs, 125, 126
Photophosphorylation, 123, *123*
  cyclic, 123, *123*
  noncyclic, 123, *123*
Photosynthesis, 125–126
  algal, 15, 311, 313
  bacterial, 13, 296–297
  carbon cycle and, 17, 703
  chloroplasts and, 107
  cytochromes and, 117
  microorganisms and, 5
  oxygenic, 126
Photosynthetic bacteria
  gas vacuoles in, 96
  photosynthesis in, 125–126
  use as food, 746
Phototrophs, 125, 295–297
Phylum, 13
Phytophthora infestans, 310
Phytoplankton, 713–714
Pia mater, 549
Picornaviruses
  multiplication of, 362
  portal of entry of, *401*
  properties of, *346*
  WIN 51, 711 and 616–617
PID; *see* Pelvic inflammatory disease
Piedras, 537
Pili, 282
  E. coli, 402, *402*
  procaryotic, 84–86, *86*
  sex, 86, 282
Pilin, 84
Pimples, 426, 525, 528
Pinkeye, 539–541, *542*
Pinocytosis, 103, 358
Pinworms, 329–330
Piperazine, *512*
Pityrosporum spp., on skin, 527
PKU; *see* Phenylketonuria
Placental transfer, 439
Plague, 400, 585–587, *600*
  arthropod vectors and, *390*
  geographic distribution of, *586*
  vaccine for, *463*
Plankton, 155
Planktonic algae, 311, *312*, 313–314
Plant viruses, 340–341, 370–371
  classification of, *370*
Plantae, 13, 255
Plaque-forming units, 349
Plaque method, 349, *349*
Plaques, 349, *349*
Plasma, 575
Plasma cells, 445
Plasma membrane
  antibiotics and, 500
  antimicrobial agents and, 185
  eucaryotic, *100*, 103
  phenols and, 195
  procaryotic, *90*, 91–92
Plasmid DNA, preparation of, Microview 2–2
Plasmids, 95–96, 236–237, 240–241
  bacteriocinogenic, 241
  conjugative, 240
  dissimilation, 241
  hereditary drug resistance and, 514
Plasmodial slime molds, 317, *317*
Plasmodium falciparum, malaria and, 596
Plasmodium malariae, malaria and, 596
Plasmodium ovale, malaria and, 596
Plasmodium spp., Microview 1–3, 317, 321–324
  portal of entry of, *401*
  reproduction of, 332

Plasmodium vivax
  life cycle of, *321*
  malaria and, 596
  trophozoites of, Microview 1–4
Plasmolysis, 154, *154*
Platelets, 575
Platyhelminthes, 325–329
Pleura, 611
Pneumococcal meningitis, 550, *551*, 553
Pneumococcal pneumonia, 621–622, *632*
  immunity and, 438
  leukocyte count in, 421
  vaccine for, 463
Pneumocystis pneumonia, 482, 501, 630–631, *633*
  pentamidine and, 509
Pneumonia, 235, 255, 284, 285, 381, 384, 400, 411, 612
  acquired immune deficiency syndrome and, 4
  bacterial, 403, 621–625
  fungal, *633*
  immunity and, 438
  Klebsiella, 622, *632*
  L. pneumophila, 3
  leukocyte count in, 421
  lobar, 402
  measles and, 535
  mycoplasmal, 622–623, *632*
    erythromycin and, 506
    tetracyclines and, 506
  pneumococcal, 621–622, *632*
  Pneumocystis, 482, 501, 630–631, *633*
  predisposing factors in, 410
  primary atypical, 289, 622–623
  vaccine for, 463
  viral, 625, *632*
Pneumonic plague, *386*, 586–587, 609
Pneumovax, 622
Point mutations, 226–228, *227*
Poison ivy, 478
Polar molecules, 35–36
Poliomyelitis, 392, 400, 411, 558–560, *568*
  immunity and, 438
  newborn and, 439
  vaccines for, 10, 462, *464*
  WIN 51, 711 and, 616
Polioviruses, 343, 559, *559*
  gastroenteritis and, 660
  multiplication of, 362
  uncoating of, 358
Pollution
  chemical, 714–716
  effects of, 714–716
Polyacrylamide gel electrophoresis, 264–265, *265*
Polyenes, 508, *512*
  mycetomas and, 538
Polyhedral viruses, 342–343, *343*
Polymers, 42
Polymorphonuclear leukocytes, 421, 426–427
Polymyxin
  cystitis and, 678
  pyelonephritis and, 678
Polymyxin B, 506–507, *512*
  plasma membrane and, 500
  structure of, *507*
Polyoma viruses, oncogenic, 368
Polypeptides, 49, 506–507, *511*
Polyribosomes, 221
Polysaccharides, 43
  biosynthesis of, 140
  granules, 96
Portals of entry, 400, *401*
Portals of exit, 411–412
Positive selection, 231
Potassium
  ion of, 29
  valence of, 28
Potassium chloride, infant diarrhea and, 654
Potassium sorbate, mold growth and, 200
Potato blight, 310
Potato spindle tuber viroid, 371, *371*
Potato yellow dwarf virus, 370
Potatoes, viral infections of, 370–371
Potential energy, 35
Pour plate method, *166*, 167
Poxviruses
  multiplication of, 358
  oncogenic, 368
  portal of entry of, *401*
  properties of, *346*
  uncoating of, 358
PPD; *see* Purified protein derivative
Praziquantal, schistosomiasis and, 511
Precipitation reactions, 452–453, *452*
Precipitin ring test, 452, *452*
Precipitins, 452
Predisposing factors, 410
Prepodyne solution, 197
Pressure cooker, 188
Primary atypical pneumonia, 622–623
Primary cell lines, 351
Primary infection, 392
Primary treatment, of sewage, *720*, 721
Prions, 15, 369–370
  autoimmunity and, 569
Privileged tissues, 480
Priviliged sites, 480
Procaine penicillin, 502
Procaryotae, 13, 255
  divisions and classes in, *259*
Procaryotes, Microview 1–3, 13
  alkaline environments and, 38
  chromosomes of, 212, *213*
  classification of, 255
  origins of, 80–81
  oxidative phosphorylation in, 123
  photosynthesis in, 125–126, *126*
Procaryotic cells, 81–100
  axial filaments of, 84
  cell wall of, 86–90
    structure of, *83*
  characteristics of, 81
  cytoplasm of, 95–96
  endospores of, 96–100
  eucaryotic cells compared, *108*
  flagella of, 84
  glycocalyx of, 82–84
  pili of, 84–86
  plasma membrane of, 91–92
    transport across, 92–95
  size, shape, and arrangement of, 81–82
Prodromal period, 411
Profundal zone, 713, *714*
Proglottids
  T. solium, Microview 1–4
  tapeworm, 327, *327*
Prokaryocytes, 365
Promoter sites, 218, 225
Prontosil, 494
Properdin, host defense and, 429–432
Prophages, 355
Propionibacterium acnes, 291
  acne and, 531
  natural environment of, 38
  on skin, 527, *527*
Propionibacterium spp.
  cheesemaking and, 740
  hair follicles and, 525–527
Propionic acid, 525–527
Propylene oxide, gaseous sterilization and, 202
Propylparaben, mold growth and, 200
Prostaglandins
  anaphylaxis and, 473, *473*
  vasodilation and, 426
Proteases, 139
Protein analysis, microorganism classification and, 264–265
Proteinase, 585
Proteins, 45–51
  antimicrobial agents and, 185
  carrier, 94
  catabolism of, 139–140
  commercial production of, 751
  denaturation of, 118, *119*
    alcohols and, 198
  formation of, anabolism and, 32
  in plasma membrane, 91
  structure of, *50*, 51
  synthesis of
    antimicrobial agents and, 499–500, *499*
    RNA and, 218–222
Protein X
  isolating gene for, Microview 2–2
  producing, Microview 2–3
Proteus mirabilis
  cells of, Microview 1–2
  septicemia and, 577
Proteus spp., 284
  endotoxins of, 407
  in large intestine, 642
  nalidixic acid and, 502
  nosocomial infections and, *388*
  pyelonephritis and, 678
  urinary tract infections and, 678
Proteus vulgaris
  cystitis and, 678
  flagella of, *85*
Protista, 13, 15, 254–255, 317–325
  cytoplasmic inclusions of, 107
Protons, 25
Protoplasmic streaming, 317
Protoplasts, 89–90, *89*
  fusion of, 748, *749*

Prototheca spp., 409
Protozoan diseases
  of cardiovascular system, 593–599
  of digestive system, 660–662, *667*
  of genital system, *691*, 692
  of lower respiratory system, 630–631, *633*
  of lymphatic system, 593–599
  water supply contamination and, 712
Protozoans, 15, 317–325
  characteristics of, *302*
  cilia of, 101
  encystment of, 319
  medically important, 319–325
  metachromatic granules in, 96
  nutrition of, 318
  parasitic, *324*
  pathogenicity of, 408–410
  pellicle of, 102
  pneumonia and, 621
  portal of entry of, *401*
  reproduction of, 318–319
Proviruses, 364, 368
Pseudobacteremia, 197
Pseudohyphae, *303*, 304
Pseudomonas aeruginosa, 280, *280*
  colonies of, Microview 1–2
  cystitis and, 678
  flagella of, *85*
  gentamicin and, 505
  meningitis and, 551
  nosocomial infections and, 388, *388*
  pathogenicity of, 531
  peritonitis and, 197
  phenol coefficient test and, 194
  pneumonia and, 621
  pyelonephritis and, 678
  pyocins and, 642
  septicemia and, 577
Pseudomonas cepacia, 387
Pseudomonas fluorescens, 704
Pseudomonas spp., 280
  anaerobic respiration and, 135
  denitrification and, 707
  Entner-Doudoroff pathway and, 131
  fresh water and, 713
  opportunistic, 183
  polymyxin B and, 507
  quaternary ammonium compounds and, 200
  skin infections and, 530–531
  urinary tract infections and, 678
Pseudopods, 15, *102*, 103, 317–318, *318*
Psittacosis, 289, *386*, 624, *632*
Psychogenic diseases, 383
Psychrophiles, 152–153
Psychrotrophs, 153
Puerperal fever, 384, 577
Puerperal sepsis, 577–579, *599*
Purified protein derivative, 619
Purines, biosynthesis of, 143, *143*
Purple nonsulfur bacteria, 296
Purple sulfur bacteria, 296
  photosynthesis in, 125–126
Pus, 427
Pyelonephritis, 678, *690*
Pyocins, 642
Pyocyanin, 531
Pyogenicity, 528
Pyrimethamine
  malaria and, 597
  toxoplasmosis and, 595
Pyrimidines, biosynthesis of, 143, *143*
Pyruvic acid
  acetyl coenzyme A and, *131*
  formation of, 122

Quaternary ammonium compounds
  microbial control and, 200
  structure of, *200*
Quellung reaction, 83–84, *84*, 621, *622*
Q fever, 287, *386*, 624–625, *632*
  transmission of, 389
Quinacrine, giardiasis and, 509, 661
Quinine, *512*
  malaria and, 11, 494, 509, 597
Quinones, 133–134

Rabbit fever, 281
Rabies, 189, *386*, 560–565, *568*
  dumb, 561
  furious, 561
  immune sera and, 439
  reported cases of, *565*
  vaccine for, 351, 462, *464*
Rabies immune globulin, 562
Rabiesvirus, 411, 560–561, *561*
  culturing of, 351
  multiplication of, 362
  Negri bodies and, 365
Radiant energy, 35, *191*
Radiation
  food preservation and, 737–738
  ionizing, 191–192, 230
  microbial control and, 191–192
  mutagenesis and, 230
  nonionizing, 192
Radioimmunoassay, 459, *460*
  hepatitis A virus and, 657
Rapid plasma reagin card test, syphilis and, 685
Rat bite fever, 585, *600*
Rat flea, *333*
  endemic typhus and, 588
  plague and, 585
Reaction rates, 34
Recalcitrance, 711
Receptors
  antigen, 438
  pathogenic adherence and, 402
Recipient cells
  competence of, 236
  genetic transfer and, 234
Recombinant DNA and RNA, Microview 2–1, 18–19, 241–244
  cloning, Microview 2–3
  constructing, Microview 2–2, *242*
  vaccine development and, 10–11
Recombinants, 234
Red algae, 311–313, *312*
Red blood cells, 575
Redi's experiment, 6, *6*
Rediae, 326
Red tides, 314
Reducing media, 161
Reduction reactions, 122
Refractive index, in microscopy, 61
Refrigeration
  microbial growth and, 152–153, 183
  salmonellosis and, 650
Regulator genes, 225
Relapsing fever, 276, 587, *600*
  arthropod vectors and, *390*
Release, viral multiplication and, 354–355
Rennin, 740
Reoviruses
  gastroenteritis and, 660
  multiplication of, 362
  particles in, *348*
  properties of, *347*
Replica plating, 231–233, *232*
Replication; *see* DNA, replication of
Repression, enzyme, 224
Repressors, 225
Reproductive system; *see* Genital system
Reservoirs of infection, 385
Resistance, 19, 417
  antimicrobial agents and, 497
  factors lowering, 432–433
  innate, 438
  nonspecific, 417, 430, 524
  specific, 417
Resistance factors, 241, 514–516
Resistance transfer factor, 241
Resistant mutants, 514
Resolution, in microscopy, 61
Resolving power, in microscopy, 61
Respiration, 131–135
  aerobic, 131–135
  anaerobic, 135, 280
  carbon cycle and, 17
  cytochromes and, 117
  fermentation and, *139*
Respirators, nosocomial infections and, 385
Respiratory syncytial viruses
  common cold and, 615
  pneumonia and, 625
Respiratory system
  diseases associated with, *632*
  lower, 611
    bacterial diseases of, 615–625, *632*
    fungal diseases of, 627–630, *633*
    helminthic diseases of, 630–631, *633*
    protozoan diseases of, 630–631, *633*
    structures of, *610*
    viral diseases of, 625–627, *632*
  normal flora of, 611
  structure and function of, 609–611
  upper, 609–611
    bacterial diseases of, 611–615, *632*
    structures of, *609*
    viral diseases of, 615, *632*
Respiratory tract, microbial invasion through, 400, *401*
Restricted transduction, 239
Restriction enzymes, 241
  DNA viruses and, 352
Reticulate bodies, chlamydial, 287, 624

Retorts, 188
Retroviruses
  acquired immune deficiency syndrome and, 362, 368, 481, 483–484
  B-type, *348*
  multiplication of, 362–364, *363*
  oncogenic, 368
  properties of, *347*
Reverse transcriptase, 362, *362*
Reversible reactions, 33
Reye's syndrome, 533–534
R factors, *240*, 241, 388, 514–516
R genes, 241
Rhabdoviruses
  multiplication of, 362
  portal of entry of, *401*
  properties of, *347*
Rh blood group system, 487–488
Rheumatic fever, 579–580, *599*
  autoimmunity and, 479
  human leukocyte antigen and, *480*
  streptococcal pharyngitis and, 612
Rheumatoid arthritis, 383, 477
  slow viruses and, 568
Rh factor, 487
Rh incompatibility, 485–488
Rhinoviruses
  common cold and, 615
  WIN 51, 711 and, 616
Rhizobium japonicum, *709*
Rhizobium spp., 281, 297
  Entner-Doudoroff pathway and, 131
  genetic transformation in, 236
  nitrogen fixation and, 157
  symbiosis and, 709
Rhizopus nigricans, 257
  spores of, 307, *307*
Rhizopus spp., 309
  spores of, 306–307
Rhizosphere, 707
Rhodospirillum spp., 707
Riboflavin, commercial production of, 749
Ribonucleic acid; *see* RNA
Ribose
  adenosine triphosphate and, 52
  formula for, 42
Ribosomal RNA, 52, 211, 221
Ribosomes, 96
  antibiotics and, 499
  eucaryotic, *104*, 105
  procaryotic, 105
Ribulose 1, 5-diphosphate carboxylase, 96
Rickets, 382
Rickettsial diseases
  of cardiovascular system, 587–590
  of lymphatic system, 587–590
Rickettsial infections, tetracyclines and, 506
Rickettsia prowazekii, 287
  epidemic typhus and, 587
Rickettsia rickettsii, 287
  portal of entry of, *401*
  Rocky Mountain spotted fever and, 589
  stain of, Microview 1–1
Rickettsia spp., 287–289, *288*
  artificial media and, 163
  biochemical properties of, *287*
Rickettsia typhi, 287
  endemic murine typhus and, 588
Rifampin, 507, *512*
  isoniazid and, 501
  legionellosis and, 624
  leprosy and, 558
  N. meningitidis and, 552
  tuberculosis and, 619
Rifamycins, 507, *512*
  nucleic acid synthesis and, 500
Riftia spp., endosymbiosis of, 156
RIG; *see* Rabies immune globulin
Rimantadine, 508
Ring stage, sporozoan, 324
Ringworm, *386*, 537, *539*, *541*
  griseofulvin and, 508
RNA, 52; *see also specific type*
  in endospore core, 98
  nitrogen and, 16
  protein synthesis and, 218–222
  synthesis of
    DNA strands and, 218–221
    nitrogen and, 156
  viral, 15
RNA polymerase, 218
  viral multiplication and, 354, 362
RNA primer, 216
RNA viruses, 341–342
  multiplication of, 359–362, *360*
Rock particles, in soil, 701
Rocky Mountain spotted fever, 287, *386*, 589–590, *600*
  arthropod vectors and, *390*
  pathways of, *590*
  reported cases of, *589*
  transmission of, 412
Root nodules, 709
  formation of, *708*
  soybean, *709*
Roundworms, 15; *see also* Aschelminthes
RTF; *see* Resistance transfer factor
Rubella, 535–536, 609
  immune sera and, 439–440
  newborn and, 439
  rash of, *536*
  vaccine for, 10, 462–463, *464*
Rubella virus, 535
Rubeola, 534–535, *541*
  reported cases of, *534*

Sabin polio vaccine, 462, 560
Saccharomyces cerevisiae, 7, *304*
  use as food, 746
Saccharomyces exiquus, 7
Saccharomyces spp., 304
Safranin, 69
St. Louis encephalitis, 565
Saliva
  constituents of, 642
  host defense and, 419
  pathogens in, 411
Salk vaccine, 462, 560
Salmonella cholerae-suis, portal of entry of, *401*
Salmonella dublin, 282, 650
Salmonella eastbourne, 650
Salmonella enteritidis, portal of entry of, *401*
Salmonella newport, 496
Salmonella paratyphi, portal of entry of, *401*
Salmonella spp., 282–284
  gastroenteritis and, 649–650
  histidine auxotrophs of, Ames test and, 233–234, *233*
  host cells and, 405
  identification of, 263
  isolation of, 164
  traveler's diarrhea and, 653
Salmonella typhi, 282
  endotoxins of, 407
  phenol coefficient test and, 194
  portal of entry of, 400, *401*
  typhoid fever and, 650
  Widal test and, 454
Salmonella typhimurium, 282, 650
  portal of entry of, *401*
Salmonellosis, 282, *386*, 649–650
  milk-borne, 649
  transmission of, 389
Salpingitis, 686
  gonorrhea and, 681
Salterns, Microview 1–3
Salts, 36–37, *37*
  dissociation of, *37*
  water as solvent for, 36, *36*
Salvarsan, syphilis and, 11
Sand filtration, water treatment and, 719
Sanitization, *184*
Saprophytes, 128
Sarcinae, 81
Sarcodina, 317, 319
Sarcodines, plasma membrane of, 102
Sarcomas, 366, *486*
Sarcoptes scabiei, scabies and, 538
Sargassum spp., 313
Scabies, 538
Scalded skin syndrome, 290, 528
  exotoxins and, *406*
Scanning electron micrograph, 67, *68*
Scanning electron microscope, *66*, 67
Scanning electron microscopy, 67
Scar tissue, 427
Scarlet fever, 384, 411, 612, *632*
  erythrogenic toxin and, 529
  exotoxins and, *406*
  predisposing factors in, 410
  transmission of, 389
Schaeffer-Fulton endospore stain, 75
Schick test, 457
Schistosoma spp., 327
  life cycle of, *598*
  schistosomiasis and, 599
Schistosomiasis, 327, 598–599, *600*
  antimonial agents and, 511
  water supply contamination and, 712
Schizogony
  plasmodial, 321–324

protozoan, 318
Schwartzman phenomenon, 551
Scientific nomenclature, 255
Scientific notation, 39
Sclerotia, 410, *410*
home canning and, 736
Scolex, tapeworm, Microview 1–4, 327, *327*
SCP; *see* Single-cell protein
Scrapie, 370
prions and, 15
Scurvy, 382
Seawater, flora of, 713–714
Sebum, 525, 531
host defense and, 419–420
Secondary infection, 392
Secondary treatment, of sewage, *720*, 722–723
Secretors, 487
Secretory component, 444
Secretory granules, 105
Secretory IgA, 444
Selective media, 163–164, *165*
Semipermeability, of plasma membrane, 91
Senile dementia, 370
Sense codons, 222
Sense strands, 218
picornavirus, 362
Sepsis, 183
Septa, 302
Septate hyphae, 302, *303*
Septic arthritis, 285
Septic tanks, 725, *725*
Septicemia, 285, 392, 528, 577
Serial dilution, 173, *174*
Serology, 451–462
hepatitis A virus and, 657
leptospirosis and, 679
lymphogranuloma venereum and, 687
microorganism classification and, 263–264
syphilis and, 685
Serovars, 282
Serratia marcescens, 284
septicemia and, 577
Serratia spp., 284
Serum, 439
separation into protein components, *439*
Serum hepatitis, 394, 657–660
transmission of, 412
Serum sickness, 555
Sewage treatment, 720–726, *720*
microorganisms and, 17–18
Sex hormones, vaginal flora and, 676–677
Sex pili, 86, 282
Sexual intercourse, disease transmission and, 388
Sexually transmitted diseases, 389, 679–688
Sexual reproduction, eucaryotic, *240*
Sexual spores, of fungi, 306–307, *306*
Sheathed bacteria, 293
Sheep scrapie, 568
Shick test, 613
Shigella boydii, 651
Shigella dysenteriae, 651
Shigella sonnei, 651
Shigella spp., 284
bacillary dysentery and, 651
host cells and, 405
identification of, 263
portal of entry of, *401*
traveler's diarrhea and, 653
Shigellosis, 284, 400, 411, 651
transmission of, 389, 390
Shingles, 369, 533–534, *541*
lesions of, *533*
Shivering, 428
Shock, anaphylactic, 473–474
Sickle-cell anemia, hemoglobin molecule of, 51
predisposing factors in, 410
Signs, 392
Silver, oligodynamic action of, 199, *199*
Silver nitrate
antimicrobial activity of, 199
neonatal gonorrheal ophthalmia and, 541, 681
Silver-sulfadiazine
burns and, 501
pseudomonads and, 531
Simian virus 40, oncogenic, 368
Simple diffusion, 93, *93*, 103
Simple lipids, 44
Simple proteins, 51
Simple stains, 69
Single-cell protein, 18, 742–746
Singlet oxygen, 158
Sinusitis, 611
Skin
bacterial diseases of, 527–531, *542*
cell-mediated hypersensitivity reactions of, 477–478
diseases associated with, *541*
function of, 525
fungal diseases of, 537–539
host defense and, 417–420
keratin in, *417*
microbial invasion through, 400, *401*
normal flora of, 525–527, *527*
structure of, 525, *526*
viral diseases of, 531–537, *542*
Skin infections
pseudomonad, 530–531
staphylococcal, 527–529
streptococcal, 529–530
Skin testing
allergy and, 475
schedule for children, *464*
tuberculin, 619
Slab gel electrophoresis, Microview 2–3
Slide agglutination test, 263, *263*
Slime layer, 83
Slime molds, 317
cellular, *316*, 317
life cycle of, *316*
plasma membrane of, 102
plasmodial, 317, *317*
Slow viral infection, 369, *370*
Slow virus diseases, 567–569
Sludge digestion, 723–725, *724*
Slug, 317
Small intestine, flora of, 642
Smallpox, 10, 400, 411, 532–533, 535, *541*
eradication of, 532
lesions of, *532*
transmission of, 389
Smallpox virus, 532
inclusion bodies and, 365
Smears
preparation of, *70*
staining, 69
Sodium, ionic bonding of, 28–29
Sodium benzoate
antifungal action of, 200
food preservation and, 738
Sodium bicarbonate, infant diarrhea and, 654
Sodium chloride
gastroenteritis and, 653
infant diarrhea and, 654
Sodium hydroxide, dissociation of, *37*
Sodium hypochlorite, disinfection and, 198
Sodium ions, 29
Sodium nitrate, food preservation and, 738–739
Sodium nitrite, food preservation and, 738–739
Soil, components of, 701–703
Soil microbiology, 701–711
Solubility, antimicrobial agents and, 496
Solutes, 36
Solutions, 94
Solvents, water as, 36, *36*
Somatostatin, bacterial synthesis of, 243
Sorbic acid
food preservation and, 738
mold growth and, 200
Spastic paralysis, 555
Specialized transduction, 239, 356, *357*
Species, 257
Specific epithet, 12, 255
Specific resistance, 417
Specimens, preparation for light microscopy, 68–75
Spectinomycin, 505
gonorrhea and, 516, 681
Sphaerotilus natans, 293, *294*
sewage treatment and, 723
Spherical particles, 658, *658*
Spheroplasts, 89–90
Spikes, viral, 342, *342*
Spiramycin, cryptosporidiosis and, 662
Spirilla, 13, 81, *82*
Spirillum ostreae, flagella of, *85*
Spirillum spp.
animal bites and, 585
lipid inclusions in, 96
Spirillum volutans, 278
Spirochetes, 81, *82*, 276–278
axial filaments of, 84, *86*, *279*
syphilis, *683*
Spirulina spp., use as food, 746
Spontaneous generation, 6–8
Pasteur's experiment and, *9*

Spontaneous mutations, 228
Sporadic disease, 392
Sporangiophores, 306
Sporangiospores, *305*, 306
Sporangium, 306
Spores
  airborne transmission of, 389
  coat, 98, *99*
  of fungi, 305–307
  septum, 98, *99*
Sporogenesis, 98, *99*
Sporosarcina ureae, *80*
  transverse septum in, *92*
Sporozoa, 318, 321–325
Sporozoites
  plasmodial, 596
  sporozoan, *321*, 324
Sporulation, 98, *99*
Spotted fevers, 589; *see also* Rocky Mountain spotted fever
Spread plate method, *166*, 167
Stabilization ponds, sewage treatment and, 725–726
Staining, 69
  of endospores, 74–75, *74*
  of flagella, *74*, 75
  microorganism classification and, 262
  negative, 74, *74*
  procedures for, *70*
Stains, Microview 1–1
  acid-fast, 73
  differential, 70–73
  Gram, 70–73, 87
  simple, 69
  special, 74–75
Standard exponential notation, 39
Standard plate count, 173–175, *174*
Staphylococci, 81, 289–291
  arrangements of, *80*
  bacitracin and, 507
  beta toxin of, 404
  coagulases of, 404
  enterotoxins of, 405–406
  food poisoning and, 646–649, 739
  genetic transformation in, 236
  hemolysins of, 404
  leukocidins and, 404
  of nasal cavity, 611
  penicillinases of, 504
  pyelonephritis and, 678
  septicemia and, 577
  on skin, 527, *527*
  skin infections and, 527–529
  subacute bacterial endocarditis and, 579
  urinary tract infections and, 678
  vancomycin and, 507
Staphylococcus aureus, *80*, 289–290, *290*, 495
  acute bacterial endocarditis and, 579
  antibiotic resistance in, 516
  clusters of, Microview 1–1
  enterotoxins of, 646–647
  exotoxins of, 405–406
  eye infections and, 539
  food poisoning and, 646–648
  gastric juice and, 420
  hospital environment and, 528
  influenza and, 627
  intoxications and, 646
  isolation of, 164
  name derivation of, 12
  of nasal cavity, 611
  nosocomial infections and, 388, *388*
  otitis media and, 615
  P. chrysogenum and, *11*
  pathogenicity of, 527–528
  phage typing of, *264*
  phenol coefficient test and, 194
  pneumonia and, 621
  toxic shock syndrome and, 3, 290, 406
  toxins of, 290
Staphylococcus epidermidis
  eye infections and, 539
  subacute bacterial endocarditis and, 579
Staphylokinase, 404
Starch, formation of, anabolism and, 32
Stationary phase, of bacterial growth, *171*, 172
STDs; *see* Sexually transmitted diseases
Stem cells, 444–445
Stereoisomers, 46
Sterilization, *184*
  dry heat, 188
  gaseous chemosterilizers and, 201–202
  hot air, 188
  radiation and, 191–192
  times for liquid solutions, *188*
  ultraviolet light and, 192
Steroids, 45
  commercial production of, 750
  sterol conversion into, *750*
Sterols, 45, 88, 500
  conversion into steroids, *750*
  in eucaryotic plasma membrane, 103
Stibophen, schistosomiasis and, 511
Sties, 528
Stipes, algal, 313
Stomach, flora of, 642
Streak plate method, 167–168, *167*
Streptobacilli, 81, *82*
  animal bites and, 585
Streptococci, 81, 289–291, *291*
  alpha-hemolytic, 290, 529
  arrangements of, *80*
  bacitracin and, 507
  beta-hemolytic, 290, 529, *529*, 612, 679
  genetic transformation in, 236
  hemolysins of, 404
  homofermentative, 138
  human carriers and, 385
  hyaluronidase and, 404
  leukocidins and, 404
  on mucosa of cheek, 642
  non-hemolytic, 290
  pharyngitis and, 612, *632*
  skin infections and, 529–530
  sore throat and, 612
  rheumatic fever and, 579
  subacute bacterial endocarditis and, 579
  of throat, 611
  urethral, 676
  urinary tract infections and, 678
  yogurt and, 741
Streptococcus faecalis, 131
Streptococcus mutans, *80*, *291*, *644*
  adherence of, 402
  dental caries and, 644
Streptococcus pneumoniae, 290, *291*, 381, *622*
  acute bacterial endocarditis and, 579
  influenza and, 627
  meningitis and, 550, *551*, 553
  mucous membranes and, 418
  otitis media and, 615
  phagocytosis and, 424
  pneumonia and, 621
  polysaccharide capsule of, 235
  portal of entry of, *401*
  quellung reaction and, 83–84, *84*
  sinusitus and, 611
  of throat, 611
  virulence of, 402
Streptococcus pyogenes, 384
  blood agar and, 164
  colony growth of, Microview 1–2
  ethanol and, *198*
  glomerulonephritis and, 679
  M protein of, 403
  phagocytosis of, *102*
  pharyngitis and, 612
  puerperal sepsis and, 577
  pyelonephritis and, 678
  rheumatic fever and, 579
  scarlet fever and, 612
  skin infection and, 529
Streptococcus salavarius, 642
Streptococcus sanguis, 642
Streptokinase, 290–291, 404, 529
Streptolysins, 290
Streptomyces griseus, 505
Streptomyces spp., 297–298, *298*
  antibiotics and, 495
  polyene antibiotics and, 508
  sterol conversion and, *750*
Streptomycin, *511*
  bacterial endocarditis and, 517
  brucellosis and, 582
  discovery of, 505
  granuloma inguinale and, 688
  isoniazid and, 501
  penicillin and, 517
  plague and, 587
  protein synthesis and, 499–500
  resistance to, 514
  ribosomal protein synthesis and, 96
  tuberculosis and, 619
  tularemia and, 581
Stress
  disease and, 383
  genital herpes and, 688
  shingles and, 534
String test, 661
Stroma, 427

Structural analogs, 501
Structural genes, 224
Subacute disease, 392
Subacute sclerosing panencephalitis, 369, 535
  slow viruses and, 569
Subarachnoid space, 549
Subclinical infection, 392
Subcutaneous mycoses, 308, 537–538
Substrate-level phosphorylation, 123
Substrates, 34, 115
  enzymatic activity and, 118–119
Sucrose
  conversion to dextran, *645*
  dental caries and, 644
  formation of, 42
  formula for, 42, 43
Sudan dyes, lipid inclusions and, 96
Sulfa drugs, 11, 494, 501
  enzyme inhibition and, 500–501
  resistance to, 514
Sulfadiazine, toxoplasmosis and, 595
Sulfadoxine, malaria and, 597
Sulfamethoxazole; *see also* Trimethoprim-sulfamethoxazole
  sites of action of, *502*
Sulfanilamide, 494
  enzyme inhibition and, 500–501
  para-aminobenzoic acid and, 120
  structure of, *501*
Sulfhydryl groups, in amino acids, 46
Sulfolobus spp., 295
Sulfonamides, 494–495, 501, *511*
  cystitis and, 678
  N. meningitidis and, 552
  trachoma and, 541
Sulfones
  enzyme inhibition and, 501
  leprosy and, 558
Sulfur
  in complex lipids, 44
  microbial growth and, 156–157
  in proteins, 45
Sulfur bacteria, 296
Sulfur granules, 96
Sulfuric acid, bacterial production of, 38
Sulfur ions, 29
Sulfur-reducing bacteria, 286
Sunlight
  antimicrobial effects of, 192
  cancer and, 383
  mutagenesis and, 230
  S. aureus and, 290
Superficial mycoses, 309, 537
Superinfection, 497, 506
Superoxide dismutase, 158
Superoxide free radicals, 158
Suppressor T cells, 451
  desensitization and, 475
Suramin, trypanosomiasis and, 567
Surface-active agents, antimicrobial activity of, 199–200, *203*
Surface phagocytosis, 424
Surfactants, 199–200, *203*
Surgery
  antiseptic, 183
  nosocomial infections and, 385–387
  resistance and, 432
Susceptibility, 417
Sweat ducts, microbial invasion through, 400
Sweat glands, 525
  host defense and, 419–420
Swimmer's ear, 531
Swine flu, 627
  vaccine for, 463
Sylvatic plague, 586
Symbiosis, 157, 381, 709
Symbiotic bacteria, 709
Symptoms, 392
Synapsis, 234
Syndrome, 392
Synergism, 501, 517
Synthesis reactions, 32
Synthetic chemicals, degradation of, 711
Synthetic drugs, 494, 501–502, *511*
Syphilis, 64, 163, 276, 384, 392, 400, 579, 682–686, *691*
  congenital, 685
  incidence and distribution of, *683*
  latent period of, 684
  lesions of, *684*
  primary stage of, 683, *684*
  salvarsan and, 11
  secondary stage of, 683–684, *684*
  spirochetes of, *683*
  tertiary stage of, 684–685
  tetracyclines and, 506
  transmission of, 389
  Wasserman test and, 457
Systemic anaphylaxis, 473–474
Systemic infections, 392
Systemic lupus erythematosus, 383, 476–477
Systemic mycoses, 308

Taenia saginata, 327, 664
  life cycle of, *648*, *664*
Taenia solium, 664
  proglottid of, Microview 1–4
  scolex of, Microview 1–4
Tampons, toxic shock syndrome and, 3, 528–529
T antigens, 368
Tapeworm, 327–329, 631
  anatomy of, *327*
  beef, 327–328, *386*
    life cycle of, *664*
  infestations, 662–665
  life cycle of, *648*
  niclosamide and, 509
Taxa, 257
Taxonomy, 254
  bacterial, 275–276
  numerical, *268*, 269
Tay-Sachs disease, 383
T cells, 438, 575, 577
  acquired immune deficiency syndrome and, 483
  activation of, *450*
  B cells compared, *451*
  cell-mediated hypersensitivity and, 472
  cell-mediated immunity and, 447–451
  cytotoxic, 447–450
    tumor cells and, 485, *485*
  delayed hypersensitivity, 450–451
  differentiation of, *445*
  helper, 450–451
  stem cells and, 444
  suppressor, 451
    desensitization and, 475
  tuberculosis and, 619
TDP; *see* Thermal death point
TDT; *see* Thermal death time
Tears, 525
  host defense and, 418–419
Teeth, flora of, 642, *643*
Teichoic acids, peptidoglycan and, 88
Temperate phages, 355
Temperature
  enzymatic activity and, 118
  microbial control and, 183, 189
  microbial growth and, 152–153
Temperature abuse, 647
Terminator sites, 221
Terramycin; *see* Oxytetracycline
Tertiary treatment, of sewage, *720*, 726
Testes, 676, *677*
Tetanospasmin, 554
Tetanus, 291, 384, 390, 394, 405, 554–555, *568*, 613, 703
  exotoxins of, *406*
  toxins of, 405
  toxoid, 554
  vaccines for, *463*
    anamnestic response and, 447
Tetracyclines, 505–506, *511*
  animal feed and, 496
  Asiatic cholera and, 653
  beta-hemolytic streptococci and, 529
  brucellosis and, 582, 584
  endemic murine typhus and, 588
  epidemic typhus and, 588
  gonorrhea and, 682
  granuloma inguinale and, 688
  inclusion conjunctivitis and, 542
  listeriosis and, 584
  lymphogranuloma venereum and, 687
  P. multocida and, 585
  penicillin and, 517
  periodontal disease and, 646
  plague and, 587
  primary atypical pneumonia and, 623
  protein synthesis and, 499–500
  psittacosis and, 624
  pyelonephritis and, 678
  resistance to, 514
  ribosomal protein synthesis and, 96
  Rocky Mountain spotted fever and, 589
  streptococcal pharyngitis and, 612
  structure of, *505*
  superinfection and, 506
  syphilis and, 685

trachoma and, 541
tularemia and, 581
V. parahaemolyticus gastroenteritis and, 653
whooping cough and, 617
Tetrads, 81
arrangements of, *80*
Tetrahedral compounds, 40, *40*
Tetrapeptide side chains, peptidoglycan and, 87
Tetroses, 42
Thallus, 302
algal, 313
lichen, *315*, 316
Thermal death point, 186
Thermal death time, 186
Thermoduric organisms, 189
Thermophilic bacteria, 152–153
anaerobic spoilage and, 735, *735*
Thermostable nuclease, 648
Thiabendazole, trichinosis and, 667
Thiobacillus ferrooxidans
energy sources for, 126
mineral extraction from ore and, *716*
Thiobacillus spp., 295
carboxysomes in, 96
denitrification and, 707
energy sources for, 126
sulfur granules in, 96
Thiobacillus thiooxidans, 38
Threonine, commercial production of, 749
Throat, 609
normal flora of, 611
Thrombocytes, 575
Thrush, 309, 539, *540*, 689
Thylakoids, 107, *107*, 126
Thymidine kinase, acyclovir and, 509
Thymine
dimers, ultraviolet light and, 192
DNA and, 51
nucleotides, 51
Ticks, *333*
Lyme disease and, 587
Q fever and, 624
Tinctures, 196
Tinea capitis, 537
griseofulvin and, 508
Tinea cruris, 537
Tinea nigra, 537, *541*
Tinea pedis, 537, *539*
Tineas, 537, *541*
Tissue cultures, virus cultivation in, 350
Tissue repair, 427
TMP-SMZ; *see* Trimethoprim-sulfamethoxazole
Tobacco mosaic virus, 342, *343*
Togaviruses
German measles and, 535
multiplication of, 362
portal of entry of, *401*
properties of, *347*
Tolerance, immunologic, 478–479
Tolnaftate, 508
tinea infections and, 537
Toluol, hydroxy group in, 40
Tomatoes, viral infections of, 370–371
Tomato spotted wilt virus, 370
Tonsillitis, 611
Tonsils, 610
Tooth decay, 643–645
Toxemia, staphylococcal, 528
Toxemia of pregnancy, urinary tract infections and, 677
Toxicity, 497
Toxic shock syndrome, 3, 528–529
exotoxins and, *406*
S. aureus and, 3, 290, 406
Toxigenicity, 405
Toxin neutralization, 457
Toxins, 379
botulinum, 405, 556
diphtheria, 405
fungal, 410
pathogenicity and, 405
S. aureus, 290
tetanus, 405
Toxoids, 405, 439
Toxoplasma gondii, 324–325
toxoplasmosis and, 593
trophozoites of, *594*
Toxoplasmosis, *386*, 593–595, *600*
acquired immune deficiency syndrome and, 482
meningitis and, 551
Trace elements, 157
Trachea, 611
mucous membrane of, *419*
Trachelomonas spp., flagellum of, *101*
Trachoma, 287, 541, *542*
Transamination, 143
Transcription, *219*
messenger RNA and, 218–221
Transduction
bacterial, 237–239, *239*
generalized, 239, 356
oncogenic, 369
specialized, 239, 356, *357*
Transfer RNA, 52, 221, *222*
Transformation, *235*
bacterial, 234–236, *267*
mechanism of, *236*
viral, 408, *409*
Transfusion reactions, 485–488
Translation, 221–222
Translocation, oncogenic, 369
Transmissible mink encephalopathy, 568
Transmission electron micrograph, 67, *68*
Transmission electron microscope, *66*, 67
Transmission electron microscopy, 67
Transverse fission, 168–169, *170*
Transverse septum, bacterial cell division and, 92, *92*
Trauma, resistance and, 432
Traveler's diarrhea, 282, 284, 653–654
exotoxins and, *406*
Trematodes, 325–327
Treponema pallidum, *82*, 276, *279*, 384, 400
axial filaments of, 84
darkfield microscopy and, 64
mucous membranes and, 418
portal of entry of, *401*
syphilis and, 682, *683*
Treponema spp., 276
Triatoms, *333*
Tricarboxylic acid cycle; *see* Krebs cycle
Trichinella spiralis, 331
life cycle of, *665*
portal of entry of, *401*
trichinosis and, 665–666
Trichinosis, 191, 331, *386*, 665–667
low-temperature food preservation and, *736*
Trichomonas vaginalis, 320, *320*, *692*
metronidazole and, 509
trichomoniasis and, 692
vaginitis and, 686
Trichomoniasis, *691*, 692
Trichophyton spp., 537
Trickling filters, sewage treatment and, 723, *724*
Triclocarban, gram-positive bacteria and, 199
Trifluorothymidine, herpetic keratitis and, 537, 542
Triglycerides, 44
Trimethoprim
enzyme inhibition and, 501
nucleic acid synthesis and, 500
resistance to, 514
sites of action of, *502*
structure of, *503*
Trimethoprim-sulfamethoxazole, 501
bacillary dysentery and, 651
chancroid and, 687
sites of action of, *502*
traveler's diarrhea and, 653
typhoid fever and, 651
Trioses, 42
Tripeptides, 49
Trisulfapyrimidines, toxoplasmosis and, 595
Trophozoites, 319, *320*
B. coli, 321
G. lamblia, *661*
P. vivax, Microview 1–4
T. gondii, *594*
Trypanomastigotes, *T. brucei*, Microview 1–4
Trypanosoma brucei, 320
trypanomastigotes of, Microview 1–4
Trypanosoma brucei gambiense, 566
Trypanosoma brucei rhodensiense, 566
Trypanosoma cruzi, 320, 595, *595*
Trypanosoma spp., 320, 322–323
Trypanosomiasis, 322–323
African, 566–567
meningitis and, 551
American, 595, *600*
arthropod vectors and, *390*
Tsetse fly, African trypanosomiasis and, 320, 322, 566
TSS; *see* Toxic shock syndrome
TSTA; *see* Tumor-specific transplantation antigen
Tube agglutination test, 454–455, *454*
Tube dilution test, 513–514, *514*
Tubercles, 618
Tuberculin skin test, 619

coccidioidomycosis and, 628
Tuberculosis, 45, 67, 183, 292, 384, 392, 400, 404, 411, 557, 618–621, *632*
bovine, 620
geographic distribution of, *620*
immunity and, 447
leukocyte count in, 421
miliary, 619
rifamycins and, 507
streptomycin and, 505
transmission of, 389
vaccine for, *463*
Tuberculous cavities, 618
Tularemia, 281, *386*, 580–581, *599*
geographic distribution of, *581*
transmission of, 412
Tumors, 365–366, 383
benign, 365, 383
cells of, transformation in, 366–368
cytotoxic T cells and, 485, *485*
malignant, 365, 383
Tumor-specific transplantation antigen, 368
Turbidity, 176, *177*
Turnover number, 115
Typhoid bacterium, isolation of, 163–164
Typhoid fever, 282, 392, 400, 411, 650–651, 739
body temperature and, 427–428
chloramphenicol and, 506
endotoxins and, 407
human carriers and, 385
leukocyte count in, 421
predisposing factors in, 410
transmission of, 390
vaccine for, *463*
water supply and, 712
Widal test and, 454
Typhoid Mary, 650
Typhus, 287, *386*, 587–589, *600*
endemic murine, 287, 588–589, *600*
arthropod vectors and, *390*
cycle of, *588*
epidemic, 287, 587–588, *600*
arthropod vectors and, *390*
vaccine for, *463*

UDP-N-acetyl-glucosamine, 140
UDPG; *see* Uridine diphosphoglucose
Ulcers, 383
Ultracentrifuge, Microview 2–2
Ultraviolet light
cancer and, 383
enzyme denaturation and, 118
microbial control and, 192
mutagenesis and, 230, *231*
phage DNA and, 355
Uncoating, of animal viruses, 358
Undecylenic acid, tinea infections and, 537
Undulant fever, 581–582
Undulating membrane, 319, *320*
Upper respiratory system, 609–611
bacterial diseases of, 611–615, *632*
structures of, *609*
viral diseases of, 615, *632*
Uracil, in RNA, 52
Ureaplasma urealyticum, 686
Ureidopenicillins, 504
Ureteritis, 678
Ureters, 675, *675*
Urethra, 675, *675*
normal flora of, 676
Urethritis, 678
chlamydial, tetracyclines and, 506
nongonococcal, 287, 686, *691*
nonspecific, 686
Uridine diphosphoglucose, 140
Urinary bladder, 675, *675*
Urinary obstructions, urinary tract infections and, 677
Urinary system
bacterial diseases of, 677–679, *690*
normal flora of, 676–677
organs of, *675*
structure and function of, 675
Urinary tract infections, 285, 381
endotoxins and, 407
Urine
host defense and, 419
pathogens in, 411
Use-dilution test, 194
Uterus, 675, *676*

Vaccination, 10–11
public health and, 462–464
smallpox and, 532
Vaccines, 10, 439, 462–464
bacterial diseases and, *463*
brucellosis and, 584
coccidioidomycosis and, 629
commercial production of, 750
diphtheria and, 613
genetic engineering and, 244
hepatitis B and, 482, 659
influenza and, 626–627
measles and, 535
multivalent, 627
mumps and, 655
pertussis and, 613, 617–618
pneumonia and, 622
poliomyelitis and, 559–560
rabies and, 351, 562, *562*
recombinant DNA and, 19
tetanus and, 613
varicella and, 534
viral diseases and, *464*
Vaccinia virus, inclusion bodies and, 365
Vacuoles, 107
digestive, *102*, 103, 424, *425*
gas, 96, 296
protozoan, 318, *318*
Vagina, 675, *676*
normal flora of, 676–677
Vaginitis, 286, 381, 677
C. albicans, 538, 689–692
G. vaginalis, 686–687, *691*
metronidazole and, 509
Valence, 28
antigenic, 441
Vancomycin, 507
glomerulonephritis and, 679
peptidoglycan synthesis and, 499
pyelonephritis and, 678
Variable surface glycoprotein, 323
Varicella, 369, 533–534, *541*
Varicella-zoster virus, latency of, 369
Variola, 532–533, *541*
major, 532
minor, 532
Vasodilation, 425–426
VDRL slide flocculation test, syphilis and, 685
VDs; *see* Venereal diseases
Vectors, arthropod, 389–390, *390*
Vegetative mycelium, 302
Veins, 575
Venereal diseases, 679–688
Venezuelan equine encephalitis, 565
Ventricles
of brain, 549
of heart, 575
Venules, 575
Vesicles, 711
of Golgi complex, 105, *105*
Vesicular-arbuscular mycorrhizae, *710*, 711
V factor, 285
Vibrio cholerae, *82*, 285, *652*
Asiatic cholera and, 651–652
enterotoxin of, 406
portal of entry of, *401*
Vibrio fischeri, Microview 1–2
Vibrio parahaemolyticus, 285
gastroenteritis and, 653
Vibrio spp., 81, *82*, 285
enterotoxins of, 406
Vibrioids, 278–279
Vibrionaceae, 285
Vidarabine, 508
herpesvirus infections and, 537
herpes zoster and, 534
herpetic keratitis and, 537, 542
Viral diseases
of cardiovascular system, 590–592
of digestive system, 654–660, *667*
of genital system, 688–689, *691*
of lower respiratory system, 625–627, *632*
of lymphatic system, 590–592
of skin, 531–537
slow, 567–569
of upper respiratory system, 611–615, *632*
Viral infections
of eye, 539–542
latent, 369
slow, 369, *370*, 567–569
Viremia, 532, 559
Virions, 340
adenovirus, *365*
assembly of, 354
enzymes of, 352
pinocytosis and, 358
Viroids, 15
plant viruses and, 370–371
recombinant RNA and, 19
slow virus diseases and, 569
Virulence, 399–400
capsules and, 83, 402–403
Viruses, 15, 339–371

animal, 340–341, *346*
growth in laboratory, 350–352
host cells and, 365
multiplication of, 357–364
artificial media and, 163
bacterial, 340–341
biochemical properties of, *287*
boiling and, 186
cancer and, 365–369, 383
characteristics of, 340–341
chemical resistance of, 184
classification of, 260–261, 344–347, *345*, *346*
complex, 343, *345*
cytopathic effects of, 351, *351*, 407, *409*
cytotoxic T cells and, 447
dessication and, 190
DNA-containing, *348*
biosynthesis of, 358–359
ECHO, 381
effects on cells, *409*
enveloped, 342, *342*, 343, *344*
budding of, 364, *364*
enzymes of, 352
filterable, 189, 339
gastroenteritis and, 660
genetic engineering and, 244
helical, 342, *343*
hemagglutination and, 455, *455*
hemorrhagic fevers and, 592
host ranges of, 340–341
inclusion bodies and, 365, *365*
interferon and, 428–429
isolation, cultivation, and identification of, 347–352
morphology of, 342–343
mouse mammary tumor, *348*
multiplication of, 352–364
naked, 342, *342*, 364
neutralization of, 457–458
noncytocidal, 407
oncogenic, 368
pathogenicity of, 407–408
plant, 340–341, 370–371
classification of, *370*
pneumonia and, 621, 625, *632*
polyhedral, 342–343, *343*
recombinant RNA and, 19
RNA-containing, *348*
multiplication of, 359–362, *360*
sizes of, 341, *341*
structure of, 341–343, *342*
upper respiratory system and, 611
vaccines for, 10–11, *464*

Vitamin B12, tapeworms and, 664
Vitamin C deficiency, 382
Vitamin D deficiency, 382
Vitamin K, synthesis of, 381
Vitamins
functions of, *116*
synthesis of, bacteria and, 642
Volutin, 96
Vomiting, intoxications and, 646
Vulvovaginal candidiasis, 689–692, *691*

Wandering macrophages, 423
Warts, 531–532, *541*
Wasserman test, 457
Water, 35–36
boiling point of, 36
chemical reactions and, 36
hydrogen bond formation in, *32*
recycling of, 17–18
in soil, 701
as solvent, 36, *36*
temperature buffering and, 36
Water pollution
chemical, 712–713
effects of, 715–716
Water purity, tests for, 716–718
Water treatment, 718–720, *721*
Weil-Felix test, 588
Wescodyne, 196
Western equine encephalitis, *386*, 565
Wheals, 475
Whey, 740, *740*
White blood cells, 575
lysosomes in, 106
Whiteheads, 531
Whooping cough, 281, 392, 411, 612, 615–618, *632*
catarrhal stage of, 616
convalescence stage of, 616
paroxysmal stage of, 616
vaccine for, 10, *463*
Widal test, 454
WIN 51, 711, 616–617
Woolsorter's disease, 583–584
Wound botulism, 556
Wound tumor virus, 370

Xenografts, 480, *481*
Xenopsylla cheopis, *333*
endemic murine typhus and, 588
plague and, 585
X factor, 285
X rays, 191
cancer and, 383
mutagenesis and, 230

Yeast infections
acquired immune deficiency syndrome and, 482
miconazole and, *500*
Yeasts, 13, 302–304
alcoholic beverages and, 741–742, *742*
cell wall of, 102
fermentation and, 6
genetic engineering and, 748
osmotic pressure and, 190
single-cell protein from, 745
Yellow fever, *386*, 592, *600*
arthropod vectors and, *390*
endemic, *593*
immunity and, 438
transmission of, 412
vaccine for, 462, *464*
Yersinia enterocolitica
food preservation and, 736
gastroenteritis and, 654
Yersinia pestis, 284
plague and, 585
portal of entry of, *401*
Yersinia spp., 284–285
Yogurt, lactic acid bacteria and, 741

Zephiran; *see* Benzalkonium chloride
Ziehl-Neelson technique, 618
Zinc, microbial growth and, 157
Zinc chloride, antimicrobial activity of, 199
Zinc ions, enzyme-substrate complexes and, 117
Zinc oxide, antifungal activity of, 199
Zinc peroxide, antimicrobial activity of, 202
Zinc undecylenate
athlete's foot and, 199
tinea infections and, 537
Zone of equivalence, 452
Zoogloea spp., 281, *723*
sewage treatment and, 722
Zoonoses, 385, *386*
Zoster, 541
portal of entry of, *401*
Zygomycota, 307
characteristics of, *310*
spores of, 306
Zygospores, 306, *306*
Zymogen, 107
Zymomonas spp., ethanol and, 749